Multiple Access Communications

IEEE PRESS
445 Hoes Lane, PO Box 1331
Piscataway, NJ 08855-1331

Multiple Access Communications

Foundations for Emerging Technologies

Edited by

Norman Abramson

University of Hawaii at Manoa

IEEE PRESS

A Selected Reprint Volume

IEEE Information Theory Society, *Sponsor*

The Institute of Electrical and Electronics Engineers, Inc., New York

This book may be purchased at a discount from the publisher
when ordered in bulk quantities. For more information contact:

IEEE PRESS Marketing
Attn: Special Sales
PO Box 1331
445 Hoes Lane
Piscataway, NJ 08855-1331
Fax: (908) 981-8062

Printed in the United States of America

10 9 8 7 6 5 4 3 2

ISBN 0-87942-292-0

IEEE Order Number: PC0287-3

Library of Congress Cataloging-in-Publication Data

Multiple access communications: foundations for emerging technologies / edited by Norman Abramson.

p. cm.

"A selected reprint volume."

"IEEE Information Theory Society, sponsor."

Includes bibliographical references and index.

ISBN 0-87942-292-0

1. IEEE order number: PC0287-3. 2. Digital communications. 3. Spread spectrum communications. 4. Local area networks (Computer networks) I. Abramson, Norman. II. IEEE Information Theory Society.

TK5103.7.M85 1993

621.382—dc20

92-23864
CIP

Contents

Part 1
Introduction

1.1 Scope of the Collection

THE TERM *multiple access communications* may be interpreted in a variety of ways in different areas of communications. One interpretation that encompasses certain theoretical questions dealing with channel capacity is common among workers in information theory. In the area of spread spectrum, multiple access usually refers to the ability of certain kinds of signals to coexist in the same frequency and time space with an acceptable level of mutual interference. A third area in telecommunications research that is concerned with another view of multiple access involves (usually digital) networks composed of large numbers of intermittent transmitters trying to share a common communication channel, often based on variations of a contention protocol. And, finally, since 1970 those involved in satellite communications have devoted considerable effort to questions related to the use of more than one satellite earth station accessing a single transponder by means of frequency division or time division (FDMA or TDMA) techniques.

In this collection of papers on multiple access communications I have made an editorial decision to include the first three areas listed above but not the fourth. The three areas included have the common characteristic that they all deal with or are directly related to statistical properties of communication channels. The papers in this collection then are connected by that common theme; and in that common theme the reader can find connections from one area to another, and even insights that are of value in one area derived from results published in another. The canonical problem that served as a guide to the selection of papers in this collection is shown in Fig. 1.

The *n* transmitters on the left of Fig. 1 might represent *n* users in a packet radio LAN (local area network) trying to communicate with the host receiver on the right. The *n* transmitters might also represent *n* separate very small aperture terminals (VSATs) transmitting packet data in a satellite network. Finally, the transmitters might provide a useful model for the kind of digital cellular personal communication networks (PCNs) that have been referred to as *third-generation wireless networks*. In all of these cases there are certain common characteristics of the network model I would like to emphasize.

1. The network is composed of independent users attempting to communicate with a single common channel or perhaps even multiple channels—but the number of channels is much smaller than the number of possible transmitters.

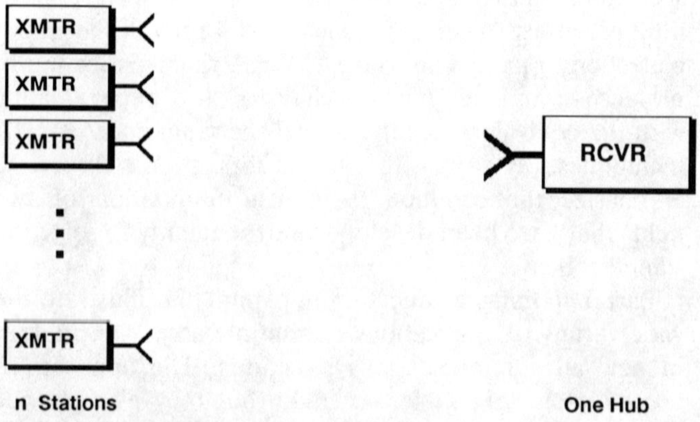

Fig. 1. Model of a multiple access channel.

2. The data from each of the users accessing the network is bursty—that is, the transmissions from a single user are separated by periods of silence (usually of durations much greater than the periods of transmission).
3. The number of users active on the network at any given time is a random variable, usually unknown to the receiver.

Not all of the papers in this collection deal with models that are consistent with all three of the assumptions listed above; but each of the papers provides insight that can teach us something about such models. And a better understanding of these models is essential for the development of the packet radio LANs, VSAT networks, and digital cellular networks that have assumed increasing importance in the telecommunications of today.

1.2. Organization

The same objectives that dictated the scope of this volume serve as a guide to its organization. My objectives are to provide a collection of papers that can be of value to those involved in the study of the theory of multiple access channels as well as to those interested in the design and analysis of real networks, based on multiple access channels. In Part 2, then, I have placed a collection of papers dealing with the fundamental properties of signals used in multiple access channels and in the fundamental information theoretic properties that limit the operation of those channels.

The multiple access capability of real networks exhibiting the three characteristics listed in section 1.1 has ordinarily been implemented by means of either a spread-spectrum–based protocol or an ALOHA-based protocol. Recent results, however, have shown that these two methods, with different origins and a different literature, are in fact simply two ways of viewing the same basic signals. Part 3 contains a set of papers dealing with spread spectrum, emphasizing the use of spread spectrum for multiple access communications. Part 4 provides a similar collection of papers on using ALOHA channels for multiple access, and this part concludes with a paper dealing with the equivalence of the spread spectrum and ALOHA techniques. My hope is that this juxtaposition will serve to emphasize the common theoretical foundation of two fields that have been developed independently to solve the same problems.

Part 5 contains a collection of papers that illustrate the wide variety of applications of multiple access technology in new telecommunications structures. The applications are of interest in their own right, but they also provide valuable feedback to the theorist interested in this field, because the most effective way to generate communication network models that are relevant to real communication networks is to build a familiarity with the real-world problems that must be analyzed. The papers in Part 5 can provide valuable insight into the motivation leading to the theoretical results in Parts 2, 3, and 4.

1.3. Comments and Additional References

1.3.1 General References

There are several previous volumes in the IEEE PRESS Selected Reprint Series as well as special issues of IEEE journals that deal with some of the topics treated in this volume. Among those that are closest in coverage to the topics covered in this volume are

1. Robert C. Dixon (Ed.), *Spread Spectrum Techniques.* New York: IEEE PRESS, 1976.
2. Special Issue on Computer Communications, R. L. Pickholtz (Ed.), *IEEE Trans. Communications,* vol. COM-25, no. 1, Jan. 1977.
3. Special Issue on Spread Spectrum Communications, L. A. Gerhardt and R. C. Dixon (Eds.), *IEEE Trans. Communications,* vol. COM-25, no. 8, Aug. 1977.
4. Special Issue on Spread-Spectrum Communications, C. E. Cook, F. W. Ellersick, L. B. Milstein, and D. L. Schilling (Guest Eds.), *IEEE Trans. Communications,* vol. COM-30, no. 5, Pt. 1, May 1982.
5. C. E. Cook, F. W. Ellersick, L. B. Milstein, and D. L. Schilling (Eds.), *Spread Spectrum Communications.* New York: IEEE PRESS, 1983.
6. Special Issue on Random-Access Communications, James L. Massey (Guest Ed.), *IEEE Trans. Information Theory,* vol. IT-31, no. 2, March 1985.
7. Special Issue on Packet Radio Networks, B. M. Leiner, D. L. Nielson, and F. A. Tobagi (Guest Eds.), *Proceedings of the IEEE,* vol. 75, no. 1, Jan. 1987.
8. Special Issue on Performance Evaluation of Multiple Access Networks, Victor O. K. Li (Guest Ed.), *IEEE J. Selected Areas Communications,* vol. SAC-5, no. 6, July 1987.
9. Special Issue on Spread Spectrum Communications, D. L. Schilling, R. L. Pickholtz, and L. B. Milstein (Guest Eds.), *IEEE J. Selected Areas Communications,* Parts I and II, vol. 8, nos. 4 and 5, May and June 1990.

Another reference that has a good deal of overlap with this volume is the proceedings of a NATO Advanced Study Institute held in August, 1980.

J. K. Skwirzynski (Ed.), *New Concepts in Multi-User Communications.* Rockville, MD: Sitjhoff and Nordhoff, 1981.

In May, 1991, the Institute of Electronics, Information and Communication Engineers of Japan published a special issue of the *IEICE Transactions on Communications* on spread spectrum. The papers in this special issue are printed in English in order to disseminate Japanese work in this field to an international audience.

Special Issue on Spread Spectrum Techniques and Its Applications, Gen Marubayashi (Ed.), *IEICE Trans. Communications,* vol. J74-B-I, no. 5, May 1991.

1.3.2 General Channel Considerations

The material in Part 2 addresses problems in multiple access channels, multiple access protocols, and multiple access signals. The papers dealing with the capacity of multiple access channels in this part are concerned primarily with multiple access in the additive Gaussian noise channel. The Gaussian channel model has received considerable attention in the multiple access literature because of the importance of this model in the case of single-transmitter single-receiver channels.

In spite of this importance, the Gaussian model has some shortcomings that have not yet been adequately covered in the literature on multiple access channels. In most of the applications described in Part 5, the number of active users on the network is a random variable, since any single network transmitter may or may not transmit at any given time. This means that we would like to be able to handle the model where the number of users on the network at any one time is a random variable (characteristic (3) of section 1.1) and where any given user transmits a Gaussian signal, but only conditioned on the fact that it is transmitting some signal. Perhaps this kind of model can be analyzed by generalizing the more common Gaussian assumption for channels with a single transmitter. Related to the question of the suitability of the Gaussian model for the kind of networks covered in this volume is the question of address information within the network. For if we really do want to analyze a network where the number of possible transmitters is large and the number of active transmitters is much less than the number of possible transmitters, then the question of protocol information within the network would seem to be relevant. The paper by Gallager in Part 2 addresses part of this issue, but considerable work remains to be done.

An extensive bibliography on information theory and multiple access channels covering the period through 1975 may be found in:

E. C. van der Meulen, "A survey of multi-way channels in information theory: 1961–1976," *IEEE Trans. Information Theory,* vol. IT-23, no. 2, Jan. 1977.

A shorter bibliography covering work from 1976 to 1980 is given at the end of the Gamal and Cover paper in Part 2.

1.3.3 Spread Spectrum Multiple Access (CDMA)

Spread spectrum multiple access communications is the subject of the articles selected for Part 3. This form of multiple access is often referred to as CDMA (code division multiple access), especially when different spreading sequence codes are assigned to different users. Recent results (see Part 4) have shown that different spreading sequences may not always be necessary or even desirable for spread spectrum multiple access. Although there is no clear agreement on the nomenclature in the technical literature, the term CDMA is preferred only for those cases where different spreading sequences are present.

Spread spectrum has its origins in a variety of military applications, and much of the early work in spread spectrum deals with the transmission of continuous signals rather than the bursty signals of interest in this volume. Again, the reader should note the same caution about many of the papers in the spread spectrum literature that was mentioned in section 1.3.2 when dealing with information theoretic models. When spread spectrum is used for multiple access of continuously transmitting information sources, a channel model where the number of transmitters and the identity of transmitters is known is an appropriate model. But if the number of transmitters is not known and if the address information of the transmitting sources is significant, then this model of spread spectrum should be reexamined.

Several textbooks cover spread spectrum in detail, although many aspects of the general spread spectrum problem are of only peripheral interest for applications in multiple access networks. In particular, synchronization, anti-jam applications, and frequency hopping are not emphasized in this selection of papers. The comprehensive text by Simon et al. includes a wealth of historical material collected by Robert Scholtz and Robert Price that is rarely seen in discussions of technical topics. A similar, but smaller, history of spread spectrum communications in Japan may be found in an article by Mitsuo Yokoyama in the special issue of the *IEICE Transactions on Communications* referenced at the end of Section 1.3.1.

1. Jack K. Holmes, *Coherent Spread Spectrum Systems.* New York: John Wiley and Sons, 1982.
2. Robert C. Dixon, *Spread Spectrum Systems*, 2nd Edition. New York: John Wiley and Sons, 1984.
3. Rodger E. Ziemer and Roger L. Peterson, *Digital Communications and Spread Spectrum Systems.* New York: Macmillan Publishing Company, 1985.
4. Marvin K. Simon, Jim K. Omura, Robert A. Scholtz, and Barry K. Levitt, *Spread Spectrum Communications*, Vols. 1, 2, and 3. Rockville, MD: Computer Science Press, 1985.
5. George R. Cooper and Clare D. McGillem, *Modern Communications and Spread Spectrum*. New York: McGraw-Hill Book Company, 1986.
6. R. Skaug and J. F. Hjelmstad, *Spread Spectrum in Communication.* London: Peter Peregrinus, Ltd., 1985.

1.3.4 ALOHA Multiple Access

The first use of ALOHA channels for multiple access took place in the ALOHANET local area packet radio network at the University of Hawaii in 1971. The original paper analyzing the ALOHA channel capacity in 1970 used a model that requires some comment in two separate areas: collision intervals and channel power constraints.

In the original ALOHA model it is assumed that two packets that overlap even a small amount result in two lost packets that must then be retransmitted over the channel. Later it was realized that the criterion of overlap in the channel was unduly restrictive and that the key question of correct packet reception was more productively addressed in terms of the overlap of packets at the output of a receiver. This distinction is particularly important in the case of packet receivers that compress the received packets in some way, since the total period of packet vulnerability can be much less than the period of possible overlap in the channel. The difference then led to the introduction of spread ALOHA–a form of ALOHA multiple access equivalent to direct-sequence spread spectrum multiple access, but which can be implemented with a simpler receiver.

In the original ALOHA model, the throughput of the channel was normalized with respect to the throughput of a transmitter sending data at the same rate as the packet-burst data rate. This normalization, therefore, compared the throughput of an ALOHA channel to the throughput of a channel transmitting continuously, even though the continuous channel might transmit more power (typically 10 dB or more) than the intermittent ALOHA transmitter. For the case of the original ALOHANET, where the constraint on the transmitter's power was a peak power constraint, such a normalization was appropriate. For other applications, however, such as VSAT networks or battery-operated transmitters in a PCN, a more appropriate normalization would compare the throughput of the ALOHA channel to that of a channel with the same average power. And as might be expected, when proper allowances are made for a 10-dB or more power mismatch, the ALOHA channel throughput looks considerably more attractive.

There are a number of Ph.D. theses and reports dealing with ALOHA channels that could not be included in this collection because of space limitations. The first careful analysis of multi-channel ALOHA networks was an informal ALOHA system report by George Turin in 1976. This work was subsequently expanded upon in the Berkeley thesis of Wing-Po Yung.

Wing-Po Yung, "Analysis of multi-channel ALOHA systems," Ph.D. thesis, University of California at Berkeley, Nov. 1978.

The same general topic was treated in a different and imaginative way by George Thomas of the Indian Space Research Organization. An abbreviated version of the results by Thomas is included in Part 4, but the interested reader might wish to consult the complete report.

George Thomas, "Multiclass spread-spectrum slotted ALOHA: A new approach for high efficiency in very large packet networks," Tech. Rep SCPO-TR-3, Satellite Communications Programmes Office, Indian Space Research Organisation Headquarters, Dept. of Space, Government of India, Bangalore 560 009, India.

Finally, note that the question of the spatial capacity of ALOHA transmitter networks is included in several of the papers in Part 4. But considerable additional results in this area are available in the reports listed below.

1. John Silvester, "On the spatial capacity of packet radio networks," UCLA, Computer Science Dept. Rep. No. UCLA-ENG-8021, May 1980.
2. Deepak Sant, "Some models of packet broadcasting networks," ALOHA Systems Tech. Rep. B81-1, University of Hawaii, Honolulu, Feb. 1981.
3. Hideaki Takagi, "Analysis of throughput and delay for single and multi-hop packet radio networks," UCLA, Computer Science Dept. Rep. No. CSD-830523, May 1983.

1.3.5 Applications

The general architecture of the worldwide telephone network assumed its present form in North America at the end of the 19th century. This architecture is based upon the use of point-to-point telecommunications channels, connected by a hierarchy of switches to provide for interconnection of large numbers of voice users. During the last two decades this telecommunications architecture has been augmented by an increasingly diverse set of special purpose telecommunications structures, motivated in many but not all cases by the need for a new telecommunications architecture better suited to digital traffic. Many of these new structures, such as local area networks (LANs), packet radio networks, VSAT networks, and digital personal communication networks (PCNs), are based on the use of a multiple access channel. Part 5 contains a collection of papers that looks at some of these new telecommunications structures with particular emphasis on the use of multiple access and this new telecommunications architecture.

The definitive reference on the early days of modern computer networks is the Harvard Ph.D. thesis of Robert Metcalfe, published as a Project MAC report at MIT.

Robert Melancton Metcalfe, "Packet communication," Rep. MAC TR-114, Project MAC, Massachusetts Institute of Technology, Cambridge MA, Dec. 1973.

After the demonstration of the effectiveness of a multiple access communications architecture in the ALOHANET in the early 1970s, the commercial exploitation of this technology was retarded by the limitations imposed by conventional frequency allocation procedures. With the exception of the use of multiple access architectures employed in the international MARISAT system in 1976 and in several VSAT networks during the late 1980s, the only possibility for commercial exploitation of this technology was in the use of multiple access procedures in cable-based local area networks. But in 1983 the ability to fully develop the commercial potential of multiple access channels was provided with an important boost by the decision of the United States Federal Communications Commission to authorize three separate multiple access frequency bands for shared use by commercial interests.

Finally, it should be noted that the large-scale investments in telecommunications based on conventional point-to-point architectures common in North America, Europe, and Japan do not exist, or exist in a much more limited form, in most of the rest of the world. The consequence of this fact is that in many of the developing countries of the world, the new telecommunications structures, based on multiple access channels, assume an even greater importance than they do in developed countries. Thus digital cellular radio, VSATs, and packet radio local area networks may provide an option to telecommunication planners in developing countries that will allow them to develop new, appropriate telecommunications structures more suited to their needs than those structures that are the norm in developed countries. The general background for these developments is addressed in:

Norman Abramson, "Satellite data networks for national development," *Telecommunications Policy*, vol. 8, no. 1, March 1984.

One specific proposal to apply this technology to meet the needs of developing countries by means of a digital network of 100,000 VSATs is described in:

Mahesh Kumar Goel, "Some studies on a low cost VSAT satellite mesh network," Ph.D. Thesis in the Dept. of Electrical Engineering, Indian Institute of Technology, Delhi, New Delhi 110 016, India, Dec. 1989.

1.4 Acknowledgment

The editor is grateful for the assistance of Robert Griffin in the collection and evaluation of relevant papers for an early version of this collection.

Part 2
General Channel Considerations

Poisson, Shannon, and the Radio Amateur*

J. P. COSTAS†, SENIOR MEMBER, IRE

Summary—**Congested band operation as found in the amateur service presents an interesting problem in analysis which can only be solved by statistical methods. Consideration is given to the relative merits of two currently popular modulation techniques, SSB and DSB. It is found that in spite of the bandwidth economy of SSB this system can claim no over-all advantage with respect to DSB for this service. It is further shown that there are definite advantages to the use of very broadband techniques in the amateur service.**

The results obtained from the analysis of the radio amateur service are significant, for they challenge the intuitively obvious and universally accepted thesis that congestion in the radio frequency spectrum can only be relieved by the use of progressively smaller transmission bandwidths obtained by appropriate coding and modulation techniques. In order to study the general problem of spectrum utilization, some basic results of information theory are required Some of the significant work of Shannon is reviewed with special emphasis on his channel capacity formula. It is shown that this famous formula, in spite of its deep philosophical significance, cannot be used meaningfully in the analysis and design of practical, present day communications systems. A more suitable channel capacity formula is derived for the practical case.

The analytical results thus obtained are used to show that broadband techniques have definite merit for both civil and military applications. Furthermore, such techniques will result in far more efficient spectrum utilization in many applications than any practical narrow-band, frequency-channelized approach. Thus broad-band techniques can, in many cases, increase the number of available "channels." With regard to military communications it is shown that the ability of a communication system to resist jamming varies in direct proportion to the transmission bandwidth for a given data rate. Thus narrow-band techniques lead progressively to more expensive communications systems and less expensive jammers. It is concluded that in the military field broad-band techniques are not only desirable but also often mandatory.

I. INTRODUCTION

MOST common usage of the radio frequency spectrum involves operation at specified frequencies as assigned by the appropriate regulatory agencies in the various countries. In contrast, the radio amateur service is assigned various bands of frequencies and properly licensed stations are permitted to operate at any frequency within these bands. This freedom of choice of frequency is necessitated by the obviously impossible administrative problem of assigning specific frequencies to specific stations and, furthermore, the available bandwidths fall short by several orders of magnitude of providing exclusive channels to each authorized station. Thus, as one might suspect, the situation in the amateur bands is a chaotic one in terms of mutual interference. There is very little tendency to "channelize" for several reasons. The crowded conditions normally leave no empty spaces in frequency so that a station starting operation has no choice but to transmit "in between" two strong stations or on top of a weaker station. Furthermore, at the higher HF frequencies, the ionospheric "skip" makes it impossible to choose a good operating frequency by listening, since the signal situation will be radically different between two points spaced many miles apart. Thus, the very nature of the amateur service would lead one to expect that any meaningful analysis of this problem must be based on a statistical approach.

A mathematical study of amateur radio communications can be of use in other important areas. Consider, for example, military communications where allocation of frequencies cannot possibly prevent interference due to the use of the same frequencies by the opposing forces. It is not hard to imagine that under such conditions each operator will shift frequency and take other appropriate action in order to get his message through. Thus, in a combat area we might well expect to find the very same chaos in the communications services that we observe in the amateur bands today. Certainly in such situations interference cannot be eliminated by allocation; interference will exist and we must simply learn to live with it. We are not speaking here of intentional jamming but rather of the casual interference which is inevitable when two opposing military forces (which today depend heavily on radio) attempt to operate independently and use the same electromagnetic spectrum. The problem of intentional jamming will be treated in detail in Section VI.

In the analysis of the radio amateur problem which follows, three modes of operation are compared. It is first assumed that all stations employ suppressed-carrier single-sideband (SSB). Then exclusive use of suppressed-carrier AM (DSB) is assumed. Finally, a frequency diversity system is examined in which each station transmits a large number of identical signals at randomly selected frequencies in the band. Intuitively we might suspect that SSB would be superior to DSB because of the two-to-one difference in signal bandwidths. The frequency diversity system is intuitively ridiculous because it apparently "wastes" bandwidth rather indiscriminantly. As we shall see, intuition is a poor guide in these matters. The feeling that we should always try to "conserve bandwidth" is no doubt caused by an environment in which it has been standard practice to share the RF spectrum on a frequency basis. Our emotions do not alter the fact that bandwidth is but one dimension of a multidimensional situation.

* Original manuscript received by the IRE, April 21, 1959; revised manuscript received, June 13, 1959.
† General Electric Co., Syracuse, N. Y.

Reprinted from *Proc. IRE*, vol. 47, no. 12, pp. 2058–2068, Dec. 1959.

II. Congested Band Analysis

SSB Case

We shall first consider the case of exclusive use of SSB. The spectral situation is shown in Fig. 1 as it might

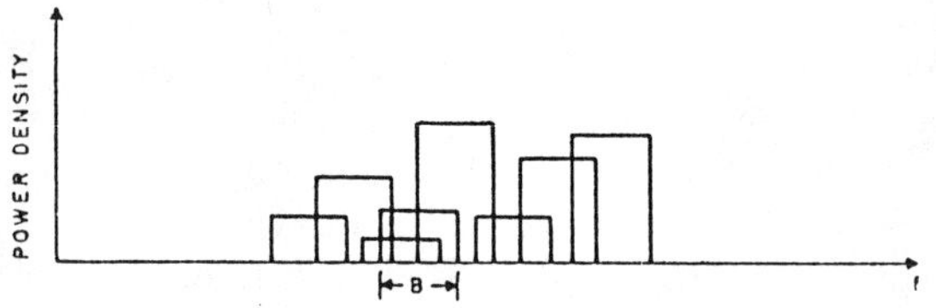

Fig. 1—Power density spectra—SSB case.

appear to a particular receiver. Each signal occupies a bandwidth B (equal to the baseband bandwidth for SSB), has a location in frequency independent of all other signal locations, and has an amplitude of power density independent of all other signal amplitudes. The signal amplitudes will have a probability distribution which will be specified at a later time. While the frequency locations of the various signals are distributed at random, it can be said that, on the average, there are a given number of signals per given unit of bandwidth. Thus, we may specify the density of loading of the band by a quantity k which represents the average number of signals per unit bandwidth. It happens that we shall need to know the probability of having a given number of signals ν falling in a bandwidth B. This, of course, is given for the conditions specified by the celebrated distribution of Poisson as

$$P(\nu, B) = \frac{(kB)^\nu}{\nu!} e^{-kB}, \qquad (1)$$

where $P(\nu, B)$ is the probability of having ν signals in the bandwidth B if there are k signals per unit bandwidth on the average.

The choice of the distribution function for the signal power densities is somewhat arbitrary and, as far as the final results are concerned, apparently not particularly critical. It is physically reasonable and mathematically convenient to choose the chi-squared distribution[1]

$$p_\nu(x) = \frac{x^{\nu/2-1}e^{-x/2}}{2^{\nu/2}\Gamma(\nu/2)} \qquad (x \geq 0), \qquad (2)$$

where $p_\nu(x)$ is the probability density function of the spectral amplitude which results from the summation of ν independent signals. For $\nu=1$ the distribution has a mean of unity. This specifies that the average signal strength at the receiver is unity which results in no loss of generality for this application.

For convenience only, we shall assume that we are receiving a signal of average strength and want to find the probability that the Signal-to-Noise Ratio at the receiver output will equal or exceed a specified value. For SSB operation, the SNR at RF is the same as the SNR at the receiver output. We shall estimate the effective noise level at the receiver input by noting the interference level at the center of the pass band. We shall now determine the probability that the interference level will be less than or equal to J, which means that the SNR at the receiver output will be equal to or greater than $1/J$, since the desired signal is assumed to be of average strength of unity. Let $P_{SSB}(\text{SNR} \geq 1/J)$ be this probability. Then

$$P_{SSB}(\text{SNR} \geq 1/J) = P(0, B) + P(1, B)\int_0^J p_1(x)dx + P(2, B)\int_0^J p_2(x)dx + P(3, B)\int_0^J p_3(x)dx + \cdots, \qquad (3)$$

which states that the event will occur if there are no signals in B, if there is one signal in B with amplitude less than J, if there are two signals in B the sum of whose amplitudes is less than J, etc. It should be clear that if an interfering signal is to contribute to the measurement of interference, its lowest frequency must fall somewhere within a frequency band extending from the center of the pass band to B cycles below. It is to this event that the terms $P(\nu, B)$ in (3) refer. Substituting (1) and (2) into (3) one obtains

$$P_{SSB}(\text{SNR} \geq 1/J) = e^{-kB}\left[1 + \sum_{\nu=1}^{\infty} \frac{(kB)^\nu}{\nu!}\int_0^J \frac{x^{\nu/2-1}e^{-x/2}}{2^{\nu/2}\Gamma(\nu/2)}dx\right]. \qquad (4)$$

Evaluation of (4) for a fixed J and variable k will give the probability of exceeding a certain receiver output SNR as a function of band loading. For example, for $J=1$ the expression gives the probability of exceeding a 0-db SNR when receiving a signal of average strength, or of exceeding a +3-db SNR when receiving a signal of twice (power) average strength, etc. Fortunately, the integral function in (4) is tabulated[2] and the series converges rather rapidly, so that the numerical work involved in evaluating (4) is not too difficult.

DSB Case

As might be suspected, the analysis of the case involving exclusive use of DSB is quite similar to the SSB analysis. There are two important differences to be noted. First, since all transmitted signals have twice the baseband bandwidth it is to be expected for a given band loading there will be more interfering signals involved than in the case of SSB. In the DSB analysis then, we will be concerned with the probability of having ν interfering signals in a bandwidth $2B$, using the same estimate of effective receiver input noise level as before.

[1] H. Cramer, "Mathematical Methods of Statistics," Princeton University Press, Princeton, N. J., ch. 18; 1946.

[2] C. D. Hodgman, "Mathematical Tables," Chemical Rubber Publishing Co., Cleveland, Ohio, p. 257; 1946.

Thus the Poisson distribution $P(\nu, 2B)$ must be used in the equation equivalent to (3) for the DSB analysis. This represents a loss caused by increased transmission bandwidth; there is a compensating gain as will be seen. The second difference between the SSB and DSB analysis involves the relationship between the predetector and postdetector SNR's. In SSB these two ratios are the same. In DSB the postdetector SNR is 3 db better than the predetector value. This difference arises because of the coherent addition of upper and lower sideband components of the signal and incoherent addition of the corresponding interference components in the synchronous detector. Thus, for identical output SNR's the interference power density will be two times as great relative to desired signal density in DSB as compared to SSB. Consequently, in the equation equivalent to (3) the upper limit on all integrals must be changed from J to $2J$ in order that J have the same meaning in both cases.

When the two changes discussed above are made, the probability of exceeding an output SNR of $1/J$ for a desired signal of mean strength (unity) becomes

$$P_{\mathrm{DSB}}(\mathrm{SNR} \geq 1/J)$$

$$= e^{-2kB}\left[1 + \sum_{\nu=1}^{\infty} \frac{(2kB)^{\nu}}{\nu!} \int_0^{2J} \frac{x^{\nu/2-1}e^{-x/2}}{2^{\nu/2}\Gamma(\nu/2)}\, dx\right]. \quad (5)$$

A comparison of (4) and (5) shows that the increased bandwidth of DSB has in some ways been detrimental ($2kB$ in place of kB in the Poisson distribution), and in other ways beneficial ($2J$ in place of J in the integral expression). As later calculations show, the increased bandwidth of DSB does not affect the relative congested band performance as compared to SSB in any significant manner. We might begin to suspect that the efficient use of broader bandwidths in a congested operating band is not necessarily a bad idea. The broader bandwidth signals will increase the tendency of frequency overlap and tend, in a sense, to cause more interference. This is obvious. *What is not so obvious is the fact that the increased bandwidth gives to the receiving system an increased ability to discriminate between the desired signal and the interference.* In order to investigate further the effects of increasing transmission bandwidth, a rather simple form of broad-band technique will now be analyzed.

Frequency Diversity Case

For this example we shall use the SSB mode of transmission (although the DSB mode would yield identical results), in a somewhat unusual manner. Each station will transmit not one but M (where M is a large number) identical signals at randomly chosen frequencies in the congested band. The receiver must know these frequency locations so that all M signals may be received, detected, and added coherently to produce the receiver output signal. With each station transmitting M identical signals, the interference spectrum amplitude will, with nearly unit probability, be very nearly equal to a constant value at all frequencies for sufficiently large M. This value may be determined quite easily by inspection.

Consider first the normal SSB situation without diversity. The received signals are distributed in amplitude of power density about a mean of unity. Thus, the average received power is B watts per station. Since there are k stations per cycle on the average, the mean interference power density will be kB watts per cycle. Going from one transmission to M transmissions per station (assuming the power of each station is now split evenly between the M signals) does not alter the value of the average interference power density. In the diversity case this *average* value will be very nearly the *actual* value of interference density level which will exist at all frequencies and at all times. The diversity receiver output SNR may now be easily calculated.

Each of the M signals will have a power B/M (for the average signal strength case) and the noise power accepted in receiving each of the M signals will be kB^2. The RF SNR at each of the M frequencies will be $1/MkB$ and coherent addition of M such signals will yield an output SNR of $1/kB$. So then

$$(\mathrm{SNR})_{\mathrm{Div}} = \frac{1}{kB} \quad (6)$$

on a power basis for a desired signal of mean strength. Note that in (6) we are able to specify the precise SNR, while in the SSB and DSB cases of (4) and (5) we can only predict the probability or the percentage time the SNR will exceed a given value.

III. Results and Discussion—Congested Band

The results represented by (4)–(6) may be interpreted in many different ways. For the purposes of this discussion let us assume that voice communications is involved and that message reception will be considered successful if the receiver output SNR equals or exceeds unity or 0 db. Keep in mind that this is not a commerical service but rather a service where the operator is willing to exert some effort in order to understand what is being said. Thus, the 0-db choice is probably reasonable with regard to sentence intelligibility where the interference is of an incoherent nature. The three equations will then be used to calculate the circuit reliability for signals at 0, +3, +6, and +9 db relative to mean signal strength as a function of kB, the band loading expressed in average number of stations per audio bandwidth. The resulting graphs are shown in Figs. 2 through 5. Turning first to Fig. 2, which assumes a received signal of mean strength, we note that the circuit reliability drops rather rapidly with band loading for both SSB and DSB. SSB shows some advantage, but of a small amount, at loadings which result in a reasonable reliability percentage. An estimate of the increased number of users for the same performance which results from SSB use may be obtained by drawing a line horizonally from any given

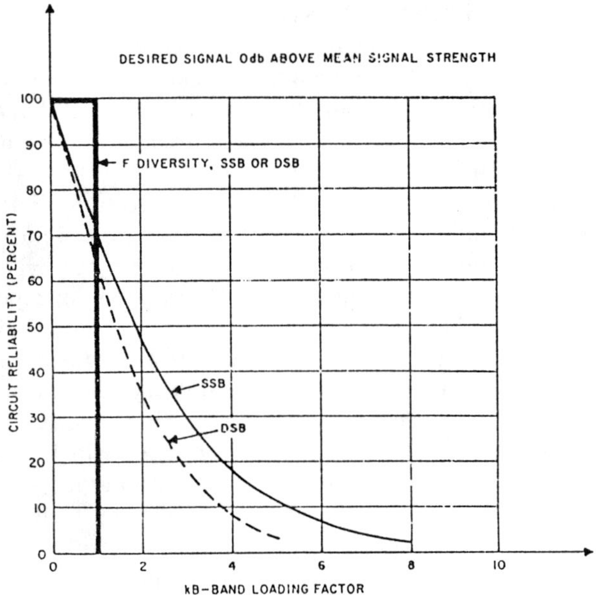

Fig. 2—Per cent circuit reliability vs band loading.

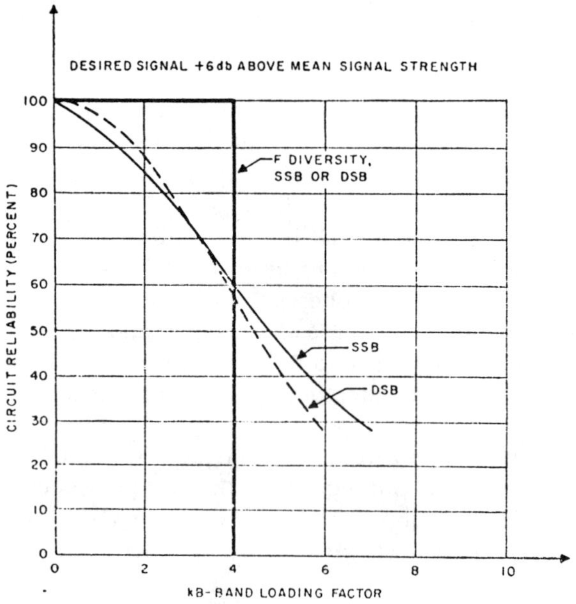

Fig. 4—Per cent circuit reliability vs band loading.

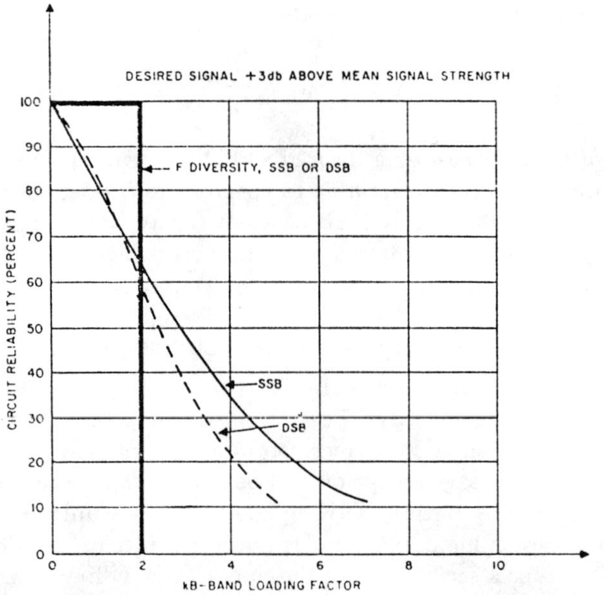

Fig. 3—Per cent circuit reliability vs band loading.

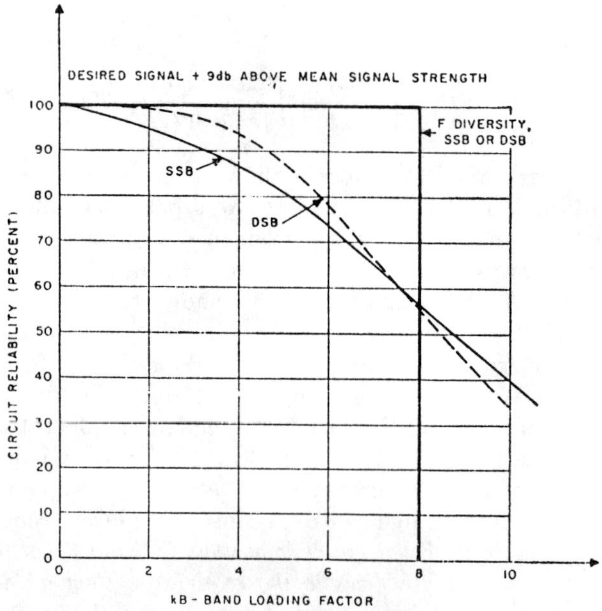

Fig. 5—Per cent circuit reliability vs band loading.

ordinate value and noting the kB values at which this line intersects the SSB and DSB curves. The two-to-one increase in loading which we might at first expect from SSB certainly does not materialize, except at values of circuit reliability which are so low as to be meaningless. Thus, the randomness of band occupancy has a significant effect on performance, and any intuitive conclusions based on orderly channel assignments are subject to considerable error. Note in particular that the circuit reliability for SSB at $kB=1$ is 70 per cent. At this loading there are enough channels to satisfy all needs, and 100 per cent reliability could be had if some organization could be obtained. About the only conclusions to be drawn from Fig. 2 are that SSB and DSB give nearly the same performance and that it is usually hopeless to try to communicate with a station whose signal strength is only average at times when the band is crowded. This last conclusion will come as no surprise to the experienced operator.

As the strength of the desired signal increases above the mean value the situation improves rather rapidly, as shown by Figs. 3–5. The SSB and DSB curves now "cross over" and both curves tend to stay at higher reliability values as kB is increased, which is to be expected. Note in Fig. 5 that DSB shows a slight advantage over SSB for the lower loading values and the cross-over occurs when the reliability factor is 63 per cent. In total these results show the futility of claiming any advantage for either SSB or DSB in this service. If one is insistent upon claiming an advantage, the specific conditions under which the comparisons are made must be given.

In our attempt to determine the sensitivity of the calculated results with respect to the choice of the amplitude distribution function, an exponential distribution was tried in place of the chi-squared. The exponential distribution gave more weight to signals above the mean than did the chi-squared. However, the final results were very nearly the same. A further calculation involving a delta-function distribution (all interfering signals of the same strength) showed no significant differences. Thus, one is led to believe that the results obtained are not particularly sensitive to the choice of any reasonable distribution function for the signal strengths.

The performance of the frequency diversity system shows up in a rather unusual manner in the graphs. This is due in part to the way in which we chose to interpret the results, and in part to the fact that in this case the interference is not random but constant. In the narrow-band cases the interference level changes considerably in short periods of time because of the random appearances and disappearances of signals close to the operating frequency. In the broad-band case, the interference observed is the net result of *nearly all* the stations on the band so that the actions of *any one* station have a negligible effect on the interference level at the output of an appropriate broad-band receiving system. Thus, for a given loading, the interference level stays fixed and only the signal strengths of the various stations to which the receiver is "tuned" will be found to vary. Some signals will be sufficiently above the noise to be understood all of the time, while others will be below the noise and will not be heard at all. We have made a rather interesting trade in going from narrow- to broad-band operation. In narrow-band operation, we can copy a strong signal most of the time and a weak signal just part of the time. In broad-band operation, we can copy a strong signal all of the time but a weak signal cannot be copied at all. The reason for the shape of the frequency diversity curves should now be clear, and the nature of the "trade-off" may be evaluated by an examination of Figs. 2–5.

Amateur band operation with broad-band systems will prove to be somewhat different in certain respects. There will be fewer stations with which contact may be established (since the weaker signals which were formerly heard intermittently will now not be heard at all), but once contact is established the conversation can be expected to continue without interruption for a considerable period of time. Since the amateur is not normally concerned with communicating with a specific person, the exchange of some freedom of choice of possible contacts for reliability of communications will probably be welcomed.

In the case of military communications, the problem is more difficult, since specific messages must be transmitted to specific stations. If the signal strengths are weak, the narrow-band approach certainly offers no solution since, as we have seen, the circuit reliability will be poor. The message will have to be repeated over and over again before it is received with any reasonable degree of completeness and accuracy. Thus, under such adverse conditions *we have been forced to lower the data rate* because the necessity for repetition has increased the time required for the transmission of a given message. Broad-band operation under the same adverse conditions will suffer the same fate, but to a lesser degree. The data rate will have to be lowered (this can be done without decreasing the bandwidth) but since the interference level will be fixed at some average value we can lower the rate by just the amount necessary to keep the error rate below the acceptable maximum. With narrow-band operation, practical considerations will no doubt force us to reduce the data rate to a value determined by the *maximum* interference level. Thus, for congested-band operation, broad-band systems appear to offer a more orderly approach to the problem and a potentially higher average traffic volume than narrow-band systems.

Nothing that has been said so far should be construed as meaning that broad-band systems will always give us the traffic volume we would like to have, or feel we must have to support operations. As the congestion becomes worse it will be impossible to avoid reducing the data rate per circuit. The important point here is that the broad-band philosophy *accepts interference as a fact of life* and an attempt is made to do the best that is possible under the circumstances. The narrow-band philosophy essentially denies the existence of interference since there is an implied assumption that the narrow-band signals can be placed in non-overlapping frequency bands and thereby prevent interference. It is perhaps redundant to state that the realities of most practical military situations almost completely destroy the validity of such reasoning.

At this point we shall leave the problem of the radio amateur and turn our attention to other communications areas. We have seen that the operating environment of the amateur is not unique to his service but that in other services, especially the military, conditions in actual practice will quite often degenerate to the congested situation of the amateur service. Under such conditions we have shown the necessity for a statistical approach to the problem. It has been further demonstrated that the efficient use of additional transmission bandwidth does not constitute a "waste" in the basic sense of the word. The policy of "conserving bandwidth" is not based on sound physical principles but is based rather on a very common but still myopic view of communications. Such a policy will, in many situations, conserve only the opportunity to communicate as efficiently as might otherwise be possible. Even worse, this point of view quite often leads to the design of systems which have little or no true military capability because of extreme sensitivity to intentional interference. These and other matters will be discussed in Sections VI and VII in more detail. First, it is necessary to derive some rather simple results from information theory.

IV. Information Theory

Consider the problem of data transmission by electrical means. We transmit pulses over a noisy circuit and the pulses, together with the noise, are received and interpreted. Errors in interpretation of the message occur because of this noise. If the error rate is too high and the transmitter power is fixed, we have traditionally lowered the data rate in order to reduce the errors. This has always worked and the reason given was very simple. A lower data rate means that the pulse lengths can be increased, which in turn allows narrower bandwidths to be used, thereby reducing the amount of noise accepted by the receiver. Thus, it became axiomatic that lower error rates could be obtained only by corresponding decreases in bandwidth and data rate. To almost everyone in the communications art the validity of this axiom was unquestioned since there was a great deal of experience in support and none in contradiction. It remained for Shannon to show that systems could be constructed, in theory at least, which would behave quite differently from what our previous experience would lead us to expect. First of all, he showed that the data rate could be held at a constant value (provided this value were below a certain maximum) and at the same time the error rate could be reduced to arbitrarily small values. As for the general belief that one should always use the minimum possible bandwidth in order to reduce the noise accepted by the receiver, Shannon showed that in the ideal case, with a white-noise background, the system bandwidth should be increased to the point where the accepted noise power is at least equal to the signal power.[3] This new theory presented a radically different picture of the limiting behavior of communications systems.

A very superficial study of Shannon-type systems will now be made in the belief that many readers, who are not specialists in information theory, might find a practical discussion of this topic interesting and perhaps useful. Fig. 6 shows a form of communications system suggested by Shannon's work. The channel has a bandwidth W and average (white) noise power N. The transmitter is limited to an average power P. Consider a white-noise generator having a bandwidth W. We record M different samples of the generator output, each sample having a duration of T seconds. These waveforms are now designated $f_1(t)$, $f_2(t)$, $f_3(t)$, $\cdots$, $f_k(t)$, $\cdots$, $f_M(t)$ and are made available as transmitted symbols, as indicated in the figure. Copies of each of the M waveforms are made and placed at the receiver in the corresponding operator units O_1, O_2, $\cdots$, O_k, $\cdots$, O_M. In operation, one of the M waveforms (say the kth one) is selected for transmission. Waveform $f_k(t)$ plus channel noise is received by each of the operator units. The operator units subtract the waveform stored within each unit from the received signal, square this difference, integrate the square for T seconds (which is the duration of the symbols), and indicate this mean-square value as shown. If T is sufficiently large, each meter (except for the kth one) will with almost unit probability read very nearly a value corresponding to $2P+N$, which is the average power of the difference voltage in each case. The kth meter will give a reading corresponding to very nearly N (again with almost unit probability), since the $f_k(t)$ portion of the received signal is completely removed in the subtraction process and only the channel noise remains. Thus, by noting which meter has the lowest reading we can identify which of the M symbols was transmitted. Of course, because of the channel noise we will make an occasional error and identify the wrong symbol.

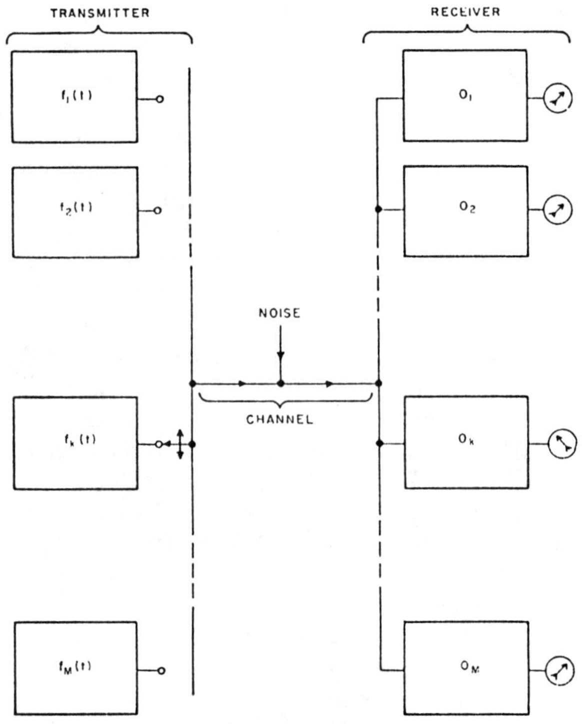

Fig. 6—An ideal communications system.

Before investigating the problem of errors we should examine the relationships between data rate R, symbol duration T, and number of symbols M. Assume that in each T seconds of time the system receives S binary digits (0, 1) to transmit. R will then be S/T bits per second. Since our symbol length is T, we must be prepared to indicate a choice of one out of 2^S possibilities with each symbol transmitted, since this is the number of different sequences of S binary digits. Then clearly

[3] C. E. Shannon, "Communication in the presence of noise," Proc. IRE, vol. 37, pp. 10–21; January, 1949.

$M = 2^S$ and

$$R = \frac{S}{T} = \frac{\log_2 M}{T} \text{ bits per second.} \tag{7}$$

Note that if the symbol length T is increased, the number of symbols M used must increase *exponentially* with T in order to keep a constant data rate. Thus, if T is *doubled*, M will have to be *squared* for the same data rate. In general terms

$$M = A^T \tag{8}$$

and

$$R = \log_2 A. \tag{9}$$

Returning again to Fig. 6, assume the kth symbol has been transmitted. Thus, we look to see if the kth meter gives the lowest reading. If this is so, there is no error. If any one of the other meters gives a lower reading, an error in selection will occur. The probability that any meter will read less than the kth one can be made progressively smaller by increasing T, which increases the integration time in the operator units. However, this is only part of the story. As T is increased to lower the probability of any one meter indicating lower than the kth, the number of such comparisons needed rises according to (8) in order that the data rate remain fixed. Thus, we have two conflicting trends as T is increased. The probability of error per comparison drops, but the number of comparisons necessary to arrive at a selection rises with increasing T. Shannon shows that we can always reduce the over-all probability of error in selection to as small a value as we may choose by letting T become large, *provided* that M does not increase with T faster than

$$M = \left(1 + \frac{P}{N}\right)^{TW}. \tag{10}$$

This maximum permissible rate of increase of M with T determines the maximum data rate which can be supported with an arbitrarily small error rate. This maximum rate is known as the channel capacity C and is obtained by substituting (10) into (7) to obtain

$$C = W \log_2 (1 + P/N). \tag{11}$$

Of course, we do not have to send data at the rate given by (11). We may send slower, and enjoy arbitrarily low, error rates. We may even send faster than C, but then we must accept a certain irreducible error rate.

As remarkable as (11) may be, the engineer concerned with practical system design needs more information than has been given thus far. We now know that multi-symbol systems of the type shown in Fig. 6 are capable, practical considerations aside, of making the most efficient possible use of the communications channel. There are two engineering constraints which must be considered carefully. First, there is an inherent delay of $2T$ seconds involved in data transmission because a T-second length sample of input binary data must be available before choice of transmitted symbol may be made, and another T seconds is required for processing at the receiver before identification may be made. What will be the order of magnitude of this transmission delay? Secondly, how many different symbols M will be required in a given situation? This last consideration is of special importance because it determines, rather directly, system complexity. We might suspect that any attempt to operate at or very near the rate C would require intolerably large T and M since this rate represents a limiting condition. Similarly, large T and M would be expected to result at operating rates lower than C if the error rate is specified at a very small value. What we really need to know is the behavior of T and M for a practical error rate, say 10^{-5}, as the data rate is varied from zero to 100 per cent of capacity. Rice, in an excellent paper,[4] gives us a good indication of the orders of magnitude involved. Rice assumed an SNR of 10 and an error rate of 10^{-5}. He then determined the number of bits per symbol S which would be necessary for various values of the ratio of actual data rate to channel capacity. The results are shown plotted in Fig. 7. Notice

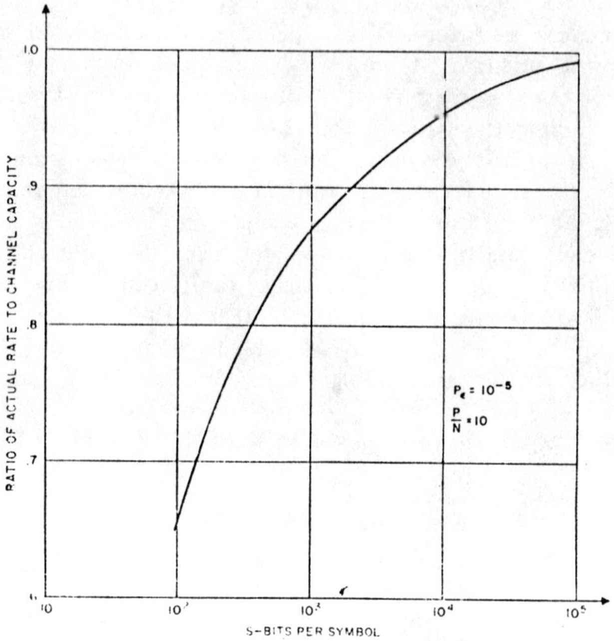

Fig. 7—Curve from Rice showing approach to capacity.

that the numbers S of bits per symbol are quite large, and keep in mind that the number of symbols M is 2^S. We need no numerical examples to conclude that the number of symbols needed will be fantastically large and that it is completely impractical to attempt to build systems which operate at rates close to the Shannon capacity under the conditions assumed above. (An

[4] S. O. Rice, "Communication in the presence of noise," *Bell Sys. Tech. J.*, vol. 29, pp. 60–93; January, 1950.

interesting piece of work by Stutt[5] shows that the situation is not quite so unreasonable if the SNR is low and the symbol waveforms are chosen systematically rather than at random.)

In brief retrospect, we (as communications engineers) have been shown by Shannon that there is an upper limit to what we can do no matter how hard we may try or how ingenious we may be. That it may be extremely difficult to achieve or even approach this upper limit in practice can hardly be blamed on Shannon. He has located the top of our mountain; the problem of reaching the peak is ours, not his.

V. A Practical System of High Efficiency

It is quite clear that any analysis of a communications problem which uses the capacity formula without careful qualification may give results of doubtful practical value. If a system of high efficiency and of reasonable complexity could be found, perhaps problem analysis could be carried out with results which would be significant in practice. Consider once more the system of Fig. 6, but now let there be only two symbols used, $f_1(t)$ and $f_2(t)$. Shannon's idea of using noise-like symbols is quite intriguing. This will be retained except that $f_2(t)$ will be the negative of $f_1(t)$ instead of being chosen at random as before. Thus, $f_1(t)$ is now transmitted for mark (or binary 1) and $-f_1(t)$ for space (or binary 0). For obvious reasons we shall refer to this two-symbol system as the binary system.

In the analysis of this binary system it is convenient to recall one form of the sampling theorem which states that a time function of T-seconds duration and of W-cycles bandwidth is completely specified by $2TW$ equally-spaced sample values of the funtion. Thus, we will represent the function $f_1(t)$ by the sequence of numbers $\{x_1, x_2, \cdots, x_{2TW}\}$, which are the values of the function at the sampling times. The function $f_1(t)$ will be noise-like except that we shall adjust the function so that we obtain the exact relationship

$$\frac{1}{2TW} \sum_{1}^{2TW} x_j^2 = P, \tag{12}$$

where P is the average transmitter power. In a like manner the channel noise, which has an average power N, will be represented by the sequence of numbers $\{n_1, n_2, \cdots, n_{2TW}\}$, where the n_j are independent normal variables with zero mean and variance N. If one performs the operations described for Fig. 6 one obtains the following for the probability of error P_ϵ:

$$P_\epsilon = \text{Prob.}\left[\frac{1}{2TW} \sum_{1}^{2TW} x_j n_j < -P\right]. \tag{13}$$

The summation term may be shown to be Gaussian with zero mean and variance $PN/2TW$. If operating conditions yield a low error probability, then

$$P_\epsilon = \frac{e^{-\gamma}}{2\sqrt{\pi}(\gamma)^{1/2}}, \tag{14}$$

where

$$\gamma = \frac{P}{N} TW. \tag{15}$$

A plot of $\log_{10} P_\epsilon$ as a function of γ is shown in Fig. 8.

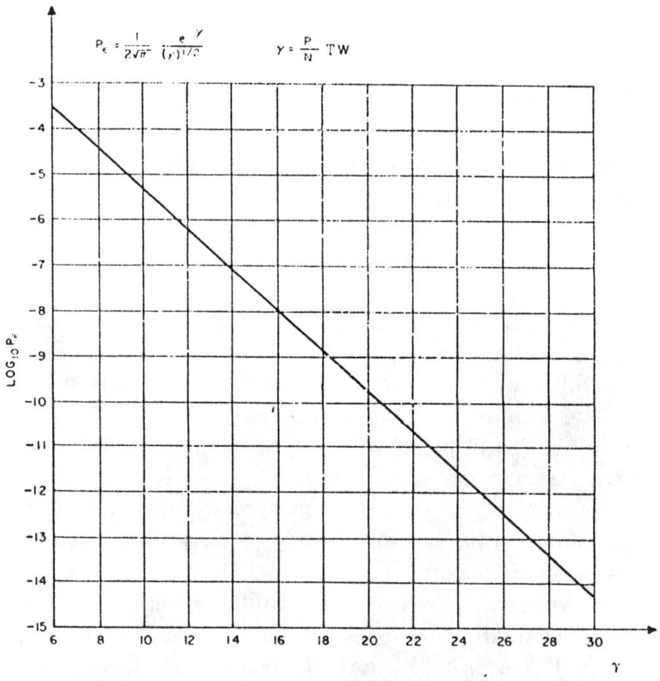

Fig. 8—Plot of $\log_{10} P_\epsilon$ vs γ for the binary system.

Note that once the system error probability is fixed, the relationship between SNR, bandwidth, and data rate ($1/T$ bits per second) is immediately determined.

We might inquire now as to how good our binary system is. It is certainly as good as any two-symbol system can be. Better results can be obtained only by increasing the number of symbols. The gain in doing this, however, does not generally appear to be worth the effort. For example, Stutt[5] shows that for a P/N of 1/10 and error probabilities in the neighborhood of 10^{-4} to 10^{-6}, the most efficient symbol choice requires the use of about 100 symbols in order to increase the data rate over binary by a factor of five. Note, however, that at a fixed error rate the data rate of the binary system may be made 5 times as large by inceasing transmitter power by 7 db. Thus, we must evaluate the relative costs of a 7-db transmitter power increase vs the increase in symbols from two (actually one in terms of equipment complexity) to 100. We must conclude, therefore, that our binary system performance represents about the best that can be done in practice. Better results may be obtained by using more symbols but the rate of improve-

[5] C. A. Stutt, "Regular Polyhedron Codes," Research Laboratory, General Electric Co., Schenectady, N. Y., Tech. Rept. No. 59-RL-2202; February, 1959.

ment will generally prove to be low.[6]

This places us in a good position to derive expressions for the channel capacity in the practical case from (14) and (15). Before doing so we should understand that for high SNR's these equations may yield rates in bits per second far in excess of the bandwidth in cycles per second. There is a mathematical limitation which prevents this. This limitation will not be discussed except to mention the fact that in theory the binary system is limited to a maximum rate $2W$ regardless of SNR.[7] In practice it is usually quite difficult to achieve even W as a rate; we shall choose this as our limiting value.

Since the rate R is $1/T$, the channel capacity C_p in the practical case may be obtained from (15) as

$$C_p = \frac{W}{\gamma} \frac{P}{N}, \tag{16}$$

$$C_p \leq W, \tag{16a}$$

where γ is fixed by the desired error probability according to (14). For an error probability of 10^{-5}, γ is approximately 9.2.

Admittedly, the result (16) is not as elegant as (11). Keep in mind, however, that the concise nature of the capacity formula (11) is made possible by a limiting process in mathematics which cannot be duplicated in practice. A valid objection could also be raised to the application of the term "capacity" to the rate indicated by (15) and expressed in (16). From purely theoretical considerations, such an objection is certainly justified. From a practical point of view, (16) does, in a sense, qualify as a capacity since the performance indicated may only be approached by the most efficient use of modulation and processing techniques. It is quite doubtful that there exist at present any operating systems which perform as well as (16) indicates is obtainable. The main point to remember is that for many years to come (16) will represent a sensible, realizable (but not easily realizable) design goal for the communications engineer; the capacity formula (11) can never serve this purpose. As processing and storage techniques improve, it is to be expected that at some future time multisymbol systems may be built whose performance will exceed that indicated by (16). This does not in any way lessen the utility of (16) as a frame of reference.

VI. Jamming

From the work of the previous section, we now derive some rather simple results which are well-known to information theory specialists the world over, but whose significance is apparently not appreciated by many engineers, at least in this country.

Consider first the performance of the binary system in a white-noise background having a density of n_0 watts per cycle. The effective noise power N will then be

$$N = n_0 W, \tag{17}$$

and (15) will read in this case

$$\gamma = \frac{PT}{n_0} = \frac{E}{n_0}, \tag{18}$$

where E is the energy per transmitted symbol. We now have derived the well-known result that for binary systems of this type operating against flat channel noise, the error probability is independent of bandwidth and is a function only of energy per symbol and noise power density. Thus, for fixed average signal power and fixed data rate, the error probability does not change as the system bandwidth is increased. It is clear that as the bandwidth increases, the noise power accepted by the receiver increases, and for large bandwidths the received noise power becomes quite large compared to received signal power. Thus, systems of this type can operate with SNR's far below unity, or put another way, these systems can be made to operate satisfactorily in the presence of very large amounts of noise power. One might begin to suspect that a broad-band system would be fairly immune to intentional jamming, since in normal operation it is contending (satisfactorily) with such large amounts of natural noise that the additional noise contributed by the jammer would be insignificant by comparison. That this is precisely the case will be made more definite in what follows.

Consider a binary communications system designed to operate in a white-noise background of power density n_0 watts per cycle. Let practical considerations demand that the error probability be kept at or below a critical value P_{e0} corresponding to a γ value of γ_0. Then the channel capacity will be from (16):

$$C_p = \frac{W}{\gamma_0} \frac{P}{n_0 W} \tag{19}$$

as far as natural noise is concerned. For the sake of argument, we shall choose to operate at a data rate R corresponding to one-half capacity. Then,

$$R = \frac{1}{T} = \frac{C_p}{2} = \frac{P}{2\gamma_0 n_0}. \tag{20}$$

Consider now the appearance of a jamming signal of average power J in the channel and let us investigate the effect of J on γ, since this factor must be kept above the assumed critical value of γ_0. The noise term N in (15) will now be

$$N = n_0 W + J, \tag{21}$$

and when (20) and (21) are substituted into (15) we obtain

$$\gamma = \gamma_0 \frac{2n_0}{n_0 + J/W}. \tag{22}$$

[6] This, like all generalizations, will have exceptions. One can conceive of situations in which the multisymbol system would have sensible application. In such cases the work of Stutt, *ibid.*, should prove quite useful.

[7] See discussion of sampling theorem which precedes (12).

This last equation tells a very interesting story. First of all, note the appearance of the J/W term in the denominator. This indicates that the effectiveness of a given average jamming power varies *inversely* as system bandwidth. The broader the bandwidth the less effective will be the resultant jamming. In particular, in order to disrupt the circuit ($\gamma = \gamma_0$) one needs an amount of average jamming power J_0 equal to

$$J_0 = n_0 W \tag{23}$$

under the conditions specified. Thus, the relationships between bandwidth and jamming power become quite clear and may be summarized as follows: *If the most efficient system design is assumed for a fixed data rate in each case, the necessary power required to jam the ciruit varies in direct proportion to system bandwidth. The broader the bandwidth the more difficult it will be to jam the circuit. Conversely, the narrower the bandwidth the easier it becomes to jam the circuit.*

It should be quite clear that if intentional jamming is a consideration, one must of necessity choose a broad-band technique. The narrow-band approach can only lead to eventual disaster.

VII. The Question of Channels

The well-known, but not necessarily sufficiently appreciated, relationship between jamming immunity and system bandwidth discussed above leads to a natural concern over loss of channels if broad-band techniques are employed, as obviously they must be in many applications. It is the purpose of this section to discuss the general problem of "channels" somewhat more thoroughly than before, through use of the practical channel capacity formula (16).

Consider the following problem. Communications service must be provided which requires that a total of K stations be permited to transmit messages at *any* time. Let α be the average fraction of time each station is active. The average signal strength (power) at a particular receiver will be denoted by $\overline{P}$ and it is assumed that Ω cycles of total bandwidth are allocated to this service. Background or natural noise will be ignored. Thus:

- Ω = Total bandwidth allocated to service.
- K = Number of stations, each of which must be permitted to transmit at any time.
- α = Average fraction of time each station is actually transmitting.
- $\overline{P}$ = Mean signal power at a receiving site (one station).
- C_{pN}, C_{pB} = Practical channel capacity per circuit in narrow- and broad-band operation, respectively.

We now wish to inquire as to the relative merits of narrow- and broad-band techniques for this service.

First let us assume an environment in which *all* stations are under the complete control of a central authority. Under this special condition, frequency division will result in circuit bandwidths of Ω/K and, since there will be no interference and background noise is ignored, the capacity per circuit will be, using (16a),

$$C_{pN} = \frac{\Omega}{K}\cdot \tag{24}$$

By comparison, the broad-band approach would yield a circuit bandwidth of Ω and a noise power N of $\alpha K\overline{P}$, which, if an average strength signal were being received, would result in a capacity per circuit of

$$C_{pB} = \frac{\Omega}{\gamma\alpha K}, \tag{25}$$

using (16). Comparing (25) and (24) we see that if such a well-disciplined environment can be found, the narrow-band system would be superior *provided* that the duty cycle factor α is kept high. For example, if $\alpha = 1$ (each station transmitting continuously) the narrow-band system appears to offer about a ten-to-one data rate advantage (for $\gamma = 10$). If, due to operational considerations, the average duty cycle is low (say, 10 per cent or even 1 per cent or less as may quite often be the case), then the broad-band system, even under such ideal conditions, becomes superior.

The reasons for this are quite clear. If the duty cycle is low, the narrow-band system wastes spectrum since most of the allocated channels in Ω will be idle at any time. This cannot be avoided since each of the K stations must have access to communications at any time. The broad-band system takes immediate advantage of a low-duty cycle since this keeps the "noise" level at low values and increases the per-circuit capacity. The narrow-band approach guarantees complete elimination of interference between stations (orthogonality, as the specialist would say), while in the broad-band case each station appears as noise to the others. Thus, at high duty cycles the narrow-band system is superior because it avoids this "noise" problem completely. We must conclude then that the narrow-band system has sensible application under very special conditions (such as in radio broadcasting), but that even where complete control of all transmitters is possible, the broad-band system can easily prove to be the more efficient user of spectrum.

We shall now consider the same communications service problem as before, except that we shall abandon any hope of a disciplined use of the bandwidth Ω. In most military applications, a congested band assumption is much more realistic for several reasons. Certainly two opposing military forces will have planned their spectrum usage independently. Under such conditions interference will be the expected rather than the unusual event. If narrow-band systems have been chosen, it is quite likely that each operator will shift frequency when severe interference is encountered in an attempt to maintain service. This is only the natural and sensible thing to do. Furthermore, under conditions where signals propagate over distances of many thousands of miles, interference will no doubt be quite common even between stations that are a part of the same military force. It seems unrealistic to expect that interference can be prevented by administrative means when the total number of users is large and when the geographic distances between groups of users is great. It must be presumed then that, in spite of careful allocation attempts, the narrow-band approach will not prevent interference and that congested operating conditions will certainly prevail.

Consider the problem that an operator faces when trying to clear messages in a congested band using narrow-band systems. As we have shown, the SNR in such a case is a statistical quantity varying from very good at one time to hopelessly poor minutes or even seconds later. If the data rate is set too high (based on those times when the SNR is good), much of the traffic will be lost and repetition will be necessary. In order to know what messages or parts of messages were lost, a return link is required, but this will also suffer from interference. Such operation is quite inefficient and it would soon be discovered that the data rate would have to be determined by the *least favorable* SNR anticipated during the operating period. Thus, the per-circuit channel capacity in this case may be approximated roughly by

$$C_{pN} = \left(\frac{P}{N}\right)_{\min} \frac{\Omega}{\gamma K}, \tag{26}$$

using (16).

The assumption of congestion does not alter the performance of the broad-band system so that (25) still holds. A rough estimate of the relative performance of these two approaches to the problem of congested band operation may now be obtained by taking the ratio of (25) to (26):

$$\frac{C_{pB}}{C_{pN}} = \frac{1}{\alpha(P/N)_{\min}}. \tag{27}$$

As rough as this approximation may be it still seems rather certain that in a congested band the broad-band system will normally far outperform the narrow-band system.

Eq. (25) may then be taken as the average capacity per circuit of a congested band. In slightly modified form we have:

$$C_{pB} = \frac{1}{\gamma k} \text{ bits per second}, \tag{28}$$

where k is the average number of actual users per cycle of bandwidth and γ is determined by the required error probability according to (14).

It might be mentioned at this point that with broad-band operation the DSB and SSB methods of modulation give identical results for the same transmission bandwidth. As a practical matter, the DSB system offers a two-to-one increase in transmission bandwidth in the modulation process over and above the bandwidth increase obtained by coding processes at baseband. This may sound strange to engineers accustomed to design work aimed at conserving bandwidth. It is still true that there are practical difficulties involved in designing equipment which uses more bandwidth *efficiently* and that the bandwidth doubling which may be obtained in the modulation process with DSB will prove quite helpful in general.

VIII. Conclusions

Since the invention, many years ago, of the frequency-selective filter, it has been common practice to share the inherent capacity of the RF spectrum among users on the basis of frequency allocations. As the number of users increased, methods were found for reducing transmission bandwidths so that new services could be accommodated in the existing spectrum. Extrapolating the past into the future has led to the natural attempt to continue this evolutionary process of seeking methods for the further narrowing of transmission bandwidths, thus providing service for the increasing user population.

This philosophy of spectrum usage is based on a particular course of development which the radio art happened to take, rather than on any fundamental physical principles. The inherent communication capacity of the spectrum can be shared in ways other than by frequency allocation and for many applications the frequency division approach represents a very poor choice indeed. In the field of military communications in particular, the tendency to follow the trends of the past quite often leads to systems having negligible military capability although good intentions may be to the contrary.

This is not to say that broad-band systems have been completely ignored in the past. It could safely be said, however, that the magnitude of the effort thus far expended on the broad-band approach is far out of proportion to the importance of this technique.

Acknowledgment

The author wishes to express appreciation to his many colleagues at the General Electric Company who assisted in the work upon which this paper is based. Special thanks are due to J. C. Kovarik who contributed much original material for the congested-band analysis and to S. Applebaum and Dr. H. D. Friedman for their constructive discussions of information theory problems.

Cooperative Broadcasting

PATRICK P. BERGMANS, MEMBER, IEEE, AND THOMAS M. COVER, FELLOW, IEEE

Abstract—**This paper shows that several transmitters operating in an additive white Gaussian noise environment can send at rates strictly dominating time-multiplex and frequency-multiplex rates by use of a superposition scheme that pools the time, bandwidth, and power allocations of the transmitters. This pooling can be achieved without cooperative action, except for agreement on the actual rate of transmission each transmitter will allow itself. The superposition scheme involves subtraction from the received signal of the estimated signals sent by the other transmitters, followed by decoding of the intended signal. This scheme has been shown to be optimal. We conclude that present methods of allocating different frequency bands to different transmitters are necessarily suboptimal.**

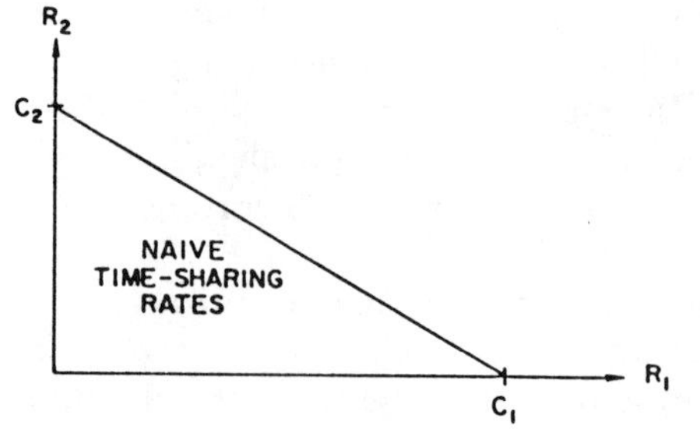

Fig. 1. Rates achievable by naive time sharing.

I. Introduction

CONSIDER two radio transmitters with total allotted power P and total available bandwidth W. Suppose that transmitter i, $i = 1,2$, is transmitting to receiver i in the presence of additive white Gaussian noise of one-sided power spectral density N_i, $N_1 \leq N_2$. The capacity of each channel operating *alone* (utilizing all of the available power P and the total bandwidth W) is then given by (Shannon [1], [2])

$$C_i = W \ln\left(1 + \frac{P}{N_i W}\right), \qquad \text{nats/s, } i = 1,2. \tag{1}$$

However, if one transmitter uses all the power and all the bandwidth, the other channel is being used at zero rate. We ask what set of rates (R_1,R_2) are simultaneously achievable. Clearly, the previous comments imply that $(C_1,0)$ and $(0,C_2)$ are achievable.

The first logical candidate for a scheme of channel sharing is that of time sharing. That is, channel 1 is used at full power P and full bandwidth W, a proportion τ_1 of the time, and channel 2 a proportion $\tau_2 = 1 - \tau_1$ of the time. This allows any rate pair

$$R_1 = \tau_1 C_1$$
$$R_2 = \tau_2 C_2 \tag{2}$$

to be achieved, as shown in Fig. 1. We shall refer to this as the naive time-sharing scheme, as opposed to the variable-power time-sharing scheme we will discuss in Section IV.

The second logical approach is the standard frequency division or band allocation approach. Here, let transmitter i operate with power P_i in a band of width W_i, $P_1 + P_2 = P$, $W_1 + W_2 = W$. The set of achievable rates is then given by

$$R_1 = W_1 \ln\left(1 + \frac{P_1}{N_1 W_1}\right)$$
$$R_2 = W_2 \ln\left(1 + \frac{P_2}{N_2 W_2}\right) \tag{3}$$

where P_i, W_i vary within the constraints just given. Fig. 2 depicts the set of (R_1,R_2) generated for a fixed frequency allocation W_1, W_2 as the power $P_1 = P - P_2$ is allowed to vary in $0 \leq P_1 \leq P$. As will be proved in Section II, allocation of power proportional to bandwidth achieves a point on the naive time-sharing line. Also, a certain range of power allocations dominates naive time sharing. Finally, the envelope of these curves, as $W_1 = W - W_2$ varies over $0 < W_1 < W$, strictly dominates the naive time-sharing bound as shown in Fig. 3. We show this in Section IV.

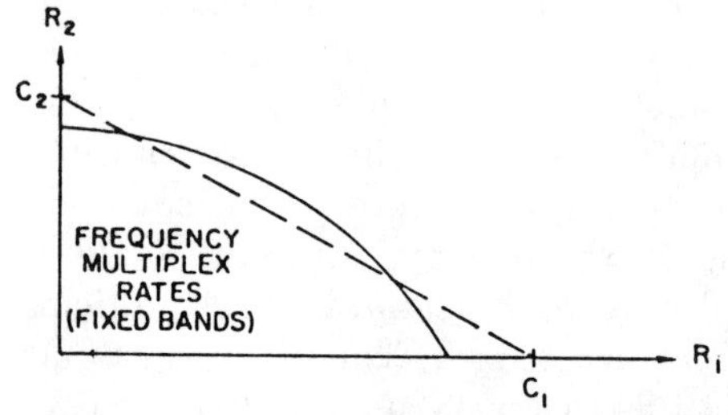

Fig. 2. Rates achievable for fixed frequency division and variable power allocation.

We now turn our attention to superposition schemes in which the transmitters make independent use of the entire band allocated for both.

The obvious decoding scheme, which we shall call naive superposition, decodes each signal separately as if the other signals were noise with respect to it. This yields rates

$$\tilde{R}_1 = W \ln\left(1 + \frac{P_1}{WN_1 + P_2}\right)$$
$$\tilde{R}_2 = W \ln\left(1 + \frac{P_2}{WN_2 + P_1}\right) \tag{4}$$

Manuscript received February 26, 1973; revised December 5, 1973. This work was supported in part by Stanford University, Stanford, Calif., under Contract AF-49(638)1517 and the National Science Foundation under Grants GK-34363 and GK-32240. Portions of this work were done at the Massachusetts Institute of Technology, Cambridge, Mass., and Harvard University, Cambridge, Mass.
P. P. Bergmans is with Cornell University, Ithaca, N.Y. 14850.
T. M. Cover is with Stanford University, Stanford, Calif. 94305.

Reprinted from *IEEE Trans. Inform. Theory*, vol. IT-20, no. 3, pp. 317–324, May 1974.

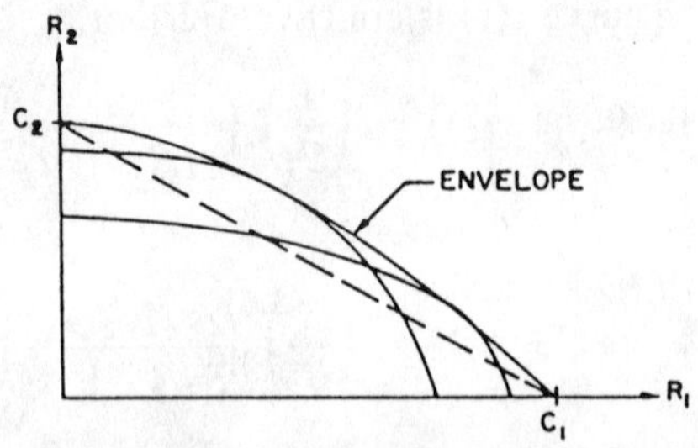

Fig. 3. Rates achievable by frequency division and variable power time sharing.

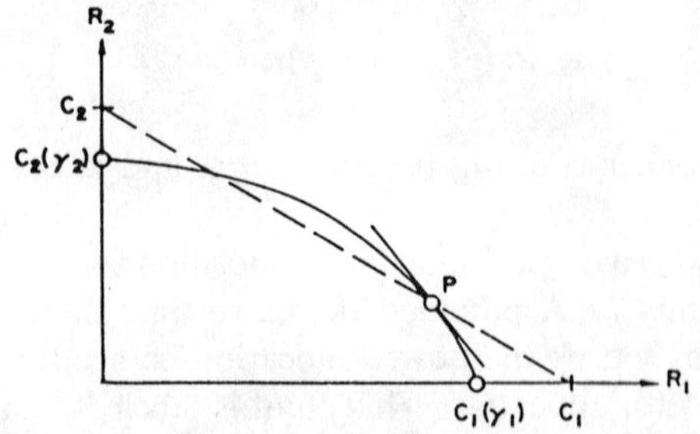

Fig. 4. Dominance of frequency division over naive time division.

as achievable for power allocation (P_1,P_2), $P_1 + P_2 = P$. See, for example, Stefanyuk [3], [4] and Costas [5]. However, this decoding scheme is strictly dominated by a scheme that subtracts out estimates of the other signal. Thus we are led to superposition codes achieving rates

$$R_1^* = W \ln\left(1 + \frac{P_1}{WN_1}\right)$$

$$R_2^* = W \ln\left(1 + \frac{P_2}{WN_2 + P_1}\right) \tag{5}$$

for $N_1 < N_2$. Note that (5) dominates (4) in the sense that $R_2^* = \tilde{R}_2$ and $R_1^* > \tilde{R}_1$. The achievability of this rate pair is shown in Cover [6] and is a special case of the subsequent results on continuous degraded channels in Bergmans [7], [8]. The optimality is proved in Bergmans [10].

II. The Achievability of the Superposition Bound

A heuristic motivation of the achievability of (5) is now given. Since $N_1 < N_2$, user 1 can also receive correctly all the information transmitted to user 2. Consequently, user 1 can subtract out that information from the received message, and decode its own information as if no transmission to user 2 was present. This justifies the expression for R_1^*. User 2, however, cannot receive all the information intended for user 1, since $N_2 > N_1$. The power P_1 of the first transmission is only noise to it, and must be added to the noise power WN_2 in its own channel. Hence the expression for R_2^*. A more thorough discussion can be found in [6]–[8].

III. Applications

We envision the use of this idea in several ways. For example, suppose that a transmitter wishes to send a stereo audio signal if the reception is good (i.e., the noise power is N_1), and wishes to send monaural if the reception is poor (i.e., the noise power is $N_2 > N_1$). Then, if the noise power is less than or equal to N_1, stereo information can be received at the rate $R_1^* + R_2^*$ given in (5). If, however, the noise power is between N_1 and N_2, only monaural information will be received—at rate R_2^*. Finally, if the noise power exceeds N_2, no information is received.

Another example is that of TV information transmission, in which black-and-white information can be sent at rate R_2^*, and color can be sent at the rate $R_1^* + R_2^*$, in the noise regions where $N_1 \leq N \leq N_2$ and $N \leq N_1$, respectively.

As a final example, suppose that a deep-space probe must transmit all of its information before it plunges into some foreign planet. Suppose also that there is some probability π_1 that the additive Gaussian noise will be low ($N = N_1$) and some probability $\pi_2 = 1 - \pi_1$ that the noise will be high ($N = N_2 > N_1$). Then (5) implies that an expected rate $\pi_1 R_1^* + R_2^*$ can be achieved.

In all three examples just cited, these rates are strictly greater than the theoretically achievable rates achievable by frequency band allocation.

IV. Frequency Multiplexing

In this section, we formalize the results on frequency multiplexing presented in the introduction. Let $\gamma_i = W_i/W$, $i = 1,2$, be the fraction of the total bandwidth allocated to each transmitter, and $\alpha_i = P_i/P$, $i = 1,2$, the fraction of the total power. Let $\boldsymbol{\alpha} = (\alpha_1,\alpha_2)$ and $\boldsymbol{\gamma} = (\gamma_1,\gamma_2)$ be two points in $\mathscr{S}_2 = \{(s_1,s_2): s_i \geq 0, s_1 + s_2 = 1\}$. All rate points of the form

$$R_1 = \gamma_1 W \ln\left(1 + \frac{\alpha_1 P}{\gamma_1 W N_1}\right)$$

$$R_2 = \gamma_2 W \ln\left(1 + \frac{\alpha_2 P}{\gamma_2 W N_2}\right) \tag{6}$$

can be achieved by simple frequency multiplexing. Without loss of generality, we can set P and W equal to 1, to simplify the notation. This is really equivalent to substituting R_i for R_i/W and N_i for N_iW/P. The capacity C_i of each channel is now given by $C_i = \ln(1 + 1/N_i)$, and (6) becomes

$$R_1 = \gamma_1 \ln\left(1 + \frac{\alpha_1}{\gamma_1 N_1}\right) = R_1(\gamma_1,\alpha_1)$$

$$R_2 = \gamma_2 \ln\left(1 + \frac{\alpha_2}{\gamma_2 N_2}\right) = R_2(\gamma_2,\alpha_2) \tag{7}$$

with

$$\boldsymbol{\alpha},\boldsymbol{\gamma} \in \mathscr{S}_2 = \{(s_1,s_2): s_i \geq 0, s_1 + s_2 = 1\}.$$

By varying the power distribution between transmitters 1 and 2, we generate a curve $C(\gamma)$ as illustrated in Fig. 4. The curve intersects the axes at the points $(C_1(\gamma_1),0)$ and $(0,C_2(\gamma_2))$, corresponding to the cases where, with the given bandwidth partition, the full power is allocated to transmitter 1 or transmitter 2. The "reduced-bandwidth" capacities $C_1(\gamma_1)$ and $C_2(\gamma_2)$ are defined by

$$C_1(\gamma_1) = R_1(\gamma_1,1) = \gamma_1 \ln\left(1 + \frac{1}{\gamma_1 N_1}\right)$$

$$C_2(\gamma_2) = R_2(\gamma_2,1) = \gamma_2 \ln\left(1 + \frac{1}{\gamma_2 N_2}\right). \tag{8}$$

$C_i(\gamma_i)$ is an increasing function of γ_i, and is equal to C_i for $\gamma_i = 1$.

We now prove the following propositions.

Proposition 1: A point on the naive time-sharing line is achieved by letting the power allocation be proportional to the bandwidth allocation. This point is labeled P on Fig. 4.

Proposition 2: Points above the naive time-sharing line are achieved for a certain range of power allocations.

To prove Propositions 1 and 2, we define

$$e = \frac{R_1}{C_1} + \frac{R_2}{C_2} - 1. \tag{9}$$

The quantity e is proportional to the distance from a given rate point (R_1,R_2) to the time-sharing line; e is positive for points above the time-sharing line, and negative for points below it.

Let $\alpha = \gamma$. Thus

$$\begin{aligned} e &= \frac{R_1(\gamma_1,\gamma_1)}{C_1} + \frac{R_2(\gamma_2,\gamma_2)}{C_2} - 1 \\ &= \frac{\gamma_1 \ln\left(1 + \frac{1}{N_1}\right)}{\ln\left(1 + \frac{1}{N_1}\right)} + \frac{\gamma_2 \ln\left(1 + \frac{1}{N_2}\right)}{\ln\left(1 + \frac{1}{N_2}\right)} - 1 \\ &= \gamma_1 + \gamma_2 - 1 = 0. \end{aligned} \tag{10}$$

Hence, the point $(R_1(\gamma_1,\gamma_1),R_2(\gamma_2,\gamma_2))$ is on the naive time-sharing line.

To prove that there are points represented by (7) above the naive time-sharing line, we first prove that $e(\gamma)$ is concave. We then differentiate e with respect to α_1 and α_2, subject to the constraint $\alpha_1 + \alpha_2 = 1$. We shall find that the distance e is not maximized for the rate point achieving time-sharing rates, and that there must be a range of the α giving points above the naive time-sharing line.

We have

$$\frac{dR_1}{d\alpha_1} = \frac{\gamma_1}{1 + \alpha_1/\gamma_1 N_1} \cdot \frac{1}{\gamma_1 N_1} = \frac{\gamma_1}{\gamma_1 N_1 + \alpha_1}$$

$$\frac{dR_2}{d\alpha_1} = \frac{-\gamma_2}{\gamma_2 N_2 + (1 - \alpha_1)}$$

$$\frac{dR_2}{dR_1} = -\frac{\gamma_1}{\gamma_2} \cdot \frac{\gamma_1 N_1 + \alpha_1}{\gamma_2 N_2 + (1 - \alpha_1)}$$

$$\begin{aligned} \frac{d^2 R_2}{dR_1^2} &= \frac{d}{d\alpha_1}\left(\frac{dR_2}{dR_1}\right) \cdot \frac{d\alpha_1}{dR_1} \\ &= -\frac{\gamma_2}{\gamma_1} \cdot \frac{\gamma_2 N_2 + (1 - \alpha_1) + (\gamma_1 N_1 + \alpha_1)}{(\gamma_2 N_2 + (1 - \alpha_1))^2} \\ &\quad \cdot \frac{\gamma_1 N_1 + \alpha_1}{\gamma_1} \\ &= -\frac{\gamma_2}{\gamma_1} \cdot \frac{\gamma_2 N_2 + \gamma_1 N_1 + 1}{(\gamma_2 N_2 + (1 - \alpha_1))^2} \cdot \frac{\gamma_1 N_1 + \alpha_1}{\gamma_1} \\ &< 0. \end{aligned} \tag{11}$$

Therefore, the curve $e(\gamma)$ is concave. Further,

$$\frac{\partial}{\partial \alpha_1}[e + \lambda(\alpha_1 + \alpha_2)] = \left(\frac{1}{C_1}\right)\left(\frac{\gamma_1}{\gamma_1 N_1 + \alpha_1}\right) + \lambda = 0. \tag{12}$$

$$\frac{\partial}{\partial \alpha_2}[e + \lambda(\alpha_1 + \alpha_2)] = \left(\frac{1}{C_2}\right)\left(\frac{\gamma_2}{\gamma_2 N_2 + \alpha_2}\right) + \lambda = 0. \tag{13}$$

Equating (12) and (13), we find

$$\gamma_2(\gamma_1 N_1 + \alpha_1) \ln\left(1 + \frac{1}{N_1}\right) = \gamma_1(\gamma_2 N_2 + \alpha_2) \ln\left(1 + \frac{1}{N_2}\right). \tag{14}$$

If (14) were satisfied for $\alpha_1 = \gamma_1$ and $\alpha_2 = \gamma_2$, we would have

$$(N_1 + 1) \ln\left(1 + \frac{1}{N_1}\right) = (N_2 + 1) \ln\left(1 + \frac{1}{N_2}\right) \tag{15}$$

which is true if and only if $N_1 = N_2$. This is in accordance with general conclusions obtained in Bergmans [7] on the impossibility of dominating time sharing when the channels are identical.

In general, (14) will not be satisfied for $\alpha = \gamma$, and the stationary point of e will be reached when e is positive, since $e = 0$ when $\alpha = \gamma$. This, together with the concavity of $e(\gamma)$, allows us to conclude that there is a nontrivial portion of that curve above the time-sharing line.

It is possible to show that

$$\left.\frac{de}{d\alpha_1}\right|_{\alpha_1 = \gamma_1} < 0 \tag{16}$$

and hence that the point corresponding to $\alpha = \gamma$ is the point P on Fig. 4. Rates dominating time-sharing rates will be achieved for

$$\begin{aligned} \delta_1 < \alpha_1 < \gamma_1 \\ 1 - \delta_1 = \delta_2 > \alpha_2 > \gamma_2 \end{aligned} \tag{17}$$

where δ is the second solution of the equation $e(\alpha) = 0$ (for a given γ). In conclusion, we should give the noisier channel a fraction of the total power which is slightly greater than the fraction of the total bandwidth it was allocated. Also, as a consequence of the fact that time sharing cannot be dominated for equal channels, it is evident that the closer the S/N ratios of the two channels, the smaller the interval $[\delta_1,\gamma_1]$ will be.

The curve of Fig. 4 was drawn for a given frequency partition γ. By letting γ vary, we shall generate a continuous set of similar curves. The envelope of all these curves strictly dominates the time-sharing line, since all the curves will have some portion of them above the time-sharing line.

Points on the envelope will be achieved for an optimal proportion of power as a function of bandwidth. There will be some tradeoff between power and bandwidth for rate points below the envelope.

V. Duality of Time and Frequency

The dominance of frequency multiplex over naive time multiplex has been established. However, in the naive approach we fixed the power of transmitter i to be a constant P over a certain proportion τ_i of the available time and let the power be zero for the remaining proportion of the time. It is clear that, for transmitter i, the average power over the *entire* time is actually $\tau_i P$ instead of P. An equivalence of the time and frequency rate curves can be established when the power is calculated with respect to the entire time interval. However, this equivalence is illusory from a practical standpoint, because it can only be achieved by lengthy power deviations from the average power constraints in the time-varying case.

To see this consider the frequency multiplex case, where the set of achievable rates is given by

$$R_1 = \gamma_1 W \ln\left(1 + \frac{\alpha_1 P}{\gamma_1 W N_1}\right)$$

$$R_2 = \gamma_2 W \ln\left(1 + \frac{\alpha_2 P}{\gamma_2 W N_2}\right), \qquad \boldsymbol{\alpha}, \boldsymbol{\gamma} \in \mathscr{S}_2 \tag{18}$$

where $\boldsymbol{\alpha}$ is the power division and $\boldsymbol{\gamma}$ is the frequency division. Similarly the set of time-sharing rates is given by

$$R_1 = \tau_1 W \ln\left(1 + \frac{P_1}{W N_1}\right)$$

$$P_2 = \tau_2 W \ln\left(1 + \frac{P_2}{W N_2}\right), \qquad \boldsymbol{\tau} \in \mathscr{S}_2 \tag{19}$$

where $\boldsymbol{\tau}$ is the division in time, and P_i is the power of transmission during communication to user i. Clearly $\tau_1 T P_1 + \tau_2 T P_2 = TP$ where T is the total duration of the waveform. For the two transmitters we let the power $P_1(t)$ and $P_2(t)$ be

$$P_1(t) = \begin{cases} P_1, & 0 \le t \le \tau_1 T \\ 0, & \tau_1 T < t \le T \end{cases}$$

and

$$P_2(t) = \begin{cases} 0, & 0 \le t \le \tau_1 T \\ P_2, & \tau_1 T < t \le T. \end{cases} \tag{20}$$

Letting $\tau_i P_i / P = \lambda_i$, $\boldsymbol{\lambda} \in \mathscr{S}_2$, we obtain

$$R_1 = \tau_1 W \ln\left(1 + \frac{\lambda_1 P}{\tau_1 W N_1}\right)$$

$$R_2 = \tau_2 W \ln\left(1 + \frac{\lambda_2 P}{\tau_2 W N_2}\right), \qquad \boldsymbol{\tau}, \boldsymbol{\lambda} \in \mathscr{S}_2 \tag{21}$$

which is, indeed, completely equivalent to the set of rates achievable by frequency multiplex. One major objection is the following. To achieve these rates in a coding sense, the time T must tend to infinity. Although it is still true that over the *total* time T the average power of the transmitter is P, over one of the two intervals the power P_i is larger than P. This may be inadmissible because of power limitations in the transmitter. An obvious solution to this problem would be to divide the total length of time T into n subintervals, and to let $n \to \infty$ as $T \to \infty$.

The argument against this is that the signal which results from this scheme does not have a vanishing power content outside $(-0, W)$ when $T \to \infty$, because of the switching between signals. If there are only two intervals ($\tau_1 T$ and $\tau_2 T$), the fraction of the total power of the resulting signal outside $(-0, W)$ will go to zero when $T \to \infty$, provided the original signals in $\tau_1 T$ and $\tau_2 T$ satisfy this condition.

In conclusion, time sharing is equivalent to frequency multiplex only if very long deviations from the average power P are allowed.

The situation is very different in frequency multiplex, because in this case, the spectrum of the transmitter need not be flat; i.e., the (single sided) power spectral density $P_i(f)$ is given by

$$P_1(f) = \begin{cases} \dfrac{\alpha_1 P}{\gamma_1 W}, & 0 \le f \le \gamma_1 W \\ 0, & \gamma_1 W < f \le W \end{cases}$$

and

$$P_2(f) = \begin{cases} 0, & 0 \le f \le \gamma_1 W \\ \dfrac{\alpha_2 P}{\gamma_2 W}, & \gamma_1 W < f \le W. \end{cases} \tag{22}$$

However, there will not be a *time* discontinuity in power. Since the transmitter is operating in the time domain, this is the important factor. One can consider time-sharing equivalent (dual) to frequency multiplex only if time variations in input power are allowed, just as spectral variations of power are allowed in the frequency multiplex case.

VI. Superposition Coding

With the simplified notation introduced in Section II, the rates achieved by superposition coding are given by

$$R_1^* = \ln\left(1 + \frac{\alpha_1^*}{N_1}\right)$$

$$R_2^* = \ln\left(1 + \frac{\alpha_2^*}{N_2 + \alpha_1^*}\right)$$

$$\boldsymbol{\alpha}^* \in \mathscr{S}_2. \tag{23}$$

It can be easily shown that the curve described by (23) is concave, and that it dominates time sharing. The question, of course, is whether it dominates the envelope of frequency multiplexing curves described in the Section IV. This section establishes this dominance.

Superposition coding will dominate frequency multiplexing if, for any given $\boldsymbol{\alpha}, \boldsymbol{\gamma} \in \mathscr{S}_2$, we can always find an $\boldsymbol{\alpha}^* \in \mathscr{S}_2$ such that

$$\ln\left(1 + \frac{\alpha_1^*}{N_1}\right) \ge \gamma_1 \ln\left(1 + \frac{\alpha_1}{\gamma_1 N_1}\right)$$

$$\ln\left(1 + \frac{\alpha_2^*}{N_2 + \alpha_1^*}\right) \ge \gamma_2 \ln\left(1 + \frac{\alpha_2}{\gamma_2 N_2}\right) \tag{24}$$

or, equivalently

$$\left(1 + \frac{\alpha_1^*}{N_1}\right) \ge \left(1 + \frac{\alpha_1}{\gamma_1 N_1}\right)^{\gamma_1}$$

$$\left(1 + \frac{\alpha_2^*}{N_2 + \alpha_1^*}\right) \ge \left(1 + \frac{\alpha_2}{\gamma_2 N_2}\right)^{\gamma_2}. \tag{25}$$

The proof of inequality (25) is deceptively difficult, and we are grateful to D. Hughes-Hartogs for his contribution of the following argument. We shall need the following lemmas, which we give without their straightforward algebraic verification.

Lemma I:

$$\max_{\gamma \in [0,1]} \left(1 + \frac{a}{\gamma}\right)^{\gamma} = 1 + a, \qquad a > 0.$$

The maximum is achieved for $\gamma = 1$.

Lemma II:

$$\max_{\gamma \in [0,1]} \left(1 + \frac{a}{\gamma}\right)^{\gamma} \left(1 + \frac{b}{1-\gamma}\right)^{1-\gamma} = 1 + a + b, \qquad a, b > 0.$$

The maximum is achieved for $\gamma = a/(a + b)$.

Proposition 3—(Hughes-Hartogs): Given $\alpha, \gamma \in \mathscr{S}_2$, i.e., $\alpha_1 + \alpha_2 = 1; \gamma_1 + \gamma_2 = 1; \alpha_i, \gamma_i \geq 0$; and $N_2 > N_1 > 0$; then there exists an $\alpha^* \in \mathscr{S}_2$ such that

$$\left(1 + \frac{\alpha_1^*}{N_1}\right) \geq \left(1 + \frac{\alpha_1}{\gamma_1 N_1}\right)^{\gamma_1} \tag{26}$$

$$\left(1 + \frac{\alpha_2^*}{N_2 + \alpha_1^*}\right) \geq \left(1 + \frac{\alpha_2}{\gamma_2 N_2}\right)^{\gamma_2}. \tag{27}$$

Proof: Using condition (26), which is then automatically satisfied, set

$$\alpha_1^* = N_1 \left[\left(1 + \frac{\alpha_1}{\gamma_1 N_1}\right)^{\gamma_1} - 1\right]. \tag{28}$$

Since $\alpha_1 \geq 0$, $\gamma_1 \geq 0$, and $N_1 \geq 0$

$$\left(1 + \frac{\alpha_1}{\gamma_1 N_1}\right)^{\gamma_1} \geq 1$$

and

$$\alpha_1^* \geq 0. \tag{29}$$

By Lemma I

$$\alpha_1^* \leq N_1 \left[\left(1 + \frac{\alpha_1}{N_1}\right) - 1\right] = \alpha_1 \leq 1. \tag{30}$$

Let $\alpha_2^* = 1 - \alpha_1^*$, and hence $\alpha^* \in \mathscr{S}_2$.

We now verify condition (27)

$$1 + \frac{\alpha_2^*}{N_2 + \alpha_1^*} = \frac{N_2 + \alpha_1^* + \alpha_2^*}{N_2 + \alpha_1^*}$$

$$= \frac{N_2 + 1}{(N_2 - N_1) + N_1 \left(1 + \dfrac{\alpha_1}{\gamma_1 N_1}\right)^{\gamma_1}}$$

$$= \frac{N_2 + 1}{(N_2 - N_1)\left(1 + \dfrac{\alpha_2}{\gamma_2 N_2}\right)^{\gamma_2} + N_1 \left(1 + \dfrac{\alpha_1}{\gamma_1 N_1}\right)^{\gamma_1} \left(1 + \dfrac{\alpha_2}{\gamma_2 N_2}\right)^{\gamma_2}} \left(1 + \frac{\alpha_2}{\gamma_2 N_2}\right)^{\gamma_2}$$

$$\geq \frac{N_2 + 1}{(N_2 - N_1)\left(1 + \dfrac{\alpha_2}{N_2}\right) + N_1 \left(1 + \dfrac{\alpha_1}{N_1} + \dfrac{\alpha_2}{N_2}\right)} \left(1 + \frac{\alpha_2}{\gamma_2 N_2}\right)^{\gamma_2}. \tag{31}$$

In (31), we invoked both Lemma I and Lemma II and the fact that $N_2 > N_1 > 0$. We now have

$$1 + \frac{\alpha_2^*}{N_2 + \alpha_1^*} \geq \frac{N_2 + 1}{N_2 - N_1 + \alpha_2 - \dfrac{\alpha_2 N_1}{N_2} + N_1 + \alpha_1 + \dfrac{\alpha_2 N_1}{N_2}} \cdot \left(1 + \frac{\alpha_2}{\gamma_2 N_2}\right)^{\gamma_2}$$

$$= \left(1 + \frac{\alpha_2}{\gamma_2 N_2}\right)^{\gamma_2}. \tag{32}$$

Equation (32) proves that condition (27) is also satisfied. The inequalities in the proof are strict for nondegenerate cases, i.e., for $\gamma_i \neq 0$.

A set of $e(\gamma)$ curves, together with the superposition coding curve, is given in Fig. 5.

We now investigate the performance of multiband mixed modes of operation, and show that any such mode of operation is dominated by a superposition coding using the full available bandwidth. The following propositions are proved in the Appendix.

Proposition 4: If a continuous AWGN channel is used to transmit information from a single source, the best performance is achieved by using a single code operating in the full available band, rather than by partitioning the total band in smaller bands and using separate codes in the smaller bands. We shall refer to this last mode of operation as the multiband mode for simple channels. Let

$$\mathscr{S}_n = \left\{(s_1, s_2, \cdots, s_n): s_i \geq 0, \sum_{i=1}^{n} s_i = 1\right\} \tag{33}$$

and $\alpha, \gamma \in \mathscr{S}_n$ be the power and frequency proportion in subband i. The total rate R_T is given by

$$R_T = \sum_{i=1}^{n} \gamma_i \ln \left(1 + \frac{\alpha_i}{\gamma_i N}\right). \tag{34}$$

The Appendix shows that $R_T \leq C = \ln(1 + 1/N)$, with equality iff $\alpha = \gamma$.

Proposition 5: If a continuous AWGN broadcast channel is used to transmit information to two users using superposition codes exclusively, the best performance is achieved

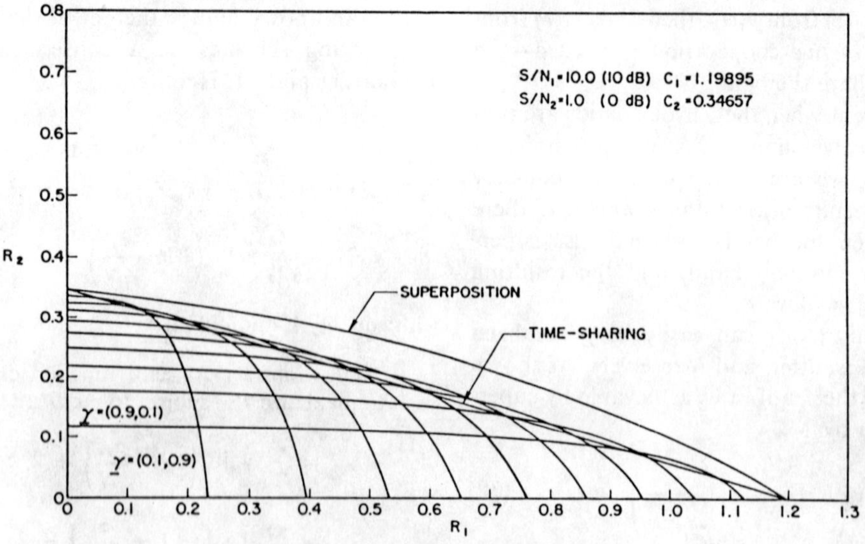

Fig. 5. Cooperative broadcasting.

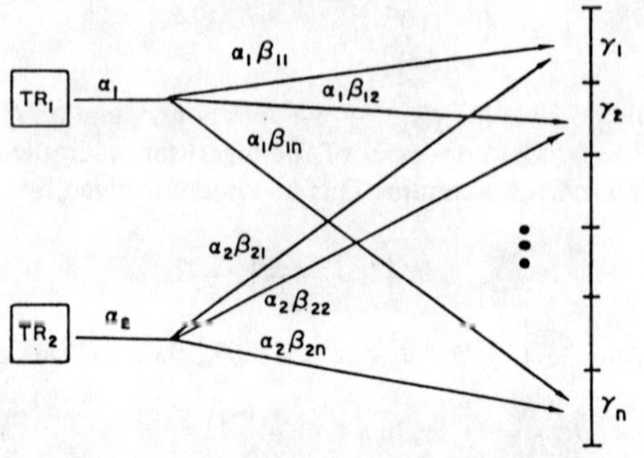

Fig. 6. Arbitrary frequency division.

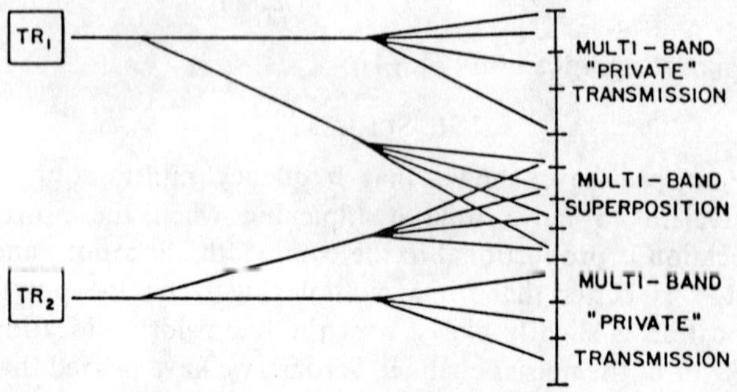

Fig. 7. Mixed multiband channel.

by using a superposition code operating in the full available band, rather than by using separate superposition codes operating in disjoint smaller subbands (multiband superposition operation).

To see this, let $\gamma \in \mathscr{S}_n$ represent the (relative) width of the ith subband, and $\alpha \in \mathscr{S}_2$ be the power repartition between transmitter 1 and transmitter 2. Finally, the repartition of the power for transmitter j ($j = 1,2$) in each subband is done proportionally to $\beta_j \in \mathscr{S}_n$, $j = 1,2$ (Fig. 6). We have

$$R_{1T} = \sum_{i=1}^{n} \gamma_i \ln\left(1 + \frac{\alpha_1\beta_{1i}}{\gamma_i N_1}\right)$$

$$R_{2T} = \sum_{i=1}^{n} \gamma_i \ln\left(1 + \frac{\alpha_2\beta_{2i}}{\gamma_i N_2 + \alpha_1\beta_{1i}}\right). \tag{35}$$

The Appendix shows that there exists an $\alpha^* \in \mathscr{S}_2$ such that

$$R_{1T} \le \ln\left(1 + \frac{\alpha_1^*}{N_1}\right)$$

$$R_{2T} \le \ln\left(1 + \frac{\alpha_2^*}{N_2 + \alpha_1^*}\right) \tag{36}$$

with equality iff $\beta_1 = \beta_2 = \gamma$.

The following procedure describes a method for finding a superposition transmission mode which uses the full available bandwidth and which dominates any multiband mixed mode of operation. In Fig. 7, we represent a mode of operation involving both frequency multiplex (in so-called "private" bands) and superposition (in the common band). Moreover, each mode is multiband, as defined previously.

Step 1: Using Proposition 4, we first replace the various private bands for each transmitter by a single private band.

Step 2: Using Proposition 3, we pool the powers and bandwidths of the two private bands and replace them by one single band with a superposition code over the width of the two private bands.

Step 3: We are now operating with multiband superposition coding, and using Proposition 5, we replace this by a single superposition code using the full available bandwidth.

At each step, the achievable rates have been increased (except in trivial cases). This finally proves the dominance of superposition coding.

For Gaussian channels, the signals can be "mixed in the air." That is

$$s(t) = s_1(t) + s_2(t)$$

$$y_i(t) = s(t) + n_i(t), \qquad i = 1,2. \tag{37}$$

User 1 first determines $s_2(t)$ from $y_1(t)$, then finds $s_1(t)$ from $y_1(t) - s_2(t)$. Thus no *active* cooperation is needed—the channels only have to share the band.

The situation is different when the various bands are preassigned, because of receiver limitations or regulations. If the bands are disjoint, we are in the case of frequency multiplexing, and the results of Section IV apply. If there is some overlap between the bands, we shall use superposition coding in the common band, and the resulting maximization problem is awkward.

The arguments of this paper can easily be generalized to a situation with N transmitters and N receivers. As shown in [8], the boundary of the set of rates achievable by superposition coding is given by

$$R_i^* = \ln\left(1 + \frac{\alpha_i^*}{N_i + \sum_{j<i} \alpha_j^*}\right) \tag{38}$$

for $\alpha^* \in \mathscr{S}_n$, as opposed to

$$\tilde{R}_i = \ln\left(1 + \frac{\alpha_i^*}{N_i + \sum_{j\neq i} \alpha_j^*}\right) \tag{39}$$

for naive superposition [3], [4].

VII. Summary

First, we have shown that frequency multiplexing is equivalent to naive time multiplexing when the power allocation is proportional to the bandwidth allocation, and that it is better than time multiplexing when the power allocation is slightly biased over the bandwidth allocation in favor of the noisier channel. Second, we have proved that superposition coding (or cooperative broadcasting) dominates frequency multiplexing, unless very long deviations from average power are allowed. Finally, we have concluded that cooperative broadcasting strictly dominates any type of mixed-mode transmission, involving both frequency multiplexing (possibly multiband) and superposition coding (possibly multiband). The absolute optimality of the cooperative broadcasting scheme achieving rates as given by (38) is established in Bergmans [10].

Moreover, we have observed that the cooperation needed for superposition codes need not be active—the signals can be independently generated by the two transmitters and "mixed in the air." These results hold for Shannon-type encoding. It is to be hoped that the general result that "superposition dominates frequency multiplexing which in turn dominates time multiplexing" also holds for modulation schemes and that improvements on existing transmission methods may be achieved by using clever modulation schemes with pooled powers and bandwidths.

Acknowledgment

We wish again to thank D. Hughes-Hartogs for his contribution of the inequality of (20). We have also benefited from the suggestions of C. Keilers on schemes for sending information over Gaussian channels. We would also like to thank J. Thorp for helpful suggestions concerning some of the inequalities. We were led to discuss the duality and possible lack thereof of the time and frequency multiplexing schemes as a consequence of remarks by T. Gaarder and M. Hellman.

Appendix

Proposition 4: If $\alpha, \gamma \in \mathscr{S}_n$ then

$$R_T = \sum_{i=1}^{n} \gamma_i \ln\left(1 + \frac{\alpha_i}{\gamma_i N}\right) \leq C = \ln\left(1 + \frac{1}{N}\right) \tag{40}$$

with equality if and only if $\alpha = \gamma$.

Proof: Consider R_T as a function of α for a fixed γ. We wish to maximize R_T subject to $\alpha \in \mathscr{S}_n$. Hence, define

$$J = \sum_{i=1}^{n} \gamma_i \ln\left(1 + \frac{\alpha_i}{\gamma_i N}\right) + \lambda \sum_{i=1}^{n} \alpha_i$$

$$= \sum_{i=1}^{n} \left(\gamma_i \ln\left(1 + \frac{\alpha_i}{\gamma_i N}\right) + \lambda\alpha_i\right)$$

$$\frac{\partial J}{\partial \alpha_i} = \frac{\gamma_i}{1 + \frac{\alpha_i}{\gamma_i N}} \cdot \frac{1}{\gamma_i N} + \lambda = \frac{\gamma_i}{\gamma_i N + \alpha_i} + \lambda = 0. \tag{41}$$

This is only possible if $\gamma_i/(\gamma_i N + \alpha_i)$ does not depend on i, which requires $\alpha = \gamma$. The convexity of the logarithm guarantees that the extremum is a maximum. This maximum is given by

$$\sum_{i=1}^{n} \gamma_i \ln\left(1 + \frac{\gamma_i}{\gamma_i N}\right) = C. \tag{42}$$

Proposition 5: If $\gamma, \beta_1, \beta_2 \in \mathscr{S}_n$, $\alpha \in \mathscr{S}_2$, then the rate point

$$R_{1T} = \sum_{i=1}^{n} \gamma_i \ln\left(1 + \frac{\alpha_1\beta_{1i}}{\gamma_i N_1}\right) \tag{43a}$$

$$R_{2T} = \sum_{i=1}^{n} \gamma_i \ln\left(1 + \frac{\alpha_2\beta_{2i}}{\gamma_i N_2 + \alpha_1\beta_{1i}}\right) \tag{43b}$$

is strictly dominated by the rate point

$$R_1^* = \ln\left(1 + \frac{\alpha_1^*}{N_1}\right) \tag{44a}$$

$$R_2^* = \ln\left(1 + \frac{\alpha_2^*}{N_2 + \alpha_1^*}\right) \tag{44b}$$

for some $\alpha^* \in \mathscr{S}_2$, unless $\beta_1 = \beta_2 = \gamma$, in which case we have equality.

Proof: We equate (43b) and (44b) to solve for α_1^*

$$\left(1 + \frac{\alpha_2^*}{N_2 + \alpha_1^*}\right) = \prod_{i=1}^{n} \left(1 + \frac{\alpha_2\beta_{2i}}{\gamma_i N_2 + \alpha_1\beta_{1i}}\right)^{\gamma_i}$$

$$\alpha_1^* = (N_2 + 1) \prod_{i=1}^{n} \left(1 + \frac{\alpha_2\beta_{2i}}{\gamma_i N_2 + \alpha_1\beta_{1i}}\right)^{-\gamma_i} - N_2. \tag{45}$$

From (43b), and by Proposition 4

$$0 \leq R_{2T} \leq \sum_{i=1}^{n} \gamma_i \ln\left(1 + \frac{\alpha_2\beta_{2i}}{\gamma_i N_2}\right) \leq \ln\left(1 + \frac{1}{N_2}\right) = C_2$$

and, since R_2^* is a continuous function of α_2^*, with $R_2^* = 0$ for $\alpha_2^* = 0$ and $R_2^* = C_2$ for $\alpha_2^* = 1$, equating (43b) and (44b) yields $\alpha^* \in \mathscr{S}_2$. Substituting α_1^* from (45) in exp (R_1^*), yields

$$\exp(R_1^*) = 1 + \frac{\alpha_1^*}{N_1} = 1 - \frac{N_2}{N_1} + \left(\frac{N_2+1}{N_1}\right)\prod_{i=1}^{n}\left(1 + \frac{\alpha_2\beta_{2i}}{\gamma_i N_2 + \alpha_1\beta_{1i}}\right)^{-\gamma_i}. \tag{46}$$

To find a lower bound for $\exp(R_1^*)$, we minimize

$$\prod_{i=1}^{n}\left(1 + \frac{\alpha_2\beta_{2i}}{\gamma_i N_2 + \alpha_1\beta_{1i}}\right)^{-\gamma_i} = \exp\left(\sum_{i=1}^{n}\gamma_i \ln\left(\frac{\gamma_i N_2 + \alpha_1\beta_{1i}}{\gamma_i N_2 + \alpha_1\beta_{1i} + \alpha_2\beta_{2i}}\right)\right)$$

subject to $\sum \beta_{2i} = 1$. Thus, taking the logarithm

$$\frac{\partial}{\partial\beta_{2i}}\sum_{i=1}^{n}\left(\gamma_i \ln\left(\frac{\gamma_i N_2 + \alpha_1\beta_{1i}}{\gamma_i N_2 + \alpha_1\beta_{1i} + \alpha_2\beta_{2i}}\right) + \lambda\beta_{2i}\right)$$

$$= \gamma_i \frac{\gamma_i N_2 + \alpha_1\beta_{1i} + \alpha_2\beta_{2i}}{\gamma_i N_2 + \alpha_1\beta_{1i}} \cdot \frac{-\alpha_2(\gamma_i N_2 + \alpha_1\beta_{1i})}{(\gamma_i N_2 + \alpha_1\beta_{1i} + \alpha_2\beta_{2i})^2} + \lambda = 0$$

or

$$\frac{\gamma_i\alpha_2}{\gamma_i N_2 + \alpha_1\beta_{1i} + \alpha_2\beta_{2i}} = \lambda \tag{47}$$

which requires

$$\beta_{2i} = \frac{\gamma_i - \alpha_1\beta_{1i}}{\alpha_2}. \tag{48}$$

Since $\ln(k_1/(k_2 + x))$ is convex for $x \geq 0$, $\boldsymbol{\beta}_2$ of (48) achieves a minimum, given by

$$\prod_{i=1}^{n}\left(1 + \frac{\alpha_2\beta_{2i}}{\gamma_i N_2 + \alpha_1\beta_{1i}}\right)^{-\gamma_i} \geq \prod_{i=1}^{n}\left(\frac{\gamma_i N_2 + \alpha_1\beta_{1i}}{\gamma_i(N_2+1)}\right)^{\gamma_i} \tag{49}$$

and

$$\exp(R_1^*) \geq 1 - \frac{N_2}{N_1} + \left(\frac{N_2+1}{N_1}\right)\prod_{i=1}^{n}\left(\frac{\gamma_i N_2 + \alpha_1\beta_{1i}}{\gamma_i(N_2+1)}\right)^{\gamma_i} = 1 - \frac{N_2}{N_1} + \frac{N_2}{N_1}\prod_{i=1}^{n}\left(1 + \frac{\alpha_1\beta_{1i}}{\gamma_i N_2}\right)^{\gamma_i}. \tag{50}$$

Finally,

$$\begin{aligned}\exp(R_1^*) - \exp(R_{IT}) &\geq 1 - \frac{N_2}{N_1} + \frac{N_2}{N_1}\prod_{i=1}^{n}\left(1 + \frac{\alpha_1\beta_{1i}}{\gamma_i N_2}\right)^{\gamma_i} - \prod_{i=1}^{n}\left(1 + \frac{\alpha_1\beta_{1i}}{\gamma_i N_1}\right)^{\gamma_i}\\ &\geq \frac{1}{N_1}\left[N_2\left(\prod_{i=1}^{n}\left(1 + \frac{\alpha_1\beta_{1i}}{\gamma_i N_2}\right)^{\gamma_i} - 1\right) - N_1\left(\prod_{i=1}^{n}\left(1 + \frac{\alpha_1\beta_{1i}}{\gamma_i N_1}\right)^{\gamma_i} - 1\right)\right]\\ &= \frac{1}{N_1}[f(N_2) - f(N_1)]\end{aligned} \tag{51}$$

where

$$f(x) \triangleq x\left[\prod_{i=1}^{n}\left(1 + \frac{\alpha_1\beta_{1i}}{\gamma_i x}\right)^{\gamma_i} - 1\right] = \prod_{i=1}^{n}\left(x + \frac{\alpha_1\beta_{1i}}{\gamma_i}\right)^{\gamma_i} - x. \tag{52}$$

We now show that $f(x)$ is strictly increasing in x, unless $\boldsymbol{\beta}_1 = \boldsymbol{\gamma}$, in which case $f(x)$ is constant. We have

$$\begin{aligned}f'(x) &= \sum_{i=1}^{n}\gamma_i\left(x + \frac{\alpha_1\beta_{1i}}{\gamma_i}\right)^{\gamma_i - 1}\prod_{j\neq i}\left(x + \frac{\alpha_1\beta_{1j}}{\gamma_j}\right)^{\gamma_j} - 1\\ &= \prod_{j=1}^{n}\left(x + \frac{\alpha_1\beta_{1j}}{\gamma_j}\right)^{\gamma_j}\sum_{i=1}^{n}\gamma_i\left(x + \frac{\alpha_1\beta_{1i}}{\gamma_i}\right)^{-1} - 1\\ &\geq \prod_{j=1}^{n}\left(x + \frac{\alpha_1\beta_{1j}}{\gamma_j}\right)^{\gamma_j}\prod_{i=1}^{n}\left(\left(x + \frac{\alpha_1\beta_{1i}}{\gamma_i}\right)^{-1}\right)^{\gamma_i} - 1\\ &= \left(\prod_{j=1}^{n}\left(x + \frac{\alpha_1\beta_{1j}}{\gamma_j}\right)^{\gamma_j}\right)\left(\prod_{i=1}^{n}\left(x + \frac{\alpha_1\beta_{1i}}{\gamma_i}\right)^{\gamma_i}\right)^{-1} - 1\\ &= 0.\end{aligned} \tag{53}$$

The inequality in (53) is a consequence of the theorem of the arithmetic and geometric means ([9], pp. 16 et seq.). We shall have equality in (53) iff $(x + \alpha_1\beta_{1i}/\gamma_i)^{-1}$ is not a function of i, i.e., iff $\alpha_1\beta_{1i}$ is proportional to γ_i, which implies $\boldsymbol{\beta}_1 = \boldsymbol{\gamma}$, since $\boldsymbol{\beta}_1, \boldsymbol{\gamma} \in \mathscr{S}_n$. Hence $f(x)$ is increasing in x (unless $\boldsymbol{\beta}_1 = \boldsymbol{\gamma}$), and $f(N_2) \geq f(N_1)$. In conclusion

$$\exp(R_1^*) - \exp(R_{IT}) \geq \frac{1}{N_1}[f(N_2) - f(N_1)] \geq 0 \tag{54}$$

and the proof of the existence of a dominating point is completed. Finally, the inequalities in this proof are strict, unless $\boldsymbol{\gamma} = \boldsymbol{\beta}_1$ and (48) are satisfied, in which case $\boldsymbol{\gamma} = \boldsymbol{\beta}_1 = \boldsymbol{\beta}_2$.

References

[1] C. E. Shannon, "A mathematical theory of communication," *Bell Syst. Tech. J.*, vol. 27, pp. 379–423, 623–656, 1948; also, reprinted in C. E. Shannon and W. Weaver, *The Mathematical Theory of Communication.* Urbana, Ill.: Univ. of Illinois Press, 1949.

[2] ——, "Probability of error for optimal codes in a Gaussian channel," *Bell Syst. Tech. J.*, vol. 38, pp. 611–656, 1959.

[3] V. L. Stefanyuk, "Kratkie Soobshcheniia o 'Vzaimopomoshchi' v Kollektivic Radiostantsii," *Probl. Peredach. Inform.*, vol. VII, pp. 103–107, 1971.

[4] V. L. Stefanyuk and M. L. Tsetlin, "Power regulation in a group of radio stations," *Probl. Peredach. Inform.*, vol. 3, pp. 49–57, 1967.

[5] J. P. Costas, "Poisson, Shannon, and the radio amateur," *Proc. IRE*, vol. 47, pp. 2058–2068, Dec. 1959.

[6] T. M. Cover, "Broadcast channels," *IEEE Trans. Inform. Theory*, vol. IT-18, pp. 2–14, Jan. 1972.

[7] P. P. Bergmans, "Random coding theorem for broadcast channels with degraded components," *IEEE Trans. Inform. Theory*, vol. IT-19, pp. 197–207, Mar. 1973.

[8] ——, "Degraded broadcast channels," Ph.D. dissertation, Dep. Elec. Eng., Stanford University, Stanford, Calif., 1972.

[9] G. H. Hardy, J. E. Littlewood, and G. Polya, *Inequalities.* New York: Cambridge, 1934.

[10] P. P. Bergmans, "A simple converse for broadcast channels with additive white Gaussian noise," *IEEE Trans. Inform. Theory* (Corresp.), vol. IT-20, pp. 279–280, Mar. 1974.

Multiple User Information Theory

ABBAS EL GAMAL, MEMBER, IEEE, AND THOMAS M. COVER, FELLOW, IEEE

Invited Paper

Abstract—A unified framework is given for multiple user information networks. These networks consist of several users communicating to one another in the presence of arbitrary interference and noise. The presence of many senders necessitates a tradeoff in the achievable information transmission rates. The goal is the characterization of the capacity region consisting of all achievable rates. The focus is on broadcast, multiple access, relay, and other channels for which the recent theory is relatively well developed. A discussion of the Gaussian version of these channels demonstrates the concreteness of the encoding and decoding necessary to achieve optimal information flow. We also offer speculations about the form of a general theory of information flow in networks.

I. Introduction

THE SHANNON theory of channel capacity has been extended successfully to many interesting communication networks in the past 10 years. We shall attempt to achieve three goals in our exposition of this theory: (i) make the theory accessible to researchers in communication theory, (ii) provide conditionally novel proofs of the theory for researchers in information theory, and (iii) present an overview of the basic problems in constructing a theory of information flow in networks.

The primary ideas can be obtained by reading the introduction and the sections on the Gaussian examples, the Shannon's theorem, and the summary. The heretofore unpublished information theoretic proofs are those for the sections on the multiple access channel, Slepian–Wolf data compression, and the degraded broadcast channel. All proofs, both new and old, are based on the idea of jointly typical sequences.

No claim for comprehensive coverage is given. For that the reader is referred to van der Meulen [1]. Rather, we are concerned with providing a unified approach to the theory. This leads naturally to a discussion of some of the major results. We begin by discussing some of the building blocks for networks.

Suppose m ground stations are simultaneously communicating to a common satellite as in Fig. 1. This is known as the multiple access channel. What are the achievable rates of communication? Does the total amount of information flow tend to infinity with the number of stations—or does the interference put an upper limit on the total communication? Does the signaling strategy change with m? Here the theory is completely known (Ahlswede [2] and Liao [3]), and all of these questions have quite satisfying answers (Section V).

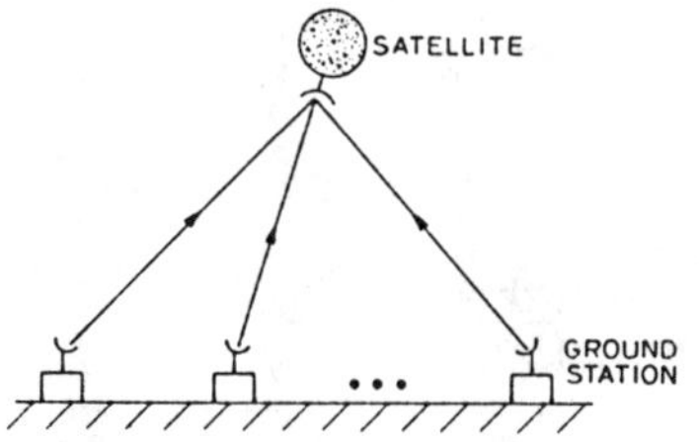

Fig. 1. Multiple access network.

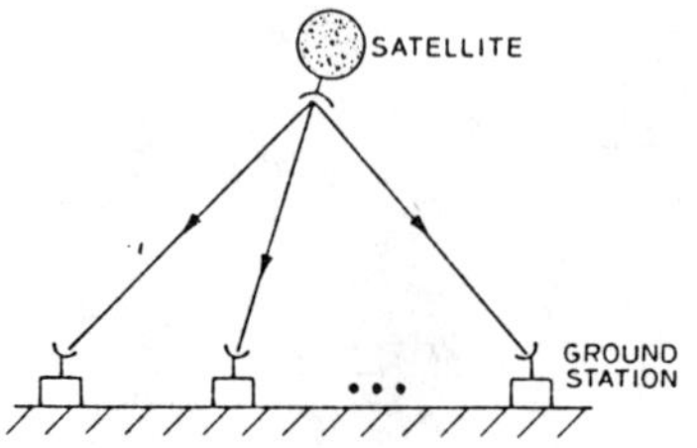

Fig. 2. Broadcast network.

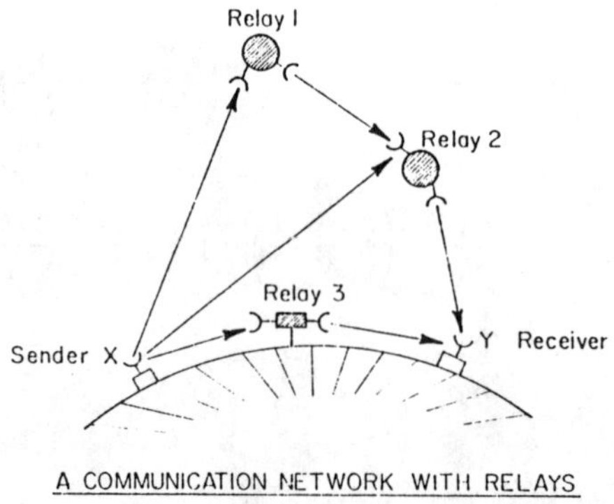

Fig. 3. Relay network.

In contrast, we can reverse the network, and consider one satellite broadcasting simultaneously to m stations as shown in Fig. 2. This is the broadcast channel. Here the achievable set of rates is not known except in special cases (Section VII). Yet another example consists of only one sender and one receiver, but includes extra channels serving as relays. This is the relay channel shown in Fig. 3.

In general, the underlying goal of work on these special channels is a theory for networks of the general form given in Fig. 4.

The interpretation of Fig. 4 is that at each instant of time the ith node sends a symbol x_i that depends on the messages

Manuscript received June 2, 1980; revised July 25, 1980. The work of A. El Gamal was partially supported by the Joint Services Electronics Program through the Air Force Office of Scientific Research (AFSC) under Contract F44620-76-C-0061 and NSF ENG 79-08948. The work of T. M. Cover was partially supported by the Joint Services Electronics Program DAAG-29-79-C-0047 and by NSF Grant ECS78-23334.

A. El Gamal is with the Department of Electrical Engineering, Stanford University, Stanford, CA 94305.

T. M. Cover is with the Departments of Electrical Engineering and Statistics, Stanford University, Stanford, CA 94305.

Reprinted from *Proc. IEEE*, vol. 68, no. 12, pp. 1466–1483, December 1980.

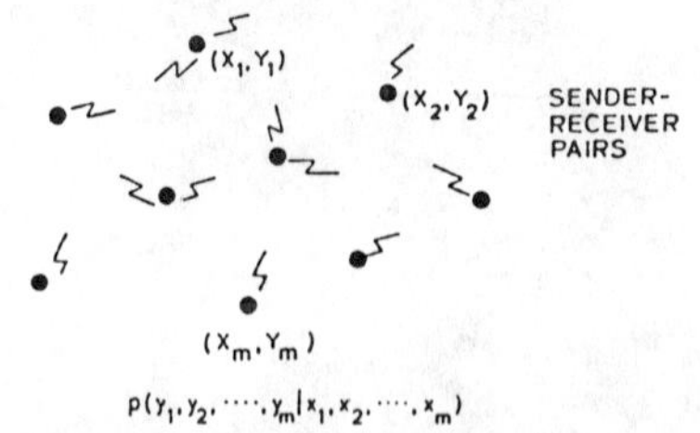

Fig. 4. General multiple user network.

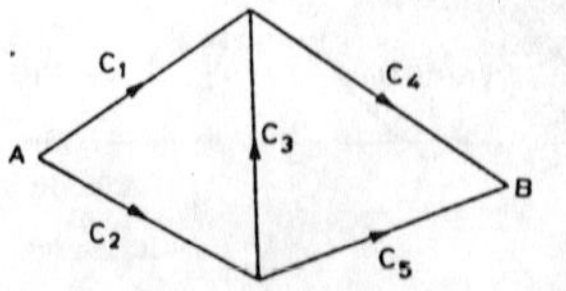

Fig. 5. Capacity is min $\{C_1 + C_2, C_1 + C_3 + C_5, C_2 + C_4, C_4 + C_5\}$. The maximum flow minimum cut theorem.

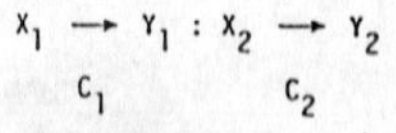

Fig. 6.

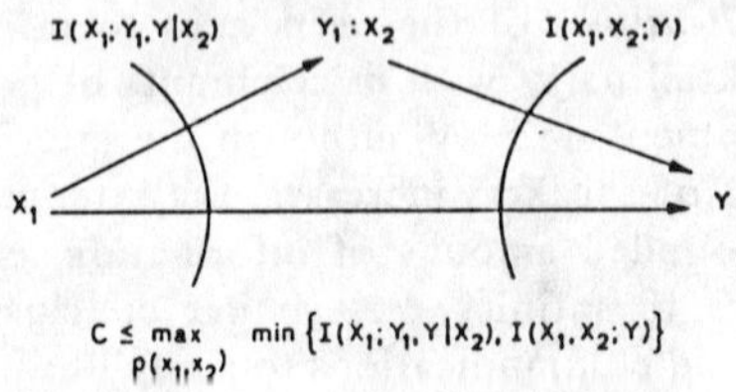

Fig. 7. Degraded and deterministic relay channel capacity.

he wishes to send and (perhaps) his past received y_i symbols. We assume that the result of the simultaneous transmission of $(x_1, x_2, \cdots, x_m)$ is a random collection of received symbols $(Y_1, Y_2, \cdots, Y_m)$ drawn according to a conditional probability $p(y_1, \cdots, y_m | x_1, \cdots; x_m)$, where $p(\cdot|\cdot)$ describes all of the effects of interference and noise in the network.

In a more restricted domain, such as the flow of water in networks of pipes, the existing theory is very satisfying. For example, in the single source, single sink network in Fig. 5, the maximum flow from A to B is easily computed from the maximum flow minimum cut theorem of Ford and Fulkerson [4].

Assume that the edges have capacities C_i as shown. Clearly, the maximum flow across any cutset is no greater than the sum of the capacities of the cut edges. Thus minimizing the maximum flow over cutsets yields an upper bound to the total flow from A to B. This flow can actually be achieved, as the Ford and Fulkerson theorem demonstrates.

However, the information flow problem involves "soft" quantities rather than "hard" commodities. A choice of node symbols x results in a random response Y, and it is difficult to see how to choose as many distinguishable x's as possible in this random environment. Consequently, it is gratifying to find that information problems like the relay channel and cascade channel admit min flow max cut interpretations. For example, the informally defined cascade network in Fig. 6, has capacity $C = \min\{C_1, C_2\}$, where C_i denotes the Shannon capacity of the ith channel. Also, for the degraded or deterministic relay channel (Section VII) we have a similar max flow min cut interpretation as shown in Fig. 7.

The structure of this paper is presented in miniature in Section II. In that section, we use Gaussian channels to run through the major results that will be given in greater generality and detail in the subsequent sections. The physically motivated Gaussian channel lends itself to concrete and easily interpreted answers. Some preliminary technical details on the properties of joint typicality are given in Section III, followed by a simple proof in Section IV of Shannon's original capacity theorem. Then treated are the multiple access channel (Section V), Slepian–Wolf data compression theorem (Section VI), the combination of both (Section VII), the broadcast channel (Section VIII), and the relay channel (Section IX).

The final summary (Section X) is a recapitulation of the paper paralleling Section II, this time in greater generality.

II. Gaussian Multiple User Channels

We shall begin our treatment of multiple user information theory by investigating Gaussian multiple user channels. This allows us to give concrete descriptions of the coding schemes and associated capacity regions. The proofs of capacity for the discrete memoryless counterparts of these channels will be given in later sections.

The basic discrete time additive white Gaussian noise channel with input power P and noise variance N is modeled by

$$Y_i = x_i + Z_i, \qquad i = 1, 2, \cdots$$

where Z_i are independent identically distributed Gaussian random variables with mean zero and variance N. The signal $x = (x_1, x_2, \cdots, x_n)$ has a power constraint

$$1/n \sum_{i=1}^{n} x_i^2 \leq P.$$

The Shannon capacity C, obtained by maximizing $I(X; Y)$ over all random variables X such that $EX^2 \leq P$ is given by

$$C = (1/2) \log (1 + P/N) \text{ bits/transmission.} \tag{2.1}$$

The continuous time Gaussian channel capacity is simply related to the discrete time capacity. If the signal $x(t)$, $0 \leq t \leq T$, has power constraint P and bandwidth constraint W, and the white noise $Z(t)$, $0 \leq t \leq T$, has power spectral density N, then the capacity of the channel $Y(t) = x(t) + Z(t)$ is given by

$$C = W \log (1 + P/NW) \text{ bit/s.} \tag{2.2}$$

The relationship between (2.1) and (2.2) can be seen informally by replacing the continuous time processes by $n = 2TW$ independent samples from the process and calculating the noise variance per sample. The full theory establishing (2.2) can be found in Wyner [5], Gallager [6], and Slepian and Pollack [7].

Having said this, we restrict our treatment to time discrete Gaussian channels.

Random Codebook: Shannon observed in 1948 that a randomly selected codebook is good (with high probability) when the rate R of the codebook is less than the channel capacity $C = \max I(X; Y)$. As mentioned above, for the Gaussian channel the capacity is given by $C = (1/2) \log (1 + P/N)$ bits per transmission.

We now set up a codebook that will be used in all of the multiple user channel models below. The codewords comprising the codebook are vectors of length n and power P. To generate such a random codebook, simply choose 2^{nR} independent identically distributed random n-vectors $\{x(1), x(2),$

$\cdots, x(2^{nR})\}$, each consisting of n independent Gaussian random variables with mean zero and variance P. The rate R will be specified later. Sometimes we will need two or more independently generated codebooks.

In the continuous channel case, one simply lets the white noise generator of power P and bandwidth W run for T seconds. Every T seconds, a new codeword is generated and we list them until we fill up the codebook.

Now we analyze the Gaussian channels shown in Fig. 8.

A. The Gaussian Channel

Here $Y = x + Z$. Choose an $R < C = 1/2 \log(1 + P/N)$. Choose any index i in the set 2^{nR}. Send the ith vector $x(i)$ from the codebook generated above. The receiver observes $Y = x(i) + Z$, then finds the index $\hat{i}$ of the closest codeword to Y. If n is sufficiently large, the probability of error $P(i \neq \hat{i})$ will be arbitrarily small. As will be seen in the definitions on joint typicality, this minimum distance decoding scheme for the Gaussian channel is essentially equivalent to finding the codeword in the codebook that is jointly typical with the received vector Y.

B. The Multiple Access Channel with m Users

We consider m transmitters, each of power P. Let

$$Y = \sum_{1}^{m} x_i + Z.$$

Specializing the results of Section IV to the Gaussian channel shows that the achievable rate region for the Gaussian channel takes on the simple form given in the following equations:

$$\begin{aligned} R_i &< C(P/N) \\ R_i + R_j &< C(2P/N) \\ R_i + R_j + R_k &< C(3P/N) \\ &\ \ \vdots \\ \sum_{1}^{m} R_i &< C(mP/N) \end{aligned} \tag{2.3}$$

where

$$C(x) = (1/2) \log(1 + x) \tag{2.4}$$

denotes the capacity of the Gaussian channel with signal to noise ratio x. When all the rates are the same, the last inequality dominates the others.

Here we need m codebooks, the ith codebook having 2^{nR_i} codewords of power P. Transmission is simple. Each of the independent transmitters chooses whatever codeword he wishes from his own codebook. The users simultaneously send these vectors. The receiver sees the codewords added together with the Gaussian noise Z.

Optimal decoding consists of looking for the m codewords, one from each codebook, such that the vector sum is closest to Y in Euclidean distance. The set of m codewords achieving the minimum distance to Y corresponds to the hypothesized collection of messages sent.

If $(R_1, R_2, \cdots, R_m)$ is in the capacity region given above, then the probability of error goes to zero as n tends to infinity.

Remarks: It is exciting to see in this problem that the sum of the rates $C(mP/N)$ of the users goes to infinity with m. Thus in a cocktail party with m celebrants of power P in the presence of ambient noise N, although the interference grows as the number of speakers increases, the intended listener receives an unbounded amount of information as the number of people goes to infinity. A similar conclusion holds of course for ground communications to a satellite.

It is also interesting to note that the optimal transmission scheme here does not involve time division multiplexing. In fact, each of the transmitters utilizes the entire time to send his message.

A practical consideration for ground transmission to a satellite involves the possible inability of the ground communicators to synchronize their transmissions. Nonetheless, it can be shown that the capacity is unchanged when there is a lack of synchronization [8].

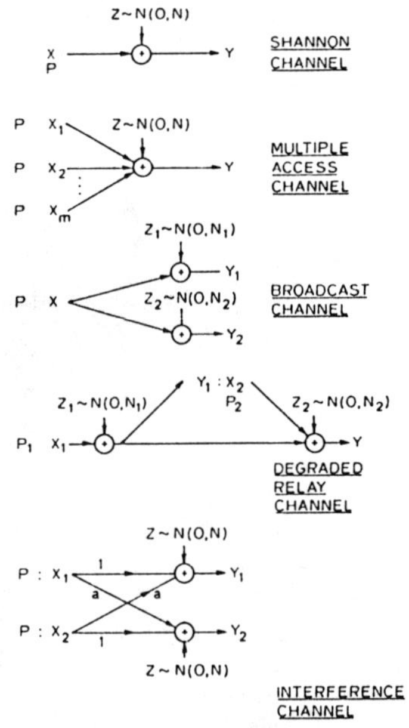

Fig. 8. Gaussian multiple user channels.

C. The Broadcast Channel

Here we assume that we have a sender of power P and two distant receivers, one with noise spectral density N_1 and the other with noise spectral density N_2. Without loss of generality, assume $N_1 < N_2$. Thus, in some sense receiver, Y_1 is better than receiver Y_2. The model for the channel is $Y_1 = x + Z_1$ and $Y_2 = x + Z_2$, where Z_1 and Z_2 are arbitrarily correlated Gaussian random variables with variances N_1 and N_2, respectively. The sender wishes to send independent messages at rates R_1 and R_2 to receivers Y_1 and Y_2, respectively.

Fortunately, all Gaussian broadcast channels belong to the class of degraded broadcast channels which will be discussed in Section VIII. Specializing that work, we find the following capacity region for the Gaussian broadcast channel:

$$\begin{aligned} R_1 &< C(\alpha P/N_1) \\ R_2 &< C(\bar{\alpha} P/(\alpha P + N_2)) \end{aligned} \tag{2.5}$$

where $0 \leqslant \alpha \leqslant 1$, $\bar{\alpha} = 1 - \alpha$, may be arbitrarily chosen to trade off rate R_1 for rate R_2 as the transmitter wishes.

To encode the messages, the transmitter generates two codebooks, one with power αP at rate R_1, and another codebook with power $\bar{\alpha} P$ and rate R_2. He has chosen R_1 and R_2 to satisfy the equation above. Then, to send an index $i \in \{1, 2, \cdots, 2^{nR_1}\}$ and $j \in \{1, 2, \cdots, 2^{nR_2}\}$ to Y_1 and Y_2, respectively, he takes codeword $x(i)$ from the first codebook and codeword $x(j)$ from the second codebook and computes the sum. He then sends the sum over the channel.

Two receivers must now do the decoding. First consider the bad receiver Y_2. He merely looks through the second codebook for the closest codeword to his received vector Y_2. His effective signal-to-noise ratio is $\bar{\alpha}P/(\alpha P + N_2)$ since Y_1's message acts as noise to Y_2. The good receiver Y_1 first decodes Y_2's codeword, which he can do because of his lower noise N_1. He subtracts this codeword $\hat{x}_2$ from Y_1. This leaves him with a channel of power αP and noise N_1. He then looks for the closest codeword in the first codebook to $Y_1 - \hat{x}_2$. The resulting probability of error can be made as low as wished.

A nice dividend of optimal encoding for degraded broadcast channels is that the better receiver Y_1 always knows the message intended for receiver Y_2 in addition to the extra information intended for himself.

D. The Relay Channel

For the relay channel, we have a sender X_1 and an ultimate intended receiver Y. Also present, however, is the relay channel intended solely to help the sender. The channel is given by

$$Y_1 = x_1 + Z_1$$
$$Y_2 = x_1 + Z_1 + x_2 + Z_2 \tag{2.6}$$

where Z_1, Z_2 are independent zero mean Gaussian random variables with variance N_1, N_2, respectively. The allowed encoding by the relay is the causal sequence

$$x_{2i} = f_i(y_{11}, y_{12}, \cdots, y_{1i-1}). \tag{2.7}$$

The sender X_1 has power P_1 and the relay X_2 has power P_2. The results of Section IX yield the capacity

$$C = \max_{0<\alpha<1} \min \left\{ C\left(\frac{P_1 + P_2 + 2\sqrt{\bar{\alpha}P_1P_2}}{N_1 + N_2}\right), C\left(\frac{\alpha P_1}{N_1}\right)\right\}$$

where

$$\bar{\alpha} = 1 - \alpha.$$

Note that if

$$P_2/N_2 \geqslant P_1/N_1$$

it can be seen that $C^* = C(P_1/N_1)$. (This is achieved by $\alpha = 1$.) The channel appears to be noise-free after the relay, and the capacity $C(P_1/N_1)$ from x_1 to the relay can be achieved. Thus the rate without the relay $C(P_1/(N_1 + N_2))$ is increased by the relay to $C(P_1/N_1)$. For large N_2, and for $P_2/N_2 \geqslant P_1/N_1$, we see that the increment in rate is from $C(P_1/(N_1 + N_2)) \approx 0$ to $C(P_1/N_1)$.

Encoding of Information: Two codebooks are needed. The first codebook has 2^{nR_1} words of power αP_1. The second has 2^{nR_0} codewords of power $\bar{\alpha}P_1$. We shall use words from these codebooks successively in order to create the opportunity for cooperation by the relay. We start by sending a codeword from the first codebook. The relay now knows the index of this codeword since $R_1 < C(\alpha P_1/N_1)$, but the intended receiver does not. However, the intended receiver has a list of possible codewords of size $2^{n(R_1 - C(\alpha P_1/N_1 + N_2))}$. The list calculation involves a result on list codes.

In the next block the relay and the transmitter would like to cooperate to resolve the receiver's uncertainty about the previously sent codeword on the receiver's list. Unfortunately, they cannot quite be sure what this list is. They do not know the received signal Y. Thus they randomly partition the first codebook into 2^{nR_0} cells with an equal number of codewords in each cell. The relay, the receiver, and the transmitter agree on this partition. The relay and the transmitter find the cell of the partition in which the codeword from the first codebook lies and cooperatively send the codeword from the second codebook with that index. That is, both X_1 and X_2 send the same designated codeword. The relay, of course, must scale this codeword so that it meets his power constraint P_2. They now simultaneously transmit their codewords. An important point to note is that the cooperative information sent by the relay and the transmitter is sent coherently. So the power of the sum as seen by the receiver Y is $(\sqrt{\bar{\alpha}P_1} + \sqrt{P_2})^2$.

However, this does not exhaust what the transmitter does in the second block. He also chooses a fresh codeword from his first codebook, adds it "on paper" to the cooperative codeword from his second codebook, and sends this sum over the channel.

The reception by the ultimate receiver Y in the second block involves first finding the cooperative index from the second codebook by looking for the closest codeword in the second codebook. He subtracts it off, and then calculates a list of indices of size 2^{nR_0} corresponding to all transmitted words from the first book which might have been sent in that second block.

Now it is time for the intended receiver Y to finish computing the codeword from the first codebook sent in the first block. He takes his list of possible codewords that might have been sent in the first block and intersects it with the cell of the partition that he has learned from the cooperative relay transmission in the second block. The rates and powers have been chosen so that it is highly probable that there will be only one codeword in this intersection. This is Y's guess about the information sent in the first block.

We are now in steady state. In each new block, the transmitter and the relay cooperate to resolve the list uncertainty from the previous block. In addition, the transmitter superimposes some fresh information from his first codebook to his transmission from the second codebook and transmits the sum.

The receiver is always one block behind, but for sufficiently many blocks, this does not affect his overall rate of reception.

E. The Interference Channel

The interference channel has two senders and two receivers. Sender 1 wishes to send information to receiver 1. He does not care what receiver 2 receives or understands. Similarly, with sender 2 and receiver 2. Each channel interferes with the other. It is not quite a broadcast channel because there is only one intended receiver for each sender, nor is it a multiple access channel because each receiver is only interested in what is being sent by the corresponding transmitter.

This channel has not been solved in general, even in the Gaussian case. But remarkably, in the case of high interference, Carleial [9] has shown that the solution to this channel is the same as if there were no interference whatsoever. To achieve this, generate two codebooks, each with power P and rate $C(P/N)$. Each sender independently chooses a word from his book and sends it. Now, if the interference a satisfies $C(a^2P/(P + N)) > C(P/N)$, the first receiver perfectly understands the index of the second transmitter. He finds it by the usual technique of looking for the closest codeword to his received signal. Once he finds this signal, he subtracts it from his received waveform. Now there is a clean channel between him and his sender. He then searches his sender's codebook to find the closest codeword and declares that codeword to be the one sent.

III. THE ASYMPTOTIC EQUIPARTITION THEOREM AND SHANNON CHANNEL CAPACITY

Every sequence of n fair coin flips has equal probability $(1/2)^n$. If the coin has bias p, all of the sequences having roughly np heads are nearly equally probable and exhaust almost all of the probability. A formalization of this idea for arbitrary random variables is known as the asymptotic equipartition property (AEP) (Shannon [10], MacMillan [11], and Breiman [12]).

Consider a sequence of independent identically distributed random variables $X = (X_1, X_2, \cdots, X_n)$, where X_i is drawn according to probability mass function $p(x)$. We are interested in defining a set A of possible outcomes, each roughly equally probable, such that $P(X \in A) \approx 1$. Toward this end, we shall say that a given sequence $x = (x_1, x_2, \cdots, x_n)$ is *ε-typical* if

$$|-(1/n) \log p(X) - H(X)| \leqslant \epsilon \tag{3.1}$$

where

$$H(X) = -\sum p(x) \log p(x) \tag{3.2}$$

is the Shannon *entropy* of $p(\cdot)$. Let us define the typical set A_ϵ to be the set of all ε-typical n-sequences x.

By the law of large numbers, for a random i.i.d. sequence X

$$-(1/n) \log p(X) = (1/n) \sum_{i=1}^{n} -\log p(X_i)$$

$$\rightarrow H(X), \text{ with probability one.} \tag{3.3}$$

Consequently, the following results are true.

1) $P(A_\epsilon) \rightarrow 1, \quad$ as $n \rightarrow \infty$.
2) $x \in A_\epsilon$ implies $2^{-n(H+\epsilon)} \leqslant p(x) \leqslant 2^{-n(H-\epsilon)}$.
3) The number of elements $\|A_\epsilon\|$ in A_ϵ is $\leqslant 2^{n(H+\epsilon)}$.

The third assertion follows from

$$1 \geqslant \sum_{A_\epsilon} p(x) \geqslant \sum_{A_\epsilon} 2^{-n(H+\epsilon)} = \|A_\epsilon\| \, 2^{-n(H+\epsilon)}. \tag{3.4}$$

Roughly speaking, a sequence is typical if the proportion of occurrences of each of its symbols is close to the true probability of occurrence.

An immediate application of the AEP yields Shannon's source coding theorem:

Theorem 1 (Shannon 1948): Given an independent identically distributed source $X_1, X_2, \cdots$ with entropy $H(X)$, there exists, for every $\epsilon > 0$, an integer n and an encoding i: $\mathfrak{X}^n \rightarrow \{1, 2, \cdots, 2^{n(H+\epsilon)}\}$ and decoding rule g: $\{1, 2, \cdots, 2^{n(H+\epsilon)}\} \rightarrow \mathfrak{X}^n$, such that

$$P\{g(i(X)) \neq X\} \leqslant \epsilon.$$

Proof: Simply index the elements x in the typical set A_ϵ.

Remark: Since the index set has $\leqslant 2^{n(H+\epsilon)}$ elements, we see that only $(H + \epsilon)$ bits/symbol are necessary to describe x.

Now, to make progress with multiple user information theory, the idea of *joint typicality* is needed. A pair of sequences x and y are said to be jointly ε-typical if x is individually ε-typical, i.e.,

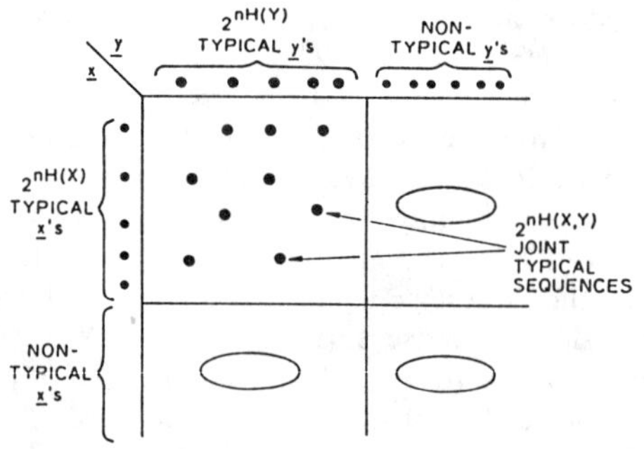

Fig. 9. Jointly typical sequences.

$$|-(1/n) \log p(x) - H(X)| \leqslant \epsilon$$

y is individually ε-typical, and (x, y) is ε-typical, i.e.,

$$|-(1/n) \log p(x, y) - H(X, Y)| \leqslant \epsilon.$$

The picture of jointly typical pairs is given in Fig. 9.

The dots in the matrix denote jointly typical pairs. It can be shown by the method of (3.4) that there are $\leqslant 2^{nH(Y|X)}$ dots in each row and $\leqslant 2^{nH(X|Y)}$ dots in each column.

The definitions and proofs of the needed results now follow. Let $\{X^{(1)}, X^{(2)}, \cdots, X^{(k)}\}$ denote a finite collection of discrete random variables with some fixed joint distribution $p(x^{(1)}, x^{(2)}, \cdots, x^{(k)})$, $(x^{(1)}, x^{(2)}, \cdots, x^{(k)}) \in \mathfrak{X}^{(1)} \times \mathfrak{X}^{(2)} \times \cdots \times \mathfrak{X}^{(k)}$. Let S denote an ordered subset of these r.v.'s, and consider n independent copies of S. Thus

$$\Pr\{S = s\} = \prod_{i=1}^{n} \Pr\{S_i = s_i\}, \qquad s \in \mathcal{S}^n. \tag{3.5}$$

For example, if $S = (X^{(j)}, X^{(l)})$, then

$$\Pr\{S = s\} = \Pr\{(X^{(j)}, X^{(l)}) = (x^{(j)}, x^{(l)})\}$$

$$= \prod_{i=1}^{n} p(x_i^{(j)}, x_i^{(l)}). \tag{3.6}$$

By the law of large numbers, for any subset S of random variables,

$$-(1/n) \log p(S_1, S_2, \cdots, S_n) = -(1/n) \sum_{i=1}^{n} \log p(S_i) \rightarrow H(S). \tag{3.7}$$

Convergence in (3.7) take place simultaneously with probability one for all 2^k subsets

$$S \subseteq \{X^{(1)}, X^{(2)}, \cdots, X^{(k)}\}.$$

Definition: The set A_ϵ of ε-typical n-sequences $(x^{(1)}, x^{(2)}, \cdots, x^{(k)})$ is defined by

$$A_\epsilon(X^{(1)}, X^{(2)}, \cdots, X^{(k)}) = A_\epsilon$$
$$= \{(x^{(1)}, x^{(2)}, \cdots, x^{(k)}):$$
$$|-(1/n) \log p(s) - H(S)| < \epsilon,$$
$$\forall S \subseteq \{X^{(1)}, X^{(2)}, \cdots, X^{(k)}\}\}. \tag{3.8}$$

Let $A_\epsilon(S)$ denote the restriction of A_ϵ to the coordinates corresponding to S. Thus if $S = (X^{(1)}, X^{(2)})$ we have

$$A_\epsilon(X^{(1)}, X^{(2)}) = \{(x^{(1)}, x^{(2)}): |-1/n \log p(x^{(1)}, x^{(2)}) - H(X^{(1)}, X^{(2)})| < \epsilon$$
$$|-1/n \log p(x^{(1)}) - H(X^{(1)})| < \epsilon$$
$$|-1/n \log p(x^{(2)}) - H(X^{(2)})| < \epsilon\}. \quad (3.9)$$

Notation: We shall use the notation

$$a_n \doteq 2^{nb} \quad (3.10)$$

to mean

$$(1/n) \log a_n \to b. \quad (3.11)$$

Lemma 1: For any $\epsilon > 0$, for sufficiently large n,

(i) $P\{A_\epsilon(S)\} \geq 1 - \epsilon, \quad \forall S \subseteq \{X^{(1)}, \cdots, X^{(k)}\}$
(ii) $\|A_\epsilon(S)\| \doteq 2^{nH(S)}$
(iii) $s \in A_\epsilon(S) \Rightarrow p(s) \doteq 2^{-n(H(S)-\epsilon)}$
(iv) Let $S_1, S_2 \subseteq \{X^{(1)}, X^{(2)}, \cdots, X^{(k)}\}$.

If

$$(s_1, s_2) \in A_\epsilon(S_1, S_2)$$

then

$$p(s_1|s_2) \doteq 2^{-n(H(S_1|S_2)-2\epsilon)}.$$

Proof: (i) follows from the convergence of the random variables in the definition of $A_\epsilon(S)$. (ii) follows immediately from the definition of $A_\epsilon(S)$. (iii) follows from

$$1 \geq \sum_{s \in A_\epsilon(S)} p(s) \overset{(ii)}{\geq} \sum_{A_\epsilon(S)} 2^{-n(H(S)+\epsilon)}$$
$$= \|A_\epsilon(S)\| 2^{-n(H(S)+\epsilon)} \quad (3.12)$$

and

$$(1-\epsilon) \leq \sum_{s \in A_\epsilon(S)} p(s) \overset{(ii)}{\leq} \|A_\epsilon(S)\| 2^{-n(H(S)-\epsilon)} \quad (3.13)$$

Finally, $(s_1, s_2) \in A_\epsilon(S_1, S_2)$ implies

$$p(s_1, s_2) \doteq 2^{-n(H(S_1,S_2)-\epsilon)}$$

and

$$p(s_2) \doteq 2^{-(H(S_2)-\epsilon)}.$$

(iv) follows from

$$p(s_1|s_2) = p(s_1, s_2)/p(s_2).$$

Lemma 2: Let S_1, S_2 be two subsets of $X^{(1)}, \cdots, X^{(k)}$. For any $\epsilon > 0$, define $\|A_\epsilon(S_1|s_2)\|$ to be the number of s_1 sequences, jointly ϵ-typical with a particular s_2 sequence. If $s_2 \in A_\epsilon(S_2)$, then for arbitrarily large n

$$\|A_\epsilon(S_1|s_2)\| \leq 2^{n(H(S_1|S_2)+2\epsilon)}.$$

Moreover,

$$(1-\epsilon) 2^{n(H(S_1|S_2)-2\epsilon)} \leq \sum_{s_2} p(s_2) \|A_\epsilon(S_1|s_2)\|.$$

Proof: Similar to part (ii) of Lemma 1.

In achievability proofs, we shall need to know the probability that conditionally independent sequences are jointly typical. Let S_1, S_2, and S_3 be three subsets of $\{X^{(1)}, X^{(2)}, \cdots, X^{(k)}\}$.

Lemma 3:

(a) Let

$$P(S_1' = s_1, S_2' = s_2, S_3' = s_3) = \prod_{i=1}^{n} p(s_{1i}|s_{3i}) p(s_{2i}|s_{3i}) p(s_{3i}).$$

Then,

$$P\{S_1', S_2', S_3') \in A_\epsilon\} \doteq 2^{-nI(S_1;S_2|S_3)}.$$

Proof:

$$P\{(S_1', S_2', S_3') \in A_\epsilon\}$$
$$= \sum_{(s_1,s_2,s_3) \in A_\epsilon} p(s_3) p(s_1|s_3) p(s_2|s_3)$$
$$\doteq \|A_\epsilon(S_1, S_2, S_3)\| 2^{-nH(S_3)} 2^{-nH(S_1|S_3)} 2^{-nH(S_2|S_3)}$$
$$\doteq 2^{nH(S_1,S_2,S_3)} 2^{-n(H(S_3)+H(S_1|S_3)+H(S_2|S_3))}$$
$$= 2^{-nI(S_1;S_2|S_3)}.$$

IV. Shannon's Capacity Theorem

We begin the theoretical discussion by reproving Shannon's original channel capacity theorem for the discrete memoryless channel (Fig. 10). When Shannon's work was published in 1948, the proofs were considered to be only heuristic—mere plausibility arguments. The first rigorous proof by Feinstein [13] 8 years later was completely different from Shannon's random coding idea. Further proofs by Wolfowitz, Fano, Gallager, and many others, were also different. The rigorous proof we give here supports the idea of Shannon's outline. It can be found as a problem in Gallager [6] and in Forney's unpublished class notes [14]. This proof technique will come up many times in subsequent sections. The fact that Shannon's original proof is the natural proof for multi-user networks is really quite remarkable.

The basic Shannon model for a communication channel is the discrete memoryless channel $(\mathcal{X}, p(y|x), \mathcal{Y})$ consisting of two alphabets—input alphabet $\mathcal{X}$ and output alphabet $\mathcal{Y}$—and a channel probability matrix $p(y|x)$. The interpretation is that if x is sent, then the received symbol Y is drawn according to probability mass function $p(y|x)$. *Discrete memoryless* means if a sequence $x_1, x_2, \cdots, x_n$ is sent, then the received sequence $y = (y_1, y_2, \cdots, y_n)$ is drawn according to $\prod_{i=1}^n p(y_i|x_i)$.

A $(2^{nR}, n)$ code for a channel consists of a set of integers $[1, 2^{nR}]$ called the message set, an *encoding function*

$$x: [1, 2^{nR}] \to \mathcal{X}^n$$

and a *decoding function*

$$g: \mathcal{Y}^n \to [1, 2^{nR}].$$

If the message $w \in [1, 2^{nR}]$ is sent, let

$$\lambda(w) = P\{g(Y) \neq w | w \text{ sent}\} \quad (4.1)$$

denote the conditional probability of error. Define the average probability of error of the code assuming a uniform distribution over the set of messages $[1, 2^{nR}]$, as

$$\bar{P}_e^n = \frac{1}{2^{nR}} \sum_w \lambda(w). \quad (4.2)$$

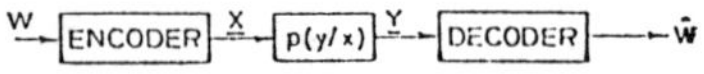

Fig. 10. The discrete memoryless channel.

The rate R of an $(2^{nR}, n)$ code is said to be *achievable* by the discrete memoryless channel if, for any $\epsilon > 0$, there exists a $(2^{nR}, n)$ code for all n sufficiently large such that $\bar{P}_e^n < \epsilon$.

The *capacity* C of the discrete memoryless channel is the supremum of the set of achievable rates. In 1948, Shannon showed that the capacity C of the discrete memoryless finite alphabet channel is given by the following.

Theorem 2 (Shannon): The capacity of the discrete memoryless channel $(\mathcal{X}, p(y|x), \mathcal{Y})$ is given by

$$C = \sup_{p(x)} I(X; Y). \tag{4.3}$$

Proof: The proof that C is the capacity of a channel involves proving:

(i) *achievability* of any $R < C$, i.e., that there exists a sequence of $(2^{nR}, n)$ codes such that $\bar{P}_e^n \to 0$, and

(ii) a *converse* showing that given any sequence of $(2^{nR}, n)$ codes with $\bar{P}_e^n \to 0$ then $R < C$.

We shall only prove the achievability part of the theorem. The converses are discussed in the following sections of this paper.

Achievability: Assume that the maximization in (4.3) is achieved for a distribution $p^*(x)$ on $\mathcal{X}$.

Random code: Choose 2^{nR} i.i.d. x sequences each with probability

$$p(x) = \prod_{i=1}^{n} p^*(x_i).$$

Decoding: Given y, choose the message w such that

$$(y, x(w)) \in A_\epsilon(X, Y)$$

is such a $w \in [1, 2^{nR}]$ exists and is unique, otherwise declare an error.

By symmetry of the random code construction, the probability of error (averaged over the random code) is independent of the index w sent. Thus without loss of generality assume that 1 is sent. Consider the events

$$E_w = \{(X(w), Y) \in A_\epsilon\}, \; w \in [1, 2^{nR}].$$

Then by the union of events bound

$$P_n = P\left(\bigcup_{w \neq 1} E_w \cup E_1^c\right)$$

$$\leq P(E_1^c) + \sum_{w \neq 1} P(E_w).$$

From Lemma 1, it follows that

$$P(E_1^c) \to 0.$$

From Lemma 3,

$$P(E_w) \doteq 2^{-nC}, \qquad \forall w \neq 1.$$

Therefore, if $R < C$, then

$$\sum_{w \neq 1} P(E_w) \doteq 2^{n(R-C)} \to 0.$$

If the error averaged over the random code tends to zero, then there must exist a sequence of codes with $\bar{P}_e^n \to 0$, and achievability is proved.

V. The Multiple Access Channel

The most completely understood multiple user channel is the multiple access channel shown in Fig. 11.

Many senders each simultaneously attempt to communicate a message to a common receiver. The most common example of this is a satellite receiver with many independent ground stations. After reception, the satellite will broadcast the information back to the ground, but that is the subject of the section on broadcast channels.

We see that the senders must contend not only with the receiver noise but with interference from each other as well.

The discrete memoryless multiple access channel $(\mathcal{X}_1 \times \mathcal{X}_2, p(y|x_1, x_2), \mathcal{Y})$ consists of three alphabets $\mathcal{X}_1, \mathcal{X}_2$, and $\mathcal{Y}$ and a probability transition matrix $p(y|x_1, x_2)$.

A $((2^{nR_1}, 2^{nR_2}), n)$ code for the multiple access channel consists of two sets of integers $\mathcal{M}_1 = [1, 2^{nR_1}]$, $\mathcal{M}_2 = [1, 2^{nR_2}]$ called the message sets, two *encoding functions*

$$X_1 : \mathcal{M}_1 \to \mathcal{X}_1^n$$

$$X_2 : \mathcal{M}_2 \to \mathcal{X}_2^n$$

and a decoding function

$$g : \mathcal{Y}^n \to \mathcal{M}_1 \times \mathcal{M}_2.$$

Assuming uniform distribution over the product message sets $\mathcal{M}_1 \times \mathcal{M}_2$ i.e., that the messages are independent and equally likely, we define the *average probability of error* for a $((2^{nR_1}, 2^{nR_2}), n)$ code to be

$$\bar{P}_e^n = \frac{1}{2^{n(R_1+R_2)}} \cdot \sum_{(w_1, w_2) \in \mathcal{M}_1 \times \mathcal{M}_2} P\{g(Y) \neq (w_1, w_2) | (w_1, w_2) \text{ sent}\}. \tag{5.1}$$

A rate pair (R_1, R_2) is said to be *achievable* for the multiple access channel if there exists a sequence of $((2^{nR_1}, 2^{nR_2}), n)$ codes with $\bar{P}_e^n \to 0$.

The *capacity region* of the multiple access channel is the closure of the set of all achievable (R_1, R_2) rate pairs.

The capacity region of the multiple access channel was established by Ahlswede [2] and Liao [3]. The following proof is different from theirs.

Theorem 3: Multiple access capacity.

The capacity of the multiple access channel $(\mathcal{X}_1 \times \mathcal{X}_2, p(y|x_1, x_2), \mathcal{Y})$ is given by the convex hull of the union of the sets

$$C(p_1 p_2) = \{(R_1, R_2) : R_1 \leq I(X_1; Y|X_2)$$
$$R_2 \leq I(X_2; Y|X_1)$$
$$R_1 + R_2 \leq I(X_1, X_2; Y)$$

where

$$p(x_1, x_2) = p_1(x_1) p_2(x_2)\} \tag{5.2}$$

over all input product probability distributions on $\mathcal{X}_1 \times \mathcal{X}_2$.

Proof:

(i) Achievability.

Random coding: Fix $p_1(x_1)$, $p_2(x_2)$. Let $p(x_1, x_2) = p_1(x_1) p_2(x_2)$. Choose a random code of 2^{nR_1} x_1's $\in \mathcal{X}_1^n$

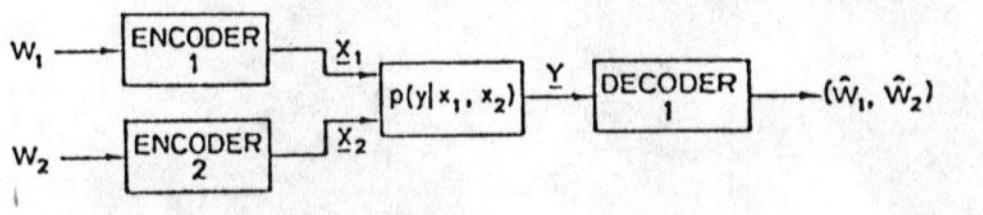

Fig. 11. The multiple access channel.

i.i.d $\sim \Pi_{i=1}^{n} p_1(x_{1i})$, and independently choose 2^{nR_2} x_2's $\in \mathcal{X}_2^n$ i.i.d $\sim \Pi_{i=1}^{n} p_2(x_{2i})$. Index these sequences as $x_1(i)$, $x_2(j)$, $i \in [1, 2^{nR_1}]$, $j \in [1, 2^{nR_2}]$.

Decoding: For decoding, given y, simply choose the pair (i, j) such that

$$(x_1(i), x_2(j), y) \in A_\epsilon$$

if such an $(i, j) \in [1, 2^{nR_1}] \times [1, 2^{nR_2}]$ exists and is unique—otherwise declare an error.

By symmetry of the random code construction, the probability of error (averaged over the random code) is independent of the index (i, j) sent. Thus without loss of generality assume that $(i, j) = (1, 1)$ is sent. Consider the events

$$E_{ij} = \{(X_1(i), X_2(j), Y) \in A_\epsilon^n\}.$$

Then, by the union of events bound

$$P_n = P\left(E_{11}^c \cup \bigcup_{(i,j)\neq(1,1)} E_{ij}\right)$$

$$\leq P(E_{11}^c) + \sum_{i\neq 1, j=1} P(E_{i1}) + \sum_{i=1, j\neq 1} P(E_{1j}) + \sum_{i\neq 1, j\neq 1} P(E_{ij}). \tag{5.3}$$

From Lemma 1, $P(E_{11}^c) \to 0$.

Next, for $i \neq 1$,

$$P(E_{i1}) = \sum_{(x_1, x_2, y)\in A_\epsilon} p(x_1)\, p(x_2, y)$$

$$\doteq \|A_\epsilon\| 2^{-nH(X_1)} 2^{-nH(X_2, Y)}$$

$$\doteq 2^{-nI(X_1; Y|X_2)}$$

where the first equality follows from $i \neq 1$, which implies the independence of X_1 from (X_2, Y). The second and third inequalities follow from the definition of A_ϵ, Lemma 1, and from $I(X_1; X_2, Y) = I(X_1; Y|X_2)$ when X_1 and X_2 are independent.

Similarly, for $j \neq 1$,

$$P(E_{1j}) \doteq 2^{-nI(X_2; Y|X_1)}$$

and for $i \neq 1, j \neq 1$,

$$P(E_{ij}) \doteq 2^{-nI(X_1; X_2; Y)}.$$

If follows that

$$P_n \leq P(E_{11}^c) + 2^{nR_1} 2^{-nI(X_1; Y|X_2)} + 2^{nR_2} 2^{-nI(X_2; Y|X_1)} + 2^{n(R_1+R_2)} 2^{-nI(X_1, X_2; Y)}. \tag{5.4}$$

Thus the conditions of the theorem cause each term to tend to zero as $n \to \infty$.

Time-sharing allows any (R_1, R_2) in the convex hull to be achieved, and the theorem is proved.

We now prove the converse to Theorem 3. This should provide the reader with some of the basic converse proof techniques.

(ii) Converse.

Given a $((2^{nR_1}, 2^{nR_2}), n)$ code for the MAC, the empirical probability distribution on $\mathcal{W}_1 \times \mathcal{W}_2 \times \mathcal{X}_1^n \times \mathcal{X}_2^n \times \mathcal{Y}^n$ is of the form

$$p(w_1, w_2, x_1, x_2, y) = \frac{1}{2^{n(R_1+R_2)}} \cdot p(x_1|w_1)\, p(x_2|w_2)\, p(y|x_1, x_2). \tag{5.5}$$

Fano's inequality [6] requires that

$$H(W_1, W_2|Y) \leq n(R_1 + R_2)\bar{P}_e^n + h(\bar{P}_e^n) \triangleq n\epsilon_n. \tag{5.6}$$

The assumption that $P_e^n \to 0$ implies that $\epsilon_n \to 0$. Consequently,

$$H(W_1 \mid Y) \leq n\epsilon_n$$

$$H(W_2 \mid Y) \leq n\epsilon_n. \tag{5.7}$$

Now consider,

(i)
$$nR_1 = H(W_1) = I(W_1; Y) + H(W_1|Y) \leq I(W_1; Y) + n\epsilon_n. \tag{5.8}$$

By the data processing inequality it follows that

$$nR_1 \leq I(W_1; Y) + n\epsilon_n \leq I(X_1; Y) + n\epsilon_n.$$

But since X_1 and X_2 are independent, we have

$$nR_1 \leq I(X_1; Y|X_2) + n\epsilon_n. \tag{5.9}$$

By symmetry it can be shown that

(ii)
$$nR_2 \leq I(X_2; Y|X_1) + n\epsilon_n. \tag{5.10}$$

Finally,

(iii)
$$n(R_1 + R_2) \leq I(W_1, W_2; Y) + n\epsilon_n. \tag{5.11}$$

The data processing theorem yields

$$I(W_1, W_2; Y) \leq I(X_1, X_2; Y). \tag{5.12}$$

Now the discrete memorylessness of the channel is easily applied to yield

$$R_1 \leq (1/n)\, I(X_1; Y|X_2) + \epsilon_n \leq (1/n) \sum_{i=1}^{n} I(X_{1i}; Y_i|X_{2i}) + \epsilon_n$$

$$R_2 \leq (1/n)\, I(X_2; Y|X_1) + \epsilon_n \leq (1/n) \sum_{i=1}^{n} I(X_{2i}; Y_i|X_{1i}) + \epsilon_n$$

and

$$R_1 + R_2 \leq (1/n)\, I(X_1, X_2; Y) + \epsilon_n \leq (1/n) \sum_{i=1}^{n} I(X_{1i}, X_{2i}; Y_i) + \epsilon_n \tag{5.13}$$

and the converse is proved.

VI. The Slepian-Wolf Source Coding Theorem

We know how to encode a source X. A rate $R > H(X)$ is sufficient. Now suppose that there are two sources $(X, Y) \sim p(x, y)$. A rate $R > H(X, Y)$ is sufficient. But what if the X-source and the Y-source must be separately described for some user who wishes to reconstruct X and Y? Clearly, by separately encoding X and Y, it is seen that a rate $R = R_x + R_y > H(X) + H(Y)$ is sufficient. However, in the surprising

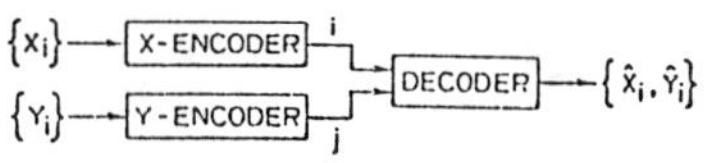

Fig. 12. Correlated source coding.

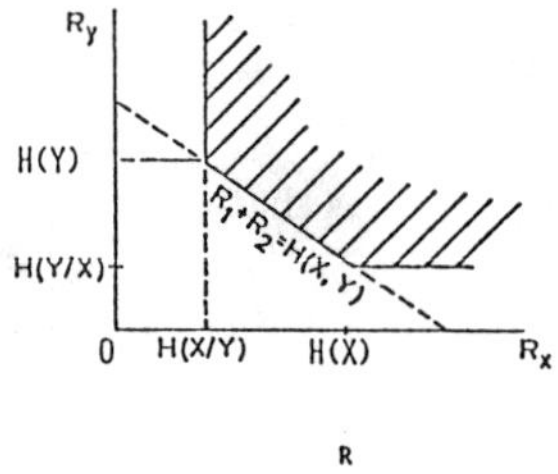

Fig. 13. Rate region for Slepian-Wolf data compression.

and fundamental paper of Slepian and Wolf [15], it is shown that a total rate $R = H(X, Y)$ suffices.

Example: Let X be a sequence of 1's and 0's generated by flipping a fair coin. Let Z be a sequence of 0's and 1's generated by flipping a coin with $\Pr(Z = 1) = p = 0.11$. Let $Y = X \oplus Z$. Note that Y is also a fair coin sequence.

Suppose that $R_y = 1$, so that Y is perfectly described. Then the information rate needed to describe X is not $R_x = 1$ but is $R_x = H(X|Y) = h(p) = \frac{1}{2}$. This is true despite the fact that the describer of X does not know the Y on which his description will be conditioned.

We now give a formal description of the Slepian-Wolf problem depicted in Fig. 12. The proof is different from that in [15] and [16] in that no time sharing is required.

A sequence $\{(X_i, Y_i)\}_{i=1}^n$ of independent copies of the pair (X, Y) of discrete random variables is to be encoded by two separate encoders; an X-encoder that observes the $\{X_i\}_{i=1}^n$ sequence and maps it into an integer $i \in [1, 2^{nR_x}]$, and a Y-encoder that observes the $\{Y_i\}_{i=1}^n$ sequence and maps it into the integer $j \in [1, 2^{nR_y}]$. The integers (i, j) are then communicated to a common decoder that tries to reproduce the $\{(X_i, Y_i)\}_{i=1}^n$ sequence. A $((2^{nR_x}, 2^{nR_y}), n)$ compression scheme for this problem consists of an integer n, two encoding functions

$$i : \mathcal{X}^n \rightarrow [1, 2^{nR_x}]$$

$$j : \mathcal{Y}^n \rightarrow [1, 2^{nR_y}]$$

and a decoding function

$$g : [1, 2^{nR_x}] \times [1, 2^{nR_y}] \rightarrow \mathcal{X}^n \times \mathcal{Y}^n.$$

The average probability of incorrect reproduction is given by

$$\bar{P}_e^n \triangleq P\{g(i(X), j(Y)) \neq (X, Y)\}.$$

A rate pair (R_x, R_y) is said to be *achievable* if there exist a sequence of compression schemes with $\bar{P}_e^n \rightarrow 0$. The problem is to find the set R of *achievable* compression rates (R_x, R_y). The rate region R (shown in Fig. 13) was established by Slepian and Wolf [15].

Theorem 4 (Slepian-Wolf): The set R of all achievable rates is given by

$$R = \{(R_x, R_y) : R_x > H(X|Y)$$

$$R_y > H(Y|X)$$

$$R_x + R_y > H(X, Y)\}. \tag{6.1}$$

The idea is to divide the $\mathcal{X}^n$-space into 2^{nR_x} bins and the $\mathcal{Y}^n$-space into 2^{nR_y} bins as shown in Fig. 14.

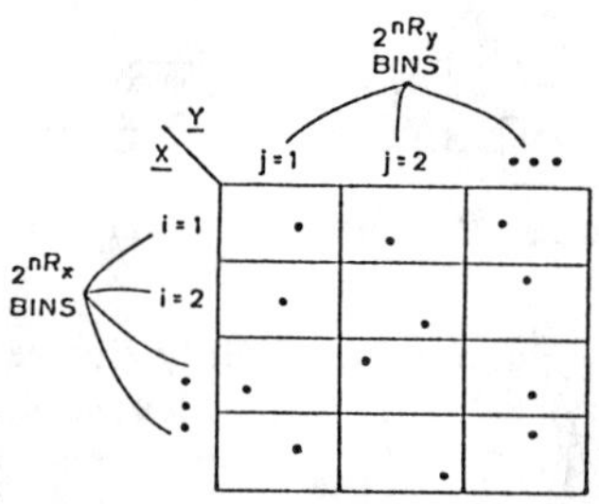

Fig. 14. Bins for Slepian and Wolf.

The dots in this figure correspond to jointly typical (x, y) pairs, and they ϵ-exhaust the probability. If R_x and R_y are large enough, there will be no more than one dot per product bin. Thus the name i of the x bin and the name j of the y bin will uniquely define the sequence (x, y) that falls in the product bin (i, j). We now prove this result.

Proof:

Encoding: Randomly assign every $x \in \mathcal{X}^n$ to one of 2^{nR_x} bins, according to a uniform distribution over the integers $[1, 2^{nR_x}]$. More precisely, for every $x \in \mathcal{X}^n$, let $p(i(x) = i) = 2^{-nR_x}$ for $i \in [1, 2^{nR_x}]$. Similarly, randomly assign every $y \in \mathcal{Y}^n$ to one of 2^{nR_y} bins, such that $p(j(Y) = j) = 2^{-nR_y}, j \in [1, 2^{nR_y}]$.

Decoding: Given (i_0, j_0) declare $(\hat{x}, \hat{y}) = (x, y)$ was sent if there is one and only one pair of sequences (x, y) such that

$$i(x) = i_0 \qquad j(y) = j_0$$

and

$$(x, y) \in A_\epsilon(X, Y).$$

Otherwise declare an error. To bound $\bar{P}_e^n$, define the events

$$E_0 = \{(X, Y) \notin A_\epsilon(X, Y)\}$$

$$E_1 = \{\exists \text{ some } (x', y') \neq (X, Y), (x', y') \in A_\epsilon,$$

$$\text{such that } i(x') = i_0, j(y') = j_0\}.$$

Using the union of events bound, we obtain

$$\bar{P}_n = P(E_0 \cup E_1) \leq P(E_0) + P(E_1). \tag{6.2}$$

The first term in (6.1) $\rightarrow 0$ as $n \rightarrow \infty$ by typicality (Lemma 1). Notice that the event E_1 is equal to the union of the events

$$E_{11} = \{\exists \text{ some } x' \text{ such that } x' \neq X$$

$$\text{and } i(x') = i_0, (x', y) \in A_\epsilon, y = Y\}$$

$$E_{12} = \{\exists \text{ some } y' \text{ such that } y' \neq Y \text{ and } j(y') = j_0,$$

$$(x, y') \in A_\epsilon, \text{and } x = X\}$$

and

$$E_{13} = \{\exists \text{ some } (x', y') \in A_\epsilon \text{ such that } x' \neq X \text{ and } y' \neq Y$$

$$\text{and } i(x') = i_0 \text{ and } j(y') = j_0\}.$$

First consider the probability of E_{11}. By the union of events bound

$$P(E_{11}) \leq \sum_{\substack{x' \neq x \\ (x', y) \in A_\epsilon(X, Y)}} P\{i(x') = i_0\}$$

$$= 2^{-nR_x} \|A_\epsilon(X|y)\|.$$

But from Lemma 2

$$\|A_\epsilon(X|y)\| \leq 2^{n(H(X|Y)+2\epsilon)}.$$

Therefore,

$$P(E_{11}) \to 0 \quad \text{as} \quad n \to \infty$$

if

$$R_x > H(X|Y).$$

Similarly, the probability of the events E_{12} and $E_{13} \to 0$ as $n \to \infty$ if $R_y > H(Y|X)$ and $R_x + R_y > H(Y, X)$, respectively.

Finally, the probability of error is averaged over random partitions. Thus there exists maps $i^*(\cdot), j^*(\cdot)$ such that $\bar{P}_e \to 0$.

The Slepian-Wolf theorem presented in this section is historically the first fundamental result in multiple user source coding theory. For a comprehensive overview of that theory, the reader is referred to Berger [17].

VII. The Multiple Access Channels with Correlated Sources

From Slepian-Wolf, we know how to give efficient separate descriptions of correlated sources. From the multiple access channel, we know how to send two separate independent descriptions over a noisy channel. Apparently we can weld these two problems together to obtain a theory of sending correlated sources over a multiple access channel (Fig. 15).

We do so in this section and find that the combined theory involves a new ingredient—correlation of the inputs for the multiple access channel. As a byproduct, we shall find a common proof of the Slepian-Wolf theorem and the multiple access channel capacity region [18].

Assume we have two information sources $U_1, U_2, \cdots$ and $V_1, V_2, \cdots$ generated by repeated independent drawings of a pair of discrete random variables U and V from a given bivariate distribution $p(u, v)$.

A block code for the channel consists of an integer n, and two encoding functions

$$x_1^n : \mathcal{U}^n \to \mathcal{X}_1^n \qquad x_2^n : \mathcal{V}^n \to \mathcal{X}_2^n$$

assigning codewords to the source outputs, and a decoding function

$$d^n : \mathcal{Y}^n \to \mathcal{U}^n \times \mathcal{V}^n.$$

The probability of error is given by

$$P_n = P\{(U^n, V^n) \neq d^n(Y^n)\}$$

$$= \sum_{(u,v) \in \mathcal{U}^n \times \mathcal{V}^n} p(u^n, v^n)\, P\{d^n(y^n) \neq (u^n, v^n) | (U^n, V^n)$$

$$= (u^n, v^n)\} \tag{7.1}$$

where the joint probability mass function is given, for a code assignment $\{x_1(u^n), x_2(v^n)\}$, by

$$p(u, v, y) = \prod_{i=1}^{n} p(u_i, v_i)\, p(y_i | x_{1i}(u^n), x_{2i}(v^n)). \tag{7.2}$$

Definition: The source (U, V) can be reliably transmitted over the multiple access channel $(\mathcal{X}_1 \times \mathcal{X}_2, \mathcal{Y}, p(y|x_1, x_2))$ if there exists a sequence of block codes $(x_1^n(u^n), x_2^n(v^n), d^n(y^n))$ such that

$$P_n = P\{d^n(Y^n) \neq (U^n, V^n)\} \to 0.$$

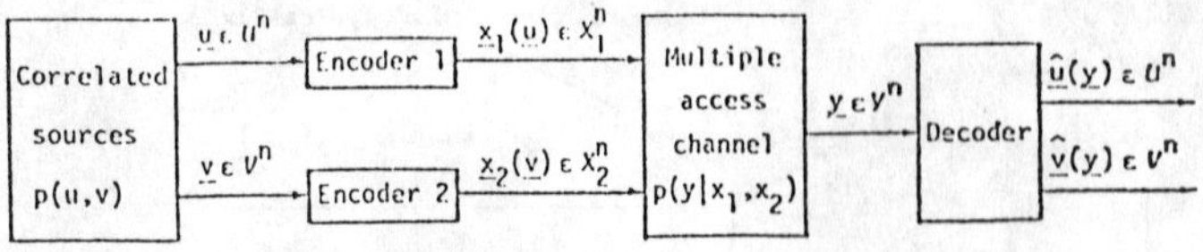

Fig. 15. The multiple access channel with arbitrarily correlated sources.

Example: Consider the transmission of the correlated sources (U, V) with the joint distribution $p(u, v)$ given by

$\mathcal{V} \backslash \mathcal{U}$	0	1
0	$\frac{1}{3}$	$\frac{1}{3}$
1	0	$\frac{1}{3}$

over the multiple access channel defined by

$$\mathcal{X}_1 = \mathcal{X}_2 = \{0, 1\}$$

$$\mathcal{Y} = \{0, 1, 2\}$$

$$Y = X_1 + X_2.$$

Here $H(U, V) = \log 3 = 1.58$ bits. On the other hand, if X_1 and X_2 are independent,

$$\max_{p(x_1)p(x_2)} I(Y; X_1, X_2) = 1.5 \text{ bits}.$$

Thus $H(U, V) > I(Y; X_1, X_2)$ for all $p(x_1)\,p(x_2)$. Consequently there is no way, even with the use of Slepian-Wolf data compression on U and V, to use the standard multiple access channel capacity region to send U and V reliably to Y. However, it is easy to see that with the choice $X_1 \equiv U$, and $X_2 \equiv V$, error-free transmission of the source over the channel is possible. This example shows that the separate source and channel coding described above is not optimal—the partial information that each of the random variables U and V contains about the other is destroyed in this separation.

To allow partial cooperation between the two transmitters, we allow our codes to depend statistically on the source outputs. This induces dependence between codewords.

We shall outline here a proof of a special case of Theorem 1 [18], in which U and V have no common part. In this case, we must show that U and V can be reliably sent to Y if, for $p(u, v)\, p(x_1|u)\, p(x_2|v)\, p(y|x_1, x_2)$,

$$H(U|V) < I(X_1; Y|X_2, V)$$

$$H(V|U) < I(X_2; Y|X_1, U) \tag{7.3}$$

$$H(U, V) < I(X_1, X_2; Y). \tag{7.4}$$

The proof will employ random coding. We first describe the random code generation and encoding-decoding schemes, and then analyze the probability of error.

Generating Random Codes: Fix $p(x_1|u)$ and $p(x_2|v)$; for each $u \in \mathcal{U}^n$ generate *one* x_1 sequence drawn according to $\prod_{i=1}^n p(x_{1i}|u_i)$ and for each $v \in \mathcal{V}^n$ generate *one* x_2 sequence drawn according to $\prod_{i=1}^n p(x_{2i}|v_i)$. Call these sequences $x_1(u)$ and $x_2(v)$, respectively.

Encoding: Transmitter 1, upon observing u at the output of source 1, transmits $x_1(u)$ and transmitter 2, after observing v at the output of source 2, transmits $x_2(v)$. Assume the maps $x_1(\cdot), x_2(\cdot)$ are known to the receiver.

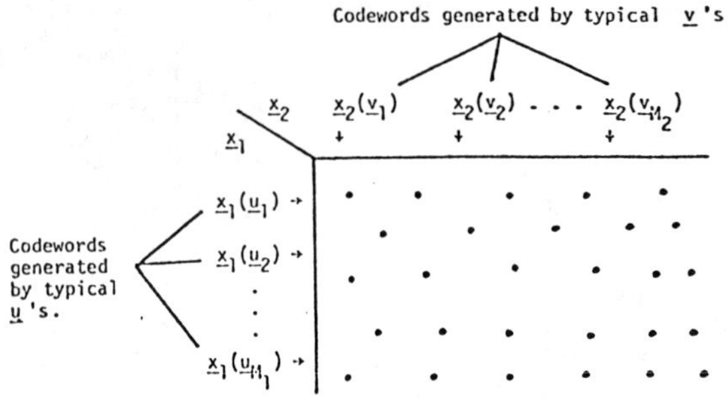

Fig. 16. *Picture of joint typicality for multiple access channel.* The dots correspond to jointly typical (X_1, X_2) pairs. Note that only $2^{nH(U,V)}$ $(x_1(u), x_2(v))$ pairs are likely to occur.

Decoding: Upon receiving y, the decoder finds the only (u, v) pair such that $(u, v, x_1(u), x_2(v), y) \in A_\epsilon$, where A_ϵ is the set of jointly ϵ-typical sequences. If there is no such (u, v) pair, or there exists more than one such pair, the decoder declares an error. A helpful picture is given in Fig. 16.

Error: Suppose (U_0, V_0) is the source output, and define the events

$$E_{(u,v)} = \{(u, v, x_1(u), x_2(v), y) \in A_\epsilon\}.$$

The probability of error averaged over (U_0, V_0) and the random codes is given by

$$P_n = P\left\{E^c_{(U_0,V_0)} \cup \bigcup_{(u,v)\neq(U_0,V_0)} E_{(u,v)}\right\}$$

$$\leqslant P\{E^c_{(U_0,V_0)}\} + P\left\{\bigcup_{(u,v)\neq(U_0,V_0)} E_{(u,v)}\right\}.$$

The first term $\to 0$ by Lemma 1. The second term is bounded above by

$$\sum_{(u_0,v_0)\in A_\epsilon} p(u_0, v_0) \cdot \sum_{\substack{u'\neq u_0\\ v'=v_0}} P\{(u', v', X_1(u'), X_2(v'), Y) \in A_\epsilon | (u_0, v_0)\}$$

$$+ \sum_{(u_0,v_0)\in A_\epsilon} p(u_0, v_0) \sum_{\substack{u'=u_0\\ v'\neq v_0}} P\{\cdot\}$$

$$+ \sum_{(u_0,v_0)\in A_\epsilon} p(u_0, v_0) \sum_{\substack{u'\neq u_0\\ v'\neq v_0}} P\{\cdot\}$$

$$\leqslant 2^{n(H(U|V)+\epsilon)} 2^{-n(I(X_1;Y|X_2,V)-\epsilon)}$$

$$+ 2^{n(H(V|U)+\epsilon)} 2^{-n(I(X_2;Y|X_1,U)-\epsilon)}$$

$$+ 2^{nH(U,V)} 2^{-n(I(X_1,X_2;Y)-\epsilon)}. \tag{7.5}$$

Consequently, $P_n \to 0$ if the conditions in (7.3) are satisfied.

VIII. The Broadcast Channel

The broadcast channel [19] is a communication network with one transmitter and many receivers.

The basic problem is to find the set of simultaneously achievable rates (R_0, R_1, R_2). To date this problem has not been solved. The special case of sequentially degraded channels has been solved by Bergmans [20] and Gallager [21]. An achievable rate region for the general broadcast channel has been put forth by Marton [22], but is not known to be the capacity region.

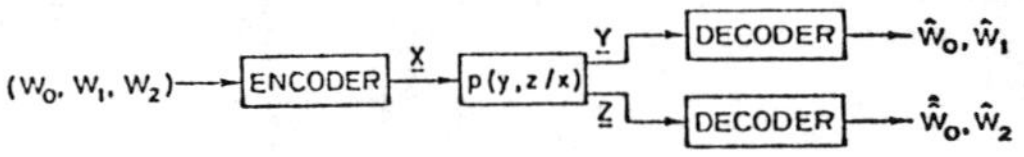

Fig. 17. The discrete memoryless broadcast channel.

We now give the formal definition of the two-receiver broadcast channel problem. A discrete memoryless broadcast channel $(\mathfrak{X}, p(y, z|x), \mathcal{Y} \times \mathcal{Z})$ as depicted in Fig. 17 consists of three finite alphabets $\mathfrak{X}$, $\mathcal{Y}$, $\mathcal{Z}$ and a probability transition matrix $p(y, z|x)$.

An $((2^{nR_0}, 2^{nR_1}, 2^{nR_2}), n)$ code for a broadcast channel consists of three sets of integers

$$\mathfrak{W}_0 = [1, 2^{nR_0}], \quad \mathfrak{W}_1 = [1, 2^{nR_1}], \quad \text{and} \quad \mathfrak{W}_2 = [1, 2^{nR_2}]$$

an *encoding function*

$$X: \mathfrak{W}_0 \times \mathfrak{W}_1 \times \mathfrak{W}_2 \to \mathfrak{X}^n$$

and two *decoding functions*

$$g_1: \mathcal{Y}^n \to \mathfrak{W}_0 \times \mathfrak{W}_1$$

$$g_2: \mathcal{Z}^n \to \mathfrak{W}_0 \times \mathfrak{W}_2.$$

The integer w_0 has the interpretation of the common part of the message, while the integers w_1, w_2 are called the independent parts of the message. Assuming uniform distribution on the set of messages $\mathfrak{W}_0 \times \mathfrak{W}_1 \times \mathfrak{W}_2$, define

$$\bar{P}^n_e = \frac{1}{2^{n(R_0+R_1+R_2)}} \sum_{(w_0,w_1,w_2)\in \mathfrak{W}_0\times\mathfrak{W}_1\times\mathfrak{W}_2} P\{g_1(Y) \neq (w_0, w_1) \text{ or } g_2(Z) \neq (w_0, w_2) | (w_0, w_1, w_2) \text{ sent}\} \tag{8.1}$$

to be the *average probability of error* of the code.

The rate triple (R_0, R_1, R_2) is said to be *achievable* by a broadcast channel if there exists a sequence of $((2^{nR_0}, 2^{nR_1}, 2^{nR_2}), n)$ codes with $\bar{P}^n_e \to 0$. The *capacity region C* for the broadcast channel is the closure of the set of all achievable rate triples (R_0, R_1, R_2). We now look at a family of broadcast channels in which one receiver can be considered worse than the other. The mathematical definition corresponding to this physical idea is as follows.

Definition: The broadcast channel $(\mathfrak{X}, p(y, z|x), \mathcal{Y} \times \mathcal{Z})$ is called *degraded* if there exists a probability transition matrix $p(z|y)$ such that for all $z \in \mathcal{Z}$ and $x \in \mathfrak{X}$

$$p(z|x) = \sum_{y\in\mathcal{Y}} p(y, z|x) = \sum_{y\in\mathcal{Y}} p(z|y) p(y|x). \tag{8.2}$$

Theorem 5: The capacity region of the degraded broadcast channel $(\mathfrak{X}, p(y, z|x), \mathcal{Y} \times \mathcal{Z})$ is given by

$$C = \{(R_0, R_1, R_2): R_0 + R_2 < I(U; Z),\ R_1 < I(X; Y|U)$$

for some $p(u)\, p(x|u)$,

$$\|\mathfrak{U}\| \leqslant \min\{\|\mathfrak{X}\|, \|\mathcal{Y}\|, \|\mathcal{Z}\|\}\}. \tag{8.3}$$

Proof: Fix $p(u), p(x|u)$.

a) Achievability: The natural coding idea necessary to solve this channel is that of superposition coding [19], [20]. We first find a low rate code for the worst receiver Z. Since this rate is below the capacity from X to Z, it is possible to further randomize the codewords chosen for Z into a small cloud of satellite codewords intended for the good receiver Y. Receiver Z then decodes the cloud center. Receiver Y first decodes the cloud center, which he can do because he can "see" at least as much as Z_1. Receiver Y then decodes the satellite codeword.

(i) Random Code: First generate $2^{n(R_2+R_0)}$ i.i.d. sequences u, each with probability

$$p(u)=\prod_{i=1}^{n} p(u_k).$$

Label these $u(i), i \in [1, 2^{n(R_0+R_2)}]$.

For every $u(i)$, generate 2^{nR_1} conditionally independent x sequences, each with probability

$$p(x \mid u(i))=\prod_{k=1}^{n} p(x_k \mid u_k(i)).$$

Label these $x(i,j), j \in [1, 2^{nR_1}]$.

Here $u(i)$ plays the role of the cloud center understandable to both Y and Z, while $x(i,j)$ is the jth satellite codeword in the ith cloud.

(ii) Decoding: If y is received, declare $(\hat{i},\hat{j})=(i,j)$ was sent if there is one and only one pair $(i,j) \in [1, 2^{n(R_0+R_2)}] \times [1, 2^{nR_1}]$, such that $(u(i), x(i,j), y) \in A_\epsilon(U,X,Y)$. If z is received, declare $\hat{\hat{i}}=i$ was sent if there is one and only one i such that $(u(i), z) \in A_\epsilon(U,Z)$. Let I, J be independent r.v.'s drawn according to uniform distributions on $[1, 2^{n(R_0+R_2)}] \times [1, 2^{nR_1}]$. Let the code be chosen randomly according to the encoding description. Then the probability of error averaged over I, J and the random code is given by

$$P_n=P\{(\hat{I},\hat{J}) \neq (I,J) \ \text{ or } \ \hat{\hat{I}} \neq I\}. \tag{8.4}$$

By the symmetry induced by the random coding, we see that each transmitted message (i,j) yields the same probability of error. Thus, setting $(i,j)=(1,1)$, we have

$$P_n=P\{(\hat{I},\hat{J}) \neq (1,1) \ \text{ or } \ \hat{\hat{I}} \neq 1 \mid (I,J)=(1,1)\}.$$

Define the events

$$E_{Zi}=\{(U(i), Z) \in A_\epsilon(U,Z)\}$$
$$E_{Yi}=\{(U(i), Y) \in A_\epsilon(U,Y)\}$$
$$E_{Yij}=\{(U(i), X(i,j), Y) \in A_\epsilon(U,X,Y)\}.$$

Then

$$P_n=P\{E_{Z1}^c \cup E_{Y11}^c \bigcup_{i\neq 1} E_{Zi} \bigcup_{i\neq 1} E_{Yi} \bigcup_{j \neq 1} E_{Y1j}\} \leqslant P\{E_{Z1}^c\}$$
$$+P\{E_{Y11}^c\}+\sum_{i\neq 1} P\{E_{Zi}\}+\sum_{i \neq 1} P\{E_{Yi}\}+\sum_{j\neq 1} P\{E_{Y1j}\}.$$

The first two terms correspond to the event that the correct codeword does not belong to the decoding set. The last terms correspond to the event that some incorrect codeword belongs to the decoding set. From Lemma 1, it follows that

$$P\{E_{Z1}^c\} \to 0, \quad P\{E_{Y11}^c\} \to 0.$$

Consider the event E_{Zi}. We observe, for $i \neq 1$, that $U(i)$ and Z are independent. Thus, by Lemma 3, for $i \neq 1$

$$P\{E_{Zi}\} \doteq 2^{-nI(U;Z)}.$$

Consequently,

$$\sum_{i\neq 1} P\{E_{Zi}\} \doteq 2^{-n(I(U;Z)-(R_0+R_2))}.$$

This term goes to zero if

$$R_0+R_2<I(U;Z).$$

Similarly, for the event $E_{Yi}, i \neq 1$

$$P\{E_{Yi}\} \doteq 2^{-nI(U;Y)}.$$

and

$$\sum_{i \neq 1} P\{E_{Yi}\} \to 0, \quad \text{if } R_0+R_2<I(U;Y) \tag{8.5}$$

But since by the degradedness of the broadcast channel $I(U;Y) \leqslant I(U;Z)$, the condition (8.5) is redundant.

Next, consider the event E_{Y1j}. From Lemma 3, it follows that

$$P\{E_{Y1j}\} \doteq 2^{-nI(X;Y|U)}.$$

Thus the term

$$\sum_{j\neq 1} P\{E_{Y1j}\} \to 0, \quad \text{if } R_1<I(X;Y \mid U). \tag{8.6}$$

b) Converse: (See Gallager [21] and Bergmans [23].)

The capacity region for the general broadcast channel is still an open problem. Several special cases have been recently solved (e.g., degraded message sets [24], more capable class [25], deterministic class [26], [27], and parallel degraded channels [28]). The most general inner bound to the capacity region of the broadcast channel is that given by Marton [22]. In the following theorem, we outline a proof of a special case of Marton's general result where it is assumed that no common part is decoded. However, this special case isolates a new coding idea. This, together with superposition, yields Marton's theorem. The proof of the following theorem is due to El Gamal and van der Meulen [29].

Theorem: Let

$$\mathcal{R}_0=\{(R_1,R_2): R_1, R_2 \geqslant 0,$$
$$R_1 \leqslant I(U;Y),$$
$$R_2 \leqslant I(V;Z),$$
$$R_1+R_2 \leqslant I(U;Y)+I(V;Z)-I(U;V),$$
$$\text{for some } p(u,v,x) \text{ on } \mathcal{U} \times \mathcal{V} \times \mathcal{X}\} \tag{8.7}$$

Then any rate pair $(R_1,R_2) \in \mathcal{R}_0$ is achievable for

$$(\mathcal{X}, p(y,z|x), \mathcal{Y} \times \mathcal{Z}).$$

Proof: (Outline)

Fix $p(u,v)$, $p(x|u,v)$. The channel $p(y,z|x)$ is given. The idea is to send u to y and v to z.

Random Coding: Generate $2^{nI(U;Y)}$ typical u's in U^n. Generate $2^{nI(V;Z)}$ typical V's in V^n. Now randomly throw the u's into 2^{nR_1} bins and the v's into 2^{nR_2} bins. For each product bin, find a *jointly* typical (u,v) pair. This can be done, as shown by the circled dots in Fig. 18, if

$$R_1+R_2<I(U;Y)+I(V;Y)-I(U;V). \tag{8.8}$$

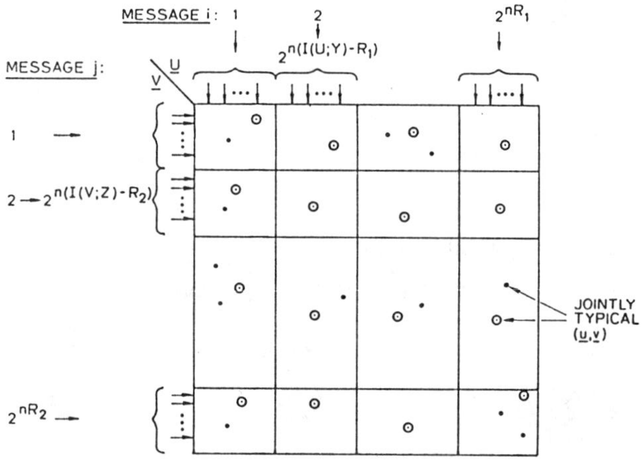

Fig. 18. Coding for Marton's Theorem.

To see this, we recall from Lemma 3 that independent choices of u and v result in a jointly typical (u, v) with probability $2^{-nI(U;V)}$. Now there are $2^{n(I(U;Y)-R_1)}$ U's in any U-bin, and $2^{n(I(V;Z)-R_2)}$ V's in any V-bin. Thus the expected number of jointly typical (U, V) pairs in a given product $U \times V$ bin is

$$2^{n(I(U;Y)-R_1)}2^{n(I(V;Z)-R_2)}2^{-nI(U;V)}. \tag{8.9}$$

The desired jointly typical (u, v) pair can be found if this expected number is much greater than 1. This is guaranteed if (8.8) holds.

Continuing with the coding, for each $U \times V$ bin and its designated jointly typical (u, v) pair, generate $x(u, v)$ according to the conditional distribution $\Pi_{k=1}^{n} p(x_k|u_k, v_k)$.

Encoding: To send i to Y and j to Z, send $x(u, v)$, where (u, v) is the designated pair in the product bin (i, j).

Decoding: Receiver Y, upon receiving y, finds the u such that (u, y) is jointly typical. Thus it is necessary that $R_1 < I(U; Y)$. He then finds the index i of the bin in which u lies. Receiver Z finds the v such that (v, z) is jointly typical. Thus we need $R_2 < I(V; Z)$. He then finds the index j of the bin in which v lies.

A Note on Feedback: In [30], it was shown that feedback cannot increase the capacity of the physically degraded broadcast channel, i.e., broadcast channels for which $p(y, z|x) = p(y|x)p(z|y)$. Surprisingly, it was later shown by Ozarow [31] and Dueck [32] that feedback can in fact increase the capacity of general broadcast channels.

IX. The Relay Channel

The discrete memoryless relay channel denoted by $(\mathcal{X}_1 \times \mathcal{X}_2, p(y, y_1|x_1, x_2), \mathcal{Y} \times \mathcal{Y}_1)$ consists of four finite sets $\mathcal{X}_1, \mathcal{X}_2, \mathcal{Y}, \mathcal{Y}_1$ and a collection of probability distributions $p(\cdot, \cdot|x_1, x_2)$ on $\mathcal{Y} \times \mathcal{Y}_1$ one for each $(x_1, x_2) \in \mathcal{X}_1 \times \mathcal{X}_2$. The interpretation is that x_1 is the input to the channel and y is the output, y_1 is the relay's output and x_2 is the input symbol chosen by the relay as shown in Fig. 19. The problem is to find the capacity of the channel between the sender x_1 and the receiver y.

The relay channel was introduced by van der Meulen [33]. The following discussion is based on Cover and El Gamal [34].

A $(2^{nR}, n)$ code for the relay channel consists of a set of integers $\mathcal{W}$ $[1, 2^{nR}]$, an encoding function

$$x_1: [1, 2^{nR}] \to \mathcal{X}_1^n$$

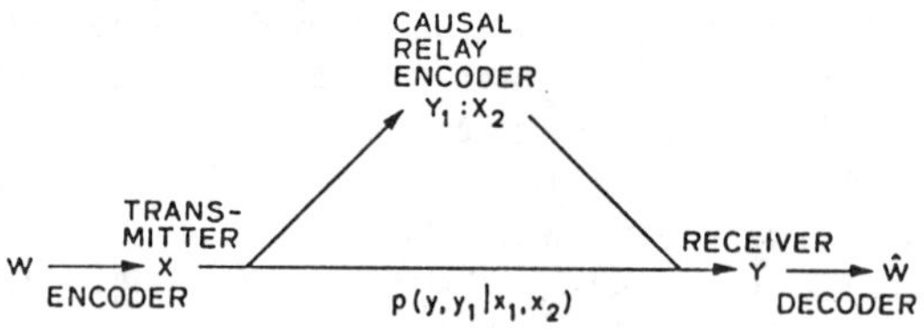

Fig. 19. The relay channel.

a set of relay functions $\{f_i\}_{i=1}^{n}$ such that

$$x_{2i} = f_i(Y_{11}, Y_{12}, \cdots, Y_{1i-1}), \qquad 1 \leqslant i \leqslant n$$

and a decoding function

$$g: \mathcal{Y}^n \to [1, 2^{nR}].$$

Note that the allowed encoding functions actually form part of the definition of the relay channel because of the nonanticipatory relay condition. The relay channel input x_{2i} is allowed to depend only on the past $y_{11}, y_{12}, \cdots, y_{1i-1}$. This is the definition used by van der Meulen [33]. The channel is memoryless in the sense that (y_i, y_{1i}) depends on the past only through the current transmitted symbols (x_{1i}, x_{2i}). Thus for any choice $p(w)$, $w \in \mathcal{W}$, and code choice $x_1: [1, 2^{nR}] \to \mathcal{X}_i^n$ and relay functions $\{f_i\}_{i=1}^{\infty}$, the joint probability mass function on $\mathfrak{M} \times \mathcal{X}_1^n \times \mathcal{X}_2^n \times \mathcal{Y}^n \times \mathcal{Y}_1^n$ is given by

$$p(w, x_1, x_2, y, y_1) = p(w) \prod_{i=1}^{n} p(x_{1i}|w)p(x_{2i}|y_{11}, y_{12}, \cdots, y_{1i-1})p(y_i, y_{1i}|x_{1i}, x_{2i}). \tag{9.1}$$

If the message $w \in [1, 2^{nR}]$ is sent, let

$$\lambda(w) = \Pr\{g(Y) \neq w | w \text{ sent}\}$$

denote the conditional probability of error. We define the average probability of error of the code as

$$\bar{P}_e^n = 2^{-nR} \sum_{w} \lambda(w).$$

The probability of error is calculated under uniform distribution over the codewords $w \in [1, 2^{nR}]$. The rate R is said to be *achievable* by a relay channel if there exists a sequence of $(2^{nR}, n)$ codes with $\bar{P}_e^n \to 0$. The *capacity* C of the relay channel is the supremum of the set of achievable rates.

We first give an upper bound to the capacity of any relay channel [34]:

Theorem 9.1: For any relay channel $(\mathcal{X}_1 \times \mathcal{X}_2, p(y, y_1|x_1, x_2), \mathcal{Y} \times \mathcal{Y}_1)$ the capacity C is bounded above by

$$C \leqslant \sup_{p(x_1, x_2)} \min\{I(X_1, X_2; Y), I(X_1; Y, Y_1|X_2)\} \tag{9.2}$$

This upper bound has a nice max flow min cut interpretation. The first term in (9.2) upper bounds the rate of information flow from senders X_1 and X_2 to receiver Y. The second term bounds the rate of transmission from X_1 to Y and Y_1.

We now consider a family of relay channels in which the relay receiver Y_1 is better than the ultimate receiver Y in the sense defined below. Here the max flow min cut upper bound in (9.2) is achieved.

Definition: The relay channel $(\mathcal{X}_1 \times \mathcal{X}_2, p(y, y_1|x_1, x_2), \mathcal{Y} \times \mathcal{Y}_1)$ is said to be *degraded* if $p(y, y_1|x_1, x_2)$ can be written in the form

$$p(y, y_1|x_1, x_2) = p(y_1|x_1, x_2)p(y|y_1, x_2). \tag{9.3}$$

Thus Y is a random degradation of the relay signal Y_1.

For the degraded relay channel, the capacity is given by the following.

Theorem 9.2: The capacity C of the degraded relay chanel is given by

$$C = \sup_{p(x_1, x_2)} \min \{I(X_1, X_2; Y), I(X_1; Y_1 | X_2)\} \quad (9.4)$$

where the supremum is over all joint distributions $p(x_1, x_2)$ on $\mathcal{X}_1 \times \mathcal{X}_2$.

Proof: Converse.

The converse follows from Theorem 9.1 and by degradedness. Thus

$$I(X_1; Y, Y_1 | X_2) = I(X_1; Y_1 | X_2). \quad (9.5)$$

This capacity region proof involves 1) random coding, 2) list codes, 3) Slepian-Wolf partitioning, 4) coding for the cooperative multiple access channel, 5) superposition coding, and 6) block Markov encoding at the relay and transmitter.

Achievability (Outline): We consider B blocks of transmission, each of n symbols. A sequence of $B - 1$ indices $W_i \in [1, 2^{nR}]$, $i = 1, 2, \cdots, B - 1$ will be sent over the channel in nB transmissions. (Note that as $B \to \infty$, for fixed n, that the rate $R(B - 1)/B$ is arbitrarily close to R.)

In each n-block $b = 1, 2, \cdots, B$, we shall use the same doubly indexed set of codewords

$$C = \{x_1(w|s), x_2(s)\}, w \in [1, 2^{nR}], s \in [1, 2^{nR_0}], x_1(\cdot|\cdot) \in \mathcal{X}_1^n, x_2(\cdot) \in \mathcal{X}_2^n. \quad (9.6)$$

We shall also need a partition

$$\mathcal{S} = \{S_1, S_2, \cdots, S_{2^{nR_0}}\} \text{ of } \mathcal{W} = \{1, 2, \cdots, 2^{nR}\}$$

into

$$2^{nR_0} \text{ cells}, S_i \cap S_j = \phi, i \neq j, \cup S_i = \mathcal{W}.$$

The partition $\mathcal{S}$ will allow us to send side information to the receiver in the manner of Slepian and Wolf [15].

Random Coding: First generate at random 2^{nR_0} independent identically distributed n-sequences in $\mathcal{X}_2^n$, each drawn according to $p(x_2) = \prod_{i=1}^n p(x_{2i})$. Index them as $x_2(s)$, $s \in [1, 2^{nR_0}]$. For each $x_2(s)$, generate 2^{nR} conditionally independent n-sequences $x_1(w|s)$, $w \in [1, 2^{nR}]$ drawn according to $p(x_1|x_2(s)) = \prod_{i=1}^n p(x_{1i}|x_{2i}(s))$. This defines a random codebook $\mathcal{C} = \{x_1(w|s), x_2(s)\}$.

The random partition $\mathcal{S} = \{S_1, S_2, \cdots, S_{2^{nR_0}}\}$ of $\{1, 2, \cdots, 2^{nR}\}$ is defined as follows. Let each integer $w \in [1, 2^{nR}]$ be assigned independently, according to a uniform distribution over the indices $s = 1, 2, \cdots, 2^{nR_0}$, to cells S_s.

Encoding: Let $w_i \in [1, 2^{nR}]$ be the new index to be sent in block i, and assume that $w_{i-1} \in S_{s_i}$. Then the encoder sends $x_1(w_i|s_i)$. The relay has an estimate $\hat{\hat{w}}_{i-1}$ of the previous index w_{i-1}. (This will be made precise in the decoding section). Assume that $\hat{\hat{w}}_{i-1} \in S_{\hat{\hat{s}}_i}$. Then the relay encoder sends the codeword $X_2(\hat{\hat{s}}_i)$ in block i.

Decoding: We assume that at the end of block $(i - 1)$ the receiver knows $(w_1, w_2, \cdots, w_{i-2})$ and $(s_1, s_2, \cdots, s_{i-1})$ and the relay knows $(w_1, w_2, \cdots, w_{i-1})$ and consequently $(s_1, \cdots, s_i)$.

The decoding procedures at the end of block i are as follows.

1) Knowing s_i, and upon receiving $y_1(i)$, the *relay* receiver estimates the message of the transmitter $\hat{\hat{w}}_i = w$ iff there exists a *unique* w such that $(x_1(w|s_i), x_2(s_i), y_1(i))$ are jointly ϵ-typical. Using Lemma 3 it can be shown that $\hat{\hat{w}}_i = w_i$ with arbitrarily small probability of error if

$$R < I(X_1; Y_1 | X_2). \quad (9.7)$$

and n is sufficiently large.

2) The *receiver* declares that $\hat{s}_i = s$ was sent iff there exists one and only one s such that $(x_2(s), Y(i))$ are jointly ϵ-typical. From Lemma 3 we know that s_i can be decoded with arbitrarily small probability of error if

$$R_0 < I(X_2; Y) \quad (9.8)$$

and n is sufficiently large.

3) Assuming that s_i is decoded successfully at the receiver, then $\hat{w}_{i-1} = w$ is declared as the index sent in block $(i - 1)$ iff there is a unique $w \in S_{s_i} \cap \mathcal{L}(Y(i - 1))$ where $\mathcal{L}(Y(i - 1))$ is the list of indices w that the receiver y considered to be "typical" with $Y(i - 1)$ in the $(i - 1)$th block. If n is sufficiently large and if

$$R < I(X_1; Y | X_2) + R_0 \quad (9.9)$$

then $\hat{w}_{i-1} = w_{i-1}$ with arbitrarily small probability of error. Combining the two constraints (9.8) and (9.9) R_0 drops out, leaving

$$R < I(X_1; Y | X_2) + I(X_2; Y) = I(X_1, X_2; Y). \quad (9.10)$$

For a detailed analysis of the probability of error, the reader is referred to [34].

Theorem 9.1 can also be shown to be the capacity for the following classes of relay channels.

(i) Reversely degrades, i.e.,

$$p(y, y_1 | x_1, x_2) = p(y_1 | x_1, x_2) p(y | y_1, x_2).$$

(ii) Relay channel with feedback.

(iii) Deterministic relay channel [35]

$$y_1 = f(x_1, x_2) \qquad y = g(x_1, x_2).$$

A general lower bound to the capacity of any relay channel can be found in [34].

X. Summary and Open Problems

Now that we have presented the Gaussian multiple user channels in Section II and proved some of the results in detail in subsequent sections, it is time to abstract the salient points of the theory. We do so by paralleling the discussion of Section II for the channels shown in Fig. 20.

1) The Shannon Channel: The codewords are n-vectors $x(1), x(2), \cdots, x(2^{nR})$. First suppose $R < I(X; Y)$. The intuitive idea is that 2^{nR} x sequences, each independent identically distributed according to $\prod_{i=1}^n p(x_i)$ will be mutually far apart in the sense that if one sends an $x(i)$ and receives a Y, then, looking back from the Y, one finds a unique x in the codebook that is jointly typical with Y. This is the correct $x(i)$.

2) List Codes: Suppose that 2^{nR} codewords are generated as above for the Shannon channel, but $R > I(X; Y)$. Then there will be exponentially many codewords jointly typical with Y. In fact, the number of codewords on the list associated with Y will be $2^{n(R-I)}$. Thus the cutoff at capacity is very sharp. One goes from one codeword in the inverse fan for $R < I$ to an exponential number of codewords for $R > I$.

3) The Multiple Access Channel: Again, random coding works. Fix p_1, p_2. Choose 2^{nR_1} x_1 sequences according to

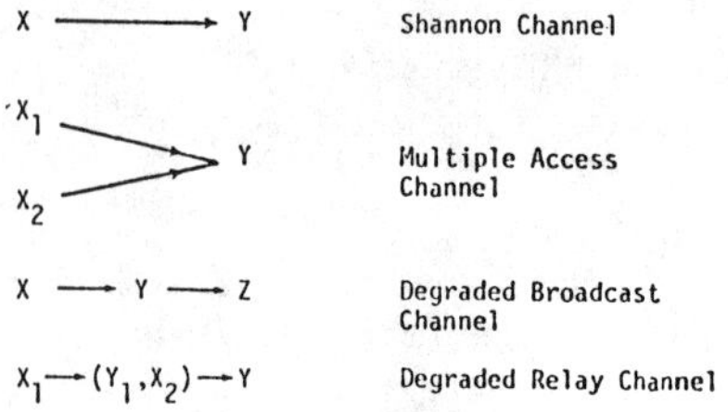

Fig. 20. Multiple user channels.

$\Pi p_1(x_{1i})$ and choose 2^{nR_2} x_2 sequences according to $\Pi p_2(x_{2i})$. Now send one of the codewords from the first codebook, and one of the codewords from the second codebook. These two together generate a random response Y drawn according to the conditional probability distribution of Y given the two sequences. With high probability, Y will be jointly typical with these two sequences. If R_1 and R_2 satisfy $R_1 < I(X_1;Y|X_2)$, $R_2 < I(X_2;Y|X_1)$, $R_1 + R_2 < I(X_1, X_2; Y)$ for some $p_1(x_1)p_2(x_2)$, then the receiver Y will see that, although there are exponentially many x_1's jointly typical with Y and exponentially many x_2's jointly typical with Y, there will be only one (x_1, x_2) *pair* in the codebook that is jointly typical with Y. Thus no error will be made in the decoding.

4) The Degraded Broadcast Channel: We assume that X sends to two receivers Y and Z, where Z is stochastically noisier than Y. Here, the idea of superposition is needed. We first introduce an auxiliary sender U and an associated channel from U to X. Thus the overall channel becomes $U \to X \to Y \to Z$. We now use a random code for the channel $U \to Z$ with 2^{nR_2} codewords u independently drawn according to $\Pi_{i=1}^{n} p(u_i)$. For each such codeword u, we generate 2^{nR_1} codewords x according to $\Pi_{i=1}^{n} p(x_i|u_i)$. If $R_2 < I(U;Z)$, then Z perceives u correctly. And if $R_1 < I(X;Y|U)$, then receiver Y perceives x correctly after decoding u. In summary, we send crude codewords to the poor receiver Z, but each of these crude codewords is really made up of small satellite codewords distinguishable by Y.

5) Degraded Relay Channel: Random coding works, but here we also need the Slepian–Wolf theorem. The idea is that we randomly choose codewords which are distinguishable by the relay, but not by the ultimate receiver. Nonetheless, by the argument for list codes, the number of codewords on the ultimate receiver's list is given by $2^{n(R - I(X_1;Y|X_2))}$. Then the relay, in cooperation with the sender, attempts to send the name of the codeword on the (unknown) list of known size $2^{n(R - I(X_1;Y|X_2))}$. He does this by randomly partitioning all the codewords into 2^{nR_0} sets each of equal size. This is the Slepian–Wolf step. He then sends the index of that set in cooperation with the transmitter. They do so by superimposing this information in the manner suggested by the degraded broadcast channel. Yet one more ingredient is necessary. The resolution of the index of the sent codeword on Y's list does not take place until the next block transmission. Thus we have a block Markov encoding of the resolution information, while superimposing the fresh information for the relay.

Summing up, we have the following points.

1) We use a random code to send information to the relay.

2) We use the list coding idea to see how many codewords are left to be resolved by the ultimate receiver Y.

3) We have a block Markov encoding scheme in which the resolution of the information is sent in the next block.

4) We use superposition to put cooperative information on top of the new information to the relay.

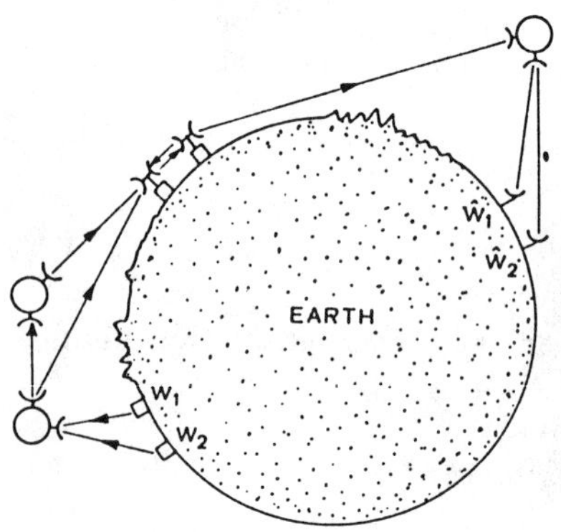

Fig. 21. Communication system.

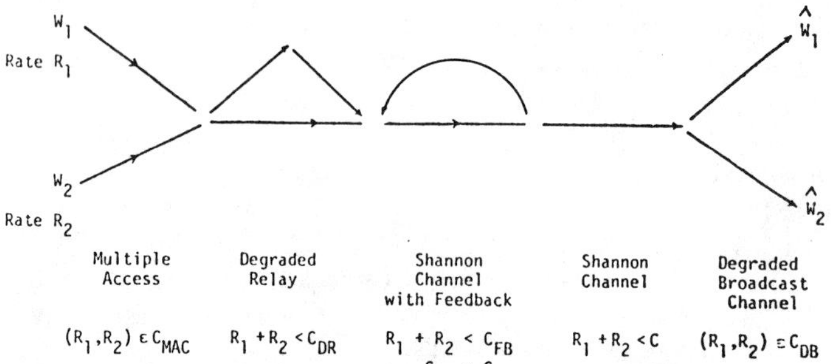

Fig. 22. Communication network with known capacity.

When we put all of these ideas together, we get the achievable rate region for the degraded relay channel. Moreover, a converse, which we do not prove here, shows that this is in fact the capacity for the degraded relay channel.

We can obviously combine these building blocks to solve networks like that in Fig. 21 and schematized in Fig. 22.

We have suppressed the input and output labels at the nodes of this graph. In this network, sender 1 wishes to send an index W_1 at rate R_1 and sender 2 wishes to send an index W_2 at rate R_2 to the receivers respectively designated as $\hat{W}_1$ and $\hat{W}_2$ with overall probability of error $P\{(\hat{W}_1, \hat{W}_2) \neq (W_1, W_2)\}$ tending to zero. This can be accomplished if and only if R_1 and R_2 simultaneously satisfy all of the constraints in Fig. 22. This overall capacity region is the convex subset of the (R_1, R_2) plane given by the intersection of the individual capacity regions. Thus max flow min cut holds for the capacity regions in this network.

Optimism is created each time a new capacity theorem is obtained, but a general theory does not yet exist for information flow in networks. Even for the building blocks of networks, some problems remain open. The multiple access capacity is known, but the capacity of the multiple access channel with feedback is still open. The capacity of the degraded broadcast channel is known, but the capacity of the general broadcast channel remains unknown. The capacity of the general relay channel is still unknown. Progress on Shannon's two-way channel [36] and on the interference channel [37] has just begun. A general theory for networks is years away. Indeed there may not be a mathematically nice answer. Even if there is a nice answer, it may not yield a practical implementation. But in communication theory, as in physics, one must have some faith. If a theory is found, it will have enormous qualitative theoretical implications. Indeed, even if optimal implementations prove to be too complex, the theory will tell the communication designer when he is close to optimality. Finally, new coding schemes will be suggested from inspection of the proofs. Fortunately, all of the results achieved to date are consistent with an information theoretic description solely in terms of quantities like $I(X_1, X_2; Y_3|X_3)$, where $I(\cdot;\cdot|\cdot)$ denotes conditional mutual information. Whether this is a result of limited technique or is true of the entire theory is yet to be revealed.

Acknowledgment

The authors are thankful to Chris Heegard for many valuable remarks during the preparation of this paper.

References

[1] E. C. van der Meulen, "A survey of multi-way channels in information theory: 1961-1976," *IEEE Trans. Inform. Theory*, vol. IT-23, no. 2, Jan. 1977.

[2] R. Ahlswede, "Multi-way communication channels," in *Proc. 2nd Int. Symp. Inform. Theory* (Tsahkadsor, Armenian S.S.R.), pp. 23-52, 1971. (Publishing House of the Hungarian Academy of Sciences, 1973.)

[3] H. Liao, "A coding theorem for multiple access communications," presented at the Int. Symp. Information Theory, Asilomar, 1972. Also Ph.D. dissertation, "Multiple Access Channels," Dep. Elec. Eng., Univ. Hawaii, Honuolulu, 1972.

[4] L. R. Ford and D. R. Fulkerson, *Flows in Networks*. Princeton, NJ: Princeton Univ. Press, 1962.

[5] A. D. Wyner, "The capacity of the band-limited Gaussian channel," *Bell Syst. Tech. J.*, vol. 45, pp. 359-371, Mar. 1965.

[6] R. G. Gallager, *Information Theory and Reliable Communication*. New York: Wiley, 1968.

[7] D. Slepian and H. O. Pollak, "Prolate spheroidal wave functions, Fourier analysis, and uncertainty-I," *Bell Syst. Tech. J.*, vol. 40, pp. 43-64. (Also see Landau and Pollak for Parts II and III.)

[8] T. Cover, R. J. McEliece, and E. Posner, "Asymchronous multiple access channel capacity," Tech. Rep. 35, Dep. Statistics, Stanford Univ., Stanford, CA, Dec. 1978. (To appear in *IEEE Trans. Inform. Theory.*)

[9] A. B. Carleial, "A case where interference does not reduce capacity," *IEEE Trans. Inform. Theory*, vol. IT-21, pp. 569-570, Sept. 1975.

[10] C. E. Shannon, "A mathematical theory of communication," *Bell Syst. Tech. J.*, vol. 27, pp. 379-423 and 623-656, July and Oct. 1948.

[11] B. McMillan, "The basic theorems of information theory," *Ann. Math. Stat.*, vol. 24, pp. 196-219, June 1953.

[12] L. Breiman, "The individual ergodic theorem of information theory," *Ann. Math. Stat.*, vol. 28, pp. 809-811, 1960.

[13] A. Feinstein, "A new basic theorem of information theory," *IRE Trans. Inform. Theory*, vol. IT-4, pp. 2-22, Sept. 1954.

[14] G. D. Forney, "Information theory," unpublished course notes.

[15] D. Slepian and J. W. Wolf, "Noiseless coding of correlated information sources," *IEEE Trans. Inform. Theory*, vol. IT-19, pp. 471-480, July 1973.

[16] T. M. Cover, "A proof of the data compression theorem of Slepian and Wolf for ergodic sources," *IEEE Trans. Inform. Theory*, vol. IT-22, pp. 226-278, Mar. 1975.

[17] T. Berger, "Multi-terminal source coding," Lecture notes presented at the 1977 CISM Summer School, Udine, Italy, July 18-20, 1977.

[18] T. Cover, A. El Gamal, and M. Salehi, "Multiple access channels with arbitrarily correlated sources," to appear in the *IEEE Trans. Inform. Theory.*

[19] T. Cover, "Broadcast channels," *IEEE Trans. Inform. Theory*, vol. IT-18, pp. 2-14, Jan. 1972.

[20] P. P. Bergmans, "Random coding theorem for broadcast channels with degraded components," *IEEE Trans. Inform. Theory*, vol. IT-19, pp. 197-207, Mar. 1973.

[21] R. G. Gallager, "Capacity and coding for degraded broadcast channels," *Problemy Peredaci Informaccii*, vol. 10, no. 3, pp. 3-14, 1974.

[22] K. Marton, "A coding theorem for the discrete memoryless broadcast channel," *IEEE Trans. Inform. Theory*, vol. IT-25, pp. 306-311, May 1979.

[23] P. Bergmans, "A simple converse for broadcast channels with additive white Gaussian noise," *IEEE Trans. Inform. Theory*, p. 279, Mar. 1974.

[24] J. Körner and K. Marton, "General broadcast channels with degraded message sets," *IEEE Trans. Inform. Theory*, vol. IT-23, no. 1, Jan. 1977.

[25] A. El Gamal, "The capacity of a class of broadcast channels," *IEEE Trans. Inform. Theory*, vol. IT-25, pp. 166-169, Mar. 1979.

[26] S. Gelfand, "Capacity of one broadcast channel," *Problems of Information Transmission*, vol. 13, no. 3, 1977, pp. 240-242 (English translation).

[27] M. Pinsker, "The capacity region of noiseless broadcast channels," *Problems of Information Transmission*, vol. 14, no. 2, pp. 97-102, 1978 (English translation).

[28] A. El Gamal, "The capacity of the product and sum of two inconsistently degraded broadcast channels," *Problems of Information Transmission*, vol. 1, 1980.

[29] A. El Gamal and E. van der Meulen, "A proof of Marton's coding theorem for the discrete memoryless broadcast channel," to appear in *IEEE Trans. on Inform. Theory.*

[30] A. El Gamal, "The feedback capacity of degraded broadcast channels," *IEEE Trans. Inform. Theory*, vol. IT-24, no. 3, May 1978.

[31] L. Ozarow, Ph.D. thesis, Massachusetts Institute of Technology, Cambridge, MA, June 1979.

[32] G. Dueck, "Semi-feedback for multiple-user channels," submitted for publication.

[33] E. C. van der Meulen, "Three-terminal communication channels," *Adv. Appl. Prob.*, vol. 3, pp. 120-154, 1971.

[34] T. Cover and A. El Gamal, "Capacity theorems for the relay channel," *IEEE Trans. Inform. Theory*, vol. IT-25, no. 5, Sept. 1979.

[35] A. El Gamal, "The capacity of the deterministic relay channel," presented at the 1979 IEEE Int. Symp. Information Theory.

[36] C. E. Shannon, "Two-way communication channels," in *Proc. 4th Berkeley Symp. Math. Stat. Prob.*, 1, Univ. California Press, pp. 611-644, 1961.

[37] R. Ahlswede, "The capacity region of a channel with two senders and two receivers," *Ann. Probability*, vol. 2, pp. 805-814, Oct. 1974.

Bibliography

This bibliography of multiple user information theory papers since 1975 is not purported to be complete. Material in this area through 1975 is thoroughly characterized in van der Meulen's 1977 survey paper.

R. Ahlswede and G. Dueck, "The best known codes are highly probable and can be produced by a few permutations," submitted to *IEEE Trans. Inform. Theory.*

R. Ahlswede and T. S. Han, "On source coding with side information via a multiple-access channel and related problems," submitted to *IEEE Trans. Inform. Theory.*

R. Ahlswede and I. Csiszar, "To get a bit of information may be as hard as to get full information," submitted to *IEEE Trans. Inform. Theory.*

R. Ahlswede, "A method of coding and its application to arbitrarily varying channels," to appear in *J. Comb., Inform. Syst. Sci.*

—, "An elementary proof of the strong converse theorem for the multiple-access channel," submitted to *J. Comb., Inform. Syst. Sci.*, 1980.

—, "Coloring hypergraphs: A new approach to multi-user source coding, Part I," *J. Comb., Inform. Syst. Sci.*, vol. 4, no. 1, pp. 76-115, 1979. Part II, to appear.

—, "Elimination of correlation in random codes for arbitrarily varying channel," *Z. Wahrscheinlichkeitstheorie u. Verw. Geb.*, vol. 44, pp. 159-175, 1978.

R. Ahlswede and G. Dueck, "Every bad code has a good subcode: A local converse to the coding theorem," *Z. Wahrscheinlichkeitstheorie u. Verw. Geb.*, vol. 34, pp. 179-182, 1976.

R. Ahlswede and P. Gacs, "Spreading of sets in product spaces and hypercontraction of the Markov operator," *Ann. Probability*, vol. 4, no. 6, pp. 925-939, 1976.

R. Ahlswede, P. Gacs, and J. Körner, "Bounds on conditional probabilities with applications in multi-user communication," *Z. Wahrscheinlichkeitstheorie u, Verw. Geb.*, vol. 34, pp. 157-177, 1976.

R. Ahlswede and P. Gacs, "Two contributions to information theory," *Colloquia Mathematica Societatis Janos Bolyai*, vol. 16. (Topics in Information Theory, Keszthely (Hungary), 1975.)

R. Ahlswede and J. Körner, "Source coding with side information and a converse for degraded broadcast channel," *IEEE Trans. Inform. Theory*, vol. IT-21, pp. 629-637, Nov. 1975.

—, "On common information and related characteristics of correlated information sources," presented at the 7th Prague Conf. Information Theory, Stat. Dec. Fct's, and Rand. Proc., Sept. 1974, to be published in the *IEEE Trans. Inform. Theory.*

M. Aref, "Information flow in relay networks," Ph.D. dissertation, Stanford Univ., Stanford, CA, Oct. 1980.

T. Basar and T. Basar, "Optimum linear coding in continuous-time communication systems with noisy side information at the decoder," *IEEE Trans. Inform. Theory*, vol. IT-25, pp. 612-616, Sept. 1979.

R. Benzel, "The capacity region of a class of discrete additive degraded

interference channels," *IEEE Trans. Inform. Theory*, vol. IT-25, pp. 228-231, Mar. 1979.
T. Berger, K. Housewright, J. Omura, S. Tung, and J. Wolfowitz, "An upper bound on the rate distortion function for source coding with partial side information at the decoder," *IEEE Trans. Inform. Theory*, vol. 25, pp. 664-666, Nov. 1979.
T. Berger, "Multiterminal source coding," *The Information Theory Approach to Communications*, G. Longo, Ed. (CISM Courses and Lectures, No. 229), New York: Springer Verlag, 1978, pp. 171-231.
T. Berger and M. Chang, "Rate-distortion with a fully informed decoder and partially informed encoder," *Abstract, 1977 IEEE Int. Symp. Information Theory* (Cornell Univ., Ithaca, NY, Oct. 1977).
T. Berger and S. Tung, "Encoding of correlated analog sources," in *Proc. 1975 IEEE-USSR Joint Workshop on Information Theory* (December 1975, Moscow, USSR), Piscataway, NJ: IEEE Press, pp. 7-10.
P. Bergmans, "The Gaussian interference channel," presented at the 1976 IEEE Int. Symp. Information Theory, Ronneby, Sweden, June 1976.
M. Bierbaum and H. Wallmeier, "A note on the capacity region of the multiple-access channel," *IEEE Trans. Inform. Theory*, vol. IT-25, p. 484, July 1979.
A. Carleial, "Interference channels," *IEEE Trans. Inform. Theory*, vol. IT-24, pp. 60-70, Jan. 1978.
A. Carleial and M. Hellman, "A note on Wyner's wiretap channel," *IEEE Trans. Inform. Theory*, vol. IT-23, pp. 387-390, May 1977.
A. Carleial, "A case where interference does not reduce capacity," *IEEE Trans. Inform. Theory*, vol. IT-21, pp. 569-570, Sept. 1975.
S. Chang and E. Weldon, "Coding for T-user multiple-access channels," *IEEE Trans. Inform. Theory*, vol. IT-25, pp. 684-691, Nov. 1979.
T. Cover, R. McEliece, and E. Posner, "Asynchronous multiple access channel capacity," to be published in *IEEE Trans. Inform. Theory*, July 1981.
T. Cover and S. K. Leung, "A rate region for the multiple access channel with feedback," to be published in *IEEE Trans. Inform. Theory*, Mar. 1981.
T. Cover, A. El Gamal, and M. Salehi, "Multiple access channels with arbitrarily correlated sources," *IEEE Trans. Inform. Theory*, vol. IT-26, Nov. 1980.
T. Cover and A. El Gamal, "Capacity Theorems for the Relay Channel," *IEEE Trans. Inform. Theory*, vol. IT-25, pp. 572-584, Sept. 1979.
T. Cover, "Some advances in broadcast channels," ch. 4 in *Advances in Communication Systems, Vol. 4, Theory and Applications*, A. Viterbi, Ed. New York: Academic Press, 1975.
—, "An achievable rate region for the broadcast channel," *IEEE Trans. Inform. Theory*, vol. IT-21, pp. 399-404, July 1975.
—, "A proof of the data compression theorem of Slepian and Wolf for ergodic sources," *IEEE Trans. Inform. Theory*, vol. IT-21, no. 2, pp. 226-228, Mar. 1975. Reprinted in *Ergodic and Information Theory*, L. Davisson and R. Gray, Eds., (Benchmark Papers in Electrical Engineering and Computer Science). Stroudsburg, PA: Dowden, Hutchinson, and Ross, Part V, pp. 305-307.
I. Csiszar and J. Körner, "Towards a general theory of source networks," *IEEE Trans. Inform. Theory*, vol. IT-26, pp. 155-166, Mar. 1980.
—, *Information Theory: Coding Theorems for Discrete Memoryless Systems*. New York: Academic Press, to be published in 1981.
—, "Graph decomposition: A new key to coding theorems," in *Preprint of Math. Inst. Hungarian Acad. Sci.* (Budapest), 1979.
—, "Broadcast channels with confidential messages," *IEEE Trans. Inform. Theory*, vol. IT-24, pp. 339-348, May 1978.
—, "Source networks with unauthorized users," *J. Combinatorics, Inform., and Syst. Sci.*, vol. 1, no. 1, pp. 25-40, 1976.
M. Deaett and J. Wolf, "Some very simple codes for the nonsynchronized two-user multiple access adder channel with binary inputs," *IEEE Trans. Inform. Theory*, vol. IT-24, pp. 635-636, Sept. 1978.
C. Downey and J. Karlof, "Group codes for the Gaussian broadcast channel with two receivers," *IEEE Trans. Inform. Theory*, vol. IT-26, pp. 406-412, July 1980.
G. Dueck, "Semi-feedback for multiple-user channels," *Preprint*. Submitted for publication.
—, "The strong converse of the coding theorem for the multiple access channel," to be published in *J. Comb., Inform. Syst. Sci.*
—, "Partial feedback for two-way and broadcast channels," to be published in *Inform. Contr.*
—, "A note on the multiple access channel with correlated sources," to be published in *IEEE Trans. Inform. Theory*, vol. IT-27, Mar. 1981.
—, "The capacity region of the two-way channel can exceed the inner bound," *Inform. Contr.*, vol. 40, pp. 258-266, 1979.
—, "Maximal error capacity regions are smaller than average error capacity regions for multi-user channels," *Prob. Cont. Inform. Theory*, vol. 7, pp. 11-19, 1978.
A. El Gamal and T. Cover, "Information theory of joint descriptions," submitted to *IEEE Trans. Inform. Theory*.
A. El Gamal, "The capacity of the deterministic relay channel," to be published in *IEEE Trans. Inform. Theory*.
—, "The capacity of the physically degraded Gaussian broadcast channel with feedback," to be published in *IEEE Trans. Inform. Theory*.
A. El Gamal and E. van der Meulen, "A proof of Marton's coding theorem for the discrete memoryless broadcast channel," to be published in *IEEE Trans. Inform. Theory*, vol. IT-27, Jan. 1981.
A. El Gamal, "The capacity of the product and sum of two inconsistently degraded broadcast channels," *Probl. Peredachi Inform.*, vol. 1, 1980.
—, "The capacity of a class of broadcast channels," *IEEE Trans. Inform. Theory*, vol. IT-25, pp. 166-169, Mar. 1979.
—, "The feedback capacity of degraded broadcast channels," *IEEE Trans. Inform. Theory*, vol. IT-24, pp. 379-381, May 1978.
A. El Gamal and C. Heegard, "On the capacity of a computer memory with defects and noise," submitted for publication in *IEEE Trans. Inform. Theory*.
N. Gaarder and J. Wolf, "The capacity region of a multiple-access discrete memoryless channel can increase with feedback," *IEEE Trans. Inform. Theory*, vol. IT-21, pp. 100-102, Jan. 1975.
S. Gelfand and M. Pinsker, "Coding for channel with random parameters," *Probl. Contr. Inform. Theory*, vol. 9, no. 1, pp. 19-31, 1980.
—, "Capacity of a broadcast channel with one deterministic component," *Probl. Inform. Transmission*, vol. 16, no. 1, 1980.
S. Gelfand, "Capacity of one broadcast channel," *Probl. Inform. Transmission*, vol. 13, no. 3, pp. 240-242 (English transl.), 1977.
R. Gray and P. Shields, "The maximum mutual information between two random processes," *Inform. Contr.*, vol. 33, no. 4, pp. 273-280, Apr. 1977.
B. Hajek and M. Pursley, "Evaluation of an achievable rate region for the broadcast channel," *IEEE Trans. Inform. Theory*, vol. IT-25, pp. 36-46, Jan. 1979.
T. S. Han, "Slepian-Wolf-Cover theorem for networks of channels," Res. Memo. RMI 79-1, Univ. Tokyo. (Submitted for publication to *Inform. Contr.*)
T. S. Han and K. Kobayashi, "A new achievable rate region for the interference channel and its evaluation," Res. Memo. RMI 79-9, Univ. Tokyo. (To be published in *IEEE Trans. Inform. Theory*, vol. IT-27, Jan. 1981.)
T. S. Han, "Deterministic broadcast channels with a common message," *IEEE Trans. Inform. Theory*, vol. IT-26, Nov. 1980.
T. S. Han and R. Ahlswede, "On source coding with side information via a multiple-access channel and related problems," submitted to *IEEE Trans. Inform. Theory*.
T. S. Han and K. Kobayashi, "An achievable rate region for noiseless source coding in a general multiterminal situation," *IEEE Trans. Inform. Theory*, vol. IT-26, May 1980.
T. S. Han, "Source coding with cross observations at the encoders," *IEEE Trans. Inform. Theory*, vol. IT-25, pp. 360-361, May 1979.
—, "The capacity region of general multiple-access channel with certain correlated sources," *Inform. Contr.*, vol. 40, no. 1, pp. 37-60, Jan. 1979.
C. Heegard, H. DePedro, and J. Wolf, "Permutation codes for the Gaussian broadcast channel with two receivers," *IEEE Trans. Inform. Theory*, vol. IT-24, no. 5, pp. 569-578, Sept. 1978.
G. D. Hu and S. Y. Shen, "Some new results of Shannon theory," Nankai University, Tianjin, China, Mar. 1980.
D. Hughes-Hartogs, "The capacity of the degraded spectral Gaussian broadcast channel," Tech. Rep. 7002-2, Information Systems Lab., Stanford Univ., Stanford, CA. Also Ph.D. dissertation, July 1975.
K. Jahn, "Coding of arbitrarily varying multiuser channels," Stanford Univ., Stanford, CA, Tech. Rep., 1979.
—, "Coding with side information and a rate-distortion theorem for arbitrarily varying correlated sources," Stanford Univ., Stanford, CA, Tech. Rep., 1979.
K. Jarett and T. Cover, "Asymmetries in relativistic information flow," to be published in *IEEE Trans. Inform. Theory*, vol. IT-27, 1981.
R. Kahn and M. Hellman, "On the wiretap channel with feedback," to be published in *IEEE Trans. Inform. Theory*.
T. Kasami and S. Lin, "Decoding of linear δ-decodable codes for a multiple access channel," *IEEE Trans. Inform. Theory*, vol. IT-24, pp. 633-635, Sept. 1978.
—, "Bounds on the achievable rates of block coding for a memoryless multiple access channel," *IEEE Trans. Inform. Theory*, vol. IT-24, pp. 187-197, Mar. 1978.
—, "Coding for a multiple access channel," *IEEE Trans. Inform. Theory*, vol. IT-22, pp. 129-137, Mar. 1976.
A. Kaspi and T. Berger, "Rate-distortion for correlated sources with partially separated encoders," submitted to *IEEE Trans. Inform. Theory*.
C. Keilers, "The capacity of the spectral Gaussian multiple-access channel," Ph.D. dissertation Dep. Elec. Eng., Stanford Univ.,

Stanford, CA, Apr. 1976.
J. Kieffer, M. Pursley, M. Wallace, and J. Wolfowitz, "Coding for correlated sources with unknown parameters," in *Proc. 1980 Conf. Information Sciences and Systems* (Princeton, NJ). pp. 168–171, Mar. 1980.
R. King, "Multiple access channels with generalized feedback," Ph.D. dissertation, Stanford Univ., Stanford, CA, ISL Rep. 6152-1, Mar. 1978.
J. Körner, "Source networks and new information quantities," in *Coll. Int. CNRS*, no. 276 (Theorie de l'Information), pp. 235–242, Paris, France, 1977.
—, "Some methods in multi-user communication," in *Information Theory, New Trends and Open Problems*, G. Longo, Ed. Vienna, Austria: Springer Verlag, 1976, pp. 172–176.
J. Körner and K. Marton, "How to encode the mod 2 sum of two binary sources," *IEEE Trans. Inform. Theory*, vol. IT-25, pp. 219–221, Mar. 1979.
—, "Images of a set via two channels and their role in multi-user communication," *IEEE Trans. Inform. Theory*, vol. IT-23, pp. 751–761, Nov. 1977.
—, "Comparison of two noisy channels," *Coll. Math. Soc. F. Boliya*, vol. 16, in *Topics in Information Theory*, Amsterdam, The Netherlands: North Holland, 1977, pp. 411–423.
—, "General broadcast channels with degraded message sets," *IEEE Trans. Inform. Theory*, vol. IT-23, pp. 60–64, Jan. 1977.
J. Körner and A. Sgarro, "Universally attainable error exponents for broadcast channels with degraded message sets," *IEEE Trans. Inform. Theory*, vol. IT-25, Nov. 1980.
V. Koshelev, "On a problem of separate coding for two dependent sources," *Probl. Trans. Inform.*, vol. 13, pp. 26–32, Jan.–Mar. 1977.
S. K. L-Y. Cheong and M. Hellman, "The Gaussian wire-tap channel," *IEEE Trans. Inform. Theory*, vol. IT-24, pp. 451–456, July 1978.
S. Lu, "On secrecy systems with side information about the message available to a cryptanalyst," *IEEE Trans. Inform. Theory*, vol. IT-25, pp. 472–475, July 1979.
K. Mackenthun, "An evaluation of the Wyner–Ziv rate region," *IEEE Trans. Inform. Theory*, vol. IT-26, pp. 347–354, May 1980.
K. Marton, "A coding theorem for the discrete memoryless broadcast channel," *IEEE Trans. Inform. Theory*, vol. IT-25, pp. 306–311, May 1979.
—, "The capacity region of deterministic broadcast channels," in *Trans. Int. Symp. Inform. Theory* (Paris-Cachan, France) 1977.
L. Ozarow, "Coding and capacity for additive white Gaussian noise multi-user channels with feedback," Ph.D. dissertation, Dep. Elec. Eng. and Comput. Sci., M.I.T., Cambridge, MA.
R. Peterson and D. Costello, "Comments on 'Convolutional tree codes for multiple access channels,'" *IEEE Trans. Inform. Theory*, vol. IT-26, pp. 627–628, Sept. 1980.
M. Pinsker, "The capacity region of noiseless broadcast channels," *Probl. Inform. Transmission*, vol. IT-14, pp. 97–102, 1978 (English transl.).
G. Sh. Poltyrev, "Capacity of parallel broadcast channels with degraded subchannels," *Probl. Trans. Inform.*, vol. 13, no. 2, pp. 23–36, Apr.–June 1977 (English transl.).
—, "Capacity for a sum of broadcast channels," *Probl. Inform. Trans*, vol. 15, no. 2, pp. 111–114, Apr.–June 1979 (English transl.).
H. Sato, "A study on Gaussian two-user channel," submitted to *IEEE Trans. Inform. Theory*.
—, "The capacity region of two-user channels—Method of extension channels," in *Abstract Proc. 1979 IEEE Int. Symp. Inform. Theory*, July 1979.
H. Sato and M. Tanabe, "A discrete two-user channel with strong interference," *Trans. IECE Japan*, vol. E61, no. 11, p. 880, Nov. 1978.
H. Sato, "On degraded Gaussian two-user channels," *IEEE Trans. Inform. Theory*, vol. IT-24, p. 637, Sept. 1978.
—, "An outer bound to the capacity region of broadcast channels," *IEEE Trans. Inform. Theory*, vol. IT-24, no. 3, p. 374, May 1978.
—, "On the capacity region of a discrete two-user channel for strong interference," *IEEE Trans. Inform. Theory*, vol. IT-24, p. 377, May 1978.
—, "Two-user communication channels," *IEEE Trans. Inform. Theory*, vol. IT-23, p. 295, May 1977.
—, "Information transmission through a channel with relay," Tech. Report ALOHA SYSTEM B-76-7, Mar. 1976.
M. Ulrey, "The capacity region of a channel with s senders and r receivers," *Inform. Contr.*, vol. 29, no. 3, pp. 185–203, Nov. 1975.
E. C. van der Meulen, "A survey of multi-way channels in information theory: 1961–1976," *IEEE Trans. Inform. Theory*, vol. IT-23, pp. 1–37, Jan. 1977.
—, "Random coding theorems for the general discrete memoryless broadcast channel," *IEEE Trans. Inform. Theory*, vol. IT-21, pp. 180–190, Mar. 1975.
E. Verriest and M. Hellman, "Convolutional encoding for Wyner's wiretap channel," *IEEE Trans. Inform. Theory*, vol. IT-25, pp. 234–236, Mar. 1979.
J. Willems, "The feedback capacity region of a class of discrete memoryless multiple access channels," *Preprint*, Univ. Leuven, Belgium, 1980.
H. Witsenhausen, "Indirect rate distortion problems," *IEEE Trans. Inform. Theory*, vol. IT-26, Sept. 1980.
—, "On source networks with minimal breakdown degradation," *Bell Syst. Tech. J.*, vol. 59, pp. 1083–1087, July–Aug. 1980.
—, "Some aspects of convexity useful in information theory," *IEEE Trans. Inform. Theory*, vol. IT-26, pp. 265–271, May 1980.
—, "The zero-error side-information problem and chromatic numbers," *IEEE Trans. Inform. Theory*, vol. IT-22, pp. 592–593, Sept. 1976.
—, "Values and bounds for the common information of two discrete random variables," *SIAM J. Appl. Math.*, vol. 31, pp. 313–333, Sept. 1976.
H. Witsenhausen and A. Wyner, "A conditional entropy bound for a pair of discrete random variables," *IEEE Trans. Inform. Theory*, vol. IT-21, pp. 493–501, Sept. 1975.
H. Witsenhausen, "On sequences of pairs of dependent random variables," *SIAM J. Appl. Math.*, vol. 28, pp. 100–113, Jan. 1975.
J. Wolf, A. Wyner, and J. Ziv, "Source coding for multiple descriptions," *Bell Syst. Tech. J.*, vol. 59, Oct. 1980.
J. Wolf, "Multi-user communication networks," in *Commun. Syst. and Random Process Theory*, J. K. Skwirzynski, Ed. Alphen aaan den Rijn-The Netherlands: Sijthoff & Noordhoff, 1978, pp. 37–53.
—, "The AEP property of random sequences and applications to information theory: Part I basic principles," *Inform. Theory, New Trends and Open Problems*, G. Longo, Ed., (CISM Courses and Lectures No. 219). New York: Springer-Verlag, 1975.
J. Wolfowitz, *Coding Theorems of Information Theory*, 3rd ed. New York: Springer-Verlag, 1978.
A. Wyner, "The rate-distortion function for source coding with side-information at the Decoder II: General sources," *Inform. Contr.*, vol. 38, pp. 60–80, July 1980.
A. Wyner and J. Ziv, "The rate-distortion function for source coding with side-information at the decoder," *IEEE Trans. Inform. Theory*, vol. IT-22, pp. 1–10, Jan. 1976.
A. Wyner, "The wire-tap channel," *Bell Syst. Tech. J.*, vol. 54, pp. 1355–1387, Oct. 1975.
—, "On source coding with side information at the decoder," *IEEE Trans. Inform. Theory*, vol. IT-21, pp. 294–300, May 1975.
—, "The common information of two dependent random variables," *IEEE Trans. Inform. Theory*, vol. IT-21, pp. 163–170, Mar. 1975.

Feedback Can at Most Double Gaussian Multiple Access Channel Capacity

JOY A. THOMAS

Abstract—The converse for the discrete memoryless multiple access channel is generalized and is used to derive strong bounds on the total capacity (sum of the rates of all the senders) of an m-user Gaussian multiple access channel in terms of the input covariance matrix. These bounds are used to show that the total capacity of the channel with feedback is less than twice the total capacity without feedback. The converse for the general multiple access channel is also used to show that for any m-user multiple access channel, feedback cannot increase the total capacity by more than a factor of m.

I. Introduction

The simplest communication situation is when we have a single sender trying to send information to a single receiver. In many practical situations, however, we have two-way links—the receiver can also send back information to the sender (for example, telephone links). Although feedback is very common in practical channels, it is still only imperfectly understood and a large number of problems remain open on the capacity of channels with feedback. In this report, we establish bounds relating this capacity to the capacity without feedback for a class of multiple access channels. Our objective is to show that feedback cannot help very much in increasing the capacity of many practical channels.

The most important and rather surprising result in this area is due to Shannon [1], who established that feedback cannot increase the capacity of single-user discrete memoryless channels. This was later extended to continuous channels by Kadota, Zakai, and Ziv [2]. Dobrushin [3] gave examples of channels where feedback does indeed increase channel capacity. Various people have proposed schemes to use feedback to devise better codes for memoryless channels; an important example is the Kailath–Schalkwijk [4] scheme for the additive white Gaussian channel which achieves a doubly exponential decay of error probability with blocklength. The capacity of Gaussian colored noise channels with feedback is not known; however, Ebert [5] and Pombra and Cover [6] have derived bounds on the capacity increase with feedback. In particular, they have shown that for any Gaussian colored noise channel, feedback does not increase the capacity by more than a factor of two.

In situations like satellite communication, we have many senders trying to communicate to a single receiver—a multiple access channel. Unlike the case of the simple discrete memoryless channel, feedback in the multiple access channel can increase total capacity even when the channel is memoryless. This is because feedback would enable the senders to cooperate with each other to a greater extent than is possible without feedback. This was first demonstrated by Gaarder and Wolf [7]. Cover and Leung [8] proved an achievable rate region for the multiple access channel with feedback. Later, Willems [9] proved that the Cover–Leung region is indeed that capacity region for a certain class of channels including the binary adder channel. Ozarow [10] has found the capacity region for the two-user Gaussian multiple access channel, using a modification of the Kailath–Schalkwijk scheme for simple Gaussian channels. However, his method does not generalize easily to three or more users. The problem of determining the capacity region for a multiple access channel with feedback still remains open.

In this correspondence, we shall prove bounds on the capacity increase with feedback for multiple access channels. As we shall show in Section II, feedback does not increase the maximum rate of any individual sender; it only helps the senders to cooperate and keep out of each other's way. Hence the only increase will be in the sum of the rates. Throughout this correspondence, we will use the total capacity of the multiple access channel to mean the maximum achievable sum of rates of all the senders.

Manuscript received April 9, 1986; revised December 8, 1986. The material in this paper was presented at the IEEE International Symposium on Information Theory, Ann Arbor, MI, October 5–9, 1986. This work was supported in part by the National Science Foundation under Grant ECS82-11568 and by the Defense Advanced Research Projects Agency under Contract N00039-84-C-0211.

The author is with the Department of Electrical Engineering, Stanford University, Stanford, CA 94305, USA.

IEEE Log Number 8613577.

II. General Converse for Multiple Access Channels

In this section we will extend the converse due to Ozarow [10] and Gaarder and Wolf [7] to the m-user case and use it to show that total capacity cannot be increased by a factor of more than m by feedback.

The general multiple access channel is characterized by an input alphabet $(\mathscr{X}_1 \times \mathscr{X}_2 \times \cdots \times \mathscr{X}_m)$, a probability transition matrix $p(y|x_1, x_2, \cdots, x_m)$, and an output alphabet $(\mathscr{Y})$. (See Fig. 1.)

To simplify notation, let S denote any arbitrary subset of $\{1,2,\cdots,m\}$ and let $\bar{S}$ denote its complement. Let X_S denote the set $\{X_j: j \in S\}$ (for example, if $S = \{1,3\}$, then $X_S = \{X_1, X_3\}$). Let $W_1, W_2, \cdots, W_m$ denote the input messages, each W_j uniformly distributed in $(1, 2^{nR_j})$ and independent of the other messages. Let W_S denote the set $\{W_j: j \in S\}$. Since we have feedback, the input symbol X_{ji} of sender j at time i is a function of the message at that sender W_j and the past values of the output, $Y_1, Y_2, \cdots, Y_{i-1}$.

Consider any $\{2^{nR_1}, 2^{nR_2}, \cdots, 2^{nR_m}, n, P_e^{(n)}\}$ code for the

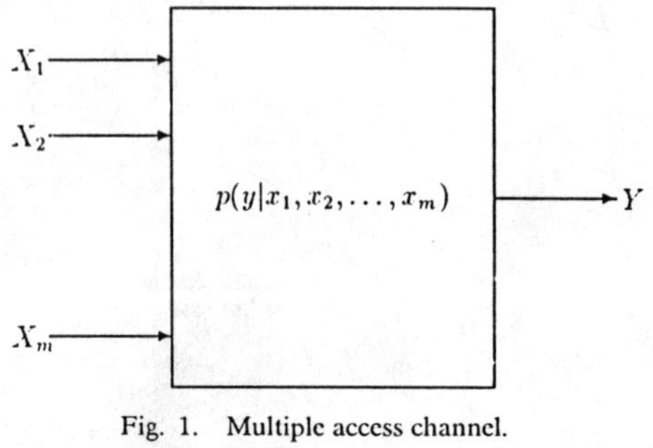

Fig. 1. Multiple access channel.

Reprinted from *IEEE Trans. Inform. Theory*, vol. IT-33, no. 5, pp. 711–716, Sept. 1987.

multiple access channel with feedback. Then by using Fano's inequality, we find

$$\sum_{j \in S} nR_j = H(W_S) \tag{1}$$

$$\overset{(a)}{\leq} I(W_S; Y^n) + n\epsilon_n \tag{2}$$

$$= H(W_S) - H(W_S|Y^n) + n\epsilon_n \tag{3}$$

$$\overset{(b)}{\leq} H(W_S|W_{\bar{S}}) - H(W_S|Y^n, W_{\bar{S}}) + n\epsilon_n \tag{4}$$

$$= I(W_S; Y^n|W_{\bar{S}}) + n\epsilon_n \tag{5}$$

$$\overset{(c)}{=} \sum I(W_S; Y_i|Y^{i-1}, W_{\bar{S}}) + n\epsilon_n \tag{6}$$

$$\overset{(d)}{=} \sum I(W_S, X_{Si}; Y_i|Y^{i-1}, W_{\bar{S}}, X_{\bar{S}i}) + n\epsilon_n \tag{7}$$

$$\overset{(e)}{\leq} \sum I(X_{Si}; Y_i|X_{\bar{S}i}) + n\epsilon_n. \tag{8}$$

(a) follows from Fano's inequality ($\epsilon_n \to 0$ as $P_e^{(n)} \to 0$). (b) is true because W_S and $W_{\bar{S}}$ are independent. (c) is the chain rule and we get (d) from the fact that $X_{Si} = f(W_S, Y^{i-1})$ and $X_{\bar{S}} = g(W_{\bar{S}}, Y^{i-1})$. Finally, since the channel is memoryless, $(Y^{i-1}, W_S, W_{\bar{S}}) \to (X_{Si}, X_{\bar{S}i}) \to Y_i$ forms a Markov chain and hence we get inequality (e).

All the mutual informations in (8) are concave functions of the per-letter joint probability distribution $p(x_1, x_2, \cdots, x_m)$. Hence by Jensen's inequality, the average of these mutual informations over time 1 to n is less than the mutual information evaluated at the average of the probability distributions. Hence after dividing by n and letting $n \to \infty$, we have the desired converse

$$\sum_{j \in S} R_j \leq I(X_S; Y|X_{\bar{S}}) \tag{9}$$

for some joint distribution $p(x_1, x_2, \cdots, x_m)p(y|x_1, x_2, \cdots, x_m)$, for all subsets S. A similar converse can be derived for an arbitrary network of nodes connected by memoryless channels. Since we have not placed any restrictions on the joint distribution, the converse is not tight in general. The basic problem in determining multiple access channel capacity with feedback is to find the class of joint distributions achievable using feedback.

The capacity of a discrete memoryless multiple access channel without feedback was first determined by Ahlswede [11] and Liao [12]. It corresponds to the convex closure of regions bounded by (9) for arbitrary input product distributions $p(x_1)p(x_2)\cdots p(x_m)p(y|x_1, x_2, \cdots, x_m)$. A related result was derived by Han [13] for a general multiple access network with correlated sources.

We will derive a few simple consequences of this converse. We will first show that feedback cannot increase the maximum individual rates of any of the senders. We will illustrate it for the case $m = 2$. Let $R_{1,f,m}$ be the maximum achievable rate with feedback from sender 1 to the receiver. Let $R_{1,nf,m}$ be the corresponding rate without feedback. From the converse for the channel with feedback, we get

$$R_{1,f,m} \leq \max_{p(x_1, x_2)} I(X_1; Y|X_2) \tag{10}$$

$$= \max_{p(x_1, x_2)} \sum_{x_2} P(X_2 = x_2) I(X_1; Y|X_2 = x_2) \tag{11}$$

$$\leq \max_{p(x_1, x_2)} \max_{x_2} I(X_1; Y|X_2 = x_2) \tag{12}$$

$$= \max_{p(x_1)p(x_2)} I(X_1; Y|X_2) \triangleq R_{1,nf,m} \tag{13}$$

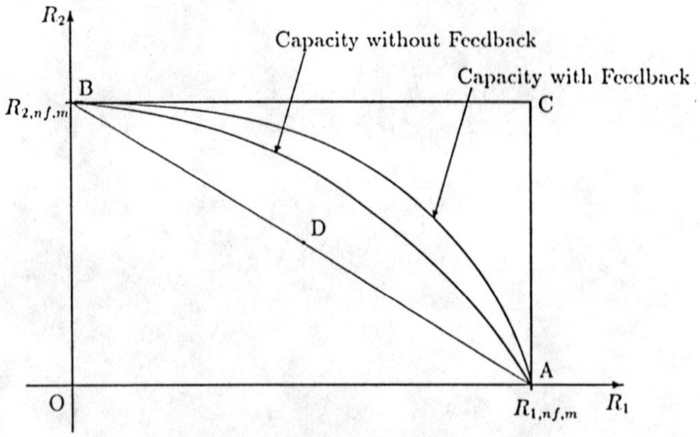

Fig. 2. Capacity region for two user multiple access channel.

since the maximum is achieved by a degenerate product distribution obtained by setting X_2 to the value that best opens up the channel between X_1 and Y. Hence the maximum rate from X_1 to Y is not increased by feedback. This can be easily justified by the fact that since X_2 does not get any better look at X_1 than does Y, he cannot help any better than by keeping quiet. We can similarly define $R_{2,nf,m}$ as the maximum rate of transmission from X_2 to Y.

We will now use a simple geometric argument to show that feedback cannot increase the total capacity of any m-user multiple access channel by more than a factor of m. The capacity regions for the case $m = 2$ are illustrated in Fig. 2. Ahlswede's results imply that the rate pairs $(R_{1,nf,m}, 0)$ and $(0, R_{2,nf,m})$ (points A and B in Fig. 2) are achievable without feedback. By time-sharing, one can achieve all points on the line joining A to B.

With feedback, the converse implies that the individual rates are less than $R_{1,nf,m}$ and $R_{2,nf,m}$, respectively. Therefore the capacity region with feedback lies within the rectangle defined by the points A, B, and O. The maximum sum of rates within this rectangle is at C, which is less than twice the sum of the rates at D, the midpoint of the main diagonal. Since D is achievable without feedback, the maximum achievable sum of rates with feedback is less than twice the maximum achievable sum of rates without feedback.

One can easily generalize this argument to m-users to obtain a factor of m as the bound on the increase in total capacity using feedback.

III. Gaussian Multiple Access Channels

There are m senders $X_1, X_2, \cdots, X_m$ all sending to the single receiver Y. The received signal at the time i is

$$Y_i = \sum_{j=1}^{m} X_{ji} + Z_i \tag{14}$$

where the Z_i are a sequence of independent identically distributed, zero-mean Gaussian noise variables with variance N (see Fig. 3). We will initially assume that there is the same power constraint P on all the senders; that is, for all senders and messages, we must have

$$E\left[\frac{1}{n}\sum_{i=1}^{n} X_{ji}^2(W_j, Y^{i-1})\right] \leq P. \tag{15}$$

We will later generalize to the case of unequal powers.

The capacity of the Gaussian multiple access channel without feedback was first found by Cover [14] and Wyner [15]. For the m-user case with equal powers, the maximum achievable sum of

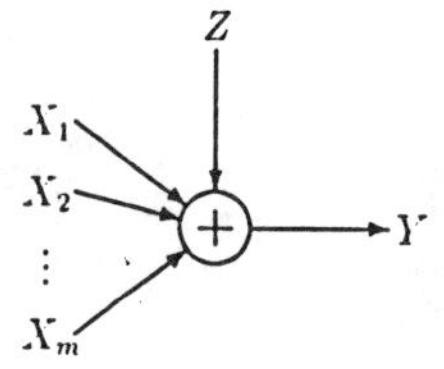

Fig. 3. Gaussian multiple access channel.

rates T is

$$T \leq C\left(m\frac{P}{N}\right) \tag{16}$$

where

$$C(x) \triangleq \frac{1}{2}\log(1+x). \tag{17}$$

Note that the total capacity with complete cooperation is $C(m^2P/N)$, which could be much larger than $C(mP/N)$. The main objective of this report is to show that even with feedback, the Gaussian multiple access channel cannot attain the cooperative bound in general.

Ozarow has determined the feedback capacity region for the two-user case, using a very interesting modification of the Kailath–Schalkwijk scheme. His expression for the capacity region in the case of equal powers is

$$T \leq 2C\left((1-\rho^2)\frac{P}{N}\right) \tag{18}$$

$$T \leq C\left(2(1+\rho)\frac{P}{N}\right) \tag{19}$$

for all $0 \leq \rho \leq 1$. It is interesting to note that Ozarow's scheme achieves the same capacity region as would be obtained by allowing all possible joint distributions in (9)—the converse is tight in this case.

IV. Factor of Two Bound

We will first extend the converse of Section II to the Gaussian case. We will use a simple lemma.

Lemma 1: Let $X_1, X_2, \cdots, X_k$ be an arbitrary set of zero-mean random variables with covariance matrix K. Let S be any subset of $\{1,2,\cdots,k\}$ and $\bar{S}$ be its complement. Then

$$h(X_S|X_{\bar{S}}) \leq h(X_S^*|X_{\bar{S}}^*) \tag{20}$$

where $(X_1^*, X_2^*, \cdots, X_k^*) \sim N(0, K)$.

Proof: We use the nonnegativity of the conditional Kullback–Liebler distance

$$D(p(x|y)\|q(x|y)) \triangleq \int p(x,y)\log\left(\frac{p(x|y)}{q(x|y)}\right).$$

Taking $p(\cdot,\cdot) = f_{X_S, X_{\bar{S}}}(\cdot,\cdot)$ and $q(\cdot,\cdot) = f_{X_S^*, X_{\bar{S}}^*}(\cdot,\cdot)$, we get

$$0 \leq D\left(f_{X_S|X_{\bar{S}}}(x_S|x_{\bar{S}})\|f_{X_S^*|X_{\bar{S}}^*}(x_S|x_{\bar{S}})\right) \tag{21}$$

$$= \int f_{X_S, X_{\bar{S}}}(x_S, x_{\bar{S}})\log\left(f_{X_S|X_{\bar{S}}}(x_S|x_{\bar{S}})\right) - \int f_{X_S, X_{\bar{S}}}(x_S, x_{\bar{S}})\log\left(f_{X_S^*|X_{\bar{S}}^*}(x_S|x_{\bar{S}})\right) \tag{22}$$

$$\overset{(a)}{=} -h(X_S|X_{\bar{S}}) - \int f_{X_S^*, X_{\bar{S}}^*}(x_S, x_{\bar{S}})\log\left(f_{X_S^*|X_{\bar{S}}^*}(x_S|x_{\bar{S}})\right) \tag{23}$$

$$= -h(X_S|X_{\bar{S}}) + h(X_S^*|X_{\bar{S}}^*) \tag{24}$$

where (a) is true because $\log(f_{X_S^*|X_{\bar{S}}^*}(x_S|x_{\bar{S}}))$ is a quadratic form in K. The proof of the lemma is complete.

Starting from the general converse and using the fact that for a power-limited Gaussian channel, all the differential entropies are finite, we get

$$\sum_{j\in S} R_j \leq I(X_S; Y|X_{\bar{S}}) \tag{25}$$

$$= h(Y|X_{\bar{S}}) - h(Y|X_S, X_{\bar{S}}) \tag{26}$$

$$= h(Y|X_{\bar{S}}) - h(Z) \tag{27}$$

$$= h(Y|X_{\bar{S}}) - \frac{1}{2}\log((2\pi e)N) \tag{28}$$

$$\leq h(Y^*|X_{\bar{S}}^*) - \frac{1}{2}\log((2\pi e)N) \text{ (by lemma)} \tag{29}$$

where $X_S^*, X_{\bar{S}}^*$ are Gaussian random variables with the same covariance matrix as $X_S, X_{\bar{S}}$. Since this covariance matrix is the average of covariance matrices for each time instant, its diagonal elements should be less than or equal to P to satisfy the power constraint (15).

Hence, dropping the asterisk for simplicity, we get

$$\sum_{j\in S} R_j \leq h(Y|X_{\bar{S}}) - \frac{1}{2}\log((2\pi e)N) \tag{30}$$

$$= \frac{1}{2}\log((2\pi e)\operatorname{var}(Y|X_{\bar{S}})) - \frac{1}{2}\log((2\pi e)N) \tag{31}$$

$$= \frac{1}{2}\log\left(\frac{\operatorname{var}(Y|X_{\bar{S}})}{N}\right). \tag{32}$$

Now let $X = (X_1, X_2, \cdots, X_m)'$ be the vector of inputs. Let K be the covariance matrix of X. We shall first show that one of the possible values of K that maximizes the total capacity is the highly symmetric form

$$K = \begin{pmatrix} P & \rho P & \rho P & \cdots & \rho P \\ \rho P & P & \rho P & \cdots & \rho P \\ \rho P & \rho P & P & \cdots & \rho P \\ \vdots & \vdots & \vdots & \ddots & \vdots \\ \rho P & \rho P & \rho P & \cdots & P \end{pmatrix} \tag{33}$$

where ρ is the correlation coefficient and $-1/(m-1) \leq \rho \leq 1$.

This is because of the symmetry of the channel. Let us assume that there is some other form of K that maximizes that sum of the rates. By appropriately re-labeling the rows and columns of K, we have a new covariance matrix which has the same total capacity. By time-sharing between the two forms of K, we can obtain a more symmetric form. Proceeding in this way, by time-sharing between all possible re-labelings of the rows and columns of K (corresponding to all possible re-labelings of the senders), we can obtain the symmetric form above. Hence, we can restrict our attention to this form of K, and we will obtain our bounds using it.

Let $e = (1,1,1,\cdots,1)$. Divide the matrix K into submatrices corresponding to S and $\bar{S}$:

$$K = \begin{matrix} & S & \bar{S} \\ S & A & B \\ \bar{S} & B' & C \end{matrix} \tag{34}$$

where A includes the rows and columns corresponding to S and C to $\bar{S}$.

The variance of Y, given $X_{\bar{S}}$, is

$$\operatorname{var}(Y|X_{\bar{S}}) = \operatorname{var}(Z) + \operatorname{var}\left(\sum_{j\in S} X_j|X_{\bar{S}}\right). \tag{35}$$

Let $V = \sum_{j\in S} X_j$ and let $\hat{V}$ be the least mean square (lms) estimate of V, given $X_{\bar{S}}$. Then

$$\operatorname{var}(V|X_{\bar{S}}) = E_{x_{\bar{S}}} \operatorname{var}(V|X_{\bar{S}} = x_{\bar{S}}) \quad (36)$$

$$\le E_{x_{\bar{S}}} E_{x_S|x_{\bar{S}}} (V - \hat{V})^2 \quad (37)$$

$$= \text{lms error in estimating } V \text{ from } X_{\bar{S}} \quad (38)$$

$$= R_{VV} - R_{VX_{\bar{S}}} R^{-1}_{X_{\bar{S}}X_{\bar{S}}} R_{X_{\bar{S}}V}$$

(see for example [16, p. 11]) (39).

$$= e(A - BC^{-1}B')e. \quad (40)$$

Using this bound for the symmetric form of K and all possible subsets S, after some algebra we obtain the following bounds on the total capacity $T = mR$:

$$T \le C\left(m(1+(m-1)\rho)\frac{P}{N}\right) \quad (41)$$

$$T \le \frac{m}{m-1} C\left((1-\rho)(m-1)(1+(m-1)\rho)\frac{P}{N}\right) \quad (42)$$

$$T \le \frac{m}{m-2} C\left((1-\rho)(m-2)\frac{1+(m-1)\rho}{1+\rho}\frac{P}{N}\right) \quad (43)$$

$$\vdots$$

$$T \le mC\left((1-\rho)\frac{1+(m-1)\rho}{1+(m-2)\rho}\frac{P}{N}\right). \quad (44)$$

In general, when S has $m-l$ elements,

$$T \le \frac{m}{m-l} C\left((1-\rho)(m-l)\frac{1+(m-1)\rho}{1+(l-1)\rho}\frac{P}{N}\right). \quad (45)$$

These bounds have various interesting properties:

- the bounds reduce to Ozarow's capacity region for $m = 2$;
- since the first bound is less than $C(mP/N)$ for negative ρ, we can restrict our attention to positive ρ;
- the first bound is monotonically increasing with ρ, the last one monotonically decreasing with ρ, and all the others first rise to a maximum, then decrease with ρ falling to 0 when $\rho = 1$.

Since each of these functions is a concave function of ρ, the pointwise minimum is also a concave function and has a unique maximum value.

To prove the factor of two bound for T, we need to use only one of the bounds corresponding to $l \approx m/2$. We will first illustrate it in the case of even m.

$$T \le \frac{m}{m-l} C\left((1-\rho)(m-l)\frac{1+(m-1)\rho}{1+(l-1)\rho}\frac{P}{N}\right) \triangleq T_l(\rho). \quad (46)$$

On differentiating and setting to 0, we find that the value of ρ that maximizes $T_l(\rho)$ is

$$\rho_c = \frac{\sqrt{l(m-l)(m-1)} - (m-1)}{(m-1)(l-1)}. \quad (47)$$

Substituting this value of ρ_c and using $m = 2l$, we get

$$T \le 2C\left(\left(\frac{l}{l-1}\right)^2 (2l - 2\sqrt{2l-1})\frac{P}{N}\right). \quad (48)$$

But since

$$\left(\frac{l}{l-1}\right)^2 (2l - 2\sqrt{2l-1}) = 2\left(\frac{l^2}{(l-1)^2}\right)\left(\frac{l^2-(2l-1)}{l+\sqrt{2l-1}}\right) \quad (49)$$

$$= 2l\left(\frac{l}{l+\sqrt{2l-1}}\right) \quad (50)$$

$$\le 2l, \quad (51)$$

we find

$$T \le 2C\left(2l\frac{P}{N}\right) = 2C\left(m\frac{P}{N}\right) = 2T_{nf}. \quad (52)$$

We proceed in the same manner for odd m. In this case we use $m = 2l+1$. Substituting ρ_c and $m = 2l+1$, we get

$$T \le \frac{2l+1}{l+1} C\left(\left(\frac{l\sqrt{2} - \sqrt{l+1}}{l-1}\right)^2 (l+1)\frac{P}{N}\right). \quad (53)$$

Now

$$\left(\frac{l\sqrt{2} - \sqrt{l+1}}{l-1}\right)^2 (l+1) \le 2l+1 \quad (54)$$

if

$$2l^3 - 3l^2 + 1 \ge 0. \quad (55)$$

But (55) is true for all positive l because the minimum of the left side for positive l is 0 occurring at $l = 1$. Hence

$$T \le \frac{2l+1}{l+1} C\left(2l+1\frac{P}{N}\right) \le 2C\left(m\frac{P}{N}\right) = 2T_{nf}. \quad (56)$$

We have thus proved that the total capacity can at most be doubled using feedback for both odd and even m.

V. Case of Unequal Powers

So far we have been dealing only with the case when all the transmitters have the same power constraints. Now let us assume that the powers are $P_1, P_2, \cdots, P_m$.

1) Without feedback: The dominating constraint on the sum of the rates is

$$T \le C\left(\frac{P_1 + P_2 + \cdots + P_m}{N}\right). \quad (57)$$

Defining $P = (1/m)\Sigma P_i$, then

$$T \le C\left(m\frac{P}{N}\right). \quad (58)$$

2) With feedback: Let $S(P_1, P_2, \cdots, P_m)$ be the total capacity with feedback. By the symmetry of the problem, S is a symmetric function of its arguments. By time-sharing, we can easily show that $S(\cdot,\cdot,\cdots,\cdot)$ is a concave function. Hence by the properties of symmetric concave functions [17, p. 104], we have

$$S(P_1, P_2, \cdots, P_m) = \frac{1}{m} S(P_1, P_2, \cdots, P_m) + \frac{1}{m} S(P_2, P_3, \cdots, P_m, P_1) + \cdots + \frac{1}{m} S(P_m, P_1, \cdots, P_{m-1}) \quad (59)$$

$$\le S\left(\frac{\Sigma P_i}{m}, \frac{\Sigma P_i}{m}, \cdots, \frac{\Sigma P_i}{m}\right) \quad (60)$$

$$= S(P, P, \cdots, P) \tag{61}$$

$$\leq 2C\left(m\frac{P}{N}\right)(\text{result for equal powers}) \tag{62}$$

$$= 2S_{nf}(P_1, P_2, \cdots, P_m). \tag{63}$$

Hence, even with different powers at the different transmitters, the total capacity with feedback is less than twice the total capacity without feedback.

VI. Conclusion

We have shown that the total capacity of any multiple access channel with white Gaussian noise can at most be doubled using feedback. Though we have not said anything about achievability, one would suspect that there exists a generalization of Ozarow's method or some other method that would show that the bounds derived in this report are achievable.

We have also shown that feedback does not help by more than a factor of m for any m-user multiple access channel. This result is weak, and we conjecture that feedback does not help by more than a factor of two even for the general multiple access channel.

Acknowledgment

The author is grateful to Prof. Thomas M. Cover for suggesting the problem and for many useful discussions.

References

[1] C. E. Shannon, "The zero error capacity of a noisy channel," *IRE Trans. Inform. Theory*, vol. IT-2, pp. 8–19, Sept. 1956.

[2] T. T. Kadota, Moshe Zakai, and Jacob Ziv, "Capacity of a continuous memoryless channel with feedback," *IEEE Trans. Inform. Theory*, vol. IT-17, pp. 372–378, July 1971.

[3] R. L. Dobrushin, "Information transmission in a channel with feedback," *Theory of Prob. and Applications*, vol. 34, pp. 367–383, Dec. 1958.

[4] J. P. M. Schalkwijk and T. Kailath, "A coding scheme for additive noise channels with feedback—Part 1: No bandwidth constraints," *IEEE Trans. Inform. Theory*, vol. IT-12, pp. 172–182, Apr. 1966.

[5] P. M. Ebert, "The capacity of the Gaussian channel with feedback," *Bell Syst. Tech. J.*, vol. 49, pp. 1705–1712, Oct. 1972.

[6] T. M. Cover and S. Pombra, "Gaussian feedback capacity," submitted to *IEEE Trans. Inform. Theory*. Also, Stanford University Statistics Tech. Rep., no. 63, 1987.

[7] N. T. Gaarder and J. K. Wolf, "The capacity region for a multiple access discrete memoryless channel can increase with feedback," *IEEE Trans. Inform. Theory*, vol. IT-21, pp. 100–102, Jan. 1975.

[8] T. M. Cover and C. S. K. Leung, "An achievable rate region for the multiple access channel with feedback," *IEEE Trans. Inform. Theory*, vol. IT-27, pp. 292–298, May 1981.

[9] F. M. J. Willems, "The feedback capacity of a class of discrete memoryless multiple access channels," *IEEE Trans. Inform. Theory*, vol. IT-28, pp. 93–95, Jan. 1982.

[10] L. H. Ozarow, "The capacity of the white Gaussian multiple access channel with feedback," *IEEE Trans. Inform. Theory*, vol. IT-30, pp. 623–629, July 1984.

[11] R. Ahlswede, "Multi-way communication channels," in *Proc. 2nd. Int. Symp. Inform. Theory*, Tsahkadsor, 1971. Publishing House of the Hungarian Academy of Sciences, pp. 23–52, 1973.

[12] H. Liao, "A coding theorem for multiple access communications," presented at *Int. Symp. on Inform. Theory*, Asilomar, 1972.

[13] T. S. Han, "Slepian–Wolf–Cover theorem for networks of channels," *Information and Control*, vol. 47, pp. 67–83, 1980.

[14] T. Cover, "Some advances in broadcast channels," in *Advances in Communication Systems*, Vol. 4, A. Viterbi, Ed. San Francisco: Academic, 1975.

[15] A. D. Wyner, "Recent results in the Shannon theory," *IEEE Trans. Inform. Theory*, vol. IT-20, pp. 2–10, Jan. 1974.

[16] T. Kailath, *Lectures on Weiner and Kalman filtering*. New York: Springer-Verlag, 1981.

[17] M. Marcus and H. Minc, *A Survey of Matrix Theory and Matrix Inequalities*. Boston: Allyn and Bacon, 1964.

A Study of Data Multiplexing Techniques and Delay Performance

By J. F. HAYES and D. N. SHERMAN

(Manuscript received May 25, 1972)

This paper is concerned with the study of data multiplexing techniques which provide local distribution for user populations whose source characteristics may be categorized as Inquiry/Response. The techniques studied are Polling, Random Access, and a Loop System. In each method a group of users is multiplexed onto the same line which is connected to a Central Processor. The Central Processor forms the interface between the local system and long haul facilities, but does no computation on arriving messages other than to route them appropriately. The advantage of such systems is flexibility of operation and economy through sharing of equipment at the Central Processor. We focus our attention on the average round-trip message delay and use this as a measure for comparisons of the three techniques.

Source models, in terms of calls per busy hour and number of bits per message for users as well as for computer responses, are developed which are appropriate in an Inquiry/Response context. Other factors taken into consideration are transmission rates, synchronization delay time, and allocation of available capacity to the transmission of overhead information, such as polling messages and acknowledgment traffic.

The results of the study are presented in graphic form, where the average delay due to traffic in the system is plotted as a function of the number of stations. Principal conclusions can be summarized as follows:

Polling—It is found that this system is sensitive to the synchronization delay which takes place each time a user station transmits to the central facility. For reasonable choices of system parameters as many as 100 stations can share the Central Processor without exceeding an average of 1-second round trip message delay.

Random Access—This system is not sensitive to synchronization delay. Over most of the range of load parameters, the Random Access system shows lower average delay than the other two systems.

Loop System—The study revealed that this system has average delay performance comparable to the other two. The Loop system is not sensitive to synchronization delay.

I. INTRODUCTION

One of the major functions of data communication networks is providing means by which users can access a distant computer in an *interactive* manner. Time-sharing, inquiry-response and credit checking are examples of applications which have a common communications requirement. They require rapid set up times for access to their respective computers and, as a consequence of their interactiveness, often require high-capacity return channels for return traffic.

Providing access to a distant computer involves providing two major data transmission media, viz. long haul facilities and local distribution. (See Figs. 1 and 2.) While there are important problems associated with each part, we shall focus our attention on local distribution techniques. We are motivated in this direction by the relatively

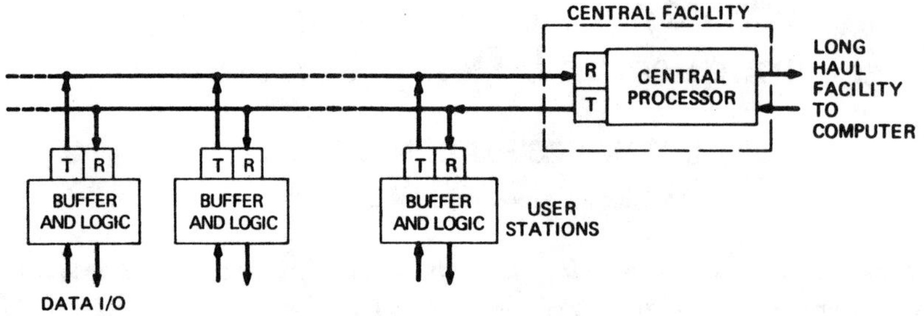

Fig. 1—Polling and random access systems.

high cost of providing local distribution compared to the total cost of transmission.

The results presented in the sequel are most appropriate to the situation where users desiring access to distant computers are geographically clustered. In this instance, traffic from several users is concentrated at what are designated as User Stations. We shall consider multiplexing techniques whereby traffic from User Stations is conveyed to a central point which we call the Central Processor. The role of the Central Processor is to route or direct message flow to and from User Stations and computers.

This paper is devoted to an analysis of the roundtrip delay of three techniques which multiplex traffic onto common facilities. The analysis is based on models of user traffic and computer responses to user messages. In the applications we consider, traffic from the user is bursty, i.e., short periods of activity followed by relatively long idle periods. Each message from a user elicits a response from the computer.

In the sequel we shall analyze the three alternatives for local distribution:

(*i*) Polling—(See Fig. 1) Stations are polled by a Central Processor and transmit over a common line. Return traffic is multiplexed over a second common line by the Central Processor.

(*ii*) Random Access—(See Fig. 1) Here User Stations transmit, at will on a common line, all messages that are generated. Positive acknowledgments are issued by the Central Processor upon error-free reception of the messages. Until an acknowledgment is received, the message is held in the station's buffer. Again, return messages travel over a second common line.

(*iii*) Loop System[1]—(See Fig. 2) User Stations share the same line. Traffic already on the line has priority, and newly generated messages are multiplexed into gaps in the line traffic.

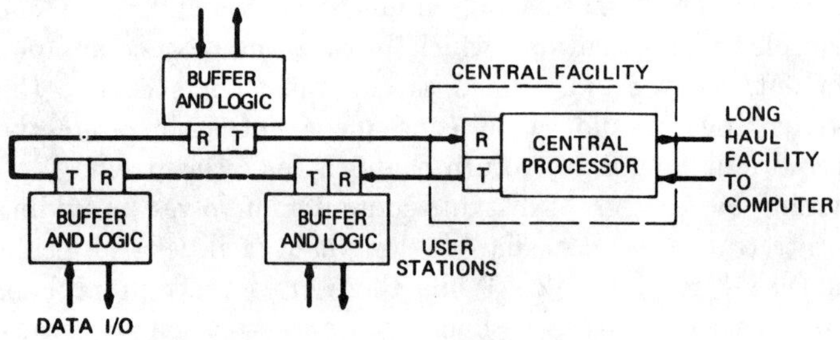

Fig. 2—Loop system.

In each of these techniques, traffic to and from different User Stations share common facilities. The basic difference among the techniques lies in the means of controlling traffic accessing the system through the User Station. In the Polling System, the Central Processor exercises tight control over entering messages. As we shall see, in order to do this, overhead is incurred which causes an increased delay. In contrast, entering traffic in the Random Access System is loosely controlled by the Central Processor and delay attributable to overhead is reduced considerably. In the Loop System, as in the Random Access System, messages are multiplexed on the line at the User Station without direct control by the Central Processor.

Our analysis of performance concentrates on buffering or queueing delay. The results of this analysis will be used to calculate the round-trip queueing delay encountered by a user-generated message. A message going to the computer from a particular station is competing for the line with messages from other stations. Consequently it encounters delay before it has sole access to shared facilities. Further, messages returning from the computer may be delayed in a queue at the Central Processor before being transmitted over the common line. We shall designate the sum of these two delays as the roundtrip delay. Of course, a message will encounter other delays such as propagation delay from the Central Processor to a distant computer and service time in the computer. However, these delays are independent of the local distribution system delays and can simply be added to the roundtrip delay that we calculate.

A common thread running through each of these systems is interactive queues. At each of the User Stations storage is assigned, if necessary, to queue up messages. However, since all the queues share the same server, the queues are not independent of one another. The exact treatment of interactive queues is mathematically difficult, therefore, in carrying out the analysis, certain approximations have been made.

The three systems described above are compared on the basis of average roundtrip delay. Since we have computed the moment generating functions of forward and return delay, higher order moments can be found easily enough. From these higher order moments one can calculate other measures of performance. In so doing, the question of correlation between forward and roundtrip delay must be considered. We did not concern ourselves with this correlation since it has no bearing on the average roundtrip delay.

In developing the models for the systems, we have attempted to take into account the constraints encountered in real communications systems. In the concluding section of the paper, examples using parameters of the Digital Data System[2] are presented.

II. GENERAL CONCLUSIONS

At the conclusion of the analytical investigations, we spend the final section of this paper in describing the delay performance of the systems. There also we compare, where possible, the advantages and disadvantages of various possible implementations. While the details appear later, it is worthwhile to give some general conclusions about the systems' behavior here.

For a given number of stations (N) and call rate (λ), the average message delay in the Polling System depends strongly on how quickly receivers can synchronize (s) as well as the available channel capacity. For instance, one can imagine that in the Polling System, polls take place over a separate channel of capacity δ. Now let us suppose that two separate systems could be designed, the first having zero synchronization time and the other taking 10 ms for the Central Processor to synchronize the polls. We would find as the number of stations increased, that the 10 ms spent at each station ultimately degrades the performance substantially. This effect can be seen by comparing Figs. 3 and 4 where we have depicted the above situation. For a large number of stations ($\sim$ 100) the two lower right families of curves rotate counterclockwise, indicating increasing delay. Note that even if we poll with capacity $\delta = 100$ bits/s (the upper right families of curves in Figs. 3 and 4), the delay is almost independent of the synchronization time and call rate. This is primarily due to the fact that we are forced to poll fewer stations (to maintain reasonable delay) and thus incur a smaller overhead induced delay.

If the aim of the system is to serve a small number of stations each of which has a high calling rate, the synchronization time is relatively unimportant. Thus it may be desirable to use a low speed line with a small synchronization time. In Fig. 5, we depict this situation and note that it is possible in this case to achieve similar performance for a small number of stations ($\leqq 25$) as is achieved with high speed line (Fig. 3).

Since the Central Processor in the Random Access technique exercises loose control over traffic flow, there is a much smaller effect of overhead and correspondingly smaller delay (see Figs. 6 and 7).

In the applications we consider, the volume of return traffic is greater than the volume of forward traffic, therefore, the first station in a Loop system suffers the largest delay. In considering the Loop system we focus on this station. Analytical results obtained in the sequel indicate that the Loop System compares favorably with each of the previous techniques. In the Loop System, such quantities as synchronization time and retransmission delay do not arise. What is relevant in this system is the processing delay, which involves examining the addresses of packets passing through the station.

III. GLOSSARY OF IMPORTANT QUANTITIES

N—number of User Stations

λ—arrival rate of messages to the User Station (messages/s or calls/busy hour)

B_F—message length generated at User Stations (bits)

C_F—capacity of forward link (bits/s)

m—time duration of message while on forward link (s)

B_R—return message length (bits)

C_R—capacity of return link (bits/s)

M—time duration of message returning from the Central Processor (s)

w—minimum time between polls in Polling System and minimum retransmission time in Random Access System (s)

$\bar{t}_c$, $\overline{t_c^2}$—mean and second moment of the time required to poll and readout messages in Polling System

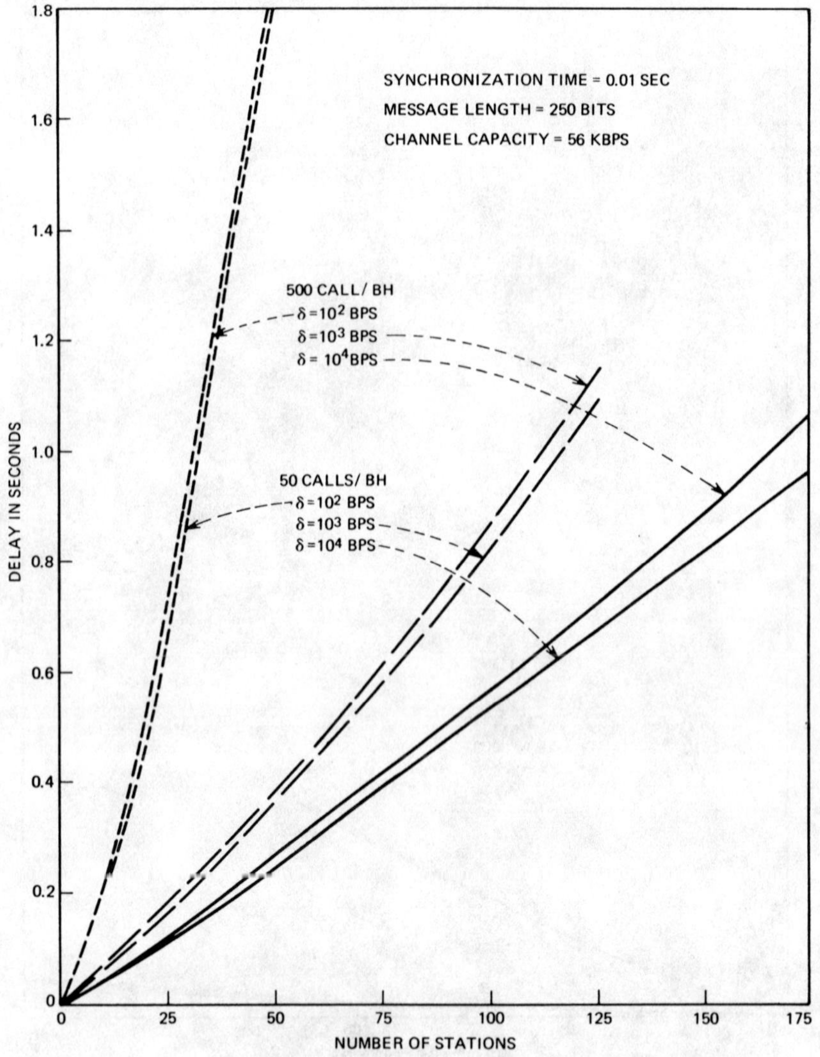

Fig. 3—Polling system delay in forward direction.

ρ—intensity of arriving traffic
d_{RA}—inbound message delay in Random Access System (s)
d_L—inbound message delay in Loop System (s)
d_P—inbound message delay in Polling System (s)
d_R—message delay on return link (s)
δ—capacity of return link allocated to polling messages in Polling System and to positive acknowledgements in Random Access System (bits/s)
s—synchronization time (s).

IV. SOURCE MODELS

As indicated in the introduction we are primarily interested in the bursty, interactive user. Such users are encountered, for example, in credit checking and Inquiry-Response applications. The models that we shall develop are also applicable to time-sharing computer systems where users access a distant computer. Traffic characteristics of such users have been studied,[3,4] so that estimates of parameters characterizing the source (holding time, rate, etc.) are available.

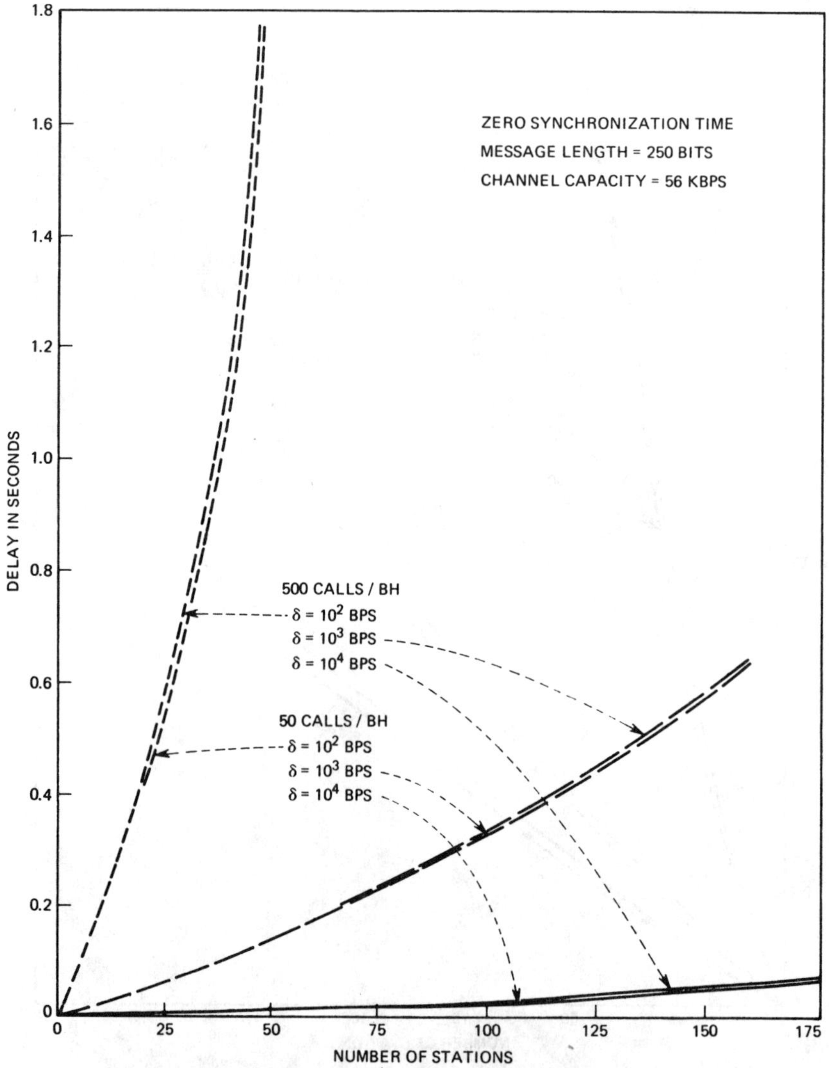

Fig. 4—Polling system delay in forward direction.

We model user traffic as consisting of Poisson arrivals of short, fixed-length messages. The arrival rate is designated as λ either in calls per busy hour or in messages per second. Each message consists of B_F bits. If the line serving a User Station has a capacity of C_F bits per second then $m = B_F/C_F$ seconds are required to transmit each message.

In our analysis we shall assume that each User Station has unlimited buffer capacity so that buffer overflow never occurs. From the delay formulas that we develop, one can also derive buffer occupancy statistics, and thus imply the buffer size required for small overload probability. For the message arrival rates that we consider, the station buffers need not be very large in terms of message lengths to achieve adequately small overflow probability.

User messages arriving at the Central Processor are routed to a computer. Each message elicits a response from the computer. All of these return messages pass back again through the Central Processor. If messages arrive at the N user stations at average rate λ messages per second, then the average rate of return message flow through the

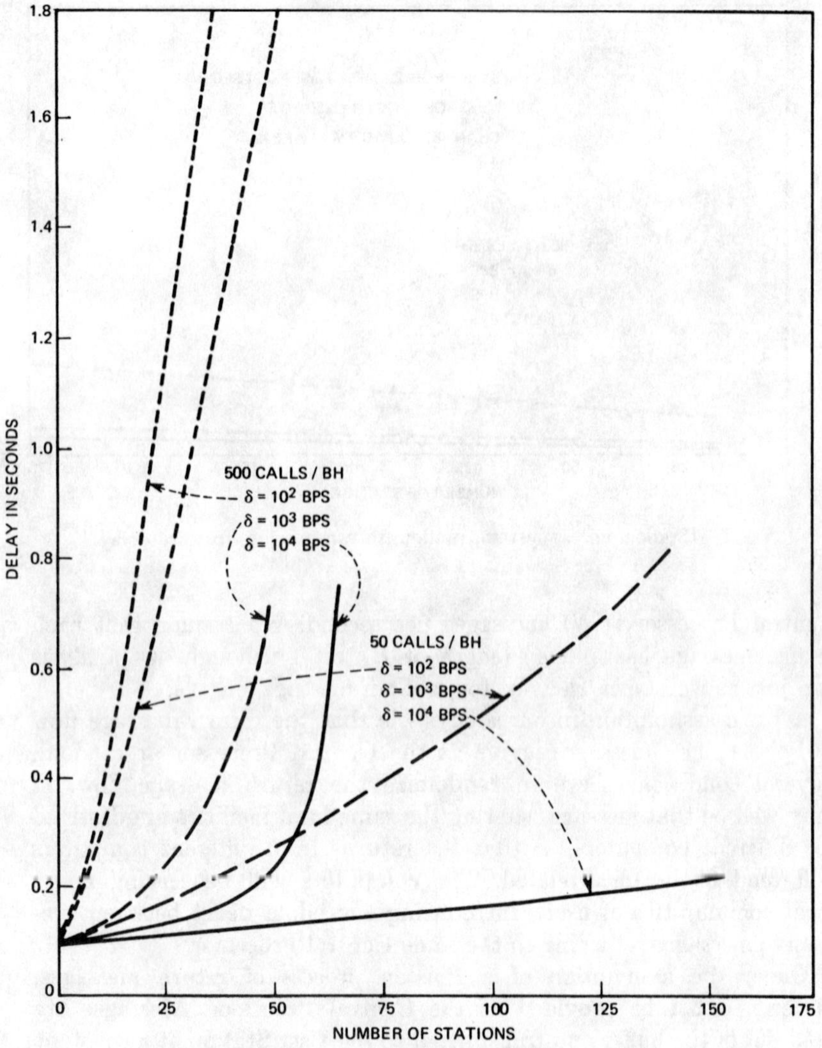

Fig. 5—Polling system delay in forward direction.

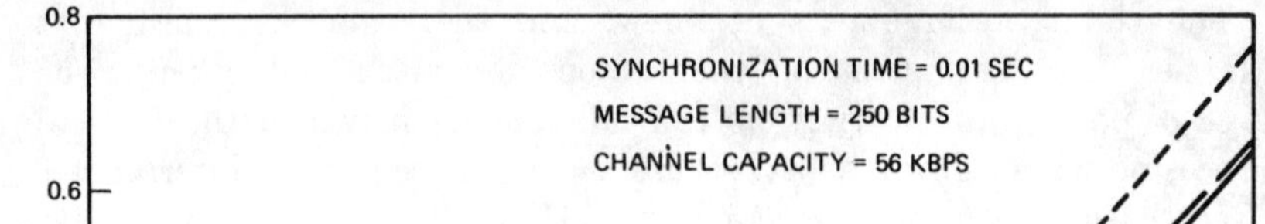

Fig. 6—Random access system, delay in forward direction, fixed-time retransmission.

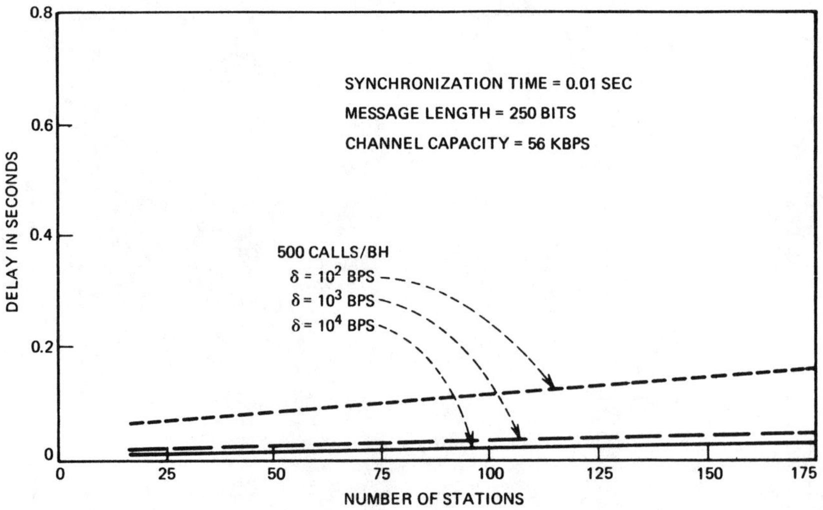

Fig. 7—Random access system, random retransmission, forward delay.

Central Processor is $N\lambda$ messages per second. We assume that each return message has a fixed length of B_R bits, although our analysis can just as well be carried out for random message lengths.

A basic assumption in our analysis is that the return message flow is Poisson, i.e., messages arrive at the Central Processor at random. Several conditions serve to randomize the return message flow. It may well be that messages sharing the same local facilities are destined for different computers, so that the returns from different computers will tend to be uncorrelated. The computers will be serving many local communities of users, introducing a random delay between successive messages returning to the same Central Processor.

Under the assumption of a Poisson process of return messages, buffering must be provided at the Central Processor. Messages are read out of the buffer and transmitted to the User Station at a constant bit rate.

V. POLLING SYSTEM

The polling technique is not new, and the reader is referred to Martin[5,6] for discussions on the present approach and others. The type of polling under study is 'roll-call' polling in which the Central Processor works down a list of users and, in sequence, interrogates User Stations sharing the same line. Upon being polled, the User Station may transmit messages. Since the generation of messages is random with respect to the time of polling, messages must queue in a buffer located at the User Station until a polling message arrives. As indicated earlier, we focus our attention on the average delay encountered by a message.

The arrangement of lines and User Stations for the Polling System is shown in Fig. 1. User Stations are bridged across common lines. Notice that, in contrast with the star connection, only one receiver and one transmitter is required at the Central Processor. The Polling System can be implemented using a "daisy chain" configuration in which the line between two user stations loops back to the Central Processor.

Since polling messages are transmitted over a common line, User Stations must be capable of recognizing messages addressed to it. The same line is used in common by messages returning from the computer. Again the User Station must be able to recognize return messages addressed to it. As we shall see, there is an interplay between forward and return message delay, in as much as polling messages and return traffic share the same physical line.

The first component of the round-trip message delay that we consider is the delay at the User Station that a message encounters before it is put on the line. The calculation of this delay is complicated by the fact that the number of messages queueing at a particular station strongly depends upon the number of messages stored in all of the other stations. Since arrivals to any station are at a Poisson rate, the less frequently a station is polled, the more messages accumulate. Moreover, as messages accumulate, the polling cycle lengthens.

The polling problem has been analyzed by Leibowitz[7] by assuming a form of independence between queues.* The results of his analysis using this assumption compares well with the results of a rigorous analysis for a two-queue network. The nature of the independence assumption is such that the largest error occurs in a two-queue system.

Of primary importance in the calculation of delay in the forward direction is a parameter that we designate as the 'walk-time' between stations. In the situation where a station has no message to transmit, the Central Processor must still spend time interrogating it. Time is required to transmit a polling message. The speed with which polling messages can be transmitted is limited, since messages returning from the computer share the same line. Further, after polling a station, the Central Processor must wait for a possible response. This time involves, for example, establishing synchronization between the User Station and the Central Processor. We shall return to the walk-time after we have considered its effect on delay.

In calculating the delay of a message in this system, it is necessary to characterize the time between polls of a particular station. From the moment generating function derived in Appendix A we obtain the mean and the mean square values of the cycle time

$$\bar{t}_c = \frac{Nw}{1 - N\rho} \tag{1a}$$

$$\overline{t_c^2} = \frac{N(N-1)(w + \rho\bar{t}_c)^2 + N[w^2 + 2w\bar{t}_c\rho + \bar{t}_c m\rho]}{1 - N\rho^2}. \tag{1b}$$

In writing (1a) and (1b), we take advantage of the fact that the walk-time is a constant value, w, and the message length is a constant value $m = B_F/C_F$. Also we define $\rho = \lambda\bar{m}$.

Messages that arrive at a User Station are buffered, awaiting the arrival of the poll. When the poll arrives, messages are read out of the buffer on a first come first served basis. Thus a message must wait until the station is polled as well as until messages that have arrived before it are read out of the buffer. In Appendix A, the generating function of delay is derived. From this generating function it can be

* A rigorous solution of the polling problem has recently been found by Eisenberg.[8] However, in our application, which may involve a large number of stations, Leibowitz's approach leads to more tractable results.

shown that the average delay is given by

$$\bar{d}_P = \frac{\overline{t_c^2}}{2\bar{t}_c}(1 + \rho) + m \qquad (2)$$

where $\bar{t}_c$ and $\overline{t_c^2}$ are given by eqs. (1a and b).

Equations (1a) and (1b) show that the walk time, w, is an important parameter in the calculation of delay. We turn now to assigning values to it. Since the polling message must contain the address of the station being polled, it must be at least $\log_2 N$ bits long. Further, since it shares the line from the processor to the station with messages returning from the computer, not all of the return line capacity is available to transmit polling messages. We treat this situation by splitting the capacity of the return line. Suppose that of the C_R bit-per-second capacity of the return line, δ bits per second, interleaved with the message bits, are allocated entirely for polling messages. The minimum time required to transmit a polling message is then

$$r = \{[\log_2 N] + 1\}/\delta \qquad (3)$$

where $[x]$ is the largest integer less than x. The remaining $C_R - \delta$ bit-per-second capacity of the return line is allocated to return message traffic. This allocation may be effected by either FDM or TDM.

Other methods may be used to share the return line between polling and return messages. For example, messages may be transmitted over the same channel with priority given to polls or user traffic. The difficulty with this approach is that there may be a large difference in the lengths of the two kinds of messages. Thus, even though a polling message is short and has priority, it may wait a relatively long time until a transmission of a return message is completed.

Since User Stations continuously receive from the same point, the Central Processor, no time is required to establish synchronization for return or polling messages. This may not be true in the reverse direction since the Central Processor is receiving from a different station on each poll. Thus in the calculation of walk-time it may be necessary to allow time for the establishment of synchronization between the User Station and the Central Processor. Let this time be denoted s. The walk-time between stations is then the sum.

$$w = s + \{[\log_2 N] + 1\}/\delta. \qquad (4)$$

The effect achieved by splitting return line capacity on return message delay can now be calculated. Return messages are B_R bits in duration and arrive at the central processor at a Poisson rate of $N\lambda$ messages per second. The time required to transmit each message is $B_R/(C_R - \delta)$ seconds. The delay of a message in the Central Processor may be found from the analysis of an M/G/1 queue.[9] It can be shown that the average return message delay is

$$\bar{d}_R = \frac{N\lambda[B_R/(C_R - \delta)]^2}{2[1 - B_R\lambda N/(C_R - \delta)]} + B_R/(C_R - \delta). \qquad (5)$$

The average roundtrip delay is the sum of $\bar{d}_P$ and $\bar{d}_R$.

VI. RANDOM ACCESS SYSTEM

The Random Access system was suggested by the University of Hawaii's ALOHA System.[10] The configuration of this system is the same as shown for the Polling System in Fig. 1. As in the Polling System, messages arrive at each User Station at rate λ messages per second. Messages are transmitted to the Central Processor on a first come first served basis. The length of a message in bits and the line rate are such that m seconds are required to transmit each constant length message. This constant length includes overhead bits required by the Random Access technique.

Messages are transmitted at random by each station in the system. There is a probability of messages from different stations overlapping. In order to detect message overlap, parity check bits are transmitted along with each message and error detection is performed upon reception. Only if a message is received error free is an acknowledgment transmitted to the User Station. Until a User Station recognizes a positive acknowledgment addressed to it, the message is held in a buffer. If, after a specified period of time, no acknowledgment is received, the message is retransmitted. This continues until an acknowledgment for a particular message is received.

Messages returning from the computer are buffered and sent to the User Stations in the same manner as the Polling System. As in the Polling System, messages must contain addresses since messages destined for different stations share the same line.

A basic relation governing traffic in the Random Access system has been derived by Abramson.[10,11] In this analysis, messages are assumed to enter the common line at Poisson rate. Suppose that a particular message is transmitted at time t_o. This message will encounter no interference if no other message is put on the line in the time interval $t_o \pm m$. We have then

$$Pr\,[\text{message retransmission}] = 1 - \exp(-2Rm) \qquad (6)$$

where R is the total transmission rate on the line, including retransmissions. The rate of retransmission on the line is $R(1 - e^{-2Rm})$.

Now messages arrive at all user stations at rate λN messages per section. If there is no continuous buildup of messages at User Stations, we must have the relationship.

$$R = N\lambda + R(1 - \exp(-2Rm)).$$

This reduces to

$$N\lambda m = Rm \exp(-2Rm). \qquad (7)$$

An examination of eq. (7) discloses that the maximum information transmission rate is $N\lambda m = 1/2e \approx 0.18$. As one attempts to exceed this, the retransmission rate increases so fast that less information gets through.

The basic assumption in this analysis is that line traffic, including retransmissions, forms a Poisson process. Simulation studies[12] show that eq. (7) holds up to $N\lambda m \cong 0.1$. For higher loadings, this simulation shows a much higher retransmission rate and, presumably, the Poisson assumption does not hold.

A basic quantity in the calculation of delay is the time that the User Station waits before retransmission if no positive acknowledgment for a prior transmission is received. Let us consider how this 'timeout' period affects the system performance that a user might experience. If two messages do interfere, then each station must initiate a retransmission. If the times of each retransmission are the same, then interference will persist. Thus some strategy must be used to avoid persistent interference as well as to provide for a short 'timeout' delay. To more fully appreciate what is meant by 'short' we have analyzed two strategies for retransmission.

The first strategy assigns to each station a fixed or designated 'timeout' delay. This approach has the advantage that it completely avoids persistent interference. It has the disadvantage that some stations will have large (on the order of the number of stations times a fixed delay per station) delay.

The second strategy (called randomized retransmission) attempts to make use of the fact that it is generally two stations that interfere, and that it is not necessary to distinguish between all N stations in the system. This approach asks interfering stations to select a retransmission time by selecting it from a random sequence of retransmission times. If each station has a different sequence, then there is a small probability of persistent interference. The approach has the advantage of shortening the retransmission delay, but has the disadvantage that there is a non-zero probability of repeated interference.

In the next two subsections we discuss the delay analysis for these strategies. Detailed calculations may be found in Appendix B.

6.1 *Fixed Timeout Delay*

Consider first the Random Access System implemented with fixed timeout retransmission. Suppose that a message is transmitted from a User Station at time $t = 0$. Since we must allow for synchronization and for the transmission of an acknowledgment message, the message can be acknowledged no sooner than time $t = s + \{[\log_2 N] + 1\}/\delta + m$. If no acknowledgment is received at this time, station i waits for $2im$ seconds ($i = 1, 2, \cdots, N$) and retransmits. This is repeated until a positive acknowledgment is received. If we assume that interferences on successive retransmissions are independent, we have from eq. (6)

$$Pr\,[\text{message clearance time} = w + m + kT_i]$$
$$= Pr\,[k \text{ retransmissions}] = [1 - \exp(-2Rm)]^k \exp(-2Rm) \quad (8)$$

where $w = \{[\log_2 N] + 1\}/\delta + s$ and $T_i = w + 2mi$.

Average message delay at each station as well as higher moments of delay may be calculated by calling upon results from the analysis of an M/G/1 queue.[9] Messages arrive at a Poisson rate of λ messages per second and each message has a geometrically distributed service time as given by (8). From the generating function derived in Appendix B, it can be shown that the average delay for a message at station i is

$$\bar{d}_{RA} = \frac{2[wP - 2Nm(1 - P)] - \lambda[w^2P + m^2 + (2Nm)^2(1 - P)]}{2\{1 - \lambda[wP + m - 2Nm(1 - P)]\}} + m \quad (9)$$

where $P = \exp(2Rm)$.

The return message delay in the random access system is the same as in the Polling System. Fixed length messages of M bits return at rate $N\lambda$ messages per second, with $C_R - \delta$ bits per second capacity available for transmission. The resulting message delay is given by eq. (5). The roundtrip delay is the sum of $\bar{d}_{RA}$ [eq. (9)] and $\bar{d}_R$ [eq. (5)].

6.2 *Randomized Retransmission Delay*

We turn now to consider a random retransmission technique. After an initial transmission, the User Station waits a minimum of $s + ([\log_2 N] + 1)/\delta$ seconds. If no acknowledgment has been received, the message is retransmitted after a randomly-distributed time interval. If no acknowledgment is received after this second transmission, the process is repeated. In our calculations we have taken the random timeout interval to be exponentially distributed with mean $1/\alpha$.

The probability of interference on the initial transmission is given by eq. (6). On subsequent retransmissions the probability of interference depends on α. For example, if α is very large, then retransmission for the two interfering stations occurs almost immediately after

$$t = s + \frac{[\log_2 N] + 1 + m}{\delta}$$

and the probability of interference is high. As α decreases, the probability of interference decreases. We approximate the probability of interference on retransmission as

$$Pr\,[\text{retransmission interference}] = 1 - \exp\,[-2m(R + \alpha)]. \quad (10)$$

The average time between retransmissions is

$$\bar{t}_r = s + \{[\log_2 N] + 1\}/\delta + 1/\alpha + m. \quad (11)$$

In Appendix B, the generating function of the time required to clear a message from a station's buffer is derived. From this we can show that the mean and the mean square times to clear a message from a station buffer are respectively

$$\bar{b} = w \exp\,(-2Rm) + X(1/\alpha + 2w) \exp\,\{+4m(R + \alpha)\} + m \quad (12a)$$

$$\begin{aligned} \overline{b^2} &= w^2 \exp\,(-2Rm) \\ &\quad - \frac{2Xw^2}{\alpha \exp\,\{-4m(R + \alpha)\}} + \frac{2X(1 + 2w\alpha)(1 + w\alpha + w\alpha Y)}{\alpha^2 \exp\,\{-6m(R + \alpha)\}} \\ &\quad + 2m\bar{b} + m^2 \end{aligned} \quad (12b)$$

where

$$X = [1 - \exp\,(-2Rm)] \exp\,\{-2m(R + \lambda)\}$$

and

$$Y = 1 - \exp\,\{-2m(R + \lambda)\}.$$

From the theory of the M/G/1 queue, it can be shown that the average delay is

$$\bar{d}_{RA} = \bar{b} + \frac{\alpha\overline{b^2}}{2(1 - \alpha\bar{b})} + m \quad (13)$$

where $\bar{b}$ and $\overline{b^2}$ are given by eqs. (12a) and (12b).

Notice from (10) and (11) that decreasing α reduces the probability of interference while increasing the average retransmission delay. Thus there is an optimum value of α that balances these effects, yielding a minimum average delay.

VII. LOOP SYSTEM

The configuration of User Stations and Central Processor for the Loop System is shown in Fig. 2. As in the Polling and Random Access Systems, the line between User Stations can be looped back to the Central Processor. Traffic flow on the line is in terms of fixed size message slots. Messages arriving at a User Station are multiplexed into these slots, one message to each slot. Traffic that is already on the line has priority; consequently a User Station must wait for an empty message slot. Traffic returning from the computer is addressed. At each station these addresses are examined. If a message is addressed to a particular station, it is taken off the line. Notice that in the Loop System, return and forward messages share the same line, consequently the total line length is half that required for Polling and Random Access Systems serving the same User Stations.

Messages arrive at each User Station at a Poisson rate of λ messages per second. These messages are multiplexed on a first come first served basis. If a message arrives when the buffer is empty, it still must wait until the line is free before it can be multiplexed.

Analytical and simulation results for this system have been presented in Refs. 13, 14, and 15. It was shown that the average message delay is approximated by the expression.

$$t_L = m(1 + \rho_L) + \frac{m\rho(1 + \rho_L)^2}{2[1 - \rho(1 + \rho_L)]} + \frac{\overline{l^2}}{\bar{l}} \frac{\rho_L}{2(1 + \rho_L)[1 - \rho(1 + \rho_L)]} \tag{14}$$

where ρ_L is the ratio of the average durations of line busy and idle periods. $\bar{l}$ and $\overline{l^2}$ are respectively the mean and the mean square values of the durations of the line busy periods. In deriving eq. (14) we make use of the fact that the message length is a constant value of m seconds.

A crucial quantity in the calculation of message delay is the line busy period. The characterization of the line busy period is simplified somewhat if we look at delay for what is usually the most critical station. In many applications the User Station receives more data than it transmits and as a consequence the traffic diminishes as one moves around the loop from the Central Processor.

We have modeled the traffic to the Central Processor as consisting of Poisson arrivals of fixed length messages. Messages arrive at rate $N\lambda$ messages per second and $M = B_R/C_R$ seconds are required to transmit each message. The busy period of the line out of the Central Processor is the busy period of an M/D/1 queue. It can be shown that[9]

$$\bar{l} = \frac{B_R/C_R}{1 - \rho_R} \tag{15a}$$

and

$$\overline{l^2} = \frac{(B_R/C_R)^2}{(1 - \rho_R)^3} \tag{15b}$$

where

$$\rho_R = B_R\lambda N/C_B .$$

The average duration of a line idle period is $1/N\lambda$. We have then

$$\rho_L = \frac{B_R N\lambda/C_R}{1 - \rho_R} = \frac{\rho_R}{1 - \rho_R}. \tag{16}$$

In order to provide a valid comparison with the other two systems, we must consider the processing time of a message at each station. A message, generated at the first station after the computer, passes through all of the other stations on the loop. At each station the address of the message is examined, entailing a delay of $\{[\log_2 N] + 1\}/C_R$ where $[x]$ is the largest integer less than x. Notice that since each station is receiving continuously from the same adjacent station, there is no synchronization delay. The cumulative delay of a message in going from the first User Station to the Central Processor is then

$$\bar{d}_L = \bar{l}_L + N([\log_2 N] + 1)/C_R \tag{17}$$

where $\bar{l}_L$ is given by eq. (14).

The return message delay for the Loop System is similar to the return message delay in the previous two systems. The delay in seconds is given by eq. (5) with $\delta = 0$. Thus the roundtrip delay is the sum of $\bar{d}_L$ [eq. (17)] and $\bar{d}_R$ [eq. (5)].

VIII. EXAMPLES OF SYSTEM BEHAVIOR

In this section, results of computations using the equations for *average* delay derived in the foregoing are presented. In presenting these results we were faced with the difficulty of choosing sets of parameters that provide meaningful comparisons. There is such a wide latitude in the choice of parameters for each of the three systems that one could easily bury the reader in a mass of curves and tables. Therefore, we have limited ourselves to relatively few cases illustrating system behavior. It is not difficult to supplement the results we present here since the expressions we have derived for average delay are relatively easy to evaluate.

Calculations have been made using values of user-related parameters (λ, B_F and B_R) which are appropriate in an Inquiry-Response context. Values for those parameters related directly to implementation (C_F, C_R and s) are chosen with the Digital Data System[2] in mind.

The results of the computations are shown in the form of sets of curves where average message delay is shown as a function of the number of user stations in the system. The rate at which messages arrive at User Stations, λ, is given two values, 50 calls per busy hour and 500 calls per busy hour, representing, respectively, light and heavy calling rates. In the application we have in mind, messages in the forward direction tend to be short. We have taken this message length to be $B_F = 250$ bits. Three values were used for the lengths of messages returning from the computer 2.5×10^3, 5×10^3, and 10^4 bits. We note that this latter value is approximately the number of bits required to fill a CRT display.

Since the systems we consider achieve economies by sharing, it is reasonable to put as many stations as possible in a system by choosing high line capacities C_F and C_R. In the Digital Data System, for example, 56 kbits per second are available transmit to local stations over short distances. Thus we take $C_R = C_F = 56 \times 10^3$ bits per second. This choice is tempered by the fact that, in order to achieve this high rate, synchronous operation is required. As we have noted in connection with the Polling and Random Access systems, the Central Processor receives data from different stations on each transmission. A low estimate for the time required to adjust synchronization from reception to reception at this speed is $s = 10$ ms. As we shall see, this value of s may lead to high values of delay for the Polling System. Therefore, for comparison we examine Polling Systems where transmission in the forward direction is asynchronous with $C_F = 2400$ bits per second and $s = 1/2400$ seconds. We have also examined fully synchronous operation where $s = 0$ and $C_F = 56 \times 10^3$ bits per second.

The parameter δ comes into play in the Polling and the Random Access system. Recall that δ is the portion of the return channel capacity allocated to the transmission of polling messages (Polling System) or positive acknowledgments (Random Access System). By varying this parameter, message delay in the forward direction is traded off against message delay in the reverse direction. For any particular system configuration, there is an optimum value for δ. However we have examined the effect of varying this parameter by choosing $\delta = 10^2$, 10^3 and 10^4 bits per second.

Sets of curves of delay in the forward direction for the Polling System are shown in Figs. 3, 4 and 5. In Figs. 7 and 8, return delay is shown as a function of the number of stations. As one might expect, traffic characteristics are important in judging the merits of implementations of the Polling System. For $\lambda = 50$ calls per busy hour, the 2,400 bps implementation (Fig. 5) yields lower delay than in the 56 kbps system with 0.01-second synchronization time (Fig. 3) in almost all cases. In fact, at this calling rate the fully synchronous 56 kbps system (Fig. 4) has delay which is lower than that of the 2,400 bps system by a relatively small amount. This advantage disappears under the heavy calling rate (500 calls per busy hour). The lower speed implementation is far more sensitive to calling rates.

Delay is sensitive to the parameter δ, which is that portion of the

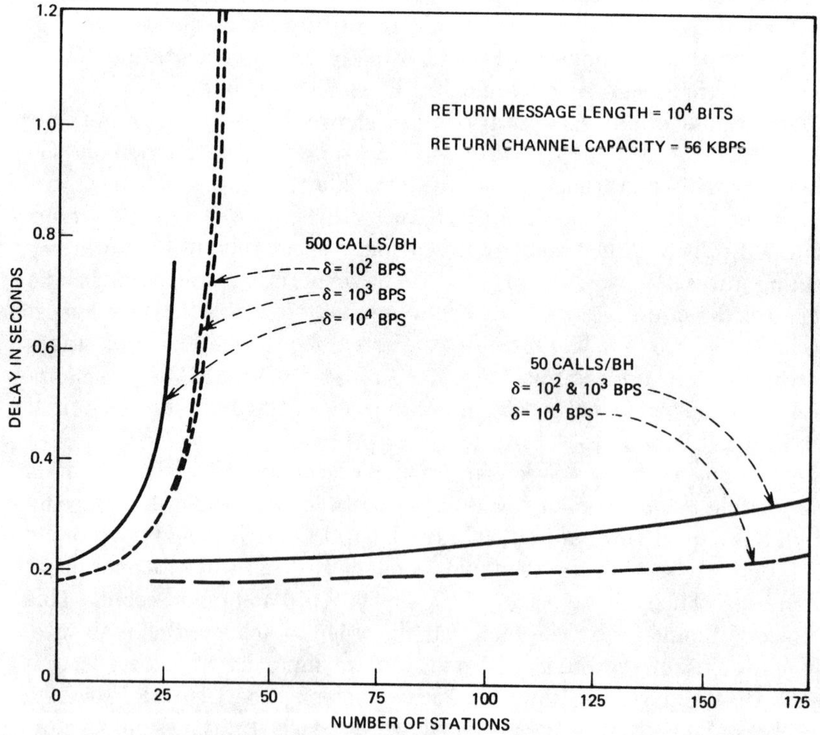

Fig. 8—Polling and random access system return delay.

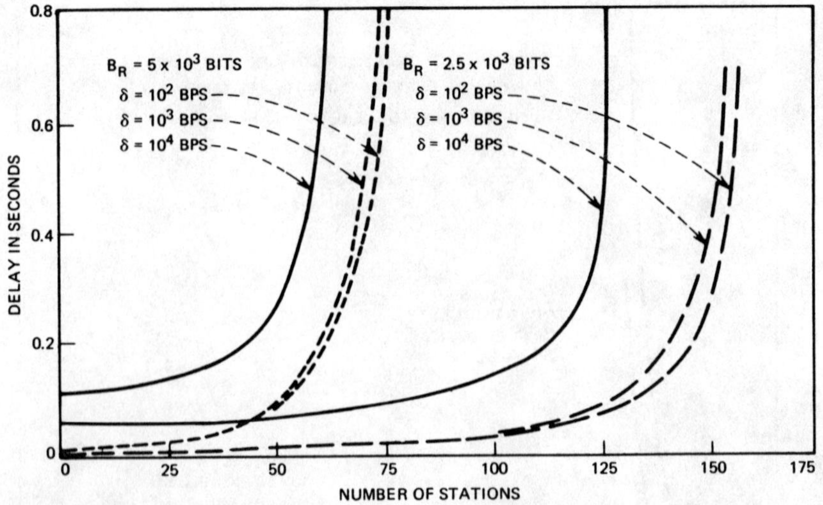

Fig. 9—Polling and random access system return delay; λ = 500 calls/BH.

return channel allocated to polling messages. It is clear from that, for the traffic we consider, $\delta = 10^2$ bps is entirely too small. By increasing δ to 10^3 bps, there is a large decrease in forward delay and a small increase in return delay. It is not clear that further advantage is obtained if δ is increased to 10^4 bps. On Figs. 8 and 9, return delay may be very large for $\delta = 10^4$ bps and, for the same number of stations, low for $\delta = 10^3$ bps (e.g., $N = 130$ stations, $M = 2.5 \times 10^3/56 \times 10^3$ seconds on Fig. 9). For any given set of traffic characteristics there is an optimum value of δ which minimizes total delay.

The results of the computation of average forward delay for the Random Access System are shown on Figs. 6 and 7. The curves of average return delay are the same as for the Polling System and are shown on Figs. 8 and 9. Unlike the Polling System, the effect of s and δ does not accumulate with the number of stations, and consequently is not very sensitive to these parameters. The Random Access System is more sensitive to calling rate than the Polling System primarily because calling rate affects the probability of a message being retransmitted [see eqs. (6) and (7)]. In Fig. 6, the results shown are for the station with the longest fixed timeout interval ($i = N$) in eq. (11).

The fixed-time retransmission implementation of the Random Access System (see Fig. 6) compares very well with the Polling System. For example, for $s = 0.01$ seconds, $\delta = 10^3$ bits per second and $\lambda = 500$ calls per busy hour, the 100 station delay in the Polling System is nearly 0.9 seconds, (see Fig. 3) while the corresponding delay in the Random Access System is less than 0.2 seconds (see Fig. 6). The Random Access System with fixed-time retransmission also performs well in comparison with the $s = 0$ implementation of the Polling System.

Recall that in the Random Access random retransmission strategy, the timeout interval is exponentially distributed with mean $1/\alpha$. By a process of trial and error we have found that $\alpha = N\lambda$ yields a rough minimum of average delay for $N > 10$ stations. This value was used to obtain the results shown in Fig. 7. As we see from Figs. 6 and 7, this realization of the Random Access System yields performance that is superior to fixed-time retransmission. Random retransmission com-

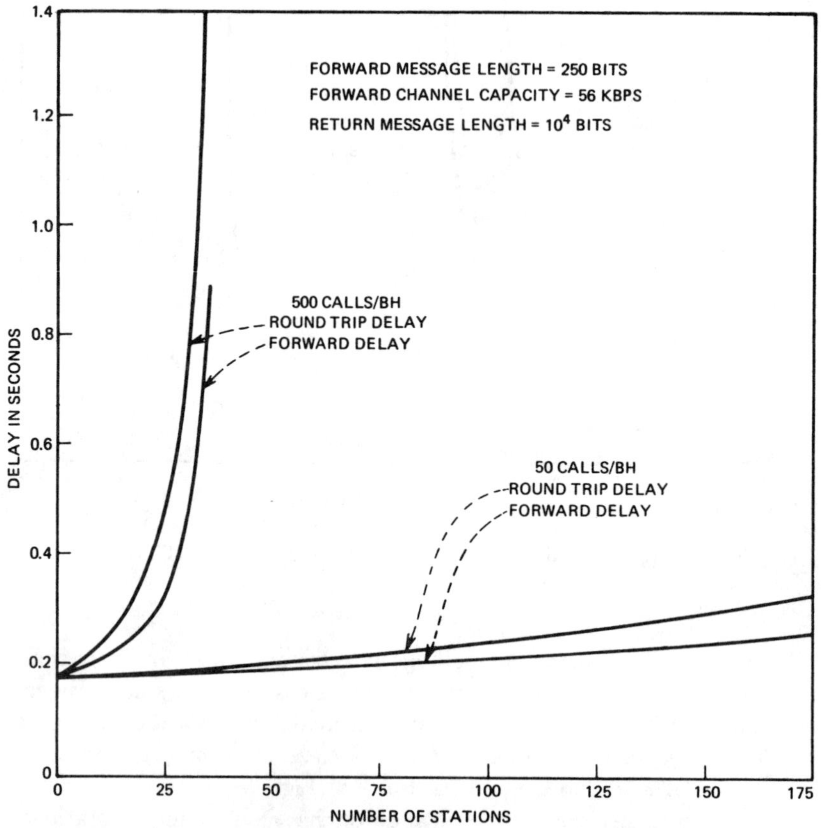

Fig. 10—Loop system forward and roundtrip delay.

pares favorably with the best implementation of the Polling System (see Fig. 4).

Our comparison may be biased somewhat in favor of the Random Access System since we have not taken into account overhead in that system. Recall that in order to detect errors, parity check bits along with information bits may be transmitted from the User Station. This will lengthen the message from the 250 bit message we have considered. A larger value of m will cause increased delay by increasing the probability of retransmission [see eqs. (6) and (7)] and by increasing the timeout interval [see eq. (11)]. However, we felt that it would not be necessary to lengthen the message very much, and consequently delay would be approximately the values we have shown.

The results of average delay calculations for the Loop System are shown on Figs. 9 and 10. Forward and roundtrip delay for the station on the line immediately after the Central Processor are shown. Roundtrip delay for the Loop System is even more sensitive to return message duration than the Polling or Random Access Systems. This is because messages going from the user terminal to the Central Processor are blocked by return messages. However, in the stable region below the knee of the delay curve, the Loop System gives performance comparable to the best implementation of the Polling and Random Access Systems. For example, for $M = 2.5 \times 10^3/5.6 \times 10^4$ seconds, $\lambda = 500$ calls per busy hour, and $N = 100$ stations, the following average roundtrip delay estimates are obtained for each of the three systems: Loop System: 0.12 seconds (Fig. 11), Polling System: 0.16 seconds

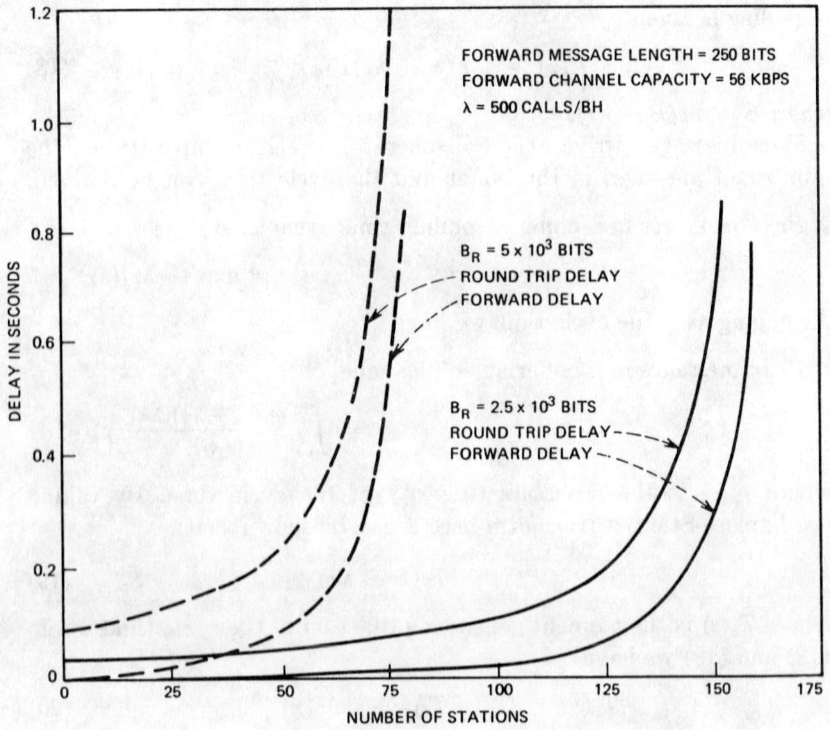

Fig. 11—Loop system forward and roundtrip delay.

(Figs. 4 and 9), and Random Access System: 0.16 seconds (Figs. 7 and 9).

IX. ACKNOWLEDGMENTS

We would like to thank R. J. Pilc for initial suggestions on the applicability of polling to local distribution, and M. Eisenberg for his assistance in developing the structure of the queueing model in the polling system.

APPENDIX A

In this appendix the moment generating function of forward delay in a Polling System is derived. Although other schedules may be analyzed, we concentrate on the situation where each station is polled once and only once in a cycle. Messages arrive for multiplexing at the User Stations at a Poisson rate of λ messages per second. In carrying out the derivation, we shall denote the generating function of the message length in seconds as $M(u)$. The generating function of the walk-time is denoted as $W(u)$. In the text, we shall apply our results to the case where message length and walk times are constants.

We consider first the "cycle time" of the polling sequence. This quantity is the time interval between polls of a particular station. As the polling sequence goes through a complete cycle, a random number of messages is encountered in each station's buffer. The analysis is simplified considerably if we assume that this number of messages is independent and identically distributed from station to station. Under this assumption, Leibowitz[7] shows that the moment generating function of the number of messages in each station's buffer at the time of polling is given by

$$P(x) = \{W(\lambda - \lambda x)P(\phi)\}^N \tag{18}$$

where $\phi = M(\lambda - \lambda x)$.

Since messages arrive at a Poisson rate, a relationship between the number of messages in the buffer and the cycle time can be derived.

$$Pr\ [n \text{ messages in a buffer at polling time cycle time } = \tau]$$
$$= \exp(-\lambda\tau)(\lambda\tau)^n/n!.$$

Averaging over the cycle time we have

$$Pr\ [n \text{ messages in a buffer at polling time}]$$
$$= \int_0^\infty \frac{\exp(-\lambda\tau)(\lambda\tau)^n}{n!} p_c(\tau)\, d\tau$$

where $p_c(\tau)$ is the probability density of the cycle time. By taking the Laplace-Stieltjes transform here it can be shown that,

$$P(x) = T_c(\lambda - \lambda x) \tag{19}$$

where $T_c(s)$ is the moment generating function of the cycle time. From (18) and (19) we have

$$T_c(u) = [W(u)T_c(\lambda - \lambda M(u))]^N. \tag{20}$$

Differentiating (20) and setting $u = 0$ yields the results on mean and mean square delay shown in eqs. (1a) and (1b) respectively.

Now a message arriving at a station must wait until the station is polled, and until all messages that have arrived before it have been multiplexed on the line. We first derive the generating function of the time the message must wait in the queue. We shall work under the assumption that messages arriving while prior messages are being multiplexed must wait until the next poll. Thus the queueing time of a customer can be written

$$d_P = \tau + \sum_{i=1}^{k} m_i \tag{21}$$

where τ is the time interval until the next poll and m_i, $i = 1, 2, \cdots, k$ are the lengths of k prior messages in the buffer. From (21) we can write the density of the delay d_P as

$$P(T) = \Pr[T < d_P \leqq T + dT] = \int_0^\infty d\tau \sum_{k=0}^{\infty} p_m^{(k)}(T - \tau)p(k, \tau) \tag{22}$$

where $p_m(s)$ is the probability density of the message length and $p(k, \tau)$ is the joint density of k and τ. It can be shown from elementary probabilistic arguments that

$$p(k, \tau) = \int_\tau^\infty dx \frac{p_c(x)}{\bar{t}_c} \frac{\lambda(x - \tau)^k}{k!} \exp\{-\lambda(x - \tau)\} \tag{23}$$

where $p_c(x)$ is the probability density of the cycle time. From (22) and (23) it can be shown that the moment generating function of the queueing delay is

$$T_{PQ}(u) = \frac{T_c\{\lambda - \lambda M(u)\} - T_c(u)}{\bar{t}_c[u - \lambda + \lambda M(u)]} \tag{24}$$

where $T_c(u)$ is the moment generating function of the cycle time. In order to find the total message delay, we must add the multiplexing time to the queueing delay. Since these quantities are independent random variables, the moment generating function of the message delay is

$$T_P(u) = T_{PQ}(u) \cdot M(u). \tag{25}$$

By differentiating $T_P(u)$ with respect to u and setting $u = 0$, the expression for delay given by eq. (2) of the text can be found.

APPENDIX B

In this appendix the generating function of message delay in the forward direction for the Random Access System is derived. Recall that after transmitting a message, a User Station waits for a positive acknowledgment from the Central Processor. The minimum time required to receive a positive acknowledgment is

$$w = \frac{[\log_2 N] + 1}{\delta} + s. \tag{26}$$

If no acknowledgment is received, the original message is retransmitted. We consider two transmission strategies; a fixed transmission time which is different for each station, and a random transmission time.

Under the fixed retransmission strategy, each station's transmission interval is different by at most $2m$ seconds, thus the retransmission interval for the ith station is $2mi$. This interval obviates the possibility of the same two stations repeatedly interfering with one another. If k retransmissions of a message are necessary to clear a message from the station's buffer, the total clearance time is

$$\tau = w + m + k(w + mi) \tag{27}$$

for the ith station.

If message flow on the line is Poisson, then the probability of two messages interfering with one another is

$$\Pr\,[\text{interference}] = 1 - \exp\,(-2Rm). \tag{28}$$

If we assume that subsequent retransmissions are independent trials, the total number of trials required to clear a message is geometrically distributed. We have

$$\Pr\,[k \text{ retransmissions}] = [1 - \exp\,(-2Rm)]^k \exp\,(-2Rm). \tag{29}$$

From (27) and (29) the moment generating function of the time required to clear a message is

$$S_{1i}(u) = E\{\exp\,(-u[w + m + kT_i])\} = \exp\,(-um)[\exp\,(uw + 2Rm) - \exp\,(2Rm - 2uim) + \exp\,(-2uim)]^{-1} \tag{30}$$

where

$$T_i = w + 2mi.$$

In the language of queueing theory, $S_{1i}(u)$ corresponds to the generating function of the service time of a customer (message). Messages arrive for multiplexing at a Poisson rate of λ messages per second.

The message delay can be found from the theory of the M/G/1 queue. We have

$$D_{RA}^{i}(u) = \frac{(1 + \lambda S_{1i}'(0))uS_{1i}(u)}{u - \lambda + \lambda S_{1i}(u)}. \tag{31}$$

By substituting (30) into (31), differentiating with respect to u and setting $u = 0$, the expression for the mean value of delay given in eq. (9) can be found.

The derivation of the generating function of message delay for random retransmission proceeds along the similar lines. The retransmission interval is a geometrically-distributed random variable with mean value $1/\alpha$ for each station. If k retransmissions are required, then the time to clear a message is

$$\tau = (k + 1)w + m + \sum_{j=1}^{k} \xi_i \tag{32}$$

where ξ_i, $j = 1, 2, \cdots, k$ are exponentially distributed random variables. The distribution of k is slightly different than in the case of fixed time-out retransmission. If two stations have interfered, there is a non-zero probability that they will interfere on the subsequent retransmission. We account for this phenomenon by approximating the probability of retransmission by the expression

$$\text{Pr [interference on retransmission]} = 1 - \exp(-2(R + \alpha)m). \tag{33}$$

The probability of no retransmissions is

$$\text{Pr [no retransmission]} = \exp(-2Rm). \tag{34}$$

The probability of k retransmissions is

$$\begin{aligned} \text{Pr } [k \text{ retransmissions}] &= [1 - \exp(-2Rm)][1 - \exp(-2(R + \alpha)m)]^{k-1} \\ &\cdot \exp(-2m(R + \alpha)); \qquad k = 1, 2, \cdots . \end{aligned} \tag{35}$$

From (31), (33) and (34) it can be shown that the generating function of message clearance is

$$\begin{aligned} S_{2i}(u) &= E\left[\exp\left(-(k + 1)w + m - \sum_{j=1}^{k} \xi_i\right)u\right] \\ &= \exp(-um)\left[\exp(-2Rm - wu) \right. \\ &\quad \left. + \frac{\lambda[1 - \exp(-2Rm)]\exp(-2m(R + \alpha))\exp(-2wu)}{\lambda + u - \lambda\exp(-wu)[1 - \exp(-2m(R + \alpha))]}\right]. \end{aligned} \tag{36}$$

The generating function of message delay can be found from the theory of the M/G/1 queue. The generating function of message delay is the same as in (31) with $S_{1i}(u)$ replaced by $S_{2i}(u)$. The first two moments of the time needed to clear a message given by (12a) and (12b) are found from successive differentiations of (36). The formula for average delay given in (13) is well known.[9]

REFERENCES

1. Pierce, J. R., "Network for Block Switching of Data," IEEE Conv. Rec. (March 1971).
2. Muench, P. E., "Bell System Private Line Digital Data Service," IEEE Conv. Rec. (March 1971).
3. Jackson, P. E., and Stubbs, C. D., "A Study of Multiaccess Computer Communications," AFIPS, Conf. Proc., *34*, p. 491.
4. Jackson, P. E., and Fuchs, E., "Estimates of Distributions of Random Variables for Certain Computer Communications Traffic Models," Proc. ACM Conf., Pine Mountain, Georgia (October 1969).
5. Martin, J., *Teleprocessing Network Organization*, Englewood Cliffs: Prentice-Hall Inc., 1970.
6. Martin, J., *Design of Real-Time Computer Systems*, Englewood Cliffs: Prentice-Hall Inc., 1967.
7. Leibowitz, M. A., "An Approximate Method for Treating a Class of Multiqueue Problems," IBM J. Res. Development, *5*, No. 3, 1961, pp. 204–209.
8. Eisenberg, M., "Queues with Periodic Service and Changeover Time," to be published, J. Oper. Res.
9. Cox, D. R., and Smith, W. L., *Queues*, Methuen and Co. Ltd., 1961, pp. 52–59.
10. "The ALOHA System," University of Hawaii Technical Report, 1969, AFOSR 70–0416 TR, AD 701417.
11. Abramson, N., "The ALOHA-System-Another Alternative for Computer Communications," University of Hawaii Technical Report B70-1, April 1970, AD 707853.
12. Bortels, W. H., "Simulation of Interference of Packets in the ALOHA Time-Sharing System, University of Hawaii Technical Report B70–2, March 1970.
13. Hayes, J. F., and Sherman, D. N., "Traffic Considerations of A Ring Switched Data Transmission System," B.S.T.J. *50*, No. 9 (November 1971), pp. 2947–2978.
14. Anderson, R. R., Hayes, J. F., and Sherman, D. N., "Simulated Performance of a Ring Switched Data Network," *COM-20*, No. 3, IEEE Trans. on Comm. Tech. (June 1972).
15. Avi-Itzhak, B., "Heavy Traffic Characteristics of a Circular Data Network," B.S.T.J. *50*, No. 8 (October 1971), pp. 2521–2549.

Basic Limits on Protocol Information in Data Communication Networks

ROBERT G. GALLAGER, FELLOW, IEEE

***Abstract*—We consider basic limitations on the amount of protocol information that must be transmitted in a data communication network to keep track of source and receiver addresses and of the starting and stopping of messages. Assuming Poisson message arrivals between each communicating source-receiver pair, we find a lower bound on the required protocol information per message. This lower bound is the sum of two terms, one for the message length information, which depends only on the distribution of message lengths, and the other for the message start information, which depends only on the product of the source-receiver pair arrival rate and the expected delay for transmitting the message. Two strategies are developed which, in the limit of large numbers of sources and receivers, almost meet the lower bound on protocol information.**

I. INTRODUCTION

A DATA communication network, for the purposes of this paper, is a finite collection of nodes; a finite collection of noiseless two-way communication links, each connecting some pair of nodes; and a finite collection of sources and receivers, each source and each receiver being connected to a node (see Fig. 1). We view each source as being paired with a receiver (typically at a different node), and the purpose of the network is to transmit messages from each source to its paired receiver. Messages are assumed to arrive from a source at randomly chosen instants of time and consist of binary strings of random length.

It seems intuitively obvious that the network must not only transmit the messages from source to receiver but must also transmit a certain amount of control information indicating, for example, the beginning, the end, and the destination of each message. It is customary in data networks to refer to such control information as *protocol information* and to refer to the conventions for representing it as *protocols.* To an information theorist, then, a protocol is a source code for representing control information.

Our major objective in this paper is to define and calculate the amount of protocol information (in the information-theoretic sense) required in the type of network described above. We shall find, rather surprisingly, that this information is related solely to the starting and stopping of messages and has nothing to do with addressing. We interpret addresses later as a particular type of source code for representing some of the information about the starting of messages. We shall use our results on protocol information to derive a lower bound on the required transmission rate of binary digits throughout the network. Finally we shall demonstrate some source coding strategies (i.e., protocols) which come very close to our lower bounds.

The following examples provide some idea of the wide variety of ways in which protocol information can be represented in different types of networks. In a *message switching network*, each message is generally preceded by a binary encoding of the source and receiver address and of the message length. Messages, with their preceding protocol bits, are queued at the individual nodes and passed from one node to another by a variety of routing algorithms which will be of no concern here. In a *packet switching network,* the messages are first divided into packets of some fixed length. Each packet is preceded by an encoding of the source and receiver address and the position of the packet within the message. The final packet, of course, might have fewer message bits than the others and this information must also be encoded. The packets are transmitted through the network independently and are reconstructed into a message at the receiver's node. The amount of protocol information is somewhat greater than for message switching, but the delay in transmission is frequently reduced.

Another possible system is to use multiplexers on each link of the network and to assign to each source-receiver pair a fixed fraction of the capacity of the links on some path from source to receiver. Frequently in such systems, there is a special string of bits called an idle character, which is used repeatedly when there is no message to be sent, and a special string of bits called a flag, which precedes and follows each message. The messages themselves are slightly encoded to prevent the occurrence of the flag character within the encoded message. The idle characters and flags should be regarded as protocol codewords which together indicate the beginning and the end of each message. Such a system is usually either very inefficient in its use of link capacities or subject to very long queues since it has no flexibility to allocate the network resources to messages as needed. If all the messages from a source are of equal length and if they arrive equally spaced in time, then the multiplexer becomes very efficient and, in fact, it is possible to eliminate *all* control characters.

In comparing the multiplexer approach with the message switching or packet switching approach, we see that the former transmits message start information but no addresses, while the latter transmits addresses but no message start information. We shall later see more clearly how message start information and addressing are related.

Manuscript received October 10, 1974; revised August 18, 1975. This work was supported in part by NASA under Grant NGL 22-099-013 and in part by the National Science Foundation under Grant GK-37582.

The author is with the Department of Electrical Engineering and Research Laboratory of Electronics, Massachusetts Institute of Technology, Cambridge, MA 02139.

Reprinted from *Trans. Inform. Theory*, vol. IT-22, no. 4, pp. 385–398, July 1976.

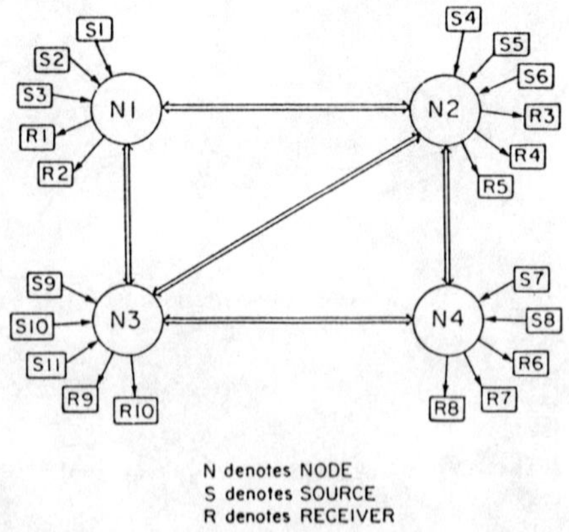

Fig. 1. Simple model of data communication network.

It should be clear that our data network model is a highly simplified version of a real data network. We have left out of consideration such questions as how new sources and receivers are connected to the network, how they are paired, and how they establish conventions for interpreting the bit-string messages between themselves. Similarly, we have ignored questions of record keeping, of privacy and security, and of removing malfunctioning nodes or links from the network and of adding new nodes and links. The control messages within the network to accomplish these functions can be, and should be, regarded as ordinary messages within the type of network we have described, with special conceptual sources and receivers within the nodes sending and receiving these messages. In data network parlance, these control functions are generally called higher level protocols, which means simply that they can be regarded as messages within the basic network described. Thus, although protocols to implement these higher level functions are important, these problems are separable from the problems of transmitting messages in our data network model.

We have also ignored the problem of errors on the communication links. These problems can be, and usually are, handled by error detection and retransmission strategies applied to individual links. Aside from some variable delay introduced into the communication links, these problems are essentially separable from those of concern here. Finally, we have excluded the possibility of a source communicating with several receivers and of a receiver getting messages from several sources. We shall see later that these latter situations can be easily incorporated into our model.

The data network model here has been chosen to focus clearly on what we view as the most important aspect of data networks, namely, the sporadic nature of the sources. This sporadic nature of data sources, generating messages occasionally but being idle most of the time, is the major reason for the practical importance of data networks (see, for example, [7] and [1]), but it is also the major cause of their complexity.

The sporadic arrivals of messages, of necessity, generate queues either within the network or at the sources awaiting message entry. These queues introduce both delay and the danger of buffer overflow. Complex flow control protocols are required to prevent buffer overflows, and frequently alternative or dynamic routing is attempted to reduce queueing delays. Alternative or dynamic routing, in turn, introduce a wide variety of protocols needed to get the information required for routing decisions and to handle the attendent problems of lost messages or packets and out-of-order arrivals at the receivers.

The sporadic arrivals of messages and the random lengths of messages also generate the need for message start–stop protocols and/or message addressing as discussed earlier. These latter protocols are the subject of this paper. We shall see that, as the number of sources at each node in the network grows (with a proportional increase in the link capacities), the queueing delays decrease and hence the need for protocols to handle queueing delays decreases. Thus our results can be interpreted as the limiting amount of protocol required for very large networks.

To our knowledge, this is the first theoretical study of protocol in data networks. We hope to convince the reader that it is indeed a subject amenable to theoretical analysis. The subject is also of practical importance. For example, the Arpanet is an example of a well engineered and researched network. A nonnegligible amount of traffic in this network comes from simple character-at-a-time transmission from interactive terminals, and the amount of protocol to handle such traffic is orders of magnitude greater than the information traffic itself. A more thorough conceptual understanding of protocol issues could undoubtedly give rise to much higher efficiencies in networks and, perhaps more importantly, to reduced complexity.

II. Summary of Results

In Section III, we start with a very simple situation, included primarily for pedogogic reasons. The network consists of only one link with K sources connected to the first node and K receivers to the second. Source k sends messages only to receiver k ($1 \le k \le K$). The sources are all synchronous. When a source is producing a message, it produces one binary digit per time unit. The lengths of all messages are independent geometrically distributed random variables with a mean denoted by $1/\epsilon$; the lengths of all idle periods between successive messages from a source are independent geometrically distributed random variables with a mean denoted by $1/\delta$. We assume that each message digit must be delivered to the appropriate receiver with a fixed delay. It is then shown that the average protocol information per message is

$$\frac{1}{\epsilon}\mathcal{H}(\epsilon) + \frac{1}{\delta}\mathcal{H}(\delta) \qquad \text{bits} \tag{1}$$

where

$$\mathcal{H}(x) = -x \log_2 x - (1-x)\log_2(1-x).$$

Furthermore, it is shown that as the number of sources K becomes large, strategies exist for which the average number of binary digits used for protocol approaches the amount in (1) as a lower bound arbitrarily closely. It is also shown that the above strategies work equally well for any distribution on the lengths of messages and idle periods which have means $1/\epsilon$ and $1/\delta$, respectively, but that the lower bound can be reduced for other distributions. In other words, the geometric distributions assumed on message lengths and idle periods are of interest not only for their simplicity but also for their extremal character.

In Section IV, we consider an arbitrary network with an arbitrary set of source-receiver pairs. For each source-receiver pair, we assume that the message arrivals are modeled by a Poisson process with rate α (α does not have to be the same for each pair). We assume that the entire message arrives at a single time and that the length of the message is again geometrically distributed with mean $1/\epsilon$. It is shown that, if the average transmission delay is given by d, the transmitted protocol information per message must be at least

$$-\log_2 (1 - e^{-\alpha d}) + \frac{1}{\epsilon} \mathcal{H}(\epsilon).$$

The first term above is interpreted as message start information and approaches zero as the allowable delay becomes larger. The second term is identified as stop information, or more appropriately as message length information and does not depend on delay.

In Section V, two strategies are developed for transmitting protocol information. Both of them consist of various ways of queueing the messages and sending them at more regular intervals, thus reducing the required information about starting time. The behavior of these strategies, for very large networks, is shown in Fig. 5.

III. Analysis of a Single Link with Identical Synchronous Sources

We start off with a very simple "network" consisting of two nodes connected by one link. There are K sources connected to the first node, and K receivers connected to the second node. Source $k, 1 \leq k \leq K$, sends messages only to receiver k. Such a network might more appropriately be considered as a simple-minded concentrator. The sources are synchronous with a given basic time unit. When a source is producing a message, it produces one binary digit per time unit; and, when a source is idle, it produces nothing for some integral number of time units. Naturally, the node must be able to distinguish a nothing output from the binary symbol outputs so that, when the source is producing nothing, we say that it produces an idle symbol i.

Assume that each source is an independent Markov source as defined in Fig. 2. The connection of sources and receivers is indicated in Fig. 3. Each transition between states of the Markov chain is described in Fig. 2, first by the probability of that transition and second by the source output corresponding to that transition; ϵ and δ are arbitrary probabilities but are the same for each source. The two states, B and I, represent the busy state and the idle state, respectively. After producing an idle symbol i the source is in the idle state, and after producing an information symbol of a message (zero or one), the source is in the busy state. The steady state probability of the busy state is $\delta/(\delta + \epsilon)$ and this is the fraction of time that the source is actually producing message symbols. A message is a string of zeros and ones with at least one idle symbol i on each side. The length of a message (the number of zeros and ones) is a geometrically distributed random variable with an expectation $1/\epsilon$. Similarly the length of each idle sequence separating messages is a geometrically distributed random variable with expectation $1/\delta$. The mean recurrence time between messages is thus $1/\epsilon + 1/\delta$. Finally, the entropy of the Markov source is easily calculated to be (see [4, sec. 3.6])

$$H_\infty(U) = \frac{\delta}{\delta + \epsilon} + \frac{\delta}{\delta + \epsilon} \mathcal{H}(\epsilon) + \frac{\epsilon}{\delta + \epsilon} \mathcal{H}(\delta) \quad \text{bits/time unit.} \tag{2}$$

Since each message symbol contains 1 bit of information and since the source is producing message symbols a fraction $\delta/(\delta + \epsilon)$ of the time, the first term on the right side of (2) may be interpreted as the entropy of the messages. Similarly, the second term may be interpreted as the entropy of the ending of the message or as the entropy in the length of the message. This information could be usefully employed by the source as part of the message to the receiver and thus could be interpreted as part of the message entropy. However, this specification of message length is traditionally regarded as protocol, and we arbitrarily define it as such here. Finally, the last term in (2) may be interpreted as the entropy in the starting time of a message or the entropy in the length of the idle sequences. Again this information can, and often does, provide useful information to the receiver, but we again define it as protocol. Note that if there is a variable delay in the transmission of messages to the receiver, then this information might be distorted at the receiver.

Next, assume that each binary digit emitted by each source is to be delivered to the corresponding receiver at some fixed delay d after the source emission time. What

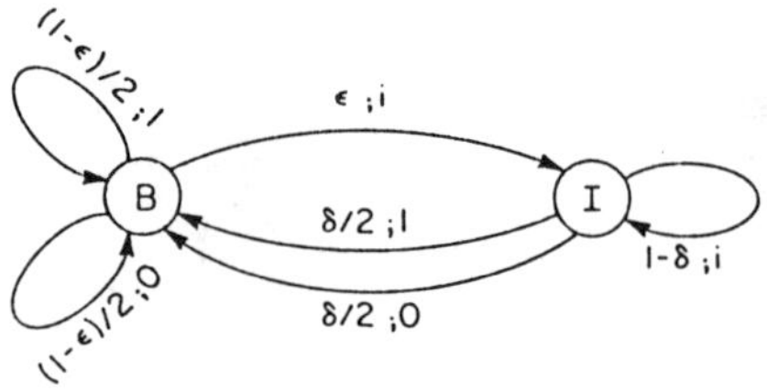

Fig. 2. Markov message source.

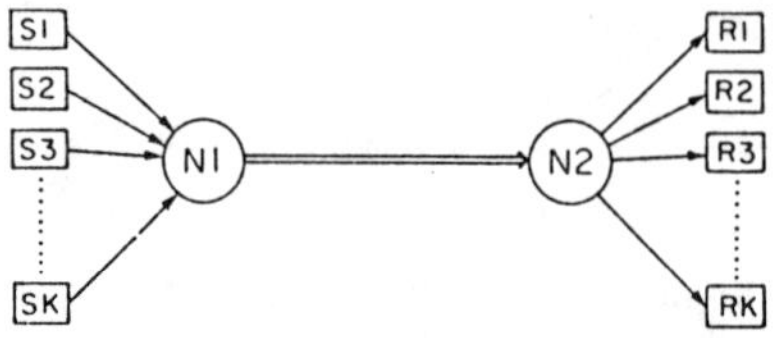

Fig. 3. Example of synchronous concentrator.

this means in effect is that in each time unit each receiver either must receive the binary digit emitted by its source d time units earlier or must receive no binary digit if none were emitted. Implicitly, therefore, whether or not the receiver is interested in the idle symbols, the idle symbols must be reconstructible from the input to the receiver. Since the output of the Markov source, idle symbols and all, must be reconstructible from the receiver input, the average mutual information between a source and its receiver is the same as the source entropy. Since we have K sources that are independent of each other, the average mutual information between the set of sources and the set of receivers is given by K times the entropy of a single source. The channel between the two nodes must have at least this much channel capacity by the converse to the coding theorem. We summarize our results thus far in the following theorem.

Theorem 1: Let K synchronous Markov sources be statistically independent of each other and each be described by Fig. 2. Let the sources be connected to a common channel with K receivers at the other end. In order for each binary digit from each source to be transmitted to the corresponding receiver with a fixed delay, it is necessary for the channel to have a capacity C satisfying

$$C \geq K H_\infty(U) \tag{3}$$

where $H_\infty(U)$ is the source entropy given in (2).

In order to interpret this result, we define the protocol information overhead ratio η as the ratio of the protocol information to the message information,

$$\eta = \mathcal{H}(\epsilon) + \frac{\epsilon}{\delta}\mathcal{H}(\delta). \tag{4}$$

For δ small, this is approximately

$$\eta \simeq \mathcal{H}(\epsilon) + \epsilon \log_2 e/\delta. \tag{5}$$

As shown in Fig. 4, a significant fraction of the channel capacity must be used for protocol information when either the expected message length is short (ϵ large) or the idle sequences are very much longer than the message sequences. We shall return to this interpretation later.

A strategy will now be described for encoding the source outputs at the transmitter node. We shall see that, in the limit of large K, the lower bound to the link capacity in (3) suffices to transmit the required information. The strategy is not intended to be practical, although practical systems could be devised using this approach. Assume that the transmitter node and receiver node each maintain a list, in numerical order, of the sources that are in the busy state at each unit of time. Each time unit, the transmitter node will send encoded protocol information to the receiver node, allowing this list at the receiver to be updated. Assume that each source emits a symbol (zero, one, or i) at each integer time j. Let $K_B(j)$ denote the number of sources in the busy state immediately after the transition at time j.

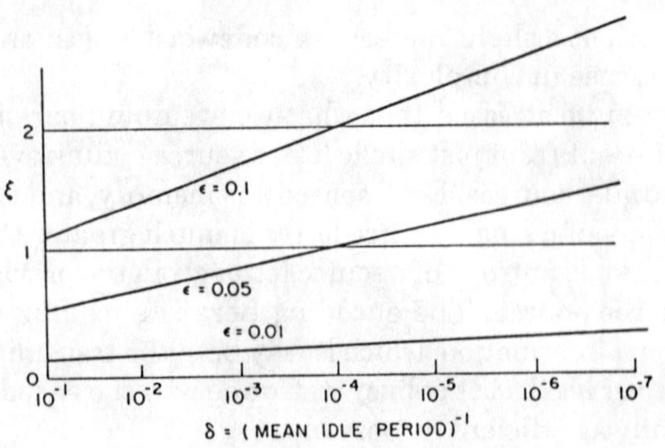

Fig. 4. Ratio of protocol to useful information as function of ϵ and δ.

At time j, the transmitter first encodes the identities of the set of sources which have gone from the busy to the idle state at time j. Each of the $K_B(j-1)$ sources in the busy state immediately preceding time j independently makes a transition to the idle state with probability ϵ. Thus this information can be represented by a binary prefix condition source code with an expected codeword length of at most $K_B(j-1)\mathcal{H}(\epsilon) + 1$ (assume a different source code for each value of K_B). Since the receiver node knows $K_B(j-1)$, it can decode this source code and remove those sources that have gone idle at time j from its busy list.

The transmitter next encodes the identities of the set of sources that have made a transition from idle to busy at time j. There are $K - K_B(j-1)$ sources in the idle state immediately before time j, and each independently has probability δ of going to the busy state. Thus the required expected length of a binary prefix condition source code here is at most $[K - K_B(j-1)]\mathcal{H}(\delta) + 1$. Upon receipt of this second codeword, the receiver adds the new busy sources to its busy list which is now updated to be the list of busy sources after the transition at time j.

Finally, the transmitter node sends the $K_B(j)$ binary digits which have been emitted by the busy sources at time j (this includes both the sources that have just become busy and those that have remained busy). These binary digits are sent in order according to the numbering of the sources ($1 \leq k \leq K$). Since the receiver node has the corresponding numbering of the receivers, it can route these bits, in order, to the receivers corresponding to the busy sources.

The steady state probability of a source being in the busy state is $\delta/(\delta + \epsilon)$, and thus the expected value of K_B in the steady state is $K\delta/(\delta + \epsilon)$. It follows then that the expected number of digits $\bar{n}$ encoded by the source node in the interval between time j and time $j + 1$ is

$$\bar{n} \leq K\frac{\delta}{\epsilon + \delta} + K\frac{\delta}{\epsilon + \delta}\mathcal{H}(\epsilon) + 1 + K\frac{\epsilon}{\epsilon + \delta}\mathcal{H}(\delta) + 1$$
$$= K H_\infty(U) + 2. \tag{6}$$

Note that, when K is large, the extra two bits that appears in this encoding are quite insignificant. Naturally, one could reduce this extra term to one bit by jointly coding the transitions into the busy state and the transitions out of it. One could also employ run length coding for the transitions, at a slight increase in codeword length and a large decrease in complexity.

It should be stressed that what we are doing here is a little different from just single letter source coding over a set of parallel sources. Each source has memory, and thus the entropy of a single source letter is much greater than the conditional entropy of a source letter given the previous state of the source. The encoding here is encoding the conditional information, which is why both the transmitter and receiver need a list of busy states and why the encoding is virtually as efficient as possible.

In showing that the source outputs can be efficiently encoded into binary digits, we have really come to grips with only half the problem. The other half is that the code is a variable length code and sending such a code over a fixed capacity channel will generate a queueing problem. Intuition suggests that, if the number of sources is great enough, the variability in codeword length should become negligible. This, of course, is the reason for sharing a channel among many sources; the channel only serves the busy sources, and with enough sources one needs little more capacity than the average rate. The following theorem, which is proved in Appendix A, bears out our intuition in this regard.

Theorem 2: Let the output from K sources, each defined by Fig. 2, be encoded by the strategy just described. Assume that at each time unit, a maximum of $KH_\infty(U) + f(K)$ bits can be transmitted where $f(K)$ satisfies

$$\lim_{K\to\infty} \frac{f(K)}{\sqrt{K}} = \infty.$$

Assume that the codewords are queued for transmission in a first-in first-out queue. Then the probability that a codeword will not be entirely transmitted within its own time unit approaches zero with increasing K.

Note that, as K gets large, the channel capacity per source required in the theorem approaches $H_\infty(U)$ which is the minimum possible as shown by Theorem 1.

In one sense this theorem is quite fragile. Real sources do not have the statistical regularity that we have assumed (for example, there are busy times of the day) and the number of sources that can share a channel is not effectively infinite. For these reasons, queuing delays are important; we shall return to the question of delay in the next section.

In a number of other senses, Theorem 2 and the coding strategy just described are very robust and depend very little on the detailed behavior of the sources and the channel. Note first that we have tacitly been assuming a noiseless binary channel. One may employ coding for error detection and retransmission or for error correction to transmit the information reliably, but this introduces delay. On the other hand, with many sources sharing one channel, many bits will be transmitted over the channel in each basic time unit, so that the delay required for reliable transmission becomes negligible as the number of sources sharing the channel increases.

Next consider what happens if the sources do not quite obey the statistical characterization of Fig. 2. As long as the fraction of sources which are busy remains close to $\delta/(\delta + \epsilon)$, and the fraction of sources making a transition to the busy state (or to the idle state) in each time interval remains close to $\delta\epsilon/(\delta + \epsilon)$, then the average codeword lengths are given approximately by (6).

One implication of this is that, if each source has a mean message length of $1/\epsilon$ and a mean idle sequence length of $1/\delta$ but an otherwise arbitrary distribution of message lengths and idle sequence lengths, the average codeword length for the strategy will not change appreciably. Since we have seen that a strategy designed for (and essentially optimal for) one probabilistic message source model works equally well for a whole class of models, we must conclude that the model of Fig. 2 has an interesting extremal property; it generates the most protocol information of any source model with the same mean message length and mean idle sequence length.

This extremal property of Fig. 2 is, in fact, easily demonstrated analytically. Consider a source model with the alphabet zero, one, and i. Assume that the expected length of runs of the idle symbol (i) is $1/\delta$ and that the expected length of runs of busy symbols (zero and one) is $1/\epsilon$. The entropy of the source can be maximized subject to these constraints, and it is seen that the maximum occurs when the idle runs and busy runs are each geometrically distributed, leading to the entropy in (2).

If one is dealing with sources which do not have the Markov property of Fig. 2, then it is possible to design more efficient source codes for those particular sources than the ones we have described here. For example, one frequently deals with sources for which the messages all have a fixed length. One could then omit the transmission of transitions from the busy to the idle state, since the receiver would already know when these transitions had to occur.

IV. THE EFFECT OF DELAY ON PROTOCOL INFORMATION

When messages can be delayed in passing through a network, there is an opportunity to send less protocol information than our results in Section III indicated. First we consider a simple example. Suppose that we modify the encoding strategy of Section III in the following way. We shall allow the transmission of messages to begin only at even numbered time intervals. If a source starts a message at an odd interval of time, then that message is delayed by one time unit before being encoded. For small δ, this reduces the average codeword length for start information (per source per time unit) from $\epsilon/(\delta + \epsilon)\mathcal{H}(\delta)$ to approximately $\epsilon/(2(\epsilon + \delta))\mathcal{H}(2\delta)$. An equivalent way to view this is that for each message the receiver has approximately one bit of uncertainty as to whether the message from the source started at an odd time or an even time; we are saving protocol by not resolving the intermessage time delay as precisely as before.

It is also instructive to look at this same type of example in a different context. Suppose we take the source model in Fig. 2 and scale down the basic time interval of synchronization while keeping the mean message *length* and the mean message interarrival *time* fixed. As the time interval shrinks, the parameter ϵ stays fixed, but the parameter δ decreases. As a result, the protocol information per message due to start information $\mathcal{H}(\delta)/\delta$ increases without bound; more and more protocol information is being used to resolve message arrival times with more and more precision.

The above situation brings out a fundamental limitation to the point of view that we adopted in Section III. In some sense, the rate at which a source produces the binary digits in a message is a detail which should not be inextricably linked to the required amount of protocol information. Our point of view in this section is that the important parameter in specifying protocol information is the average delay that messages experience. When the delay is small, then the receiver must receive considerable information about message starting times; when the delay is large, the receiver needs less information.

In order to focus on the problem of delay in the simplest context, we abandon the synchronous model of Fig. 2 and assume instead that the message arrival times of each source form a Poisson process of a given rate α. The parameter α is the expected number of arrivals per second, and $1/\alpha$ is the mean interarrival time. We assume that an entire message arrives instantaneously, rather than one bit per time unit, but it will be seen from the analysis that the results are not critically dependent upon this assumption. The message lengths have some given probability mass function $P_M(m)$, and an entropy $H(M) = \Sigma_{n=1}^{\infty} - P_M(m) \log P_M(m)$. We assume that the messages must have their identity preserved in transmission to the receiver. That is, if a source produces a message 011 followed by a message 1011, then it is not acceptable for them to be combined and presented to the receiver as a single message 0111011. This means that the protocol information about message lengths must be transmitted to the receiver. Our only problem is to find out how much information about arrival times must be transmitted.

In the analysis that follows, we consider an arbitrary network as described in the Introduction. We consider an arbitrary source receiver pair in the network and derive a lower bound on the average information per message that must be sent to the receiver in order to limit the average delay per message to a given value d. This lower bound is independent of the network topology, but in the next section we shall describe protocols that approach the bound quite closely for networks with very large numbers of source receiver pairs. This rather surprising independence from network topology will be discussed more fully in Section VI.

We assume that different source-receiver pairs may have different message arrival rates and different distributions on message length. We also generalize the model in the Introduction slightly to allow a source to send messages to several different receivers (and vice versa). We simply regard such a source as being several virtual sources, one virtual source being paired with, and sending messages to, each of the receivers.

We are now ready to analyze a given source-receiver pair. Let X_i, $i = 1, 2, \cdots$, be the message arrival times and Y_i, $i = 1, 2, \cdots$, be the times at which the messages are delivered to the receiver. Since we assume the message arrivals to be a Poisson process with rate α, the interarrival times $T_i = X_i - X_{i-1}$, $i = 1, 2, \cdots$, (with $X_0 = 0$) are independent and each have the probability density $\alpha e^{-\alpha t}$.

For any given network and network protocol, and for any given integer N, there will be a joint probability measure P_N on $X^N = (X_1, \cdots, X_N)$ and $Y^N = (Y_1, Y_2, \cdots, Y_N)$. There are two obvious constraints on the probability measure P_N: first, the marginal distribution of X^N must satisfy the Poisson process assumption, and second, for each message, the delay $D_i = Y_i - X_i$ must be nonnegative with probability one. For a given P_N, there is an average mutual information, $I_{P_N}(X^N; Y^N)$ between the sequence X^N and Y^N. This mutual information is a lower bound, given P_N, on the information provided to the receiver about message arrival times at the source. Note that we are not asserting that the receiver should be interested in this information, nor that it is in a form suitable for any particular use; we are simply asserting that the very existence of P_N and the delivery of the messages provides this information whether we want to provide it or not.

For a given P_N, there is also an expected delay per message, $\bar{D}_N$ given by

$$\bar{D}_N = \frac{1}{N} \sum_{i=1}^{N} E(D_i) \tag{7}$$

where $D_i = Y_i - X_i$ and $E(D_i)$ is the expected value of D_i for the probability measure P_N. Define $\mathcal{P}_N(d)$ as the class of probability measures P_N which both satisfy the above constraints and have an expected delay per message $\bar{D}_N \leq d$. We can obtain a lower bound on the amount of transmitted protocol information about message arrival times, subject to a constraint on expected delay, by minimizing $I_{P_N}(X^N; Y^N)$ over $P_N \in \mathcal{P}_N(d)$. This will be recognized by information theorists as a rate-distortion problem, and we define the Nth order rate-distortion function by

$$R_N(d) = \inf_{P_N \in \mathcal{P}_N(d)} \frac{1}{N} I_{P_N}(X^N; Y^N). \tag{8}$$

The rate-distortion function $R(d)$ is then defined by

$$R(d) = \lim_{N \to \infty} \inf R_N(d). \tag{9}$$

From the definitions it is clear that $R(d)$ is a lower bound to the average protocol information about message arrival times that must be transmitted to limit the expected delay per message to d. It is not clear whether, for any given network, protocols exist which actually transmit this minimum amount of information. Even if such protocols did exist, it is not clear that the information could be efficiently encoded. In other applications of rate-distortion theory, one can demonstrate the existence of efficient encodings by introducing delay and complexity; that approach does not work here since delay *is* the distortion measure.

There may be some question as to why we have used mean delay as our criterion here rather than, let us say, mean-square delay or maximum delay. Mean-square delay is both less tractable analytically and less conventional in data networks than mean delay and has no clear cut advantages. Maximum delay appears to be quite intractable and has the further disadvantage that, in a finite network with finite capacities and Poisson arrivals, there is clearly no possibility of guaranteeing a finite maximum delay for all messages, since there is no maximum on the number of message arrivals within a given interval.

The following theorem, which is proved in Appendix B, provides a lower bound to $R(d)$.

Theorem 3: $R_1(d)$, as given by (8) with $N = 1$, is a lower bound to $R_N(d)$, for all $N > 1$, to $R(d)$, and to the average protocol information per message about message arrival times between a source-receiver pair for Poisson message arrivals of rate α and expected delay d. Furthermore, $R_1(d)$ is given by

$$R_1(d) = -\log_2 (1 - e^{-\alpha d}) \quad \text{bits/message.} \tag{10}$$

The probability measure P_1 that achieves $R_1(d)$ is defined implicitly by

$$Y_1 = \max (X_1, d) + Z \tag{11}$$

where Z is a nonnegative random variable, independent of X_1, with probability density $p_Z(z) = (\alpha + \rho) \exp - (\alpha + \rho)z$, where ρ is given by

$$\rho = \frac{\alpha e^{-\alpha d}}{1 - e^{-\alpha d}}. \tag{12}$$

Note that $R_1(d)$ is a function only of the product αd (it is not hard to see that $R_N(d)$ and $R(d)$ also must be functions only of αd). The asymptotic behavior of $R_1(d)$, for small d and large d, are given by the approximations

$$R_1(d) \simeq \begin{cases} -\log_2 \alpha d, & \alpha d \ll 1 \quad (13) \\ e^{-\alpha d} \log_2 e, & \alpha d \gg 1. \quad (14) \end{cases}$$

Small αd is the case of major interest for most network applications since messages are usually relatively infrequent. One possible exception to this is that of a teletype source in which each character is modeled as a message, but even here it is often desirable to keep αd small.

Fig. 5 compares the lower bound $R_1(d)$ with the average number of binary digits used for message arrival time protocol in the strategies of Section V. It is seen that the difference between the strategies and the bound is always less than 0.6 bits, and that the ratio approaches one for small αd. We conjecture that, for large αd, $R(d)$ (as contrasted with $R_1(d)$) approaches zero as $(\alpha d)^{-2}$ rather than as $e^{-\alpha d}$, but our reason for this conjecture will be explained later.

Next we look at the total average mutual information per message between a source-receiver pair. Assume that

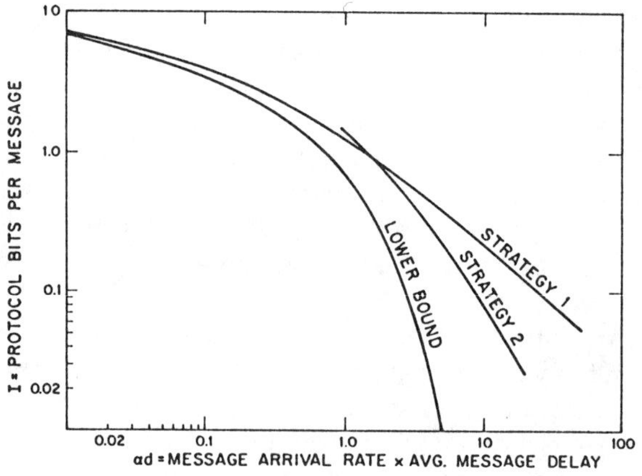

Fig. 5. Message start protocol information as function of average message delay.

the message lengths and message arrival times are statistically independent and assume that the messages are composed of independent equiprobable binary digits. Then the total mutual information, per message I, is the sum of the message length information $H(M)$, the message arrival time information, lowerbounded by $-\log_2 (1 - e^{-\alpha d})$, and the actual message information $1/\epsilon$, where $1/\epsilon$ is defined as the mean message length,

$$I \geq H(M) - \log_2 (1 - e^{-\alpha d}) + 1/\epsilon. \tag{15}$$

If the message lengths are geometrically distributed, this becomes

$$I \geq \frac{1}{\epsilon} \mathcal{H}(\epsilon) - \log_2 (1 - e^{-\alpha d}) + 1/\epsilon. \tag{16}$$

The first two terms in each of the above expressions are classified as protocol information, and the final term is classified as message information. The mutual information per unit time is, of course, just αI.

If all of the virtual sources, corresponding to the source-receiver pairs in the network, are statistically independent, then the data processing theorem asserts that the information for each source-receiver pair must be transmitted through the links of the network, thus imposing constraints on the channel capacities of the links. These capacity constraints[1] have been intensively studied in the literature as multicommodity flow problems [3]. The only difference between the classical problem and that here is that the information flow here includes the protocol terms in (15). Also, since (15) underbounds the information flow, the multicommodity flow solution will yield lower bounds on the feasible capacities.

V. Strategies for Minimizing Protocol Information with a Delay Constraint

In this section we discuss two strategies for encoding protocol information. For each of the strategies, we regard the sources and receivers in the network as being split into

[1] We are interested not in queueing delays here but only in link capacities that are capable of carrying the required traffic; we discuss the implication of queueing delays in Section VI.

virtual sources and virtual receivers such that each virtual source sends messages to only one virtual receiver and vice versa.

The set of virtual source receiver pairs are partitioned in such a way that, within any block of the partition, all the sources are connected to one node and all the receivers to some other node. Also all sources within one block of the partition have the same message arrival rate α, and the same desired average message delay d. All of the sources within one block of the partition will be encoded together into a bit stream which is then transmitted through a single path in the network.

Frequently, in order to avoid overloading particular links in the network, it is necessary to split the traffic between a given pair of nodes into several paths. In this case we simply further subdivide the blocks above, each smaller block containing the proper traffic for a given path. In the limit of an arbitrarily large number of source receiver pairs in each block, this subdivision can approximate the desired fraction of traffic for each path arbitrarily closely.

For simplicity we assume that, when the bit streams from several blocks must pass through the same link in the network, they are simply multiplexed together (see Fig. 6). From a practical point of view, such multiplexing is usually not a good idea because of the nonsteady flow on each bit stream. From a conceptual point of view, however, the argument of Theorem 2 can be applied to each of the strategies suggested here to show that, as the number of sources within a block of the partition becomes large, the encoded data from that block becomes steady. Our objective is to focus on the tradeoff between protocol information and the delay introduced by reducing protocol information. The assumption of many sources, leading to steady flow, is an artifice for studying this relationship independently of network queuing problems.

The first strategy is the simplest. The output of each source within a block is separately queued at the input to the encoder for that block. The encoder samples the source queues cyclically, sampling each queue once each $2d$ seconds. When the encoder samples a queue, it removes all the messages from the queue, encoding first the number of messages in the queue, then the length of each message, and then the messages themselves.

Consider the entropy of the contents of a queue when it is sampled. On the assumption that the message arrivals form a Poisson process with rate α, the probability that n messages arrive in the $2d$ second interval between inspections is given by the Poisson distribution

$$P_{2\alpha d}(n) = \frac{e^{-2\alpha d}(2\alpha d)^n}{n!}, \qquad n \ge 0. \tag{17}$$

Thus the entropy of the number of message arrivals in the interval is

$$H(P_{2\alpha d}) = \sum_{n=0}^{\infty} \frac{e^{-2\alpha d}(2\alpha d)^n}{n!} \log \frac{n! e^{2\alpha d}}{(2\alpha d)^n} \quad \text{bits.} \tag{18}$$

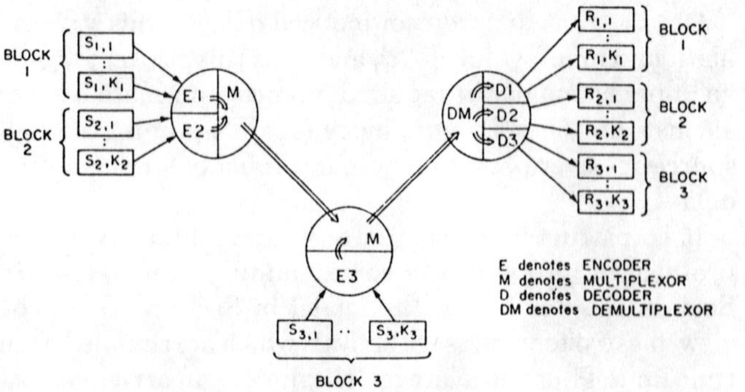

Fig. 6. Data network with source-receiver pairs partitioned into blocks.

Assuming as before that the message lengths are independent of the arrivals and are independent identically distributed random variables of mean $1/\epsilon$ and entropy $H(M)$, we see that the entropy of the message lengths in the interval is $2\alpha dH(M)$. On the assumption that the messages are composed of independent equiprobable binary digits, their entropy is $2\alpha d/\epsilon$. The entropy of the queue contents per sample is thus

$$H(P_{2\alpha d}) + 2\alpha dH(M) + 2\alpha d/\epsilon. \tag{19}$$

Assuming an arbitrarily large number of sources per encoder, the encoder can generate a bit stream with arbitrarily little more than the above number of bits per source per sample and with arbitrarily little delay. Since the message arrivals have a uniform distribution between sample points, the average delay of a message before being sampled by the encoder, and thus essentially before appearing on the bit stream going to the receiver, is d.

The number of bits transmitted per source per sample in (19) can, as usual, be separated into protocol information about message arrivals ($H(P_{2\alpha d})$), protocol information about message lengths ($2\alpha dH(M)$), and useful information ($2\alpha d/\epsilon$). In order to compare the protocol information about message arrivals with the lower bound in Section IV, we observe that the number of messages per sample approaches $2\alpha d$ over a long period of time. Thus the number of protocol bits about message arrival times transmitted per message in this strategy is

$$\frac{H(P_{2\alpha d})}{2\alpha d}. \tag{20}$$

This value is compared with the lower bound $-\log_2(1 - e^{-\alpha d})$ in Fig. 5. For small αD, the expression in (20) approaches $\log_2(e/2\alpha d)$, which is larger than the bound by $\log_2 e/2$ bits per message.

The delay d in these results ignores the propagation delay on the path through the network. Thus we call d the *mean protocol induced delay;* the total expected delay will be d plus the propagation delay. If the number of source-receiver pairs is not sufficiently large, there will, of course, also be a nonnegligible queuing delay. The delay in the lower bound of Section IV. was also a protocol-induced delay (i.e., total delay less propagation delay) as can be seen by defining time zero at the receiver in Section IV to be one propagation delay later than time zero at the source.

Observe that the protocol-induced delay in this strategy has a maximum value of $2d$, and thus this strategy yields an upper bound on the required protocol information given a constraint on maximum delay (again assuming enough source-receiver pairs to ignore improbably large queuing delays).

It is instructive to compare the message arrival time protocol in (20) with the corresponding term $\mathcal{H}(\delta)/\delta$ in Section III. For $\delta \ll \epsilon$, the model in Section III can be viewed as Poisson message arrivals which are sampled each time unit. Thus the average delay before an arriving message is sampled is ½ time unit and the arrival rate is δ, so that $\alpha d = \delta/2$. Finally, $\mathcal{H}(\delta)/\delta \approx \log_2 (e/\delta)$, for δ small, which is the same as (20) with $\alpha d = \delta/2 \ll 1$.

It is also instructive to compare the message arrival time protocol here with addressing in a message switching network. First we investigate what happens in the strategy here as αd becomes small. In this limit, using (17), the probability of finding one message in a source queue approaches $2\alpha d$ and the probability of finding more than one message is negligible. Most source queues are empty when sampled, which suggests that an efficient way to encode the presence or absence of a message is to use run length coding across the set of sources in the block. As shown in [5], the mean number of bits per message used in this encoding is very close to the entropy in (20), which is approximately $\log_2 [(e/(2\alpha d)]$.

Now let K be the number of source-receiver pairs in the block and let the source-receiver pairs be assigned addresses from zero to $K - 1$. Then an alternative strategy to the run length coding above would be to take each message as it arrives from a source, prefix it with the encoded address of the source-receiver pair and the encoded message length, and enter it directly onto the queue for transmission to the receiver node. The number of protocol bits per message for this addressing is about $\log_2 K$. Since the messages are transferred to the encoder output queue immediately upon arrival, the protocol induced delay in this addressing can be interpreted to be zero. This appears to contradict the lower bound of Section IV, but it does not for two reasons: first, the queueing delay can not be neglected here, and second, the capacity of the path from source node to receiver node must be greater than the information rate, and the idle bits when the queue is empty carry information about message arrival times.

We see from the above that addressing can be regarded as the low protocol delay, high protocol information end, of the tradeoff between delay and protocol information. The use of addressing is preferable for $2\alpha d > Ke$. The design question of where a network with given requirements should be on that tradeoff curve is difficult and unsolved except in the limit of arbitrarily large numbers of source-receiver pairs.

Most message switched and packet switched networks use separate addresses for source and receiver (one notable exception is Tymnet [8] which assigns addresses only to communicating source-receiver pairs). This requires more protocol bits than addressing within a block and, from an information-theoretic point of view, most of these extra bits are wasted. This loss of efficiency is sometimes justified because, first, separate full addresses simplifies dynamic routing, and second, the loss of efficiency is unimportant if the messages are long.

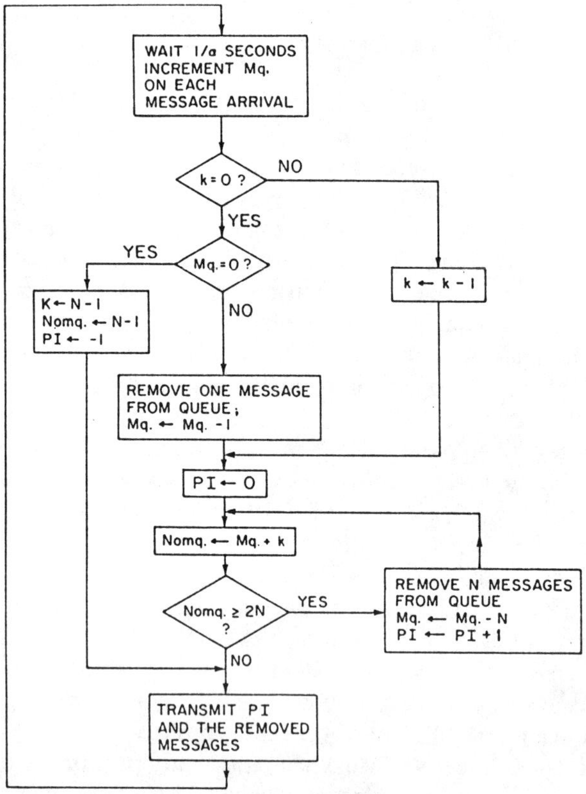

Fig. 7. Algorithm for strategy 2.

We next describe a strategy designed for large values of αd. The encoder for a given block of sources samples the source queues cyclically, as in the first strategy, but the sampling rate is $1/\alpha$. Let N be a positive integer and assume that we want to have an average message delay of about N/α. The strategy then, with two exceptions, is to remove one message from the queue at each sample time and encode it. The first exception is that if no message is in the queue then no message is taken out for N consecutive sample times. The other exception is that if more than $2N$ messages are in the queue then $N + 1$ messages are removed from the queue and encoded. This description leaves a few fine points uncovered, such as what happens during an N sample idle period if more than $2N$ messages accumulate (we transmit N of them), and what happens if, after transmitting an extra N messages, there are still more than $2N - 1$ messages on the queue (we transmit another extra N). The flow diagram in Fig. 7 describes the strategy in detail. The idea behind the strategy is to minimize the protocol information by making the two exceptional events relatively rare. Waiting for N sample times when the queue is empty, or transmitting $N + 1$ messages when the queue is too full, have the tendency of leaving the queue half full (with N messages) after each exceptional event. Since messages normally enter and leave the queue at the same rate, the exceptional events will not happen very often.

This strategy is analyzed in Appendix C. It is shown that, for any given N, the protocol information per message about message arrivals is bounded by

$$I \leq \begin{cases} 2\mathcal{H}\left(\dfrac{1}{e(e-1)}\right), & N = 1 \\ 2\mathcal{H}\left(\dfrac{1}{2N(N+4/3)}\right), & N \geq 2 \end{cases} \tag{21}$$

and the average delay per message by

$$d \leq \frac{1}{\alpha}\left[N - \frac{5N-3}{12N+16}\right]. \tag{22}$$

These expressions are compared with the lower bound in Fig. 5.

We note that these expressions have the protocol information per message going to zero with increasing d roughly as $(\alpha d)^{-2}$, whereas the lower bound approaches zero as $e^{-\alpha d}$. The standard deviation of the time required for N^2 arrivals, however, is N/α. Thus it appears that somewhere on the order of one bit of information should be required by the receiver every N^2 messages to track the arrivals within N/α seconds. Since $d \approx N/\alpha$, we conjecture that the lower bound is not tight for large αd and that this upper bound has at least the right qualitative behavior.

VI. Discussion

The preceding two sections have derived lower and upper bounds on the number of protocol bits per message required in a data network as a function of delay (excluding propagation delays). Both bounds assumed that the message arrivals from each source formed a Poisson process. The lower bound also assumed statistical independence between different source-receiver pairs and between message lengths and arrival times for each pair. The upper bound also assumed an arbitrarily large number of source-receiver pairs at each pair of nodes and assumed noiseless channels. In this section, we give a qualitative discussion of these assumptions.

The most restrictive assumption, of course, is that of an arbitrarily large number of source-receiver pairs, which was the assumption that allowed us to ignore queueing delays. Existing and planned data networks use orders of magnitude more protocol bits per message than we have here. Part of this extra protocol is devoted to higher level protocols, and part to error control on the channels, which are largely separable issues. A major part, however, is devoted to flow control and to protocols required for alternative or dynamic routing. This paper does not address these flow control and routing protocols except to show that they become wasteful in the limit of many source-receiver pairs per node pair. It is hoped, however, that the analysis given here of addressing or message arrival protocols will lay the groundwork for future analysis of flow control and routing.

The results here do suggest one connection between protocol delay and queueing delay. In existing networks, the protocol per message *increases* beyond a certain point of network loading. This causes throughput, in messages per second, to decrease beyond a certain delay, thus causing a further increase in congestion and an instability similar to that with automobile traffic. Our results here suggest that, at least as far as addressing or message arrival information is concerned, this effect can be partly counteracted in a well designed system by using less protocol per message for addressing as delay increases.

Next we look at the assumption that different source-receiver pairs are independent of each other. The major situation in which this is a poor assumption is that of interactive communication, which we would model as two source-receiver pairs going in opposite directions. In this case, a receiver will have some side information about when messages arrive at the distant source node, and the results of Section IV do not quite apply. If we assume, as we should from the standpoint of functional modularity, that the network protocols must be independent of the detailed interactive characteristics of the sources, then the results of Section IV still apply to the data through the network.

Finally, we consider the assumption of Poisson arrivals. We have not been able to establish that, for a given arrival rate, the rate-distortion function of Section IV is maximized by Poisson arrivals. However, it is not difficult to see that, for small αd, the performance of the first strategy in Section V is virtually independent of the source arrival statistics and depends primarily on the arrival rate.

Appendix A

Proof of Theorem 2

The strategy described in the text uses separate source codes for encoding transitions out of the busy state, transitions into the busy state, and binary data. The length of each codeword for the transitions can be chosen to exceed the conditional self information of the transitions by at most one, and the length of the codewords for the binary data is precisely equal to the conditional self information of the binary data. Thus, in one time unit, the length of the entire encoded sequence for that time unit exceeds the conditional self information of the outputs of the K sources by at most two. Since $KH_\infty(U) + f(K)$ bits can be transmitted within the time unit, the probability that the codeword requires more than one time unit for transmission is at most the probability that the conditional self information exceeds $KH_\infty(U) + f(K) - 2$. Unfortunately, a codeword may be short enough to be transmitted within the time unit, but still might be delayed because of previous codewords that spilled over into the present time unit (in other words, we have a queuing problem). If the combined length of the m codewords ($m \geq 1$) ending at the present time exceed $m[KH_\infty(U) + f(K)]$, then the present codeword will be delayed and, moreover, the present codeword will be delayed *only* if the above condition is satisfied for some $m \geq 1$. Let $I_{m,K}$ be the conditional self information from the K sources over the m time units up to and including the present. Then we can upper bound the probability that the current codeword is delayed P_D by

$$P_D \leq \sum_{m=1}^{\infty} P\{I_{m,K} \geq m[KH_\infty(U) + f(K) - 2]\}. \tag{A1}$$

Since the K sources are statistically independent, the conditional self information $I_{m,K}$ is the sum of K independent identically distributed random variables, namely, the conditional self information for the individual sources. Let $\mu_m(s)$ be the semi-invariant moment generating function for the conditional self information from a single source over m time units,

$$\mu_m(s) = \ln E\,[\exp(sI_{m,1})]. \qquad \text{(A2)}$$

Using the Chernoff bound, we than have, for any $s \geq 0$,

$$P_D \leq \sum_{m=1}^{\infty} \exp\{K\mu_m(s) - sm[KH_\infty(U) + f(K) - 2]\}. \qquad \text{(A3)}$$

We need the following lemma about $\mu_m(s)$; we will use the Lemma to complete the proof of the theorem and then prove the Lemma.

Lemma: For any source parameters $0 < \epsilon < 1, 0 < \delta < 1$, there exists $s_0 > 0$ and a finite number α such that for all $m \geq 1$ and all $s, 0 \leq s \leq s_0$,

$$\mu_m(s) \leq smH_\infty(U) + \alpha m s^2. \qquad \text{(A4)}$$

Substituting (A4) into (A3), we have, for $0 \leq s \leq s_0$,

$$P_D \leq \sum_{m=1}^{\infty} \exp\{\alpha m K s^2 - sm[f(K) - 2]\}. \qquad \text{(A5)}$$

Let $s = K^{-1/2}$; for large enough K, this is within the range $0 \leq s \leq s_0$. Then

$$P_D \leq \sum_{m=1}^{\infty} \exp\left[m\left(\alpha - \frac{f(K) - 2}{\sqrt{K}}\right)\right]$$

$$= \frac{\exp[\alpha - (f(K) - 2)K^{-1/2}]}{1 - \exp[\alpha - (f(K) - 2)K^{-1/2}]}. \qquad \text{(A6)}$$

The limit of the right side as $K \to \infty$ is zero, completing the proof.

Proof of Lemma: Let $A(s)$ be the 2×2 matrix with components

$$A_{ij} = P(j|i)e^{sI(j|i)}$$

in which $P(j|i)$ is the probability of a transition to state j given state i for the Markov chain defining the source, and $I(j|i)$ is the conditional self information of the source output for the same transition. Thus

$$A(s) = \begin{bmatrix} (1-\epsilon)^{1-s}\,2^s & \epsilon^{1-s} \\ \delta^{1-s}\,2^s & (1-\delta)^{1-s}\,2^s \end{bmatrix}.$$

The semi-invariant moment generating function $\mu_m(s)$ in (A2) is then given by

$$\mu_m(s) = \ln\left[\xi_0\, A^m(s)\, 1^T\right] \qquad \text{(A7)}$$

in which the row vector $\xi_0 = \delta/(\delta + \epsilon), \epsilon/(\delta + \epsilon)$ is the steady state probability assignment for the source and 1^T is the column vector of two ones. The eigenvalues of $A(s)$ are easily calculated to be

$$\lambda_i(s) = \frac{1}{2}\Big\{(1-\epsilon)^{1-s}\,2^s + (1-\delta)^{1-s}$$

$$\pm \sqrt{[(1-\delta)^{1-s} - 2^s(1-\epsilon)^{1-s}]^2 + 4(\epsilon\delta)^{1-s}\,2^s}\Big\} \qquad \text{(A8)}$$

where the plus sign corresponds to $\lambda_1(s)$ and the minus to $\lambda_2(s)$.

Expanding $A(s)$ in terms of eigenvalues and eigenvectors and multiplying the terms of (A7), we obtain

$$\mu_m(s) = \ln\{[1 - \beta(s)]\lambda_1^m(s) + \beta(s)\lambda_2^m(s)\}, \qquad \text{(A9)}$$

$$\beta(s) = \frac{1}{2}$$

$$- \frac{(\epsilon - \delta)[(1-\delta)^{1-s} - (1-\epsilon)^{1-s}\,2^s] + 2\delta\epsilon[\epsilon^{-s} + 2^s\delta^{-s}]}{2(\epsilon + \delta)\sqrt{[(1-\delta)^{1-s} - 2^s(1-\epsilon)^{1-s}]^2 + 4(\epsilon\delta)^{1-s}\,2^s}} \qquad \text{(A10)}$$

Since $|\lambda_2(s)| < \lambda_1(s)$, we can overbound (A9) by

$$\mu_m(s) \leq \ln[(1 + 2|\beta(s)|)\lambda_1{}^m(s)] \leq 2|\beta(s)| + m\ln\lambda_1(s). \qquad \text{(A11)}$$

By direct calculation, $\beta(s)$ and $d\beta(s)/ds$ have the value zero at $s = 0$. The second derivative of $\beta(s)$ is messy, but easily seen to be finite and continuous for $0 \leq s < 1$. Thus, given $0 < s_0 < 1$, there exists a_1 such that $|\beta(s)| \leq a_1 s^2$, for $0 \leq s \leq s_0$. Similarly, $\ln\lambda_1(s)$ and its first derivative have the values zero and $H_\infty(U)$, respectively, at $s = 0$. The second derivative of $\ln\lambda_1(s)$ is finite and continuous for $0 \leq s < 1$ so that, for large enough a_2, $\ln\lambda_1(s) \leq sH_\infty(U) + a_2 s^2$, for $0 \leq s \leq s_0$. Choosing $\alpha = 2a_1 + a_2$, we have (A4), completing the proof.

Appendix B

Proof of Theorem 3

Let $X^N = (X_1, X_2, \cdots, N_N)$, $Y^N = (Y_1, \cdots, Y_N)$ be a joint ensemble in which $X_n (1 \leq n \leq N)$ is the arrival time of the nth message at the source node and $Y_n (1 \leq n \leq N)$ is the time of delivery of the nth message to the destination. Let P_N be a joint probability measure on X^N, Y^N and assume that $P_N \in \mathcal{P}_N(d)$. First, we must show that the average mutual information between X^N and Y^N satisfies

$$I_{P_N}(X^N; Y^N) \geq NR_1(d) \qquad \text{(B1)}$$

and then we must calculate $R_1(d)$.

Using standard relations between average mutual informations and entropies (see [4, ch. 2]), we have

$$I_{P_N}(X^N; Y^N) = H(X^N) - H(X^N|Y^N), \qquad \text{(B2)}$$

$$H(X^N|Y^N) = H(X_1|Y^N) + \sum_{n=2}^{N} H(X_n|X_1, \cdots, X_{n-1}, Y^N)$$

$$\leq H(X_1|Y_1) + \sum_{n=2}^{N} H(X_n|X_{n-1}, Y_n)$$

$$= H(X_1|Y_1) + \sum_{n=2}^{N} H(X_n - X_{n-1}|X_{n-1}, Y_n). \qquad \text{(B3)}$$

Here we have used the chain rule, then the fact that conditioning cannot increase entropy, and finally the fact that, conditioned on X_{n-1}, the interarrival interval $X_n - X_{n-1}$ is simply a translation of X_n. Now define the random variable Z_n by

$$Z_n = Y_n - X_{n-1}. \qquad \text{(B4)}$$

Since Z_n is a deterministic function of Y_n and X_{n-1},

$$H(X_n - X_{n-1}|X_{n-1}, Y_n) = H(X_n - X_{n-1}|Z_n, X_{n-1}, Y_n)$$

$$\leq H(X_n - X_{n-1}|Z_n), \qquad \text{(B5)}$$

$$H(X^N|Y^N) \leq H(X_1|Y_1) + \sum_{n=2}^{N} H(X_n - X_{n-1}|Z_n). \qquad \text{(B6)}$$

Since the interarrival times $X_n - X_{n-1}$ are independent of each other, we also have

$$H(X^N) = H(X_1) + \sum_{n=2}^{N} H(X_n - X_{n-1}), \qquad \text{(B7)}$$

$$I_{P_N}(X^N; Y^N) \geq I(X_1; Y_1) + \sum_{n=2}^{N} I(X_n - X_{n-1}; Z_n) \qquad \text{(B8)}$$

where (B8) results from substituting (B6) and (B7) into (B2).

Now observe from (B4) that the delay for the nth message, $D_n = Y_n - X_n$, also satisfies

$$D_n = Z_n - (X_n - X_{n-1}), \qquad 2 \le n \le N. \tag{B9}$$

Let the expected value of D_n be d_n (which is determined by the joint probability measure P_N). Since $X_n - X_{n-1}$ has the same distribution as X_1 and since D_n satisfies (B9), we have

$$I(X_n - X_{n-1}; Z_n) \ge R_1(d_n), \qquad n \ge 2 \tag{B10}$$

by the definition of the function R_1. Letting $d_1 = E(Y_1 - X_1)$, and substituting (B10) into (B8), we obtain

$$I_{P_N}(X^N; Y^N) \ge \sum_{n=1}^{N} R_1(d_n) \tag{B11}$$

$$\ge NR_1\left(\frac{1}{N}\sum_{n=1}^{N} d_n\right) \tag{B12}$$

where we have used the convexity of R_1. Finally, since $P_N \in \mathcal{P}_N(d)$, $(1/N)\ \sum_{n=1}^{N} d_n \le d$. Since R_1 is a nonincreasing function, (B12) implies (B1). Since $P_N \in \mathcal{P}_N(d)$ is arbitrary, we have shown that $R_N(d) \ge R_1(d)$.

We now calculate $R_1(d)$; we use natural logarithms here and convert back to binary units in Section IV. Let X denote the arrival time of the message at the source and Y the delivery time at the destination. Let P be a probability measure on X, Y with the given marginal density on X,

$$p_X(x) = \alpha e^{-\alpha x}, \qquad x \ge 0. \tag{B13}$$

Let the distortion measure be given by

$$D(x, y) = \begin{cases} y - x, & y \ge x \\ \infty, & y < x. \end{cases} \tag{B14}$$

The infinite distortion for $y < x$ is simply an artifice to exclude probability measures that allow $Y < X$.

Define the Lagrange multiplier function

$$R_0(\rho, P) = I_p(X; Y) + \rho E[D(X, Y)], \tag{B15}$$

$$R_0(\rho) = \inf_P R_0(\rho, P). \tag{B16}$$

Because of the convexity of $R_1(d)$, we have

$$R_1(d) = \inf_{\rho \ge 0} [R_0(\rho) - \rho d]. \tag{B17}$$

Thus $R_0(\rho)$ may be interpreted as the ordinate axis intercept of the tangent of slope $-\rho$ to $R_1(d)$.

There is a well-known upper bound [6] to the function $R_0(\rho)$ given by

$$R_0(\rho) \le -\int dx\, p_X(x) \ln\left[\int dy\, w(y)\, e^{-\rho D(x,y)}\right]$$

$$= -\int dx\, p_X(x) \ln\left[e^{\rho x} \int_x^{\infty} dy\, w(y)\, e^{-\rho y}\right] \tag{B18}$$

where $w(y)$ is an arbitrary probability density. Similarly there is a well-known lower bound ([4, theorem 9.4.1] or [2, theorem 4.2.3]) given by

$$R_0(\rho) \ge \int dx\, p_X(x) \ln \frac{f(x)}{p_X(x)} \tag{B19}$$

where $f(x) > 0$ is any function satisfying the inequality

$$\int dx\, f(x)\, e^{-\rho D(x,y)} = e^{-\rho y} \int_0^y dx\, f(x)\, e^{\rho x} \le 1 \tag{B20}$$

for all $y \ge 0$. Furthermore, (B19) (and (B18)) are satisfied with equality if

$$e^{\rho x} \int_x^{\infty} dy\, w(y)\, e^{-\rho y} = \frac{p_X(x)}{f(x)} \tag{B21}$$

and, under these conditions, (B20) is also satisfied with equality for almost all y for which $w(y) > 0$.

We now find a probability density $w(y)$ and a function $f(x)$ satisfying (B20) and (B21). We note that the restriction $X \le Y$ means that small values of Y yield large mutual informations about X, and thus we hypothesize that there is some number y_0 such that $w(y) > 0$, for $y \ge y_0$, and $w(y) = 0$, for $y < y_0$. Differentiating (B20) with respect to y, and assuming equality for $y \ge y_0$, we find that $f(x) = \rho$, for $x > y_0$. Using this in (B21) and differentiating with respect to x yields

$$w(y) = \begin{cases} \dfrac{\alpha(\alpha + \rho)}{\rho} e^{-\alpha y}, & y \ge y_0 \\ 0, & y < y_0, \end{cases} \tag{B22}$$

$$y_0 = \frac{1}{\alpha} \ln \frac{\alpha + \rho}{\rho} \tag{B23}$$

where (B23) follows from the constraint that $w(y)$ is a probability density.

Substituting this value for $w(y)$ into (B21) and integrating, we have

$$f(x) = \begin{cases} \rho \exp[(\alpha + \rho)(y_0 - x)], & x \le y_0 \\ \rho, & x \ge y_0. \end{cases} \tag{B24}$$

Finally, we must demonstrate that this $f(x)$ satisfies (B20) for all $y \ge 0$. Defining $h(y)$ as the integral in (B20) and integrating, we get

$$h(y) \triangleq e^{-\rho y} \int_0^y dx\, f(x)\, e^{\rho x}$$

$$= \frac{\rho}{\alpha} e^{(\alpha+\rho)y_0 - \rho y} [1 - e^{-\alpha y}], \qquad y \le y_0.$$

Using (B23), we see that $h(y_0) = 1$. Also we see that the derivative of $h(y)$ is positive for $y < y_0$, from which it follows that $h(y) < 1$, for $y < y_0$. A similar integration verifies that $h(y) = 1$, for $y > y_0$.

Since $f(x)$ in (B24) has now been shown to satisfy (B19) with equality, we can integrate (B19) to obtain

$$R_0(\rho) = \frac{\alpha + \rho}{\alpha} \ln \frac{\alpha + \rho}{\rho} + \ln \frac{\rho}{\alpha}. \tag{B25}$$

Minimizing over ρ in (B17), we get

$$d = \frac{1}{\alpha} \ln \frac{\alpha + \rho}{\rho} = y_0, \qquad \rho = \frac{\alpha e^{-\alpha d}}{1 - e^{-\alpha d}} \tag{B26}$$

$$R_1(d) = -\ln(1 - e^{-\alpha d}). \tag{B27}$$

The transition probability density that achieves $R_1(d)$ is related to $w(y)$ and $f(x)$ by (see [4] or [2])

$$p(y|x) = \frac{w(y) f(x) e^{-\rho D(x,y)}}{p_X(x)} \tag{B28}$$

$$p(y|x) = \begin{cases} (\alpha + \rho) e^{-(\alpha+\rho)(y-x)}, & y \ge x \ge d \\ (\alpha + \rho) e^{-(\alpha+\rho)(y-d)}, & y \ge d > x \\ 0, & \text{elsewhere.} \end{cases}$$

This is equivalent to (12), completing the proof.

APPENDIX C

Analysis of Strategy 2

Observe from the flow diagram describing strategy 3 that, at the end of each sample, *Mq* is the number of messages on the queue and *Nomq* is the nominal number of messages on the queue; *Nomq* is normally the same as *Mq*, but, on an idle sequence, it is equal to the sum of *Mq* and the number of samples for which the queue remains in the idle state. Note that *Nomq*, from one sample to the next, is incremented by the number of message arrivals and decremented by one. If the above operation would leave *Nomq* negative, *Nomq* is set to $N-1$; whereas if it would leave *Nomq* greater than $2N-1$, then *Nomq* is successively decremented by N until it is less than or equal to $2N-1$

With these observations, we can describe *Nomq* by a Markov chain. The probability of i arrivals during a sample period is $(ei!)^{-1}$, and thus the transition probabilities for *Nomq* are given by

$$\begin{aligned} P(i|j) &= [e(i-j+1)!]^{-1}, \qquad N-1>i-j\geq -1, i\leq N-1 \\ &= e^{-1}+(eN!)^{-1}, \qquad i=N-1, j=0 \\ &= \sum_{l=0}^{\infty}[e(i-j+1+lN)!]^{-1}, \qquad i\geq j-1, i\geq N \\ &= \sum_{l=1}^{\infty}[e(i-j+1+lN)!]^{-1}, \qquad i<j-1, i\geq N \\ &= 0, \qquad \text{otherwise.} \end{aligned}$$

The above chain is clearly ergodic, and we let $Q_N(i)$ denote the steady state probability of state $i(0\leq i\leq 2N-1)$ for a given N. These equations are not as difficult to solve as they appear. The simplification comes from the fact that state 0 can only be reached from itself and state 1, leading to the equation

$$Q_N(1)=(e-1)Q_N(0).$$

Similarly, state 1 can be reached only from 0, 1 and 2, so that

$$Q_N(2)=(e-1)Q_N(1)-\frac{1}{2}Q_N(0).$$

Similar equations can be written for each state $i<N-1$, leading to the following lemma.

Lemma: Let the sequence of numbers $C_0, C_1, \cdots$ be defined by the recurrence

$$C_i=(e-1)C_{i-1}-\sum_{j=2}^{i}\frac{C_{i-j}}{j!}, \qquad i\geq 1 \tag{C1}$$

with the initial condition $C_0=1$. Then, for all $N\geq 1$ and all $i\leq N-1$,

$$Q_N(i)=Q_N(0)C_i. \tag{C2}$$

The numbers C_i in (C1) can be calculated for any given i, but it is useful to have an asymptotic expression for C_i. To obtain this, let $g(x)$ be the generating function for the sequence,

$$g(x)=\sum_{i=0}^{\infty}C_ix^i. \tag{C3}$$

By multiplying both sides of (C1) by x^i and summing over i, we find that

$$g(x)=\frac{1}{e^x-xe}. \tag{C4}$$

Regarding $g(x)$ as a function of a complex variable, we see that it is analytic within and on the unit circle except for a second order pole at $x=1$. Evaluating the residues of the poles, we find

$$g(x)=\frac{2}{e(x-1)^2}-\frac{2}{3e(x-1)}+g_1(x) \tag{C5}$$

where $g_1(x)$ is analytic within and on the unit circle. Let $g_1(x)=\Sigma_{i=0}^{\infty}\epsilon_ix^i$. Because of the analyticity, $\lim_{i\to\infty}\epsilon_i=0$. Expanding the other two terms in (C5), we find that

$$C_i=\frac{2i}{e}+\frac{8}{3e}+\epsilon_i. \tag{C6}$$

Numerically, ϵ_i converges remarkably quickly. We find that $\epsilon_2=4\times 10^{-5}$, and, for $i>2$, ϵ_i is even smaller.

We now return to the transition probabilities on the Markov chain for *Nomq*. Note that, for all $i\leq N-1$ and all j,

$$P(i|j)+P(N+i|j)=\sum_{l=0}^{\infty}\frac{1}{e[(i-j+1) \bmod N+lN]!}.$$

This quantity remains the same if j is increased or decreased by N. This means that a Markov chain also describes (*Nomq*) mod N, and, because of the modular symmetry,

$$Q_N(i)+Q_N(i+N)=\frac{1}{N}, \qquad 0\leq i\leq N-1. \tag{C7}$$

This equation is used below in finding the steady state probability of state $N-1$. We have

$$\begin{aligned} Q_N(N-1) &= \frac{1}{e}Q_N(N)+\frac{1}{e}Q_N(0)+\sum_{j=0}^{N-1}Q_N(j)\frac{1}{e(N-j)!} \\ &= \frac{1}{Ne}+\sum_{j=0}^{N-1}Q_N(j)\frac{1}{e(N-j)!} \end{aligned}$$

$$\frac{1}{N}=Q_N(N-1)(e-1)-\sum_{j=0}^{N-2}Q_N(j)\frac{1}{(N-j)!}. \tag{C8}$$

Substituting (C2) into this equation and comparing with (C1), we find that

$$\frac{1}{N}=C_NQ_N(0). \tag{C9}$$

This equation, (C1), (C2), and (C7) allow us to calculate the steady state probabilities. These in turn allow us to calculate the protocol information required by the receiver about message arrivals. The protocol information is denoted by *PI* in the flow diagram for the algorithm. When an idle sequence starts, $PI=-1$ is sent. This happens when the queue is in the state $Nomq=0$ and no messages enter for a sample interval. Thus

$$\Pr(PI=-1)=\frac{Q_N(0)}{e}=\frac{1}{eNC_N}. \tag{C10}$$

For $N\geq 2$, this is given with negligible error from (C6) as

$$\Pr(PI=-1)\approx\frac{1}{2N(N+4/3)}. \tag{C11}$$

Special protocol information is also required when extra messages are sent for a particular sample. We have $PI=l$ if Nl extra messages are sent at the sample time. Since the number of message arrivals divided by the number of sample times approaches one over a long interval of time, we expect the number of extra batches of N messages to balance the number of idle periods yielding

$$\Pr(PI=-1)=\sum_{l=1}^{\infty}l\Pr(PI=l). \tag{C12}$$

To derive this formally, observe that, for *Nomq* to make a transition from j to i with $PI = l$, we must have $j - i + 1 + lN$ arrivals in the intersample interval. Thus

$$\Pr(PI = l) = \sum_{j=0}^{2N-1} Q_N(j) \sum_{i=N}^{2N-1} \frac{1}{e(i - j + 1 + lN)!}. \quad \text{(C13)}$$

Next observe that

$$\sum_{j=0}^{2N-1} \sum_{i=0}^{2N-1} Q_N(j) P(i|j)(i - j) = 0. \quad \text{(C14)}$$

This equation simply states that the mean value of *Nomq* is constant in the steady state and follows from the equation $Q_N(i) = \Sigma_j Q_N(j)P(i|j)$. Substituting the equation for $P(i|j)$ into (C14), we get

$$\sum_{j=0}^{2N-1} Q_N(j) \sum_{i=\max(j-1,0)}^{2N-1} \frac{(i-j)}{e(i-j+1)!} - \frac{(N-1)Q_N(0)}{e} + \sum_{j=0}^{2N-1} Q_N(j) \sum_{i=N}^{2N-1} \sum_{l=1}^{\infty} \frac{i-j}{e(i-j+1+lN)!} = 0.$$

If we extend the first sum on i to start at $j - 1$, we include the additional term $Q_N(0)/e$. We can also rewrite the final sum over i and l by the change of variable $k = i + lN$.

$$\sum_{j=0}^{2N-1} Q_N(j) \sum_{i=j-1}^{2N-1} \frac{i-j}{e(i-j+1)!} - \frac{NQ_N(0)}{e} + \sum_{j=0}^{2N-1} Q_N(j) \sum_{l=1}^{\infty} \sum_{k=N+lN}^{2N-1+lN} \frac{k-j-lN}{e(k-j+1)!} = 0.$$

Combining part of the final sum with the first sum, we find

$$\sum_{j=0}^{2N-1} Q_N(j) \sum_{i=j-1}^{\infty} \frac{i-j}{e(i-j+1)!} - \frac{NQ_N(0)}{e} - \sum_{j=0}^{2N-1} Q_N(j) \sum_{i=N}^{2N-1} \sum_{l=1}^{\infty} \frac{lN}{e(i-j+1+lN)!} = 0. \quad \text{(C15)}$$

The first sum over i is clearly zero for every j, and, comparing the final term with (C13), we see that (C15) is equivalent to (C12).

It is rather messy to calculate $\Pr(PI = l)$, for each l. On the other hand, given a random variable *PI* which takes on integer values from -1 to ∞ and which satisfies (C12), it is easy to obtain an upper bound on the entropy of *PI*. We simply maximize the entropy of *PI* subject to the constraint in (C12), getting $2\mathcal{H}(\Pr(PI = -1))$. The entropy of *PI* is the protocol information about message arrivals, and using the value for $\Pr(PI = -1)$ in (C10) and (C11), we have completed the derivation of (21).

We turn now to the average message delay. We split this delay into two terms: first the delay of each message before the first sample time following the message arrival, and second the integral number of sample times that the message waits until the sample when it is transmitted. Since the message arrivals are Poisson, the average delay before the next sample is half a sample time or $1/2\alpha$. For the second component of average delay, we use the law of large numbers. The accumulated delay in this component over all the messages that arrive before the Lth sample time is simply $1/\alpha$ times the sum of the number of messages in the queue over the first $L - 1$ sample times. By the law of large numbers, this sum divided by L approaches the mean length of the queue with probability one. Similarly the number of message arrivals divided by L approaches one with probability one. Thus the average message delay d is given by

$$d = \frac{1}{2\alpha} + \frac{1}{\alpha} E(Mq) \quad \text{(C16)}$$

when Mq is the length of the queue at the end of a sample.

The length of the queue is related to the nominal queue length *Nomq*. Whenever $PI = -1$ occurs, *Nomq* jumps from zero to $N - 1$ and Mq remains at zero. On each subsequent sample time, the difference between *Nomq* and Mq is reduced by one until the difference is zero. Thus the mean value of *Nomq* exceeds the mean value of Mq by $\Pr(PI = -1)N(N - 1)/2$,

$$d = \frac{1}{2\alpha} - \frac{1}{\alpha} \Pr(PI = -1) \frac{N(N-1)}{2} + \frac{1}{\alpha} E(Nomq), \quad \text{(C17)}$$

$$E(Nomq) = \sum_{j=0}^{2N-1} j Q_N(j).$$

Using the fact that $Q_N(j) + Q_N(j + N) = 1/N$, for $0 \le j \le N$,

$$E(Nomq) = N + \frac{N-1}{2} - N \sum_{j=0}^{N-1} Q_N(j).$$

From (C9) and (C2),

$$Q_N(j) = \frac{C_j}{NC_N}.$$

Using (C6) and overbounding by ignoring the error terms ϵ_i, we obtain

$$E(Nomq) \le N - 2\left(\frac{N+1}{3N+4}\right). \quad \text{(C18)}$$

Substituting this into (C17) and simplifying, we have

$$d \le \frac{1}{\alpha}\left[N - \frac{5N-3}{12N+16}\right]. \quad \text{(C19)}$$

REFERENCES

[1] N. Abramson, "Packet switching with satellites," 1973 National Computer Conf., *AFIPS Conference Proceedings*, vol. 42, 1973.

[2] T. Berger, *Rate Distortion Theory, A Mathematical Basis for Data Compression*. Englewood Cliffs, N. J.: Prentice Hall, 1971.

[3] L. R. Ford and D. R. Fulkerson, *Flows in Networks*. Princeton, N. J.: Princeton University Press, 1962.

[4] R. G. Gallager, *Information Theory and Reliable Communication*. New York: Wiley, 1968.

[5] R. G. Gallager and D. C. Van Voorhis, "Optimal source codes for geometrically distributed alphabets," *IEEE Trans. Inform. Theory*, vol. IT-21, pp. 228–230, Mar. 1975.

[6] B. Haskell, "The computation and bounding of rate distortion functions," *IEEE Trans. Inform. Theory*, vol. IT-15, pp. 525–531, Sept. 1969.

[7] R. Kahn, "Resource-sharing computer communication networks," *Proc. IEEE*, vol. 60, Nov. 1972.

[8] L. Tymes, "Tymnet—A terminal oriented communications network," *AFIPS Conf. Proc.*, vol. 38, 1971.

GROUP SYNCHRONIZING OF BINARY DIGITAL SYSTEMS

R. H. Barker

Signals Research and Development Establishment, Christchurch England

SUMMARY

A sequence of binary digits has very little meaning unless the significance of the individual digits is known. If one digit can be labelled in some way others may be identified by counting. It is shown that transmission of a prearranged synchronizing pattern can perform such a labelling operation and guidance is given as to the length of pattern necessary under various specified conditions of operation. In particular, the probabilities are calculated of failure to synchronize and of false synchronizing when some of the digits are received wrongly by reason of noise or interference. Certain patterns are shown to be specially suitable for the purpose.

INTRODUCTION

All digital systems for the transmission of information depend to a greater or less extent for their efficiency upon correct synchronization. In general two distinct synchronizing operations must be performed. First of all there is digit synchronizing which can be said to establish equal time scales at the two ends of the link. In practice it means the generation at the receiver of a regular train of pulses having the same frequency as and a known phase relationship to a similar train of pulses at the transmitter. It enables a distorted signal to be regenerated with a low probability of error.

Secondly there is group synchronization which effectively pin-points an origin of time. If full advantage is to be taken of the binary codes, each digit of a group is differently weighted, and it is essential to be able to assign these weights correctly. It is equivalent to tying a label to one digit, after which any other may be related to it by counting. In a system where the number of digits in the group is small and constant, a convenient method of synchronizing is to provide circuits which can decide on a time average basis whether or not synchronization is correct. If not, the group phase at the receiver is altered by one digit, the process being repeated as many times as is necessary.

An alternative method—the one to be considered in more detail—is to transmit a special pattern of digits which is unambiguously recognized at the receiver. It is undoubtedly the better method when the group is a long one. In the extreme the whole message may constitute the group, in which case each message would be prefixed by the synchronizing pattern and the system would be of the start–stop variety. One might perhaps suppose that transmission of the first digit is an adequate indication of the beginning of a message, but a good receiver, intended to operate on weak and perhaps noisy signals, will do its best to generate binary digits on all occasions, and there may be no obvious way to distinguish between a true message and a random binary sequence produced by a noise only input. The present approach assumes that the synchronizing signal will consist of a pattern of digits which is sufficiently unlikely to occur by chance in a random sequence or in a previous message.

A logical starting point in assessing the necessary length of the synchronizing signal is to decide how much information is conveyed by the act of synchronizing. This quantity depends upon one's *a priori* knowledge of when the labelled digit is most likely to be received. For example, there may be an agreement between the operators concerned to commence working at

Reprinted with permission from *Communication Theory*, pp. 273–287, 1953, Butterworth-Heinemann Limited.

9.00 a.m. and to allow for errors in timing, a tolerance of $\pm$ 30 sec is permitted. If the digit rate is r per sec, then the labelled digit is one of a set of $60r$. It is of course most likely that this one will be nearer to the middle of the possible set than the edge, but to simplify the problem we shall for the moment assume that all positions of the set are equally likely. The synchronizing operation is then equivalent to specifying one of the $60r$ equally likely numbers and the information conveyed is

$$H = \log_2 m \text{ bits} \qquad \ldots.(1)$$

where m is the number of equally likely choices, in this case $60r$.

It might appear, therefore, that the necessary information could be conveyed by H binary digits. This is false, however, since it assumes that the digits have a weighted significance whereas at the receiver all digits must initially have equal significance.

PROBABILITY OF SYNCHRONIZING CORRECTLY

Suppose a synchronizing pattern n digits long is to be used to label one out of a possible choice of m digits. That is to say, the pattern may be sent so as to terminate at any one of m different instants. To identify the synchronizing instant correctly the receiver must be able to examine a total of $m + n - 1$ digits. If $n \ll m$ the probability of the pattern occurring by chance within a random set $m + n - 1$ is $(m + n - 1)2^{-n}$. When the pattern is received, therefore, the probability that it is actually being transmitted is $1 - (m + n - 1)2^{-n}$. It is possible to ensure that this is sufficiently near to unity for practical purposes, even though n may be very much smaller than m. When n is comparable with m the true expression becomes very complicated and does in fact depend upon the particular pattern chosen for synchronizing. Certainty is of course realized when $n = m$.

The situation is more complicated if the communication link is noisy so that the regenerated digits are not always correct. There is then no certainty of being able to recognize the synchronizing pattern when it is sent, and the receiver circuits must be designed to permit a suitable margin for errors. The best that can be done is to determine the probabilities of the various alternatives.

P_1 = probability that synchronizing pattern is sent and is recognized;

P_2 = probability that synchronizing pattern is sent but not recognized;

P_3 = probability that synchronizing pattern is not sent but is recognized by chance coincidence;

P_4 = probability that synchronizing pattern is not sent and is not recognized.

The sum of these, $P_1 + P_2 + P_3 + P_4 = 1$.

The following symbols will be used in evaluating these probabilities:

S = the *a priori* probability that the synchronizing signal was sent;

h = probability of any digit being received in error;

n = the number of digits in the synchronizing pattern;

y = the number, out of the last n digits received, which agree with the expected synchronizing pattern;

$x = n - y$;

k = maximum number of errors permitted by the pattern recognizer;

$\varepsilon = nh$.

P_1: *Probability that pattern is sent and recognized*

If the pattern is known to be sent, the probability that it is received with x errors is $C^n_x\,(1-h)^{n-x}h^x$ and the probability of recognition is

$$\sum_{x=0}^{k} C^n_x\,(1-h)^{n-x}h^x$$

Hence

$$P_1 = S\sum_{x=0}^{k} C^n_x\,(1-h)^{n-x}h^x \qquad \ldots.(2)$$

P_2: *Probability that pattern is sent but not recognized*

If the pattern is known to be sent, the probability that it is received with more than k errors is

$$\sum_{x=k+1}^{n} C^n_x\,(1-h)^{n-x}h^x$$
$$= 1 - \sum_{x=0}^{k} C^n_x\,(1-h)^{n-x}h^x$$

Hence

$$P_2 = S[1 - \sum_{x=0}^{k} C^n_x\,(1-h)^{n-x}h^x] \qquad \ldots.(3)$$

When $h \ll 1$ the probability that the pattern, if sent, will be received with ust x errors is given approximately by the Poisson law:

$$p(x) = \frac{e^{-\varepsilon}\varepsilon^x}{x!} \qquad \text{where } \varepsilon = nh$$

The probability that it is received with more than k errors is

$$\sum_{k+1}^{\infty} p(x) = \Pi(k+1,\,\varepsilon), \text{ say, so that}$$
$$P_2 = \Pi(k+1,\,\varepsilon) \qquad \ldots.(4)$$

and

$$P_1 = S[1 - \Pi(k+1,\,\varepsilon)] \qquad \ldots.(5)$$

The advantage of equations 4 and 5 is that the probability is now basically a function of two variables, k and ε, instead of three. Furthermore the Π function is to be found ready tabulated[1].

P_3: *Probability that pattern is recognized by chance*

Knowing that the pattern is not sent, there is still some chance of its being generated by a random series of digits. The probability of its occurring with x errors is $2^{-n}C^n_x$ and of its occurring with k errors or less is $2^{-n}\sum_{x=0}^{k} C^n_x$

Hence

$$P_3 = (1-S)2^{-n}\sum_{x=0}^{k} C^n_x \qquad \ldots.(6)$$

P_4: *Probability that pattern was not sent or recognized*

This follows simply from equation 6

$$P_4 = (1-S)\,[1 - 2^{-n}\sum_{x=0}^{k} C^n_x] \qquad \ldots.(7)$$

The situation which arises after the reception of each successive digit is as follows: Of the previous n digits, y agree with the synchronizing pattern and x do not. Is the last digit y received to be labelled as the synchronizing instant? We make the arbitrary choice to do so if $x \leq k$. The probability that this will happen is $P_1 + P_3$. If it does so happen, what is the probability that our interpretation is correct? This is the *a posteriori* probability that the synchronizing signal was sent. Of more direct interest is the probability that it was not sent, since this is the probability of a false synchronizing operation occurring. Let it be denoted by $F(k)$. By application of Bayes's theorem, we obtain

$$F(k) = \frac{P_3}{P_1 + P_3}$$

$$= \frac{(1 - S)2^{-n} \sum^{k} C_x^n}{S[1 - \Pi(k, \varepsilon)] + (1 - S)2^{-n} \sum^{k} C_x^n} \qquad \ldots (8)$$

If a system is to work satisfactorily in practice, the conditions must be such that both the probability of a failure to synchronize to a signal which is sent, and the probability of synchronizing to a signal not sent are small. Both $F(k)$ and Π are small in these circumstances and equation 8 reduces to the approximate expression

$$SF(k) = 2^{-n} \sum_{x=0}^{k} C_x^n \qquad \ldots (9)$$

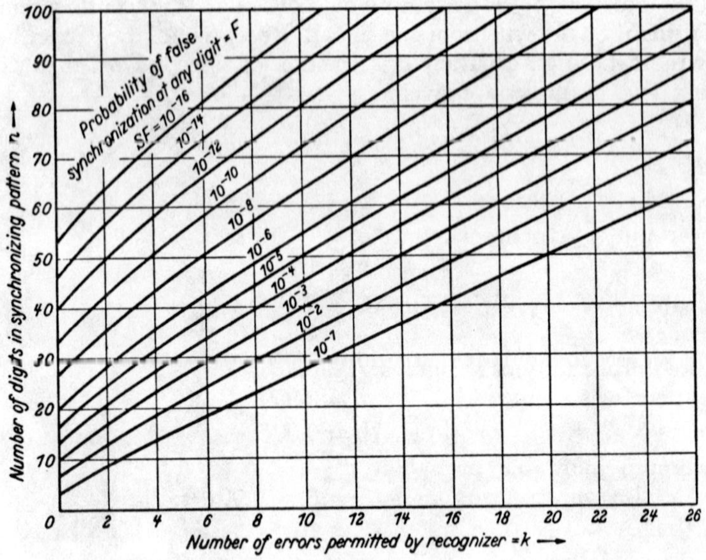

Figure 1

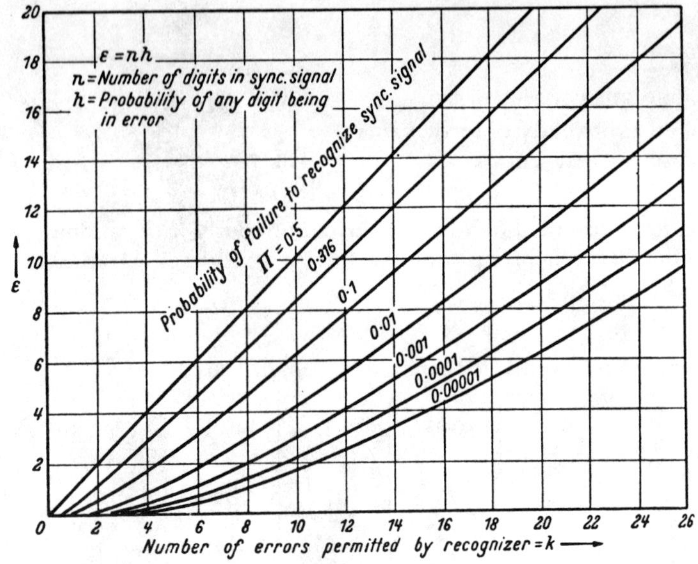

Figure 2

The right hand side is plotted in *Figure 1* for various values of k.

The probability of false synchronizing, $F(k)$, and the probability of a failure to synchronize to a signal that is actually sent are the most important factors to be considered. The latter, from equation 4, is simply Π. This is plotted in *Figure 2* as a function of k and ε.

Sufficient assumptions must be made concerning the conditions of operation of a system to enable upper limits to be set to $F(k)$ and Π. *Figures 1* and 2 and a little trial and error then enable suitable values of n and k to be determined. These are the important parameters of the synchronizing system. The longer the pattern is, the less the chance of its occurring out of a random sequence. The larger the margin of error, k, the less is the probability of a failure to recognize a synchronizing signal actually sent. A compromise must be made by balancing good performance against complexity of equipment.

Equation 9 shows that it is desirable for S to be as large as possible. It may be possible in some applications to know quite accurately when the synchronizing signal is to be expected. The probability that it is sent at that time is then quite high. For example on a 'Press to Transmit' system the first digits might be the synchronizing signal. Reception of the carrier would probably indicate to an accuracy of a fraction of a second when to expect the signal. Two examples are given.

Example 1

No. of digits per second $= 5 \times 10^5$

Uncertainty of synchronizing time $= \pm 0{\cdot}1$ sec

Hence $S = 10^{-5}$

Probability of any digit being in error $= h = 0{\cdot}05$

Desired probability of recognizing pattern when sent $= 0{\cdot}999$

$\Pi = 0{\cdot}001$

Maximum allowable probability of a false synchronizing $= F = 0{\cdot}001$ (say)

$SF = 10^{-8}$

A short trial and error process with *Figures 1* and *2* shows that approximately these probabilities are obtained when $n = 64$ and $k = 10$.

Example 2

No. of digits per second $= 28000$

Desired probability of recognizing pattern when sent $= 0{\cdot}99$

$\Pi = 0{\cdot}01$

The system is assumed to be operating continuously with random synchronizing. One false synchronization per hour is considered acceptable.

$$F(k) = \frac{1}{28 \times 10^3 \times 3600} \approx 10^{-8}$$

Suitable values of n and k, for different values of h are:

h	0·1	0·05	0·02	0·01	0
n	55	38	27	24	17
k	12	6	3	2	0

EQUIVOCATION

The equivocation, as defined by SHANNON, measures the uncertainty about the input when the output is known. In this application the output is known to the extent that y out of n digits agree with the synchronizing pattern and there is a corresponding uncertainty as to whether the synchronizing signal was sent or not. By arguments similar to those outlined above, we can calculate the *a posteriori* probability that the signal received with x errors is false. It is

$$F(x) = \frac{(1 - S)2^{-n}}{(1 - S)2^{-n} + Sh^{x}(1 - h)^{n-x}} \qquad \ldots .(10)$$

The expectation of x is hn. Substituting this in equation 10 gives F_l, the most likely *a posteriori* probability of false synchronizing. With the simplifying assumption that $S \ll 1$, this is

$$F_l = \frac{1}{1 + Sh_0^n} \qquad \ldots .(11)$$

where

$$h_0 = 2h^h(1 - h)^{1-h} \qquad \ldots .(12)$$

Since h_0 is never less than unity, increasing the number of digits, n, in the synchronizing pattern always decreases the probability of false synchronization, except in the special case of $h = 0{\cdot}5$, when no information can be transmitted.

The actual equivocation in any circumstance depends upon the amount of agreement between the digits actually received and those of the pattern expected, but equation 11 may be used to calculate the most likely value

$$R_l = -[F_l \log_2 F_l + (1 - F_l) \log_2 (1 - F_l)] \qquad \ldots .(13)$$

If F_l is small an approximate expression is

$$\log_{10} R_l = \log_{10} F_l + \log_{10} (1{\cdot}44 - 3{\cdot}32 \log_{10} F_l) \qquad \ldots .(14)$$

Figure 3 shows the way in which it depends upon h and n. The dependence upon the *a priori* probability of synchronizing, S, can be taken into account by a correction added to the ordinate. The actual number of digits, n, is given by

$$n = n' - c \log_{10} S \qquad \ldots .(15)$$

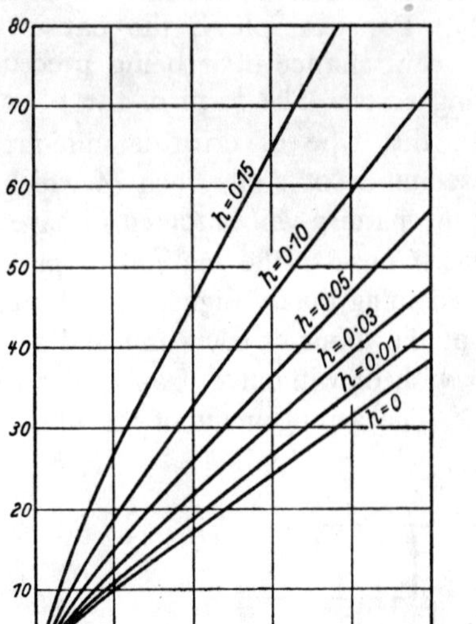

Figure 3

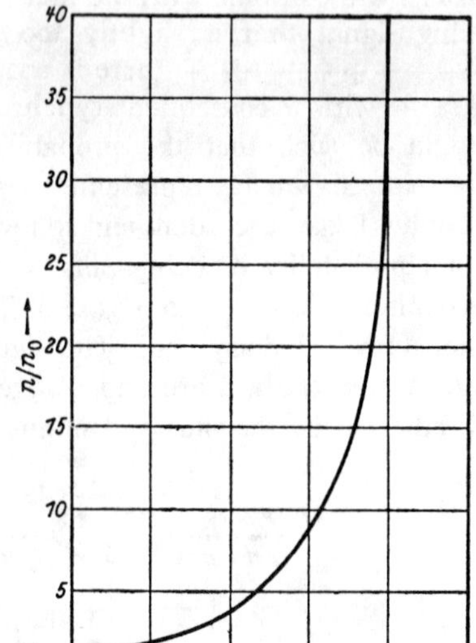

Figure 4

where $c = 1/\log_{10}h_0$ and is tabulated below.

h	0	0·01	0·03	0·05	0·10	0·15
c	3·32	3·61	4·15	4·90	6·25	8·77

Suppose that F_l, the most likely *a posteriori* probability of false synchronizing, and S the *a priori* probability of synchronizing are arbitrarily specified. The number of digits of the pattern can then be expressed in terms of the error rate. If $F_l \ll 1$ it follows from equation 11 that

$$\frac{n}{n_0} = \frac{1}{\log_2 h_0} \qquad \ldots.(16)$$

where n_0 is the value of n when there are no errors ($h = 0$). This function is plotted in *Figure 4* and shows the way in which the number of digits must be increased in order to convey the same amount of information in the presence of errors. The same relation, of course, holds good for the transmission of a message. There is, however, a big difference between the use of the redundancy in the two cases.

In message transmission it is theoretically possible by suitable encoding to utilize the redundancy so as to make the probability of correct interpretation as near to unity as may be desired. This encoding is accompanied by delay. The same amount of redundancy applied to the synchronizing operation results in an *a posteriori* probability of synchronizing, the most probable value of which is as given in equation 11. It is not possible by any rearrangement of the digits or other encoding to improve this probability.

For this reason the number of digits needed to secure a useful *a posteriori* probability of synchronizing increases more rapidly with transmission errors than would be the case for a message.

THE SYNCHRONIZING PATTERN

The arguments given above are concerned with the selection of a suitable length for the synchronizing pattern and providing $m \gg n$ they are not in the least affected by the nature of the pattern. A further important factor, however, is that there is a possibility that the random digits preceding the pattern will combine with the first part of the pattern to produce a synchronizing signal that is slightly too early. For example if the pattern is ++++++++++ there is a 50 per cent chance of it being preceded by a + with a consequent synchronizing error. The form of the pattern should be such that the probability of this type of error is minimized. The pattern will be represented by a sequence of digits each of which is ± unity. Digits preceding and following the pattern will be taken to have an equal probability of being plus or minus. Consider the method of pattern recognition illustrated in *Figure 5*. The incoming binary digits are fed into a chain of $n - 1$ delay units. The signals at the input of each unit and at the output from the last are added together with or without a change of phase, as indicated by the plus and minus signs. The arrangement of the plus and

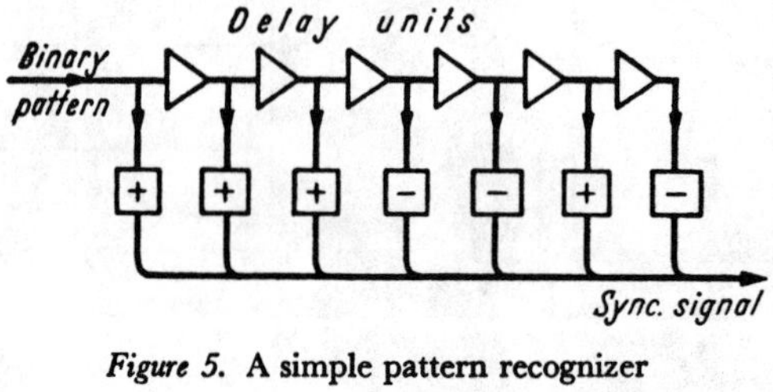

Figure 5. A simple pattern recognizer

minus signs used for this purpose is identical with the pattern used for synchronizing, but reversed in order. In the example of *Figure 5* the latter would be −+−−+++.

Imagine that the single digit + 1, preceded and followed by zeros, is fed into the recognizer. The output is the sequence

$$1, 1, 1, -1, 1, -1$$

so that the pulse transfer function[2] is

$$1 + z^{-1} + z^{-2} - z^{-3} - z^{-4} + z^{-5} - z^{-6}$$

A synchronizing signal timed to terminate at $t = 0$ can be represented by the sequence

$$-z^{6} + z^{5} - z^{4} - z^{3} + z^{2} + z + 1$$

where z is the operator of the sequence transform.

The output from the recognizer with this as the input is the product of these two expressions, namely

$$-z^{6} - z^{4} - z^{2} + 7 - z^{-2} - z^{-4} - z^{-6}$$

This is in fact the auto-correlation function of the pattern, and may be alternatively written as

$$-1, 0, -1, 0, -1, 0, 7, 0, -1, 0, -1, 0, -1$$

It is the output sequence obtained when the synchronizing signal is preceded and followed by zeros. In the binary system under consideration the two possible states are represented by + 1 and − 1, so that this is impossible. The result does, however, have a useful meaning since random elements preceding and following cause deviations which are equally likely to be positive or negative. The output of + 7 units is obtained when the pattern is exactly contained within the delay line store. The large pulse obtained at this instant is the label used for marking the instant of synchronization. It is not possible for the full amplitude to be reached by any random digits in combination with one or more of the pattern. This is illustrated in *Figure 6* in which the maximum possible amplitudes are shown around the synchronizing instant, together with the probabilities that they shall occur by chance. In this illustration the reception of the synchronizing digits has been assumed free from error. Any errors have the effect of decreasing the amplitude of the wanted pulse and of permitting an increase in that of unwanted ones.

It is possible to show that none of the 0 or − 1 terms in the above auto-correlation sequence can be made more negative without some other of these terms becoming more positive. The pattern suggested is in this sense an ideal one, and it would appear desirable to search for other patterns, preferably with more digits, having similar properties. The following conclusions have been reached as a result of attempts in this direction.

(1) The number of digits must be one less than an integral multiple of four.

(2) If digits are paired counting from both ends, the odd pairs must be different, even pairs alike.

(3) The simplest patterns are

$n = 3$ ++−

$n = 7$ +++−−+−

$n = 11$ +++−−−+−−+−

(4) No longer patterns have been found and it is considered unlikely that they exist, although this has not been proved.

It is possible to obtain longer patterns which are very nearly ideal and many such may be found by trial and error. Patterns whose auto-correlation

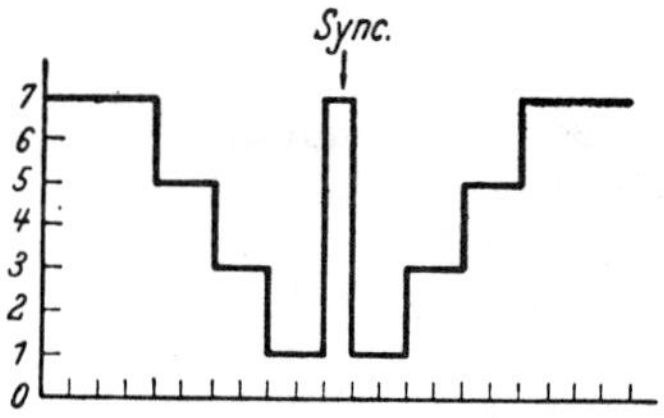

Figure 6a. Maximum possible output near Sync. digit

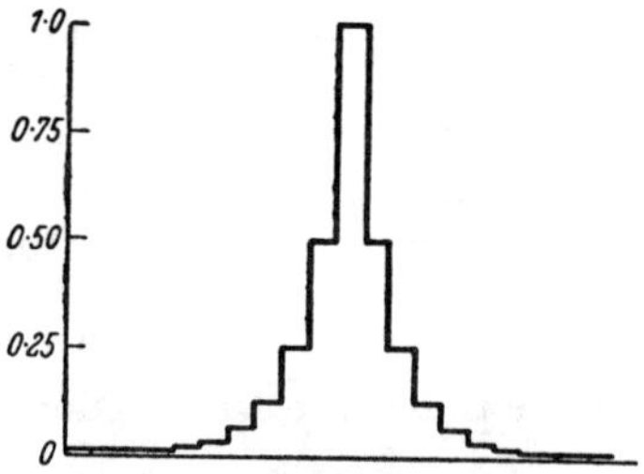

Figure 6b. Probability that maximum output is obtained

sequence terms do not exceed + 1 (except at the centre) can, however, be constructed by combining ideal patterns in ideal groups. The meaning will be made clear by an example. The above pattern of three digits (++−) can be written down twice in the same sense followed by once in the opposite sense, ++−++−−−+. This has the auto-correlation sequence

$$1, 0, -3, 0, 1, 0, -3, 0, 9, 0, -3, 0, 1, 0, -3, 0, 1$$

The other possible combinations using the above ideal patterns have the numbers of digits $n = 21, 33, 49, 77, 121$.

The instrumentation for the reception and recognition of these long sequences must necessarily be somewhat complicated. It may be accomplished by means of two delay lines, the first dealing with individual digits and the second with groups of digits. For the sequence of 21 digits

$$+++--+-+++--+----++-+$$

the arrangement would be as in *Figure 7*.

TWO-STAGE SYNCHRONIZING

It may well be desirable to utilize patterns which fall short of those discussed so far, but which may be generated and recognized with less complicated equipment. Such patterns will have to be longer if the standard of performance is to be maintained. One method is to perform the synchronizing operation in two stages. The first stage is to label one digit of a repeated frame, and the second is to label one particular frame of the sequence.

Suppose the first stage consists of sending a repetition of an ideal pattern of n elements and that a recognition unit such as that in *Figure 5* is available. The output, if there are no errors, will be the sequence

$$-----1, -1, n, -1, -1 --- -1, n, -1, -1,$$
$$--- n, -1, ---, \text{etc.}$$

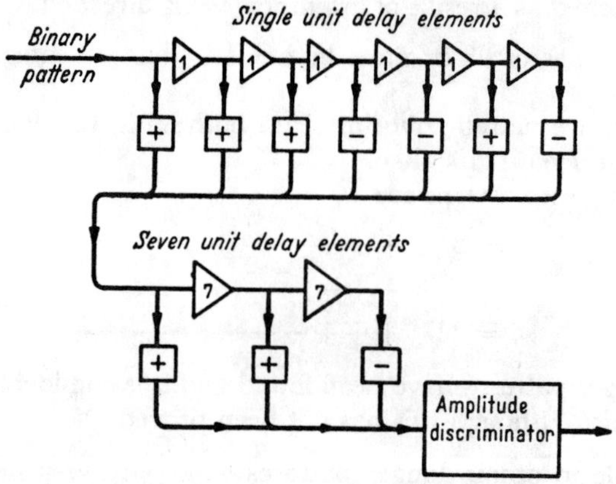

Figure 7. Recognizer for 21-digit pattern

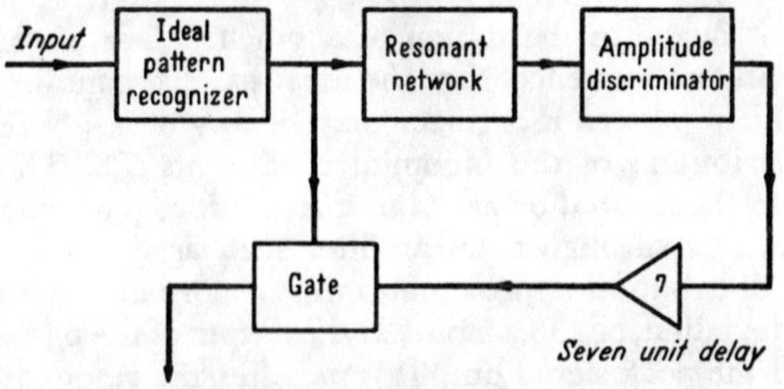

Figure 8. A two-stage recognizer

in which the positive pulses of amplitude n are separated by $n - 1$ pulses of amplitude -1. This signal may be fed into a resonant system so that the repeated pulses build up to large amplitude. A simple device would be a delay line open at the end to yield an in-phase reflection delayed by n digits and fed from a high impedance circuit to give repeated reflections. If each reflection is attenuated by d relative to the initial input, the output pulses will build up in amplitude to yield finally

$$n(1 + d + d^2 + \ldots) = \frac{n}{1 - d}$$

if the summation is continued to infinity. Errors occur at random times and therefore do not add in phase.

The second stage, that of labelling one frame, can be undertaken when the ideal pattern has been repeated a sufficient number of times to build up a reasonable amplitude of pulse. A suitable method of labelling is to transmit a frame in which the sense (polarity) of the pattern is reversed. The regular sequence of pulses built up in the first stage may then be used to define the instants at which the output of the ideal pattern recognizer is tested for polarity. *Figure 8* shows a suitable arrangement.

The operation may be understood more readily by reference to the waveforms of *Figure 9* based upon the use of an ideal pattern of seven digits.

At *Figure 9a* is the synchronizing signal consisting of the ideal 7-digit pattern repeated four times in the positive sense and once in the negative sense. The output of the recognizer is shown at b and the build up of the pulses in the resonant network at c. When the pulse amplitude has built up sufficiently to operate the amplitude discriminator, pulses are fed to the gate circuit with a delay of 7 digits. This delay is to ensure that a pulse is available to operate the gate when the sense of the pattern is reversed. If the decrement of the resonant network is sufficiently small the stored energy would serve this purpose and this unit could be dispensed with. The gate is of a type such that a pulse is generated by the combination of a negative pulse on its input terminal and a positive pulse on its control terminal.

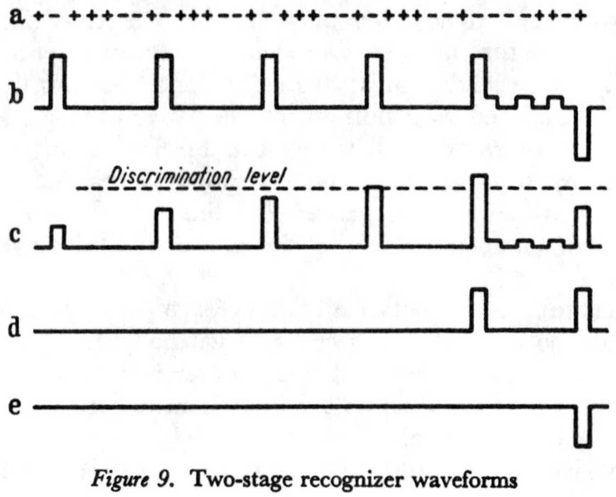

Figure 9. Two-stage recognizer waveforms

An important factor in the design of a synchronizer of this type is the response to a random sequence. Let the ideal pattern contain n digits. The output of the ideal pattern recognizer may be any of $n + 1$ levels with the probability distribution of the binomial coefficients C^n_x. This distribution approximates to the normal or gaussian distribution, particularly when n is large. After passing through a linear filter such as the resonant network, the amplitude distribution approximates much more closely to the normal. The standard deviation, or r.m.s. noise, at the output of the pattern recognizer is $\sigma = \sqrt{n}$ and the peak signal output is n. After the resonant network the peak signal output is $n/(1 - d)$ and the mean square noise is $\sigma^2/(1 - d^2)$. The signal–noise ratio is therefore improved by the factor $\sqrt{[(1 + d)/(1 - d)]}$, and for the same peak signal output the standard deviation of the noise is correspondingly reduced.

$$\frac{\text{Peak signal}}{\text{r.m.s. noise}} \approx \sqrt{\left[\frac{n(1+d)}{1-d}\right]}$$

and the probability that the noise builds up to exceed a fraction α of the peak signal is given approximately by the error function

$$\frac{1}{2} \operatorname{erf} \sqrt{\left[\frac{n(1+d)}{1-d}\right]}$$

The approximation is very inaccurate when α is close to unity. The formula yields a finite probability of the noise building up to exceed the peak signal, which is, of course, physically impossible. A better approximation for this condition is probably to use the binomial distribution based upon an equivalent number of digits n_0 where $n_0 = n(1 + d)/(1 - d)$. This is the number of digits which would have to be used for a simple ideal pattern to give the same performance. *Figures 1* and *2* may be used and the value of k so obtained applied to determine the level of amplitude discrimination.

REPEATED SYNCHRONIZING

The length of the synchronizing pattern may be reduced if the *a priori* probability of synchronizing is increased. One method of doing this is to send the synchronizing pattern at regular intervals. The synchronizing signal is then only accepted as correct if it is received at the instant it is expected. This method is obviously not capable of yielding absolute synchronization, but only of marking repeatedly the beginning of a group. It is suitable for most types of time multiplex systems. *Figure 10* illustrates one way in which it may be applied.

A simple ideal pattern is inserted into the transmission at regular intervals of m digits. At the receiver an ideal pattern recognizer, such as in *Figure 5*, generates a pulse whenever this pattern is received, either because it was inserted or by chance. If no other pulse has been received for some time previously, it will pass through gate 1 and will reset the counter of modulus m. The action of resetting sends a pulse to a coincidence detector unit. Because the resetting was initiated by a pulse from the pattern recognizer, this unit will perceive coincidence and will trigger the flip-flop circuit so cutting off further counter resetting pulses. This condition is maintained until the flip-flop restores, which will happen after a time which may be controlled arbitrarily though not accurately. We will assume that this time corresponds to M digits, where $M > m$.

The counter restores automatically after m pulses. If synchronism is correct the pulse so generated will coincide with a pulse from the pattern recognizer and any interference with the counter is held up for another M digits. In other words, as long as the system stays synchronized no false pulses from the pattern recognizer can get through to upset it. If the system is not synchronized, each pulse from the recognizer is assumed to be a

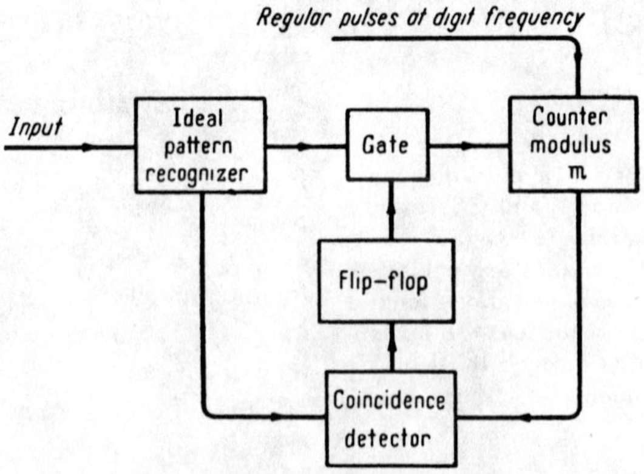

Figure 10. System for repeated synchronizing

synchronizing pulse until subsequent lack of coincidence with the counter output pulse show it to be wrong.

It will be seen that the method bears some resemblance to the search method mentioned earlier, but it has the advantage that the search does not extend to all digits of the group. It includes only those in which the combination of n immediately preceding happens to be identical with the chosen pattern (or nearly so if errors are allowed for) and the chance of this is relatively small.

CONCLUSIONS

The minimum number of digits for the synchronizing pattern is a function of four variables:

(1) The *a priori* probability of synchronizing, S.
(2) The probability of a transmission error, h.
(3) The required probability of correct operation when the pattern is sent, $(1 - \Pi)$.
(4) The maximum allowable probability of false synchronization due to chance coincidence, F_l.

Graphs have been prepared which enable this number to be calculated, and also the best margin to allow for transmission errors. A study of the properties of various patterns as revealed by their auto-correlation sequences shows that a few particular patterns are specially suitable. Furthermore equipment for generating and identifying them need not be unduly complicated.

Although only binary systems have been considered it is apparent that these same 'ideal' patterns could advantageously be applied to the synchronization of multi-level systems and in fact to continuous systems in which a peak power limitation becomes effective before a mean power limitation.

ACKNOWLEDGEMENTS

Acknowledgement is made to the chief Scientist, Ministry of Supply and to the Controller of H.M. Stationery Office for permission to publish this paper.

REFERENCES

[1] FRY, T. C., *Probability and its Engineering Uses*, New York, 1928
[2] BARKER, R. H., "The Pulse Transfer Function and its Application to Sampling Servo Systems", *Inst. elect. Engrs Monograph* No. 43, July 1952.

Generalized Barker Sequences

S. W. GOLOMB AND R. A. SCHOLTZ, MEMBER, IEEE

Abstract—A generalized Barker sequence is a finite sequence $\{a_r\}$ of complex numbers having absolute value 1, and possessing a correlation function $C(\tau)$ satisfying the constraint $|C(\tau)| \leq 1, \tau \neq 0$. Classes of transformations leaving $|C(\tau)|$ invariant are exhibited. Constructions for generalized Barker sequences of various lengths and alphabet sizes are given. Sextic Barker sequences are investigated and examples are given for all lengths through thirteen. No theoretical limit to the length of sextic sequences has been found.

I. Introduction

A BARKER SEQUENCE [1] is a sequence $\{a_r\}$ of $+1$'s and -1's of some finite length k such that the correlation function $C(\tau)$ defined by $C(\tau) = \sum_{r=1}^{k-\tau} a_r a_{r+\tau}^*$ satisfies $|C(\tau)| \leqq 1$ for $\tau \neq 0$. Consider "generalized Barker sequences" $\{a_r\}$ of finite length k, where the terms a_r are allowed to be complex numbers of absolute value 1, where correlation is now the Hermitian dot product $C(\tau) = \sum_{r=1}^{k-\tau} a_r a_{r+\tau}^*$ (z^* denotes the complex conjugate of z) and the same restriction $|C(\tau)| \leqq 1$ for $\tau \neq 0$ is imposed. We will exhibit a class of transformations which leave the absolute value of the correlation function unaltered, so that, in particular, generalized Barker sequences are changed into other generalized Barker sequences. We examine the effect of these transformations on the original (± 1) Barker sequences and on four-valued (± 1, $\pm i$) Barker sequences, and then exhibit constructions for generalized Barker sequences of a variety of lengths and all possible alphabet sizes. Finally, we observe some very interesting facts about sextic Barker sequences, including the fact that they occur for all lengths tested (i.e., for $k \leqq 13$), and could conceivably occur for all possible sequence lengths.

II. Barker-Preserving Transformations

Let $\{u_r\}$ be a complex-valued sequence of length k, and let its auto-correlation function $C_u(\tau) = \sum_{r=1}^{k-\tau} u_r u_{r+\tau}^*$. We define a new complex-valued sequence $\{v_r\}$ of length k as follows:

$$v_r = u_r e^{2\pi i r/m} \tag{1}$$

where m is any nonzero integer. Then we observe that the auto-correlation function $C_v(\tau)$ of $\{v_r\}$ satisfies

$$C_v(\tau) = \sum_{r=1}^{k-\tau} v_r v_{r+\tau}^* = \sum_{r=1}^{k-\tau} u_r e^{2\pi i r/m} u_{r+\tau}^* e^{-2\pi i(r+\tau)/m}$$

$$= \sum_{r=1}^{k-\tau} u_r u_{r+\tau}^* e^{-2\pi i \tau/m} = e^{-2\pi i \tau/m} C_u(\tau) \tag{2}$$

for all τ. In particular, since $|e^{-2\pi i \tau/m}| = 1$, we have $|C_v(\tau)| = |C_u(\tau)|$ for all τ, as well as $|v_r| = |u_r|$ by virtue of (1). As a special case, if $|u_r| = 1$ for all r, the transformation (1) takes a generalized Barker sequence of length k into another generalized Barker sequence of length k.

Further, we observe that the somewhat more general transformation

$$v_r = u_r e^{2\pi i (r+\alpha)/x} \tag{3}$$

where α and x are any real numbers, $x \neq 0$, also preserves $|v_r| = |u_r|$ and $|C_v(\tau)| = |C_u(\tau)|$. Finally, the transformation $v_r = u_{k-r+1}$, which merely runs the sequence backwards, clearly preserves the Barker property, as does the transformation $w_r = u_r^*$ (where the *star* again denotes complex conjugation).

In summary, there is a group of $4m^2$ transformations, each of which takes m-phase Barker sequences into m-phase Barker sequences. As generators of this group, we may take the two order-2 transformations of *time reversal* and *complex conjugation*, and the two order-m transformations of *constant multiplication* by ζ^α and *progressive multiplication* by successive powers of ζ where $\zeta = e^{2\pi i/m}$.

III. Specific Examples

If we take $m = 2$ in (1), then the sequence $(u_1, u_2, u_3, u_4, \cdots, u_k)$ is changed into $(u_1, -u_2, u_3 - u_4, \cdots, (-1)^{k-1} u_k)$. In particular, this transformation changes ordinary binary Barker sequences into other such sequences, though nothing very startling is obtained in this way. Taking $m = 4$, the sequence $(u_1, u_2, u_3, u_4, \cdots, u_k)$ is changed into $(u_1, iu_2 - u_3, -iu_4, \cdots, i^{k-1} u_k)$. In particular, this transformation sets up a one-to-one correspondence between ordinary (binary) Barker sequences and four-symbol Barker sequences (the symbols being ± 1, $\pm i$) in which real and imaginary terms alternate. This correspondence is illustrated in Table I for binary Baker sequences of all known lengths. (By the results of Storer and Turyn [2], there are no other binary Barker sequences of *odd* length, and the possibility of other such sequences of *even* length appears quite remote.) Another description of the correspondence is that between successive terms in the binary sequence, there is a transition of either 0° or 080°, while between successive terms in the alternating quaternary sequence, there is a transition of either 90° or 270°, and we let 0° transitions correspond to 90° transitions, while 180° transitions correspond to 270° transitions, respectively.

It is interesting to note that the quaternary alternating Barker sequences of odd length are all palindromic (i.e., read the same forward and backward), thus evidencing

Manuscript received March 12, 1965; revised July 9, 1965. This paper was prepared by the Electronic Sciences Lab., University of Southern California, Los Angeles, supported by the Jet Propulsion Lab./NASA, under Contract 951076.

The authors are with the Dept. of Electrical Engineering, University of Southern California, Los Angeles, Calif.

Reprinted from *IEEE Trans. Inform. Theory*, vol. IT-11, no. 4, pp. 533–537, Oct. 1965.

TABLE I

CORRESPONDENCE BETWEEN BINARY AND ALTERNATING QUATERNARY BARKER SEQUENCES

Length k	Binary Sequence
1	$+1$
2	$+1, +1$
3	$+1, +1, -1$
4	$+1, +1, +1, -1$
5	$+1, +1, +1, -1, +1$
7	$+1, +1, +1, -1, -1, +1, -1$
11	$+1, +1, +1, -1, -1, -1, +1, -1, -1, +1, -1$
13	$+1, +1, +1, +1, +1, -1, -1, +1, +1, -1, +1, -1, +1$

	Quaternary Barker Sequence
1	$+1$
2	$+1, +i$
3	$+1, +i, +1$
4	$+1, +i, -1, +i$
5	$+1, +i, -1, +i, +1$
7	$+1, +i, -1, +i, -1, +i, +1$
11	$+1, +i, -1, +i, -1, -i, -1, +i, -1, +i, +1$
13	$+1, +i, -1, -i, +1, -i, +1, -i, +1, -i, -1, +i, +1$

TABLE II

TRANSFORMATIONS ON A QUATERNARY BARKER SEQUENCE

Original Sequence	$+1,$	$-1,$	$+1,$	$+i,$	$-i,$	$-1,$	$+i,$	$-i,$	$-i,$	$+1,$	$+i,$	$+i,$	$+1,$	$+1,$	$+1$
Tranoform with $m = 2$	$+1,$	$+1,$	$+1,$	$-i,$	$-i,$	$+1,$	$+i,$	$+i,$	$-i,$	$-1,$	$+i,$	$-i,$	$+1,$	$-1,$	$+1$
Transform with $m = 4$	$+1,$	$-i,$	$-1,$	$+1,$	$-i,$	$-i,$	$-i,$	$-1,$	$-i,$	$+i,$	$-i,$	$+1,$	$+1,$	$+i,$	-1

TABLE III

CONSTRUCTION FOR GENERALIZED BARKER SEQUENCES FOR ALL LENGTHS UP TO 5, AND ALL ALPHABET SIZES GREATER THAN 1

k	sequence	correlation function	which n?
1	1	1	all $n \geqq 1$
2	$1, 1$	$2, 1$	all $n \geqq 1$
3	$1, 1, \eta$	$3, 1 + \eta, \eta$	all $n \geqq 2$
4	$1, 1, \eta, 1$	$4, 1 + \eta + \eta^*, 1 + \eta, 1$	all $n \geqq 2$
5	$1, 1, 1, -1, 1$	$5, 0, 1, 0, 1$	all *even* n } all $n \geqq 2$
	$1, 1, \rho, \eta, \rho$	$5, 1 + \rho + \eta + \eta^*, 1 + \rho + \eta, \rho + \eta, \rho$	all n except 1, 2, 4 } all $n \geqq 2$

TABLE IV

k	sequence	correlation function	which n?
7	$1, 1, 1, -1, -1, 1, -1$	$7, 0, -1, 0, -1, 0, -1$	all even n
	$1, 1, \beta, \beta, 1, \beta, 1$	$7, 2(1 + u), 2 + u + \beta,$ $2 + u, 1 + u, 1 + \beta, 1$ where $u = \beta + \beta^*$.	all odd $n > 1$ except 5 & 7, as well as large even n.
9	$1, 1, \epsilon, \epsilon, \epsilon^*, -\epsilon, \epsilon, -1, 1$	$9, \epsilon^{*2} - \epsilon^2 + \epsilon - \epsilon^*, \epsilon^{*2} + \epsilon^2 + 2(\epsilon + \epsilon^*) - 1, 0, 1, 0,$ $\epsilon + \epsilon^* - 1, 0, 1.$	all even n except 2, 4, & 10
	$1, 1, \delta, \delta, 1, 1,$ $\delta^*, 1, \delta^*$	$9, -\delta, \delta^*, 0, -\delta, \delta^*,$ $0, -\delta, \delta^*.$	all multiples of 3.

a symmetry which is obscured in the binary representation.

Next, we consider a quaternary Barker sequence of length 15 [3] which is *not* equivalent to a binary Barker sequence. In Table II, we see such a sequence, together with its transforms corresponding to $m = 2$ and to $m = 4$.

We observe that the effect of the $m = 2$ transformation is to replace the original terms by their complex conjugates in reverse order. Since reversal of order and conjugation of elements are both elementary symmetries which preserve Barker correlation, nothing essentially new has been obtained. However, the transformation with $m = 4$ yields a very different sequence from the original. For example, whereas the original sequence had one run of length 3 and two runs of length 2 each, the transformed sequence has one run of length 3 but only one run of length 2. Also, the original sequence has its run of length 3 at the end, whereas the transformed sequence has its longest run near the middle.

It may be mentioned in passing that despite its title, [4] contains no examples of quaternary Barker codes.

IV. General Alphabets

Let $\rho = e^{2\pi i/n}$, and define the alphabet A_n to be the set $\{1, \rho, \rho^2, \rho^3, \cdots, \rho^{n-1}\}$. For subsequent convenience, we also define $\eta = \rho^{[(n+1)/2]}$, the lowest power of ρ whose phase angle is at least 180°.

We ask the general question on the existence of Barker sequences as follows: "For given sequence length k, what are all the values of n such that a Barker sequence of length k can be constructed from the alphabet A_n?"

In Table III, we answer this question completely for $k = 1, 2, 3, 4, 5$ by exhibiting constructions which work for all $n \geqq 2$. A couple of simple observations will assist the reader in verifying that the out-of-phase correlations are indeed $\leqq 1$, as required.

a) The sum of 2 unit vectors lies within the unit circle if and only if the angle between those vectors is between 120° and 240°. (In particular, $1 + \eta$ lies within or on the unit circle for $n \geqq 2$, the case $n = 3$ being the tightest squeeze, while $\rho + \eta$ lies within or on the unit circle except when $n = 1, 2, 4$, the cases $n = 3$ and 6 being tightest.)

b) The sum of 3 unit vectors lies within the unit circle if and only if there is no semicircle properly containing all three vectors. (In particular, since the sets $1, \eta, \eta^*$ and $1, \rho, \eta$ each defy containment within a semicircle, the vectors $1 + \eta + \eta^*$ and $1 + \rho + \eta$ do not extend beyond the unit circle.) Finally, we may observe that $1 + \rho + \eta + \eta^*$ lies *outside* the unit circle for $n = 1, 2, 4$; *on* the unit circle for $n = 3, 5, 6$; and *within* the unit circle for $n \geqq 7$. (In fact, this vector sum decreases in magnitude to zero as n increases to infinity.)

A simple necessary and sufficient condition for the sum of four unit vectors to lie within the unit circle has not been found. However, the following two necessary conditions are sharp.

1) The four vectors must not be contained within an arc of less than $2\cos^{-1}(\frac{1}{4}) \approx 151°$.

2) No three of the four vectors may be contained within an arc of less than $180° - \cos^{-1}(\frac{1}{4}) \approx 104.5°$.

For the case $k = 6$, it has long been known that there is no binary Barker sequence. An exhaustive search has shown that there is also no generalized Barker sequence of length 6 for alphabet sizes of 3, 4, or 5. However, the following example using the alphabet A_6 has been found:

$$1,\ 1,\ \epsilon,\ -1,\ 1,\ -\epsilon$$

where $\epsilon = e^{2\pi i/6}$. The correlation values for this sequence are $6, \epsilon^2, \epsilon^1, \epsilon^3, \epsilon^5, \epsilon^4$.

Except for the Barker-preserving transformations of Sec. II, this example is unique. However, in one of its transformed avatars, it takes the form

$$1,\ \epsilon,\ -1,\ 1,\ -\epsilon,\ -1$$

for which the correlation values are $6, -1, -1, 1, -1, -1$.

The case of sequence length 7 has been settled definitively, as shown in Table IV, which augments Table III. Here β is an element of A_n with a phase angle between 120° and π-$\cos^{-1}(\frac{3}{4}) \approx 138\frac{1}{2}°$. Not only are there no 5th or 7th roots of unity with phase angles in this range, but an exhaustive search has shown that there are no Barker sequences of length 7 over A_5 or A_7.

The case of sequence length 9 is not entirely settled, but important partial results are listed in Table IV. Since a ternary sequence exists, length 9 sequences can be constructed for nth roots of unity when n is a multiple of three, as indicated in Table IV, where $\delta = e^{2\pi i/3}$. For even n, the sequence listed works when ϵ is a root of unity with phase angle in the range (38.2°, 68.7°). Since it has been shown by other investigators that Barker sequences of length 9 do not exist for $n = 2$ and 4, the only case in question for even n is $n = 10$.

V. The Sextic Alphabet

There is strong evidence to indicate that a Barker sequence of length 6 can *only* be achieved over an alphabet A_n for which n is a multiple of 6. Moreover, the sixth roots of unity (with zero adjoined) have the useful property of being partially closed under addition, as well as closed under multiplication.

In Table IV, we see the present state of knowledge concerning binary, ternary, quaternary, and sextic Barker sequences.[1] In Table V a *blank* indicates that no such sequence exists; a *question mark* means that the question has not yet been exhaustively searched. We see that for lengths $k = 6, 8, 10$, and 12, there are *no* examples of binary, ternary, or quaternary sequences, but that sextic examples exist.

Examples of length 6, 8, and 10 were first discovered by the authors. Examples of length 12 were found in a computer search conducted by Harold Fredricksen. Of

[1] The appropriate earlier references for these cases are binary [2], ternary [5], and quaternary [3]. However, these cases have all been rechecked.

TABLE V

COMPARISON OF THE SEQUENCES AVAILABLE FROM THE ALPHABETS A_2, A_3, A_4, AND A_6

k	n = 2	3	4	6
1	√	√	√	√
2	√	√	√	√
3	√	√	√	√
4	√	√	√	√
5	√	√	√	√
6				√
7	√	√	√	√
8				√
9		√		√
10				√
11	√		√	√
12				√
13	√		√	√
14				?
15			√	?

TABLE VI

TABLE OF SEXTIC BARKER SEQUENCES

Length k = 3			
1)	0 0 2	P, T	2 1 k 5 4
2)	0 0 3	P, B	3 z k z 3
Length k = 4			
1)	0 0 1 4		4 z 1 k 5 z 2
2)	0 0 1 5		5 0 0 k 0 0 1
3)	0 0 2 0	P, T	0 1 z k z 5 0
4)	0 0 3 0	B	0 z 3 k 3 z 0
Length k = 5			
1)	0 0 0 2 5		5 z 0 1 k 5 0 z 1
2)	0 0 0 3 0	B	0 z 0 z k z 0 z 0
3)	0 0 0 3 1		1 2 1 5 k 1 5 4 5
4)	0 0 1 4 2	P	2 3 1 z k z 5 3 4
5)	0 0 2 4 2	P, T	2 3 z 2 k 4 z 3 4
6)	0 0 2 5 3		3 4 1 3 k 3 5 2 3
7)	0 0 3 1 3		3 2 1 3 k 3 5 4 3
Length k = 6			
1)	0 0 1 3 0 4	P	4 5 3 1 2 k 4 5 3 1 2
Length k = 7			
1)	0 0 0 2 5 4 2		2 3 4 5 2 5 k 1 4 1 2 3 4
2)	0 0 0 3 3 0 3	B	3 z 3 z 3 z k z 3 z 3 z 3
3)	0 0 1 0 3 4 1		1 z 4 2 z 1 k 5 z 4 2 z 5
4)	0 0 1 0 4 5 2		2 z 5 4 5 0 k 0 1 2 1 z 4
5)	0 0 1 4 5 2 0	P	0 1 5 z 4 2 k 4 2 z 1 5 0
6)	0 0 2 1 5 0 4		4 5 0 4 z 5 k 1 z 2 0 1 2
7)	0 0 2 2 0 1 5		5 0 1 z 4 0 k 0 2 z 5 0 1
8)	0 0 2 2 0 2 0	P, T	0 1 z 0 1 z k z 5 0 z 5 0
9)	0 0 2 2 5 1 5		5 0 z 4 2 2 k 4 4 2 z 0 1

TABLE VI (*Continued*)

Length k = 8			
1)	0 0 0 0 3 2 5 1		1 0 1 2 4 1 0 *k* 0 5 2 4 5 0 5
2)	0 0 0 1 3 0 5 3		3 4 5 *z* 0 2 0 *k* 0 4 0 *z* 1 2 3
3)	0 0 0 1 3 5 3 1		1 2 *z* 4 3 1 1 *k* 5 5 3 2 *z* 4 5
4)	0 0 0 2 3 0 3 1		1 2 1 4 2 1 2 *k* 4 5 4 2 5 4 5
5)	0 0 0 2 4 1 0 4		4 5 0 1 *z* 3 *z* *k* *z* 3 *z* 5 0 1 2
6)	0 0 0 2 5 2 1 5		5 0 1 2 0 1 4 *k* 2 5 0 4 5 0 1
7)	0 0 1 0 4 2 3 0		0 *z* 3 1 2 *z* 5 *k* 1 *z* 4 5 3 *z* 0
8)	0 0 1 0 5 3 0 2		2 1 1 4 0 0 5 *k* 1 0 0 2 5 5 4
9)	0 0 1 1 5 4 1 3		3 2 2 5 5 2 0 *k* 0 4 1 1 4 4 3
10)	0 0 2 2 1 5 1 5	*P*	5 0 *z* 5 3 1 5 *k* 1 5 3 1 *z* 0 1
11)	0 0 2 2 4 2 3 1		1 2 3 *z* 2 1 1 *k* 5 5 4 *z* 3 4 5
Length k = 9			
1)	0 0 0 0 2 4 0 4 2	*T*	2 3 *z* 4 5 *z* 0 1 *k* 5 0 *z* 1 2 *z* 3 4
2)	0 0 0 0 3 2 5 0 3		3 *z* 5 *z* 0 2 3 0 *k* 0 3 4 0 *z* 1 *z* 3
3)	0 0 0 0 3 2 5 1 3		3 2 *z* 2 1 *z* *z* 1 *k* 5 *z* *z* 5 4 2 4 3
4)	0 0 0 1 3 5 2 0 4	*P*	4 5 *z* *z* *z* *z* 1 2 *k* 4 5 *z* *z* *z* *z* 1 2
5)	0 0 0 1 3 5 2 1 5		5 0 1 *z* 2 0 3 1 *k* 5 3 0 4 *z* 5 0 1
6)	0 0 0 1 4 0 3 2 0		0 1 2 *z* 2 0 5 *z* *k* *z* 1 0 4 *z* 4 5 0
7)	0 0 0 1 4 0 3 2 5		5 *z* 3 3 1 5 0 2 *k* 4 0 1 5 3 3 *z* 1
8)	0 0 0 2 4 1 4 3 1		1 2 3 4 3 1 *z* 3 *k* 3 *z* 5 3 2 3 4 5
9)	0 0 0 2 5 1 4 3 1		1 2 3 4 1 1 5 3 *k* 3 1 5 5 2 3 4 5
10)	0 0 1 0 4 3 0 2 5		0 1 *z* 0 3 2 2 4 *k* 2 4 4 3 0 *z* 5 0
11)	0 0 1 1 5 4 1 3 0		0 *z* *z* *z* 0 *z* *z* *z* *k* *z* *z* *z* 0 *z* *z* *z* 0
12)	0 0 1 1 5 4 1 3 1		1 2 1 1 *z* 4 1 5 *k* 1 5 2 *z* 5 5 4 5
13)	0 0 1 1 5 4 1 4 0		0 5 5 4 5 0 1 *z* *k* *z* 5 0 1 2 1 1 0
14)	0 0 1 1 5 4 2 5 1		1 0 0 4 4 1 2 5 *k* 1 4 5 2 2 0 0 5
15)	0 0 1 1 5 5 2 4 2		2 3 2 2 4 5 5 0 *k* 0 1 1 2 4 4 3 4
16)	0 0 1 2 0 0 3 5 3		3 4 3 *z* 2 0 5 1 *k* 5 1 0 4 *z* 3 2 3
17)	0 0 1 2 0 5 3 5 3		3 4 3 4 3 5 1 5 *k* 1 5 1 3 2 3 2 3
18)	0 0 2 2 0 0 4 0 4	*P*, *T*	4 5 *z* 4 5 *z* 4 5 *k* 1 2 *z* 1 2 *z* 1 2
Length k = 10			
1)	0 0 0 0 2 4 0 3 1 5		5 0 *z* 0 3 *z* 0 0 2 *k* 4 0 0 *z* 3 0 *z* 0 1
2)	0 0 0 1 0 4 3 0 2 5		5 *z* 0 3 5 *z* 1 0 5 *k* 1 0 5 *z* 1 3 0 *z* 1
3)	0 0 0 1 4 4 1 3 0 4		4 5 4 2 5 2 4 5 2 *k* 4 1 2 4 1 4 2 1 2
4)	0 0 0 1 5 3 4 1 1 4		4 *z* 1 2 5 *z* 3 4 0 *k* 0 2 3 *z* 1 4 5 *z* 2
5)	0 0 0 2 2 0 0 3 0 4		4 5 4 2 *z* 0 1 2 5 *k* 1 4 5 0 *z* 4 2 1 2
6)	0 0 1 1 2 5 1 5 4 2		2 3 5 1 5 3 1 1 0 *k* 0 5 5 3 1 5 1 3 4
7)	0 0 1 1 3 0 1 5 4 2		2 3 5 1 5 1 1 3 0 *k* 0 3 5 5 1 5 1 3 4
8)	0 0 1 1 4 4 1 3 0 4		4 5 3 3 0 *z* 3 2 2 *k* 4 4 3 *z* 0 3 3 1 2
9)	0 0 1 1 4 4 1 4 0 4		4 5 4 4 5 4 2 1 2 *k* 4 5 4 2 1 2 2 1 2
10)	0 0 1 1 5 5 2 4 1 5		5 0 4 4 0 0 *z* 5 *z* *k* *z* 1 *z* 0 0 2 2 0 1
11)	0 0 1 2 5 5 3 5 2 0		0 1 5 4 5 3 3 *z* 1 *k* 5 *z* 3 3 1 2 1 5 0
Length k = 11			
1)	0 0 0 0 2 2 5 0 3 0 4		4 5 4 5 0 3 0 1 1 1 *k* 5 5 5 0 3 0 1 2 1 2
2)	0 0 0 2 3 0 1 5 3 0 4		4 5 4 *z* *z* 0 1 4 2 3 *k* 3 4 2 5 0 *z* *z* 2 1 2
3)	0 0 1 2 3 1 0 3 0 1 5		5 0 0 *z* 5 4 0 1 2 1 *k* 5 4 5 0 2 1 *z* 0 0 1
4)	0 0 2 2 3 1 0 3 0 1 5		5 0 1 4 4 5 *z* *z* 2 *z* *k* *z* 4 *z* *z* 1 2 2 5 0 1
5)	0 0 2 3 4 3 2 5 2 4 2	*P*	2 3 *z* *z* 0 4 *z* 3 1 1 *k* 5 5 3 *z* 2 0 *z* *z* 3 4
6)	0 0 2 3 4 3 2 5 3 4 2		2 3 4 4 1 *z* 0 2 2 0 *k* 0 4 4 0 *z* 5 2 2 3 4
Length k = 12			
1)	0 0 1 0 5 2 2 4 0 4 1 3		3 2 2 5 4 3 1 0 3 1 2 *k* 4 5 3 0 5 3 2 1 4 4 3
2)	0 0 1 1 5 5 2 3 0 3 0 4		4 5 3 4 5 2 2 0 *z* 5 3 *k* 3 1 *z* 0 4 4 1 2 3 1 2
3)	0 0 1 2 3 1 1 5 0 3 0 4		4 5 3 *z* 0 *z* 5 3 2 5 1 *k* 5 1 4 3 1 *z* 0 *z* 3 1 2

these, only S_6 is unique up to Barker-preserving transformations. This is one of several facts supporting the highly tentative conjecture that sextic Barker sequences exist for *all* sequence lengths, in sharp contrast to the binary case [2].

The results of an exhaustive computer search for sextic Barker sequences through length twelve are given in Table VI. All sequences listed are distinct under the Barker preserving transformations and are lexicographically the smallest of their respective equivalence classes. Numerals in Table VI denote powers of the primitive sixth root of unity, k denotes sequence length, and z denotes a zero; sequences marked P may be transformed to palindromic sequences; B and T indicate binary and ternary sequences, respectively.

A Barker sequence is called *palindromic* if the sequence reads the same forward and backward, except for at most a real constant factor. Clearly the correlation function of a palindromic sequence is real. We have observed that there are sextic palindromic sequences of every length through 9, none of length 10, one of length 11, and none of lengths 12 through 17.

References

[1] R. H. Barker, "Group synchronizing of binary digital systems," in *Communication Theory*. London: Butterworth, 1953, pp. 273–287.

[2] R. Turyn and J. Storer, "On Binary Sequences," *1961 Proc. AMS*, vol. 12, pp. 394–399.

[3] G. Carley, unpublished report. M.I.T., 1963.

[4] G. R. Welti, "Quaternary Codes for Pulsed Radar," *IRE Trans. on Information Theory*, vol. IT-6, pp. 400–408, June 1960.

[5] D. F. de Long Jr., "Three-phase codes," M.I.T. Lincoln Lab., Lexington, Mass., Group Rept. 47-28, July 24, 1959.

Optimal Binary Sequences for Spread Spectrum Multiplexing

ROBERT GOLD

MAGNAVOX RESEARCH LABORATORIES, TORRANCE, CALIFORNIA 90503

I. Introduction

Linear shift register sequences (see Zierler[1] and Gold[2]) have found extensive applications in spread spectrum communication systems. The binary sequences generated by shift register devices serve as the encoding mechanism of such systems which, when added to the baseband information, results in a wideband low-power-density signal which has statistical properties similar to noise. The casual listener is thus denied access to the baseband information which can be recovered from the wideband signal only through correlation with a stored reference sequence in the receiver which is an exact replica of the original encoding sequence.

The usefulness of the maximal linear sequences in spread spectrum communications depends in large part on their ideal autocorrelation properties. The autocorrelation function of a binary sequence h is defined as $\theta_h(\tau)$ = (number of agreements − number of disagreements) when the sequence h is compared with a cyclic shift of itself. It is well known that for maximal linear sequences $\theta_h(0)$ = period of the sequence and $\theta_h(\tau) = -1$ for $\tau \neq 0$. The detection by the receiver of the high in-phase correlation value $\theta_h(0)$ determines the synchronization between transmitter and receiver necessary for the removal of the encoding sequence and the recovery of the baseband information. In multiplexing applications many systems will be operating in the same neighborhood and each communication link will employ a different maximal encoding sequence. In general, the cross-correlation function between different maximal sequences may be relatively large. Thus different systems operating in the same environment can interfere with the successful attainment and maintenance of proper synchronization by having the receiver of one communication link lock onto the cross-correlation peaks obtained by correlating with the encoding sequence of a different communication link. Thus the successful use of spread spectrum communication systems in multiplexing applications depends upon the construction of large families of encoding sequences with uniformly low cross-correlation values. In this paper we present an analytical technique for the construction of such families of linear binary encoding sequences.

II. Notation

Following the notation of Zierler[1] we denote by $V(f)$ the vector space of linear sequences generated by the recursion relation corresponding to the polynomial f of degree n and further identify the members of $V(f)$ with binary $2^n - 1$ tuples. If f is a primitive polynomial over the field $K = \{0, 1\}$ then $h \in V(f)$ implies h is a maximal linear sequence. We denote by $||h||$ the number of ones in the sequence h and by $\bar{h}$ the sequence such that $\bar{h}(i) = h(i) + 1$. The correlation function θ of two binary sequences a, b has been defined as

$$\theta(\tau) = \sum_{i=0}^{2^n-1} \chi a(i) \chi b(i+\tau)$$

where χ is the unique isomorphism of the additive group $\{0, 1\}$ onto the multiplicative group $\{1, -1\}$. We note that $\theta(a, b)(\tau)$ is simply described as the number of agreements − number of disagreements of the sequences a and b for each τ and that

$$\theta(a, b) = 2^n - 1 - 2\,||a + b||.$$

In what follows α will always denote a primitive $2^n - 1$ root of unity in a splitting field of $x^{2^n-1} + 1$ and the minimal polynomial of α^i will be denoted by f_i. Finally, we note the following result of Bose and Chaudhuri.[3]

Theorem 1

Let α be any primitive element of the splitting field of $x^{2^n-1} + 1$. Let f_i be the minimal polynomial of α^i. Let

$$g = \frac{x^{2^n-1} + 1}{\operatorname{lcm}\{f_1, f_2, \cdots f_{2k}\}}$$

Then $a, b \in V(g)$ implies $||a + b|| > 2k$.

III. Statement and Proof of Result

Our techniques for the construction of large families of encoding sequences with uniformly low cross-correlation values is based on the following result.

Theorem 2

Let α be any primitive element of $GF(2^n)$. Let f_1 be the minimal polynomial of α. Let f_t be the minimal polynomial of α^t where

$$t = \begin{cases} (2^{(n+1)/2}) + 1 & (n \text{ odd}) \\ (2^{(n+2)/2}) + 1 & (n \text{ even}). \end{cases}$$

Then $a \in V(f_1)$ and $b \in V(f_t)$ implies $|\theta(a, b)| \leq t$.

The significance of this theorem is that it tells how to select shift register tap connections which will generate maximal linear sequences with a known bound on the cross-correlation function. Since α is primitive, the sequence generated by the shift register corresponding to f_1 is maximal. Since

$$2^{(n+1)/2} + 1 \quad \text{and} \quad 2^{(n+2)/2} + 1$$

are both relatively prime to $2^n - 1$ for $n \not\equiv 0$ mod the 4 sequence corresponding to the polynomial f_t is also maximal in these cases. Theorem 2 thus permits the selection of pairs of maximal sequences with known bound on the cross-correlation function. This result is of practical importance since, for example, for $n = 13$ there are 630 maximal sequences and there exist pairs of these sequences

Manuscript received July 22, 1966; revised February 4, 1967.

Reprinted from *IEEE Trans. Inform. Theory*, vol. IT-13, no. 4, pp. 619–621, Oct. 1967.

whose correlation values are as high as $\theta = 703$ while Theorem 2 guarantees the selection of pairs of sequences such that $|\theta| \leq 129$.

This result is a special case of the more general theorem stated in the following which has been obtained independently by Gold[5] and Kasami,[4] and is related to the weight distribution of error-correcting codes.

Theorem

Let a and b be maximal linear sequences given by

$$a(i) = T(\alpha^{-i}) \quad \text{and} \quad b(i) = T((\alpha^{2^l+1})^{-i})$$

where α is a primitive $2^n - 1$ root of unity (n odd), l is any integer such that $(l, k) = 1$, and T is the trace of $GF(2^n)$. Then $\theta(a, b)(n) = -1$ when $a(\tau) = 0$ and

$$\theta(a, b)(\tau) = \begin{cases} -(2^{(n+1)/2} + 1) & \\ \text{or} & \text{when } a(\tau) = 1. \\ (2^{(n+1)/2} - 1) & \end{cases}$$

The proof of this theorem is contained in Gold[5].

In the remainder of this section, we proceed to the proof of Theorem 2 by means of a series of lemmas.

Lemma 1

Let α be any primitive $2^n - 1$ root of unity in a splitting field of

$$x^{2^n-1} + 1.$$

Let

$$g_k = \frac{x^{2^n-1} + 1}{\text{lcm}\,\{f_1, f_2, \cdots f_k\}}$$

where f_i is the minimal polynomial of α^i. Let f be an irreducible polynomial of degree n. Let A_f be the conjugate class of roots of f. Let $m_f = \min\,\{i \mid \alpha^i \in A_f\}$, the class leader of A_f. Then $m_f > k$ implies f is a factor of g_k.

Proof: f irreducible of degree n implies $f \mid x^{2^n-1} + 1$ implies $f \mid g_k$ lcm $\{f_1, f_2, \cdots f_k\}$, and $m_f > k$ implies $f_i \mid$ lcm $\{f_1 \cdots f_k\}$ implies $f \mid g_k$.

Lemma 2

Let α be any primitive $2^n - 1$ root of unity in a splitting field of

$$x^{2^n-1} + 1.$$

Let $u = 2^{n-1} - 1$. Let

$$v = \begin{cases} 2^{n-1} - 1 - 2^{n-1/2} & \text{for } n \text{ odd} \\ 2^{n-1} - 1 - 2^{n/2} & \text{for } n \text{ even.} \end{cases}$$

Let f_u be the minimal polynomial of α^u. Let f_v be the minimal polynomial of α^v. Then $m_{f_u} = u$ and $m_{f_v} = v$.

Proof: α^r and α^s belong to the same conjugate class of $GF(2^n)$ if, and only if, there exists an integer k such that $\alpha^r = (\alpha^s)^{2^k}$ if, and only if, $r = s \cdot 2^k$ modulo $2^n - 1$ if, and only if, there exists a cyclic permutation p such that $[r(0), r(1), \cdots r(n-1)] = [s(p(0)), s(p(1)) \cdots s(p(n-1))]$ where

$$r = \sum_{i=0}^{n-1} r(i)2^i \quad \text{and} \quad s = \sum_{i=0}^{n-1} s(i)2^i.$$

Now $u = 2^n - 1 = \sum_{i=0}^{n-1} u(i)\, 2^i$ where $[u(0), u(1), \cdots u(n-1)] = [1\ 1 \cdots 1\ 0]$. Clearly any permutation of $[1\ 1 \cdots 1\ 0]$ corresponds to a larger integer and hence $m_{f_u} = u$. Now

$$v = 2^{n-1} - 1 - 2^{n-1/2} = \sum_{i=0}^{n-1} v(i)2^i$$

where

$$\left[v(0), v(1), \cdots v\left(\frac{n-3}{2}\right), v\left(\frac{n-1}{2}\right), \cdot v\left(\frac{n+1}{2}\right) \cdots v(n-2)v(n-1)\right]$$

$$= [\underbrace{1, 1, \cdots 1}_{n-1/2 \text{ ones}},\ 0,\ \underbrace{1 \cdots 1}_{n-1/2 \text{ ones}},\ 0].$$

Again it is clear that any cyclic permutation of $[v(0), \cdots v(n-1)]$ will result in a larger integer and hence $m_{f_v} = v$. A similar argument holds when $v = 2^{n-1} - 1 - 2(n/2)$.

Lemma 3

f_u and f_v are factors of g_{v-1} where u and v are as in Lemma 2 and

$$g_{v-1} = \frac{x^{2^n-1} + 1}{\text{lcm}\,\{f_1, f_2, \cdots f_{v-1}\}}$$

then $f_v \mid g_{v-1}$.

Proof:

$$m_{f_u} = u \text{ (by Lemma 2)} = 2^n - 1$$

$> 2^{n-1} > v - 1$ implies $f_u \mid g_{v-1}$ (by Lemma 1)

$$m_{f_v} = v \text{ (by Lemma 2)} > v - 1 \text{ implies } f_v \mid g_{v-1}.$$

Lemma 4

Let f_i denote the minimal polynomial of α^i. Let

$$g_k = \frac{x^{2^n-1} + 1}{\text{lcm}\,\{f_1, f_2, \cdots f_k\}}.$$

Then $a, b \in V(g_k)$ implies $|\theta(a, b)| < 2^n - 1 - 2k$.

Proof: $a, b \in V(g_k)$ implies $a + b \in V(g_k)$ implies $||a + b|| > k$ by Theorem 1. Since 1 is not a primitive root of unity, $1 + x$ is clearly a factor of g, and hence $V(1 + x) \subset V(g_k)$. Thus the constant sequence of ones is a member of $V(g_k)$, and hence $V(g_k)$ is closed with respect to the operation of complementation, i.e., $h \in V(g_k)$ implies $\tilde{h} \in V(h_k)$ where $\tilde{h}(i) = 1 + h(i)$.

Thus $a, b \in V(g_k)$ implies $a + b \in V(g_k)$ implies $\widetilde{a + b} \in V(g_k)$ implies $||\widetilde{a + b}|| = 2^n - 1 - ||a + b|| > k$ implies $k < ||a + b|| < 2^n - 1 - k$ implies $|\theta(a, b)| < (2^n - 1) - 2k$.

Proof of Theorem 2: Let $\beta = \alpha^{-2}$. Then β is clearly a primitive $2^n - 1$ root of unity and

$$\alpha = \alpha^{-2}(2^{n-1} - 1) = \beta^{2^{n-1}} - 1 = \beta^u$$

$$\alpha^t = \begin{cases} \alpha^{(2^n+1)/2} + 1 & (\text{for } n \text{ odd}) = \\ \alpha^{(2^n+2)/2} + 1 & \text{for } n \text{ even;} \end{cases}$$

$$\left.\begin{cases} \alpha^{-2}[2^{n-1} - 1 - 2^{(n-1)/2}] = \beta^{2^{n-1}} - 1 - 2^{n-1/2} \\ \alpha^{-2}[2^{n-1} - 1 - 2^{n/2}] = \beta^{2^{n-1}} - 1 - 2^{n/2} \end{cases}\right\} = \beta^v.$$

Now f_1 is the minimal polynomial of $\alpha = \beta_u$ and f_t is the minimal polynomial of $\alpha^t = \beta_v$. By Lemma 2 $m_{f_1} = u > v - 1$ and $m_{f_t} = v > v - 1$. Thus by Lemma 1 f_1 and f_t are factors of g_{v-1}. Hence, $V(f_1) \subset V(g_{v-1})$ and $v(f_t) \subset V(g_{v-1})$. Thus $a \in V(f_1)$ and $b \in V(f_t)$ implies a and $b \in (g_{v-1})$. Thus by Lemma 4

$$|\theta(a, b)| < 2^n - 1 - 2(v - 1) =$$

$$\begin{cases} 2^n - 1 - 2[2^{n-1} - 2 - 2^{n-1/2}] = 2^{(n+1)/2} + 3 \\ 2^n - 1 - 2[2^{n-1} - 2 - 2^{n/2}] = 2^{(n+2)/2} + 3. \end{cases}$$

Since the value of the cross-correlation function is always odd we have

$$|\theta(a, b)| \leq 2^{n+1/2} + 1 \quad \text{for} \quad n \text{ odd}$$

$$|\theta(a, b)| \leq 2^{n+2/2} + 1 \quad \text{for} \quad n \text{ even}.$$

For 13-stage shift registers there are pairs of maximal sequences with cross-correlation peaks as high as $\theta(\tau) = 703$ while proper selection of shift registers in accordance with the above theorem guarantees sequences whose cross correlation satisfies the inequality

$$|\theta(\tau)| \leq 2^{(13+1)/2} + 1 = 129.$$

We note further that for purely random sequences of length $2^{13} - 1 = 8191$ we would expect the cross-correlation function to exceed $2\sigma = 2\sqrt{8192} \sim 180$ for 5 percent of the correlation values and hence linear sequences chosen in accordance with our technique perform better with respect to their cross-correlation properties than purely random sequences.

IV. Construction of Encoding Families

In this section we show how to provide large families of encoding sequences each of period $2^n - 1$ and such that the cross-correlation function of any pair of sequences of the family has a cross-correlation function θ which satisfies the inequality

$$|\theta(\tau)| \leq 2^{(n+2)/2} + 1.$$

In a spread spectrum multiplexing application such families form ideal codes which minimizes interlink interference. Instead of having each communication link employ a different maximal sequence we assign to each link a member of the encoding family to be constructed below. These are nonmaximal linear sequences, and hence their autocorrelation function will not be two-valued; however, the out-of-phase value of the autocorrelation function will satisfy the above inequality. Thus by slightly relaxing the conditions on the autocorrelation function we obtain a family of encoding sequences with the high cross-correlation peaks eliminated.

The procedure for generating these encoding families is embodied in the following theorem.

Theorem

Let f_1 and f_t be a preferred pair of primitive polynomials of degree n whose corresponding shift registers generate maximal linear sequences of period $2^n - 1$ and whose cross-correlation function θ satisfies the inequality.

$$|\theta| \leq t = \begin{cases} 2^{(n+1)/2} + 1 & \text{for} \quad n \text{ odd} \\ 2^{(n+2)/2} + 1 & \text{for} \quad n \text{ even} \quad n \not\equiv \bmod 4. \end{cases}$$

Then the shift register corresponding to the product polynomial $f_1 \cdot f_t$ will generate $2^n + 1$ different sequences each period $2^n - 1$ and such that the cross-correlation function θ of any pair of such sequences satisfies the above inequality.

Proof: $a \in V(f_1 \cdot f_t) = V(f_1) + V(f_t)$ implies $a = b + c$ where $b \in V(f_1)$ and $c \in V(f_t)$. Period $(b + c) = \text{lcm}\{\text{period } b, \text{period } c\} = 2^n - 1$. Since degree $f_1 \cdot f_t = 2n$, there are $(2^{2n} - 1)/(2^n - 1) = 2^n + 1$ essentially different sequences in $V(f_1 \cdot f_t)$. Finally $a, b \in V(f_1 \cdot f_t)$ implies

$$\begin{cases} a = a_1 + a_t \\ b = b_1 + b_t \end{cases}$$

where $a_1, b_1 \in V(f_1)$, and $a_t, b_t \in V(f_t)$. $|\theta(a, b)| = |\theta(a_1 + b_1, a_t + b_t)| \leq t$ by Theorem 2 since $a_1 + b_1 \in V(f_1)$ and $a_1 + b_1 \in V(f_t)$.

Thus, by way of illustration, if we consider the pair of polynomials, $f_1(x) = 1 + x + x^2 + x^3 + x^7$ and $f_t(x) = 1 + x + x^2 + x^3 + x^4 + x^5 + x^7$ then the product polynomial is $f_1(x) f_2(x) = 1 + x^2 + x^6 + x^8 + x^{11} + x^{12} + x^{14}$ and the corresponding 14-stage shift register will generate 129 different linear sequences of period 127. The cross-correlation function θ of any pair of such sequences will satisfy the inequality $|\theta(\tau)| \leq 17$.

References

[1] N. Zierler, "Linear recurring sequences," *J. SIAM*, vol. 7, March 1959.
[2] R. Gold, "Characteristic linear sequences and their coset functions," (accepted for publication *J. Soc. Ind. Appl. Math.*, May 1965).
[3] W. W. Peterson, *Error Correcting Codes*. New York: Wiley, 1961.
[4] T. Kasami, "Weight distribution formula for some class of cyclic codes," University of Illinois, Urbana, Rept. R-265, April 1966.
[5] R. Gold, "Maximal recursive sequences with 3-valued recursive cross-correlation functions" (submitted for publication, January 1967).

Lower Bounds on the Maximum Cross Correlation of Signals

L. R. WELCH

Abstract—Some communication systems require sets of signals with impulse-like autocorrelation functions and small cross correlation. There is considerable literature on signals with impulse-like autocorrelation functions but little on sets of signals with small cross correlation. A possible reason is that designers put too severe a restriction on cross correlation magnitudes. This correspondence establishes lower bounds on how small the cross correlation and autocorrelation can simultaneously be.

INTRODUCTION

When a receiver uses matched filter detection, the output of the detector is the correlation function of the input signal and the desired signal. When several communicators are using the same channel without phase coherence, it is desirable that the cross correlation functions be small in absolute value. In addition, an impulse-like autocorrelation function is needed for some applications such as radar. There is considerable literature on single signals with good autocorrelation functions, e.g., [1]–[15], but little on sets of signals with good cross correlation [18], [19]. This correspondence establishes bounds on how small the cross correlation and autocorrelation values can be. The bounds on periodic correlation show that the reported sets of signals in [18] are near optimal and, conversely, the reported sets of signals show the bounds to be reasonably tight in the applicable cases.

LOWER BOUNDS

Although, in practice, signals defined for continuous time are needed for nonsynchronous communication systems, the application of a sampling theorem reduces the signal design problem to one of selecting sets of vectors $\{(a_1^\nu \cdots a_L^\nu): \nu = 1,\cdots,M\}$ whose components a_i^ν are complex numbers with

$$c_{\nu\nu}(0) = \sum_{i=1}^{L} |a_i^\nu|^2 = 1.$$

We begin with a theorem on inner products among a set of M vectors of length L and norm 1.

Define

$$c_{\nu\lambda} = \sum_{i=1}^{L} a_i^\nu \bar{a}_i^\lambda$$

and

$$c_{\max} = \max_{\nu \neq \lambda} |c_{\nu\lambda}|.$$

Manuscript received October 3, 1973; revised November 12, 1973. This work was supported by the Army Research Office, Durham, N.C., under Grant DA-ARO-D-31-124-73-G-153 and by the National Science Foundation under Grant GK-36532.

The author is with the Department of Electrical Engineering, University of Southern California, Los Angeles, Calif. 90007.

Theorem (*Inner Products*): Let k be a positive integer. Then

$$(c_{\max})^{2k} \ge \frac{1}{M-1}\left[\frac{M}{\binom{L+k-1}{k}} - 1\right].$$

Proof: Obviously

$$M(M-1)c_{\max}^{2k} + M \ge \sum_{\nu,\lambda} |c_{\nu,\lambda}|^{2k} = B_k.$$

We have

$$B_k = \sum_{\nu,\lambda} (c_{\nu\lambda})^k (\bar{c}_{\nu\lambda})^k = \sum_{\nu,\lambda}\left(\sum_{i=1}^{L} a_i^\nu \bar{a}_i^\lambda\right)^k \left(\sum_{j=1}^{L} \bar{a}_j^\nu a_j^\lambda\right)^k$$

$$= \sum_{\nu,\lambda} \sum_{\substack{t_1\cdots t_k=1\\ \tau_1\cdots\tau_k=1}}^{L} \prod_{i=1}^{k} a_{t_i}^\nu \bar{a}_{\tau_i}^\nu \bar{a}_{t_i}^\lambda a_{\tau_i}^\lambda.$$

Interchanging orders of summation produces

$$B_k = \sum_{\substack{t_1\cdots t_k=1\\ \tau_1\cdots\tau_k=1}}^{L} \left|\sum_{\nu} \prod_{i=1}^{k} a_{t_i}^\nu \bar{a}_{\tau_i}^\nu\right|^2.$$

Various choices of $(t_1 \cdots t_k)$ and $(\tau_1 \cdots \tau_k)$ give rise to the same product of the a and $\bar{a}$; the number of choices can be expressed as multinomial coefficients. This observation can be used to rearrange the sum in the form

$$B_k = \sum_{\substack{b_1\cdots b_L\\ d_1\cdots d_L}} \binom{k}{b}\binom{k}{d} \left|\sum_{\nu=1}^{M} \prod_{i=1}^{L} (a_i^\nu)^{b_i} (\bar{a}_i^\nu)^{d_i}\right|^2$$

where $b_i, d_i \ge 0, \sum_{i=1}^{L} b_i = \sum_{i=1}^{L} d_i = k$, and

$$\binom{k}{b} = \frac{k!}{\prod_i (b_i!)} \qquad \binom{k}{d} = \frac{k!}{\prod_i (d_i!)}.$$

Since the summands are nonnegative, the terms with

$$(b_1,\cdots,b_L) \neq (d_1,\cdots,d_L)$$

may be dropped to yield

$$B_k \ge \sum_{b_1\cdots b_L} \left[\binom{k}{b} \sum_{\nu=1}^{M} \prod_{i=1}^{L} |a_i^\nu|^{2b_i}\right]^2.$$

Cauchy's inequality for sums of squares then gives

$$B_k \ge \frac{\left[\sum_{b_1\cdots b_L} \binom{k}{b} \sum_{\nu=1}^{M} \prod_{i=1}^{L} |a_i^\nu|^{2b_i}\right]^2}{\sum_{b_1\cdots b_L} 1}$$

where all sums have as their range $\{(b_1 \cdots b_L): b_i \ge 0, \sum b_i = k\}$. Interchanging orders of summation and applying the multi-

Reprinted from *IEEE Trans. Inform. Theory*, vol. IT-20, no. 3, pp. 397–399, May 1974.

nomial expansion theorem gives

$$B_k \geq \frac{\left(\sum_{\nu=1}^{M}\left(\sum_{i=1}^{L}|a_i^{\nu}|^2\right)^k\right)^2}{\binom{L+k-1}{k}}.$$

Since all vectors are assumed to have norm 1, the innermost sum equals one and the expression reduces to

$$B_k \geq \frac{M^2}{\binom{L+k-1}{k}}.$$

From a previous inequality this yields

$$M(M-1)(c_{\max})^{2k} + M \geq \frac{M^2}{\binom{L+k-1}{k}}$$

and the theorem follows.

The vector $(a_k \cdots a_L, a_1 \cdots a_{k-1})$ is called the kth cyclic shift of $(a_1 \cdots a_L)$. If all cyclic shifts of members of a set S are added that set to give a set C, then the periodic auto- or cross correlation values of pairs from S are inner products of pairs from C. If there are M members of S, there are $M \cdot L$ members of C, and the previous theorem translates to the following.

Theorem (Periodic Correlation): Define

$$c_{\nu\lambda}(\tau) = \sum_{i=1}^{L-\tau} a_i^{\nu}\bar{a}_{i+\tau}^{\lambda} + \sum_{i=1}^{\tau} a_{i+L-\tau}^{\nu}\bar{a}_i^{\lambda}$$

$$c_1 = \max_{\nu \neq \lambda}\ \max_{0 \leq \tau < L} |c_{\nu\lambda}(\tau)|$$

$$c_2 = \max_{\nu}\ \max_{1 \leq \tau < L} |c_{\nu\nu}(\tau)|$$

and

$$c_{\max} = \max(c_1, c_2).$$

Then

$$c_{\max}^{2k} \geq \frac{1}{ML-1}\left[\frac{ML}{\binom{L+k-1}{k}} - 1\right].$$

Next suppose $L = 2N - 1$ and $a_i^{\nu} = 0$ for $N < i \leq L$; then the periodic correlations of the L-dimensional vectors become the aperiodic correlations of the N-dimensional vectors, and the previous theorem becomes the following.

Theorem (Aperiodic Correlation):

$$c_{\max}^{2k} \geq \frac{1}{M(2N-1)-1}\ \frac{M(2N-1)}{\binom{2N+k-2}{k}} - 1.$$

The preceding results are, of course, trivial if the right side of the inequality is negative. However, in the correlation theorems the case with $k = 1$ and $M > 1$ is nontrivial and gives the following results.

Corollary (Periodic):

$$c_{\max} \geq \sqrt{\frac{M-1}{ML-1}}.$$

Corollary (Aperiodic):

$$c_{\max} \geq \sqrt{\frac{M-1}{M(2N-1)-1}}.$$

In [15], Golay introduced the idea of a signal consisting of two sequences which are to be transmitted in separate channels each of which has a matched filter, and in which the outputs of the two filters are to be added for detection. This idea was generalized in [16], [17] to K sequences transmitted in separate channels. In [17], for certain choices of K and N, it was shown that there are K signals each consisting of K sequences of length N and K detectors whose outputs are zero unless the correct signal is present with the correct timing. In other words, K communicators can share K channels in such a way that they do not interfere with one another but each has the advantage of an impulse-type detector response. The next theorem shows what happens when more than K communicators try to share K channels.

Let S be a set of M signals, each consisting of K sequences of length N, such that

$$\sum_{i=1}^{K}\sum_{t=1}^{N} |a_i^{\nu}(t)|^2 = 1$$

where $a_i^{\nu}(t)$ is the tth term of the ith sequence of the νth signal. Define $c_{\nu,\lambda}(i,\tau)$ to be the aperiodic correlation function between the ith sequences of the νth and λth signal and

$$c_{\nu,\lambda}(\tau) = \sum_{i=1}^{K} c_{\nu,\lambda}(i,\tau).$$

Further, define

$$c_1 = \max_{\nu \neq \lambda}\ \max_{\tau} |c_{\nu,\lambda}(\tau)|$$

$$c_2 = \max_{\nu}\ \max_{\tau \neq 0} |c_{\nu,\nu}(\tau)|$$

and

$$c_{\max} = \max(c_1, c_2).$$

Theorem (Multichannel Aperiodic Correlation):

$$c_{\max}^{2k} \geq \frac{1}{M(2N-1)-1}\left[\frac{M(2N-1)}{\binom{2NK-K+k-1}{k}} - 1\right].$$

Proof: Define b_{iK+j}^{ν} to be the $(i+1)$st term of the jth sequence of the νth signal for $0 \leq i < N$ and define $b_i^{\nu} = 0$ for $KN < i \leq K(2N-1) = L$. To the set of vectors $\{(b_1^{\nu} \cdots b_L^{\nu})\}$ adjoin all lth cyclic shifts for all l which are multiples of K. It is easily seen that there are $M(2N-1)$ sequences in the resulting set and every inner product equals $c_{\nu,\lambda}(\tau)$ for some choice of parameters. Applying the inner product theorem completes the proof.

Corollary (Multichannel):

$$c_{\max} \geq \sqrt{\frac{M/K-1}{M(2N-1)-1}}.$$

Proof: Set $k = 1$ in the previous theorem.

Concluding Remarks

All of the corollaries are obtained by setting $k = 1$ in the appropriate theorems. If M is less than L this is the only choice of k giving nontrivial results. However, the theorems may be used to give asymptotic results. In particular, if, as L tends to infinity, M grows as the tth power of L, k may be set equal to t in the inner product theorem to show that the bound is asymptotically

$$c_{\max} \geq \frac{1}{\sqrt{L}}(t! - 1)^{1/2t}.$$

In [20, p. 129], Van Lint shows that the duals of certain BCH codes have L^t codewords of length $L - 1$ and have inner products not exceeding $2(t - 1)\sqrt{L}$. From these observations it can be concluded that when M grows as the tth power of L

$$c_{\max} \geq a_t/\sqrt{L}$$

where

$$(t! - 1)^{1/2t} \leq a_t \leq 2(t - 1).$$

Since BCH codes are cyclic, this result extends to periodic correlation bounds by replacing t by $t + 1$.

The application of these bounds to binary error-correcting codes is not straightforward. If a binary error-correcting code of length L has minimum distance D, and the inner product between codewords a and b is defined as

$$c_{ab} = \frac{1}{L} \sum_{t=1}^{L} (-1)^{a_t + b_t}$$

then

$$\max_{a \neq b} c_{ab} = L - 2D.$$

However, the preceding theorems give bounds on $|c_{ab}|$, which in general is not equal to $L - 2D$. It is possible, in the binary case, to modify the preceding approach to give bounds on D; the results are nearly as good as the Gilbert existence bound for D near $L/2$. This result will be proven in a future correspondence.

References

[1] R. H. Barker, "Group sync of binary digital systems," in *Communication Theory*. London: Butterworth, 1953.

[2] S. W. Golomb and R. A. Scholtz, "Generalized Barker sequences," *IEEE Trans. Inform. Theory*, vol. IT-11, pp. 533–537, Oct. 1965.

[3] R. A. Scholtz, "Search for digital radar ranging signals," in *Hawaii Int. Conf. System Sciences*, Jan. 1971.

[4] R. Turyn and J. Storer, "On binary sequences," *Proc. Amer. Math. Soc.*, vol. 12, pp. 394–399, 1961.

[5] G. R. Welti, "Quaternary codes for pulsed radar," *IRE Trans. Inform. Theory*, vol. IT-6, pp. 400–408, June 1960.

[6] D. A. Huffman, "The generation of impulse-equivalent pulse trains," *IRE Trans. Inform. Theory* (Int. Symp.), vol. IT-8, pp. S10–16, Sept. 1962.

[7] M. H. Ackroyd, "Synthesis of efficient Huffman sequences," *IEEE Trans. Aerosp. Electron. Syst.*, vol. AES-8, pp. 2–8, Jan. 1972.

[8] J. Injeyan, "Generation of Huffman codes," Canadian General Electric Co., Tech. Memo RQ65EE2, Oct. 1965.

[9] C. E. Cook and M. Bernfeld, *Radar Signals*. New York: Academic Press, 1967, ch. 8.

[10] A. Boehmer, "Binary pulse compression codes," *IEEE Trans. Inform. Theory*, vol. IT-13, pp. 156–166, Apr. 1967.

[11] D. C. Chu, "Polyphase codes with good periodic correlation properties," *IEEE Trans. Inform. Theory* (Corresp.), vol. IT-18, pp. 531–532, July 1972.

[12] R. Turyn, "Sequences with small correlation," in *Error Correcting Codes*, H. B. Mann, Ed. New York: Wiley, 1968.

[13] R. L. Frank, "Polyphase codes with good nonperiodic correlation properties," *IEEE Trans. Inform. Theory*, vol. IT-9, pp. 43–45, Jan. 1963.

[14] R. L. Frank, S. A. Zadoff, and R. C. Heimiller, "Phase shift pulse codes with good periodic correlation properties," *IRE Trans. Inform. Theory* (Corresp.), vol. IT-8, pp. 381–382, Oct. 1962.

[15] M. J. E. Golay, "Complementary series," *IRE Trans. Inform. Theory*, vol. IT-7, pp. 82–87, Apr. 1961.

[16] C. L. Liu and C. Tseng, "Complementary sets of sequences," *IEEE Trans. Inform. Theory*, vol. IT-18, pp. 644–651, Sept. 1972.

[17] B. P. Schweitzer, "Generalized complementary code sets," Ph.D. dissertation, University of California, Los Angeles, 1971.

[18] R. M. Lerner, "Signals having good correlation functions," in *WESCON Conv. Rec.*, 1961.

[19] R. A. Scholtz and L. R. Welch, "Group characters: Sequences with good correlation properties," unpublished.

[20] J. H van Lint, *Coding Theory*. New York: Springer, 1971.

Bounds on the Number of Signals with Restricted Cross Correlation

HARRISON E. ROWE, FELLOW, IEEE

Abstract—**We seek the maximum number of users possible for multiuser, spread-spectrum radio systems, as the signal set is varied, under the following ideal conditions: 1) no fading; 2) synchronous system; 3) orthogonal signal set for each user; 4) matched-filter individual receivers; 5) no additive noise.**

Welch has given an upper bound on the number of vectors (or equivalently signals) w possible in a d-dimensional space, given the maximum cross-correlation magnitude C_{max} between any vector pair. However, the error rate for the systems under study may depend on a suitably defined root-mean-square cross correlation C_{rms}, rather than on the maximum cross correlation C_{max} of a vector set.

We extend Welch's results to d-dimensional vector sets of size w divided into alphabets of size A, the vectors of each alphabet being strictly orthogonal, and give the maximum size of such a vector set as function of C_{rms}.

We demonstrate that as Welch's limit is approached, all cross correlations must approach C_{max}, and given precise limits on the way in which this must occur. Similarly, close to the present bounds (in terms of C_{rms}) all rms cross correlations must be close to C_{rms}.

Particular examples are given of signal sets close to the present bounds.

SYSTEM MODEL

A COMMON channel is shared by U spread-spectrum transmitters. Each has an orthogonal alphabet of A waveforms, each of duration no greater than T seconds and confined to a band of width B Hz. We represent these $A \cdot U$ waveforms by the same number of vectors [1], in a space of $d \equiv 2TB$ dimensions.

$$i\text{th symbol, } k\text{th user's alphabet} \Leftrightarrow V_i^k; \quad 1 \leqslant k \leqslant U, \quad 1 \leqslant i \leqslant A. \tag{1}$$

All signals have unit energy:

$$|V_i^k|^2 = 1. \tag{2}$$

Each alphabet is strictly orthogonal:

$$V_i^k \cdot V_{i'}^k = 0, \qquad i \neq i'. \tag{3}$$

Define

$$C_{max} \equiv \max_{i,i',k \neq k'} |V_i^k \cdot V_{i'}^{k'}|. \tag{4}$$

For $C_{max} \ll 1$ different alphabets are approximately orthogonal.

A related system uses biorthogonal alphabets; each consists of the above A orthogonal signals and their negatives, a total of $2A$ waveforms. We use the convention

$$V_{-i}^k \equiv -V_i^k. \tag{5}$$

In (1) the index i takes on the values $-A \leqslant i \leqslant -1$ and $1 \leqslant i \leqslant A$, and (3) is modified to

$$V_i^k \cdot V_{i'}^k = 0, \qquad i = \pm i'. \tag{6}$$

Obviously from (5), $V_i^k \cdot V_{-i}^k = -1$.

U receivers correspond to these transmitters. Each consists of A filters matched to the alphabet of its corresponding transmitter. These filters are numbered from 1 to A, the impulse response of the ith filter corresponding to the vector V_i^k. (For the biorthogonal case the ith filter gives +1 output for V_i^k, −1 output for V_{-i}^k.) Synchronous operation is assumed; for a typical time slot the (common) receiver input is

$$V = \sum_{k=1}^{U} V_{i_k}^k \tag{7}$$

where the kth transmitter has sent symbol i_k from its alphabet; either $1 \leqslant i_k \leqslant A$ (orthogonal), or $-A \leqslant i_k \leqslant -1$ or $1 \leqslant i_k \leqslant A$ (biorthogonal). The output of the ith matched filter of the kth receiver is from (2), (3) or (6), and (7)

$$V_i^k \cdot V = \begin{cases} I_i^k, & i \neq \pm i_k, \\ \pm 1 + I_i^k, & i = \pm i_k, \end{cases} \qquad 1 \leqslant i \leqslant A \tag{8}$$

where the ± applies to the biorthogonal case; the interference parameters I_i^k are given by

$$I_i^k = \sum_{\substack{k'=1 \\ k' \neq k}}^{U} V_i^k \cdot V_{i_{k'}}^{k'}, \qquad 1 \leqslant i \leqslant A. \tag{9}$$

I_i^k, of course, depends on the set $\{i_{k'}: k' \neq k\}$ of symbols sent by every other transmitter. The $|I_i^k|$ will be small if C_{max} is sufficiently small.

Each receiver examines its A matched filter outputs, selects the one with the largest magnitude, and declares its corresponding symbol to be the one transmitted. Error will occur for a typical symbol i of receiver k if and only if at least one other filter has a larger output:

$$|\pm 1 + I_i^k| \leqslant |I_{i'}^k|, \qquad i' \neq i, \quad 1 \leqslant i, \quad i' \leqslant A \tag{10}$$

the ± again applying to the biorthogonal case.

Manuscript received July 9, 1981; revised January 14, 1982. This paper was presented at the National Telecommunications Conference, New Orleans, LA, December 1981.

The author is with Bell Laboratories, Holmdel, NJ 07733.

Reprinted from *IEEE Trans. Commun.*, vol. COM-30, no. 5, pp. 966–974, May 1982.

Rather than characterizing the interference properties of a particular signal set by its I_i^k, we consider the statistical model described in the following section.

STATISTICAL MODEL

Assume each transmitter sends independent, equally likely symbols from its alphabet, independently from all other transmitters. Thus, for orthogonal alphabets the kth transmitter sends symbols $V_1{}^k, V_2{}^k, \cdots, V_A{}^k$, each with probability $1/A$; for biorthogonal alphabets it sends $\pm V_1{}^k, \pm V_2{}^k, \cdots, \pm V_A{}^k$, each with probability $1/(2A)$.

The probability of error for each symbol of a given signal set can in principle be determined for this statistical model, by (10) and (9). To do this we would require the complete probability distribution for the interference parameters of each alphabet, i.e., the joint distribution for $\{I_i^k\}$ for each k. While this is not possible in general, the means and variances of the I_i^k are readily found.

For this purpose, it is useful to denote the average output of the kth transmitter as $\langle V^k \rangle$, where we drop the subscript inside the $\langle\ \rangle$ in accord with a convention used throughout, that parameters averaged over are omitted.

$$\langle V^k \rangle = \begin{cases} \dfrac{1}{A} \sum_{i=1}^{A} V_i{}^k, & \text{orthogonal} \\ 0, & \text{biorthogonal.} \end{cases} \tag{11}$$

The first and second moments of the output interference quantities I_i^k of (9) are given as follows:

$$\langle I_i^k \rangle = V_i^k \cdot \sum_{\substack{k'=1 \\ k' \neq k}}^{U} \langle V^{k'} \rangle. \tag{12}$$

Define

$$\delta I_i^k \equiv I_i^k - \langle I_i^k \rangle. \tag{13}$$

Then

$$\langle (\delta I_i^k)^2 \rangle = \sum_{\substack{k'=1 \\ k' \neq k}}^{U} \{\langle (V_i^k \cdot V^{k'})^2 \rangle - (V_i^k \cdot \langle V^{k'} \rangle)^2\}. \tag{14}$$

Consider first the biorthogonal case, where each alphabet contains $2A$ vectors, consisting of the original A orthogonal vectors and their negatives (5). Substituting (11) into (12) and (14),

$$\langle I_i^k \rangle = 0$$

$$\langle (I_i^k)^2 \rangle = \langle (\delta I_i^k)^2 \rangle = \sum_{\substack{k'=1 \\ k' \neq k}}^{U} \langle (V_i^k \cdot V^{k'})^2 \rangle. \tag{15}$$

The mean-square interference to each signal is simply the sum of the mean-square interferences due to the other $U-1$ users.

This result (15) will not be strictly true for orthogonal alphabets, since $\langle V^k \rangle$ is no longer 0 as in the biorthogonal case (11); rather, for orthogonal alphabets containing A vectors

$$|\langle V^k \rangle| = \frac{1}{\sqrt{A}}. \tag{16}$$

Nevertheless, in many cases (15) remains approximately true for the orthogonal case as well. To see this, examine the k'th term of (14). By the Schwarz inequality this term $\{\ \} \geqslant 0$. The limiting value 0 is attained only when $V_i^k \cdot V_{i'}^{k'}$ = constant, independent of i'; i.e., the interference in the ith filter of the kth receiver must be identical for every member of the k'th alphabet. Now every biorthogonal alphabet has 2^A corresponding orthogonal alphabets; only a single one of these could satisfy such a requirement (i.e., interfere with a given filter of another alphabet independently of which symbol of the interfering alphabet was transmitted). This argument suggests that (15) may hold approximately for most signal sets with orthogonal alphabets.

Moreover, frequency-hop signals are frequently used in applications, consisting of contiguous sine waves of different frequencies ("chips"); if there is no phase coherence between signals of different users at the receiver, there can be no dc interference components. A rough analog here would have each transmitter use a biorthogonal alphabet of $2A$ symbols, but have each receiver identify every vector and its negative (5) as a single symbol. Then (15) will hold exactly.

The detector of (10) is optimum for Gaussian interference, i.e., when the I_i^k are independent, identically distributed, zero-mean, Gaussian random variables. For a large number of users, $U \gg 1$, the I_i^k will be approximately Gaussian. In cases of interest the interference will have zero mean, as discussed above; otherwise the detector must account for the average values of the interference outputs, and if the user set changes, do this adaptively. However, we have no information about the correlation between interference outputs of different filters, i.e., $\langle I_i^k I_{i'}^k \rangle$, and therefore cannot know for certain that different (Gaussian) I_i^k are approximately independent, although this may be so in many cases.

PERFORMANCE CRITERIA

We adopt the mean-square interference outputs of the receiver filters (15) as a measure of performance. The U active users may be chosen from a possibly much larger set of M potential users, $M \geqslant U$, each possible set U satisfying the conditions of (1)-(3) (and (5), (6) for biorthogonal signals); $C_{\max}$ is now defined by (4) with $1 \leqslant k, k' \leqslant M$, i.e., $C_{\max}$ is the largest cross-correlation magnitude over the entire signal set of $w \equiv A \cdot M$ vectors. Now define $C_{\mathrm{rms}}{}^2$ as the mean-square interference per interferer in the worst case:

$$C_{\mathrm{rms}}{}^2 \equiv \max_{i,k,\{k_l\}} \frac{1}{A(U-1)} \sum_{k' \in \{k_l\}} \sum_{i'=1}^{A} (V_i^k \cdot V_{i'}^{k'})^2 \tag{17}$$

where

$$\{k_l\} \equiv \{k_1, k_2, \cdots, k_{U-1}\}, \qquad k_l \neq k, \quad 1 \leqslant k_l \leqslant M. \tag{18}$$

Thus, C_{rms}^2 is the largest mean-square interference per interferer, for any signal and any of the corresponding $(M-1)!/[(U-1)!(M-U)!]$ possible set of interferers. Equations (17) and (18) imply that

$$\frac{1}{A}\sum_{\substack{k'=1\\k'\neq k}}^{M}\sum_{i'=1}^{A}(V_i^k\cdot V_{i'}^{k'})^2\leqslant(M-1)C_{rms}^2,\qquad \text{all } i,k. \quad (19)$$

We find the maximum number of users (or equivalently signal vectors) as the signal set is varied, for a given C_{rms}; the interference powers are then limited by $\langle (I_i^k)^2\rangle \leqslant (U-1)C_{rms}^2$. As noted above, the interference outputs I_i^k will be approximately Gaussian. We place no constraint on the correlation between different interference outputs $\langle I_i^k I_{i'}^k\rangle$. Correlation can only increase the maximum symbol error rate in a biorthogonal system over that corresponding to independent, Gaussian interference. Since, as shown below, most $\langle (I_i^k)^2\rangle$ are close to their upper limit $(U-1)C_{rms}^2$, for correlation to *decrease* the maximum error rate in an orthogonal system, most $\langle I_i^k I_{i'}^k\rangle$ would have to be positive, a possibility we ignore. Consequently, we assume the maximum symbol error rate will equal or exceed that corresponding to independent Gaussian interference with power $(U-1)C_{rms}^2$ [1, pp. 257-263].

UPPER BOUNDS—C_{max}

Welch has shown for a general set of w unit vectors in d dimensions (i.e., *not* divided into orthogonal or biorthogonal alphabets, as above), denoted by V_ν, $1\leqslant\nu\leqslant w$, that [2]

$$\frac{w^2}{d}\leqslant\sum_{\nu=1}^{w}\sum_{\nu'=1}^{w}(V_\nu\cdot V_{\nu'})^2\leqslant w(w-1)C_{max}^2+w \quad (20)^1$$

where he derived the left-hand member by a sequence of inequalities and the right-hand member by inspection. The outer inequality then yields Welch's result (in different notation and modified form), for the maximum number of vectors with maximum cross correlation C_{max}:

$$\frac{w-1}{d-1}\leqslant\frac{1}{1-C_{max}^2 d}. \quad (21)$$

The equality of (21) is shown as the curve of Fig. 1.

J. E. Mazo has suggested that signal sets close to Welch's limit (21) must have most cross correlations close to C_{max}. Now equality in (21) implies equality between the outer members of (20), which in turn requires

$$\sum_{\nu=1}^{w}\sum_{\substack{\nu'=1\\ \nu\neq\nu'}}^{w}(V_\nu\cdot V_{\nu'})^2=w(w-1)C_{max}^2. \quad (22)$$

1 It is interesting to observe that the lower bound is attained by a vector set having w/d vectors along each of the d coordinate axes.

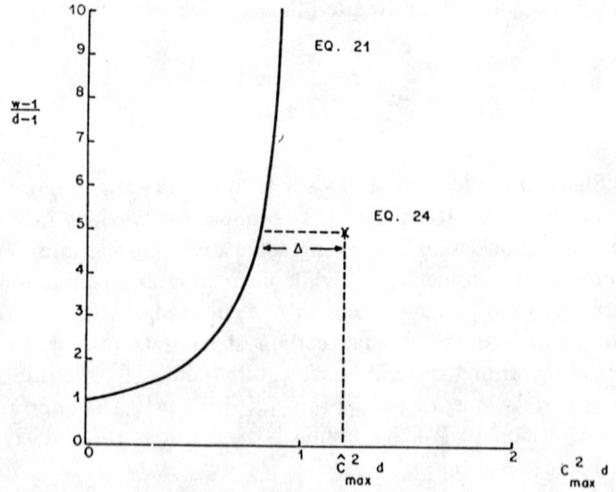

Fig. 1. Welch's bound, and a typical signal set lying below it.

Since $|V_\nu\cdot V_{\nu'}|\leqslant C_{max}$ for $\nu\neq\nu'$ and since the $\Sigma\Sigma$ in (22) has $w(w-1)$ terms, equality in (21) therefore implies that

$$|V_\nu\cdot V_{\nu'}|=C_{max},\qquad \nu\neq\nu'. \quad (23)$$

Any signal set at Welch's limit must have all cross correlations equal to C_{max}.

More generally, consider possible signal sets close to, but not at, Welch's limit (21). Fig. 1 shows (21), together with a point representing a particular signal set having correlation limit $\hat{C}_{max}$, lying a distance Δ to the right of this curve:

$$\frac{w-1}{d-1}=\frac{1}{1-\hat{C}_{max}^2 d+\Delta},\qquad 0\leqslant\Delta. \quad (24)$$

Substituting (24) into (20)

$$\hat{C}_{max}^2\left(1-\frac{\Delta}{\hat{C}_{max}^2 d}\right)\leqslant\frac{1}{w(w-1)}$$

$$\cdot\sum_{\nu=1}^{w}\sum_{\substack{\nu'=1\\ \nu\neq\nu'}}^{w}(V_\nu\cdot V_{\nu'})^2\leqslant\hat{C}_{max}^2. \quad (25)$$

From (25), no more than $w(w-1)/K$ of the $(V_\nu\cdot V_{\nu'})^2$ can be smaller than $\hat{C}_{max}^2-K\Delta/d$, for $K\geqslant 1$ (see the Appendix). Alternatively expressing this result in terms of the horizontal-axis variable $C_{max}^2 d$ of Fig. 1, no greater fraction than $1/K$ of the $w(w-1)$ quantities $(V_\nu\cdot V_{\nu'})^2 d$ can lie to the left of $\hat{C}_{max}^2 d-K\Delta$. Clearly, this result is useful only if the point (23) representing the particular signal set lies close to the curve (21) representing Welch's limit; however, this point can lie slightly to the right of $C_{max}^2 d=1$, where (21) has a pole.

A case of particular interest is

$$\hat{C}_{max}^2 d=1. \quad (26)$$

Here no more than $w(w-1)/K$ of the $w(w-1)$ cross correlations (i.e., no greater fraction than $1/K$) can satisfy

$$(V_\nu \cdot V_{\nu'})^2 d \leqslant 1 - K\frac{d-1}{w-1}. \tag{27}$$

Signal sets close to Welch's limit must have C_{rms} close to C_{max}. However, it is plausible to suppose that we may modify such a vector set by increasing some cross correlations (and therefore increasing C_{max}) while decreasing others, maintaining C_{rms} (and perhaps the error rate) unchanged. This is consistent with the fact that certain signal sets proposed for spread-spectrum systems have quite unevenly distributed cross correlations, i.e., $C_{rms} \ll C_{max}$ [3], [4]. The appropriate extension to Welch's bound is straightforward and given in the next section.

UPPER BOUNDS—C_{rms}

We retain the left-hand inequality of (20), given by Welch [2], we modify the central expression ($\Sigma\Sigma$) to a form appropriate for orthogonal alphabets of size A, and finally we use the restriction of (19) to give a new right-hand inequality:

$$\frac{w^2}{d} \leqslant \sum_{k,k'=1}^{M} \sum_{i,i'=1}^{A} (V_i^k \cdot V_{i'}^{k'})^2 \leqslant wA(M-1)C_{rms}^2 + w$$

$$= w(w-A)C_{rms}^2 + w. \tag{28}$$

The outer inequality of (28) yields the appropriate modification of Welch's result (21), for the maximum number of vectors divided into sets of orthogonal alphabets of size A, with maximum mean-square interference per interfering alphabet C_{rms}^2:

$$\frac{w-A}{d-A} \leqslant \frac{1}{1-C_{rms}^2 d}. \tag{29}$$

The equality of (29) is shown as the curve of Fig. 2. Note that this curve is identical to the curve of Fig. 1 [the equality of (21)] with redefined axes. For biorthogonal alphabets of size $2A$, the corresponding bound for w is twice as large.

We find analogous results to those of the previous section, for signal sets close to the limit (29). Equality in (29) implies equality in (28), which requires

$$\sum_{k=1}^{M} \sum_{i=1}^{A} \sum_{\substack{k'=1\\k'\neq k}}^{M} \sum_{i'=1}^{A} (V_i^k \cdot V_{i'}^{k'})^2 = wA(M-1)C_{rms}^2. \tag{30}$$

The inner pair of $\Sigma\Sigma$ in (30) must be $\leqslant A(M-1)C_{rms}^2$ by (19) for every i, k. Therefore, (30) implies equality in (19); this result and (17) require

$$\frac{1}{A}\sum_{i=1}^{A} (V_i^k \cdot V_{i'}^{k'})^2 = C_{rms}^2, \quad \text{all } i, k, \quad k' \neq k; \quad M > U. \tag{31}$$

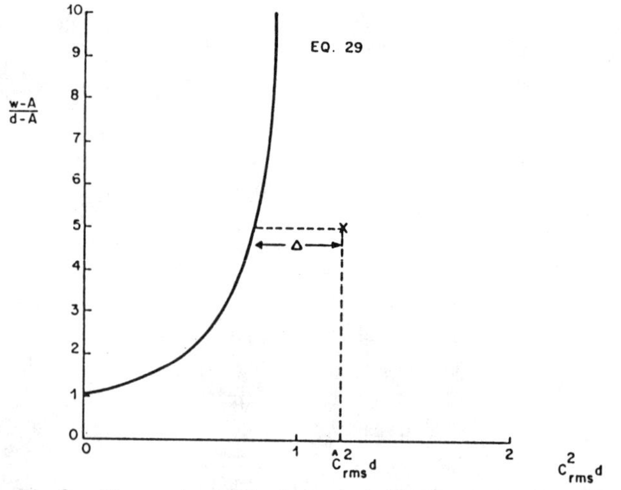

Fig. 2. The rms bound, and a typical signal set lying below it.

Any signal set at the limit (i.e., with equality) in (29) must possess the property (31); the rms interference to every signal from every other alphabet is equal to C_{rms}. Note that (31) is not necessarily satisfied for $M = U$; here

$$\frac{1}{A(U-1)} \sum_{\substack{k'=1\\k'\neq k}}^{U} \sum_{i'=1}^{A} (V_i^k \cdot V_{i'}^{k'})^2$$

$$= C_{rms}^2, \qquad \text{all } i, k; \quad M = U. \tag{32}$$

The total interference to every signal is identical (rather than the individual interferences from each of the other alphabets); however, only this result is needed in the $M = U$ case.

More generally, consider possible signal sets close to, but not at, the limit (29), a typical example being illustrated in Fig. 2 for a signal set with $\hat{C}_{rms}^2$ [defined by (17)] lying a distance Δ to the right of the curve (29). As above, several different cases must be considered (see the Appendix). For $M = U$, no greater fraction than $1/K$ of the w quantities

$$\frac{d}{A(U-1)} \sum_{\substack{k'=1\\k'\neq k}}^{U} \sum_{i'=1}^{A} (V_i^k \cdot V_{i'}^{k'})^2$$

can lie to the left of $\hat{C}_{rms}^2 d - K\Delta$ on Fig. 2. For $M > U$, if the additional condition

$$\frac{1}{A}\sum_{i'=1}^{A} (V_i^k \cdot V_{i'}^{k'})^2 \leqslant \hat{C}_{rms}^2, \qquad \text{all } i, k, \ k' \neq k \tag{33}$$

is imposed, no greater fraction than $1/K$ of the $w(w-A)/A$ quantities $(d/A)\ \Sigma_{i'=1}^{A}\ (V_i^k \cdot V_{i'}^{k'})^2$ can lie to the left of $\hat{C}_{rms}^2\ d - K\Delta$ on Fig. 2. If (33) is abandoned, this result is only slightly degraded for $M \gg U$; in general no greater fraction than $1/[K(1 - U/M)]$ of the $w(w-A)/A$ quantities $(d/A)\ \Sigma_{i'=1}^{A}\ (V_i^k \cdot V_{i'}^{k'})^2$ can lie to the left of $\hat{C}_{rms}^2\ d - K\Delta$ on Fig. 2.

Any signal set close to the limit (29) must have uniform interference properties, the exact nature of the uniformity depending on the case (e.g., $M = U, M \gg U$). However, such signal sets can differ radically from those close to Welch's limit (21) of the preceding section, having quite unevenly distributed cross correlations, i.e., $C_{\mathrm{rms}} \ll C_{\max}$.

LOWER BOUNDS

Consider first a signal set in d dimensions composed of two sets of d vectors each. The first set consists of d unit vectors along the d coordinate axes; the second set consists of d unit vectors with components $\pm 1/\sqrt{d}$, the $\pm$ signs chosen according to a Sylvester-type Hadamard matrix [5].

$$\begin{aligned}
V_1 &= [1\ \ 0\ \ 0 \cdots 0\ \ 0] \\
V_2 &= [0\ \ 1\ \ 0 \cdots 0\ \ 0] \\
&\ \ \vdots \\
V_d &= [0\ \ 0\ \ 0 \cdots 0\ \ 1] \\
V_{d+1} &= \frac{1}{\sqrt{d}}[1\ \ 1\ \ 1 \cdots 1\ \ 1] \\
V_{d+2} &= \frac{1}{\sqrt{d}}[1\ \ -1\ \ 1 \cdots 1\ \ -1] \\
&\ \ \vdots \\
V_{2d} &= \frac{1}{\sqrt{d}}[1\ \ -1\ \ -1 \cdots \pm 1].
\end{aligned} \tag{34}$$

By the Hadamard property, d must be a power of 2; we take $M = U$. Since this set of $2d$ vectors contains U alphabets of size A, $U \cdot A = 2d$; consequently, U and A must both be powers of 2. From (17) and (18) with $M = U$, and from (34)

$$C_{\mathrm{rms}}^2 d = \frac{1}{2\left(1 - \dfrac{A}{2d}\right)}; \qquad C_{\max}^2 d = 1. \tag{35}$$

Since $w = 2d$

$$\frac{w-A}{d-A} = \frac{2 - \dfrac{A}{d}}{1 - \dfrac{A}{d}}. \tag{36}$$

A/d is restricted to the values

$$A/d = 2/d, \cdots, 1/16, 1/8, 1/4, 1/2, 1. \tag{37}$$

The corresponding values for (35) and (36) are

$$\begin{aligned}
C_{\mathrm{rms}}^2 d &= \frac{1}{2}\,\frac{1}{1 - \dfrac{1}{d}}, \cdots, 16/31, 8/15, 4/7, 2/3, 1. \\
\frac{w-A}{d-A} &= 2\,\frac{1 - \dfrac{1}{d}}{1 - \dfrac{2}{d}}, \cdots, 31/15, 15/7, 7/3, 3, \infty.
\end{aligned} \tag{38}$$

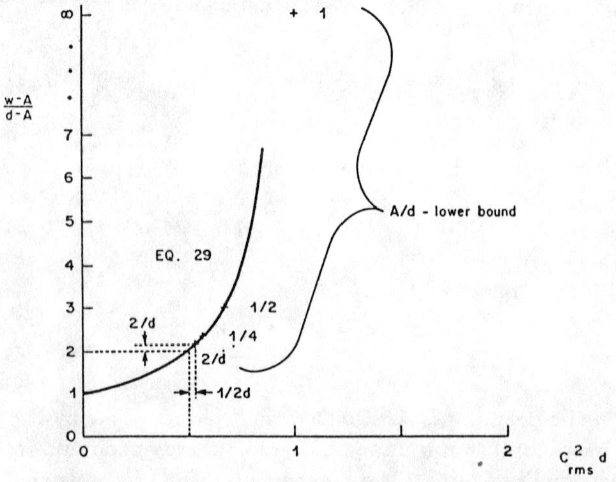

Fig. 3. The rms upper bound with semi-Hadamard lower bound. $M = U, d = 2^{\text{integer}}, U = 2^{\text{integer}}, A \equiv 2d/U = 2^{\text{integer}}$

Solving (35) for A/d and substituting into (36)

$$\frac{w-A}{d-A} = \frac{1}{1 - C_{\mathrm{rms}}^2 d}. \tag{39}$$

The values (38) for the signal set (34) lie precisely on the upper bound (29), plotted in Fig. 2. We replot this upper bound, together with the present lower bounds (38) indicated by +'s lying on the upper bound, on Fig. 3. Note that the point $C_{\mathrm{rms}}^2\, d = 1$, $(w - A)/(d - A) = \infty$ corresponds to the case $U = 2, A = d$ for the signal set of (34), and that the point for the case $U = d, A = 2$ (binary alphabets) lies close to the point $C_{\mathrm{rms}}^2\, d = 0.5$, $(w - A)/(d - A) = 1$, when the dimensionality is high, $d \gg 1$, as indicated by the dotted lines in Fig. 3.

Finally, we consider an abstract description of a frequency-hop signal set discussed by Einarsson [6]. Let ϕ_c, $c = 1, 2, \cdots, d$ represent unit vectors along the d coordinate axes. Divide this set of unit vectors into L groups, each with A unit vectors:

$$d = AL. \tag{40}$$

Each signal vector is composed of one vector from each group. Using Einarsson's notation, the address of the kth user (alphabet) is the L-dimensional vector (in a space unrelated to that of the signal vectors V_i^k)

$$a_k = [a_{k1}\ \ a_{k2} \cdots a_{kL}]. \tag{41}$$

The a_k belong to a finite field of A elements $GF(A)$; A is a prime number or a power of a prime number.

$$0 \leqslant a_{k1}, a_{k2}, \cdots, a_{kL} \leqslant A - 1. \tag{42}$$

Then, using the notation of (1), the first member of the kth alphabet is

$$V_1^k = \frac{1}{\sqrt{L}}\left(\phi_{a_{k1}} + \phi_{A + a_{k2}} + \cdots + \phi_{(L-1)A + a_{kL}}\right). \tag{43}$$

The other members of alphabet k are defined by

$$V_i^k = \frac{1}{\sqrt{L}} (\phi_{y_{k1}} + \phi_{A+y_{k2}} + \cdots + \phi_{(L-1)A+y_{kL}}) \tag{44}$$

where

$$y_k = a_k + (i-1)\mathbf{1}, \qquad i = 1, 2, \cdots, A \tag{45}$$

and $\mathbf{1}$ is the all-ones vector

$$\mathbf{1} = (1 \quad 1 \cdots 1). \tag{46}$$

Arithmetic operations (addition and multiplication) are mod A if A is prime; if A is a power of a prime these operations are more complicated [6]. The components of y_k are restricted the same as those of a_k in (42).

The vectors of each alphabet are obtained by cyclic permutation of the unit coordinate vectors within each group of L. To complete the signal set design we must assign the addresses a_k of (41). Einarsson does this as follows [6].

$$a_k = [(k-1) \quad (k-1)\beta \quad (k-1)\beta^2 \cdots (k-1)\beta^{L-1}], \qquad k = 1, 2, \cdots, A$$

A = prime, or power of prime

β = primitive element of $GF(A)$ (47)

$2 \leqslant L \leqslant A - 1.$

The final relation of (47) and (40) implies that

$$\sqrt{d + 1/4} + 1/2 \leqslant A \leqslant d/2. \tag{48}$$

There are A alphabets, each with A symbols; hence

$$M = A, \qquad w = MA = A^2. \tag{49}$$

Einarsson shows that vectors from different alphabets (i.e., having different addresses a_k, $a_{k'}$) can have no more than one coordinate axis vector ϕ_c in common. Therefore, for any given vector i of alphabet k, the cross correlations with the symbols of another alphabet k' are distributed as follows:

$$V_i^k \cdot V_{i'}^{k'} = \begin{cases} \dfrac{1}{L} & \text{for } L \text{ vectors of } k' \\ 0 & \text{for } A - L \text{ vectors of } k'. \end{cases} \tag{50}$$

Consequently

$$C_{\max}{}^2 = \frac{1}{L^2}. \tag{51}$$

This signal set satisfies (33), which is more restrictive than (17) or (19), with

$$C_{\rm rms}{}^2 = \frac{1}{AL} = \frac{1}{d}. \tag{52}$$

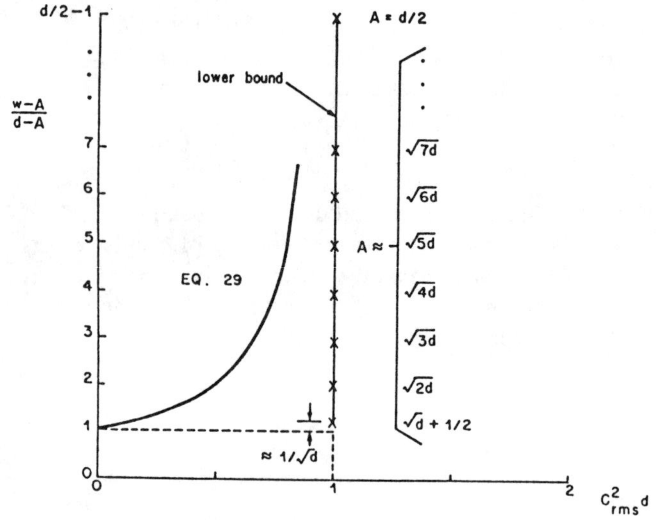

Fig. 4. The rms upper bound, with Einarsson lower bound. $U \leqslant M = A = (\text{prime})^{\text{integer}}$, d/A = integer', $\sqrt{d + 1/4} + 1/2 \leqslant A \leqslant d/2$.

Therefore

$$\left(\frac{C_{\rm rms}}{C_{\max}}\right)^2 = \frac{d}{A^2}. \tag{53}$$

From (48) and (53)

$$\frac{4}{d} \leqslant \left(\frac{C_{\rm rms}}{C_{\max}}\right)^2 \leqslant \frac{1}{1 + \dfrac{1}{2d} + \sqrt{\dfrac{1}{d} + \dfrac{1}{4d^2}}} \approx 1 - \frac{1}{\sqrt{d}}. \tag{54}$$

For large alphabets the cross correlations are quite unevenly distributed; i.e., $(C_{\rm rms}/C_{\max})^2 \ll 1$.

By (52) all signal sets described by (40)-(47) must have $C_{\rm rms}{}^2 d = 1$. From (49)

$$\frac{w - A}{d - A} = \frac{A(A-1)}{d - A}. \tag{55}$$

For different A's satisfying (48) we plot the corresponding vertical coordinate (55) at the horizontal coordinate $C_{\rm rms}{}^2 d = 1$, in Fig. 4. For minimum and maximum alphabet and signal set size we have respectively

$$A = \sqrt{d + 1/4} + 1/2 \approx \sqrt{d} + 1/2;$$

$$\frac{w - A}{d - A} = \frac{1}{1 - \sqrt{\dfrac{1}{d} - \dfrac{1}{4d^2}} - \dfrac{1}{2d}} \approx 1 + \frac{1}{\sqrt{d}}$$

$$A = d/2; \tag{56}$$

$$\frac{w - A}{d - A} = \frac{d}{2} - 1.$$

In general

$$A \approx \sqrt{d}\sqrt{\frac{w-A}{d-A}}, \qquad A \ll d. \tag{57}$$

Note that for U active users the interference power to each is $(U-1)/d$; if all alphabets are used ($U = M$) the interference power is $(A-1)/d$.

DISCUSSION

We have given an upper bound, shown in Fig. 2, for the number of users in a multiuser spread-spectrum system with orthogonal (or biorthogonal) alphabets, given the allowable interference power. This interference power will largely determine the error rate for most practical cases.

The descriptive parameter C_{rms}, defined in (17), is the worst rms interference per interfering alphabet. C_{rms} depends on the particular signal set, with M alphabets; if the number of active users U is less than M, C_{rms} may or may not depend on U as well. In the Hadamard example (34), (35) applies *only* for $U = M$; e.g., for $U \leqslant M/2 + 1$, $C_{rms}{}^2 d = 1$. In contrast, the Einarsson signal set (40)-(47) has $C_{rms}{}^2$ given by (52) for every $U \leqslant M = A$ (49).

Our bound [Fig. 2 or (29)] shows the maximum number of vectors w or equivalently alphabets $M = w/A$ possible for a given $C_{rms}{}^2$. Any signal set close to this limit must have uniform interference properties, which we have described in a precise way. At the limit:

1) If $U = M$, the total interference to every signal is the same.

2) If $U < M$, the interference to every signal from every other alphabet is the same (however, the interference from different members of the interfering alphabet may vary).

The Hadamard example illustrates case 1 (Fig. 3). However, the Einarsson signal set (Fig. 4) does not lie at the limit, although it does satisfy the condition (33) that the interference to every signal from every other alphabet is the same. A signal set at Welch's limit [Fig. 1 or (21)] must have $C_{rms} = C_{max}$, and the same interference between every pair of signals.

We next inquire how tight is the upper bound of (29) and Fig. 2. The point at $C_{rms}{}^2 d = 0$, $(w-A)/(d-A) = 1$ is easily attained by a signal set consisting of d unit vectors along the d coordinate axes; these may be divided into alphabets of any size. Next, the Hadamard signal set of Fig. 3 attains the upper bound for $U = M$ for $C_{rms}{}^2 d$ ranging from a little greater than 1/2 to 2/3 for alphabet sizes varying from $A = 2$ to $A = d/2$. This set also has a point at $C_{rms}{}^2 d = 1$, $(w-A)/(d-A) = \infty$. Finally, the Einarsson set shown on Fig. 4 is confined to the vertical line $C_{rms}{}^2 d = 1$, with $(w-A)/(d-A)$ varying from a little greater than 1 to $d/2 - 1$ for alphabet size varying from $A \approx \sqrt{d}$ to $A = d/2$.

Our upper bounds thus are tight in the sense that signal sets with some alphabet size A can be found that attain the limit for values of $C_{rms}{}^2 d$ near either 0 or 1, or in the middle of its range. However, we desire more lower bounds, for signal sets with various restrictions.

1) Can the bound be attained in the mid-range ($C_{rms}{}^2 d \approx 0.5$) by signal sets with $U < M$, for large and small alphabet size A?

2) At the pole $C_{rms}{}^2 d = 1$, the Einarsson set (for which we can have $U < M$ as well as $U = M$) attains a maximum value of $(w-A)/(d-A) = d/2 - 1$, for large alphabet size $A = d/2$. How large can $(w-A)/(d-A)$ be for small alphabet size, both for $U = M$ and for $U < M$?

3) How close can we come to the rms bound with signal sets for which $C_{rms} = C_{max}$, i.e., having equal interference between every pair of signals from different alphabets?

The present C_{rms} bound (Fig. 2) was obtained by modifying Welch's analysis for his C_{max} bound (Fig. 1). Welch also presented higher order bounds for $C_{max}{}^2 d > 1$ [2]; similar results for $C_{rms}{}^2 d > 1$ are not available.

The present rms bound yields limits on the number of users possible in spread-spectrum systems under various conditions (e.g., [7]).

APPENDIX

SIGNALS SETS CLOSE TO WELCH'S AND rms LIMITS

Consider Welch's C_{max} bound (21), and a particular signal set close to it (24), illustrated in Fig. 1. Subtracting w from the outer members of (20), deleting the $\nu = \nu'$ terms from the $\Sigma\Sigma$ [which sum to w by (2)], dividing by $w(w-1)$, and rearranging the left-hand member, yields

$$\frac{1}{d}\left(1-\frac{d-1}{w-1}\right) \leqslant \frac{1}{w(w-1)} \sum_{\substack{\nu,\nu'=1 \\ \nu \neq \nu'}}^{w} (V_\nu \cdot V_{\nu'})^2 \leqslant C_{max}{}^2. \tag{58}$$

The descriptive parameters of the particular signal set under consideration are related by (24); substituting (24) in the left-hand member of (58), and changing the right-hand member to $\hat{C}_{max}{}^2$, yields (25) after minor rearrangement, where the Σ includes $w(w-1)$ terms.

Now if

$$0 \leqslant x_i \leqslant X \tag{59}$$

then necessarily

$$\frac{1}{N}\sum_{i=1}^{N} x_i \leqslant X. \tag{60}$$

If further

$$X - \Delta \leqslant \frac{1}{N}\sum_{i=1}^{N} x_i \tag{61}$$

then it is obvious that no more than N/K of the x_i can be $\leqslant X - K\Delta$. Note that the limiting case—exactly N/K of the $x_i = X - K\Delta$—is attained when the rest of the x_i, $N(1 - (1/K))$ in number, have their maximum value X. Applying this result to (25) yields directly the bounds on cross correlations stated immediately following (25).

Consider finally the C_{rms} bound (29), and a signal set close to it, illustrated in Fig. 2. This signal set has parameters satisfying

$$\frac{w-A}{d-A}=\frac{1}{1-\hat{C}_{rms}^{\,2}d-\Delta} \tag{62}$$

where $\hat{C}_{rms}^{\,2}$ is defined by (17) for the particular signal set under consideration, and, as noted in the Discussion, may depend on U if $U < M$. Following the above analysis for Welch's C_{max} bounds, we observe from (3) that for $k = k'$ in (28) the terms inside the $\Sigma\Sigma$ are 0 for $i \neq i'$, 1 for $i = i'$; therefore, the $k = k'$ terms sum to w. Subtracting w from the outer members of (28) and deleting the $k = k'$ terms in the $\Sigma\Sigma$, we have after rearrangement

$$\frac{1}{d}\left(1-\frac{d-A}{w-A}\right) \leqslant \frac{1}{w(w-A)}\sum_{k=1}^{M}\sum_{i=1}^{A}\sum_{\substack{k'=1\\k'\neq k}}^{M}\sum_{i'=1}^{A}(V_i^k\cdot V_{i'}^{k'})^2 \leqslant C_{rms}^{\,2}. \tag{63}$$

Since the signal set under discussion satisfies (62), substituting (62) in the left-hand member of (63) and changing $C_{rms}^{\,2}$ in the right-hand member to $\hat{C}_{rms}^{\,2}$ yields

$$\hat{C}_{rms}^{\,2}-\frac{\Delta}{d} \leqslant \frac{1}{w(w-A)}\sum_{k=1}^{M}\sum_{i=1}^{A}\sum_{\substack{k'=1\\k'\neq k}}^{M}\sum_{i'=1}^{A}(V_i^k\cdot V_{i'}^{k'})^2 \leqslant \hat{C}_{rms}^{\,2} \tag{64}$$

corresponding to (25) for Welch's C_{max} bounds.

The results (31) and (32) suggest that the $M = U$ and the $M > U$ cases must be treated separately. Consider first the $M = U$ case. Rewrite (64) as

$$\hat{C}_{rms}^{\,2}d-\Delta \leqslant \frac{1}{w}\sum_{k=1}^{U}\sum_{i=1}^{A}\frac{d}{U-1}\cdot\left\{\sum_{\substack{k'=1\\k'\neq k}}^{U}\frac{1}{A}\sum_{i'=1}^{A}(V_i^k\cdot V_{i'}^{k'})^2\right\} \leqslant \hat{C}_{rms}^{\,2}d. \tag{65}$$

The quantity inside { } of (65) represents the total interference to V_i^k; there are $w = AU$ such quantities. By definition [(17) and (18) with $M = U$] each is $\leqslant (U-1)\hat{C}_{rms}^{\,2}$. By (59)-(61) no greater fraction than $1/K$ of the w quantities

$$\frac{d}{A(U-1)}\sum_{\substack{k'=1\\k'\neq k}}^{U}\sum_{i'=1}^{A}(V_i^k\cdot V_{i'}^{k'})^2$$

can be less than $\hat{C}_{rms}^{\,2}d-K\Delta$.

We treat the $M > U$ case in two steps. First, we impose the condition (33), which places a uniform limit on the rms interference to any signal V_i^k from any other alphabet k'. This condition is satisfied by the Einarsson signal set (40)-(47), but not by the Hadamard signal set (34), for example; (33) is stronger than (17) and (18). Rewrite (64) as

$$\hat{C}_{rms}^{\,2}d-\Delta \leqslant \frac{A}{w(w-A)}\sum_{k=1}^{M}\sum_{i=1}^{A}\sum_{\substack{k'=1\\k'\neq k}}^{M} d\cdot\left\{\frac{1}{A}\sum_{i'=1}^{A}(V_i^k\cdot V_{i'}^{k'})^2\right\} \leqslant \hat{C}_{rms}^{\,2}d. \tag{66}$$

The quantity inside { } of (66) is the interference to V_i^k from the kth alphabet, limited as noted above by (33). Then (59)-(61) yield directly the result that no greater fraction than $1/K$ of the $w(w-A)/A = MA(M-1)$ quantities $(d/A)\ \Sigma_{i'=1}^{A}\ (V_i^k\cdot V_{i'}^{k'})^2$ can be less than $\hat{C}_{rms}^{\,2}\ d-K\Delta$. However, if we abandon (33), the interference now being limited only by (17) and (18), some of the quantities inside { } of (66) can now exceed $\hat{C}_{rms}^{\,2}$, and so the fraction of $(d/A)\ \Sigma_{i'=1}^{A}\ (V_i^k\cdot V_{i'}^{k'})^2$ less than $\hat{C}_{rms}^{\,2}\ d-K\Delta$ can now be greater than $1/K$.

We therefore consider this problem for $M \gg U$, with $\hat{C}_{rms}^{\,2}$ of (62) defined by (17) and (18) and subject to no other restrictions. For each i, k reorder the terms inside the $\Sigma_{k'}$, i.e., $d\{\ \}$, such that their magnitudes are nondecreasing; then by (17) and (18) the first $(U-1)$ such terms must sum to $\leqslant (U-1)\hat{C}_{rms}^{\,2}\ d$. No more than the first $(U-2)$ such terms can exceed $\hat{C}_{rms}^{\,2}\ d$, and if any do, all successive terms must be less than $\hat{C}_{rms}^{\,2}\ d$.

First, it is clear that to maximize the number of terms $d\{\ \}$ in (66) that are $\leqslant \hat{C}_{rms}^{\,2}\ d-K\Delta$, we must replace the $\leqslant$ by =. Second, divide such terms into two sets:

1) those belonging to the first $(U-1)$ terms of the reordered $\Sigma_{k'}$ and

2) those belonging to the remaining terms of the reordered $\Sigma_{k'}$.

The largest number in the second category is the same as given following (66) for the case where the additional constraint (33) was imposed, i.e., $MA(M-1)/K$. The number in the first category will be maximized when those in the second category are confined to as few i, k as possible; then for each of the $[MA(M-1)/K]/[K(M-U)]$ such i, k we can have no more than $(U-2)$ additional $d\{\ \}$ at the minimum value $\hat{C}_{rms}^{\,2}\ d-K\Delta$. Combining these, an upper bound on the number of $d\{\ \} \leqslant \hat{C}_{rms}^{\,2}\ d-K\Delta$ is $[MA(M-1)/K]$ [1 +

$(U-2)/(M-U)] < [w(w-A)/A]/[K(1-U/M)]$, as stated following (33).

To summarize the above three paragraphs, if we impose a uniform limit (33) on the rms interference to any signal V_i^k from any other alphabet k', the fraction of the $w(w-A)/A$ quantities $(d/A)\ \Sigma_{i'=1}^{A}\ (V_i^k \cdot V_{i'}^{k'})^2$ that can be less than $\hat{C}_{rms}{}^2\, d - K\Delta$ is less than $1/K$. If we abandon this uniform limit (33), this fraction is less than $1/[K(1-U/M)]$, a very small increase for $U \ll M$.

ACKNOWLEDGMENT

I would like to thank J. E. Mazo for several helpful discussions, and in particular for bringing Welch's bounds to my attention. I am also grateful to D. J. Goodman, B. G. King, W. L. Mammel, V. K. Prabhu, W. R. Young, and O. C. Yue for helpful discussions.

REFERENCES

[1] J. M. Wozencraft and I. M. Jacobs, *Principles of Communication Engineering*. New York: Wiley, 1965, ch. 4.

[2] L. R. Welch, "Lower bounds on the maximum cross correlation of signals," *IEEE Trans. Inform. Theory*, vol. IT-20, pp. 397–399, May 1974.

[3] A. J. Viterbi, "A processing satellite transponder for multiple access by low-rate mobile users," presented at the *Digital Satellite Commun. Conf.*, Montreal, P.Q., Canada, Oct. 23–25, 1978.

[4] D. J. Goodman, P. S. Henry, and V. K. Prabhu, "Frequency-hopped multilevel FSK for mobile radio," *Bell Syst. Tech. J.*, vol. 59, pp. 1257–1275, Sept. 1980.

[5] M. Harwit and N. J. A. Sloane, *Hadamard Transform Optics*. New York: Academic, 1979, pp. 17,210.

[6] G. Einarsson, "Address assignment for a time-frequency-coded, spread-spectrum system," *Bell Syst. Tech. J.*, vol. 59, pp. 1241–1255, Sept. 1980.

[7] H. E. Rowe, "Bounds on the number of users in spread-spectrum systems," in *Proc. Nat. Telecommun. Conf.*, Nov. 30–Dec. 4, 1980, pp. 69.6.1–69.6.6.

Part 3
Spread Spectrum (CDMA)

The Spread Spectrum Concept

ROBERT A. SCHOLTZ, SENIOR MEMBER, IEEE

Abstract—This paper describes an idealized spread-spectrum communication system. The processing gain concept is developed as a measure of a well-designed system's robust performance against independent wide-sense stationary interference. Multipath and repeater jammer rejection, partial correlation problems, and security requirements are related to spread-spectrum code properties.

1. INTRODUCTION

THOUGH well known for more than twenty years, the subject of spread spectrum (SS) communications has not enjoyed the limelight until recently.[1-3] Undoubtedly the low profile of SS techniques has been due to its distinct military (classified) advantages which are based on a robust immunity to interference and jamming. The recent upsurge in open discussions of SS techniques has probably been caused in part by applications of the concept to multiple-user communication situations where large amounts of interference are encountered, and in part by a rapidly advancing technology which is making more intricate signal processing feasible.

The objective of this tutorial paper is to explain how immunity to interference is achieved through use of SS techniques and to present a reasonably rigorous spectral analysis of the SS system. In the process, several major technical problems confronting the system engineer will be revealed.

2. SPREAD SPECTRUM SIGNALS

Roughly speaking, a spread spectrum signal is generated by modulating a data signal onto a wideband carrier so that the resultant transmitted signal has bandwidth which is much larger than the data signal bandwidth and which is relatively insensitive to the data signal. A "mathematical" block diagram of the transmitter is shown in Figure 1. In the notation used here, u(e.g., $d(u, t)$) simply indicates that the quantity involved should be viewed as being random in some way. Double lines in block diagrams indicate in-phase and quadrature channel signals with the in-phase channel carrying the real part of the indicated complex signal, and the quadrature channel carrying the imaginary part. Thus the transmitted signal $s(u, t)$ is viewed as the real part of the product of three complex random signals:

$$s(u, t) = \text{Re}\,[d(u, t)c(u, t)e^{j(\omega_0 t+\varphi(u))}]\,. \tag{1}$$

The actual mechanization of the transmitter may differ considerably from that indicated in Figure 1 (e.g., single channel processing at an i.f. or r.f. frequency may be used to replace two channel processing at baseband), but the mathematical model of the signal given in (1) generally will be applicable.

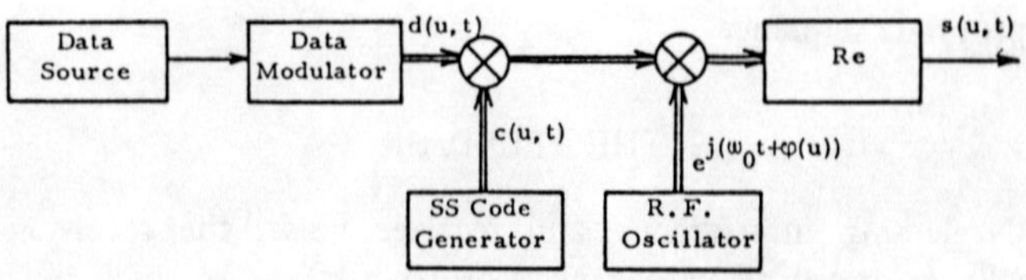

Figure 1. A Spread Spectrum Transmitter.

The expression for the SS transmitted signal $s(u, t)$ indicated in (1) would be the same as for non-SS modulation schemes if it were not for the extra factor $c(u, t)$ which we will refer to as the *spread spectrum code* signal. To simplify the demodulation process, SS codes usually are frequency or phase modulated, and not amplitude modulated, i.e.,

$$|\,c(u, t)\,| = 1, \tag{2}$$

the one obvious exception to this being time-hopping SS codes in which $|\,c(u, t)\,|$ takes on the values 0 or 1, turning the transmitter off and on in an irregular manner. By far the two most widely discussed code signals are the following.

a) *Direct sequence (DS) Signal with chip time T_c:*

$$c(u, t) = \sum_n a_n \boxed{T_c}\,(t - nT_c) \tag{3}$$

where $|\,a_n\,| = 1$ for all n. This signal contains no random parameters.

b) *Noncoherent Frequency Hopping (FH) Signal with hop-time T_h:*

$$c(u, t) = \sum_n e^{j(\omega_n t+\varphi_n(u))}\,\boxed{T_h}\,(t - nT_h) \tag{4}$$

where $\varphi_n(u)$ is a sequence of independent random phase variables, uniform on $(-\pi, \pi)$.

The notation $\boxed{T}(t)$ is used here to denote a square T-second pulse of unit amplitude, centered at the time origin, and hence it is easily verified that both the DS and FH signals satisfy the constant power condition (2).

The complex sequence $\{a_n\}$ in the DS case, or the frequency sequence $\{\omega_n\}$, in the FH case, must be agreed upon in advance by transmitter and receiver, and in fact have a status similar to that of a key in a cryptographic system. That is, with knowledge of the appropriate sequence, demodulation is possible and without knowledge of that sequence, demodulation is extremely difficult. From a cryptographic viewpoint it would be nice to make the SS code sequences purely random with no mathematical structure. However since all

Manuscript received February 19, 1976; revised March 4, 1977. This work was supported in part by the Army Research Office under Grants DA-ARO-D-31-124-73-G153 and DAAG29-76-G-0246.

The author is with the Department of Electrical Engineering, University of Southern California, Los Angeles, CA 90007, and with the Missile Systems Group, Hughes Aircraft Company, Canoga Park, CA 91304.

Reprinted from *IEEE Trans. Commun.*, vol. COM-25, no. 8, pp. 748–755, August 1977.

systems have a finite memory constraint, all practical SS code sequences have some periodic structure, i.e.,

$$a_n = a_{n+N} \text{ or } \omega_n = \omega_{n+N} \tag{5}$$

for all n where N will be used to denote the period of the appropriate sequence.

3. THE RECEIVER

Neglecting interference and receiver noise, the receiver ideally is presented with a waveform $r(u, t)$ from which the data modulation $d(u, t)$ must be extracted. We assume that

$$r(u, t) = \text{Re}\,\{c(u, t - \tau(u))d(u, t - \tau(u)) \cdot e^{j((\omega_0 + \omega_d(u))t + \theta(u))}\}, \tag{6}$$

i.e., the channel inserts a random delay and Doppler shift. This simple model is sufficient to illustrate the demodulation difficulties which the receiver encounters.

A mathematical block diagram of an SS receiver is shown in Figure 2. Again in-phase and quadrature baseband signaling has been chosen for tutorial simplicity. The indicated multiplications (mixing operations) are the receiver's attempt to first reduce the received signal to baseband and then strip the SS code from the data signal. The baseband filter can be considered to be the basic data detection filter (e.g., a matched filter in the digital signal case), possessing a bandwidth comparable to the bandwidth of the data modulation. Assuming that the RF filter passes $r(u, t)$ without distortion, the output of the baseband filter is

$$v(u, t) = \int_{-\infty}^{\infty} 2h(t - \alpha) r(u, \alpha) c_r^*(u, \alpha - \hat{\tau}) \cdot e^{-j((\omega_0 + \hat{\omega}_d)\alpha + \hat{\theta})}\, d\alpha \tag{7}$$

where $h(t)$ is the impulse response of the baseband filter, $c_r(u, t)$ is a receiver generated replica of the transmitter's SS code, and $(\ )^*$ denotes conjugation. Assuming that second harmonics of the carrier frequency are eliminated and ideal mixing takes place, $v(u, t)$ in (7) can be evaluated for the signal (6) to give

$$v(u, t) = \int_{-\infty}^{\infty} h(t - \alpha) c(u, \alpha - \tau(u)) d(u, \alpha - \tau(u)) \cdot c_r^*(u, \alpha - \hat{\tau}) e^{j((\omega_d(u) - \hat{\omega}_d)\alpha + \theta(u) - \hat{\theta})}\, d\alpha. \tag{8}$$

Considerable analytical effort is required to determine the effects of the nonideal RF filters and the nonideal mixers in real systems.

The values $\hat{\omega}_d$, $\hat{\theta}$, $\hat{\tau}$ are provided by synchronization tracking loops (note shown in Figure 2) in an attempt to align the receiver VCO and SS code generator with the corresponding received signal components. With regard to terminology:

$$\hat{\tau} \approx \tau(u) \Rightarrow \text{SS code sync} \tag{9a}$$

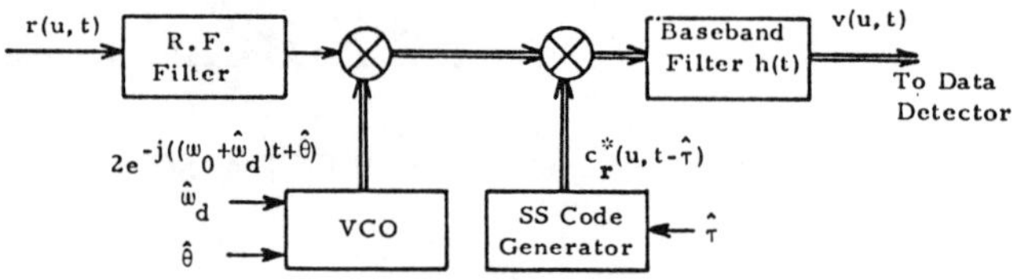

Figure 2. A Spread Spectrum Receiver.

$$\hat{\omega}_d \approx \omega_d(u) \Rightarrow \text{frequency lock} \tag{9b}$$

$$\hat{\theta} \approx \theta(u) \Rightarrow \text{phase lock.} \tag{9c}$$

When the receiver is perfectly locked (exact equalities in (9)) and the receiver SS code signal $c_r(u, t)$ is an exact replica of a constant power (2) transmitted SS code, the output of the baseband filter is

$$v(u, t) = \int_{-\infty}^{\infty} h(t - \alpha) d(u, \alpha - \hat{\tau})\, d\alpha \quad \text{(perfect lock).} \tag{10}$$

Obviously in this "perfect" case, an ideal equivalent complex baseband channel from the data modulator in the transmitter to the baseband filter in the receiver has been created.

In many cases the receiver cannot reproduce a perfect replica of the transmitted SS code $c(u, t)$ due to random parameters in the code. For example the noncoherent FH code contains random phase jumps inserted by the transmitter when frequency changes occur. Not knowing these phase jumps *a priori* and not making an attempt to learn them, the receiver will output the signal

$$v(u, t) = \int_{-\infty}^{\infty} h(t - \alpha) d(u, \alpha - \hat{\tau}) \sum_n e^{j\varphi_n'(u)} \cdot \boxed{T_h}\,(\alpha - nT_h - \hat{\tau})\, d\alpha \quad \text{(phase incoherent)} \tag{11}$$

where $\{\varphi_n'(u)\}$ is a sequence of independent random phases. In effect here we have assumed ideal SS code sync and frequency lock, but no phase lock. Hence the baseband equivalent channel from data modulator to baseband filter now contains random phase jumps every T_h seconds. These jumps must be considered part of the data signal [as in (11)] and have the effect of increasing the bandwidth of the baseband filter $h(t)$ to the order of $1/T_h$ and requiring that envelope detection and possibly postdetection integration techniques be employed in data demodulation.

4. PRIVACY VS. FLEXIBILITY

If the SS system is built so that the SS generator-modulator in the transmitter is operated independently of the data modulator, then it may be possible to use various modulation formats with the same SS code system and thereby build flexibility into the system. This is possible provided that the code synchronizer (which provides τ in the receiver) is compatible with a variety of data modulation formats, and provided that the data demodulators can all be built to cope with any uncertainties (e.g., phase jumps) left in the signal by the SS code modulator-demodulator system.

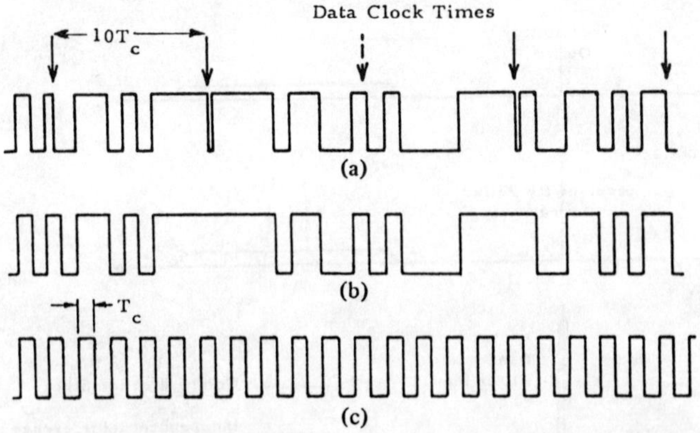

Figure 3. SS Transmitted Modulation with (a) "independent" and (b) "coincident" data and SS code clocks. (c) SS Code Clocks.

One problem with making the data modulator and SS code generator independent is that it may be possible for anyone to read the data directly from a clean copy of the received signal. For example, suppose the data and a DS-SS code are both biphase modulated onto the carrier, with the SS code bit rate at about 10 times the data bit rate, as shown in Figure 3. (The code rate used in this illustration is extremely low. Code bit rates are normally on the order of 1000 or more times higher than the data rate.) If one first estimates the SS code clock [Figure 3 (c)], then it is an easy matter to determine unscheduled phase shifts [solid arrows in Figure 3 (a)] in the transmitted modulation which must be due to phase modulation of the data. It is then possible in Figure 3 (a) to determine the data clock pulses (solid and dashed arrows) and determine the sequence of data bit changes. However if the data clock is divided down from the SS clock so that possible phase change times in the data modulation line up with phase change times in the SS code modulation, no unscheduled phase shifts occur [see Figure 3(b)]. Hence making the SS code and data clocks coincident means that the data cannot be read unless the SS code is known by the receiver.

Systems which have coincident data and SS code clocks are often said to have a data *privacy* feature. In fact systems with privacy features are also simpler to build, with much of the modulation and demodulation equipment shared by the SS code and data signal. Typical privacy system transmitted modulations are the following.

a) *DD-SS System with PSK Data Modulation:*

$$d(u, t)c(u, t) = \sum_n a_n d_{[n/M]}(u) \boxed{T_c} (t - nT_c) \tag{12}$$

where $\{a_n\}$ is the SS code, $\{d_m(u)\}$ is the data sequence, and $[n/M]$ denotes the integer part of n/M. Hence the data clock rate is $1/M^{\text{th}}$ of the SS code clock rate. This difference in clock rates (M is large) is necessary to produce spread spectrum effects.

b) *Noncoherent FH-SS System with FSK Data Modulation:*

$$d(u, t)c(u, t) = \sum_n e^{j((\omega_n + d_{[n/M]}(u))t + \varphi_n(u))} \boxed{T_h} (t - nT_n) \tag{13}$$

where $\{\omega_n\}$ is the SS code and $\{d_n(u)\}$ is the data sequence. Again the data clock rate is $1/M^{\text{th}}$ of the SS clock rate. However spectral spreading can be achieved by use of a variety of frequencies $\{\omega_n\}$, and M is sometimes 1 in this system. Since one data symbol time MT_h contains $M-1$ internal phase jumps, the data demodulator will have to employ coherent detection within T_h second internals, followed by postdetection integration over M hop times.

Both of these privacy systems have combined the SS code and data modulators into single units.

5. WIDE-SENSE STATIONARY INDEPENDENT INTERFERENCE

Assuming that all tracking loops are operating in a stable fashion with $\hat{\tau}$, $\hat{\omega}_d$, and $\hat{\theta}$ relatively constant, we can ignore feedback effects through these loops and treat the signal processing from receiver input to baseband filter output as linear processing. Hence it is possible to separately analyze the effects of various components of the input signal on the output. Here we consider any wide-sense stationary random process $N(u, t)$ appearing at the output of the RF filter

$$N(u, t) = \text{Re}\{n(u, t)e^{j\omega_0 t}\} \tag{14}$$

where the spectral relationship between the real valued RF signal $N(u, t)$ and the complex valued baseband signal $n(u, t)$ is given by

$$S_N(f) = \tfrac{1}{4}[S_n(f - f_0) + S_n(-f - f_0)] \tag{15}$$

with $f_0 = \omega_0/2\pi$. Hence the baseband power spectral density $S_n(f)$ cannot have bandwidth exceeding B_{RF}, the bandwidth of the RF filter which determines the maximum spectral width of $S_N(f)$. We further assume that $n(u, t)$ is independent of all other receiver inputs and random parameters within the receiver. Certainly the usual receiver noise, some types of jamming, and unrelated interference can be modeled in this fashion. Since the processing is linear, we can normalize the signal so that $n(u, t)$ has unit average power.

$$E\{|n(u, t)|^2\} = 1 = \int_{-\infty}^{\infty} S_n(f)\, df \tag{16}$$

We shall now determine the amount of power in the baseband filter output $v_n(u, t)$ caused by $n(u, t)$.

Following the processing indicated in Figure 2, the baseband filter output is related to $n(u, t)$ by

$$v_n(u, t) = \int_{-\infty}^{\infty} h(t - \alpha) n(u, \alpha) e^{-j(\hat{\omega}_d \alpha + \hat{\theta})} c_r^*(u, \alpha - \hat{\tau})\, d\alpha. \tag{17}$$

After a change of variables, the expected squared value of this output under the stationarity and independence assumptions is easily shown to be:

$$E\{|v_n(u,t)|^2\} = \iint_{-\infty}^{\infty} h(\alpha)h^*(\beta)R_n(\beta-\alpha)$$
$$\cdot e^{j\hat{\omega}_d(\alpha-\beta)}E\{c_r^*(u,t-\hat{\tau}-\alpha)$$
$$\cdot c_r(u,t-\hat{\tau}-\beta)\}\,d\alpha\,d\beta \qquad (18)$$

where $R_n(\tau)$ is the ensemble autocorrelation of $n(u, t)$, i.e. the Fourier transform of $S_n(f)$. The result of this computation depends on $t - \hat{\tau}$, and hence the resultant output interference power is a periodic function of t, synchronized with the period of the SS receiver code.

By settling for a time-averaged value of the output power (18), it is possible to significantly simplify both the computation and the interpretation of results. Denoting time average by $\langle\,\rangle$, time averaging (18) gives

$$\sigma_{v_n}{}^2 \triangleq \langle E\{|v_n(u,t)|^2\}\rangle = \iint_{-\infty}^{\infty} h(\alpha)h^*(\beta)R_n(\beta-\alpha)$$
$$\cdot e^{j\hat{\omega}_d(\beta-\alpha)}R_{c_r}(\alpha-\beta)\,d\alpha\,d\beta \qquad (19)$$

where

$$R_{c_r}(\tau) \triangleq \langle E\{c_r(u,t+\tau)c_r^*(u,t)\}\rangle \qquad (20)$$

is no longer a function of t. The Fourier transform of $R_{c_r}(\tau)$ is the power spectral density $S_{c_r}(f)$ of the SS code signal replica. In terms of Fourier transforms (19) reduces to

$$\sigma_{v_n}{}^2 = \int_{-\infty}^{\infty} |H(f)|^2[S_{c_r}(f) * S_n(f-\hat{f}_d)]\,df \qquad (21)$$

where ( )*( ) denotes convolution, $H(f)$ is the system function of the baseband filter, and $\hat{f}_d = \hat{\omega}_d/2\pi$. Hence the *interference power spectral density $S_n(f - \hat{f}_d)$ is spread by the convolution with the code power spectral density $S_{c_r}(f)$ and then reduced by the baseband filter.*

A typical sequence of power spectral densities for the processing of a SS signal and narrowband interference is shown in Figure 4. The key operation is obviously the mixing process with the SS code which compresses the desired signal into the bandwidth of the baseband filter and simultaneously spreads the interference power.

By interchanging the order of convolution and integration over f in (21), it is possible to perform a worst case analysis:

$$\sigma_{v_n}{}^2 = \int_{-\infty}^{\infty} S_n(\alpha-\hat{f}_d)g(\alpha)\,d\alpha \qquad (22a)$$

where

$$g(\alpha) \triangleq \int_{-\infty}^{\infty} |H(f)|^2\,S_{c_r}(f-\alpha)\,df. \qquad (22b)$$

Since $n(u, t)$ has unit power [see (16)], the highest possible

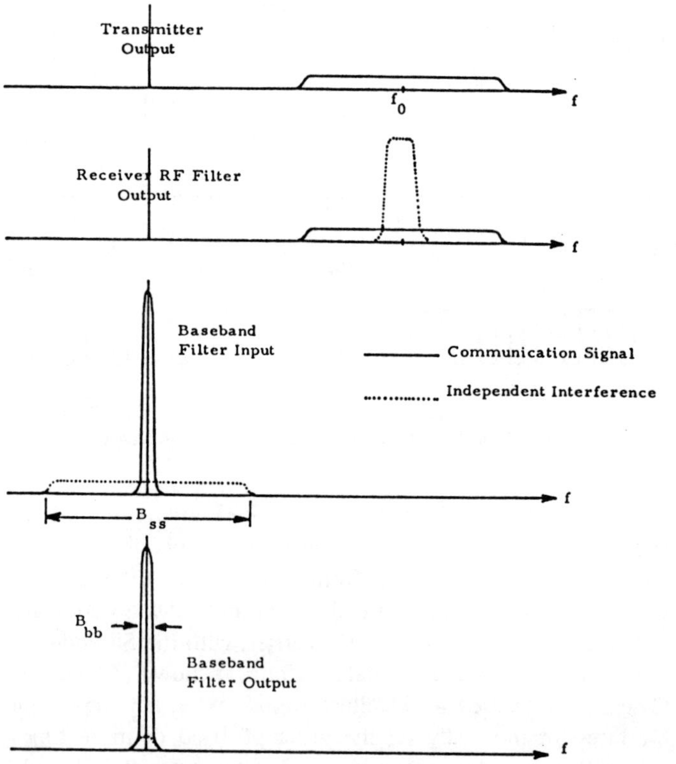

Figure 4. Power Spectral Density Plots for Signal and Interference; Signal-to-Interference Ratio = 1 at RF filter output; Processing Gain = 16.

value of $\sigma_{v_n}{}^2$ occurs when $S_n(\alpha - \hat{f}_d)$ is concentrated at the maximum of $g(\alpha)$. Hence

$$\sigma_{v_n}{}^2 \leqslant \max_{\alpha} \int_{-\infty}^{\infty} |H(f)|^2 S_{c_r}(f-\alpha)\,df. \qquad (23)$$

To minimize this bound by good SS code selection, one must make the power spectral density $S_{c_r}(f)$ as flat as possible over the bandwidth B_{ss} of the SS code, i.e., the SS bandwidth of the system. In other words, since $c_r(u, t)$ has unit power in bandwidth B_{ss}, $\max_\alpha g(\alpha)$ can be underbounded by

$$\max_{\alpha} g(\alpha) \geqslant \frac{1}{B_{ss}}\int_{-\infty}^{\infty} |H(f)|^2\,df = \frac{B_{bb}}{B_{ss}} \qquad (24)$$

with equality possible for ideal SS code design. The last equality in (24) is made by assuming that the baseband filter has a maximum gain of unity, and using B_{bb} for the noise bandwidth of the baseband filter. Furthermore *for the ideal flat SS code spectrum, $g(\alpha)$ is approximately B_{bb}/B_{ss} for all α of interest and the baseband filter power output is approximately independent of the shape of the interference's spectral composition.*

$$\sigma_{v_n}{}^2 \approx \frac{B_{bb}}{B_{ss}} \qquad \text{(for ideal SS code).} \qquad (25)$$

The synchronous receiver, as described in this section and the previous section recovers the data modulation in the baseband filter without attenuation, i.e., with unit gain. On the other hand, wide-sense stationary independent interference modulation is reduced on the average by the factor in (25). Hence the receiver provides a

$$\text{Processing Gain} \triangleq \frac{B_{ss}}{B_{bb}} \tag{26}$$

advantage to the desired data signal.

One byproduct of SS system design with high processing gain is the *inherent nonobservability of the transmitted signal.* Suppose for example that an SS system with a 30 dB processing gain is operating with a 10 dB signal-to-noise energy ratio at the output of the baseband filter. This implies that the signal-to-noise energy ratio in the RF portion of the receiver is −20 dB. Another receiver, with an identical antenna and RF section but not containing the SS code multiplier, would have an extremely difficult time determining the presence of the RF signal at −20 dB SNR. Somehow the surreptitious listener would have to develop a 20 dB advantage over the SS receiver by reducing the length of the propagation path, using a higher gain antenna, a cooler receiver, etc. Even if this were possible and the listener could detect the SS signal, he could not demodulate data in the privacy format without first knowing the SS code.

6. CODE-RELATED INTERFERENCE

Many types of interference cannot be modeled by wide-sense stationary independent random processes. Two examples of such interference are multipath and repeater jamming. In both cases the interfering signal is modulated with the proper SS code but is received with a greater delay than the direct path and possibly at a different carrier frequency. Hence an adequate model for this interference is

$$m(u, t) = \mathrm{Re}\,\{c(u, t - \tau(u) - \tau_M) d(u, t - \tau(u) - \tau_M) \cdot e^{j((\omega_0 + \omega_d(u) + \omega_M)t + \theta_M(u))}\} \tag{27}$$

where as in Section 4, $\tau(u)$ and $\omega_d(u)$ are the direct path parameters, and now τ_M and ω_M are differential parameters indicating the added time delay and frequency shift of the multipath or jamming relative to the direct path. The modulation $d(u, t)$ can be viewed as data modulation, jammer modulation, or both as the case may be.

Assuming that the receiver is properly synchronized in frequency and time to the direct path parameters, the component of the output signal due to $m(u, t)$ is given by

$$v_m(u, t) = \int_{-\infty}^{\infty} h(t - \alpha) c(u, \alpha - \hat{\tau} - \tau_M) c_r^*(u, \alpha - \hat{\tau}) \cdot d(u, \alpha - \hat{\tau} - \tau_M) e^{j(\omega_M \alpha + \theta_M(u))}\, d\alpha. \tag{28}$$

There are two approaches to output power computations depending on the relation between the "data" modulation $d(u, t)$ and the SS code modulation. In either case, ensemble averages will indicate that $E\{|v_m(u, t)|^2\}$ is a function, not only of τ_M and f_M, but also of $t - \hat{\tau}$. Again as in the case of non-code related interference, the expected power in the output interference is a periodic function of t, relative to the epoch of the SS receiver code. Hence we will compute appropriate time averaged values of the expected output power.

(a) Case I: Wide-Sense Stationary Independent Modulation

Assume $d(u, t)$ is a wide-sense stationary random process, independent of $c(u, t)$ and $c_r(u, t)$. Then

$$\sigma_m^2(\tau_M, \omega_M) \triangleq \langle E\{|v_m(u, t)|^2\}\rangle = \iint_{-\infty}^{\infty} h(\alpha) h^*(\beta) R_{cc_r^*(\tau_M)}(\beta - \alpha) \cdot R_d(\beta - \alpha) e^{-j\omega_M(\beta - \alpha)}\, d\alpha\, d\beta \tag{29}$$

where $R_d(\tau)$ is the correlation function of $d(u, t)$ and

$$R_{cc_r^*}(\tau) = \langle E\{c(u, t + \tau - \tau_M) c_r^*(u, t + \tau) c^*(u, t - \tau_M) \cdot c_r(u, t)\}\rangle \tag{30}$$

is the correlation function of the SS code product $c(u, t - \tau_M) c_r^*(u, t)$. Defining $S_{cc_r^*(\tau_M)}(f)$ and $S_d(f)$ to be the spectral densities of the corresponding correlation functions, (29) can be simplified to:

$$\sigma_m^2(\tau_M, \omega_M) = \int_{-\infty}^{\infty} |H(f)|^2\, [S_{cc_r^*(\tau_M)}(f) * S_d(f + f_M)]\, df. \tag{31}$$

Notice that equations (31) and (21) are formally the same and thus the analysis techniques of the previous section are now applicable. Hence by analogy a processing gain advantage of K/B_{bb} against a properly SS coded signal with relative delay τ_M can be guaranteed if the power spectral density of the code product signal $c(u, t - \tau_M) c_r^*(u, t)$ (a unit power signal) is no higher than $1/K$. Even with a flat power spectral density this requires that the code product signal have bandwidth of at least K.

(b) Case II: Modulation in the Privacy Format

Analyses involving privacy formats can become notationally complicated if done in great generality. Here we shall analyze a DS-SS system carefully, indicating the types of manipulations which lead to simplified results in both DS and FH cases.

The output (28) of a DS-SS privacy system with multipath interference at the input is:

$$v_m(u, t) = \int_{-\infty}^{\infty} h(t - \alpha) \left[\sum_n a_n d_{[n/M]}(u)\, \boxed{T_c}\, (\alpha - \hat{\tau} - \tau_M - nT_c)\right] \left[\sum_m a_m\, \boxed{T_c}\, (\alpha - \hat{\tau} - mT_c)\right]^* \cdot e^{j(\omega_M \alpha + \theta_M(u))}\, d\alpha. \tag{32}$$

We assume that the baseband filter-sampler is actually mechanized as an integrate and dump circuit with integration time equal to the duration of a data bit, namely MT_c. (In a noncoherent FH system the integration time is limited by phase jumps to T_h.) Hence we assume in (32) that when $v_m(u, t)$ is sampled at $t = \hat{\tau} + (iM - \frac{1}{2})T_c$, then the baseband filter integrates over the $m = (i-1)_M, (i-1)M+1, \cdots, iM-1$ pulses in the SS code pulse train. After inserting a scale factor to give the filter a maximum gain of unity, we have

$$h(t) = \frac{1}{MT_c} \boxed{MT_c}\left(t - \frac{MT_c}{2}\right). \tag{33}$$

Another simplification comes from the fact that in (32), each pulse in the replica code $c_r(u, \alpha - \hat{\tau})$ is overlapped by at most two pulses in the multipath pulse train. The overlapping pulses are easily determined mathematically by first computing

$$\tau_M = K_M T_c + \tau_0, \qquad 0 \leqslant \tau_0 < T_c \tag{34}$$

which indicates that τ_M is a delay of K_M (integer) chip times plus an additional delay τ_0 of less than a chip time. The result of these simplifications to (32) is:

$$\begin{aligned} v_i(u) &\triangleq v_m(u, \hat{\tau} + (iM - \tfrac{1}{2})T_c) \\ &= e^{j(\theta_M(u) + \omega_M \hat{\tau})}[\psi(i, M, K_M, \omega_M T_c)\chi(\tau_0, \omega_M) \\ &\quad + \psi(i, M, K_M + 1, \omega_M T_c)\chi(\tau_0 - T_c, \omega_M)] \end{aligned} \tag{35}$$

where

$$\psi(i, M, K, \gamma) = \frac{1}{M} \sum_{m=(i-1)M}^{iM-1} d_{\lfloor \frac{m-K}{M} \rfloor}(u) a_{m-K} a_m^* e^{jm\gamma} \tag{36}$$

$$\chi(\tau, \omega) = \begin{cases} e^{j\omega\tau/2}\left(1 - \dfrac{|\tau|}{T_c}\right) \dfrac{\sin(\omega(T_c - |\tau|)/2)}{\omega(T_c - |\tau|)/2}, & \tau \leqslant T_c \\ 0, & \tau > T_c. \end{cases} \tag{37}$$

The well-versed reader will recognize $\chi(\tau, \omega)$ as the radar ambiguity function of a T_c second square pulse, and, since

$$|\chi(\tau, \omega)| \leqslant \chi(0, 0) = 1 \tag{38}$$

it is a simple matter to work out relatively tight bounds on $|v_j(u)|$ in terms of the $\psi(i, M, K, \gamma)$ sums by themselves.

Though the precise computations are straightforward, let us further reduce the amount of bookkeeping by specializing to the case $\tau_0 = 0$ and using the bound

$$\begin{aligned} |v_i(u)| &\leqslant |\psi(i, M, K_M, \omega_M T_c)| \\ &= \Bigg| d_{i-2-A}(u) \frac{1}{M} \sum_{m=(i-1)M}^{(i-1)M+B-1} a_{m-K_M} a_m^* \\ &\quad \cdot e^{jm\omega_M T_c} + d_{i-1-A}(u) \frac{1}{M} \\ &\quad \cdot \sum_{m=(i-1)M+B}^{iM-1} a_{m-K_M} a_m^* e^{jm\omega_M T_c} \Bigg|. \end{aligned} \tag{39}$$

Here A and B are integers determined using the equation

$$K_M = AM + B, \qquad 0 \leqslant B < M. \tag{40}$$

Since the data variables are outside the sums in (39), it is now a simple matter to compute moments of the data filter output.

If data bits are uncorrelated with unit variance and if M and N are relatively prime, then the time (i) and ensemble (u) averaged, code-related interference level is given by

$$\begin{aligned} \langle E\{|v_i(u)|^2\}\rangle = \frac{1}{N} \sum_{n=0}^{N-1} & [|f(n, B, K_M, \omega_M T_c)|^2 \\ & + |f(n, M-B, K_M, \omega_M T_c)|^2] \end{aligned} \tag{41}$$

where

$$f(n, B, K_M, \omega_M T_c) = \frac{1}{M} \sum_{m=0}^{B-1} a_{m+n-K_M} a_{m+n}^* e^{jm\omega_M T_c}. \tag{42}$$

It generally is very difficult to evaluate (42) without resorting to a computer, and in cases where the code period is large, even complete computer results may be impossible to obtain.

When $\{a_n\}$ is a maximum length shift register (MLSR) sequence[4] and $K_M \neq 0$, then $\{a_{n-K_M} a_n^*\}$ is another MLSR sequence. Further assuming that $\omega_M = 0$, (41) can be evaluated using the results of Lindholm[12] to give

$$\begin{aligned} \langle E\{|v_i(u)|^2\}\rangle = \frac{1}{M}\Bigg[& \frac{B}{M}\left(1 - \frac{B-1}{N}\right) \\ & + \left(1 - \frac{B}{M}\right)\left(1 - \frac{M-B-1}{N}\right)\Bigg] \end{aligned} \tag{43}$$

which is on the order of $1/M$. Hence in this case with $K_M \neq 0$ mod N, a processing gain of M is possible. This again illustrates the fact that code related interference rejection requires that $\{a_{n+K_M} a_n^*\}$ also have the properties of SS code.

7. DIRECT SEQUENCE SPECTRAL COMPUTATIONS

Evaluation of processing gain against various types of interference requires that certain SS code power spectral density computations be performed. In the DS case the transmitted code $c(u, t)$ and the receiver's replica $c_r(u, t)$ are identical periodic coherent waveforms having no random parameters. Hence the SS code correlation function required in (19) is given by

$$\begin{aligned} R_{c_r}(\tau) &= \frac{1}{NT_c} \int_0^{NT_c} c_r(u, t+\tau) c_r^*(u, t)\, dt \\ &= \left(1 - \frac{\tau_0}{T_c}\right) R_a(K) + \frac{\tau_0}{T_c} R_a(K+1) \end{aligned} \tag{44a}$$

where

$$\tau = KT_c + \tau_0, \qquad 0 \leqslant \tau_0 < T_c \tag{44b}$$

and

$$R_a(K) = \frac{1}{N} \sum_{m=0}^{N-1} a_{m+K} a_m^*. \tag{44c}$$

It is apparent from (44) that $R_{c_r}(\tau)$ is a periodic correlation function with period NT_c in the variable τ, since $\{a_n\}$ is a periodic sequence.

The periodic form of $R_{c_r}(\tau)$ implies that the power spectral density of $c_r(u, t)$ must be a line spectral density with

$$S_{c_r}(f) = \sum_n P_n \delta\left(f - \frac{n}{NT_c}\right) \tag{45}$$

where

$$P_n = \frac{1}{NT_c} \int_0^{NT_c} R_{c_r}(\tau) e^{-j2\pi n(\tau/NT_c)}\, d\tau$$

$$= S_a(n) \left(\frac{\sin (\pi n/N)}{\pi n/N} \right)^2 \tag{46}$$

Here $S_a(n)$ is the discrete Fourier transform of one period of the SS code sequence:

$$S_a(n) = \frac{1}{N} \sum_{K=0}^{N-1} R_a(K) e^{j2\pi n(K/N)}. \tag{47}$$

The spectral density $S_{c_r c_r^*(\tau_M)}(f)$ required to evaluate the effects of code related interference can be evaluated by applying these same techniques to the new sequence $\{a_{n-K_M} a_n^*\}$ when $\tau_M = K_M T_c$.

Since $S_{c_r}(f)$ is a line spectral density, i.e., max $S_{c_r}(f) = \infty$, it would seem possible that the transmitted signal is in fact observable by a surreptitious listener and also that independent wide-sense-stationary interference may cause significantly more trouble than is predicted in Section 5. Such is not the case if the following precautions are observed: (a) *Make sure that the data modulation bandwidth is much larger than* $1/NT_c$ (the reciprocal of the SS code modulation's period). This insures that the width of $|H(f)|^2$ is much greater than the line-spacing in $S_{c_r}(f)$ and hence that $g(\alpha)$ in (22b) is a smooth function of α. Hence independent interference cannot cause unexpected difficulties. In this case data modulation will also spread the lines in the SS code's spectral density to yield a smooth transmitted signal power spectral density, devoid of lines. (b) If it is impossible to guarantee data modulation of sufficient bandwidth, then *make sure that NT_c is very large,* implying very close line-spacing in $S_{c_r}(f)$. Several periods of the SS code will be required to surreptitiously detect the SS transmitted signal using a filter narrow enough to isolate a spectral line and gain an advantageous signal-to-noise ratio for detection. If NT_c is large enough, this type of detection will be impossible due to oscillator drifts, Doppler shift variations, etc.

8. NONCOHERENT FREQUENCY HOPPING SPECTRAL COMPUTATIONS

The independent uniformly distributed random phase jumps which occur in noncoherent FH signals lead to significant mathematical simplifications and eliminate the possibility of a line structure to the spectral density. Computation of the ensemble autocorrelation gives

$$E\{c_r(u, t+\tau) c_r^*(u, t)\}$$

$$= \sum_n e^{j\omega_n \tau} \boxed{T_h - |\tau|}\, (t - nT_h + \tau/2). \tag{48}$$

Here it is understood that $\boxed{X}\,(t)$ is identically zero when X is negative. The function in (48) is periodic in t since the hopping sequence is periodic, but it is not periodic in τ. Let us denote the set of frequencies present in the FH sequence $\{\omega_n\}$ by Ω and let $k(\omega)$ be the number of times that the frequency ω occurs in one period of the FH sequence. Then the time average of (48) is easily computed:

$$R_{c_r}(\tau) = \begin{cases} \left(1 - \dfrac{|\tau|}{T_h}\right) \displaystyle\sum_{\omega' \epsilon \Omega} \frac{k(\omega')}{N} e^{j\omega' \tau}, & |\tau| < T_h \\ 0, & |\tau| \geqslant T_h. \end{cases} \tag{49}$$

Fourier transforming gives the spectral density:

$$S_{c_r}(f) = T_h \sum_{\omega' \epsilon \Omega} \frac{k(\omega')}{N} \left| \frac{\sin (\pi(f - f')T_h)}{\pi(f - f')T_h} \right|^2 \tag{50}$$

where $f' = \omega'/2\pi$. Notice that this power spectral density does not depend on the order of the frequencies in the FH sequence, but only on the relative usage $k(\omega')/N$ of frequencies $\omega' \epsilon \Omega$.

As in the case of DS-SS codes, the product of a FH-SS code signal with a $K_M T_h$ shifted version of itself is a new FH-SS signal. When $\tau_M = K_M T_h$, then the spectral density $S_{c_r c_r^*(\tau_M)}(f)$ required in the study of code-related interference can be determined by applying the above computational techniques to the sequence $\{\omega_{n-K_M} - \omega_n\}$. One conclusion about FH code design which can be drawn at this point is that codes with ω_n increasing or decreasing linearly with n are to be avoided if code-related interference is expected.

9. CLOSING COMMENTS

Even in the simple SS system models discussed here, there are two major classes of theoretical problems which we have not yet touched: (1) the synchronization problems involving phase, frequency and SS code acquisition and tracking, and (2) the problems of SS code design. The following are just a few of the problems related specifically to the SS concept.

In an SS system, the signal energy in any portion of the RF spectrum is dominated by the noise energy. This fact makes it extremely difficult to first lock up the RF carrier tracking loop which supplies $\hat{\omega}_d$ and possibly $\hat{\theta}$ in the receiver, even using sophisticated loops.[13] It seems that *SS code sync must be established first* to concentrate the received energy in a narrow band and allow suppressed carrier tracking to be established. The current literature[13-15] contains several techniques at various levels of sophistication for handling special cases of this problem.

Since an SS system uses bandwidth resources extravagantly, the frequency spectrum generally must be shared with other systems, leading to a (SS) code division multiple access (CDMA) mode of operation. A well known CDMA design for DS-SS codes is the Gold codes[5, 6] whose cross-correlation properties are nearly optimal.[7] Good FH-SS code sets have been proposed by Lempel and Greenberger[8] and by Solomon.[9] Unfortunately these code sets and the simple MLSR sequences for single DS-SS systems *are not cryptographically secure.* The structure of a complete code sequence can be determined from an observation of a small segment of the sequence.[11] It is conceivable that an intelligent listener with an operating gain advantage can use this fact either to read the data or jam the system.

Another CDMA problem is the *near-far problem,* which occurs when the power levels of the undesired signals of the other users are very large compared to the power level of the desired signal. When the processing gain of the system is not large enough to counteract this power imbalance, then time hopping modes of operation may be required to eliminate all but occasional interference. One possible time hopping code is discussed by Cohen *et al.*[10]

While this article has not touched on hardware design problems, it already should be obvious that significant problems must be solved in any system design which hopes to achieve the significant advantages of the SS concept.

REFERENCES

(1) R. C. Dixon, *Spread Spectrum Systems,* John Wiley and Sons, 1976.

(2) "Spread Spectrum Communications," AGARD Lecture series no. 58, National Technical Information Service AD-766-914, July 1973.

(3) *Proceedings of the 1973 Symposium on Spread Spectrum Communications,* Naval Electronics Laboratory Center, San Diego, California, March 13-16, 1973.

(4) S. W. Golomb, *Shift Register Sequences,* Holden-Day, 1967, part II.

(5) R. Gold, "Optimum Binary Sequences for Spread Spectrum Multiplexing," *IEEE Trans. on Information Theory,* Oct. 1967, pp. 619-621.

(6) R. Gold, "Maximal Recursive Sequences with 3-Valued Recursive Crosscorrelation Functions," *IEEE Trans. on Information Theory,* January, 1968, pp. 154-156.

(7) L. R. Welch, "Lower Bounds on the Maximum Cross Correlation Between Signals," *IEEE Trans. on Information Theory,* May, 1974.

(8) A. Lempel and H. Greenberger, "Families of Sequences with Optimal Hamming-Correlation Properties," *IEEE Trans. on Information Theory,* Jan. 1974, pp. 90-94.

(9) G. Solomon, "Optimal Frequency Hopping Sequences for Multiple Access," in volume 1 of reference (3).

(10) A. R. Cohen, J. A. Heller, and A. J. Viterbi, "A New Coding Technique for Asynchronous Multiple Access Communication," *IEEE Trans. on Communication Technology,* October 1971, pp. 849-855.

(11) W. H. Mills, "Continued Fractions and Linear Recurrences," *Mathematics of Computation,* January, 1975, pp. 173-180.

(12) J. H. Lindholm, "An Analysis of the Pseudo-Randomness Properties of Subsequences of Long m-Sequences," *IEEE Trans. on Information Theory,* July 1968, pp. 569-576.

(13) W. C. Lindsey, *Synchronization Systems in Communication and Control,* Prentice-Hall, 1972, Section 3.8.

(14) J. J. Stiffler, *Theory of Synchronous Communications,* Prentice-Hall, 1971, Chapter 6.

(15) C. J. Kilgus, "Pseudo-Noise Code Acquisition Using Majority Logic Decoding," *Proc. of ICC,* June 1973, pp. 23.16-23.17.

Theory of Spread-Spectrum Communications—A Tutorial

RAYMOND L. PICKHOLTZ, FELLOW, IEEE, DONALD L. SCHILLING, FELLOW, IEEE, AND LAURENCE B. MILSTEIN, SENIOR MEMBER, IEEE

Abstract—**Spread-spectrum communications, with its inherent interference attenuation capability, has over the years become an increasingly popular technique for use in many different systems. Applications range from antijam systems, to code division multiple access systems, to systems designed to combat multipath. It is the intention of this paper to provide a tutorial treatment of the theory of spread-spectrum communications, including a discussion on the applications referred to above, on the properties of common spreading sequences, and on techniques that can be used for acquisition and tracking.**

I. INTRODUCTION

SPREAD-spectrum systems have been developed since about the mid-1950's. The initial applications have been to military antijamming tactical communications, to guidance systems, to experimental antimultipath systems, and to other applications [1]. A definition of spread spectrum that adequately reflects the characteristics of this technique is as follows:

> "Spread spectrum is a means of transmission in which the signal occupies a bandwidth in excess of the minimum necessary to send the information; the band spread is accomplished by means of a code which is independent of the data, and a synchronized reception with the code at the receiver is used for despreading and subsequent data recovery."

Under this definition, standard modulation schemes such as FM and PCM which also spread the spectrum of an information signal do not qualify as spread spectrum.

There are many reasons for spreading the spectrum, and if done properly, a multiplicity of benefits can accrue simultaneously. Some of these are

- Antijamming
- Antiinterference
- Low probability of intercept
- Multiple user random access communications with selective addressing capability
- High resolution ranging
- Accurate universal timing.

Manuscript received December 22, 1981; revised February 16, 1982.

R. L. Pickholtz is with the Department of Electrical Engineering and Computer Science, George Washington University, Washington, DC 20052.

D. L. Schilling is with the Department of Electrical Engineering, City College of New York, New York, NY 10031.

L. B. Milstein is with the Department of Electrical Engineering and Computer Science, University of California at San Diego, La Jolla, CA 92093.

The means by which the spectrum is spread is crucial. Several of the techniques are "direct-sequence" modulation in which a fast pseudorandomly generated sequence causes phase transitions in the carrier containing data, "frequency hopping," in which the carrier is caused to shift frequency in a pseudorandom way, and "time hopping," wherein bursts of signal are initiated at pseudorandom times. Hybrid combinations of these techniques are frequently used.

Although the current applications for spread spectrum continue to be primarily for military communications, there is a growing interest in the use of this technique for mobile radio networks (radio telephony, packet radio, and amateur radio), timing and positioning systems, some specialized applications in satellites, etc. While the use of spread spectrum naturally means that each transmission utilizes a large amount of spectrum, this may be compensated for by the interference reduction capability inherent in the use of spread-spectrum techniques, so that a considerable number of users might share the same spectral band. There are no easy answers to the question of whether spread spectrum is better or worse than conventional methods for such multiuser channels. However, the one issue that is clear is that spread spectrum affords an opportunity to give a desired signal a power advantage over many types of interference, including most intentional interference (i.e., jamming). In this paper, we confine ourselves to principles related to the design and analysis of various important aspects of a spread-spectrum communications system. The emphasis will be on direct-sequence techniques and frequency-hopping techniques.

The major systems questions associated with the design of a spread-spectrum system are: How is performance measured? What kind of coded sequences are used and what are their properties? How much jamming/interference protection is achievable? What is the performance of any user pair in an environment where there are many spread spectrum users (code division multiple access)? To what extent does spread spectrum reduce the effects of multipath? How is the relative timing of the transmitter-receiver codes established (acquisition) and retained (tracking)?

It is the aim of this tutorial paper to answer some of these questions succinctly, and in the process, offer some insights into this important communications technique. A glossary of the symbols used is provided at the end of the paper.

II. SPREADING AND DIMENSIONALITY—PROCESSING GAIN

A fundamental issue in spread spectrum is how this technique affords protection against interfering signals with

Reprinted from *IEEE Trans. Commun.*, vol. COM-30, no. 5, pp. 855–884, May 1982.

finite power. The underlying principle is that of distributing a relatively low dimensional (defined below) data signal in a high dimensional environment so that a jammer with a fixed amount of total power (intent on maximum disruption of communications) is obliged to either spread that fixed power over all the coordinates, thereby inducing just a little interference in each coordinate, or else place all of the power into a small subspace, leaving the remainder of the space interference free.

A brief discussion of a classical problem of signal detection in noise should clarify the emphasis on finite interference power. The "standard" problem of digital transmission in the presence of thermal noise is one where both transmitter and receiver know the set of M signaling waveforms $\{S_i(t), 0 \leqslant t \leqslant T;\ 1 \leqslant i \leqslant M\}$. The transmitter selects one of the waveforms every T seconds to provide a data rate of $\log_2 M/T$ bits/s. If, for example, $S_j(t)$ is sent, the receiver observes $r(t) = S_j(t) + n_w(t)$ over $[0, T]$ where $n_w(t)$ is additive, white Gaussian noise (AWGN) with (two-sided) power spectral density $\eta_0/2$ W/Hz.

It is well known [3] that the signal set can be completely specified by a linear combination of no more than $D \leqslant M$ orthonormal basis functions (see below), and that although the white noise, similarly expanded, requires an infinite number of terms, only those within the signal space are "relevant" [3]. We say that the signal set defined above is D-dimensional if the minimum number of orthonormal basis functions required to define all the signals is D. D can be shown to be [3] approximately $2B_DT$ where B_D is the total (approximate) bandwidth occupancy of the signal set. The optimum (minimum probability of error) detector in AWGN consists of a bank of correlators or filters matched to each signal, and the decision as to which was the transmitted signal corresponds to the largest output of the correlators.

Given a specific signal design, the performance of such a system is well known to be a function only of the ratio of the energy per bit to the noise spectral density. Hence, against white noise (which has infinite power and constant energy in every direction), the use of spreading (large $2B_DT$) offers no help. The situation is quite different, however, when the "noise" is a jammer with a fixed *finite* power. In this case, the effect of spreading the signal bandwidth so that the jammer is uncertain as to where in the large space the components are is often to force the jammer to distribute its finite power over many different coordinates of the signal space.

Since the desired signal can be "collapsed" by correlating the signal at the receiver with the known code, the desired signal is protected against a jammer in the sense that it has an effective power advantage relative to the jammer. This power advantage is often proportional to the ratio of the dimensionality of the space of code sequences to that of the data signal. It is necessary, of course, to "hide" the pattern by which the data are spread. This is usually done with a pseudonoise (PN) sequence which has desired randomness properties and which is available to the cooperating transmitter and receiver, but denied to other undesirable users of the common spectrum.

A general model which conveys these ideas, but which uses random (rather than pseudrandom) sequences, is as follows. Suppose we consider transmission by means of D equiprobable and equienergy orthogonal signals imbedded in an n-dimensional space so that

$$S_i(t) = \sum_{k=1}^{n} S_{ik}\phi_k(t); \qquad 1 \leqslant i \leqslant D; \qquad 0 \leqslant t \leqslant T$$

where

$$S_{ik} = \int_0^T S_i(t)\phi_k(t)\,dt$$

and where $\{\phi_k(t); 1 \leqslant k \leqslant n\}$ is an orthonormal basis spanning the space, i.e.,

$$\int_0^T \phi_l(t)\phi_m(t)\,dt = \delta_{lm} \triangleq \begin{cases} 1 & l = m \\ 0 & l \neq m. \end{cases}$$

The average energy of each signal is

$$\int_0^T \overline{S_i^2(t)}\,dt = \sum_{k=1}^{n} \overline{S_{ik}^2} \triangleq E_s; \qquad 1 \leqslant i \leqslant D \tag{1}$$

(the overbar is the expected value over the ensemble).

In order to hide this D-dimensional signal set in the larger n-dimensional space, choose the sequence of coefficients S_{ik} independently (say, by flipping a fair coin if a binary alphabet is used) such that they have zero mean and correlation

$$\overline{S_{ik}S_{il}} \triangleq \frac{E_s}{n}\delta_{kl}; \qquad 1 \leqslant i \leqslant D. \tag{2}$$

Thus, the signals, which are also assumed to be known to the receiver (i.e., we assume the receiver had been supplied the sequences S_{ik} before transmission) but denied to the jammer, have their respective energies uniformly distributed over the n basis directions as far as the jammer is concerned.

Consider next a jammer

$$J(t) = \sum_{k=1}^{n} J_k\phi_k(t); \qquad 0 \leqslant t \leqslant T \tag{3}$$

with total energy

$$\int_0^T J^2(t)\,dt = \sum_{k=1}^{n} J_k^2 \triangleq E_J \tag{4}$$

which is added to the signal with the intent to disrupt communications. Assume that the jammer's signal is independent of the desired signal. One of the jammer's objectives is to devise a strategy for selecting the components J_k^2 of his fixed total energy E_J so as to minimize the postprocessing signal-to-noise ratio (SNR) at the receiver.

The received signal

$$r(t) = S_i(t) + J(t) \tag{5}$$

is correlated with the (known) signals so that the output of the ith correlator is

$$U_i \triangleq \int_0^T r(t)S_i(t)\,dt = \sum_{k=1}^{n} (S_{ik}^2 + J_k S_{ik}). \tag{6}$$

Hence,

$$E(U_i \mid S_i) = \sum_{k=1}^{n} \overline{S_{ik}^2} = E_s \tag{7}$$

since the second term averages to zero. Then, since the signals are equiprobable,

$$E(U_i) = \frac{E_s}{D}. \tag{8}$$

Similarly, using (1) and (2),

$$\text{var}(U_i \mid S_i) = \sum_{k,l} J_k J_l \overline{S_{ik} S_{il}}$$

$$= \sum_{k=1}^{n} J_k^2 \overline{S_{ik}^2}$$

$$= \frac{E_s}{n} E_J \tag{9}$$

and

$$\text{var}\, U_i = \frac{E_s}{nD} E_J. \tag{10}$$

A measure of performance is the signal-to-noise ratio defined as

$$\text{SNR} = \frac{E^2(U)}{\text{var}(U)} = \frac{E_s}{E_J} \cdot \frac{n}{D}. \tag{11}$$

This result is *independent of the way that the jammer distributes his energy*, i.e., regardless of how J_k is chosen subject to the constraint that $\Sigma_k J_k^2 = E_J$, the postprocessing SNR (11) gives the signal an advantage of n/D over the jammer. This factor n/D is the *processing gain* and it is exactly equal to the ratio of the dimensionality of the *possible* signal space (and therefore the space in which the jammer must seek to operate) to the dimensions needed to actually transmit the signals. Using the result that the (approximate) dimensionality of a signal of duration T and of approximate bandwidth B_D is $2B_D T$, we see the processing gain can be written as

$$G_P = \frac{n}{D} \cong \frac{2B_{ss}T}{2B_D T} = \frac{B_{ss}}{B_D} \tag{12}$$

where B_{ss} is the bandwidth in hertz of the (spread-spectrum) signals $S_i(t)$ and B_D is the minimum bandwidth that would be required to send the information if we did not need to imbed it in the larger bandwidth for protection.

A simple illustration of these ideas using random binary sequences will be used to bring out some of these points. Consider the transmission of a single bit $\pm\sqrt{E_b/T}$ with energy E_b of duration T seconds. This signal is one-dimensional. As shown in Figs. 1 and 2, the transmitter multiplies the data bit $d(t)$ by a binary ± 1 "chipping" sequence $p(t)$ chosen randomly at rate f_c chips/s for a total of $f_c T$ chips/bit. The dimensionality of the signal $d(t)p(t)$ is then $n = f_c T$. The received signal is

$$r(t) = d(t)p(t) + J(t), \qquad 0 \leqslant t \leqslant T, \tag{13}$$

ignoring, for the time being, thermal noise.

The receiver, as shown in Fig. 1, performs the correlation

$$U \triangleq \sqrt{\frac{E_b}{T}} \int_0^T r(t)p(t)\,dt \tag{14}$$

and makes a decision as to whether $\pm\sqrt{E_b/T}$ was sent depending upon $U \gtrless 0$. The integrand can be expanded as

$$r(t)p(t) = d(t)p^2(t) + J(t)p(t) = d(t) + J(t)p(t), \tag{15}$$

and hence the data bit appears in the presence of a code-modulated jammer.

If, for example, $J(t)$ is additive white Gaussian noise with power spectral density $\eta_{0J}/2$ (two-sided), so is $J(t)\ p(t)$, and U is then a Gaussian random variable. Since $d(t) = \pm\sqrt{E_b/T}$, the conditional mean and variance of U, assuming that $\pm\sqrt{E_b/T}$ is transmitted, is given by E_b and $E_b(\eta_{0J}/2)$, respectively, and the probability of error is [3] $Q(\sqrt{2E_b/\eta_{0J}})$ where $Q(x) \triangleq \int_x^\infty (1/\sqrt{2\pi})e^{-y^2/2}\,dy$. Against white noise of unlimited power, spread spectrum serves no useful purpose, and the probability of error is $Q(\sqrt{2E_b/\eta_{0J}})$ regardless of the modulation by the code sequence. White noise occupies all dimensions with power $\eta_{0J}/2$. The situation is different, however, if the jammer power is limited. Then, not having access to the random sequence $p(t)$, the jammer with available energy E_J (power E_J/T) can do better than to apply this energy to one dimension. For example, if $J(t) = \sqrt{E_J/T}$, $0 \leqslant t \leqslant T$, then the receiver output is

$$U = E_b + \sqrt{E_b E_J}\, \frac{1}{n} \sum_{i=1}^{n} X_i \tag{16}$$

where the X_i's are i.i.d.[1] random variables with $P(X_i = +1) = P(X_i = -1) = \frac{1}{2}$. The signal-to-noise ratio (SNR) is

$$\frac{E^2(U)}{\text{var}(U)} = \frac{E_b}{E_J} n. \tag{17}$$

Thus, the SNR may be increased by increasing n, the process-

[1] Independent identically distributed.

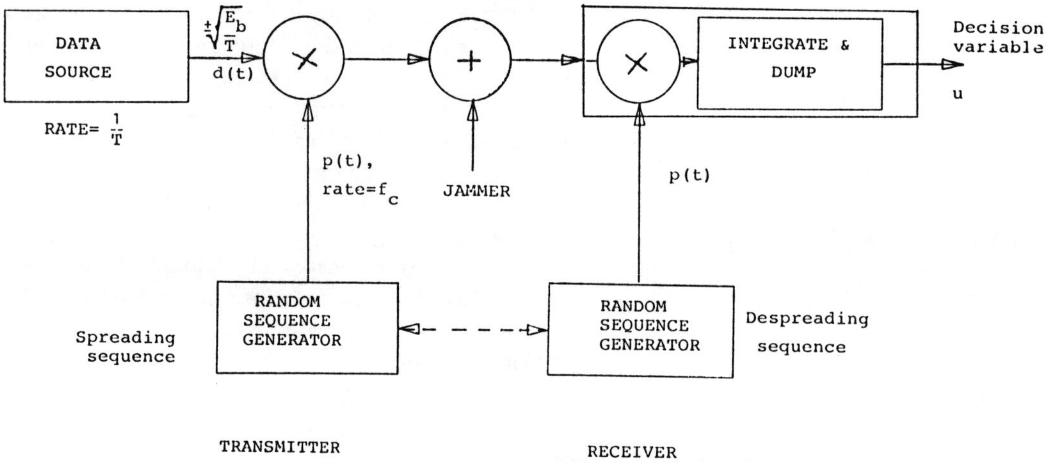

Fig. 1. Direct-sequence spread-spectrum system for transmitting a single binary digit (baseband).

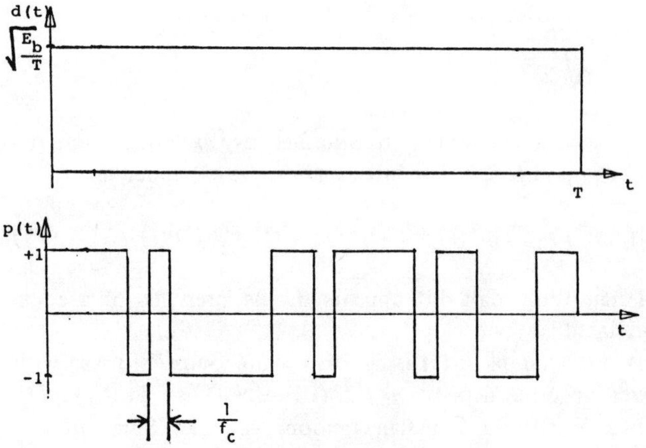

Fig. 2. Data bit and chipping sequence.

ing gain, and it has the form of (11). As a further indication of this parameter, we may compute the probability P_e that the bit is in error from (16). Assuming that a "minus" is transmitted, we have

$$\begin{aligned} P_e &= P(U > 0) \\ &= P(Z_n > \alpha n) \\ &= \begin{cases} \dfrac{1}{2^n} \displaystyle\sum_{[\alpha n]}^{n} \binom{n}{k}; & \dfrac{E_b}{E_J} < 1 \\ 0; & \dfrac{E_b}{E_J} \geqslant 1 \end{cases} \end{aligned} \tag{18}$$

where

$Z_n \triangleq \frac{1}{2} \sum_{i=1}^{n} (1 + X_i)$ is a Bernoulli random variable with

mean $\frac{n}{2}$ and variance $\frac{n}{4}$,

$$\alpha \triangleq \frac{1}{2}\left(1 + \sqrt{\frac{E_b}{E_J}}\right),$$

and $[X]$ is defined as the integer portion of X. The partial binomial sum on the right-hand side of (18) may be upper bounded [2] by

$$P_e \leqslant \frac{1}{2^n}\left(\frac{1}{1-\alpha}\right)^n \left(\frac{1-\alpha}{\alpha}\right)^{\alpha n}; \qquad \tfrac{1}{2} < \alpha \leqslant 1$$

or

$$P_e \leqslant 2^{-n[1-H(\alpha)]}; \qquad \tfrac{1}{2} < \alpha \leqslant 1 \tag{19}$$

where $H(\alpha) \triangleq -\alpha \log_2 \alpha - (1-\alpha)\log_2 (1-\alpha)$ is the binary entropy function. Therefore, for any $\alpha > \frac{1}{2}$ (or $E_b \neq 0$), P_e may be made vanishingly small by increasing n, the processing gain. (The same result is valid even if the jammer uses a chip pattern other than the constant, all-ones used in the example above.) As an example, if $E_J = 9E_b$ (jammer energy 9.5 dB larger than that of the data), then $\alpha = 2/3$ and $P_e \leqslant 2^{-0.085n}$. If $n = 200$ (23 dB processing gain), $P_e < 7.6 \times 10^{-6}$.

An approximation to the same result may be obtained by utilizing a central limit type of argument that says, for large n, U in (16) may be treated as if it were Gaussian. Then

$$P_e = P(U < 0) \cong Q\left(\sqrt{\frac{E_b}{E_J} n}\right) \tag{20}$$

and, if $E_b/E_J = -9.5$ dB and $n = 200$ (23 dB), $P_e \cong Q(\sqrt{22}) \simeq 1.5 \times 10^{-6}$. The processing gain can be seen to be a multiplier of the "signal-to-jamming" ratio E_b/E_J.

A more traditional way of describing the processing gain, which brings in the relative bandwidth of the data signal and that of the spread-spectrum modulation, is to examine the power spectrum of an *infinite sequence* of data, modulated by the rapidly varying random sequence. The spectrum of the random data sequence with rate $R = 1/T$ bits/s is given by

$$S_D(f) = T\left(\frac{\sin \pi f T}{\pi f T}\right)^2 \tag{21}$$

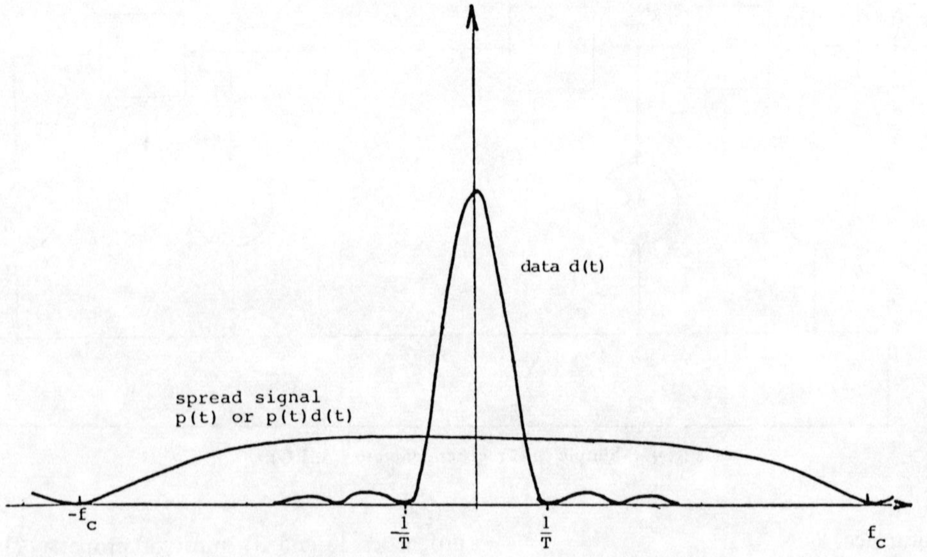

Fig. 3. Power spectrum of data and of spread signal.

and that of the spreading sequence [and also that of the product $d(t)\,p(t)$] is given by

$$S_{ss}(f) = \frac{1}{f_c}\left(\frac{\sin \pi f/f_c}{\pi f/f_c}\right)^2. \tag{22}$$

Both are sketched in Fig. 3. It is clear that if the receiver multiplies the received signal $d(t)p(t) + J(t)$ by $p(t)$ giving $d(t) + J(t)p(t)$, the first term may be extracted virtually intact with a filter of bandwidth $1/T \triangleq B_D$ Hz. The second term will be spread over at least f_c Hz as shown in Fig. 3. The fraction of power due to the jammer which can pass through the filter is then roughly $1/f_cT$. Thus, the data have a power advantage of $n = f_cT$, the processing gain. As in (12), the processing gain is frequently expressed as the ratio of the bandwidth of the spread-spectrum waveform to that of the data, i.e.,

$$G_p \triangleq \frac{B_{ss}}{B_D} = f_cT = n. \tag{23}$$

The notion of processing gain as expressed in (23) is simply a power improvement factor which a receiver, possessing a replica of the spreading signal, can achieve by a correlation operation. It must not be automatically extrapolated to anything else. For example, if we use frequency hopping for spread spectrum employing one of N frequencies every T_H seconds, the total bandwidth must be approximately N/T_H (since keeping the frequencies orthogonal requires frequency spacing $\approx 1/T_H$). Then, according to (12), $G_p = (N/T_H)/B_D$. Now if we transmit 1 bit/hop, $T_HB_D \simeq 1$ and $G_P = N$, the number of frequencies used. If $N = 100$, $G_p = 20$ dB, which seems fairly good. But a single spot frequency jammer can cause an average error rate of about 10^{-2}, which is not acceptable. (A more detailed analysis follows in Section IV below.) This effectiveness of "partial band jamming" can be reduced by the use of coding and interleaving. Coding typically precludes the possibility of a small number of frequency slots (e.g., one slot) being jammed causing an unacceptable error rate (i.e., even if the jammer wipes out a few of the code symbols, depending upon the error-correction capability of the code, the data may still be recovered). Interleaving has the effect of randomizing the errors due to the jammer. Finally, an analogous situation occurs in direct sequence spreading when a pulse jammer is present.

In the design of a practical system, the processing gain G_p is not, by itself, a measure of how well the system is capable of performing in a jamming environment. For this purpose, we usually introduce the *jamming margin* in decibels defined as

$$M_J = G_P - \left(\frac{E_b}{\eta_{0J}}\right)_{\min} - L. \tag{24}$$

This is the residual advantage that the system has against a jammer after we subtract both the minimum required energy/bit-to-jamming "noise" power spectral density ratio $(E_b/\eta_{0J})_{\min}$ and implementation and other losses L. The jamming margin can be increased by reducing the $(E_b/\eta_{0J})_{\min}$ through the use of coding gain.

We conclude this section by showing that regardless of the technique used, spectral spreading provides protection against a broad-band jammer with a finite power P_J. Consider a system that transmits R_0 bits/s designed to operate over a bandwidth B_{ss} Hz in white noise with power density η_0 W/Hz. For any bit rate R,

$$\left(\frac{E_b}{\eta_0}\right)_{\text{actual}} = \frac{P_s}{\eta_0 R} = \frac{P_s}{P_N}\frac{B_{ss}}{R} \tag{25}$$

where

$$P_s \triangleq E_bR = \text{signal power}$$

$$P_N \triangleq \eta_0 B_{ss} = \text{noise power}.$$

Then for a specified $(E_b/\eta_0)_{\min}$ necessary to achieve mini-

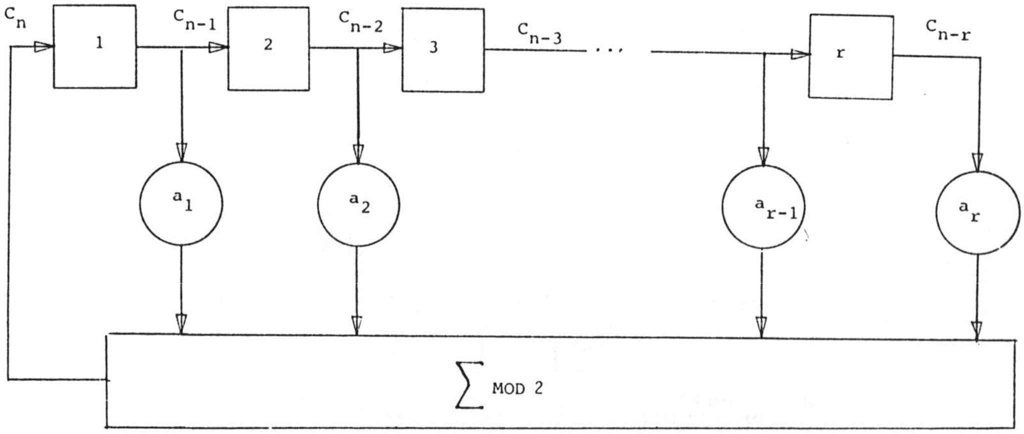

Fig. 4. Simple shift register generator (SSRG).

mum acceptable performance,

$$R \leqslant \frac{P_s}{P_N} \frac{B_{ss}}{(E_b/\eta_0)_{\min}} \triangleq R_0. \tag{26}$$

If a jammer with power P_J now appears, and if we are already transmitting at the maximum rate R_0, then (25) becomes

$$\left(\frac{E_b}{\eta_0}\right)_{\text{actual}} = \frac{P_s}{P_N + P_J} \frac{B_{ss}}{R_0}$$

$$= \left(\frac{E_b}{\eta_0}\right)_{\min} \frac{P_N}{P_N + P_J}$$

or

$$\left(\frac{E_b}{\eta_0}\right)_{\text{actual}} = \left(\frac{E_b}{\eta_0}\right)_{\min} \frac{\eta_0}{\eta_0 + P_J/B_{ss}}. \tag{27}$$

Thus, if we wish to recover from the effects of the jammer, the right-hand side of (27) should be not much less than $(E_b/\eta_0)_{\min}$. This clearly requires that we increase B_{ss}, since for any finite P_J, it is then possible to make the factor $\eta_0/(\eta_0 + P_J/B_{ss})$ approach unity, and thereby retain the performance we had before the jammer appeared.

III. PSEUDORANDOM SEQUENCE GENERATORS

In Section II, we examined how a purely random sequence can be used to spread the signal spectrum. Unfortunately, in order to despread the signal, the receiver needs a replica of the transmitted sequence (in almost perfect time synchronism). In practice, therefore, we generate pseudorandom or pseudonoise (PN) sequences so that the following properties are satisfied. They

1) are easy to generate
2) have randomness properties
3) have long periods
4) are difficult to reconstruct from a short segment.

Linear feedback shift register (LFSR) sequences [4] possess properties 1) and 3), most of property 2), but not property 4). One canonical form of a binary LFSR known as a simple shift register generator (SSRG) is shown in Fig. 4. The shift register consists of binary storage elements (boxes) which transfer their contents to the right after each clock pulse (not shown). The contents of the register are linearly combined with the binary (0, 1) coefficients a_k and are fed back to the first stage. The binary (code) sequence C_n then clearly satisfies the recursion

$$C_n = \sum_{k=1}^{r} a_k C_{n-k} \,(\text{mod } 2); \qquad a_r = 1. \tag{28}$$

The periodic cycle of the states depends on the initial state and on the coefficients (feedback taps) a_k. For example, the four-stage LFSR generator shown in Fig. 5 has four possible cycles as shown. The all-zeros is always a cycle for any LFSR. For spread spectrum, we are looking for *maximal length* cycles, that is, cycles of period $2^r - 1$ (all binary r-tuples except all-zeros). An example is shown for a four-state register in Fig. 6. The sequence output is 1 0 0 0 1 1 1 1 0 1 0 1 1 0 0 ··· (period $2^4 - 1 = 15$) if the initial contents of the register (from right to left) are 1000. It is always possible to choose the feedback coefficients so as to achieve maximal length, as will be discussed below.

If we do have a maximal length sequence, then this sequence will have the following pseudorandomness properties [4].

1) There is an approximate balance of zeros and ones (2^{r-1} ones and $2^{r-1} - 1$ zeros).

2) In any period, half of the runs of consecutive zeros or ones are of length one, one-fourth are of length two, one-eighth are of length three, etc.

3) If we define the ±1 sequence $C_n' = 1 - 2C_n$, $C_n = 0, 1$, then the autocorrelation function $R_{C'}(\tau) \triangleq 1/L \, \Sigma_{k=1}^{L} C_k' C_{k+\tau}'$ is given by

$$R_{C'}(\tau) = \begin{cases} 1, & \tau = 0, L, 2L \cdots \\ -\dfrac{1}{L}, & \text{otherwise} \end{cases} \tag{29}$$

where $L = 2^r - 1$. If the code *waveform* $p(t)$ is the "square-wave" equivalent of the sequences C_n', if $L \gg 1$, and if we

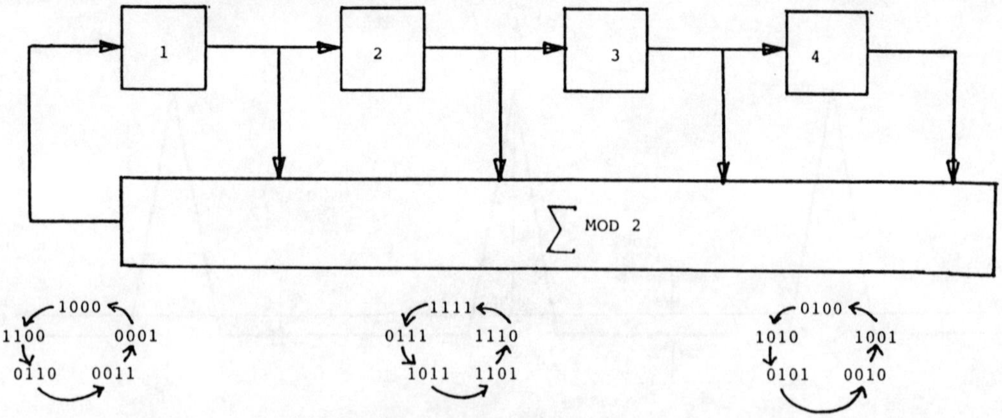

Fig. 5. Four-stage LFSR and its state cycles.

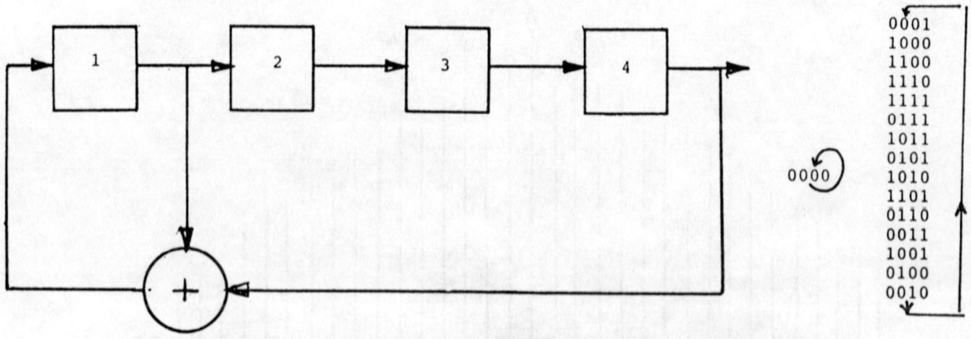

Fig. 6. Four-stage maximal length LFSR and its state cycles.

define

$$q(\tau) \triangleq \begin{cases} 1 - |\tau| f_c; & |\tau| \leqslant \dfrac{1}{f_c} \\ 0; & \text{otherwise} \end{cases}$$

then

$$R_p(\tau) \simeq \sum_i q\left(\tau - \frac{iL}{f_c}\right). \tag{30}$$

Equation (29), and therefore (30), follow directly from the "shift-and-add" property of maximal length (ML) LFSR sequences. This property is that the chip-by-chip sum of an MLLFSR sequence C_k and any shift of itself $C_{k+\tau}$, $\tau \neq 0$ is the *same* sequence (except for some shift). This follows directly from (28), since

$$(C_n + C_{n+\tau}) = \sum_{k=1}^{L} a_k (C_{n-k} + C_{n+\tau-k}) \pmod 2. \tag{31}$$

The shift-and-add sequence $C_n + C_{n+\tau}$ is seen to satisfy the same recursion as C_n, and if the coefficients a_k yield maximal length, then it must be the same sequence regardless of the initial (nonzero) state. The autocorrelation property (29) then follows from the following isomorphism:

$$(\{0, 1\}, +) \leftrightarrow (\{1, -1\}, \times).$$

Therefore,

$$C_k + C_{k+\tau} \leftrightarrow C_k' C_{k+\tau}'$$

and if C_k' is an MLLFSR ± 1 sequence, so is $C_k' C_{k+\tau}'$, $\tau \neq 0$. Thus, there are 2^{r-1} 1's and $(2^{r-1} - 1)$ −1's in the product and (29) follows. The autocorrelation function is shown in Fig. 7(a).

Property 3) is a most important one for spread spectrum since the autocorrelation function of the code sequence *waveform* $p(t)$ determines the spectrum. Note that because $p(t)$ is *pseudorandom*, it is periodic with period $(2^r - 1) \cdot 1/f_c$, and hence so is $R_p(\tau)$. The spectrum shown in Fig. 7(b) is therefore the line spectrum

$$S_p(f) = \left[\sum_{\substack{m=-\infty \\ m \neq 0}}^{\infty} \delta(f - mf_0)\right] \frac{L+1}{L^2} \left(\frac{\sin \pi f/f_c}{\pi f/f_c}\right)^2 + \frac{1}{L^2}\,\delta(f) \tag{32}$$

where

$$f_0 = \frac{f_c}{2^r - 1}.$$

If $L = 2^r - 1$ is very large, the spectral lines get closer together, and for practical purposes, the spectrum may be viewed as being continuous and similar to that of a purely random binary waveform as shown in Fig. 3. A different, but commonly used implementation of a linear feedback shift register is the modular shift register generator (MSRG) shown in Fig. 8. Additional details on the properties of linear feedback shift registers are provided in the Appendix.

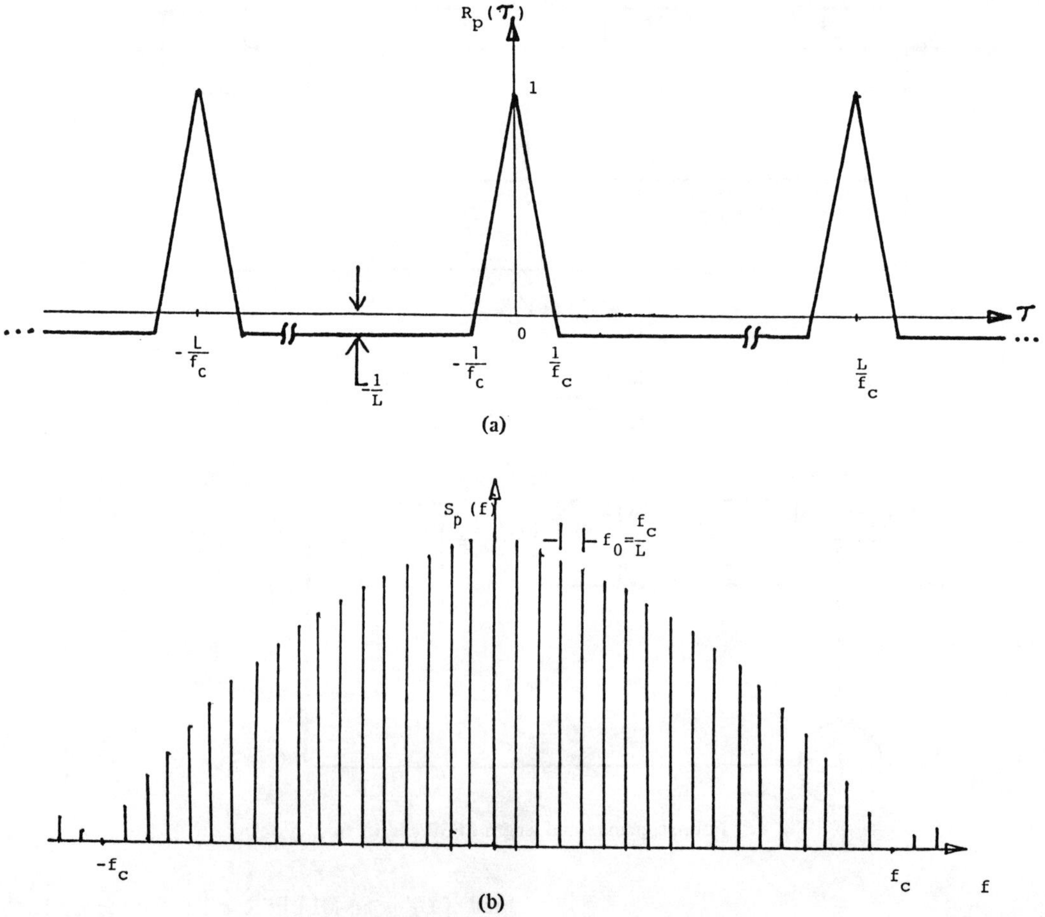

Fig. 7. Autocorrelation function $R_p(\tau)$ and power spectral density of MLLFSR sequence waveform $p(t)$. (a) Autocorrelation function of $p(t)$. (b) Power spectral density of $p(t)$.

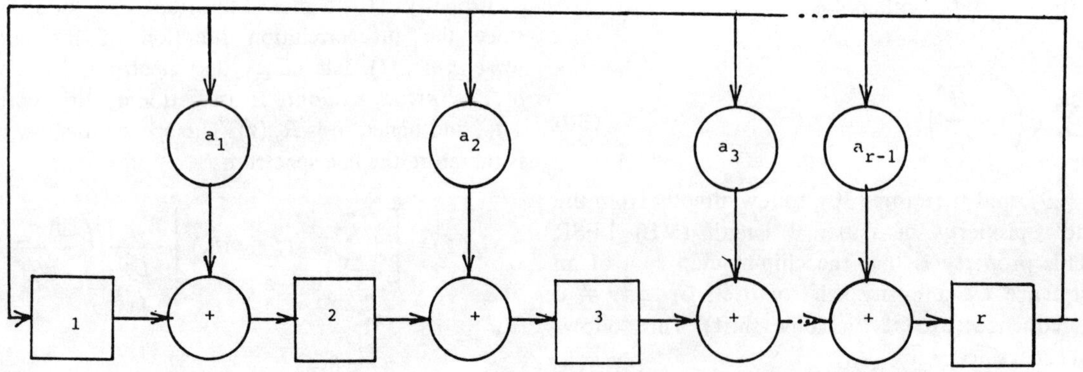

Fig. 8. Implementation as a modular shift register generator (MSRG).

For spread spectrum and other secure communications (cryptography) where one expects an adversary to attempt to recover the code in order to penetrate the system, property 4) cited in the beginning of this section is extremely important. Unfortunately, LFSR sequences do not possess that property. Indeed, using the recursion (28) or (A8) and observing only $2r-2$ consecutive bits in the sequence C_n allows us to solve for the $r-2$ middle coefficients and the r initial bits in the register by linear simultaneous equations. Thus, even if $r = 100$ so that the length of the sequence is $2^{100}-1 \simeq 10^{30}$, we would be able to construct the entire sequence from 198 bits by solving 198 linear equations (mod 2), which is neither difficult nor that time consuming for a large computer. Moreover, because the sequence C_n satisfies a recursion, a very efficient algorithm is known [7], [8] which solves the equations or which equivalently synthesizes the shortest LFSR which generates a given sequence.

In order to avoid this pitfall, several modifications of the LFSR have been proposed. In Fig. 9(a) the feedback function is replaced by an arbitrary Boolean function of the contents of the register. The Boolean function may be implemented by ROM or random logic, and there are an enormous number of these functions (2^{2^r}). Unfortunately, very little is known [4] in the open literature about the properties of such non-

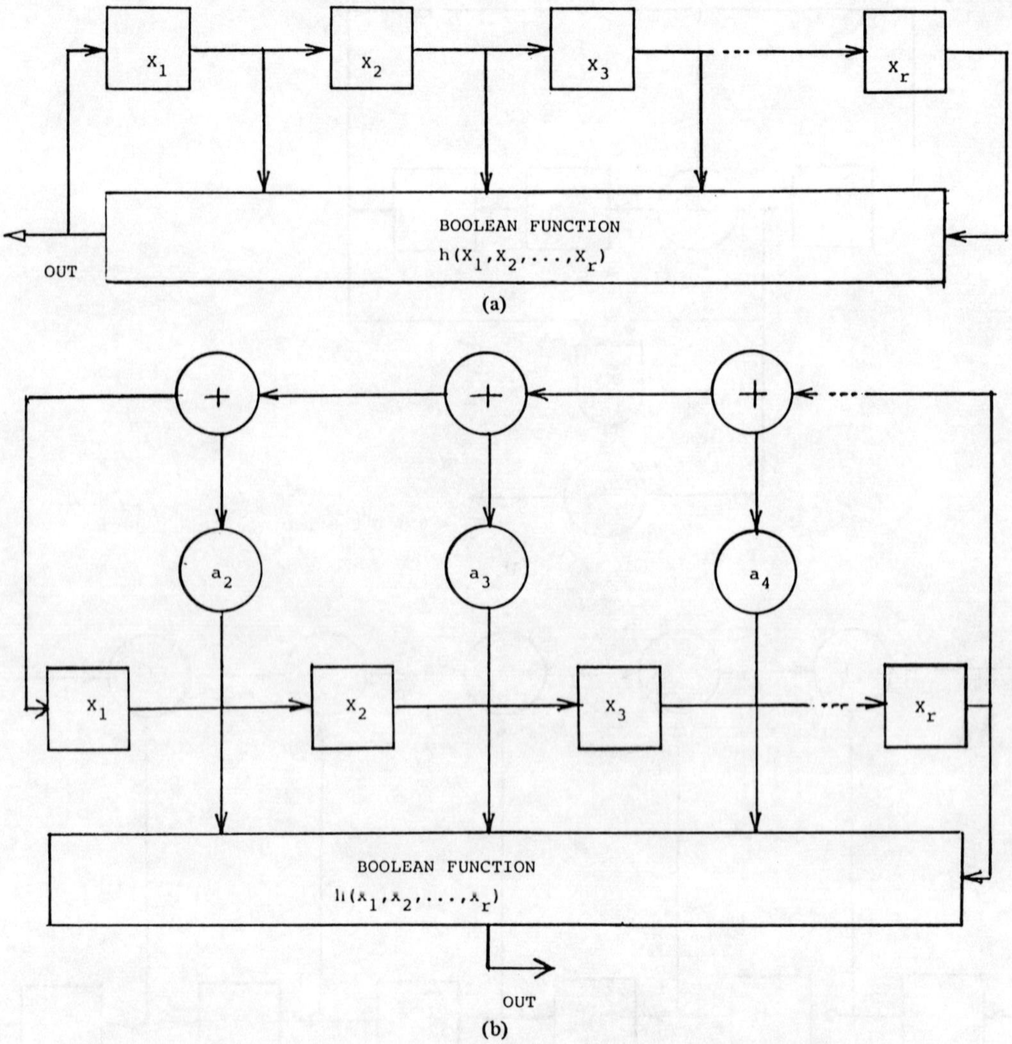

Fig. 9. Nonlinear feedback shift registers. (a) Nonlinear FDBK. Number of Boolean functions = 2^{2^r}. (b) Linear FSR, nonlinear function of state, i.e., nonlinear output logic (NOL).

linear feedback shift registers. Furthermore, some nonlinear FSR's may have no cycles or length > 1 (e.g., they may have only a *transient* that "homes" towards the all-ones state after any initial state). Are there feedback functions that generate only *one* cycle of length 2^r? The answer is yes, and there are exactly $2^{2^{r-1}-r}$ of them [9]. How do we find them? Better yet, how do we find a subset of them with all the "good" randomness properties? These are, and have been, good research problems for quite some time, and unfortunately no general theory on this topic currently exists.

A second, more manageable approach is to use an MLLFSR with nonlinear output logic (NOL) as shown in Fig. 9(b). Some clues about designing the NOL while still retaining "good" randomness properties are available [10]-[12], and a measure for judging how well condition 4) is fulfilled is to ask: What is the degree of the shortest LFSR that would generate the same sequence? A simple example of an LFSR with NOL having three stages is shown in Fig. 10(a). The shortest LFSR which generates the same sequence (of period 7) is shown in Fig. 10(b) and requires six stages.

When using PN sequences in spread-spectrum systems, several additonal requirements must be met.

1) The "partial correlation" of the sequence C_n' over a window w smaller than the full period should be as small as possible, i.e., if

$$\rho(w;j,\tau) \triangleq \sum_{n=j}^{j+w-1} C_n' C_{n+\tau}',$$

$$\rho(w) = \max_{j,\tau} |\rho(w;j,\tau)| \tag{33}$$

should be $\ll L = 2^r - 1$.

2) Different code pairs should have uniformly low cross correlation, i.e.,

$$R_{C'C''}(\tau) \triangleq \frac{1}{L} \sum_{k=1}^{L} C_k' C_{k+\tau}'' \tag{34}$$

should be $\ll 1$ for all values of τ.

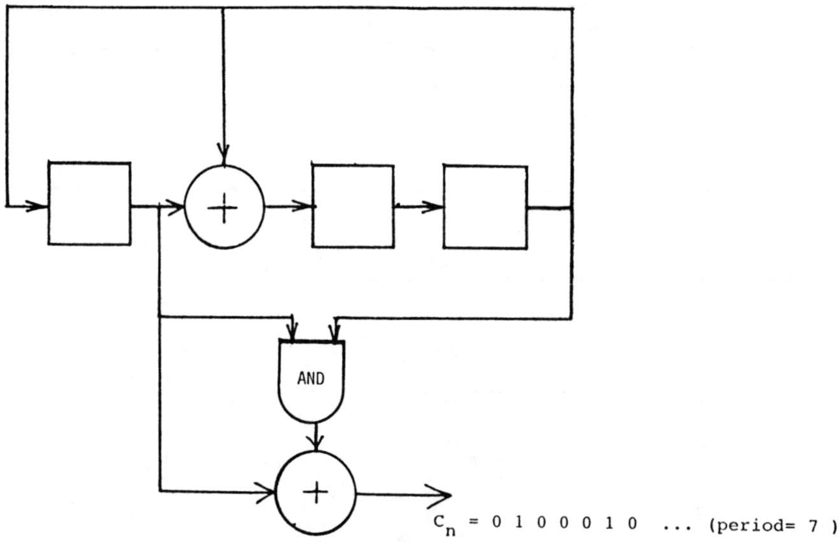

(a)

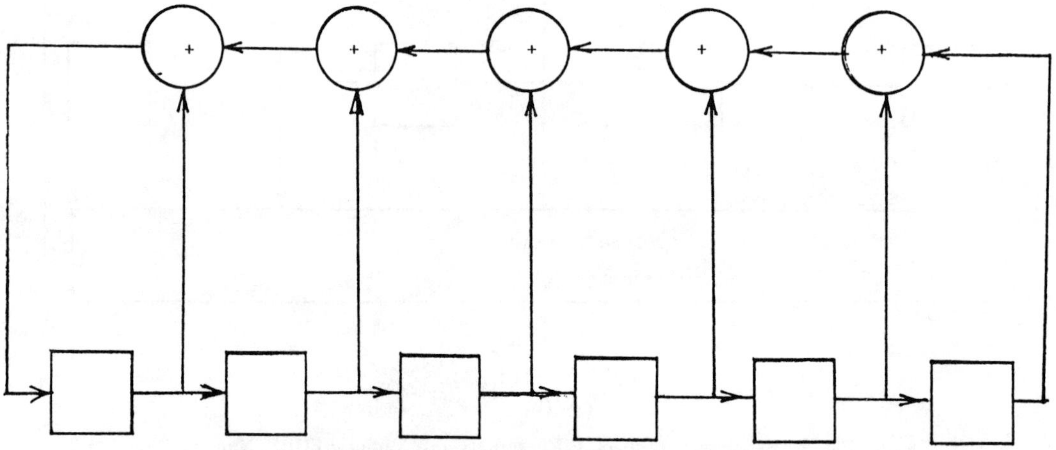

(b)

Fig. 10. LFSR with NOL and its shortest linear equivalent. (a) Three-stage LFSR with NOL. (b) LFSR with $f(x) = 1 + x + x^2 + x^3 + x^4 + x^5 + x^6$ which generates the same sequence as that of (a) under the initial state 1 0 0 0 1 0.

3) Since the code sequences are periodic with period L, there are two correlation functions (depending on the relative polarity of one of the sequences in the transition over an initial point τ on the other). If we define the finite-cross-correlation function [13] as

$$f_{C'C''}(\tau) \triangleq \frac{1}{L} \sum_{k=1}^{\tau} C_k' C_{k+\tau}'', \tag{35}$$

then the so-called even and odd cross-correlation functions are, respectively,

$$R_{C'C''}^{(e)}(\tau) = f_{C'C''}(\tau) + f_{C'C''}(L-\tau)$$

and

$$R_{C'C''}^{(o)}(\tau) = f_{C'C''}(\tau) - f_{C'C''}(L-\tau)$$

and we want

$$\max_{\tau} |R_{C'C''}^{e}(\tau)| \quad \text{and} \quad \max_{\tau} |R_{C'C''}^{(o)}(\tau)|$$

to be $\ll 1$.

The reason for 1) is to keep the "self noise" of the system as low as possible since, in practice, the period is very long compared to the integration time per symbol and there will be fluctuation in the sum of any filtered (weighted) subsequence. This is especially worrisome during acquisition where these fluctuations can cause false locking. Bounds on $\rho(w)$ [14] and averages over j of $\rho(w; j, \tau)$ are available in the literature.

Properties 2) and 3) are both of direct interest when using PN sequences for code division multiple access (CDMA) as will be discussed in Section V below. This is to ensure minimal cross interference between any pair of users of the common spectrum. The most commonly used collection of sequences which exhibit property 2) are the Gold codes [15]. These are sequences derived from an MLLFSR, but are *not* of maximal length. A detailed procedure for their construction is given in the Appendix.

Virtually all of the known results about the cross-correlation properties of useful PN sequences are summarized in [16].

As a final comment on the generation of PN sequences for spread spectrum, it is not at all necessary that feedback shift registers be used. *Any* technique which can generate "good" pseudorandom sequences will do. Other techniques are described in [4], [16], [17], for example. Indeed, the generation of good pseudorandom sequences is fundamental to other fields, and in particular, to cryptography [18]. A "good" cryptographic system can be used to generate "good" PN sequences, and vice versa. A possible problem is that the specific additional "good" properties required for an operational spread-spectrum system may not always match those required for secure cryptographic communications.

IV. ANTIJAM CONSIDERATIONS

Probably the single most important application of spread-spectrum techniques is that of resistance to intentional interference or jamming. Both direct-sequence (DS) and frequency-hopping (FH) systems exhibit this tolerance to jamming, although one might perform better than the other given a specific type of jammer.

The two most common types of jamming signals analyzed are single frequency sine waves (tones) and broad-band noise. References [19] and [20] provide performance analyses of DS systems operating in the presence of both tone and noise interference, and [21]-[26] provide analogous results for FH systems.

The simplest case to analyze is that of broad-band noise jamming. If the jamming signal is modeled as a zero-mean wide sense stationary Gaussian noise process with a flat power spectral density over the bandwidth of interest, then for a given fixed power P_J available to the jamming signal, the power spectral density of the jamming signal must be reduced as the bandwidth that the jammer occupies is increased.

For a DS system, if we assume that the jamming signal occupies the total RF bandwidth, typically taken to be twice the chip rate, then the despread jammer will occupy an even greater bandwidth and will appear to the final integrate-and-dump detection filter as approximately a white noise process. If, for example, binary PSK is used as the modulation format, then the average probability of error will be approximately given by

$$P_e = Q\left(\sqrt{\frac{2E_b}{\eta_0 + \eta_{0J}}}\right). \tag{36}$$

Equation (36) is just the classical result for the performance of a coherent binary communication system in additive white Gaussian noise. It differs from the conventional result because an extra term in the denominator of the argument of the $Q(\cdot)$ function has been added to account for the jammer. If P_e from (36) is plotted versus E_b/η_0 for a given value of P_J/P_s, where P_s is the average signal power, curves such as the ones shown in Fig. 11 result.

Expressions similar to (36) are easily derived for other modulation formats (e.g., QPSK), and curves showing the performance for several different formats are presented, for example, in [19]. The interesting thing to note about Fig. 11 is that for a given η_{0J}, the curve "bottoms out" as E_b/η_0 gets larger and larger. That is, the presence of the jammer will cause an irreducible error rate for a given P_J and a given f_c. Keeping P_J fixed, the only way to reduce the error rate is to increase f_c (i.e., increase the amount of spreading in the system). This was also noted at the end of Section II.

For FH systems, it is not always advantageous for a noise jammer to jam the entire RF bandwidth. That is, for a given P_J, the jammer can often increase its effectiveness by jamming only a fraction of the total bandwidth. This is termed *partial-band jamming*. If it is assumed that the jammer divides its power uniformly among K slots, where a slot is the region in frequency that the FH signal occupies on one of its hops, and if there is a total of N slots over which the signal can hop, we have the following possible situations. Assuming that the underlying modulation format is binary FSK (with noncoherent detection at the receiver), and using the terminology MARK and SPACE to represent the two binary data symbols, on any given hop, if

1) $K = 1$, the jammer might jam the MARK only, jam the SPACE only, or jam neither the MARK nor the SPACE;

2) $1 < K < N$, the jammer might jam the MARK only, jam the SPACE only, jam neither the MARK nor the SPACE, or jam both the MARK and the SPACE;

3) $K = N$, the jammer will always jam both the MARK and the SPACE.

To determine the average probability of error of this system, each of the possibilities alluded to above has to be accounted for. If it is assumed that the N slots are disjoint in frequency and that the MARK and SPACE tones are orthogonal (i.e., if a MARK is transmitted, it produces no output from the SPACE bandpass filter (BPF) and vice versa), then the average probability of error of the system can be shown to be given by [23], [24]

$$P_e = \frac{(N-K)(N-K-1)}{N(N-1)}\frac{1}{2}\exp\left(-\tfrac{1}{2}\,\mathrm{SNR}\right) + \frac{K(N-K)}{N(N-1)}\exp\left[-\frac{1}{\dfrac{2}{\mathrm{SNR}}+\dfrac{1}{\mathrm{SJR}}}\right] + \frac{K(K-1)}{N(N-1)}\frac{1}{2} \cdot \exp\left[-\frac{1}{2}\,\frac{1}{\dfrac{1}{\mathrm{SNR}}+\dfrac{1}{\mathrm{SJR}}}\right] \tag{37}$$

where SNR is the ratio of signal power to thermal noise power at the output of the MARK BPF (assuming that a MARK has been transmitted) and SJR is the ratio of signal

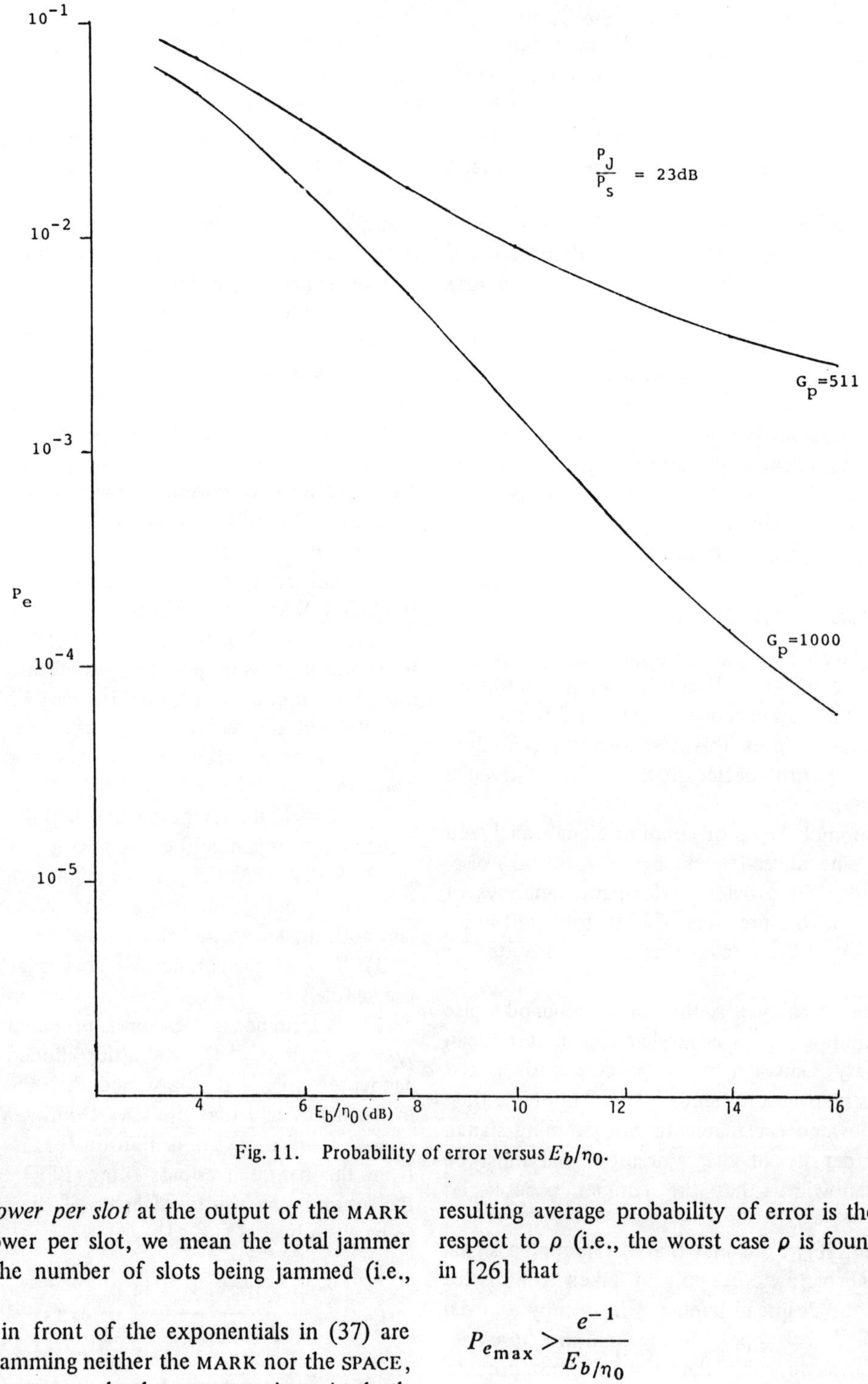

Fig. 11. Probability of error versus E_b/η_0.

power to *jammer power per slot* at the output of the MARK BPF. By jammer power per slot, we mean the total jammer power divided by the number of slots being jammed (i.e., $\mathrm{SJR} = P_s/(P_J/K)$).

The coefficients in front of the exponentials in (37) are the probabilities of jamming neither the MARK nor the SPACE, jamming only the MARK or only the SPACE, or jamming both the MARK and the SPACE. For example, the probability of jamming both the MARK and the SPACE is given by $K(K-1)/N(N-1)$. In Fig. 12, the P_e predicted by (37) is plotted versus SNR for $K = 1$ and $K = 100$ for a P_J/P_s of 10 dB. These two curves are labeled "uncoded" on the figure.

Often, a somewhat different model from that used in deriving (37) is considered. This latter model is used in [26], and effectively assumes that either MARK and SPACE are simultaneously jammed or that neither of the two is jammed. For this case, a parameter ρ, where $0 < \rho \leqslant 1$, representing the fraction of the band being jammed, is defined. The resulting average probability of error is then maximized with respect to ρ (i.e., the worst case ρ is found), and it is shown in [26] that

$$P_{e_{\max}} > \frac{e^{-1}}{E_b/\eta_0}$$

where e is the base of the natural logarithm. It can be seen that partial band jamming affords the jammer a strategy whereby he can degrade the performance significantly (i.e., P_e can be forced to be inversely proportional to E_b/η_0 rather than exponential).

For tone jamming, the situation becomes somewhat more complicated than it is for noise jamming, especially for DS systems. This is because the system performance depends upon the location of the tone (or tones), and upon whether the period of the spreading sequence is equal to or greater than the duration of a data symbol. Oftentimes the effect

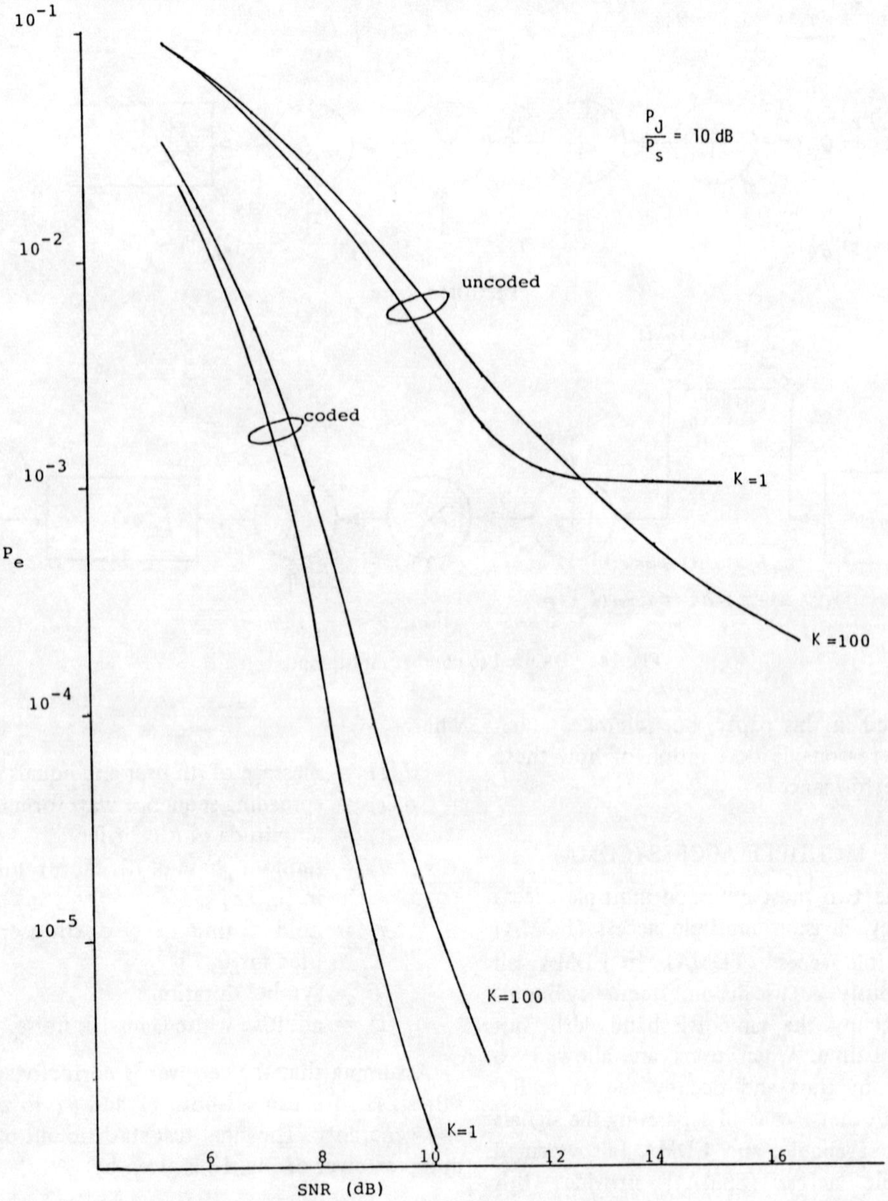

Fig. 12. Probability of error versus SNR.

of a despread tone is approximated as having arisen from an equivalent amount of Gaussian noise. In this case, the results presented above would be appropriate. However, the Gaussian approximation is not always justified, and some conditions for its usage are given in [20] and [27].

The situation is simpler in FH systems operating in the presence of partial-band tone jamming, and as shown, for example, in [24], the performance of a noncoherent FH–FSK system in partial-band tone jamming is often virtually the same as the performance in partial-band noise jamming. One important consideration in FH systems with either noise or tone jamming is the need for error-correction coding. This can be seen very simply by assuming that the jammer is much stronger than the desired signal, and that it chooses to put all of its power in a single slot (i.e., the jammer jams one out of N slots). The $K = 1$ uncoded curve of Fig. 12 corresponds to this situation. Then with no error-correction coding, the system will make an error (with high probability) every time it hops to a MARK frequency when the corresponding SPACE frequency is being jammed or vice versa. This will happen on the average one out of every N hops, so that the probability of error of the system will be approximately $1/N$, independent of signal-to-noise ratio. This is readily seen to be the case in Fig. 12. The use of coding prevents a simple error as caused by a spot jammer from degrading the system performance. To illustrate this point, an error-correcting code (specifically a Golay code [2]) was used in conjunction with the system whose uncoded performance is shown in Fig. 12, and the performance of the coded system is also shown in Fig. 12. The advantage of using error-correction coding is obvious from comparing the corresponding curves.

Finally, there are, of course, many other types of common jamming signals besides broad-band noise or single frequency tones. These include swept-frequency jammers, pulse-burst jammers, and repeat jammers. No further discussion of these

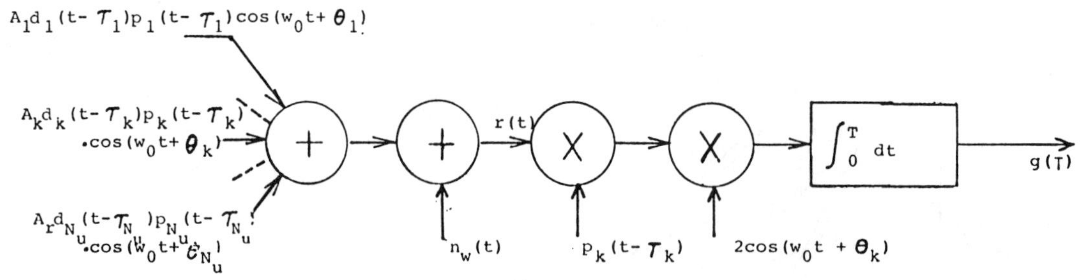

Fig. 13. DS CDMA system.

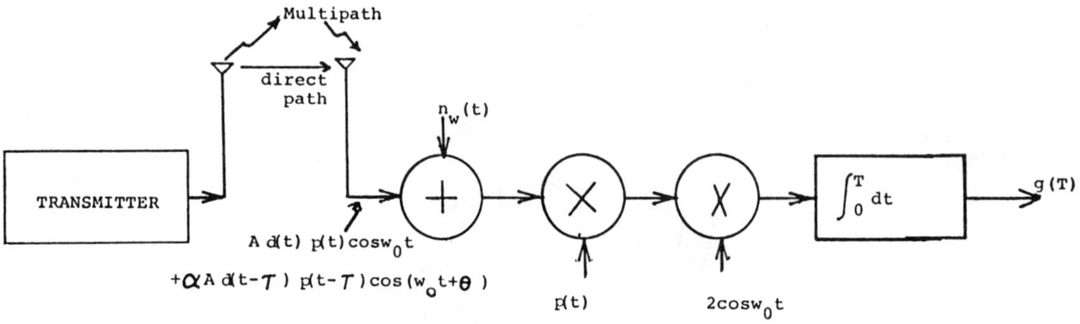

Fig. 14. DS used to combat multipath.

jammers will be presented in this paper, but references such as [28]-[30] provide a reasonable description of how these jammers affect system performance.

V. CODE DIVISION MULTIPLE ACCESS (CDMA)

As is well known, the two most common multiple access techniques are frequency division multiple access (FDMA) and time division multiple access (TDMA). In FDMA, all users transmit simultaneously, but use disjoint frequency bands. In TDMA, all users occupy the same RF bandwidth, but transmit sequentially in time. When users are allowed to transmit simultaneously in time and occupy the same RF bandwidth as well, some other means of separating the signals at the receiver must be available, and CDMA [also termed spread-spectrum multiple access (SSMA)] provides this necessary capability.

In DS CDMA [31]-[33], each user is given its own code, which is approximately orthogonal (i.e., has low cross correlation) with the codes of the other users. However, because CDMA systems typically are asynchronous (i.e., the transition times of the data symbols of the different users do not have to coincide), the design problem is much more complicated than that of having, say, N_u spreading sequences with uniformly low cross correlations such as the Gold codes discussed in Section III and in the Appendix. As will be seen below, the key parameters in a DS CDMA system are both the cross-correlation and the partial-correlation functions, and the design and optimization of code sets with good partial-correlation properties can be found in many references such as [16], [34], and [35].

The system is shown in Fig. 13. The received signal is given by

$$r(t) = \sum_{i=1}^{N_u} A_i d_i(t-\tau_i) p_i(t-\tau_i) \cos(\omega_0 t + \theta_i) + n_w(t) \quad (38)$$

where

$d_i(t)$ = message of ith user and equals ± 1
$p_i(t)$ = spreading sequence waveform of ith user
A_i = amplitude of ith carrier
θ_i = random phase of ith carrier uniformly distributed in $[0, 2\pi]$
τ_i = random time delay of ith user uniformly distributed in $[0, T]$
T = symbol duration
$n_w(t)$ = additive white Gaussian noise.

Assuming that the receiver is correctly synchronized to the kth signal, we can set both τ_k and θ_k to zero without losing any generality. The final test statistic out of the integrate-and-dump receiver of Fig. 14 is given by

$$g(T) = A_k + \frac{1}{T} \sum_{\substack{i=1 \\ i \neq k}}^{N_u} A_i \int_0^T d_i(t-\tau_i) \cdot p_i(t-\tau_i) p_k(t) \cos(\theta_i)\, dt + \frac{2}{T} \int_0^T n_w(t) p_k(t) \cos(\omega_0 t)\, dt \quad (39)$$

where double frequency terms have been ignored.

Consider the second term on the RHS of (39). It is a sum of $N_u - 1$ terms of the form

$$A_i \cos(\theta_i) \int_0^T d_i(t-\tau_i) p_i(t-\tau_i) p_k(t)\, dt.$$

Notice that, because the ith signal is not, in general, in sync with the kth signal, $d_i(t-\tau_i)$ will change signs somewhere in the interval $[0, T]$ 50 percent of the time. Hence, the above integral will be the sum of two partial correlations of $p_i(t)$ and $p_k(t)$, rather than one total cross correlation. Therefore, (39) can be rewritten

$$g(T) = A_k + \sum_{\substack{i=1 \\ i \neq k}}^{N_u} A_i[\pm\rho_{ik}(\tau_i) \pm \hat{\rho}_{ik}(\tau_i)] \cos(\theta_i) + n(T) \quad (40)$$

where

$$\rho_{ik}(\tau_i) \triangleq \frac{1}{T}\int_0^{\tau_i} p_i(t-\tau_i)p_k(t)\, dt$$

$$\hat{\rho}_{ik}(\tau_i) \triangleq \frac{1}{T}\int_{\tau_i}^{T} p_i(t-\tau_i)p_k(t)\, dt$$

and

$$n(T) \triangleq \frac{2}{T}\int_0^T n_w(t)p_k(t)\cos\omega_0 t\, dt.$$

Notice that the coefficients in front of $\rho_{ik}(\tau_i)$ and $\hat{\rho}_{ik}(\tau_i)$ can independently have a plus or minus sign due to the data sequence of the ith signal. Also notice that $\rho_{ik}(\tau_i) + \hat{\rho}_{ik}(\tau_i)$ is the total cross correlation between the ith and kth spreading sequences. Finally, the continuous correlation functions $\rho_{ik}(\tau) \pm \hat{\rho}_{ik}(\tau)$ can be expressed in terms of the discrete even and odd cross-correlation functions, respectively, that were defined in Section III.

While the code design problem in CDMA is very crucial in determining system performance, of potentially greater importance in DS CDMA is the so-called "near-far problem." Since the N_u users are typically geographically separated, a receiver trying to detect the kth signal might be much closer physically to, say, the ith transmitter rather than the kth transmitter. Therefore, if each user transmits with equal power, the signal from the ith transmitter will arrive at the receiver in question with a larger power than that of the kth signal. This particular problem is often so severe that DS CDMA cannot be used.

An alternative to DS CDMA, of course, is FH CDMA [36]-[40]. If each user is given a different hopping pattern, and if all hopping patterns are orthogonal, the near-far problem will be solved (except for possible spectral spillover from one slot into adjacent slots). However, the hopping patterns are never truly orthogonal. In particular, any time more than one signal uses the same frequency at a given instant of time, interference will result. Events of this type are sometimes referred to as "hits," and these hits become more and more of a problem as the number of users hopping over a fixed bandwidth increases. As is the case when FH is employed as an antijam technique, error-correction coding can be used to significant advantage when combined with FH CDMA.

FH CDMA systems have been considered using one hop per bit, multiple hops per bit (referred to as fast frequency hopping or FFH), and multiple bits per hop (referred to as slow frequency hopping or SFH). Oftentimes the characteristics of the channel over which the multiple users transmit play a significant role in influencing which type of hopping one employs. An example of this is the multipath channel, which is discussed in the next section.

It is clearly of interest to consider the relative capacity of a CDMA system compared to FDMA or TDMA. In a perfectly linear, perfectly synchronous system, the number of orthogonal users for all three systems is the same, since this number only depends upon the dimensionality of the overall signal space. In particular, if a given time-bandwidth product G_P is divided up into, say, G_P disjoint time intervals for TDMA, it can also be "divided" into N binary orthogonal codes (assume that $G_P = 2^m$ for some positive integer m).

The differences between the three multiple-accessing techniques become apparent when various real-world constraints are imposed upon the ideal situation described above. For example, one attractive feature of CDMA is that it does not *require* the network synchronization that TDMA requires (i.e., if one is willing to give up something in performance, CDMA can be (and usually is) operated in an asynchronous manner). Another advantage of CDMA is that it is relatively easy to add additional users to the system. However, probably the dominant reason for considering CDMA is the need, in addition, for some type of external interference rejection capability such as multipath rejection or resistance to intentional jamming.

For an asynchronous system, even ignoring any near-far problem effects, the number of users the system can accommodate is markedly less than G_P. From [31] and [35], a rough rule-of-thumb appears to be that a system with processing gain G_P can support approximately $G_P/10$ users. Indeed, from [31, eq. (17)], the peak signal voltage to rms noise voltage ratio, averaged over all phase shifts, time delays, and data symbols of the multiple users, is approximately given by

$$\overline{\text{SNR}} \simeq \left[\frac{N_u - 1}{3G_P} + \frac{\eta_0}{2E_b}\right]^{-1/2}$$

where the overbar indicates an ensemble average. From this equation, it can be seen that, given a value of E_b/η_0, $(N_u - 1)/G_P$ should be in the vicinity of 0.1 in order not to have a noticeable effect on system performance.

Finally, other factors such as nonlinear receivers influence the performance of a multiple access system, and, for example, the effect of a hard limiter on a CDMA system is treated in [45].

VI. MULTIPATH CHANNELS

Consider a DS binary PSK communication system operating over a channel which has more than one path linking the transmitter to the receiver. These different paths might consist of several discrete paths, each one with a different attenuation and time delay relative to the others, or it might consist of a continuum of paths. The RAKE system described in [1] is an example of a DS system designed to operate effectively in a multipath environment.

For simplicity, assume initially there are just two paths, a direct path and a single multipath. If we assume the time delay the signal incurs in propagating over the direct path is smaller than that incurred in propagating over the single multipath, and if it is assumed that the receiver is synchronized to the time delay and RF phase associated with the direct path, then the system is as shown in Fig. 14. The received signal is given by

$$r(t) = Ad(t)p(t) \cos \omega_0 t + \alpha Ad(t-\tau)p(t-\tau) \cdot \cos(\omega_0 t + \theta) + n_w(t) \quad (41)$$

where τ is the differential time delay associated with the two paths and is assumed to be in the interval $0 \leqslant \tau \leqslant T$, θ is a random phase uniformly distributed in $[0, 2\pi]$, and α is the relative attenuation of the multipath relative to the direct path. The output of the integrate-and-dump detection filter is given by

$$g(T) = A + [\pm \alpha A\rho(\tau) + \alpha A\hat{\rho}(\tau)] \cos \theta \quad (42)$$

where $\rho(\tau)$ and $\hat{\rho}(\tau)$ are partial correlation functions of the spreading sequence $p(t)$ and are given by

$$\rho(\tau) \triangleq \frac{1}{T}\int_0^{\tau} p(t)p(t-\tau)\,dt \quad (43)$$

and

$$\hat{\rho}(\tau) \triangleq \frac{1}{T}\int_{\tau}^{T} p(t)p(t-\tau)\,dt. \quad (44)$$

Notice that the sign in front of the second term on the RHS of (42) can be plus or minus with equal probability because this term arises from the pulse preceding the pulse of interest (i.e., if the ith pulse is being detected, this term arises from the $i-1$th pulse), and this latter pulse will be of the same polarity as the current pulse only 50 percent of the time. If the signs of these two pulses happen to be the same, and if $\tau > T_c$ where T_c is the chip duration, then $\rho(\tau) + \hat{\rho}(\tau)$ equals the autocorrelation function of $p(t)$ (assuming that a full period of $p(t)$ is contained in each T second symbol), and this latter quantity equals $-(1/L)$, where L is the period of $p(t)$. In other words, the power in the undesired component of the received signal has been attenuated by a factor of L^2.

If the sign of the preceding pulse is opposite to that of the current pulse, the attenuation of the undesired signal will be less than L^2, and typically can be much less than L^2. This is analogous, of course, to the partial correlation problem in CDMA discussed in the previous section.

The case of more than two discrete paths (or a continuum of paths) results in qualitatively the same effects in that signals delayed by amounts outside of $\pm T_c$ seconds about a correlation peak in the autocorrelation function of $p(t)$ are attenuated by an amount determined by the processing gain of the system.

If FH is employed instead of DS spreading, improvement in system performance is again possible, but through a different mechanism. As was seen in the two previous sections, FH systems achieve their processing gain through interference avoidance, not interference attenuation (as in DS systems). This same qualitative difference is true again if the interference is multipath. As long as the signal is hopping fast enough relative to the differential time delay between the desired signal and the multipath signal (or signals), all (or most) of the multipath energy will fall in slots that are orthogonal to the slot that the desired signal currently occupies.

Finally, the problems treated in this and the previous two sections are often all present in a given system, and so the use of an appropriate spectrum-spreading technique can alleviate all three problems at once. In [41] and [42], the joint problem of multipath and CDMA is treated, and in [43] and [44], the joint problem of multipath and intentional interference is analyzed. As indicated in Section V, if only multiple accessing capability is needed, there are systems other than CDMA that can be used (e.g., TDMA). However, when multipath is also a problem, the choice of CDMA as the multiple accessing technique is especially appropriate since the same signal design allows both many simultaneous users and improved performance of each user individually relative to the multipath channel.

In the case of signals transmitted over channels degraded by both multipath and intentional interference, either factor by itself suggests the consideration of a spectrum-spreading technique (in particular, of course, the intentional interference), and when all three sources of degradation are present simultaneously, spread spectrum is a virtual necessity.

VII. ACQUISITION

As we have seen in the previous sections, pseudonoise modulation employing direct sequence, frequency hopping, and/or time hopping is used in spread-spectrum systems to achieve bandwidth spreading which is large compared to the bandwidth required by the information signal. These PN modulation techniques are typically characterized by their very low repetition-rate-to-bandwidth ratio and, as a result, synchronization of a receiver to a specified modulation constitutes a major problem in the design and operation of spread-spectrum communications systems [46]-[50].

It is possible, in principle, for spread-spectrum receivers to use matched filter or correlator structures to synchronize to the incoming waveform. Consider, for example, a direct-sequence amplitude modulation synchronization system as shown in Fig. 15(a). In this figure, the locally generated code $p(t)$ is available with delays spaced one-half of a chip ($T_c/2$) apart to ensure correlation. If the region of uncertainty of the code phase is N_c chips, $2N_c$ correlators are employed. If no information is available regarding the chip uncertainty and the PN sequence repeats every, say, 2047 chips, then 4094 correlators are employed. Each correlator is seen to examine λ chips, after which the correlator outputs $V_0, V_1, \cdots, V_{2N_c-1}$ are compared and the largest output is chosen. As λ increases, the probability of making an error in synchronization decreases; however, the acquisition time increases. Thus, λ is usually chosen as a compromise between the probability of a synchronization error and the time to acquire PN phase.

A second example, in which FH synchronization is employed, is shown in Fig. 15(b). Here the spread-spectrum signal

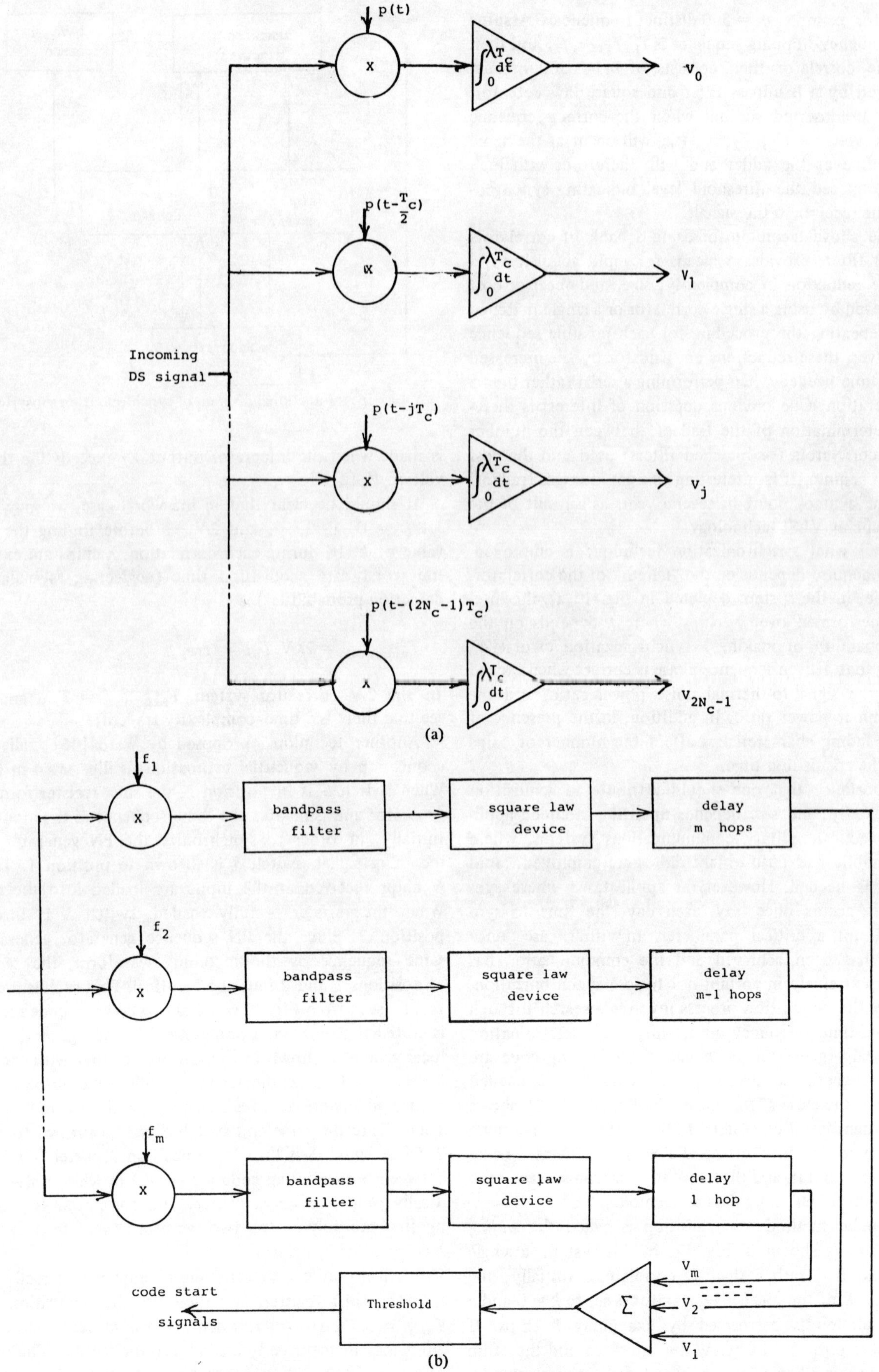

Fig. 15. (a) Direct sequence acquisition using $2N_c$ correlators. (b) Passive correlator structure for a frequency-hopping coarse acquisition scheme.

hops over, for example, $m = 500$ distinct frequencies. Assume that the frequency-hopping sequence is $f_1, f_2, \cdots, f_m$ and then repeats. The correlator then consists of $m = 500$ mixers, each followed by a bandpass filter and square law detector. The delays are inserted so that when the correct sequence appears, the voltages $V_1, V_2, \cdots, V_m$ will occur at the same instant of time at the adder and will, therefore, with high probability, exceed the threshold level indicating synchronization of the receiver to the signal.

While the above technique of using a bank of correlators or matched filters provides a means for rapid acquisition, a considerable reduction in complexity, size, and receiver cost can be achieved by using a single correlator or a single matched filter and repeating the procedure for each possible sequence shift. However, these reductions are paid for by the increased acquisition time needed when performing a serial rather than a parallel operation. One obvious question of interest is therefore the determination of the tradeoff between the number of parallel correlators (or matched filters) used and the cost and time to acquire. It is interesting to note that this tradeoff may become a moot point in several years as a result of the rapidly advancing VLSI technology.

No matter what synchronization technique is employed, the time to acquire depends on the "length" of the correlator. For example, in the system depicted in Fig. 15(a), the integration is performed over λ chips where λ depends on the desired probability of making a synchronization error (i.e., of deciding that a given sequence phase is correct when indeed it is not), the signal-to-thermal noise power ratio, and the signal-to-jammer power ratio. In addition, in the presence of fading, the fading characteristics affect the number of chips and hence the acquisition time.

The importance that one should attribute to acquisition time, complexity, and size depends upon the intended application. In tactical military communications systems, where users are mobile and push-to-talk radios are employed, rapid acquisition is needed. However, in applications where synchronization occurs once, say, each day, the time to synchronize is not a critical parameter. In either case, once acquisition has been achieved and the communication has begun, it is extremely important not to lose synchronization. Thus, while the acquisition process involves a search through the region of time-frequency uncertainty and a determination that the locally generated code and the incoming code are sufficiently aligned, the next step, called *tracking*, is needed to ensure that the close alignment is maintained. Fig. 16 shows the basic synchronization system. In this system, the incoming signal is first locked into the local PN signal generator using the acquisition circuit, and then kept in synchronism using the tracking circuit. Finally, the data are demodulated.

One popular method of acquisition is called the *sliding correlator* and is shown in Fig. 17. In this system, a single correlator is used rather than L correlators. Initially, the output phase k of the local PN generator is set to $k = 0$ and a partial correlation is performed by examining λ chips. If the integrator output falls below the threshold and therefore is deemed too small, k is set to $k = 1$ and the procedure is repeated. The determination that acquisition has taken place

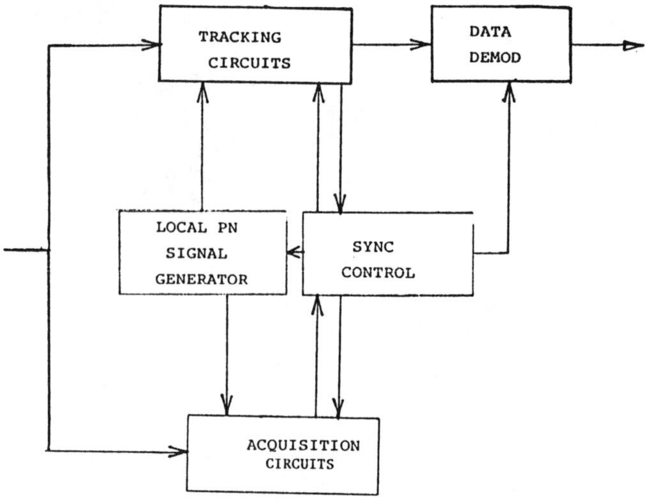

Fig. 16. Functional diagram of synchronization subsystem.

is made when the integrator output V_I exceeds the threshold voltage $V_T(\lambda)$.

It should be clear that in the worst case, we may have to set $k = 0, 1, 2, \cdots$, and $2N_c - 1$ before finding the correct value of k. If, during each correlation, λ chips are examined, the worst case acquisition time (neglecting false-alarm and detection probabilities) is

$$T_{\text{acq,max}} = 2\lambda N_c T_c. \tag{45}$$

In the $2N_c$-correlator system, $T_{\text{acq,max}} = T_c\lambda$, and so we see that there is a time-complexity tradeoff.

Another technique, proposed by Ward [46], called rapid acquisition by sequential estimation, is illustrated in Fig. 18. When switch S is in position 2, the shift register forms a PN generator and generates the same sequence as the input signal. Initially, in order to synchronize the PN generator to the incoming signal, switch S is thrown to position 1. The first N chips received at the input are loaded into the register. When the register is fully loaded, switch S is thrown to position 2. Since the PN sequence generator generates the same sequence as the incoming waveform, the sequences at positions 1 and 2 must be identical. That such is the case is readily seen from Fig. 19 which shows how the code $p(t - jT_C)$ is initially generated. Comparing this code generator to the local generator shown in Fig. 18, we see that with the switch in position 1, once the register is filled, the outputs of both mod 2 adders are *identical*. Hence, the bit stream at positions 1 and 2 are the same and switch S can be thrown to position 2. Once switch S is thrown to position 2, correlation is begun between the incoming code $p(t - jT_c)$ in white noise and the locally generated PN sequence. This correlation is performed by first multiplying the two waveforms and then examining λ chips in the integrator.

When no noise is present, the N chips are correctly loaded into the shift register, and therefore the acquisition time is $T_{\text{acq}} = NT_c$. However, when *noise is present*, one or more chips may be incorrectly loaded into the register. The resulting waveform at 2 will then not be of the same phase as the sequence generated at 1. If the correlator output after λT_c ex-

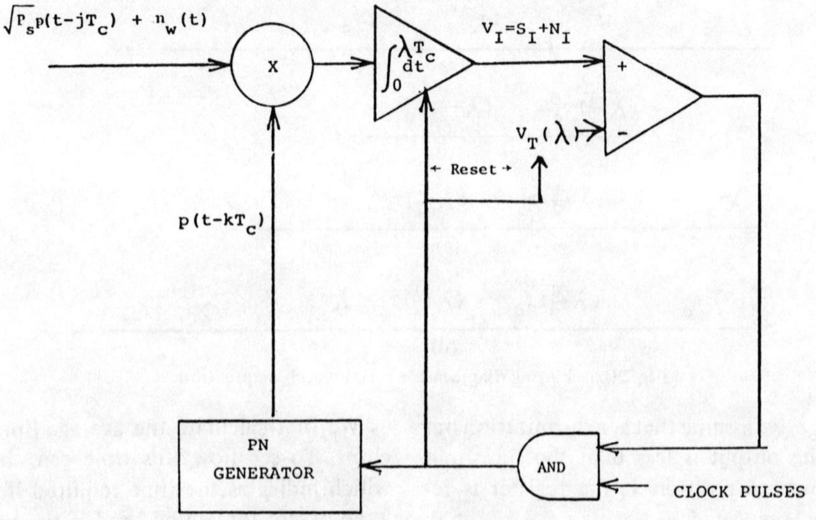

Fig. 17. The "sliding correlator."

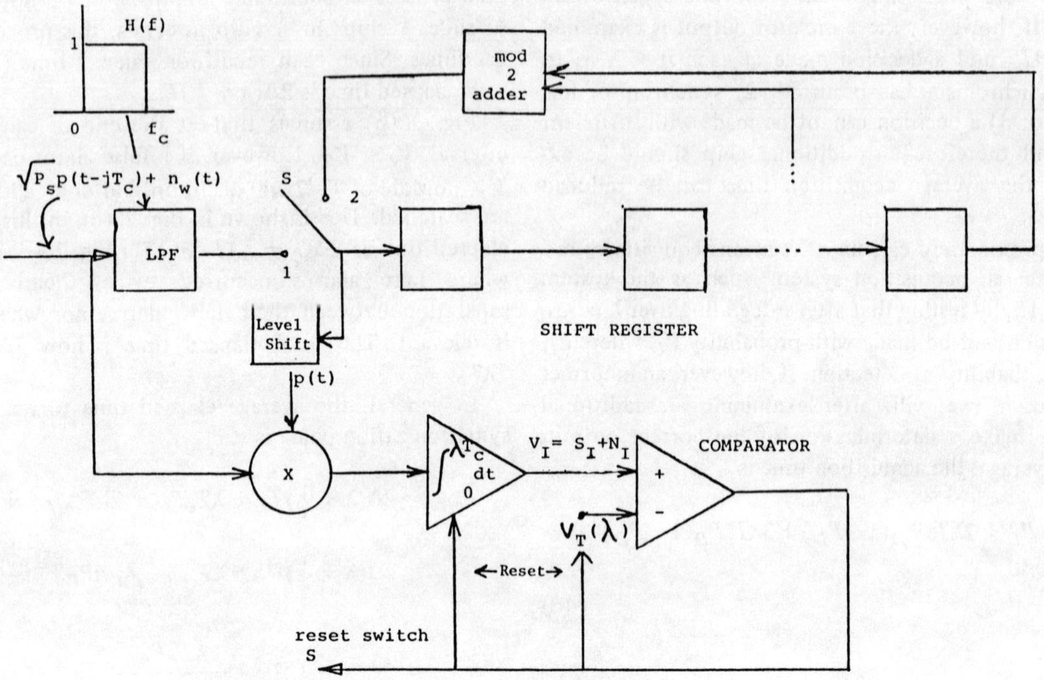

Fig. 18. Shift register acquisition circuit.

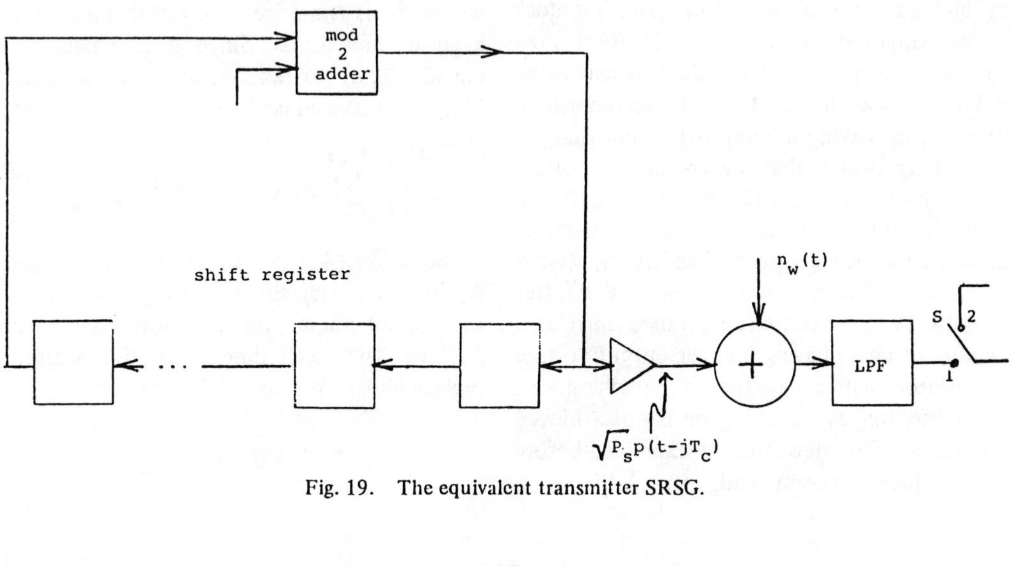

Fig. 19. The equivalent transmitter SRSG.

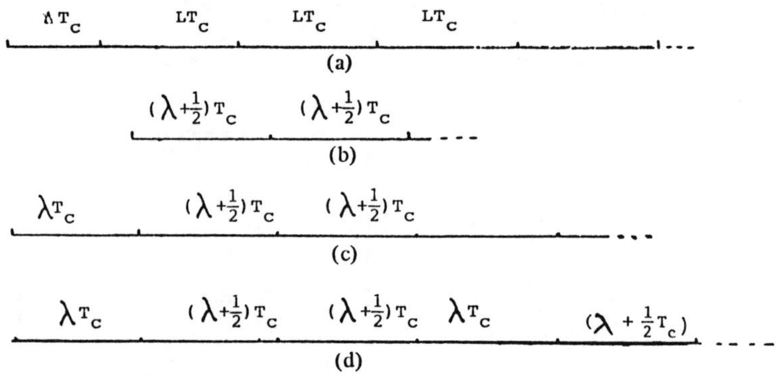

Fig. 20. Timing diagram for serial search acquisition.

ceeds the threshold voltage, we assume that synchronization has occurred. If, however, the output is less than the threshold voltage, switch S is thrown to position 1, the register is reloaded, and the procedure is repeated.

Note that in both Figs. 17 and 18, correlation occurs for a time λT_c before predicting whether or not synchronism has occurred. If, however, the correlator output is examined after a time nT_c and a decision made at each $n \leqslant \lambda$ as to whether 1) synchronism has occurred, 2) synchronism has not occurred, or 3) a decision cannot be made with sufficient confidence and therefore an additional chip should be examined, then the average acquisition time can be reduced substantially.

One can approximately calculate the mean acquisition time of a parallel search acquisition system, such as the system shown in Fig. 15, by noting that after integrating over λ chips, a correct decision will be made with probability P_D where P_D is called the probability of detection. If, however, an incorrect output is chosen, we will, after examining an additional λ chips, again make a determination of the correct output. Thus, on the average, the acquisition time is

$$\bar{T}_{\mathrm{acq}} = \lambda T_c P_D + 2\lambda T_c P_D(1-P_D) + 3\lambda T_c P_D(1-P_D)^2 + \cdots$$
$$= \frac{\lambda T_c}{P_D} \tag{46}$$

where it is assumed that we continue searching every λ chips even after a threshold has been exceeded. This is not, in general, the way an actual system would operate, but does allow a simple approximation to the true acquisition time.

Calculation of the mean acquisition time when using the "sliding correlator" shown in Fig. 17 can be accomplished in a similar manner (again making the approximation that we never stop searching) by noting that we are initially offset by a random number of chips Δ as shown in Fig. 20(a). After the correlator of Fig. 17 finally "slides" by these Δ chips, acquisition can be achieved with probability P_D. (Note that this P_D differs from the P_D of (46), since the latter P_D accounts for false synchronizations due to a correlator matched to an incorrect phase having a larger output voltage than does the correlator matched to the correct phase.) If, due to an incorrect decision, synchronization is not achieved at that time, L additional chips must then be examined before acquistion can be achieved (again with probability P_D).

We first calculate the average time needed to slide by the Δ chips. To see how this time can change, refer to Fig. 20(b) which indicates the time required if we are not synchronized. λ chips are integrated, and if the integrator output $V_I < V_T$ (the threshold voltage), a $\frac{1}{2}$ chip delay is generated, and we then process an additional λ chips, etc. We note that in order to slide Δ chips in $\frac{1}{2}$ chip intervals, this process must occur 2Δ times. Since each repetition takes a time $(\lambda + \frac{1}{2})T_c$, the total elapsed time is $2\Delta(\lambda + \frac{1}{2})T_c$.

Fig. 20(b) assumes that at the end of each examination interval, $V_I < V_T$. However, if a false alarm occurs and $V_I > V_T$, no slide of $T_c/2$ will occur until after an additional λ chips are searched. This is shown in Fig. 20(c). In this case, the total elapsed time is $2\Delta(\lambda + \frac{1}{2})T_c + \lambda T_c$. Fig. 20(d) shows the case where false alarms occurred twice. Clearly, neither the separation between these false alarms nor where they occur is relevant. The total elapsed time is now $2\Delta(\lambda + \frac{1}{2})T_c + 2\lambda T_c$.

In general, the average elapsed time to reach the correct synchronization phase is

$$\bar{T}_{s/\Delta} = 2\Delta(\lambda + \tfrac{1}{2})T_c + \lambda T_c P_F + 2\lambda T_c P_F^2 + \cdots$$
$$= 2\Delta(\lambda + \tfrac{1}{2})T_c + \lambda T_c P_F \sum_{n=1}^{\infty} n P_F^{n-1}$$
$$= 2\Delta(\lambda + \tfrac{1}{2})T_c + \frac{\lambda T_c P_F}{(1-P_F)^2} \tag{47}$$

where P_F is the false alarm probability. Equation (47) is for a given value of Δ. Since Δ is a random variable which is equally likely to take on any integer value from 0 to $L-1$, $\bar{T}_{s/\Delta}$ must be averaged over all Δ. Therefore,

$$\bar{T}_s \triangleq \frac{1}{L}\sum_{\Delta=0}^{L-1} \bar{T}_{s/\Delta} = L(\lambda + \tfrac{1}{2})T_c + \frac{\lambda T_c P_F}{(1-P_F)^2}. \tag{48}$$

Equation (48) is the average time needed to slide through Δ chips. If, after sliding through Δ chips, we do not detect the correct phase, we must now slide through an additional L chips. The mean time to do this is given by (47), with Δ replaced by L. We shall call this time $\bar{T}_{s/L}$:

$$\bar{T}_{s/L} = 2L(\lambda + \tfrac{1}{2})T_c + \frac{\lambda T_c P_F}{(1-P_F)^2}. \tag{49}$$

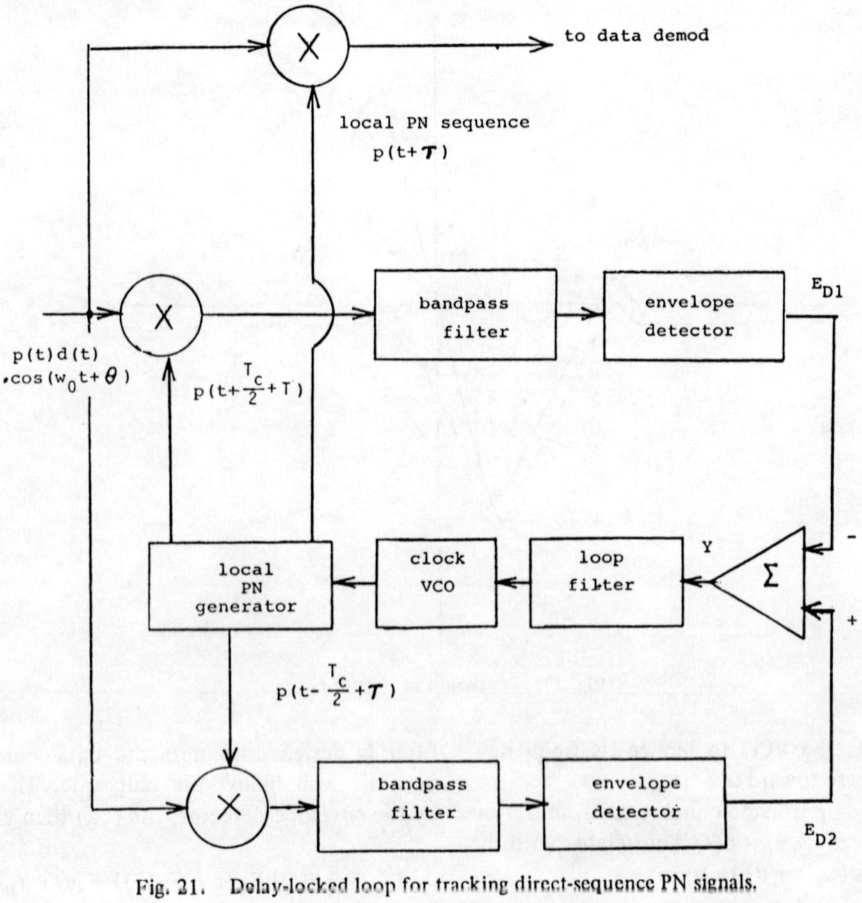

Fig. 21. Delay-locked loop for tracking direct-sequence PN signals.

The mean time to acquire a signal can now be written as

$$\bar{T}_{acq} = \bar{T}_s + \bar{T}_{s/L}[P_D(1-P_D) + 2P_D(1-P_D)^2 + \cdots]$$

$$\bar{T}_s + \frac{1-P_D}{P_D}\bar{T}_{s/L}$$

or

$$\bar{T}_{acq} = \left[L(\lambda + \tfrac{1}{2})T_c + \frac{\lambda T_c P_F}{(1-P_F)^2}\right] + \frac{1-P_D}{P_D}\left[2L(\lambda + \tfrac{1}{2})T_c + \frac{\lambda T_c P_F}{(1-P_F)^2}\right]. \qquad (50)$$

VIII. TRACKING

Once acquisition, or coarse synchronization, has been accomplished, tracking, or fine synchronization, takes place. Specifically, this must include chip synchronization and, for coherent systems, carrier phase locking. In many practical systems, no data are transmitted for a specified time, sufficiently long to ensure that acquisition has occurred. During tracking, data are transmitted and detected. Typical references for tracking loops are [51]-[54].

The basic tracking loop for a direct-sequence spread-spectrum system using PSK data transmission is shown in Fig. 21. The incoming carrier at frequency f_0 is amplitude modulated by the product of the data $d(t)$ and the PN sequence $p(t)$. The tracking loop contains a local PN generator which is offset in phase from the incoming sequence $p(t)$ by a time τ which is less than one-half the chip time. To provide "fine" synchronization, the local PN generator generates two sequences, delayed from each other by one chip. The two bandpass filters are designed to have a two-sided bandwidth B equal to twice the data bit rate, i.e.,

$$B = 2R = 2/T. \qquad (51)$$

In this way the data are passed, but the product of the two PN sequences $p(t)$ and $p(t \mp T_c/2 + \tau)$ is *averaged*. The envelope detector eliminates the data since $|d(t)| = 1$. As a result, the output of each envelope detector is approximately given by

$$E_{D1,2} \cong \left|\overline{p(t)p\left(t \pm \frac{T_c}{2} + \tau\right)}\right| = \left|R_p\left(\tau + \frac{T_c}{2}\right)\right| \qquad (52)$$

where $R_p(x)$ is the autocorrelation function of the PN waveform as shown in Fig. 7(a). [See Section III for a discussion of the characteristics of $R_p(x)$.]

The output of the adder $Y(t)$ is shown in Fig. 22. We see from this figure that, when τ is positive, a positive voltage, proportional to Y, instructs the VCO to increase its frequency, thereby forcing τ to decrease, while when τ is negative, a

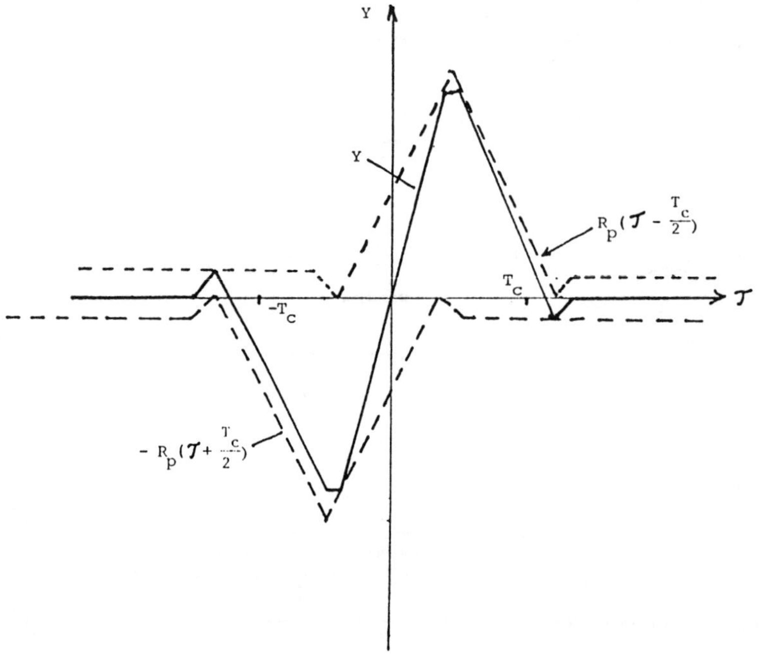

Fig. 22. Variation of Y with τ.

negative voltage instructs the VCO to reduce its frequency, thereby forcing τ to increase toward 0.

When the tracking error τ is made equal to zero, an output of the local PN generator $p(t+\tau) = p(t)$ is correlated with the input signal $p(t) \cdot d(t) \cos(\omega_0 t + \theta)$ to form

$$p^2(t)d(t) \cos(\omega_0 t + \theta) = d(t) \cos(\omega_0 t + \theta).$$

This despread PSK signal is inputted to the data demodulator where the data are detected.

An alternate technique for synchronization of a DS system is to use a tau-dither (TD) loop. This tracking loop is a delay-locked loop with only a single "arm," as shown in Fig. 23(a). The control (or gating) waveforms $g(t)$, $\bar{g}(t)$, and $g'(t)$ are shown in Fig. 23(b), and are used to generate both "arms" of the DLL even though only one arm is present. The TD loop is often used in lieu of the DLL because of its simplicity.

The operation of the loop is explained by observing that the control waveforms generate the signal

$$V_p(t) = g(t)p(t+\tau - T_c/2) + \bar{g}(t)p(t+\tau + T_c/2). \quad (53)$$

Note that either one or the other, but not both, of these waveforms occurs at each instant of time. The voltage $V_p(t)$ then multiplies the incoming signal

$$d(t)p(t) \cos(\omega_0 t + \theta).$$

The output of the bandpass filter is therefore

$$E_f(t) = d(t)g(t)\,|p(t)p(t+\tau + T_c/2)|$$
$$+ d(t)g(t)\,|p(t)p(t+\tau - T_c/2)| \quad (54)$$

where, as before, the average occurs because the bandpass filter is designed to pass the data and control signals, but cuts off well below the chip rate. The data are eliminated by the envelope detector, and (54) then yields

$$E_d(t) = g(t)\,|R_p(\tau + T_c/2)| + g(t)|R_p(\tau - T_c/2)|. \quad (55)$$

The input $Y(t)$ to the loop filter is

$$Y(t) = E_d(t)g'(t)$$
$$= g(t)\,|R_p(\tau - T_c/2)| - \bar{g}(t)\,|R_p(\tau - T_c/2)| \quad (56)$$

where the "—" sign was introduced by the inversion caused by $g'(t)$.

The narrow-band loop filter now "averages" $Y(t)$. Since each term is zero half of the time, the voltage into the VCO clock is, as before,

$$V_c(t) = |R_p(\tau - T_c/2)| - |R_p(\tau + T_c/2)|. \quad (57)$$

A typical tracking system for an FSK/FH spread-spectrum system is shown in Fig. 24. Waveforms are shown in Fig. 25. Once again, we have assumed that, although acquisition has occurred, there is still an error of τ seconds between transitions of the incoming signal's frequencies and the locally generated frequencies. The bandpass filter BPF is made sufficiently wide to pass the product signal $V_p(t)$ when $V_1(t)$ and $V_2(t)$ are at the same frequency f_i, but sufficiently narrow to reject $V_p(t)$ when $V_1(t)$ and $V_2(t)$ are at different frequencies f_i and f_{i+1}. Thus, the output of the envelope detector $V_d(t)$, shown in Fig. 24, is unity when $V_1(t)$ and $V_2(t)$ are at the same frequency and is zero when $V_1(t)$ and $V_2(t)$ are at different frequencies. From Fig. 25, we see that $V_g(t) = V_d(t)\ V_c(t)$ and is a three-level signal. This three-level signal is filtered to form a dc voltage which, in this case, presents a negative voltage to the VCO.

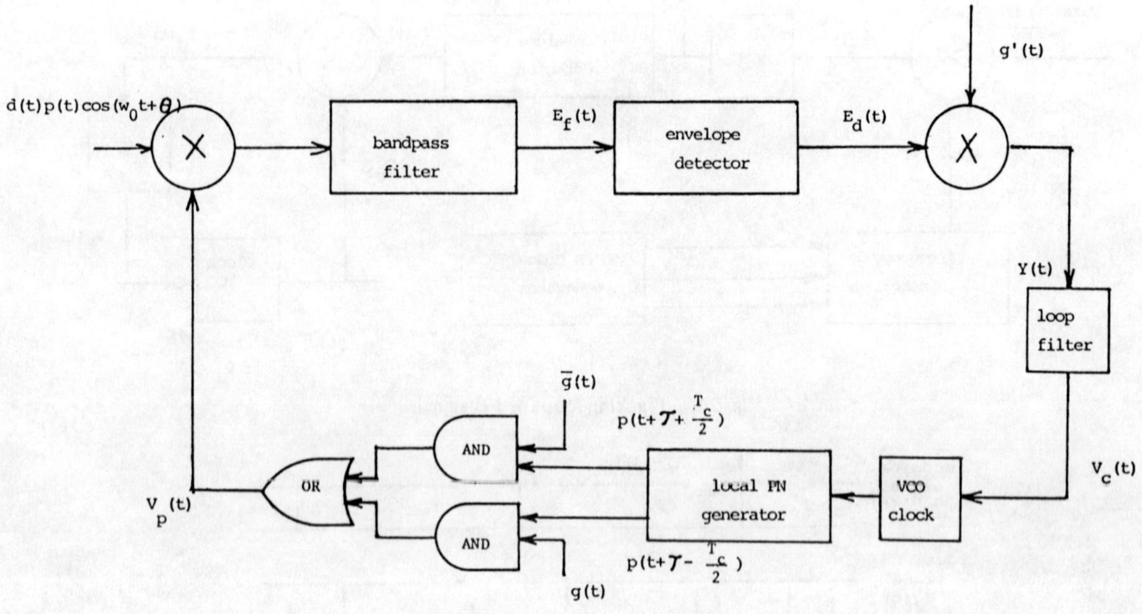

(a)

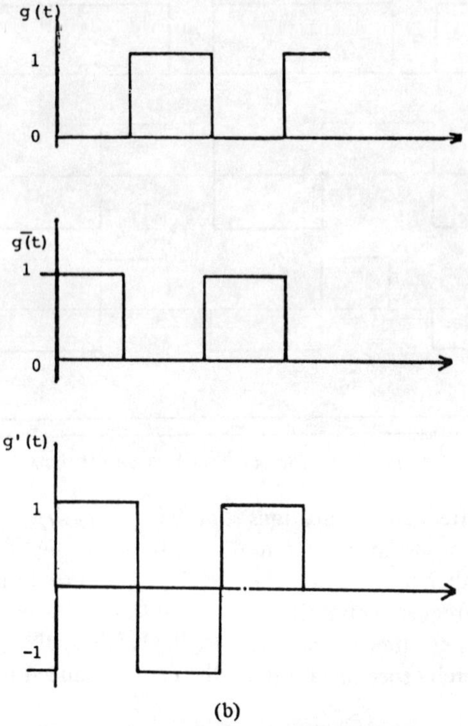

(b)

Fig. 23. The tau-dither loop. (a) Block diagram. (b) Control waveforms.

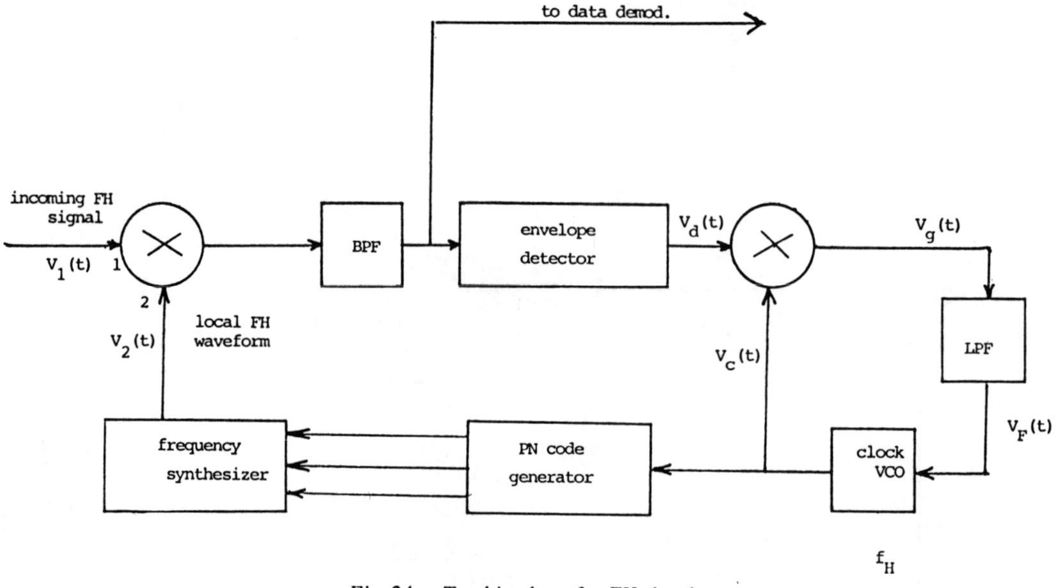

Fig. 24. Tracking loop for FH signals.

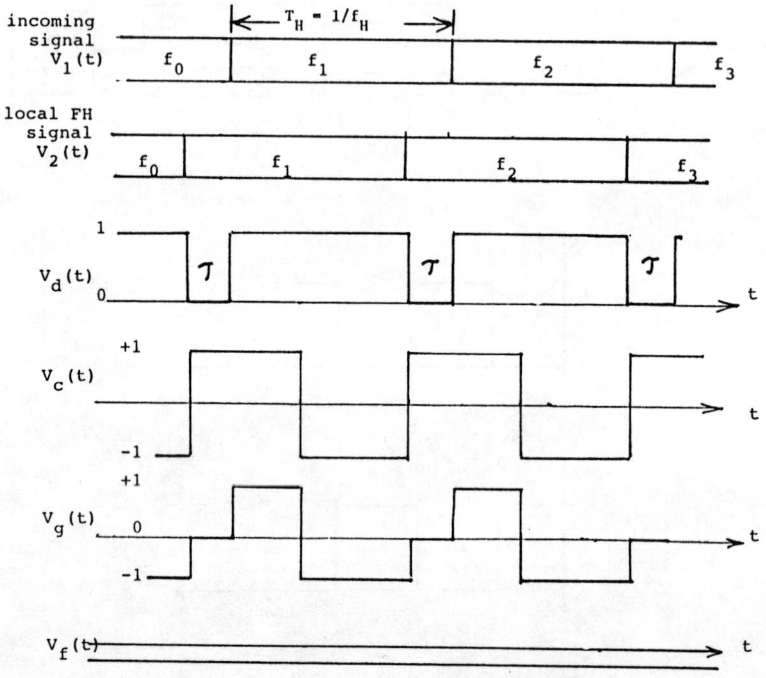

Fig. 25. Waveforms for tracking an FH signal.

It is readily seen that when $V_2(t)$ has frequency transitions which precede those of the incoming waveform $V_1(t)$, the voltage into the VCO will be negative, thereby delaying the transition, while if the local waveform frequency transitions occur after the incoming signal frequency transitions, the voltage into the VCO will be positive, thereby speeding up the transition.

The role of the tracking circuit is to keep the offset time τ small. However, even a relatively small τ can have a major impact on the probability of error of the received data. Referring to the DS system of Fig. 21, we see that if τ is not zero, the input to the data demodulator is $p(t)p(t+\tau)d(t)$ cos $(\omega_0 t + \theta)$ rather than $p^2(t)d(t)$ cos $(\omega_0 t + \theta) = d(t)$ cos $(\omega_0 t + \theta)$. The data demodulator removes the carrier and then averages the remaining signal, which in this case is $p(t)p(t+\tau)d(t)$. The result is $\overline{p(t)p(t+\tau)d(t)}$. Thus, the amplitude of the data has been reduced by $\overline{p(t)p(t+\tau)} = R_p(\tau) \leqslant 1$. For example, if $\tau = T_c/10$, that data amplitude is reduced to 90 percent of its value, and the power is reduced to 0.81. Thus, the probability of error in correctly detecting the data is reduced from

$$P_e = Q\left(\sqrt{\frac{2E_b}{\eta_0}}\right)$$

to

$$P_e(\tau = T_c/10) = Q\left(\sqrt{\frac{1.62E_b}{\eta_0}}\right),$$

and at an E_b/η_0 of 9.6 dB, P_e is increased from 10^{-5} to 10^{-4}.

IX. CONCLUSIONS

This tutorial paper looked at some of the theoretical issues involved in the design of a spread-spectrum communication system. The topics discussed included the characteristics of PN sequences, the resulting processing gain when using either direct-sequence or frequency-hopping antijam considerations, multiple access when using spread spectrum, multipath effects, and acquisition and tracking systems.

No attempt was made to present other than fundamental concepts; indeed, to adequately cover the spread-spectrum system completely is the task for an entire text [55], [56]. Furthermore, to keep this paper reasonably concise, the authors chose to ignore both practical system considerations such as those encountered when operating at, say, HF, VHF, or UHF, and technology considerations, such as the role of surface acoustic wave devices and charge-coupled devices in the design of spread-spectrum systems.

Spread spectrum has for far too long been considered a technique with very limited applicability. Such is not the case. In addition to military applications, spread spectrum is being considered for commercial applications such as mobile telephone and microwave communications in congested areas.

The authors hope that this tutorial will result in more engineers and educators becoming aware of the potential of spread spectrum, the dissemination of this information in the classroom, and the use of spread spectrum (where appropriate) in the design of communication systems.

APPENDIX

ALGEBRAIC PROPERTIES OF LINEAR FEEDBACK SHIFT REGISTER SEQUENCES

In order to fully appreciate the study of shift register sequences, it is desirable to introduce the polynomial representation (or generating function) of a sequence

$$C(x) = \sum_{i=0}^{\infty} C_i x^i \leftrightarrow (C_0, C_1, C_2, \cdots). \tag{A1}$$

If the sequence is periodic with period L, i.e.,

$$C_0, C_1, C_2, \cdots, C_{L-1} C_0 C_1, \cdots, C_{L-1}, C_0, \cdots,$$

then since $x^L C(x) \leftrightarrow (0, 0, \cdots, 0, C_0, C_1, C_2, \cdots)$,

$$C(x)(1 - x^L) = \sum_{i=0}^{L-1} C_i x^i \triangleq R(x) \tag{A2}$$

with $R(x)$ the (finite) polynomial representation of one period.

Thus, for any periodic sequence of period L,

$$C(x) = \frac{R(x)}{1 - x^L}; \qquad \deg R(x) < L. \tag{A3}$$

Next consider the periodic sequence generated by the LFSR recursion. Multiplying each side by x^n and summing gives

$$\sum_{n=0}^{\infty} C_n x^n = \sum_{k=1}^{r} a_k \sum_{n=0}^{\infty} C_{n-k} x^n$$

$$= \sum_{k=1}^{r} a_k \sum_{l=0}^{k-1} C_{l-k} x^l + \sum_{k=1}^{r} a_k x^k \left(\sum_{n=0}^{\infty} C_n x^n \right).$$

The left-hand side is the generating function $C(x)$ of the sequence. The first term on the right is a polynomial of degree $<r$, call it $g(x)$, which depends only on the initial state of the register $C_{-1}, C_{-2}, C_{-3}, \cdots, C_{-r}$. Thus, the basic equation of the register sequence may be written as

$$C(x) = \frac{g(x)}{f(x)}; \qquad \deg g(x) < r \tag{A4}$$

where $f(x) \triangleq 1 - \sum_{k=1}^{r} a_k x^k$ is the characteristic polynomial[2] (or connection polynomial) of the register. Since $C(x)$ is the generating polynomial of a sequence of period $L = 2^r - 1$, it can be shown from (A3) and (A4) that $f(x)$ must divide $1 - x^L$. This is illustrated in the following example.

Example

The three-stage binary maximal length register with $f(x) = 1 + x + x^3$ has period 7. If the initial contents of the register are $C_{-3} = 1$, $C_{-2} = 0$, $C_{-1} = 0$, then $g(x) = a_3 = 1$ and $C(x) = 1/(1 + x + x^3)$. Long division (modulo 2) yields

$$C(x) = 1 + x + x^2 + x^4 + x^7 + x^9 + \cdots$$

which is the generating function of the periodic sequence

$$1\,1\,1\,0\,1\,0\,0 \,\vdots\, 1\,1\,1\,0\,1 \cdots,$$

and which is precisely the sequence generated by the corresponding recursion

$$C_n = C_{n-1} + C_{n-3} \pmod 2.$$

Observe that

$$(1 + x + x^3)(1 + x + x^2 + x^4) = 1 + x^7$$

so that $f(x)$ divides $1 + x^7$. Also, we may write

$$C(x) = \frac{1}{1 + x + x^3} \cdot \frac{1 + x + x^2 + x^4}{1 + x + x^2 + x^4} = \frac{1 + x + x^2 + x^4}{1 + x^7}$$

which is in the form of (A3).

[2] For binary sequences, all sums are modulo two and minus is the same as plus. The polynomials defining them have 0, 1 coefficients and are said to be polynomials over a finite field with two elements. A field is a set of elements, and two operations, say, + and ·, which obey the usual rules of arithmetic. A finite field with q elements is called a Galois field and is designated as $GF(q)$.

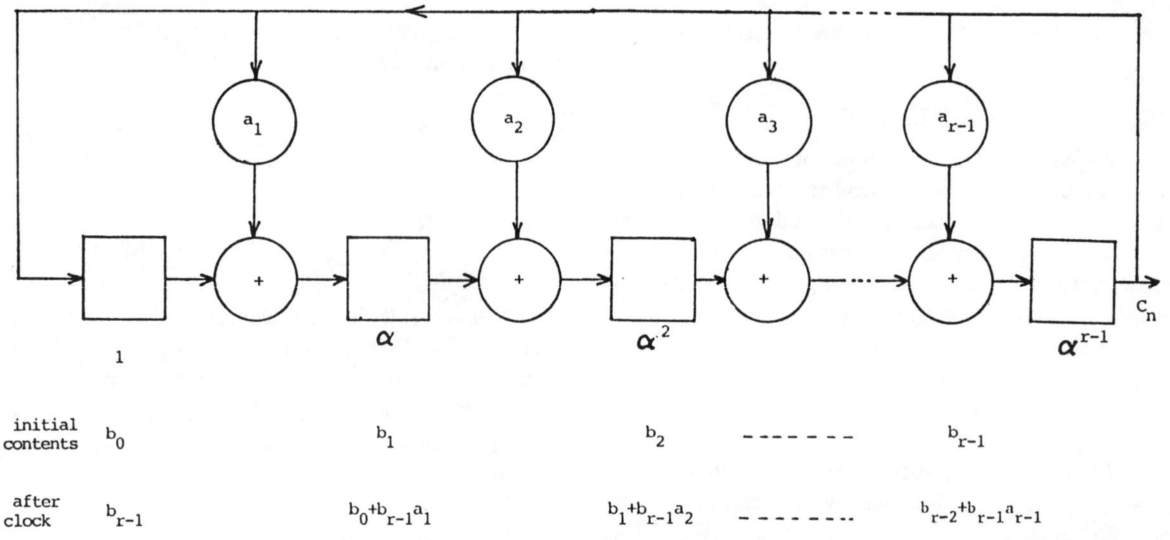

Fig. 26. Binary modular shift register generator with polynomial $f_M(x) = 1 + a_1x + a_2x^2 + \cdots + a_{r-1}x^{r-1} + x^r$.

It is easy to see (by multiplying and equating coefficients of like powers) that if

$$\frac{1}{f(x)} = \frac{1}{1 + a_1x + a_2x^2 + \cdots + a_rx^r}$$

$$= C_0 + C_1x + C_2x^2 + \cdots = C(x)$$

then

$$C_n = \sum_{k=1}^{r} a_k C_{n-k},$$

so that (except for initial conditions) $f(x)$ completely describes the maximal length sequence. Now what properties must $f(x)$ possess to ensure that the sequence is maximal length? Aside from the fact that $f(x)$ must divide $1 + x^L$, it is necessary (but not sufficient) that $f(x)$ be irreducible, i.e., $f(x) \neq f_1(x) \cdot f_2(x)$. Suppose that $f(x) = f_1(x)\, f_2(x)$ with $f_1(x)$ of degree r_1, $f_2(x)$ of degree r_2, and $r_1 + r_2 = r$. Then we can write, by partial fractions,

$$\frac{1}{f(x)} = \frac{\alpha(x)}{f_1(x)} + \frac{\beta(x)}{f_2(x)}; \qquad \deg \alpha(x) < r_1 \quad \deg \beta(x) < r_2.$$

The maximum period of the expansion of the first term is $2^{r_1}-1$ and that of the second term is $2^{r_2}-1$. Hence, the period of $1/f(x) \leqslant$ least common multiple of $(2^{r_1}-1, 2^{r_2}-1) < 2^r-3$. This is a contradiction, since if $f(x)$ were maximal length, the period of $1/f(x)$ would be 2^r-1. Thus, a *necessary* condition that the LFSR is maximal length is that $f(x)$ is irreducible.

A *sufficient* condition is that $f(x)$ is *primitive*. A primitive polynomial of degree r over $GF(2)$ is simply one for which the period of the coefficients of $1/f(x)$ is 2^r-1. However, additional insight can be had by examining the roots of $f(x)$. Since $f(x)$ is irreducible over $GF(2)$, we must imagine that the roots are elements of some larger (extension) field. Suppose that α is such an element and that $f(\alpha) = 0 = \alpha^r + a_{r-1}\alpha^{r-1} + \cdots + a_1\alpha + 1$ or

$$\alpha^r = a_{r-1}\alpha^{r-1} + \cdots + a_1\alpha + 1. \tag{A5}$$

We see that *all* powers of α can be expressed in terms of a linear combination of $\alpha^{r-1}, \alpha^{r-2}, \cdots, \alpha, 1$ since any powers larger than $r-1$ may be reduced using (A5). Specifically, suppose we have some power of α that we represent as

$$\beta \triangleq b_0 + b_1\alpha + b_2\alpha^2 + \cdots + b_{r-1}\alpha^{r-1}. \tag{A6}$$

Then if we multiply this β by α and use (A5), we obtain

$$\beta\alpha = b_{r-1} + (b_0 + b_{r-1}a_1)\alpha + (b_1 + b_{r-1}a_2)\alpha^2 + \cdots + (b_{r-2} + b_{r-1}a_{r-1})\alpha^{r-1} \tag{A7}$$

The observations above may be expressed in another, more physical way with the introduction of an LFSR in modular form [called a modular shift register generator (MSRG)] shown in Fig. 26. The feedback, modulo 2, is between the delay elements. The binary contents of the register at any time are shown as $b_0, b, \cdots, b_{r-1}$. This vector can be identified with β as

$$\beta = b_0 + b_1\alpha + \cdots + b_{r-1}\alpha^{r-1} \leftrightarrow [b_0, b_1, \cdots b_{r-1}],$$

the contents of the first stage being identified with the coefficient of α^0, those of the second stage with the coefficient of α^1, etc. After one clock pulse, it is seen that the register contents correspond to

$$\beta\alpha = b_{r-1} + (b_0 + b_{r-1})\alpha + \cdots + (b_{r-2} + b_{r-1}a_{r-1})\alpha^{r-1}$$

$$\leftrightarrow [b_{r-1}, \cdots, b_{r-2} + b_{r-1}a_{r-1}].$$

Thus, the MSRG is an α-multiplier. Now if $\alpha^0, \alpha, \alpha^2, \alpha^3, \cdots, \alpha^{L-1}$, $L = 2^r-1$ are all *distinct*, we call α a *primitive* element of $GF(2^r)$. The register in Fig. 26 cycles through all states (starting in any nonzero state), and hence generates a maximal length sequence. Thus, another way of describing that the polynomial $f_M(x)$ is primitive (or maximal length) is that it has a primitive element in $GF(2^r)$ as a root.

There is an intimate relationship between the MSRG shown in Fig. 26 and the SSRG shown in Fig. 4. From Fig. 26 it is easily seen that the output sequence C_n satisfies the recursion

$$C_n = \sum_{k=0}^{r-1} a_k C_{n-r+k}. \tag{A8}$$

Multiplying both sides by x^n and summing yields

$$C(x) \triangleq \sum_{n=-\infty}^{\infty} C_n x^n = \sum_{k=0}^{r-1} a_k \sum_{n=0}^{\infty} C_{n-r+k} x^n$$

$$= \sum_{k=0}^{r-1} a_k x^{r-k} \sum_{l=-r+k}^{-1} C_l x^l$$

$$+ x^r \sum_{k=0}^{r-1} a_k x^{-k} \left(\sum_{n=0}^{\infty} C_n x^n \right) \tag{A9}$$

or

$$C(x) = g_M(x) + x^r \sum_{k=0}^{r-1} a_k x^{-k} C(x). \tag{A10}$$

$g_M(x)$ is the first term on the right-hand side of (A9) and is a polynomial of degree $< r$ which depends on the initial state. Then we have

$$C(x) = \frac{g_M(x)}{f_M(x)} \tag{A11}$$

where

$$f_M(x) = 1 - \{a_0 x^r + a_1 x^{r-1} + a_2 x^{r-2} + \cdots + a_{r-1} x\}$$

(recall that in $GF(2)$, minus is the same as plus) is the characteristic (or connection) polynomial of the MSRG. Since the sequence C_n [of coefficients of $C(x)$] when $f_M(x)$ is primitive depends *only* on $f_M(x)$ (discounting phase), the relationship between the SSRG and the MSRG which generates the *same* sequence is

$$f(x) = x^r f_M\left(\frac{1}{x}\right). \tag{A12}$$

$f_M(x)$ is called the *reciprocal* polynomial of $f(x)$ and is obtained from $f(x)$ by reversing the order of the coefficients.

There are several good tables of irreducible and primitive polynomials available [2], [5], [6], and although the tables do not list all the primitive polynomials, algorithms exist [7] which allow one to generate all primitive polynomials of a given degree if one of them is known. The number $\lambda(r)$ of primitive polynomials of degree r is [4]

$$\lambda(r) = \frac{\phi(2^r - 1)}{r} \tag{A13}$$

where $\phi(m)$ is the number of integers less than m which are relatively prime to m (Euler totient function). The growth of this number with r is shown in Table I.

TABLE I
THE NUMBER OF MAXIMAL LENGTH LINEAR SRG SEQUENCES OF DEGREE $r = \lambda(r) = \phi(2^r - 1)/r$

r	2^r-1	$\lambda(r)$
1	1	1
2	3	1
3	7	2
4	15	2
5	31	6
6	63	6
7	127	18
8	255	16
9	511	48
10	1,023	60
11	2,047	176
12	4,095	144
13	8,191	630
14	16,383	756
15	32,767	1,800
16	65,535	2,048
17	131,071	7,710
18	262,143	8,064
19	524,287	27,594
20	1,048,575	24,000
21	2,097,151	87,672
22	4,194,303	120,032

The algebra of LFSR's is useful in constructing codes with uniformly low cross correlation known as Gold codes. The underlying principle of these codes is based on the following *theorem* [15].

If $f_1(x)$ is the minimal polynomial of the primitive element $\alpha \in GF(2^r)$ and $f_t(x)$ is the minimal polynomial of α^t, where both $f_1(x)$ and $f_t(x)$ are of degree r and

$$t = \begin{cases} 2^{\frac{r+1}{2}} + 1, & r \text{ odd} \\ 2^{\frac{r+2}{2}} + 1, & r \text{ even,} \end{cases}$$

then the product $f(x) \triangleq f_1(x) f_t(x)$ determines an LFSR which generates $2^r + 1$ different sequences (corresponding to the $2^r + 1$ states in distinct cycles) of period $2^r - 1$, and such that for any pair C' and C'',

$$L\,|R_{C'C''}(\tau)| < t.$$

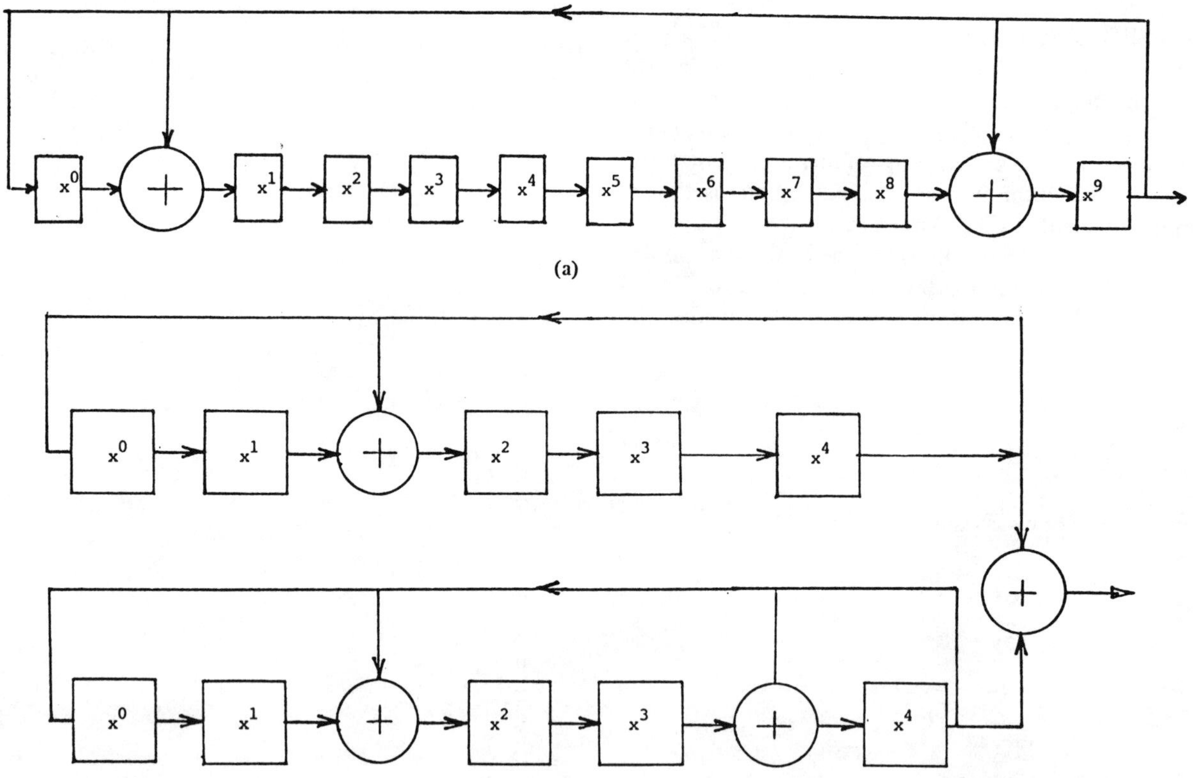

(b)

Fig. 27. Two implementations of LFSR which generate Gold codes of length $2^5 - 1 = 31$ with maximum cross correlation $t = 9$. (a) LFSR with $f(x) = 1 + x + x^3 + x^9 + x^{10}$. (b) LFSR which generates sequences corresponding to $f(x) = (1 + x^2 + x^5)\cdot(1 + x^2 + x^4 + x^5) = 1 + x + x^3 + x^9 + x^{10}$.

Futhermore, $R_{C'C''}(\tau)$ is only a three-valued function for any integer τ.

A minimal polynomial of α is simply the smallest degree monic[3] polynomial for which α is a root. With the help of a table of primitive polynomials, we can identify minimal polynomials of powers of α and easily construct Gold codes. For example, if $r = 5$ and $t = 2^3 + 1 = 9$, using [2] we find that $f_1(x) = 1 + x^2 + x^5$ and $f_9(x) = 1 + x^2 + x^4 + x^5$. Then $f(x) = 1 + x + x^3 + x^9 + x^{10}$. The two ways to represent this LFSR (in MSRG form) are shown in Fig. 27. Fig. 27(a) shows one long nonmaximal length register of degree 10 which generates sequences of period $2^5 - 1 = 31$. Since there are $2^{10} - 1$ possible nonzero initial states, the number of initial states that result in distinct cycles is $(2^{10} - 1)/(2^5 - 1) = 2^5 + 1 = 33$. Each of these initial states specifies a different Gold code of length 31. Fig. 27(b) shows how the same result can be obtained by adding the outputs of the two MLFSR's of degree 5 together modulo two. This follows simply from the observation that the sequence(s) generated by $f(x)$ are just the coefficients in the expansion of $1/f(x) = 1/f_1(x)\cdot f_9(x)$. By using partial fractions, one can see that the resulting coefficients are the (modulo two) sum of the coefficients of like powers in the expansion of $1/f_1(x)$ and $1/f_9(x)$. Naturally, the sequence resulting will depend on the relative *phases* of the two degree-5 registers. As before, there are $(2^{10} - 1)/(2^5 - 1) = 2^5 + 1 = 33$ relative phases which result in 33 different sequences satisfying the cross-correlation bound given by the theorem.

[3] A monic polynomial is one whose coefficient of its highest power is unity.

GLOSSARY OF SYMBOLS

a_n	{0, 1} feedback taps for LFSR.
B_D	One-sided bandwidth (Hz) for data signal(s).
B_{ss}	One-sided bandwidth (Hz) of baseband spread-spectrum signal.
$C(x)$	Generating function of C_n; $C(x) = \Sigma_{n=0}^{\infty} C_n x^n$.
C_n	{0, 1} LFSR sequence.
C_n' or C_n''	{1, −1} LFSR sequence.
D	Dimensionality of underlying signal space.
$d(t)$	Data sequence waveform.
Δ	Initial offset, in chips, of incoming signal and locally generated code.
DS	Direct sequence.
E_b	Energy/information bit.
E_J	Jammer energy over the correlation interval.
E_s	Energy/symbol.
$f(x)$	Characteristic (connection) polynomial of an LFSR, $f(x) = 1 + a_1 x + \cdots + a_{r-1} x^{r-1} + x^r$.
f_c	Chip rate; $T_c = 1/f_c$.
FH	Frequency hopping.
G_P	Processing gain.
$J(t)$	Jammer signal waveform.

K	Number of frequencies jammed by partial-band jammer.
$L = 2^r - 1$	Period of PN sequence.
L	Implementation losses.
λ	Number of chips examined during each search in the process of acquisition.
$\lambda(r)$	Number of binary maximal length PN codes of degree r (length $L = 2^r - 1$).
M	Signal alphabet size.
M_J	Jamming margin.
n	Number of chips/bit or number of dimensions of spread signal space.
N	Number of frequencies in FH.
N_c	Number of chips in uncertainty region at start of acquisition.
N_u	Number of users in CDMA system.
η_0	One-sided white noise power spectral density (W/Hz).
η_{0J}	$P_J/2f_c$ = power density of jammer.
$n_w(t)$	Additive white Gaussian noise (AWGN).
$p(t)$	Spreading sequence waveform.
P_D	Probability of detection.
P_e	Probability of error.
P_F	Probability of false alarm.
P_J	Jammer power.
P_N	Noise power.
P_s	Signal power.
PN	Pseudonoise sequence.
r	Number of stages of shift register.
$r(t)$	Received waveform.
R	Data rate (bits/s).
$R_p(\tau)$	Autocorrelation function.
$R_{C'C''}(\tau)$	Cross-correlation function of two (periodic) ± 1 sequences C_n', C_n''.
$\rho(\tau)$	$(1/T) \int_0^\tau p(t)\, p(t-\tau)\, dt$ (partial correlation function).
$\hat{\rho}(\tau)$	$(1/T) \int_\tau^T p(t)\, p(t-\tau)\, dt$.
$S(t)$	Transmitted signal waveform.
$S_p(f)$	Power spectral density of spreading sequence waveform [also denoted $S_{ss}(f)$].
SNR	Signal-to-noise power ratio.
SJR	Signal-to-jammer power ratio.
T	Signal or symbol duration.
T_c	Chip duration.
T_H	Time to hop one frequency; $1/T_H$ = hopping rate.
V	Correlator output voltage.

ACKNOWLEDGMENT

The authors wish to thank the anonymous reviewers for their constructive suggestions in the final preparation of this paper.

REFERENCES

[1] R. A. Scholtz, "The origins of spread-spectrum communications," this issue, pp. 822–854.

[2] W. W. Peterson and E. J. Weldon, Jr., *Error Correcting Codes*, 2nd ed. Cambridge, MA: M.I.T. Press, 1972.

[3] J. M. Wozencraft and I. M. Jacobs, *Principles of Communication Engineering*. New York: Wiley, 1965.

[4] S. W. Golomb, *Shift Register Sequences*. San Francisco, CA: Holden Day, 1967.

[5] R. W. Marsh, *Table of Irreducible Polynomials over GF(2) Through Degree 19*. Washington, DC: NSA, 1957.

[6] W. Stahnke, "Primitive binary polynomials," *Math. Comput.*, vol. 27, pp. 977–980, Oct. 1973.

[7] E. R. Berlekamp, *Algebraic Coding Theory*. New York: McGraw-Hill, 1968.

[8] J. L. Massey, "Shift-register synthesis and BCH decoding," *IEEE Trans. Inform. Theory*, vol. IT-15, pp. 122–127, Jan. 1969.

[9] N. G. deBruijn, "A combinatorial problem," in *Koninklijke Nederlands Akademie Van Wetenschappen Proc.*, 1946, pp. 758–764.

[10] E. J. Groth, "Generation of binary sequences with controllable complexity," *IEEE Trans. Inform. Theory*, vol. IT-17, pp. 288–296, May 1971.

[11] E. L. Key, "An analysis of the structure and complexity of nonlinear binary sequence generators," *IEEE Trans. Inform. Theory*, vol. IT-22, pp. 732–736, Nov. 1976.

[12] H. Beker, "Multiplexed shift register sequences," presented at CRYPTO '81 Workshop, Santa Barbara, CA, 1981.

[13] J. L. Massey and J. J. Uhran, "Sub-baud coding," in *Proc. 13th Annu. Allerton Conf. Circuit and Syst. Theory*, Monticello, IL, Oct. 1975, pp. 539–547.

[14] J. H. Lindholm, "An analysis of the pseudo randomness properties of the subsequences of long *m*-sequences," *IEEE Trans. Inform. Theory*, vol. IT-14, 1968.

[15] R. Gold, "Optimal binary sequences for spread spectrum multiplexing," *IEEE Trans. Inform. Theory*, vol. IT-13, pp. 619–621, 1967.

[16] D. V. Sarwate and M. B. Pursley, "Cross correlation properties of pseudo-random and related sequences," *Proc. IEEE*, vol. 68, pp. 598–619, May 1980.

[17] A. Lempel, M. Cohn, and W. L. Eastman, "A new class of balanced binary sequences with optimal autocorrelation properties," IBM Res. Rep. RC 5632, Sept. 1975.

[18] A. G. Konheim, *Cryptography, A Primer*. New York: Wiley, 1981.

[19] D. L. Schilling, L. B. Milstein, R. L. Pickholtz, and R. Brown, "Optimization of the processing gain of an *M*-ary direct sequence spread spectrum communication system," *IEEE Trans. Commun.*, vol. COM-28, pp. 1389–1398, Aug. 1980.

[20] L. B. Milstein, S. Davidovici, and D. L. Schilling, "The effect of multiple-tone interfering signals on a direct sequence spread spectrum communication system," *IEEE Trans. Commun.*, vol. COM-30, pp. 436–446, Mar. 1982.

[21] S. W. Houston, "Tone and noise jamming performance of a spread spectrum *M*-ary FSK and 2, 4-ary DPSK waveforms," in *Proc. Nat. Aerosp. Electron. Conf.*, June 1975, pp. 51–58.

[22] G. K. Huth, "Optimization of coded spread spectrum systems performance," *IEEE Trans. Commun.*, vol. COM-25, pp. 763–770, Aug. 1977.

[23] R. H. Pettit, "A susceptibility analysis of frequency hopped *M*-ary NCPSK—Partial-band noise on CW tone jamming," presented at the Symp. Syst. Theory, May 1979.

[24] L. B. Milstein, R. L. Pickholtz, D. L. Schilling, "Optimization of the processing gain of an FSK-FH system," *IEEE Trans. Commun.*, vol. COM-28, pp. 1062–1079, July 1980.

[25] M. K. Simon and A. Polydoros, "Coherent detection of frequency-hopped quadrature modulations in the presence of jamming—Part I: QPSK and QASK; Part II: QPR class I modulation," *IEEE Trans. Commun.*, vol. COM-29, pp. 1644–1668, Nov. 1981.

[26] A. J. Viterbi and I. M. Jacobs, "Advances in coding and modulation for noncoherent channels affected by fading, partial band, and multiple access interference," in *Advances in Communication Systems*, vol. 4. New York: Academic, 1975.

[27] J. M. Acin and R. D. Turner, "Effect of co-channel interference on CPSK carriers," *IEEE Trans. Commun.*, vol. COM-21, pp. 783–790, July 1973.

[28] R. H. Pettit, "Error probability for NCFSK with linear FM jamming," *IEEE Trans. Aerosp. Electron. Syst.*, vol. AES-8, pp. 609–614, Sept. 1972.

[29] A. J. Viterbi, "Spread spectrum communications—Myths and realities." *IEEE Commun. Mag.*, pp. 11–18, May 1979.

[30] D. J. Torrieri, *Principles of Military Communication Systems*. Dedham, MA: Artech House, 1981.

[31] M. B. Pursley, "Performance evaluation for phase-coded spread spectrum multiple-access communication—Part I: System analysis," *IEEE Trans. Commun.*, vol. COM-25, pp. 795–799, Aug. 1977.
[32] K. Yao, "Error probability of asynchronous spread-spectrum multiple access communication systems," *IEEE Trans. Commun.*, vol. COM-25, pp. 803–809, Aug. 1977.
[33] C. L. Weber, G. K. Huth, and B. H. Batson, "Performance considerations of code division multiple access systems," *IEEE Trans. Veh. Technol.*, vol. VT-30, pp. 3–10, Feb. 1981.
[34] M. B. Pursley and D. V. Sarwate, "Performance evaluation for phase-coded spread spectrum multiple-access communication—Part II. Code sequence analysis," *IEEE Trans. Commun.*, vol. COM-25, pp. 800–803, Aug. 1977.
[35] M. B. Pursley and H. F. A. Roefs, "Numerical evaluation of correlation parameters for optimal phases of binary shift-register sequences," *IEEE Trans. Commun.*, vol. COM-25, pp. 1597–1604, Aug. 1977.
[36] G. Solomon, "Optimal frequency hopping sequences for multiple access," in *Proc. 1973 Symp. Spread Spectrum Commun.*, vol. 1, AD915852, pp. 33–35.
[37] D. V. Sarwate and M. B. Pursley, "Hopping patterns for frequency-hopped multiple-access communication," in *Proc. 1978 IEEE Int. Conf. Commun.*, vol. 1, pp. 7.4.1–7.4.3.
[38] P. S. Henry, "Spectrum efficiency of a frequency-hopped-DPSK spread spectrum mobile radio system," *IEEE Trans. Veh. Technol.*, vol. VT-28, pp. 327–329, Nov. 1979.
[39] O. C. Yue, "Hard-limited versus linear combining for frequency-hopping multiple-access systems in a Rayleigh fading environment," *IEEE Trans. Veh. Technol.*, vol. VT-30, pp. 10–14, Feb. 1981.
[40] R. W. Nettleton and G. R. Cooper, "Performance of a frequency-hopped differentially modulated spread-spectrum receiver in a Rayleigh fading channel," *IEEE Trans. Veh. Technol.*, vol. VT-30, pp. 14–29, Feb. 1981.
[41] D. E. Borth and M. B. Pursley, "Analysis of direct-sequence spread-spectrum multiple-access communication over Rician fading channels," *IEEE Trans. Commun.*, vol. COM-27, pp. 1566–1577, Oct. 1979.
[42] E. A. Geraniotis and M. B. Pursley, "Error probability bounds for slow frequency-hopped spread-spectrum multiple access communications over fading channels," in *Proc. 1981 Int. Conf. Commun.*
[43] L. B. Milstein and D. L. Schilling, "Performance of a spread spectrum communication system operating over a frequency-selective fading channel in the presence of tone interference," *IEEE Trans. Commun.*, vol. COM-30, pp. 240–247, Jan. 1982.
[44] ——, "The effect of frequency selective fading on a noncoherent FH-FSK system operating with partial-band interference," this issue, pp. 904–912.
[45] J. M. Aein and R. L. Pickholtz, "A simple unified phasor analysis for PN multiple access to limiting repeaters," this issue, pp. 1018–1026.
[46] R. B. Ward, "Acquisition of pseudonoise signals by sequential estimation," *IEEE Trans. Commun. Technol.*, vol. COM-13, pp. 474–483, Dec. 1965.
[47] R. B. Ward and K. P. Yiu, "Acquisition of pseudonoise signals by recursion-aided sequential estimation," *IEEE Trans. Commun.*, vol. COM-25, pp. 784–794, Aug. 1977.
[48] P. M. Hopkins, "A unified analysis of pseudonoise synchronization by envelope correlation," *IEEE Trans. Commun.*, vol. COM-25, pp. 770–778, Aug. 1977.
[49] J. K. Holmes and C. C. Chen, "Acquisition time performance of PN spread-spectrum systems," *IEEE Trans. Commun.*, vol. COM-25, pp. 778–783, Aug. 1977.
[50] S. S. Rappaport, "On practical setting of detection thresholds," *Proc. IEEE*, vol. 57, pp. 1420–1421, Aug. 1969.
[51] J. J. Spilker, Jr., "Delay-lock tracking of binary signals," *IEEE Trans. Space Electron. Telem.*, vol. SET-9, pp. 1–8, Mar. 1963.
[52] P. T. Nielson, "On the acquisition behavior of delay lock loops," *IEEE Trans. Aerosp. Electron. Syst.*, vol. AES-12, pp. 415–523, July 1976.
[53] ——, "On the acquisition behavior of delay lock loops," *IEEE Trans. Aerosp. Electron. Syst.*, vol. AES-11, pp. 415–417, May 1975.
[54] H. P. Hartman, "Analysis of the dithering loop for PN code tracking," *IEEE Trans. Aerosp. Electron. Syst.*, vol. AES-10, pp. 2–9, Jan. 1974.
[55] R. C. Dixon, *Spread Spectrum Systems*. New York: Wiley, 1976.
[56] J. K. Holmes, *Coherent Spread Spectrum Systems*. New York: Wiley, 1982.

Throughput Analysis for Code Division Multiple Accessing of the Spread Spectrum Channel

JOSEPH Y. N. HUI, MEMBER, IEEE

Abstract—We consider the use of error correction codes of rate r on top of pseudonoise (PN) sequence coding for code division multiple accessing of the spread spectrum channel. The channel is found to have a maximum throughput of 0.72 and 0.36 based on the evaluation of channel capacity and cutoff rate, respectively. More generally, these two values are derived for given bandwidth expanding n/r versus n/N where n is the length of the PN sequence and N is the number of simultaneous users. It is found that to achieve the maximum throughput, n should be small. This implies that coding schemes with short PN sequences and low rate codes are superior in terms of throughput or antijam capability. The extreme case of $n=1$ corresponds to using a very low rate code with no PN sequence coding. Convolutional codes are recommended and analyzed for their error rate and decoding complexity.

I. Introduction

CODE division multiple accessing [1], [6] is attractive for mobile communication for the following reasons. First, a large number of users may subscribe to such a system while system performance (error rate) depends only on the number of simultaneous users. The channel degrades gracefully as the number of active users increases. Second, each user may transmit asynchronously and no scheduling of the channel is required. This uniformity makes entry of new subscribers easy. For such systems, the signal of the other active users becomes interference (having statistics that are approximately Gaussian) to a receiver. To reduce the effect of this interference, each user redundantly encodes its message and spreads the coded signal over a large spectral bandwidth. The obvious question then, is how many users (each with fixed data rate) may transmit simultaneously and reliably for a given spectral bandwidth? Is there a capacity analogous to the e^{-1} throughput of the slotted ALOHA system? Before answering this question, Section I describes the coding scheme and subsequently models the channel. A Gaussian approximation is used in the process. Section II then evaluates the total capacity and cutoff rate for the active users. For convolutional codes and sequential decoding, it is shown that the maximum throughput is $(\log_2 e)/4 = 0.36$. A similar result using a different model is derived by Viterbi and reported in [6]. Section III gives an analysis for the bit error probability and decoding complexity as a function of the constraint length and rate of the convolutional code, as well as the length of the PN sequence.

Manuscript received July 28, 1983; revised February 8, 1984. This work was supported by the National Science Foundation under Grant NSF/ECS 78-19880 and the Advanced Project Agency under Grant ONR-N00014-75-C1183.

The author is with Bell Communications Research, Murray Hill, NJ 07974.

II. Modeling

Let N be the number of active users. Each user i, $1 \leqslant i \leqslant N$, is assigned an n bit PN sequence $\{c_{ij}\}_{j=1}^{n}$, $c_{ij} \in \{\pm 1\}$. A chip is defined as an interval of length d. The PN sequence carrier is the function

$$c_i(t) = \sum_{j=1}^{n} c_{ij} s(t - jd)$$

in which $s(t)$ equals 1 if $0 \leqslant t \leqslant d$, and 0 elsewhere. Each user encodes the binary (0 or 1) data stream $\{\cdots u_{i,-1} u_{i,0} u_{i,1} \cdots\}$ into the antipodal (1 or -1) code stream $\{\cdots x_{i,-1} x_{i,0} x_{i,1} \cdots\}$. The rate of encoding is r ($\leqslant 1$). A coding scheme using a convolutional code is shown in Fig. 1. The signal sent by each user[1] is

$$x_i(t) = \sqrt{P_i} \sum_{k=-\infty}^{\infty} x_{i,k} c_i(t - knd + D_i)$$

in which P_i is the power for the ith transmitter, and D_i is the delay for the ith user, which is randomly distributed in $[0, nd]$. The channel output is

$$y(t) = \sum_{i=1}^{N} x_i(t).$$

The ith receiver first acquires synchronization with the ith transmitter by estimating D_i, which we assume can be estimated accurately by some means. Demodulation is performed by match filtering and the values

[1]Viterbi's derivation considers BPSK with a random phase $\cos\theta_i$ for each user. Consequently, the results for sum-capacity and sum-cutoff rates (1.44 and 0.72, respectively) are twice that of our model.

Reprinted from *IEEE J. Selected Areas Commun.*, vol. SAC-2, no. 4, pp. 482–486, July 1984.

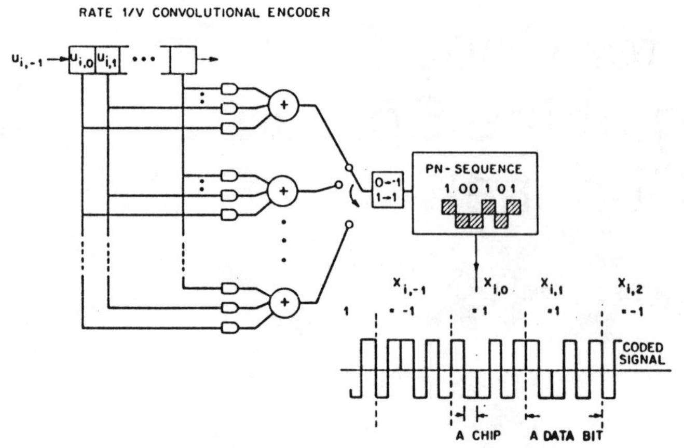

Fig. 1. Coding scheme for the spread spectrum channel.

$$y_{i,k} = \left(nd^2 \sum_{\substack{j=1 \\ j \neq i}}^{N} P_j \right)^{-1/2} \int_{-D_i + knd}^{-D_i + (k+1)nd} y(t) \cdot c_i(t - knd + D_i)\, dt$$

are obtained. The $y_{i,k}$'s are subsequently quantized. We shall assume either hard quantization or no quantization. The channel, as viewed by the ith user, can be characterized by $P(y_{i,k}/x_{i,k})$. Assuming that the PN sequences are generated randomly with equiprobable use of 1 and -1,[2] it follows readily that the random variable $y_{i,k}$ has mean

$$E(y_{i,k}/x_{i,k} = m) = m \left(nP_i \Big/ \sum_{\substack{j=1 \\ j \neq i}}^{N} P_j \right)^{1/2}$$

and variance

$$E\left((y_{i,k} - \bar{y}_{i,k})^2 / x_{i,k} = m \right) = 1$$

for $m = 1$ or $= 1$. Since $y_{i,k}$ is the sum of a large number of random variables, the statistics of $y_{i,k}$ are approximately Gaussian by the law of large numbers. Hence, by defining

$$N_{o,i}/2 = \sum_{\substack{j=1 \\ j \neq i}} P_j d$$

which has the interpretation of equivalent one-sided noise spectral density [5], and dropping the subscript k[3] we have the approximation

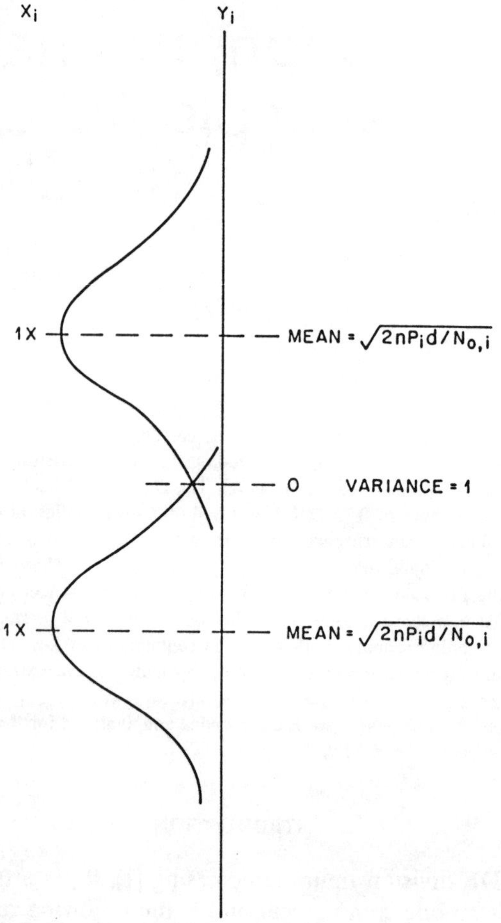

Fig. 2. The binary input Gaussian output channel.

$$P(y_i/x_i = m) = \exp\left(-\left(y_i - m\sqrt{2nP_i d/N_{o,i}} \right)^2 / 2 \right) / \sqrt{2\pi}.$$

The channel (actually the channel plus quantizer) can be modeled by the channel transition diagram in Fig. 2 and Fig. 3 for the cases of no quantization and hard quantization, respectively. Fig. 2 is essentially a binary input Gaussian output channel, whereas Fig. 3 is a binary symmetric channel (BSC) with crossover probability $p = Q(\sqrt{2nP_i d/N_{o,i}})$ in which

$$Q(z) = \int_{-\infty}^{-z} e^{-x^2/2}\, dx / \sqrt{2\pi}.$$

Consider the case of no coding on top of the PN sequence (i.e., $x_{i,k} = 1$ if $u_{i,k} = 1$ and $x_{i,k} = -1$ if $u_{i,k} = 0$). The channel is therefore the BSC of Fig. 3. Assuming equal power for all users, the bit error probability is upper bounded by

$$Q\left(\sqrt{2nP_i d/N_{o,i}} \right) = Q\left(\sqrt{n/(N-1)} \right) \cong Q(S^{-1/2})$$

in which S is the total throughput (N/n bits/chip). This upper bound is shown in Fig. 4. It is readily seen that the throughput has to be reduced substantially for tolerable error probability. An immediate conclusion is that coding should be used on top of PN sequence coding to achieve a lower bit error probability.

[2] There is the question of whether specifically designed PN sequences (such as the Gold code [2]) may achieve a higher capacity than PN sequences that are generated randomly. The answer is no, due to the asynchronism of the transmitters. The proof of this statement follows from a coding theorem for the asynchronous multiple access channel [3].

[3] The $y_{i,k}$'s may not be independent for different k's. This dependence, however, is weak and diminishes as the number of users N becomes large. Exploiting this dependence for achieving lower error probability or higher throughput requires a more sophisticated decoding algorithm. We restrict ourselves to simple decoder structures. The issue of channel capacity with restrictions on the structures of the decoder is treated in [3].

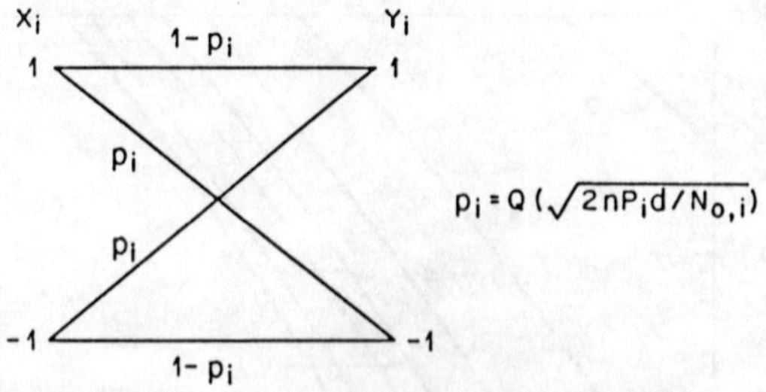

Fig. 3. The binary symmetric channel.

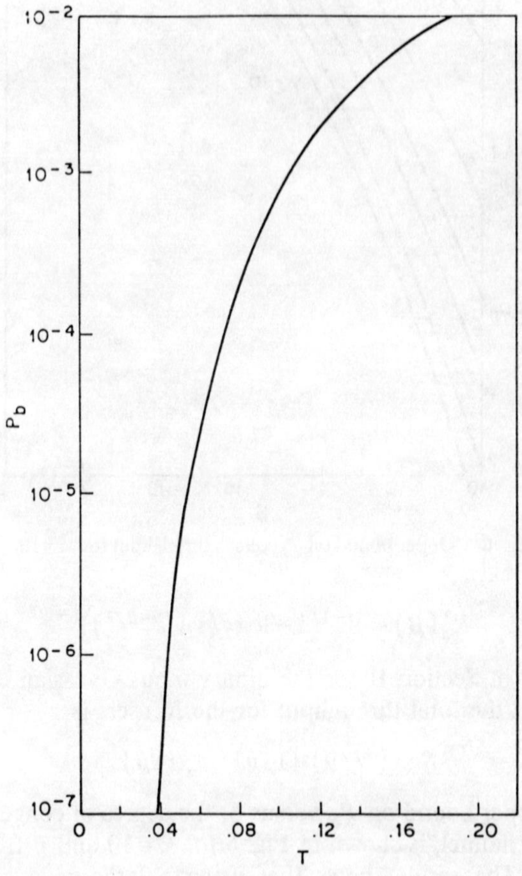

Fig. 4. Error probability versus throughput with no error correction code.

III. Capacity and Cutoff Rates

The capacity of a channel is the upper bound on the rate of reliable communication. The cutoff rate is the value of the random coding exponent function with $\rho = 1$ [4]. When sequential decoding is used, the cutoff rate is the maximum rate of communication over the channel with a bounded average number of computations for decoding an information bit. For the BSC of Fig. 3, the cutoff rate and capacity (in bits/code symbol, i.e., x_i) are derived in [4], giving

$$R_{o,i} = 1 - \log_2\left(1 + \sqrt{4p_i(1-p_i)}\right)$$
$$C_i = 1 - H(p_i)$$

in which

$$p_i = Q\left(\sqrt{2nP_i d/N_{o,i}}\right)$$

and H is the binary entropy function. For the binary input Gaussian output channel, we have [4]

$$R_{o,i} = 1 - \log_2\left(1 + e^{-nP_i d/N_{o,i}}\right)$$
$$C_i = \log 2\pi e - \int_{-\infty}^{\infty} P^i(y) \log_2 P^i(y)\, dy$$

in which

$$P^i(y) = \left(P^i_{-1}(y) + P^i_1(y)\right)/2$$

and

$$P^i_m(y) = P(Y_i = y_i / x_i = m)$$
$$= \exp\left(-\left[y_i - m\sqrt{2nP_i d/N_{o,i}}\right]^2/2\right)/\sqrt{2\pi}.$$

We shall assume henceforth that the transmitting power is equal for all users. Define $\beta = n/N$, the ratio of the length of the PN sequence to the total number of active users. Summing over the N users and normalizing by n, the number of chips in a code symbol x_i, the sum of the capacity and cutoff rate (in bits/chip) for the BSC is

$$R^T_o(\beta) = \frac{N}{n} R_{o,i} = \beta^{-1}\left(1 - \log_2\left(1 + \sqrt{4p(1-p)}\right)\right)$$
$$C^T(\beta) = \frac{N}{n} C_i = \beta^{-1}(1 - H(p))$$

in which

$$p = Q\left(\sqrt{n/(N-1)}\right) \approx Q(\beta^{1/2}) \quad \text{for large } N.$$

For the binary input Gaussian channel [4],

$$R^T_o(\beta) = \beta^{-1}\left(1 - \log_2\left(1 + e^{-\beta/2}\right)\right)$$
$$C^T(\beta) = \beta^{-1}\left(-\log_2 2\pi e - \int_{-\infty}^{\infty} P(y) \log_2 P(y)\, dy\right)$$

in which

$$P(y) = \left(P_1(y) + P_{-1}(y)\right)/2$$

and

$$P_m(y) = \exp\left(-\left[y - m\beta^{1/2}\right]^2/2\right)/\sqrt{2\pi}.$$

These four functions of β are shown in Fig. 5. All four functions are observed to be monotonically decreasing in β. It can be readily shown that all four converge to the function $1/\beta$ for large values of β. Asymptotic evaluation of these functions for vanishing values of β gives

$$R^T_o = \frac{1}{2\pi} \log_2 e = 0.2296$$
$$C^T = \frac{1}{\pi} \log_2 e = 0.4592$$

for the BSC, and

$$R^T_o = \frac{1}{4} \log_2 e = 0.3607$$
$$C^T = \frac{1}{2} \log_2 e = 0.7213$$

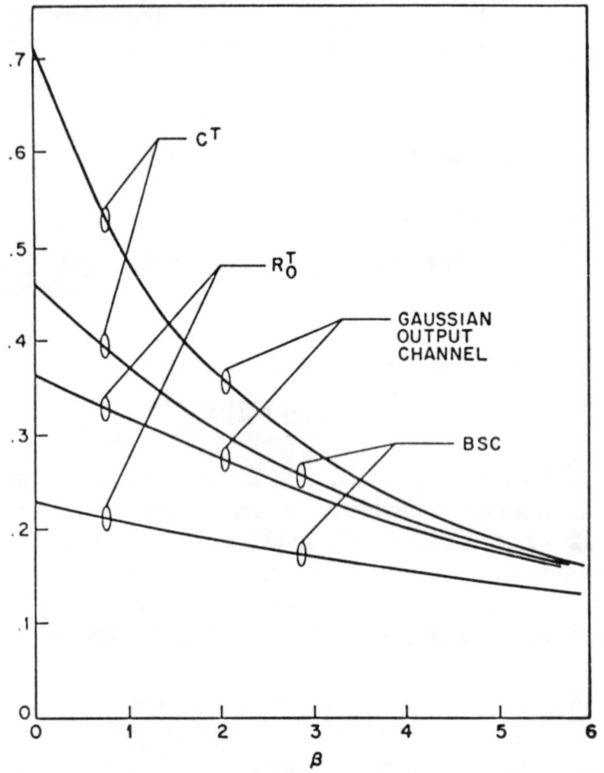

Fig. 5. Capacity and cutoff rates.

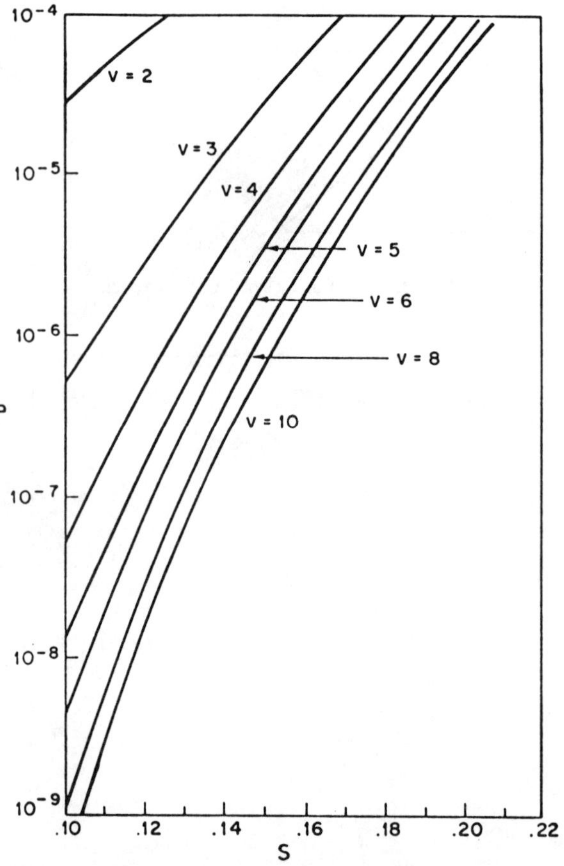

Fig. 6. Upper bound on P_b versus throughput for $k=10$.

for the binary input Gaussian channel. (It is well known that hard quantization reduces the capacity by a factor of $\pi/2$ [4].) It is noteworthy that small values of β correspond to very noisy channels in Figs. 2 and 3, when the cutoff rate can be shown to be half the capacity. In fact, both R_o^T and C^T for diminishingly small β can be derived alternatively using a very noisy channel model [4]. The asymptotic result for large β suggests that using long PN sequences decreases the sum of the capacity and cutoff rate. The lesson is that we should use very low rate encoders and short PN sequences so that β can be made small. The smallest value of β is achieved for $n=1$ which corresponds to using a very low rate code with no PN sequence coding.

IV. Error Probability and Decoding Complexity

This section studies the upper bound on error probability as a function of complexity. For constraint length k and rate $1/v$ convolutional codes, the ensemble average of expected number of bit errors for an error event starting at a given time [4] is upper bounded by

$$\begin{aligned} P_b &< 2^{-kvR_{o,i}}/[1-2^{-(vR_{o,i}-1)}]^2 \\ &= 2^{-kv\beta R_o^T(\beta)}\Big/\Big[1-2^{-(v\beta R_o^T(\beta)-1)}\Big]^2 \\ &= \left[(1+e^{-1/2Sv})/2\right]^{vk}\Big/\left[1-2\left((1+e^{-1/2Sv})/2\right)^{v}\right]^2. \end{aligned} \tag{1}$$

The last equality follows from

$$R_o^T(\beta)=\beta^{-1}\left(1-\log_2\left(1+e^{-\beta/2}\right)\right)$$

derived in Section II for the binary input Gaussian channel, and the total throughput for the N users is

$$S=(N/n)\cdot(1/v)=1/(v\beta).$$

This upper bound on P_b versus S, the degree of congestion of the channel, is shown in Fig. 6 for $k=10$ and v from 2 to 10. The result shows that using a large v gives far superior error probability for given k. Consequently, using a small β also improves the antijam capability of a user. By setting to zero the denominator of the upper bound on P_b, we have

$$S=-1/\left(2v\ln\left[2(1/2)^{1/v}-1\right]\right).$$

This value of S versus v is given in Fig. 7 together with the corresponding value of $\beta(=1/(Sv))$. It is observed that a large v (consequently, a small β) increases the value of S. A value of v around 5 would give a value of S close to the maximum R_o^T.

From the bound on P_b in (1) the error probability is decreasing at a rate that is exponential in $nvkR_o^T(\beta)/N$. It is noteworthy that nv is the number of chips for the output on a transition of the trellis. Consider the use of Viterbi decoding, for which decoding complexity is measured by k (the trellis therefore has 2^{k-1} states), for fixed nv and N. Consider changing the value of β. (Since $\beta=n/N$ and nv is fixed, a smaller β would imply a smaller n and a larger

V	β	MAX S (BITS/CHIP)
2	1.765	.284
3	1.066	.313
4	0.766	.326
5	0.599	.334
6	0.493	.339
7	0.418	.342
8	0.363	.344

Fig. 7. Maximum throughput for rate $1/v$ code.

v.) Due to the fact that $R_o^T(\beta)$ is maximized by small values of β, the complexity k is minimized for small values of β at a fixed error probability.

Using small β, however, may require a decoder complexity larger than that reflected by the value of k due to two reasons. First, decoding can be performed in units of code symbols since circuits on a chip are available for demodulating the entire PN sequence carrier (which is considered as a code symbol in the channel transition diagram). A small β would require a large v, the number of code symbols on each transition of the trellis. Second, the use of small β requires finer quantization for decoding since the channel (Fig. 2) becomes more noisy. The merging of the four functions in Fig. 5 in fact shows that rough quantization may be used with little effect on R_o^T or C^T for large β.

Since user interference is severe in such a system, using a long constraint length code is crucial for low error probability. Unfortunately, Viterbi decoding is too complex to implement for constraint lengths beyond 10. We suggest searching for good convolutional codes to decrease error probability and increase total throughput. (Our previous bound on P_b is a random coding bound for an ensemble of codes chosen at random.) A directory of good encoders would be built up from which each subscriber would choose an encoder. The maximum size of the directory, which determines the number of subscribers for the system, depends on the density of good codes times the volume of the code space. We believe that searching for good convolutional encoders [4] is a better and more tractable way to improve code orthogonality than trying to design PN sequences with good autocorrelation and low cross correlation, which is difficult due to code asynchronism.

A long PN sequence may also make synchronization at the receiver easier to achieve. It should be noted that two synchronizations have to be performed, namely the synchronization of the PN sequence carrier and the synchronization of the convolutional encoder, up to the code symbols. For the case of $n=1$, there is no need for PN sequence synchronization. In practice, a synchronization signal can be sent on top of the coded signal for easy code synchronization.

Priorities among the users can be set up by the authorization of transmitting power levels for the users. Using a higher transmitting power, a user may choose to increase its data rate or decrease its error rate. The analysis in this paper can be carried further to show that the capacity for each user is directly proportional to its transmitting power. Fading, due to distance and poor weather conditions, can also cause variations of signal power and result in dramatically deteriorated system performance. The system, in order to operate reliably in most conditions, would have to include a power margin that reduces the total throughput of the system. Also, power monitoring for fairness is difficult in practice.

V. Conclusion

It has been suggested [5] that CDMA for the spread spectrum channel should operate at a throughput of about 10 percent for reasonable error performance. The analysis in this paper gives a maximum throughput of 36 percent based on the cutoff rate. In practice, the throughput would have to be reduced for a smaller decoding complexity achieving a tolerable bit error rate. We have shown that a throughput of 20 percent is feasible in practice.

References

[1] J. P. Costas, "Poisson, Shannon, and the radio amateur," *Proc. IRE*, vol. 47, pp. 2058–2068, Dec. 1959.

[2] R. Gold, "Optimal binary sequences for spread spectrum multiplexing," *IEEE Trans. Inform. Theory*, vol. IT-13, pp. 619–621, 1967.

[3] J. Y. Hui, "Fundamental issues of multiple accessing," Ph.D. dissertation, Massachusetts Inst. Technol., Cambridge, MA, 1983.

[4] A. J. Viterbi and J. K. Omura, *Principles of Digital Communication and Coding*. New York, NY: McGraw-Hill, 1979.

[5] R. L. Pickholtz, D. L. Schilling, and L. B. Milstein, "Theory of spread-spectrum communications—A tutorial," *IEEE Trans. Commun.*, vol. COM-30, pp. 855–884, May 1982.

[6] O. Yue, "Spread spectrum mobile radio, 1977–1982," *IEEE Trans. Veh. Technol.*, vol. VT-32, no. 1, Feb. 1983.

RECENT PROGRESS IN MULTIUSER DETECTION

Sergio Verdú
Dept. Electrical Engineering
Princeton University
Princeton, NJ 08544

I. Introduction

Multiuser Detection is an active research area in Detection Theory. Its objective is the study of strategies to demodulate the digital information sent simultaneously by several transmitters who share a multiple-access channel. Common channels that are encompassed by this general model include up-link satellite channels, local-area networks and radio networks.

Static strategies, such as Frequency-Division Multiple-Access (FDMA) and Time-Division Multiple-Acess (TDMA), whereby the multiple-access channel is effectively partitioned into independent single-user sub-channels are common in practice. The analysis of those approaches lies fully whithin the domain of classical single-user communication, and offers few, if any, conceptual challenges. Moreover, from the practical viewpoint, static strategies tend to be wasteful in applications where most users actively send information during a very small percentage of time. Current dynamic channel-sharing strategies, which allow the active users a larger share of the channel while they are transmitting, fit into two categories: random-access communication and simultaneous transmission systems. In random-access communication it is assumed that the receiver is not capable of demodulating more than one simultaneous transmission, and so the problem is to design protocols to schedule channel access at nonoverlapping times, and if collisions between messages occur (as they do when the protocol operates in a decentralized fashion) to ensure that those messages are eventually retransmitted successfully. Simultaneous transmission systems differ from static strategies and random-access protocols in that users are allowed to transmit simultaneously, asynchronously and through the same channel. The destination receives a noisy version of the superposition of all the transmitted waveforms and it is the task of the receiver to demodulate all (as in the case of the satellite channel) or a subset (as in multipoint-to-multipoint topologies) of the transmitted messages. This is the communication problem that is the subject of multiuser detection and multiuser information theory. In practice, the major multiaccess strategy using the simultaneous transmission philosophy is Code-Division Multiple-Access (CDMA). In this technique, each transmitter is assigned a fixed, distinct signature waveform which he uses to modulate his message as in single-user communication. Then the information sent by each user can be demodulated by correlating the received signal with each of the signature waveforms. This demodulator, whose use is widespread in practice, is referred to as the conventional single-user detector. As is well-known, when the channel output is corrupted by additive white Gaussian noise, the conventional single-user detector minimizes the probability of error in a single-user channel, i.e., in the absence of interfering users. The fact that this is no longer true in the multiple-access channel is the raison d'être of the area of multiuser detection.

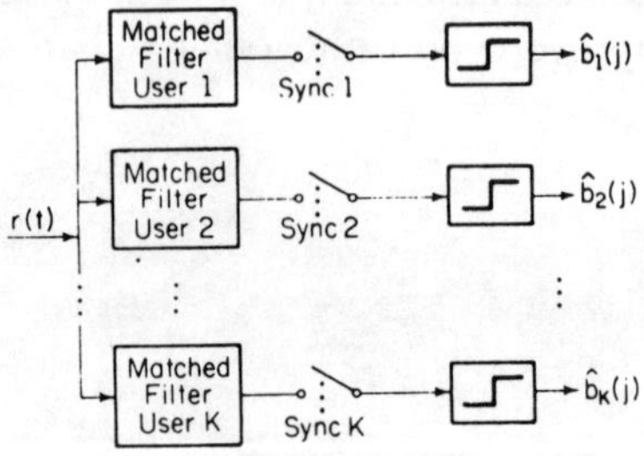

Fig. 1. Conventional multiuser detector.

The performance of the conventional single-user detector is acceptable provided that the energies of the received signals are not too dissimilar and that the signature waveforms are designed so that their crosscorrelations are low enough (this depends on the desired maximum number of simultaneous users). In practice, low crosscorrelations are usually achieved employing Spread-Spectrum Pseudonoise sequences of long periodicity. If the received signal energies are indeed dissimilar, i.e., some users are very weak in comparison to others, then the conventional single-user detector is unable to recover the messages of the weak users reliably, even if the signature waveforms have very low crosscorrelations. This is known as the ***near-far problem*** and is the main shortcoming of

Reprinted with permission from *Advances in Communication and Signal Processing*, © 1989 Springer-Verlag Berlin-Heidelberg.

currently operational Direct-Sequence Spread-Spectrum Multiple-Access systems.

Due to the reduced multiple-access capability and the increased vulnerability to hostile sources caused by the near-far problem, its solution or alleviation had been a target of researchers in the area for several years. Before the emergence of solutions based on multiuser detection, success had been very limited and, essentially, the only remedies available were power control and the design of signals with even more stringent crosscorrelation properties. Unfortunately, power control (i.e., the adaptive adjustment of transmitter power depending on its location and on the received powers of the other users) dictates significant reductions in the transmitted powers of the strong users in order for the weaker users to achieve reliable communication. Thus, power control can become self-defeating since it actually decreases the overall multiple-access and antijamming capabilities of the system. Furthermore, more and more complex signature waveforms lead to rapid increases in system cost and bandwidth, and, as we have noted, do not eliminate the near-far problem. For these reasons, it can be seen why the solution to the near-far problem has been highlighted as the main achievement of multiuser detection.

The chief reason why multiuser detection did not develop until relatively recently was the belief shared by many a worker in Spread Spectrum that multiuser interference is accurately modeled as a white Gaussian random process, and thus the conventional detector is essentially optimum. This serves as a good illustration of A. N. Kolomogorov's assertion that "the Central Limit Theorem is a dangerous tool in the hands of amateurs," and it ranks along with the stability of the original ALOHA algorithm, and the independence assumption in queueing models of store-and-forward communication networks as one of the maximum exponents of the area of "voodoo multiuser communication." While it is not difficult to build an infinite population multiuser signal model which can be rigorously shown to be asymptotically Gaussian as the individual amplitudes go to zero with the appropriate speed, the number of transmitters, signature waveforms, and power levels encountered in many practical situations (e.g. in near-far environments) render the Gaussian approximation completely useless. This was recognized by H. V. Poor who proposed in 1980 [9] the use of techniques from both minimax robustness and non-Gaussian signal detection to improve the performance of the conventional single-user detector in multiuser channels. Several earlier works had already investigated receivers that used the knowledge of the interfering signature waveforms to improve the performance of Code-Division Multiplexing systems; Timor [15, 16] showed that it was possible to double the maximum number of simultaneous users achievable with the conventional noncoherent demodulator of Frequency-Hopped FSK systems, Horwood and Gagliardi [3] and Schneider [12, 13] considered multiuser receivers for synchronous systems. In particular, [12] claimed that an appropriately chosen linear transformation of a bank of matched filter outputs results in optimum decisions. Although that claim turned out to be erroneous, such a receiver and its generalization to the asynchronous case do have very desirable properties as we discuss in the sequel. In the same work, Schneider briefly considered the asynchronous channel arguing that it could be modeled as a finite state machine and conjectured that Viterbi's convolutional decoding algorithm could be used to demodulate the data. Those ideas were closely related to earlier work to combat crosstalk in single-user multichannel communication systems by Shnidman [14], Kaye and George [4] and Van Etten [1, 2] among others. Those works dealt with PAM systems through multi-input multi-output dispersive channels, which are no more than a vector generalization of the conventional scalar intersymbol interference model. In contrast, in the asynchronous multiuser channel each signal overlaps with two consecutive signals of each of the interferers; thereby introducing memory in the channel in a way which (as we will see in Section II) is akin to a scalar periodically time-varying generalization of the conventional intersymbol interference model.

Section II reviews the main strategies proposed so far in multiuser detection, and Section III summarizes the main results on the performance of the various receivers. For the sake of clarity and in order to highlight the fundamental features of the approaches reviewed in this paper, it is convenient to circumscribe the discussion to the two-user case, and the reader is referred to the corresponding treatment with an arbitrary number of users in each of the referenced sources.

II. Multiuser Detectors

The two-user white Gaussian asynchronous multiple-access channel considered in this paper is

$$y(t) = \sum_{i=-M}^{M} b_1(i)\, s_1(t-iT-\tau_1) + \sum_{i=-M}^{M} b_2(i)\, s_2(t-iT-\tau_2) + n(t), \tag{1}$$

where $n(t)$ is white Gaussian noise with power spectral density equal to σ^2. The unit-energy signature waveforms assigned to both users are $s_1(t)$ and $s_2(t)$ and have duration T. The symbol streams $\{b_1(i)\}$ and $\{b_2(i)\}$ take values on $\{-\sqrt{w_1},+\sqrt{w_1}\}$ and $\{-\sqrt{w_2},+\sqrt{w_2}\}$, respectively, where w_k denotes the received energy per bit of user k. The delays $\tau_1 \in [0,T]$ and $\tau_2 \in [0,T]$ account for the asynchronism between both transmitted streams.

The objective is to find the optimum demodulator for the streams of data transmitted by both users. This would appear to apply only to multipoint-to-point topologies where detection is centralized (e.g. the up-link satellite channel). However, in situations where each location is interested in demodulating only one of the transmitted streams, the same strategy can be used as long as the receiver knows the signature waveforms of the interfering users. (The derivation of optimum detectors without knowledge of the interfering signature waveforms is discussed later in this section.) A key point in the derivation of the optimum receiver [20] for the asynchronous channel (1) is that in order to make optimum decisions on any particular bit it is necessary to use the observation of the entire received waveform rather than just the received waveform on the interval of that particular bit. In other words, due to the asynchronism between the users, we face a problem of sequence detection rather than one-shot detection. It is easy to show [20] that the sequence of outputs of matched filters for $\{s_1(t)\}$ and $\{s_2(t)\}$ is a set of sufficent statistics to demodulate the data streams $\{b_1(i)\}$ and $\{b_2(i)\}$. Those sufficient statistics can be obtained by a receiver that knows the signature waveforms of both users upon acquisition of the timing of the bit-epochs of each stream. Assuming without loss of generality that $\tau_1 \le \tau_2$, the sufficient statistics admit the following expression in terms of the data

$$\begin{bmatrix} y_1(i) \\ y_2(i) \end{bmatrix} = \begin{bmatrix} 0 & \rho_{21} \\ 0 & 0 \end{bmatrix} \begin{bmatrix} b_1(i-1) \\ b_2(i-1) \end{bmatrix} + \begin{bmatrix} 1 & \rho_{12} \\ \rho_{12} & 1 \end{bmatrix} \begin{bmatrix} b_1(i) \\ b_2(i) \end{bmatrix} + \begin{bmatrix} 0 & 0 \\ \rho_{21} & 0 \end{bmatrix} \begin{bmatrix} b_1(i+1) \\ b_2(i+1) \end{bmatrix} + \begin{bmatrix} n_1(i) \\ n_2(i) \end{bmatrix} \tag{2}$$

for $-M \le i \le M$ and where

$$\rho_{12} = \int_0^T s_1(t)\, s_2(t+\tau_1-\tau_2)\, dt, \qquad \rho_{21} = \int_0^T s_1(t)\, s_2(t+T+\tau_1-\tau_2)\, dt \tag{3}$$

are the crosscorrelations between the signature waveforms. The discrete-time random process $\{[n_1(i)\ n_2(i)]^T\}$ in (2) is Gaussian with zero-mean and covariance matrix:

$$E\left[\begin{bmatrix} n_1(i) \\ n_2(i) \end{bmatrix} [n_1(j)\ \ n_2(j)]\right] = \sigma^2\, \mathbf{H}(i-j)$$

where $\mathbf{H}(i) = 0$ if $|i| > 1$, and $\mathbf{H}(1)$, $\mathbf{H}(0)$ and $\mathbf{H}(-1)$ are the matrices appearing in (2), i.e.,

$$\mathbf{H}(0) = \begin{bmatrix} 1 & \rho_{12} \\ \rho_{12} & 1 \end{bmatrix}, \qquad \mathbf{H}(1) = \mathbf{H}^T(-1) = \begin{bmatrix} 0 & \rho_{21} \\ 0 & 0 \end{bmatrix}.$$

Since the noise in (1) is white and Gaussian, maximum likelihood decisions are obtained by selecting the symbol sequences $\{b_1(i)\}$ and $\{b_2(i)\}$ that best explain the observations in a mean-square sense, i.e., assuming that the optimum detector knows the received energies, it should select the argument achieving

$$\min_{\substack{b_1(i) \in \{-\sqrt{w_1}, +\sqrt{w_1}\} \\ b_2(i) \in \{-\sqrt{w_2}, +\sqrt{w_2}\} \\ i=-M, \cdots M}} \int [y(t) - \sum_{k=1}^{2} \sum_{i=-M}^{M} b_k(i)\, s_k(t-iT-\tau_k)]^2\, dt. \tag{4}$$

Before considering how to obtain a detector that solves for the minimum in (4) it is instructive to consider the synchronous ($\tau_1 = \tau_2$) version of the problem for which a one-shot approach suffices:

$$\min_{\substack{b_1(0) \in \{-\sqrt{w_1}, +\sqrt{w_1}\} \\ b_2(0) \in \{-\sqrt{w_2}, +\sqrt{w_2}\}}} \int_0^T [y(t) - \sum_{k=1}^{2} b_k(0)\, s_k(t)]^2\, dt = \max_{\mathbf{b} \in \{-\sqrt{w_1}, \sqrt{w_1}\} \times \{-\sqrt{w_2}, \sqrt{w_2}\}} 2\mathbf{b}^T\mathbf{y} - \mathbf{b}^T\mathbf{H}\mathbf{b} \tag{5}$$

where $\mathbf{y} = [y_1(0), y_2(0)]^T$, $\mathbf{b} = [b_1(0), b_2(0)]^T$ and $\mathbf{H} = \begin{bmatrix} 1 & \rho \\ \rho & 1 \end{bmatrix}$, $\rho = \int_0^T s_1(t) s_2(t)\, dt$.

The solution to the right-hand side of (5) is easy; the function therein can be computed for each of the four possible values of $\mathbf{b}$ and the maximizing vector gives the solution to the optimum demodulation problem. In the K-user case, there are 2^K possible values of $\mathbf{b}$ and no algorithm is known to solve (5) in a number of steps which is polynomial in the number of users. In fact, it has been shown that (5) is an NP-hard combinatorial optimization problem [25], and, therefore, it can admit a polynomial solution only if the same is true for longstanding combinatorial problems such as the *travelling salesman* and *integer linear programming* problems.

In the asynchronous case, it is possible to follow the same approach by viewing each symbol as transmitted by a different fictitious user over the interval $[-MT, MT+2T]$. In the K-user case, this is equivalent to having a synchronous channel with $(2M+1)K$ fictitious users. Thus, the complexity of the above brute-force method is out of the question because of the large values of M one expects in applications. Fortunately, the asynchronous problem admits a much more efficient algorithm whose complexity per demodulated bit is very similar to that of the synchronous case (exponential in K but independent of M), making it practical for situations where the number of simultaneous active users does not exceed the range 10-15. The reason for the existence of such an algorithm is that the $(2M+1)K \times (2M+1)K$ matrix $\mathbf{H}$ of crosscorrelations between each possible pair of symbols has a great deal of structure, as each symbol only overlaps with two consecutive symbols of the other users. For example, in the two-user case it is equal to

$$\mathbf{H} = \begin{bmatrix} 1 & \rho_{12} & & & & & \\ \rho_{12} & 1 & \rho_{21} & & & & \\ & \rho_{21} & 1 & \rho_{12} & & & \\ & & & \cdot & & & \\ & & & & \cdot & & \\ & & & & & \cdot & \\ & & & & \rho_{21} & 1 & \rho_{12} \\ & & & & & \rho_{12} & 1 \end{bmatrix} \tag{6}$$

This implies that the objective function in the right-hand side of (5) admits the following additive decomposition

$$\begin{aligned}2\mathbf{b}^T\mathbf{y} - \mathbf{b}^T\mathbf{H}\mathbf{b} = {} & \lambda_{-M}^{(1)}[0, b_1(-M)] + \lambda_{-M}^{(2)}[b_1(-M), b_2(-M)] + \lambda_{-M+1}^{(1)}[b_2(-M), b_1(-M+1)] + \cdots \\ & + \lambda_0^{(1)}[b_2(-1), b_1(0)] + \lambda_0^{(2)}[b_1(0), b_2(0)] + \lambda_1^{(1)}[b_2(0), b_1(1)] + \cdots \\ & + \lambda_{M-1}^{(2)}[b_1(M-1), b_2(M-1)] + \lambda_M^{(1)}[b_2(M-1), b_1(M)] + \lambda_M^{(2)}[b_1(M), b_2(M)], \end{aligned} \tag{7}$$

where

$$\lambda_i^{(k)}[a, b] = 2by_k(i) - w_k - 2ab\,\rho_{jk}, \qquad j \neq k.$$

The interesting thing to note on (7) is that the dependence on the components of **b** is sequential; each of the functions $\lambda_i^{(k)}$ depends on only two symbols and every symbol appears only in two consecutive functions. Therefore, the maximization of (7) is equivalent to finding the longest path in a layered directed graph (trellis) where there are two nodes in each layer (Fig. 2). (The K-user algorithm in [20] has 2^{K-1} nodes per layer and each node is connected to two nodes in the following layer.) The natural solution to this combinatorial problem is dynamic programming, which can be implemented in real time by running it in forward fashion and with decisions that look ahead over a finite window of stages, rather than waiting for all the data to be received (the Viterbi algorithm). Hence, the structure of the optimum multiuser receiver is now complete: a matched filter front-end followed by a Viterbi algorithm (Fig. 3). Due to the interference between the users, maximum likelihood is not the unique optimality criterion even if the transmitted streams are assumed independent and identically distributed. Minimum probability of error decisions can be accomplished by a backward-forward dynamic programming algorithm [19, 23]; however, the bit-error-rate of the maximum likelihood receiver turns out to be indistinguishable from the minimum achievable bit-error-rate in the region of signal-to-noise ratios of interest, so for most applications it is not worth incurring the additional implementation cost required by the backward-forward algorithm over the Viterbi algorithm.

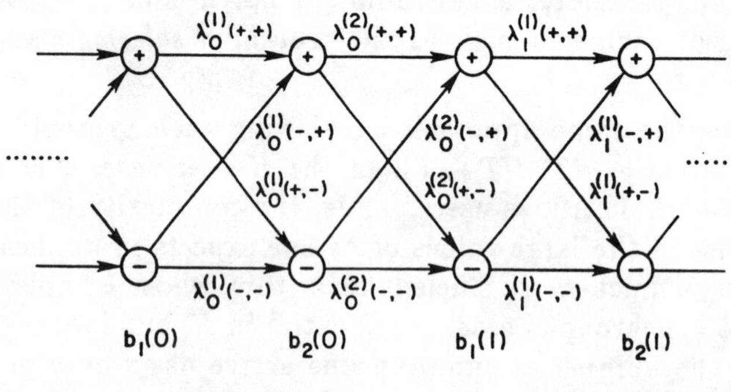

Figure 2. Trellis for two-user asynchronous demodulation.

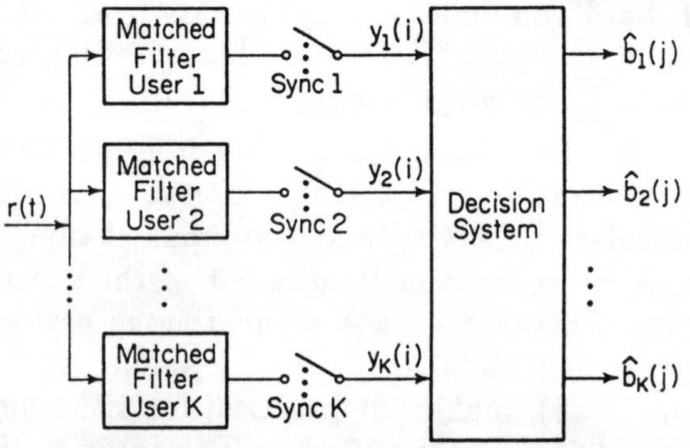

Figure 3. Optimum multiuser detector.

In the special case when the energies of the users are the same and $\rho_{12} = \rho_{21}$ (i.e. **H** is Toeplitz), the functions $\lambda_i^{(k)}$ in (7) no longer depend on k and i (except through the data $y_k(i)$) and they become identical to the metric functions obtained by Ungerboeck [17] for maximum likelihood sequence detection in the single-user intersymbol interference channel. This is because if the energies and signature waveforms of both users are identical and $\tau_2 - \tau_1 = T/2$, then the multiuser channel (1) is entirely equivalent to an intersymbol interference channel where each symbol overlaps with two symbols. Thus, the decomposition in (7) is akin to a periodically time-varying generalization of Ungerboeck's decomposition [17]. Similar decompositions [21] of the likelihood function are also feasible in asynchronous multiple-access channels which are much less structured than (1).

As we discuss in Section III the optimum detector affords important performance gains over the conventional single-user detector, and, in particular, it solves the near-far problem. However, the price for this is exponential complexity in the number of users, since the combinatorial optimization problem of selecting the most likely transmitted bits given the matched filter outputs is inherently hard [25]. This has spurred several works that propose lower complexity multiuser detectors whose performance lies somewhere in the big gap between optimum and conventional bit-error-rate. A natural alternative to explore is the replacement of the Viterbi algorithm by a sequential decoding algorithm. This is done in [11] where a metric is proposed for a stack sequential decoder based on the additive decomposition of the loglikelihood function in (7). At each point in the search in the binary tree of the sequential decoding algorithm the metric is the sum of two parts: one that the depends on "past" symbols exclusively and the other that depends on "future" unexplored symbols. The metric proposed in [11] takes the decisions of the conventional detector in lieu of those unexplored symbols. As usual, results on the computational complexity of the resulting sequential decoding algorithm are hard since the number of steps depends on the noise realization and there is the possibility of buffer overflow. But it appears that unless the signal-to-noise ratios are extremely poor such a strategy does indeed afford important computational savings over the Viterbi algorithm when the number of users is large.

The approach taken in [5-8] to find low-complexity multiuser detectors that exhibit good performance, and, in particular, are near-far resistant, is to restrict attention to linear transformations of the bank of matched filter outputs. As in the discussion of the maximum likelihood detector, it is beneficial to consider first the synchronous version of the problem. The maximization in (5) is over a four-element set because the receiver is assumed to know the energies of both users. If those energies were unknown and we were interested in obtaining maximum likelihood estimates of the energies in addition to the maximum likelihood decisions on the transmitted streams, then the maximization in (5) would be over the whole plane, and the solution would be $\mathbf{H}^{-1}\mathbf{y}$, i.e., the sign of the components of $\mathbf{H}^{-1}\mathbf{y}$ gives the most likely transmitted bit and their absolute values give the maximum-likelihood amplitude estimates. (Note that better amplitude estimates would be possible with a sequence approach rather than a one-shot approach.) This boils down to using a simple modification of the conventional detector: instead of correlating the received signal with $s_1(t)$ and comparing the output to zero to determine the bit transmitted by user 1, we correlate the received signal with $s_1(t) - \rho\, s_2(t)$. In the hypothetical case when there is no background Gaussian noise ($\sigma^2 = 0$), we have $\mathbf{y} = \mathbf{Hb}$, and the bits selected by this receiver are $sgn(\mathbf{H}^{-1}\mathbf{y}) = sgn(\mathbf{b})$, i.e., the true data, regardless of the values of the received energies. A nice property, which augurs well for the performance of this receiver (at least in the high SNR region). Indeed, it is the lack of this feature that makes the conventional receiver so sensitive to the near-far problem. In the general K-user case, the essence of the $sgn(\mathbf{H}^{-1}\mathbf{y})$ receiver is that instead of correlating with $s_k(t)$ it correlates with the projection of $s_k(t)$ on the subspace orthogonal to the subspace spanned by the other signature waveforms $\{s_j(t),\ j \neq k\}$. Thus, it effectively tunes out the multiuser interference, and hence its name: *decorrelating detector*.

The foregoing derivation of the decorrelating detector has highlighted that if no prior information on the transmitted energies is available, maximum likelihood decisions are obtained by a linear receiver. Note that Schneider [12] sought to minimize error probability in a synchronous multiuser channel with equal-energy signals and, erronously, arrived at the same receiver $sgn(\mathbf{H}^{-1}\mathbf{y})$. At this point, it is natural to ask whether it is possible to achieve better performance by choosing a different linear transformation when the received energies are known. The answer is affirmative; for example, the optimum linear receiver (for high SNR's) for user 1 correlates the received signal with [6]:

$$\begin{cases} s_1(t) - \sqrt{\dfrac{w_2}{w_1}}\, sgn(\rho)\, s_2(t) & \text{if } \dfrac{w_2}{w_1} \leq \rho^2 \\ s_1(t) - \rho\, s_2(t) & \text{otherwise} \end{cases}$$

However, as we discuss in Section III, the decorrelating receiver provides the same degree of near-far resistance as the optimum multiuser receiver [20] (which uses knowledge of the received energies) and it is much easier to compute than the optimum linear receiver in the K-user case.

How do we generalize the decorrelating receiver to the asynchronous case? We can take the same approach as in the derivation of the optimum receiver: each of the $(2M+1)K$ symbols can be viewed as transmitted by a different user. However, the resulting detector would be impossible to compute as it would involve the inversion of a crosscorrelation matrix of dimension $(2M+1)K$. Fortunately, as $M \rightarrow \infty$ the rule $sgn(\mathbf{H}^{-1}\mathbf{y})$ is equivalent to passing the sufficient statistics $\mathbf{y}(i) = [y_1(i)\ y_2(i)]^T$ through a two-input two-output discrete-time linear time invariant system with transfer function [7]

$$[\mathbf{H}^T(1)\, z + \mathbf{H}(0) + \mathbf{H}(1)\, z^{-1}]^{-1} = \frac{1}{1-\rho_{12}^2-\rho_{21}^2-\rho_{12}\rho_{21}z-\rho_{12}\rho_{21}z^{-1}} \begin{pmatrix} 1 & -\rho_{12}-\rho_{21}z^{-1} \\ -\rho_{12}-\rho_{21}z & 1 \end{pmatrix} \quad (8)$$

Notice that the matrix in (8) performs the same function as the decorrelating detector in the synchronous case, namely, it substracts from each matched filter output (say, the one corresponding to the ith bit of user 1) the matched filter outputs corresponding to the two overlapping bits of the other user (bits i−1 and i of user 2). However, now this is not enough because this operation introduces intersymbol interference between the symbols of user 1 as the matched filter outputs of the overlapping bits (i−1 and i of user 2) are contaminated by bits i−1 and i+1 of user 1. This is the origin of the recursive part of the filter (8) which acts as a linear equalizer for the intersymbol interference introduced by the nonrecursive part. Note that the impulse response of the recursive part is of the form $[\alpha(\rho_{12},\rho_{21})]^{|n|}$, i.e., it is non-causal, and therefore, in practice, it is necessary to approximate it by truncation.

An alternative to the decorrelating detector for asynchronous channels, which is presented here for the first time, can be obtained by taking a one-shot approach where each symbol interval is considered separately. Let us focus attention on bit 0 of user 1, which occupies the interval [0, T] (assuming without loss of generality $\tau_1 = 0$). This bit overlaps with bit −1 of user 2, over the interval $[0, \tau_2]$ and with bit 0 of user 2 over the interval $[\tau_2, T]$. We can view this situation as a 3-user synchronous channel (see Fig. 4) with unit-energy signature waveforms $\tilde{s}_1(t) = s_1(t)$; $\tilde{s}_2(t) = e_2^{-1/2} s_2^L(t)$; $\tilde{s}_3(t) = (1-e_2)^{-1/2}\, s_2^R(t)$, where

$$s_2^L(t) = \begin{cases} s_2(t+T-\tau_2) & 0 \le t \le \tau_2 \\ 0 & \tau_2 \le t \le T \end{cases} \qquad s_2^R(t) = \begin{cases} 0 & 0 \le t \le \tau_2 \\ s_2(t-\tau_2) & \tau_2 \le t \le T \end{cases}$$

and $e_2 = \int_0^{\tau_2} s_2^2(t+T-\tau_2)dt$ $(0<e_2<1)$ is the partial energy of the interfering signal over the left overlapping interval. We can now solve for a 3-user decorrelating detector with

$$\mathbf{H} = \begin{bmatrix} 1 & e_2^{-1/2}\rho_{12} & (1-e_2)^{-1/2}\rho_{21} \\ e_2^{-1/2}\rho_{12} & 1 & 0 \\ (1-e_2)^{-1/2}\rho_{21} & 0 & 1 \end{bmatrix}, \quad (9)$$

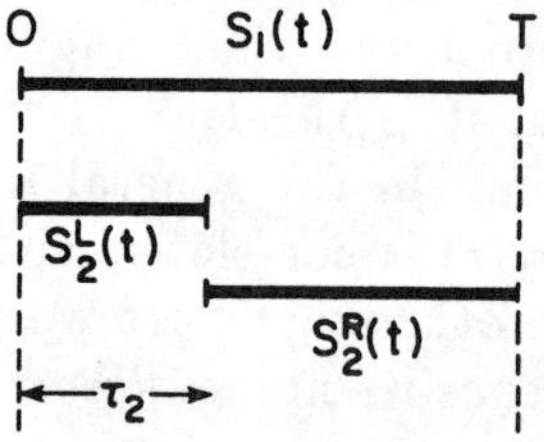

Figure 4. Intervals for one-shot decorrelating detector.

resulting in a detector for user 1 that correlates the received signal with

$$[s_1(t) - \frac{\rho_{12}}{e_2}\, s_2^L(t) - \frac{\rho_{21}}{1-e_2}\, s_2^R(t)].$$

This one-shot decorrelating detector has lower complexity than the asynchronous decorrelating detector described before at the expense of some performance degradation.

Another approach to obtain suboptimum low-complexity multiuser signal detectors is proposed in [18] by Varanasi and Aazhang. Again, it is best to consider first the synchronous case. If we had an estimate, say $\hat{b}_2$, of the symbol transmitted by user 2 in the interval $[0, T]$, then we could subtract $\hat{b}_2\, s_2(t)$ from the received signal and process the remaining signal as if $b_2 = \hat{b}_2$, i.e., correlate it with $s_1(t)$ and find the polarity of the output. The detector proposed in [18] is assumed to know the received energies, and the tentative decisions are obtained with the conventional single-user detector, i.e., $\hat{b}_2$ is $+\sqrt{w_2}$ or $-\sqrt{w_2}$ depending on the polarity of the output of the matched filter of the signature waveform of user 2 (Fig. 5). If the signal of user 2 is comparatively strong, then the conventional receiver will select $\hat{b}_2 = b_2$ with high probability, in which case the decision on the bit transmitted by user 1 will noticeably improve the (poor) decision made by the conventional receiver. On the other hand, the decision on the bit of user 2

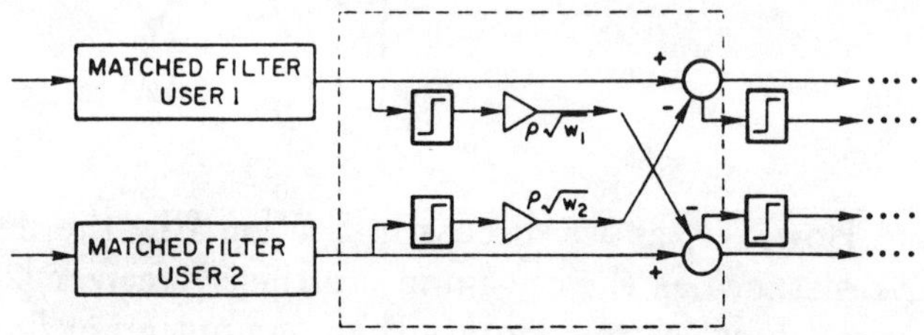

Figure 5. Multi-stage detector for two synchronous users.

will not improve the one obtained with the conventional detector since $b_1 = \hat{b}_1$ does not occur with high enough probability. However, we can now repeat the same procedure as many times as desired using the tentative decisions of the previous stage in lieu of those obtained by the conventional detector. For example, at the output of the second stage, the decisions on user 2 will be near-optimal because the available estimate of the bit of user 1 is excellent, whereas the decisions on user 1 will not improve those obtained in the first stage. Roughly speaking, the decisions put out by even [resp. odd]-numbered stages will be good for the powerful [resp. weak]-transmitters. When the energies are similar, the prerequisite for adequate operation is that the crosscorrelations and noise level be such that the initial decisions put out by the conventional detector have moderate bit-error-rate. In the asynchronous case, the principle of operation is very similar except that now we need to subtract two terms corresponding to the two symbols that overlap with each symbol. Therefore, at the output of each stage we need to store a sliding window of decisions. For example, if there is only one stage we need to store two consecutive outputs of each conventional single-user detector, whereas if there are two stages, we need two consecutive outputs of the first stage and hence, a window of three consecutive conventional decisions at the beginning. In general, if there are m stages, the kth stage needs to store $m - k + 1$ consecutive outputs for each user. As in the case of the one-shot decorrelating detector, this detector can be simplified using (partial) correlations with $s_2^L(t)$ and $s_2^R(t)$.

We now turn our attention to the derivation of demodulators for decentralized (multipoint-to-multipoint) topologies, where each receiver is interested in demodulating the information transmitted by only one of the active users. Obviously, all the foregoing detectors can still be used provided the signature waveforms of the interferers are known and provided that synchronism with each of the transmitted streams is acquired. The decorrelating detector is especially suited for this, as we can implement the linear transformation dictated by each row of $\mathbf{H}^{-1}$ or $[\mathbf{H}^T(1)\, z + \mathbf{H}(0) + \mathbf{H}(1)\, z^{-1}]^{-1}$ separately, thus resulting in a simple decentralized version of the detector. The rest of the detectors discussed above do not lend themselves to such decentralized implementations, but in many cases it is possible to lower their complexity substantially, without compromising performance, by ignoring the comparatively weak interferers. In any case, it is important to investigate which detection structures result when the signature waveforms of the interferers are not known or when the receiver acquires synchronization only with the user of interest. The problem becomes one of single-user signal detection in additive colored non-Gaussian noise, since the equivalent noise seen by each signal is the sum of the multiple-access interference and the additive background Gaussian noise.

The solution to this signal detection problem is found in [10] and [19, Chap. 5] both in the case where the interfering signature waveforms are known (but no synchronization is available) and in the case where the modulation format is Direct Sequence Spread Spectrum and the interfering signature sequences are not known. Even though the unavailability of the timing epochs of the interfering users results in very complex nonlinear detectors, it is shown in [10] that important reductions in complexity are possible in several asymptotic cases such as when the number of chips per symbol is large and the background Gaussian noise level is low. Also, when the interfering users are weak the locally optimum detector reduces to a modification of the conventional single-user detector, whereby the received signal is correlated with a smoothed replica of the signature waveform of the user of interest. This is because for the purposes of locally optimum detection it is necessary to take care only of the nonwhiteness of the multiple-access noise.

III. Performance Analysis

We focus attention now on the bit-error-rate achieved by the detectors reviewed in Section II. We denote by $P_k(\sigma)$ the bit-error-rate of the kth user when the white Gaussian noise level is σ^2. Of particular interest is the bit-error-rate in the region of low σ^2, because that quantifies the inherent performance degradation due to the presence of other users in the channel. A convenient way to quantify bit-error-rate in the region of high signal to noise ratios is the *asymptotic efficiency* which is defined as follows [20, 22]. Define the *effective energy* of user k, $e_k(\sigma)$, as the energy that that user would require to achieve bit-error-rate $P_k(\sigma)$ in the same Gaussian channel but without interfering users, i.e., $P_k(\sigma) = Q(e_k^{1/2}(\sigma)/\sigma)$. Thus, the *efficiency* or ratio between the effective and actual energies $e_k(\sigma)/w_k$ is an alternative way to characterize $P_k(\sigma)$. The *asymptotic efficiency* is defined as $\eta_k = \lim_{\sigma \to 0} e_k(\sigma)/w_k$ and measures the slope with which $P_k(\sigma)$ goes to 0 in the high signal-to-noise ratio region. A third useful performance measure is the *near-far resistance*, which is defined as the minimum asymptotic efficiency over the relative energies of all the other users. This measure quantifies the robustness of a receiver against the near-far problem, which as we argued in the introduction is the major shortcoming of coherent CDMA systems.

The minimum probability of error achievable in a Gaussian asynchronous multiple-access channel was obtained in [20]. Prior to the appearance of that paper, works devoted to the analysis of error probability in multiuser channels had focused attention exclusively on the bit-error-rate of the conventional single-user receiver. For fixed, arbitrary offset between the signals, it is easy to see that the two-user asynchronous error probability of a conventional receiver for user 1 is

$$P_1^c(\sigma) = \frac{1}{4}\, Q\left((\sqrt{w_1} + \sqrt{w_2}\,\rho_{12} + \sqrt{w_2}\,\rho_{21})/\sigma\right) + \frac{1}{4}\, Q\left((\sqrt{w_1} + \sqrt{w_2}\,\rho_{12} - \sqrt{w_2}\,\rho_{21})/\sigma\right)$$
$$+ \frac{1}{4}\, Q\left((\sqrt{w_1} - \sqrt{w_2}\,\rho_{12} + \sqrt{w_2}\,\rho_{21})/\sigma\right) + \frac{1}{4}\, Q\left((\sqrt{w_1} - \sqrt{w_2}\,\rho_{12} - \sqrt{w_2}\,\rho_{21})/\sigma\right) \qquad (10)$$

In the region of low background Gaussian noise the expression in (10) is dominated by the Q-function with the smallest argument. Hence, according to the foregoing definition, the asymptotic efficiency is

$$\eta_1^c = \max{}^2\{0,\ 1 - \sqrt{\frac{w_2}{w_1}}\,(\,|\rho_{12}| + |\rho_{21}|)\} \qquad (11)$$

If the interferer is very weak $w_2 << w_1$, then, η_1^c is close to unity (the conventional receiver is quasi-optimal because very little penalty is paid for ignoring user 2). As w_2/w_1 grows, η_1^c decreases monotonically, until it becomes zero for

$$\frac{w_2}{w_1} \geq \frac{1}{[\,|\rho_{12}| + |\rho_{21}|\,]^2},$$

in which case, the probability of error is worse than 0.25 regardless of the background noise level. Therefore, the near-far resistance is equal to zero and there is always an interference level that renders the conventional receiver useless no matter how good the signal design.

The minimum probability of error derived in [20] exhibits a very different behavior. The main contribution of that paper was the introduction of the so-called *method of indecomposable sequences* to obtain upper bounds on the probability of error of m-ary hypothesis testing problems in terms of the error probability of binary problems. This method has been successfully used in a wide variety of channels: both single-user [24] and multiple-access channels with both synchronous and asynchronous users, and with both Gaussian and Poisson [21] observation models. Space limitations prevent us from an exposition of the method of indecomposable sequences and its application to code division multiple-access channels. The main conclusion from that analysis is that the probability of error of the maximum likelihood receiver is optimum in the region of moderate to large signal-to-noise ratios and it behaves as that of a single transmitter with reduced energy. This reduction in effective energy is quantified by the optimum asymptotic efficiency [22]

$$\eta_1 = 1 - \frac{\sqrt{w_2}}{\sqrt{w_1}}\left[\left(2\,|\rho_{12}| - \frac{\sqrt{w_2}}{\sqrt{w_1}}\right)^+ + \left(2\,|\rho_{21}| - \frac{\sqrt{w_2}}{\sqrt{w_1}}\right)^+\right] \qquad (12)$$

As in the case of the conventional receiver, η_1 is close to unity if $w_2/w_1 << 1$. However, unlike

(11), η_1 is not monotonic in w_2/w_1. Actually, if

$$\frac{w_2}{w_1} \geq 4 \max\{\rho_{12}^2, \rho_{21}^2\}, \tag{13}$$

then $\eta_1 = 1$. Therefore, as long as the energy of user 2 exceeds the threshold given by (13) the bit-error-rate of user 1 is equivalent to the single-user case where user 2 is not active. The explanation of this behavior of the optimum receiver is that if the interfering user is sufficiently powerful, then the primary source of errors committed in the optimum demodulation of user 1 is the background Gaussian noise, rather than the randomness of the information carried by the interfering signal. Note that according to the threshold in (13), an interferer who is 3 dB weaker than the user of interest has no appreciable effect on the bit-error-rate as long as the maximum crosscorrelation is below 0.35 (a mild condition on the signal design). Interestingly, the same is true if the relative energy of the interferer is *higher* than −3 dB. The minimization of (12) with respect to w_2/w_1 yields

$$1 - \max{}^2\{\,|\rho_{12}|,\ |\rho_{21}|,\ (\,|\rho_{12}| + |\rho_{21}|)/\sqrt{2}\},$$

which implies that there is a nonzero level of guaranteed asymptotic efficiency regardless of the received energies (unless both signature signatures are identical and there is no offset between them). A more stringent measure of near-far resistance is obtained by minimizing the asymptotic efficiency over the energies of all the users letting those energies be time-varying; the rationale for this being that the near-far problem occurs primarily in networks with dynamically changing topologies. In such case, the near-far resistance is given by

$$\overline{\eta}_1 = \sqrt{[1 - (\rho_{12} + \rho_{21})^2][1 - (\rho_{12} - \rho_{21})^2]}. \tag{14}$$

In the general K-user case, it can be shown that $\overline{\eta}_1$ is equal to the smallest energy of the signals in Ξ_1, which is the set of multiuser signals such that $b_1(0) = 1$, i.e., $\Xi_1 = \{\sum_i \sum_{k=1}^{K} b_k(i) s_k(t - iT - \tau_k)$, $b_1(0) = 1$, $b_k(i) \in R$, $(k,i) \neq (1,0)\}$. In the synchronous case, $\overline{\eta}_k$ can be shown [6] to equal $1/(\mathbf{H}^{-1})_{kk}$, whereas in the asynchronous case [7] we have

$$\overline{\eta}_k = \left[\frac{1}{\pi}\int_0^{\pi} [\mathbf{H}(1)^T e^{j\omega} + \mathbf{H}(0) + \mathbf{H}(1)\, e^{-j\omega}]_{kk}^{-1}\, d\omega\right]^{-1}.$$

The decorrelating detector eliminates the multiuser interference from the decision statistic. Therefore, its bit-error-rate has the very desirable property that it is independent of the energy of the interfering users. It can be shown that it is equal to (in the K-user case) [7]:

$$P_k^d(\omega) = Q\left(\frac{\sqrt{w_k \overline{\eta}_k}}{\sigma}\right) \tag{15}$$

which implies that the effective energy is $e_k(\omega) = w_k \overline{\eta}_k$, and therefore, the efficiency is independent of the noise level as well as of the energies and is equal to the near-far resistance of the optimum receiver. Hence, we arrive at the interesting conclusion that the decorrelating detector achieves optimum near-far resistance. Figure 6 illustrates the near-far behavior of the conventional, optimum and decorrelating detectors, comparing their asymptotic efficiencies as a function of the relative energies. Note that there is a gap between the minimum efficiency achieved by the optimum receiver and the efficiency of the conventional detector. This is because the asymptotic efficiency of the optimum detector is the function in (12) where the energies are assumed time-invariant. The minimum of the optimum asymptotic efficiency over the amplitudes of each of the symbols does indeed coincide with the decorrelating asymptotic efficiency.

The bit-error-rate of the one-shot decorrelating detector for asynchronous channels introduced in Section II is also independent of the energy of the interferers. Its efficiency (and near-far resistance) is given by the inverse of the first diagonal element of the inverse of (9), i.e.,

$$\eta_1^{od} = 1 - \frac{\rho_{12}^2}{e_2} - \frac{\rho_{21}^2}{1 - e_2}.$$

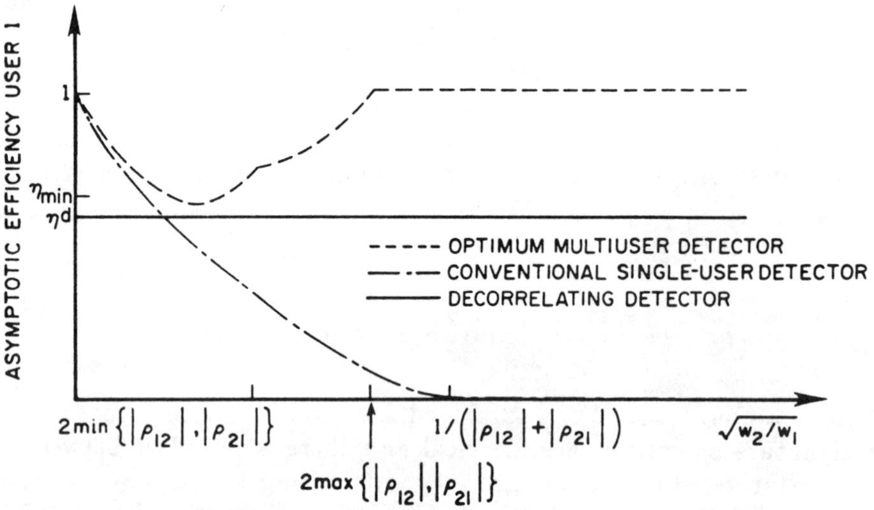

Figure 6. Asymptotic efficiencies for $\rho_{12} = 0.3$, $\rho_{21} = 0.5$. $\eta_{\min} = 0.68$ and $\eta^d = 0.59$.

The performance analysis of the remaining detectors reviewed in Section II does not lend itself to similar analytical results due to the nonlinear nature of those detectors. A numerical computation of the receiver proposed in [18] shows that in the two-user case a single stage detector achieves noticeable improvements over the conventional receiver especially when the interferer is very strong and the signature waveforms have good crosscorrelation properties. No bit-error-rate analysis of the nonlinear single-user detectors obtained in [10] has been undertaken to date. Nevertheless, [10] shows the interesting result that in the absence of background Gaussian noise the transmitted bits can be demodulated perfectly even if the signature waveforms of the interferers are not known to the receiver, thus hinting that some degree of robustness to the near-far effect is also possible in that case. However, no practically implementable receiver that ignores the interfering signature waveforms is known to be near-far resistant.

Acknowledgment

This work was supported by the U.S. Army Research Office under contract DAAL03-87-K-0062.

References

1. W. Van Etten, "An Optimum Linear Receiver for Multiple Channel Digital Transmission Systems," *IEEE Trans. on Communications*, pp. 828-834, Aug 1975.
2. W. Van Etten, "Maximum Likelihood Receiver for Multiple Channel Transmission Systems," *IEEE Trans. Commun.*, vol. COM-24, no. 2, pp. 276-283, February 1976.
3. D. Horwood and R. Gagliardi, "Signal Design for Digital Multiple Access Communications," *IEEE Trans. Commun.*, vol. COM-23, pp. 378-383, March 1975.
4. A.R. Kaye and D.A. George, "Transmission of multiplexed PAM signals over multiple channel and diversity systems," *IEEE Trans. on Communication Technology*, vol. COM-18, pp. 520-525, Oct 1970.

5. R. Lupas-Golaszewski and S. Verdu, "Asymptotic Efficiency of Linear Multiuser Detectors," *Proc. 25th Conf. on Decision and Control*, pp. 2094-2100, Athens, Dec. 1986.

6. R. Lupas and S. Verdu, "Linear Multiuser Detectors for Synchronous Code-Divsion Multiple-Access Channels," *IEEE Trans. Information Theory*, vol. IT-34. to appear.

7. R. Lupas and S. Verdu, "Near-Far Resistance of Multiuser Detectors in Asynchronous Channels," *IEEE Trans. Communications*, vol. COM-37. to appear

8. R. Lupas and S. Verdu, "Optimum Near-Far Resistance of Linear Detectors for Code-Division Multiple-Access Channels," *Abs. 1988 IEEE Int. Symp. Information Theory*, p. 14, Kobe, Japan, June 1988.

9. H.V. Poor, "Signal detection in multiple access channels," U.S. Army Research Office Proposal (Contract DAAG29-81-K-0062 to Coordinated Science Laboratory, University of Illinois),, 1980.

10. H. V. Poor and S. Verdu, "Single-User Detectors for Multiuser Channels," *IEEE Trans. Communications*, vol. 36, pp. 50-60, Jan. 1988.

11. C. Rushforth and Z. Xie, "Multiuser Signal Detection using Sequential Decoding," *1988 IEEE Military Communications Conf.*, San Diego, Oct. 1988.

12. K.S. Schneider, "Optimum Detection of Code Division Multiplexed Signals," *IEEE Trans. Aerosp. Electron. Syst.*, vol. AES-15, no. 1, pp. 181-185, January 1979.

13. K.S. Schneider, "Crosstalk Resistant Receiver for *M*-ary Multiplexed Communications," *IEEE Trans. on Aerospace and Electronic Systems*, vol. AES-16, pp. 426-433, Jul 1980.

14. D.A. Shnidman, "A Generalized Nyquist Criterion and an Optimum Linear Receiver for a Pulse Modulation System," *The Bell System Technical Journal*, pp. 2163-2177, Nov 1967.

15. U. Timor, "Improved Decoding Scheme for Frequency-Hopped Multilevel FSK System," *The Bell System Technical Journal*, vol. 59, pp. 1839-1855, December 1980.

16. U. Timor, "Multistage Decoding of Frequency-Hopped FSK System," *The Bell System Technical Journal*, vol. 60, pp. 471-483, April 1981.

17. G. Ungerboeck, "Adaptive Maximum Likelihood Receiver for Carrier-Modulated Data Transmission Systems," *IEEE Trans. Commun.*, vol. COM-22, no. 5, pp. 624-636, May 1974.

18. M. K. Varanasi and B. Aazhang, "Near Optimum demodulation for Coherent Communications in Asynchronous Gaussian CDMA Channels," *Proc. 22nd Princeton Conf. on Information Sciences and Systems*, pp. 832-839, Mar. 1988.

19. S. Verdu, "Optimum multi-user signal detection," Ph. D. Dissertation, Department of Electrical and Computer Engineering, University of Illinois at Urbana-Champaign. Report T-151 Coordinated Science Laboratory, Urbana, IL, Aug. 1984.

20. S. Verdu, "Minimum Probability of Error for Asynchronous Gaussian Multiple-Access Channels," *IEEE Trans. on Information Theory*, vol. IT-32, pp. 85-96, Jan. 1986.

21. S. Verdu, "Multiple-Access Channels with point-process observations: Optimum demodulation," *IEEE Trans. Information Theory*, vol. IT-32, pp. 642-651, September 1986.

22. S. Verdu, "Optimum multiuser asymptotic efficiency," *IEEE Trans. Communications*, vol. COM-34, pp. 890-897, Sept. 1986.

23. S. Verdu and H. Vincent Poor, "Abstract Dynamic Programming Models under Commutativity Conditions," *SIAM J. Control and Optimization*, vol. 24, pp. 990-1006, Jul. 1987.

24. S. Verdu, "Maximum Likelihood Sequence Detection for Intersymbol Interference Channels: A New Upper Bound on Error Probability," *IEEE Trans. Information Theory*, vol. IT-33, pp. 62-68, Jan. 1987.

25. S. Verdu, "Computational Complexity of Optimum Multiuser Detection," *Algorithmica*, vol. 4, to appear.

Pseudo-Noise Sequences and Interference Cancellation Techniques for Spread Spectrum Systems —Spread Spectrum Theory and Techniques in Japan—

Ryuji KOHNO†

SUMMARY Spread spectrum theory and techniques have been studied and developed for commercial applications in Japan, rather than for military use as in USA. Main purpose of the study and development is not to prevent jamming or intentional interference, but to carry out more efficient utilization of available frequency spectra. This paper introduces some spread spectrum theory and techniques which have been studied for the past several years in Japan for improving capacity of transmission. Co-channel interference in code-division multiple access (CDMA) based on spread spectrum technique is the most dominant factor in the limitation of capacity. Various classes of pseudo-noise (PN) sequences or spreading sequences have been proposed in order to reduce co-channel interference or achieve fast acquisition and high security. On the other hand, co-channel interference can be eliminated by digital signal processing approaches, such as adaptive digital filtering and beam-forming. This paper gives an overview of these PN sequences and interference cancellation techniques which appear to be most promising for use in commercial spread spectrum systems.

1. Introduction

Spread spectrum (SS) techniques are used for communicating and ranging where the transmission bandwidth is typically much larger than that which is required to transmit the information over the channel. This spreading of the signal spectrum gives us many advantages such as robustness against interference and noise, low probability of intercept, realization of code-division multiple access (CDMA), high accuracy ranging and so on[(1)-(4)]. These advantages depend on modulation method of spreading the spectrum such as direct-sequence (DS) and frequency hopping (FH) spectrum spreading, and a pseudo-noise (PN) sequence that is of such a wide band that it spreads the spectrum of a signal and identifies each user in CDMA.

SS systems have been studied and developed primarily for military use in USA, because an SS signal resists jamming and is hard for an enemy to intercept. Techniques for transmission security such as anti-jamming were chiefly taken into account[(5),(6)]. However, recently, research and development of SS systems have increased for commercial applications such as mobile communication[(7),(8)].

In Japan, civil and commercial communications are the most important applications of SS techniques[(9)]. Since the radio wave is a common resource for every user, its usage is strictly regulated by law. Efficient utilization of available frequency spectra is of major importance. In order to substitute SS systems for conventional ones, it is necessary for SS systems to improve in the use efficiency of frequency spectra while retaining the above-mentioned advantages. Research and development on SS techniques for consumer communication is more popular than that for common carrier communication in Japan. Consumer communication, that is, personal communication controlled by users according to their individual purposes and circumstances is characterized as a Japanese market of SS techniques[(10),(11)]. Since consumer communication usually needs a low cost system and employs a poor quality channel such as a low power radio or power line, advantages of SS techniques such as asynchronous CDMA, robustness against interference and noise, and less interference to other systems are useful.

Many SS techniques and applications have been proposed and discussed at IEICE technical meetings on SS systems which have been held three or four times a year since 1986 in Japan. However, most of those papers were written in Japanese and unknown to researchers outside Japan. Therefore, in particular, useful techniques which improve transmission capacity in SS systems are introduced.

CDMA is necessary for SS communications to achieve capacity comparable to conventional narrow band communications. In a CDMA system, users having different PN sequences serving as identifying codes can transmit SS signals at the same time and in the same frequency band, receive the sum of SS signals, and recover only the desired signal by using the same PN sequence employed in transmission. However, since PN sequences have non-zero crosscorrelation, there is interference among SS signals of users simultaneously accessing the channel, that is called co-channel interference. Co-channel interference restricts the num-

Manuscript received February 6, 1991.
† The author is with the Faculty of Engineering, Yokohama National University, Yokohama-shi, 240 Japan.

Reprinted from *IEICE Trans. Commun.*, vol. J74-B-I, no. 5, pp. 1083–1092, May 1991.

ber of available users who can simultaneously access a channel or the capacity of CDMA (unlike FDMA and TDMA capacities which are primarily bandwidth limited).

In order to reduce co-channel interference and improve capacity, two types of approaches have been considered. First, various kinds of PN sequences having good correlation properties have been proposed. PN sequences play an important role in achieving fast acquisition and high security as well. Since coding theory is useful for generating and analyzing PN sequences, coding theorists approach this subject. Second, digital signal processing approaches are proposed to eliminate co-channel interference, because co-channel interference is distinguishable from noise and narrow band interference. Adaptive digital filters and array antenna can cancel co-channel interference in temporal and spatial domains.

Although there are many other advanced SS theory and techniques in Japan, an overview of just PN sequences and interference cancellation techniques is provided in the following sections.

2. CDMA System Based on Spread Spectrum Techniques

First of all, in order to clarify a problem in CDMA based on SS techniques, a model of CDMA is described in this section. For the sake of simplicity, I describe a CDMA system using DS spread spectrum modulation, i.e., DS/SSMA, where SSMA is spread spectrum multiple access.

Figure 1 shows a model of a DS/SSMA system, where M users are individually transmitting spread spectrum signals $S_m(t)$, $m=1, 2, \cdots, M$ at the same time and in the same frequency band.

In order to spread a signal spectrum in the transmitter of the mth user, the message sequence signal $d_m(t)$ is directly multiplied by pseudo-noise (PN) sequence signal $C_m(t)$ which has a much higher symbol rate than $d_m(t)$. $S_m(t)$ can be written as

$$S_m(t)=\mathrm{Re}\{d_m(t)\,C_m(t)\exp[j\omega_c t+\theta_m]\} \tag{1}$$

where $j=\sqrt{-1}$, ω_c and θ_m are the carrier frequency and the phase offset, respectively. $d_m(t)$ and $C_m(t)$ are expressed by

$$d_m(t)=\sum_{k=-\infty}^{\infty} d_{m,k}U_{T_d}(t-kT_d) \tag{2}$$

$$C_m(t)=\sum_{j=-\infty}^{\infty} C_{m,j}U_T(t-jT) \tag{3}$$

and

$$U_T(t)=\begin{cases}1 & \text{if } 0\leqq t<T\\ 0 & \text{otherwise}\end{cases} \tag{4}$$

where $d_{m,k}$ is a symbol of the message sequence $\{d_{m,k}\}$ of the mth user at instant kT_d, $C_{m,j}$ is a symbol of the PN sequence $\{C_{m,j}\}$ assigned to the mth user at instant jT. In the case of BPSK, $d_{m,k}$, $C_{m,j}\in\{+1,-1\}$. In the cases of QPSK and polyphase-shift-keying, $C_{m,j}$ is a complex number. Here, T is the duration of a 'chip' in the PN sequences, while T_d is that of a bit of message sequences. We assume that T_d is a multiple of T and define $N=T_d/T$.

In a receiver, multiplexed SS signals from all users are received and multiplied again by the same PN sequence used in transmitter j in order to transform the spread spectrum of the desired SS signal j into its original narrow band. The received signal $r(t)$ can be represented as

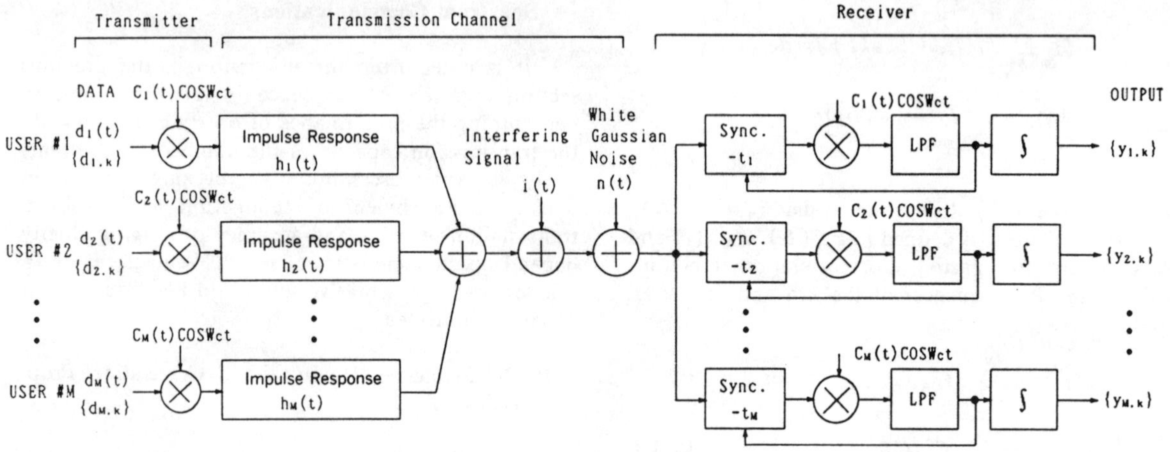

Fig. 1 A model of a DS/SSMA system.

$$r(t)=\sum_{m=1}^{M}\int_{-\infty}^{\infty}h_m(\tau)S_m(t-\tau-t_m)d\tau+i(t)+n(t) \tag{5}$$

where $h_m(t)$ is the impulse response of the transmission channel from the mth user to the receiver, $i(t)$ is an interfering or jamming signal, $n(t)$ is channel noise, and t_m is the phase difference between the received and transmitted SS signals of the mth user.

SS systems are usually applied in channels where the average power of noise $\bar{N}$ and that of interference $\bar{I}$ are much larger than that of a desired signal, $\bar{S}$. For instance, $\bar{S}/(\bar{N}+\bar{I})$ is often less than -20dB. In such a bad channel, although it is impossible to directly detect each chip of the transmitted sequence, SS systems can recover the transmitted message sequences by using correlation detection. In SS systems, a correlator that obtains correlation between received signals and the local PN sequence is used in order to achieve acquisition of the PN sequence and demodulate message sequences.

All the multiplexed SS signals except the desired SS signal interfere with the desired SS signal, due to the crosscorrelation of the different PN sequences assigned to individual users, and is described as co-channel interference.

The output signal of the correlator for the nth user can be derived by integrating $r(t)C_n(t)$ over $T_d=NT$. When PN sequence acquisition is performed, the sample value $y_{n,k}$ of the output at instant kT_d, can be represented in the form

$$\begin{aligned} y_{n,k}=&\,h_{n,0}d_{n,k}\\ &+\sum_{i\neq 0}\sum_{j=0}^{N-1}h_{n,j+iN}\Theta_{n,n}(j)\,d_{n,k-i}\\ &+\sum_{i}\sum_{j=0}^{N-1}\Bigg[\sum_{m=0}^{n-1}h_{m,j+iN}\{R_{n,m}(j+j_m)\,d_{m,k-i-1}\\ &+\hat{R}_{n,m}(j+j_m)\,d_{m,k-i}\}\\ &+\sum_{m=n+1}^{M}h_{m,j+iN}\{R_{n,m}(j+j_m)\,d_{m,k-i}\\ &+\hat{R}_{n,m}(j+j_m)\,d_{m,k-i+1}\}\Bigg]\\ &+i_{n,k}+n_{n,k} \end{aligned} \tag{6}$$

where $h_{m,j}=h_m(jT)$. It is assumed that $j_m=(t_n-t_m)/T \bmod N$, where t_m is defined in Eq.(5). $R_{n,m}(j)$ and $\hat{R}_{n,m}(j)$ are the partial crosscorrelation functions between the PN sequences of the nth and mth user, given by

$$R_{n,m}(j)=\frac{1}{N}\sum_{i=N-j}^{N-1}C_{n,i}C^{*}_{m,i+j} \tag{7}$$

$$\hat{R}_{n,m}(j)=\frac{1}{N}\sum_{i=0}^{N-j-1}C_{n,i}C^{*}_{m,i+j} \tag{8}$$

where* identifies the complex conjugate. Also, the periodic or even correlation function $\Theta_{n,m}(j)$ and odd correlation function $\hat{\Theta}_{n,m}(j)$ are defined by

$$\Theta_{n,m}(j)=R_{n,m}(j)+\hat{R}_{n,m}(j) \tag{9}$$

$$\hat{\Theta}_{n,m}(j)=R_{n,m}(j)-\hat{R}_{n,m}(j). \tag{10}$$

In Eq.(6), the first term of the right hand side indicates the desired baseband message sequences, the second term, the intersymbol interference of the signal from the desired user, the third, the co-channel interference from the undesired users, the fourth, the interfering signal component, and the fifth, the noise component. The magnitude of the co-channel interference is determined by the periodic or even crosscorrelation function $\Theta_{n,m}(j)$ or the odd crosscorrelation function $\hat{\Theta}_{n,m}(j)$ depending on whether $d_{m,k}=d_{m,k-1}$ or $d_{m,k}=-d_{m,k-1}$.

Even in a channel of considerably low $\bar{S}/(\bar{N}+\bar{I})$, the correlator can recover the transmitted message sequence when the ratio of spread spectrum bandwidth to baseband bandwidth, i.e.,

$$G_p=\frac{1/T}{1/T_d}=\frac{T_d}{T}=N \tag{11}$$

is large. G_p represents the capability of suppressing interference and noise, and is called the 'processing gain' of the system.

In the presence of co-channel interference, the crosscorrelation between the PN sequences of desired and undesired users prevents acquisition and tracking of the PN sequence and restricts the number of users simultaneously accessing the channel. This is because side lobes of the output signal of the correlator hamper the detection of the main lobe of autocorrelation and the side lobes increase with the number of simultaneous users.

3. Studies on Pseudo-Noise Sequences for Spread Spectrum Communications

It is noted from the discussion in the previous section that the PN sequence is very important in determining the performance of SS systems. Not only the transmission capacity or the number of simultaneously accessing users but also the power spectrum density, establishment of acquisition and tracking, transmission security, and accuracy of ranging chiefly depend on characteristics of the PN sequence[(1)-(4)]. In this section, representative studies on PN sequences in Japan are outlined.

3.1 PN Sequences Having Good Correlation Properties

Generally speaking, correlation properties of PN sequences are the most dominant factors in determining the capacity and other performances in SSMA. The required correlation properties of the PN sequence for DS/SSMA systems are both sharp autocorrelation and low crosscorrelation.

In order to uniformly spread a signal spectrum to a considerably wide band, the chip rate of a sequence should be much higher than the bit rate of the message sequences. Furthermore, autocorrelation of the sequence should have a sharp main peak and low side lobes like white noise to achieve stable acquisition and tracking. This is the main reason why a PN sequence is used as a spreading sequence. In particular, this property is important for features in SS techniques such as low probability of interception and less interference to other systems.

Crosscorrelation between two arbitrary PN sequences that are assigned to users in SSMA should be small in an absolute value sense for any phase shift or quasi orthogonal. This is because the crosscorrelation prevents achieving acquisition and tracking of the sequence and restricts the number of simultaneously accessing users as mentioned in the previous section.

It is well known that linear feedback shift register (LFSR) sequences such as the maximum length (M) sequence and Gold sequence have good correlation properties and have been used in practical SS systems[12]. However, some other classes of PN sequences possessing better correlation properties have been proposed.

3. 1. 1 Polyphase Sequences

Although binary PN sequences have been used in SSMA because of less complexity in implementation, there are polyphase sequences with better correlation properties exist. A periodic sequence is called an orthogonal sequence when the periodic autocorrelation function is zero in every term except for period-multiple-shift terms, i.e., $\Theta_{n,n}(j)=0$ for $j\neq 0 \bmod N$. It was shown that a polyphase periodic sequence whose absolute value of any symbol is constant can be transformed into an orthogonal sequence by discrete Fourier transform (DFT)[13]. Maximum absolute value of the crosscorrelation function between orthogonal sequences will satisfy the mathematical lower bound if any absolute value of the crosscorrelation is constant[14]. Therefore, such orthogonal sequences are good for a DS/SSMA system.

When every row of a product matrix of a generalized DFT matrix and a diagonal matrix whose absolute value of every diagonal element is constant is combined in series, it is a class of orthogonal sequences satisfying the mathematical lower bound of crosscorrelation[15]. In an SSMA system using such a class of orthogonal sequences, co-channel interference can be eliminated by using its orthogonality. In a similar manner, other classes of polyphase orthogonal sequences based on DFT coefficients have been proposed[16].

In general, a binary sequence can be transformed to a polyphase sequence if spectra of the binary sequence are shifted on a frequency axis. Two sequences whose frequency spectra hardly overlap each other usually have good crosscorrelation property. Therefore, even if spectra of binary sequences are imbalanced, we can obtain various classes of polyphase PN sequences with good crosscorrelation by shifting spectra of the binary sequences[17]. The shift operation of sequence spectra results in efficient utilization of available frequency spectra in SSMA.

Moreover, aperiodic or finite length sequences can be used in DS systems as well as periodic sequences. Even-shift orthogonal sequences and their related N-shift cross-orthogonal sequences have good correlation properties[18],[19]. The latter consists of N sequences with a length of N^n. Its autocorrelation function is zero for N-multiple-shift terms except zero shift and its crosscorrelation function is zero for N-multiple-shift terms. Acquisition of these PN sequences can be achieved by a matched filter.

3. 1. 2 Real-Valued Sequences

Polyphase shift keying such as QPSK and 8PSK must be used in order to spread signal spectra by using polyphase sequences, while amplitude shift keying (ASK) may be employed to do it by real-valued sequences. DS system using real valued sequences can be implemented with less complexity than that using polyphase sequences. We can obtain some classes of orthogonal sequences in which every symbol is a real number, i.e., real-valued orthogonal sequence[20]-[23]. Absolute values of symbols of some real-valued orthogonal sequences have an upper bound of $1/\sqrt{N}$ or $2/\sqrt{N}$ where N is the period of the sequence. These classes of real-valued orthogonal sequences are represented by Fourier series in general. The nth sequence $\{C_{n,i}\}$ among a set of sequences having a period of N can be written as

$$C_{n,i}=\frac{1}{\sqrt{N}}\sum_{k=0}^{N-1} W^{ki+\phi_{n,k}}, \tag{12}$$

where $W=\exp(j\cdot 2\pi/N)$ and $\{\phi_{n,k}\}$ are named phase coefficients. If the phase coefficients are not conditioned at all, the sequence $\{C_{n,i}\}$ will have complex values. If the phase coefficients satisfy with

$$\phi_{n,0}=0 \tag{13}$$

$$\phi_{n,k}=-\phi_{n,-k}=-\phi_{n,N-k}, \quad k=1, 2,\cdots, N-1, \tag{14}$$

$\{C_{n,i}\}$ will be read-valued sequence. We can generate various classes of real-valued sequences by determining specific phase coefficients.

3. 1. 3 PN Sequences on a Finite Field

In a similar manner to real-valued orthogonal sequences defined on a complex number field, we can derive an orthogonal sequence on a finite field. If a primitive element of $GF(q)$ is used as W in Eq. (12) instead of $\exp(j\cdot 2\pi/N)$, the sequence defined on $GF(q)\{C_{n,i}\}$ will be an orthogonal sequence(24).

LFSR sequences such as M and Gold are generally defined on a finite field. In particular, the periodic correlation function of the Kasami and bent sequence(12),(25) satisfies Welch lower bound(26). A bent sequence which is generated by a bent function is a nonlinear binary sequence and is suitable for DS/SSMA, because it has low correlation values, no difference between the number of 1's and that of 0's in a period, that is, "balanced", and large linear complexity or linear span(27),(30).

As shown in Eq. (6), co-channel interference is determined not only by periodic or even crosscorrelation function $\Theta_{n,m}(j)$ but also by odd crosscorrelation function $\hat{\Theta}_{n,m}(j)$. Both the even and the odd crosscorrelations are important, because the probabilities that $d_{m,k}=d_{m,k-1}$ and $d_{m,k}=-d_{m,k-1}$ are equal. In order to guarantee low odd as well as even crosscorrelation, a class of sequences produced by alternately inverting values in PN sequences was proposed(31). The class of sequences $\{A_i\}$ is derived from a biphase PN sequence $\{C_k\}$ with an odd period of N, $C_k\in\{+1,-1\}$ by

$$A_i=(-1)^i C_k \quad \text{for } k=i \bmod N, \tag{15}$$

where the period is $2N$. Such inversion is considered to be a kind of spectra shift operation, which can improve odd correlation properties of polyphase PN sequences(17).

A DS/SS system using QPSK is more robust against interference and jamming than one using BPSK in general. Some quadriphase PN sequences with low periodic correlation have been proposed by combining two biphase PN sequences, such as

$$B_k=\frac{1}{\sqrt{2}}(C_{1,k}+jC_{2,k}), \tag{16}$$

where $\{B_k\}$ is a quadriphase sequence, $\{C_{1,k}\}$ and $\{C_{2,k}\}$ are biphase ones. Quadriphase sequences which consist of binary M sequence at in-phase and the shifted sequence at quadrature-phase, i.e., 2-tuple M sequences can reduce total crosscorrelation in a DS/SS-QPSK system(32),(33). Moreover, if a binary PN sequence with a good periodic correlation and its alternately inverted sequence by Eq. (15) are used at in-phase and quadrature-phase respectively, the produced quadriphase PN sequence will have better crosscorrelation properties(34).

In order to investigate more precise correlation properties of PN sequences, we should consider the high-order moment of the PN sequences, by which performance of acquisition is evaluated. The third and fourth order moments provide natural measures of skewness and flattening, respectively. The third order moment was used in selecting proper sets of truncated M sequences for Tracking and Data Relay Satellite (TDRS) and its related systems(3).

3. 2 PN Sequences for Fast Acquisition

As the processing gain of SS systems is large, advantageous properties of the SS systems such as robustness against interference, low probability of interception, and accuracy of ranging improve. A PN sequence with a long period is required to obtain large processing gain. However, it usually takes a long time to achieve acquisition of PN sequences with a long period. Some classes of PN sequences which are synthesized by elemental PN sequences with short periods have been proposed in order to achieve fast acquisition of long period PN sequences.

A generator of synthesized PN sequences is shown in Fig. 2, where $\{X_k^{(i)}\}$ for $i=1,2,\cdots,M$ are elemental PN sequences with periods of N_i, and $\{Z_k\}$ is a synthesized PN sequence produced by operation W(36). $\{Z_k\}$ can be written as

$$Z_k=W(X_k^{(1)}, X_k^{(2)}, \cdots, X_k^{(M)}). \tag{17}$$

Gold and Kasami sequences can be represented as synthesized PN sequences of elemental biphase M sequences with the same period in the product operation W in which each symbol is a product of symbols of every elemental PN sequence.

If periods of elemental PN sequences N_i are relatively prime, we will obtain a synthesized PN sequence with a long period equal to the product of the periods of elemental ones, i.e., $N=N_1N_2\cdots N_M$. In particular, when elemental PN sequences are biphase M sequences with the same chip duration and the operation is multiplication, synthesized PN sequence is called JPL sequence which is applied to accurate ranging(1)-(3). Concatenated PN sequence is synthesized by the multiplication operation such that a period of

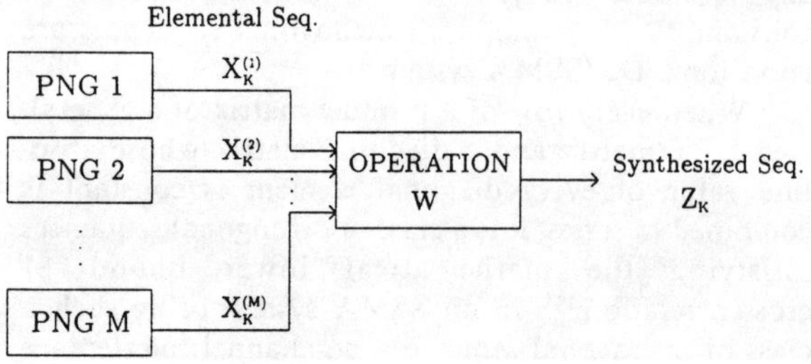

Fig. 2 A generator of a synthesized PN sequence.

one elemental PN sequence may correspond to a chip duration of another elemental one(37). Since a synthesized PN sequence usually has large correlation with elemental sequences, we can achieve fast acquisition if crosscorrelations between the synthesized and elemental sequences are calculated in order of period.

In these PN sequences, the synthesizing operation is linear. If nonlinear operation is used for synthesizing PN sequences, security or complexity of estimating the sequence will improve. Two classes of nonlinear synthesized PN sequences have been proposed.

3. 2. 1 Majority Logic Combination Sequence

Let elemental sequences $\{X_k^{(i)}\}$ for $i=1, 2, \cdots, M$ be biphase PN sequences with relatively prime periods in Eq. (17). Majority logic combination sequence is defined by

$$Z_k = \mathrm{Sign}(X_k^{(1)} + X_k^{(2)} + \cdots + X_k^{(M)}), \tag{18}$$

where M is odd number and Sign ( ) is the sign function, i.e., $\mathrm{Sign}(X)=1$ for $X>0$ and $\mathrm{Sign}(X)=-1$ for $X<0$.

It has been found that majority logic combination sequence using M sequences as elemental ones have a larger correlation peak at instant of synchronization than that of JPL and concatenated sequences even in the case of low SN ratio reception(38).

3. 2. 2 Generalized Geffe Sequence

Let element sequences $\{X_k^{(i)}\}$ for $i=1, 2, \cdots, M-1$ be biphase PN sequences and $\{X_k^{(M)}\}$ be a multi-level PN sequence whose periods are relatively prime. Generalized Geffe sequence is defined by

$$Z_k = \sum_{i=1}^{M} X_k^{(i)} \delta(i, X_k^{(M)}) = X_k^{(X_k^{(M)})}, \tag{19}$$

where $X_k^{(M)} \in \{1, 2, \cdots, M\}$ and

$$\delta(i, j) = \begin{cases} 1 & \text{if } i=j \\ 0 & \text{if } I \subset + j \end{cases} \tag{20}$$

A symbol of the synthesized sequence is selected from symbols of elemental sequences $\{X_k^{(i)}\}$ for $i=1, 2, \cdots, M-1$ corresponding to a symbol of $\{X_k^{(M)}\}$(36). Since symbols of elemental sequences which are not selected disappear, the order of symbols of elemental sequences in the synthesized sequence is different from that of the original elemental sequences. If the operation is modified such that the order can be the same as that of the original ones, correlation between the synthesized and elemental sequences will improve sharply. Therefore, required time to achieve acquisition will be shortened(39). Moreover, these classes of synthesized PN sequences have such large linear complexity compared to LFSR sequences such as M and Gold sequences that greater security of SS systems can be designed.

If these classes of synthesized sequences are applied to DS/SSMA systems, crosscorrelation between synthesized sequences should be kept to a minimum. It was clarified that using balanced PN sequences such as bent sequences as the elemental ones is good to suppress crosscorrelation(40).

4. Interference Cancellation Techniques in SSMA

If both even and odd crosscorrelations of PN sequences are not completely zero at any instant in DS/SSMA systems, co-channel interference (that is, an interfering SS signal from an undesired user due to crosscorrelation among PN sequences assigned to individual users in SSMA) will remain and limit the number of simultaneously accessing users or the transmission capacity. Co-channel interference results in the "near-far" problem that is related to the problem of very strong signals from undesired users at a receiver swamping out the weaker signals from desired user(1)-(3). Moreover, if enough inherent processing gain of an SS system cannot be obtained (e.g., due to restricted transmission bandwidth in a channel), co-channel interference will graw larger as the period of the PN sequences or the processing gain decreases.

To further aid the SS receiver in suppressing co-channel interference, various digital signal processing techniques can be employed. In this section, techniques for cancelling co-channel interference on the temporal and the spatial domains are introduced. Since the chip rate of the PN sequence must be much larger than the data rate, very high speed signal processing capabilities are required. Fortunately, these techniques have become feasible because of the recent development of high-speed digital signal processing devices.

4. 1 Temporal Domain Approach

As shown in Fig. 1, received SS signals include interfering signals from the other systems and noise as well as co-channel interfering SS signals within the DS/SSMA system. The interfering signals except those from co-channel interference might arise from intentional jamming, multipath, and conventional narrow-band communications coexisting in the same frequency band. Such narrow-band interfering signals are not inherent in SS systems and can be rejected by means of adaptive notch filters(5). This paper shall I focus on cancellation of wide-band co-channel interference in DS/SSMA.

From Eq. (6), note that co-channel interference can be regenerated in the receiver, unlike thermal noise because it is due to SS signals of simultaneously accessing other users. Therefore, it can be cancelled in the receiver by generating a replica if PN sequences of the users and channel characteristics are known(41).

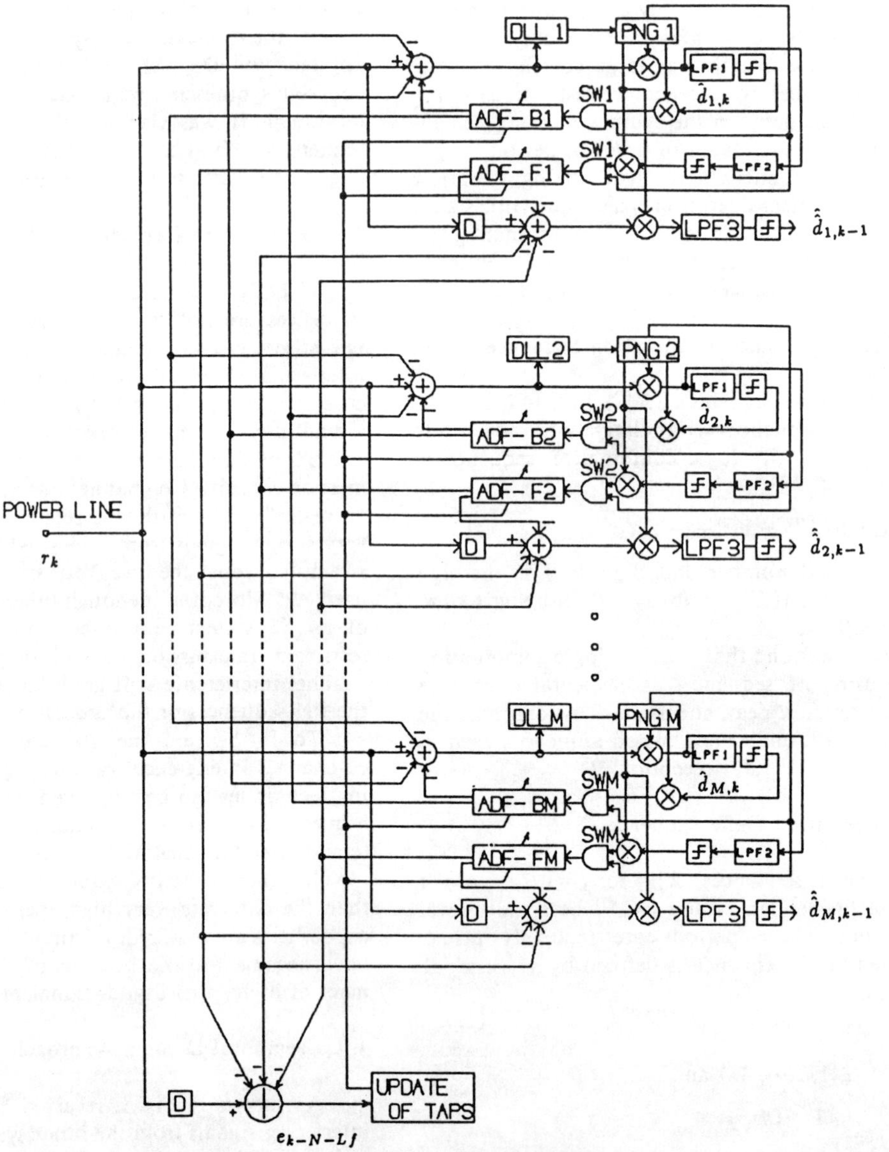

Fig. 3 The receiver for every user including the proposed adaptive canceller of interference in DS/SSMA.

In the temporal domain, an adaptive digital filter (ADF) can be employed to adaptively identify the time-varying impulse response of the channel and cancel intersymbol interference as well as co-channel interference in DS/SSMA system without excessive noise enhancement.

Figure 3 shows the receiver for every user including the proposed adaptive canceller using ADFs[42]. In Fig. 3, carrier demodulation is omitted. The receiver demodulates a received signal $r_k=r(kT)$ in order to estimate data from all the users $d_{n,k}(n=1, 2,\cdots, M)$, where the number of users accessing the channel is M. The receiver consists of demodulators for all the users and a circuit to update ADFs. The demodulator for the nth user consists of a DLL(DLLn), a PN sequence generator (PNGn), multipliers, integrators (LPF1, LPF2, LPF3), threshold decision devices, a gate switch (SWn), ADFs (ADF-Bn, ADF-Fn) and a delay (D). Here three sets of both a multiplier and an integrator constitute three correlators. ADF-Bn and ADF-Fn act as the canceller of intersymbol and co-channel interference.

In the demodulator for the nth user, ADF-Bn and ADF-B$i(i\neq n)$ cancel postcursors of the intersymbol and co-channel interference component in a received signal, respectively. DLLn achieves acquisition and

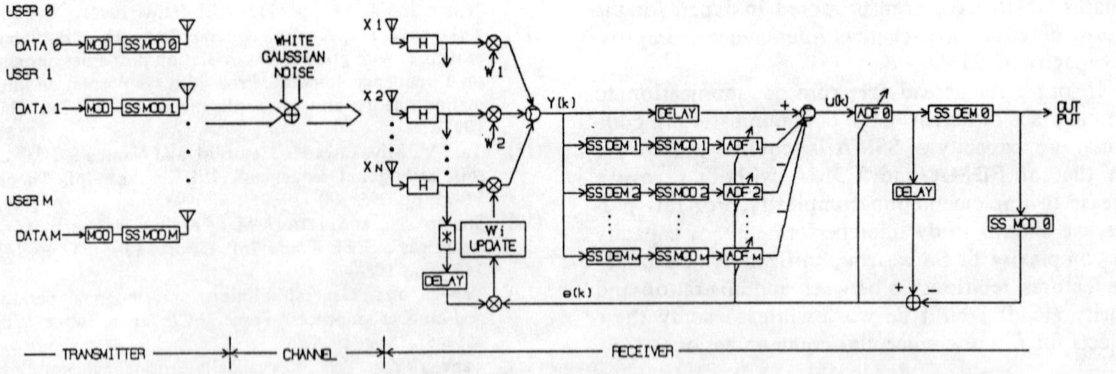

Fig. 4 A DS/SSMA system with an adaptive array antenna including a canceller of interference.

tracking of the PN sequence. After the cancellation and acquisition, the signal is despread by the correlator which consists of a multiplier and LPF1 which is an integrator with an integrating interval of L_1T ($L_1<N$). Then $\hat{d}_{n,k}$, an estimate of $d_{n,k}$, is obtained by the first decision device. $\hat{d}_{n,k}$ is multiplied again by the PN sequence in order to obtain a respread spectrum signal $\hat{S}_{n,k}$, which is used by ADF-Bn and ADF-Fn in order to generate the replica of intersymbol and co-channel interference.

The adaptive canceller is useful in order to operate with stable synchronization and adequate number of users in a channel where sufficient processing gain cannot be obtained due to restriction on bandwidth, as found in a power line.

4.2 Spatial Domain Approach

In radio SSMA communication, received signals with different arrival angles can be distinguished by beam-forming of antenna. An adaptive array antenna is useful in suppressing interfering signals, because it can adaptively control directivity of the antenna even if the desired SS signal's arrival angle is unknown[43],[44]. An adaptive array antenna can be considered to be an adaptive spatial filter, while an adaptive channel equalizer, an adaptive temporal filter. The weights of antenna elements can be updated by using a reference signal determined by partially demodulating the SS signal in a way simular to the tap coefficients of an adaptive equalizer. However, if there is a high-level interfering SS signal from an undesired user with the same arrival angle as that of the desired user, an array antenna cannot suppress it.

In order to solve this problem, combination of an adaptive array antenna in the spatial domain and an adaptive interference canceller in the temporal domain was proposed[45]. The proposed system can suppress interfering SS signals, i.e., co-channel interference, with arrival angles different from that of a desired user by using a null steering array antenna and eliminate by means of a canceller the residual interference and co-channel interference having an arrival angle the same as that of the desired SS signal.

Figure 4 shows the structure of the DS/SSMA system based on combination of adaptive spatial and temporal filtering, which consists of array antenna and canceller using ADFs. The adaptive array antenna suppresses co-channel interference having arrival angles different from that of a desired user. When the interfering signals from M undesired users remain at the output of the array antenna, the adaptive canceller can eliminate them in the similar manner mentioned in Sect.4.1. H for each antenna element in Fig. 4 represents a quadrature hybrid splitting the received signal into quadrature components. In the canceller, the interfering SS signal from the ith user ($i=1, 2, \cdots, M$) is demodulated and respread by SSDEMi and SSMODi, respectively. ADFi ($i=1, 2, \cdots, M$) is used to identify the entire channel characteristics of both the channel for the ith user and the array and to generate a replica of the distorted interference component in the array output. Then every replica of interference is subtracted from the delayed output signal of the array. The output signal from the canceller is then fed to ADF0, which compensates for the distortion of the desired SS signal from the 0th or the desired user, and demodulated by SSDEM0. The final output data are respread by SSMOD0 in order to produce the reference signal for the array and the ADFs.

The proposed system can achieve stable demodulation and improve the error rate of decoded data even in a heavy interference channel where a conventional array antenna system cannot achieve acquisition.

5. Concluding Remarks

I have introduced a variety of PN sequences and filtering techniques on the spatial and the temporal domains which have been proposed in Japan for the purpose of reducing co-channel interference or improving capacity of SSMA.

In order for spread spectrum communication to take the place of narrow-band communication in commercial use, capacity of SSMA is required to improve over that of FDMA and TDMA without excessive increase in implementation complexity. For this purpose, we should study total performance of capacity and complexity of SS systems, utilization of remarkable features, relationship between administration and security etc. It would be worthwhile to study these subjects for future commercial communications.

References

(1) Simon M. K., Omura J. K., Scholtz R. A. and Levitt B. K.: "Spread Spectrum Communications", Computer Science Press (1985).

(2) Dixon R. C.: "Spread Spectrum Systems", John Wiley & Sons (1976).

(3) Holmes J. K.: "Coherent Spread Spectrum Systems", John Wiley & Sons (1982).

(4) Yokoyama M.: "Spread Spectrum Communication Systems", Kagakugijutsu-shuppan (1988).

(5) Milstein L. B. and Iltis R. A.: "Signal processing for interference rejection in spread spectrum communications", IEEE ASSP Magazine, **3**, 2, pp. 18-31 (April 1986).

(6) Nicholson D. L.: "Spread Spectrum Signal Design LPE & AJ Systems", Computer Science Press (1988).

(7) Cooper G. R. and Netteleton R. W.: "A spread-spectrum technique for high capacity mobile communications", IEEE Trans. Veh. Technol., **VT-27**, pp. 264-275 (1978).

(8) Gilhousen K. S., Jacobs I. M., Padovani R., Viterbi A. J., Weaver L. A. and Wheatley C. E.: "On the capacity of a cellular CDMA system", IEEE Trans. Veh. Technol. (to be appeared in 1991).

(9) Marubayashi G., Nakagawa M. and Kohno R.: "Research and development activities on spread spectrum communication systems", J. IEICE, **73**, 5, pp. 580-592 (May 1989).

(10) Kohno R.: "A trend of research and development concerning low power radio systems for consumer communications", IEICE Technical Report, **SSTA89**-1 (March 1989).

(11) Nakagawa M.: "Consumer communication and spread spectrum communication", IEICE Technical Report **SSTA90**-18 (June 1990).

(12) Sarwate D. V. and Pursley M. B.: "Crosscorrelation properties of pseudorandom and related sequences", Proc. IEEE, **68**, pp. 593-619 (1980).

(13) Suehiro N. and Hatori M.: "Polyphase periodic sequences without crosscorrelation and their application to asynchronous SSMA systems", Trans. IEICE Japan, **J68-A**, 10, pp. 1087-1093 (Oct. 1986).

(14) Suehiro N.: "Polyphase periodic sequences with lowest peak of crosscorrelation", Trans. IECE Japan, **J66-A**, 5, pp. 402-407 (May 1983).

(15) Suehiro N.: "Modulatable orthogonal sequences and their application to SSMA systems", IEEE Trans. Inf. Theory, **IT-34**, 1, pp. 93-100 (1988).

(16) Suehiro N.: "Cross-correlation functions for quadratic-phase sequences with different periods", Trans. IECE Japan, **J66-A**, 11, pp. 1115-1121 (Nov. 1983).

(17) Fukumasa H., Kohno R. and Imai H.: "Polyphase sequences with good cross-correlation properties designed on a frequency domain", Proc. 13th symposium on information theory and its applications, pp. 705-710 (Jan. 1991).

(18) Taki Y., Miyakawa H., Hatori M. and Namba S.: "Even-shift orthogonal sequences", IEEE Trans. Inf. Theory, **IT-15**, pp. 295-300 (1969).

(19) Suehiro N. and Hatori M.: "N-shift cross-orthogonal sequences", IEEE Trans. Inf. Theory, **IT-34**, 1, pp. 143-146 (Jan. 1988).

(20) Tanada Y.: "General solution for orthogonal periodic real-number sequences", Trans. IECE Japan, **J68-A**, 5, pp. 451-457 (May 1985).

(21) Tanada Y.: "Sets of cyclically orthogonal multilevel sequences derived from monomial phases", Trans. IECE Japan, **J69-A**, 1, pp. 106-113 (Jan. 1986).

(22) Tanada Y.: "Cyclically orthogonal sequences with low multiplicity of values derived from quadratic residues", Trans. IECE Japan, **J69-A**, 8, pp. 967-973 (Aug. 1986).

(23) Tanada Y.: "Synthesis of cyclically orthogonal multilevel sequences with compound-number length", Trans. IECE Japan, **J69-A**, 9, pp. 1117-1123 (Sept. 1986).

(24) Tanada Y., Yoshida H. and Nagasawa Y.: "Cyclically orthogonal pseudonoise sequences over a finite field", IEICE Technical Report, **SSTA90**-28 (June 1990).

(25) Kasami T.: "Weight distribution formula for some class of cyclic codes", Coordinated Science Lab., Illinois. Urban, Tech. Rep., R-285 (April 1966).

(26) Welch L. R.: "Lower bounds on the maximum cross correlation of signals", IEEE Trans. Inf. Theory, **IT-20**, pp. 397-399 (1974).

(27) Koga N., Matsufuji S., Imamura K. and Soejima S.: "Correlation function of bent function sequences", Proc. the 11th symposium on information theory and its applications, pp. 537-542 (Dec. 1988).

(28) Matsufuji S., Imamura K. and Soejima S: "Balanced binary pseudorandom sequences with low periodic correlation", Abstracts of IEEE 1990 international symposium on information theory, p. 149 (Jan. 1990).

(29) Matsufuji S. and Imamura K.: "Generator of binary sequences with low periodic correlation", IEICE Technical Report, **SSTA90**-35 (Oct. 1990).

(30) Matsufuji S. and Imamura K.: "Real-valued bent function and its application to the design of balanced quadriphase sequences with optimal correlation properties", Abstracts of 8-th international conference on applied algebra, algebraic algorithms and error-correcting codes, p. 29 (Aug. 1990).

(31) Kohno R., Tanaka T. and Imai H.: "On odd-correlation functions for the class of sequences produced by inversion of alternate values in pseudonoise sequences", Trans. IECE Japan, **J65-A**, 10, pp. 1029-1030 (Oct. 1982).

(32) Kohno R. and Imai H.: "On pseudo-noise sequences for direct-sequence QPSK spread spectrum communication systems", Trans. IECE Japan, **J65-A**, 1, pp. 69-76 (Jan. 1982).

(33) Kohno R.: "Pseudo-noise sequences and its applications", Proc. 1986 Workshop on Coding Theory and Its Applications, pp. 31-33 (Aug. 1986).

(34) Matsufuji S., Hayashi K. and Soejima S.: "A construction of quadriphase sequences with low periodic correlation", IEICE Technical Report, **SSTA89**-38 (Nov. 1989).

(35) Fukumasa H., Kohno R. and Imai H.: "Pseudo-noise sequences for tracking and data relay satellite and related systems", IEICE Trans., **E74**, 5, pp. 1137-1144 (May

1991).
(36) Okada T., Kohno R., Imai H. and Hatori M.: "A note on a generalized Geffe pseudo noise sequence", IEICE Technical Report, **SS88**-17(April 1988).
(37) Maskara S. L. and Das J.: "Concatenated sequences for spread spectrum systems", IEEE Trans. Aerosp. & Electron. Syst., **AES-17**, pp. 342-351 (May 1981).
(38) Tachikawa S. and Marubayashi G.: "Majority logic combination sequences using M-Sequences for rapid acquisition in direct sequence/spread spectrum communication system", Trans. IEICE, **J73-B-I**, 12, pp. 902-912 (Dec. 1990).
(39) Fukumasa H., Kuroiwa N., Sugita T., Kohno R. and Imai H.: "A note on methods to generalize Geffe pseudo noise sequence", IEICE Technical Report, **SSTA89**-23 (Aug. 1989).
(40) Matsufuji S., Imamura K., Mizoguchi T. and Soejima S.: "Sequences constructed by component sequences with relatively prime period", IEICE Technical Report, **SS88**-37 (Nov. 1988).
(41) Kohno R., Imai H. and Hatori M.: "Cancellation technique of co-channel interference in asynchronous spread spectrum multiple access systems", Trans. IECE Japan, **J65-A**, 5, pp. 416-423 (May 1983).
(42) Kohno R., Imai H., Hatori M. and Pasupathy S.: "An adaptive canceller of cochannel interference for spread-spectrum multiple-access communication networks in a power line", IEEE J. Selected Areas in Commun., **JSAC8**, 4, pp. 691-698 (May 1990).
(43) Compton R. T., Jr.: "An adaptive array in a spread-spectrum communication system", Proc. IEEE, **66**, pp. 289-298 (March 1978).
(44) Compton R. T., Jr.: "Adaptive antennas concepts and performance", Prentice-Hall (1988).
(45) Kohno R., Imai H., Hatori M. and Pasupathy S.: "Combination of an adaptive array antenna and a canceller of interference for direct-sequence spread-spectrum multiple-access system", IEEE J. Selected Areas in Commun. **JSAC8**, 4, pp. 675-682 (May 1990).

The Role of Spread Spectrum in Packet Radio Networks

MICHAEL B. PURSLEY, FELLOW, IEEE

Invited Paper

This paper is devoted to an examination of the key features of spread-spectrum signaling in packet radio networks. The multiple-access capability, capture, and anti-multipath capability of spread spectrum are the principal topics, and the basic features of spread spectrum that enable it to provide these capabilities are illustrated. The interaction between the spread-spectrum signaling and the network protocols is discussed. Methods for performance evaluation are reviewed, and analytical results on the multiple-access capability of spread spectrum are presented.

I. Introduction

There are several properties of spread spectrum that can be exploited in a packet radio network. These properties are derived from the signal structures used in spread spectrum and from the processing that takes place in the receivers. In this paper we consider four desirable characteristics of radios in a packet radio network: signal capture, multiple access, anti-multipath, and narrow-band interference rejection. Each of these four features can be obtained via the proper use of spread spectrum. The emphasis in this paper is on how these features are derived from the spread-spectrum signaling and what impact they have on the performance of the packet radio network.

It is advantageous to begin with a cursory definition of each of the four characteristics. Capture (or capture effect) refers to the ability of the radio to demodulate at least one of a number of overlapping packets, all of which are addressed to it. The situation envisioned here is that two or more radios transmit packets to another radio at about the same time and one packet arrives at the receiver before the reception of another is complete. This is depicted in Fig. 1 (all of the transmissions are intended for radio *B*). If the radio cannot demodulate more than one packet at a time—and this is typical for ground-based, mobile, spread-spectrum radios—the goal is to provide the capability for the radio to receive one of the overlapping packets. This packet is said to have captured the receiver in question. Usually, the packet that captures the receiver corresponds to either the first-arriving signal or the strongest signal. Spread-spectrum techniques with good capture capability can significantly improve delay and throughput performance of a packet radio network [8], [20].

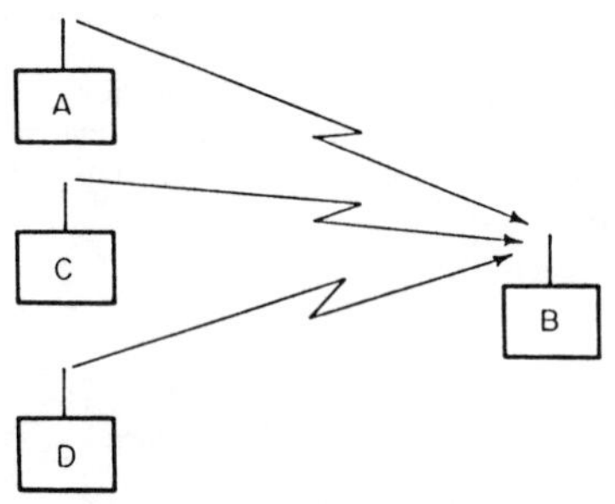

Fig. 1. Access to a single receiver.

Our convention is that capture refers to the ability to demodulate one packet among a subset of packets which are all intended for the radio under consideration (e.g., radio *B* in Fig. 1). In this situation, capture is achieved by distinguishing between the packets on the basis of their power levels or arrival times. Contrast this with multiple-access capability, which, according to the convention followed in this paper, refers to the ability of a radio to receive a packet that is addressed to it but is subjected to the interference caused by the transmission of packets addressed to other radios. More precisely, multiple-access capability is the ability of two or more radios to simultaneously receive packets from different transmitters, even though the transmissions overlap in time and each of the transmitting radios is within range of each of the receivers. This latter situation is depicted in Fig. 2 where we illustrate the simultaneous transmission of one packet addressed to radio *B* and one to radio *D*.

Unlike capture capability, multiple-access capability is a measure of the performance in a subset of the network that includes multiple receivers as well as multiple transmitters.

Manuscript received March 10, 1986; revised July 31, 1986. This research was supported by the Army Research office under Contract DAAG-28-84-K-0088, The Defense Advanced Research Projects Agency under Contract MDA-903-82-C-0086, and the Joint Service Electronics Program under Contract N00014-84-C-0149.

The author is with the Coordinated Science Laboratory, University of Illinois, Urbana, IL 61801, USA.

Reprinted from *Proc. IEEE*, vol. 75, no. 1, pp. 116–134, Jan. 1987.

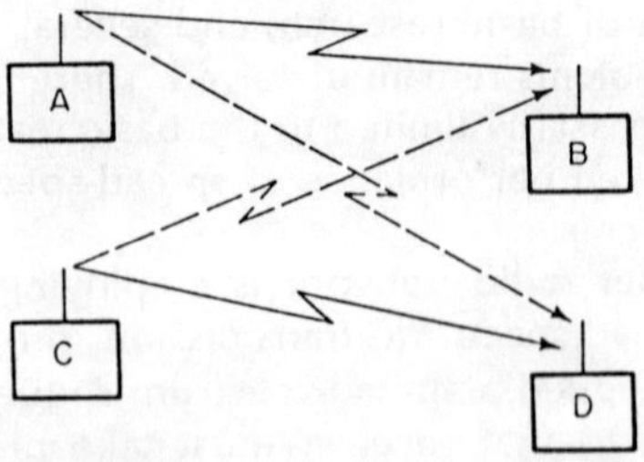

Fig. 2. Multiple access.

Also, the discrimination between packets in a spread-spectrum multiple-access situation is primarily on the basis of their spreading sequences rather than their arrival times. In an ideal multiple-access system, packets can have identical power levels and arrival times, yet they can be received correctly provided the spread-spectrum signals are carefully selected.

A key observation is that these two properties, capture and multiple access, deal with multiple-terminal communication channels. Specifically, the channel in question has two or more inputs which correspond to the transmitting radios that are within range of a given receiver. The channel can have any number of outputs (receivers); the number depends on the number of radios within range of the transmitters under consideration.

Anti-multipath capability refers to the ability to communicate reliably over a link which has multiple transmission paths. This is illustrated in Fig. 3. Without spread-spectrum signaling or special processing at the receiver, multipath channels may not provide reliable communication between radios because of the fading and intersymbol interference that result.

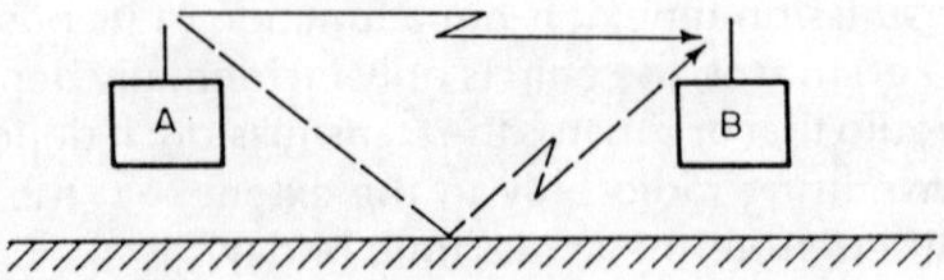

Fig. 3. Multipath.

The final characteristic is the radio's ability to receive a packet in the presence of narrow-band interference. This interference may be due to hostile jamming or to other radios (not part of the packet radio network) that are transmitting in the same frequency band. This is illustrated in Fig. 4.

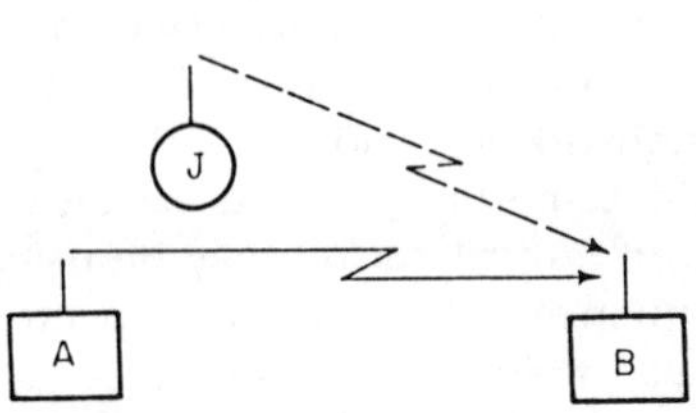

Fig. 4. Narrow-band interference.

The coverage given to these four topics is not uniform in this paper. This is, in part, due to the differences in the roles that these features play in determining network performance and in the degree of interplay between the individual characteristics and the protocols employed in the packet radio network. Because the focus of this paper is on the special role that spread spectrum plays in packet radio networks, we are particularly concerned with the radio characteristics that have greatest influence on network performance and network protocol selection and design. This leads us to a consideration of the *capture* and *multiple-access* capabilities as the two most important characteristics, because of their strong influence on the throughput-delay performance of the network and on the design of channel-access protocols for the packet radio network.

We will also devote some attention to the benefits of spread spectrum for communication over multipath channels. This is primarily because the features of spread spectrum that provide good capture also provide good anti-multipath performance. Another reason is that the DARPA spread-spectrum packet radio network operates at frequencies that are subject to severe multipath distortion. On the other hand, we do not consider narrow-band interference rejection, primarily because this important feature of spread spectrum plays no special role in packet radio networks.

Clearly, all four characteristics mentioned above can provide improved performance on the individual links in the network, and this certainly enhances the overall performance of the packet radio network. One key difference is that the design of spread-spectrum signals for channels with multipath and narrow-band interference has been accomplished via the theory of single-link communication (i.e., a single transmitter and a single receiver). The fact that there may be a number of simultaneously active links in the network has not played a major role. By contrast, the capture and multiple-access features are fundamentally multiple-transmission issues. Without the possibility that two or more transmitting radios are within range of the same receiving radio, there is no need for the radio to have capture or multiple-access capabilities.

This list of four capabilities that can be derived from spread-spectrum signaling is by no means exhaustive; in fact, we have not even touched upon one of the primary reasons for the use of spread spectrum: the difficulty of detection by an unauthorized receiver. Moreover, we make no claim that the relative emphasis given the individual properties in this paper is proportional to their relative importance. Rather, we are governed by the extent of their interactions with the network protocols and other network-oriented issues.

The subject of Section II is the interplay between general properties of spread spectrum and the protocols for packet radio networks. Spread-spectrum transmission protocols are emphasized, but some attention is given to channel-access and acknowledgment protocols. Only those protocol issues that are strongly coupled with spread-spectrum signaling are considered.

Sections III through V are devoted to a more detailed discussion of the properties of spread spectrum and its role in packet radio networks. This presentation is divided into two parts. The first part (Sections III and IV) deals with the direct sequence form of spread-spectrum modulation in which the spreading code is modulated directly onto the carrier along with the digital data signal. In fact, in direct-sequence spread-spectrum modulation, we can combine the digital spreading code and the digital data (e.g., modulo-2 addition) before they are modulated onto the RF carrier. The second part (Section V) deals with the frequency hopping form of spread-spectrum modulation. In particular, we will concentrate on slow-frequency-hop spread spectrum in which the carrier frequency is changed at a rate that is slow in comparison with the data rate.

II. Spread-Spectrum Signaling and Packet Radio Network Protocols

There are some general properties shared by both direct-sequence and frequency-hop spread spectrum that have tremendous impact on the design and implementation of the protocols in a packet radio network. Among the protocols that are most strongly influenced are the spread-spectrum transmission protocol, the channel-access protocol, and the acknowledgment protocol. The nature of the coupling between these protocols and the spread-spectrum modulation is discussed in this section. Only those aspects of the protocols that are strongly tied to the properties of spread-spectrum signaling are addressed here; more complete discussions of network protocols can be found in [23] and [72].

Considerations of the specific properties of direct sequence and frequency hopping and their influence on network protocols are included in Sections III–V. It should be kept in mind that some of the protocols are more suitable for one form of spread spectrum than the other. Even the qualitative statements made in this section about the feasibility or the performance of various protocols are usually not *equally* applicable to both forms.

A. Spread-Spectrum Transmission Protocols

The spreading sequences permit the receivers to distinguish one spread-spectrum transmission from another when both use the same modulation type and format (e.g., two direct-sequence PSK signals). For direct-sequence spread spectrum the spreading sequence is called the *signature sequence* (also known as the pseudorandom sequence, PN sequence, chip sequence, or spreading code); for frequency-hop spread spectrum it is known as the *frequency-hopping pattern*. For convenience, we refer to the elements of the spread-spectrum sequence as *chips* for both frequency-hop and direct-sequence spread-spectrum modulation, even though this term is more commonly reserved for direct sequence.

For many spread-spectrum radios it is possible to change the spreading sequence from one transmission to the next. The choice of the sequence for a particular transmission will strongly affect the probability of success for that transmission. *Spread-spectrum transmission protocols* are the rules that govern the selection of the spreading sequence to be used in the transmission of a particular packet. This selection can be influenced by the identity of the transmitter, the identity of the intended receiver (or receivers), the time of day, the starting time for the transmission, the priority assigned to the message, and several other factors.

Although a number of different spread-spectrum transmission protocols have been discussed in the literature (e.g., see [48] and [67]), there are not many performance results available. The design and analysis of these protocols is still an active area of basic research, and several of the most important problems remain unsolved. Therefore, our discussion is necessarily limited to the basic features rather than the detailed performance of spread-spectrum transmission protocols.

If the packet radio network is employing a *receiver-oriented* spread-spectrum transmission protocol, each receiver is assigned a spread-spectrum sequence and all transmissions to that receiver must take place on that receiver's sequence. The performance of such a protocol also depends on the timing mechanism for generating the sequence and on the time offsets between different packets that are transmitted to the same receiver. For example, the sequence timing for a given packet can be derived from the time of day or from the position in the packet, and either slotted or unslotted transmission is possible. The protocol may also call for the transmitters to delay some or all of the transmissions by random amounts.

An important part of the transmission protocol is the timing mechanism for the sequence. An infinite-length spreading sequence $x = \cdots, x_{-1}, x_0, x_1, x_2, \cdots$ has infinitely many shifts or phases. Each of the chips x_i could be selected as the first chip in the packet. If the jth chip is selected, the spreading sequence for the packet is $x_j, x_{j+1}, \cdots, x_{j+J-1}$, where J is the number of chips per packet. This is the jth starting phase for the sequence x. One function of the timing mechanism is to determine which phase is to be employed for each packet that is to be transmitted.

The first timing mechanism that we consider employs time of day as the reference. The chip that is used at a particular point in the packet is a function of the sequence and the transmission time; it is not a function of its position in the packet. In fact, the chip is only incidentally dependent on the radio that originates the transmission: it depends on the transmitting radio only to the extent that the radio's clock differs from the actual time of day.

When two or more packets using the same phase of the same spreading sequence arrive at a given receiver at nearly the same time, there will be insufficient time offset between the two sequences to permit the receiver to distinguish between them. As a result, it is highly probable that none of the packets will be received correctly. This is because spread-spectrum demodulation provides no discrimination between such signals, so there will be a large number of errors in the received packets (too many for practical error-correcting codes to handle). We refer to this event as a *collision*.

A collision in a spread-spectrum system has about the same effect as a collision in a narrow-band radio. The packets involved in the collision are lost and they must be retransmitted at a later time (thereby increasing the delay). An important difference that arises in a spread-spectrum packet radio network is the possibility that two packets collide, yet, at the same time, a third packet is successfully received (e.g., the third packet may be using a different spreading sequence). Thus if spread-spectrum transmission is employed, collision events are associated with individual packets rather than with time slots or with individual receivers. Moreover, the error mechanism that leads to collisions should not be confused with the random errors that might occur when multiple packets are transmitted simultaneously, perhaps to different receivers, using different sequences or time-offset versions of the same sequence.

The motivation for the use of a receiver-oriented transmission protocol and the use of time of day to generate the phases for the spreading sequences is the reduced time required by the receiver to acquire synchronism with the spreading sequence on an incoming packet. In addition, slotted transmission is known to provide increased throughput (at least in narrow-band radio networks), and time slotting will also help reduce acquisition times. However, based on the above considerations, it is clear that a spread-spectrum packet radio network with small differential propagation delays that uses such a protocol and timing mechanism will experience a large number of collisions (at least at high traffic levels). In terms of our four features of spread spectrum outlined in Section I, we say that such a system does not provide good capture capability. One alternative is to have the sequences keyed to the transmission start time, rather than to time of day, and this brings us to the second scheme.

A timing mechanism for the spreading sequences that gives better capture properties is to use the position in the packet as a reference (e.g., x_0 is the first chip in the packet, x_1 the second, etc.). For such an approach, either the sequences are reused or there must be a method whereby they are changed periodically. For a military application, periodic changing of the sequence is necessary in order to prevent interception and other undesirable consequences. For a slotted packet radio network, the sequence changing might be accomplished by associating a different starting point in the original infinite-length sequence for each packet slot. As a result, all packet transmissions that begin in a given packet slot use the same sequence.

In order to avoid collisions, the transmitters must incorporate randomized transmission delays within the slot. Even if the receiver-oriented transmission protocol is employed, two packets with transmission start times that are sufficiently far apart will arrive at the receiver as time-shifted versions of the spreading sequence. Providing that initial acquisition has been accomplished, they will not collide: they will suffer random errors only.

The disadvantage with this approach is that, although the receiver knows the starting point for the sequence before the first packet arrives, there is a large uncertainty in the actual timing for the sequence. The uncertainty that must be resolved in the acquisition process is greater than the transmitter clock offsets and the differential propagation times, because it also includes the randomized transmission delays. This protocol is receiver oriented in the sense that the transmissions all use the receiver's sequence, but it is transmitter-oriented in the sense that the transmitters randomly select the starting time of the sequence. Thus although the capture properties are better than for the first method, the acquisition times and acquisition probabilities are worse.

The third approach has the potential to combine the best features of the first two, but it requires better network timing than is needed for slotted transmission alone. The timing mechanism in this approach is based on dividing the packet slot into mini-slots. The transmissions initiated in different mini-slots utilize independent sequences or at least different starting points in the same sequence. This can be thought of as a variation of the second scheme in the sense that it is a method for periodic updating of the spreading sequence that also provides a certain degree of quantization for the acquisition process. Within each mini-slot, the receiver is searching for arrivals on a specific code, and the uncertainties in the arrival times are just the transmitter clock errors and differential propagation times.

With either the second or third timing mechanisms, multiple transmissions to the same radio can take place in the same packet slot without necessarily resulting in a collision. Simultaneous transmissions on independent sequences produce random errors (albeit not necessarily independent errors) rather than packet collisions. With proper use of error-control coding, it is possible to correct the errors if the number of interfering packets is not too large. In this way, the receiver may be able to capture one of the packets being sent to it and demodulate it correctly. Unless the receivers are able to simultaneously despread and demodulate *multiple* packets—and this is not usually the case—the correct reception of one packet per packet slot by a given receiver is the most that can be achieved with any protocol and timing mechanism.

The degree to which this capture capability can actually be achieved depends on the details of the spread-spectrum modulation and the processing in the receiver. In order to carry the discussion further, we must be more specific about the modulation type. Consequently, we postpone further consideration of receiver-oriented transmission protocols to Sections III–V.

Another way to provide capture in a spread-spectrum packet radio network is to employ a *transmitter-oriented* spread-spectrum transmission protocol. For this protocol, packets transmitted by different radios will be on different spreading sequences, no matter when the transmissions were originated. Even if the transmissions begin at exactly the same time, they will not collide. There will, of course, be random errors due to the mutual interference between the two transmissions.

This protocol has better capture properties than receiver-oriented protocols. However, the disadvantage with a transmitter-oriented protocol is that the receiver does not know which spreading sequence will be used on an incoming packet. It must search over the set of spreading sequences of all radios within range. This assumes, of course, that the receiver has the information and equipment needed to generate the spreading sequences used by these radios. Clearly, transmitter-oriented protocols place a considerable burden on the receiver. In fact, such protocols have the worst acquisition performance of all the spread-spectrum transmission protocols described in this paper.

The simplest spread-spectrum transmission protocol to implement is one that uses a common spreading sequence for all transmissions. Discrimination between two or more incoming packets must be made on the basis of time of arrival of the packets and time offsets among the received versions of the sequences. Clearly, this protocol has even worse capture properties than the receiver-oriented protocol: if a common sequence is used for all receivers, collisions can even take place between packets transmitted to different receivers.

The use of a common sequence does, however, facilitate the transmission of broadcast message because all radios are "tuned" to this sequence at all times. Transmitter-oriented transmission protocols are also suitable for broadcasting, but receiver-oriented protocols are very poor, because a separate transmission is required for each receiver to whom the broadcast message is to be sent. An acceptable compromise between good capture, rapid acquisition, and low delay for broadcast messages may be to designate a small fraction of the packet slots as broadcast slots and use a common spreading sequence in these slots. Transmission in the remaining slots could be according to a receiver-oriented protocol.

There are some combinations of the three spread-spectrum transmission protocols discussed above. Perhaps the most commonly suggested combination is based on the use of a common spreading sequence for the synchronization preamble and address portion of the packet. The data portion of the packet can be either transmitter-oriented or receiver-oriented. Because chip synchronism has already been accomplished by the time the data portion of the packet has arrived (otherwise, the packet is lost anyway), switching to a receiver-oriented protocol causes no extra burden on the receiver. Switching to a transmitter-oriented protocol is not much more complicated; the only additional requirement is that the receiver be able to generate the sequences used by all transmitters within range.

It is not clear how much interference protection (i.e., capture and discrimination against multiple-access interference) is provided by the use of a common spreading sequence for the preamble and address. A rigorous quantitative analysis has not been carried out, but the vulnerability of the preamble to collisions with other preambles suggests that use of a common spreading sequence for the header will not perform well in a slotted system unless randomized delays are employed. If mini-slotting is used, the length of the mini-slots should be at least as long as the common-sequence portion of the packet. In a dense network or in a network with heavy traffic, considerably better performance can be obtained by using a receiver-oriented transmission protocol for the synchronization preamble, and a transmitter-oriented for the message portion of the packet.

We close this subsection with a few comments about the use of acknowledgment protocols in a spread-spectrum radio network. More on this subject can be found in [72]. Passive acknowledgments are commonly considered for use in packet radio networks due to their potential for increased efficiency. In a relay situation, the acknowledgment process is carried out without the need for a special transmission, because the acknowledgment is obtained from the relayed packet. This has the potential for reducing delay and decreasing network traffic. The mechanism is fairly simple for narrow-band modulation, because all radios are tuned to all transmissions.

Passive acknowledgments can also be used in conjunction with spread-spectrum modulation, provided that compatible transmission and routing protocols are employed. For both transmitter-oriented protocols and those that use common spreading sequences, the passive acknowledgment protocol can work as effectively with spread-spectrum radios as with narrow-band radios. This is not necessarily true for receiver-oriented protocols, however. The radios would have to have the information and equipment needed to generate spreading sequence used by the relay radio; otherwise, separate acknowledgment packets are required.

B. Channel-Access Protocols

The characteristics of spread spectrum influence the selection of a channel-access protocol in two major ways. First, because of the multiple-access capability of spread spectrum, not only can simultaneous transmissions to different receivers be tolerated, but the number of such transmissions should be large in order to achieve high network throughput. Unlike for narrow-band radio networks, an ideal channel-access protocol in a spread-spectrum packet radio network would have multiple transmissions taking place at all times. Thus the goal of channel sensing in a spread-spectrum radio network is to determine if the intended receiver is busy rather than to determine if there are other packets being transmitted within range of the intended receiver.

Second, the type of channel sensing that is used to eliminate conflicts in narrow-band radio networks is virtually useless in a spread-spectrum packet radio network. For example, if the transmitter-oriented transmission protocol is employed, the transmitters do not know which spreading sequences to look for to see if the intended receiver is busy. By their very nature, spread-spectrum transmissions are very difficult to detect by receivers that do not know the spreading sequence. If a common spreading sequence is used, a transmitter that wishes to see if the intended receiver is busy must check all packets that are being transmitted. This includes acquiring synchronization and demodulating the address portion of the packet. Even for receiver-oriented protocols, the hidden terminal problem will limit the effectiveness of channel sensing.

Because of these two considerations, and because of the interference rejection capability of spread spectrum, it is typically more efficient to use ALOHA or slotted ALOHA than to employ channel-access protocols that require channel sensing. There are some exceptions, but most of these are for frequency-hop spread spectrum, so they will be discussed in Section V. Quantitative results on the throughput of fully connected packet radio networks with ALOHA and slotted ALOHA transmission are given in [49] and [56].

Another alternative to protocols that require channel sensing is the busy-tone channel-access protocol. In narrow-band packet radio networks, this protocol is referred to as busy-tone multiple access (BTMA) [23]. In the most general form, a subset of the radios that hear that a given receiver is busy transmit a busy tone. The subset may include the receiver itself, provided that the radios can simultaneously receive data and transmit a busy tone at a different frequency (most radios do not have this capability). All radios that hear the busy tone refrain from transmitting.

Again, in making the transition from protocols for narrow-band radios to spread-spectrum radios, we must keep in mind that we are not trying to eliminate multiple transmissions, as long as different receivers are involved. Thus BTMA has a different objective in a spread-spectrum packet radio network: the goal is to reduce the number of transmissions to a busy receiver. In a spread-spectrum version of BTMA, the busy "tone" should be a busy "signal" with at least as much spread-spectrum protection as the spread-spectrum data signal. The reason is that interference on the busy-signal channel is potentially more disruptive than interference on the data channel. Generation of the busy signal by an unauthorized transmitter can be even more disruptive (all radios assume the channel is busy, so no packets are transmitted).

III. Basic Features of Direct-Sequence Spread Spectrum

This section is devoted to a tutorial discussion of the capture, multiple-access, and anti-multipath characteristics of direct-sequence spread spectrum. We illustrate how these properties are derived from the signal structure and receiver processing. This is carried out in the simplest situation of interest: only two signals arrive at the receiver. To simplify the presentation, we restrict attention to *binary* direct-sequence spread-spectrum signals in both this section and the next. All of the important principles can be explained in the context of this signal structure.

Binary direct-sequence spread-spectrum radios employ signals of the form

$$s(t) = A\, a(t)\, b(t) \cos\,[2\pi f_c t + \phi] \qquad (3.1)$$

where $b(t)$ is a binary baseband data signal and $a(t)$ is a baseband spectral-spreading signal. The latter signal usually has a bandwidth that is much greater than required to transmit the data; that is, the bandwidth of $a(t)$ is much larger than the bandwidth of $b(t)$. If the basic pulse shape is rectangular for both $b(t)$ and $a(t)$, the signal defined by (3.1) is mathematically equivalent to binary phase-shift-key (PSK) modulation [42, p. 796], and even though $s(t)$ is expressed as amplitude modulation in (3.1), we usually refer to it as binary PSK direct-sequence spread spectrum. The natural generalization to quaternary direct-sequence modulation is obtained by adding to $s(t)$ a binary direct-sequence spread-spectrum signal that is orthogonal to $s(t)$; this is described in [45].

The data signal $b(t)$ consists of a sequence of positive and negative rectangular pulses, so it can be written as

$$b(t) = \sum_n b_n p_T(t - nT) \qquad (3.2)$$

where $p_T(t)$ is the rectangular pulse of duration T, Σ_n denotes the sum over all integers n that correspond to elements in the packet (in this case, the elements are data symbols), and the data symbols are from the data sequence

$$(b_n) = \cdots, b_{-1}, b_0, b_1, b_2, \cdots.$$

The binary digit b_n is either +1 or −1, depending on the data symbol to be sent in the nth time interval. The spectral-spreading signal consists of a sequence of positive and negative pulses, but the pulse shape (i.e., the chip waveform) need not be rectangular. If the chip waveform is denoted by $\psi(t)$, the spectral-spreading signal can be written as

$$a(t) = \sum_j a_j \psi(t - jT_c). \qquad (3.3)$$

The chip waveform $\psi(t)$ is assumed to be a time-limited pulse of duration T_c. The sequence

$$(a_j) = \cdots, a_{-1}, a_0, a_1, a_2, \cdots$$

is called the signature sequence or spreading code.

For convenience, we follow the usual convention whereby the signal is normalized such that

$$T_c^{-1} \int_0^T [\psi(t)]^2\, dt = 1.$$

It is also convenient to define the signal baseband $v(t)$ by

$$v(t) = a(t)\, b(t)$$

the product of the spectral-spreading signal and the data signal. It is important to keep in mind that the sign of the data signal $b(t)$ can change at integer multiples of the data pulse duration T.

We consider direct-sequence spread-spectrum signals for which the data pulse duration is an integer multiple of the chip duration. If $T = NT_c$ for some integer N, there are N chips per data pulse, and the bandwidth of the spread-spectrum signal $v(t)$ is roughly N times the data rate. For some systems, the signature sequence is periodic with period equal to N; in other cases, the period is much greater than N.

The optimum receiver for a binary direct-sequence spread-spectrum system in which white Gaussian noise is the only interference is the correlation receiver shown in Fig. 5. The received signal $Y(t)$ is multiplied by

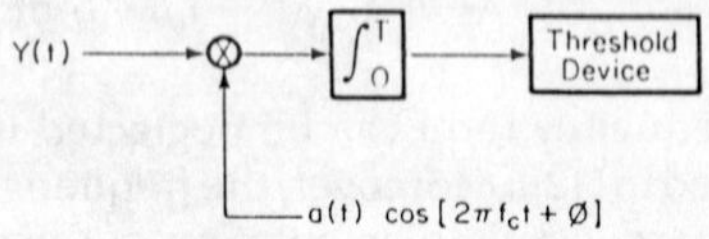

Fig. 5. Coherent correlation receiver for binary direct-sequence spread spectrum.

$$a(t) \cos\,(2\pi f_c t + \phi)$$

and the product is integrated over an interval of length T.

There is no improvement in performance when we use direct-sequence spread spectrum on a channel in which the only interference is white Gaussian noise. However, for channels with other forms of interference, spread spectrum offers many improvements over narrow-band modulation. These improvements are derived from the fact that the spectral-spreading signal can be selected to give the $s(t)$ several desirable properties which can be exploited to enable the receiver to discriminate against these other forms of interference. For example, if there is an additive interference signal $Z(t)$, the received signal can be written as $Y(t) = s(t) + Z(t) + n(t)$ where $n(t)$ is the white Gaussian noise process. The effects of this interference can be reduced by choosing the spectral-spreading signal to minimize

$$I = \int_0^T Z(t)\, a(t) \cos\,(2\pi f_c t + \phi)\, dt. \qquad (3.4)$$

All four characteristics discussed in the Introduction can be modeled as additive interference. In particular, in order to consider the capture phenomenon, we let the interference be a delayed version of the signal $s(t)$. This is also an appropriate model for a specular multipath channel with two propagation paths. For either phenomenon, however, the amplitude of the interference signal may differ from that of $s(t)$. Thus we let

$$Z(t) = A'v'(t - t_d) \cos (2\pi f_c t + \theta) \tag{3.5}$$

where A' is the amplitude of the interference, t_d is the time delay of the interference relative to $s(t)$, and θ is the phase of the interference signal.

To provide sufficient generality to include both multipath and capture in the same discussion, we let $v'(t) = a(t)\, b'(t)$ where $b'(t)$ is a data signal which is possibly different than $b(t)$. The data signal $b'(t)$ corresponds to a data sequence (b'_n) that may differ from (b_n). In particular, for the capture problem, (b'_n) is the data sequence from a different radio.

Note that for the multipath situation, the delay t_d corresponds to the relative lengths of the two paths. The time delay t_d produces a corresponding carrier phase shift of $-2\pi f_c t_d$ for the interference signal relative to the signal $s(t)$, and this phase shift is included in the phase θ of the interference signal. If, instead, the interference is due to overlapping packet transmissions, there is a component of θ that accounts for the phase difference of the RF carriers generated by the two transmitters.

The effect of the interference signal on the output of the correlation receiver is determined by the quantity I defined in (3.4). From (3.4) and (3.5) we see that if the double frequency term is ignored,

$$I = A' \cos [\phi - \theta] \int_0^T v(t - t_d)\, v(t)\, dt. \tag{3.6}$$

The double frequency term can be neglected if $f_c T_c \gg 1$ (this is discussed in [12]). Moreover, the frequency response of a practical hardware implementation of the correlation receiver is negligible for frequencies near $2f_c$.

The phase difference $\phi - \theta$ and the delay t_d are usually not under our control, so we would like to have the interference small for all values of ϕ, θ, and t_d. Therefore, we wish to choose the spectral-spreading signal $a(t)$ in a way that makes the *magnitude* of

$$R_v(t_d) = \int_0^T v(t - t_d)\, v(t)\, dt \tag{3.7}$$

small for all values of the delay. We may, without loss of generality, assume that a positive pulse was sent in the first interval (i.e., $b_0 = +1$), so that $b(t) = +1$ for $0 \le t < T$. This allows us to write

$$\begin{aligned} R_v(t_d) &= \int_0^T v(t - t_d)\, a(t)\, dt \\ &= b'_{-1} \int_0^{t_d} a(t - t_d)\, a(t)\, dt \\ &\quad + b'_0 \int_{t_d}^T a(t - t_d)\, a(t)\, dt \end{aligned} \tag{3.8}$$

but this is related to a certain autocorrelation function for the signal $a(t)$. More precisely, if the data symbols b'_0 and b'_{-1} are equal, the right-hand side of (3.8) is the *sum* of two partial-period autocorrelation functions for the spectral-spreading signal $a(t)$; if the data symbols are not equal, then it is the *difference* of two such functions. The implications of this observation are discussed in [31], [51], [44], and [60].

Clearly, it is not possible to make $|R_v(t_d)|$ small for very small values of t_d, because

$$|R_v(0)| = \int_0^T [a(t)]^2\, dt = NT_c = T.$$

Notice that this is an in-phase autocorrelation. However, proper selection of the signature sequence will make $R_v(t_d)$ small for all values of delay in the range $T_c \le t_d \le T - T_c$ if the sequence has period N (in this case, $|R_v(0)| = |R_v(T)|$). Values of t_d in this range produce out-of-phase autocorrelations. If the sequence is not periodic or has a period that is longer than the packet, the interference can be made small for all $t_d \ge T_c$.

There are two different time offsets that are important. The first is the difference between the arrival times of the two packets (the interarrival time), and the second is the time offset between the two versions of the signature sequence (the sequence offset time). A large interarrival time facilitates the acquisition of time synchronism and phase coherence with the received signal, but there is no firm constraint. Depending on the system, sequence, prior timing information available to the receiver, clock stability, and other factors, it may be possible to acquire one of two packets that arrive at precisely the same time. On the other hand, there is a firm constraint on the sequence offset time: it must be large enough to give an out-of-phase autocorrelation in order to permit discrimination between the two packets. Without this much offset, which amounts to the duration of a chip or perhaps slightly less, acquisition is impossible. If this constraint is met and acquisition is achieved, the resulting error probability for the packet will depend primarily on the mutual interference $|R_v(t_d)|$. The error probability decreases as $|R_v(t_d)|$ is decreased, and this is accomplished by selecting the signature sequence (a_j) to give small out-of-phase autocorrelation values. The same correlation property is needed for good anti-multipath capability.

The situation is quite similar for multiple access in a direct-sequence spread-spectrum packet radio network. In such a system, the spread-spectrum signals from other transmitters are a form of interference to a receiver which attempts to demodulate the signal $s(t)$. Let $s'(t)$ denote one of these other signals, and denote its amplitude by A', its spectral-spreading signal by $a'(t)$, its signature sequence by (a'_j), its data signal by $b'(t)$, and its data sequence by (b'_n). Also define $v'(t) = a'(t)\, b'(t)$. If t_d denotes the time offset between the two signals (due to offsets in the transmitters' clocks and differential propagation delays) and ϕ' denotes the relative phase, the interference is

$$\int_0^T v'(t - t_d)\, a(t)\, dt \cos (\phi'). \tag{3.9}$$

Since this should be small for all values of the phase angles, we must consider the magnitude of

$$\begin{aligned} R_{v',v}(t_d) &= b_0 \int_0^T v'(t - t_d)\, a(t)\, dt \\ &= b_0 b'_{-1} \int_0^{t_d} a'(t - t_d)\, a(t)\, dt \\ &\quad + b_0 b'_0 \int_{t_d}^T a'(t - t_d)\, a(t)\, dt. \end{aligned} \tag{3.10}$$

Notice that this is a crosscorrelation, rather than an autocorrelation as in (3.8). The goal in selecting signature sequences to provide good multiple-access capability is to make this crosscorrelation small for all values of the time offset t_d.

By allowing the possibility that $a'(t)$ might be the same as $a(t)$, we can unify the analysis of capture, multipath, and multiple access. From (3.9), (3.10), and the observation that it is the *magnitude* of the correlation that we wish to minimize, we conclude that we must make the magnitudes of both

$$R(t_d) = \int_0^{t_d} a'(t - t_d)\, a(t)\, dt \qquad (3.11)$$

and

$$\hat{R}(t_d) = \int_{t_d}^{T} a'(t - t_d)\, a(t)\, dt \qquad (3.12)$$

small for all appropriate values of t_d (e.g., $t_d = 0$ is not an "appropriate" value for the multipath problem). These are the widely studied continuous-time partial correlation functions for the spectral spreading signals $a'(t)$ and $a(t)$ (see [42] and [45] for further details). The functions are autocorrelation functions if $a(t) = a'(t)$; they are crosscorrelation functions otherwise.

The selection of signature sequences with small correlation is the subject of several journal articles that have been published during the last decade. The papers [30] and [60] discuss the most commonly considered classes of sequences, the so-called pseudorandom sequences and their close relatives (e.g., Gold sequences). The importance of these sequences for spread-spectrum systems is described in [60], and data on correlation values for some specific sets of sequences are given in several references [11], [51], [60], [63], [66].

In summary, we have seen that the multiple-access capability is derived from the crosscorrelation properties of the signature sequences. If the signature sequences have small crosscorrelation values, reliable reception of simultaneous transmissions may be possible. The capture and anti-multipath capabilities are derived from the out-of-phase autocorrelation values for the signature sequences. If the signature sequences are carefully selected, the network will have the property that time offsets of at least one chip duration will produce sufficiently small interference to permit reliable demodulation provided that initial synchronization has been achieved.

IV. Error Probabilities for Direct-Sequence Spread Spectrum in a Packet Radio Network

Information on the basic features of direct-sequence spread spectrum can be found in [7], [19], [22], [36], [45], and [86]. Papers which emphasize multiple access in direct-sequence spread-spectrum systems include [2], [13], [42], [53], and [81]; and discussions of the performance of direct-sequence spread spectrum on a multipath channel can be found in [15], [26], [31], [44], and [76]. Material on the important topic of acquisition and tracking in spread-spectrum receivers, a subject that is not covered here, can be found in [7], [19], [37]–[39], and [86].

In order to describe the basic features of direct-sequence spread spectrum in Section III, we restricted attention to simple examples with two signals only. In practice, of course, there are several signals rather than two, and the receiver must process the composite signal to extract the desired information. The success of the receiver depends on the sequences used in the direct-sequence modulation, the number of interfering signals and their relative power levels, and the type of processing employed. In this section, we discuss these factors, describe analytical methods for performance evaluation, and present numerical results for the error probabilities in spread-spectrum packet radio networks.

We begin by discussing receivers for specular multipath channels because of the dichotomy that exists in the receiver design philosophy for such channels. One type of receiver for this channel fits into a general framework for which precise analytical methods of performance evaluation exist, but the other type does not. In order to describe the two approaches, we consider a specular multipath situation with no multiple-access interference. In general, there are several multipath components (as many as six to ten is not uncommon [34], [35], [76], [79]), and there are two fundamentally different approaches to the demodulation of the received signal.

The first approach is to attempt to combine the multipath components in some way. For convenience, we refer to a receiver that is based on this approach as a *multipath-combining receiver.* In a multipath-combining receiver, the energy in all of the multipath components is used to extract the data. In the second approach, the receiver attempts to isolate a single component and extract the data from that one component only. Such a receiver is just a correlation receiver matched to the single component, so it does not take advantage of the energy in the other multipath components. However, it does discriminate against these components in order to eliminate (or at least significantly reduce) both intrasymbol and intersymbol interference.

The multipath-combining receiver is more complex than the correlation receiver, but the former provides better performance on a pure multipath channel. Perhaps the most famous of the multipath-combining receivers is the RAKE system due to Price and Green [40]; however, several other types of multipath-combining receivers have been employed [10], [69], [76]. The type of multipath-combining receiver employed in the DARPA low-cost packet radio is based on a "coherent recursive integrator," which is closely related to the adaptive RAKE receiver and post-detection integrator. Most of the recent research on the performance of multipath-combining receivers is due to Turin, who has published both analytical and experimental results [74]–[79]. Other results can be found in [25] and [26]. These analyses are typically based on approximations, such as the approximation of the distribution of the interference signals by a Gaussian distribution.

When other interference is present in addition to the multipath signals, both the operation and the analysis of the multipath-combining receiver are more complicated than for the pure multipath channel. A practical problem that arises in a multipath-combining receiver when both multipath and multiple-access interference are present is the difficulty of distinguishing between autocorrelation peaks (needed for the multipath combining process) and cross-correlation peaks that arise from the multiple-access interference. There are several analytical problems that complicate the performance evaluation for multipath-combining receivers. In particular, the problem of determining the performance of such a receiver for channels with narrow-band interference (e.g., jamming) or multiple-access interference is very difficult. This is especially true when the interference distribution is not accurately approximated by a Gaussian distribution, which is the case, for example, if the number of interfering signals is large compared with the number of chips per symbol. A more subtle problem that arises in the performance analysis of the multipath-combining receiver is due to the strong statistical dependence that occurs between the signals that are being combined.

The correlation receiver is inferior to the multipath-combining receiver on channels with a large number of equal-strength multipath components. However, the operation of the correlation receiver is simpler, and it will give satisfactory performance if one of the multipath components is strong enough to provide an adequate signal-to-noise ratio. This is often the case when one of the multipath components is significantly larger than the others, such as when one component is due to a direct line-of-sight path and the others are due to reflections. For many applications, the correlation receiver may be used, even though it is suboptimum, because it is fairly robust against other forms of interference.

In contrast to the situation for the multipath-combining receiver, very accurate methods for the evaluation of the bit error probability are available for systems with a correlation receiver. The first successful attempt at a precise calculation of the error probability for a direct-sequence spread-spectrum multiple-access communication system with an additive white Gaussian channel (no multipath) was published by Yao (e.g., [92]) about a decade ago. Although various approximations based on Gaussian models and signal-to-noise ratio calculations were available much earlier, the majority of these did not account for the correlation properties of the sequences used in the system (among the few exceptions are [2] and [41]). Moreover, there were no results on their accuracy at that time (several cases have since been discovered for which these approximations are not reliable). During the past six years, many new methods for the evaluation of error probabilities have been developed for both coherent and noncoherent demodulation of direct-sequence spread spectrum for specular multipath channels [15], [16] and multiple-access channels [9], [13], [24], [25], [27], [53], [82]–[85].

Some of the work has also been extended to include channels with both narrow-band interference and multiple-access interference [9]. There are also many recent results on the performance of direct-sequence spread spectrum for channels with narrow-band interference only [32], [33], [65], [86]. Although none of these apply to specular multipath channels, there are some results for narrow-band interference on channels with diffuse frequency-selective fading [33].

An additional complication that arises for direct-sequence spread spectrum is the fact that the symbol errors are not independent for the kinds of channels encountered in a packet radio network. This further complicates the performance analysis of a direct-sequence spread-spectrum system that uses error-control coding. Even if we can compute the symbol error probability on the channel, we are not necessarily able to evaluate the error probability at the output of the decoder. Moreover, whether the system uses error-control coding or not, it is not an easy task to compute the packet error probability. In general, we must know more about the channel than its symbol error probability in order to evaluate either the bit error probability or packet error probability at the decoder output.

A. Model for Multiple-Access and Multipath Interference

In order to present some of the performance results for direct-sequence spread-spectrum packet radio, we first describe the basic model for multiple-access and multipath interference in such a system. All results presented are for the correlation receiver. There are three basic types of interference: wide-band interference due to other direct-sequence spread-spectrum signals generated by the network (e.g., multiple-access interference and multipath), wide-band Gaussian noise (e.g., thermal noise and broad-band jamming), and narrow-band interference (e.g., narrow-band transmitters in the same band and hostile narrow-band jamming).

Suppose that radio A transmits the signal $s_1(t)$ to radio B. The composite signal seen by B is given by

$$Y(t) = y(t) + X(t) + J(t) \tag{4.1}$$

where the term $X(t)$ is a random process representing noise and interference that can be modeled as Gaussian (e.g., thermal noise, broad-band Gaussian jamming, and possibly certain other types of RF interference) and the term $J(t)$ represents non-Gaussian interference (e.g., tone jamming). The term $y(t)$ represents the composite spread-spectrum signal; that is,

$$y(t) = \sum_{k=1}^{K} s_k(t - T_k) \tag{4.2}$$

where

$$s_k(t) = A_k a_k(t)\, b_k(t) \cos\,[2\pi f_c t + \theta_k]. \tag{4.3}$$

The variables A_k and T_k are the amplitude and time delay, respectively, for the kth signal.

We begin by considering the interference from the other direct-sequence spread-spectrum signals (i.e., $s_k(t)$ for $k > 2$). For now, assume that the spectral-spreading signals $a_k(t)$, $1 \le k \le K$, are derived from *deterministic* sequences such as maximal-length linear-feedback shift-register sequences (m-sequences). Results for random signature sequences are given later in this section.

Assume that B's receiver has acquired phase coherence and time synchronism with the first signal ($k = 1$), and let the arrival of this signal serve as the time reference. Thus we can let $T_1 = 0$ so that T_k represents the relative time delay (relative to the first signal) encountered by the kth signal. The time delay T_k accounts for the differential propagation time and the differences (if any) between the clocks of A and the radio that transmitted the kth signal.

Assume that B's receiver is in the process of demodulating $s_1(t)$ to extract the data from $b_1(t)$. The other $K - 1$ signals are therefore treated as sources of interference. The kth signal may represent a packet being transmitted by another radio, or it may be a multipath component of A's transmission. The correlation receiver processes these two types of interference in exactly the same way (the multipath-combining receiver processes them quite differently).

If the kth signal was transmitted by a radio other than A, the ability of the receiver to discriminate against the signal is usually referred to as capture capability if the signal was intended for B or multiple-access capability if it was intended for a radio other than B. The difference is that $a_1(t)$ and $a_k(t)$ may be derived from the same signature sequence if the two signals are intended for the same receiver, but $a_1(t)$ and $a_k(t)$ are usually derived from different signature sequences if the two signals are intended for different receivers. This is true, for example, if a receiver-oriented protocol is employed to determine the signature sequence to be used on a given packet transmission. For such a protocol, all packets to be transmitted to a given receiver must use the sequence assigned to that receiver.

If the kth signal is a multipath component of A's transmission, the ability of the receiver to discriminate against the signal is its antimultipath capability. In this situation, $a_1(t)$ and $a_k(t)$ are derived from the same signature sequence. The similarity between the antimultipath and capture capabilities is that discrimination against the interference is based on the autocorrelation properties of the signature sequence used to generate $a_1(t)$ and $a_k(t)$. The multiple-access capability is derived from the crosscorrelation properties of the two signature sequences used to generate $a_1(t)$ and $a_k(t)$.

Let the data sequence for $b_k(t)$ be denoted by

$$(b_{k,n}) = \cdots, b_{k,-1}, b_{k,0}, b_{k,1}, b_{k,2}, \cdots.$$

In order to evaluate the bit error probability for B's receiver, consider the output of B's correlator corresponding to the data bit $b_{1,0}$. The interference at the output of this correlator due to the presence of the kth signal depends on two of the data symbols from the sequence $(b_{k,n})$. Without loss of generality, we can assume these are the symbols $b_{k,-1}$ and $b_{k,0}$, and we denote this pair of symbols by $\mathbf{b}_k$. The interference at the correlator output also depends on the time delay T_k and the phase θ_k. Specifically, it is given by

$$J_k(\mathbf{b}_k, T_k, \theta_k) = \tfrac{1}{2}A_k T\, I_k(\mathbf{b}_k, T_k, \theta_k)$$

where the normalized interference I_k is defined in [45] and several other references [13], [50]. An important feature is that this normalized interference depends on the correlation between the signals $a_k(t)$ and $a_1(t)$, and this correlation is the partial-period (aperiodic) correlation defined in (3.11) and (3.12), rather than a full-period (periodic) correlation.

We now consider the combined effects of the interfering spread-spectrum signals and Gaussian interference. The appropriate model for the received signal in this case is $Y(t) = y(t) + X(t)$ where $y(t)$ is given by (4.2) and $X(t)$ is a stationary Gaussian random process. The output of the correlation receiver can be written as

$$Z = W + \tfrac{1}{2}A_1 T\left\{ b_{1,0} + \sum_{k=2}^{K} e_k I_k(\mathbf{b}_k, T_k, \theta_k)\right\} \tag{4.4}$$

where $e_k = A_k/A_1$ and W is the random variable representing the portion of the output that is due to the Gaussian process. The parameters T_k and θ_k are assumed to be random variables with T_k (modulo T) uniformly distributed on $[0, T]$ and θ_k (modulo 2π) uniformly distributed on $[0, 2\pi]$. Moreover, the set consisting of the time delays T_k, $1 \le k \le K$, and phase angles θ_k, $1 \le k \le K$, is a set of $2K$ mutually independent random variables. In the most general situation, the amplitudes A_k, $1 \le k \le K$, are assumed to be mutually independent and independent of the time delays and phase angles; this is the model analyzed in [15] for the multipath channel. In what follows, we consider only the special case in which these amplitudes are deterministic constants. Finally, the data sequences are assumed to be independent, identically distributed random variables, the values $+1$ and -1 are equally probable, the data sequences are independent of all the other parameters, and data sequences originating at different sources are assumed independent of each other.

Because of the symmetry of the distributions of the random phase angles and data symbols, the conditional probability of error given $b_{1,0}$ does not depend on the value of $b_{1,0}$, so we take it to be $+1$. Similarly, the normalized interference

$$I(\mathbf{b}, \mathbf{T}, \mathbf{\theta}) = \sum_{k=2}^{K} e_k I_k(\mathbf{b}_k, T_k, \theta_k) \tag{4.5}$$

has a symmetrical distribution about zero. The vector $\mathbf{\theta}$ is the $(K-1)$-tuple of phase angles $(\theta_2, \cdots, \theta_K)$. Similarly, $\mathbf{b}$ and $\mathbf{T}$ are vectors of all of the relevant data symbols and time delays, respectively, for the signals $s_k(t)$, $2 \le k \le K$. For convenience, we assume in all that follows that the amplitudes A_k (and hence and coefficients e_k) are deterministic. An analysis for random amplitudes is given in [15] for the multipath channel.

B. Evaluation of the Error Probability

The probability of error, averaged over all time delays, phase angles, and data bits, can be written as

$$P_e = E\{Q([1 + I(\mathbf{b}, \mathbf{T}, \mathbf{\theta})]/\sigma)\} \tag{4.6}$$

where E denotes the mathematical expectation and Q is the complementary Gaussian distribution function

$$Q(x) = \int_x^{\infty} \exp(-y^2/2)\, dy.$$

The parameter σ is the standard deviation of the normalized noise; that is, $\sigma^2 = \text{var}\{W'\}$ where W' is defined by

$$W' = (\tfrac{1}{2}A_1 T)^{-1} W. \tag{4.7}$$

If the wide-band Gaussian noise is white with two-sided spectral density $N_0/2$, then $\sigma^2 = (2E_s/N_0)^{-1}$, where E_s denotes the energy per symbol. As a check, note that if there is no interference, (4.6) reduces to

$$P_e = Q([2E_s/N_0]^{1/2})$$

which is the correct result for antipodal signaling on an additive white Gaussian noise channel with no other sources of interference.

For simplicity, we denote $I(\mathbf{b}, \mathbf{T}, \mathbf{\theta})$ by I, but it must be remembered that I is a random variable that depends on the data symbols, time delays, and phase angles. The error probability can be expressed as [85]

$$P_e = \tfrac{1}{2} E\{Q([1 + I]/\sigma)\}$$

$$= \tfrac{1}{2} E\{Q([1 + |I|]/\sigma)\} + \tfrac{1}{2} E\{Q([1 - |I|]/\sigma)\} \quad (4.8)$$

or [13]

$$P_e = \tfrac{1}{2} - \tfrac{1}{2} P(-1 < W' + I < +1). \quad (4.9)$$

Straightforward evaluation of either of these expressions requires a multidimensional integration, and the number of computations is proportional to $(4Nn_T n_\theta)^{K-1}$, where n_T and n_θ are the numbers of points needed in one-dimensional numerical integration with respect to the time delay and phase angle, respectively. The major difficulty revealed by this consideration is that the amount of computation is exponential in the number of interfering signals. Moreover, the calculation must be carried out for each value of σ of interest (i.e., for each choice of E_s/N_0) [13]. Fortunately, more efficient methods exist for the evaluation of (4.8) and (4.9).

Yao [85] applied moment-space bounding techniques to (4.8) to obtain upper and lower bounds on P_e. We give a brief description of this approach. Let β be the largest value I can attain for the particular sequences under consideration, and let f_I be the density function for the random variable I. Let g_1 and g_2 be bounded continuous functions on the interval $[-\beta, \beta]$, and define the moment

$$m_i(F) = \int_{-\beta}^{\beta} g_i(u)\, dF(u).$$

The probability distribution function F is arbitrary except that it must satisfy $F(-\beta^-) = 0$ and $F(\beta) = 1$; let D_β be the set of all such distribution functions, and define

$$M = \{(m_1(F), m_2(F)) : F \in D_\beta\}.$$

Also define the space of function values by

$$C = \{(g_1(u), g_2(u)) : -\beta \le u \le \beta\}$$

which actually represents a curve in two-dimensional Euclidean space as u varies from $-\beta$ to β.

The moment-space bounds are based on the fact that M is the convex hull of the curve C, and so one can determine results about moments from knowledge of the points that make up the set C, the latter being much easier to compute than the former. In particular, if $g_2(u) = Q([1 + u]/\sigma)$, then $P_e = E\{g_2(I)\}$. The function g_1 is selected to satisfy two opposing constraints. The function g_1 should be such that the curve C is fairly "straight" in order that the moment space is thin and a tight bound is obtained (if C is a straight line, the bound is exact), but g_1 must also be selected to make it feasible to compute the moment $m_1 = E\{g_1(I)\}$. Upper and lower bounds on P_e are then obtained from a "slice" of M; that is, the upper bound is

$$P_e^U = \max \{y: (m_1, y) \in M\}$$

and the lower bound is

$$P_e^L = \min \{y: (m_1, y) \in M\}.$$

Examples of reasonable choices for $g_1(u)$ are u^2 (second-moment bounds) and $\exp[c(1 + u)]$ for some constant c (exponential bounds).

Generalizations to multiple exponential bounds are necessary to give tight bounds for large values of K and N, and the resulting software needed to evaluate the bounds is more complex. In general, it may be necessary to solve a set of nonlinear equations in order to compute the bounds [85]. Nevertheless, the moment-space bounding method does produce tight upper and lower bounds, and this method was used to generate the first numerical results for the probability of error in direct-sequence spread-spectrum multiple-access communications.

Chronologically, the next results were obtained by exploiting the convexity of $Q(x)$ for $x > 0$. Both upper and lower bounds were developed in the late 1970s and later published in [53]. The original results were for systems in which the maximum multiple-access interference does not exceed the desired signal; that is, the maximum value of I over all choices of $\mathbf{b}$, $\mathbf{T}$, and $\mathbf{\theta}$ must be less than 1. Because of this restriction, the method of [53] cannot be applied to systems in which the ratio K/N is large (e.g., for $N = 31$ we were unable to find sequences for which the condition $I < 1$ is satisfied for $K \ge 5$). Later, this approach was extended to systems in which the maximum interference exceeds the desired signal ($I > 1$), and it was employed to analyze the effects of multiple-tone interference as well as multiple-access interference [9].

Results obtained in [53] by applying this bounding method for $N = 31$ are shown in Fig. 6 for three equal-power spread-spectrum signals. The two sets of bounds correspond to

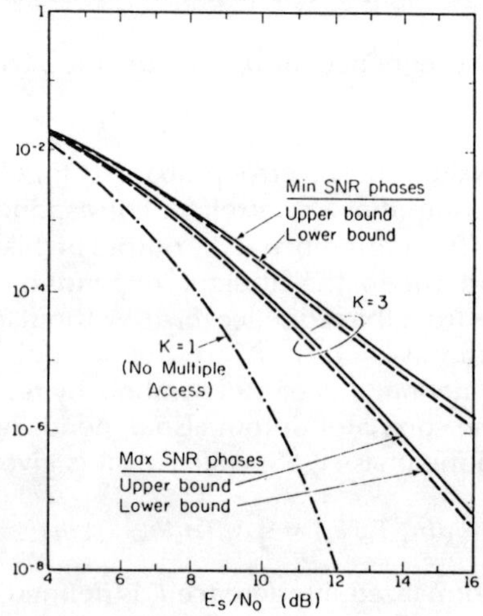

Fig. 6. Upper and lower bounds for two sets of sequences with different phases. ($N = 31$.)

two different orientations (starting phases) for the three nonreciprocal m-sequences of period 31. These phases, which satisfy the condition $\max\{I\} < 1$, were selected according to the signal-to-noise ratio parameter defined in [42] and studied further in [11], [45], and [50]. Also shown in Fig. 6 is the error probability for a single binary direct-sequence spread-spectrum signal (no multiple access). Comparisons with this curve show the cost of permitting three simultaneous transmissions. This cost is measured in terms of the required increase in transmitter power to achieve a given value of P_e; alternatively, it can be measured in terms of the increase in error probability for a fixed power level. For this figure (and Figs. 7–9 as well) the axis labeled E_s/N_0 (dB) should be interpreted as $-10\log_{10}(2\sigma^2)$ if the wideband Gaussian noise process $X(t)$ does not have a flat spectral density.

One method for evaluating P_e that is based on an approximation rather than bounds is the characteristic-function method that has been applied to both the multiple-access and multipath problems [13], [15]. Like the moment-space bounding technique, the characteristic-function method had been used earlier for calculating the error probability in systems with intersymbol interference. As adapted to coherent spread-spectrum communications, it provides a very effective method for approximating the symbol error probability.

The characteristic-function method is based on the characteristic function for the interference on the channel. The multidimensional integration required to evaluate the convolutions (the density function for a sum of independent random variables) is replaced by the multiplication of the characteristic functions. Consider (4.9) and let ϕ_I and $\phi_{W'}$ be the characteristic functions for the interference I and the channel noise W', respectively. It follows from (4.5) that

$$\phi_I(u) = E\{\exp(juI)\}$$
$$= \prod_{k=2}^{K} E\{\exp(jue_k I_k(b_k, T_k, \theta_k))\} \quad (4.10)$$

and from the fact that W' is a zero-mean Gaussian random variable with variance σ^2 that

$$\phi_{W'}(u) = \exp\{-\tfrac{1}{2}\sigma^2u^2\} \quad (4.11)$$

where j denotes the complex square root of -1. If the wideband noise has constant spectral density $N_0/2$, σ^2 can be replaced by $(2E_s/N_0)^{-1}$ in (4.11).

The terms of the product in (4.10) are the characteristic functions of the multiple-access interference components. It is shown in [13] that these can be expressed in terms of a single integration from 0 to $\pi/2$ with respect to the phase variable θ, and that (4.9) is equivalent to

$$P_e = Q(\sigma^{-1}) + \int_0^{\infty} [(\sin u)/u]\phi_{W'}(u)[1 - \phi_I(u)]\, du/\pi. \quad (4.12)$$

The integral in (4.12) can be truncated to have a finite upper limit Ln, which makes it possible to evaluate the integral numerically. It is shown in [13] that for a maximum error p in the error probability, it is sufficient to choose L such that

$$L > Q^{-1}(\sigma p[\pi/8]^{1/2})/(\sigma\pi).$$

The amount of computation required to evaluate P_e via (4.10)–(4.12) for a binary PSK system is proportional to $n_\theta n_u N(K-1)$ where n_θ and n_u are the numbers of points needed for the numerical integration with respect to the phase and the numerical integration of (4.11), respectively. Of course, n_θ and n_u will depend on the desired accuracy, the value of σ, and perhaps, to some extent, on N and K. However, the amount of computation is approximately proportional to the product NK, as opposed to the exponential dependence on K required by direct numerical integration and many of the other techniques. This makes it feasible to evaluate the performance of direct-sequence spread-spectrum multiple access for a fairly dense packet radio network. For example, results on P_e are given in [13] for values of K as large as 24; this corresponds to a packet radio network in which 23 radios within range of radio B (see Fig. 2) are transmitting packets on other spreading sequences at the same time that radio A is transmitting a packet to B.

Results obtained from the characteristic-function method for $N = 127$ are shown in Fig. 7 for several values of K between 1 to 12 (the signals have the same power). Note that

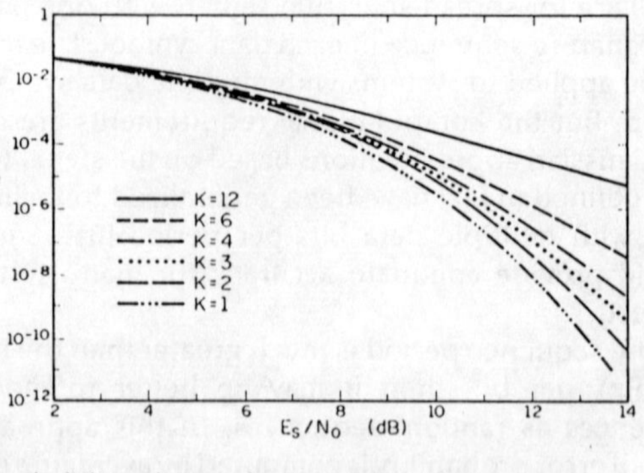

Fig. 7. Probability of error for $N = 127$ and $K = 1, 2, 3, 4, 6$, and 12.

in order to provide a symbol error probability of 10^{-4} in a packet radio network in which we can expect as many as eleven other packet transmissions within range of B, more than 2 dB of additional transmitter power is required to compensate for the multiple-access interference. The penalty is less than 1 dB, however, if there will be only two or three packet transmissions within range of B. Alternatively, we see that if E_s/N_0 is fixed at 10 dB, the range of values of P_e is about 10^2 for K ranging between 1 and 12.

Although we have concentrated on binary PSK modulation in the discussion above, the characteristic-function method has been applied to several other modulation formats such as offset quadriphase shift keying (OQPSK) and minimum shift keying (MSK). The basic principles are the same as for binary modulation. The results show that although analyses based on signal-to-noise ratio calculations and Gaussian approximations conclude that, for a given bandwidth and data rate, binary PSK, QPSK, and OQPSK all have the same error probability, the actual error probabilities are slightly different for these three systems. The details of the analysis and some numerical results for these forms of direct-sequence spread-spectrum modulation are given in [13]. Several other approaches have been developed or adapted for use in the evaluation of the symbol error probability in direct-sequence systems. There are series expansion methods [13], [82]–[84] and methods based on Gauss-quadrature integration formulas [24]. All of the methods mentioned above are potentially suitable for application to packet radio network performance analysis. The choice as to which method is best depends on the specific parameters of the modulation and the network, and it also depends heavily on the software that is available.

Because of the near–far problem that arises in packet radio networks, the fact that some of the methods cannot easily handle problems with unequal-power signals eliminates them from consideration. For instance, the characteristic-function method is superior in this regard to the bounding method derived from convexity that was employed to generate the results of Fig. 6. Unlike some of the methods discussed in this paper, it is straightforward to apply the characteristic-function method to situations in which not all of the signals have the same power. Some results for such an application are presented below for random signature sequences.

Thus far we have assumed the signature sequences are deterministic, and the numerical results presented (Figs. 6 and 7) are for spread-spectrum signals with one period of the signature sequence in each data symbol. The methods can be applied to systems with multiple data symbols per period, but the computational requirements are greater. The Gaussian approximations based on the signal-to-noise ratio defined in [42] have been generalized to include systems with multiple data bits per period [61], and these should provide adequate accuracy for many systems of interest.

If the sequence period is much greater than the number of chips per bit, then it may be better to model the sequences as random sequences. In this approach, the symbol error probability is computed by averaging over the ensemble of sequences in addition to averaging over the phase angles and time delays. This is a particularly good approach for the DARPA low-cost packet radio (LPR) in its "bit-to-bit code changing" mode. In this mode, a new pseudo-random sequence selection is made for each successive data symbol. Thus to evaluate the average symbol error probability for a large number of symbols (e.g., a 2000-symbol packet), it makes sense to average over the corresponding signature sequences. For the LPR, this corresponds to a time-average probability of error. The mathematical model we adopt for a random signature sequence is a sequence of independent binary random variables, each of which takes values −1 or +1 with equal probability. The signature sequences for signals originating from different transmitters are assumed to be statistically independent.

Results obtained by applying the characteristic-function method to three different systems which employ random signature sequences with 127 chips per data symbol are shown in Fig. 8. The important common feature is that the

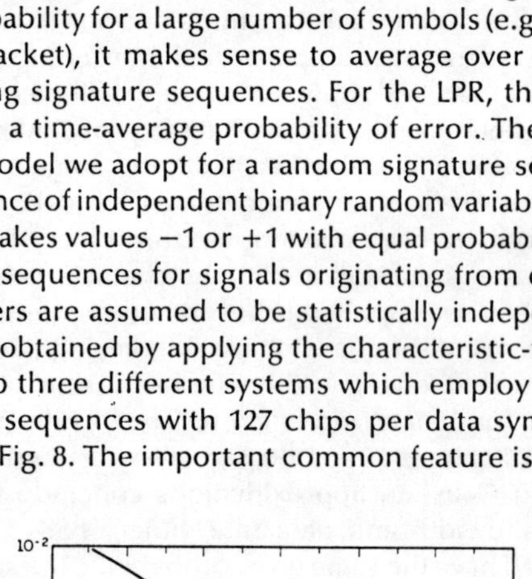

Fig. 8. Error probability for unequal-power signals. (Binary DS, N = 127.)

total interference power is the same for all three systems. However, the power is distributed among a different number of interfering signals in the three systems. The first system has eight interfering signals in addition to the signal that is intended for the receiver under consideration, and all nine signals have the same power. The voltage ratios appearing in (4.4) and (4.5) are given by $e_k = 1, 2 \le k \le 9$, for this system. In the second system, there are four interfering signals, each of which has twice as much power as the intended signal ($e_k = \sqrt{2}, 2 \le k \le 5$). Finally, the third system has only two interfering signals, but each has four times as much power as the intended signal ($e_2 = e_3 = 2$). As is clearly shown in Fig. 8, the error probability is not a function of the total interference power only, it depends on the distribution of the power among the interfering signals. This rules out the possibility of accurately calculating P_e using simple signal-to-interference power ratios.

Rather than work with the characteristic function for the interference, one can work directly with the probability density functions for the interference components. This is the method employed in [25], [27] where tight upper and lower bounds are obtained for the symbol error probability in multiple-access systems which employ random signature sequences. The bounds can be made as tight as desired by increasing the amount of computation, and the computations can be carried out efficiently on an array processor. The upper and lower bounds obtained from this method are shown in Fig. 9 for equal-power signals, N = 127, and three choices of K. Notice that the bounds are considerably tighter for K = 3 than for the larger values of K. This is fairly typical of most bounding techniques.

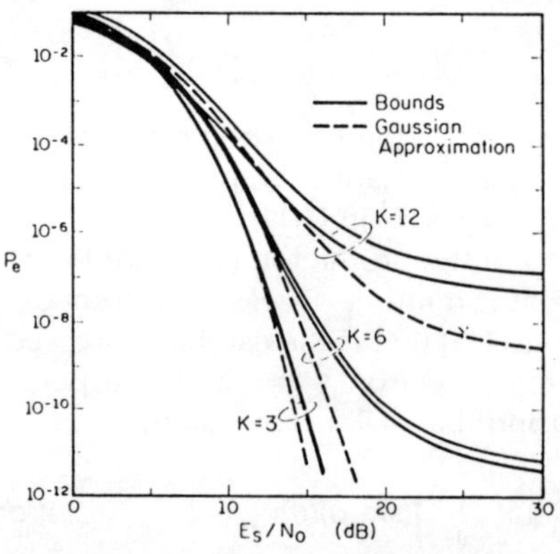

Fig. 9. Error probability bounds and the Gaussian approximation for random signature sequences with N = 127.

Also shown in Fig. 9 are the results of applying the approximation

$$P_e \approx P_G = Q(\mathrm{SNR}) \tag{4.13}$$

where, for random sequences, the signal-to-noise ratio SNR is given by

$$\mathrm{SNR} = \{\sigma^2 + [(K-1)/3N]\}^{-1/2}. \tag{4.14}$$

This approximation, which was suggested in [42], is based on the mean and variance of the output of the correlation receiver. For the derivation of these moments see [57] or [58]. As illustrated by the curves in Fig. 9, the approximation given in (4.13) is accurate if the ratio K/N is not too large. One might think that a larger number of interfering signals makes the interference more nearly Gaussian, and so the approximation will be better for large values of K. The flaw in this argument is the fact that the interference is not normalized as required by the central limit theorem. In order that the distribution of the sum of the $K-1$ independent random variables representing the interference terms approach the Gaussian distribution, we would have to have N increasing as well, which is not the case in Fig. 9.

Thus far, we have dealt with the symbol error probability. Unfortunately, in direct-sequence spread-spectrum packet radio systems, the computation of the packet error probability cannot be based on the symbol error probability alone. This is in large part due to the fact that the symbol errors in a direct-sequence spread-spectrum multiple-access system are not independent. Rather, they are highly coupled through the phase angles and time delays (and signal amplitudes also, if we treat them as random).

The problem is even more difficult if error-control coding is employed. Even if the errors on the channel were independent, the errors at the output of the decoder are highly dependent. Moreover, in typical packet radio networks, the problem is not alleviated by interleaving. In part, this is because packet lengths are usually not long enough to provide the amount of separation between symbols that is necessary to make the errors independent.

The only coding system for which results on the packet error probability have been published are for convolutional coding with Viterbi decoding [71], [55], [56]. These results are upper bounds that are based on the fact that for equal-power binary PSK direct-sequence spread-spectrum signals, the largest symbol error probability occurs when all of the interfering signals have the same phase as the signal being demodulated and have a time offset that is a multiple of the chip duration (this result is proven in [56]). Fortunately, because the sequences are random, this upper bound corresponds to a model in which the symbol errors on the channel are independent. Unfortunately, as stated above, this is not enough to guarantee that the errors at the decoder output are independent (indeed, they are not).

The approach developed in [71] is utilized [55] and [56] to account for the effects of the Viterbi decoder and derive an upper bound on the packet error probability. These results were used to obtain the results shown in Fig. 10 for the convolutional code of constraint length 7 and rate $\frac{1}{2}$ in a direct-sequence spread-spectrum system with 127 chips per symbol. The two curves shown are for systems with three interfering signals ($K = 4$) and twelve interfering signals ($K = 13$). The packet length is 1000 information bits (2000 symbols) in both cases.

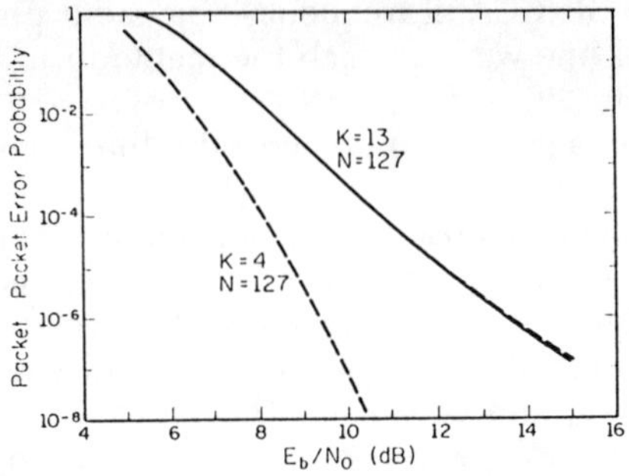

Fig. 10. Upper bounds on the packet error probability. (Convolutional coding, constraint length 7, rate $\frac{1}{2}$, random signature sequences.)

Although the numerical results given in this section are for multiple-access interference, the evaluation of the error probability can be carried out for multipath interference in exactly the same way. Details of the analysis and numerical results can be found in [44] and [15] for spread-spectrum receivers with a correlator. Efficient evaluation of error probabilities for multipath-combining receivers is, to a large extent, still an open problem.

Another open problem is the determination of acquisition and capture performance for spread-spectrum packet radio networks. Generally speaking, we expect that a reasonably well-designed spread-spectrum system that has good multiple-access performance will also have good anti-multipath and capture performance if the spreading sequences have been selected to have good autocorrelation properties. However, this assumes that reliable acquisition can be accomplished, and there are virtually no available analytical results on the acquisition performance of direct-sequence spread-spectrum receivers in a multiple-access environment. The situation for frequency-hop packet radio networks is somewhat better.

We close this section with a brief summary of the major points and conclusions. Direct-sequence spread spectrum with proper sequence selection is an effective means of combatting multipath interference and multiple-access interference, both of which arise in packet radio networks (such as the DARPA low-cost packet radio network) that operate in the upper end of the UHF band. Several methods are available for analytical and numerical evaluation of bit error probabilities for direct-sequence spread-spectrum multiple-access communication systems that employ correlation receivers, and the choice of which method to use depends on the required accuracy and the range of parameters (number of simultaneous transmissions and number of chips per symbol). The situation is not quite as good for the evaluation of the performance of direct-sequence spread spectrum for multipath channels if a combining receiver is employed (most of the available methods involve Gaussian approximations).

If the multiple-access signals are of unequal power, the resulting performance is not only degraded (for the weaker signals) but it is also more difficult to evaluate in the sense that some of the available methods do not apply. However, the characteristic-function method is applicable, and the results shown in Fig. 8 demonstrate that the total interference power is not a good predictor of performance at low error probabilities; the distribution of the power among the interfering signals is also important. This also points out a limitation of the Gaussian approximation based on signal-to-interference ratios only. Finally, some extensions to the evaluation of packet error probabilities in systems with convolutional coding were discussed.

V. Performance of Frequency-Hop Packet Radio Networks

The reader not familiar with the basic features of frequency-hop spread-spectrum communications may find it advantageous to consult any of a number of recent books and papers on the subject such as [7], [14], [36], or [86]. More specialized results on the performance of frequency-hop spread-spectrum communications on channels with narrow-band and partial-band interference are given in [46],

[54], [65], [68], and [80]. In addition, a recent paper [49] describes the role of frequency-hop transmission in packet radio and gives results on error probabilities and throughput for frequency-hop packet radio networks.

In a frequency-hop spread-spectrum packet radio network, the total RF bandwidth is divided into q subbands called frequency slots, and there is one carrier frequency available in each of these slots. The RF signal from a given transmitter is hopped from slot to slot by changing the carrier frequency. The sequence of carrier frequencies used by a signal is known as its frequency-hopping pattern. The interval of time between two consecutive hop epochs is called the hop interval, and the length of this interval is called the hop period. If all hop intervals are of length T_h, the hopping rate is $R_h = T_h^{-1}$. In this paper, only *slow* frequency hopping will be considered (i.e., the hopping rate is less than the data rate).

The requirement for *synchronous* frequency hopping is that the hop intervals for all transmissions must be aligned at all receivers. For many applications, this requirement cannot be met, and only *asynchronous* hopping is feasible [49]. For very low hopping rates and networks with small differential delays, synchronous hopping may be possible, but we will take a conservative approach and use the characteristics of frequency hopping in a way that gives an upper bound on the error probability for both synchronous and asynchronous frequency-hop packet radios.

We assume that time slotting is employed; that is, time is divided into intervals called *packet slots*, and each transmission must take place wholly within a packet slot. The assumption of time slotting is not really critical to any of the discussion (unslotted frequency-hop packet radio networks are considered in [49]), but it does simplify the presentation. The conditions required for slotted transmission are much less strict than the requirements for synchronous frequency hopping. In particular, we assume that a guard time of a few percent of the packet length is sufficient to permit slotted operation in the packet radio network. For a 4000-symbol packet at 400 000 symbols per second, a 1-percent guard time requires 50 μs, which can accommodate a differential propagation distance of about 15 km.

Whenever two or more symbols from different radios are transmitted simultaneously in the same frequency slot, we say a *hit* occurs. Suppose two packets are transmitted in the same packet slot on two different asynchronous hopping patterns, and consider a single symbol from one of them. The probability that the other signal is transmitted at the same frequency at any time during the transmission interval for that symbol is called the probability of a hit and is denoted by P_h.

For the model of a frequency-hop radio network adopted in [49], each transmission takes place on a random hopping pattern, which is just a random sequence of carrier frequencies. The hopping patterns are identically distributed, and different hopping patterns are statistically independent. Typical models for the random hopping patterns (e.g., memoryless or Markovian [14]) are such that $P_h \leq 2/q$.

The same four characteristics—capture, multiple access, anti-multipath, and narrow-band interference rejection—described in Sections III and IV for direct-sequence spread spectrum are possessed to varying degrees by frequency-hop spread spectrum as well. For specular multipath, the only form of multipath considered in this paper, frequency-hop spread spectrum does not provide very much protection. Only the multipath components with delays on the order of the hop period are rejected by frequency hopping alone. Some form of equalization or hybrid modulation (direct sequence and frequency hopping) is necessary if there are significant multipath components with delays less than T_h. With respect to the other three characteristics, however, frequency hopping can be superior to direct sequence, depending on the detailed features of the channel and the interference.

A. Capture and Spread-Spectrum Transmission Protocols

A general discussion of the connection between spread-spectrum signaling, transmission protocols, and capture is given in Section II-A. Frequency-hop spread spectrum has some special properties that provide better capture properties than direct sequence for certain transmission protocols. Suppose that a packet is transmitted on a particular hopping pattern and the intended receiver has acquired the necessary timing information. Once this initial acquisition has been accomplished, transmissions on other hopping patterns or delayed versions of the same hopping pattern will not disrupt synchronization or demodulation. The interference due to these transmissions will cause random errors rather than collisions. In fact, unlike the situation for direct-sequence spread spectrum, subsets of these errors will be conditionally independent given the number and power levels of the interfering transmissions.

The capture interval for a frequency-hop packet radio depends on the spread-spectrum transmission protocol and the timing mechanism for the hopping pattern generator. The situation is as described in Section II-A, except that there are some special considerations for minislotting for frequency-hop radios. It is often convenient to let the minislot length be the hop period, but, if the hopping rate is low, a smaller interval can be used. If there are a large number of hop intervals per packet slot, a reasonable choice of channel-access protocol will permit high throughput in the frequency-hop packet radio network with an acceptably low probability of packet collision. Moreover, the implementation is not particularly complex if the minislots are not too small. The SINCGARS frequency-hop radio provides an example of one way in which the method can be implemented [28], [29], [64].

The advantage in this approach is that the receiver knows exactly which hopping pattern will be used for packets that arrive in a given minislot. A search over a set of hopping patterns is not required. This greatly reduces acquisition time and increases the probability of successful acquisition. Moreover, this protocol and timing mechanism provide good capture capability. Theoretically, the capture interval is the hop period or smaller (i.e., the length of the minislot).

B. Multiple-Access Capability

Multiple access, as depicted in Fig. 2, is concerned with multiple transmissions on different hopping patterns. A measure of the multiple-access capability is the number of simultaneous transmissions as a function of the packet error probability. As the number of transmissions is increased, the number of hits increases, and this increases the packet error probability. Of course, a larger packet error probability implies that a larger number of retransmissions will be required, and this implies increased delay.

The multiple-access capability of frequency-hop packet radio networks is considerably enhanced if the receiver can determine which symbols in the packet have been hit. This information, typically called side information, is used in the demodulator and decoder to improve error control. A key feature of slow-frequency-hop radios is the relative ease with which side information can be developed. For example, it can be obtained from the hopping pattern generators in frequency-hop radios that track multiple hopping patterns. Other methods for developing side information include the transmission of test symbols during each hop period, exploitation of the asynchronism of the frequency hopping in the network, or comparisons between the outputs of the matched filters.

A very powerful error-control technique is to erase the symbols that have been hit and employ the features of a suitable code (such as the Reed–Solomon code) to correct the symbols that have been erased. A given packet is received correctly if the number of hits does not exceed the code's erasure-correcting capability. Even if the side information is not perfect, this approach has been shown to work very well, and it is being employed in the packet radio version of the SINCGARS frequency-hop radio [28], [29], [64].

Suppose we take as one measure of performance the probability of error P_E for an individual transmission (or a retransmission) of a packet from one radio to its neighbor. This measure is closely coupled to the delay as mentioned above. For a fully connected network employing a slotted ALOHA channel-access protocol, it is known that, for any given error probability less than about 10^{-1}, the throughput per unit bandwidth for frequency-hop radios is much greater than for narrow-band radios [49]. It is also known that the maximum (unconstrained) throughput per unit bandwidth is as large for frequency hopping with long block codes as it is for ALOHA with narrow-band radios (e.g., see the report by Hajek in [62]).

At error probabilities less than 10^{-2} the difference between the throughput of a frequency-hop packet radio network and a narrow-band packet radio network is very dramatic. In order to give a quantitative measure of the difference, it is necessary to define network throughput carefully. We restrict attention to fully connected networks with fixed-length packets and slotted transmission. If the network is not fully connected, the measure we define is actually a measure of local throughput [49]. The definition can easily be applied to unslotted systems [49, sec. V], and it can be modified to handle variable-length packets.

The throughput is defined as the average number of successful packets per packet slot. If μ denotes the average number of transmissions attempted per slot, and $p_e(\mu)$ denotes the probability of error for a particular packet when the average number of packets per slot is μ, the throughput in packets per packet interval is

$$s(\mu) = \mu\,[1 - p_e(\mu)].$$

Notice that this throughput measures does not distinguish between packets according to their destinations. It is, as desired, a measure of the multiple-access *capability* of the modulation and coding technique which is divorced from the particular protocols being used in the network. In a sense, $s(\mu)$ is the average number of receivable packets. A packet is receivable if the number of hits it encounters is within the correction capability of the code. The question of the existence of a radio that is ready to receive this packet is another matter. Similarly, a packet that is intended for multiple destinations is counted as one successful transmission only, even if it is actually received successfully by several radios.

The best the spread-spectrum system can do is provide the capability for successful simultaneous transmissions within the network. Whether this multiple-access capability is actually achieved depends on the radios, the transmission protocol, and the channel-access protocol. If each radio can receive multiple packets simultaneously and if it can receive and transmit simultaneously, then the multiple-access capability is always achieved. If the radios do not have these features—and most do not—the multiple-access capability is achieved only if no two packets are intended for the same receiver and none of the receivers are busy transmitting. The goal of the network protocols is to minimize the difference between $s(\mu)$ and the average number of packets that are delivered to their destinations.

The maximum value of $s(\mu)$ is the maximum local throughput, which is denoted by S_{max}. The value of μ for which $s(\mu) = S_{max}$ is denoted by μ_m. For the codes considered in [49], $s(\mu)$ is a strictly increasing function for values of μ not greater than μ_m. For traffic levels above μ_m the throughput decreases, the required number of retransmissions increases, and the average delay becomes very large. For Poisson traffic and slotted ALOHA without frequency hopping $\mu_m = 1$, the throughput at this traffic level is e^{-1}, and the resulting error probability is $1 - e^{-1}$ which is approximately 0.632. For frequency-hop radio networks with error-control coding, the value of μ_m depends on the code and the bandwidth used by the spread-spectrum system, but it is typically much larger than one.

The error probability $p_e(\mu_m)$ at traffic level μ_m is typically 0.1 or greater; it can be as large as 0.632 if the bandwidth or coding are inadequate. This means at least 10 percent of the packets must be retransmitted, which is too large for many applications [29]. Consequently, an important performance measure is the maximum throughput subject to a constraint on the probability of error

$$S^*(P_E) = \max\,\{s(\mu) : p_e(\mu) \le P_E\}.$$

In order to make valid comparisons between frequency-hop and narrow-band radio networks, it is necessary to normalize $s(\mu)$ and $S^*(P_E)$ to give throughput per unit bandwidth. If the code has rate r and the spread-spectrum modulation produces bandwidth expansion β, the normalized throughput at traffic level μ is

$$w(\mu) = rs(\mu)/\beta$$

and the maximum normalized throughput at error probability P_E is

$$W^*(P_E) = rS^*(P_E)/\beta.$$

For frequency-hop packet radio networks, the bandwidth expansion is approximately q, the number of frequency slots.

Some values of the normalized throughput are given in Table 1 for two values of P_E. The codes are (n, k) extended

Table 1 Normalized Throughput for Several Reed–Solomon Codes of Block Length 64 ($q = 100$)

k	W^* (10^{-2})	W^* (10^{-6})
32	0.0917	0.0372
24	0.1056	0.0503
20	0.1078	0.0545
16	0.1056	0.0561
12	0.0978	0.0541

Reed–Solomon codes of block length $n = 64$, the number of frequency slots is 100, and the traffic has Poisson distribution. The parameter k is the number of information symbols (the code rate is $r = k/n$).

The limiting value for the normalized throughput as both the average traffic and the number of frequency slots get large (i.e., $\mu \to \infty$ and $q \to \infty$ with $\mu/q = t$ held constant) is given by [18], [49]

$$v(t) = rtF[e; n, 1 - \exp(-2t)].$$

The function $F(x;n,p)$ is the probability of x or fewer successes in n independent trials when the probability of success on a given trial is p (i.e., F is the binomial distribution function). The limiting throughputs for P_E equal to 10^{-2} and 10^{-6} are given in Table 2. The limiting values for the throughput for narrow-band radios with these same error probabilities are given in Table 3.

Table 2 Limiting Values of the Normalized Throughput for Several Reed–Solomon Codes of Block Length 64 with Large q

k	W^* (10^{-2})	W^* (10^{-6})
32	0.1126	0.0668
24	0.1243	0.0783
20	0.1245	0.0802
16	0.1199	0.0785
12	0.1091	0.0722

Table 3 Normalized Throughput for Narrow-Band Radios

W^* (10^{-2})	W^* (10^{-6})
0.0099	1.0000E − 6

Throughput can also be determined for direct-sequence spread-spectrum packet radio networks, but the process is more involved. This is largely due to the statistical dependence between symbol errors in such a system. Some comparisons between the throughput for direct-sequence and frequency-hop packet radio networks are given in [56].

VI. Concluding Remarks

We have described several ways in which spread-spectrum signaling can enhance the performance of a packet radio network, and we have also pointed out that network protocol considerations impose constraints on the spread-spectrum modulation. There are many important topics that could not be included in this paper, but an attempt was made to emphasize the features of spread spectrum that have the greatest interaction with network protocols and network operation. A bibliography is included which gives several publications which provide deeper coverage of the topics considered in this paper as well as a number of publications that deal with topics such as acquisition and anti-jam communications that were not covered here.

Acknowledgment

The author wishes to thank E. A. Geraniotis, J. S. Lehnert, W. E. Stark, and D. J. Taipale for the numerical results presented in Figs. 6–10, and the reviewers and editors of this special issue for their comments on the original draft of the paper.

References and Selected Bibliography

[1] J. M. Aein and J. W. Schwartz, Eds., "Multiple access to a communication satellite with a hard-limiting repeater—Vol. II: Proceedings of the IDA multiple access summer study," Institute for Defense Analysis, Rep. R-108, 1965.

[2] D. R. Anderson and P. A. Wintz, "Analysis of a spread-spectrum multiple-access system with a hard limiter," *IEEE Trans. Commun. Technol.*, vol. COM-17, pp. 285–290, Apr. 1969.

[3] D. E. Borth and M. B. Pursley, "Analysis of direct-sequence spread-spectrum multiple-access communication over Rician fading channels," *IEEE Trans. Commun.*, vol. COM-27, pp. 1566–1577, Oct. 1979.

[4] W. C. Fifer and F. J. Bruno, "The low-cost packet radio," this issue, pp. 33–42.

[5] E. L. Caples, K. E. Massad, and T. R. Minor, "A UHF channel simulator for digital mobile radio," *IEEE Trans. Vehic. Technol.* vol. VT-29, pp. 281–289, May 1980.

[6] C. E. Cook, F. W. Ellersick, L. B. Milstein, and D. L. Schilling, Eds., *IEEE Trans. Commun.* (Special Issue on Spread-Spectrum Communications), vol. COM-30, May 1982.

[7] G. R. Cooper and C. D. McGillem, *Modern Communications and Spread Spectrum.* New York, NY: McGraw-Hill, 1986.

[8] D. H. Davis and S. A. Gronemeyer, "Performance of slotted ALOHA random access with delay capture and randomized time of arrival," *IEEE Trans. Commun.*, vol. COM-28, pp. 703–710, May 1980.

[9] R.-.H. Dou and L. B. Milstein, "Error probability bounds and approximations for DS spread-spectrum communication systems with multiple tone or multiple access interference," *IEEE Trans. Commun.*, vol. COM-32, pp. 493–501, May 1984.

[10] S. C. Fralick, "An improved packet radio demodulator for mobile operation," Temporary Packet Radio Note 132, Mar. 1975.

[11] F. D. Garber and M. B. Pursley, "Optimal phases of maximal-length sequences for asynchronous spread-spectrum multiplexing," *Electron. Lett.*, vol. 16, pp. 756–757, Sept. 1980.

[12] —, "Performance of offset quadriphase spread-spectrum multiple-access communications," *IEEE Trans. Commun.*, vol. COM-29, pp. 305–314, Mar. 1981.

[13] E. A. Geraniotis and M. B. Pursley, "Error probability for direct-sequence spread-spectrum multiple-access communications—Part II: Approximations," *IEEE Trans. Commun.*, vol. COM-30, pp. 985–995, May 1982.

[14] —, "Error probabilities for slow-frequency-hopped spread-spectrum multiple-access communications over fading channels," *IEEE Trans. Commun.*, vol. COM-30, pp. 996–1009, May 1982.

[15] —, "Performance of coherent direct-sequence spread-spectrum communications over specular multipath fading channels," *IEEE Trans. Commun.*, vol. COM-33, pp. 502–508, June 1985.

[16] —, "Performance of noncoherent direct-sequence spread-spectrum communications over specular multipath fading channels," *IEEE Trans. Commun.*, vol. COM-34, pp. 219–226, March 1986.

[17] L. A. Gerhardt and R. C. Dixon, Eds., *IEEE Trans. Commun.*, (Special Issue on Spread Spectrum Communications), vol. COM-25, Aug. 1977.

[18] B. Hajek, "Recursive retransmission control—Application to a frequency-hopped spread-spectrum system," in *Proc. 1982 Conf. on Information Science and Systems* (Princeton Univ.,

Princeton, NJ), pp. 116–120, Mar. 1982.

[19] J. K. Holmes, *Coherent Spread Spectrum Systems*. New York, NY: Wiley, 1982.

[20] R. E. Kahn, "The organization of computer resources into a packet radio network," *IEEE Trans. Commun.*, vol. COM-25, pp. 169–178, Jan. 1977.

[21] R. E. Kahn, S. A. Gronemeyer, J. Burchfiel, and R. C. Kunzelman, "Advances in packet radio technology," *Proc. IEEE*, vol. 66, pp. 1468–1496, Nov. 1978.

[22] J. Kaiser, J. W. Schwartz, and J. M. Aein, "Multiple access to a communication satellite with a hard-limiting repeater—Vol. I: Modulation techniques and their applications," Institute for Defense Analysis, Rep. R-108, 1965.

[23] L. Kleinrock, *Queueing Systems, Vol. II: Computer Applications*. New York, NY: Wiley, 1976.

[24] D. Laforgia, A. Luvison, and V. Zingarelli, "Bit error rate evaluation for spread-spectrum multiple-access systems," *IEEE Trans. Commun.*, vol. COM-32, pp. 660–668, June 1984.

[25] J. S. Lehnert, "Direct-sequence spread-spectrum signaling with applications to packet radio systems," Rep. T-153, Coordinated Science Lab., University of Illinois, Urbana, Nov. 1984.

[26] J. S. Lehnert and M. B. Pursley, "Multipath diversity reception of coherent direct-sequence spread-spectrum communications," in *Proc. 1983 Conf. on Information Sciences and Systems* (Johns Hopkins Univ., Baltimore, MD), pp. 770–775, Mar. 1983.

[27] —, "Error probabilities for binary direct-sequence spread-spectrum communications with random signature sequences," *IEEE Trans. Commun.*, vol. COM-35, no. 1, pp. 85–96, Jan. 1987.

[28] J. R. McChesney, "System applications for SINCGARS VHF tactical radio," in *Proc. Workshop on Research Trends in Spread Spectrum Systems* (sponsored by the U.S. Army Research Office and the University of Illinois, Aug. 4–7, 1985).

[29] J. R. McChesney and M. B. Pursley, "Channel-access and multiple-access performance for SINCGARS packet-switched radio," in *Proc. AFCEA–IEEE Conf. on Tactical Communications* (Fort Wayne, IN, Apr. 1986).

[30] F. J. MacWilliams and N. J. A. Sloane, "Pseudo-random sequences and arrays," *Proc. IEEE*, vol. 64, pp. 1715–1729, Dec. 1976.

[31] J. L. Massey and J. J. Uhran, "Sub-baud coding," in *Proc. 13th Ann. Allerton Conf. on Circuit and System Theory* (Univ. Illinois, Urbana), pp. 539–547, Oct. 1975.

[32] L. B. Milstein, S. Davidovici, and D. L. Schilling, "The effect of multiple-tone interfering signals on a direct sequence spread spectrum communication system," *IEEE Trans. Commun.*, vol. COM-30, pp. 436–446, Mar. 1982.

[33] L. B. Milstein and D. L. Schilling, "Performance of a spread spectrum communication system operating over a frequency-selective fading channel in the presence of tone interference," *IEEE Trans. Commun.*, vol. COM-30, pp. 240–247, Jan. 1982.

[34] D. L. Nielson, "Microwave propagation and noise measurements for mobile digital radio application," SRI Tech. Rep., Packet Radio Note 4, Jan. 1975.

[35] —, "Microwave propagation measurements for mobile digital radio application," *IEEE Trans. Vehic. Technol.*, vol. VT-27, pp. 117–132, Aug. 1978.

[36] R. L. Pickholtz, D. L. Schilling, and L. B. Milstein, "Theory of spread-spectrum communications—A Tutorial," *IEEE Trans. Commun.*, vol. COM-30, pp. 855–884, May 1982.

[37] A. Polydoros and C. L. Weber, "A unified approach to serial search spread-spectrum code acquisition—Part I: General theory," *IEEE Trans. Commun.*, vol. COM-32, pp. 542–549, May 1984.

[38] —, "A unified approach to serial search spread-spectrum code acquisition—Part II: A matched-filter receiver," *IEEE Trans. Commun.*, vol. COM-32, pp. 550–560, May 1984.

[39] —, "Analysis and optimization of correlative code-tracking loops in spread-spectrum systems," *IEEE Trans. Commun.*, vol. COM-33, pp. 30–43, Jan. 1985.

[40] R. Price and P. E. Green, Jr., "A communication technique for multipath channels," *Proc. IRE*, vol. 46, pp. 555–570, Mar. 1958.

[41] M. B. Pursley, "Evaluating performance of codes for spread-spectrum multiple-access communications," in *Proc. 12th Ann. Allerton Conf. on Circuit and Systems Theory* (Univ. Illinois, Urbana) pp. 765–774, Oct. 1974.

[42] —, "Performance evaluation for phase-coded spread-spectrum multiple-access communication—Part I: System analysis," *IEEE Trans. Commun.*, vol. COM-25, pp. 795–799, Aug. 1977.

[43] M. B. Pursley, Ed., *IEEE Trans. Vehic. Technol.*, (Special Issue on Mobile Spread-Spectrum Communications), vol. VT-30, Feb. 1981.

[44] M. B. Pursley, "Effects of specular multipath fading on spread-spectrum communications," in *New Concepts in Multi-User Communications* (NATO Advanced Study Institute—Series E), J. K. Skwirzynski, Ed. Sijthoff and Nordhoof, The Netherlands: Alphen Van den Rijn, 1981, pp. 481–505.

[45] —, "Spread-spectrum multiple-access communications," in *Multi-User Communication Systems*, G. Longo, Ed. Vienna and New York: Springer-Verlag, 1981, pp. 139–199.

[46] —, "Coding and diversity for channels with fading and pulsed interference," in *Proc. 1982 Conf. on Information Sciences and Systems* (Princeton Univ., Princeton, NJ), pp. 413–418, Mar. 1982.

[47] —, "Throughput of frequency-hopped spread-spectrum communications for packet radio networks," in *Proc. 1983 Conf. on Information Sciences and Systems* (Johns Hopkins Univ., Baltimore, MD), pp. 550–556, Mar. 1983.

[48] —, "Adaptive spread spectrum radio networks," Coordinated Science Laboratory Rep. T-131, Univ. Illinois, Urbana, Dec. 1983 (reprinted in [62]).

[49] —, "Frequency-hop transmission for satellite packet switching and terrestrial packet radio networks," *IEEE Trans. Inform. Theory*, vol. IT-32, pp. 652–667, Sept. 1986; also Coordinated Science Laboratory Rep. T-144, Univ. Illinois, Urbana, June 1984.

[50] M. B. Pursley, F. D. Garber, and J. S. Lehnert, "Analysis of generalized quadriphase spread-spectrum communications," in *Conf. Rec., IEEE Int. Conf. on Communications*, vol. 1, pp. 15.3.1–15.3.6, June 1980.

[51] M. B. Pursley and H. F. A. Roefs, "Numerical evaluation of correlation parameters for optimal phases of binary shift-register sequences," *IEEE Trans. Commun.*, vol. COM-27, pp. 1597–1604, Oct. 1979.

[52] M. B. Pursley and D. V. Sarwate, "Performance evaluation for phase-coded spread-spectrum multiple-access communication—Part II: Code sequence analysis," *IEEE Trans. Commun.*, vol. COM-25, pp. 800–803, Aug. 1977.

[53] M. B. Pursley, D. V. Sarwate, and W. E. Stark, "Error probability for direct-sequence spread-spectrum multiple-access communications—Part I: Upper and lower bounds," *IEEE Trans. Commun.*, vol. COM-30, pp. 975–984, May 1982.

[54] M. B. Pursley and W. E. Stark, "Performance of Reed-Solomon coded frequency-hop spread-spectrum communications in partial-band interference," *IEEE Trans. Commun.*, vol. COM-33, pp. 767–774, Aug. 1985.

[55] M. B. Pursley and D. J. Taipale, "Error probabilities for spread-spectrum packet radio with convolutional codes and Viterbi decoding," in *IEEE Military Communications Conf. Rec.*, vol. 2, pp. 438–441, Oct. 1985.

[56] M. B. Pursley and D. J. Taipale, "Error probabilities for spread-spectrum packet radio with convolutional codes and Viterbi decoding," *IEEE Trans. Commun.*, vol. COM-35, no. 1, pp. 1–12, Jan. 1987.

[57] H. F. A. Roefs and M. B. Pursley, "Correlation parameters of random and maximal length sequences," in *Proc. IEEE Canadian Conf. on Communications and Power* (Montreal, Que., Canada), pp. 141–143, Oct. 1976.

[58] —, "Correlation parameters of random binary sequences," *Electron. Lett.*, vol. 13, no. 16, pp. 488–489, Aug. 4, 1977.

[59] D. V. Sarwate, "Bounds on the crosscorrelation and autocorrelation of sequences," *IEEE Trans. Inform. Theory*, vol. IT-25, pp. 720–724, Nov. 1979.

[60] D. V. Sarwate and M. B. Pursley, "Crosscorrelation properties of pseudorandom and related sequences," *Proc. IEEE*, vol. 68, pp. 593–619, May 1980.

[61] D. V. Sarwate, M. B. Pursley, and T. U. Basar, "Partial correlation effects in direct-sequence spread-spectrum multiple-access communication systems," *IEEE Trans. Commun.*,

vol. COM-32, pp. 567–573, May 1984.

[62] P. Sass, Ed., "Adaptive spread spectrum network research," U.S. Army Research and Development Tech. Rep. CECOM-TR-84-4, Apr. 1984.

[63] M. S. Schmidt, "Sequence selection for spread-spectrum packet radio systems," M.S. thesis, Dep. Elect. Eng., Univ. Illinois, Urbana (also Coord. Sci. Lab. Tech. Rept. T-121), Jan. 1983.

[64] A. Shohara *et al.*, "Design Plan: SINCGARS packet switch overlay," Preliminary Draft, SRI Project 1244, under Contract DAAB07-85-C-K581 (U.S. Army CECOM), Apr. 24, 1985.

[65] M. K. Simon, J. K. Omura, R. A. Scholtz, and B. K. Levitt, *Spread Spectrum Communications* (3 vols.). Rockville, MD: Computer Sci. Press, 1985.

[66] R. Skaug, "Numerical evaluation of the nonperiodic autocorrelation parameter for optimal phases of maximal length sequences," *Proc. Inst. Elec. Eng.*, pt. F, vol. 127, pp. 230–237, June 1980.

[67] E. S. Sousa and J. A. Silvester, "A spreading code protocol for a distributed spread spectrum packet radio network," in *Proc. IEEE Global Communications Conf.*, pp. 481–486, Nov. 1984.

[68] W. E. Stark, "Coding for frequency-hopped spread-spectrum channels with partial-band interference," Coord. Sci. Lab. Tech. Rep. R-945, University of Illinois, Urbana, July 1982.

[69] S. Stein, *Communication Systems and Techniques*, pt. III. New York, NY: McGraw-Hill, 1966.

[70] J. S. Storey and F. A. Tobagi, "Throughput performance of a direct sequence CDMA packet radio network," SEL Tech. Rep. 85-277 (also SURAN Temporary Note 20), Stanford Univ., Stanford, CA, June 1985.

[71] D. J. Taipale, "Direct-sequence spread-spectrum packet radio performance," Rep. T-155, Coord. Sci. Lab., Univ. Illinois, Urbana, Nov. 1984.

[72] F. A. Tobagi, "Modeling and performance analysis of multihop packet radio networks," this issue, pp. 135–155.

[73] G. L. Turin, "An introduction to digital matched filters," *Proc. IEEE*, vol. 64, pp. 1092–1112, July 1976.

[74] —, "Antimultipath techniques for packet radio systems," Packet Radio Temporary Note 262, Nov. 22, 1978.

[75] —, "Antimultipath techniques for packet radio systems II: Summary of the work of M. A. Kamil and L.-F. Wei," Packet Radio Temporary Note 282, Dec. 17, 1979.

[76] —, "Introduction to spread-spectrum antimultipath techniques and their application to urban digital radio," *Proc. IEEE*, vol. 68, pp. 328–353, Mar. 1980.

[77] —, "Antimultipath techniques for packet radio systems III: The M-ary case," Packet Radio Temporary Note 296, Oct. 17, 1980.

[78] —, "Antimultipath techniques for packet radio systems IV: Feasibility Study by L.-F. Wei and G. L. Turin," Packet Radio Temporary Note 299, Mar. 16, 1981.

[79] G. L. Turin *et al.*, "A statistical model of urban multipath propagation," *IEEE Trans. Vehic. Technol.*, vol. VT-21, pp. 1–9, Feb. 1972.

[80] A. J. Viterbi and I. M. Jacobs, "Advances in coding and modulation for noncoherent channels affected by fading, partial band, and multiple-access interference," in *Advances in Communication Systems*, vol. 4. New York, NY: Academic Press, 1975.

[81] C. L. Weber, G. K. Huth, and B. H. Batson, "Performance considerations of code division multiple-access systems," *IEEE Trans. Vehic. Technol.*, vol. VT-30, pp. 3–10, Feb. 1981.

[82] K.-T. Wu, "Direct sequence spread-spectrum communications: Applications to multiple access and jamming resistance," Ph.D. dissertation, Univ. Michigan, Ann Arbor, 1981.

[83] —, "Average error probability for DS-SSMA communications: The Gram-Charlier expansion approach," in *Proc. 19th Ann. Allerton Conf. on Communications, Control, and Computing* (Univ. Illinois, Urbana), pp. 237–246, Sept. 1981.

[84] K.-T. Wu and D. L. Neuhoff, "Average bit error probability for direct sequence spread-spectrum multiple-access communication systems," in *Proc. 18th Ann. Allerton Conf. on Communications, Control, and Computing* (Univ. Illinois, Urbana), pp. 359–368, Oct. 1980.

[85] K. Yao, "Error probability of asynchronous spread spectrum multiple access communication systems," *IEEE Trans. Commun.*, vol. COM-25, pp. 803–809, Aug. 1977.

[86] R. E. Ziemer and R. L. Peterson, *Digital Communications and Spread Spectrum Systems*. New York, NY: Macmillan, 1985.

Slotted Random Access Spread-Spectrum Networks: An Analytical Framework

ANDREAS POLYDOROS, MEMBER, IEEE, AND JOHN SILVESTER, SENIOR MEMBER, IEEE

Abstract—An analytic framework is proposed for the study of single-hop spread-spectrum networks using random access and packet switching under various network topologies and channel conditions. The key feature of the theory is the identification of a set of probabilistic parameters, which, based on a symmetry argument, serve to efficiently summarize the effect on performance of various network considerations such as transmitter–receiver configuration, spread-spectrum code allocation, error correction and detection mechanisms, spreading format, jamming conditions, etc. Examples investigating capture effects, coding tradeoffs, and scheduling optimizations are presented. Various previously known results are shown to be special cases of the framework that we describe.

I. Introduction

MULTIPLE-ACCESS radio for data communication presents many research and implementation challenges. The first such operational example was the ALOHA network at the University of Hawaii [1], subsequently analyzed in detail in [16]. Since that time, much work has been done on proposing novel access schemes and investigating network performance. More recently, we have seen the introduction of spread-spectrum as a modulation and access tool in these networks [15], motivated mainly by military applications that take advantage of its inherent antijam (AJ) and low probability of intercept (LPI) capabilities. In addition, spread-spectrum techniques can be used to provide (simultaneous) multiuser capability. Most of the spread-spectrum multiple-access (SSMA) analysis thus far has focused on the physical level of communication, with only a few papers treating link-level issues and even fewer addressing topological and other network-level issues. Thus, putting aside the impact of the bursty traffic and access protocol on the overall system (retransmissions, etc.), authors have investigated (and strongly debated) the efficiency of SSMA as a signaling alternative in benign (nonjammed) environments [5]–[7], [10]–[12], [14], [18], [21], [23], [26]–[30], [40]–[44], where the figure of merit has been taken to be the number of simultaneously active users supportable at a specified bit error rate (BER) (this is elaborated on in Section II).

Manuscript received March 31, 1986; revised January 20, 1987. This work was supported by the Army Research Office (ARO) under Contract DAAG29-84-K-0084. This paper was presented in part at IEEE MILCOM 85, Boston, MA, October 1985.

The authors are with the Communication Sciences Institute, Department of Electrical Engineering—Systems, University of Southern California, Los Angeles, CA 90089-0272.

IEEE Log Number 8714854.

Although this is a rather simplified and incomplete figure of merit in a topologically complex multiple-access environment, it nonetheless provides a useful yardstick for comparisons. Furthermore, the above results can be appropriately folded into a more general analytical model in order to quantify the impact of the physical level in a parametric fashion (discussed further in Section V).

On the link-level aspects of SSMA signaling, notable papers are those by Davis and Gronemeyer [8], which looks at capture issues in a centralized spread-spectrum network, and by Raychaudhuri [31], which analyzes paired SSMA links in a common environment; they have provided much of the impetus for our present work (see also [13] for a related study, as well as the more recent [24], [27], [32]).

Finally, on the network level (end-to-end considerations), we mention the papers by Sousa and Silvester [35], [36], which look at spreading code assignment strategies and other issues such as the effect of half-duplex operation, and the work in [3], [38], [39], which has extended Boorstyn's packet radio network performance models [2] to the SSMA environment. These last references study true multihop throughput performance for some access protocols; delay results in this area are lacking, however, and we must turn to simple approximate models such as [19], [33], [37].

As indicated, the purpose of this paper is to develop a *generic model* for the study of local throughput in slotted SSMA networks, motivated by the fact that past results were constrained to specific (individual) system models and assumptions, thus making it rather difficult to make meaningful comparisons. The many parameters that need to be considered are explicitly included, and by modification of these parameters one can investigate many interesting submodels, including those mentioned above.

The paper is structured as follows. In Section II, the notions of throughput, utilization, and topology are reviewed, and their interrelationship is examined. In Section III, a general throughput expression is established, and the interpretation of associated parameters is undertaken in Section IV. The factorization of one such crucial parameter, the receiver's probability of acceptance, is analyzed in Section V, with the capture and retention mechanisms elaborated upon further. Numerical results for some examples and discussions are also offered in this section to illuminate various tradeoffs, optimizing design choices, etc. Section VI provides some closing remarks.

Reprinted from *IEEE J. Selected Areas Commun.*, vol. SAC-5, no. 6, pp. 989–1002, July 1987.

II. Throughput, Utilization, and Topology

The fundamental measure of performance encountered in the literature on random access networks is the throughput–delay tradeoff [1], [16]. For slotted systems, the throughput β is measured in packets per slot and the delay is measured in slot times. In order to account for the broader spectrum used for SSMA, it is meaningful from a system viewpoint to talk about the *utilization* or *channel efficiency* (cf. [31], [41]) measured in information bits per Hertz of available bandwidth, u_{ib}, or in packets per second per Hertz, u_p. Alternatively, the closely related measure of the number of users (all active) that can be supported at a specified BER can be used. The number of users is commonly expressed as a fraction of the bandwidth expansion factor with regard to a single user's requirements.

We shall deal here exclusively with slotted systems mainly because of their ease of analysis. It should be pointed out, however, that many military SSMA systems employ slotted systems since security considerations (such as periodic code changes) imply common clocks and timing. Furthermore, the "graceful degradation" feature of SSMA systems supports the general conjecture that the difference in efficiency between slotted and unslotted systems is much less significant than in narrow-band ALOHA (where there is a factor of 2) since collisions are catastrophic; recent results for particular systems [27] corroborate this judgement.

We let $T_{\text{slot}} = T_{\text{packet}}$ denote the slot or packet duration (i.e., one packet per slot) and assume that each slot is subdivided into L bauds (modulated symbol intervals) with $T_{\text{baud}} = T_{\text{slot}}/L$. In addition, each packet occupies a bandwidth $B_{\text{packet}} \leq W_{\text{RF}}$ Hertz where W_{RF} is the total available RF bandwidth. If the baud symbols belong to an M-ary alphabet and are encoded using a rate r random-error-correcting code, then each packet carries $rL \log_2 M$ information bits; furthermore, for unspread systems, B_{packet} is proportional to $1/T_{\text{baud}}$ and (perhaps) M. The constant of proportionality depends on the particular modulation technique used [25]; for instance, in phase-coherent modulation, the product $G \triangleq T_{\text{baud}} B_{\text{packet}}$ is about 1, for noncoherent MFSK, $G_{\text{MFSK}} = M$, etc. Thus, in the unspread case, B_{packet} is directly related to the baud time-width via G_{unspread}. On the other hand, if spread-spectrum modulation is used, B_{packet} incurs an expansion proportional to the *spreading ratio* η and is no longer determined by T_{baud}; rather, the bandwidth involved in transmitting a whole packet should be observed, hence the terminology. The exact expansion factor depends on the specific design of the spread-spectrum system (DS versus FH, overlapping versus nonoverlapping frequency sets, slow versus fast hopping, etc.) as well as on the precise definition of η (the definition of the spreading ratio in the open literature is far from unanimous). We can generally write that $G_{\text{spread}} = C\eta G_{\text{unspread}}$ where the constant C can be appropriately determined. For instance, if we *agree* to define $\eta \triangleq B_{\text{packet}}/R_b$, i.e., as the ratio of the used bandwidth per packet over the information bit rate, then $\eta = R_c/R_b$ in DS/BPSK with a code rate R_c, while for FH with Q_{FH} orthogonal frequencies and a frequency separation interval Δf, $\eta = Q_{\text{FH}} \Delta f / R_b$. For fast FH with an R_H hopping rate, $\Delta f = R_H$, while in the slow FH case, Δf is a function of the modulation employed per hop. In all cases, the constant C can be related to the impact of coding and the "unspread" part of the modulation as $C = R_b / B^{\text{unspread}}_{\text{packet}}$.

Let β_{RF} denote the total throughput in packets per slot over the available RF bandwidth. This β_{RF} is a function of the random access mechanism, the modulation/coding/spreading format, and the topology. We can then express the total system utilization u_{ib} in information bits per second per Hertz as

$$u_{ib} = \frac{rL \log_2 M \, \beta_{\text{RF}}(r) \, 1/T_{\text{slot}}}{W_{\text{RF}}} = \log_2 M \frac{r\beta_{\text{RF}}(r)}{W_{\text{RF}} T_{\text{baud}}} \tag{1a}$$

where the dependence of β_{RF} on r is explicitly shown. Let us now suppose that $B_{\text{packet}} \leq W_{\text{RF}}$, so that there exist $Z \triangleq W_{\text{RF}}/B_{\text{packet}}$ orthogonal "subcarriers" or spectral "groups" (for simplicity, we assume that Z is an integer), each of which can operate independently of the others.[1] In this case, the denominator of (1a) can be written as

$$W_{\text{RF}} T_{\text{baud}} = ZG \tag{1b}$$

where $G = B_{\text{packet}} T_{\text{baud}}$ is the time bandwidth product per packet as discussed previously for both spread and unspread systems. We can therefore write

$$u_{ib} = \log_2 M \frac{r\beta_{\text{RF}}(r)}{ZG} \triangleq \log_2 M \frac{r\beta(r)}{G} \tag{2}$$

where $\beta(r) \triangleq \beta_{\text{RF}}(r)/Z$ is the throughput per subcarrier or group. We can make the following observations.

1) u_{ib} is independent of T_{slot} and Z, indicating that the total utilization is independent of the actual bit or symbol rate and the resulting number of orthogonal subcarriers for a fixed total bandwidth; therefore, spectral groupings and subscriber subdivision into independent groups (a simultaneous frequency division and random multiple-access scheme) is not expected to affect system efficiency to the extent that this grouping does not alter the statistics of the random access scheme (e.g., the same arrival model should still apply, such as Poisson for an infinite population).

2) The expression is general and holds for spread and unspread systems. In particular, if we wish to demonstrate explicitly the dependence of the throughput β on the coding rate and the spreading ratio, we can adopt the no-

[1] Such spectral groups are assumed to operate at different frequencies, spaced sufficiently apart so that crosstalk is negligible (orthogonality assumption). Quadrature modulations at the same frequency are not treated here as "subcarriers," but rather as modulation alternatives. This is necessary in view of the fact that packet collisions and other channel disturbances destroy orthogonality between the quadrature components of a subcarrier.

tation

$$u_{ib} = \begin{cases} C_{\text{unspread}}\, r\beta(r) & \text{for unspread} \\ C_{\text{spread}}\, r\beta(r, \eta)/\eta & \text{for spread} \end{cases} \tag{3}$$

where the constants C_{spread} and C_{unspread} are independent of r and η. It is, therefore, important to determine the (implicit) dependence of β on the particular modulation/coding/spreading formats and the topology.

Our development will be fairly general in that it accounts for the possible statistical dependence among users beyond the standard like-user interference ("noise effect"). Certain "symmetry" or "homogeneity" conditions will be imposed on the model so that compact closed-form expressions can be derived and analyzed. The concepts do allow for generalization to asymmetric models in a fairly straightforward manner, however. Regarding acknowledgments, we assume that they are almost (costless from a capacity viewpoint), fully reliable, and instantaneous. These are reasonable assumptions in light of the fact that the one bit of acknowledgment information can be heavily coded for reliability. For a more detailed study of the impact of acknowledgments in SSMA, see [20].

III. A General Throughput Expression

We shall derive here a general expression for the throughput–delay characteristic of an arbitrary topology involving N_T transmitters and N_R receivers under a symmetry condition, namely, that each transmitter (TR) face the same probabilistic circumstances in the channel; an analogous symmetry holds from each receiver's (RCVR's) viewpoint. This will become clearer below when we elaborate on the notion of "success" as it is perceived from the transmitter's, the receiver's, and the topology's viewpoints. We shall see, in particular, that the receiver's conditional "success," (conditioned on the random number of transmitters in a slot, which brings about the topological dependence), is strongly related to the physical-level parameters (modulation, spreading, coding, etc.), while the topology's notion of "success" can be determined from networking and routing considerations. It will follow that the fundamental parameters $p_{s|m}$ (defined below), which quantify "success" at a link or local level, constitute the virtual analytic bond between the link level and the physical and network levels.

To be more specific, certain important parameters are defined in Table I, some of which are fixed (deterministic) and some of which are random variables (r.v.'s). Certain identities result immediately from these definitions, such as $0 \le M_T \le N_T$; $0 \le M_R \le N_R$; $0 \le B \le N_T$; and $M_T = M_B + M_0$. Certain other equalities follow from the specification of the particular scenario; for instance, for *paired-off* users [31], $U/2 = N_T = N_R$, $M_T = M_R$, while, in a more general configuration, there can be N_T fixed transmitters competing for N_R fixed receivers ($U = N_R + N_T$, $M_R = N_R$, $N_R \neq N_T$). A special case of this is the star topology of [8], where N_T transmitters compete for the attention of a single receiver ($N_R = 1$). An even more general situation occurs when $U = N_T = N_R$, i.e., each unit can be either transmitting or receiving, with the restriction that $M_T + M_R = U$. Let us note here that the term *topology* does not so much refer to the spatial aspects of networking[2] since we are really concerned with "local" scenarios, whereby each of the units U under consideration is potentially accessible by any other in a single slot. Rather, among other things, it pertains to the *connectivity* between active transmitters and receivers arising from two issues: *availability* (in half-duplex operation) and *spreading-code distribution* between users (transmitter-based versus receiver-based versus common code). The examples studied so far, i.e., star topology [8] and paired-off users [13], [31] are the simplest from a topological viewpoint. The first implies a common code between all transmitters (that is intelligible to the central receiver), while the second assumes a unique code assignment to each transmitter–receiver pair and complete receiver availability whenever the corresponding transmitter is active. Therefore, no topological "competition" takes place between active transmitters in this latter case, and the only disruptive interaction between pairs is that of multiuser interference.

TABLE I
KEY PARAMETERS

U	Total number of radio units in the local channel (fixed)
N_T	Maximum number of potential transmitters in a slot (fixed)
N_R	Maximum number of potential receivers in a slot (fixed)
M_T	Number of active transmitters in a particular slot (r.v.)
M_R	Number of active receivers in a particular slot (r.v.)
B	Number of backlogged users at the beginning of a slot (r.v.)
M_B	Number of backlogged users retransmitting in a slot (r.v.)
M_0	Number of new packets transmitted in a slot (r.v.)
S	Number of "channel successes" in a slot (r.v.)
$f_{M_T}(m)$	Prob $\{M_T = m\}$, unconditional pdf
$\Delta(\xi, \Xi, p)$	Pr$\{\xi$ successes in Ξ trials$\}$, binomial with parameter p

A full-fledged quantification of all possible topological scenarios would require a joint probabilistic description of the (random) number of active transmitters and receivers per slot, further subdivided into the random group sizes of units with the *same* code (in a transmitter-based or receiver-based system). This is quite an elaborate task, for which it is not clear that obvious answers exist (see [35], [36] for some of the combinational aspects of this problem). Our purpose here is to provide a framework that generalizes upon previously considered topologies, yet remains simple enough to be insightful. As such, we have focused on the case where the number of active transmitters per slot suffices to determine the conditional success rate. A *common-code* network with a fixed but arbitrary number of receivers ($M_r = N_r$), fits this model (it can be viewed as a generalization of the star topology), as does the paired-off structure. We therefore choose this fixed N_R scenario as our case study. An immediate generalization is to interpret N_R as M_R, view the throughput expression

[2]However, the spatial distribution of units does affect more subtle aspects of the network operation, for example, the statistical dependence among receivers in a stressed environment (see Section IV, Example 2).

as conditional (given M_R), and obtain the final answer through appropriate averaging whenever such a procedure is applicable.[3] The implication of $M_T \neq M_R$ is that "successes" for different transmitters will, in general, be statistically *dependent*, unlike the model in [31] whose throughput/delay results were derived with an independence assumption regarding the "success" of different transmitter–receiver pairs.

We note here that the concept of a single transmitter or receiver does not necessarily coincide with the true (physical) distribution of modems, but rather it relates to their functional properties from an analytic viewpoint. Thus, a single physical receiver with multicode receiving capability is, as far as the present model is concerned, equivalent to as many receivers as packets that can simultaneously be captured. Stated differently, a single receiver in the present context can capture one and only one packet (if any at all).

The performance expressions developed below are expressible in terms of the following three important parameters:

$$p_S^T(m) \triangleq \text{Prob}\left\{\text{success from a particular TR's viewpoint given another } (m-1) \text{ packets in the channel}\right\},$$

$$p_{s|m} \triangleq \text{Prob}\left\{S = s \text{ channel successes, given } M_T = m \text{ users in the slot}\right\},$$

$$p_A^R(m) \triangleq \text{Prob}\left\{\text{a particular RCVR } \textit{accepts} \text{ a packet, given } m \text{ packets in the slot}\right\}.$$

It is standard to define the throughput β in packets per slot as the expected number of successes per slot [8], [16], [31]:

$$\beta \triangleq E\{S\}$$

where $E\{S\}$ denotes the unconditional expectation over all possible underlying uncertainties. We then have the following general proposition (Proposition 1) for the case where a single packet can result in *at most one* success; this corresponds to the case where even if several receivers successfully decode the same packet, it is only counted as a *single* success. Alternatively, we could conceive of scenarios where multiple receptions of the same packet are desirable (broadcasting applications), in which case we might count each successful reception as a contribution to the throughput (see Section IV).

Proposition 1: Under the aforementioned assumptions of symmetry and at most one success per packet, β can be uniquely determined through the quantities in (2) as

$$\beta = \sum_{m=1}^{N_T} m p_S^T(m) f_{M_T}(m) \tag{4a}$$

where

$$p_S^T(m) = \frac{1}{m} \sum_{s=1}^{m} s\, p_{s|m}. \tag{4b}$$

Furthermore, $f_{M_T}(m)$ is a uniquely determined function of the set of parameters $\{p_{s|m}\}$. For the sake of continuity, we give the proof of only (4) here; the statement regarding $f_{M_T}(m)$ is documented in Appendix A.

[3] It can be argued that a variety of interesting models, such as independent or linearly dependent M_T and M_R, can be incorporated here.

Proof of Proposition 1: Through the chain rule of expectations,

$$\beta = E\{S\} = E_{M_T}\{E\{S|M_T\}\}$$
$$= \sum_{m=1}^{N_T} E\{S|M_T = m\} f_{M_T}(m)$$
$$= \sum_{m=1}^{N_T} \left(\sum_{s=1}^{m} s\, p_{s|m}\right) f_{M_T}(m).$$

Note that the limits of the inner summation are from 1 to m, according to the assumption that there can be no more "successes" than packets; a more liberal interpretation is discussed later.

We shall now show (4b). Let S_m^s be the set of $s \leq m$ successful transmitters out of m packets attempting in a slot. The probability that a particular transmitter T_i belongs to S_m^s is

$$\text{Prob}\left\{T_i \in S_m^s \,\middle|\, \begin{matrix} m \text{ attempts,} \\ s \text{ successes} \end{matrix}\right\} = \frac{\binom{m-1}{s-1}}{\binom{m}{s}} = \frac{s}{m}$$

where the symmetry condition has been taken into account. Thus, from the above and the definition of $p_S^T(m)$,

$$p_S^T(m) = \sum_{s=1}^{m} \Pr\{T_i \in S_m^s | m \text{ attempts, } s \text{ successes}\}\, p_{s|m}$$
$$= \sum_{s=1}^{m} \frac{s}{m} p_{s|m} = \frac{1}{m} \sum_{s=1}^{m} s\, p_{s|m}.$$

Certain comments can now be made.

Comment 1: The above proof incorporates all statistical dependence among users. Thus, it accounts for both the multiuser "noise-like" interference effects during a packet as well as the topological effect of a transmitter's attempt to secure attention from any potential receiver. Removing the latter topological uncertainty (and setting $M_R = M_T$) results in the special case of independent (or paired-off) users, as examined in [31]. Only then is it true that

$$p_{s|m} = \Delta(s, m, p_S^T(m)) = \Delta(s, m, p_A^R(M)) \tag{5}$$

in which case a very specific form of symmetry is trivially satisfied and Proposition 1 follows easily. Note that the paired-off model is, from a topological viewpoint, a rather stringent one, for which the condition $M_T = M_R$ is necessary but not sufficient.

Comment 2: In certain cases, it is easier to work directly with $p_{s|m}$ or $p_S^T(m)$, while in others, the determination of $p_A^R(m)$ is preferable; examples of both will be given later.

Comment 3: It was stated in Proposition 1 that $f_{M_T}(m)$ is a function of $\{p_{s|m}\}$. This makes intuitive sense since the composite traffic should depend on the rate at which retransmissions must be made, which is itself a function of the channel-success rate. However, as shown in Appendix A, $f_{M_T}(m)$ is *independent* of $\{p_{s|m}\}$ whenever the probability of a new transmission is equal to the probability of a retransmission ($p_0 = p_r$). For this special case, we can state a throughput expression in a more general form:

$$\beta = \sum_{m=1}^{N_T} \left(\sum_s s\, p_{s|m} \right) f_{M_T}(m) \tag{6}$$

where the limits of the inner summation are unspecified, implying that "success" could be interpreted in a more general sense (i.e., more than one success per packet); the following examples should illustrate the concept. Note, however, that (4b) would *not* be meaningful in this case since the transmitter's and topology's notions of success differ.

Comment 4: One can *define* the probability of success per transmitted packet as $x \triangleq \beta / y$ where y is the average offered traffic [see (13)]. However, the determination of this probability necessarily requires evaluating β as per Proposition 1.

IV. Interpretations of Success and Determination of $p_{s|m}$

In order to proceed and evaluate the probability of s successes given m active transmitters, $p_{s|m}$, we must specify the configuration and what counts as a successful transmission. As a simple case, we consider the following.

Example 1: This is the case of N_T transmitters addressing a single receiver. Using the previous definitions, it follows that

$$p_{s|m} = \begin{cases} 1 - p_A^R(m) & s = 0 \\ p_A^R(m) & s = 1 \\ 0 & s \geq 2, \end{cases} \tag{7a}$$

in which case $\Sigma_{s=1}^m s p_{s|m} = p_A^R(m)$, and by Proposition 1,

$$p_S^T(m) = \frac{p_A^R(m)}{m} \tag{7b}$$

so that

$$\beta = \sum_{m=1}^{N_T} p_A^R(m) f_{M_T}(m). \tag{7c}$$

In view of the symmetry conditions, (7b) simply states that a particular transmitter has probability $1/m$ of being the "successful" one. Here, transmitter and channel (or topology) success are identical notions. Note that the standard ALOHA throughput [16] is derived from (7c) by setting $p_A^R(1) = 1$; $p_A^R(m) = 0$ for $m \geq 2$. Similarly, we can model the scenario where N_T transmitters compete for two *independently* operating receivers. "Success" can be defined here such that a packet is successful whenever it is accepted by one of the two receivers; however, if the *same* packet is accepted by both receivers, it only counts as one success (at most one success per packet). We can calculate the relevant $p_{s|m}$ as

$$p_{1|m} = 2p_A^R(m)\,(1 - p_A^R(m)) + \frac{1}{m}\,(p_A^R(m))^2 \tag{8a}$$

$$p_{2|m} = \left(\frac{m-1}{m}\right)(p_A^R(m))^2, \tag{8b}$$

from which

$$\sum_{s=1}^{2} s p_{s|m} = 2p_A^R(m) - \frac{(p_A^R(m))^2}{m} \tag{8c}$$

and the associated throughput follows from (4) and (8c).

The above procedure could be extended to $N_R \geq 3$ through the evaluation of $p_{s|m}$ and the use of (4); however, such a procedure would become increasingly cumbersome. Instead, by employing the notion of $p_S^T(m)$ and assuming independently operating receivers, a rather simple result is derived below.

Proposition 2: If N_T transmitters compete for N_R *independently operating* receivers and success per packet is at most one, then

$$p_S^T(m) = 1 - \left(1 - \frac{p_A^R(m)}{m}\right)^{N_R} \tag{9a}$$

so that

$$\beta(N_T, N_R) = \sum_{m=1}^{N_T} m \left[1 - \left(1 - \frac{p_A^R(m)}{m}\right)^{N_R} \right] f_{M_T}(m). \tag{9b}$$

Proof: By the definition of success given above, a packet is successful if at least one receiver accepts it, thus contributing one success count to the throughput. Therefore, from a "typical" transmitter's TR_i viewpoint,

$$\begin{aligned} p_S^T(m) &= 1 - \text{Prob}\{\text{TR}_i\text{'s packet not accepted}\} \\ &= 1 - \prod_{j=1}^{N_R} \text{Prob}\{\text{RCVR}_j \text{ does not accept TR}_i\text{'s packet}\} \\ &= 1 - \prod_{j=1}^{N_R} [1 - \text{Prob}\{\text{RCVR}_j \text{ accepts TR}_i\text{'s packet}\}] \\ &= 1 - \left[1 - \frac{p_A^R(m)}{m}\right]^{N_R} \end{aligned}$$

i.e., (9a), where the symmetry condition has been used from both the TR_i's and RCVR_j's viewpoint. (As a check, (7) and (8) can be verified against Proposition 2.)

As an alternative to Proposition 2, consider the case where the number of successes equals the number of accepted packets, *regardless of their source*. In other words, a packet originating from a single transmitter and simultaneously accepted by N_R receivers would contribute $s = N_R$ successes to the throughput. This definition would be useful whenever a routing protocol required multiple copies of the same packet to propagate in different directions, either because of a desired distribution of information (say, from a control center to adjacent nodes) or because this would increase the reliability of final delivery at a remote destination (multihop routing). Then, the appropriate expression for β is (6) where $p_{s|m}$ is best evaluated from the receiver's viewpoint. Indeed, if N_T transmitters compete for N_R independent receivers in this latter fashion, symmetry yields the binomial expression

$$p_{s|m} = \Delta(s, N_R, p_A^R(m)) \tag{10a}$$

which results in

$$\sum_s s p_{s|m} = N_R \, p_A^R(m) \tag{10b}$$

and

$$\beta(N_T, N_R) = N_R \sum_{m=1}^{N_T} p_A^R(m) f_{M_T}(m) \tag{10c}$$

(multisuccess/packet).

Note that (10c) is distinctly different from (9b) since the topological notions of success are quite different in the two cases. Furthermore, by comparison we conclude that

$$\left[\begin{array}{c}\beta(N_T, N_R)\\ \text{(multisuccess/packet)}\end{array}\right] = N_R \cdot \left[\begin{array}{c}\beta(N_T, 1)\\ \text{(single-success/packet)}\end{array}\right]. \tag{11}$$

Thus, in the multisuccess/packet case, throughput increases linearly with the number of available receivers N_R, while it is a more complicated function of N_R in the single-success/packet case. For the latter, we note from (9b) that

$$\beta(N_T, 1) = \sum_{m=1}^{N_T} p_A^R(m) f_M(m) \tag{12}$$

while

$$\lim_{N_R \to \infty} \beta(N_T, N_R) = \sum_{m=1}^{N_T} m f_{M_T}(m) \triangleq y \tag{13}$$

where y is, by definition, the average offered traffic in packets per slot. Furthermore, the derivative $d\beta/dN_R < 0$, which implies that $\beta(N_T, N_R)$ is monotonically increasing to its asymptotic limit y. The interpretation of (13) is that, as $N_R \to \infty$, there are "enough" receivers to successfully absorb all offered traffic, despite any arbitrarily large percentage of failures (thus, independently of $p_A^R(m)$ as long as it is not exactly zero).

We have seen so far that the assumption of independently operating receivers allows the derivation of performance expressions that are rather compact and easy to handle. A wide variety of scenarios supports this assumption quite well, either through strong topological arguments (paired-off users, independent thermal noise and/or jamming per receiver, etc.) or as a convenient analytical approximation whenever the presence of many users with random spreading codes and appropriate receiver processing (coding, interleaving, etc.; see Section V) serve to sufficiently "randomize" interference among receivers.[4] On the other hand, there exist cases where receiver independence is patently not arguable and its use would cause misleading results. It should be stressed that the present analytical framework is quite functional and valid even then, except that the burden of evaluating $p_{s|m}$ shifts from the receiver's to the transmitter's viewpoint. This is best illustrated by means of an example.

Example 2: Consider two transmitters addressing two receivers in a paired-off fashion. Spreading is achieved by means of random frequency hopping patterns and no channel coding is employed, so that a single collision ("hit"), i.e., simultaneous occupancy of a frequency bin by the two transmitters results in the most severe destruction (total loss of packet). Clearly, a collision-free packet received by $RCVR_1$ immediately implies the same for $RCVR_2$, i.e., total statistical dependence. For simplicity, let us assume no other form of interference other than collisions (generalizations to independent noise/jamming per receiver are immediate). Also, let Pr [no hit] denote the probability that the two random (or pseudorandom) patterns do not collide at all within the packet. This probability can be evaluated given the specifics of the spreading system (see [13], for example). It is then straightforward to evaluate $p_{s|m}$ and, via Proposition 1, $p_S^T(m)$ as

$$p_S^T(m) = \begin{cases} 0 & m = 0 \\ 1 & m = 1 \\ \Pr[\text{no hit}] & m = 2 \end{cases} \quad \text{(paired-off)}$$

without invoking any receiver arguments. Furthermore, if the topology is changed to competitive (common code, say) with at most one success per packet, then a simple enumeration yields

$$p_S^T(m) = \begin{cases} 0 & m = 0 \\ 1 & m = 1 \\ \frac{3}{4}\Pr[\text{no hit}] & m = 2 \end{cases} \quad \text{(competitive)}$$

which clearly depicts the impact of topology as well as of the computational viewpoint (transmitter or receiver).

Not much more can be said about throughput unless the various probabilities involved are specified somewhat further. This is pursued in the following section.

[4]See supportive evidence in, for instance, [17] for a related numerical comparison between exact and approximate models for a specific scenario.

V. Factorization of $p_A^R(m)$ and the Capture Effect

An important factor needed to calculate $\{p_{s|m}\}$ (which uniquely specify performance as explained above) is the probability that an intended receiver will accept the packet, $p_A^R(m)$. In [31], $p_A^R(m)$ was identified with $p_c(m)$, i.e., the probability that the packet is correctly decoded given m users in the slot, while in [8] it was identified with the probability that the receiver captures the first packet out of m contending ones under a random time-of-arrival assumption. Thus, in [31], capture was assumed to occur with probability 1, whereas in [8] it was assumed that the packet is always decoded correctly once captured. In reality, both of these events occur in a nonperfect manner and should be accounted for in the factorization of $p_A^R(m)$. Such a factorization implies that there exists at least one packet with the correct code for the intended receiver, which possibly competes against other packets during the synchronization (capture) phase and, once captured, it must be correctly decoded against noise, jamming, and other packets then appearing as like-user interference. Thus, considerations of topology, traffic, etc., do not enter into the calculation of $p_A^R(m)$ since they have already been taken into account in the determination of $p_{s|m}$.

If we let $p_{\text{capture}}^R(m)$ denote the conditional capture probability and $p_{\text{retain/capture}}^R(m)$ denote the conditional probability that the receiver will *retain* the packet after decoding (typically associated with the event that no remaining errors are detected after error correction), given m users in the slot, we can write p_A^R as

$$p_A^R(m) = p_{\text{capture}}^R(m)\, p_{\text{retain/capture}}^R(m). \qquad (14)$$

In the following two sections, we investigate the impact of these two parameters in more detail.

A. Capture Models

Capture occurs because of 1) random arrival delays (delay-capture), which is improved by using randomization of the start-transmit time [8], and 2) different receive powers (power-capture), which depends on node location and the propagation laws [21], [22]. In the literature, it is common to assume that only the first packet to be received is eligible for capture, and for the purposes of this paper, we shall also make this assumption. This corresponds to a system that, upon detecting an acquisition sequence at the beginning of a packet, switches to a prolonged acquisition/tracking mode and is therefore not able to lock onto any later arriving packets even if the first is not acquired (because of, for example, other packets arriving before the tracking and gain-control loops have locked onto the first packet).

For illustrative purposes, let us first assume perfect retention and consider some simple capture models, allowing us to investigate system performance as a function of N_R via (9b). We model the capture probability p_{capture}^R and hence the acceptance probability p_A^R by: 1) the purely exponential model

$$p_A^R(m) = C^m \qquad m \ge 1 \qquad (15a)$$

and 2) the semiexponential model

$$p_A^R(m) = \begin{cases} 1 & m = 1 \\ C^m & m \ge 2 \end{cases} \qquad (15b)$$

where C is an appropriate constant. Reference [8] proved and adopted the model of (15b), while (15a) can be thought of as a mathematically convenient approximation to (15b). Furthermore, the standard Poisson composite traffic is assumed, i.e.,

$$f_{M_T}(m) = e^{-y}\frac{y^m}{m!} \qquad (16)$$

where y is the average offered traffic of (13). From (12), (14), and (16), it follows that

$$\beta(\infty, 1) = e^{-y}\sum_{m=1}^{\infty}\frac{(Cy)^m}{m!} = e^{-y}[e^{Cy} - 1] \qquad (17)$$

($N_T = \infty$ is implicit in the Poisson model). Equation (17) along with (13) constitutes the boundary values of

$$\beta(\infty, N_R) = ye^{-y}\sum_{m=1}^{\infty}\frac{y^{m-1}}{(m-1)!}\left[1 - \left(1 - \frac{C^m}{m}\right)^{N_R}\right]. \qquad (18)$$

In Figs. 1–5, we investigate the behavior of this equation as we experiment with y, C, and N_R. In Fig. 1(a), we

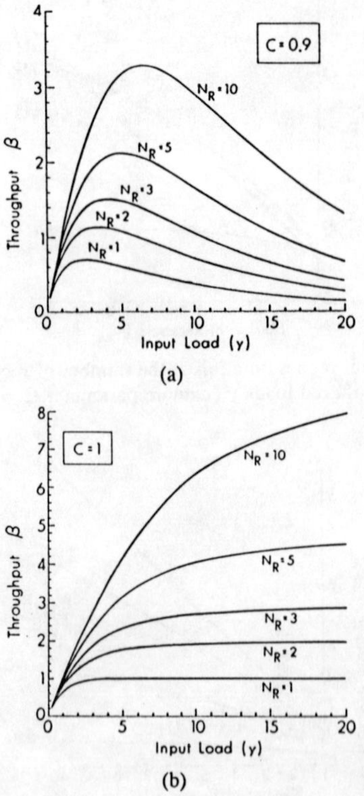

Fig. 1. Throughput β as a function of the Poisson offered traffic y for different number of receivers N_R. (a) Capture probability $C = 0.9$. (b) Capture probability $C = 1.0$.

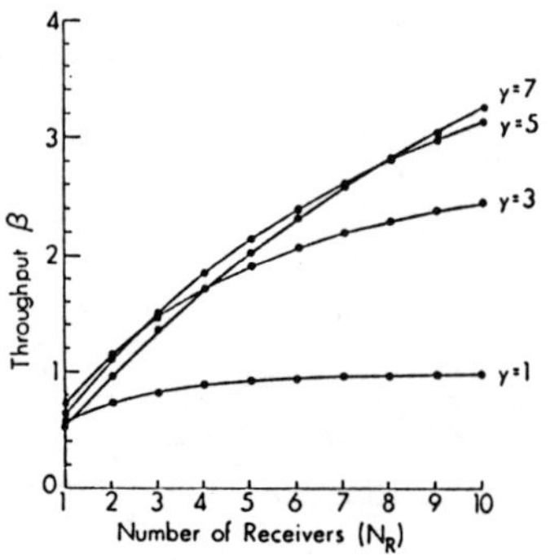

Fig. 2. Throughput β as a function of the number of receivers N_R for different offered loads y (capture parameter $C = 0.9$).

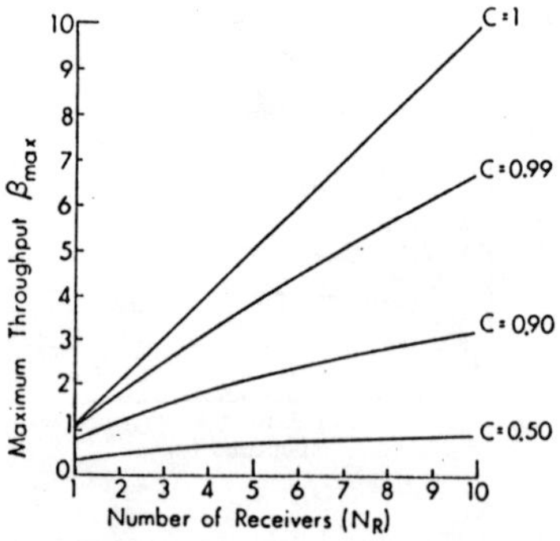

Fig. 3. Maximum throughput $\beta_{\max}$ as a function of the number of receivers N_R for different capture parameters.

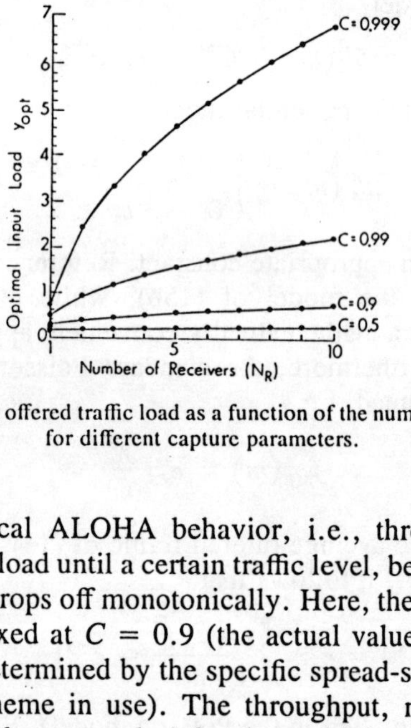

Fig. 4. Optimal offered traffic load as a function of the number of receivers for different capture parameters.

see the typical ALOHA behavior, i.e., throughput increases with load until a certain traffic level, beyond which throughput drops off monotonically. Here, the capture parameter is fixed at $C = 0.9$ (the actual value of this parameter is determined by the specific spread-spectrum acquisition scheme in use). The throughput, measured in packets per slot, exceeds 1 for the multiple receiver case due to the increased bandwidth and multiuser capability of the spread-spectrum system. Let us note that the results are presented in terms of throughput (as opposed to normalized throughput) since the connection between C and the spreading ratio cannot be assessed unless a particular acquisition/capture scheme is invoked. When interpreting and comparing results with the different values of C, care should be exercised.

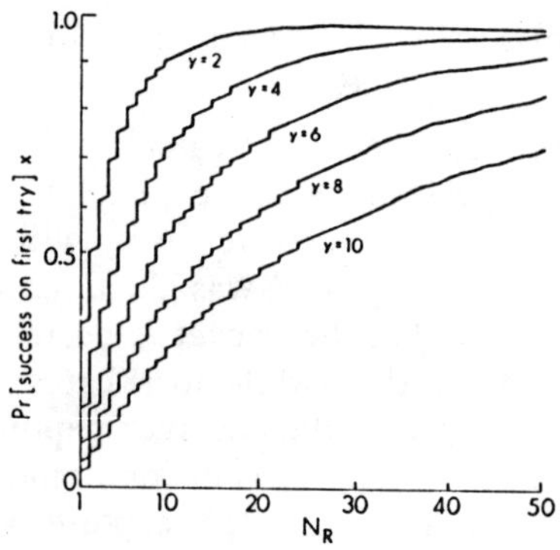

Fig. 5. Probability of success on first try x as a function of the number of receivers N_R for different traffic loads.

In Fig. 1(b), we show a similar set of curves for a capture probability of $C = 1$. We note that for this (ideal) case there is no reduction in throughput as the load is increased since one of the packets will always be captured. Asymptotically, as $y \to \infty$, the throughput tends to the number of available receivers since the probability that two receivers happen to capture the same packet goes to zero for high load (this can also be seen in the $C = 1$ curve of Fig. 3). Fig. 2 shows how the throughput is affected by the number of receivers for a fixed loading factor. We note that the optimal load (to maximize throughput) differs, depending on the number of receivers in use. Figs. 3 and 4 show the maximum throughput and the associated optimal load as a function of the number of receivers for different capture parameters. As noted above, the curve for $C = 1$ in Fig. 3 is a straight line and corresponds to an infinite load. Another parameter of interest is the probability of succeeding on the first attempt (important for delay-constrained systems) as defined in Comment 4, which is plotted in Fig. 5. Note that, for Poisson traffic, x is the same for all packets (backlogged or new).

Returning to the semiexponential model (15b), we can study the star topology (single receiver) of [8] under the Poisson model (16) by evaluating (12):

$$\beta(\infty, 1) = f_{M_T}(1) + \sum_{m=2}^{\infty} C^m f_{M_T}(m)$$
$$= (1 - C) y e^{-y} - e^{-y} + e^{(C-1)y},$$

or letting

$$Q = 1 - C, \qquad (19)$$

we conclude that

$$\beta = (Qy - 1) e^{-y} + e^{-Qy}. \qquad (20)$$

The performance dictated by this equation is shown in Fig. 6. The curve for $C = 0$ corresponds to the usual (nonspread) ALOHA model.

For different values of Q (corresponding to the particular receiver and acquisition models), (20) can be optimized with respect to y. Setting $d\beta/dy = 0$, we find that y_{opt} is the solution of

$$1 + Q^{-1} - y_{\text{opt}} = e^{(1-Q) y_{\text{opt}}}. \qquad (21a)$$

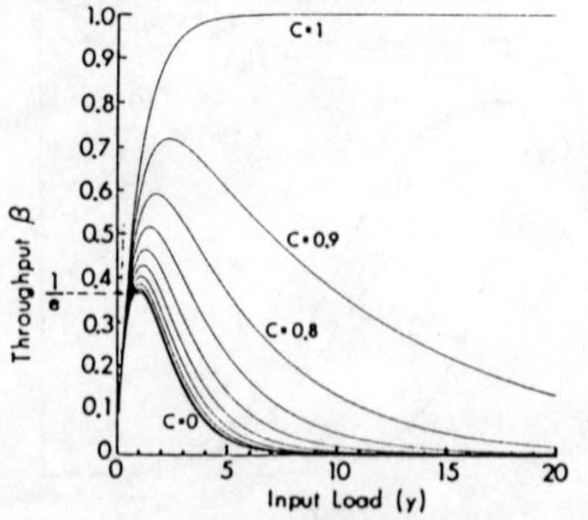

Fig. 6. Throughput β versus load y for the semiexponential model and a star topology with a single receiver for different values of the capture parameter C.

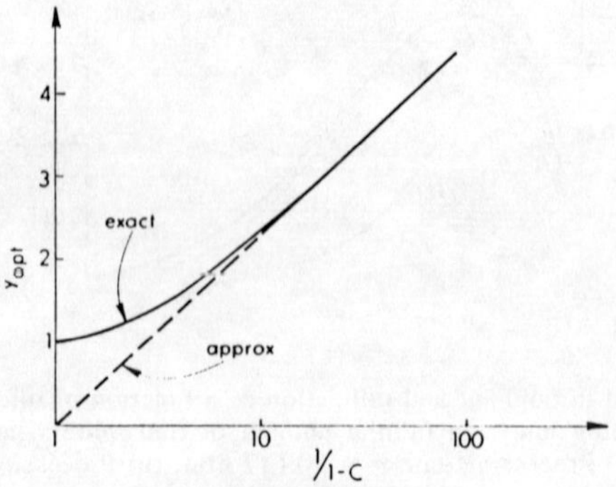

Fig. 7. Exact and approximate optimal traffic load y_{opt} for the semiexponential model.

Clearly, y_{opt} must satisfy $0 < y_{opt} < 1 + Q^{-1}$, but no closed-form solution exists. Under the assumption that $Q \approx 0$, which implies $C \approx 1$ (strong capture affect), (21a) can be approximated by

$$e^{y_{opt}} + y_{opt} \approx Q^{-1}, \tag{21b}$$

which is easier to solve numerically. In fact, for very small Q, y_{opt} is asymptotically close to

$$y_{opt} \approx -\ln Q \tag{21c}$$

as shown in Fig. 7. We now turn our attention to the retention mechanism once capture has been assured.

B. Retention Models

In both [8] and [31], it is assumed that a perfect error detection capability is available and that this capability consumes very little channel capacity (almost zero overhead). In certain applications, such as packet voice, packets containing a few bit errors might be acceptable, however. For now, we concentrate on $p_{correct}(m)$, which is the probability that no detectable errors are present after error correction. This depends on the error-correction mechanism, the spreading format, the forward-error-correction (FEC) rate r, etc. A good FEC scheme might use concatenated codes [15], with a combination of convolutional encoding and Viterbi or sequential decoding for the inner or (channel) code, and of Reed–Solomon or BCH encoding and algebraic decoding for the outer (or packet) code. In other words, the inner code attempts to lower the BER from the channel, while the outer code attempts to correct as many errors as possible on a packet basis. Depending on the particular spreading format, interpacket or intrapacket interleaving might be required in order to randomize the error locations from the channel and the packet decoding, respectively. This is particularly true for DS multiple-access schemes with slotted operation since the channel errors occur in bursts [30]. On the other hand, spreading formats with inherent frequency or time diversity, such as fast FH or TH, offer protection against bursty channel behavior (neglecting bursty jamming activity). However, even if the channel noise (multiple-access or other) is truly random at the input to the channel decoder, the output bits (which are fed into the packet decoder) do suffer from bursty error behavior (as, for instance, in Viterbi decoding of convolutional codes). If the average length of such bursts is larger than the correcting capability of the outer code, interpacket interleaving will be necessary. This, of course, assumes buffering capability at the transmitter and receiver. If no buffers are present, such interleaving is not possible, which should be taken into account when selecting the spreading format.

Assuming an e-error-correcting outer code with either interleaving or other diversity that renders independent bit errors at the decoder input, we can employ the standard conservative bound [25]:

$$p_{correct}(m) \geq \sum_{j=0}^{e} \binom{L}{j} p_{b,in}^{j}(m)\left(1 - p_{b,in}(m)\right)^{L-j} \tag{23}$$

where L is the packet (codeword for block coding) length, e is the number of correctable errors, and $p_{b,in}(m)$ is the bit error probability at the input to the decoder. For any particular class of codes, such as Reed–Solomon, more precise expressions can be derived based on the symbol error probability [4].

In certain schemes, the despreading and inner decoding parts are performed simultaneously; an example is the tree-coded multiple-access scheme using orthogonal convolutional codes [5]. Then, $p_{b,in}(m)$ is the output BER of such a decoder in the presence of m users. As another example, we can consider multiple frequency shift keying frequency hopping (MFSK/FH) with Reed–Solomon codes using M-ary symbols. We can conceptualize the inner code as the grouping of the input bits to M-ary symbols before they are fed into the Reed–Solomon encoder.[5]

It follows from the above that the modulation/spreading combination and channel conditions manifest themselves via $p_{b,in}(m)$. This is a function of the channel or raw $E_b/N_{0,eq}$ where E_b is the energy per bit and $N_{0,eq}$ is the "equivalent" noise power spectral density, accounting both for thermal noise and multiuser interference. For instance, for a slotted DS system with m users of equal

[5]In this case, however, the above bound (23) will be of no use since the word (packet) error probability is directly expressible in terms of the M-ary channel-symbol error probability, the latter being a function of the multiuser interference (see [13], [27]).

power, [9], [26], [43] provide an expression for $E_b/N_{0,\mathrm{eq}}$ as

$$\frac{E_b}{N_{0,\mathrm{eq}}} = \left[\left(\frac{E_b}{N_{0,\mathrm{th}}}\right)^{-1} + \frac{(m-1)\alpha_{\mathrm{ma}}}{\eta}\right]^{-1} \tag{25}$$

where α_{ma} is a coefficient depending on the cross correlation between the particular multiple-access codes in use, $m - 1$ is the number of interfering users, $E_b/N_{0,\mathrm{th}}$ is the bit signal-to-thermal-noise ratio, and $\eta \triangleq R_c/R_b$ is the spreading ratio (code rate over bit rate). Thus, for a coherent BPSK/DS system with no inner code, $p_{b,\mathrm{in}}(m)$ is simply given by

$$p_{b,\mathrm{in}} = Q\left(\sqrt{\frac{2E_b}{N_{0,\mathrm{eq}}}}\right) \tag{26}$$

where $Q(x)$ is the integral of the Gaussian pdf tail.

In Figs. 8 and 9, we investigate the effect of coding rate on system performance for such a coherent system with hard decisions at the symbol level and BCH coding. The bound (for L = 1024)

$$r \leq 1 - \frac{11e}{1024}$$

is used to determine the coding rates corresponding to a value for e used in (23). In this set of figures it is assumed that the capture probability is $C = 1$, (so that the emphasis is upon the coding effects) and the composite arrival is Poisson (infinite transmitter population). Furthermore, we set $\alpha_{\mathrm{ma}} = 1$.[6] In Fig. 8(a), we plot performance for an SNR of 10 dB, a processing gain of 50 (17 dB), and a single receiver. Performance is measured in scaled throughput as well as utilization. We see that for any particular loading there is an optimal coding rate and that for this range of parameters the overall optimum is for $r = 0.89$. Fig. 8(b) shows a similar set of curves for a smaller processing gain (13 dB as opposed to 17 dB). We observe similar behavior, but note that performance is much lower and drops to zero faster due to the reduced processing gain (the horizontal scales of the two figures are different). As expected, coding is much more important in the lower processing gain case (compare the difference between the $r = 1$ and $r = 0.89$ curves for the two figures.) In Fig. 9, we increase the number of receivers and observe the corresponding throughput increase. Also notice that coding is again important, even though the processing gain is 17 dB, since we must overcome the higher multiuser interference caused by the higher traffic levels needed to load up the multiple receivers.

Figs. 10 and 11 consider the finite population case where the transmission probability of new traffic, p_0, and that of retransmission traffic, p_r, must be specified. The Markov chain for the number of backlogged users is now solved to find system performance, as discussed in Appendix A. In Fig. 10, we study the binomial arrival model

[6]Any other appropriate value of α_{ma} can be consolidated into η [see (25)].

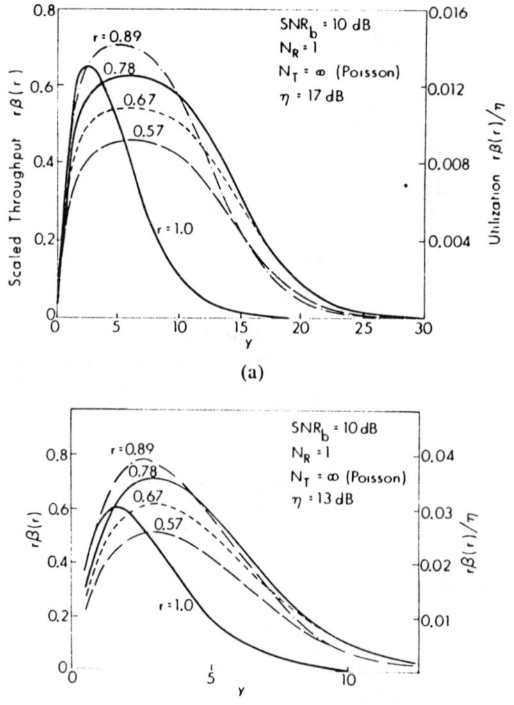

Fig. 8. Scaled throughput and utilization as a function of offered load y, various coding rates, an infinite number of transmitters, and a single receiver. (a) Processing gain $\eta = 50$ (17 dB). (b) Processing gain $\eta = 20$ (13 dB).

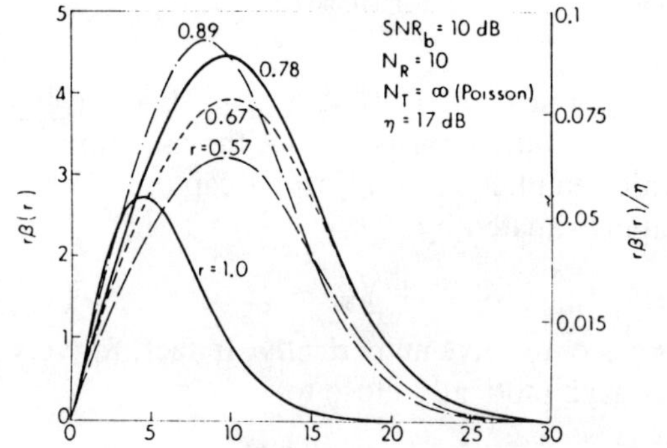

Fig. 9. Scaled throughput and utilization as a function of offered load, various coding rates, an infinite number of transmitters, and multiple receivers ($N_R = 10$) with a processing gain of 50 (17 dB).

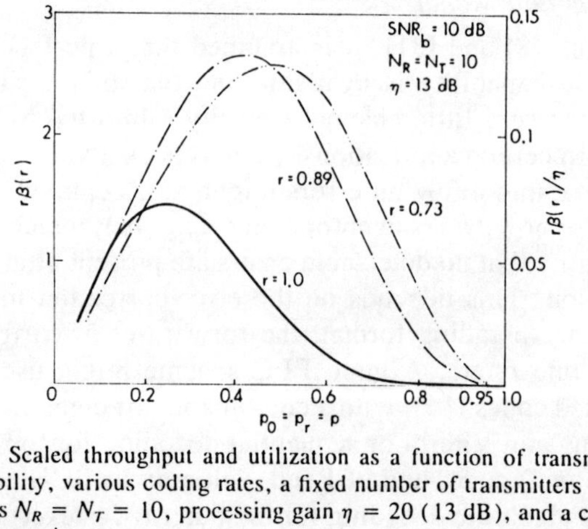

Fig. 10. Scaled throughput and utilization as a function of transmission probability, various coding rates, a fixed number of transmitters and receivers $N_R = N_T = 10$, processing gain $\eta = 20$ (13 dB), and a competitive scenario.

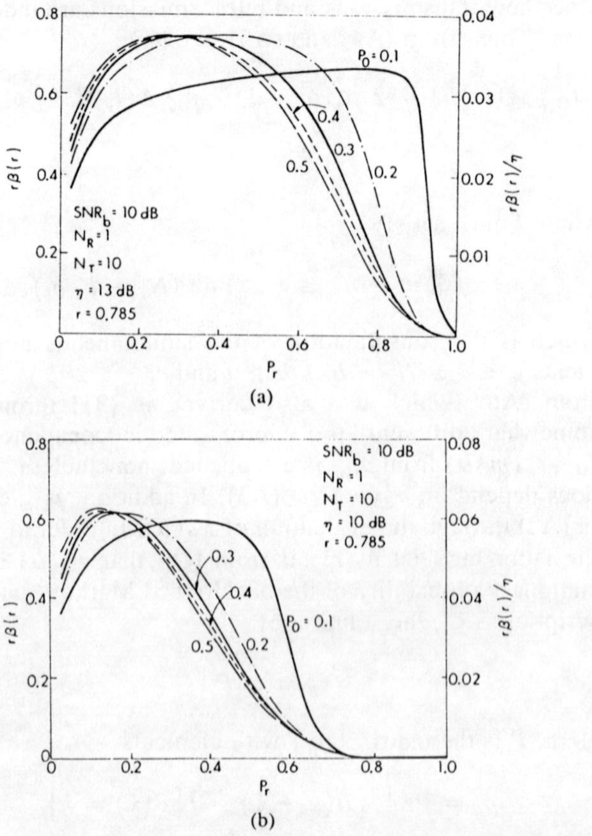

Fig. 11. Scaled throughput and utilization as a function of the retransmission probability, various new transmission probabilities, ten transmitters, and a single receiver. (a) Processing gain $\eta = 20$ (13 dB). (b) Processing gain $\eta = 10$ (10 dB).

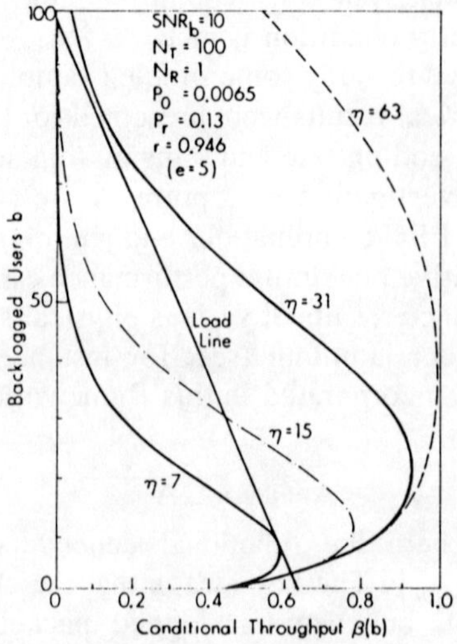

Fig. 12. Throughput versus backlog size; ten transmitters, various processing gains.

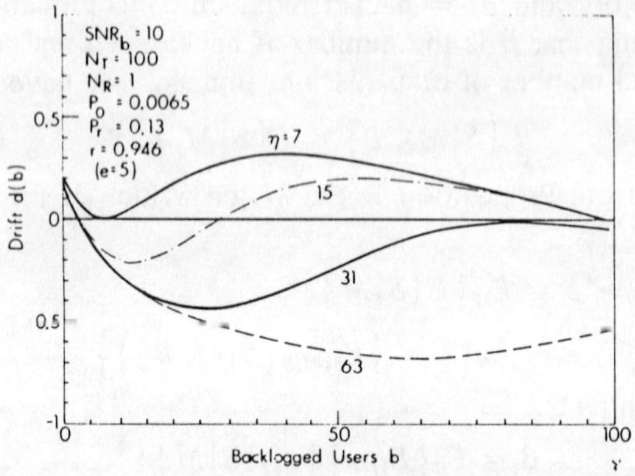

Fig. 13. Drift versus backlog size; ten transmitters, various processing gains.

($p_0 = p_r = p$) for $N_T = N_R = 10$ in a competitive (non-paired-off) scenario. For the set of parameters used here, the optimal coding rate is relatively insensitive to load since we have a high processing gain for the number of users. Fig. 11(a) and (b) investigate the p_0 versus p_r trade-off for ten transmitters and one receiver for a processing gain of 13 dB [Fig. 11(a)] and 10 dB [Fig. 11(b)]. Clearly, selection of the right retransmission rate is more critical for the low processing gain system.

C. Stability Considerations

Instability is a well-recognized problem with narrow-band random access systems, and we can anticipate similar problems to occur for the SSMA case. We investigated the behavior of SSMA systems using an approach based on [16]. In Fig. 12, we present throughput versus backlog size for various processing gains where we have selected the retransmission probability to be large enough that it can potentially create instability. We also plot the "load line" to show the operating points. The behavior is shown more clearly in Fig. 13, which shows the drift (expected increase or decrease in the backlog size) as a function of the backlog. We see that the processing gain is one additional parameter (in addition to the number of users and retransmission probability) that determines the stability of the system. Thus, for low values of the processing gain ($\eta < 7$), we have a single operating point, corresponding to low throughput and high backlog (an overloaded system). For higher values of the processing gain ($\eta > 31$), we again have a single operating point corresponding to high throughput and low backlog. Intermediate values lead to systems with three equilibrium points (two stable and one unstable) for the range of parameters considered. Provided sufficient processing gain is available, the more forgiving nature of SSMA systems (soft degradation) results in a system that has robust performance as a function of the backlog (see the $\eta = 63$ curve).

VI. Conclusions

In this paper, we have presented an analytic framework for the study of random access packet-switched spread-spectrum links under various network topologies and channel conditions. The key feature of the theory is the identification of a set of probabilistic parameters which serve to efficiently summarize the effect of various network considerations on the link performance, such as transmitter–receiver configuration, spread-spectrum code allocation, error correction and detection mechanisms, spreading format, jamming condition, etc., provided a certain symmetry condition is met.

We proceeded to study some simple examples using this theory. Numerical results were presented for the selection of the optimal coding rate and transmission strategies, as well as the investigation of capture. More sophisticated spreading and FEC combinations and jamming conditions require alternative but similar performance estimates. The BER performance results of various physical spread-spectrum links under jamming (see, for instance, [34]) can thus be nicely incorporated in this framework for further study.

Appendix A

We derive here the functional dependence between $f_M(m)$ and $\{p_{s|m}\}$. The key idea is that the channel does not distinguish between backlogged packets and new packets, it only counts the total number of packets in the slot, m. First, however, we define the new and retransmission probabilities: p_0 = new packet transmission probability and p_r = packet retransmission probability. Recalling that B is the number of backlogged and M_T is the total number of transmissions in a slot, we have

$$E\{S|M_T, B\} = E\{S|M_T\}. \tag{A1}$$

Thus, from Proposition 1, (let M_T be written M for simplicity)

$$\beta = E_M\{E\{S|M\}\}$$
$$= E_B\{E_{M|B}\{E_{S|M,B}\{S|M, B\}\}\},$$

or from (A1),

$$\beta = E_B\{E_{M|B}\{E_{S|M}\{S|M\}\}\}$$
$$= \sum_{b=0}^{N_T} \pi_B(b)\,\beta(b) \tag{A2}$$

where $\pi_B(b)$ is the stationary pdf of the "backlog" r.v. B and $\beta(b)$ is the conditional throughput

$$\beta(b) = E_{M|B=b}\{E_{S|M}\{S|M\}\}$$
$$= \sum_{m=0}^{N_T} m p_S^T(m)\, f_{M|B=b}(m|b). \tag{A3}$$

In (A3), $f_{M|B=b}(m|b)$ is the conditional distribution of M given $B = b$. Presently, we show that this conditional pdf does not depend on $p_S^T(m)$ (or $p_{s|m}$). Indeed, from the definition of $F_{M|B=b}(M|b)$, we have

$$f_{M|B=b}(m|b) = \sum_{l=0}^{N_T-b} \text{Prob}\{m \text{ transmissions} \,|\, B = b,\, l \text{ new packets}\}\ \text{Prob}\{l \text{ new packets}\}$$
$$\triangleq \sum_{l=0}^{N_T-b} \text{Prob}\{m \text{ transmissions} \,|\, b, l\} \cdot \Delta(l, N_T - b, p_0).$$

But

$$\text{Prob}\{m \text{ transmissions} \,|\, b, l\} = \text{Prob}\{(m - l) \text{ retransmissions} \,|\, b, l\}$$
$$= \text{Prob}\{(m - 1) \text{ retransmissions} \,|\, b\}$$
$$= \begin{cases} 0 & m < l \text{ or } \{m \ge l \text{ and } m - l > b\} \\ \Delta(m - l, b, p_r) & \text{otherwise} \end{cases} \tag{A5}$$

since new transmissions and retransmissions are independent. Thus, from (A4) and (A5), we have

$$f_{M|B=b}(m|b) = \sum_{l} \Delta(m - l, b, p_r)\,\Delta(l, N_T - b, p_0) \tag{A6}$$

where l must satisfy

$$\max(0, m - b) \le l \le \min(N_T - b, m),$$

which is the consolidation of the simultaneous requirements $0 \le l \le N_T - b$, $\{m \ge l \text{ and } m - l \le b\}$. Thus, from (A6) (which was also derived in [31] through a somewhat different procedure) the independence of $f_{M|B=b}(m|b)$ from $p_{s|m}$ is established; nonetheless, $\beta(b)$ does depend on $p_S^T(m)$ via (A3). In addition, $p_{s|m}$ enters in (A2) through the evaluation of $\pi_B(b)$. In order to show the latter fact, let us recall from [16] that $\pi_B(b)$ is the stationary probability of the backlogged Markov state description, i.e., the solution of

$$\pi_B = \pi_B P. \tag{A7}$$

Here, P is the matrix $\{p_{ij}\}$ with elements

$$p_{ij} = \text{Prob}\{B(k + 1) = j \,|\, B(k) = i\} \tag{A8}$$

where i, j are the numbers of backlogged users in the successive slots $(k, k + 1)$, respectively. Let m_b (out of i) and m_0 (out of $N_T - i$) be the numbers of backlogged and new users who do transmit in the kth slot, respectively, so that $m_b + m_0 = m$ (total transmitting users in kth slot). The number of potential backlogged users in $(k + 1)$st slot is $(i + m_0)$ if none succeeds. This number will be reduced by the number of successes s out of m possible. Thus, at the end of the kth slot, or the beginning of the $(k + 1)$st, we have $j = i + m_0 - s$ backlogged users. Thus, from (A8) and the above discussion,

$$p_{ij} = \sum_{m_b=0}^{i} \sum_{m_0=0}^{N_T-i} \text{Prob}\{B(k + 1) = j \,|\, B(k) = i, M_0 = m_0, M_B = m_b\} \cdot \Delta(m_b, i, p_r)\,\Delta(m_0, N_T - i, p_0) \tag{A9}$$

where

$$\text{Prob}\left\{B(k+1) = j \mid B(k) = i, M_0 = m_0, M_B = m_B\right\}$$
$$= \text{Prob}\left\{s = i + m_0 - j \text{ successes} \mid (m_0 + m_b) \text{ trials}\right\}. \quad \text{(A10)}$$

Since, however, $0 \le s \le m_0 + m_b$, it follows that

$$\Pr\left\{B(k+1) = j \mid B(k) = i, m_0, m_b\right\} = \begin{cases} 0 & \text{if } (j - i) > m_0 \\ & \text{or } (i - j) > m_b \\ p_{i+m_0-j \mid m_0+m_b} & \text{otherwise.} \end{cases} \quad \text{(A11)}$$

It can thus be concluded that

$$p_{ij} = \sum_{m_0 = \max(0, j-i)}^{N_T - i} \Delta(m_0, N_T - i, p_0) \cdot \sum_{m_b = \max(0, i-j)}^{i} \Delta(m_b, i, p_r) p_{i+m_0-j \mid m_0+m_b}, \quad \text{(A12)}$$

which specifies the dependence of P, and thus of π_B, upon the "success" parameters $p_{s|m}$. Let us note that (A12) is a more compact and topologically more general expression than those found in [8], [16], [22], [31] since the only requirements imposed herein are the symmetry conditions. It is useful whenever $p_0 \neq p_r$; indeed, it is a straightforward exercise to show that, when $p_0 = p_r = p$, (A6) reduces to

$$f_{M|B=b}(m \mid b) = f_M(m) = \Delta(m, N_T, p), \quad \text{(A13)}$$

independent of b and the channel parameters $p_{s|m}$. It depends only on p, a communicator's design choice.

References

[1] N. Abramson, "The ALOHA system—Another alternative for computer communications," in *AFIPS Conf. Proc. 1970 Fall Joint Comput. Conf.*, vol. 37, pp. 281–285, 1970.

[2] R. Boorstyn and A. Kershenbaum, "Throughput analysis of multihop packet radio networks," in *Proc. ICC*, Seattle, WA, June 1980.

[3] M.-S. Chen, R. Boorstyn, and A. Kershenbaum, "Throughput analysis of multiple hop packet radio networks using a code division multiple access (CDMA) protocol," in *Proc. Allerton Conf.*, 1984, pp. 20–29.

[4] G. C. Clark and J. B. Cain, *Error Correction Coding for Digital Communication*, New York: Plenum, 1982.

[5] A. Cohen, J. A. Heller, and A. J. Viterbi, "A new coding technique for asynchronous multiple access communications," *IEEE Trans. Commun. Technol.*, vol. COM-19, Oct. 1971.

[6] G. C. Cooper and R. W. Nettleton, "A spread-spectrum technique for high-capacity mobile communications," *IEEE Trans. Vehicular Technol.*, vol. VT-27, pp. 264–275, Nov. 1978.

[7] G. C. Cooper, R. W. Nettleton, and D. P. Grybos, "Cellular land-mobile radio: Why spread spectrum?" *IEEE Commun. Mag.*, vol. 17, pp. 17–24, Mar. 1979.

[8] D. H. Davis and S. A. Gronemeyer, "Performance of slotted ALOHA random access with delay capture and randomized time of arrival," *IEEE Trans. Commun.*, vol. COM-28, pp. 703–710, May 1980.

[9] R. Gagliardi, *Satellite Communications*. New York: Reinhold, 1984, ch. 7.

[10] E. A. Geraniotis and M. B. Pursley "Error probability for direct-sequence spread-spectrum multiple-access communications, Part II: Approximations," *IEEE Trans. Commun.*, Special Issue on Spread Spectrum Communications, vol. COM-30, pp. 985–995, May 1982.

[11] E. A. Geraniotis, Coding for BPSK SFH/SSMA communications over fading channels," in *Proc. MILCOM 82*, Boston, MA, 1982, pp. 35.1.1–35.1.5.

[12] D. J. Goodman, P. S. Henry, and V. K. Prabhu, "Frequency-hopped multilevel FSK for mobile radio," *Bell Syst. Tech. J.*, vol. 59, no. 7, pp. 1257–1275, Sept. 1980.

[13] B. Hajek, "Recursive retransmission control—Application to a frequency hopped spread-spectrum system," in *Proc. 1982 Conf. Inform. Sci. Syst.*, Baltimore, MD, Mar. 1982, pp. 116–120.

[14] J. Y. N. Hui, "Throughput analysis for code-division multiple accessing of the spread spectrum channel," *IEEE J. Select. Areas Commun.*, vol. SAC-2, pp. 482–486, July 1984.

[15] R. Kahn, S. Gronemeyer, J. Burchfiel, and R. Kunzeman, "Advances in packet radio technology," *Proc. IEEE*, vol. 66, pp. 1468–1496, Nov. 1978.

[16] L. Kleinrock and S. Lam, "Packet switching in a multiaccess broadcast channel: Performance evaluation," *IEEE Trans. Commun.*, vol. COM-23, pp. 410–423, Apr. 1975.

[17] T. Ketseoglo, "Spread spectrum random access communications with multiple reception capability," M. S. thesis, Department of Electrical Engineering, University of Maryland, College Park, Dec. 1986.

[18] J. S. Lehnert and M. B. Pursley, "Error probabilities for binary direct sequence spread spectrum communications with random signature sequences," *IEEE Trans. Commun.*, vol. COM-35, pp. 87–98, Jan. 1987.

[19] I. Lee and J. A. Silvester, "An iterative scheme for performance modelling of slotted ALOHA packet radio networks," in *Proc. ICC*, Philadelphia, PA, June 1982, pp. 6G.1–6G.6.

[20] S. S. Lee and J. A. Silvester, "Analysis of the effect of acknowledgement traffic on the performance of spread spectrum packet radio networks," in *Proc. ICC*, Toronto, Ont., Canada June 1986, pp. 57.4.1–57.4.8.

[21] S. L. Musa and W. Wasylkiwskyj, "Co-channel interference of spread spectrum systems in a multiple user environment," *IEEE Trans. Commun.*, vol. COM-26, pp. 1405–1413, Oct. 1978.

[22] C. Namislo, "Analysis of mobile radio slotted ALOHA networks," *IEEE J. Select. Areas Commun.*, vol. SAC-2, pp. 583–588, July 1984.

[23] J. K. Skwirzynski, Ed. "New concepts in multi-user communications," in *Proc. NATO Conf.*, The Netherlands: Sizthoff and Noordhoff, 1981.

[24] A. Polydoros and J. A. Silvester, "An analytical framework for slotted random access spread spectrum networks," in *Proc. MILCOM 85*, Boston, MA, Oct. 1985, pp. 25.5.1–25.5.7.

[25] J. G. Proakis, *Digital Communications*. New York: McGraw-Hill, 1983.

[26] M. B. Pursley, "Performance evaluation for phase-coded spread-spectrum multiple access communications—Part I: System analysis," *IEEE Trans. Commun.*, vol. COM-25, pp. 795–799, Aug. 1977.

[27] ——, "Frequency-hop transmission for satellite packet switching and terrestrial packet radio networks," *IEEE Trans. Inform. Theory*, vol. IT-32, pp. 652–667, Sept. 1986.

[28] M. B. Pursley and D. V. Sarwate, "Performance evaluation for phase coded spread spectrum multiple-access communication, Part II: Code sequence analysis," *IEEE Trans. Commun.*, vol. COM-25, pp. 800–803, Aug. 1977.

[29] M. B. Pursley, D. V. Sarwate, and W. E. Stark, "Error probability for direct sequence spread spectrum multiple access communications—Part I: Upper and lower bounds," *IEEE Trans. Commun.*, vol. COM-30, May 1982.

[30] M. B. Pursley and D. J. Taipale, "Error probabilities for spread-spectrum packet radio with convolutional codes and Viterbi decoding," *IEEE Trans. Commun.*, vol. COM-35, pp. 1–12, Jan. 1987.

[31] D. Raychaudhuri, "Performance analysis of random access packet switched code division multiple access systems," *IEEE Trans. Commun.*, vol. COM-29, pp. 895–901, June 1981.

[32] A. R. K. Sastry, "Effect of acknowledgment traffic on the performance of slotted ALOHA/CDMA systems," *IEEE Trans. Commun.*, vol. COM-22, pp. 1219–1222, Nov. 1984.

[33] J. A. Silvester and I. Lee, "Throughput/delay performance of multihop slotted ALOHA networks," Commun. Sci. Inst., Dep. Elec. Eng., Univ. Southern Calif., Los Angles, CA, CSI Rep. 83-09-01 (revised), June 1984.

[34] M. K. Simon, J. K. Omura, R. A. Scholtz, and B. K. Levitt, *Spread Spectrum Communications*. Rockville, MD: Computer Science, 1985.

[35] E. S. Sousa and J. A. Silvester, "A spreading code protocol for a

distributed spread spectrum packet radio network," in *Proc. GLOBECOM 84*, Atlanta, GA, Dec. 1984.
[36] —, "A code switching technique for distributed spread spectrum packet radio networks," in *Proc. Int. Conf. Commun.*, Chicago, IL, June 1985.
[37] S. L. Su and V. O. K. Li, "An iterative model to analyze multi-hop packet radio networks," in *Proc. Allerton Conf.*, Oct. 1985, pp. 545–554.
[38] F. A. Tobagi and J. S. Storey, "Improvements in throughput of a CDMA packet radio network due to a channel load sense access protocol," in *Proc. Allerton Conf.*, 1984, pp. 40–49.
[39] —, "Throughput performance of a direct sequence CDMA packet radio network," Comput. Syst. Lab., Stanford Univ., Stanford, CA, SEL Tech. Rep. 85-277, June 1985.
[40] A. J. Viterbi, "A processing satellite transponder for multiple access by low-rate mobile users," in *Proc. Digital Satellite Commun. Conf.*, Montreal, P.Q., Canada, Oct. 1978, pp. 166–174.
[41] —, "When not to spread spectrum—A sequel," *IEEE Commun. Mag.*, vol. 23, pp. 12–17, Apr. 1985.
[42] C. L. Weber, G. K. Huth, and B. H. Batson, "Performance considerations of frequency-hopping multiple access," in *Proc. Nat. Telecommun. Conf.*, Seattle, WA, Dec. 1980, pp. 69.4.1–69.4.5.
[43] —, "Performance considerations of code division multiple access systems," *IEEE Trans. Vehicular Technol.*, vol. VT-30, Feb. 1981.
[44] K. Yao, "Error probability of asynchronous spread spectrum multiple access communications systems," *IEEE Trans. Commun.*, vol. COM-25, pp. 803–809, Aug. 1977.

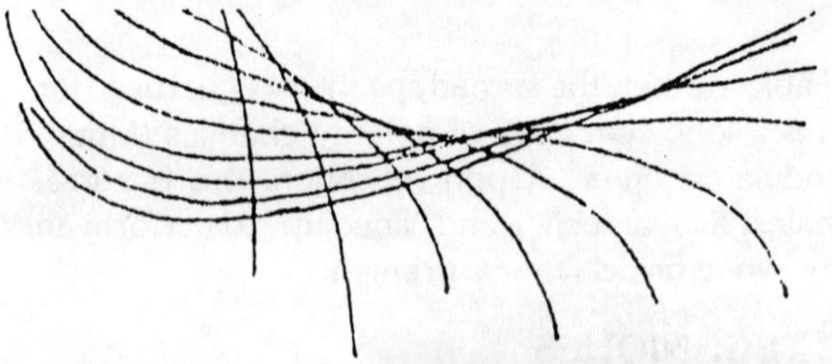

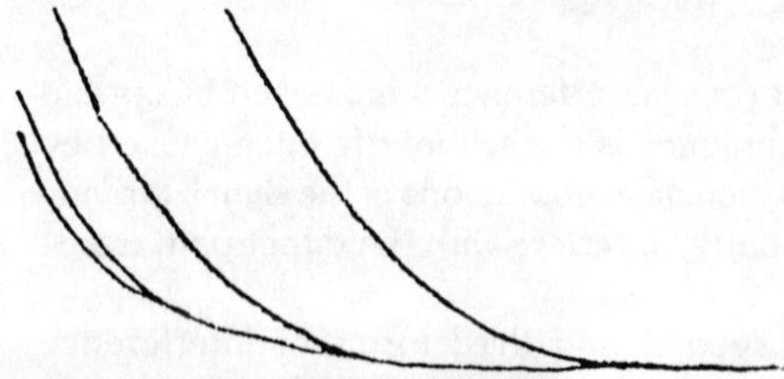

SPREAD SPECTRUM COMMUNICATIONS— MYTHS AND REALITIES

Andrew J. Viterbi

Coding is always beneficial and sometimes crucial for the suppression of interference in spread spectrum communications.

INTRODUCTION

Spread spectrum communication techniques date back to the early fifties. Since the earliest applications, system improvements have been more evolutionary than revolutionary. Like most improvements in electronic systems, these are due primarily to the availability of ever higher speed integrated circuit components, which translate in this case to wider spread spectra. In three decades the achievable spreading factor has grown by about three orders of magnitude[1] to the point that we are now limited more by bandwidth allocations than by technology limitations. Before we examine the quantitative effects of spreading, let us catalog briefly the multiple purposes of spread spectrum communications.

First, we note that spreading here refers to expansion of the bandwidth well beyond what is required to transmit digital data. Thus, a system transmitting data at a rate (R) of 100 Mbits/s using approximately 100 MHz of bandwidth (W) is not spread at all, while a system transmitting at 100 bits/s spread over a spectrum of about 100 MHz has a factor $W/R = 10^6$, or 60 dB of so-called *processing gain*.

The author is with the Linkabit Corporation, San Diego, CA 92121.

[1]Which parallels the evolution of data rate capabilities of digital communications.

PURPOSES

The purpose and applicability of spread spectrum techniques is threefold:

- Interference Suppression
- Energy Density Reduction
- Ranging or Time Delay Measurement

Foremost among these is the suppression of interference which may be characterized as any combination of the following:

1) Other Users: intentional (hostile or unintentional),
2) Multiple Access: spectrum sharing by "coordinated" users,
3) Multipath: self-jamming by delayed signal.

Protection against in-band interference is usually called anti-jamming (A/J). This is the single most extensive application of spread spectrum communication. A similar application is that of multiple access by numerous users who share the same spectrum in a coordinated manner, in that each employs signaling characteristics or parameters (often referred to as codes) which are distinguishable from those of all other users. One reason for using this shared spectrum, so-called code-division multiple access (CDMA), is that by distinguishing signals in this way, separation in the more common dimensions of frequency or time is not required, and hence the usual transmission tolerances need not be imposed on these parameters.

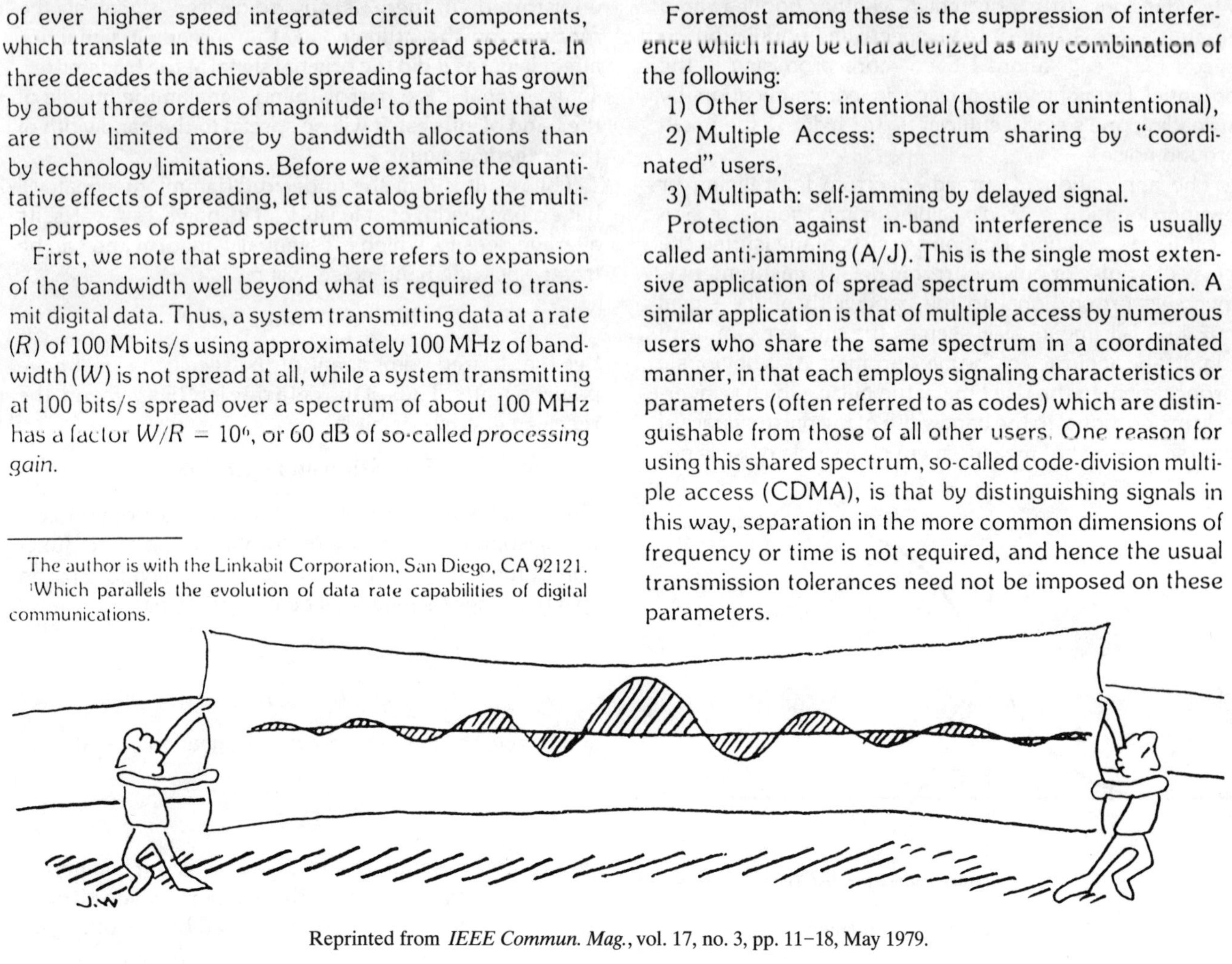

Reprinted from *IEEE Commun. Mag.*, vol. 17, no. 3, pp. 11–18, May 1979.

The third form of interference suppressed by spread spectrum techniques is the self interference caused by multipath in which delayed versions of the signal, arriving via alternate paths, interfere with the direct path transmission.

While the second and third forms of interference would appear more benign than that of a hostile emitter, the technique and effect are the same. What makes the intentional interference more challenging is the game aspect of the problem and the fact that the interfering source is generally granted much more power than the communicator, which is usually not the case for cooperating users and even less so for multipath interference.

The second class of applications centers about the reduction of the energy density of the transmitted signal. This, too, has a threefold purpose:

1) to meet international allocations regulations,
2) to minimize detectability,
3) for privacy.

Downlink transmissions from satellites must meet international regulations on the spectral density of the signals received on earth. By spreading this energy over a wider bandwidth, total transmitted power can be increased, and hence performance improved. Spreading also decreases the detectability of a signal by a regulatory body which employs spectral analysis to monitor or regulate emissions. (It is not known whether bootleg radio amateurs are using spread spectrum modulation to evade FCC regulations.) Even more promising is the potential for achieving privacy in communication by spreading one's signal sufficiently to "hide" in the background noise.

The application of spread spectrum for ranging or position location is rapidly gaining in importance. In simplest terms, position location consists of measuring the delay of a pulse or pulses. Error in delay measurement is inversely proportional to the bandwidth of the signal pulse. This is most easily seen by the simple example of Fig. 1. The accuracy of the measurement Δt is obviously proportional to the rise time of the pulse, which is inversely proportional to the bandwidth of the pulse signal. Of course, a one-shot measurement on a single pulse is not very reliable. Rather, the spread spectrum signal used for ranging is a long sequence of polarity changes (binary PSK-modulated signal). Upon reception, this is correlated against a local replica and "lined up" to perform an accurate range or delay measurement.

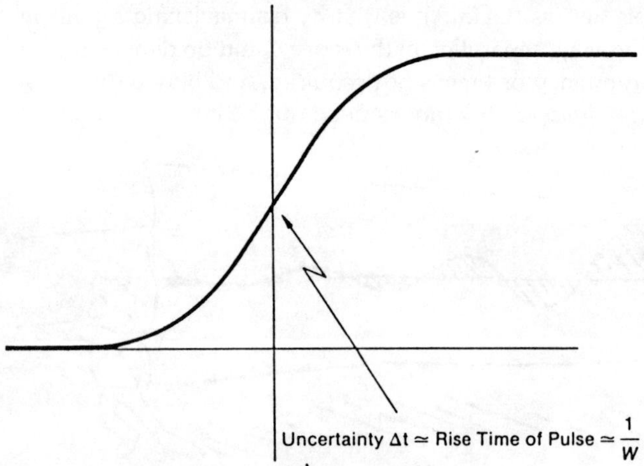

Fig. 1. Time delay measurement.

BASIC TECHNIQUES

Having outlined the multiple uses of spectrum spreading, we must examine at least a superficial description of the concept before we can proceed to dispel myths and uncover realities about this increasingly popular technique. Fig. 2 is an all-purpose diagram to describe spread spectrum modulation. Multiplication of two unrelated signals produces a signal whose spectrum is the *convolution* of the spectra of the two component signals. Thus, if the digital data (binary) signal is relatively narrow-band compared to the spreading signal, the product signal will have nearly the spectrum of the wider (spreading) signal. So much for the modulator. At the demodulator, the received signal is multiplied by exactly the same spreading signal. Now if the spreading signal, locally generated at the receiver, is lined up (synchronized) with the received spread signal, the result is the original signal plus, possibly, some spurious higher frequency components outside the band of the original signal, and hence easily filtered to reproduce the original data essentially undistorted. If there is any undesired signal at the receiver, on the other hand, the spreading signal will affect it just as it did the original signal at the transmitter. Thus, even if it is a narrow-band signal in the middle of the band of interest, it will be spread to the bandwidth of the spreading signal.

The result is that the undesired (jamming) signal will have a bandwidth of at least W. If its power is J watts, its average density, which is essentially uniform and can be treated as wide-band noise, will be

$$N_0 = J/W \text{ watts/Hz.}$$

Let the desired component of the received signal have power S watts. Thus, if the data rate is R bits/second, the received energy per bit is

$$E_b = S/R \text{ watts} \cdot \text{second.}$$

Now it is generally recognized that digital communication system bit error rate performance is a direct function of the dimensionless ratio E_b/N_0, which for spread spectrum signals may thus be expressed as

$$\frac{E_b}{N_0} = \frac{S}{J}\frac{W}{R}$$

and hence, the jamming power-to-signal power ratio is

$$\frac{J}{S} = \frac{W/R}{E_b/N_0}. \quad (1)$$

This establishes that if E_b/N_0 is the *minimum bit energy-to-noise density ratio* needed to support a given bit error rate, *and if* W/R is the ratio of spread bandwidth to the

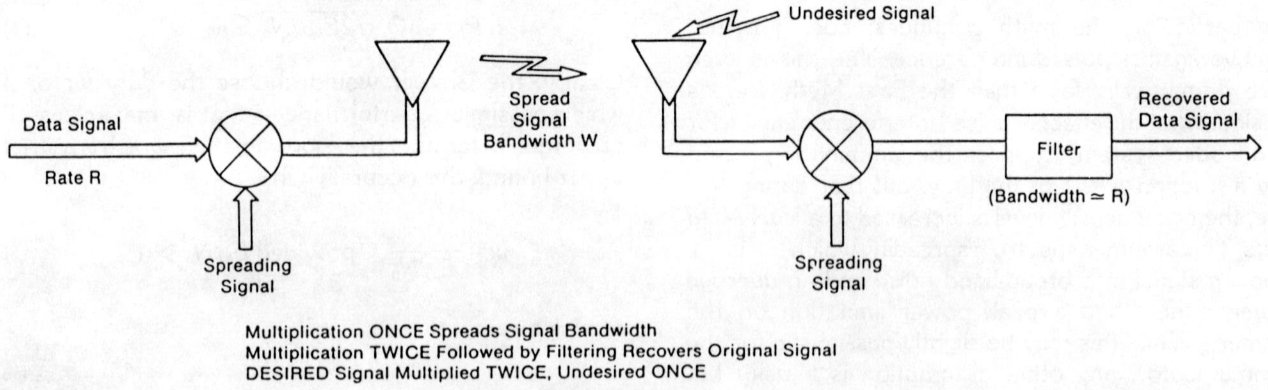

Fig. 2. Basic spread-spectrum techniques.

original data bandwidth, also called the *processing gain*, then J/S is the *maximum tolerable jamming power-to-signal power ratio*, also known as the *jamming margin*.

We have come this far without even specifying the characteristics of the spreading signal. There are, in fact, two distinct classes of spreading techniques. The first is called *direct sequence* or *pseudonoise* (PN) spread spectrum. Here the spreading is achieved by multiplication by a binary pseudorandom sequence whose symbol (switching) rate is many times the binary data bit rate. The spreading sequence symbol rate is sometimes called the *chip rate*.

The second class utilizes a *frequency hopping* carrier. Here the spreading signal remains at a given frequency for each bit or even for several bits. Thus, locally it is no wider than the data signal, but when it hops to a new frequency, it may be anywhere within the "spreading" bandwidth W.

One fundamental difference between the two techniques is that direct sequence PN spread signals can be coherently demodulated. With frequency hopped signals, on the other hand, phase coherence is difficult to maintain when the signal frequency is hopped over a wide range; hence, this modulation is usually demodulated noncoherently.

We are now ready to explore several firmly entrenched items of common wisdom regarding the relative desirability of various features of spread spectrum systems. Often these attitudes are misguided, as we shall presently show. In all cases, the ideas hold for both classes of spread spectrum techniques, but for all but the last concept the arguments are somewhat simpler for direct sequence spreading, which we shall therefore consider.

We are ready now to reveal the first of four myths.

First Myth:

Error-correcting coding requires redundancy, which spreads bandwidth and thus reduces available processing gain for the available bandwidth.

Coding does not reduce the *effective* processing gain in a spread spectrum system.

Reality is, in fact, just the opposite. To see that coding does not reduce the *effective* processing gain, let us rewrite jamming margin (1) in terms of the *symbol*[2] *rate* R_s and the *symbol energy* E_s. These are related to the bit rate and the bit energy through the code rate r, defined as the number of data bits per transmitted symbol, or the inverse of the coding expansion factor. (For example, a rate 1/2 coded system transmits two code symbols for each data bit.) It follows that

$$R_s = R/r \qquad \text{and} \qquad E_s = E_b r.$$

Now if we repeat the previous dimensional argument replacing bits by symbols everywhere, we have

$$J/S = \frac{W/R_s}{E_s/N_0},$$

but substituting the preceding definitions for symbol rate and energy, we obtain for the maximum tolerable J/S ratio

$$J/S = \frac{W}{R/r}\frac{N_0}{E_b r} = \frac{W/R}{E_b/N_0}$$

which gets us back to (1). This may seem like sleight of hand, but it really is not. Moreover, although it will take some further reading to be convinced, we are really ahead of the game. For with coding, the required E_b/N_0 for a given level of performance (bit error rate) is actually *reduced*. Thus, for a given processing gain (W/R) the jamming margin is further *increased* by coding.

For those who are satisfied that spectrum spreading (especially direct sequence PN) techniques make the noise look "white" while the signal energy, without or with coding, can be fully recovered by the receiver's "correlating" multiplier, the dispelling of the First Myth will come as no surprise. Yet it is often this sophisticated group who will fall prey to the

---

Second Myth:

Error-correcting coding is effective only against uniform interference.

---

[2]Symbol rate refers to the *code symbol* of the error-correcting code—not that of the PN spreading code, which is usually called chip rate.

In particular, the myth continues; coding is not effective against *pulsed interference*. Yet, this is even more dramatically false than the First Myth. Let us consider what the effect of pulsed interference can be for an uncoded system. Suppose the jamming is present only a fraction $\rho < 1$ of the time, but that during this time, the noise density level is increased to a level N_0/ρ watts. This assumes spectrum spreading which turns the jamming signal into broad-band noise and an *average power* rather than a peak power limitation on the jammer. (While this may be slightly pessimistic for the communicator, any other assumption is a risky bet against technological progress.) Now it is well known that with coherent demodulation an *uncoded* BPSK modulated system produces a bit error rate P_b related to E_b/N_0 as

$$P_b = Q(\sqrt{2E_b/N_0}) \lesssim e^{-E_b/N_0} \tag{2}$$

where

$$Q(x) = \frac{1}{\sqrt{2\pi}} \int_x^\infty e^{-u^2/2}\, du.$$

But if the noise is intermittent, and hence only with probability ρ corrupts a given transmitted bit[3] with the higher noise density N_0/ρ, the resulting bit error rate becomes

$$P_b = \rho Q\sqrt{2E_b\rho/N_0} \lesssim \rho e^{-\rho E_b/N_0}. \tag{2'}$$

Clearly, the jammer would choose the duty factor ρ which pessimizes performance—that is, maximizes bit error rate. In terms of the approximation, which is a strict upper bound, this occurs when

$$\rho = \frac{1}{E_b/N_0} \quad \text{provided } E_b/N_0 > 1$$

at which value

$$\max_{0<\rho<1} P_b \lesssim \frac{e^{-1}}{E_b/N_0}. \tag{3}$$

(Note that although we worked with the approximation for its simplicity, had we used the exact (Q-error-function) expression, the worst case ρ would be nearly the same and the maximum bit error rate would not be significantly lower.)

The result is quite dramatic. Pulse jamming—with spread spectrum but without coding—changes an exponential relation into an inverse-linear one. Numerically, if we desire bit error rate performance on the order of $P_b \simeq 10^{-5}$, stationary noise (or jamming) requires only $E_b/N_0 \simeq 10$ dB, while with pulse jamming we must have $E_b/N_0 \simeq 45$ dB, an increase in *required signal power of over three orders of magnitude!*

Amazingly, coding can almost fully restore this deplorable situation, but before we can explain why, we must briefly explore a summary of the general capabilities of coded systems.

[3]We assume for simplicity that a given interference pulse corrupts an integral number of bits. This is a reasonable assumption if the pulse width is many times the bit duration. Otherwise, the situation is actually less favorable to the jammer.

Coding Fundamentals

When a binary data stream must be transmitted over a noisy channel with a troublesome bit error rate, coding can be used to significantly reduce the error rate incurred by the message.

In **block coding** schemes, the message bit stream is partitioned into blocks of k bits, where k is the **block length.** Each such message block is replaced with an n bit code word (n is bigger than k) which is transmitted in its place. Thus, every n bit transmitted "contains" only k message bits so that the **rate** r of the code is k/n bits per code symbol.

A common noisy channel model is the so-called **Gaussian channel** where each bit, viewed as a square pulse of amplitude ± 1, is independently subjected to additive noise and an error occurs when the noise alters the pulse polarity.

As a result of errors, the received n-bit block can be any of 2^n possible words. Since there are only 2^k different code words that could have been transmitted (one for each k-bit message block) and 2^k is typically much less than 2^n, the number of possible received words 2^n is much greater than the number of code words 2^k. For each received code word, the decoder decides what was the most likely code word that was transmitted, and the receiver then identifies the corresponding k-bit message block. In this way error correction can be achieved.

A notably different approach is **convolutional coding**. Here the incoming message bit stream is applied to a K-stage shift register which is shifted b bits at a time. For each K message bits stored in the register, there are n linear logic circuits which operate on the register contents to produce n code bits of the encoded output stream. For each shift of the register, b new message bits are inserted and n code bits are delivered, so that the rate is b/n information bits per code symbol. In this case, a particular code bit depends on K message bits where K is called the **constraint length** of the code. Note also that a particular message bit remains in the register for K/b shifts, and thus influences the value of nK/b code bits.

Unlike block coding, the optimal decoding operation for convolutional codes requires a memory that stores, in effect, a function of the entire past history of the received bit stream. The performance (as measured by error rate) of a convolutional coding system improves as the complexity (i.e., memory) allowed for the decoder is increased. Several methods of decoding convolutional codes have been developed. The optimal (maximum likelihood) scheme is generally known as the Viterbi algorithm. Viterbi decoding for reasonably short constraint lengths is feasible to implement and high decoding speeds are achievable. For extremely low error probabilities, a large constraint length K is required. The computational complexity of Viterbi decoding for large K makes this approach impractical. Another approach, **sequential decoding,** then becomes more attractive. A third technique, **feedback decoding,** though inferior in performance against random errors, is particularly well suited to correcting systematic error bursts which may occur in fading channels.

Both fading and pulse jamming introduce memory in the channel and further modify the channel statistics. Yet the same coding techniques as used for the Gaussian channel are at least as effective here, provided interleaving is employed to reduce or eliminate this memory.

ERROR CORRECTING CODING FUNDAMENTALS

All that we need to know about coding for the present purpose is that for practically any memoryless channel, there are many good binary codes of rate r bits/symbol for which the bit error rate is upper bounded [1] by either

$$P_b < 2^{-K(\alpha-1)}, \quad \text{if it is a } \textit{block} \text{ code of block length } K \tag{4}$$

or

$$P_b < \frac{2^{-K\alpha}}{[1-2^{-(\alpha-1)}]^2}, \quad \text{if it is a } \textit{convolutional} \text{ code of constraint length } K. \tag{5}$$

In either case

$$\alpha = \frac{r_0}{r} > 1 \tag{6}$$

provided the code rate $r < r_0 < 1$.

Performance then depends strongly on the value of the parameter[4] r_0. This parameter, and consequently α, is increased if the decoder is furnished with everything the receiver "knows" about the channel; that is, for binary symbols, not only the receiver's "belief" that the transmitted symbol was a "zero" or a "one," but how strongly the receiver believes this. This confidence in the decision is called a "soft decision," "quality information," or simply a "metric."

Now for a uniform Gaussian channel, with soft decisions furnished to the decoder, the all-important parameter r_0 is a function of only the symbol energy-to-noise density. That is,

$$r_0 = 1 - \log_2 [1 + e^{-E_s/N_0}] \tag{7}$$

where, as previously defined,

$$E_s = E_b r. \tag{8}$$

Combining (6), (7), and (8), we can relate E_b/N_0 to α and r_0 as

$$\frac{E_b}{N_0} = -\alpha \frac{\ln (2^{1-r_0} - 1)}{r_0}. \tag{9}$$

This quantity expressed in decibels ($10 \log E_b/N_0 - 10 \log \alpha$) is plotted in Fig. 3 as a function of r_0. Of course, Fig. 3 or (9) is meaningful only when taken together with (4) or (5). The interpretation is that for a given acceptable complexity of implementation, which is roughly proportional to 2^K for either class of codes, and a given code rate r, we need to select a value of α to guarantee the required P_b, according to (4) or (5). This establishes r_0 according to (6), and finally we obtain E_b/N_0 by adding $10 \log \alpha$ to the ordinate of Fig. 3 for the given r_0. If the resulting r_0 is greater than 1, we must choose a smaller code rate r for which $r_0 < 1$ for the required α. Although

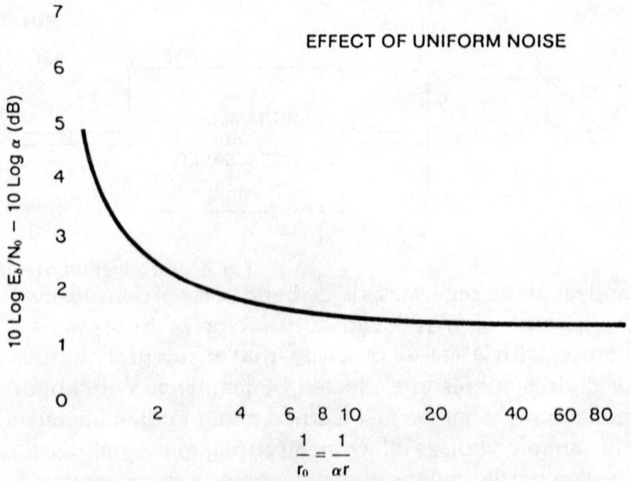

Fig. 3. E_b/N_0 **requirement in additive uniform noise.**

(4) and (5) are only upper bounds, and hence pessimistic estimates of the value of coding, they are sufficient to establish its merit, and in fact are reasonably accurate. As a practical matter, commonly used and commercially available convolutional decoders require an E_b/N_0 which is about 0.5-2 dB above the curve of Fig. 3 to achieve $P_b \simeq 10^{-5}$.

For specific comparison, we have from (2) that to achieve $P_b = 10^{-5}$ on the uniform Gaussian channel without coding requires $E_b/N_0 = 9.6$ dB. With rate 1/2 convolutional codes, practical soft-decision decoders require between 3.5 and 4.5 dB for the same performance. Rate 3/4 codes require approximately 1 dB more. If the decoder is provided with only hard-decision inputs, it requires approximately 2 dB higher E_b/N_0.

Thus far we have considered only stationary wideband noise, whether of thermal origin, or so rendered by the direct sequence spread spectrum technique we have investigated. We now return to the pulse jammer and show how coding remedies the deplorable situation which we left before the present digression.

SECOND MYTH REVISITED—APPROPRIATELY ARMED

Suppose now that we code as before, but for the nonuniform (pulse) jammer. Spreading causes this to

[4] This parameter is also the so-called computational cutoff rate beyond which the sequential-decoding mean computational load becomes unbounded.

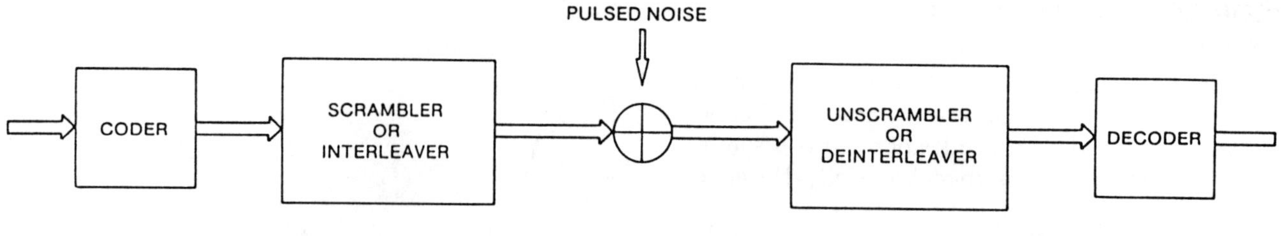

Fig. 4. Introduction of interleaving for disposing bursts.

appear at the receiver as wide-band noise of density level N_0/ρ, but for a reduced duty factor ρ. Suppose, as before, with little loss of reality, that an integral number of code symbols are affected by jamming. We cannot quite apply what we just learned about coding because the jamming pulses affect many contiguous symbols; so we can hardly call the channel memoryless as required. But this is easily remedied. Suppose we construct a device which randomly scrambles the order of the symbols prior to transmission, but after coding, and puts them back in the right order after reception, but before decoding (Fig. 4). (Scramblers and unscramblers are more commonly called interleavers and deinterleavers.) But the unscrambler which restores the transmitted symbols to their right place in order actually scrambles the regular jamming pulses into random patterns.[5] Scrambling or interleaving thus makes our system memoryless again and we can apply our new-found coding knowledge.

Without belaboring the exact details, arguing intuitively and believingly on the basis of (2) and (2′), let us replace e^{-E_s/N_0} by $\rho e^{-\rho E_s/N_0}$ in all the formulas of the previous (uniform noise) section. Thus, (7) is replaced by

$$r_0 = 1 - \log_2 [1 + \rho e^{-\rho E_s/N_0}]. \qquad (7')$$

Combining (7′) with (6) and (8), we get

$$\frac{E_b}{N_0} = -\frac{\alpha}{\rho r_0} \ln \left[\frac{2^{1-r_0} - 1}{\rho}\right] \qquad (9')$$

which obviously reduces to (9) for uniform jamming ($\rho = 1$).

This is maximized by a jammer with duty cycle

$$\rho = (2^{1-r_0} - 1)e \qquad (10)$$

provided $r_0 > 1 - \log_2 (1 + e^{-1}) = 0.548$ for which

$$\max_{0<\rho<1} \frac{E_b}{N_0} = \frac{\alpha e^{-1}}{r_0(2^{1-r_0} - 1)} \quad \text{for } r_0 > 0.548. \qquad (11)$$

Even the severe impact of pulsed jamming can be entirely contained by suitable coding.

If $r_0 \leqslant 0.548$, $\rho = 1$ maximizes E_b/N_0 and (9′) reduces to (9). This says that if r_0 is small enough, *no penalty* is paid to a pulse jammer. Even for $r_0 > 0.548$, the penalty is small, as seen in Fig. 5, which shows the new pulsed noise case and reproduces the uniform noise case from Fig. 3.

Thus, even more amazing than the original 35 dB loss to pulsed noise, *coding recovers it all and then some.* In spite of this, there still are skeptics who believe in the

Third Myth:

Interleaving destroys memory. Memory can be exploited to correct errors. Hence, interleaving is bad.

The discussion leading to the system design of Fig. 3 should suffice to dispel such misgivings, particularly when we recall that soft decisions contain about all the information available about the channel (granted, of course, that memory may be exploited to extract such quality information). To definitely put to rest all discussion of the matter, let us consider a burst interference phenomenon which always affects B symbols, and let there be no overlaps of bursts (overlaps can only help the communicator). In this case, the key parameter of (7′) becomes

$$r_0 = 1 - \frac{1}{B} \log_2 \{(1 - \rho) + \rho[1 + e^{-\rho E_s/N_0}]^B\} \qquad (7'')$$

which causes (9′) to be replaced by

$$\frac{E_b}{N_0} = -\frac{\alpha}{\rho r_0} \ln \left\{\left[\frac{2^{(1-r_0)B} - (1 - \rho)}{\rho}\right]^{1/B} - 1\right\} \qquad (9'')$$

which leads to the plots of Fig. 6. Clearly (7″) and (9″) reduce to (7′) and (9′) for $B = 1$, which is the case when interleaving is employed.

In fact, the situation is even worse than shown in Fig. 6.

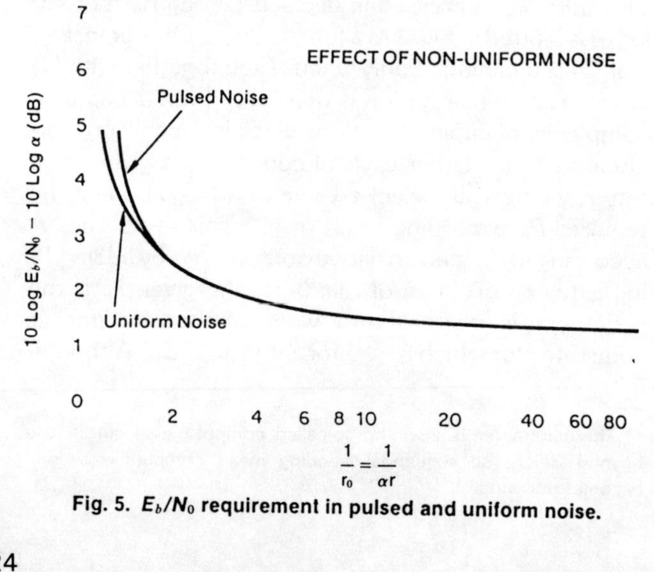

Fig. 5. E_b/N_0 requirement in pulsed and uniform noise.

[5]Note the parallel with spectrum spreading; the second multiplier unspread the signal and spread the interference; here the second device unscrambles the desired code sequence and scrambles the undesired pulsed interference.

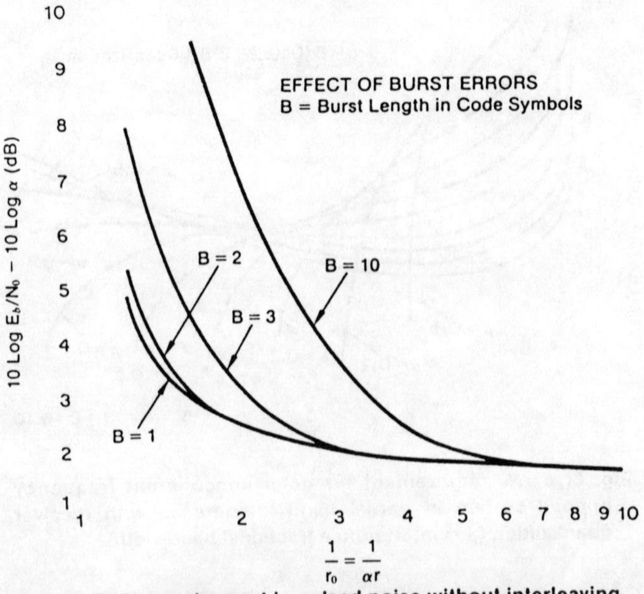

Fig. 6. E_b/N_0 **requirement in pulsed noise without interleaving.**

For as r decreases, the number of symbols per bit and, for a constant duration interference pulse, the number of symbols per pulse (B) increases. Thus, keeping B fixed, as is done for convenience in Fig. 6, gives misleadingly favorable results.

Our final "myth" happens fortuitously to coincide with reality. We call it, therefore, a "Folk Theorem." It concerns an interesting comparison of direct sequence (coherent) spreading with frequency hopping (noncoherent) spreading:

Fourth Folk Theorem: (Myth = Reality)

Performance of frequency-hopped spread spectrum is 3 dB worse than that of direct sequence (PN) spread spectrum.

The commonly invoked "mythical" argument is that noncoherent systems can utilize at best orthogonal signals (e.g., binary FSK modulation) instead of antipodal signals (binary PSK modulation) and this accounts for the 3 dB. The trouble with this argument is that it ignores the possiblity of higher signaling alphabets (such as MFSK) and, worse still, the real possibility that frequency-hopped systems may be more vulnerable to nonuniform interference.

We note, in fact, that in frequency-hopped systems, the jammer need not pay the cost of a higher peak power signal, for if he jams just a fraction of the band,[6] $\rho < 1$, with power density N_0/ρ, he will appear to the receiver just as a partial time jammer. Note also that if the hopping rate is at least as great as the symbol rate, interleaving is unnecessary.

For alphabets of size q, (4) and (5) must be modified to become

$$P_b < q^{-K(\alpha-1)} \quad \text{block codes} \qquad (4')$$

and

$$P_b < \frac{(q-1)q^{-K\alpha}}{[1-q^{-(\alpha-1)}]^2} \quad \text{convolutional codes} \qquad (5')$$

which reduce to (4) and (5) when $q = 2$. There is nothing to gain by using multiple signal alphabets for coherent, direct sequence systems, but with frequency-hopped systems, we can show [2] that for the worst case partial-band jammer

$$\max_{0<\rho<1} \frac{E_b}{N_0} = \alpha \frac{(q-1)4e^{-1}}{(\log_2 q)r_0(q^{1-r_0}-1)}$$

provided

$$\frac{E_b}{N_0} \geq \frac{3}{r \log_2 q}.$$

For asymptotically large q, this approaches

$$\frac{E_b}{N_0} \simeq (4 \ln 2)\alpha$$

which is exactly a factor of 2 (3 dB) above the minimum of Fig. 3 which occurs as $r_0 \longrightarrow 0$. This comparison is shown in Fig. 7, which also shows the diminishing returns of using alphabet size $q > 8$. The asymptotic minimum is virtually reached for $q > 32$.

Notice that noncoherent frequency-hopped systems exhibit a minimum, while coherent direct sequence systems improve monotonically as $1/r_0$ increases. The explanation of this behavior is better understood by examining Fig. 8 which is a more detailed and more realistic examination of performance for an octal alphabet. Here the assumptions are more realistic. Specifically, channel quality information is limited to two bits (four levels) out of each of the q matched filters. This also allows for a practical automatic gain control (AGC) technique, which has not been mentioned up to this point.

The curve for $\rho = 1$ is, of course, for uniform noise. The increase at high rates (r and r_0 close to 1) is due to the lack of coding redundancy. The increase at low rates is due to the higher loss, characteristic of noncoherent combining of symbols in high diversity (here low rate) noncoherent communication systems. As ρ, the fraction of the interference bandwidth decreases, performance gets increasingly worse at high rates since the diversity is lacking to overcome the strong jammer. But as r_0 decreases, diversity becomes sufficient to fully defeat the low ρ jammer, as shown by the family of curves of Fig. 8.

[6] Possibly varying this by hopping himself in order to defeat the obvious communicator strategy of determining the jammed region and staying out of it.

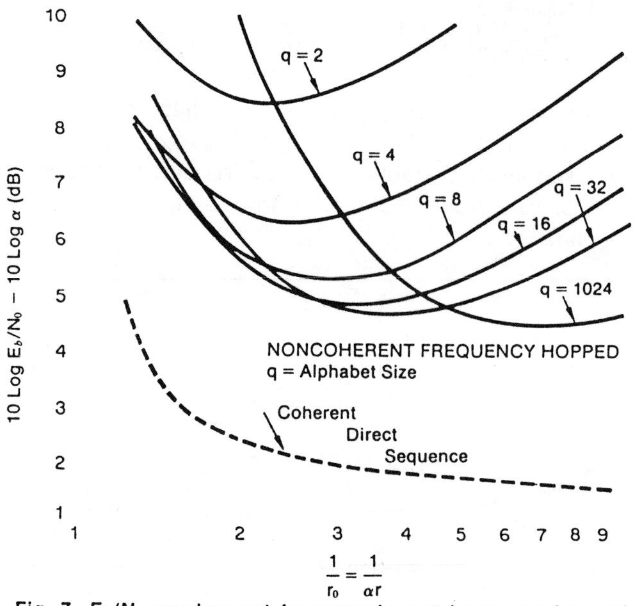

Fig. 7. E_b/N_0 requirement for noncoherent frequency hopped systems.

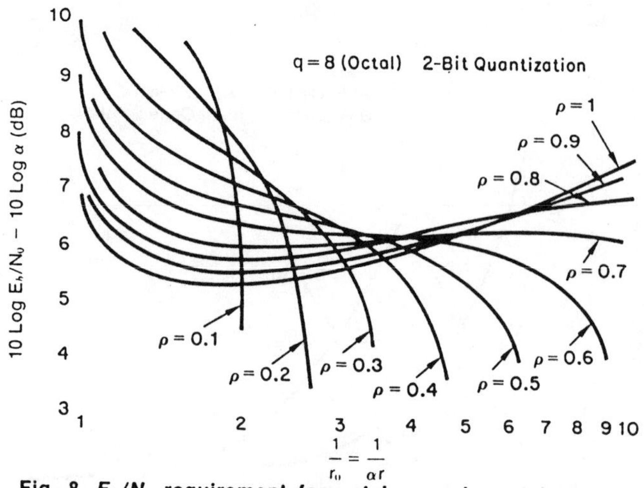

Fig. 8. E_b/N_0 requirement for octal noncoherent frequency hopped system in partial band interference with receiver quantization (ρ = interference fractional bandwidth).

SUMMARY

Beyond cataloging the many uses of spread spectrum communication, we have made no attempt to be uniform in our treatment of this extensive and many-faceted field. We have concentrated rather on its application for the suppression of interference, and have made three main points:

- Coding is always useful, and it may be critical to adequate performance of spread spectrum systems, particularly when the nature of the interference is partial-time, or in the case of frequency-hopped spreading, partial band. Proper interfacing of the decoder to the demodulator, in utilizing quality (soft decision) information, is important to ensure maximum benefit from coding.
- Interleaving or scrambling may be equally essential in the presence of burst interference.
- Direct sequence spread spectrum efficiency is about double that for frequency hopping. This is tantamount to doubling the processing gain W/R. However, frequency-hopping technology may have an edge in achievable band spreading of one or more orders of magnitude over direct sequence spreading technology which greatly overshadows the "system edge" of the latter.

Although the efficiency of direct sequence spread spectrum is about double that of frequency hopping, this advantage is overshadowed by the greater band spreading achievable with frequency hopping technology.

REFERENCES

[1] A. J. Viterbi and J. K. Omura, *Principles of Digital Communication and Coding.* New York: McGraw-Hill, 1979.
[2] A. J. Viterbi and I. M. Jacobs, "Advances in coding and modulation for noncoherent channels affected by fading, partial-band and multiple-access interference," in *Advances in Communication Systems,* vol. 4. New York: Academic, 1975.

Wireless Digital Communication: A View Based on Three Lessons Learned

Shannon's information theory is applied to PCN, with examples of improvements in satellite communication, magnetic recording and modem technologies that have approached information-theoretic limits.

Andrew J. Viterbi

The first two installments of "Technologies on the Horizon" in September and November, 1990, featured personal viewpoints by Ray Steele and Don Cox on the emerging area of personal communication networks (PCNs). Both authors addressed key issues that need to be resolved to expedite large scale deployment of low power, low cost, low complexity wireless access technologies.

My request for comments or reactions from members of the communications community drew a passionate and informed response from a third leading contributor to recent advances in personal communications: Andrew J. Viterbi, Chief Technical Officer of QUALCOMM, Inc., took exception to Don Cox's view that variations in power control and soft handoff imperfections will cause loss of system capacity in PCNs based on spread-spectrum techniques.

As Viterbi noted in his response to me, ". . . Cox faults spread-spectrum digital techniques for excessive complexity, but at the same time claims they will be ineffective in 'real-world' situations." Viterbi then adds, "If he were right on the latter, there would be no need to argue the former. He is, I contend, wrong on both, because the complexity, which is exclusively in the digital processing, can be made as cheap and virtually as power thrifty, as needed by market-demand influence on economies of scale."

At my invitation, Andy Viterbi contributes his views this month on PCN design and complexity to inform readers of IEEE Communications Magazine about recent advances in spread-spectrum approaches to wireless digital communications. Viterbi outlines the implications of Shannon's information theory for PCN design, providing examples of improvements in satellite communication, magnetic recording and modem technologies that have increasingly approached information-theoretic limits. He emphasizes the application of Shannon's work to control of four types of interference: multi-user, multiple cell, multipath and multiple media. Viterbi then closes by applying "Shannon's third lesson" to a comparison of code-division multiple access (CDMA), time-division multiple access (TDMA) and frequency-division multiple access (FDMA) techniques in personal communications.

This month's column, together with earlier "Technologies on the Horizon" contributions by Ray Steele and Don Cox, and references provided by all three authors, constitute a useful tutorial for interested, nonspecialist readers on the major technical alternatives currently being pursued as personal communication networks move from the laboratory to field experiments and initial service applications. Of course, the real feasibility test for any proposed approach to PCNs will ultimately be demonstrated in sizable field experiments involving large numbers of users, cells and very imperfect, "real-world" transmission environments. Such field experiments are currently under way, and we can expect to learn about real-world performance, complexity and cost in the next year or so, as field results for alternative wireless communications systems are reported.

With thanks to increasingly responsive and enthusiastic readership, I invite additional comments and suggestions for this column beyond the ones currently on my horizon. All correspondence should be sent to: Howard L. Lemberg, Bellcore, 445 South Street, Room 2M289, Morristown, NJ 07962-1910

ANDREW VITERBI is Vice Chairman, and Chief Technical Officer, QUALCOMM, Inc.

Technologies on the Horizon is edited by Howard Lemberg of Bellcore.

It is a virtual certainty that wireline and fiber communication will be fully digital by the end of the century. It is almost as likely that the same will be true for most wireless communication. Beginning with broadcast of high quality digital audio, already underway in Japan and Europe, and with high definition digital video broadcasting on the horizon, it appears that by the end of the decade digital modulation will have superseded analog in virtually all forms of broadcasting. Digital communication provides excellent reproduction and greatest efficiency of transmission bandwidth and power through effective utilization of two fundamental techniques: source compression coding to greatly reduce the transmission rate for a given degree of fidelity; and error control coding to further reduce signal-to-noise and bandwidth requirements.

While digital modulation is also certain to become the universal choice for two-way personal and mobile communication, and while the two techniques just mentioned will be equally important here for the same reasons, there is another even

Reprinted from *IEEE Commun. Mag.*, vol. 29, no. 9, pp. 33–36, September 1991.

Shannon's theory can be summarized in three lessons, which are transforming all communication products and services that surround us.

more critical need for such applications: to afford many users simultaneous and equal access to the network. Whichever multiple access technique is employed, the ultimate performance limitation is the system's susceptibility to interference. In narrowband multiple access techniques, whether frequency division or time division, this interference manifests itself as inter-user, intersymbol and multipath interference. Mitigation of the detrimental effects of these disturbances is limited to equalization and to isolation by frequency reuse with sufficient spatial separation. The first may be difficult to implement, particularly in mobile vehicles, while the second reduces spectrum efficiency.

Historical Precedents

Before delving deeper into these issues and the underlying lessons which lead to their resolution, it is useful to trace the origins of moder electronics from the discovery of the transistor in 1948. This singular event opened the solid state electronics era, which revolutionized business styles and lifestyles universally. The most visible fruits of the dramatic progress in solid state integration over the past two decades has been the ubiquitous application of computer power descending from large office applications to the most modest of home and personal users. A natural extension, with almost as great an impact, is to liberate the individual communicator from the tether, whether through cordless telephony, mobile cellular telephony or the emerging generic concept of the personal communication system. Equally natural and correct is the view that digital techniques, having evolved to such enormous speed and memory capabilities through computer applications of solid state technology (and to a lesser extent signal processing applications), will produce equally economical and powerful personal communication products.

The pitfall of the computer-communication analogy [1] is the exporting of not only hardware capabilities, but of system concepts as well, from the computer to the communication arena. Thus, for example, the concept of OSI layering, which is a logical extension of software hierarchies in computers, presents a deceptively over simplified view of communication systems and networks. The design of the lowest (or physical) layer has a much more profound impact on the feasibility of communication processes at higher layers than can be explained solely by issues of speed and memory. For an understanding of this key point, it is necessary to explore more fully the fundamentals of communications which by a remarkable coincidence were revealed in a research journal published in 1948, the same year in which the discovery of the transistor was made, and at the same research organization, the Bell Telephone Laboratories in Murray Hill, New Jersey.

Shannon's Three Lessons

This journal, "The Mathematical Theory of Communication" by Claude Shannon [2], more popularly referred to as Shannon's Information Theory, provided in a relatively abstract form the foundations for design of efficient wireless communication systems, including those seeking multiple access to a common medium. Of course it took longer than a decade before a sufficient number of communication engineers had been trained to understand and envision the ultimate practical embodiments of this theory, and about two decades beyond that before the above mentioned solid state evolution of the transistor reached the technological and economic level to implement these embodiments.

The essence of the Shannon theory, which pertains to digital communications, can be summarized in the form of three lessons, learned and applied over nearly a half century. Ranging in order from intuitive to somewhat surprising, these lessons are as follows:

- Never discard information prematurely that may be useful in making a decision until after all decisions related to that information have been completed.
- Completely separate techniques for digital source compression from those for channel transmission, even though the first removes redundancy and the second inserts it.
- In the presence of interference or jamming, intentional or otherwise, the communicator, through signal processing at both transmitter and receiver, can ensure that performance degradation due to the interference will be no worse than that caused by Gaussian noise at equivalent power levels. This implies that the jammer's optimal strategy is to produce Gaussian noise interference. Against such interference, the communicator's best waveform should statistically appear as Gaussian noise. Thus, the "minimax" solution to the contest is that signals and interference should all appear as noise which is as wideband as possible. This is a particularly satisfying solution when, as we shall see, one user's signal is another user's interference.

These three lessons are transforming all the communication products and services that surround us. The first lesson has radically changed the design approach to satellite communication over the past two decades. It has also had a major impact on the wireline modem industry, and is just beginning to be felt as an economic factor in the recording industry as well. Each field has applied Shannon's first lesson and assigned its own jargon to its implementation. In satellite communication they are called "forward error correction (FEC) with soft decision decoding." [3] In high quality wireline modems the techniques are known as "maximum-likelihood sequence estimation" [4] and "trellis coding." [5] In magnetic recording it is called "partial response maximum-likelihood (PRML) detection." [6]

More interesting is the impact of these techniques:

- In recording, they are replacing the very inefficient "peak detector" and increasing recording densities several fold.
- In wireline modems, adaptive linear equalization in the seventies raised the data rate

of a 3 KHz leased line from 1200 baud to 9600 baud; enhancement by trellis coding in the eighties has produced transmission rates as high as 19.2 Kbaud.

- For satellites, convolutional codes with soft decision decoders have reduced the link budget requirement for digital communication by 6 dB, thus ensuring the feasibility [7] of a new billion dollar industry, namely, VSAT's providing up to 64 kbps service using only 1.2m to 2.4m dishes. The impact of this technology on satellite digital TV broadcasting is just over the horizon. Mobile satellite communication in the heretofore underused Ku-band spectrum is even more impressive. A one-watt transmitter provides both position-location and messaging communications over one transponder of a conventional non-processing satellite [8] for tens of thousands of trucks on North American highways today; the same technology will be available in Europe, and possibly Japan, in the near future.

The second lesson of information theory is just now being learned. Digital compression of speech and video has been a research topic for at least three decades. Bit rate reductions of almost two orders of magnitude are now the norm. High definition television (HDTV) broadcasting is now a major field of research and development at the world's largest electronic manufacturers. Until recently an all-digital approach, which completely separates video-source compression coding from transmission coding, was scoffed at as hopelessly impractical, not because of the feasibility of implementing such compression factors, but because of the difficulty in transmitting 20 to 60 Mbps of digital information, and especially because of the cost of consumer receivers for such.

In 1989, the U.S. Defense Advanced Research Project Agency (DARPA) awarded three development contracts for HDTV signal processing technology, including both image compression and transmission systems. A team headed by Sarnoff Laboratories and a team headed by MIT each received an award, with the third award going to QUALCOMM, Inc. [9] Of these, the first two proposed analog-digital hybrid methodologies, while the third was the only one advocating an all-digital approach. [10]

Since then all known participants have joined the "all digital bandwagon," including General Instruments, NBC Thomson S.A., Philips N.V., Sarnoff, and, more recently, the AT&T-Zenith joint adventure and the MIT-General Instruments "American Television Alliance." [11] Some are concentrating on DBS while others concentrate on terrestrial or cable delivery. All seem to be learning the second lesson.

Multiple Access: A More Subtle Lesson

Which brings us finally to Shannon's third lesson, not yet as widely accepted as the first two. For nearly half a century, military communication specialists, who deal in particularly pernicious and malignant interference, have intuitively understood that by proper processing at both transmitter and receiver, these could be tamed into behaving like the most benign of interference, thermal (White Gaussian) noise. (It is, in fact, no coincidence that just prior to the research that led to his 1948 publication, Shannon's primary preoccupation was military communication and secrecy.) [12] But personal, wireless and cellular communication also must deal with a variety of interference forms, primarily "the four multiples": multiple-user access, multiple cell-sites, multipath and multiple media.

Through the use of spread spectrum techniques, the detrimental effect of all four can be mitigated, and in some cases (multiple cell-sites and multipath in particular), can even be used to improve communication performance. For example, with wideband spread spectrum modulation, the multiple paths of multipath propagation can be isolated and through diversity combining can be used to advantage. Actually, this is but one of several diversity techniques utilized in a well-designed spread spectrum cellular system. Another is cell-site antenna diversity combining, with a third being the combining of signals from and to two or more cell sites through a technique known as "softhandoff." Equally important unique features of spread spectrum CDMA are interference reduction through "voice activity gating" and "cell-site sectorization gain." What is left, after these mitigating techniques have been exhausted, is the benign additive Gaussian noise, which is itself mitigated by FEC coding.

Most important is the gain in "Reuse Factor," a cellular parameter that is inversely proportional to the number of different frequency assignments necessary to guarantee that neighboring cells are assigned disjoint frequency bands. Reuse factors of 1/19, 1/12, 1/7, 1/4 and 1/3 have been used or proposed for progressively more optimistic designs of FDMA and TDMA systems. [13] CDMA systems employ ubiquitous frequency reuse and thus have a factor of 1. [14] Ubiquitous and universal frequency reuse applies to all users in the assigned spectrum or channel, to all cell sites, and to all media, terrestrial and satellite, both geostationary and low earth orbit. In fact, with CDMA, seamless handover between media becomes possible.

There is, however, one major impediment to universal reuse that must be overcome, particularly in terrestrial systems. Known as the "Near-Far" problem, this refers to the condition in which some users are much closer to the base station than others, thus introducing excessive interference. This has been solved in cellular systems by implementing several levels of rapid, tight, power control. Experimental systems have demonstrated control of power up to 100 dB in fractions of a second, with a total variation in controlled power on the order of + 1.5 dB.

This is not the forum for detailed analyses. Such have been presented by my associates and myself in other venues [15] and publications. [16] To provide some measure of the capacity of CDMA, we quote the current goal of the QUALCOMM CDMA system:

$$\text{CAPACITY (CDMA)} \approx 1\ \text{Bit/Sec/Hz/Cell}$$

This presupposes a Voice Activity factor of 1/2 and Sectorization Gain on the order of 4 to 6

Advances in digital communication in the latter half of this century were guided by the lessons of information theory but fueled by the progress in solid state electronics.

Whichever multiple access technique is employed, the ultimate performance limitation is the system's susceptibility to interference.

dB. (This is by no means a physical limitation; increases of 2 to 4 or even more may well be feasible in second generation systems.)

By comparison, the current equivalent capacity of analog AMPS (assuming the same 10 Kbit/sec voice quality) is

$$\text{CAPACITY (AMPS)} \approx 1/21 \text{ Bit/Sec/Hz/Cell}$$

(Narrow-band AMPS may, at best, double this capacity given that it triples the calls per current channel, but it will undoubtedly require a lower reuse factor.) North American TDMA will provide at best

$$\text{CAPACITY(N.A.—TDMA)} \approx 1/7 \text{ BIT/SEC/HZ/CELL}$$

assuming the current reuse factor 1/7, provided the current C/I -18 dB requirement can be tolerated. European (GSM) TDMA is designed to

$$\text{CAPACITY (GSM-TDMA)} \approx 1/10 \text{ BIT/SEC/Hz/cELL}$$

assuming a reuse factor of 1/4.

What is it about TDMA that prevents it from achieving a higher reuse factor? In TDMA, users in the same cell are kept from interfering (much) with one another by assigning them disjoint frequency and time slots. Once all frequency slots assigned to a given cell have been allocated, the neighboring cells must be assigned another set of frequency slots and so forth, until we reach a distant enough cell that the original frequency can be reused. What keeps this from happening too soon (and therefore with higher reuse factor) is that both the multi-user interference and the multipath interference (that which has not been neutralized by equalization) are seriously detrimental, unlike the benign noise interference characteristic of CDMA.

In short, the principal drawback of TDMA (and FDMA) is that while its proponents have learned the first two lessons of Information Theory (they do employ soft decision decoding and equalization, and they do separate voice compression Vocoding from channel transmission processing and FEC soft-decision decoding), they have missed the third and most important lesson on rendering the interference benign. This is only possible with CDMA and spread spectrum, the logical choice for personal, mobile and wireless digital access.

Economic Considerations

Though capacity and call quality may be the primary concern, other desirable features in personal and mobile communication include: transmitter power requirements of subscriber units; cell-site costs, which are dominated by RF and analog circuitry; transition plan for gradual and profitable conversion from analog cellular and coexistence with existing systems; and security and privacy.

CDMA excels over all other multiple access techniques in providing the best solution for all of the above. Without embarking on a detailed exploration of why and how, suffice it to say again that all are facilitated by the wideband benign interference properties of CDMA of which we have spoken, and that universal frequency reuse avoids the complicated issue of frequency-management planning when additional cells are introduced.

No assessment is complete without a comparison of implementation complexity and cost. Doubtless, CDMA is conceptually more difficult to understand. But difficulty of concept should not be confused with difficulty of implementation. Were this so, the CD-player would never have become the popular low priced consumer product it is. And this brings me back to my original premise, that advances in digital communication in the latter half of this century were guided by the lessons of information theory but fueled by the spectacular progress in solid state electronics. Specifically, the "smart" and "difficult" algorithms are relegated to the solid state circuitry, where levels of integration and speed double every two years, making yesterday's hopelessly complex and costly implementation into the high volume low cost microchips and ASIC implementations of today. What differentiates CDMA from other multiple access techniques is mostly contained in these chips. At the same time, the conventional analog and RF circuitry in the cell site is greatly reduced because it is shared among more users, with user separation performed at baseband, again in the same powerful chips. The subscriber equipment complexity is also reduced, primarily because of reduced transmitter power requirements.

Let me conclude with a personal observation. I have been privileged to participate in the communication engineering profession and the telecommunication industry over the majority of the period during which we have learned how to apply the key lessons of Shannon's Information Theory. I firmly believe that the final decade of the half century will bring this knowledge to complete fruition in the form of ubiquitous digital communication products and services undreamed of as I entered the field.

References

[1] K. Kobayashi, Computers and Communications: A Vision of C & C, MIT Press, Cambridge, MA, 1986.

[2] C. E. Shannon, "A Mathematical Theory of Communication," Bell System Technical Journal, Vol. 27, pp. 379-423 and 623-656, Ju and October, 1948.

[3] J. A. Heller and I. M. Jacobs, "Viterbi Decoding for Satellite and Space Communication," IEEE Transactions on Communications Technology, Vol. COM-l9, pp. 835 848, October, 1971.

[4] G. D. Forney, Jr., "Maximum-Likelihood Sequence Estimation of Digital Sequences in the Presence of Intersymbol Interference, "IEEE Transactions on Information Theory," Vol. IT-18, pp. 363-378, May, 1972.

[5] G. Ungerboeck, "Channel Coding with Multilevel/Phase Signals," IEEE Transactions on Information Theory, Vol. IT-28, pp. 55-67, January, 1982.

[6] R. Wood, "Magnetic Megabits," IEEE Spectrum, Vol. 27, pp. 32-38, May, 1990.

[7] Having previously doubled the communication ranges of space missions such as Voyager and Galileo.

[8] I. M. Jacobs et al., "The Application of a Novel Two-Way Mobile Satellite Communications and Vehicle Tracking System to the Transportation Industry," IEEE Transactions on Vehicular Technology, Vol. 40, pp. 57-63, February, 1991.

[9] QUALCOMM, Inc., "Technical Proposal for an HDTV Receiver/Processor," submitted to DARPA, Arlington, VA, February, 1989.

[10] United Press International National Wire Service, "Pentagon Names More HDTV Contractors," October, 26, 1989.

[11] New York Times, "Advanced TV Testing Set Amid Tumult on Technology" pp. Cl and C6, November 15,1990 and "AT&T and Zenith in Venture Plan to Jointly Build all Digital System for Advanced TV," pp. Cl and C18, December 18,1990.

[12] R. Price (Ed.), "A Conversation with Claude Shannon," IEEE Communications Magazine, Vol. 22, pp. 123-126, May, 1984.

[13] W. C-Y Lee, Mobile Cellular Telecommunications Systems, McGraw-Hill, New York, 1989.

[14] This factor is effectively reduced by the increase in interference from neighboring cells, but still remains greater than 3/5. See also Reference [16]

[15] IEEE GLOBECOM Workshop on CDMA in Satellite and Terrestrial Applications, San Diego, CA, December, 1990.

[16] K.S. Gilhousen, et al., "On the Capacity of a Cellular CDMA System," IEEE Transactions on Vehicular Technology, Vol. VT-40, May, 1991.

Part 4
ALOHA Multiple Access

The Throughput of Packet Broadcasting Channels

NORMAN ABRAMSON, FELLOW, IEEE

Abstract—Packet broadcasting is a form of data communications architecture which can combine the features of packet switching with those of broadcast channels for data communication networks. Much of the basic theory of packet broadcasting has been presented as a byproduct in a sequence of papers with a distinctly practical emphasis. In this paper we provide a unified presentation of packet broadcasting theory.

In Section II we introduce the theory of packet broadcasting data networks. In Section III we provide some theoretical results dealing with the performance of a packet broadcasting network when the users of the network have a variety of data rates. In Section IV we deal with packet broadcasting networks distributed in space, and in Section V we derive some properties of power-limited packet broadcasting channels, showing that the throughput of such channels can approach that of equivalent point-to-point channels.

I. INTRODUCTION

A. Packet Switching and Packet Broadcasting

THE transition of packet-switched computer networks from experimental [1] to operational [2] status during 1975 provides convincing evidence of the value of this form of communications architecture. Packet switching, or statistical multiplexing [3], can provide a powerful means of sharing communications resources among large number of data communications users when those users can be characterized by a high ratio of peak to average data rates. Under such circumstances, data from each user are buffered, address and control information is added in a "header," and the resulting bit sequence, or "packet," is routed through a shared communications resource by a sequence of node switches [4], [5].

Packet-switched networks, however, still employ point-to-point communication channels and large multiplexing switches for routing and flow control in a fashion similar to conventional circuit switched networks. In some situations [6]-[10] it is desirable to combine the efficiencies achievable by a packet communications architecture with other advantages obtained by use of broadcast communication channels. Among these advantages are elimination of routing and network switches, system modularity, and overall system simplicity. In addition, certain kinds of channels available to the communications systems designer, notably satellite channels, are basically broadcast in their structure. In such cases use of these channels in their natural broadcast mode can lead to significant system performance advantages [11], [12].

B. Outline of Results

Packet broadcasting is a form of data communications architecture which can combine the features of packet switching with those of broadcast channels for data communication networks. Much of the basic theory of packet broadcasting has been presented as a byproduct in a sequence of papers with a distinctly practical emphasis. In this paper we provide a unified presentation of packet broadcasting theory.

In Section II we introduce the theory of packet broadcasting as implemented in the ALOHA System at the University of Hawaii; also in Section II we explain a modification of the basic ALOHA method, called slotting. In Section III we provide some theoretical results dealing with the performance of a packet broadcasting channel when the users of the channel have a variety of data rates. In Section IV we deal with packet broadcasting networks distributed in space, and present some incomplete results on the theoretical properties of such networks. Finally, in Section V we derive some properties of power limited packet broadcasting channels showing that the throughput of such channels can approach that of equivalent point-to-point channels. This result is of importance in satellite systems using small earth stations since it implies that the multiple access capability and the complete connectivity (in the topological sense) of packet broadcasting channels can be obtained at no price in average throughput.

II. PACKET BROADCASTING CHANNELS

A. Operation of a Packet Broadcasting Channel

Consider a number of widely separated users, each wanting to transmit short packets over a common high-speed channel. Assume that the rate at which users generate packets is such that the average time between packets from a single user is much greater than the time needed to transmit a single packet. In Fig. 1 we indicate a sequence of packets transmitted by a typical user.

Conventional time or frequency multiplexing methods (TDMA or FDMA) or some kind of polling scheme could be employed to share the channel among the users. Some of the disadvantages of these methods for users with high peak-to-average data rates are discussed by Carleial and Hellman [13]. In addition, under certain conditions polling may require unacceptable system complexity and extra delay.

In a packet broadcasting system the simplest possible solution to this multiplexing problem is employed. Each user transmits its packets over the common broadcast channel in a completely unsynchronized (from one user to another) manner. If each individual user of a packet broadcasting chan-

Manuscript received January 19, 1976; revised June 11, 1976. This work was supported by The ALOHA System, a research project at the University of Hawaii which is supported by the Advanced Research Projects Agency of the Department of Defense and monitored by NASA Ames Research Center under Contract NAS2-8590. The views and conclusions contained in this paper are those of the author and should not be interpreted as necessarily representing the official policies, either expressed or implied, of the Advanced Research Projects Agency of the United States Government.

The author is with The ALOHA System, University of Hawaii, Honolulu, HI 96822.

Reprinted from *IEEE Trans. Commun.*, vol. COM-25, no. 1, pp. 117–128, January 1977.

Fig. 1. Packets from a typical user.

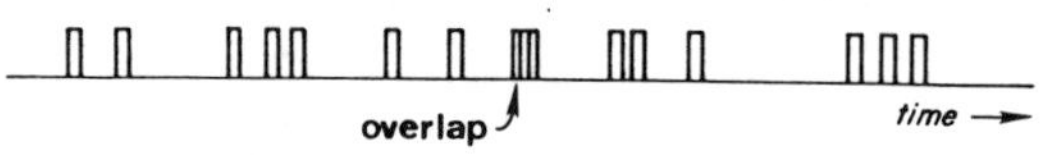

Fig. 2. Packets from several users on an ALOHA channel.

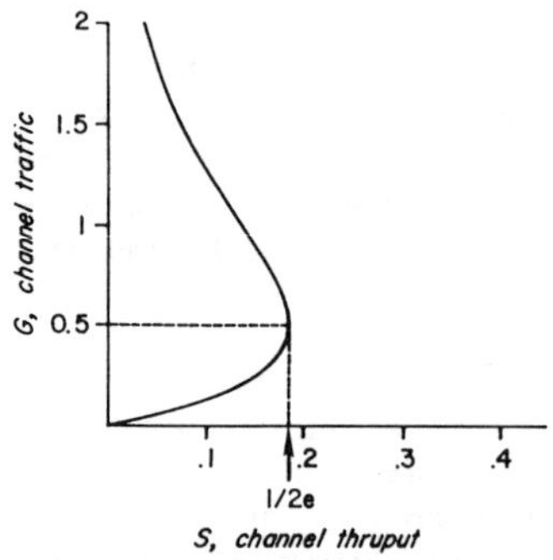

Fig. 3. Channel throughput versus channel traffic for an ALOHA channel.

nel is required to have a low duty cycle, the probability of a packet from one user interfering with a packet from another user is small as long as the total number of users on the common channel is not too large. As the number of users increases, however, the number of packet overlaps increases and the probability that a packet will be lost due to an overlap also increases. The question of how many users can share such a channel and the analysis of various methods of dealing with packets lost due to overlap are the primary concerns of this paper. In Fig. 2 we show a packet broadcasting channel with two overlapping packets. Since the first packet broadcasting channel was put into operation in the ALOHA System radio-linked computer network at the University of Hawaii [6], they have been referred to as ALOHA channels.

B. ALOHA Capacity

A transmitted packet can be received incorrectly or lost completely because of two different types of errors: 1) random noise errors and 2) errors caused by packet overlap. In this paper we assume that the first type of error can be ignored, and we shall be concerned only with errors caused by packet overlap. In Section II-D we describe several methods of dealing with the problem of packets lost due to overlap, but first we derive the basic results which tell us how many packets can be transmitted with no overlap.

Assume that the start times of packets in the channel comprise a Poisson point process with parameter λ packets/second. If each packet lasts τ seconds, we can define the normalized channel traffic G where

$$G = \lambda\tau. \tag{1}$$

If we assume that only those packets which do not overlap with any other packet are received correctly, we may define $\lambda' < \lambda$ as the rate of occurrence of those packets which are received correctly. Then we define the normalized channel thruput S by

$$S = \lambda'\tau. \tag{2}$$

The probability that a packet will not overlap a given packet is just the probability that no packet starts τ seconds before or τ seconds after the start time of the given packet. Then, since the point process formed from the start times of all packets in the channel was assumed Poisson, the probability that a packet will not overlap any other packet is $e^{-2\lambda\tau}$, or e^{-2G}. Therefore

$$S = Ge^{-2G} \tag{3}$$

and we may plot the channel throughput versus channel traffic for an ALOHA channel (Fig. 3).

From Fig. 3 we see that as the channel traffic increases, the throughput also increases until it reaches its maximum at $S = 1/2e = 0.184$. This value of throughput is known as the capacity of an ALOHA channel, and it occurs for a value of channel traffic equal to 0.5. If we increase the channel traffic above 0.5, the throughput of the channel will decrease.

C. Application of an ALOHA Channel

In order to indicate the capabilities of such a channel for use in an interactive network of alphanumeric computer terminals, consider the 9600 bits/s packet broadcasting channel used in the ALOHA System. From the results of Section II-B we see that the maximum average throughput of this channel is 9600 bits/s times $1/2e$, or about 1600 bits/s. If we assume the conservative [14] figure of 5 bits/s as the average data rate (including overhead) from each active[1] terminal in the network, this channel can handle the traffic of over 300 active terminals and each terminal will operate at a peak data rate of 9600 bits/s. Of course, the total number of terminals in such a network can be much larger than 300 since only a fraction of all terminals will be active and a terminal consumes no channel resources when it is not active.

D. Recovery of Lost Packets

Since the packet broadcasting technique we have described will result in some packets being lost due to packet overlaps, it is necessary to introduce some technique to compensate for this loss. We may list four different packet recovery techniques for dealing with the problem of lost packets. The first three make use of a feedback channel to the packet transmitter and the repetition of lost packets, while the fourth is based on coding.

1) Positive Acknowledgments (POSACKS): Perhaps the most direct way to handle lost packets is to require the receiver of the packet to acknowledge correct receipt of the packet. Each packet is transmitted and then stored in the transmitter's buffer until a POSACK is received from the receiver. If a POSACK is not received in a given amount of time, the transmitter can repeat the transmission and continue to repeat until a POSACK is received or until some other criterion is met. The POSACK can be transmitted on a sepa-

[1] A terminal is defined as active from the time a user transmits an attempt to log on until he transmits a log off message.

rate channel (as in the ALOHANET [6]) or transmitted on the same channel as the original packets (as in the ARPA packet radio system [15]). An error detection code and a packet numbering system can be used to increase the reliability of this technique.

2) Transponder Packet Broadcasting: Certain communication channels—notably communication satellite channels—transmit packets on one frequency to a transponder which retransmits the packets on a second frequency. In such cases all units in a packet broadcasting network can receive their own packet retransmissions, determine whether a packet overlap has occurred, and repeat the packet if necessary. This technique has been employed in ATS-1 satellite experiments in the Pacific Educational Computer Network (PACNET) [16] and in the ARPA Atlantic INTELSAT IV packet broadcasting experiments [17].

3) Carrier Sense Packet Broadcasting: For ground-based packet broadcasting networks where the signal propagation time over the furthest transmission path is much less than the packet duration, it is feasible to provide each transmission unit with a device to inhibit packet transmission while another unit is detected transmitting. A carrier sense capability can increase the channel throughput, even if these conditions are not met, when used in conjunction with other packet recovery methods. Carrier sense systems have been analyzed by Tobagi [18] and by Kleinrock and Tobagi [19]. A comprehensive yet compact analysis of such systems is provided in [42].

4) Packet Recovery Codes: When a user employs a packet broadcasting channel to transmit long files by breaking them into large numbers of packets, it is possible to encode the files so that packets lost due to broadcasting overlap can be recovered. It is clear that some of the existing classes of multiple burst error-correcting codes [20] and cyclic product codes [21] can be used for packet recovery in transmissions of long files. It is also clear that these codes are not as efficient as possible for packet recovery and that considerable work remains to be done in this area.

E. Slotted Channels

It is possible to modify the completely unsynchronized use of the ALOHA channel described above in order to increase the maximum throughput of the channel. In the pure ALOHA channel each user simply transmits a packet when ready without any attempt to coordinate his transmission with those of other users. While this strategy has a certain elegance, it does lead to somewhat inefficient channel utilization. If we establish a time base and require each user to start his packet only at certain fixed instants, it is possible to increase the maximum value of the channel thruput. In this kind of channel, called a slotted ALOHA channel, a central clock establishes a time base for a sequence of "slots" of the same duration as a packet transmission [41]. Then when a user has a packet to transmit, he synchronizes the start of his transmission to the start of a slot. In this fashion, if two messages conflict they will overlap completely, rather than partially.

To analyze the slotted ALOHA channel, define G_i as the probability that the ith user will transmit a packet in some slot. Assume that each user operates independently of all other users, and that whether or not a user transmits a packet in a given slot does not depend upon the state of any previous slot. If we have n users, we can define the normalized channel traffic for the slotted channel G where

$$G = \sum_{i=1}^{n} G_i. \tag{4}$$

Note that G may be greater than 1.

As before, we can also consider the rate at which a user sends packets which do not experience an overlap with other user packets. Define $S_i \leqslant G_i$ as the probability that a user sends a packet and that this packet is the only packet in its slot. If we have n users, then we define the normalized channel throughput for the slotted channel S where

$$S = \sum_{i=1}^{n} S_i. \tag{5}$$

Note that S is less than or equal to 1 and $S \leqslant G$.

For the slotted ALOHA channel with n independent users, the probability that a packet from the ith user will not experience an interference from one of the other users is

$$\prod_{\substack{j=1\\ j\neq i}}^{n} (1 - G_j).$$

Therefore we may write the following relationship between the message rate and the traffic rate of the ith user:

$$S_i = G_i \prod_{\substack{j=1\\ j\neq i}}^{n} (1 - G_j). \tag{6}$$

If all users are identical, we have

$$S_i = \frac{S}{n} \tag{7}$$

and

$$G_i = \frac{G}{n} \tag{8}$$

so that (6) can be written

$$S = G\left(1 - \frac{G}{n}\right)^{n-1} \tag{9}$$

and in the limit as $n \to \infty$, we have

$$S = Ge^{-G}. \tag{10}$$

Equation (10) is plotted in Fig. 4 (curve labeled "slotted ALOHA"). Note that the message rate of the slotted ALOHA channel reaches a maximum value of $1/e = 0.368$, twice the capacity of the pure ALOHA channel.

This result for slotted ALOHA channels was first derived by Roberts [41] using a different method.

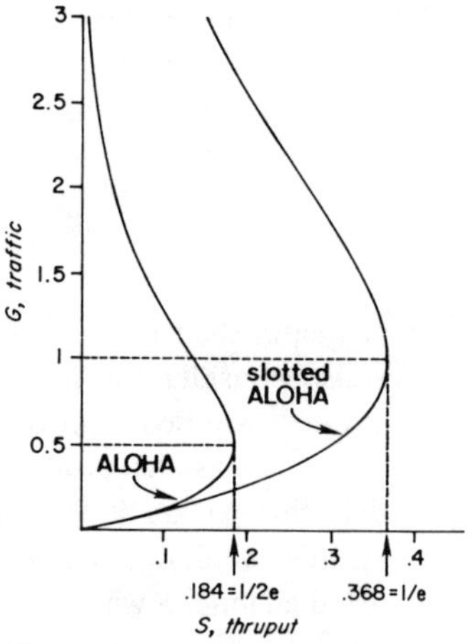

Fig. 4. Traffic versus throughput for an ALOHA channel and a slotted ALOHA channel.

III. PACKET BROADCASTING WITH MIXED DATA RATES

A. Unslotted Case: Variable Packet Lengths

In Section II we were concerned with the analysis of ALOHA channels carrying a homogeneous mix of packets. If some channel users have a higher average data rate than others, however, the high rate users must either transmit packets more frequently or transmit longer packets. In this section we shall analyze the unslotted ALOHA channel when carrying packets of different lengths, and we shall analyze the slotted ALOHA channel when the probability of transmitting in a given slot varies from user to user.

Let us assume an unslotted ALOHA channel with two different possible packet durations, τ_2 and τ_1. Assume $\tau_2 \geqslant \tau_1$, and therefore we refer to the two different length packets as long packets and short packets, respectively. Assume also the start times of the long packets and short packets form two Poisson point processes with parameters λ_2 and λ_1 packets/second, and that the two Poisson point processes are mutually independent. Then we can define the normalized channel traffic for those packets of duration τ_i:

$$G_i = \lambda_i \tau_i, \qquad i = 1, 2. \tag{11}$$

Again assume that only those packets which do not overlap with any other packet are received correctly and define $\lambda_i' < \lambda_i$ as the rate of occurrence of those packets of duration τ_i which are received correctly. Define the normalized throughput of packets of duration τ_i as

$$S_i = \lambda_i' \tau_i, \qquad i = 1, 2. \tag{12}$$

Since we assumed two independent Poisson point processes, the probability that a short packet will be received correctly is

$$\exp\left[-\lambda_1(2\tau_1) - \lambda_2(\tau_1 + \tau_2)\right] \tag{13}$$

and if we define

$$G_{12} \triangleq \lambda_1 \tau_2 \tag{14a}$$

$$G_{21} \triangleq \lambda_2 \tau_1 \tag{14b}$$

(13) becomes

$$\exp\left[-2G_1 - G_{21} - G_2\right]. \tag{15}$$

Therefore

$$S_1 = G_1 \exp\left[-2G_1 - G_{21} - G_2\right] \tag{16a}$$

and, by a similar argument, the throughput of long packets is

$$S_2 = G_2 \exp\left[-G_{12} - G_1 - 2G_2\right]. \tag{16b}$$

For any given values of λ_1 and λ_2 we may calculate G_1, G_2, G_{12}, and G_{21}; substitution of these values into (16a) and (16b) will allow calculation of the throughputs S_1 and S_2. Therefore (16a) and (16b) may be used to define an allowable set of throughput pairs (S_1,S_2) in the (S_1,S_2) plane.

To determine the boundary of this region we define

$$\alpha \triangleq \frac{\tau_2}{\tau_1}. \tag{17}$$

Note that $\alpha \geqslant 1$. We may rewrite (16a) and (16b) in terms of α, the ratio of long packet duration to short packet duration:

$$S_1 = G_1 \exp\left[-2G_1 - \left(1 + \frac{1}{\alpha}\right) G_2\right] \tag{18a}$$

$$S_2 = G_2 \exp\left[-(1+\alpha)G_1 - 2G_2\right]. \tag{18b}$$

The boundary of the set of allowable (S_1,S_2) pairs in the (S_1,S_2) plane is defined by setting the Jacobian

$$J = \left| \frac{\partial S_i}{\partial G_j} \right|, \qquad i,j = 1, 2 \tag{19}$$

equal to zero. A simple calculation shows that the Jacobian is zero when

$$G_2 = \frac{1 - 2G_1}{\left(\dfrac{(\alpha - 1)^2}{\alpha} G_1 + 2\right)}. \tag{20}$$

Note that this checks for $G_1 = 0$ and for $\alpha = 1$.

We need only substitute this expression for G_2 into (18a) and (18b) to obtain two equations for S_1, the short packet throughput, and S_2, the long packet throughput, in terms of the single parameter G_1; and as G_1 varies from 0 (all long packets) to 1/2 (all short packets), we will trace out the boundary of the achievable values of throughput in the (S_1, S_2) plane. These achievable throughput regions are indicated for several values of α in Fig. 5.

The basic conclusion of this analysis is that the total channel throughput can undergo a significant decrease if all packets are not of the same length. Thus if the two different

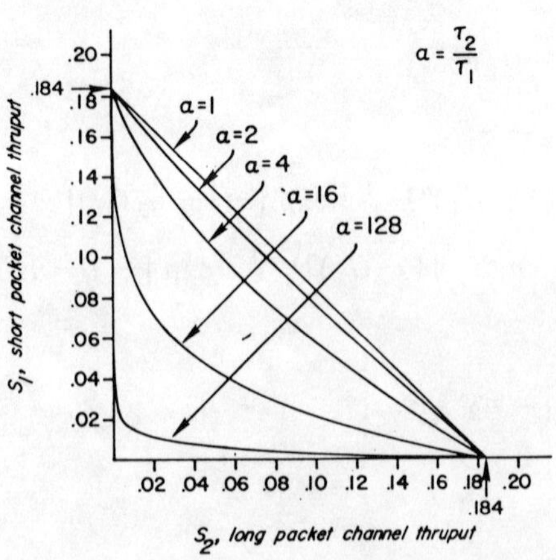

Fig. 5. Achievable throughput regions in an unslotted ALOHA channel.

packet lengths differ by a large factor, it is often preferable to break up long packets into many shorter packets as long as the overhead necessary to transmit the text in each packet is small. Ferguson [23] has generalized these results to show that channel throughput is maximized over all possible packet length distributions with fixed length packets.

In view of this discouraging result, we might conclude that an inhomogeneous mix of users inevitably leads to a decrease in the maximum value of channel throughput. Surprisingly, this conclusion is not warranted, and we shall show in Section III-B that a mix of users of varied data rates can lead to an increase in the maximum values of channel throughput.

B. Slotted Case: Variable Packet Rates

In the section we shall consider a slotted ALOHA channel used by n users, possibly with different values of channel traffic G_i. From (6) we have a set of n nonlinear equations relating the channel traffics and the channel throughputs for these n users:

$$S_i = G_i \prod_{\substack{j=1 \\ j \neq i}}^{n} (1 - G_j), \qquad i = 1, 2, \cdots, n. \tag{21}$$

Define

$$\alpha = \prod_{j=1}^{n} (1 - G_j); \tag{22}$$

then (21) can be written

$$S_i = \frac{G_i}{1 - G_i}\, \alpha, \qquad i = 1, 2, \cdots, n. \tag{23}$$

For any set of n acceptable traffic rates $G_1, G_2, \cdots, G_n$, these n equations define a set of channel throughputs $S_1, S_2, \cdots, S_n$ or a region in an n-dimensional space whose coordinates are the S_i. In order to find the boundary of this region, we calculate the Jacobian:

$$J = \left| \frac{\partial S_j}{\partial G_k} \right|, \qquad j, k = 1, 2, \cdots, n. \tag{24}$$

Since

$$\frac{\partial S_j}{\partial G_k} = \begin{cases} \prod\limits_{\substack{i=1 \\ i \neq j}}^{n} (1 - G_i), & j = k \\ -G_j \prod\limits_{\substack{i=1 \\ i \neq j,k}}^{n} (1 - G_i), & j \neq k \end{cases} \tag{25}$$

after some algebra we may write the Jacobian as

$$J = \alpha^{n-2} \begin{vmatrix} (1 - G_1) & -G_1 & -G_1 & \\ -G_2 & (1 - G_2) & -G_2 & \cdots \\ -G_3 & -G_3 & (1 - G_3) & \\ & \vdots & & \end{vmatrix}$$

$$= \alpha^{n-2} [1 - G_1 - G_2 - \cdots - G_n]. \tag{26}$$

Thus the condition for maximum channel throughputs is

$$\sum_i G_i = 1. \tag{27}$$

This condition can then be used to define a boundary to the n-dimensional region of allowable throughputs $S_1, S_2, \cdots, S_n$.

Consider the special case of two classes of users with n_1 users in class 1 and n_2 users in class 2:

$$n_1 + n_2 = n. \tag{28}$$

Let S_1 and G_1 be the throughputs and traffic rates for users in class 1, and let S_2 and G_2 be the throughputs and traffic rates for users in class 2. Then the n equations (21) can be written as the two equations

$$S_1 = G_1 (1 - G_1)^{n_1 - 1} (1 - G_2)^{n_2} \tag{29a}$$

$$S_2 = G_2 (1 - G_2)^{n_2 - 1} (1 - G_1)^{n_1}. \tag{29b}$$

For any pair of acceptable traffic rates G_1 and G_2, these two equations define a pair of channel throughputs S_1 and S_2 or a region in the (S_1, S_2) plane.

From (27) we know that the boundary of this region is defined by the condition

$$n_1 G_1 + n_2 G_2 = 1. \tag{30}$$

We can use (30) to substitute for G_1 in (29a) and (29b) and obtain two equations for S_1 and S_2 in terms of a single parameter G_2. Then as G_2 varies from 0 to 1, the resulting (S_1, S_2) pairs define the boundary of the region we seek. These achievable regions are indicated for various values of n_1 and n_2 in Figs. 6 and 7.

The important point to notice from Figs. 6 and 7 is that in a lightly loaded slotted ALOHA channel, a single large user can transmit data at a significant percentage of the total channel data rate, thus allowing use of the channel at rates well above the limit of $1/e$ or 37 percent obtained when all users have the same message rate. A throughput data rate above the $1/e$ limit

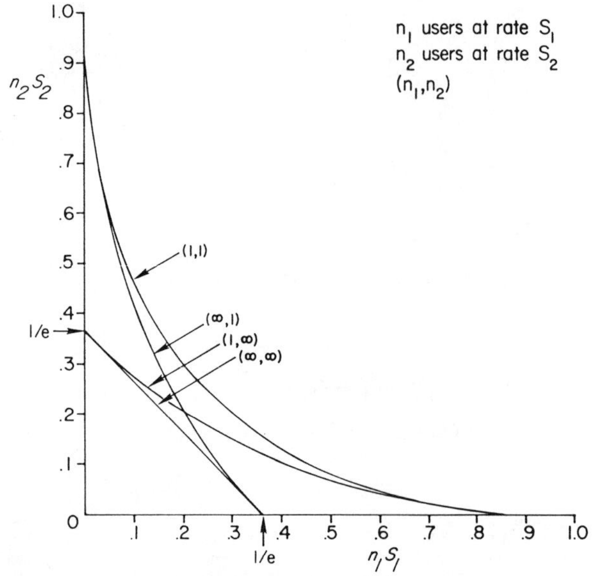

Fig. 6. Allowable channel throughputs.

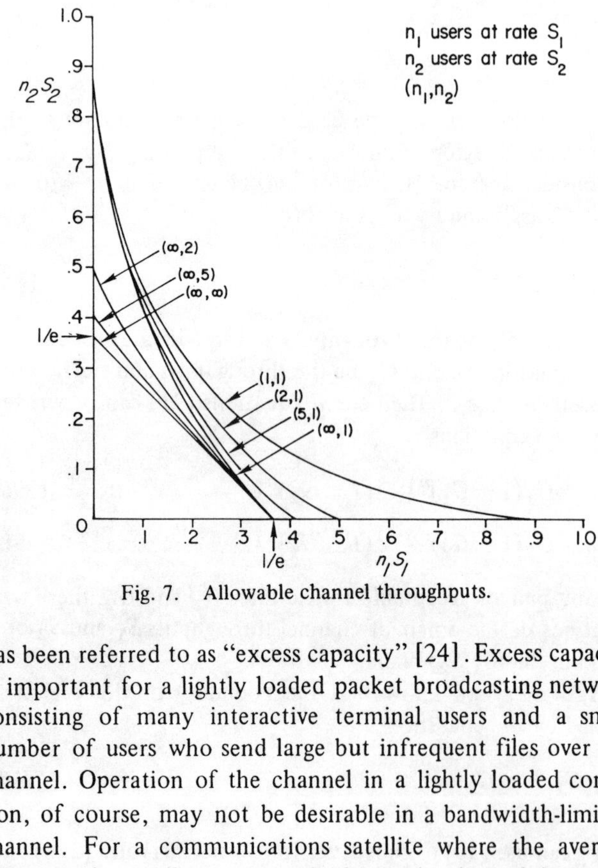

Fig. 7. Allowable channel throughputs.

has been referred to as "excess capacity" [24]. Excess capacity is important for a lightly loaded packet broadcasting network consisting of many interactive terminal users and a small number of users who send large but infrequent files over the channel. Operation of the channel in a lightly loaded condition, of course, may not be desirable in a bandwidth-limited channel. For a communications satellite where the average power in the satellite transponder limits the channel, however, operation in a lightly loaded packet-switched mode is an attractive alternative. Since the satellite will transmit power only when it is relaying a packet, the duty cycle in the transponder will be small and the average power used will be low (see Section V).

Finally, we note that it is possible to deal with certain limiting cases in more detail, to obtain equations for the boundary of the allowable (S_1,S_2) region.

1) For $n_1 = n_2 = 1$:

Upon using (30) in (29), we obtain

$$S_1 = G_1{}^2 \tag{31a}$$

$$S_2 = (1 - G_1)^2. \tag{31b}$$

2) For $n_2 \to \infty$:

$$S_1 = G_1(1 - G_1)^{n_1 - 1} \cdot \exp\left[-(1 - n_1 G_1)\right] \tag{32a}$$

$$S_2 = (1 - n_1 G_1)(1 - G_1)^{n_1 - 1} \cdot \exp\left[-(1 - n_1 G_1)\right]. \tag{32b}$$

3) For $n_1 = n_2 \to \infty$:

$$S_1 = \frac{G_1}{e} \tag{33a}$$

$$S_2 = \frac{1 - G_1}{e}. \tag{33b}$$

Additional details dealing with excess capacity and the delay experienced with this kind of use of a slotted ALOHA channel may be found in [11] and [25]. A different view of the use of a slotted packet broadcasting for different sources may be found in [43].

IV. SPATIAL PROPERTIES OF PACKET BROADCASTING NETWORKS

A. Packet Repeaters

In this section we deal with certain spatial properties of packet broadcasting networks. Not long after the initial units of the ALOHA System went into operation, it was realized that the range of the network could be extended beyond the range of a single radio link in the network (about 200 km) by the use of packet repeaters. A packet repeater operates in much the same manner as a conventional radio repeater with one major exception. Since radio transmission in a packet broadcasting network is intermittent, a packet repeater can receive a packet and retransmit that packet in the same frequency band by turning off its receiver during a retransmission burst. Thus a packet repeater can sidestep many of the frequency allocation and spatial cell problems [26] of conventional land-based repeater networks.

The use of packet repeaters leads to the consideration of packet broadcasting networks with more than one central station distributed over very large areas. Users transmit a packet, and if the packet cannot be received directly by its destination, it is forwarded to its destination by one or more packet repeaters according to some routing algorithm [27]. The study of such networks has led to the analysis of two communication theory issues related to the performance of the networks: 1) capture effect and 2) the distribution of packet traffic and packet throughput in space.

B. Capture Effect

Up to this point we have analyzed packet broadcasting channels under the pessimistic assumption that if two packets overlap at the receiver, both packets are lost. In fact, this assumption provides a lower bound to the performance of real packet broadcasting channels, since in many receivers the stronger of two overlapping packets may capture the receiver and may be received without error. Metzner [40] has used this fact to derive an interesting result, showing that by dividing users into two groups—one transmitting at high power and the other at low power—the maximum throughput can be increased by about 50 percent. This result is of importance for packet broadcasting networks with a mixture of data and packetized speech traffic.

In order to include the effect of capture in a packet broadcasting network, we consider a distribution of packet generators over a two-dimensional plane and a single packet broadcasting receiver which receives packets from these generators [41]. The receiver then may be viewed as a "packet sink" and the packet generators as a distribution of "packet sources" in the plane. We assume that the rate of generation of packets in a given area depends only on r, the distance from the packet sink, and is independent of direction θ.

Then we may define a traffic density and a throughput density analogous to the normalized traffic G and normalized thoughput S defined in Section II-B.

$G(r)$ = normalized packet traffic per unit area at a distance r.

$S(r)$ = normalized packet thruput per unit area at a distance r.

The traffic due to all packet generators in a differential ring of width dr at a radius r is

$$G(r)\,2\pi r\,dr. \tag{34}$$

We assume that packets from different users are generated so that the packet starting times of all packets generated in the differential ring constitute a Poisson point process. Then since the sum of two independent Poisson processes is a Poisson point process, if users in different rings are independent, the start times of all packets generated in a circle of radius r also constitute a Poisson point process, and the total traffic generated by all users within a distance r of the center is

$$\int_0^r G(x)2\pi x\,dx. \tag{35}$$

If we assume that a packet from a user at a distance r from the center will be received correctly unless it is overlapped by a packet sent from a user at a distance ar or less ($a \geqslant 1$), then using the results of Section II-B the probability that such a packet will be received correctly is

$$\exp\left[-4\pi \int_0^{ar} G(x)x\,dx\right]. \tag{36}$$

Any packet generated from a packet source in the circle of radius ar shown in Fig. 8 will interfere with packets generated from a source in the circle of radius r. A packet generated outside the circle of radius ar will not interfere with packets generated from a source in the circle of radius r.

We can relate the normalized packet throughput to the normalized packet traffic in the usual way:

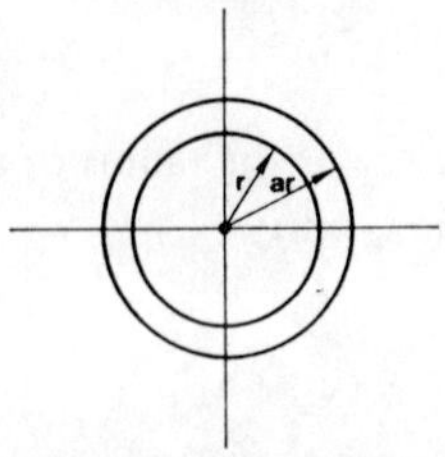

Fig. 8. Regions of interfering packets.

$$2\pi r S(r)\,dr = 2\pi r G(r) \exp\left[-4\pi \int_0^{ar} G(x)x\,dx\right] dr$$

or

$$S(r) = G(r) \exp\left[-4\pi \int_0^{ar} G(x)x\,dx\right]. \tag{37}$$

If we take a derivative of (37) with respect to r and use (37) to substitute for the exponential, we get

$$S'(r)G(r) = G'(r)S(r) - 4\pi r a^2 S(r)G(r)G(ar). \tag{38}$$

We have not found a general solution of (38) for relating $S(r)$ to $G(r)$ in the presence of capture. We have been able to analyze two special cases, however.

C. Two Solutions

In the first of these special cases we assume a constant traffic density $G(r)$. We can then show that the throughput density $S(r)$ has a Gaussian form, due to the fact that those packets generated further from the receiver will be received correctly less frequently than those packets generated close to the receiver.

In the second special case analyzed we assume a constant packet throughput density $S(r)$ and perfect capture ($a = 1$). Under these assumptions, the packet traffic density will increase as the distance from the receiver increases. We show that there exists a radius r_0 such that the packet traffic density is finite within a circle of radius r_0 around the receiver, while the packet traffic density becomes unbounded on the circle of radius r_0.

For the important case of a packet broadcasting channel distributed over some geographical area and using a packet retransmission policy (Section II-D), this result has an interesting interpretation. In such a situation any packet transmitted from a terminal located within the circle of radius r_0 will be received correctly with probability one (after a finite number of retransmissions), while the expected number of retransmissions required for a packet transmitted from a terminal further from the center than r_0 will be unbounded. Thus there exists a circle of radius r_0 such that terminals transmitting from within this circle can get their packets into the central receiver, while terminals transmitting from outside this circle spend all their time retransmitting their packets in vain. We call r_0 the Sisyphus distance of the ALOHA channel.

1) Constant Packet Traffic Density: Assume the density of normalized packet traffic is constant over the plane

$$G(r) = G_0 \tag{39}$$

and define the distance r_1 as the radius of a circle within which the total packet traffic is unity:

$$\pi r_1^2 G_0 \triangleq 1. \tag{40}$$

Then (38) reduces to

$$S'(r) = -\frac{4ra^2}{r_1^2} S(r) \tag{41a}$$

with the boundary condition

$$S_0 = G_0 \tag{41b}$$

so that the packet throughput density is

$$S(r) = G_0 \exp\left[-2a^2\left(\frac{r}{r_1}\right)^2\right] \tag{42}$$

and the total normalized packet thruput from a circle of radius r is

$$S = \int_0^r S(r')2\pi r'\, dr' = \frac{1}{2a^2}\left\{1 - \exp\left[-2\left(\frac{ar}{r_1}\right)^2\right]\right\} \tag{43}$$

and

$$\lim_{r \to \infty} S = \frac{1}{2a^2}. \tag{44}$$

Note that a total throughput which can be supported by a single packet sink with "perfect capture" ($a = 1$) is equal to one half.

2) Constant Packet Throughput Density: Another case of interest where we have found a solution for (38) is that of constant packet throughput density in the plane. Assume

$$S(r) = S_0 \tag{45}$$

over the region in the plane where $S(r)$ and $G(r)$ are bounded.

Then (38) becomes

$$G'(r) = 4\pi ra^2 G(r)G(ar). \tag{46}$$

For the case of $a = 1$ (perfect capture), (46) becomes

$$G'(r) = 4\pi r G^2(r) \tag{47}$$

with the boundary condition

$$G(0) = S_0 \tag{48}$$

so that

$$G(r) = \frac{S_0}{1 - 2\pi r^2 S_0} \tag{49}$$

Note that the normalized packet traffic per unit area is finite

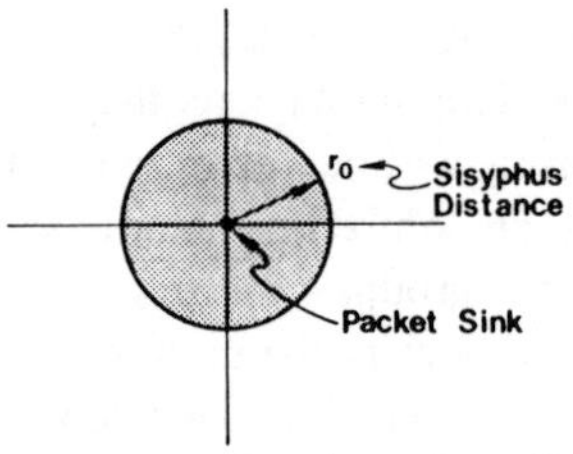

Fig. 9. Region of constant packet throughput S_0 for a single packet sink.

for

$$0 \leq r < r_0 \tag{50}$$

where

$$r_0 \triangleq [2\pi S_0]^{-1/2} \tag{51}$$

and r_0 is the Sisyphus distance mentioned in Section IV-C. Note that the Sisyphus distance also has the property that

$$\pi r_0^2 S_0 = \frac{1}{2}. \tag{52}$$

As in the previous case, the total packet throughput which can be supported by a single packet sink operating with perfect capture is one half.

V. PACKET BROADCASTING WITH AVERAGE POWER LIMITATIONS

A. Satellite Packet Broadcasting

In previous sections we have analyzed the performance of packet broadcasting channels and compared the performance of these channels to that of conventional point-to-point channels operating at the same peak data rate. Such a comparison is of interest in the case of channels limited by multiple access interference rather than noise, since an increase in the transmitted power of such channels will not lead to improved performance. But just as the average data rate of a packet broadcasting channel can be well below its peak data rate when it is operated at a low duty cycle, the average transmitted power of a packet broadcasting channel can be well below its peak transmitted power.

In this section we analyze the throughput of a packet broadcasting channel when compared to that of a conventional point-to-point channel of the same average power. This analysis is of interest in the case of satellite information systems employing thousands of small earth stations. For a satellite system the fundamental limitation in the downlink is the average power available in the satellite transponder rather than the peak power. Our results show that in the limit of large numbers of small earth stations, the packet throughput approaches 100 percent of the point-to-point capacity. Thus the multiple access capability and the complete connectivity (in the topological sense) of an ALOHA channel can be obtained at no price in average throughput. Furthermore, since our results suggest the use of higher peak power in the satellite transponder (while the average power is kept constant), the small earth stations may use smaller antennas and simpler receivers and modems than would be necessary in a conventional system.

In existing satellite systems the TWT output power in each transponder cannot be varied dynamically. In such systems the advantages implied by our analysis may be realized by frequency-division sharing a single transponder among several voice users and a single channel, operating in an ALOHA mode or some other burst mode, and occupying a frequency band equivalent to one or more voice users. The type of operation implied by our analysis also suggests investigation of high peak power satellite burst transponders (perhaps employing power devices similar to those used in radar systems) for use in information systems composed of large numbers of ultra-small earth stations.

B. Burst Power and Average Power

The capacity of a satellite channel can be calculated by the classical Shannon equation

$$C = W \log \left(1 + \frac{P}{N}\right) \tag{53}$$

where C is the capacity in bits (if the log is a base two logarithm), W is the channel bandwidth, P is the average received signal power at the earth station, and N is the average noise power at the earth station. Equation (53) expresses the capacity of the satellite channel under the assumption that the transponder transmits continuously.

If the channel is used in burst mode the transponder will emit power only when a data burst occurs, and the average power out of the transponder will be less than the burst power. Let D be the ratio of the average power transmitted to the power transmitted during a data burst. For a linear transponder D will equal the channel traffic G, and for a hard-limiting transponder D will equal the duty cycle of the channel. For both the unslotted and slotted ALOHA channel the duty cycle is $1 - e^{-G}$. Thus for a linear transponder[2]

$$D = G, \tag{54a}$$

while for a hard-limiting transponder

$$D = 1 - e^{-G}. \tag{54b}$$

Note that in the case of a hard-limiting transponder with small values of channel traffic, the duty cycle approaches that of a linear transponder.

If we retain P as the notation for the average signal power received at the earth station, the power received during a data burst will be P/D. Thus (53) should be modified in two ways.

[2] Our analysis is of significance only for $G < 1$. The analysis is formally correct, however, for all G, even though the designation of the power transmitted during bursts as "peak power" becomes inappropriate for the linear transponder case when $G > 1$. (In such a situation the "peak power" is less than the average power.)

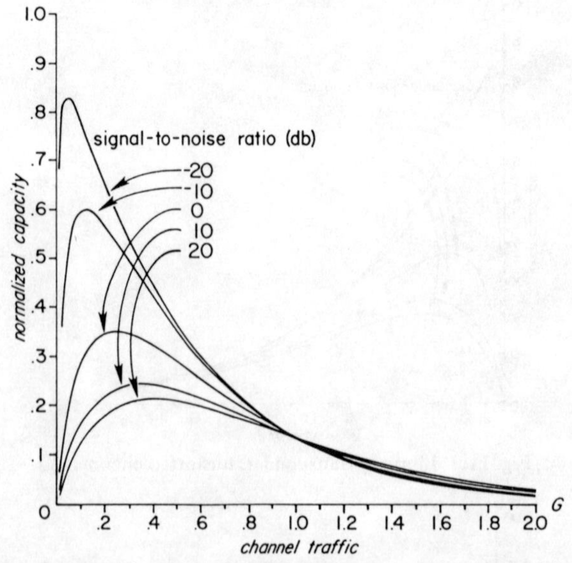

Fig. 10. Linear transponder; unslotted channel.

1) We replace W by SW to account for the fact that the channel is only used intermittently.

2) We replace P in (53) by P/D to keep the average power of the channel fixed at P.

We should note that when we make these changes, we are assuming that the packet length of the system is long enough so that the asymptotic assumptions which are used to derive (53) still apply. In practice, this is not a problem.

With these two changes then, we have four different cases.

1) Unslotted channel, linear transponder:

$$C_1 = Ge^{-2G} W \log \left(1 + \frac{P}{GN}\right). \tag{55a}$$

2) Unslotted channel, limiting transponder:

$$C_2 = Ge^{-2G} W \log \left(1 + \frac{P}{(1 - e^{-g})N}\right). \tag{55b}$$

3) Slotted channel, linear transponder:

$$C_3 = Ge^{-G} W \log \left(1 + \frac{P}{GN}\right). \tag{55c}$$

4) Slotted channel, limiting transponder:

$$C_4 = Ge^{-G} W \log \left(1 + \frac{P}{(1 - e^{-G})N}\right). \tag{55d}$$

We have calculated the normalized capacities C_i/C for $i =$ 1, 2, 3, 4 for different values of P/N, the signal-to-noise ratio of the earth station when the transponder operates continuously. The normalized capacities are plotted in Figs. 10, 11, 12, and 13 for P/N equal to –20, –10, 0, 10, and 20 dB. Of particular interest in these curves is the fact that the highest values of C_i/C occur just where we would want them to occur—for small values of channel traffic (G) and for small earth stations (low P/N). In the limit we have (for a fixed value of G)

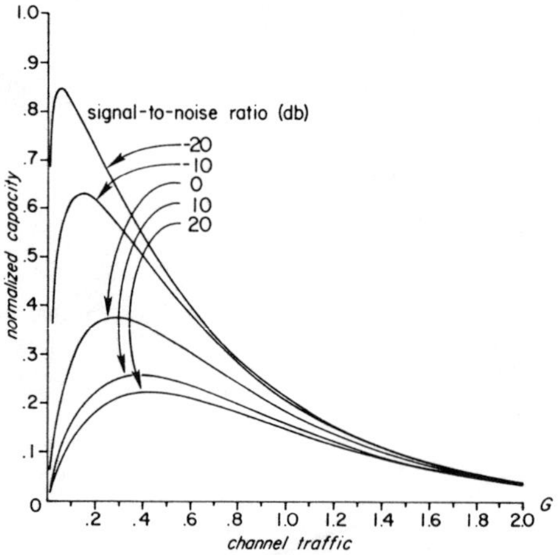

Fig. 11. Limiting transponder; unslotted channel.

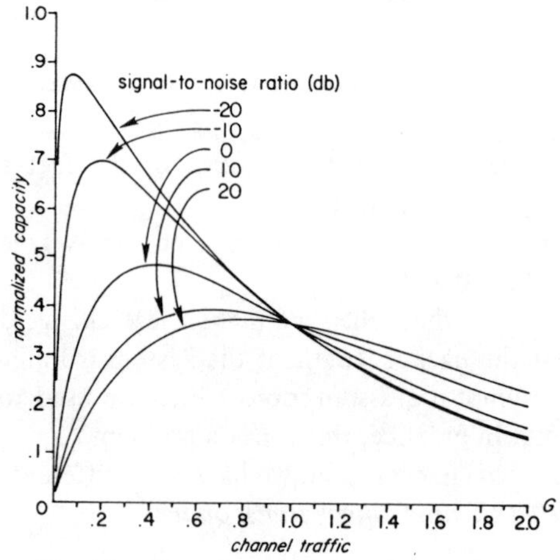

Fig. 12. Linear transponder; slotted channel.

$$\lim_{\frac{P}{N}\to 0} \frac{C_i}{C} = \frac{S}{D}, \qquad i = 1, 2, 3, 4 \tag{56}$$

so that

1) unslotted channels, linear transponder

$$\lim_{\frac{P}{N}\to 0} \frac{C_1}{C} = e^{-2G} \tag{57a}$$

2) unslotted channels, limiting transponder

$$\lim_{\frac{P}{N}\to 0} \frac{C_2}{C} = \frac{Ge^{-2G}}{(1 - e^{-G})} \tag{57b}$$

3) slotted channel, linear transponder

$$\lim_{\frac{P}{N}\to 0} \frac{C_3}{C} = e^{-G}$$

4) slotted channel, limiting transponder

$$\lim_{\frac{P}{N}\to 0} \frac{C_4}{C} = \frac{Ge^{-G}}{(1 - e^{-G})} \tag{57d}$$

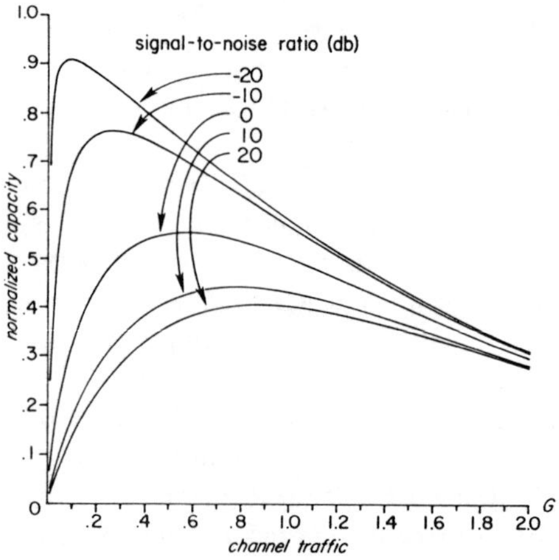

Fig. 13. Limiting transponder; slotted channel.

and in all cases

$$\lim_{G\to 0} \lim_{\frac{P}{N}\to 0} \frac{C_i}{C} = 1. \tag{58}$$

Thus this multiplexing technique allows a network of small inexpensive earth stations to achieve the maximum value of channel capacity, at the same time providing complete connectivity and multiple access capability.

VI. BACKGROUND AND ACKNOWLEDGMENT

The term "packet broadcasting" was first coined by Robert Metcalfe in his Ph.D. dissertation [28]. As is often the case with simple ideas, the concept of combining burst transmission and Poisson user statistics to provide random access to a channel has occurred independently to a number of investigators. The first attempt at an analysis of such a system of which I am aware is contained in an internal Bell Laboratories memorandum by Schroeder [29], suggested by an earlier paper by Pierce and Hopper [30]. Two other early related papers were written by Costas [31] and Fulton [32]. Of course, a theoretical analysis is not necessary in order to build such a system, and anyone who has sat in a taxi listening to the staccato voice bursts of a radio dispatcher and a set of taxi drivers sharing a single voice channel will recognize the operation of a voice packet broadcasting channel using a carrier sense protocol. And even after an analysis is available, the concept of packet broadcasting may be suggested without reference to the theory [33].

The first papers analyzing packet broadcasting in the form implemented in the ALOHA System [6] assumed fixed packet throughput and a retransmission protocol as described in Section II-D-1). This approach leads to a number of questions involving optimum retransmission policy [28], the behavior of the channel with a finite number of users [39], stability of the channel [13], and transmission of long files by means of various reservation schemes [34]; [44]. A comprehensive treatment of these as well as other interesting packet broadcasting questions may be found in Kleinrock [42]. In this paper we have taken a different approach by assuming a given packet traffic rather than throughput. With such a starting point, the questions mentioned above do not assume key importance in the theory, although their practical importance is not diminished.

Much of the theory of packet broadcasting was developed in two working groups sponsored by the Advanced Research Projects Agency of the Department of Defense. These groups circulated a private series of working papers—the ARPANET Satellite System notes (ASS notes) and the Packet Radio Temporary notes (PRT notes)—where many of the theoretical results described or referenced in this paper appeared for the first time. Unfortunately, the several references to ASS notes in papers subsequently published in the open literature may have produced some confusion in the minds of those trying to trace the references. Among the most significant of the ASS note and PRT note results was the first derivation of the capacity of a slotted ALOHA channel and the first analysis of the use of the capture effect in packet broadcasting, both by Larry Roberts. That note has since been republished in the open literature [41].

The results of Section III-A dealing with two different packet lengths were suggested by an ASS note written by Tom Gaarder, and the results of Section III-B dealing with the excess capacity of a slotted channel were suggested by an ASS note written by Randy Rettberg. Other problems which were first analyzed in ASS notes or PRT notes but not emphasized in this paper include various packet broadcasting reservation systems [22], [35], [36], carrier sense packet broadcasting [18], [19], and questions dealing with packet routing and protocol issues in a network of repeaters [37]. The reader interested in theoretical network protocol questions should also see Gallagher [38], although this work did not originate in an ASS note or PRT note.

The first system to employ packet broadcasting techniques was the ALOHA System computer network at the University of Hawaii in 1970. Subsequently, packet repeaters were added to the network and packet broadcasting by satellite was demonstrated in the system. Some of the people involved in the implementation and development of the system were Richard Binder, Chris Harrison, Alan Okinaka, and David Wax.

The historical relevance of [29] and [32] was pointed out to me by Joe Aein, to whom I am indebted, in spite of my embarassment at having forgotten I was thesis supervisor on the second of these papers.

REFERENCES

[1] L. G. Roberts and B. D. Wessler, "Computer network development to achieve resource sharing," in *1970 Spring Joint Comput. Conf., AFIPS Conf. Proc.*, vol. 36. Montvale, NJ: AFIPS Press, 1970, pp. 543–549.

[2] L. G. Roberts, "Data by the packet," *IEEE Spectrum*, vol. 11, pp. 46–51, Feb. 1974.

[3] W. W. Chu, "A study of asynchronous time division multiplexing for time-sharing computer systems," in *1969 Spring Joint Comput. Conf., AFIPS Conf. Proc.*, vol. 35. Montvale, NJ: AFIPS Press, 1969, pp. 669–678.

[4] F. Heart, R. Kahn, S. Ornstein, W. Crowther, and D. Walden, "The interface message processor for the ARPA computer network," in *1970 Spring Joint Comput. Conf., AFIPS Conf. Proc.*, vol. 36. Montvale, NJ: AFIPS Press, 1970, pp. 551–567.

[5] H. Frank, I. Frisch, and W. Chou, "Topological considerations in the design of the ARPA computer network," in *1970 Spring Joint Comput. Conf., AFIPS Conf. Proc.*, vol. 36. Montvale, NJ: AFIPS Press, 1970, pp. 581–587.

[6] N. Abramson, "The ALOHA system—Another alternative for computer communications," in *1970 Fall Joint Comput. Conf., AFIPS Conf. Proc.*, vol. 37. Montvale, NJ: AFIPS Press, 1970, pp. 281–285.

[7] R. E. Kahn, "The organization of computer resources into a packet radio network," in *Nat. Comput. Conf., AFIPS Conf. Proc.*, vol. 44, May 1975, pp. 177–186.

[8] N. Abramson and E. R. Cacciamani, Jr., "Satellites: Not just a big cable in the sky," *IEEE Spectrum*, vol. 12, pp. 36–40, Sept. 1975.

[9] I. T. Frisch, "Technical problems in nationwide networking and interconnection," *IEEE Trans. Commun.*, vol. COM-23, pp. 78–88, Jan. 1975.

[10] R. M. Metcalfe and D. R. Boggs, "ETHERNET: Distributed packet switching for local computer networks," *Commun. Ass. Comput. Mach.*, to be published.

[11] N. Abramson, "Packet switching with satellites," in *Nat. Comput. Conf., AFIPS Conf. Proc.*, vol. 42, 1973, pp. 695–702.

[12] R. D. Rosner, "Optimization of the number of ground stations in a domestic satellite system," in *EASCON'75 Rec.*, Sept. 29–Oct. 1, 1975, pp. 64A–64J.

[13] A. B. Carleial and M. E. Hellman, "Bistable behavior of ALOHA-type systems," *IEEE Trans. Commun.*, vol. COM-23, pp. 401–410, Apr. 1975.

[14] P. E. Jackson and C. D. Stubbs, "A study of multiaccess computer communications," in *1969 Spring Joint Comput. Conf., AFIPS Conf. Proc.*, vol. 34. Montvale, NJ: AFIPS Press, 1969, pp. 491–504.

[15] T. J. Klein, "A tactical packet radio system," in *Proc. Nat. Telecommun. Conf.*, New Orleans, LA, Dec. 1975.

[16] K. Ah Mai, "Organizational alternatives for a Pacific educational computer-communication network," ALOHA Syst. Tech. Rep. CN74-27, Univ. Hawaii, Honolulu, May 1974.

[17] R. Binder, R. Rettberg, and D. Walden, *The Atlantic Satellite Packet Broadcast and Gateway Experiments.* Cambridge, MA: Bolt Beranek and Newman, 1975.

[18] L. Kleinrock and F. A. Tobagi, "Packet switching in radio channels: Part I—Carrier sense multiple-access modes and their throughput-delay characteristics," *IEEE Trans. Commun.*, vol. COM-23, pp. 1400–1416, Dec. 1975.

[19] F. A. Tobagi and L. Kleinrock, "Packet switching in radio channels: Part II—The hidden terminal problem in carrier sense multiple-access and the busy-tone solution," *IEEE Trans. Commun.*, vol. COM-23, pp. 1417–1433, Dec. 1975.

[20] R. T. Chien, L. R. Bahl, and D. T. Tang, "Correction of two erasure bursts," *IEEE Trans. Inform. Theory*, vol. IT-15, pp. 186–187, Jan. 1969.

[21] N. Abramson, "Cyclic code groups," *Problems of Inform. Transmission*, Acad. Sci. USSR, Moscow, vol. 6, no. 2, 1970.

[22] L. G. Roberts, "Dynamic allocation of satellite capacity through packet reservation," in *Nat. Comput. Conf. AFIPS Conf. Proc.*, vol. 42, June 1973, pp. 711–716.

[23] M. J. Ferguson, "A study of unslotted ALOHA with arbitrary message lengths," in *Proc. 4th Data Commun. Symp.*, Quebec, Canada, Oct. 7–9, 1975, pp. 5-20–5-25.

[24] R. Rettberg, "Random ALOHA with slots-excess capacity," ARPANET Satellite Syst. Note 18, NIC Document 11865, Stanford Res. Inst., Menlo Park, CA, Oct. 11, 1972.

[25] N. Abramson, "Excess capacity of a slotted ALOHA channel (continued)," ARPANET Satellite Syst. Note 30, NIC Document 13044, Stanford Res. Inst., Menlo Park, CA, Dec. 6, 1972.

[26] L. Schiff, "Random-access digital communication for mobile radio in a cellular environment," *IEEE Trans. Commun.*, vol. COM-22, pp. 688–692, May 1974.

[27] H. Frank, R. M. Van Slyke, and I. Gitman, "Packet radio network design—System considerations," in *Nat. Comput. Conf., AFIPS Conf. Proc.*, vol. 44, May 1975, pp. 217–232.

[28] R. M. Metcalfe, "Packet communication," Rep. MAC TR-114, Project MAC, Massachussetts Inst. Technol., Cambridge, July 1973.

[29] M. R. Schroeder, "Nonsynchronous time multiplex system for speech transmission," Bell Lab. Memo., Jan. 19, 1959.

[30] J. R. Pierce and A. L. Hopper, "Nonsynchronous time division with holding and random sampling," *Proc. IRE*, vol. 40, pp. 1079–1088, Sept. 1952.

[31] J. P. Costas, "Poisson, Shannon, and the radio amateur," *Proc. IRE*, vol. 47, p. 2058, Dec. 1959.
[32] F. F. Fulton, Jr., "Channel utilization by intermittent transmitters," Tech. Rep. 2004-2, Stanford Electron. Lab., Stanford Univ., Stanford, CA, May 12, 1961.
[33] K. D. Levin, "The overlapping problem and performance degradation of mobile digital communication systems," *IEEE Trans. Commun.*, vol. COM-23, pp. 1342–1347, Nov. 1975.
[34] R. Binder, "A dynamic packet-switching system for satellite broadcast channels," in *Conf. Rec., Int. Conf. Commun.*, vol. III, June 1975, pp. 41-1–41-5.
[35] S. S. Lam and L. Kleinrock, "Packet switching in a multiaccess broadcast channel: Dynamic control procedures," *IEEE Trans. Commun.*, vol. COM-23, pp. 891–904, Sept. 1975.
[36] L. Kleinrock and S. S. Lam, "Packet switching in a multiaccess broadcast channel: Performance evaluation," *IEEE Trans. Commun.*, vol. COM-23, pp. 410–423, Apr. 1975.
[37] I. Gitman, "On the capacity of slotted ALOHA networks and some design problems," *IEEE Trans. Commun.*, vol. COM-23, pp. 305–317, Mar. 1975.
[38] R. G. Gallager, "Basic limits on protocol information in data communication networks," *IEEE Trans. Inform. Theory*, vol. IT-22, pp. 385–398, July 1976.
[39] M. J. Ferguson, "On the control, stability, and waiting time in a slotted ALOHA random-access system," *IEEE Trans. Commun.*, vol. COM-23, pp. 1306–1311, Nov. 1975.
[40] J. J. Metzner, "On improving utilization in ALOHA networks," *IEEE Trans. Commun.*, vol. COM-24, pp. 447–448, Apr. 1976.
[41] L. G. Roberts, "ALOHA packet system with and without slots and capture," *Comput. Commun. Rev.*, vol. 5, pp. 28–42, Apr. 1975.
[42] L. Kleinrock, *Queueing Systems, Volume 2: Computer Applications.* New York: Wiley, 1976, pp. 360–407.
[43] I. Gitman, R. M. Van Slyke, and H. Frank, "On splitting random accessed broadcast communication channels," in *Proc. 7th Hawaii Int. Conf. Syst. Sci.–Suppl. on Comput. Nets*, Western Periodicals, Jan. 1974, pp. 81–85.
[44] W. Crowther, R. Rettberg, D. Walden, S. Ornstein, and F. Heart, "A system for broadcast communication: Reservation ALOHA," in *Proc. 6th Hawaii Int. Conf. Syst. Sci.*, Western Periodicals, Jan. 1973, pp. 371–374.

Aloha Packet System with and Without Slots and Capture

LAWRENCE G. ROBERTS

CAPTURE

FM receivers will track the strongest of many signals if the power of the next-to-strongest is down by 1.5–3 dB, a specification called the *capture ratio.* Two major factors affect the power received: the $1/r^2$ distance factor and random individual power variations. The distance factor is major, whereas the individual power levels are usually less than 1.5–3 dB. Since variation will help guarantee that one signal is captured, and since individual variations will probably not help much, we will look only at distance variation.

Define

Capture Ratio: $CR = 10 \log (P_1/P_2)$, P_1 is stronger signal
Power Ratio: $\beta = P_2/P_1$, $\quad 0 \le \beta \le 1$
Distance factor: $r_2^2 = \beta r_1^2$ $\quad \beta^{1/2} = r_2/r_1$,

Assuming the receiver is surrounded by an equal density population of consoles within a circle of radium R, the only consoles that can interfere with a given station at radius r are the stations within the circle of radium $\beta^{-1/2} r$.

Define

$\gamma =$ The total bit rate being transmitted by all stations as a fraction of the total bandwidth.
$\sigma =$ The actual bit rate fraction being received correctly.
$q =$ Probability of packet being received correctly.

Thus $\sigma = \gamma q$.

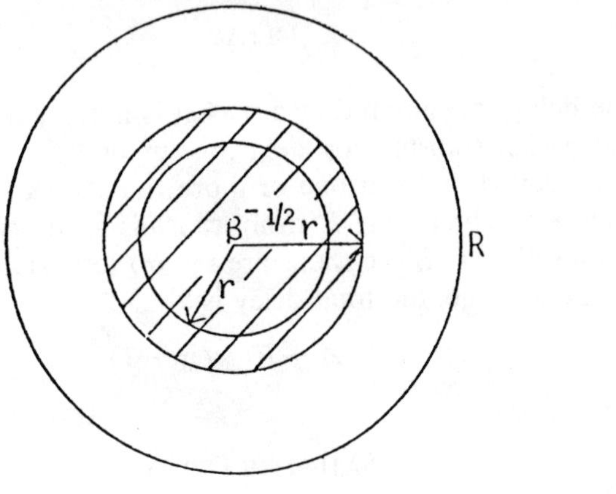

To compute q for the average station given a transmit rate γ is the task.

There are two cases, asynchronous and synchronous. In both cases we assume a Poisson point process for the totality of stations wanting to start packets of length T seconds at random time with a rate $\lambda = \gamma/T$. Thus, the probability of no conflict within a time (mT) is:

$$q_o = e^{-\gamma m}$$

However, if only a fraction k of the stations are in the proper position to prohibit your being captured, the Poisson rate is reduced by k. At a given radius r, the interfacing stations are those in the area $\pi r^2/\beta$ whereas the total area is πR^2. Thus, the fraction of stations that would cause conflict is the area ratio. Thus:

$$q(r) = e^{-k\gamma m}, \qquad k = \frac{r^2}{\beta R^2}$$

The number of stations at any radius r increases linearly with r as the circumference increases. The probability density is therefore:

$$p(r) = \frac{2\pi r}{\pi R^2}, \qquad \int_0^R p(r)\, dr = 1$$

Now, to compute the average q we must integrate the product $p(r)q(r)$ over r out to $R\sqrt{\beta'}$ and $p(r)q_o$ from $R\sqrt{\beta'}$ to R, since here all stations interfere.

$$q = \int_0^{R\sqrt{\beta'}} p(r)q(r)\, dr + \int_{R\sqrt{\beta'}}^R p(r)q_o\, dr$$

$$q = \int_0^{R\sqrt{\beta'}} \frac{2r}{R^2} e^{-\frac{r^2\gamma m}{\beta R^2}}\, dr + \int_{R\sqrt{\beta'}}^R \frac{2r}{R^2} e^{-\gamma m}\, dr$$

$$q = \frac{\beta}{\gamma m}(1 - e^{-\gamma m}) + (1 - \beta)e^{-\gamma m}$$

Having found the average q, the $\sigma = \gamma_q$ relationship will give us σ:

$$\sigma = \frac{\beta}{m}(1 - e^{-\gamma m}) + (1 - \beta)\gamma e^{-\gamma m}$$

Note that this is the linear combination of two solutions, the solution for no capture (when $\beta = 0$) and the solution for full capture, where one packet is always received correctly.

Now it is necessary to look at the two cases to determine m.

Reprinted from *Computer Communications Review*, vol. 5, no. 2, pp. 28–42, April 1975. Reprinted with Permission.

Asynchronous

A given packet will come into conflict with packets starting within a period $2T$ since each packet has length T.

$\leftarrow T \rightarrow$ packet

$| \leftarrow 2T \rightarrow |$ conflict period

Thus, as was derived by University of Hawaii, $m = 2$.

$$\sigma_a = \frac{\beta}{2}(1 - e^{-2\gamma a}) + (1 - \beta)\gamma_a e^{-2\gamma_a}$$

Synchronous Slots

Packets are always put into slots. Each station keeps a clock that counts off slot times and is resynchronized by observing other packets. The slots must have a slightly longer T to prevent errors but this factor will be considered layer. The period of packet starting times giving rise to a transmission in a given slot is just the previous slot time. Thus, the Poisson period is $T(m = 1)$.

$\leftarrow T \rightarrow$ slot

$| \leftarrow T \rightarrow |$ conflict period

$$\sigma_s = \beta(1 - e^{-\gamma_s}) + (1 - \beta)\gamma_s e^{-\gamma_s}$$

Note that these two formulas are identical under the following substitutions:

$$\sigma_a = \tfrac{1}{2}\sigma_s, \qquad \gamma_a = \tfrac{1}{2}\gamma_s$$

Thus, slot solutions are good for both, keeping in mind that for asynchronous channel use the rates are 1/2 that of the synchronous case. That is, for any given channel bandwidth and capture ratio the peak rate achievable, or the rate at a given delay (dependent on γ/σ), will be only half as good in the asynchronous case as if slots are defined. Slot padding will mean some of this advantage is lost, but unless the slots become twice the packet size the slot technique is significantly better.

Other Population Densities

If the population density of stations is not uniform within a circle, the next most realistic ground-based case is that of a long narrow strip of land or a circular situation with population density decreasing with radius (like most cities). Both cases result in the same distribution, so consider the long strip of land:

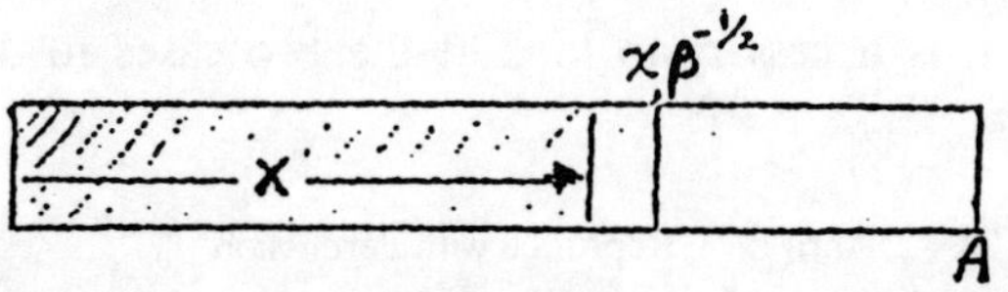

Probability of x:

$$p(x) = 1/A$$

Fraction of stations to cause conflict:

$$k = \frac{x}{A\sqrt{\beta'}}$$

$$q = \int_0^{a\sqrt{\beta'}} \frac{1}{A} e^{-k\gamma m}\, dx + \int_{A\sqrt{\beta'}}^{A} \frac{1}{\beta} e^{-\gamma m}\, dx$$

$$q = \frac{\sqrt{\beta'}}{\gamma m}(1 - e^{-\gamma m}) + \left(1 - \sqrt{\beta'}\right)e^{-\gamma m}$$

$$\sigma = \frac{\sqrt{\beta'}}{m}(1 - e^{-\gamma m}) + \left(1 - \sqrt{\beta'}\right)\gamma e^{-\gamma m}$$

Therefore, the solution is identical except that $\sqrt{\beta'}$ is substituted for β, improving the situation since the capture ratio required for a given performance can be twice that of the previous case.

The worst case is typified by a satellite situation where all stations are almost equidistant from each other (via the satellite). In the satellite case capture can be presumed impossible ($\beta = 0$), unless capture ratios considerably better than 1.5 dB can be achieved (1.5 dB = the power ratio of the extreme case).

Maximum σ for Slots

Except for the unrealizable case where the capture ratio $= 0$ ($\beta = \infty$) in which the maximum $\sigma = 1$ at $\gamma = \infty$, there is a true maxima for σ.

$$\sigma \max = \beta + (1 - \beta)e^{-\gamma_1}$$

where

$$\gamma_1 = \frac{1}{1 - \beta}$$

Delay

The delay encountered by a packet is initially the channel and packet transmission delay, assuming no conflict. Call this delay A. Next, if one or more retransmissions occur, each takes some time R from transmission to subsequent retransmission. Since there are (γ/σ) total transmissions on the average, the total delay is:

$$D = A + R(\gamma/\sigma - 1)$$

Satellite Delay

For a satellite the initial delay, A, is the up ≠ down satellite delay ($C = .27$ s) plus the packet time, T.

$$A = C + T$$

Retransmission occurs after seeing your own down signal

being in conflict, and waiting randomly about 10 packet times to avoid direct reconflict.

$$R = C + 6T$$

Thus

$$D = (C = 6T)\frac{\gamma}{\sigma} - 5T$$

Typical times for a 50KB channel, 1400 bit packets are:

$$D = .438\frac{\gamma}{\sigma} - .14\,\mathrm{s}$$

Ground Radio Delay

Here the propagation delay, C, is much less but the same rule holds for A.

$$A = C + T$$

The retransmission can only occur after acknowledgment. However, this can always be sent within one packet time (ave $= 1/2\ T$). Two transmission delays occur, however, and a random wait over nine-packet times is assumed.

$$R = 2C + 7T$$

Thus

$$D = (2C - 7T)\frac{\gamma}{\sigma} - (C + 6T)$$

Typical time for 50KB, 1400 bit packets at distances under 100 miles are

$$D = .196\frac{\gamma}{\sigma} - .168\,\mathrm{s}$$

Operating Point

When σ is pushed to the maximum for any variant, the system is dynamically unstable, as it is if γ is somehow pushed beyond the second solution of the σ, γ equation. Thus, an operating point must be chosen below the maximum σ point, which is dynamically stable and has a reasonable delay. A convenient point to pick, both for operation and for discussing the capacity of the system, is the point where $q = 1/2$ and $\gamma = 2\sigma$. This means the delay is fixed if comparisons between variants are considered and that for actual use there are only two transmissions on the average. This also constrains the maximum delay which can grow quickly if $(1 - q)$ gets closer to unity.

At $q = 1/2$:

$$\beta = \frac{\gamma m(.5 - e^{-8m})}{1 - (1 + \gamma m)e^{-\gamma m}}, \qquad \sigma = .5\gamma$$

where $m = 1$ for slots and $m = 2$ for asynchronous.

	Capture Ratio	β	σ_a	σ_s
Infeasible	0	1.0	.40	.80
Good FM	1.5 dB	.707	.30	.60
Mod. FM	3.0 dB	.500	.25	.50
Poor FM	6.0 dB	.250	.205	.41
Satellite	0	0	.173	.347

Slot Padding

For satellite operation, each station knows its delay to the satellite and the padding can be minimal unless very high bandwidths are used. For UHF radio with a 25-mile radius the padding is negligible at 50 KB, 10% at 500 KB, and 50% at 5 MB for 1200 bit packets. This is assuming no delay correction computations as with satellites. Beyond 5 MB either delay correction must be incorporated or the asynchronous system used if 1200 bit packets are still desired. Of course, it is at this same bandwidth that multiple receivers begin to help sort out the conflicts—so delay correction may be unnecessary to achieve a consistently high channel utilization.

Conclusion

For satellite use, slots will improve the capacity by a factor of two to .347. For ground radio, a good FM receiver would per mit a further improvement to 60% of the full bandwidth. The delay in this case is only eight packet times. slots clearly need not be used at very low channel utilizations since the asynchronous system will work just as well at low rates. Thus, the resynchronization of the slot clock is not a problem. If no packets come by and the clock drifts, no harm is done—we just have the asynchronous case.

ALOHA SLOT PACKET SYSTEM

Maximum Channel Utilization vs. FM Capture Ratio
for evenly populated area around station

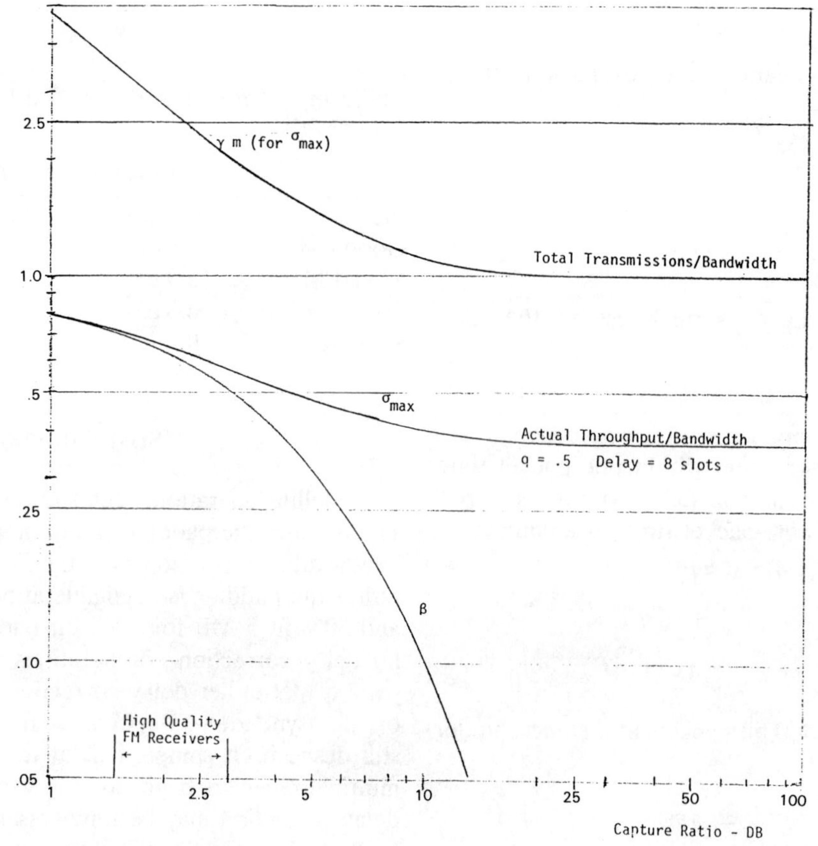

SATELLITE DELAY

Asynchronous and Slot Systems
for 50 KB Using 1400 Bit Packets

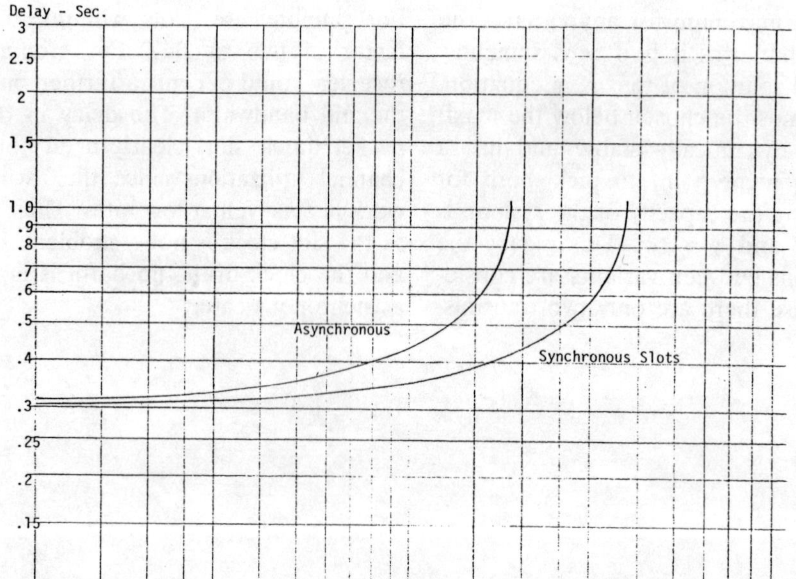

ALOHA SLOT SYSTEM

Ground Use Delay for 1400 Bits/50 KB

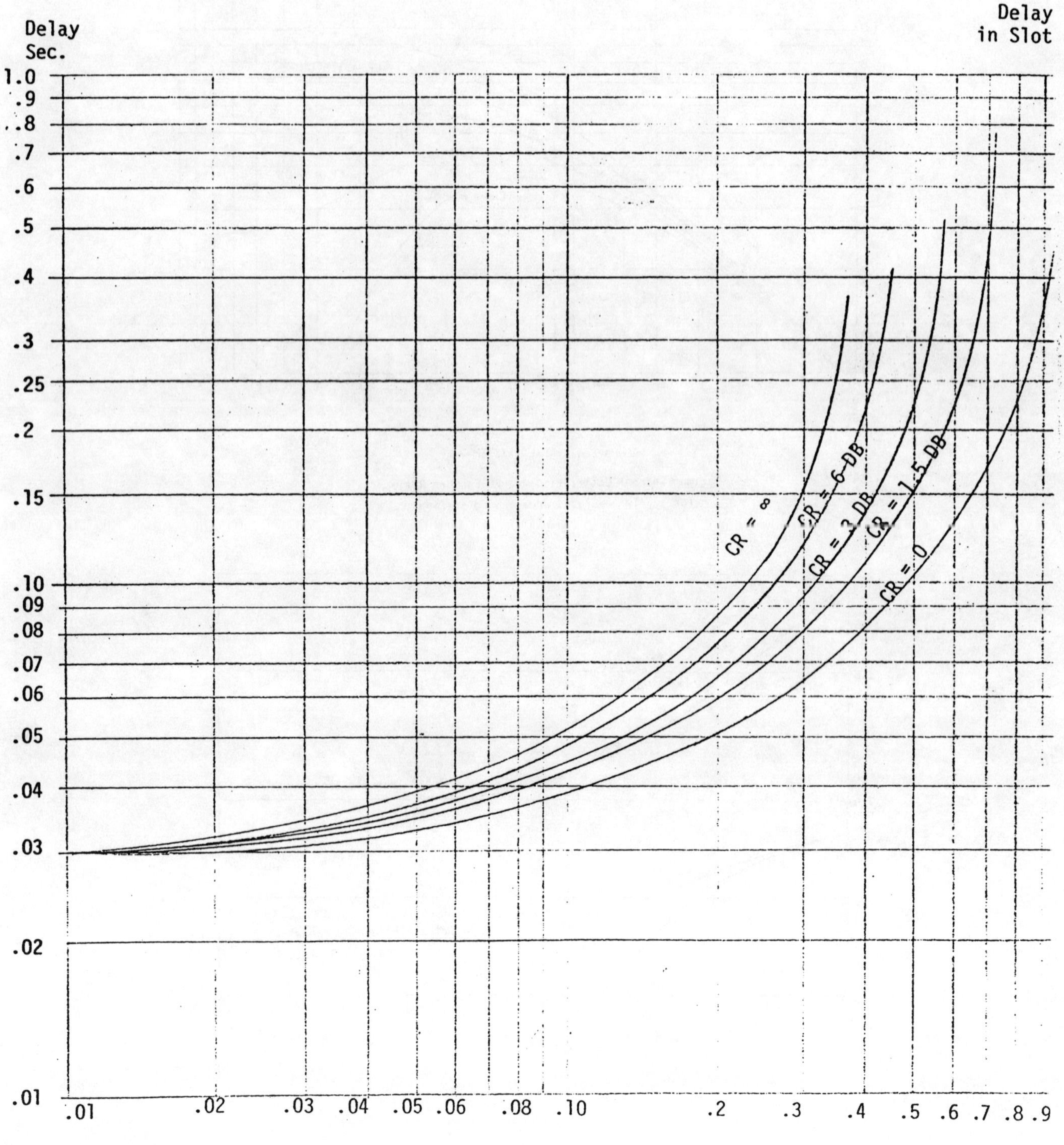

ALOHA SLOT SYSTEM

σ vs. γ

For Capture Ratios from 0-∞

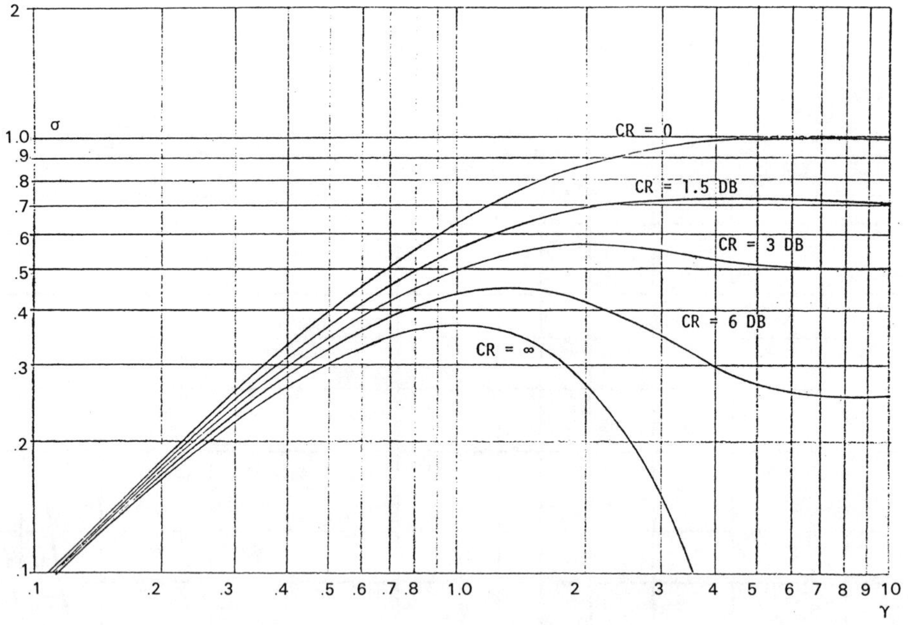

ALOHA SLOT PACKET SYSTEM

Maximum Channel Utilization vs. FM Capture Ratio

for evenly populated area around station

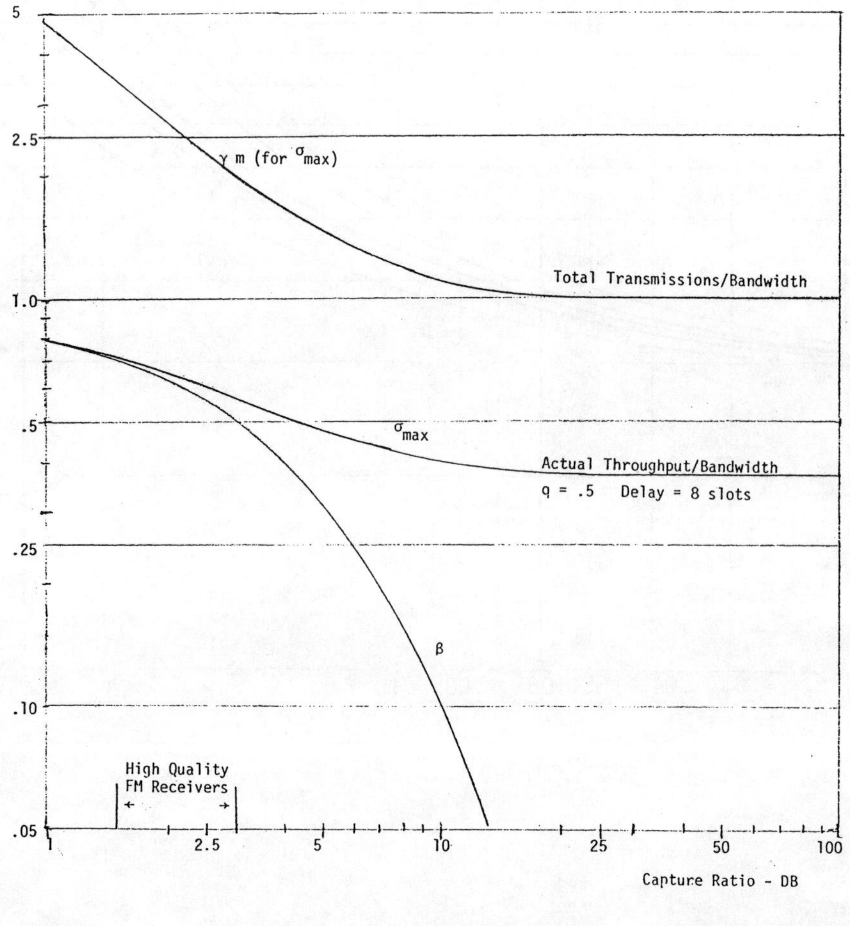

SATELLITE DELAY

Asynchronous and Slot Systems

for 50 KB Using 1400 Bit Packets

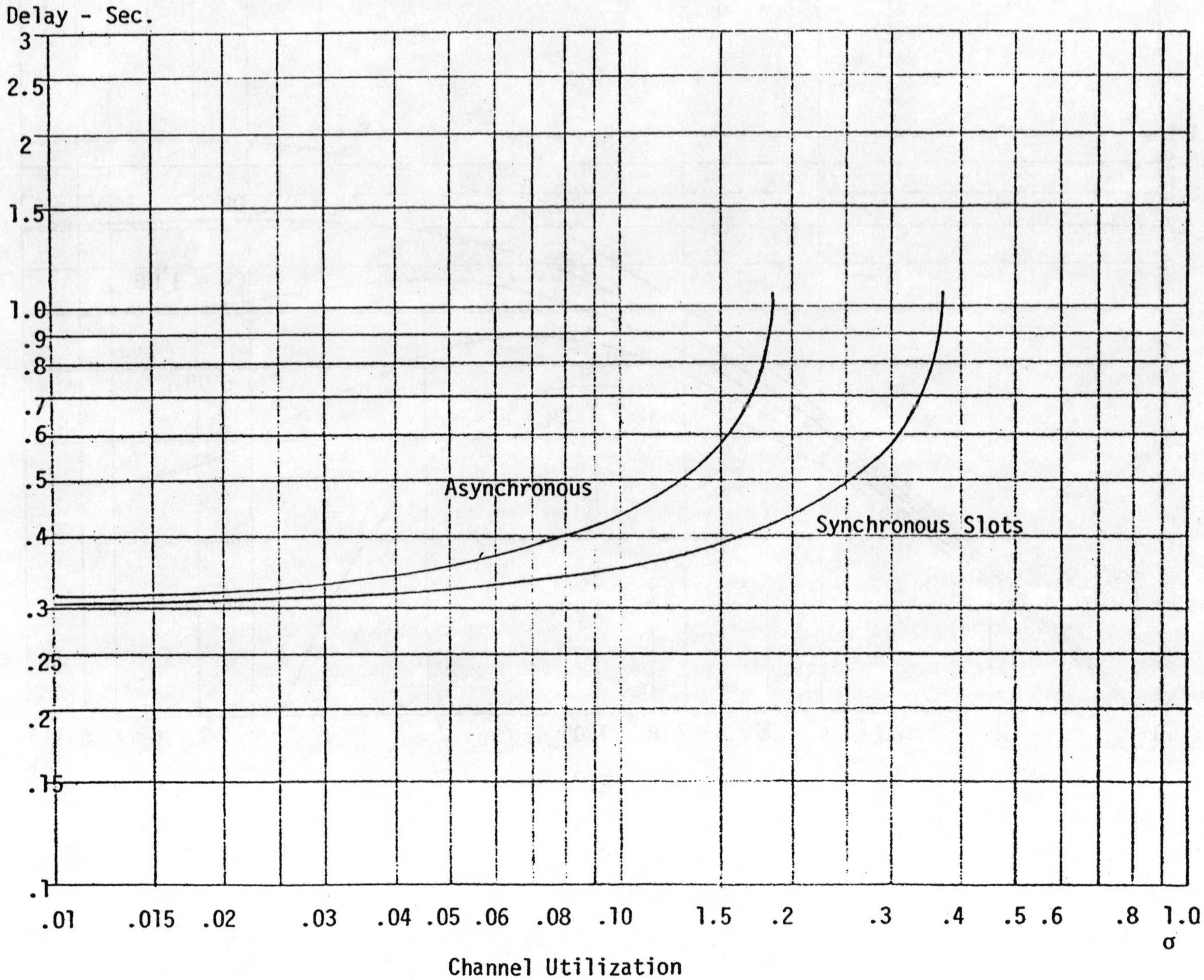

ALOHA SLOT SYSTEM

σ vs. γ

For Capture Ratios from 0-∞

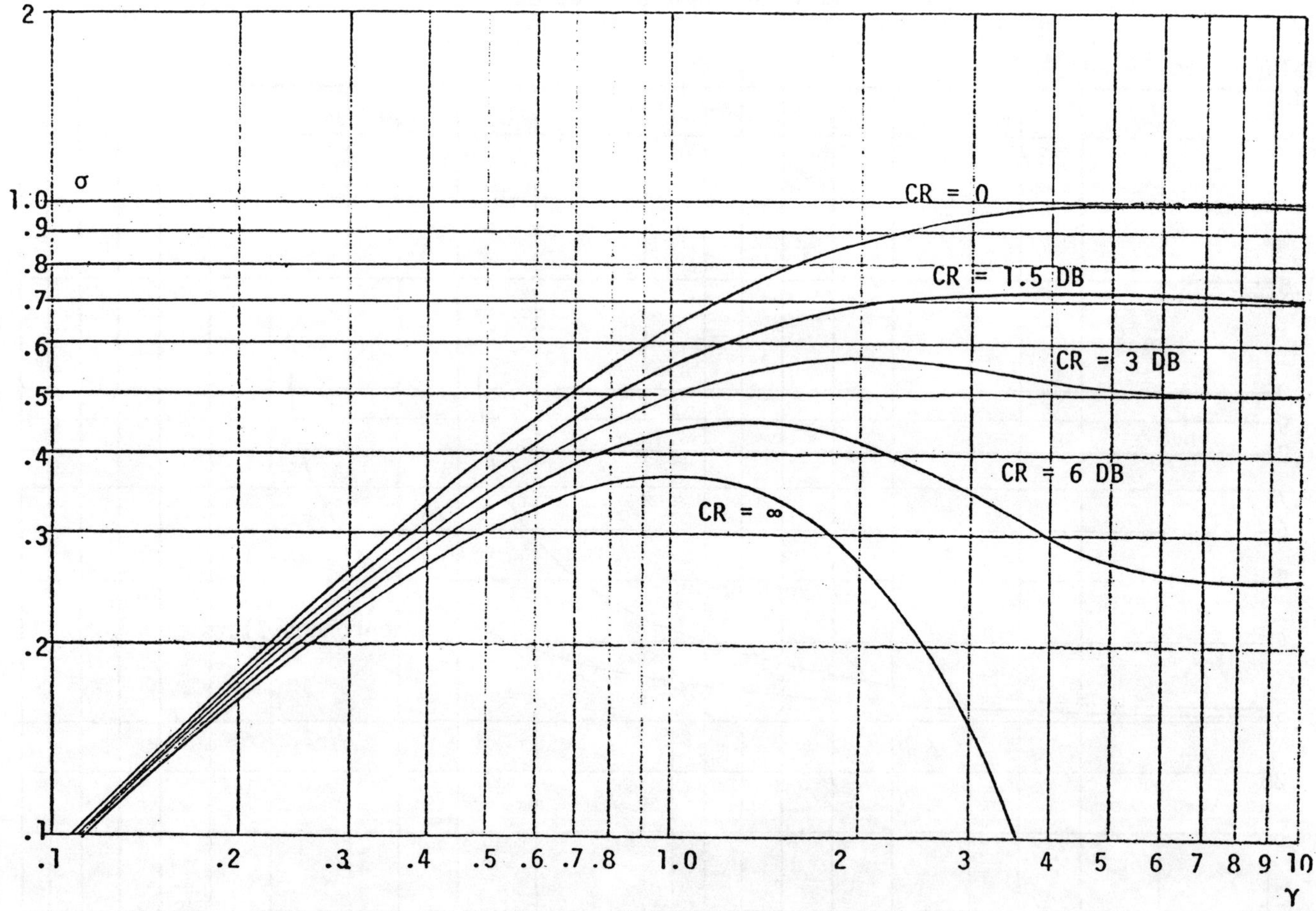

Bistable Behavior of ALOHA-Type Systems

AYDANO B. CARLEIAL, STUDENT MEMBER, IEEE, AND MARTIN E. HELLMAN, MEMBER, IEEE

Abstract—**Packet switching has found widespread application in computer communications because of its ability to efficiently handle high ratios of peak-to-average data rate. Packet radio is the application of packet switching techniques to radio channels. The resultant multiple-access problem requires novel approaches. Such approaches have been developed by others and have primarily been analyzed in steady-state behavior. This paper demonstrates and analyzes an important aspect of the dynamic characteristics of packet radio, namely, that of bistable behavior. That is, the system possesses two statistically stable equilibrium points, one in a desirable low-delay region, and the other in an undesirable high-delay region. Since the stability is only statistical in nature, the system oscillates between these two points. Even if the resultant steady-state behavior is very poor, this dynamic analysis frequently shows that system performance will be acceptable. This is due to quiet periods (such as at night) which allow the system to recover.**

I. INTRODUCTION

IN THIS paper we are concerned with communication systems using digital data packets in a multiple-access radio channel. Multiple-access communication problems arise when several sources of information must reach a single destination over a common channel. Examples of this rather common situation are many earth stations simultaneously transmitting to a satellite, or newsmen in a press conference attempting to address the President of the United States. Techniques have been developed to achieve reliable communications in a multiple-access situation. Besides physically separating signals (directive antennas, cables, etc.), one can separate them by partitioning time, or the frequency spectrum, among the transmitters. The resultant time-division multiple-access (TDMA) and frequency-division multiple-access (FDMA) systems are efficient and practical in most situations. More sophisticated schemes have been devised to obtain higher rates.[1] Not all multiple-access situations warrant the use of such techniques, however; important considerations often include expandability and flexibility, suitability to time-varying load, simplicity of implementation, time delays, etc.

Computer communications using digital data packets are characterized by short high-rate bursts occurring at random points in time. The statistical nature of the users' demand and the high peak-to-average ratio of data rates make user-assigned TDMA or FDMA inadequate. Consider a system with 400 terminals of which 100, on the average, are active. Let each of these generate an average of one data packet each 100 s, to be sent to the central receiver. Each packet has 1000 bits, and the system's assigned frequency band permits a flow of 10^4 bits/s into the receiver. Conventional TDMA or FDMA can be used by assigning time or frequency slots to the terminals, but long delays will be incurred. It takes 40 s just to transmit a packet (at $10^4/400 = 25$ bits/s). At the cost of sign-on procedures and added complexity, one could improve on this performance using some form of demand assignment, but even then the delay will be at least 10 s.

ALOHA systems [1] were proposed to cope with this type of multiple-access problem and make much shorter delays possible. They are random time-division multiple-access (RTDMA) systems because terminals make use of the channel at random times. There are no user-assigned time slots.

System users generate information according to a random process. The data are formatted into standard-size packets, each carrying the terminal's address, parity-check bits for error detection, and other required information. In an ALOHA system, the channel is available to any terminal whenever it has a packet ready for transmission. The entire frequency band is made available, allowing transmission in a relatively short burst. Provided no other terminal attempts to use the channel during the same time, the communication is successfully accomplished. Sometimes, however, overlapping transmission will occur due to the independent, random origination of packets at the terminals. It is assumed that interference will cause overlapping packets to be lost by the receiver, which therefore does not acknowledge them. The receiver detects (and ignores) damaged packets using the parity-check bits. Each terminal stores its last transmitted packet. If an acknowledgment is not received, the terminal enters the retransmission mode and retransmits the packet until it gets through to the receiver. Retransmission times are randomized to prevent recurrent interference among the same terminals. New messages can be generated only

Paper approved by the Associate Editor for Computer Communication of the IEEE Communications Society for publication without oral presentation. Manuscript received July 25, 1974; revised September 30, 1974. The work of A. B. Carleial was supported in part by the U. S. Air Force Office of Scientific Research under Contract F44620-73-C-0065. The work of M. E. Hellman was supported by the Advanced Research Projects Agency under Contract DAHC 15-73-C-0187.

A. B. Carleial is with the Instituto de Pesquisas Espaciais (INPE), São José dos Campos, São Paulo, Brazil, and with the Department of Electrical Engineering, Stanford University, Stanford, Calif. 94305.

M. E. Hellman is with the Department of Electrical Engineering, Stanford University, Stanford, Calif. 94305 and is a consultant to SRI, Menlo Park, Calif.

[1] It has recently been shown [2] that the highest communication rates over a multiple-access channel without feedback are achieved through a scheme of superimposed signals, with all transmitters operating all the time and using the entire frequency band. The key to this method is the receiver's ability to reliably decode one of the signals in the presence of interference and noise. After subtracting out this signal, it can proceed with similar decoding steps on the remaining waveform.

Reprinted from *IEEE Trans. Commun.*, vol. COM-23, no. 4, pp. 401–410, April 1975.

when the terminal is cleared. We say that it has then returned to the origination mode.

The channel is not fully utilized, since it is idle part of the time and at times it is subjected to destructive interference. It has been shown [1], nevertheless, that such a system can operate well in practice, provided it is not too heavily loaded. It matches the randomness of user demand in a simple and natural way.

The delay involved in successful transmission of a packet depends on how many times it needs to be retransmitted and how long the terminal must wait to retransmit. A good system should not require too many retransmissions. The average wait time for a retransmission should be short, typically much shorter than the average time between new packet originations. This is consistent with a situation where, at any given time, most of the terminals are in the origination mode. Backlogged packets will normally get through after a short delay, provided there are only a few terminals in the retransmission mode. If, however, a large number of interfering transmissions occur, the number of retransmission-mode terminals rises and a runaway effect may take place. Due to the high repetition rate of retransmissions, interference becomes so frequent that fewer and fewer transmissions are successful. Still more users are placed in the retransmission mode, more interference takes place, and so on.

Based on the above heuristics, we conjectured that systems would exhibit a bistable behavior. One possible stable condition (with short delays and reasonably good throughput) would have most terminals in the origination mode. The other stable condition would have most terminals in the retransmission mode, obstructing the channel with prevailing interference. Transitions from one condition to the other should occasionally take place, due to statistical fluctuations. Analysis and simulations have confirmed this conjecture. The objective of this paper is to examine bistable behavior and conditions for successful operation of ALOHA systems.

We shall demonstrate bistable RTDMA systems of both the synchronous and nonsynchronous type. In a synchronous (or slotted) system, transmissions may start only at precise clock times. A common time reference must be supplied to all terminals, and different propagation delays have to be compensated for. The clock period is chosen such that each time slot can accommodate one packet transmission. The analysis of synchronous systems can proceed in discrete time. In a nonsynchronous system, terminals transmit packets without a clock reference. Partially overlapping transmissions interfere, and communication capacity is consequently smaller than in the synchronous case. Simplicity is the great advantage. Synchronous operation may not be feasible in systems with variable propagation delays (e.g., mobile terminals).

We shall model the generation of new packets by an active terminal as a Poisson process for a nonsynchronous system because this is a good characterization of a thought process and also simplifies analysis. Furthermore, for a large number of users, the analysis becomes quite insensitive to the exact distribution of individual terminal traffic generation. The overall traffic generation process tends to a Poisson process. Let the time rate of new packet origination per terminal be denoted by λ_o. Then the average time a user waits before transmitting while in the origination mode is $1/\lambda_o = T_o$.

As previously mentioned, packet generation is inhibited when the terminal is in the retransmission mode. Retransmission of the backlogged packet will then be attempted, also at random points in time,[2] until successful reception takes place. This artificial process of repetitions has rate λ_r. Practical system designs will tend to have λ_r much greater than λ_o in order to obtain short delays. We will show that bistable behavior results for a range of parameter values under this condition. This model is reasonable if the transmission time for a packet is much smaller than $T_o = 1/\lambda_o$ and $T_r = 1/\lambda_r$. These conditions are satisfied in practice for all systems with more than a few terminals.

In the case of synchronous systems, it is convenient to use a discrete-time model. An origination-mode terminal will transmit a new packet in any given time slot with probability p_o, whereas retransmissions take place with probability p_r. If τ is the duration of a time slot, p_o and p_r correspond to $\tau\lambda_o$ and $\tau\lambda_r$ in the continuous-time model, respectively. The important feature of either characterization is the lack of memory, given the terminal's current mode.

Unstable behavior of synchronous systems was originally predicted by Rettberg [3] using a deterministic model. Metcalfe [4] used a steady-state analysis to demonstrate the existence of two stable equilibria in synchronous systems, and discussed control strategies. Further analysis of slotted systems was made by Lu [5], with a derivation of steady-state behavior averaged over the dynamics of the system. A model similar to Lu's was developed more fully by Kleinrock and Lam [6], [7] to include effects of system dynamics and control strategies; the reader is also referred to [8] for details. We have independently developed a similar model which is discussed in Section II of this paper.

Exact analysis of slotted systems is simpler, and we develop analytic results for this case first. Then we analyze the behavior of unslotted systems with the aid of simulations. A comparison shows that the two types of system behave qualitatively the same. This equivalence is further explored in Section IV, which treats the limiting behavior as the number of users tends to inifinity. It is found that, except for a scaling of the data rates, the analysis is almost identical.

[2] Although a real system will probably use a bounded randomized wait time, we model this time as an exponential random variable, with mean value $1/\lambda_r$. Again the analysis is not very sensitive to the actual distribution, provided there are many terminals.

II. SYNCHRONOUS RTDMA SYSTEMS

We now proceed with analysis of slotted ALOHA systems, as characterized in the last section. Consider N terminals, each emitting packets at random time slots. Transmission is successful if and only if a single terminal attempts communication during the time slot. An unsuccessful attempt puts the terminal temporarily in the retransmission mode.

The channel may be viewed as a discrete-time system with $N + 1$ possible states, corresponding to the number $X(t)$ of terminals in the retransmission mode at the beginning of time slot $t = 1,2,3,\cdots$. The sequence of system states forms a discrete-time Markov chain with state-transition matrix $P = [p_{nm}]$ where $p_{nm} = P_r[X(t+1) = m \mid X(t) = n]$. In one step, the system may move down one state (a successful retransmission); stay at the same state (no transmissions, or one successful new transmission, or an interference event involving retransmissions only); or move higher in the chain by one or more states (interference event involving one or more fresh packets). We thus find for $n = 0,1,2,\cdots,N$,

$$\begin{aligned}
p_{n,m} &= 0, \qquad \text{for } m \leq n-2 \\
p_{n,n-1} &= (1-p_o)^{N-n} n p_r (1-p_r)^{n-1} \\
p_{n,n} &= (1-p_o)^{N-n}(1 - n p_r (1-p_r)^{n-1}) \\
&\quad + (N-n) p_o (1-p_o)^{N-n-1} (1-p_r)^n \\
p_{n,n+1} &= (N-n) p_0 (1-p_0)^{N-n-1} [1 - (1-p_r)^n] \\
p_{n,m} &= \binom{N-n}{m-n} p_o^{m-n} (1-p_o)^{N-m}, \qquad \text{for } m \geq n+2.
\end{aligned} \tag{1}$$

For $0 < p_o < 1$ and $0 < p_r < 1$, the Markov chain is irreducible, aperiodic, and positive-recurrent, with a vector of long-term state occupation probabilities $\boldsymbol{\mu} = (\mu_0,\mu_1,\cdots,\mu_N)^T$ satisfying the equation $\boldsymbol{\mu}^T = \boldsymbol{\mu}^T P$ [9, pp. 247–253].

The system's expected packet flow per time slot (effective throughput) in a given state n is found to be $f_n = (1-p_o)^{N-n-1}(1-p_r)^{n-1}[(N-n)p_o(1-p_r) + n(1-p_o)p_r]$ for $n = 0,1,2,\cdots,N$. In steady state, the system's throughput is

$$\bar{f} = \boldsymbol{\mu}^T f = \sum_{n=0}^{N} \mu_n f_n,$$

with $f = (f_0,f_1,\cdots,f_N)^T$. If $p_o = p_r = p$, the expected packet flow is the same in all states, $f_n = Np(1-p)^{N-1} = \bar{f}$. We note here that this throughput is maximized by choosing $p = 1/N$, in which case it converges (from above) to $\bar{f} = \exp(-1) \doteq 0.368$ as $N \to \infty$.

Statistics on the transitions from each state are given by the state-transition matrix P. As an indicator of system dynamics, it is interesting to compute the expected drift from state $n = 0,1,2,\cdots,N$. This is simply

$$\bar{d}_n = \sum_{m=0}^{N} (m-n) p_{n,m}. \tag{2}$$

A graph of expected drift versus present state gives a good idea of how the system tends to move in its random walk over the state space. Clearly, $\bar{d}_0$ is positive and $\bar{d}_N$ is negative, so that there is at least one neighborhood of states where the expected drift is near zero and has negative slope. This constitutes an equilibrium point for the system. Since $\bar{d}_n$ changes from positive to negative in such a region, we call it a stable equilibrium point: small excursions result in statistical drifts tending to restore the equilibrium. The system, therefore, lingers in the region. If $p_o = p_r$, for example, $\bar{d}_n$ is monotone decreasing in n (and approximately linear); there is a single equilibrium point, and it is stable. It is indeed verified that states in this neighborhood have the highest occurrence probabilities. See Fig. 1.

With $p_r > p_o$, as in a practical implementation, the system can exhibit different types of behavior. Fig. 2 shows graphs of expected drift and state probabilities for different values of the parameters. Consider the case of Fig. 2(b). Starting from state zero, the system will normally drift toward the first stable point, above which $\bar{d}_n$ becomes negative. At this point, the quick dispatch of backlogged packets prevents a rapid rise in state number. Note, however, that the concept of stability is employed only in a statistical sense in the present context. Beyond this stable point lies a second equilibrium region, this one unstable (because $\bar{d}_n$ is changing from negative to positive here). If the system passes this point, it is likely to drift to a still higher, stable region of states, characterized by recurrent interferences, few successful transmissions, and long delays. It will usually not return to the more desirable, lo : operating point for a long time.

Statistical noise will cause the system to oscillate between these two stable points. The average time spent in one stable region before moving to the other depends on the depth of the "potential well" associated with each region and on the magnitude of the statistical noise in each region (the variance of the distance moved, for example).

Note that the steady-state behavior of a system will depend crucially on the proportion of time spent in each region. For example, if the average time to move from the "good" region to the "bad" region is 1 year and the average time to move in the opposite direction is 100 years then, in steady-state only about 1 percent of the system's time is spent in the good region. However, if the system is lightly loaded at night and is fully utilized for only 8 h a day, then the mean time between failures due to overloading will be about 1000 days. When such a failure occurs, users will leave the system (probably cursing), thereby allowing it to restart in a good initial state the next day. This example shows that steady-state behavior can be a poor indicator of system performance.

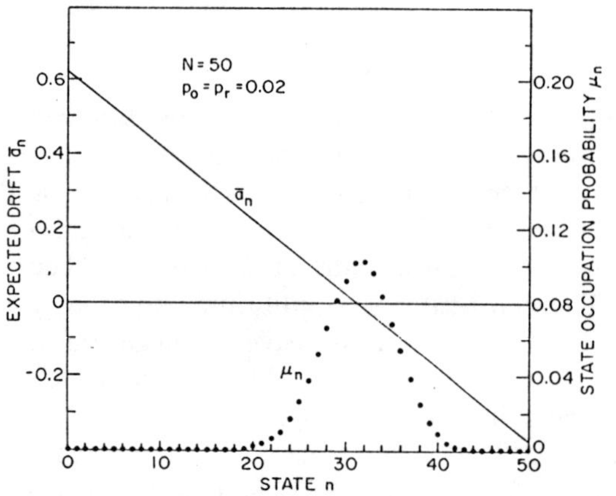

Fig. 1. Expected drift and state occupation probability for a synchronous system with 50 terminals and $p_o = p_r = 0.02$.

We are interested in operating the system in a region of states with reasonably high packet flow and short delay. Over a period of time $[1,T]$, define the average delay D_T (in time slots per packet) to be

$$D_T = \frac{\sum_{t=1}^{T} X(t)}{\sum_{t=1}^{T} \psi(t)} \tag{3}$$

where $X(t)$ is the state and $\psi(t)$ is the number of packets successfully transferred (zero or one) at time slot t. D_T is a random variable whose distribution depends on the path taken by the system's random walk. Letting $\bar{D} = \lim_{T\to\infty} D_T$, we see that

$$\bar{D} = \lim_{T\to\infty} \frac{[\sum_{t=1}^{T} X(t)]/T}{[\sum_{t=1}^{T} \psi(t)]/T} \tag{4}$$

and therefore,

$$\bar{D} = \frac{\sum_{n=0}^{N} \mu_n n}{\sum_{n=0}^{N} \mu_n f_n} = \frac{\bar{n}}{\bar{f}} \quad \text{with probability one.} \tag{5}$$

$\bar{D}$ is thus equal to the expected value of the delay for transmission of a packet in steady state.[3]

$\bar{D}$ can be considerably larger than typical values of D_T obtained while the system remains in low-numbered states (i.e., paths of the random walk that do not visit the higher stable region in the bistable case). It is important to investigate the possibility of exploring transient operation

[3] Radio wave propagation delay is not considered. Only the waiting time resulting from the need to retransmit is included. Accordingly, a packet received on the first attempt has delay zero.

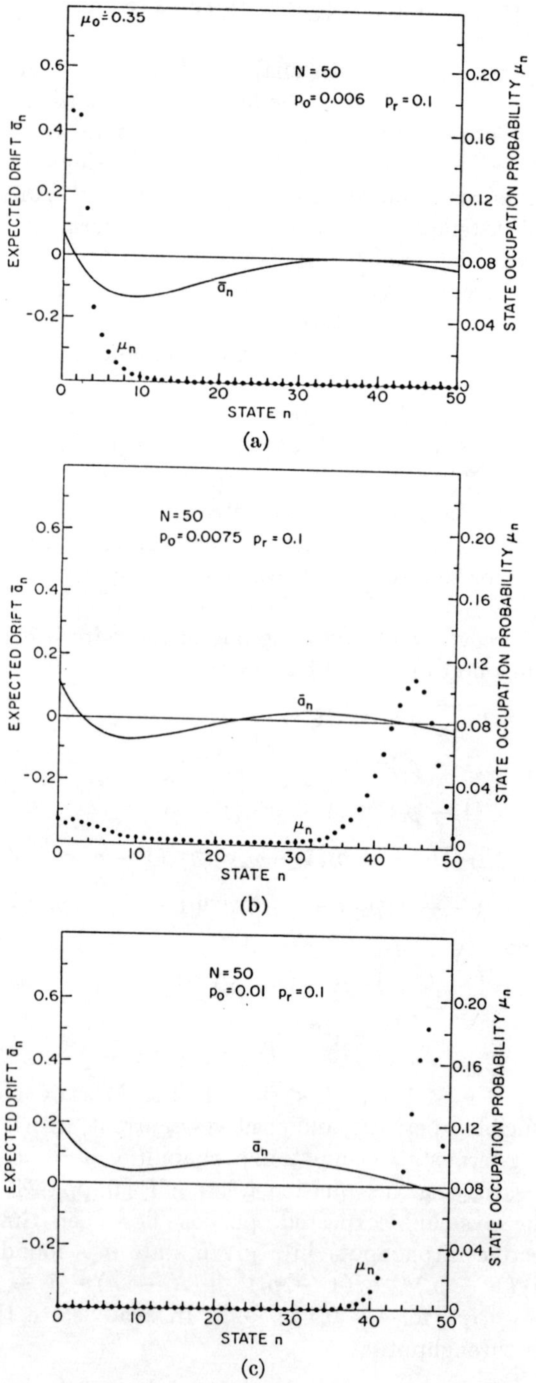

Fig. 2. Expected drift and state occupation probability for three synchronous systems with 50 terminals and $p_r > p_o$. (a) $p_o = 0.006$, $p_r = 0.1$. (b) $p_o = 0.0075$, $p_r = 0.1$. (c) $p_o = 0.01$, $p_r = 0.1$.

of the system in the favorable states, in particular by computing statistics on transition times between the two stable regions.

Define the random variable t_{nm} as the time (number of time slots) it takes the system to reach state m for the first time, starting from state n. A method described in [5, pp. 241–246] can be used to determine the characteristic function $\phi_{nm}(u)$ of this random variable and thereby find its distribution. A modified Markov chain is employed where state m is changed into an absorbing state. The other states then become nonrecurrent. The analysis gives correct results because, up to the first visit to state m, the modified chain behaves the same as the original chain.

The characteristic function is defined as $\phi_{nm}(u) = E[\exp(iut_{nm})]$. We take $t_{mm} = 0$ and thus $\phi_{mm}(u) = 1$ For $n \neq m$, we can write

$$\begin{aligned}\phi_{nm}(u) &= \sum_{l=0}^{N} p_{nl}E[\exp(iut_{nm}) \mid \text{first transition is } n \to l] \\ &= p_{nm}\exp(iu1) + \sum_{l\neq m} p_{nl}E\{\exp[iu(1+t_{lm})]\} \\ &= \exp(iu)[p_{nm} + \sum_{l\neq m} p_{nl}\phi_{lm}(u)].\end{aligned} \quad (6)$$

If we now define the N-dimensional vectors

$$\begin{aligned}\boldsymbol{\phi}_{(m)} &= (\phi_{0m}\,\phi_{1m},\cdots,\phi_{m-1,m},\phi_{m+1,m},\cdots,\phi_{Nm})^T \\ \boldsymbol{p}_{(m)} &= (p_{0m},p_{1m},\cdots,p_{m-1,m},p_{m+1,m},\cdots,p_{Nm})^T\end{aligned} \quad (7)$$

and the $N \times N$ matrix $Q_{(m)}$, obtained from P by eliminating its mth row and its mth column, the following equation holds:

$$\exp(-iu)I\boldsymbol{\phi}_{(m)} = \boldsymbol{p}_{(m)} + Q_{(m)}\boldsymbol{\phi}_{(m)}. \quad (8)$$

The solution

$$\boldsymbol{\phi}_{(m)} = (\exp(-iu)I - Q_{(m)})^{-1}\boldsymbol{p}_{(m)} \quad (9)$$

provides the characteristic function of t_{nm} for each initial state $n \neq m$. Note that $\boldsymbol{p}_{(m)}$ and $Q_{(m)}$ depend on the choice of final state. The probability mass function of t_{nm} is found by expressing ϕ_{nm} in the power series form

$$\phi_{nm} = \sum_{t=1}^{\infty} \pi_{nm}(t)\exp(iut). \quad (10)$$

We have $\Pr\{t_{nm} = t\} = \pi_{nm}(t)$.

Computing the distribution of t_{nm} for a large system may be quite laborious. While all the moments can be found by differentiation of its characteristic function, the following formula for the mean is obtained directly by the same method:

$$\begin{bmatrix}\bar{t}_{0m}\\ \cdot\\ \cdot\\ \cdot\\ \bar{t}_{m-1,m}\\ \bar{t}_{m+1,m}\\ \cdot\\ \cdot\\ \cdot\\ \bar{t}_{Nm}\end{bmatrix} = (I - Q_{(m)})^{-1}\begin{bmatrix}1\\ 1\\ 1\\ \cdot\\ \cdot\\ \cdot\\ 1\\ 1\end{bmatrix}. \quad (11)$$

If n is in a stable region of states and m is in the other stable region, $\bar{t}_{nm}$ and $\bar{t}_{mn}$ are good measures of the mean time for transitions from one region to the other.

A computationally straightforward technique for obtaining partial information on the statistics of t_{nm} is also possible. Take the state transition matrix P and modify it by changing m into an absorbing state. Then compute powers of this modified matrix $P_{(m)}$ and observe how the (n,m)th element grows to unity. With a sequence of k matrix-squaring operations, one can compute $P^2_{(m)}, P^4_{(m)}, P^8_{(m)}, \cdots, P^{2^k}_{(m)}$ and thereby find the probability that state m is reached by the system after 2^k steps or less. Fig. 3 shows the result of such a computation for the bistable system of Fig. 2(b) with $n = 0$ and $m = 44$. Assuming T_o to be 1 min, the duration of a time slot is $\tau = T_o p_o = 0.0075$ min $= 0.45$ s. The graph shows that the probability of reaching state 44 (in the high delay region) within one hour (8000 time slots) of operation is about 55 percent. This is unacceptable system performance. However, any practical implementation of an ALOHA system will probably involve many more than 50 users since the advantage over TDMA and FDMA grows with N. It is easier to obtain acceptable system performance for a larger value of N because the probability of failure within a given time period (e.g., 1 h) goes to zero rapidly in N, provided the system is scaled appropriately to handle the larger number of users. This effect is discussed in Section IV.

III. NONSYNCHRONOUS RTDMA SYSTEMS

A nonsynchronous RTDMA (unslotted ALOHA) system cannot be described solely by a Markov chain. Rather, the appropriate stochastic process model is that of a semi-Markov process [10]. If $\{X_i\}_{i=1}^{\infty}$ is a Markov chain, then $X(t)$ is a derived semi-Markov process if

$$X(t) = X_i, \qquad \sum_{j=1}^{i-1} T_j \le t < \sum_{j=1}^{i} T_j \quad (12)$$

where $\{T_j\}_{j=1}^{\infty}$ is a sequence of random variables with the following properties.

1) Given $\{X_i\}_{i=1}^{\infty}$, the random variables $\{T_j\}_{j=1}^{\infty}$ are independent.

2) Given $\{X_i\}_{i=1}^{\infty}$, T_j and T_k have the same distribution if $X_j = X_k$ and $X_{j+1} = X_{k+1}$.

In other words, the length of time the semi-Markov process spends in a particular state depends on that state and the next state of the process, and given those two states, the time of residence is independent of all other times of residence and states occupied.

For an unslotted ALOHA system, let $X(t)$ equal the number of users in the retransmission mode just after the last time the channel was quiet. Fig. 4 shows a typical sample function from a process with five users. The pulses indicate transmissions, and the number just over a pulse is the identity (1–5) of the user transmitting. It is assumed that any overlap of pulses (interference) causes all affected packets to fail. The height of a pulse is immaterial, and the only reason for the differing heights is to distinguish pulses. The intervals between transitions will be called epochs. By definition, the state of the system is constant during an epoch. The first epoch begins at time $t_1 = 0$.

We now indicate why $X(t)$ as defined in the preceding paragraph is a semi-Markov process. Letting

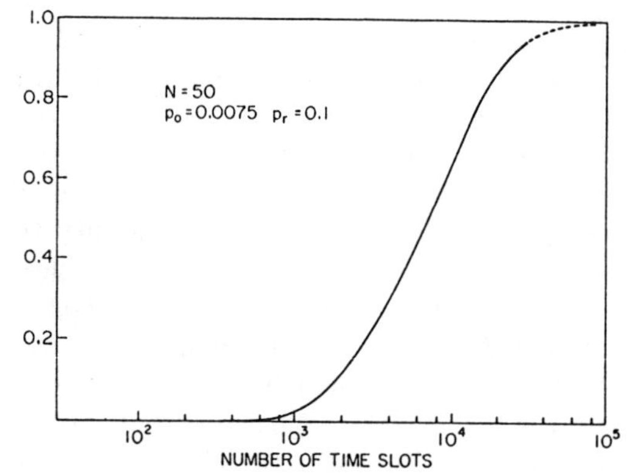

Fig. 3. Probability of reaching state 44 (in high-delay stable region) within the number of time slots indicated for the bistable system of Fig. 2(b) starting from state zero.

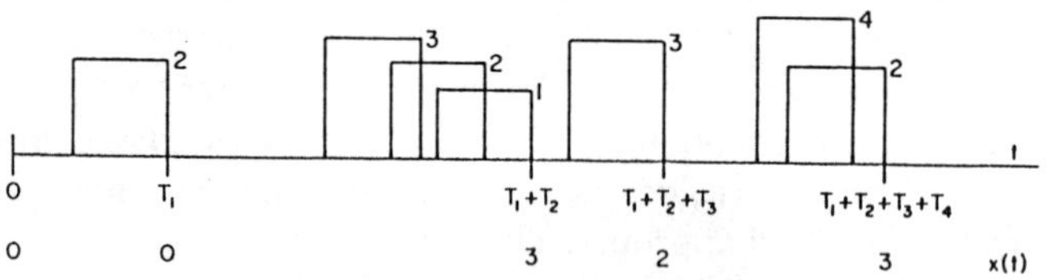

Fig. 4. An example of the packet transmission process in a nonsynchronous system with five terminals.

$$t_i = \sum_{j=1}^{i-1} T_j$$

denote the start of the ith epoch, we have $X_i = X(t_i)$ users in the retransmission mode and the remaining $N - X_i$ users in the origination mode, and no terminals transmitting at time t_i. Due to the memoryless nature of the exponential distribution, the additional time from t_i until a particular user starts transmitting is still exponentially distributed (with parameter λ_o if the user is in the origination mode, and with parameter λ_r if it is in the retransmission mode). Therefore, the statistical behavior of the system after t_i is completely determined by X_i and is independent of the manner in which this state was reached. It follows that the discrete-time process $\{X_i\}_{i=1}^{\infty}$ is a Markov chain. Some additional thought shows that the sequence $\{T_i\}_{i=1}^{\infty}$ satisfies the conditions for a semi-Markov process.

While the nature of this process makes the analysis of unslotted ALOHA systems more cumbersome than that of slotted systems, considerable information can be obtained by considering only the associated Markov chain

$$X_i = X(\sum_{j=1}^{i-1} T_j). \qquad (13)$$

Its statistics are completely specified by the initial state $X_1 = X(0)$, usually taken as 0, and its state transition matrix $P = [p_{ij}], 0 \leq i,j < N$. The elements p_{ij} are defined by

$$p_{ij} = \Pr(X_{n+1} = j \mid X_n = i). \qquad (14)$$

For example, $p_{i,i-1}$ is the probability that during an epoch a user in the retransmission mode transmits and is not interfered with; $p_{i,i+2}$ is the probability that two users in the origination mode interfere with each other and/or with any number of users in the retransmission mode; etc. Except for certain special values of i and j, the p_{ij} are difficult to compute analytically. However, they are easily obtained by simulation and can then be used to predict system behavior. This is easier than estimating system behavior directly. Basically, what we have done is to parameterize the estimation problem.

Once the p_{ij} are known within a reasonable accuracy, gross behavior of the system can be inferred from

$$\bar{a}_n = \sum_{j=0}^{N} (j - n) p_{nj}. \qquad (15)$$

The quantity $\bar{a}_n$ is the expected distance moved in a transition from state n. As for slotted systems, the only stable regions are those in which $\bar{a}_n$ changes sign from positive to negative as n increases.

Fig. 5 shows $\bar{a}_n$ versus n for $N = 25$ and six values of $T_o = T_r \equiv T$. Message duration τ was normalized to be 1 s. The parameter N/T is the average number of packets generated by all users per unit time. Two hundred transitions were simulated for each point plotted. This number is large enough for inferring gross behavior, but should not be used as an indication of exact behavior. We see that when $T_o = T_r$, $\bar{a}_n$ is a decreasing function of n, just as in the slotted case. Also, as one would expect, for n fixed, $\bar{a}_n$ is a decreasing function of T. This is because T is the mean time until the first user transmits, and for N large, it is also approximately the mean time between all transmissions which occur in state n. A larger value of T/N therefore indicates a smaller probability of interference, causing the drift to be more toward lower numbered states.

Now let us turn to the more interesting and realistic case where $T_o \gg T_r$. The feedback effect is then present and various behaviors are possible. If the system is conservatively designed, then $\bar{a}_n < 0$ for most states, and even when most users are in the retransmission mode, the net drift is toward lower numbered states. For example, in Fig. 6, when $T_o = 10T_r$, we find $\bar{a}_n < 0$ for $i > 1$. Such a system spends most of its time near states 1 and 2. At the opposite extreme, a poorly designed system has $\bar{a}_n > 0$ for most states, and spends most of its time in high numbered states. In Fig. 6, $T_o = 20T_r$ is an example of such a system. It too has a single equilibrium point, in this case between states 24 and 25.

A bistable situation can also exist, as shown in Fig. 6 with $T_o = 15T_r$. The plot of $\bar{a}_n$ versus n has three crossings of the axis; the first and last are stable equilibrium points, while the middle crossing is an unstable equilibrium point.

IV. POISSON APPROXIMATION FOR LARGE VALUES OF N

When the number of users N is large, a simplifying approximation can be made that yields the expected drift values $\{\bar{a}_n\}_{n=0}^{N}$ for both synchronous and nonsynchronous ALOHA systems without recourse to simulation.

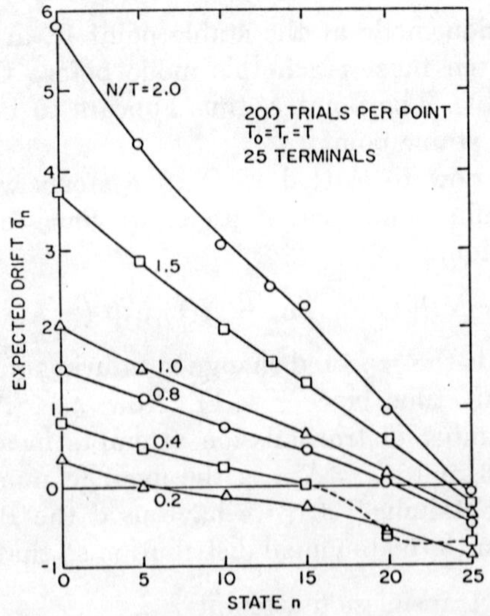

Fig. 5. Simulation estimates of expected drift per epoch for six nonsynchronous systems with 25 terminals and $T_o = T_r$.

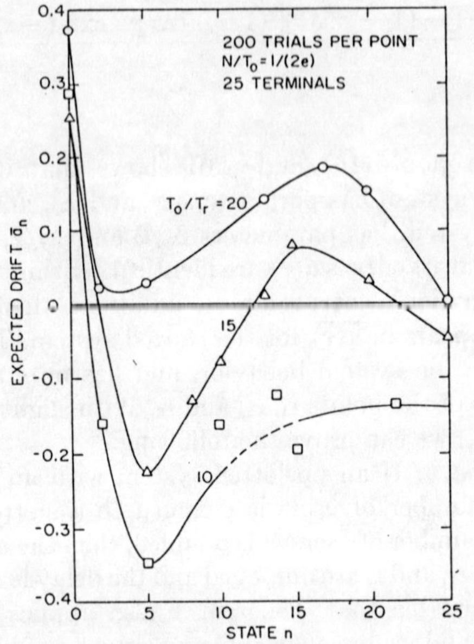

Fig. 6. Simulation estimates of expected drifts per epoch for three nonsynchronous systems with 25 terminals and $T_r < T_o$. The value of T_o is the constant, and the ratio T_o/T_r is indicated.

Considering nonsynchronous systems first, let us establish the duration of a packet transmission τ as our unit of time (i.e., $\tau = 1$). If all terminals were in the origination mode, the generation of transmissions would be a Poisson process with intensity $\Lambda_o = N\lambda_o$. If they were all in the retransmission mode, the intensity would be $\Lambda_r = N\lambda_r$. If a fraction r of the terminals are in the retransmission mode, then the intensity of the Poisson process is $\Lambda(r) = (1 - r)\Lambda_o + r\Lambda_r$. The rate at which users move from the origination mode into the retransmission mode is $(1 - r)\Lambda_o[1 - \exp(-2\Lambda(r))]$, since $(1 - r)\Lambda_o$ is the rate of generation of new transmissions and $[1 - \exp(-2\Lambda(r))]$ is the probability that a transmission is not successful [1]. Similarly, the rate at which terminals leave the retransmission mode is $r\Lambda_r \exp(-2\Lambda(r))$. The net rate of increase of users in the retransmission mode is the difference between these two quantities and is also equal to $N\overline{a(r)}$ where $a(r)$ denotes the time rate of change of r. We now regard r as the system state. It follows that

$$\overline{a(r)} = N^{-1}[(1 - r)\Lambda_o - \Lambda(r) \exp(-2\Lambda(r))]. \quad (16)$$

Note that (16) gives the true expected time rate of change or r (with τ as the unit of time) whereas in Section III we calculated $\bar{a}_n$, the expected drift per epoch.

If $\Lambda_o = \Lambda_r = \Lambda$, then (16) reduces to

$$\overline{a(r)} = N^{-1}[(1 - r)\Lambda - \Lambda \exp(-2\Lambda)], \quad (17)$$

which is linear in r and crosses the r axis at the stable point $r = 1 - \exp(-2\Lambda)$. If Λ_r is increased, then for reasonable values of Λ_o ($\Lambda_o < 0.5$, to be exact), the stable equilibrium point moves to a lower value of r. See Fig. 7. This is desirable since it results in smaller delays. If, however, Λ_r is increased beyond a certain point, two stable equilibria develop. The smaller of these, which we shall denote r_1, continues to move to smaller, more desirable values of r; but the other, denoted r_3, is in the region of high delay. Fortunately, if N is very large, then even a small stable region holds the system for a very long time because the law of large numbers has a chance to set in during the system's random walk on r.

With reference to Fig. 7, we see that if $\Lambda_o = 0.2$ and $\Lambda_r = 3.0$, the smaller stable value of r is 0.087. At this point, the system throughput is equal to the rate of generation of new packets and is thus equal to $(1 - 0.087)\Lambda_o = 0.183$. The maximum possible throughput for an unslotted ALOHA system is $1/(2e) \doteq 0.184$, and occurs for $\Lambda_o = \Lambda_r = 0.5$. Thus we pay only a slight penalty in throughput, but, as we shall see, gain significantly in delay characteristics. The expected delay for a packet when the system is in a state r is[4]

$$\overline{D(r)} = \left[\frac{1}{\exp(-2\Lambda(r))} - 1\right] T_r = [\exp(2\Lambda(r)) - 1)T_r \quad (18)$$

since the probability of success on each transmission is $\exp(-2\Lambda(r))$ and each retransmission delay has an average value of $T_r = 1/\lambda_r$. If $\Lambda_o = 0.2$, $\Lambda_r = 3.0$, and the system is in the lower stable region, the expected delay is $1.43T_r = 0.095T_o$. For $\Lambda_o = \Lambda_r = 0.5$, the delay is $1.72T_o$. We lose only about 0.5 percent in throughput, but gain a reduction in delay by a factor of 18, a very advantageous tradeoff.

If $\Lambda_o = 0.1845$ and $\Lambda_r = 30$, the lower stable equilibrium point moves to $r = 0.0088$. Under these conditions, the throughput is 0.183, the same as above, but the delay is $1.45T_r = 0.0089T_o$, about one tenth of its previous value. This process can be continued letting Λ_o approach 0.183 and Λ_r approach infinity. The throughput in the lower stable region can be kept constant, while the delay tends to zero. In fact, it can be seen that if the throughput f_{r_1}

[4] More precisely, $\overline{D(r)}$ is the expected delay given that the system's random walk does not drift away from state r during the packet's transmissions.

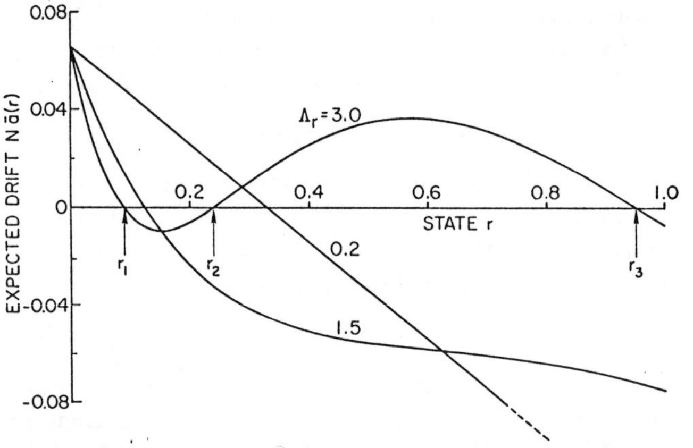

Fig. 7. Expected drift $\bar{a}_n \equiv N\overline{a(r)}$ for three nonsynchronous systems with $\Lambda_o = 0.2$ and a very large number of terminals. The values of Λ_r are 0.2, 1.5, and 3.0. This last value results in a bistable system for which the three equilibrium points r_1, r_2, and r_3 are indicated.

is kept equal to a constant η, then increasing Λ_r by a given factor (say, ten) reduces the delay by exactly the same factor at the lower stable point r_1. This is because at an equilibrium point, the throughput η is equal to the rate of generation of new packets $(1 - r)\Lambda_o$. Thus $\Lambda(r_1) = (1 - r_1)\Lambda_o + r_1\Lambda_r = \eta + r_1\Lambda_r$, and from (16)

$$\overline{a(r_1)} = N^{-1}[\eta - (\eta + r_1\Lambda_r) \exp(-2(\eta + r_1\Lambda_r))] = 0. \quad (19)$$

In general, if $a(r) = 0$ for given values of η and Λ_r, then $\overline{a(r/k)} = 0$ for the same value of η provided we now use $\Lambda_r' = k\Lambda_r$. Note that $\Lambda'(r/k)$ is the same as $\Lambda(r)$ even though the stable point moves. From (18) we thus see that the expected delay, measured in terms of T_r, is constant. But T_r is reduced by a factor of k when Λ_r is increased by a factor of k. We thus have the following.

Theorem 1: When operating at a fixed throughput η at a stable equilibrium point, an unslotted ALOHA system with an essentially infinite number of users has an expected delay which is inversely proportional to the rate of retransmissions λ_r.

It will be noted that in the above examples where η was held at 0.183 and Λ_r was increased from 3 to 30, the delay did not decrease by exactly a factor of 10. This was due to roundoff error.

The price that is paid for increasing Λ_r is that the unstable equilibrium point, denoted r_2, also moves closer to the origin. In the above examples, with $f_{r_1} \equiv \eta = 0.183$ as Λ_r increases from 3 to 30, r_2 moves from 0.243 to 0.0131. For a finite number N of terminals, there is a limit to how small r_2 can be before the stability of the equilibrium at r_1 becomes threatened. For example, if $r_1 = 0.01$ $r_2 = 0.02$, and $N = 10^6$, then 10 000 users are in the retransmission mode at the stable equilibrium point r_1, and another 10 000 would have to be placed in the retransmission mode before the unstable point r_2 is reached. This appears to be an acceptable situation. If, however, $N = 1000$ and r_1 and r_2 are still 0.01 and 0.02, only ten users are in the retransmission mode at the stable point r_1, and only an additional ten must reach this mode before the system goes unstable. The point r_1 thus appears to be a rather ephermeral stable point.

Turning now to slotted ALOHA systems with an essentially infinite number of users, we find, keeping the same notation,

$$\overline{a(r)} = N^{-1}[(1 - r)\Lambda_o - \Lambda(r) \exp(-\Lambda(r))] \quad (20)$$

where $\overline{a(r)}$ is the expected change in r during one time slot (or per unit time since $\tau = 1$). Now $\Lambda_o = Np_o$ is the average number of transmission attempts in a time slot when $r = 0$ and $\Lambda_r = Np_r$ is the average number when $r = 1$. In obtaining (20), we have used the Poisson approximation to the binomial distribution so that

$$\Pr\left\{\begin{matrix} k_1 \text{ new transmission attempts} \\ \text{and} \\ k_2 \text{ retransmission attempts}\end{matrix}\right\} = \frac{\exp[-(1-r)\Lambda_o][(1-r)\Lambda_o]^{k_1}}{k_1!} \cdot \frac{\exp(-r\Lambda_r)(r\Lambda_r)^{k_2}}{k_2!}. \quad (21)$$

Comparison of (16) and (20) shows that if a slotted ALOHA system has parameters Λ_o and Λ_r, and if an unslotted system has parameters $\Lambda_o/2$ and $\Lambda_r/2$, then their expected drift curves $\overline{a(r)}$ are identical (within the Poisson approximation) except for an additional factor of two which appears in $\overline{a(r)}$ for the slotted system. This has no real effect on system behavior, and has no effect on the locations of the points r_1, r_2, and r_3. With slight additional reasoning, we can prove the following.

Theorem 2: If an unslotted system with an essentially infinite number of users is changed to a slotted system, and the number of users N is doubled, then the equilibrium points r_1, r_2, and r_3 are unmoved and the delay is unaffected.

We thus see that Theorem 1 also applies to slotted systems, and we have a complete equivalence between the two types of system. The only real difference is that unslotted systems have half the capacity of slotted systems.

V. DISCUSSION

Previous analyses of ALOHA systems have assumed the traffic generation to be a Poisson process with constant intensity. This tacitly assumes that the average time a terminal waits before retransmitting, T_r, is equal to the average thought time to produce a packet, T_o. Since T_o is typically on the order of 1 min, this would result in unacceptably long delays. Any practical system will have a much shorter value of T_r, typically less than 1 s. This results in Λ_r/Λ_o ratios of 100 and even larger. The results presented in this paper show that such a system is capable of bistable behavior, with one stable point being in a desirable low-delay region, and the other stable point being in the undesirable high-delay region. We have given qualitative guidelines concerning system design to ensure that the undesirable stable point is not reached. In particular, most systems will have peak periods during which they are heavily loaded, followed by slack periods of low use. For example, the peak period may be 8 h and the slack period may be 16 h. If the probability of the system reaching the undesirable region within 8 h is small, then many days will elapse before the system "fails." And even when it fails, the following slack period will allow it to return to the desirable region. Thus the prediction that the system spends a long time in the undesirable region is really mathematical fiction.

It may be possible to use a dynamically adaptive strategy to reduce, or completely eliminate, even the occasional failures. If T_r were to be increased when the system was approaching the unstable region, then the position of the unstable region would be moved further from the current operating point, and a margin of safety added. As the system recovered, T_r could again be reduced. The delay would be longer during the recovery period, but at least the delay is allowing greater throughput. If a user is to suffer a given delay, it is best for him to interfere with as few other packets as possible.

There are at least two techniques which could be used to control T_r. If there is a central control station which can monitor the channel, then it can control the value of T_r through occasional commands to the terminals. Alternatively, a terminal could increase T_r whenever it had to retransmit a packet and decrease T_r whenever a packet was successfully transmitted. The increment sizes would have to be carefully chosen, but it appears that such a strategy could be effective.

We cannot consider exercising a similar control over T_o, since it is our assumption that this parameter, unlike T_r, is a characteristic of the users' thought processes. It would be possible, nonetheless, to send an inhibiting command from the central control to the terminals which would temporarily block all new packet generation (say, by locking keyboards). This would be done whenever some dangerously high state was reached, thereby assuring no further rise in state. Retransmissions would continue, however, and normal operation would resume when some specified lower state was reached. A design tradeoff results between the duration of the outages and their frequency. Note that in all centrally controlled strategies, only an estimate of the current state is available (from the intensity of traffic), but for large N and $\Lambda_r \gg \Lambda_o$, such an estimate can be quite accurate.

In summary, we have shown that bistable operation is not only possible, but is probably to be found in any practical system. It is hoped that the techniques developed in this paper will provide a basis for designing packet radio systems with smaller delays and less frequent failures.

ACKNOWLEDGMENT

The authors wish to thank Dr. S. Fralick of the Stanford Research Institute and Dr. R. Kahn of the Advanced Research Projects Agency for valuable discussions concerning this work. They also wish to thank P. Murthy, currently at the University of Hawaii, for performing the simulations of Section III. One of the reviewers helpfully contributed several of the references.

REFERENCES

[1] N. Abramson, "The ALOHA system—Another alternative for computer communications," in *1970 Fall Joint Comput. Conf., AFIPS Conf. Proc.*, vol. 37. Montvale, N. J.: AFIPS Press, 1970, pp. 281–285.

[2] A. D. Wyner, "Recent results in the Shannon theory," *IEEE Trans. Inform. Theory*, vol. IT-20, pp. 2–10, Jan. 1974.

[3] R. Rettberg, "A brief simulation of the dynamics of an Aloha system with slots," ARPA Satellite Syst. Note 11, July 31, 1972.

[4] R. Metcalfe, "Steady-state analysis of a slotted and controlled Aloha system with blocking,' presented at the 6th Hawaii Int. Conf. on Syst. Sci., Honolulu, Hawaii, 1973.

[5] S. Lu, "Dynamic analysis of slotted Aloha with blocking," ARPA Satellite Syst. Note 36, Mar. 12, 1973.

[6] L. Kleinrock and S. S. Lam, "On stability of packet switching in a random multi-access broadcast channel," presented at the 7th Hawaii Int. Conf. on Syst. Sci., Honolulu, Hawaii, Jan. 8–10, 1974; also, *Proc. Special Subconf. on Comput. Nets*, pp. 73–77.

[7] S. S. Lam, "Packet switching in a multi-access broadcast channel with application to satellite communication in a computer network," Ph.D. dissertation, Dep. Comput. Sci., Univ. California, Los Angeles, Mar. 1974.

[8] L. Kleinrock and S. S. Lam, "Packet switching in a multiaccess broadcast channel: Performance evaluation," this issue, pp. 410–423.

[9] E. Parzen, *Stochastic Processes.* San Francisco, Holden-Day, 1962.

[10] R. Pyke, "Markov renewal processes: Definitions and preliminary properties," *Ann. Math. Stat.*, vol. 32, pp. 1231–1242, 1961.

An Approximate Method for Design and Analysis of an ALOHA System

HISASHI KOBAYASHI, SENIOR MEMBER, IEEE, YOSHIKUNI ONOZATO,
AND DIEU HUYNH, MEMBER, IEEE

Abstract—We develop here an approximate method for the design and performance prediction of a multiaccess communication system which employs the ALOHA packet-switching technique. Our model is based on the use of a diffusion process approximation of an ALOHA-like system (with or without time-slotting). A simple closed-form solution for the variable $Q(t)$, a variant of the number of backlog messages at time t, is given in terms of a few system and user parameters. Final results are expressed in terms of ordinary performance measures such as throughput and average delay. Several numerical examples are given to demonstrate the usefulness of the approximation technique developed.

I. INTRODUCTION

SINCE a novel form of packet switching was introduced in the ALOHA System at the University of Hawaii [1], [2], various forms of packet communication techniques, in combined form of *random-access multiplexing* and *packet broadcasting,* have been investigated. The ALOHA packet-switching technique is applicable to any data communication system with a multiaccess channel, be it a satellite channel or ground-radio channel. Its basic operations principles will be briefly reviewed below.

A user transmits his data messages into the random-access channel in the form of packets. If a packet occupies the channel for the entire packet duration, it will be successfully received by a message-switching processor at the other end of the channel. The message-switching processor may be located on the ground (like MENEHUNE in the ALOHANET) or may be a transponder in a satellite. If more than one user transmits packets simultaneously and these transmission times overlap each other, even partially—an event called a *collision*—then all of the transmissions are considered unsuccessful. These packets must then be retransmitted after randomly chosen delay times elapse.

The original scheme implemented in the ALOHANET is completely *asynchronous* in the sense that no control-clock is required for timing purposes. All users can initiate transmission at arbitrary instants. Subsequently, a modified scheme was proposed [3] in which the time axis is segmented into contiguous sequence of *time slots.* In this slotted version, users are still allowed, as in the original ALOHA technique, to generate messages at random epochs, but a packet needs to be of fixed length τ_s (or less), and it must fall exactly in one of the time slots. The slotted ALOHA scheme certainly requires a more costly implementation than the unslotted scheme. Some studies [2], [3] have shown, however, that the slotted version reduces frequencies of collisions, and thereby increases the maximum throughput of the random-access channel. Various design and performance issues of the slotted ALOHA scheme have been addressed by a number of authors [4]-[9].

Most of the analyses reported thus far are based on Markovian models in which both retransmission delays and *system-user times* (to be defined in Section II) are independent and identically distributed (iid) random variables with exponential distributions. Kleinrock and Lam [6] considered a satellite communication system employing the slotted ALOHA technique and showed that a slotted ALOHA channel may, under some certain conditions, exhibit a sudden collapse of a sustained performance, once the system load exceeds some critical threshold level. They gave a graphical explanation (based on the fluid approximation) on how the system may fall into an *unstable* region of operating points. Carleial and Hellman [7] conducted a Markov chain analysis and discussed the *bistable* behavior of the slotted ALOHA system.

The performance analysis of the unslotted ALOHA scheme, on the other hand, has remained more or less unexplored since the original static analyses conducted by Abramson [2] and Roberts [3]. Carleial and Hellman [7] discuss an imbedded Markov chain approach defined over epochs which are just after *busy periods* of an unslotted ALOHA channel. They did not carry out a complete analysis, however, and proposed a numerical solution technqiue aided by a Monte Carlo simulation. Ferguson [10] analyzed an unslotted ALOHA system in which packet lengths are assumed to be exponentially distributed. He further assumes that both *system-user times* and *retransmission delays* are identically distributed.

In this paper we will present an approximate method for design and performance prediction of ALOHA systems of both types. Major results are highlighted below.

1) An approximation model based on the Ornstein-Uhlenbeck process approximation is introduced to characterize a random jump process $Q(t)$ (a quantity similar to the number of backlog messages at time t; to be defined in Section II).

2) An approximate but simple closed-form solution is given in terms of a few system and user parameters. Final results are expressed in terms of the ordinary performance measures such as throughput, average delay, etc.

Manuscript received March 18, 1976; revised September 15, 1976. This work was supported by the ALOHA System, a research project at the University of Hawaii, which is supported by the Advanced Research Projects Agency of the Department of Defense and monitored by NASA Ames Research Center under Contract NAS2-8590.

H. Kobayashi was with the University of Hawaii, Honolulu, HI 96822, as a Consultant to the ALOHA System Project, on leave from the IBM Thomas J. Watson Research Center, Yorktown Heights, NY 10598.

Y. Onozato is at the University of Hawaii, Honolulu, HI 96822.

D. Huynh was with the ALOHA System, University of Hawaii, Honolulu, HI 96822. He is now with IBM System Communications Division, Kingston, NY 12401.

Reprinted from *IEEE Trans. Commun.*, vol. COM-25, no. 1, pp. 148–157, Jan. 1977.

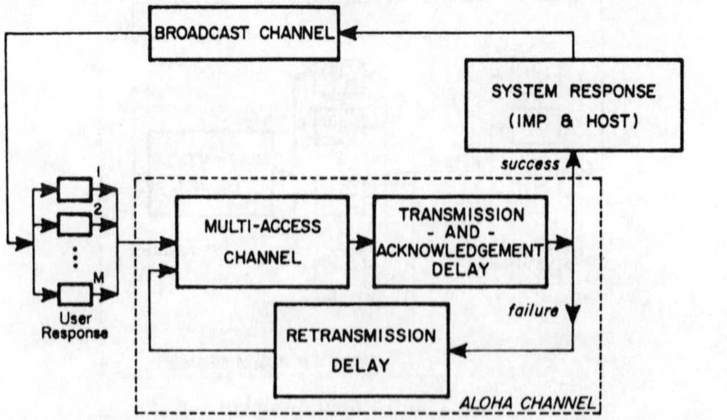

Fig. 1. Data communication system with ALOHA channel (slotted or unslotted).

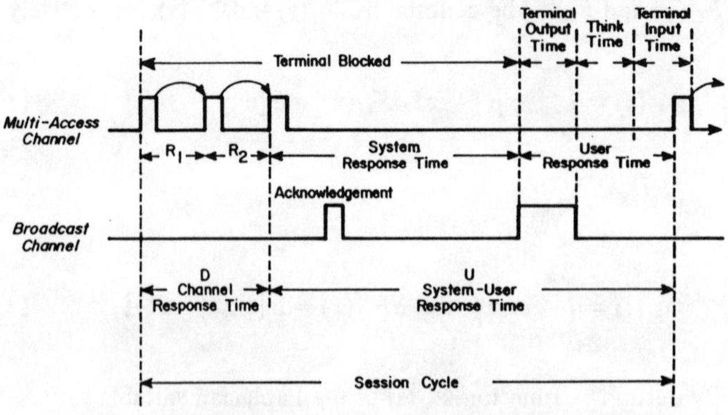

Fig. 2. Timing diagram of a typical user-system session.

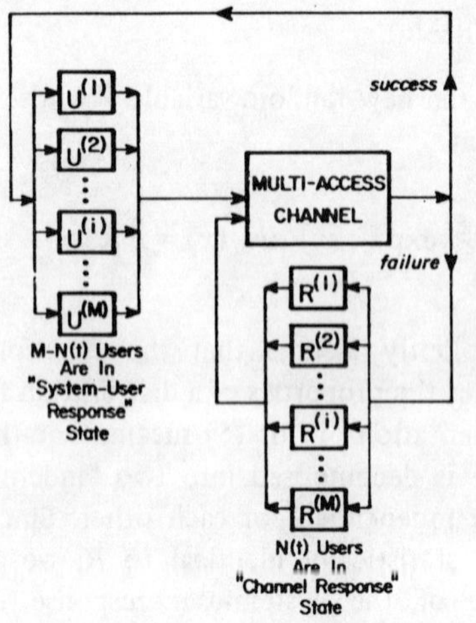

Fig. 3. Queueing model of the ALOHA system of Fig. 1.

3) Both unslotted and slotted ALOHA schemes are treated in unified manner.

4) Illustrative numerical examples are given to demonstrate usefulness of the approximation technique. The accuracy of the results obtained is assessed in the few special cases, for which there are exact solutions obtained from a Markov chain analysis.

In Section II we formulate our system model and introduce a few transformation steps. Section III discusses characterization of channel inputs and outputs, as a preliminary step toward the diffusion approximation of the ALOHA channel behavior, which is duly discussed in Section IV. Section V is devoted to numerical examples in which we demonstrate the validity and practicality of the approximation method developed.

II. MODEL FORMULATION

Let us consider a system with the ALOHA (slotted or unslotted) channel as depicted in Fig. 1, in which there are a fixed number M of users. Fig. 2 illustrates a timing diagram of a typical session between a (terminal) user and the system. The message shown in this figure collides with other packets twice in a row and hence is retransmitted twice. Here R_1 and R_2 represent random variates which are the sum of *transmission and acknowledgment delay* and *retransmission delay* in retransmission cycles 1 and 2, respectively. We define the *system-user response time* U as the duration from the moment the transmission is successfully done until the instant the same user transmits a new packet. The system-user response is thus composed of *system response time* and *user response time:* the system response time is the amount of time required for transmitting the successful message and processing it at the host system of its destination. In the ALOHANET at the University of Hawaii [2], for example, a host system may be either TSO of IBM/370 of the campus computing center, or any computer system in the ARPANET to which ALOHANET is connected. In the ALOHANET both acknowledgment and responding messages will be sent from MENEHUNE (IMP) to the originating terminal via the packet broadcasting channel. We wish to emphasize that the system response time defined above (hence, the system-user response U as well) depends not only on behavior of the message but also on the degrees of congestion of the destination host and communication links involved. The response time perceived by a user is therefore the sum of the channel response time (which we call later *delay* and denote by D), and the system response time. During this response time period we assume that the user terminal is *blocked.*

Having defined those terms given above, we are now ready to introduce a specific queueing model we are going to analyze. The model diagram is depicted in Fig. 3. In the diagram those boxes which are associated with the variates $U^{(i)}$ and $R^{(i)}$, $1 \leqslant i \leqslant M$, represent random delays, or equivalently two sets of parallel servers with no queueing. We assume that all the M users constitute a homogeneous but independent population. Thus $U^{(i)}$'s, $1 \leqslant i \leqslant M$, are statistically (iid) random variables with a common distribution $F_U(x) = P[U \leqslant x]$. Similarly $R^{(i)}$'s, $1 \leqslant i \leqslant M$ are iid random variables with distribution $F_R(x) = P[R \leqslant x]$.

Let us assume that we can decompose the random variable U into two components:

$$U = V + W \tag{1}$$

where V and W are statistically independent, and the variable W possesses the same distribution as that of R, i.e.,

$$F_W(x) = F_R(x). \tag{2}$$

In order to state the above decomposition more formally, let the Laplace-Stieltjes transform (LST) of the distributions

$F_U(x)$ and $F_R(x)$ be denoted by $\Phi_U(s)$ and $\Phi_R(s)$, respectively:

$$\Phi_U(s) = \int_0^\infty \exp\{-sx\}\, dF_U(x) = E[\exp\{-sU\}] \tag{3}$$

and

$$\Phi_R(s) = \int_0^\infty \exp\{-sx\}\, dF_R(x) = E[\exp\{-sR\}]. \tag{4}$$

By defining a function $\Phi_V(s)$ of the Laplacian variable s by

$$\Phi_V(s) = \frac{\Phi_U(s)}{\Phi_R(s)} \tag{5}$$

we introduce the new random variable V whose distribution is $F_V(s)$ such that

$$\Phi_V(s) = \int_0^\infty \exp\{-sx\}\, dF_V(x) = E[\exp\{-sV\}]. \tag{6}$$

Here we implicitly assume that the function $F_V(x)$ thus defined satisfies the properties of a distribution function.

The representation (1) or (5) means that the system-user response time is decomposed into two tandem phases V and W, which are independent of each other. Since the random variable W is statistically identical to R, we can merge the *second phase* of the system-user response time into the channel response time, obtaining the transformed model as depicted in Fig. 4, in which the *modified ALOHA channel* is defined as the multiaccess channel together with a set of random delay devices $R^{(i)}$, $1 \leq i \leq M$.

In reference to the original model, i.e., Fig. 3, let us denote by $N(t)$ the number of backlog users (or equivalently the number of outstanding messages which require retransmissions). The remaining $M - N(t)$ users are those who are either waiting for responses to their successfully transmitted packets or in the midst of preparing a new packet to be sent out. In the model of Fig. 4 we define $Q(t)$ as the number of users in the modified ALOHA channel, i.e., $Q(t) = N(t) + L(t)$ where $L(t)$ is the number of those users who are in the second phase of system-user response time. Thus we may call $Q(t)$ the number of *enlarged* backlog users.

III. CHARACTERIZATION OF CHANNEL INPUTS AND OUTPUTS

Let us choose the time origin $t = 0$ arbitrarily and let $A(t)$ be the number of packet arrivals into the modified ALOHA channel during the interval $[0,t]$. Similary let $D(t)$ denote the total departures (i.e., successfully transmitted packets) during that interval. Then the quantity $Q(t)$ defined above can be written as

$$Q(t) = Q(0) + A(t) - D(t). \tag{7}$$

Then for a given finite interval $[t, t + \Delta]$ the number of enlarged backlog users will change by the amount

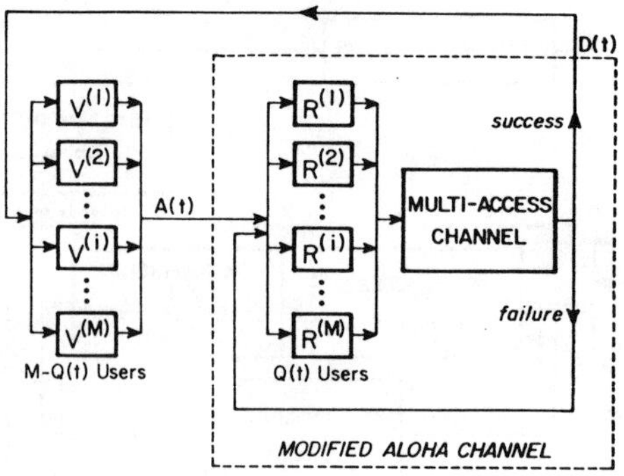

Fig. 4. Transformed model of Fig. 3.

$$Q(t + \Delta) - Q(t) = \{A(t + \Delta) - A(t)\} - \{D(t + \Delta) - D(t)\} \tag{8}$$

which we write, for notational conciseness, as

$$\Delta Q(t) = \Delta A(t) - \Delta D(t). \tag{9}$$

The modified ALOHA channel can be viewed as a single-server queueing system with random queue discipline. If Δ is sufficiently large that we observe a number of arrivals and departures during $[t, t + \Delta]$, then by virtue of the *central-limit theorem,* we approximate the integer-valued variable $\Delta A(t)$ in terms of a Gaussian variable with mean $\mu_a\Delta$ and variance $\sigma_a^2\mu_a^3\Delta$, where $1/\mu_a$ and σ_a^2 are the mean and variance of the interarrival time. For the derivation above see, for example, Cox and Miller [11]. Similarly, we approximate $\Delta D(t)$ by a Gaussian random variable to mean $\mu_d\Delta$ and variance $\sigma_d^2\mu_d^3\Delta$ where $1/\mu_d$ and σ_d^2 represent the mean and variance of the interdeparture time. Thus we can approximate $\Delta Q(t)$ by a Gaussian variable with mean

$$E[\Delta Q(t)] = (\mu_a - \mu_d)\Delta \tag{10}$$

and variance

$$\text{var}\,[\Delta Q(t)] \cong (\sigma_a^2\mu_a^3 + \sigma_d^2\mu_d^3)\Delta - 2\,\text{cov}\,[\Delta A(t), \Delta D(t)]. \tag{11}$$

The number of users who are in the first phase of the system-user time is $M - Q$.[1] Then the mean interarrival rate is

$$\mu_a = (M - Q)\mu_V \tag{12}$$

where $1/\mu_V$ is the mean value of the random variable V. Since the arrival traffic is a superposition of messages generated by $M - Q$ independent and asynchronous sources, it can be approximated by a *Poisson process* with rate μ_a of (12). Then it follows that the standard deviation of the interarrival times is

[1] We drop the time parameter t hereafter, and simply write Q instead of $Q(t)$.

$$\sigma_a = \frac{1}{\mu_a} = \frac{1}{(M - Q)\mu_V}. \tag{13}$$

If the multiaccess channel is a slotted ALOHA channel, the variables V and R are both discrete-valued variables, taking on only integral multiples of the slot size τ_s. In this case we also adopt (13) by arguing that the arrival process is approximately a *Poisson sequence*: the numbers of arrivals in successive time slots are iid with Poisson distribution of mean $\lambda = 1 - e^{-\mu_a \tau_s} \cong \mu_a \tau_s$.

Now we will investigate statistical properties of the departure process $D(t)$. We treat two cases separately, i.e., the slotted ALOHA and unslotted ALOHA schemes.

A. Slotted ALOHA

The simplest and most frequently assumed form for the retransmission delay (including acknowledgment delay) distribution is a geometric distribution. The so-called memoryless property of the geometric deistribution implies that any message in one of the R boxes of Fig. 4 attempts a transmission in the next slot time with probability p_R. If there are Q such outstanding messages in the modified ALOHA channel, the probability that we see a departure (i.e., a successful transmission) in the next slot is

$$S(Q) = Qp_R(1 - p_R)^{Q-1} \tag{14}$$

which we call *channel throughput*. The departure rate is then

$$\mu_d(Q) = \frac{S(Q)}{\tau_s} = \frac{Qp_R}{\tau_s}(1 - p_R)^{Q-1}. \tag{15}$$

Fig. 5 plots curves $S(Q)$ versus Q for various values of retransmission parameters p_R.

If we transform the parameter p_R into a new parameter μ_R by the relation

$$p_R = 1 - \exp\{-\mu_R \tau_s\} \cong \mu_R \tau_s \tag{16}$$

then $1/\mu_R$ corresponds to the mean value of the exponential distribution that fits closest to the geometric distribution. We can then write the channel throughput as

$$S(Q) = Q\mu_R \tau_s \exp\{-(Q - 1)\mu_R \tau_s\}. \tag{17}$$

We now define the *channel traffic rate* $G(Q)$

$$G(Q) = Qp_R \cong Q\mu_R \tau_s \tag{18}$$

which is the expected number of total traffic (original transmissions plus retransmissions) to be offered to the ALOHA channel in the next slot, when $Q(t) = Q$ in the present slot. By combining (17) and (18) we obtain the following relationship between the throughput and channel traffic rate:

$$S \cong G \exp\{-G\} \exp\{\mu_R \tau_s\} \cong G \exp\{-G\} \tag{19}$$

where we assume $\mu_R \tau_s \ll 1$ in obtaining the last expression. This is the formula originally obtained by Roberts [3].

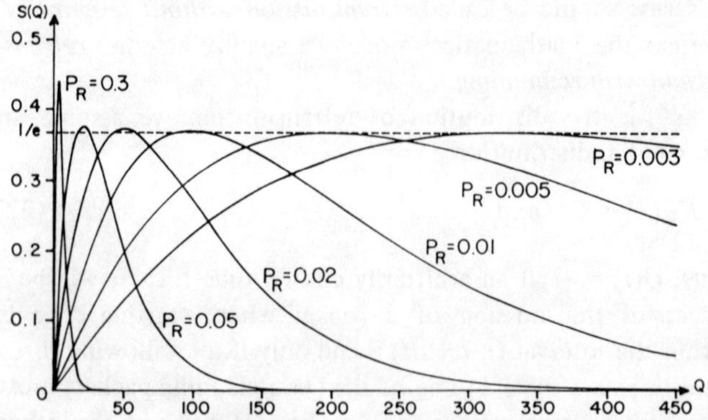

Fig. 5. Departure rate (throughput) versus Q for various retransmission parameters P_R in a slotted ALOHA System.

The output process from the ALOHA channel (and the modified ALOHA channel as well) is a 1-0 sequence, where "1" mean a successful transmission in the given time slot and "0" means no departure, that is, either no transmissions are attempted or a collision occurs in that slot. Since we can model the modified slotted ALOHA channel of Fig. 4 as a (discrete-time) *processor-sharing server,* or what we may more appropriately call a *capacity-sharing channel* with queue-dependent processing rate $S(Q)$ (see the Appendix), the departure process can be characterized by a *Bernoulli sequence* in which it takes on values "1" or "0," independently for different slots, with probabilities S and $1 - S$, respectively. An intuitive explanation follows.

It is known that in a continuous time system the output of an $M/G/1$ queueing system under the processor-sharing discipline is a Poisson process. An analogous property holds for its discrete version. Thus, when the input is a Bernoulli sequence, so is the output from the processor-shared server. For a more rigorous discussion the reader is referred to an earlier report by Kobayashi [12].

With the arguments given above, we approximate the mean and variance of interdepature times by

$$\frac{1}{\mu_d(Q)} = \frac{\tau_s}{S(Q)} \tag{20}$$

and

$$\sigma_d{}^2(Q) = \frac{1}{\mu_d{}^2(Q)}, \tag{21}$$

respectively.

B. Unslotted ALOHA

The rest of the present section is devoted to obtaining an expression similar to (17), when the random-access channel is an unslotted ALOHA channel like the one implemented in the ALOHANET. We assume that the packet length τ is a random variable with an arbitrary distribution. For the sake of mathematical tractability we assume that each time a packet attempts retransmission a new packet length is sampled. In reality, however, the packet length for a message is sampled only once, regardless of the number of collisions experienced.

This case should be called *retransmission without resampling,* whereas the mathematical model we analyze assumes *retransmission with resampling.*

As for the distribution of retransmission we assume an exponential distribution

$$F_R(x) = 1 - \exp\{-\mu_R x\}. \tag{22}$$

When $Q(t) = Q$ at an arbitrarily chosen time t, there will be a successful transmission of a packet whose starting edge is within the interval $(t, t + dt)$ if and only if the following three conditions are met: 1) one of the Q outstanding packets must start transmission within $(t, t + dt)$; 2) none of the other $(Q - 1)$ packets may start transmissions in the interval $(t, t + \tau_1)$, where τ_1 is the duration of the packet defined above (which we call the *marked packet* hereafter); and 3) there are no prior transmissions which may interfere with the marked packet. The probability of condition 1) is given by $Q\mu_R dt$; the probability of the condition 2) is $(\exp\{-\mu_R\tau_1\})^{Q-1} = \exp\{-\mu_R\tau_1(Q-1)\}$.

Condition 3) will be violated if there is a transmission of a packet of length τ starting within the interval $(t - \tau, t)$. In order to evaluate this probability we first need to know the properties of the departing process. As in the case of slotted ALOHA, we can model the channel under consideration as a processor-sharing server (see the Appendix), and we can show that the departure process $D(t)$ of Fig. 4 is Poisson, if its input process $A(t)$ is Poisson. Conversely if $D(t)$ is Poisson, so is $A(t)$ for *arbitrary* distribution of the variables $V^{(i)}$ of Fig. 4 [13], [14].

If a process is Poisson, so is its time-reversed process as discussed by Reich [15]. Thus by applying the above result to our model we can show that

$$\lim_{\Delta \to 0} \frac{P[\text{departure in } (t - \Delta, t) \mid Q(t) = Q]}{\Delta} = \lambda(Q) \triangleq Q\mu_R \tag{23}$$

where $\lambda(Q)$ is the arrival rate when $Q(t) = Q$. Thus the interval between the most recent depature and the current time t is exponentially distributed with mean $1/Q\mu_R$. Let us denote by τ_0 the packet length of the most recently attempted transmission. Then the probability of condition 3) being met is $\exp\{-\mu_R Q\tau_0\}$.

Combining the probabilities of the three conditions obtained above, we have the following expression for the departure rate $\mu_d(Q)$ of the unslotted ALOHA system;

$$\mu_d(Q) = E[Q\mu_R \cdot \exp\{-\mu_R\tau_1(Q-1)\} \cdot \exp\{-\mu_R Q\tau_0\}] \tag{24}$$

where the expectation should be taken with respect to the variables τ_0 and τ_1. Since they are statistically independent, we immediately obtain

$$\mu_d(Q) = Q\mu_R \Phi_\tau(\mu_R(Q-1))\Phi_\tau(\mu_R Q) \tag{25}$$

where $\Phi_\tau(s)$ is the LST of the distribution $F_\tau(x)$.

Example 1) Consider an unslotted ALOHA system in which packet length is constant τ_s, i.e.,

$$F_\tau(x) = \begin{cases} 0, & \tau < \tau_s \\ 1, & \tau \geqslant \tau_s . \end{cases}$$

Then by taking the LST of this distribution, we have

$$\Phi_\tau(s) = \exp\{-s\tau_s\}.$$

Therefore, the departure rate of the successful packets from the channel is

$$\mu_d(Q) = Q\mu_R \exp\{-(2Q-1)\mu_R\tau_s\}$$

when the number of enlarged backlog users is Q. By deriving the channel traffic rate G and throughput rate S by

$$G(Q) = Q\mu_R\tau_s$$

and by

$$S(Q) = \mu_d(Q)\tau_s,$$

respectively, we obtain the following relation between S and G:

$$S = G \exp\{-2G\} \exp\{\mu_R\tau_s\} \cong G \exp\{-2G\}$$

where the last expression was obtained under the assumption $\mu_R\tau_s \ll 1$. This expression agrees with the formula originally derived by Abramson [1] from a different approach.

IV. SOLUTIONS VIA THE DIFFUSION APPROXIMATION

Now let us return to (10) and (11) and proceed to obtain a system equation which characterizes the random process $Q(t)$. From the results of the previous section, we now have

$$E[\Delta Q] \cong [(M - Q)\mu_V - \mu_d(Q)]\Delta \tag{26}$$

and

$$\text{var}[\Delta Q] \cong [(M - Q)\mu_V + \mu_d(Q)]\Delta. \tag{27}$$

The last equation is obtained by assuming the correlation between ΔA and ΔD is negligibly small. We justify this assumption as follows. The number of departure ΔD in the interval $[t, t + \Delta]$ is dependent upon $Q(t')$, $t \leqslant t' \leqslant t + \Delta$. The value $Q(t')$ in turn depends on the *past* arrivals $A(s)$, $s \leqslant t'$ (as well as on the past departures). Thus if $Q(t)$ is sufficiently large compared with the additional arrivals ΔA in the following interval $[t, t + \Delta]$, then the variation in ΔD of the same period is largely a function of Q, and is weakly dependent on ΔA.

Although the values which $Q(t)$ can take on are only nonnegative integer values, we choose to approximate this *jump process* $Q(t)$ by a *continuous-path random process* $x(t)$, as has been done in the diffusion approximation of a queueing system [16]-[18]. The reason for such a substitution is simply that the system equation for $x(t)$ (to be shown later) is much more tractable than that for $Q(t)$.

In order for $x(t)$ to mimic the behavior of $Q(t)$, the incremental change process

$$dx(t) = x(t + dt) - x(t) \tag{28}$$

should possess properties similar to those of the corresponding process $dQ(t) = Q(t + dt) - Q(t)$. Based on the results of (26) and (27), we define $dx(t)$ as a Gaussian process with

$$E[dx(t)] = \beta(x)dx, \qquad \text{var}\,[dx(t)] = \alpha(x)dx \tag{29}$$

where

$$\beta(x) = (M - x)\mu_V - \mu_d(x),$$
$$\alpha(x) = (M - x)\mu_V + \mu_d(x). \tag{30}$$

The process $x(t)$ thus defined is a continuous-path Markov process and is often called a diffusion process.

Since $x(t)$ is a continuous random process, we can define the probability density function $p(x; t)$ as

$$p(x; t)dx = P[x < x(t) \leqslant x + dx]. \tag{31}$$

Let $\bar{x}$ be such a value that satisfies $\beta(\bar{x}) = 0$, i.e.,

$$(M - \bar{x})\mu_V = \mu_d(\bar{x}) \tag{32}$$

as is illustrated in Fig. 6. Then by linearizing $\mu_d(x)$ around $x = \bar{x}$, we have

$$\mu_d(x) \cong \mu_d(\bar{x}) + \mu_d{}'(\bar{x})(x - \bar{x}). \tag{33}$$

Then the function $\beta(x)$ becomes also a linear function of x:

$$\beta(x) \cong \beta_0 - \beta_1 x \tag{34}$$

where

$$\beta_1 = \mu_d{}'(\bar{x}) + \mu_V, \qquad \beta_0 = \beta_1\bar{x}. \tag{35}$$

If we consider a fairly narrow region around $x = \bar{x}$, we can approximate $\alpha(x)$ by a constant, i.e. $\alpha(\bar{x})$, assuming the first-order term $\alpha'(\bar{x})(x - \bar{x})$ is much smaller than $\alpha(\bar{x})$, so that

$$\alpha(x) \triangleq \alpha(\bar{x}) = 2(M - \bar{x})\mu_V = 2\mu_d(\bar{x}) = \alpha_0. \tag{36}$$

Then we can show that $p(x; t)$ should satisfy the following partial-differential equation (Appendix B of [26])

$$\frac{\partial p(x; t)}{\partial t} = \frac{\alpha_0}{2}\frac{\partial^2 p(x; t)}{\partial x^2} + \beta_1 \frac{\partial}{\alpha x}[(x - \bar{x})p(x; t)] \tag{37}$$

which is known by the name of the diffusion equation of Ornstein-Uhlenbeck (O-U) process [11], [19]. The O-U process is a variant of the Brownian motion process, and can

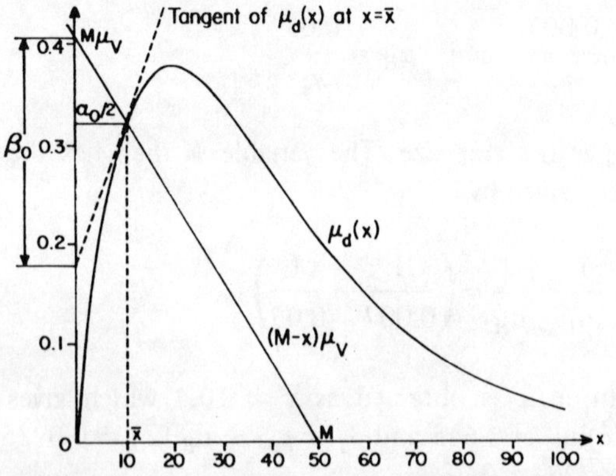

Fig. 6. A graphical interpretation of $\bar{x}$ of (32).

be viewed as a limiting form of a random walk with central restoring tendency. The diffusion approximation of this type was proposed earlier by Kobayashi [20] as a stochastic model of program behavior in virtual storage environment and was analyzed in [21].

Since the process $Q(t)$ is bounded between $Q = 0$ and $Q = M$, so is its diffusion approximation $x(t)$:

$$0 \leqslant x(t) \leqslant M, \qquad \text{for all } t \geqslant 0. \tag{38}$$

Furthermore, the following initial condition must be met

$$p(x; 0 \mid x_0; 0) = \delta(x - x_0) \tag{39}$$

where $\delta(x)$ is Dirac's delta function. The solution of the diffusion equation (37) with the boundary conditions (38) and (39) is discussed by Sweet and Hardin [22]. The equilibrium solution is a Gaussian distribution with center $\bar{x}$ and variance $\alpha_0/2\beta_1$, truncated at $x = 0$ and $x = M$, namely

$$p(x) = \lim_{t\to\infty} p(x; t \mid x_0; 0) = \begin{cases} G^{-1}\exp\left\{-\dfrac{(x - \bar{x})^2}{\sigma_x{}^2}\right\}, & 0 \leqslant x \leqslant M \\ 0, & \text{elsewhere} \end{cases} \tag{40}$$

where

$$\sigma_x{}^2 = \frac{\alpha_0}{2\beta_1} \tag{41}$$

and G is the normalization constant.

Based on the steady-state solution (40), we can derive various measures of performance pertinent to design and performance prediction of an ALOHA-like system. The following section discusses several numerical examples of such applications.

V. NUMERICAL EXAMPLES

Example 2) Let us assume a slotted ALOHA system with $M = 50$ homogeneous users. Let the random variables U and R both be exponentially distributed with parameters

$$\mu_U = \frac{0.007}{\tau_s} \quad \text{and} \quad \mu_R = \frac{0.05}{\tau_s}$$

where τ_s is the slot size. The variable V then has the mean value $1/\mu_V$ given by

$$\frac{1}{\mu_V} = \frac{1}{\mu_U} - \frac{1}{\mu_R} = \left(\frac{1}{0.007} - \frac{1}{0.05}\right)\tau_s.$$

The solution $\bar{x}$ is obtained as $\bar{x} = 10.4$ which gives $\alpha_0 = 2(M - \bar{x})\mu_V = 0.645$ and $\beta_1 = \mu_V + \mu_d'(\bar{x}) = 0.0225$, thus $\sigma_x = \sqrt{\alpha_0/2\beta_1} = 3.78$.

The normalization constant G is determined as $G \cong \sqrt{2\pi}\sigma_x[1 - \Phi(-(\bar{x}/\sigma_x))] = 9.45$. The average value of Q is given by

$$E[Q] = E[x] \cong \int_0^\infty xp(x)\,dx$$

$$= G^{-1}\sigma_x{}^3 \exp\left\{-\frac{\bar{x}^2}{2\sigma_x{}^2}\right\} + \bar{x} = 10.43.$$

The average throughput rate [per slot time] is calculated as

$$\lambda_s = \int_0^M S(x)p(x)\,dx$$

$$\cong S(\bar{x}) + (E[x] - \bar{x})S'(\bar{x}) + \frac{(E[x] - \bar{x})^2}{2} S''(\bar{x})$$

$$= 0.306 \text{ [packets/slot]} \tag{42}$$

where we use the second-order approximation of $S(x)$ around $x = \bar{x}$. By applying Little's formula the average length of time [in slot times] that a packet stays in the modified ALOHA channel of Fig. 4 is

$$T = \frac{E[Q]}{\lambda_s} = 34.1 \text{ [slots]}.$$

Therefore, the average delay D that a message experiences from the moment it is generated at the user station until it successfully goes through the channel is

$$D = T - \frac{1}{\mu_R\tau_s} = 14.1 \text{ [slots]}.$$

The average number of outstanding messages is thus given by

$$E[N] = \lambda_s D = E[Q] - \frac{\lambda_s}{\mu_R\tau_s} = 4.32. \tag{43}$$

In order to assess the accuracy of the approximate method adopted above, we evaluated the distribution of the variable N based on the Markovian assumption made for the model of Fig. 3. The model is essentially equivalent to those studied by Carleial and Hellman [7], Murthy [8], and others. The computational procedure we used is a recursive algorithm, which can avoid the inversion of an $M \times M$ matrix. The mean value of N is evaluated as $E[N] = 4.78$, which indicates that the error of our approximate solution obtained above is in the order of 10 percent. Throughput and delay were also calculated in the Markovian model. They are $\lambda_s = 0.317$ [packets/slot] and $D = 15.1$ [slots], which also confirm the validity of the approximate solutions obtained earlier.

Example 3) Stability Analysis: In general there is some difference between the center $\bar{x}$ of the distribution and the expectation $E[x]$ due to the truncation of $p(x)$ at the boundaries $x = 0$ and $x = M$. In the example shown above the difference was rather small. If we are willing to ignore this difference and accept $\bar{x}$ as a approximation to the mean value, then we are adopting the *fluid approximation* instead of the diffusion approximation. An alternative interpretation of this graphical solution is that we are neglecting variations of the variables V and R; if the variables V and R are both constants $1/\mu_V$ and $1/\mu_R$, respectively, the corresponding solution is exactly what we observe in this simple graphical analysis.

The straight line $(M - x)\mu_V$ in this graph represents the *inflow rate* [packets/second] to the modified ALOHA channel, when the number of users in that channel is x; similarly, the curve $\mu_d(x) = S(x)/\tau_s$ represents the *outflow* [packets/second] from that channel. If, for a given value of x, the inflow rate is greater than the outflow rate, then x will increase. Conversely in the region of x where the outflow rate exceeds the inflow rate, x will decrease. Thus if the curve $\mu_d(x)$ and the straight line $(M - x)\mu_V$ cross each other at $x = x_A$, x_B, and x_C as shown in Fig. 7, then the points x_A and x_C are *stable* operating points, whereas x_B is *unstable,* since x tends to diverge either towards x_A or x_C, away from the point x_B. This type of analysis and graphical interpretation are often used in an electrical circuit analysis, where the x value represents an operating *voltage* and the flow rate represents electrical *current.* The parameters M and μ_V then correspond to the source voltage and the impedance. The unstable point x_B exists in the region of negative resistance.

Variations in the flow of the particles (packets) are significant in that the actual operating points are *distributed* around $x = x_A$ and $x = x_C$. The diffusion approximation discussed in this paper is a refinement over the fluid approximation in that an approximate distribution around x_A and x_C can be obtained.

Once the system is driven into the region around x_C, its performance suddenly collapses; throughput decreases virtually to zero, and the channel delay becomes immensely large. The quantity T [time slots] defined in Example 2 is given as the intercept of the line **OC** with the horizontal line $S = 1$ as shown in Fig. 7. The average delay that a packet experiences in the ALOHA channel is given by $D = T - 1/\mu_R\tau_s$.

Graphical analyis of the sort outlined above was applied to the slotted ALOHA system by Kleinrock and Lam [5]. The difference between their result and ours here is as follows. In their model both the inflow and outflow (throughput) rates are plotted against N instead of Q (or x). As a result the throughput curve includes parameters μ_V and M as well as μ_R. In our curve, S versus Q, the throughput curve includes only

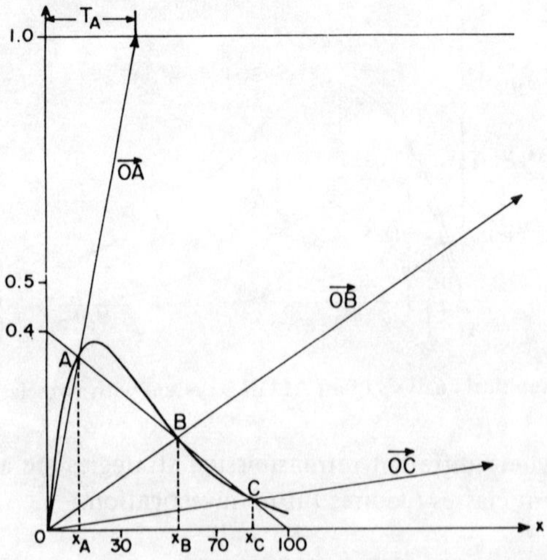

Fig. 7. Stable operating points *A* and *C* and unstable point *B* ($M = 100$, $P_R = 0.05$, $P_V = 0.004$).

one parameter, i.e., μ_R. Thus, by preparing a set of S versus Q curves for various values of the retransmission parameter μ_R, one can easily analyze the effect of such system parameters as M, μ_U (hence μ_V), and μ_R on the system's stability, throughput, and delay.

It is worthwhile here to mention that graphical analysis of a similar nature has been applied by Courtois [23] to an analysis of a multiprogrammed virtual storage system. An increase in paging rates due to either a sudden increase in system work load or poor control in memory allocation/management policy often results in a sudden increase in paging overheads, hence a sharp decline in the actual throughput. This phenomenon is often referred to as "thrashing," which corresponds in our problem at hand, to an avalanche of packet collisions in the ALOHA channel. For further discussion on this subject, see Kobayashi [24].

Example 4) Case with $\mu_V < 0$: Thus far we assumed the distributions of the random variables U and R to be such that the random variable V defined by (1) is nonnegative. Clearly the inequality $\mu_U < \mu_R$ is a necessary condition for that requirement. Now we demonstrate that we can violate this condition and still obtain a meaningful result by formally adopting the computational and graphical analysis procedure discussed in the preceding examples. Let us assume the same model as in Example 2 but we now choose a much smaller value for the mean system-user response time,

$$\mu_U = \frac{0.3}{\tau_s}.$$

Then the parameter μ_V takes on a negative value,

$$\mu_V = \frac{\mu_U \mu_R}{\mu_R - \mu_U} = -\frac{0.06}{\tau_s}$$

and the line $(M - x)\mu_V \tau_s$ versus x now possesses a positive slope as shown in Fig. 8. The cross-over point of the curve $S(x)$ with this line is at $\bar{x} = 53.2$ which is even greater than $M = 50$. The corresponding throughput value is

$$\lambda_s = S(\bar{x}) = (M - \bar{x})\mu_V \tau_s = 0.181 \text{ [packets/slot]}$$

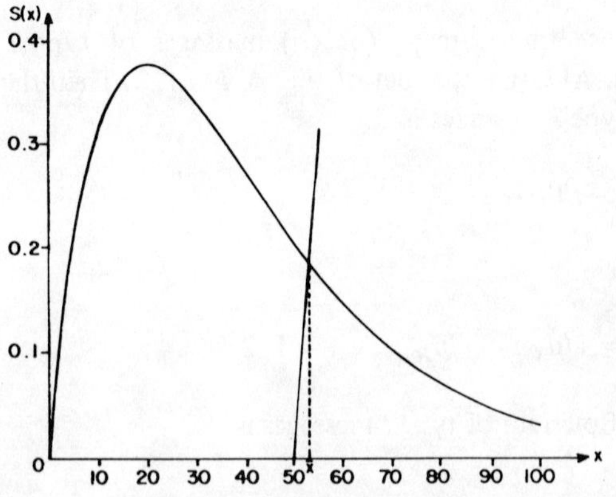

Fig. 8. Case with $\mu_v < 0$.

which gives

$$T = \frac{\bar{x}}{\lambda_s} = 293.9 \text{ [slots]}.$$

Then the average delay that a packet experiences within the ALOHA channel is

$$D = T - \frac{1}{\mu_R \mu_s} = 293.9 - 20 = 273.9 \text{ [slots]}$$

which in turn gives the average number of outstanding packets as

$$E[N] = \lambda_s D = 49.6.$$

The corresponding quantities derived via the Markovian assumption are $E[N] = 49.4$, $\lambda_s = 0.183$ [packets/slot], and $D = 269$ [slots]. The accuracy of the approximate solutions is as good as in our previous examples.

A formal proof on the generalization to cases with $\mu_V < 0$ is yet to be shown. An intuitive interpretation is that whenever V is negative it corresponds to a *time advance* (i.e., delay of negative amount). In the context of fluid approximation $\mu_V < 0$ means the direction of flow is reversed. Our conjecture is that the decomposition (1) should be treated as a special case of Cox's representation of a general distribution in terms of exponential distributions in cascade [25]. Note that in his representation the distribution parameters (such as μ_V and μ_R) can be complex numbers. When the retransmission delay variable R is not exponentially distributed, it can be further decomposed with the same spirit as that of Cox's representation. A further generalization of our method along this line is left for future investigation.

Example 5) Heterogeneous Users: The approximation method and graphical analysis discussed in the preceding examples can be extended to some types of heterogeneous population source models. Let us assume, for example, a slotted ALOHA system with *two classes* of users. The two classes are characterized by different values of μ_V which we denote by μ_{V_1} and μ_{V_2}. Let there be M_1 users of class 1 and M_2 users of class 2. We make no distinctions in the retransmission delay strategy between the two classes.

Suppose there are x_i (or Q_i) messages of type i in the modified ALOHA channel of Fig. 4, $i = 1, 2$. Then the inflow rate of type i messages is

$$(M_i - x_i)\mu_i \tau_s, \qquad i = 1, 2$$

where

$$1/\mu_i = 1/\mu_{V_i} - 1/\mu_R, \qquad i = 1, 2.$$

The outflow rate of type i messages is

$$S_i = \frac{x_i}{x} S(x), \qquad i = 1, 2$$

where $x = x_1 + x_2$ and $S(\cdot)$ is the throughput function defined earlier, which has μ_R as a parameter. Then by equating the inflow rate and outflow rate given above, we obtain

$$\frac{(M_1 - x_1)\mu_1}{x_1} = \frac{(M_2 - x_2)\mu_2}{x_2}$$

which is the necessary condition for the values $\bar{x}_1$ and $\bar{x}_2$ of the operating point in equilibrium. By equating the last expression to a variable parameter θ we write the total inflow rate as

$$\sum_{i=1}^{2} (M_i - x_i)\mu_i \tau_s = \theta(x_1 + x_2) = \theta x.$$

Based on the last equation we synthesize the inflow rate curve I as illustrated in Fig. 9; for a given value of θ, we draw a straight line which emanates from the origin 0 with slope θ. The crossing points with the two straight lines $(M - x_i)\mu_i \tau_s$, $i = 1, 2$ are denoted by points P and Q. Then we obtain a new point R by the following vector summation:

$$\boldsymbol{OR} = \boldsymbol{OP} + \boldsymbol{OQ}.$$

The curve I is obtained by changing θ over the range $0 < \theta < \infty$. The intersection of the I curve and the $S(x)$ curve gives the fluid approximation solution $\bar{x}$, the corresponding $\bar{x}_i$, and their throughput values λ_i

$$\lambda_i = (M_i - \bar{x}_i)\mu_i \tau_s, \qquad i = 1, 2.$$

The average delay D_i and the average number of outstanding messages $E[N_i]$ are thus immediately obtainable as

$$D_i = \frac{\bar{x}_i}{\lambda_i} - \frac{1}{\mu_R \tau_s}, \qquad i = 1, 2$$

and

$$E[N_i] = \lambda_i D_i = \bar{x}_i - \frac{\lambda_i}{\mu_R \tau_s}, \qquad i = 1, 2.$$

It is straightforward to extend this simple procedure to the case with more than two classes of users. A generalization to

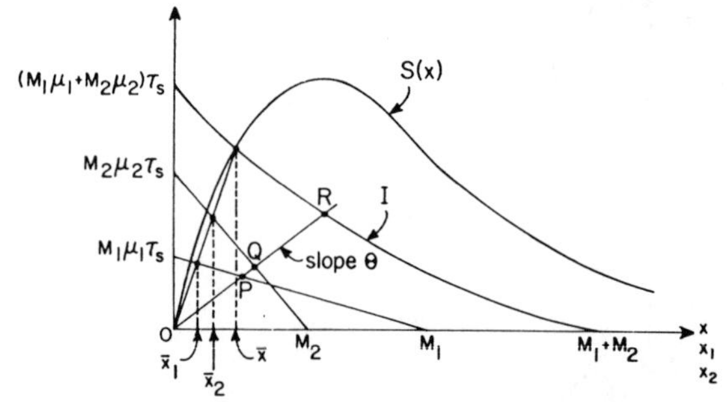

Fig. 9. Graphical analysis of an ALOHA system with two class users.

the case where different retransmission strategies are assumed for different classes requires future investigation.

VI. CONCLUSION

The major advantage of the approximation method discussed in the present paper is its *simplicity* and *generality*. By introducing the fictitious variable $Q(t)$ as a measure of system congestion, we are able to represent the dynamic behavior of an ALOHA-like system in terms of a diffusion process, known as the Ornstein–Uhlenbeck process. The use of the variable $Q(t)$ (and its diffusion process approximation $x(t)$) also enables us to separate such key parameters as M, the number of users; μ_V, the message generation rate of a user; p_R, the retransmission delay parameter, etc., in the performance equations and graphical analyses developed in the paper. Several illustrative examples of Sections III–V have clearly demonstrated the utility and accuracy of our analytic method. For more detailed discussions and additional examples, the reader is directed to our ALOHA technical report [26].

APPENDIX

PROCESSOR-SHARING SERVER AND CAPACITY-SHARING CHANNEL

The notion of *processor-sharing* (PS) was originally introduced for continuous-time system as the limiting case of the so-called *round-robin* (RR) scheduling discipline, in which the time quantum is made to approach infinitesimal zero. Under this discipline all jobs (or customers) present in the system receive service simultaneously, sharing the server equally. If there are n jobs in the system at a given time, and the total processing rate of the server is $C(n)$ [work units/second], then each job receives service with the rate of $C(n)/n$. It has been shown lately that under the PS discipline a system $M/G/1$ behaves as if it were an $M/M/1$ system; the departure process exhibits a Markovian property, i.e., a Poisson process with the same rate as the input. A proper translation of this important property for a *discrete-time system* requires the following definition of PS [12].

Definition: Processor-sharing (PS) discipline in a discrete-time queueing system is a scheduling rule which selects, in every time slot, one of n customers found in the system *randomly* and *equally* likely, and the processes it for one unit time PS discipline originally defined by Kleinrock [27].

Thus if the service requirement of a customer is $k\tau_s$, then it must be scheduled and processed k times before it is completed. Note that the PS discipline is distinct from the RR, although in the limit $\tau_s \to 0$ they converge to the continuous-time PS discipline originally defined by Kleinrock [27].

When the server is a channel, the PS rule defines a special multiplexing method, and a more appropriate name would be a *capacity-sharing channel.*

ACKNOWLEDGMENT

The authors would like to thank Professors N. Abramson and M. Ferguson for their careful reading of the manuscript and valuable suggestions on this work.

REFERENCES

[1] N. Abramson, "The ALOHA System—Another alternative for computer communications," in *Fall Joint Comput. Conf., AFIPS Conf. Proc.*, vol. 37, 1970, pp. 281–285.

[2] —, "The ALOHA System," in *Computer Communication Networks,* N. Abramson and F. F. Kuo, Eds. Englewood Cliffs, NJ: Prentice-Hall, 1973, pp. 501–517.

[3] L. G. Roberts, "ALOHA packet system with and without slots and capture," ARPA Network Information Center, Stanford Res. Inst., Stanford, CA, June 1972, ASS note 8.

[4] R. M. Metcalfe, "Steady-state analysis of a slotted and controlled ALOHA System with blocking," in *Proc. 4th Hawaii Int. Conf. System Sciences,* Western Periodicals, Jan. 1973, pp. 375–378.

[5] L. Kleinrock and S. S. Lam, "Packet-switching in a slotted satellite channel," in *Nat. Comput. Conf., AFIPS Conf. Proc.*, vol. 42, 1973, pp. 703–710.

[6] —, "Packet switching in a multiaccess broadcast channel: Performance evaluation," *IEEE Trans. Commun.*, vol. COM-23, pp. 410–423, Apr. 1975.

[7] A. B. Carleial and M. E. Hellman, "Bistable behavior of ALOHA-type systems," *IEEE Trans. Commun.*, vol. COM-23, pp. 401–410, Apr. 1975.

[8] P. Murthy, "Analysis and static control procedures for slotted ALOHA with blocking," Univ. of Hawaii, Honolulu, ALOHA System tech. rep. B75-9, Feb. 1975.

[9] I. Gitman, "On the capacity of slotted ALOHA networks and some design problems," *IEEE Trans. Commun.*, vol. COM-23, pp. 305–317, Mar. 1975.

[10] M. J. Ferguson, "A study of unslotted ALOHA with arbitrary message lengths," Univ. of Hawaii, Honolulu, ALOHA System tech. rep. B75-13, Feb. 1975.

[11] D. R. Cox and M. D. Miller, *The Theory of Stochastic Processes.* New York: Wiley, 1965.

[12] H. Kobayashi, "On discrete-time processes in a packetized communication system," Univ. of Hawaii, Honolulu, ALOHA System tech. rep. B75-28, Aug. 1975.

[13] J. L. Doob, *Stochastic Processes.* New York: Wiley, 1953, p. 405.

[14] N. M. Mirasol, "The output of a M/G/∞ queueing system is Poisson," *Oper. Res.*, vol. 11, pp. 282–284, 1963.

[15] E. Reich, "Departure Processes," in *Congestion Theory,* W. L. Smith and W. E. Wilkinson, Eds. Chapel Hill, NC: Univ. of North Carolina Press, 1965, pp. 439–457.

[16] G. F. Newell, *Applications of Queueing Theory.* London, England: Chapman and Hall, 1971.

[17] D. P. Gaver and G. S. Shedler, "Processor utilization in multiprogramming systems via diffusion approximation," *Oper. Res.*, vol. 21, pp. 569–576, 1973.

[18] H. Kobayashi, "Application of the diffusion approximation to queueing networks, parts I and II," *J. Ass. Comput. Mach.*, vol. 21, pp. 316–328, Apr. 1974, and vol. 21, pp. 459–469, July 1974.

[19] G. E. Uhlenbeck and L. S. Ornstein, "On the theory of the Brownian motion," *Phys. Rev.*, vol. 36, pp. 823–841, 1930; also in *Selected Papers on Noise and Stochastic Processes,* N. Wax, Ed. New York: Dover, 1954.

[20] H. Kobayashi, "Some recent progress in analytic studies of system performance," *Proc. 1st USA-Japan Comp. Conf.*, Tokyo, Japan, Oct. 1972, pp. 130–138.

[21] M. Z. Ghanem and H. Kobayashi, "A parametric representation of program behavior in a virtual memory system," IBM T. J. Watson Research Center, Yorktown Heights, NY, IBM res. rep. RC 4560, Oct. 1970.

[22] A. L. Sweet and J. C. Hardin, "Solution for some diffusion processes with two barriers," *J. Appl. Prob.*, pp. 423–431, 1970.

[23] P. J. Courtois, "Decomposability, instabilities and saturation in multiprogramming systems," *Commun. Ass. Comput. Mach.*, vol. 18, pp. 371–376, July 1975.

[24] H. Kobayashi, "Queueing network models of computing systems," in *Current Trends in Programming Methodology, Vol. III: Software Modeling,* K. M. Chandy and R. T. Yeh, Ed. Englewood Cliffs, NJ: Prentice-Hall, 1976.

[25] D. R. Cox, "A use of complex probabilities in theory of stochastic processes," *Proc. Cambridge Phil. Soc.*, vol. 51, pp. 313–319, 1955.

[26] H. Kobayashi, Y. Onozato, and D. Huynh, "An approximate method for design and analysis of an ALOHA System," University of Hawaii, Honolulu, ALOHA System tech. rep. B76-4, Mar. 1976.

[27] L. Kleinrock, "Time-shared systems: A theoretical treatment," *J. Ass. Comput. Mach.*, vol. 14, pp. 242–261, Apr. 1967.

Packet Switching in Radio Channels: Part I—Carrier Sense Multiple-Access Modes and Their Throughput-Delay Characteristics

LEONARD KLEINROCK, FELLOW, IEEE, AND FOUAD A. TOBAGI

Abstract—Radio communication is considered as a method for providing remote terminal access to computers. Digital byte streams from each terminal are partitioned into packets (blocks) and transmitted in a burst mode over a shared radio channel. When many terminals operate in this fashion, transmissions may conflict with and destroy each other. A means for controlling this is for the terminal to sense the presence of other transmissions; this leads to a new method for multiplexing in a packet radio environment: carrier sense multiple access (CSMA). Two protocols are described for CSMA and their throughput-delay characteristics are given. These results show the large advantage CSMA provides as compared to the random ALOHA access modes.

I. INTRODUCTION

LARGE COMPUTER installations, enormous data banks, and extensive national computer networks are now becoming available. They constitute large expensive resources which must be utilized in a cost/effective fashion. The constantly growing number of computer applications and their diversity render the problem of *accessing* these large resources a rather fundamental one. Prior to 1970, wire connections were the principal means for communication among computers and between users and computers. The reasons were simple: dial-up and leased telephone lines were available and could provide inexpensive and reasonably reliable communications for short distances, using a readily available and widespread technology. It was long recognized that this technology was inadequate for the needs of a computer-communication system which is required to handle bursty traffic (i.e., large peak to average data rates). For example, the inadequacies included the long dial-up and connect time, the minimum three-minute tariff structure, the fixed and limited data rates, etc. However, it was not until 1969 that the cost to switch communication bandwidth dropped below the cost of the bandwidth being switched [1]. At that time, the new technology of packet-switched computer networks emerged and developed a cost/effective means for connecting computers together over long-distance high-speed lines. However, these networks did not solve the *local* interconnection problem, namely, how can one efficiently provide access from the user to the network itself? Certainly, one solution is to use wire connections here also. An alternate solution is the subject of this paper, namely, ground radio packet switching.

We wish to consider broadcast radio communications as an alternative for computer and user communications. The ALOHA System [2] appears to have been the first such system to employ wireless communications. The advantages in using broadcast radio communications are many: easy access to central computer installations and computer networks; collection and dissemination of data over large distributed geographical areas independent of the availability of preexisting (telephone) wire networks; the suitability of wireless connections for communications with and among mobile users (a constantly growing area of interest and applications); easily bypassed hostile terrain; etc. Perhaps, this broadcast property is the key feature in radio communication.

The Advanced Research Projects Agency (ARPA) of the Department of Defense recently undertook a new effort whose goal is to develop new techniques for packet radio communication among geographically distributed, fixed or mobile, user terminals and to provide improved frequency management strategies to meet the critical shortage of RF spectrum. The research presented in this paper is an integral part of the total design effort of this system which encompasses many other research topics [3]–[9].

Consider an environment consisting of a number of (possibly mobile) users in line-of-sight and within range of each other, all communicating over a (broadcast) radio channel in a common frequency band. The classical approach for satisfying the requirement of two users who need to communicate is to provide a communication channel for their use so long as their need continues (line-switching). However, the measurements of Jackson and Stubbs [10] show that such allocation of scarce communication resources is extremely wasteful. Rather than providing channels on a user-pair basis, we much prefer to provide a single high-speed channel to a large number of users which can be shared in some fashion. This, then, allows us to take advantage of the powerful "large number laws" which state that with very high probability, the demand at any instant will be approximately equal to

Paper approved by the Associate Editor for Computer Communication of the IEEE Communications Society for publication after presentation at the National Computer Conference, Anaheim, Calif., 1975. Manuscript received December 5, 1974; revised June 11, 1975. This work was supported in part by the Advanced Research Projects Agency, Department of Defense under Contract DAHC15-73-C-0368.

The authors are with the Computer Science Department, University of California, Los Angeles, Calif. 90024.

Reprinted from *IEEE Trans. Commun.*, vol. COM-23, no. 12, pp. 1400–1416, December 1975.

the sum of the average demands of that population. We wish to take advantage of these gains due to resource sharing.

Of interest to this paper is the consideration of radio channels for packet switching (also called packet radio channels). A packet is merely a package of data prepared by one user for transmission to some other user in the system. As soon as we deal with shared channels in a packet-switching mode, then we must be prepared to resolve conflicts which arise when more than one demand is simultaneously placed upon the channel. In packet radio channels, whenever a portion of one user's transmission overlaps with another user's transmission, the two collide and "destroy" each other. The existence of some acknowledgment scheme permits the transmitter to determine if his transmission was successful or not. The problem we are faced with is how to control the access to the channel in a fashion which produces, under the physical constraints of simplicity and hardware implementation, an acceptable level of performance. The difficulty in controlling a channel which must carry its own control information gives rise to the so-called random-access modes. A simple scheme, known as "pure ALOHA," permits users to transmit any time they desire. If, within some appropriate time-out period, they receive an acknowledgment from the destination, then they know that no conflicts occurred. Otherwise, they assume a collision occurred and they must retransmit. To avoid continuously repeated conflicts, some scheme must be devised for introducing a *random* retransmission delay, spreading the conflicting packets over time. A second method for using the radio channel is to modify the completely unsynchronized use of the ALOHA channel by "slotting" time into segments whose duration is exactly equal to the transmission time of a single packet (assuming constant-length packets). If we require each user to start his packets only at the beginning of a slot, then when two packets conflict, they will overlap completely rather than partially, providing an increase in channel efficiency. This method is referred to as "slotted ALOHA" [11]–[13].

The radio channel as considered in this paper is characterized as a wide-band channel with a propagation delay between any source-destination pair which is very small compared to the packet transmission time.[1] This suggests a third approach for using the channel; namely, the carrier sense multiple-access (CSMA) mode. In this scheme one attempts to avoid collisions by listening to (i.e., "sensing") the carrier due to another user's transmission.[2] Based on this information about the state of the channel, one may think of various actions to be taken by the terminal. Two protocols will be described and analyzed which we call "persistent" CSMA protocols: the nonpersistent and the p-persistent CSMA. Below, we present the protocols, discuss the assumptions, and finally establish and display the throughput-delay performance for each.

II. CSMA TRANSMISSION PROTOCOLS AND SYSTEM ASSUMPTIONS

The various protocols considered below differ by the action (pertaining to packet transmission) that a terminal takes after sensing[3] the channel. However, in all cases, when a terminal learns that its transmission was unsuccessful, it reschedules the transmission of the packet according to a randomly distributed retransmission delay. At this new point in time, the transmitter senses the channel and repeats the algorithm dictated by the protocol. At any instant a terminal is called a *ready terminal* if it has a packet ready for transmission at this instant (either a new packet just generated or a previously conflicted packet rescheduled for transmission at this instant).

A terminal may, at any one time, either be transmitting or receiving (but not both simultaneously). However, the delay incurred to switch from one mode to the other is negligible. Furthermore, the time required to detect the carrier due to packet transmissions is negligible (that is a zero detection time is assumed).[4] All packets are of constant length and are transmitted over an assumed noiseless channel (i.e., the errors in packet reception caused by random noise are not considered to be a serious problem and are neglected in comparison with errors caused by overlap interference). The system assumes noncapture (i.e., the overlap of any fraction of two packets results in destructive interference and both packets must be retransmitted). We further simplify the problem by assuming the propagation delay (small compared to the packet transmission time) to be identical[5] for all source–destination pairs.

We first consider the *nonpersistent CSMA*. The idea here is to limit the interference among packets by always rescheduling a packet which finds the channel busy upon arrival. More precisely, a ready terminal senses the channel and operates as follows.

1) If the channel is sensed idle, it transmits the packet.

2) If the channel is sensed busy, then the terminal schedules the retransmission of the packet to some later time according to the retransmission delay distribution. At this new point in time, it senses the channel and repeats the algorithm described.

A slotted version of the nonpersistent CSMA can be

[1] Consider, for example, 1000-bit packets transmitted over a channel operating at a speed of 100 kbits/s. The transmission time of a packet is then 10 ms. If the maximum distance between the source and the destination is 10 mi, then the (speed of light) packet propagation delay is of the order of 54 μs. Thus the propagation delay constitutes only a very small fraction ($a = 0.005$) of the transmission time of a packet. On the contrary, when one considers satellite channels [13] the propagation delay is a relatively large multiple of the packet transmission time ($a \gg 1$).

[2] Sensing carrier prior to transmission is a well-known concept in use for (voice) aircraft communication. In the context of packet radio channels, it was originally suggested by D.Wax of the University of Hawaii in an internal memorandum dated Mar. 4, 1971.

[3] Each terminal has the capability of sensing carrier on the channel. The practical problems of feasibility and implementation of sensing, however, are not addressed here.

[4] The detection time is considered negligible for relatively wide-band channels (100 kHz). In Part II [19] the detection time on the "busy-tone" narrow-band channels (on the order of 2 kHz) will be accounted for in the analysis.

[5] By considering this constant propagation delay equal to the largest possible, one gets lower (i.e., pessimistic) bounds on performance.

considered in which the time axis is slotted and the slot size is τ seconds (the propagation delay). All terminals are synchronized[6] and are forced to start transmission only at the beginning of a slot. When a packet's arrival occurs during a slot, the terminal senses the channel at the beginning of the next slot and operates according to the protocol described above.

We next consider the *p-persistent CSMA* protocol. However, before treating the general case (arbitrary p), we introduce the special case of $p = 1$.

The *1-persistent CSMA* protocol is devised in order to (presumably) achieve acceptable throughput by never letting the channel go idle if some ready terminal is available. More precisely, a ready terminal senses the channel and operates as follows.

1) If the channel is sensed idle, it transmits the packet with probability one.

2) If the channel is sensed busy, it waits until the channel goes idle (i.e., persisting on transmitting) and only then transmits the packet (with probability one—hence, the name of 1-persistent).

A slotted version of this 1-persistent CSMA can also be considered by slotting the time axis and synchronizing the transmission of packets in much the same way as for the previous protocol.

The above 1-persistent and nonpersistent protocols differ by the probability (one or zero) of not rescheduling a packet which upon arrival finds the channel busy. In the case of a 1-persistent CSMA, we note that whenever two or more terminals become ready during a transmission period (TP), they wait for the channel to become idle (at the end of that transmission) and then they all transmit with probability one. A conflict will also occur with probability one! The idea of randomizing the starting time of transmission of packets accumulating at the end of a TP suggests itself for interference reduction and throughput improvement. The scheme consists of including an additional parameter p, the probability that a ready packet persists ($1 - p$ being the probability of delaying transmission by τ seconds). The parameter p will be chosen so as to reduce the level of interference while keeping the idle periods between any two consecutive nonoverlapped transmissions as small as possible. This gives rise to the *p-persistent CSMA*, which is a generalization of the 1-persistent CSMA.

More precisely, the protocol consists of the following: the time axis is finely slotted where the (mini) slot size is τ seconds. For simplicity of analysis, we consider the system to be synchronized such that all packets begin their transmission at the beginning of a (mini) slot.

Consider a ready terminal. If the channel is sensed idle, then: with probability p, the terminal transmits the packet; or with probability $1 - p$, the terminal delays the transmission of the packet by τ seconds (i.e., one slot). If at this new point in time, the channel is still detected idle, the same process is repeated. Otherwise, some packet must have started transmission, and our terminal schedules the retransmission of the packet according to the retransmission delay distribution (i.e., acts as if it had conflicted and learned about the conflict).

If the ready terminal senses the channel busy, it waits until it becomes idle (at the end of the current transmission) and then operates as above.

III. TRAFFIC MODEL: ASSUMPTIONS AND NOTATION

In the previous section, we identified the system protocols, operating procedures, and assumptions. Here we characterize the traffic source and its underlying assumptions.

We assume that our traffic source consists of an infinite number of users who collectively form an independent Poisson source with an aggregate mean packet generation rate of λ packets/s. This is an approximation to a large but finite population in which each user generates packets infrequently and each packet can be successfully transmitted in a time interval much less than the average time between successive packets generated by a given user. Each user in the infinite population is assumed to have at most one packet requiring transmission at any time (including any previously blocked packet).

In addition, we characterize the traffic as follows. We have assumed that each packet is of constant length requiring T seconds for transmission. Let $S = \lambda T$. S is the average number of packets generated per transmission time, i.e., it is the input rate normalized with respect to T. Under steady-state conditions, S can also be referred to as the channel throughput rate. Now, if we were able to perfectly schedule the packets into the available channel space with absolutely no overlap or gaps between the packets, we could achieve a maximum throughput equal to 1; therefore we also refer to S as the *channel utilization*. Because of the interference problem inherent in the random nature of the access modes, the achievable throughput will always be less than 1. The maximum achievable throughput for an access mode is called the *capacity* of the channel under that mode.

Since conflicts can occur, some acknowledgment scheme is necessary to inform the transmitter of its success or failure. We assume a positive acknowledgment scheme[7]: if within some specified delay (an appropriate time-out period) after the transmission of a packet, a user does not receive an acknowledgment, he knows he has conflicted. If he now retransmits immediately, and if all users behave likewise, then he will definitely be interfered with again (and forever!). Consequently, as mentioned above, each user delays the transmission of a previously collided packet by some random time whose mean is $\bar{X}$ (chosen, for example, uniformly between 0 and $X_{max} = 2\bar{X}$). The traffic

[6] In this paper, the practical problems involved in synchronizing terminals are not addressed.

[7] The channel for acknowledgment is assumed to be separate from the channel we are studying (i.e., acknowledgments arrive reliably and at no cost).

offered to the channel from our collection of users consists not only of new packets but also of previously collided packets: this increases the mean *offered* traffic rate which we denote by G (packets per transmission time T) where $G \geq S$.

Our two further assumptions are the following.

Assumption 1: The average retransmission delay $\bar{X}$ is large compared to T.

Assumption 2: The interarrival times of the point process defined by the start times of all the packets plus retransmissions are independent and exponentially distributed.

It is clear that Assumption 2 is violated in the protocols we consider. (We have introduced it for analytic simplicity.) However, in Section V, some simulation results are discussed which show that performance results based on this assumption are excellent approximations, particularly when the average retransmission delay $\bar{X}$ is large compared to T. Moreover, in the context of slotted ALOHA it was analytically shown [14] in the limit as $\bar{X} \to \infty$, that Assumption 2 is satisfied; furthermore, simulation results showed that only the first moment of the retransmission delay distribution had a noticeable effect on the average throughput-delay performance.

So far, we have defined the following important system variables: S (throughput), G (offered channel traffic rate), T (packet transmission time), $\bar{X}$ (average retransmission delay), τ (propagation delay), and p (p-persistent parameter). Without loss of generality, we choose $T = 1$. This is equivalent to expressing time in units of T. We express $\bar{X}$ and τ in these normalized time units as $\delta = \bar{X}/T$ and $a = \tau/T$.

IV. THROUGHPUT ANALYSIS

We wish to solve for the channel capacity of the system for all of the access protocols described above. This we do by solving for S in terms of G (as well as the other system parameters). The channel capacity is then found by maximizing S with respect to G. S/G is merely the probability of a successful transmission and G/S is the average number of times a packet must be transmitted (or scheduled) until success. In Section V, we discuss delay and give the throughput-delay tradeoff for these protocols.

This analysis is based on renewal theory and probabilistic arguments requiring independence of random variables provided by Assumption 2. Moreover steady-state conditions are assumed to exist. However from the (S,G) relationships found below one can see that steady state may not exist because of inherent instability of these random-access techniques. This instability is simply explained by the fact that when statistical fluctuations in G increase the level of mutual interference among transmissions, then the positive feedback causes the throughput to decrease to 0. Nevertheless, the results are useful for the following reasons.

1) They are meaningful for a finite (and possibly long) period of time. (Simulations supporting these analytic results showed no saturation over the simulated period of time when $\bar{X}$ was large enough; see Section V.)

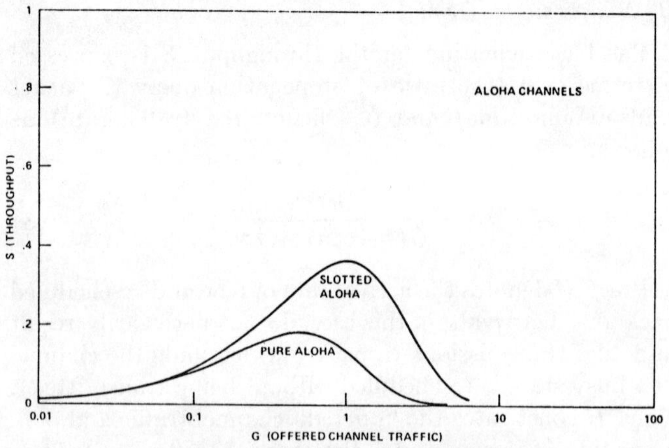

Fig. 1. Throughput in ALOHA channels.

2) In finite population cases, stable situations are possible for which steady-state results prevail over an infinite time horizon. (See [14] and [16].)

3) Control procedures have been prescribed for the slotted ALOHA random access [14] which stabilize unstable channels, achieving performance very close to the equilibrium results.

A. *ALOHA Channels*

In the *pure ALOHA* access mode, each terminal transmits its packet over the data channel in a completely unsynchronized manner. Under the system and model assumptions (mainly Assumption 2), we have

$$S = GP_s$$

where P_s is the probability that an arbitrary offered packet is successful. A given packet will overlap with another packet if there exists at least one start of transmission within T seconds before or after the start time of the given packet (i.e., over a "vulnerable" interval of length $2T$). Using the Poisson traffic assumption, Abramson [2] first showed that

$$S = Ge^{-2G}. \tag{1}$$

Thus, we see that pure ALOHA achieves a maximum throughput of $1/(2e) = 0.184$ (at $G = 1/2$).

In the *slotted ALOHA*, if two packets conflict, they will overlap completely rather than partially (i.e., a vulnerable interval only of length T). The throughput equation then becomes

$$S = Ge^{-G} \tag{2}$$

and was first obtained by Roberts [12] who extended Abramson's result in (1). With this simple change, the maximum throughput is increased by a factor of two to $1/e = 0.368$ (at $G = 1$). In Fig. 1, we plot the throughput S versus the offered traffic G for these two systems. From these results, it is all too evident that a significant fraction of the channel's ultimate capacity ($C = 1$) is not utilized with the ALOHA access modes; we recover a major portion of this loss with the CSMA protocols, as we now show.

B. Nonpersistent CSMA

The basic equation for the throughput S is expressed in terms of a (the ratio of propagation delay to packet transmission time) and G (the offered traffic rate) as follows:

$$S = \frac{Ge^{-aG}}{G(1 + 2a) + e^{-aG}}. \quad (3)$$

Proof: G denotes the arrival rate of new and rescheduled packets. All arrivals, in this case, do not necessarily result in actual transmissions (a packet which finds the channel in a busy state is rescheduled without being transmitted). Thus, G constitutes the "offered" channel traffic and only a fraction of it constitutes the channel traffic itself. Consider the time axis[8] (See Fig. 2)[9] and let t be the time of arrival of a packet which senses the channel idle and such that no other packet is in the process of transmission. Any other packet arriving between t and $t + a$ will find (sense) the channel as unused, will transmit, and hence will cause a conflict. If no other terminal transmits a packet during these a seconds (the "vulnerable" period), then the first packet will be successful.

Let $t + Y$ be the time of occurrence of the last packet arriving between t and $t + a$. The transmission of all packets arriving in $(t, t + Y)$ will be completed at $t + Y + 1$. Only a seconds later will the channel be sensed unused. Now, any terminal becoming ready between $t + a$ and $t + Y + 1 + a$ will sense the channel busy and hence will reschedule its packet. The interval between t and $t + Y + 1 + a$ is called a *transmission period* (TP). Note that there can be at most one successul transmission during a TP. Define an *idle period* to be the period of time between two consecutive TP's (also called busy periods in this simple case). A busy period plus the following idle period constitute a cycle. Let $\bar{B}$ be the expected duration of the busy period, $\bar{I}$ the expected duration of the idle period, and $\bar{B} + \bar{I}$ the expected length of a cycle. Let U denote the time during a cycle that the channel is used without conflicts. Using renewal theory arguments, the average channel utilization is simply given by

$$S = \frac{\bar{U}}{\bar{B} + \bar{I}}. \quad (4)$$

The probability that a TP is successful is simply the probability that no terminal transmits during the first a seconds of the period and is equal to e^{-aG}. Therefore

$$\bar{U} = e^{-aG}. \quad (5)$$

The average duration of an idle period is simply $1/G$. The average duration of a busy interval is $1 + \bar{Y} + a$, where $\bar{Y}$ is the expected value of Y.

[8] The reference time axis considered in this and subsequent proofs is the transmitter's time. Shifting all transmissions by τ seconds will give a description of events on the station's time axis. Any time overlap in transmission on the station's time axis results in packet interference.

[9] In this and other figures, a vertical arrow represents a terminal becoming ready.

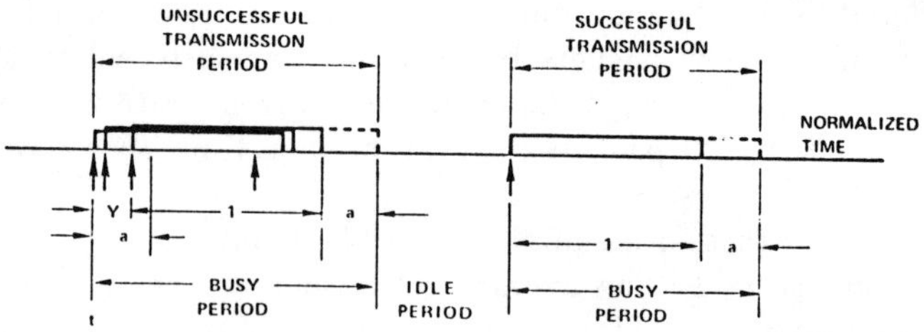

Fig. 2. Nonpersistent CSMA: Busy and idle periods.

The distribution function for Y is

$$F_Y(y) \triangleq \Pr\{Y \le y\} = \Pr\{\text{no arrival occurs in an interval of length } a - y\}$$
$$= \exp\{-G(a - y)\}, \qquad (y \le a). \quad (6)$$

The average of Y is therefore given by

$$\bar{Y} = a - \frac{1}{G}(1 - e^{-aG}). \quad (7)$$

Applying (4) and using the expressions found for $\bar{U}$, $\bar{B}$, and $\bar{I}$, we get (3). Q.E.D.

It is easy to prove that the throughput equation for the *slotted* nonpersistent CSMA is given by

$$S = \frac{aGe^{-aG}}{(1 - e^{-aG}) + a}. \quad (8)$$

Note that for both cases we have

$$\lim_{a \to 0} S = G/(1 + G). \quad (9)$$

This shows that when $a = 0$, a throughput of 1 can theoretically be attained for an offered channel traffic equal to infinity. S versus G for various values of a is plotted in Fig. 3.

C. 1-Persistent CSMA

The throughput equation for this protocol is given by

$$S = \frac{G[1 + G + aG(1 + G + aG/2)]e^{-G(1+2a)}}{G(1 + 2a) - (1 - e^{-aG}) + (1 + aG)e^{-G(1+a)}}. \quad (10)$$

Proof: Consider Fig. 4 and again let t be the time of arrival of a packet which senses the channel to be idle with no other packet in the process of transmission. In this protocol, any packet arriving in the interval $[t + a, t + Y + 1 + a]$ will sense the channel busy and hence must wait until the channel is sensed idle (at time $t + 1 + Y + a$) at which time they will *all* transmit! The number of packets accumulated at the end of TP is the number of arrivals in $1 + Y$ seconds. If this total is equal to or greater than two, then a conflict occurs in the next TP with probability 1.

Define a *busy period* to be the time between t and the end of that TP during which no packets accumulate. De-

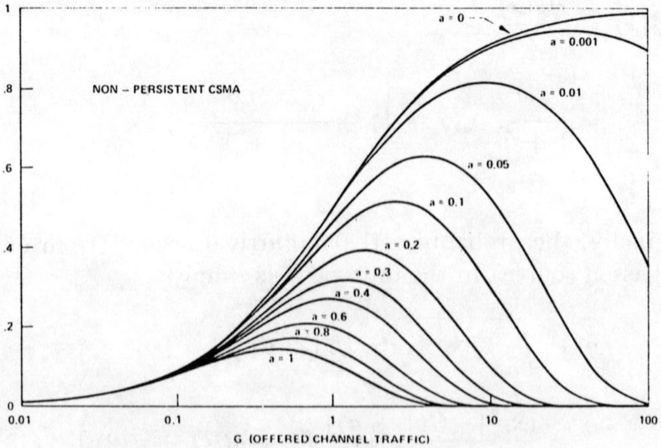

Fig. 3. Throughput in nonpersistent CSMA.

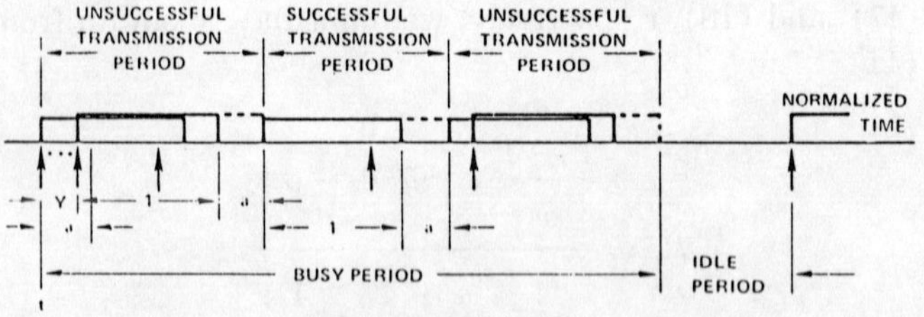

Fig. 4. 1-persistent CSMA: TP's, busy periods, and idle periods.

fine an *idle period* to be the period of time in which the channel is idle and no packets are present awaiting transmission. A busy period plus the following idle period constitute a *cycle*.

Let $\bar{B}$ be the expected duration of the busy period, $\bar{I}$ the expected duration of the idle period, and $\bar{B}+\bar{I}$ the expected length of a cycle.

Let us now consider the transmission of an arbitrary packet. Three situations must be considered.

1) If the packet arrives to an idle system, then its transmission is successful if and only if no packets arrive during its first a seconds; its probability of success is therefore e^{-aG}.

2) If the packet arrives during the first a seconds of a TP, then its probability of success is 0.

3) If the packet arrives during the channel busy period (excluding the first a seconds of the TP), then it is successful (in the next TP) if and only if it is the only packet to arrive during this TP and no packets arrive during *its* first a seconds. To calculate its probability of success, we observe that a TP is of random length equal to $1 + a + Y$ where Y is a random variable. Let B' denote the time during a cycle that the channel is in its busy period excluding the first a seconds of each TP. B' is a sequence of segments of random length $1 + Y \triangleq Z$ separated by periods of a seconds. Knowing that a packet arrives in B', this packet is more likely to arrive in a longer segment Z than in a shorter one (due to the "paradox of residual life" [17]). Let $\hat{Z}$ denote the segment in which the arrival occurred, and $\hat{q}_0$ (derived below) be the probability that no arrival occurs in $\hat{Z}$; the probability of success of the packet is therefore $\hat{q}_0 e^{-aG}$.

Only cases 1) and 3) contribute to a successful transmission. Let $\bar{B}'$ be the expected value of B'. From renewal theory arguments, the probability that an arrival finds the channel idle [case (1)] is given by $\bar{I}/(\bar{B}+\bar{I})$, and the probability that an arrival finds the channel in situation 3) is $\bar{B}'/(\bar{B}+\bar{I})$; then the probability of success of the packet is given by

$$P_s \triangleq \Pr\{\text{success}\} = \frac{\bar{I}}{\bar{B}+\bar{I}} e^{-aG} + \frac{\bar{B}'}{\bar{B}+\bar{I}} \hat{q}_0 e^{-aG}. \qquad (11)$$

The determination of $\bar{I},\bar{B},\bar{B}'$, and $\hat{q}_0$ follows.

Since the traffic is Poisson, it is clear that the average idle period is given by

$$\bar{I} = 1/G. \qquad (12)$$

For $\bar{B},\bar{B}'$, and $\hat{q}_0$ we must first obtain some intermediate results as follows. The distribution function for Y and its average are given in (6) and (7), respectively. The Laplace transform of the probability density function of Y, defined as

$$F_Y^*(s) \triangleq \int_0^\infty e^{-sy}\, dF_Y(y),$$

is given by

$$F_Y^*(s) = e^{-aG} + \frac{G(e^{-as} - e^{-aG})}{G - s}. \qquad (13)$$

Let us now find the distribution of the number of packets accumulated at the end of a TP.

Let

$$q_m(y) \triangleq \Pr\{m \text{ packets accumulated at end of TP} \mid Y = y\}$$

and

$$q_m = \int_0^a q_m(y)\, dF_Y(y).$$

Let $Q(z)$ denote the generating function of q_m defined by

$$Q(z) \triangleq \sum_{m=0}^{\infty} q_m z^m.$$

The number of packets accumulated at the end of a TP is equal to the number of packets arriving during a period of time equal to $1 + Y$. Let m_1 denote the number of packets arriving in $T = 1$, and m_2 the number of packets arriving in Y. Let $Q_1(z)$ and $Q_2(z)$ denote the generating functions of the probability distributions for m_1 and m_2, respectively. Since the arrival process is Poisson, the random variables m_1 and m_2 are independent and the generating function $Q(z)$ of q_m, where $m = m_1 + m_2$, is given by

$$Q(z) = Q_1(z)Q_2(z).$$

We have [17]

$$Q_1(z) = \exp\{G(z-1)\}$$

and

$$Q_2(z) = F_Y^*(G(1-z)).$$

From (13) we get

$$Q(z) = \exp\{G(z-1)\}\exp\{-aG\}\left[1 + \frac{\exp\{aGz\} - 1}{z}\right]. \tag{14}$$

We may invert this explicit expression for $Q(z)$; in particular we find that the probability of zero packets accumulated at the end of a TP is

$$q_0 = Q(z)\big|_{z=0} = \exp\{-G(1+a)\}[1+aG]. \tag{15}$$

To find the average busy period, we let Y_i denote the random variable Y defined above corresponding to the ith TP in a busy period. All Y_i, $i = 1,2,\cdots$, are independent and identically distributed. It is easy to see that the number of TP's in a busy period is geometrically distributed with mean $1/q_0$. Conditioned on the fact that we have exactly k TP's in the busy period and that $Y_i = y_i$ for $i = 1,2,\cdots,k$, the average busy period is

$$\bar{B}(y_1,y_2,\cdots,y_k) = k(1+a) + y_1 + y_2 + \cdots + y_k.$$

Therefore, by removing the conditions on k and Y_i, we get $\bar{B}$ as

$$\bar{B} = \cdots\int_{y_i=0}^{a}\cdots\int_{y_1=0}^{a}\sum_{k=1}^{\infty}[k(1+a) + y_1 + \cdots + y_k]$$

$$\cdot\, q_0(y_k)\prod_{i=1}^{k-1}(1-q_0(y_i))\, dF_{Y_1}(y_1)\cdots dF_{Y_i}(y_i)\cdots.$$

It is easy to see that by inverting the order of summation and integration, the contribution of the term $k(1+a)$ reduces to $(1+a)/q_0$ and the contribution of the generic term y_j simply reduces to $\bar{Y}(1-q_0)^{j-1}$. Finally, we have

$$\bar{B} = \frac{1+a}{q_0} + \sum_{j=1}^{\infty}\bar{Y}(1-q_0)^{j-1} = \frac{1+a+\bar{Y}}{q_0}. \tag{16}$$

Since the average number of TP's is $1/q_0$, from the distribution of B' we have

$$\bar{B}' = \frac{1+\bar{Y}}{q_0}. \tag{17}$$

In (11) for P_s, it remains only to compute $\hat{q}_0$. The probability density function of $Z = 1 + Y$ is easily obtained from the distribution of Y. From (6), the probability density function of Y can be expressed as

$$f_Y(y) = \exp\{-aG\}u_0(y) + G\exp\{-aG\}\exp\{Gy\}, \qquad 0 \le y \le a$$

where $u_0(y)$ is the unit impulse at $y = 0$. Thus we have

$$f_Z(x) = \exp\{-aG\}u_0(x-1) + G\exp\{-aG\} \cdot \exp\{G(x-1)\}, \qquad 1 \le x \le 1+a.$$

The probability density function of $\hat{Z}$ is given by [17]

$$f_{\hat{Z}}(x) = \frac{xf_Z(x)}{\bar{Z}} = \frac{e^{aG}}{1+\bar{Y}}u_0(x-1) + \frac{Gxe^{-aG}e^{G(x-1)}}{1+\bar{Y}}, \qquad 1 \le x \le 1+a.$$

Finally, the probability that no arrival occurs (from our Poisson source) in the interval $\hat{Z}$ is simply

$$\hat{q}_0 = \int_{x=1}^{1+a}\exp\{-Gx\}f_{\hat{Z}}(x)\,dx = \frac{\exp\{-G(1+a)\}}{1+\bar{Y}}[1 + aG(1+a/2)]. \tag{18}$$

Using our expressions for $\bar{I}$, $\bar{B}$, $\bar{B}'$, and $\hat{q}_0$ in (12), (16), (17), and (18), respectively, we immediately obtain from (11)

$$P_s = \frac{\dfrac{1+\bar{Y}}{q_0}e^{-aG}\hat{q}_0 + \dfrac{1}{G}e^{-aG}}{\dfrac{1+a+\bar{Y}}{q_0} + \dfrac{1}{G}}.$$

Substituting the expressions obtained for q_0, $\hat{q}_0$, and $\bar{Y}$, and recalling that $S = GP_s$, we have finally established (10).

Q.E.D.

Slotted 1-persistent CSMA: Let us now consider the *slotted* version of 1-persistent CSMA. The throughput equation for this case is given by

$$S = \frac{G\exp\{-G(1+a)\}[1+a-\exp\{-aG\}]}{(1+a)(1-\exp\{-aG\}) + a\exp\{-G(1+a)\}}. \tag{19}$$

Proof: In this slotted version, as in slotted ALOHA, if two packets conflict, they will overlap completely. The length of a TP is always equal to $1 + a$. (We have assumed that the packet transmission time is an integer multiple of the propagation delay.)

Since the traffic process is an independent one (Assumption 2), the number of slots in an idle period is geometrically distributed with a mean equal to $1/(1-e^{-aG})$. Thus the average idle period is given by

$$\bar{I} = \frac{a}{1-e^{-aG}}. \tag{20}$$

Using a similar argument, we find that the average busy period is given by

$$\bar{B} = \frac{1+a}{\exp\{-G(1+a)\}}. \tag{21}$$

Let $\bar{U}$ again denote the expected time during a cycle that the channel is used without conflicts. In order to find $\bar{U}$ we need to determine the probability of success over each

TP in the busy period. The probability of success over the first TP is given by

$$\Pr\{\text{success over first TP}\} = \Pr\{\text{only one packet arrives during the last slot of the preceding idle period/some arrival occurred}\}$$

$$= \frac{aGe^{-aG}}{1 - e^{-aG}}.$$

Similarly we have:

$$\Pr\{\text{success over any other TP}\} = \frac{G(1+a)\exp\{-G(1+a)\}}{1-\exp\{-G(1+a)\}}.$$

The number of TP's in a busy period is geometrically distributed with a mean equal to

$$\exp\{G(1+a)\} \triangleq 1/q_0,$$

thus

$$\bar{U} = \frac{aG\exp\{-aG\}}{1-\exp\{-aG\}} + \left(\frac{1}{q_0} - 1\right) \cdot \frac{G(1+a)\exp\{-G(1+a)\}}{1-\exp\{-G(1+a)\}}. \quad (22)$$

Applying (4) and using the expressions found for $\bar{U}$, $\bar{I}$, and $\bar{B}$, we get (19). Q.E.D.

The ultimate performance in the ideal case ($a = 0$), for both slotted and unslotted versions, is

$$S = \frac{Ge^{-G}(1+G)}{G + e^{-G}}. \quad (23)$$

For any value of a, the maximum throughput S will occur at an optimum value of G. In Fig. 5 we show S versus G for the nonslotted version of 1-persistent CSMA for various values of a.

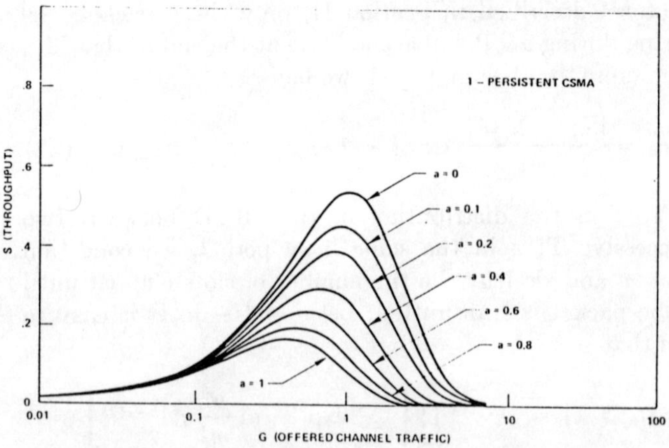

Fig. 5. Throughput in 1-persistent CSMA.

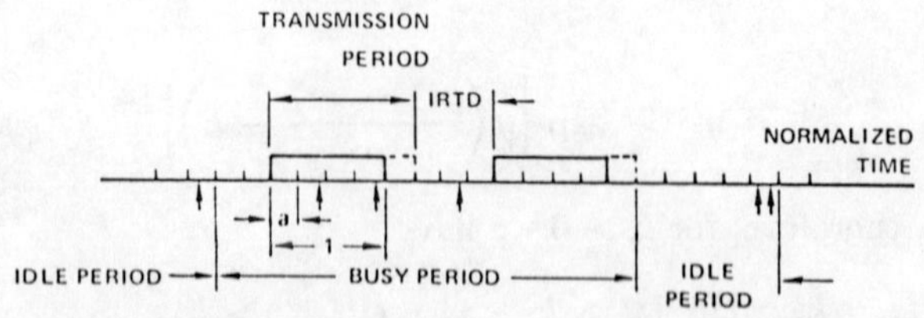

Fig. 6. p-persistent CSMA: TP's, busy periods, and idle periods.

D. p-Persistent CSMA

For a given offered traffic G and a given value of the parameter p, we can determine the throughput S as

$$S(G,p,a) = \frac{(1-e^{-aG})[P_s'\pi_0 + P_s(1-\pi_0)]}{(1-e^{-aG})[a\bar{l}'\pi_0 + a\bar{l}(1-\pi_0) + 1 + a] + a\pi_0} \quad (24)$$

where P_s', P_s, $\bar{l}'$, $\bar{l}$, and π_0 are defined in the following proof in (37), (34), (36), (30), and (25), respectively.

Proof: Consider a TP and assume that some packets arrive during the period as shown in Fig. 6. These packets sense the channel busy and accumulate at the end of the TP, at which point they randomize the starting times of their transmission according to the randomizing process described in Section II. This randomization creates a random delay before a TP starts, called the initial random transmission delay (IRTD), during which time the channel is "wasted." If, at the start of a new TP, two or more terminals decide to transmit, then a conflict will certainly occur. All other packets which have delayed their transmission by τ seconds will then sense the channel busy and will have to be rescheduled for transmission by incurring a retransmission delay δ. Thus, at the expense of creating this IRTD, we greatly improve the probability of success over a TP.

Consider Fig. 6 in which we observe two TP's separated by an IRTD. One can also define busy periods and idle periods in much the same way as before. An idle period is that period of time during which the channel is idle and no packets are ready for transmission. A busy period consists of a sequence of transmission periods such that some packets arrive during each transmission period *except* the last one. Let $\mathcal{T}_i$ denote the ith TP of a busy period. In order to find the channel utilization, we once again apply (4), which requires identifying and determining the average busy and idle periods, the gaps between TP's, as well as the condition for success over each TP. This we do as follows.

Recall that we require the system to be (mini-) slotted (the slot size equal to a, the normalized propagation delay) and all transmissions to start at the beginning of a slot. Here again we consider the transmission time of a packet to be an integer number $1/a$ slots (recall $T = 1$). Let $g = aG$; g is the average arrival rate of new and rescheduled packets during a (mini) slot.

We first determine the distribution of the number of packets accumulated at the end of a TP. Let N denote this number and let $\pi_n \triangleq \Pr\{N = n\}$. According to the protocol described in Section II, only those packets arriving during a TP will accumulate at the end of that TP. Therefore, by Assumption 1, we have

$$\pi_n = \frac{[(1+a)G]^n}{n!} \exp\{-(1+a)G\}, \qquad n \geq 0. \quad (25)$$

To find the distribution of the IRTD between two successive TP's in the same busy period, we condition $N = n$ and we let t_n be the number of slots elapsed until some packet is transmitted. Let $q = 1 - p$. It is easy to see that

$$\Pr\{t_n > k\} = q^{(k+1)n} \prod_{j=1}^{k} \left[\sum_{m=0}^{\infty} \exp\{-g\} \frac{g^m}{m!} q^{m(k-j+1)} \right]$$

$$= q^{(k+1)n} \prod_{j=1}^{k} \exp\{g(q^{k-j+1} - 1)\}$$

$$= q^{(k+1)n} \exp\left\{g\left(\frac{q(1-q^k)}{p} - k\right)\right\} \quad (26)$$

and, therefore, for $k > 0$ we have

$$\Pr\{t_n = k\} = \Pr\{t_n > k-1\} - \Pr\{t_n > k\}$$

$$= q^{kn}[1 - q^n \exp\{-g(1-q^k)\}]$$

$$\cdot \exp\left\{g\left(\frac{q(1-q^{k-1})}{p} - (k-1)\right)\right\} \quad (27)$$

and for $k = 0$,

$$\Pr\{t_n = 0\} = 1 - q^n. \quad (28)$$

The average IRTD is given by

$$\bar{t}_n = \sum_{k=0}^{\infty} \Pr\{t_n > k\}. \quad (29)$$

Removing the condition on N, we get

$$\bar{t} = \sum_{n=1}^{\infty} \bar{t}_n \frac{\pi_n}{1-\pi_0}. \quad (30)$$

$\bar{t}$ is the average gap between two consecutive TP's in a busy period.

In order to find the probability of success over a TP $\mathfrak{J}_i$ one has to distinguish two cases: $i = 1$ and $i \neq 1$. We first treat the second case, $i \neq 1$. Given $N = n$, define[10]:

$P_s(n)$	probability of success over $\mathfrak{J}_i$
L_n	the number of packets present at the starting time of $\mathfrak{J}_i$
$L_n - n$	merely the number of packets arriving during the gap t_n.

By the Poisson assumption we have

$$\Pr\{L_n = l / t_n = k\} = \frac{(kg)^{l-n}}{(l-n)!} e^{-kg}, \qquad l \geq n. \quad (31)$$

[10] The quantities $P_s(n)$, P_s, and L_n need no index i since they are identical for all $\mathfrak{J}_i$, $i \neq 1$.

Removing the condition on t_n,

$$\Pr\{L_n = l\} = \sum_{k=1}^{\infty} \frac{(kg)^{l-n}}{(l-n)!} e^{-kg} \Pr\{t_n = k\} + (1-q^n)\delta_{l,n}, \qquad l \geq n \quad (32)$$

where $\delta_{i,j}$ is the Kronecker delta. The probability of success over $\mathfrak{J}_i$ is equal to the probability that none of the L_n transmit over $\mathfrak{J}_i$:

$$P_s(n) = \sum_{l=n}^{\infty} \frac{lpq^{l-1}}{1-q^l} \Pr\{L_n = l\}. \quad (33)$$

Removing the condition on N, we get

$$P_s = \sum_{n=1}^{\infty} P_s(n) \frac{\pi_n}{1-\pi_0}. \quad (34)$$

For the probability of success over $\mathfrak{J}_1$ we note that the number of packets present at the beginning of a busy period, denoted by N', is the number of packets arriving in the last slot of the previous idle period. We then have

$$\pi_n \triangleq \Pr\{N' = n\}$$

$$= \frac{g^n}{n!} \frac{e^{-g}}{1-e^{-g}}, \qquad n \geq 1. \quad (35)$$

Given $N' = n$, let t_n' denote the first initial random transmission delay of the busy period, and $P_s'(n)$ denote the probability of success over $\mathfrak{J}_1$. The distribution of t_n' and its average $\bar{t}_n'$ are the same as for t_n [(27) and (29)]. $P_s'(n)$ is the same as $P_s(n)$ [see (33)]. Removing the condition on N', we get

$$\bar{t}' = \sum_{n=1}^{\infty} \bar{t}_n' \pi_n' \quad (36)$$

$$P_s' = \sum_{n=1}^{\infty} P_s'(n) \pi_n'. \quad (37)$$

It remains to compute $\bar{B}$, $\bar{U}$, and $\bar{I}$. It is clear that the number of TP's in a busy period is equal to m with probability $\pi_0(1-\pi_0)^{m-1}$.

Consider a busy period with m TP's. Let N_i denote the number of packets accumulated at the end of the ith TP. We know that $N_m = 0$, and that all other $N_i \geq 1$ are independent and identically distributed random variables. Conditioned on the fact that $N_i = n_i, i = 1, \cdots, m-1$, the average busy period is given by

$$\bar{B}_m(n_1, \cdots, n_{m-1}) = a\bar{t}' + \sum_{i=1}^{m-1} a\bar{t}_{n_i} + m(1+a). \quad (38)$$

The expected time, during the busy period, that the channel is used without conflicts is given by

$$\bar{U}_m(n_1, \cdots, n_{m-1}) = P_s' + \sum_{i=1}^{m-1} P_s(n_i). \quad (39)$$

On the other hand, we know that

$$\Pr\{N_i = n_i\} = \frac{\pi_{n_i}}{1 - \pi_0}, \qquad n_i \geq 1,\ i = 1,2,\cdots,m-1. \tag{40}$$

Therefore, removing the conditions $N_i = n_i$ in (38) and (39), we get

$$\bar{B}_m = a\bar{l}' + (m-1)a\bar{l} + m(1+a) \tag{41}$$

$$\bar{U}_m = P_s' + (m-1)P_s \tag{42}$$

and removing the condition on m we get

$$\bar{B} = \sum_{m=1}^{\infty} \bar{B}_m \pi_0 (1-\pi_0)^{m-1} = a\bar{l}' + \frac{a\bar{l}(1-\pi_0) + 1 + a}{\pi_0} \tag{43}$$

$$\bar{U} = P_s' + \frac{1-\pi_0}{\pi_0} P_s. \tag{44}$$

The idle period is geometrically distributed with mean $1/(1 - e^{-g})$; its average is:

$$\bar{I} = \frac{a}{1 - e^{-g}}. \tag{45}$$

Finally, using (4) and substituting for $\bar{B}$, $\bar{U}$, and $\bar{I}$ the expressions found in (43), (44), and (45), respectively, we get the throughput S; it is a function of G, p, and $a = 1/T$ and is expressed as

$$S(G,p,a) = \frac{P_s' + \dfrac{1-\pi_0}{\pi_0} P_s}{a\bar{l}' + a\bar{l}\dfrac{1-\pi_0}{\pi_0} + \dfrac{1+a}{\pi_0} + \dfrac{a}{1-e^{-g}}} \tag{46}$$

which reduces to (24). Q.E.D.

In order to evaluate $S(G,p,a)$, a PL/1 program was written and run on the IBM 360/91 of the Campus Computing Network at UCLA. For small values of p $(0.01 \leq p \leq 0.1)$, the numerical computation as suggested by (24) becomes time consuming and requires an extremely large amount of storage. Fortunately some approximations have been found useful which lead to a closed-form solution for the throughput (see the derivations of $S'(G,p,a)$ in Appendix A).

Special case $a = 0$: Let us now consider the special case $a = 0$. For finite G, $g = aG = 0$. Equation (26) becomes

$$\Pr\{t_n > k\} = q^{(k+1)n}.$$

The average IRTD is then given by (29), and is expressed as

$$\bar{t}_n = \sum_{k=0}^{\infty} \Pr\{t_n > k\} = \frac{q^n}{1-q^n}.$$

It is important to note that $\bar{t}_n$ is finite, so is $\bar{t}$. On the other hand the idle period given in (45) becomes

$$\bar{I} = \frac{1}{G}.$$

Since $\bar{l}$ and $\bar{l}'$ are finite, by letting $a \to 0$ in (46) we get

$$S(G,p,a=0) = \frac{P_s' + \dfrac{1-\pi_0}{\pi_0} P_s}{\dfrac{1}{\pi_0} + \dfrac{1}{G}}. \tag{47}$$

To compute P_s we have to get back to (31) through (34). With $a = 0$ we have

$$\Pr\{L_n = l/t_n = k\} = 1, \qquad l = n$$

and

$$\Pr\{L_n = n\} = 1.$$

Therefore

$$P_s(n) = \frac{npq^{n-1}}{1-q^n} \tag{48}$$

and

$$P_s = \sum_{n=1}^{\infty} \frac{npq^{n-1}}{1-q^n} \frac{\pi_n}{1-\pi_0} \tag{49}$$

where

$$\pi_n = \frac{G^n}{n!} e^{-G}. \tag{50}$$

By the same token, we see from (35) that

$$\pi_1' = \frac{ge^{-g}}{1-e^{-g}} \underset{g\to 0}{\longrightarrow} 1$$

and that

$$P_s' = P_s'(1) = 1.$$

With these considerations, the throughput is given by

$$S(G,p,a=0) = \frac{G[\pi_0 + (1-\pi_0)P_s]}{G + \pi_0} \tag{51}$$

where P_s and π_n are given in (49) and (50), respectively. When $p = 1$, we have, from (48),

$$P_s(1) = 1$$

$$P_s(n) = 0, \qquad n > 1$$

and therefore

$$P_s = \frac{Ge^{-G}}{1-e^{-G}}.$$

Equation (51) then becomes

$$S(G,p=1,a=0) = \frac{G(1+G)e^{-G}}{G+e^{-G}}$$

which is (and should be) identical to the 1-persistent CSMA when $a = 0$ [see (23)]. Let us now consider $p \to 0$. Since $1 - q_n \approx np$, (48) then becomes

$$P_s(n) = q^{n-1}$$

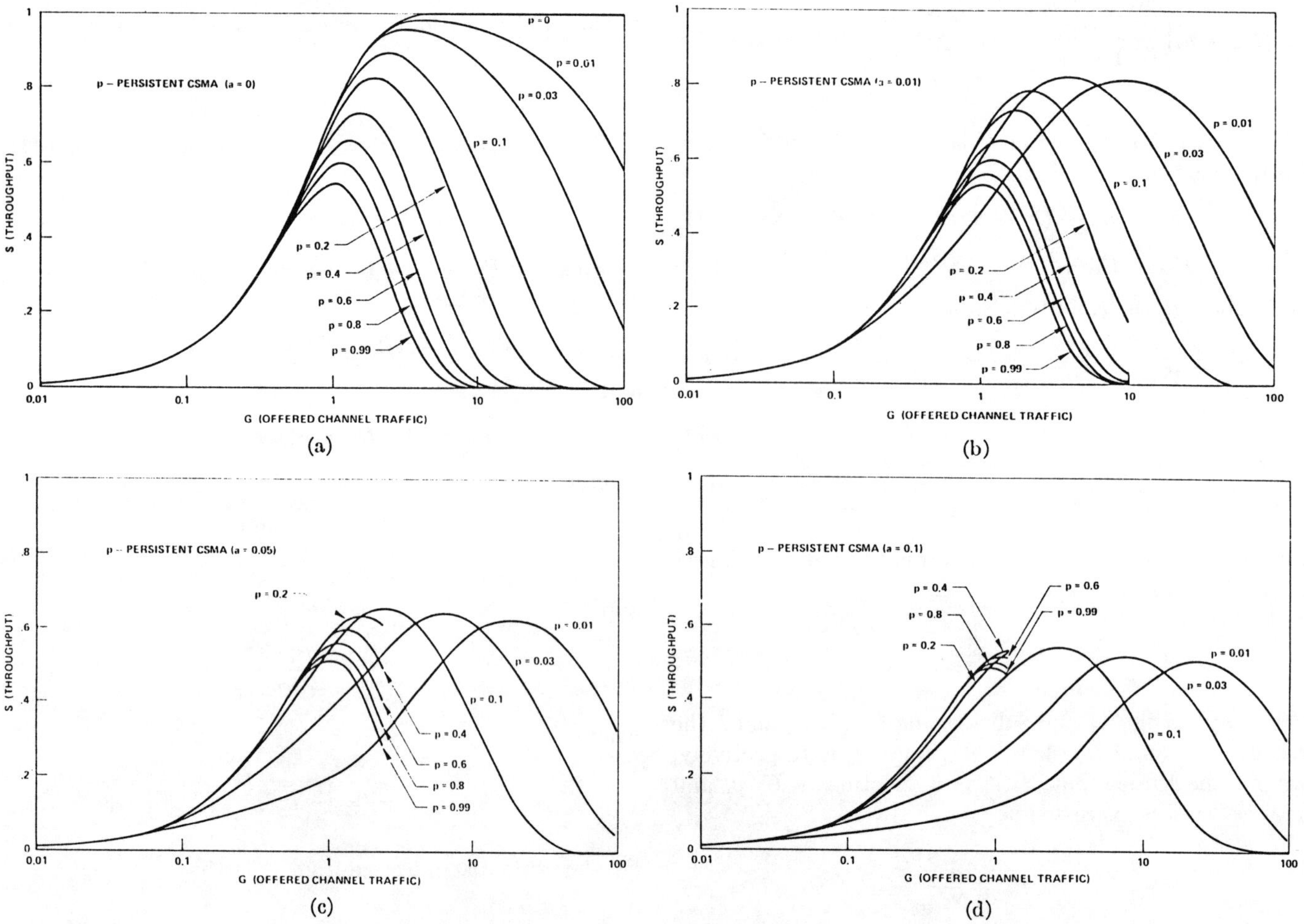

Fig. 7. Channel throughput in p-persistent CSMA. (a) $a = 0$. (b) $a = 0.01$. (c) $a = 0.05$. (d) $a = 0.1$.

and

$$P_s = \sum_{n=1}^{\infty} \frac{q^{n-1}G^n e^{-G}}{n!(1 - e^{-G})}$$

$$= \frac{qe^{-G}(e^{qG} - 1)}{1 - e^{-G}}.$$

In particular, $p \to 0$ gives $P_s(n) \to 1$, for all n, and $P_s \to 1$. In this limit the throughput is then given by

$$S(G,p \to 0,a = 0) \to \frac{G}{G + e^{-G}} \qquad (52)$$

which shows that a channel capacity of 1 can be achieved when $G \to \infty$.

For each value of a, one can plot a family of curves S versus G with parameter p [as shown in Fig. 7 (a)–(d)]. The channel capacity for each value of p can be numerically determined at an optimum value of G. In Fig. 8 we show the channel capacity as a function of p, for $a = 0$, 0.01, 0.05, and 0.1. We note that the capacity is not very sensitive to small variations of p; for $a = 0.01$, it reaches its highest value (i.e., the channel capacity for this protocol) at a value $p = 0.03$. When $p = 1$, the (slotted) p-persistent CSMA reduces to the slotted 1-persistent CSMA. Indeed we can check that, when $p = 1$, (24) reduces to (19), since P_s, $\bar{l}$, and $\bar{l}'$ then become

$$P_s = \frac{aGe^{-G}}{1 - e^{-aG}}$$

$$\bar{l}' = \bar{l} = 0.$$

E. Performance Comparison and Sensitivity of Capacity to the Parameter a

To summarize, we plot in Fig. 9 for $a = 0.01$, S versus G for the various access modes introduced so far and thus show the relative performance of each, as also indicated in Table I.

While the capacity of ALOHA channels does not depend on the propagation delay, the capacity of a CSMA channel does. An increase in a increases the vulnerable period of a packet. This also results in "older" channel state information from sensing. In Fig. 10 we plot, versus a, the channel capacity for all of the above random-access modes. We note that the capacities for nonpersistent and p-persistent CSMA are more sensitive to increases in a, as compared to the 1-persistent scheme. Nonpersistent CSMA drops below 1-persistent for larger a. Also, for large a,

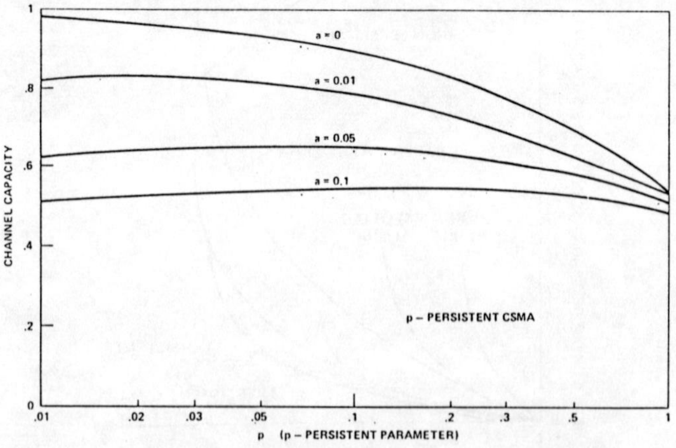

Fig. 8. p-persistent CSMA: effect of p on channel capacity.

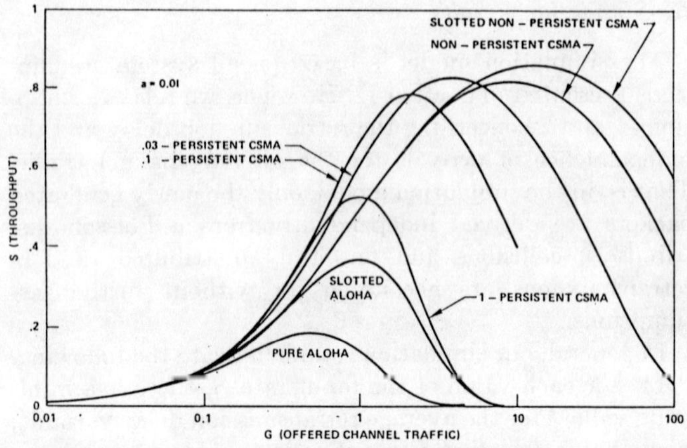

Fig. 9. Throughput for the various access modes ($a = 0.01$).

TABLE I
CAPACITY C FOR THE VARIOUS PROTOCOLS CONSIDERED ($a = 0.01$)

Protocol	Capacity C
Pure ALOHA	0.184
Slotted ALOHA	0.368
1-Persistent CSMA	0.529
Slotted 1-Persistent CSMA	0.531
0.1-Persistent CSMA	0.791
Nonpersistent CSMA	0.815
0.03-Persistent CSMA	0.827
Slotted Nonpersistent CSMA	0.857
Perfect Scheduling	1.000

slotted ALOHA (and even "pure" ALOHA) is superior to any CSMA mode since decisions based on partially obsolete data are deleterious; this effect is due in part to our assumption about the constant propagation delay. (For p-persistent, numerical results are shown only for $a \leq 0.1$. Clearly, for larger a, optimum p-persistent is lower-bounded by 1-persistent.)

V. DELAY CONSIDERATIONS

A. Delay Model

In the previous section, we analyzed the performance of CSMA modes in terms of maximum achievable throughput. We now introduce the expected packet delay D defined as the average time from when a packet is generated until it is successfully received.

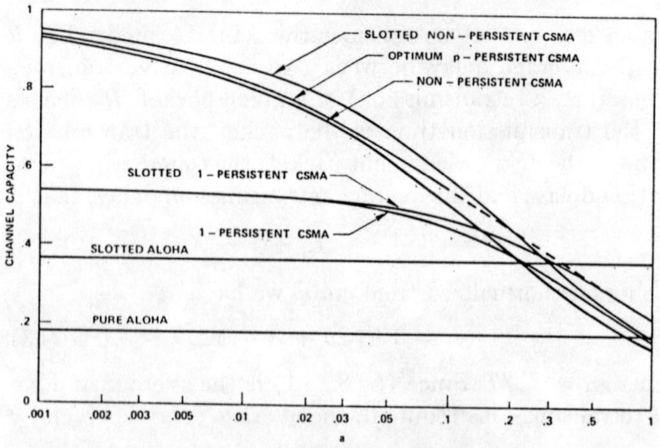

Fig. 10. Effect of propagation delay on channel capacity.

Our principal concern in this section is to investigate the tradeoff between the average delay D and the throughput S.

As we have already stated, for the correct operation of the system, a positive acknowledgment scheme is needed. If an acknowledgment is not received by the sender of a packet within a specified time-out period, then the packet is retransmitted (incurring the random retransmission delay X, introduced to avoid repeated conflicts). For the present study, we assume the following.

Assumption 3: The acknowledgment packets are always correctly received with probability one.

The simplest way to accomplish this is to create a separate channel[11] (assumed to be available) to handle acknowledgment traffic. If sufficient bandwidth is provided to this channel overlaps between acknowledgment packets are avoided, since a positive acknowledgment packet is created only when a packet is correctly received, and there will be at most one such packet at any given time. Thus, if T_a denotes the transmission time of the acknowledgment packet on the separate channel, then the time-out for receiving a positive acknowledgment is $T + \tau + T_a + \tau$, provided that we make the following assumption.

Assumption 4: The processing time needed to perform the sumcheck and to generate the acknowledgment packet is negligible.

Assumption 2 further simplifies our delay model by implicitly assuming that the probability of a packet's success is the same whether the packet is new or has been blocked, or interfered with any number of times before; this probability is simply given by the throughput equation, i.e.,

$$P_s = \frac{S}{G} = \frac{\text{throughput}}{\text{offered traffic}}.$$

Bearing these assumptions in mind, we can write the delay equations for each of the previous access modes.

[11] The reader is referred to [16] for a study of the effect of acknowledgment traffic on channel throughput when acknowledgment packets are carried by the same channel.

As an example let us consider the ALOHA mode. Let R be the average delay between two consecutive transmissions (i.e., a retransmission) of a given packet. R consists of the transmission time of the packet, the transmission time of the acknowledgment packet, the round-trip propagation delay, and the average retransmission delay, that is

$$R = T + \tau + T_a + \tau + \bar{X}.$$

Using our normalized time units, we have

$$R = 1 + 2a + \alpha + \delta \tag{53}$$

where $\alpha = T_a/T$. Since $(G/S - 1)$ is the average number of retransmissions required, the average delay is given by

$$D = \left(\frac{G}{S} - 1\right)R + 1 + a. \tag{54}$$

(Special attention must be devoted to the CSMA modes in which packets may incur pretransmission delays, and in which all arrivals do not necessarily correspond to actual transmissions. The delay equations and their derivations are given in Appendix B.)

Let us begin with some comments concerning the above delay equations. First, G/S as obtained from the throughput equations rests on two important and strong Assumptions 1 and 2; namely, that we have an independent Poisson point process and that δ is infinite, or large compared to the transmission time (in which case delays are also large and unacceptable). On the other hand, δ cannot be arbitrarily small. It is intuitively clear that when a certain backlog of packets is present, the smaller δ is, the higher is the level of interference and hence the larger is the offered channel traffic G. Thus, $G = G(S,\delta)$ is a decreasing function of δ such that the average number of transmissions per packet, $[G(S,\delta)]/S$, decreases with increasing values of δ, and reaches the asymptotic value predicted by the throughput equation. Thus, for each S, a minimum delay can be achieved by choosing an optimal δ. Such an optimization problem is difficult to solve analytically, and simulation techniques have been employed in our evaluations below.[12]

Before we proceed with the discussion of simulation results, we compare the various access modes in terms simply of the average number of transmissions (or average number of schedulings[13]) G/S. For this purpose, we plot G/S versus S in Fig. 11 for the ALOHA and CSMA modes, when $a = 0.01$. Note that CSMA modes are superior in that they provide lower values for G/S than the ALOHA modes. Furthermore, for each value of the throughput, there exists a value of p such that p-persistent is optimal. For small values of S, $p = 1$ (i.e., 1-persistent) is optimal. As S increases, the optimum p decreases.

[12] We have been able to solve the problem analytically in the case of the nonpersistent CSMA when we are in presence of a large population but with a finite number of users; all conclusions obtained from simulation in Section V-B have been verified by the analysis. For this the reader is referred to reference [16].

[13] For the nonpersistent and p-persistent CSMA, G measures the offered channel traffic and not the actual channel traffic. G/S represents, then, the average number of times a packet was scheduled for transmission before success.

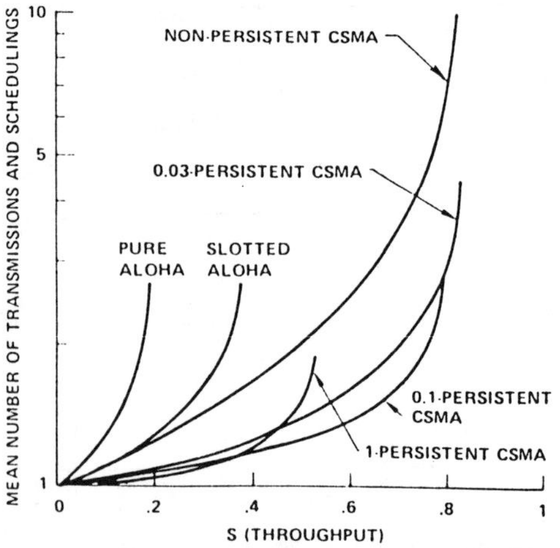

Fig. 11. G/S versus throughput ($a = 0.01$).

B. Simulation Results

The simulation model is based on all system assumptions presented in Section II. However, we relax Assumptions 1 and 2 concerning the retransmission delay and the independence of arrivals for the offered channel traffic. That is, in the simulation model, only the newly generated packets are derived independently from a Poisson distribution; collisions and uniformly distributed random retransmissions are accounted for without further assumptions.

In general, our simulation results indicate the following.

1) For each value of the input rate S, there is a minimum value δ for the average retransmission delay variable, such that below that value it is impossible to achieve a throughput equal to the input rate.[14] The higher S is, the larger δ must be to prevent a constantly increasing backlog, i.e., to prevent the channel from saturating. In other words, the maximum achievable throughput (under assumed stable conditions) is a function of δ, and the larger δ is, the higher is the maximum throughput.

2) Recall that the throughput equations were based on the assumption that $\bar{X}/T = \delta \gg 1$. Simulation shows that for finite values of δ, $\delta > \delta_0$, but not too large compared to 1, the system already "reaches" the asymptotic results ($\delta \to \infty$). That is, for some finite values of δ, Assumption 2 is excellent and delays are acceptable. Moreover, the comparison of the (S,G) relationship as obtained from simulation and the results obtained from the analytic model exhibits an excellent match. Simulation experiments were also conducted to find the optimal delay; that is, the value of $\delta(S)$ which allows one to achieve the indicated throughput with the minimum delay.

Finally, in Fig. 12[15] we give the throughput-minimum delay tradeoff for the ALOHA and CSMA modes (when $a = 0.01$). *This is the basic performance curve.* We conclude

[14] Such behavior is characteristic of random multiple-access modes. Similar results were already encountered by Kleinrock and Lam [13] when studying slotted ALOHA in the context of a satellite channel.

[15] In Fig. 12, the curve corresponding to slotted ALOHA is obtained from the analytical model developed in [13] successfully verified by simulation.

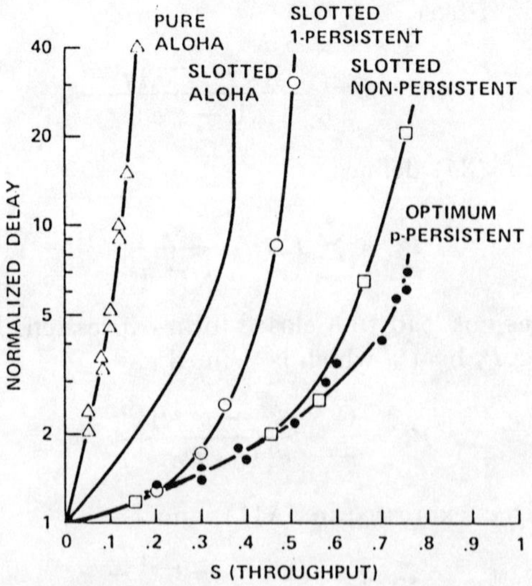

Fig. 12. Throughput-delay tradeoffs from simulation ($a = 0.01$).

that the optimum p-persistent CSMA provides us with the best performance; on the other hand the performance of the (simple) nonpersistent CSMA is quite comparable.

VI. SUMMARY AND DISCUSSION

We have introduced and evaluated the new CSMA mode and have shown it to be an efficient means for randomly accessing packet switched radio channels which have a small ratio of propagation delay to packet transmission time. Just as with most "contention" systems, these random multi-access broadcast channels (ALOHA, CSMA) are characterized by the fact that the throughput goes to zero for large values of channel traffic. At an optimum traffic level, we achieve a maximum throughput which we define to be the system capacity. This and the throughput-delay performance were obtained by a steady-state analysis under the assumption of equilibrium conditions.

However, these channels exhibit unstable behavior at most input loads as shown by Kleinrock and Lam [18]. In this last reference, the dynamic behavior and stability of an ALOHA channel are considered; quantitative estimates for the relative stability of the channel are given, which indicate the need for special control procedures to avoid a collapse. Optimal control procedures have been found [14], [15] and similar procedures are necessary for CSMA as well, since it can be shown [16] that CSMA exhibits similar unstable behavior.

Throughout the paper, it was assumed that all terminals are within range and in line-of-sight of each other. A common situation consists of a population of terminals, all within range and communicating with a single "station" (computer center, gate to a network, etc.) in line-of-sight of all terminals. Each terminal, however, may *not* be able to hear all the other terminals' traffic. This gives rise to what is called the "*hidden-terminals*" problem. The latter badly degrades the performance of CSMA as shown in Part II of this paper [19]. Fortunately, in a single-station environment, the hidden-terminal problem can be eliminated by dividing the available bandwidth into two separate channels: a busy tone channel and a message channel. As long as the station is receiving a signal on the message channel, it transmits a busy tone signal on the busy tone channel (which terminals sense for channel state information). The CSMA with a busy tone under a nonpersistent protocol has been analyzed. It is shown to provide a maximum channel capacity of approximately 0.65 when $a = 0.01$ for a channel bandwidth W of 100 kHz (modulated at 1 bit/Hz); when $W = 1$ MHz and $a = 0.01$, the channel capacity is 0.71 [19]. These values compare favorably with the capacity of 0.815 for nonpersistent CSMA with no hidden terminals.

APPENDIX A

SMALL p APPROXIMATIONS IN p-PERSISTENT CSMA

We claim, for small p, that $S(G,p,a)$ may be approximated by

$$S'(G,p,a) = \frac{(1-e^{-aG})[\hat{P}_s'\pi_0 + \hat{P}_s(1-\pi_0)]}{(1-e^{-aG})[a\hat{\hat{t}}'\pi_0 + a\hat{\hat{t}}(1-\pi_0) + 1 + a] + a\pi_0} \tag{A1}$$

where $\hat{P}_s'$, $\hat{P}_s$, $\hat{\hat{t}}$, and $\hat{\hat{t}}'$ are defined hereafter in the proof.

Proof: We show here that, with some approximations, we can get a closed-form solution for the throughput when p has *small values* ($p < 0.1$). These approximations are validated by comparing the results obtained in this section with those obtained from Section IV-D for $p = 0.1$.

For the distribution of idle time between two TP's, we have from (26)

$$\Pr\{t_n > k\} = q^{(k+1)n} \exp\left\{g\left(\frac{q(1-q^k)}{p} - k\right)\right\}. \tag{A2}$$

When p is small, we may make the following approximation (actually a lower bound):

$$q^k = (1-p)^k \simeq 1 - kp \tag{A3}$$

and therefore we may rewrite (A2) as

$$\Pr\{t_n > k\} \simeq q^{(k+1)n}e^{-kpg} = q^n[q^n e^{-pg}]^k. \tag{A4}$$

Let $t_{n>}^*(z)$ and $t_n^*(z)$ be the generating functions defined by

$$t_{n>}^*(z) \triangleq \sum_{k=0}^{\infty} \Pr\{t_n > k\} z^k \tag{A5}$$

$$t_n^*(z) \triangleq \sum_{k=0}^{\infty} \Pr\{t_n = k\} z^k. \tag{A6}$$

We have

$$t_{n>}^*(z) = q^n \sum_{k=0}^{\infty} (q^n e^{-pg} z)^k = \frac{q^n}{1 - q^n e^{-pg} z}. \tag{A7}$$

Since

$$\Pr\{t_n = k\} = \Pr\{t_n > k-1\} - \Pr\{t_n > k\}, \qquad k > 0$$

and

$$\Pr\{t_n = 0\} = 1 - \Pr\{t_n > 0\},$$

we have

$$t_n^*(z) = 1 + (z-1)t_{n>}^*(z) = 1 + \frac{q^n(z-1)}{1 - q^n e^{-pg}z}. \quad \text{(A8)}$$

The averages defined in (29) can now be written as

$$\bar{t}_n = \left.\frac{\partial t_n^*(z)}{\partial z}\right|_{z=1} = \frac{q^n}{1 - q^n e^{-pg}}. \quad \text{(A9)}$$

Equation (30), which defines $\bar{t}$ as $\sum_{n=1}^{\infty} \bar{t}_n \pi_n/(1-\pi_0)$, does not lead to a closed-form expression. Instead, we replace $\bar{t}$ by $\hat{\bar{t}}$, which is defined as

$$\hat{\bar{t}} = \frac{C}{1 - Ce^{-pg}} \quad \text{(A10)}$$

where $C = \sum_{n=1}^{\infty} q^n \pi_n/(1-\pi_0)$. ($\hat{\bar{t}}$ is smaller than $\bar{t}$ since $\bar{t}_n = q^n/(1-q^n e^{-pg})$ is a convex function of q^n.)

C can be expressed as

$$C = \frac{\exp\{-(1+a)pG\} - \pi_0}{1-\pi_0} = \frac{\pi_0^p - \pi_0}{1-\pi_0} \quad \text{(A11)}$$

and therefore,

$$\hat{\bar{t}} = \frac{\pi_0^p - \pi_0}{1 - \pi_0 - (\pi_0^p - \pi_0)e^{-pg}}. \quad \text{(A12)}$$

To find the probability of success over TP $\mathfrak{I}_i$, $i \neq 1$, we first define the following generating functions:

$$L_n^*(z) \triangleq \sum_{l=n-1}^{\infty} \Pr\{L_n = l\}z^l \quad \text{(A13)}$$

$$L_n^*(z/k) \triangleq \sum_{l=n-1}^{\infty} \Pr\{L_n = l/t_n = k\}z^l. \quad \text{(A14)}$$

It is clear that

$$L_n^*(z/k) = \exp\{kg(z-1)\}z^{n-1}. \quad \text{(A15)}$$

Removing the condition on k, we get

$$L_n^*(z) = \sum_{k=0}^{\infty} L_n^*(z/k)\cdot\Pr\{t_n = k\}$$

$$= z^{n-1}\sum_{k=0}^{\infty} \exp\{kg(z-1)\}\cdot\Pr\{t_n = k\}$$

$$= z^{n-1}t_n^*(\exp\{g(z-1)\})$$

$$= \frac{q^n(\exp\{g(z-1)\} - 1)z^{n-1}}{1 - q^n\exp\{-pg\}\exp\{g(z-1)\}} + z^{n-1}. \quad \text{(A16)}$$

The probability of success $P_s(n)$, defined in (33), is now simply expressed (since $1 - q^l \approx lp$) as

$$P_s(n) = L_n^*(q) = q^{n-1} - \frac{(1-e^{-gp})q^{2n-1}}{1 - q^n e^{-2gp}}. \quad \text{(A17)}$$

Here again, (34) defines

$$P_s = \sum_{n=1}^{\infty} P_s(n)\cdot\frac{\pi_n}{1-\pi_0}$$

which does not lead to a closed-form expression. Instead, we replace P_s by $\hat{P}_s$, which is defined as

$$\hat{P}_s = \frac{C}{q} - \frac{(1-e^{-gp})C'}{q(1-Ce^{-gp})} \quad \text{(A18)}$$

where C is as expressed in (A11), and

$$C' = \sum_{n=1}^{\infty} q^{2n}\frac{\pi_n}{1-\pi_0} = \frac{\pi_0^{1-q^2} - \pi_0}{1-\pi_0}. \quad \text{(A19)}$$

Finally, $\hat{P}_s$ can be expressed as shown in the following equation:

$$\hat{P}_s = \frac{\pi_0^p - \pi_0}{q(1-\pi_0)} - \frac{(1-\exp\{-gp\})(\pi_0^{1-q^2} - \pi_0)}{q(1-\pi_0) - q\exp\{-2gp\}(\pi_0^p - \pi_0)}. \quad \text{(A20)}$$

The quantities $\hat{\bar{t}}'$ and $\hat{P}_s'$ are readily obtained from (A12), and (A20), respectively, by replacing

$$\pi_0 \triangleq \exp\{-G(1+a)\}$$

by the quantity e^{-g}. The substitution of $\hat{P}_s$, $\hat{\bar{t}}$, $\hat{P}_s'$, and $\hat{\bar{t}}'$ for P_s, $\bar{t}$, P_s', and $\bar{t}'$, respectively, in (46) provides us with a closed-form solution for $S(G,p,a)$ when p is small.

In Table II, we compare for $p = 0.1$ the "exact" results obtained from Section IV-D to those obtained by the approximation; note that the closed-form solution is quite satisfactory for $p < 0.1$.

APPENDIX B

DELAY EQUATIONS

A. Nonpersistent CSMA

In this case, the average delay R between two successive sense points of the same packet is

$$R = \begin{cases} 1 + \alpha + 2a + \delta, & \text{if the packet is transmitted} \\ \delta, & \text{if the packet is blocked.} \end{cases} \quad \text{(B1)}$$

Let P_b be the probability that an arrival gets blocked (i.e., senses the channel busy). We have

$$1 - P_b = \frac{a + 1/G}{\bar{C}}$$

$$= \frac{1 + aG}{1 + G(1 + a + \bar{Y})}. \quad \text{(B2)}$$

Under the traffic independence assumption, the rate of

TABLE II
COMPARISON OF RESULTS FOR THROUGHPUT S OBTAINED FROM THE EXACT ANALYSIS (24) AND RESULTS OBTAINED FROM THE APPROXIMATION (APPENDIX A) WHEN $p = 0.1$

	$a = 0.01$		$a = 0.05$	
G	Exact	Approximate	Exact	Approximate
0.1	0.098	0.098	0.095	0.094
0.2	0.192	0.192	0.179	0.178
0.3	0.279	0.279	0.252	0.251
0.4	0.358	0.358	0.316	0.314
0.5	0.428	0.428	0.370	0.367
0.6	0.490	0.490	0.417	0.413
0.7	0.544	0.544	0.457	0.453
0.8	0.589	0.590	0.490	0.486
0.9	0.628	0.630	0.519	0.515
1.0	0.661	0.663	0.543	0.539
1.1	0.689	0.691	0.563	0.560
1.2	0.711	0.714	0.580	0.578
1.3	0.730	0.733	0.594	0.593
1.4	0.745	0.749	0.606	0.605
1.5	0.757	0.761	0.616	0.616
1.6	0.766	0.771	0.624	0.625
1.7	0.773	0.778	0.630	0.632
1.8	0.778	0.784	0.635	0.638
1.9	0.781	0.787	0.639	0.643
2.0	0.783	0.790	0.642	0.647
2.1	0.784	0.791	0.644	0.649
2.2	0.784	0.791	0.645	0.651
2.3	0.783	0.790	0.646	0.653

actual transmissions is given by

$$H = G(1 - P_b).$$

Since $(H/S) - 1$ represents the average number of actual retransmissions per packet, the average delay D is therefore

$$D = (H/S - 1)[1 + \alpha + 2a + \delta][(G - H)/S]\delta + 1 + a \tag{B3}$$

where G/S is given by the nonpersistent CSMA throughput equation (3).

If we choose to treat all packet arrivals in a uniform manner, we may assume that when a packet is blocked, it behaves as if it could transmit, and learned about its blocking only T_a seconds after the end of its "virtual" transmission. With this simplification, the delay equation is

$$D = (G/S - 1)(1 + 2a + \alpha + \delta) + 1 + a \tag{B4}$$

thus introducing an additional delay equal to $(GP_b/S)[1 + \alpha + 2a]$.

B. 1-Persistent CSMA

Unlike the ALOHA channel, a packet on a CSMA channel incurs an additional pretransmission delay r, if upon its arrival, that packet detects the channel busy. Recall that the probability of finding the channel busy is given by (see Section IV-C)

Pr {a packet finds the channel busy}

$$= \frac{\bar{B} - a/q_0}{\bar{B} + \bar{I}} = \frac{1 + \bar{Y}}{q_0(\bar{B} + \bar{I})} \tag{B5}$$

where $\bar{B}$, $\bar{I}$, $\bar{Y}$, and q_0 are given in (16), (12), (7), and (15), respectively.

Under the condition that the packet found the channel busy, the average waiting time until the channel is detected idle (i.e., until the end of the TP) is simply equal [17] to $\overline{Z^2}/2\bar{Z}$ by the Poisson assumption. The second moment of Z is simply given by

$$\overline{Z^2} = \overline{(1 + Y)^2} = 1 + 2\bar{Y} + \overline{Y^2}.$$

From the distribution of Y given in (6) we then have

$$\overline{Z^2} = 1 + a^2 + 2(1 - 1/G)\bar{Y}. \tag{B6}$$

Therefore the average pretransmission delay $\bar{r}_1$ can be easily expressed as

$$\bar{r}_1 = \frac{1 + a^2 + 2(1 - 1/G)\bar{Y}}{2(1 + \bar{Y})} \cdot \text{Pr}\{\text{the packet finds the channel busy}\}$$

$$= \frac{1 + a^2 + 2(1 - 1/G)\bar{Y}}{2q_0(\bar{B} + \bar{I})}. \tag{B7}$$

Finally, the expected packet delay is

$$D = (G/S - 1)(1 + 2a + \alpha + \delta + \bar{r}_1) + \bar{r}_1 + 1 + a \tag{B8}$$

where G/S is given by the 1-persistent CSMA throughput equation (10).

C. p-Persistent CSMA

Similar to the special case of 1-persistent CSMA, a packet in this general scheme incurs an initial delay which we denote by r_p. In order to compute its expected value $\bar{r}_p$, one must consider the following situations.

1) An arbitrary packet, upon arrival, will find the channel idle with probability $\bar{I}/(\bar{B} + \bar{I})$, in which case its average initial wait is $a\bar{l}'$.

2) An arbitrary packet, upon arrival, will find the channel in the first IRTD (first t' seconds) of a busy period with probability $a\bar{l}'/(\bar{B} + \bar{I})$. In this case, its average initial delay is $a\overline{l'^2}/2\bar{l}'$.

3) An arbitrary packet, upon arrival, will find the channel in the remaining part of a busy period with probability $(B - a\bar{l}')/(\bar{B} + \bar{I})$, in which case the average initial wait is $\overline{(1 + a + al)^2}/2(1 + a + a\bar{l})$.

Therefore

$$\bar{r}_p = \frac{\bar{I}}{\bar{B} + \bar{I}} a\bar{l}' + \frac{a\bar{l}'}{\bar{B} + \bar{I}} \cdot \frac{a\overline{l'^2}}{2\bar{l}'} + \frac{B - a\bar{l}'}{\bar{B} + \bar{I}} \cdot \frac{\overline{(1 + a + al)^2}}{2(1 + a + a\bar{l})}. \tag{B9}$$

Treating all transmissions and schedulings uniformly (by introducing artificial delays due to "virtual" transmissions and acknowledgment), the expected delay can simply be expressed as

$$D = (G/S - 1)[1 + 2a + \delta + \bar{r}_p] + 1 + a + \bar{r}_p \tag{B10}$$

where G/S is given by the p-persistent CSMA throughput equation (24).

REFERENCES

[1] L. G. Roberts, "Data by the packet," *IEEE Spectrum*, vol. 11, pp. 46–51, Feb. 1974.

[2] N. Abramson, "The ALOHA System—Another alternative for computer communications," in *1970 Fall Joint Comput. Conf., AFIPS Conf. Proc.*, vol. 37. Montvale, N. J.: AFIPS Press, 1970, pp. 281–285.

[3] R. E. Kahn, "The organization of computer resources into a packet radio network," in *Nat. Comput. Conf., AFIPS Conf. Proc.*, vol. 44. Montvale, N. J.: AFIPS Press, 1975, pp. 177–186.

[4] L. Kleinrock and F. Tobagi, "Random access techniques for data transmission over packet-switched radio channels," in *Nat. Comput. Conf., AFIPS Conf. Proc.*, vol. 44. Montvale, N. J.: AFIPS Press, 1975, pp. 187–201.

[5] R. Binder, N. Abramson, F. Kuo, A. Okinaka, and D. Wax, "ALOHA packet broadcasting—A retrospect," in *Nat. Comput. Conf., AFIPS Conf. Proc.*, vol. 44. Montvale, N. J.: AFIPS Press, 1975, pp. 203–215.

[6] H. Frank, I. Gitman, and R. Van Slyke, "Packet radio system—Network considerations," in *Nat. Comput. Conf., AFIPS Conf. Proc.*, vol. 44. Montvale, N. J.: AFIPS Press, 1975, pp. 217–231.

[7] S. Fralick and J. Garrett, "Technological considerations for packet radio networks," in *Nat. Comput. Conf., AFIPS Conf. Proc.*, vol. 44. Montvale, N. J.: AFIPS Press, 1975, pp. 233–243.

[8] J. Burchfiel, R. Tomlinson, and M. Beeler, "Functions and structure of a packet radio station," in *Nat. Comput. Conf., AFIPS Conf. Proc.*, vol. 44. Montvale, N. J.: AFIPS Press, 1975, pp. 245–251.

[9] S. Fralick, D. Brandin, F. Kuo, and C. Harrison, "Digital terminals for packet broadcasting," in *Nat. Comput. Conf., AFIPS Conf. Proc.*, vol. 44. Montvale, N. J.: AFIPS Press, 1975, pp. 253–261.

[10] P. E. Jackson and C. D. Stubbs, "A study of multi-access computer communications," in *1969 Spring Joint Comput. Conf., AFIPS Conf. Proc.*, vol. 34. Montvale, N. J.: AFIPS Press, 1969, pp. 491–504.

[11] N. Abramson, "Packet switching with satellites," in *Nat. Comput. Conf., AFIPS Conf. Proc.*, vol. 42. Montvale, N. J.: AFIPS Press, 1973, pp. 695–702.

[12] L. Roberts, "ARPANET Satellite System," Notes 8 (NIC Document 11290) and 9 (NIC Document 11291), available from the ARPA Network Information Center, Stanford Research Institute, Menlo Park, Calif.

[13] L. Kleinrock and S. Lam, "Packet-switching in a slotted satellite channel," in *Nat. Comput. Conf., AFIPS Conf. Proc.*, vol. 42. Montvale, N. J.: AFIPS Press, 1973, pp. 703–710.

[14] S. Lam, "Packet switching in a multi-access broadcast channel with application to satellite communications in a computer network," School of Eng. and Appl. Sci., Univ. of California, Los Angeles, rep. UCLA-ENG 7429, Apr. 1974.

[15] S. Lam and L. Kleinrock, "Dynamic control schemes for a packet switched multi-access broadcast channel," in *Nat. Comput. Conf., AFIPS Conf. Proc.*, vol. 44. Montvale, N. J.: AFIPS Press, 1975, pp. 143–153.

[16] F. Tobagi, "Random access techniques for data transmission over packet switched radio networks," Ph.D. dissertation, Comput. Sci. Dep., School of Eng. and Appl. Sci., Univ. of California, Los Angeles, rep. UCLA-ENG 7499, Dec. 1974.

[17] L. Kleinrock, *Queueing Systems, Vol. I, Theory; Vol. II, Computer Applications.* New York: Wiley Interscience, 1975.

[18] L. Kleinrock and S. S. Lam, "Packet switching in a multiaccess broadcast channel: Performance evaluation," *IEEE Trans. Commun.*, vol. COM-23, pp. 410–423, Apr. 1975.

[19] F. A. Tobagi and L. Kleinrock, "Packet switching in radio channels: Part II—The hidden terminal problem in carrier sense multiple access and the busy tone solution," *IEEE Trans. Commun.*, this issue, pp. 1417–1433.

A STUDY OF UNSLOTTED ALOHA WITH ARBITRARY MESSAGE LENGTHS

Michael J. Ferguson
University of Hawaii
Honolulu, Hawaii
and
McGill University
Montreal, Quebec, Canada

ABSTRACT

This study of unslotted ALOHA results in three major conclusions:
1. Throughput is maximized over all possible packet distributions with fixed length packets.
2. "Real" throughput (data only--excluding headers and unfilled packets) is, for exponentially distributed message lengths, almost always maximized by using variable length rather than fixed length packets in the unslotted channel.
3. "Real" throughput in the slotted channel exceeds that of the variable length packets in the unslotted channel but by less than 33% (not 100%) if the overhead is about equal to the average message length.

The model studied assumes that a terminal sends a packet with arbitrary length and then waits an exponential length time (mean λ^{-1}) before attempting to transmit again. Each time it transmits, it obtains a new independent message length according to $L_0(t)$. This model discriminates against long packets by not requiring repeated successive retransmissions in the event of failure.

1. INTRODUCTION

A previous paper [1], introduced a simple and slightly unrealistic model for variable length unslotted ALOHA. It assumes that each terminal produces a sequence of transmit and idle states with each idle state, corresponding to either a "think" time or retransmit time that is exponentially distributed with mean λ^{-1} and each transmit interval, corresponding to the transmission of a packet, was drawn independently from an exponential density with mean μ^{-1}. Here we use the same model but extend the offered packet length to be any density, $L_0(t)$.

2. ANALYSIS - THROUGHPUT AND WAITING TIME

We assume N terminals each of which alternates between idle (no transmission) and transmission. A no-transmission interval is assumed exponential with mean λ^{-1}; the transmit interval has length with density $L_0(t)$ and mean $\bar{L}_0$. We define the state of the system as the number n, that are actually transmitting. Since the exponential idle (no-transmission) time ensures the process is ergodic we have the problem of determining how many of N independent processes are "on" at any given time. The probability that any one is on is just $(\bar{L}_0/\bar{L}_0+\lambda^{-1})$ and since the processes are independent we have; P_n, the probability of n terminals transmitting as

$$P_n = \binom{N}{n}\left(\frac{\bar{L}_0\lambda}{1+\bar{L}_0\lambda}\right)^n\cdot\left(\frac{1}{1+\bar{L}_0\lambda}\right)^{N-n} \qquad (1)$$

Defining $\bar{L}_0\lambda=\rho$ we can write (1) as

$$P_n = \binom{N}{n}\frac{\rho^n}{(1+\rho)^N} \qquad (2)$$

In fact (2) is true for any terminal idle density so long as the resulting process is ergodic. This neat analysis is due to Abramson [2].

The system alternates between busy and idle periods where an idle period is defined as "no terminals transmitting" and a busy period by one or more transmitting.

Because of the no memory property of the exponential idle times, we have the idle time exponentially distributed with mean, $\bar{I}_N$,

$$\bar{I}_N = \frac{1}{N\lambda} \qquad (3)$$

Thus, if we define the busy period mean as $\bar{B}_N$, because of the ergodicity we have,

$$P_0 = \frac{\bar{I}_N}{\bar{B}_N+\bar{I}_N} \qquad (4)$$

Again, because of the exponential nature of the idle times, all system busy periods start with exactly one transmitting terminal. Thus, a busy period results in a successful transmission if and only if $L_0<I_{N-1}$ where I_{N-1} is the system idle time of the (N-1) other terminals not transmitting. This is obviously exponential with mean $((N-1)\lambda)^{-1}$. Thus if we define q_s as the probability that an idle period results in a successful transmission we have

$$q_s = \Pr(L_0<I_{N-1}) = \int_0^\infty L_0(t)e^{-(N-1)\lambda t}dt \qquad (5)$$

The distribution of successful-packet transmissions is found noting that a success occurs if and only if $L_0<I_{N-1}$. The density of L_s, the random variable representing the length of successful transmissions is

$$\Pr(L_s=t) = \Pr(L_0=t|L_0<I_{N-1})$$
$$= \frac{\Pr(L_0=t;L_0<I_{N-1})}{\Pr(L_0<I_{N-1})}$$
$$= \frac{1}{q_s}\Pr(L_0<I_{N-1}|L_0=t)\Pr(L_0=t)$$

Thus

$$\Pr(L_s=t) = \frac{e^{-(N-1)\lambda t}L_0(t)}{q_s} \qquad (6)$$

Since system busy periods are independent we have $\bar{M}_N$, the average number of successful packets per second as that proportion of busy periods resulting in success.

This paper was supported by The ALOHA System, a research project at the University of Hawaii, which is supported by the Advanced Research Projects Agency of the Department of Defense and monitored by NASA Ames Research Center under Contract No. NAS2-8590.

Reprinted from *Proc. DATACOMM*, pp. 5.20–5.25, Quebec, Canada, October 1975.

Thus, using (2,3,4) we have

$$\overline{M}_N = \frac{q_s}{\overline{I}_N+\overline{B}_N} = \frac{N\lambda}{(1+\rho)^N} q_s \tag{7}$$

The throughput, S_N, is just $\overline{M}_N \cdot \overline{L}_s$ and is, using (6) and (7)

$$S_N = \frac{N\lambda}{(1+\rho)^N} \int_0^{\infty} t\, L_0(t)\, e^{-(N-1)\lambda t} dt \tag{8}$$

3. THE OPTIMAL DISTRIBUTION OF PACKET LENGTHS

In this section we find the distribution of packet lengths which maximizes throughputs for large $N\to\infty$. We note from (8) that S_N will go to zero as $N\to\infty$ unless $(1+\rho)^N$ remains finite. The necessary and sufficient condition for this is to have $\rho \simeq 0(N^{-1})$. Thus for large N we write

$$\rho = \frac{K}{N} \tag{9}$$

Letting $N\to\infty$ in (8) and noting that $\left(1+\frac{K}{N}\right)^N \to e^K$ we have

$$S_\infty = Ke^{-K} \int_0^{\infty} \frac{t\, L_0(t)}{\overline{L}_0} e^{-(Kt/\overline{L}_0)} dt \tag{10}$$

Thus the objective is to find the $L_0(t)$ which maximizes (10) subject to $\overline{L}_0$ being a constant.

We first note that $te^{-Kt/\overline{L}_0}$ is a monotone increasing from 0 to $\overline{L}_0/K$ and concave over the range $0<t<2\overline{L}_0/K$. We assume $K<1$. Thus the peak and concave region go out beyond the mean $\overline{L}_0$. Since $\overline{L}_0$ is less than the abscissa of the peak, then it is easy to see that the entire function is concave with respect to $\overline{L}_0$. Thus if we pick any two points on either side of $\overline{L}_0$, the chord drawn lies below the function value at $\overline{L}_0$. This is shown in Figure 1 with a sketch of $te^{-Kt/\overline{L}_0}$. All that is required for the optimum $L_0(t)$ to be a delta function at $\overline{L}_0$ is for the $te^{-Kt/\overline{L}_0}$ to be concave with respect to $\overline{L}_0$. The proof consists of showing that for any density with probability mass at two points above and below $\overline{L}_0$ such that the mean is $\overline{L}_0$, a higher value of (10) will result if the mass is moved to be concentrated at $\overline{L}_0$. This is then extended to any arbitrary density completing the proof. The details are in the Appendix. The optimum result is thus $L_0(t)=\delta(t-\overline{L}_0)$ and thus S_∞ becomes

$$\max_{L_0(t)} S_\infty = Ke^{-2K} \tag{11}$$

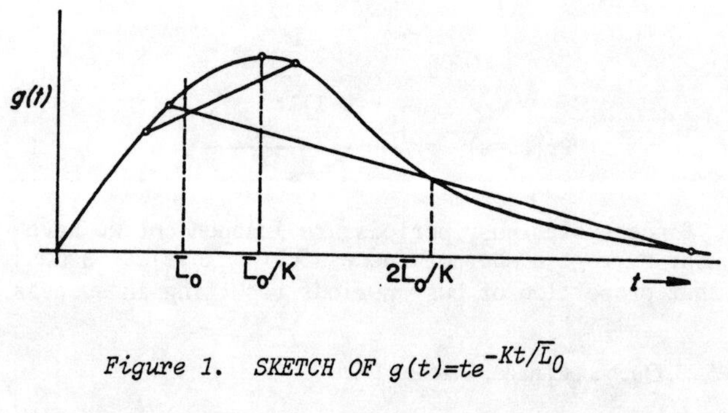

Figure 1. SKETCH OF $g(t)=te^{-Kt/\overline{L}_0}$

Maximizing (11) over K gives K=1/2 and $S_{\infty_{OPT}}$ as

$$S_{\infty_{OPT}} = \max_K \max_{L_0(t)} S_\infty = \frac{e^{-1}}{2} = .186 \tag{12}$$

This confirms the value of capacity of unslotted ALOHA found by Abramson [3]. Note that for this case, with fixed packet lengths, the model accurately reflects channel behavior and has no approximations.

4. SOME SPECIAL CASES

(i) Two Packet Lengths (Header-Only and Full Packet)

Assume we have two packet lengths a header-only packet with length H and probability α, and a full packet with length L_m.

Thus $L_0(t)=\alpha\delta(t-H)+(1-\alpha)\delta(t-L_m)$

$$\overline{L}_0 = \alpha H + (1-\alpha)L_m$$

and hence

$$S_{N_T} = \frac{(N\lambda)\{\alpha He^{-(N-1)\lambda H}+(1-\alpha)L_m e^{-(N-1)\lambda L_m}\}}{(1+\rho)^N} \tag{13}$$

Another interpretation of (13) is that the throughput is being divided between a short-packet channel and a long-packet channel. Thus we can define S_{N_H} and S_{N_m} as the short packet throughput and long packet throughput, respectively. Thus

$$S_{N_H} \triangleq \frac{\alpha N\lambda He^{-(N-1)\lambda H}}{(1+\rho)^N} \tag{14a}$$

$$S_{N_m} = \frac{N\lambda(1-\alpha)L_m e^{-(N-1)\lambda L_m}}{(1+\rho)^N} \tag{14b}$$

and

$$S_{N_T} = S_{N_H} + S_{N_m}$$

Again we let $\rho=K_T/N$ and define

$$H\lambda \triangleq \rho_H = \frac{K_H}{N}; \quad L_m\lambda \triangleq \rho_m = \frac{K_m}{N}$$

Thus

$$\rho = \alpha\rho_H + (1-\alpha)\rho_m \tag{15a}$$

$$K_T = \alpha K_H + (1-\alpha)K_m \tag{15b}$$

Hence as $N\to\infty$ we have

$$S_{\infty_H} = \alpha K_H e^{-(K_T+K_H)} \tag{16a}$$

$$S_{\infty_m} = (1-\alpha)K_m e^{-(K_T+K_m)} \tag{16b}$$

$$S_{\infty_T} = S_{\infty_H} + S_{\infty_M} \tag{16c}$$

We now wish to find the K which maximizes S_{∞_T} to see the ultimate throughput for this channel. We first define r, the ratio of the lengths of short to long packets as

$$r = H/L_m$$

Thus, noting the definitions of K_T, K_m, K_H, we first see that $K_H = rK_m$ and hence

$$K_H = \frac{rK_T}{\alpha r + (1-\alpha)} \tag{17a}$$

$$K_m = \frac{K_T}{\alpha r + (1-\alpha)} \tag{17b}$$

Substituting (17) into (16) gives S_{∞_T} as a function of K_T with α and r as parameters. The optimization over K_T was performed numerically and the following conclusions resulted.

1. The maximum throughput is not very sensitive to K_T over a wide range of say $.3 \leq K_T \leq .6$.
2. For $.2 \leq r \leq 1$ and for all α, $K_T = .5$ resulted in throughputs that were within a few percent of the optimal.
3. As K_T was increased over its "optimal" range, the short-packet channel increased throughput and the long-packet channel decreased throughput but the sum remained remarkably constant.

Figure 2 shows the variation in total throughput, and for the various channels for both $\hat{K}_T$, the optimum and $K_T = .5$. In Figure 2, r=.1 and the abscissa is the fraction of offered traffic in the short, header-only channel, α. Note the difference in total throughput is negligible but there is a definite change in the mix as shown by $\hat{S}_{\infty_H}$ and S_{∞_H}.

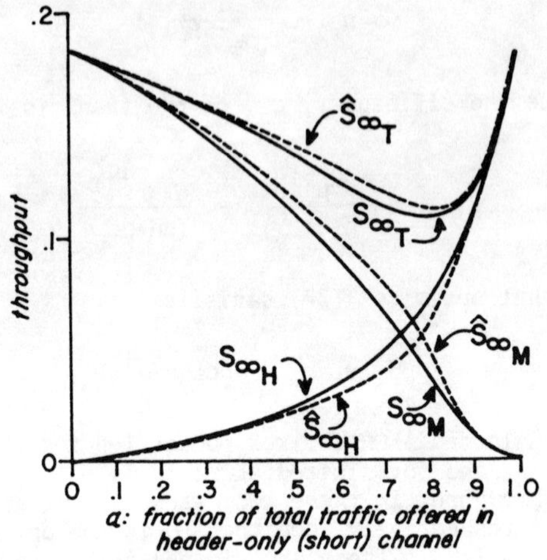

Figure 2. *THROUGHPUT IN SHORT/LONG PACKET CHANNEL*

Figure 3 shows S_{∞_T} for $K_T = .5$ and r=.1, .3, .5, .8 and 1.

(ii) Exponential Messages Plus Headers

Here we assume a header of length H. The message is exponential with mean μ^{-1}. Thus the packet has the following density:

$$L_0(t) = \begin{cases} 0 & 0 \leq t \leq H \\ \mu e^{-\mu(t-H)} & H \leq t \leq \infty \end{cases} \tag{18}$$

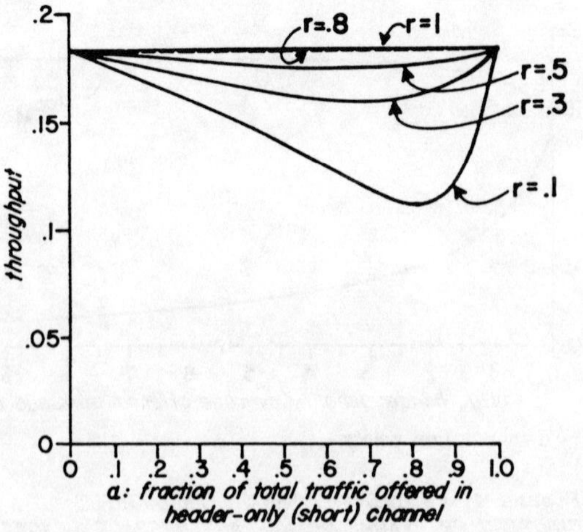

Figure 3. *TOTAL THROUGHPUT AS A FUNCTION OF SHORT TO LONG PACKET LENGTHS AND MIX*

Thus we have

$$\bar{L}_0 = H + \frac{1}{\mu} \tag{19}$$

and S_{N_M}, the throughput as

$$S_{N_M} = \frac{N\lambda}{(1+\rho)^N}\left\{\frac{He^{-(N-1)\lambda H}}{1+(N-1)\lambda/\mu} + \frac{1}{\mu}\cdot\frac{1}{(1+(N-1)\lambda/\mu)^2}e^{-(N-1)\lambda H}\right\} \tag{20}$$

Define, as before,

$$\rho = \lambda\bar{L}_0 = K/N$$

$$\rho_H = \lambda H = K_H/N \tag{21}$$

$$\rho_m = \lambda/\mu = K_m/N$$

Then using (19) and (21) we have

$$K = K_H + K_m$$

and substituting (21) into (20), letting $N \to \infty$ gives

$$S_{\infty_M} = \frac{e^{-(K_m+2K_H)}}{(1+K_m)}\left\{K_H + \frac{K_m}{(1+K_m)}\right\} \tag{22}$$

Again define the ratio of header length to average message length as $r \triangleq H\mu$. Then $K_H = \mu H K_m = rK_m$ and

$$K_H = \frac{rK}{1+r}$$

$$K_m = \frac{K}{1+r} \tag{23}$$

Thus we can substitute (23) into (22) and optimize over K. This was done, again numerically and again the result was not very sensitive to K. Figure 4 shows the throughput S_{∞_M} for K=.5 as r ranges from 0 to 10. For large r, the header totally dominates the effective packet size and the capacity approaches $.5e^{-1} = .184$. The first point computed was for r=.1 and $S_{\infty_M} = .144$ with $K_{OPT} = .406$. These values are not significantly different when K=.5.

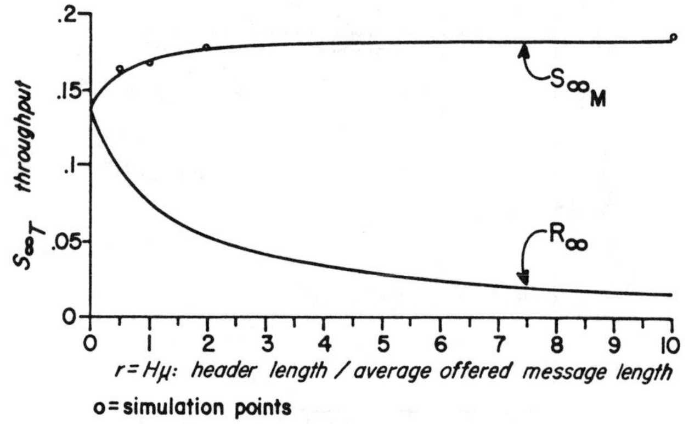

Figure 4. TOTAL AND "REAL" THROUGHPUT AS A FUNCTION OF HEADER TO AVERAGE MESSAGE LENGTH

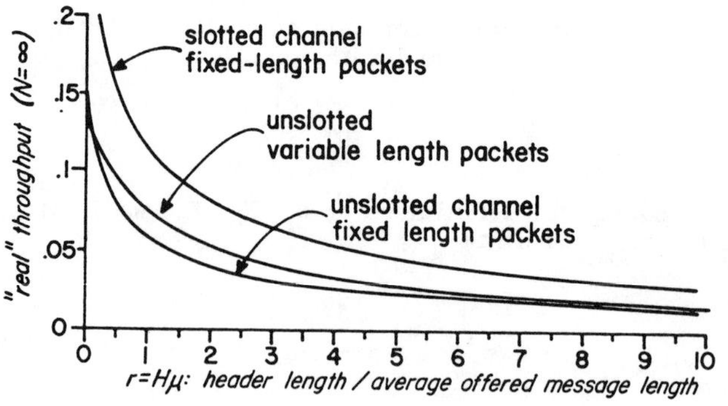

Figure 5. "REAL" THROUGHPUT FOR SLOTTED/UNSLOTTED CHANNELS AND VARIABLE-LENGTH PACKETS

We also show, in Figure 4, the results of simulations and note that the difference between the points and the curves are negligible. The simulation was of a system that included headers, acknowledgements, and repetition of actual blocked packets. Since the header size was fixed in the simulation, various traffic levels were produced by varying the number of terminals. The number used, to give a K≈.5, ranged from 75 for r=.5 to 205 for r=10. The results also confirm the lack of sensitivity to N for N large.

The results for S_{∞_M} in Figure 4 show that throughput is maximized for large r. This corresponds to the variable part of the packet being insignificant. However, most likely the variable part is our message and the header is overhead. Thus we can define the "real" throughput R_∞, as the message part of the throughput S_{∞_M}. To determine what part of the throughput is real, we note that by using (5), (6) and (18) the mean successful packet length is

$$\overline{L}_S = H + \frac{1}{\mu(1+K_m)} \qquad (24)$$

Obviously the first term in (24) is the header and the second the actual successful message length. Thus the fraction of throughput is (22) that is "real", defined as f, is

$$f = \frac{\frac{1}{\mu}\cdot\frac{1}{1+K_m}}{H+\frac{1}{\mu}\cdot\frac{1}{1+K_m}} = \frac{1}{r(1+K_m)+1}$$

$$R_\infty = f\,S_{\infty_M} = \frac{Ke^{-K}e^{-rK/(1+r)}}{(1+r)\left(1+\frac{K}{1+r}\right)^2} \qquad (25)$$

We optimize (25) over K by differentiating with respect to K and setting to zero to give

$$K_{OPT} = \frac{\sqrt{2-\frac{r^2}{(1+r)^2}}-1}{1-\frac{r^2}{(1+r)^2}} \qquad (26)$$

The optimum K in (26) ranges from .414 for r=0 to .5 for $r\to\infty$. Again the optimum is not much different than K=.5. The optimum "real" throughput is plotted as a function of r in Figure 4 using the optimum K in (26).

5. FIXED OR VARIABLE LENGTH OPTIMUM?

In this section we compare the maximum "real" throughput of fixed length packets versus variable length and as an aside also include channel slotting. To obtain the most favorable comparison we optimize the fixed length packets with respect to the average message distribution. For simplicity we assume messages of exponentially distributed lengths and mean μ^{-1}, headers of length H, and the data portion of the packet of length L_D. We assume messages are broken up into n packets each with a header. The probability that exactly n packets will be needed is, with message length L_m, just

$$Pn((n-1)L_D < L_m \le nL_D) = e^{-\mu L_D(n-1)}(1-e^{-\mu L_D}) \qquad n=1,\ldots,\infty$$

Thus the mean number of packets/message, $\overline{n}$, is

$$\overline{n} = \frac{1}{1-e^{-\mu L_D}}$$

and hence the efficiency, ζ_e, of the transmissions is

$$\zeta_e = \frac{\mu^{-1}}{\overline{n}(H+L_D)} = \frac{1-e^{-\mu L_D}}{(\mu H+\mu L_D)} \qquad (27)$$

The L_D that optimizes (27) satisfies

$$\mu H = e^{\mu L_D} - \mu L_D - 1 \qquad (28)$$

For an unslotted ALOHA, fixed packet lengths the capacity is $e^{-1}/2$ and for slotted ALOHA it is e^{-1}. Thus the "real" throughput in these two cases is just $\hat{\zeta}_e\cdot e^{-1}/2$ and $\hat{\zeta}_e e^{-1}$ respectively when the "^" is the optimum L_D chosen from (28). These are plotted in Figure 5 along with R_∞ from Section (4.ii). From this figure it can be seen that variable length packets have better real throughput in the unslotted channel as long as the header is at least 1/10 the average message length (i.e. $\mu H \ge .1$). Thus in most realistic cases we have greater "real" throughput by using variable length packets in the unslotted channel. In the slotted channel, the real throughput always seems to be better than variable lengt packets in the unslotted system. Note however that the difference is rather small (33%) if the header is about equal to the average message length--a not too uncommon situation. The major question then is whether the small increase in throughput warrants the slotting.

6. CONCLUSION

A major result of this analysis is a confirmation of previous partial analyses by Abramson [4] and Gaarder [5] that variable length packets reduce capacity in the random access channel. The optimality of fixed-length packets obtained here is not impaired by the model assumption on the behavior of retransmissions and shows that even in the optimistic environment assumed for variable length packets, fixed length packets came out best. The simulations performed show that the model is remarkably accurate in predicting throughput for large N.

However, if we measure "real" throughput, under a wide range of realistic conditions, variable length packets are maximized throughput. Finally, the improvement in "real" throughput when the channel is slotted is much less than the factor of 2 that results with fixed-packet length comparisons.

7. ACKNOWLEDGEMENT

Norman Abramson noted the essential independence assumption in the analysis and N. Thomas Gaarder acted as a sounding board for various attempts at the optimality problem.

8. REFERENCES

[1] FERGUSON, M.J.: "An Analysis of Variable Length Packets in Unslotted ALOHA," *ALOHA System Technical Report B75-7*, University of Hawaii, February 1975.

[2] ABRAMSON, N.: Private communication.

[3] ABRAMSON, N.: "The ALOHA System - Another Alternative for Computer Communications," *Proceedings of the AFIPS Fall Joint Computer Conference*, Vol. 37, November 1970, pp. 281-285.

[4] ABRAMSON, N.: "ARPANET Satellite System," *ASS Note 4 (NIC 11286)*, Stanford Research Institute, Menlo Park, California, May 1972.

[5] GAARDER, N.T.: "ARPANET Satellite System," *ASS Note 3 (NIC 11285)*, Stanford Research Institute, Menlo Park, California, April 1972.

APPENDIX: OPTIMIZATION OF PACKET LENGTHS

We wish to show that

$$\int_0^\infty f(x)g(x)dx \le g(m) \tag{A1}$$

where

$$\int_0^\infty f(x)dx = 1 \tag{A2}$$

$$f(x) \ge 0 \text{ for all } x \ge 0 \tag{A3}$$

$$\int_0^\infty xf(x)dx = m \tag{A4}$$

and for any $x_1 \le m \le x_2$ such that

$$\alpha x_1 + (1-\alpha)x_2 = m \tag{A5}$$

we have

$$\alpha g(x_1) + (1-\alpha)g(x_2) \le g(m) \tag{A6}$$

Assume also that g(x) is continous.

Case 1

Suppose $f(x)=\alpha\delta(x-x_1)+(1-\alpha)\delta(x-x_2)$ and satisfies (A2), (A3), and (A4). Therefore

$$\alpha x_1 + (1-\alpha)x_2 = m$$

and we obtain, evaluating

$$\int_0^\infty f(x)g(x)dx = \alpha g(x_1) + (1-\alpha)g(x_2) \le g(m)$$

Thus we have proved (A1) in this case. The result here implies that we should place all the probability mass at m.

Case 2

Suppose $\int_0^\infty f(t)dt$ is continuous for all t. Define ζ_1 and ζ_2, and ε_1, ε_2 small as

$$\zeta_i = \frac{1}{\varepsilon_i}\int_{x_i-\varepsilon_i/2}^{x_i+\varepsilon_i/2} f(x)dx \tag{A7}$$

and

$$\varepsilon_1\zeta_1x_1 + \varepsilon_2\zeta_2x_2 = (\varepsilon_1\zeta_1+\varepsilon_2\zeta_2)m \tag{A8}$$

If we define

$$\alpha = \frac{\varepsilon_1\zeta_1}{\varepsilon_1\zeta_1+\varepsilon_2\zeta_2} \tag{A9}$$

then we have (A8) equivalent to (A5). Basically we have chosen $x_1 \le m \le x_2$. Then we have adjusted ε_1 and ε_2, the infinitesimal interval about x_1 and x_2 such that we are extracting the correct amount of probability mass to give the same mean as f(x). We now show that this mass would be better employed at the mean. We do this by noting that

$$\int_{x_1-\varepsilon_1/2}^{x_1+\varepsilon_1/2} f(x)g(x)dx + \int_{x_2-\varepsilon_2/2}^{x_2+\varepsilon_2/2} f(x)g(x)dx$$

$$\simeq g(x_1)\int_{x_1-\varepsilon_1/2}^{x_1+\varepsilon_1/2} f(x)dx + g(x_2)\int_{x_2-\varepsilon_2/2}^{x_2+\varepsilon_2/2} f(x)dx$$

$$= \varepsilon_1\zeta_1 g(x_1) + \varepsilon_2\zeta_2 g(x_2)$$

$$= (\varepsilon_1\zeta_1+\varepsilon_2\zeta_2)[\alpha g(x_1)+(1-\alpha)g(x_2)]$$

$$\le (\varepsilon_1\zeta_1+\varepsilon_2\zeta_2)g(m) \tag{A10}$$

Thus we would be better off moving the probability mass in the intervals chosen to the mean m. The proof is completed by noting that if we start at $\bar{L}_0$ and move out in arbitrarily small steps we will successively place all of the mass at the mean. That we can do this and not run into an imbalance of mass is due to (A4) since

$$\int_0^m (x-m)f(x)dx = \int_m^\infty (m-x)f(x)dx \qquad (A11)$$

The proof in this section would go through even if f(x) were allowed δ functions. However, the definition of ζ_i in (A7) is not sufficient in this case. We must modify the problem as follows:

Step 1: Suppose we have a δ function in interval of weight f_1. We define

$$\varepsilon_{1,i}\zeta_{1,i} = \delta_i f_1 \qquad i=1,2,\ldots$$

such that $\varepsilon_{1i}\cdot\zeta_{1i}$ satisfies (A8) for the x_2 and ε_2 under consideration. We continue N times until $\sum_{i=1}^{N} \delta_i = 1$ and all of the mass has been used up.

Step 2: We then continue in the same manner as Case 2 with the "density" portion of f(x). Again we are assured that there will be no imbalance because of (A11).

Thus we have shown that for arbitrary mixed probability densities with mean m we have (A1) true. In our case

$$f(x) = L_0(x)$$

$$m = \bar{L}_0$$

$$g(x) = xe^{-K/\bar{L}_0 x} \qquad x \geq 0$$

This completes the sketch of the proof. The requirement that g(x) be continuous is not really necessary. Greater care could easily remove it.

On the Throughput of an ALOHA Channel with Variable Length Packets

SANDRO BELLINI AND FLAMINIO BORGONOVO

Abstract—**In this paper we give a new simple expression for the probability of successful transmission on an infinite population ALOHA channel with variable length packets. Expressions for the throughput and the probability density of the packet length on the channel are derived in a straightforward way along with the best and the worst length densities.**

I. INTRODUCTION

The pure random access scheme ALOHA was introduced by Abramson [1]. A summary of results on throughput with constant length packets along with an extensive set of references is given in [2]. In the same paper Abramson studied the special case of two different packet lengths. Ferguson [3], [4] gave general results for all possible packet length distributions. He showed that in the steady-state operation the length probability density $b(x)$ of the packets transmitted on the channel, is different from the length probability $a(x)$ of offered packets, i.e., the successfully transmitted ones.

In this paper it is shown that the collision probability does not depend on the packet length distribution, but only on its mean value. This important fact allows $b(x)$ and the throughput to be related to $a(x)$ directly so that the influence of $a(x)$ on capacity and other performance measures can be analyzed in a straightforward way. Thus some of the results already presented in [3], [4] are more concisely derived.

The investigations are performed under the infinite population and the steady-state assumption and are mainly devoted to the unslotted channel case. However, the slotted case can be handled in a similar way.

II. UNSLOTTED CHANNEL

In this analysis we consider the infinite population model as specified by the following assumptions.

A1) The packets are offered for the first transmission with rate λ_a and have a length drawn from the probability density $a(x)$ with mean a and moment generating function $M_a(\lambda) = E[\exp(\lambda x)]$;

A2) The transmissions occur according to a Poisson point process with rate λ_b;

A3) The transmitted packets have lengths drawn from the probability density $b(x)$ with mean b and moment generating function $M_b(\lambda)$;

Paper approved by the Editor for Computer Communication of the IEEE Communications Society for publication without oral presentation. Manuscript received July 16, 1979; revised May 21, 1980.

The authors are with the Space Communication Center, C.N.R., Istituto di Elettrotechnica ed Elettronica, Politecnico di Milano, 20133 Milan, Italy.

A4) When two or more packets interfere, they destroy each other and have to be retransmitted.

With these assumptions the following theorem holds.

Theorem 1: The probability P_s of successful transmission of a packet of length x is given by

$$P_s = \exp(-\lambda_b(b + x)). \tag{1}$$

Proof: A packet of length x is successfully transmitted if

(i) no other transmission starts during x, and

(ii) previously transmitted packets do not interfere with it.

Then

$$P_s = \Pr(i) \Pr(ii).$$

From A2) we have $\Pr(i) = \exp(-\lambda_b x)$. In order to get $\Pr(ii)$ we consider the starting point B of the packet transmission and a time interval $\overline{AB}$ of length y (Fig. 1).

According to A2) if there are N transmission starting points in the interval, then these are uniformly distributed inside it. Let Z be the distance from B to one of these points, say C, and let L be the length of the corresponding packet. Interference due to this packet occurs if $L > Z$.

The probability $P_y(ii)$ that the packets in $\overline{AB}$ do not interfere with B is

$$P_y(ii) = \sum_{n=0}^{\infty} \Pr(N = n)(1 - P_y)^n \tag{2}$$

where

$$P_y = \Pr(L > Z) = \frac{1}{y} \int_0^y \Pr(L > \alpha)\, d\alpha.$$

Since

$$\Pr(N = n) = \frac{(\lambda_b y)^n}{n!} \exp(-\lambda_b y)$$

we get from (2)

$$P_y(ii) = \exp(-\lambda_b y P_y). \tag{3}$$

Now we take the limit as $y \to \infty$.

$$\lim_{y \to \infty} yP_y = \int_0^{\infty} \Pr(L > \alpha)\, d\alpha = b \tag{4}$$

Reprinted from *IEEE Trans. Commun.*, vol. COM-28, no. 11, pp. 1932–1935, Nov. 1980.

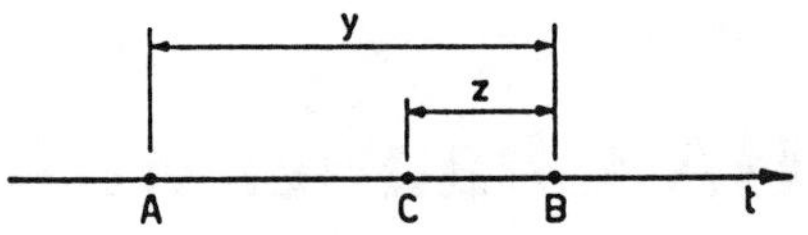

Fig. 1.

so that (3) gives

$$\Pr(ii) = \exp(-\lambda_b b). \tag{5}$$

Therefore (1) holds.

It is noteworthy that (5) depends only on the channel traffic $G = \lambda_b b$. This enables us to state the following.

Theorem 2: The steady-state operation is possible only if $a(x)$ falls off at least as fast as an exponential $\exp(-\mu x)$ and

$$\lambda_b < \mu. \tag{6}$$

The channel packet length density and the throughput are given by

$$b(x) = \frac{1}{M_a(\lambda_b)} a(x) \exp(\lambda_b x) \tag{7}$$

and

$$S = \lambda_a a = \frac{\lambda_b a}{M_a(\lambda_b)} \exp\left(-\lambda_b \frac{M_a'(\lambda_b)}{M_a(\lambda_b)}\right). \tag{8}$$

Proof: The rate at which packets of length x are successfully transmitted is P_s times the rate at which they are transmitted

$$\lambda_a a(x) = \lambda_b b(x) \exp(-\lambda_b(b+x))$$

so that (6) holds, and

$$b(x) = \frac{\lambda_a}{\lambda_b} \exp(\lambda_b b) a(x) \exp(\lambda_b x). \tag{9}$$

Observing that

$$M_b(0) = \frac{\lambda_a}{\lambda_b} \exp(\lambda_b b) M_a(\lambda_b) = 1 \tag{10}$$

we get (7), which gives the mean value

$$b = M_b'(0) = \frac{M_a'(\lambda_b)}{M_a(\lambda_b)}. \tag{11}$$

From (10) and (11) we get (8).

The following theorem gives some properties of the throughput (8) as a function of λ_b or G, respectively.

Theorem 3: The throughput S is a bell-shaped function of λ_b in $(0, \mu)$ (μ possibly infinite). It reaches its maximum value for $\lambda_b = \lambda_{bM} \leqslant 1/2b$. For any value of λ_b the maximum is obtained in the case of constant packet length. Finally, at least for $G \leqslant 0.5$

$$S \leqslant G \exp(-2G) \tag{12}$$

where the equality holds only in the constant length case.

Proof: The following inequalities will be useful:

$$M_a(\lambda) \exp(-\lambda a) \geqslant 1, \qquad \lambda \geqslant 0$$

which follows from

$$\frac{d}{d\lambda} M_a(\lambda) \exp(-\lambda a) = \int_0^\infty (x-a) \exp(-\lambda(x-a)) a(x)\, dx \geqslant 0, \qquad \lambda \geqslant 0$$

and

$$\frac{d}{d\lambda} \frac{M_a^{(n)}(\lambda)}{M_a(\lambda)} = \frac{M_a^{(n+1)}(\lambda) M_a(\lambda) - M_a^{(n)}(\lambda) M_a'(\lambda)}{M_a^2(\lambda)} \geqslant 0, \qquad n = 1, 2, \cdots$$

which can be obtained by the Holder's inequality. In fact

$$E[x^n \exp(\lambda x)] \leqslant (E[x^{n+1} \exp(\lambda x)])^{\frac{n}{n+1}} \cdot (E[\exp(\lambda x)])^{\frac{1}{n+1}}$$

$$E[x \exp(\lambda x)] \leqslant (E[x^{n+1} \exp(\lambda x)])^{\frac{1}{n+1}} \cdot (E[\exp(\lambda x)])^{\frac{n}{n+1}}.$$

In particular we note that $b = M'_a(\lambda_b)/M_a(\lambda_b)$ and $M''_a(\lambda_b)/M_a(\lambda_b)$ are nonincreasing functions of λ_b. Besides $M''_a(\lambda_b)/M_a(\lambda_b) \geqslant b^2$. It is easily seen from (8) that $dS/d\lambda_b$ has the same sign as

$$\begin{aligned}(1-\lambda_b b)^2 &- \lambda_b^2 M_a''(\lambda_b)/M_a(\lambda_b) \\ &\leqslant (1-\lambda_b b)^2 - \lambda_b^2 b^2 \\ &= 1 - 2\lambda_b.\end{aligned}$$

Therefore S has no maxima for $\lambda_b b > 0.5$.

For $0 \leqslant \lambda_b \leqslant 0.5$, $(1-\lambda_b b)^2$ decreases steadily while $\lambda_b^2 M''_a(\lambda_b)/M_a(\lambda_b)$ increases, so that S has only one maximum.

Next we show that the throughput, for a given λ_b, cannot be greater than that obtained with constant packet length

$$S_M(\lambda_b) = \lambda_b a \exp(-2\lambda_b a).$$

In fact, since $b \geqslant a$ and $M_a(\lambda_b) \exp(-\lambda_b a) \geqslant 1$ we have

$$S(\lambda_b) = \frac{\lambda_b a}{M_a(\lambda_b)} \exp(-\lambda_b b) \leqslant S_M(\lambda_b). \tag{13}$$

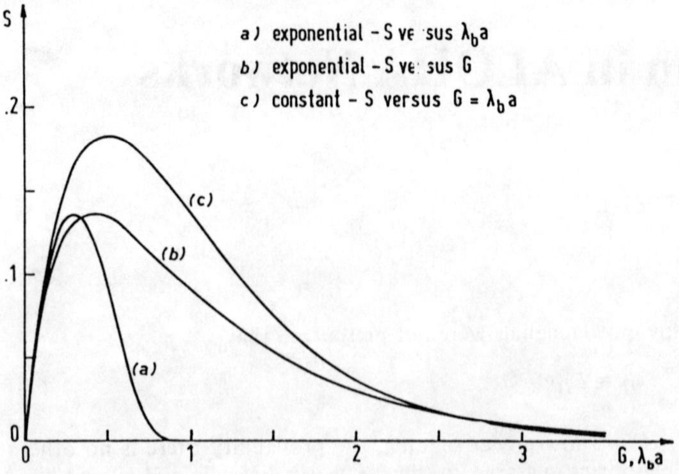

Fig. 2. Throughput with exponential and constant packet lengths.

Finally, for a given λ_b the smallest b, i.e., the smallest G, are obtained in the constant length case. The fact, combined with (13), gives (12), at least for $G \leqslant 0.5$.

We note that at large values of G, S can be greater than $G \exp(-2G)$. As an example plots of S are shown in Fig. 2 for the constant and the exponential length cases.

The worst density $a(x)$, i.e., the one which gives the lowest throughput is given in the following.

Theorem 4: Given a minimum m, a mean a, and a maximum packet size M, the throughput is minimized for any λ_b by the density which concentrates the probability at m and M.

The lowest S for a fixed G is given by the same density, at least for values of G such that $dG/dS \geqslant 0$.

Proof: It is sufficient to show that if there exists an interval (x_1, x_2) with $x_1 > m$ and $x_2 < M$, in which $a(x) \neq 0$, then it is possible to increase both b and $M_a(\lambda_b)$. In fact it can be easily seen that the density

$$a'(x) = A_1 a(x+\Delta) r(x+\Delta) + A_2 a(x-\Delta) r(x-\Delta) + a(x)(1-r(x))$$

where

$$r(x) = \begin{cases} 1, & x_1 \leqslant x \leqslant x_2 \\ 0, & \text{elsewhere} \end{cases}$$

$$0 < \Delta \leqslant \min(x_1 - m, M - x_2)$$

$$A_1 = \frac{\exp(\lambda_b \Delta) - 1}{2Sh\lambda_b\Delta} > \frac{1}{2}$$

$$A_2 = \frac{1 - \exp(-\lambda_b \Delta)}{2Sh\lambda_b\Delta} < \frac{1}{2}$$

gives the same $M_a(\lambda_b)$ as $a(x)$, a lower value of a, and a greater value of b. Now the previous mean value a can be restored by shifting to M a sufficient amount of probability mass. It is straightforward to see that this increases both $M_a(\lambda_b)$ and b. The second part of the theorem can be proved in a way similar to the last part of Theorem 3.

III. SLOTTED CHANNEL

This case differs from the previous one because the time axis is divided in slots of length τ and transmissions can start only at the beginning of a slot.

In the case in which the packet lengths are constrained to be integer multiples of τ, it is simple to show that (7) still holds, while (1) and (8) have to be replaced by

$$P_s = \exp(-\lambda_b(b - \tau + x)) \tag{1'}$$

$$S = \frac{\lambda_b a}{M_a(\lambda_b)} \exp\left(-\lambda_b\left(\frac{M_a'(\lambda_b)}{M_a(\lambda_b)} - \tau\right)\right). \tag{8'}$$

In the case of constant length $x = \tau$, (1') and (8') reduce to the well-known results [2]

$$P_s = \exp(-G)$$

$$S = G \exp(-G).$$

Similar results can be obtained for general densities. We mention only the special case of packet lengths not exceeding τ. The throughput is

$$S = \lambda_b a \exp(-\lambda_b \tau) = G \exp(-\lambda_b \tau).$$

Note that it does not depend on the actual density.

IV. CONCLUSIONS

It has been shown that the probability of successful transmission in an infinite population ALOHA channel depends upon the channel packet length density only through its mean value. This fact allows us to get simple expressions for the steady-state throughput, and to restate other known results in a more precise way. It can also simplify the transmission delay calculations.

REFERENCES

[1] N. Abramson, "The ALOHA system—Another alternative for computer communication," in *Proc. AFIPS 1970 Fall Joint Comput. Conf.*, vol. 37, pp. 281–285.

[2] ——, "The throughput of packet broadcasting channels," *IEEE Trans. Commun.*, vol. COM-25, pp. 117–128, Jan. 1977.

[3] M. J. Ferguson, "A study of unslotted ALOHA with arbitrary message lengths," DATA COMM., P.Q., Canada, Oct. 1975.

[4] ——, "An approximate analysis of delay for fixed and variable length packets in an unslotted ALOHA channel," *IEEE Trans. Commun.*, vol. COM-25, pp. 644–654, July 1977.

On Improving Utilization in ALOHA Networks

JOHN J. METZNER

Abstract—By the simple expedient of dividing users into two groups—one transmitting at high power and the other at low power—the maximum utilization of a slotted ALOHA communication system can be increased from 36.8 percent to about 53 percent. Similar comments apply to the unslotted ALOHA case. An extension to more than two power groups also is described.

There is a fairly simple method of increasing the utilization of ALOHA networks [1], [2] which I have not found mentioned in the literature.[1] In the slotted ALOHA case [2], for example, it is assumed that if two or more users occupy the same time slot, all the message packets in that time slot are lost and must be retransmitted. It has been recognized [3], [4] that more complex procedures could permit decoding of the interfering signals. However, a simpler way of gaining an improvement is to divide the users into two groups as follows. Group I users (the "loud talkers") transmit to the satellite at high power, while group II users (the "quiet talkers") send at a low power. (Power is not a very critical parameter on the earth-to-satellite link.) Whenever exactly one group I user and one or more group II users occupy the same time slot, it is not unreasonable to assume that the group I signal can be decoded, but the other signal(s) cannot. On the other hand, if two or more group I users occupy the same time slot, we assume that all the sendings in the time slot are lost. By a simple extension of prior results, it is shown below that the maximum utilization is increased from $1/e$ or 36.8 percent of the slots to about 53 percent of the slots by this two-group procedure. The same general result also holds for the unslotted ALOHA case, except for a factor of 1/2 reduction in utilization rates.

SLOTTED ALOHA CASE

Let f_H be the average number of high-power message packets per time slot, and let F_H be the average number of high-power packets plus retransmissions. Let f_L, F_L be the corresponding parameters for the low-power transmissions. As in previous analyses [1], [2], assume that there are many users and selection of packets and retransmissions is governed by a Poisson process.

The high-power group statistics are determined as if the low-power signals were not present, so that

$$f_H = F_H e^{-F_H}. \tag{1}$$

For a low-power sending, the probability there is no other low-power message in the same time slot is e^{-F_L}, and the probability there is no high-power message in the slot is e^{-F_H}. Assuming these two events are independent, the probability that a low-power message will have to be retransmitted is $1 - e^{-(F_L+F_H)}$. Then, as in [1, (14.5)] (except that the unslotted case is being treated there),

$$F_L = f_L + F_L(1 - e^{-(F_L+F_H)}). \tag{2}$$

Using (1) in (2),

$$F_L e^{-F_L} = f_L \frac{F_H}{f_H}. \tag{3}$$

Given a particular f_H (and consequent F_H), the maximum f_L (at which saturation occurs for the low-power signals) is at $F_L = 1$, for which we have

$$f_{L\,\max} = \frac{f_H}{F_H} \cdot \left(\frac{1}{e}\right). \tag{4}$$

Fig. 1 shows a plot of $f_{L\max} + f_H$ (the total average utilization) versus f_H. The maximum total average utilization of about 0.53 (exact value $e^{-(1-1/e)}$) occurs at a point where the high-power rate f_H is somewhat greater than the low-power rate f_H.

UNSLOTTED ALOHA

Analogous to [1], let $r_H\tau$ be the channel utilization and $R_H\tau$ the channel traffic in the high-power case. Let $r_L\tau$ and $R_L\tau$ be the corresponding low-power parameters. Then, by arguments similar to [1] and above,

$$r_H\tau = R_H\tau e^{-2R_H\tau}, \tag{5}$$

$$R_L = r_L + R_L(1 - e^{-2(R_L+R_H)\tau}), \tag{6}$$

and

$$R_L\tau e^{-2R_L\tau} = (r_L\tau)\frac{R_H\tau}{r_H\tau}. \tag{7}$$

Given a particular $r_H\tau$, saturation for the low-power case

Paper approved by the Associate Editor for Computer Communication of the IEEE Communications Society for publication without oral presentation. Manuscript received October 15, 1975; revised December 24, 1975. This research was sponsored by the National Science Foundation under Grant ENG 75-07839.

The author is with Wayne State University, Detroit, MI 48202.

[1] After this paper was submitted for publication, some related unpublished work by Abramson, Fralik, Gaarder, and Roberts in informal ARPA Network Information Center notes during 1972–1975 was brought to the author's attention by a referee. Its relation to the present work is discussed in the final paragraph.

Reprinted from *IEEE Trans. Commun.*, vol. COM-24, no. 4, pp. 447–448, April 1976.

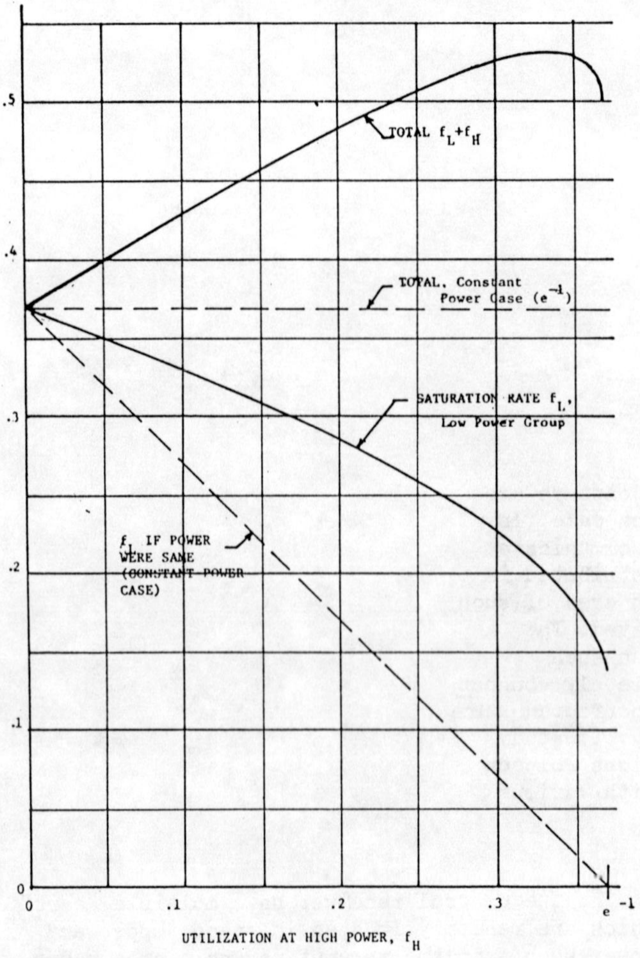

Fig. 1.

occurs when $R_L \tau = 1/2$, for which

$$(2r_L\tau)_{\max} = \left(\frac{2r_H\tau}{2R_H\tau}\right)\left(\frac{1}{e}\right). \tag{8}$$

Note that (8) and (4) are identical except for factor-of-2 scaling.

EXTENSION TO K POWER CLASSES

The following extension was proposed by M. Hellman. Suppose there are K classes of users at K different power levels (numbered $1, 2, \cdots, K$) such that if $i < j$ a class i signal always dominates a class j signal. Consider the slotted ALOHA case, and let F_i, f_i be defined analogous to F_H, f_H. Define

$$\left.\begin{aligned} c_r &= e^{-\sum_{j=1}^{r} F_j} \\ c_0 &= 1 \end{aligned}\right\} 1 \leqslant r \leqslant K. \tag{9}$$

Then, using the dominance property,

$$f_i = F_i c_i, \qquad 1 \leqslant i \leqslant K. \tag{10}$$

Let a superscript asterisk denote the maximum of a quantity. Suppose that the following is true for a given i, $1 < i \leqslant K$: given any fixed $F_1, F_2, \cdots, F_{i-1}$,

$$\left(\sum_{j=i}^{K} f_j\right)^* = [e^{-F_i^*}]\, c_{i-1}. \tag{11}$$

Then, from (9)–(11),

$$f_{i-1} + \left(\sum_{j=i}^{K} f_j\right)^* = (F_{i-1} + e^{-F_i^*})[e^{-F_{i-1}}]\, c_{i-2}. \tag{12}$$

The value of F_{i-1} which maximizes (12) is readily found to be $1 - e^{-F_i^*}$. This result is independent of $F_{i-2}, \cdots, F_2, F_1$. Thus,

$$F_{i-1}^* = 1 - e^{-F_i^*}. \tag{13}$$

Substituting (13) in (12),

$$\left(\sum_{j=i-1}^{K} f_j\right)^* = [e^{-F_{i-1}^*}]\, c_{i-2}. \tag{14}$$

Clearly, $F_K^* = 1$, and from (9) and (10),

$$f_K^* = c_K = e^{-1} c_{K-1} = [e^{-F_K^*}]\, c_{K-1}. \tag{15}$$

Thus, it follows by induction that (11) is true for all i. The total utilization is

$$\sum_{j=1}^{K} f_j^* = e^{-F_1^*}. \tag{16}$$

Because of (13), $F_1^* \to 0$ as $K \to \infty$, and perfect utilization is achieved. The convergence is rather slow, 18 classes being required to reach 90 percent efficiency for example.

This possibility of perfect utilization by having different signal powers also was observed in the aforementioned work of Abramson and others, but from the following different viewpoint. If all senders use equal power, the received power at the satellite would decrease monotonically with distance from the satellite. If the modulation-demodulation process has the property of perfect capture, then there is a heirarchy of users similar to the above assumption.

REFERENCES

[1] N. Abramson, "The Aloha system," in *Computer Communication Networks*, N. Abramson and F. Kuo, Ed. Englewood Cliffs, NJ: Prentice-Hall, 1973, ch. 14.

[2] L. Kleinrock and S. S. Lam, "Packet-switching in a slotted satellite channel," in *1973 Nat. Comput. Conf., AFIPS Conf. Proc.*, vol. 42, AFIPS Press, 1973, pp. 703–710.

[3] A. D. Wyner, "Recent results in the Shannon theory," *IEEE Trans. Inform. Theory*, vol. IT-20, pp. 2–10, Jan. 1974.

[4] A. B. Carleial and M. E. Hellman, "Bistable behavior of ALOHA-type systems," *IEEE Trans. Commun.*, vol. COM-23, pp. 401–409, Apr. 1975.

Some Models of Packet Radio Networks with Capture

L. Fratta
Politecnico di Milano, Italy
D. Sant
Univ. of Hawaii at Manoa, USA

ABSTRACT

Models for two classes of packet radio networks with capture are presented and the network thruput is evaluated in each case. In the first type of networks, mobile stations wish to communicate with a central receiver which has multiple "ears" distributed in space. Upper and lower bounds on the thruput per unit area of such a network with perfect and imperfect capture are derived. The second class of networks considered are those in which each station communicates with other stations. For a completely connected mobile network with a uniform traffic matrix and perfect capture it is shown that the maximum network thruput is unity. To study the effect of imperfect capture, a network with stations Poisson distributed in space is considered and the thruput with various traffic assignments is evaluated.

I. INTRODUCTION

The observation that the phenomenon of capture increases the thruput of an ALOHA channel was first made by Roberts /1/. This observation was further investigated by Metzner /2/, Abramson /3/ and Gaarder /4/. The results obtained were for a single receiver configuration i.e. a channel in which many transmitters desire to communicate with a single central receiver.

In this paper, we investigate how capture affects the thruput behaviour of an ALOHA network i.e. a system which has multiple transmitters and receivers. The configuration we shall study is assumed to have a random topology in which the transmitters and receivers are randomly located in a two dimensional space.

In section II we briefly review how capture has been modelled by incorporating it in a distance metric.

In section III we study an extension of the single receiver ALOHA channel. Specifically we let the central receiver have multiple "ears" which are randomly located in space. Upper and lower bounds on the thruput per unit area under perfect and imperfect capture conditions are obtained.

In section IV we consider a completely connected mobile packet radio network in which each mobile unit communicates with all the others. Only perfect capture is considered here.

The effect of imperfect capture on a network somewhat similar to section IV is considered in section V.

II. THE CAPTURE CONCEPT

Although the concept of capture is simple, it is difficult to define it in an exact and succinct model. FM receivers can correctly receive the strongest of many intefering signals if the ratio of the received power of the strongest to the received power of the next strongest exceeds a minimum value called the capture ratio (Roberts /1/).

Consider two transmitters T_1 and T_2 located at distances r_1 and r_2 from the receiver. Assume that both of them transmit with equal power and that the received power degrades monotonically with the ath power of the distance. In the event of conclict, T_1 will capture the receiver

This research was supported by NSF and NATO under grants NSF-ENG 78-12284 and NATO-1568.

Reprinted with permission from *Proc. 5th International Conference Computer Commun.*, pp. 155–161, Atlanta, Georgia, October 27–30, 1980.

when $10 \log(1/r_1^a)/(1/r_2^a) \geq CR$ i.e. when $r_2 \geq \alpha r_1$; where CR is the capture ratio in dB and $\alpha = 10^{0.1CR/a}$ is called the capture parameter. Throughout this paper we shall assume that a=2 i.e. the received power degrades monotonically with the square of the distance; even though in some practical experiments a higher value of a has been experienced.

The above says that capture can be modelled by a distance measure. Perfect capture is said to occur when the capture ratio CR is 0 dB and the capture parameter α is unity.

The above model implicitly assumes that capture is a deterministic binary event: if the ratio of the received powers exceeds the capture ratio, the stronger signal always captures the receiver. Further, capture is assumed to depend only on the ratio of the received powers and not on the magnitudes of the actual powers received. It has been pointed out (Fralick /5/) that the above is a realistic model of capture in a digital channel with ideal receivers and no noise. The capture ratio in this case is 0 dB (perfect capture) and in the event of conflict, the transmitter nearest to the receiver always succeeds in capturing the receiver. With noise, however, the situation becomes much more complicated. The capture event now becomes probabilistic and depends also on other quantities such as the magnitude of the received signal to noise ratio of each signal, the amount of transmission overlap, etc. If however the minimum received signal to noise ratio for all transmitters is given, then for any given modulation scheme we can calculate the capture ratio so that the strongest signal captures the receiver with very high probability. e.g. For a binary PSK, slotted ALOHA (complete transmission overlap) channel, if the received signal to noise ratio for all transmitters exceeds 15 dB then the capture ratio can be taken as 5 dB. Doing so would mean that capture occurs with a probability of 0.99 (Fralick /5/). For higher values of minimum received signal to noise ratio, the capture ratio should decrease.

In this paper we assume that capture can be modelled by the distance measure with α as the capture parameter. i. e. A receiver will correctly receive a packet from a transmitter located at a distance r from it, if no other transmitter within a distance αr of the receiver is also transmitting.

III. A MULTIPLE EAR ALOHA CHANNEL

It has been shown that (Gaarder /4/) an ALOHA channel with a single central receiver can support only a certain number of transmitters. This number, called the Sisyphus number, depends only on the non-zero thruput of each transmitter and not on the spatial distribution of the transmitters (sources). In order to support more than the Sisyphus number of transmitters (sources) one must have more than one receiver. This can be accomplished by letting the central receiver have multiple "ears" (sinks) which are distributed in space. The thruput of such a network will strongly depend on the relative locations of the sources and sinks. A particular spatial placement of sinks may be optimum for a particular placement of sources (each sink covering exactly the Sisyphus number of sources, no source being covered by more than one sink) but terrible for another. Moreover if the sources are mobile (which is an important case in packet radio networks), the problem of determining the optimum sink placements becomes almost impossible. We therefore choose to model the multiple "ear" receiver/ mobile transmitters network by a random placement of both sources and sinks in space.

Consider a very large area (so that we can ignore edge effects) in which the spatial distribution of sources and sinks are statistically independent two dimensional Poissons with parameters σ and ρ respectively. Assume that the channel is slotted and that p is the probability that each source is active (transmits a packet) in a given slot. The sources communicate with the central receiver via the sinks. A source is said to have a successful transmission if at least one sink receives the transmitted packet correctly. We are not concerned here with how the sinks forward the received packets to the central receiver. We just assume that it is done with no use of common channel.

Let α be the capture parameter and let R_j be the random variable which denotes the distance from a source to its jth nearest sink. It can be shown that (Abramson /6/) the joint density of $(R_1, R_2 \ldots R_j)$ is given by

$$p_{R_1,R_2,\ldots R_j}(r_1,r_2,\ldots r_j) = \begin{cases} (2\pi\rho)^j (r_1,r_2\ldots r_j) e^{-\pi\rho r_j^2}, & r_j > r_{j-1} \ldots > r_1 > 0 \\ 0, \text{ otherwise} \end{cases} \quad (1)$$

and

$$p_{R_j}(r) = \frac{2(\pi\rho r^2)^j}{r(j-1)!} e^{-\pi\rho r^2}, \quad r>0 \quad (2)$$

Probability of No Interference

Let A_j denote the event that a packet transmitted by a source is correctly heard by its jth nearest sink. Then $P(A_j | R_j = r)$ is the probability that no other source within a distance

αr of the sink is also active. The number of active sources in an area Z is a Poisson random variable with parameter $p\sigma Z$. Thus we have

$$P(A_j|R_j=r) = e^{-p\sigma\pi\alpha^2 r^2} \quad (3)$$

Now

$$P(A_j) = \int_0^\infty P(A_j|R_j=r)p_{R_j}(r)dr \quad (4)$$

which, by using (2), becomes

$$P(A_j) = \frac{1}{(1+\alpha^2 p\gamma)^j} \quad (5)$$

where $\gamma=\sigma/\rho$ is the ratio of the spatial densities of sources to sinks.

Let A denote the event that a packet transmitted by a source is correctly received by at least one sink. Writing A as the sum of mutually ex - clusive events, we have

$$P(A) = P(A_1)+P(\bar{A}_1A_2) + P(\bar{A}_1\bar{A}_2A_3) + \ldots \quad (6)$$

A simple upper bound to P(A) is the union bound which is given by

$$P(A)\underline{<}\min\left\{1, \sum_{j=1}^{\infty} P(A_j)\right\} \quad (7)$$

Taking the lower bound to P(A) as the probability of being heard by the nearest sink (the first term in (6)) we get grom (5) and (7)

$$\frac{1}{1+\alpha^2 p\gamma} \leq P(A) \leq \min\left\{1, \frac{1}{\alpha^2 p\gamma}\right\} \quad (8)$$

In the case of perfect capture (α=1) a tighter upper bound can be obtained as follows. Consider

$$P(\overline{A_1}\,\overline{A_2}\ldots\overline{A_{j-1}}A_j)=$$
$$= P(A_j|\overline{A_1}\,\overline{A_2}\ldots\overline{A_{j-1}})P(\overline{A_{j-1}}|\overline{A_1}\ldots\overline{A_{j-2}})..P(\overline{A_2}|\overline{A_1})P(\overline{A_1})$$
$$\underline{<}P(A_j|\overline{A_1}\,\overline{A_2}\ldots\overline{A_{j-1}})P(\overline{A_1}) \quad (9)$$

Now $P(A_j|\overline{A_1}\,\overline{A_2}\ldots\overline{A_{j-1}})$ will be maximum when the circle corresponding to the jth nearest sink has no area in common with any of the circles corresponding to the ith nearest sink, i<j, a situation depicted in Figure 1.

Clearly in this case because of the disjoint areas,

$$P(A_j|\overline{A_1}\,\overline{A_2}\ldots\overline{A_{j-1}})P(\overline{A_1})=P(A_j)P(\overline{A_1})=P(A_j\overline{A_1}) \quad (10)$$

Now

$$P(A_j\overline{A_1}|R_1=r_1,R_2=r_2,\ldots R_j=r_j) = = (1-e^{-p\sigma\pi r_1^2})e^{-p\sigma\pi r_j^2} \quad (11)$$

and

$$P(A_j\overline{A_1})= \int\int\ldots\int_{0<r_1<r_2\ldots<r_j} P(A_j\overline{A_1}|R_1=r_1,\ldots R_j=r_j) \quad (12)$$
$$p_{R_1,\ldots R_j}(r_1,\ldots r_j)dr_1dr_2\ldots dr_j$$

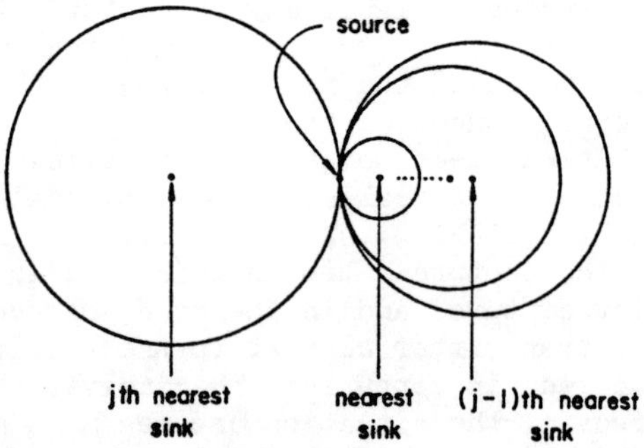

Fig. 1

which by using (1) becomes

$$P(A_j\overline{A_1}) = \frac{p\gamma}{(1+p\gamma)^j(1+2p\gamma)} \quad (13)$$

Thus from (9), (10) and (13) we have

$$P(A_j\overline{A_1}\,\overline{A_2}\ldots\overline{A_{j-1}}) \leq \frac{p\gamma}{(1+p\gamma)^j(1+2p\gamma)} \quad (14)$$

and from (6) we get

$$P(A) \leq \min\left\{1, \frac{2}{1+2p\gamma}\right\} \quad (15)$$

<u>Thruput</u>

Knowing the probability of being correctly heard by at least one sink, P(A), the thruput per source, s, can be written as

$$s = pP(A) \quad (16)$$

which from (8) and (15) can be bounded by

$$\frac{p}{1+\alpha^2 p\gamma} \underline{<} s \underline{<} \begin{cases} p\min\{1, \frac{2}{1+2p\gamma}\} & , \alpha = 1 \\ p\min\{1, \frac{1}{\alpha^2 p\gamma}\} & , \alpha > 1 \end{cases} \quad (17)$$

Now, the sources are Poisson distributed in space with parameter σ. Bounds on the average thruput per unit area, $S_u=\sigma s$ in terms of the

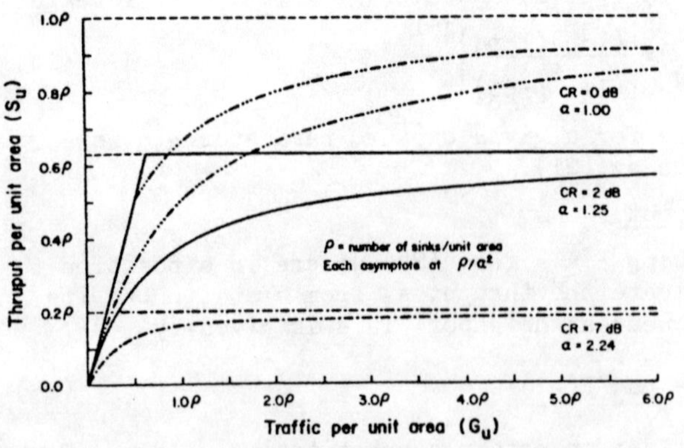

Fig. 2- Upper and lower bounds to the thruput/ area vs. traffic/area.

average traffic per unit area, $G_u = \sigma p$, will thus be given by

$$\frac{G_u}{1+\alpha^2 G_u/\rho} \le S_u \le \begin{cases} G_u \min\{1, \frac{2}{1+2G_u/\rho}\}, & \alpha = 1 \\ G_u \min\{1, \frac{1}{\alpha^2 G_u/\rho}\}, & \alpha > 1 \end{cases} \quad (18)$$

These bounds are illustrated in Figure 2 for various values of capture ratio. From (18) it can be seen that the bounds on the average thruput per unit area depend only on the average traffic per unit area and the average number of sinks per unit area. From Figure 2 we see that the capacity (maximum thruput) per unit area of our random multiple "ear" network with capture parameter α and an average of ρ "ears" per unit area equals ρ/α^2.

IV. A FULLY CONNECTED MIBILE NETWORK

In this section we investigate the thruput of a mobile packet radio network in which each mobile station wishes to communicate with all the others. Each station has both transmit and receive capabilities and further all stations transmit and receive on the same frequency. The network covers an arbitrary region in space; the only requirement being that each station remains within line of sight of all the others at all times. We assume that perfect capture occurs.

Let there be n stations in such a network. Let the channel be slotted and let the stations transmit independently of one another with p denoting the probability of a station being active in a slot. We assume a uniform traffic matrix i.e. an active station chooses to address any one of the others with probability 1/(n-1). We define a useful packet to be a packet that is correctly received by its destination.

We now evaluate the probability s_j that a station, say station j, will receive a useful packet in a slot. This event will occur if and only if the following three events occur.

i) Station j is silent (else it will receive its own transmission)
ii) At least one of the other (n-1) stations is active
iii) The active station nearest to station j chooses to address it (because perfect capture implies that a station always correctly receives only the packet transmitted by its nearest active station).

We thus have

$$s_j = \frac{(1-p)}{n-1}\left[1-(1-p)^{n-1}\right] \quad (19)$$

Clearly s_j is the thruput received by station j. Because the stations are mobile and hence identical in all respects, the average thruput of the network of n stations S_n, is given by

$$S_n = n s_j = \frac{n}{n-1}(1-p)\left[1-(1-p)^{n-1}\right] \quad (20)$$

The above is maximum when $p = 1-(\frac{1}{n})^{\frac{1}{n-1}}$ and $S_{n,max} = (\frac{1}{n})^{\frac{1}{n-1}}$. Also $\lim_{n\uparrow\infty} S_{n,max} = 1$.

If there is no capture in the channel then the thruput is given by the well known slotted ALOHA thruput equation (Abramson /3/) $S_n = np(1-p)^{n-1}$ the maximum value of which is $S_{n,max} = (1-1/n)^{n-1}$. The improvement in the network thruput obtained due to capture is shown in Figure 3. We see that as the number of stations increases, the channel bit rate is utilized completely in spite of the

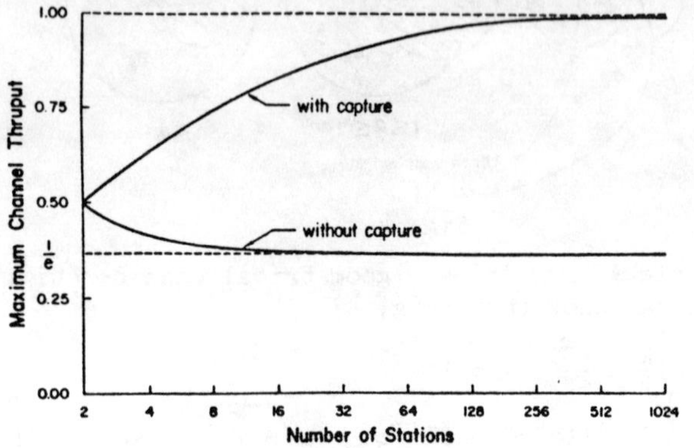

Fig. 3- Thruput performance with and without capture.

fact that stations randomly access the channel and address any other station in the network with the same probability.

V. A PARTIALLY CONNECTED NETWORK

In this section we consider a class of networks in which each station whishes to send packets to only some specific other stations. Specifically we consider an infinite plane (so that we can ignore edge effects) in which stations are randomly located according to a two dimensional Poisson distribution with parameter σ. Each station has both transmit and receive capabilities and all stations transmit and receive on the same frequency. The channel is slotted and as before p denotes the probability that a station is active in a slot. Further an active station chooses to address its jth nearest neighbor with probability a_j. Clearly we must have $\sum_j a_j=1$. Note that the above represents a non-symmetric traffic situation. In fact, station X may be the jth nearest neighbor of station Y but the reverse may not be true.

Probability of No Interference

Let A_j denote the event that a packet transmitted by a station is correctly received by its jth nearest neighbor. We wish to evaluate $P(A_j|R_j=r)$ i.e. the probability of a successful transmission to the jth nearest neighbor given that he is a distance r away. This is given by the probability that no other source within a distance αr of the destination is active, where α is the capture parameter. Referring to Figure 4, we see that this is the probability that the interior of circle B contains no active stations given that the interior of circle A has exactly (j-1) stations (excluding the source). The number of stations in B_1 is no longer Poisson distributed because we know that circle A contains precisely (j-1) stations. From the Poisson hypothesis it turns out that each of the (j-1) stations is uniformly distributed with area in circle A. Let q denote the ratio of the area B_1 to the area of

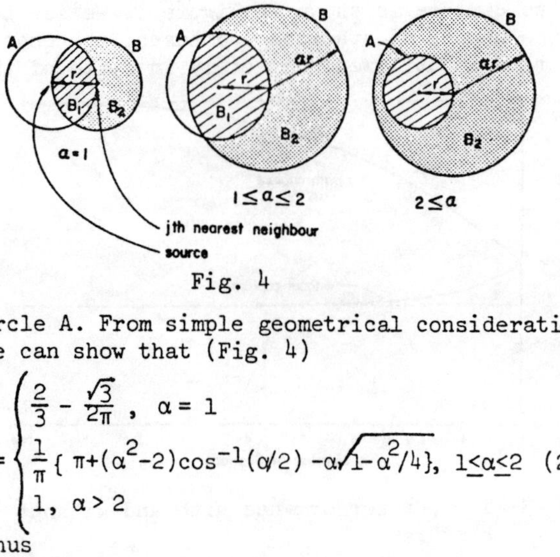

Fig. 4

circle A. From simple geometrical considerations one can show that (Fig. 4)

$$q = \begin{cases} \frac{2}{3} - \frac{\sqrt{3}}{2\pi}, & \alpha = 1 \\ \frac{1}{\pi}\{ \pi+(\alpha^2-2)\cos^{-1}(\alpha/2) -\alpha\sqrt{1-\alpha^2/4}\}, & 1\leq\alpha\leq 2 \\ 1, & \alpha > 2 \end{cases} \quad (21)$$

Thus

$$P(A_j|R_j=r)= (1-p)e^{-p\sigma B_2}\sum_{i=0}^{j-1}\binom{j-1}{i}q^i\bar{q}^{j-1-i}(1-p)^i \quad (22)$$

where

$\bar{q} = 1-q$.

The above simplifies to

$$P(A_j|R_j=r) = (1-p)e^{-p\sigma B_2}(1-pq)^{j-1} \quad (23)$$

By substituting $\pi r^2(\alpha^2-q)$ for the area B_2, using (2) for $P_{R_j}(r)$ and averaging (23) over r we get

$$P(A_j)= \frac{(1-p)(1-pq)^{j-1}}{(1+\alpha^2 p-pq)^j} \quad (24)$$

where for a given capture parameter α, q is given by (21).

Thruput

Knowing $P(A_j)$ from (24), we are in a position to evaluate the thruput s_j from a station to its jth nearest neighbor. This is given by

$$s_j = pa_j\, P(A_j) \quad (25)$$

In order to obtain a quantitative measure of the thruput in the network, we need to specify the probabilities a_j. From (24) and (25), the total thruput $S=\sum s_j$ is maximum when $a_1= 1$ and $a_j=0$ for $j\geq 2$ i.e. the maximum thruput is obtained in the trivial case when each station transmits to only its nearest neighbor. We do not expect this in practice, however we can expect that a station would send out more packets to its nearer stations than those far away. We therefore let the a_j's be geometric with parameter μ i.e. $a_j=(1-\mu)^{j-1}\mu$. In that case by using (24), (25) we can show that the total thruput per station is given by

$$S = \frac{\mu p(1-p)}{\mu+\alpha^2 p-\mu pq} \quad (26)$$

where the q for any given capture parameter α is given by (21). The maximum values of S(over p) for some values of α have been evaluated numerically and are shown in Figure 5 as functions of μ.

The average total thruput per station given by (26) is a global measure of the useful packets transmitted by a station with no regard to the individual s_j's. Now the probabilities $P(A_j)$ rapidly decrease with j and thus most of the

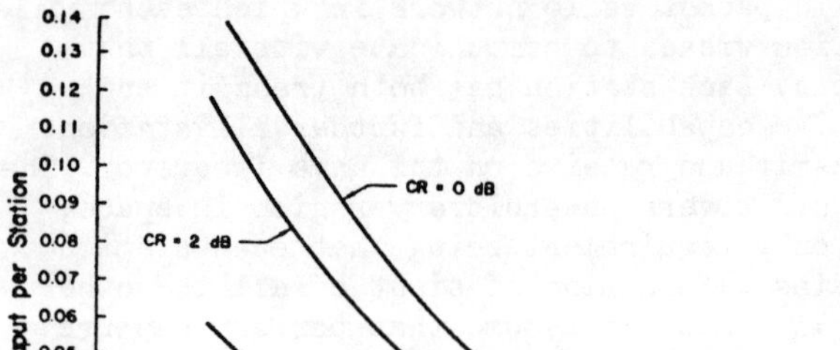

Fig. 5-Maximum thruput per station vs. 1/μ for three values of capture ratio.

thruput S is delivered to the very near stations while very little, if not zero, thruput is delivered to the far away stations. This brings out the difference in concept between s_j and a_j. In fact, a_j is a measure of the desirability of sending packets to the jth nearest neighbor while s_j is a measure of the satisfaction of this desire. Hence instead of the sum thruput $S=\sum s_j$ it may be more meaningful to study the thruput vector $\underline{S}=(s_1,s_2,\ldots s_j,\ldots)$. Due to the limited channel capacity, not all the vectors $\underline{S}$ are achievable i.e. in the $\underline{S}$ space there will be a finite region of feasible thruput vectors $\underline{S}$. We do not develop here any general treatment of feasibility but rather limit ourselves to the particular case where the thruput vector $\underline{S}$ has N equal components i.e. $s_1=s_2\ldots=s_N=s$ and $s_{N+j}=0$ for $j>0$. This case corresponds to a situation in which each station has an equal thruput s to all its N nearest neighbors. We would like to determine the maximum possible s for a given N and α. This can be done as follows From (24),(25) we see that

$$s = \frac{pa_j(1-p)(1-pq)^{j-1}}{(1+\alpha^2 p-pq)^j}, \quad j=1,2,\ldots N \tag{27}$$

By using the fact $\sum_{j=1}^{N} a_j=1$, we get

$$s = \frac{p(1-p)\left[1-\frac{1+\alpha^2 p-pq}{1-pq}\right]}{(1+\alpha^2 p-pq)\left[1-\left(\frac{1+\alpha^2 p-pq}{1-pq}\right)^N\right]} \tag{28}$$

The maximum value of s for several capture parameters α has been determined numerically and is depicted in Figure 6, where we have plotted Ns (rather than s) versus N. Taking N as a crude measure of connectivity of the network, we see that Figure 6 shows the thruput per station (Ns) vs. connectivity behaviour of the network. For N=10, the behaviour of the total thruput per station (Ns) versus the capture parameter α is shown in Figure 7.

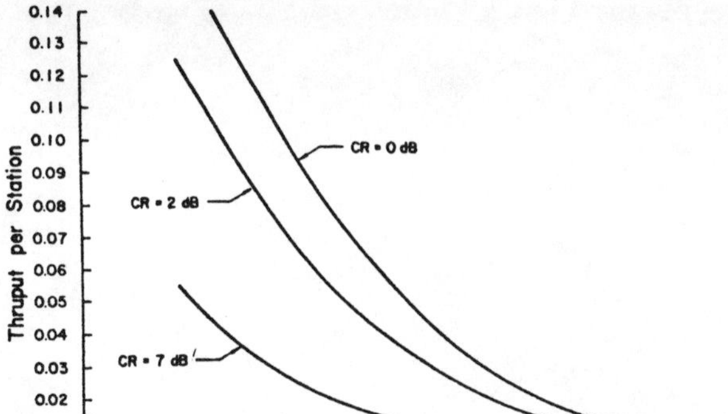

Fig. 6-Maximum Thruput per station vs. N for three values of capture ratio.

VI. CONCLUSIONS

We have presented and studied models for two classes of packet radio networks with capture.

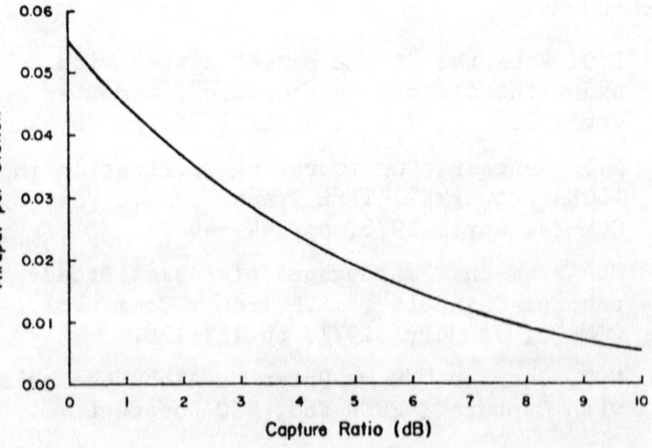

Fig. 7-Maximum thruput per station vs. capture ratio.

In the first one, stations desire to communicate with a central receiver which has multiple "ears" geographically scattered over an area. Upper and lower bounds on the thruput per unit area of such a network with perfect and imperfect capture were derived. The capacity (maximum thruput) of such a network was shown to be a ratio of the average number of "ears" per unit area and the square of the network's capture parameter.

The second class of networks considered were those in which stations wish to communicate with other stations. For a completely connected mobile network with a uniform traffic matrix and perfect capture, it was shown that the maximum achievable thruput is unity. To study the effect of imperfect capture we need to assign some topology to the network. We have considered the case where stations are Poisson distributed in space. The network thruput with various traffic assignments is evaluated. Rather than the traffic vector specification, we have considered here the more meaningful thruput vector specification. This guarantees an effective thruput to the specified destinations.

The analysis performed in this paper pertains only to slotted networks. By using renewal theory arguments /7/, it can easily be extended to unslotted networks. The only change in the results will be a scale factor of half in the thruput.

In section V we raised the question of network thruput versus connectivity. This topic is still under investigation by the authors with particular attention to its implications on network protocols.

ACKNOWLEDGEMENTS

The authors are grateful to N. Abramson for his suggestions.

REFERENCES

/1/ L.G. Roberts, "ALOHA packet system with and without slots and capture", Computer Comm. Review, Vol.5, April 1975, pp.28-42.

/2/ J.J. Metzner, "On improving utilization in ALOHA networks", IEEE Trans. Comm., Vol. COM-24, April 1976, pp. 447-448.

/3/ N. Abramson, "Throughput of Packet Broadcasting Channels", IEEE Trans. Comm. Vol. COM-25, January 1977, pp.117-128.

/4/ N.T. Gaarder, "Some Notes on Aloha Channels with Capture", PRTN #60, NIC Document #16753, Stanford Research Inst., Menlo Park, California, June 1, 1973.

/5/ S. Fralick, "Is Prometheus Really Bound", PRTN 66, NIC Document#17366, Stanford Res. Inst. Menlo Park, California, July 19, 1973.

/6/ N. Abramson, "The Statistics of Neighboring Points", ALOHA SYSTEM Technical Report #B79-3.

/7/ D. Sant, "Thruput of Unslotted Aloha Channels with Arbitrary Packet Interarrival Time Distributions", ALOHA SYSTEM Technical Report#B79-2.

Application of Multiuser Coding to Random Access Protocols

GEORGE THOMAS

Abstract—**We examine the usefulness of multiuser coding for throughput enhancement in various random access protocols with (0, 1, *e*) feedback. A scheme is proposed where the user population is divided into two or more classes and multiuser coding is used in such a manner that all collisions among at most one user per class are resolved by virtue of the multiuser code. For ALOHA, tree, and interval-splitting random access protocols, we demonstrate significant improvements in throughput when multiuser coding techniques are employed.**

I. Slotted ALOHA Throughput Improvement Due to Multiuser Coding

Assume that a finite population of N users of a slotted ALOHA channel are divided into two classes 1 and 2, such that there are m users in class 1 and n users in class 2, $m + n = N$. Let (C_1, C_2) be a pair of two-user codes. All users in the ith class collectively use code C_i for their transmissions, $i = 1, 2$. Ternary (0, 1, e) feedback is assumed to be available at the end of each slot. A collision does not occur in this case if at most one user transmits simultaneously from each class. Following [1] for notations, we can therefore describe the throughputs from the two classes as

$$S_1 = mG_1(1+G_1)^{m-1}[(1-G_2)^n + nG_2(1-G_2)^{n-1}] \quad (1)$$

$$S_2 = nG_2(1-G_2)^{n-1}[(1-G_1)^m + mG_1(1-G_1)^{m-1}]. \quad (2)$$

To determine the boundary of the set of feasible throughput pairs, we equate the Jacobian determinant $(\delta S_i/\delta G_j)$ to zero. This leads [5] to a quadratic constraint on mG_1 and nG_2 in (1) and (2), in contrast with the well-known linear constraint ([1], (27)) for the case without coding. It can also be verified [5] that for modest values of m and n, this quadratic constraint quickly converges to the positive quadrant of the unit circle (Fig. 1).

In Fig. 1, the sets A and B represent our results for two-class coding whereas C and D represent the uncoded case [1]. The linear traffic curves C and the slightly convex-down throughput curves D for the uncoded cases are seen to be transformed to sharply convex-up curves A and B with coding, with obvious improvement in throughput. As seen from Fig. 2, the total throughput can be as high as 0.587 (coded) packets per slot. Also, in contrast with the uncoded case, this maximal total throughput is attained when both classes share the channel equally. Thus, fairness in access and maximization of channel utilization are not mutually conflicting objectives in the coded case.

In the case of more than two classes, a similar analysis [5] leads to throughput formulas analogous to (1) and (2) with the class traffic variables constrained to lie on the positive "orthant" of the unit hypersphere.

II. Effective Throughputs with Multiuser Codes

Throughput calculations in the preceding section need to be discounted for the loss of rate due to coding. For this, we recall some facts from the coding theory for the noiseless, binary input, $(T + 1)$-ary output, linear multiple access channel for T independent users without feedback or interuser cooperation, abbreviated as the binary additive multiple access channel (BAMAC). The basic BAMAC model and its capacity region are well known [2]. The special case of the two-user BAMAC has been studied extensively (see [4]), and its capacity region is known to be the set of rate points (R_1, R_2) bounded by $0 < R_1 < 1$, $0 < R_2 < 1$, and $R_1 + R_2 < 1.5$. See [2, Fig. 4] for the capacity region of the three-user BAMAC.

In two-class throughput curves such as in Fig. 1 (marked B), we need to scale down the axes by the two-user code rates R_1 and R_2. For each rate point (R_1, R_2) on the outer boundary $(R_1 + R_2 < 1.5)$ of the two-user BAMAC capacity region, we can move a point (S_1, S_2) along the limiting throughput curve $(m = n = \infty)$ in Fig. 1 (marked B), and plot R_2S_2 versus R_1S_1. Such a family of throughput curves and their outer envelope (marked A) are seen in Fig. 3. It is this outer envelope A which represents the effective throughput and which should really be compared to the corresponding uncoded case [1] represented by curve B in Fig. 3. The outer envelope A in Fig. 3 can be interpreted as the boundary of the true capacity region of two-class coded ALOHA because we have used rate points from the outer boundary of the BAMAC capacity region. This capacity region is seen to be nonconvex, a feature recently encountered in other random access studies also, e.g., [3], [11], and [14] therein.

To determine the maximum effective throughput for two-class coded ALOHA we define, for any coded throughput pair (S_1, S_2) and for any two-user code rate pair (R_1, R_2) within the capacity region, the effective throughput (after discounting for the loss of rate due to coding) as the inner product of (S_1, S_2) and (R_1, R_2), i.e.,

$$S_{\text{eff}} = S_1R_1 + S_2R_2. \quad (3)$$

For each fixed point (S_1, S_2), (3) is a linear function in the rate variables R_1 and R_2. Maximization of (3) under the linear constraints which define the BAMAC capacity region now follows the well-known principles of linear programming. In particular, for each fixed (S_1, S_2) on the outer boundary of the feasible coded throughput region in Fig. 4, (3) attains its maximum value at one of the "corner points" (0.5, 1) or (1, 0.5) depending on whether $S_1 < S_2$ or $S_1 > S_2$. Since the

Paper approved by the Editor for Random Access Systems of the IEEE Communications Society. Manuscript received May 12, 1985; revised January 15, 1988. This paper was presented at the Indo-US Workshop on Systems and Signal Processing, Bangalore, India, January 1988.

The author is with INSAT Programme Office, ISRO Headquarters, Bangalore, India, 560009.

IEEE Log Number 8822466.

Reprinted from *IEEE Trans. Commun.*, vol. 36, no. 8, pp. 983–986, August 1988.

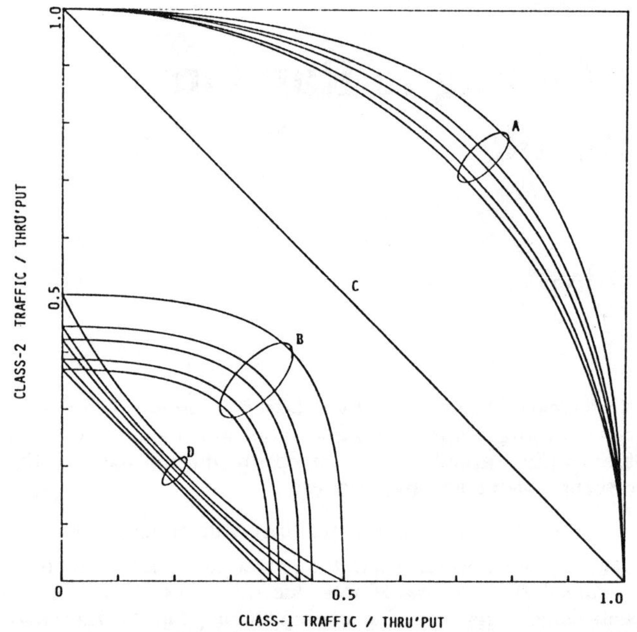

Fig. 1. Throughput and traffic in two-class ALOHA. (Each set of five curves from top down is for two equal classes of 2, 3, 4, 10, and infinite number of users per class.)

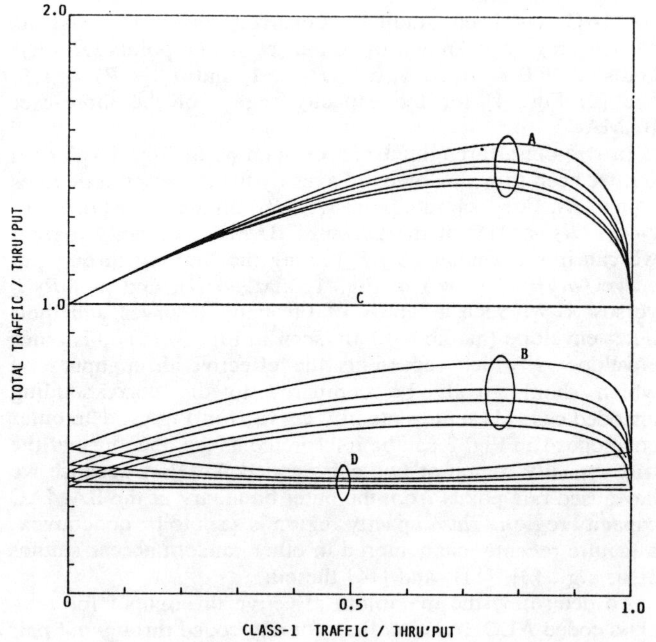

Fig. 2. Total throughput and traffic in two-class ALOHA. (Each set of five curves from top down is for two equal classes of 2, 3, 4, 10, and infinite number of users per class.)

maximizing vector is thus fixed, the inner product (3) is now proportional to the projection of the coded throughput vector (S_1, S_2) onto this fixed rate vector. The geometry in Fig. 4 indicates that the vector (S_1, S_2) which has the largest projection onto the rate vector (0.5, 1) is identified by the point on the outer boundary of the feasible (S_1, S_2) region where the tangent to this boundary curve is normal to the rate vector (0.5, 1); and similarly for the rate vector (1, 0.5) also. We can determine numerically [5] that the maximum two-class coded effective throughput for ALOHA is 0.4505, a significant improvement over the uncoded maximum of 0.367.

In [4, Table III], we have a list of the best known

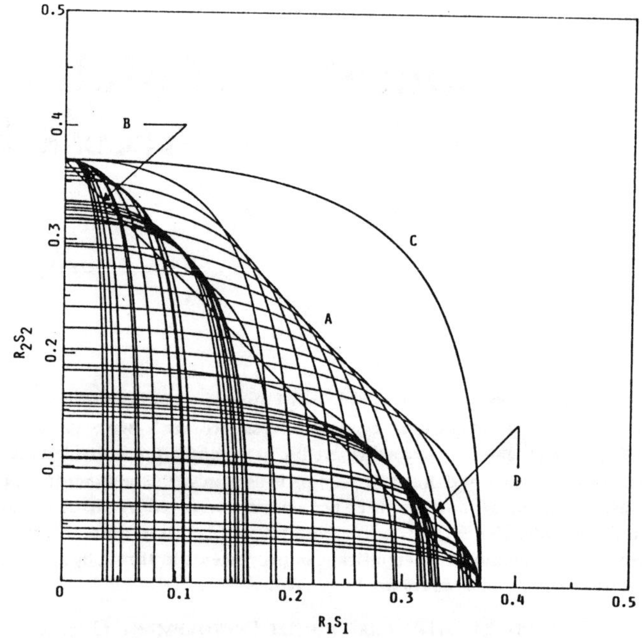

Fig. 3. Capacity region for two-class coded ALOHA.

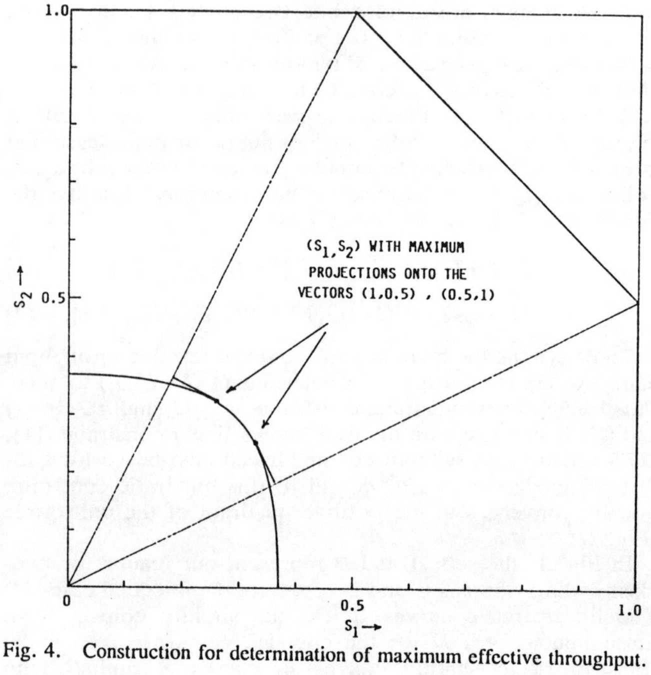

Fig. 4. Construction for determination of maximum effective throughput.

constructive code rate pairs for the two-user BAMAC. In Fig. 3, the curve D has been obtained as the outer envelope of the family od plots of (3) for these practically attainable code rate pairs. Fig. 3 shows that even for these rate points well inside the capacity region, the performance is significantly superior to that of uncoded ALOHA.

Given two two-user codes with rate pairs (u_1, u_2) and (v_1, v_2), respectively, when can we assert that one is superior to the other (with respect to maximum attainable effective throughput in two-class called ALOHA?) In [5], we have shown that a sufficient condition for a code rate pair to be superior to another is that both its sum rate and its maximal component rate be larger than the respective parameters of the other code.

Further, along the straight line joining two rate points (R_1, R_2) and (R_2, R_1), every interior point has a smaller maximal

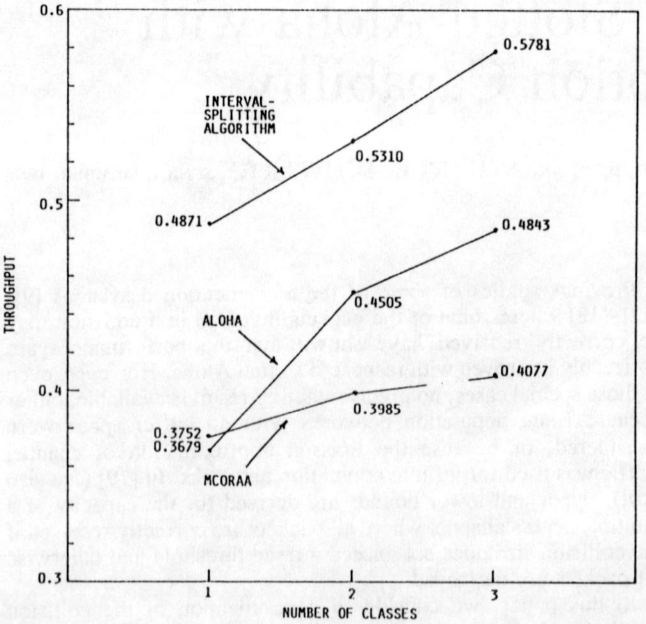

Fig. 5. Throughput improvement in multiclass coded random access protocols.

rate component than the two end points of the line. Consequently, none of these interior points can attain a higher effective throughput (3) than that attained by the end points. Thus, we conclude that in multiclass coded ALOHA, time-sharing of the codes does not accrue any advantage.

Using the same general approach as above but using the three-user BAMAC [2] instead, and with a lot more computational effort, we have determined [5] that the highest attainable effective ALOHA throughput in this case is 0.4843. Computational tedium has precluded extension of the numerical search to the case of four or more classes.

III. Collision Resolution Protocols with Multiuser Coding

We have also investigated the multiuser coding improvements in throughput in the cases of Capetanakis' tree protocols and Gallager's interval-halving algorithm, following essentially the same approach as above. The details are in [5] and the results, for two and three classes, are in Fig. 5.

IV. Conclusion

We have shown that multiuser coding affords improvements in throughputs in two- and three-class ALOHA as well as in tree- and interval-splitting random access algorithms. ALOHA and tree algorithms appear to stand little further to gain by using four or more classes, whereas any trend to such saturation in interval-splitting protocols is not apparent up to three classes. The limit to which the throughput would converge for larger number of classes is an open problem.

References

[1] N. Abramson, "The throughput of packet broadcasting channels," *IEEE Trans. Commun.*, vol. COM-25, pp. 117–126, Jan. 1977.

[2] S. C. Chang and E. J. Weldon, Jr., "Coding for *T*-user multiple access channels," *IEEE Trans. Inform. Theory*, vol. IT-25, pp. 684–689, 1979.

[3] R. G. Gallager, "A perspective on multiaccess channels," *IEEE Trans. Inform. Theory*, vol. IT-31, pp. 124–142, 1985.

[4] P.A.B.M.C. van der Braak and H.C.A. van Tilborg, "A family of good uniquely decodable code pairs for the two-access binary adder channel," *IEEE Trans. Inform. Theory*, vol. IT 31, pp. 3 9, 1985.

[5] G. Thomas, "Application of multi-user coding to random access protocols," SatCom Programmes Office, ISRO HQ, Bangalore, India, Res. Rep., July 1987.

Stability Properties of Slotted Aloha with Multipacket Reception Capability

SYLVIE GHEZ, STUDENT MEMBER, IEEE, SERGIO VERDÚ, MEMBER, IEEE, AND STUART C. SCHWARTZ, SENIOR MEMBER, IEEE

Abstract—**The stability of the Aloha random access algorithm in an infinite-user slotted channel with multipacket reception capability is considered. This channel is a generalization of the usual collision channel, in that it allows the correct reception of one or more packets involved in a collision. The number of successfully received packets in each slot is modeled as a random variable which depends exclusively on the number of simultaneous attempted transmissions. This general model includes as special cases channels with capture, noise, and code division multiplexing. It is shown by means of drift analysis that the channel backlog Markov chain is ergodic if the packet arrival rate is less than the expected number of packets successfully received in a collision of n as n goes to infinity. Finally, the properties of the backlog in the nonergodicity region are examined.**

I. Introduction

One of the main problems in random access communications is the determination of the maximum stable throughput. In particular, an important result is that the Aloha protocol is unstable [1]–[3] in an infinite-user slotted collision channel where a transmission is successful only if no other users attempt transmissions simultaneously. Several strategies have been designed to stabilize this channel, such as collision resolution algorithms (see [4], for example) where transmissions are deferred until the current conflict is solved, and more recently, Aloha-type strategies using decentralized control, where the retransmission probability is updated according to previous channel outcomes. It has been shown [5]–[7] that the maximum stable throughput achievable by such Aloha-type strategies with decentralized control is e^{-1}.

However, the collision channel model does not hold in many important practical multiuser communication systems [8]–[21] because simultaneous transmission of several packets does not necessarily result in the destruction of all the transmitted information. For instance, the capture phenomenon is common in local area radio networks [12]–[15]; if the power of one of the received packets is sufficiently large compared to the power of the other packets involved in a collision, then the strongest packet can be correctly decoded, while the other packets are lost. Other examples are multiple-access channels where several users transmit simultaneously in the same frequency band, and a multiuser detector demodulates the information transmitted by all active users (e.g., [8]–[11]). Although those systems do not necessarily require a random access protocol, it is sometimes useful to exercise some flow control through such a protocol so as to limit the maximum number of simultaneous transmitters, in order to bound the multiuser receiver complexity and guarantee lower bit-error rates.

Manuscript received July 23, 1987; revised January 8, 1988. Paper recommended by Past Associate Editor A. Ephremides. This work was supported in part by the Office of Naval Research under Contract N00014-87-k-0054 and by the Army Research Office under Contract DAAL03-87-k-0062.

The authors are with the Department of Electrical Engineering, Princeton University, Princeton, NJ 08544.

IEEE Log Number 8821359.

Previous studies of some of the aforementioned systems [9], [12]–[18] where some of the packets involved in a collision may be correctly received have shown that the performances are noticeably improved with respect to slotted Aloha. However, even in those special cases, no precise stability result is available, either because finite population networks with no buffer space were considered, or because the Poisson approximation of channel traffic was used for infinite population networks. In [19] (see also [20]), upper and lower bounds are derived for the capacity of a multiple access channel where all packets are correctly received if the collision size does not exceed a fixed threshold and otherwise all packets are destroyed.

In this paper, we consider a generalization of the collision channel, where the receiver can demodulate several packets simultaneously. It is assumed that the number of correctly demodulated packets is a random variable, which, given the number of packets simultaneously transmitted, is independent of the backlog and of the number of previous retransmission attempts. This random variable can take any integer value between zero and the collision size. Thus, the channel is described by a matrix of conditional probabilities (ϵ_{nk}) where ϵ_{nk} is the probability that k packets are correctly demodulated given that there were n simultaneous transmissions. We analyze the usual Aloha algorithm with the multipacket reception capability just described. Users are synchronized so that transmissions take place within one slot, and at the end of each slot, stations that did transmit a packet learn whether or not their transmission was successful. Unsuccessful or backlogged packets are retransmitted in each subsequent slot with probability p, $0 < p \leq 1$. It turns out that multipacket reception capability can stabilize Aloha. Our main result states that the maximum stable throughput is equal to the limit of the average number of packets correctly received in collisions of size n when n goes to infinity. To show this, we model the channel backlog as a Markov chain, and then study its properties by using some simple drift analysis techniques.

The last part of this paper is a study of the properties of the backlog in the nonergodicity region. Unlike the backlog Markov chain for slotted Aloha which is always transient [1], the backlog for our model does in general have a null recurrence region of positive length, which depends on the matrix (ϵ_{nk}) and on the retransmission probability p. However, transience in the nonergodicity region can be ensured for a large class of systems, and in particular for channels where the number of successful simultaneous transmissions is bounded.

II. Multipacket Reception Model

Let A_k be the number of new packets arriving during time slot k. Assume that $(A_k)_{k\geq 0}$ are i.i.d. random variables with probability distribution:

$$P[A_k = n] = \lambda_n \qquad (n \geq 0)$$

such that the mean arrival rate $\lambda = \sum_{n=1}^{\infty} n\lambda_n$ is finite. New packets are transmitted with probability one at the beginning of the first slot following their arrival.

Given that n packets are being transmitted in one slot, we define

Reprinted from *IEEE Trans. Automat. Contr.*, vol. 33, no. 7, pp. 640–649, July 1988.

for $n \geq 1, 0 \leq k \leq n$

$$\epsilon_{nk} = P[k \text{ packets are correctly received} \mid n \text{ are transmitted}].$$

The multipacket reception properties of the channel are summarized by the stochastic matrix

$$E = \begin{bmatrix} \epsilon_{10} & \epsilon_{11} & & \\ \epsilon_{20} & \epsilon_{21} & \epsilon_{22} & 0 \\ \cdot & \cdot & & \\ \epsilon_{n0} & \epsilon_{n1} & \epsilon_{nn} & \\ \vdots & \vdots & & \ddots \end{bmatrix}$$

which we refer to as the *reception matrix* of the channel. For instance, the reception matrix for the usual collision channel is

$$\begin{bmatrix} 0 & 1 & \\ 1 & 0 & 0 \\ 1 & 0 & \\ 1 & \cdot & \\ \vdots & \vdots & \end{bmatrix}$$

while for a system with capture it has the form

$$\begin{bmatrix} 0 & 1 & \\ 1-x_2 & x_2 & 0 \\ \cdot & \cdot & \\ 1-x_n & x_n & \\ \vdots & \vdots & \end{bmatrix}$$

where x_n is the probability of capture given that the collision size is n. The model studied in [19], [20] can be described by a reception matrix of the form

$$\begin{bmatrix} 0 & 1 & & & \\ 0 & 0 & 1 & & 0 \\ \cdot & \cdot & & & \\ 0 & 0 & & 1 & \\ 1 & 0 & & & \\ 1 & 0 & & & 0 \\ \vdots & \vdots & & & \end{bmatrix}$$

Note that by letting $\epsilon_{10} \neq 0$ our model allows not only collisions but also background noise to be a source of errors.

Denote by X_n the number of backlogged packets in the system at the beginning of slot n. The discrete-time process $(X_n)_{n \geq 0}$ is easily seen to be a homogeneous Markov chain. We define the system to be stable if $(X_n)_{n \geq 0}$ is ergodic and unstable otherwise. The average number of packets correctly received in collisions of size n is denoted by $C_n = \Sigma_{k=1}^{n} k\epsilon_{nk}$. We can now state the main result.

Theorem 1: If C_n has a limit $C = \lim_{n \to \infty} C_n$, then[1] the system is stable for all arrival distributions such that $\lambda < C$ and is unstable for $\lambda > C$. This also holds if C is infinite: if $\lim_{n \to \infty} C_n = +\infty$, then the system is always stable.

The proof is given in Section III. In the remainder of this section, we use Theorem 1 to analyze several simple random access channels that fall within the scope of the multipacket reception channel.

1) Mobile Users with Pairwise Transmissions: Consider an infinite number of transmitters $T_1, T_2, \cdots$, and an infinite number of receivers $R_1, R_2, \cdots$, whose positions in the plane are i.i.d. random variables. Suppose that transmissions are pairwise

[1]This result holds under the assumption that the Markov chain of the number of backlogged packets is irreducible and aperiodic (for details and sufficient conditions, see Section III).

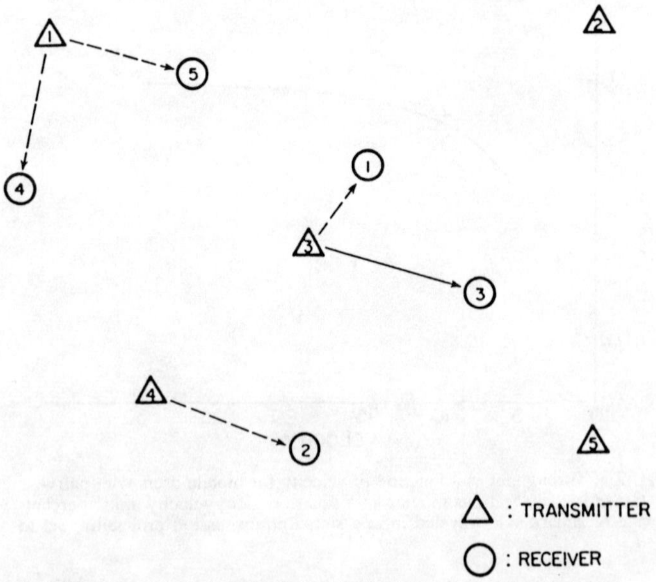

Fig. 1. Pairwise transmissions with only one success (3-3).

in the sense that transmitter T_n sends packets only to receiver R_n, and R_n is only interested in the packets sent by T_n (see Fig. 1). Assume also that each receiver can only detect correctly the packet sent by the closest transmitter (in particular, this is the case if there is perfect capture, see Example 3 below). The successes of transmissions occurring at the same time are independent, so that for $n \geq 2$

$$\epsilon_{nk} = \binom{n}{k} p(n)^k (1 - p(n))^{n-k}$$

where $p(n)$ is the probability that any given transmitter is successful in a collision of size n, which is equal to $1/n$ if we assume that all locations are memoryless, i.e., independent from slot to slot. It follows that

$$C_n = np(n) = 1$$

and the maximum throughput is 1. More generally, if because of channel noise, the message of the closest transmitter is received correctly with probability α (in other words $\epsilon_{11} = \alpha$), then the throughput is equal to α. The assumption that the locations of the stations are memoryless is equivalent to assuming that they move infinitely fast. If this simplifying assumption is dropped, then the number of successes depends not only on the current number of retransmissions, but also on the previous history of retransmissions, and thus the problem is no longer encompassed by our multipacket reception model. In Fig. 2, the result of a simulation shows that for moderate speeds, the actual throughput is well approximated by the foregoing analysis.

2) Frequency Hopping Random Access Channel: Consider a finite population of N users transmitting by frequency hopping, as in [11], [22]. For each packet he wants to transmit, a user selects with equal probability one frequency in a fixed set of q frequencies. A packet is correctly received iff no other packet is transmitted on the same frequency during the same slot. We compute $(\epsilon_{Nk})_{1 \leq k \leq N}$, and $C = \lim_{N \to \infty} C_N$. If the users have infinite buffer space, then C can be taken as a good approximation for large N of the maximum stable throughput of the system, which is unknown. If the users have no buffer space, as is often assumed, the backlog Markov chain is always ergodic, but even then, one should expect reasonable delays in large population problems only for arrival rates below C. The computation of the reception matrix of this channel is a simple combinatorial problem of random assignment of objects to cells (e.g., see [23, App. A]).

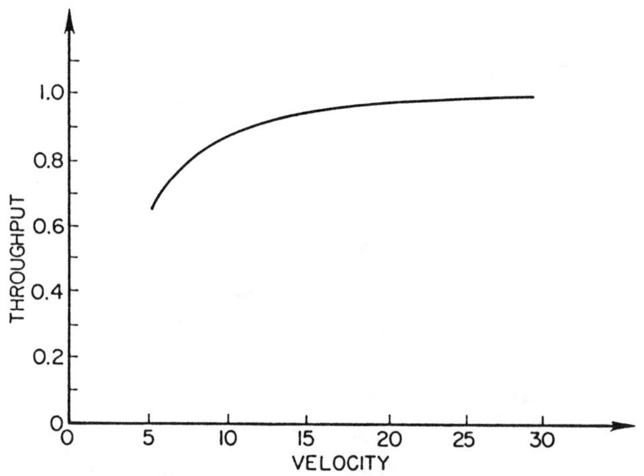

Fig. 2. Throughput as a function of velocity for mobile users with pairwise transmissions. Stations moving in a square region; velocity units: percentage of square side traveled in one slot. Retransmission probability set to 0.1.

Denote by $T_1, T_2, \cdots, T_N$ the users, all involved in the collision, and also denote by S the set of users whose packets are correctly received. Two cases need to be considered.

a) $2 \le N \le q$: We have, for $1 \le j \le N$

$$\epsilon_{Nj} = \binom{N}{j} P[S = \{T_1, T_2, \cdots, T_j\}] \tag{1}$$

and the following decomposition:

$$P[\{T_1, T_2, \cdots, T_j\} \subseteq S] = P[S = \{T_1, T_2, \cdots, T_j\}] + P\left[\bigcup_{k=j+1}^{N} \{\{T_1, \cdots, T_j, T_k\} \subseteq S\}\right]$$

easily yields the desired expression

$$P[S = \{T_1, T_2, \cdots, T_j\}] = \sum_{k=0}^{N-j} (-1)^k \cdot \binom{N-j}{k} P[\{T_1, T_2, \cdots, T_{k+j}\} \subseteq S] \tag{2}$$

where only one term is left to compute

$$P[\{T_1, T_2, \cdots, T_{k+j}\} \subseteq S] = \frac{q(q-1) \cdots (q-j-k+1)(q-j-k)^{N-j-k}}{q^N} \tag{3}$$

for $1 \le j \le N, 0 \le k \le N - j$. Putting (1), (2), and (3) together gives the result

$$\epsilon_{Nj} = \binom{N}{j} \sum_{k=0}^{N-j} (-1)^k \binom{N-j}{k} \cdot \frac{q(q-1) \cdots (q-j-k+1)(q-j-k)^{N-j-k}}{q^N} \tag{4}$$

for $1 \le j \le N$. Notice in particular that $\epsilon_{N,N-1} = 0$. Let us now compute the average number of packets correctly received in collisions of size N, $C_N = \sum_{j=1}^{N} j\epsilon_{Nj}$. By using (4) and summing at $j + k$ constant, we get

$$q^N C_N = \sum_{i=1}^{N} q(q-1) \cdots (q-i+1)(q-i)^{N-i} \cdot \sum_{n=0}^{i-1} (-1)^n \frac{N!}{n!(i-n-1)!(N-i)!}$$

which can be simplified as

$$q^N C_N = \sum_{i=1}^{N} \frac{N!}{(i-1)!(N-i)!} \cdot q(q-1) \cdots (q-i+1)(q-i)^{N-i}(1-1)^{i-1}$$

to get the final result

$$C_N = N\left(1 - \frac{1}{q}\right)^{N-1}.$$

b) $N > q$: In this case, there can be at best $q - 1$ successes in a collision of size N. The same method applies to get the following probabilities:

$$\epsilon_{Nj} = \binom{N}{j} \sum_{k=0}^{q-j-1} \binom{N-j}{k} (-1)^k \cdot \frac{q(q-1) \cdots (q-j-k+1)(q-j-k)^{N-j-k}}{q^N} \qquad (1 \le j \le q-1)$$

$$\epsilon_{Nj} = 0 \qquad (q \le j \le N)$$

resulting in the same expected number of successes as before

$$C_N = N\left(1 - \frac{1}{q}\right)^{N-1}.$$

Now we let the population size N go to infinity and we apply our result. If we let N grow to infinity while keeping q constant, we have $\lim_{N\to\infty} C_N = 0$, so the system is always unstable. On the other hand, if we let N go to infinity while keeping q equal to a fixed percentage of the population size, i.e., N/q constant, then $\lim_{N\to\infty} C_N = +\infty$, and the system is always stable. It is easily shown that to get a finite maximum stable throughput, q has to grow as $N/\ln N$.

3) Mobile Radio Network with Capture: Consider an infinite number of users independently and uniformly distributed in a circle of radius R, whose positions are independent from slot to slot. Users transmit packets to a common receiver located at the center of the network. Denote by P_1 and P_2 the received powers of the strongest and the next to strongest packets involved in a collision. Assume, as in [12]–[14], that the strongest packet is correctly received iff $P_1/P_2 > K$ (K being a system dependent constant), and that all the other packets involved in the collision are not received successfully. Assume, moreover, that the received power of a packet only depends on the distance r between the sender and the receiver

$$P = \frac{\text{constant}}{r^\alpha} \qquad (\alpha \ge 2).$$

Then there will be capture iff

$$r_2 > \beta r_1$$

where $\beta = K^{1/\alpha}$ is the capture parameter, and r_1, r_2 are the distances of the closest and the next to closest senders from the receiver.

Denote by D the distance between a given user and the receiver. It is easily shown that the pdf of D is given by

$$p_D(r)=2\frac{r}{R^2}\qquad(0\le r\le R).$$

Given N users, denote by U_N the closest from the receiver, and by D_N its distance from the receiver. Computing the cdf of D_N and taking its derivative, we obtain

$$p_{D_N}(r)=2N\frac{r}{R^2}\left[1-\left(\frac{r}{R}\right)^2\right]^{N-1}\qquad(0\le r\le R).\tag{5}$$

Given $D_N = r$, the other $N-1$ users are uniformly distributed in the annular region (r, R). So if N users collide and $D_N = r$, U_N will be correctly received iff all the other users are in the annular region $(\beta r, R)$, which is empty if $\beta r > R$. Therefore, if we denote by

$$P_N(r)=P[\text{capture}\,|\,N \text{ collide, } D_N=r]\qquad(N\ge 2)$$

we have

$$P_N(r)=\begin{cases}\left[\dfrac{R^2-\beta^2r^2}{R^2-r^2}\right]^{N-1} & \text{if } r\le\dfrac{R}{\beta}\\ 0 & \text{if } r\ge\dfrac{R}{\beta}\end{cases}.\tag{6}$$

Thus, the probability of capture in a collision of N ($N \ge 2$) is

$$\epsilon_{N1}=\int_0^{R/\beta}P_N(r)p_{D_N}(r)\,dr.$$

Using (5) and (6), and with the change of variable $x = \beta/R$, this is easily computed

$$\epsilon_{N1}=\int_0^{1/\beta}2Nx(1-\beta^2x^2)^{N-1}\,dx=\frac{1}{\beta^2}.$$

It follows that $C = 1/\beta^2$ is the maximum stable throughput. Notice, in particular, that for $\beta = 1$ (perfect capture), we have $C = 1$ and for $\beta \to \infty$ (no capture), we have $C \to 0$.

Under certain conditions, the performances of Aloha in the multipacket channel can be improved by varying the retransmission probability as a function of the channel history, and a maximum stable throughput of $\sup_{x\ge 0} e^{-x}\Sigma_{n=1}^{\infty} C_n/n!\,x^n$ can be reached (see [31]).

III. Ergodicity Region

The number of backlogged packets in the system at time n, $(X_n)_{n\ge 0}$, is a homogeneous Markov chain whose one-step transition probability matrix can be computed as a function of p, $(\lambda_k)_{k\ge 0}$, and E. Denoting by $B_i(j)$ the probability of having j retransmissions out of i backlogged packets

$$B_i(j)=\binom{i}{j}p^j(1-p)^{i-j}\tag{7}$$

we get

$$P_{00}=\lambda_0+\sum_{n=1}^{\infty}\lambda_n\epsilon_{nn}$$

$$P_{0k}=\sum_{n=0}^{\infty}\lambda_{k+n}\epsilon_{k+n,n}\qquad(k\ge 1)$$

and for $i \ge 1$

$$P_{i,i-k}=\sum_{n=0}^{\infty}\lambda_n\sum_{j=k}^{i}B_i(j)\epsilon_{n+j,n+k}\qquad(1\le k\le i)$$

$$P_{ii}=\lambda_0\left[B_i(0)+\sum_{j=1}^{i}B_i(j)\epsilon_{j0}\right]+\sum_{n=1}^{\infty}\lambda_n\sum_{j=0}^{i}B_i(j)\epsilon_{n+j,n}$$

$$P_{i,i+k}=\sum_{n=0}^{\infty}\lambda_{k+n}\sum_{j=0}^{i}B_i(j)\epsilon_{j+k+n,n}\qquad(k\ge 1).\tag{8}$$

Sufficient conditions for $(X_n)_{n\ge 0}$ to be irreducible and aperiodic are as follows:

- if $0 < p < 1$:

$$\lambda_0\ne 0\tag{9a}$$

$$\lambda_0+\sum_{n=1}^{\infty}\lambda_n\epsilon_{nn}<1\tag{9b}$$

$$\epsilon_{10}\ne 1\tag{9c}$$

- if $p = 1$:

$$\lambda_0\ne 0\tag{9a}$$

$$\lambda_0+\sum_{n=1}^{\infty}\lambda_n\epsilon_{nn}<1\tag{9b}$$

$$\text{for all } i\ge 1,\ \epsilon_{i0}\ne 1.\tag{9d}$$

These are only sufficient conditions, but they hold for almost all nontrivial systems. For example, if (9b) does not hold, then zero is an absorbing state, since the left-hand side of (9b) is equal to P_{00}. Also, (9c) simply means that the successful reception of a single packet in the absence of other active users is possible. Assume, for instance, that $0 < p < 1$ and that the arrivals are Poisson distributed. Then we only have to assume (9c), and (9b) is true unless there is perfect reception, that is $\epsilon_{nn} = 1$ for all $n \ge 1$, in which case the system would of course always be stable. The case $p = 1$ gives rise to a number of pathological situations, hence, the much stronger condition (9d). It generally turns out that either (9d) is not necessary or the stability region of the system is obvious. For instance, it is clear from the transition probabilities that slotted Aloha with $p = 1$ is always unstable. In any case, it is assumed in what follows that $(X_n)_{n\ge 0}$ is irreducible and aperiodic.

Proof of Theorem 1: The proof is based on drift analysis. Recall that in general, the drift at state i ($i \ge 0$) is defined by

$$d_i=E[X_{t+1}-X_t\,|\,X_t=i].$$

If we denote by Σ_t the number of successful transmissions in slot t, we have

$$X_{t+1}-X_t=A_t-\Sigma_t$$

and therefore

$$d_i=\lambda-E[\Sigma_t\,|\,X_t=i].\tag{10}$$

Now if R_t is the number of retransmissions in slot t, we get

$$P[\Sigma_t=k\,|\,X_t=i,\ A_t=n,\ R_t=j]=\epsilon_{n+j,k}$$

for $0 \le j \le i$, $0 \le k \le n + j$ and with the convention that $\epsilon_{00} =$

$C_0 = 0$. Thus,

$$E[\Sigma_t | X_t = i,\ A_t = n,\ R_t = j] = C_{n+j}$$

and

$$E[\Sigma_t | X_t = i] = \sum_{n=0}^{\infty} \lambda_n \sum_{j=0}^{i} B_i(j) C_{n+j}. \tag{11}$$

The value of the drifts for our model follows from (10) and (11)

$$d_i = \lambda - \sum_{n=0}^{\infty} \lambda_n \sum_{j=0}^{i} B_i(j) C_{n+j}. \tag{12}$$

The idea of the proof is to compute $\lim_{i\to\infty} d_i$ which will turn out to be a very simple expression, and then apply the results of [3] and [24] to determine the ergodicity region of $(X_n)_{n\geq 0}$. Let us first recall the two results that will be used in the sequel.

Lemma A (Pakes [24]): Let $(X_n)_{n\geq 0}$ be an irreducible and aperiodic Markov chain having as state space the nonnegative integers, denote by (P_{ij}) its transition probability matrix, and by d_i its drift at state i. Then if for all i $|d_i| < \infty$, and if $\limsup_{i\to\infty} d_i < 0$, $(X_n)_{n\geq 0}$ is ergodic.

Lemma B (Kaplan [3]): Under the assumptions of Lemma A, if for some integer $N \geq 0$ and some constants $B \geq 0$, $c \in [0, 1]$ the following two conditions hold, then $(X_n)_{n\geq 0}$ is not ergodic:

i) for all $i \geq N$, $d_i > 0$

ii) for all $i \geq N$, all $\theta \in [c, 1]$, $\theta^i - \Sigma_j P_{ij}\theta^j \geq -B(1 - \theta)$.

From (12), it can be seen that $|d_i|$ is finite since

$$|d_i| \leq \lambda + \sum_{n=0}^{\infty} \lambda_n \sum_{j=0}^{i} B_i(j) C_{n+j} \leq 2\lambda + ip.$$

Next, the drift limit is given by the following lemma.

Lemma 1: If C_n has a limit C, finite or not, then $\lim_{i\to\infty} \Sigma_{n=0}^{\infty} \lambda_n \Sigma_{j=0}^{i} B_i(j) C_{n+j} = C$.

Proof of Lemma 1: We consider two separate cases depending on whether C is finite.

1) $C = +\infty$.

Fix $A > 0$ and pick $r \geq 0$ such that $\lambda_r \neq 0$. There exists an integer M such that for all $n \geq M$, $C_n > A$. Fix such an M. Then we have for $i \geq M$

$$\sum_{n=0}^{\infty} \lambda_n \sum_{j=0}^{i} B_i(j) C_{n+j} > \lambda_r \sum_{j=0}^{i} B_i(j) C_{j+r} > \lambda_r A \sum_{j=M}^{i} B_i(j)$$

which terminates the proof, since for any fixed $M \geq 0$

$$\lim_{i\to\infty} \sum_{j=M}^{i} B_i(j) = 1. \tag{13}$$

2) $C < +\infty$.

We have for $i > M$

$$\left| \sum_{n=0}^{\infty} \lambda_n \sum_{j=0}^{i} B_i(j) C_{n+j} - C \right| \leq \sum_{j=0}^{M} B_i(j) \sum_{n=0}^{\infty} \lambda_n |C_{n+j} - C|$$

$$+ \sum_{j=M+1}^{i} B_i(j) \sum_{n=0}^{\infty} \lambda_n |C_{n+j} - C|. \tag{14}$$

Fix $\epsilon > 0$. There exists M such that for all $n > M$, $|C_n - C| < \epsilon/2$. Fix such an M. Then

$$\sum_{j=M+1}^{i} B_i(j) \sum_{n=0}^{\infty} \lambda_n |C_{n+j} - C| < \frac{\epsilon}{2}.$$

Also, if L is an upper bound for C_n

$$\sum_{j=0}^{M} B_i(j) \sum_{n=0}^{\infty} \lambda_n |C_{n+j} - C| \leq 2L \sum_{j=0}^{M} B_i(j) < \frac{\epsilon}{2}$$

for i big enough because (13) holds, which takes care of the first term in (14) and ends the proof of Lemma 1.

Putting together (12) and Lemmas A and 1, we get that 1) if $\lim_{n\to\infty} C_n = +\infty$, then $\lim_{i\to\infty} d_i = -\infty$, and $(X_n)_{n\geq 0}$ is ergodic; and 2) if $\lim_{n\to\infty} C_n = C < +\infty$, then $\lim_{i\to\infty} d_i = \lambda - C$, and $(X_n)_{n\geq 0}$ is ergodic for $\lambda < C$. If $\lambda > C$, we can apply Lemma B and conclude that $(X_n)_{n\geq 0}$ is not ergodic provided that Kaplan's condition ii) holds. This is the purpose of Lemma 2, which is the last step in the proof of Theorem 1.

Lemma 2: If for all $n \geq 1$, $C_n < L$ for some $L \in (0, \infty)$, then Kaplan's condition holds: there exists a constant B, an integer N, and a real $c \in [0, 1]$ such that

$$\theta^i - \sum_j P_{ij}\theta^j \geq -B(1-\theta) \qquad \text{all } i \geq N,\ \theta \in [c, 1].$$

Proof of Lemma 2: According to [25], it is enough to show that the downward part of the drift, defined as

$$D(i) = -\sum_{k=1}^{i} kP_{i,i-k}$$

is bounded below. From the transition probabilities (8), we get

$$D(i) = -\sum_{k=1}^{i} k \sum_{n=0}^{\infty} \lambda_n \sum_{j=k}^{i} B_i(j) \epsilon_{n+j,n+k}$$

which can also be put in the form

$$D(i) = -\sum_{j=1}^{i} B_i(j) \sum_{n=0}^{\infty} \lambda_n \sum_{k=1}^{j} k\epsilon_{n+j,n+k}$$

from which it follows that

$$D(i) \geq -\sum_{j=1}^{i} B_i(j) \sum_{n=0}^{\infty} \lambda_n C_{n+j} \geq -L.$$

□

Notice that in the proof of Theorem 1 (and this also holds for Theorem 2 below), the exact expression (7) for $B_i(j)$ is never used. The only requirements are that $(B_i(j))_{0\leq j\leq i}$ is a probability distribution, and that (13) holds. Therefore, our results are valid for a larger class of retransmission policies than was first assumed. For example, there could be K priority groups, each with a different retransmission probability.

Although Theorem 1 is quite general, in many practical cases, the reception matrix has a very simple structure and the stability region can be obtained with virtually no computations. This happens for instance in radio networks with capture where all is needed is the limit of the second column of the matrix, or also in the simple case where above a certain collision size N, the transmission is too garbled for the receiver to be able to decode anything correctly, so that $C_n = 0$ for $n > N$.

This last example is a particular case of a noteworthy feature of Theorem 1, namely that the stability region does not depend on any finite number of rows of the reception matrix. In fact, any number of modifications of the matrix that leaves $\lim_{n\to\infty} C_n$ unchanged does not affect the stability region. Although this may be surprising at first sight, it can be intuitively explained by the fundamental instability of the collision channel: unless the receiver is perfect (all ϵ_{nn} equal to 1), the backlog will eventually exceed any prefixed value with probability one, thus it is the limit of C_n that determines the stability region.

The stability region is also unchanged if the first transmission of packets is delayed. If new packets are backlogged, that is, transmitted for the first time with probability p in each slot following their arrival (this transmission rule appears in the literature as controlled-access or delayed first transmission), the drifts become $d_i = \lambda - \Sigma_{j=1}^{i} B_i(j)C_j$ for $i \geq 1$, and from Lemmas 1 and 2 the ergodicity region remains the same.

If C_n does not have a limit, Theorem 1 does not give the stable throughput of the system. Even though in almost all practical cases, and indeed in all the examples of Section II, C_n does have a limit, it is conceptually interesting to examine the case when $\liminf_{n\to\infty} C_n \neq \limsup_{n\to\infty} C_n$. It is worth pointing out that adding constraints as strong as the following on the reception matrix still does not imply that C_n has a limit:

i) $(\epsilon_{n0})_{n\geq 1}$ is nondecreasing

ii) $(\epsilon_{nk})_{n\geq k}$ is nonincreasing for all $k \geq 1$

iii) $\epsilon_{nk} \geq \epsilon_{n,k+1}$ for $n \geq 2,\ 1 \leq k \leq n-1$

although the counterexamples we have been able to build are somewhat contrived. Notice that conditions i) and ii) above imply that each column has a limit $\alpha_k = \lim_{n\to\infty} \epsilon_{nk} (k \geq 0)$, which is very likely to happen in practice. In any case, Theorem 2 below still gives some information on the stability region, although the exact result requires in general the complete knowledge of the sequence $(C_n)_{n\geq 1}$. In fact, given any nonnegative numbers $\alpha < \gamma < \beta$, one can construct a reception matrix with nth row average C_n such that:

i) $\liminf_{n\to\infty} C_n = \alpha$

ii) $\limsup_{n\to\infty} C_n = \beta$

and such that the maximum stable throughput is γ.

Theorem 2: The system is stable for $\lambda < \liminf_{n\to\infty} C_n$ and unstable for $\lambda > \limsup_{n\to\infty} C_n$.

Proof:

a) If $\lambda < \liminf_{n\to\infty} C_n$, then $(X_n)_{n\geq 0}$ is ergodic.

If $\liminf_{n\to\infty} C_n = +\infty$, then $\lim_{n\to\infty} C_n = +\infty$, and the result has already been proved, so assume that $\liminf_{n\to\infty} C_n$ is finite. From Lemma A, it is enough to prove that for all $\epsilon > 0$, there exists N such that

$$d_i < \lambda - \liminf_{n\to\infty} C_n + \epsilon \qquad \text{all } i \geq N.$$

Recall from (12) that we have

$$d_i = \lambda - \sum_{n=0}^{\infty} \lambda_n \sum_{j=0}^{i} B_i(j) C_{n+j}. \tag{15}$$

So it is only needed to prove that for all $\epsilon > 0$ there exists N such that

$$\sum_{n=0}^{\infty} \lambda_n \sum_{j=0}^{i} B_i(j) C_{n+j} > \liminf_{n\to\infty} C_n - \epsilon \qquad \text{all } i \geq N.$$

Now by definition there exists M such that for all $k \geq M$:

$$C_k > \liminf_{n\to\infty} C_n - \epsilon$$

and therefore for all $i > M$:

$$\sum_{n=0}^{\infty} \lambda_n \sum_{j=0}^{i} B_i(j) C_{n+j} > (\liminf_{n\to\infty} C_n - \epsilon) \sum_{j=M}^{i} B_i(j)$$

which completes the proof since (13) holds.

b) If $\lambda > \limsup_{n\to\infty} C_n$, then $(X_n)_{n\geq 0}$ is not ergodic.

Since λ is finite, in this case $\limsup_{n\to\infty} C_n$ is necessarily finite. Therefore, $(C_n)_{n\geq 1}$ is bounded and from Lemma 2, Kaplan's condition holds. Thus, it is enough to show that for all $\epsilon > 0$, there exists N such that

$$d_i > \lambda - \limsup_{n\to\infty} C_n - \epsilon \qquad \text{all } i \geq N.$$

From (15), we only need to show

$$\sum_{n=0}^{\infty} \lambda_n \sum_{j=0}^{i} B_i(j) C_{n+j} < \limsup_{n\to\infty} C_n + \epsilon \qquad \text{all } i \geq N.$$

Since there exists M such that for all $k \geq M$

$$C_k < \limsup_{n\to\infty} C_n + \epsilon$$

then if L is an upper bound for C_n, we have for $i \geq M$

$$\sum_{n=0}^{\infty} \lambda_n \sum_{j=0}^{i} B_i(j) C_{n+j} < L \sum_{j=0}^{M-1} B_i(j) + \limsup_{n\to\infty} C_n + \epsilon$$

from which the result follows, using (13). □

IV. Behavior of the Backlog Markov Chain in the Nonergodicity Region

In this section, we further investigate the properties of $(X_n)_{n\geq 0}$ in the case $\lambda > C$, assuming of course that $(C_n)_{n\geq 1}$ has a finite limit. It has been proved in [1] that the backlog Markov chain for the usual slotted Aloha algorithm is transient, but this result cannot be generalized to our model when $\lambda > C$. We give below an example showing that $(X_n)_{n\geq 0}$ can be null recurrent when the mean arrival rate λ belongs to an interval of positive length. The boundary between the null recurrence and the transience regions generally depends in a rather complicated manner on both the reception matrix and the retransmission probability p. We give a sufficient condition for $(X_n)_{n\geq 0}$ to be transient when $\lambda > C$, as well as bounds on the recurrence region.

Consider the reception matrix defined by

$$\epsilon_{nk} = \frac{1}{n^2} \qquad (1 \leq k \leq n)$$

$$\epsilon_{n0} = 1 - \frac{1}{n}$$

for $n \geq 1$. Then $C_n = \Sigma_{k=1}^{n} k/n^2 = (n+1)/2n$, and $C = 1/2$. Using Lemmas C and D below, we show in [26] that X_n is recurrent for $\lambda < R(p)$ and transient for $\lambda > R(p)$, where $R(p)$ is a function of the retransmission probability p and is given by

$$R(p) = \frac{1}{p} + \frac{(1-p)}{p^2} \ln(1-p) \qquad (0 < p < 1)$$

$$R(1) = 1.$$

It is easily seen that $R(p)$ is an increasing function of p for $p \in]0, 1[$ with extrema $\lim_{p\to 0} R(p) = 1/2$ and $\lim_{p\to 1} R(p) = 1$. Fig. 3 summarizes the behavior of X_n for this example.

It is somehow surprising to see that in this case, as well as in all

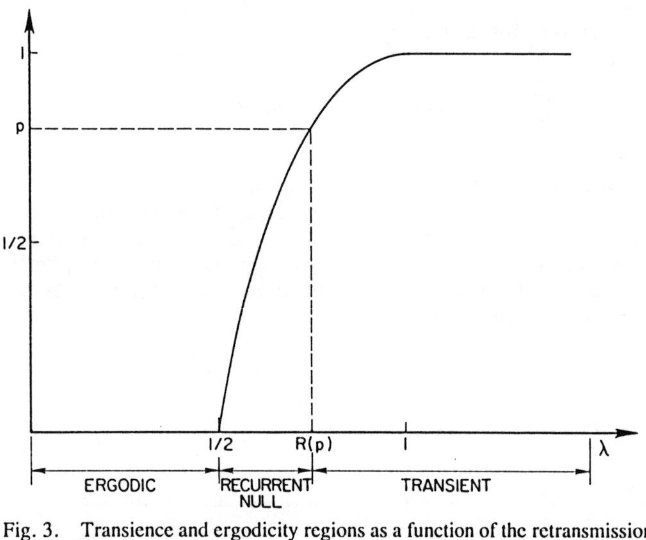

Fig. 3. Transience and ergodicity regions as a function of the retransmission probability when $\epsilon_{nk} = 1/n^2$.

the other examples we have computed, the recurrence region becomes larger as p increases. Intuitively, the recurrence of X_n when $\lambda > C$ seems to be due to the fact that transitions from any state i to 0 (or to some fixed integer k_0) are possible and that the probability of such an event, P_{i0} (or P_{ik_0}), goes to zero slowly with i. It can be checked that these probabilities are increasing functions of p when i is large enough.

Transience is ensured for $\lambda > C$ if the supremum of the elements of the kth-column goes to zero faster than k^2. This condition holds for all the examples in Section II, as well as for many real life cases, due to the practical limitations on the receiver capabilities. In particular, it is always verified if the reception matrix has only a finite number of nonzero columns (or equivalently, if the backlog Markov chain has uniformly bounded downwards transitions, as defined in [3]) which happens, for instance, if there is capture. Note that the proof of Theorem 3 below is of course valid for the conventional collision channel, and in this case becomes somewhat simpler than the proof in [1].

Theorem 3: If $\lim_{k\to\infty} k^2 \sup_{n\geq k} \epsilon_{nk} = 0$, then $(X_n)_{n\geq 0}$ is transient for $\lambda > C$.

Because of the complexity and lack of structure of the one-step transition probabilities (8), few results on the recurrence and transience of Markov chains can be applied to our model. Before proving Theorem 3, let us introduce the following two criteria from [27].

Lemma C: Let $(X_n)_{n\geq 0}$ be an irreducible and aperiodic Markov chain, having as state space the set of nonnegative integers, and with one-step transition probability matrix $P = (P_{ij})$. $(X_n)_{n\geq 0}$ is recurrent if and only if there exists a sequence $(y_n)_{n\geq 0}$ such that

1) $\lim_{n\to\infty} y_n = +\infty$

2) for some integer $N>0$ $\sum_{j=0}^{\infty} y_j P_{ij} \leq y_i$ all $i \geq N$.

We will only use the sufficiency part, which has also been proved in [24].

Lemma D: With the same assumptions as in Lemma C, $(X_n)_{n\geq 0}$ is transient if and only if there exists a sequence $(y_n)_{n\geq 0}$ such that

1) $(y_n)_{n\geq 0}$ is bounded

2) for some integer $N>0$ $\sum_{j=0}^{\infty} y_j P_{ij} \leq y_i$ all $i \geq N$

3) for some $k \geq N$ $y_k < y_0, \cdots, y_{N-1}$.

Sufficiency under the additional constraints $y_i > 0$ and $\lim_{i\to\infty} y_i = 0$ has also been proved in [28]. Also, the sufficiency parts of both lemmas are an immediate consequence of [29, Theorems 5 and 6] together with the results in [30].

Proof of Theorem 3: We use Lemma D with $y_n = 1/(n+1)^\theta$, $\theta \in]0, 1[$. We have

$$\sum_j P_{ij} y_j \leq y_i \Leftrightarrow \sum_{k=1}^{i} (y_{i-k} - y_i) P_{i,i-k} + \sum_{k=1}^{\infty} (y_{i+k} - y_i) P_{i,i+k} \leq 0 \quad (16)$$

and

$$(i+1)^{1+\theta} \sum_{k=1}^{i} (y_{i-k} - y_i) P_{i,i-k} + (i+1)^{1+\theta} \cdot \sum_{k=1}^{\infty} (y_{i+k} - y_i) P_{i,i+k} = D'(i) + U'(i) \quad (17)$$

where we have defined

$$D'(i) = (i+1)^{1+\theta} \sum_{k=1}^{i} \left[\frac{1}{(i+1-k)^\theta} - \frac{1}{(i+1)^\theta} \right] \cdot \sum_{n=0}^{\infty} \lambda_n \sum_{j=k}^{i} B_i(j) \epsilon_{n+j,n+k}$$

$$U'(i) = (i+1)^{1+\theta} \sum_{k=1}^{\infty} \left[\frac{1}{(i+1+k)^\theta} - \frac{1}{(i+1)^\theta} \right] \cdot \sum_{n=0}^{\infty} \lambda_{k+n} \sum_{j=0}^{i} B_i(j) \epsilon_{n+k+j,n}. \quad (18)$$

The drift of X_n at state i can be computed from the transition probabilities (8)

$$d_i = -\sum_{k=1}^{i} k P_{i,i-k} + \sum_{k=1}^{\infty} k P_{i,i+k} = D(i) + U(i) \quad (19)$$

where we have defined

$$D(i) = -\sum_{k=1}^{i} k \sum_{n=0}^{\infty} \lambda_n \sum_{j=k}^{i} B_i(j) \epsilon_{n+j,n+k}$$

$$U(i) = \sum_{k=1}^{\infty} k \sum_{n=0}^{\infty} \lambda_{n+k} \sum_{j=0}^{i} B_i(j) \epsilon_{j+k+n,n}. \quad (20)$$

The idea of the proof is to show that

$$\lim_{i\to\infty} [D'(i) + U'(i)] = -\theta \lim_{i\to\infty} d_i \quad (21)$$

and since it has been proved in Section III that $\lim_{i\to\infty} d_i = \lambda - C$, we will be able to conclude that $(X_n)_{n\geq 0}$ is transient for $\lambda > C$.

1) $\lim_{i\to\infty} [D'(i) + \theta D(i)] = 0$.

From (18) and (20)

$$D'(i)+\theta D(i)=(i+1)\sum_{k=1}^{i}\left[\left(\frac{i+1}{i+1-k}\right)^{\theta}-1-\frac{\theta k}{i+1}\right] \cdot \sum_{n=0}^{\infty}\lambda_n \sum_{j=k}^{i} B_i(j)\epsilon_{n+j,n+k}$$

which is more conveniently written as

$$D'(i)+\theta D(i)=(i+1)\sum_{n=0}^{\infty}\lambda_n\sum_{j=1}^{i}B_i(j)\sum_{k=1}^{j} \cdot\left[\left(\frac{i+1}{i+1-k}\right)^{\theta}-1-\frac{\theta k}{i+1}\right]\epsilon_{n+j,n+k}.$$

This expression is nonnegative since

$$\left(\frac{i+1}{i+1-k}\right)^{\theta}-1-\frac{\theta k}{i+1}>0 \qquad (1\le k\le i).$$

Define $\gamma_k = \sup_{n\ge k}\epsilon_{nk}$. Then

$$0\le D'(i)+\theta D(i)\le(i+1)\sum_{n=0}^{\infty}\lambda_n\sum_{j=1}^{i}B_i(j)\sum_{k=1}^{j} \cdot\left[\left(\frac{i+1}{i+1-k}\right)^{\theta}-1-\frac{\theta k}{i+1}\right]\gamma_{n+k}$$

$$\le(i+1)\sum_{n=0}^{\infty}\lambda_n\sum_{k=1}^{i}\left[\left(\frac{i+1}{i+1-k}\right)^{\theta}-1-\frac{\theta k}{i+1}\right]\gamma_{n+k}.$$

That is

$$D'(i)+\theta D(i)\le x_1(i)+x_2(i) \tag{22}$$

with, assuming for instance that i is odd

$$x_1(i)=(i+1)\sum_{n=0}^{\infty}\lambda_n\sum_{k=1}^{(i+1)/2}\left[\left(\frac{i+1}{i+1-k}\right)^{\theta}-1-\frac{\theta k}{i+1}\right]\gamma_{n+k}$$

$$x_2(i)=(i+1)\sum_{n=0}^{\infty}\lambda_n\sum_{k=(i+3)/2}^{i}\left[\left(\frac{i+1}{i+1-k}\right)^{\theta}-1-\frac{\theta k}{i+1}\right]\gamma_{n+k}. \tag{23}$$

We show that $x_1(i)$ and $x_2(i)$ go to zero independently. Fix $\epsilon > 0$. Define for $0 < x \le i$ the function

$$p_i(x)=\frac{i+1}{x^2}\left[\left(\frac{i+1}{i+1-x}\right)^{\theta}-1\right]-\frac{\theta}{x}.$$

It is easily proved that for each $i \ge 1$, $p_i(x)$ is a positive nondecreasing function of x. Also

$$p\left(\frac{i+1}{2}\right)=\frac{1}{i+1}\left[4(2^{\theta}-1)-2\theta\right]=\frac{A}{i+1}$$

where A is a positive constant depending only on θ. From (23)

$$x_1(i)=\sum_{n=0}^{\infty}\lambda_n\sum_{k=1}^{(i+1)/2}k^2p_i(k)\gamma_{n+k}\le\frac{A}{i+1}\sum_{n=0}^{\infty}\lambda_n\sum_{k=1}^{n+(i+1)/2}k^2\gamma_{n+k}.$$

If $\lim_{k\to\infty}k^2\gamma_k = 0$, then $\lim_{n\to\infty}1/n\Sigma_{k=1}^{n}k^2\gamma_k = 0$. So we can choose i large enough so that for $n \ge (i+1)/2$, $\Sigma_{k=1}^{n}k^2\gamma_k < n\epsilon$. Then

$$x_1(i)\le\epsilon\frac{A}{i+1}\sum_{n=0}^{\infty}\lambda_n\left(n+\frac{i+1}{2}\right)=\epsilon A\left(\frac{\lambda}{i+1}+\frac{1}{2}\right).$$

Now if we choose i big enough so that for $k > (i+3)/2$, we have $\gamma_k < \epsilon/k^2$, then

$$x_2(i)\le\epsilon\sum_{n=0}^{\infty}\lambda_n\sum_{k=(i+3)/2}^{i}(i+1) \cdot\left[\left(\frac{i+1}{i+1-k}\right)^{\theta}-1-\frac{\theta k}{i+1}\right]\frac{1}{(n+k)^2}$$

$$\le\frac{4\epsilon}{i+3}\sum_{k=(i+3)/2}^{i}\left[\left(\frac{i+1}{i+1-k}\right)^{\theta}-1-\frac{\theta k}{i+1}\right].$$

By bounding the sum in the last equation by integrals, it can be seen that it is upper bounded by a linear function of i.

2) $\lim_{i\to\infty}[U'(i)+\theta U(i)] = 0$.

From (18) and (20)

$$U'(i)+\theta U(i)=\sum_{k=1}^{\infty}(i+1)\left[\left(\frac{i+1}{i+1+k}\right)^{\theta}-1+\frac{\theta k}{i+1}\right] \cdot\sum_{n=0}^{\infty}\lambda_{k+n}\sum_{j=0}^{i}B_i(j)\epsilon_{j+k+n,n}.$$

With a change of variable

$$U'(i)+\theta U(i)=(i+1)\sum_{j=0}^{i}B_i(j)\sum_{n=1}^{\infty}\lambda_n\sum_{k=1}^{n} \cdot\left[\left(\frac{i+1}{i+1+k}\right)^{\theta}-1+\frac{\theta k}{i+1}\right]\epsilon_{n+j,n-k}.$$

By using the following inequalities:

$$0\le\frac{1}{(1+x)^{\theta}}-1+\theta x\le\theta(1+\theta)\frac{x^2}{2} \qquad (x\ge0,\ 0<\theta<1)\}$$

we get

$$0\le U'(i)+\theta U(i)\le\theta\frac{(1+\theta)}{2}(i+1)\sum_{j=0}^{i}B_i(j) \cdot\sum_{n=1}^{N}\lambda_n\sum_{k=1}^{n}\frac{k^2}{(i+1)^2}\epsilon_{n+j,n-k}$$

$$+\theta(i+1)\sum_{j=0}^{i}B_i(j)\sum_{n=N+1}^{\infty}\lambda_n\sum_{k=1}^{n}\frac{k}{i+1}\epsilon_{n+j,n-k}$$

$$\le\frac{1}{i+1}\sum_{n=1}^{N}n^2\lambda_n+\sum_{n=N+1}^{\infty}n\lambda_n.$$

Fix $\epsilon > 0$. Choose N such that $\Sigma_{N+1}^{\infty}n\lambda_n < \epsilon/2$, and then, N being fixed, choose i large enough so that $1/(i+1)\Sigma_{n=1}^{N}n^2\lambda_n < \epsilon/2$. □

It should be clear at this point that unlike the ergodicity region, the recurrence region depends in general on the elements of the reception matrix (instead of only the row averages) and on the retransmission probability p. For this reason, an exact expression for the recurrence region seems rather difficult to obtain; nonetheless, the method (see [26]) that we used to study the example in Fig. 3 can be generalized to obtain the following upper and lower bounds on the recurrence region.

Theorem 4: $(X_n)_{n\geq 0}$ is recurrent for $\lambda < L$ and transient for $\lambda > U$, with $L = \max\{l_1, \sup_{0<\theta<1} l_\theta, \sup_{0<\theta<1} l'_\theta\}$ and $U = \min\{u_1, \inf_{0<\theta<1} u_\theta, \inf_{0<\theta<1} u'_\theta\}$ where

$$l_1 = \lim_{i\to\infty} (i+1) \sum_{n=0}^{\infty} \lambda_n \sum_{j=1}^{i} B_i(j) \sum_{k=1}^{n+j} \ln\left(\frac{i+n+1}{i+n-k+1}\right) \epsilon_{n+j,k}$$

$$l_\theta = \frac{1}{\theta} \lim_{i\to\infty} (i+1)^{1-\theta} \sum_{n=0}^{\infty} \lambda_n \sum_{j=1}^{i} B_i(j) \cdot \sum_{k=1}^{n+j} [(i+n+1)^\theta - (i+n-k+1)^\theta] \epsilon_{n+j,k}$$

$$l'_\theta = \frac{1}{\theta} \lim_{i\to\infty} (i+1)\, [\ln(i+1)]^{1-\theta} \sum_{n=0}^{\infty} \lambda_n \sum_{j=1}^{i} B_i(j) \cdot \sum_{k=1}^{n+j} [[\ln(i+n+1)]^\theta - [\ln(i+n-k+1)]^\theta] \epsilon_{n+j,k}$$

and

$$u_1 = \lim_{i\to\infty} (i+1)\, [\ln(i+1)]^2 \sum_{n=0}^{\infty} \lambda_n \sum_{j=1}^{i} B_i(j) \cdot \sum_{k=1}^{n+j} \left[\frac{1}{\ln(i+n+2-k)} - \frac{1}{\ln(i+n+2)}\right] \epsilon_{n+j,k}$$

$$u_\theta = \frac{1}{\theta} \lim_{i\to\infty} (i+1)^{1+\theta} \sum_{n=0}^{\infty} \lambda_n \sum_{j=1}^{i} B_i(j) \cdot \sum_{k=1}^{n+j} \left[\frac{1}{(i+n+1-k)^\theta} - \frac{1}{(i+n+1)^\theta}\right] \epsilon_{n+j,k}$$

$$u'_\theta = \frac{1}{\theta} (i+1)\, [\ln(i+1)]^{1+\theta} \sum_{n=0}^{\infty} \lambda_n \sum_{j=1}^{i} B_i(j) \cdot \sum_{k=1}^{n+j} \left[\frac{1}{[\ln(i+n+2-k)]^\theta} - \frac{1}{[\ln(i+n+2)]^\theta}\right] \epsilon_{n+j,k}.$$

We are assuming that the limits above exist, which indeed happens in most practical cases. The theorem is valid if any of these limits is infinite. In particular, if $L = +\infty$, then X_n is always recurrent. Note that usually, it is not necessary to carry out all the computations, because one of the three terms in the definition of L is equal to one of the terms in the definition of U. In fact, in most cases, we have $\sup_{0<\theta<1} l_\theta = \inf_{0<\theta<1} u_\theta$ if $0 < p < 1$, and $u_1 = l_1$ if $p = 1$. The proof of Theorem 4 can be found in [26].

References

[1] W. A. Rosenkrantz and D. Towsley, "On the instability of the slotted Aloha multiaccess algorithm," *IEEE Trans. Automat. Contr.*, vol. AC-28, pp. 994–996, 1983.

[2] G. Fayolle, E. Gelenbe, and J. Labetoulle, "Stability and optimal control of the packet switching broadcast channel," *J. Ass. Comput. Mach.*, vol. 24, no. 3, pp. 375–386, 1977.

[3] M. Kaplan, "A sufficient condition for nonergodicity of a Markov chain," *IEEE Trans. Inform. Theory,* vol. IT-25, pp. 470–471, 1979.

[4] D. Bertsekas and R. Gallager, *Data Networks.* Englewood Cliffs, NJ: Prentice-Hall, 1987.

[5] B. Hajek and T. Van Loon, "Decentralized dynamic control of a multiaccess broadcast channel," *IEEE Trans. Automat. Contr.*, vol. AC-27, pp. 559–569, 1982.

[6] B. Hajek, "Hitting-time and occupation-time bounds implied by drift analysis with applications," *Adv. Appl. Prob.*, vol. 14, pp. 502–525, 1982.

[7] J. N. Tsitsiklis, "Analysis of a multiaccess control scheme," *IEEE Trans. Automat. Contr.*, vol. AC-32, pp. 1017–1020, 1987.

[8] S. Verdú, "Minimum probability of error for asynchronous Gaussian multiple-access channels," *IEEE Trans. Inform. Theory,* vol. IT-32, pp. 85–96, 1986.

[9] D. Raychaudhuri, "Performance analysis of random access packet-switched code division multiple access systems," *IEEE Trans. Commun.*, vol. COM-29, pp. 895–901, 1981.

[10] A. Polydoros and J. Silvester, "Slotted random access spread-spectrum networks: An analytical framework," *IEEE J. Select. Areas Commun.*, vol. SAC-5, no. 6, pp. 989–1002, 1987.

[11] M. K. Simon, J. K. Omura, R. A. Scholtz, and B. K. Levitt, *Spread-Spectrum Communications.* New York: Computer Science Press, 1985.

[12] N. Abramson, "The throughput of packet broadcasting channels," *IEEE Trans. Commun.*, vol. COM-25, pp. 117–128, 1977.

[13] D. J. Goodman and A. A. M. Saleh, "The near/far effect in local Aloha radio communications," *IEEE Trans. Vehicular Technol.*, vol. VT-36, Feb. 1987.

[14] L. G. Roberts, "Aloha packet system with and without slots and capture," *Comput. Commun. Rev.*, no. 5, pp. 28–42, 1975.

[15] C. Namislo, "Analysis of mobile radio slotted Aloha networks," *IEEE J. Select. Areas Commun.*, vol. SAC-2, pp. 583–588, 1984.

[16] D. H. Davis and S. A. Gronemeyer, "Performance of slotted Aloha random access with delay capture and randomized time of arrival," *IEEE Trans. Commun.*, vol. COM-28, pp. 703–710, 1980.

[17] J. C. Arnbak and W. van Blitterswijk, "Capacity of slotted Aloha in Rayleigh fading channels," *IEEE J. Select. Areas Commun.*, vol. SAC-5, no. 2, pp. 261–269, 1987.

[18] C. C. Lee, "Random signal levels for channel access in packet radio networks," *IEEE J. Select. Areas Commun.*, vol. SAC-5, no. 6, pp. 1026–1034, 1987.

[19] B. S. Tsybakov, V. A. Mikhailov, and N. B. Likhanov, "Bounds for packet transmission rate in a random multiple access system," *Problemy Peredachi Informatsii,* vol. 19, no. 1, pp. 61–81, 1983.

[20] N. Mehravari, "Collision resolution in random access systems with multiple reception," preprint, 1987.

[21] A. Shwartz and M. Sidi, "Erasure, capture and noise errors in controlled multiple access networks," in *Proc. 25th Conf. Decision Contr.*, Dec. 1986, pp. 1333–1334.

[22] R. E. Kahn, S. A. Gronemeyer, J. Burchfiel, and R. C. Kunzelman, "Advances in packet radio technology," *Proc. IEEE,* vol. 66, pp. 1468–1496, 1978.

[23] W. Szpankowski, "Analysis and stability considerations in a reservation multiaccess system," *IEEE Trans. Commun.*, vol. COM-31, no. 5, pp. 684–692, 1983.

[24] A. G. Pakes, "Some conditions for ergodicity and recurrence of Markov chains," *Oper. Res.*, vol. 17, pp. 1058–1061, 1969.

[25] L. I. Sennott, P. A. Humblet, and R. L. Tweedie, "Mean drifts and the non-ergodicity of Markov chains," *Oper. Res.*, vol. 31, pp. 783–789, 1983.

[26] S. Ghez, S. Verdú, and S. C. Schwartz, "Random access communication with multipacket reception capability. I: Stability of Aloha," Dep. Elect. Eng., Princeton Univ., Tech. Rep. ISS. 8748, 1987.

[27] J. F. Mertens, E. Samuel-Cahn, and S. Zamir, "Necessary and sufficient conditions for recurrence and transience of Markov chains, in terms of inequalities," *J. Appl. Prob.*, vol. 15, pp. 848–851, 1978.

[28] C. M. Harris and P. G. Marlin, "A note on feedback queues with bulk service," *J. Assoc. Mach.*, vol. 19, pp. 727–733, 1972.

[29] F. G. Foster, "On the stochastic matrices associated with certain queuing processes," *Ann. Math. Statist.*, no. 24, pp. 355–360, 1953.

[30] P. Vit, "On the equivalence of certain Markov chains," *J. Appl. Prob.*, vol. 13, pp. 357–360, 1976.

[31] S. Ghez, S. Verdú, and S. C. Schwartz, "On decentralized control algorithms for multipacket Aloha," in *Proc. 25th Allerton Conf. on Commun., Contr., and Comput.*, Oct. 1987.

Tree Algorithms for Packet Broadcast Channels

JOHN I. CAPETANAKIS

Abstract—The multiaccessing of a broadcast communication channel by independent sources is considered. Previous accessing techniques suffer from long message delays, low throughput, and/or congestion instabilities. A new class of high-speed, high-throughput, stable, multiaccessing algorithms is presented. Contentions resolving tree algorithms are introduced, and they are analyzed for specific probabilistic source models. It is shown that these algorithms are stable (in that all moments of delay exist) and are optimal in a certain sense. Furthermore, they have a maximum throughput of 0.430 packets/slot and have good delay properties. It is also shown that, under heavy traffic, the optimally controlled tree algorithm adaptively changes to the conventional time-division multiple access protocol.

I. Introduction

A. Problem Definition and Issues

THIS paper considers the time-domain multiple accessing of a broadcast channel by a set of independent sources. A broadcast channel is a communications channel in which a signal generated by one transmitter can be received by many receivers. Some examples of these channels are a satellite channel in which a satellite acts as a transponder; a ground radio network where a number of terminals employ common frequencies; and coaxial cables where a number of terminals use a single cable for communicating by time multiplexing. When a number of spatially isolated independent sources have access to such a channel, a problem usually arises in allocating the channel capacity to the various users. This is so because the only means these independent users have of communicating their requests for channel capacity is through the channel itself.

The data are assumed to be transmitted in packet form. A packet is a fixed length block of digital data that contains the information to be transmitted, along with the source and destination addresses and any other overhead information that might be necessary such as error correcting and error detecting bits. We will assume that if more than one transmitter transmits simultaneously, then they will interfere with each other and all the packets will be received incorrectly. The resolution of this contention is the heart of the multiaccessing problem, and it is here that we will focus our attention by developing and analyzing a new class of conflict-resolving algorithms.

Manuscript received May 25, 1978; revised September 22, 1978. This paper was presented in part at the IEEE International Symposium on Information Theory, Ithaca, NY, October 10–14, 1977.

The author is with Lincoln Laboratory, Massachusetts Institute of Technology, Lexington, MA 02173.

There are three main issues that are involved in multiaccessing a common channel: a) the percentage of the capacity used in accessing, b) the time it takes to access the channel, and c) the stability of the multiaccessing system, i.e., whether the system may behave so that increasingly many sources are attempting to access the channel but decreasingly few are actually succeeding. These three properties (throughput, delay, and stability) are defined more precisely as follows.

Average throughput: the ratio of the number of packets that are successfully transmitted in a very long interval to the maximum number of packets that could have been transmitted with continuous transmission on the channel.

Average delay: the ratio of the total delay of the packets in a very long interval to the number of packets in the interval. The delay is the time from the instant that a packet arrives at a source to the instant that it is successfully received.

Stability: we shall call the system kth order stable if the first k moments of the delay of a randomly chosen packet are finite.

It should be clear that a desirable multiaccessing system is one that is at least first-order stable, has high average throughput, and has low accessing delay.

B. Channel Model

The channel is assumed to be slotted; that is, the channel time is divided into equal segments called slots. The length of each slot equals that of a packet, and it is assumed that each source is synchronized to the channel and transmits a packet within only one slot. Furthermore, the channel is such that the sources can determine after transmission whether there have been zero, one, or multiple packets in the slot. Multiple packets per slot correspond to a collision, and under such circumstances no packet gets through. If, however, a packet does not collide with other packets, then it is assumed to be successfully transmitted. This last assumption allows us to focus on the multiaccessing properties of the channel.

C. Poisson Source Model

The source model that will be used here is the Poisson source model, which assumes the existence of an infinite number of independent identical sources that collectively

Reprinted from *IEEE Trans. Inform. Theory*, vol. IT-25, no. 5, pp. 505–515, September 1979.

generate k packets/slot, where k is a Poisson random variable with mean λ. Furthermore, the number of arrivals in any one slot is independent of the arrivals in other slots. Note that, since the number of sources is infinite but the new traffic is finite, the probability that any one source will have more than one message arrival in a finite interval is zero.

D. The Multiaccessing Protocol

Two of the more prominent basic past techniques for accessing a communications channel in the time domain are conventional time-division multiple access (TDMA) and the Aloha algorithm. In TDMA, contention is avoided by dedicating a portion of the channel to each of the sources. Although TDMA is effective in situations where the number of sources is small and the message lengths are long, it suffers from a large average delay when the number of sources is large and the duty cycle is short.

In the Aloha algorithm, when a source has a new packet, it transmits it and then listens to the channel to determine whether or not a collision occurred. If a collision is detected, then the source retransmits the packet at a randomly selected time. The retransmission takes place at a randomly selected time so that the conflicting packets will surely not collide again. It has been shown [1]–[6] that when the sources satisfy the Poisson source model, then the maximum throughput for the slotted Aloha system is $1/e$. However, a multiaccess Aloha type system for such a source model is unstable; eventually it reaches a situation where the number of retransmitting sources tends to infinity and the throughput tends to zero. Therefore, although the delay and throughput properties might be satisfactory in the short term, they are quite poor when observed over a long interval of time. Actually, if the sources satisfy the Poisson source model, the average delay of the Aloha system, as well as of TDMA, is infinite.

There are numerous "dynamic reservation schemes" in the literature [7]–[11]. These schemes generally use two channels: a reservation channel and a data channel. Before data can be transmitted, the data channel must be reserved through the reservation channel by the TDMA and/or Aloha techniques. It follows then, that the disadvantages of the TDMA and Aloha algorithms apply to the reservation channel.

In this paper, we introduce a new time-domain multiple access scheme, the tree algorithm, that is based on the observation that a contention among several active sources (an active source is one with a packet to transmit) is resolved if, and only if, all the sources are somehow subdivided into groups such that each group contains at most one active source. In contrast with TDMA and Aloha, the tree algorithm has a maximum average throughput of 0.430 packet/slot and is stable in that all the moments of the delay are finite. (The tree algorithm can, of course, also be used to place reservations in a reservation system.)

It should be emphasized that, depending on the message arrival and length statistics, reservation schemes are capable of throughput approaching one. One, however, should be careful not to compare the performance of the tree algorithm to the overall reservation system. The comparison should be made to the performance of only the reservation channel, for it is here where the conflict resolving tree algorithm would be used.

E. Outline and Results

In Section II the static binary tree algorithm is presented and analyzed. The main results obtained are a) the average throughput versus average delay curve shown in Fig. 5; b) the maximum average throughput is 0.347 packets/slot; and c) all the moments of the delay are finite if λ (the packet arrival rate) is less than 1/3 packets/slot. (The last result seems overly conservative, and indications are that the moments are finite for $\lambda <$ 0.347.) In Section III the optimum dynamic tree algorithm is determined and analyzed. First, it is shown that the optimum tree is binary everywhere, except for the root node whose optimum degree depends on traffic conditions (see (3.4)). Second, the E[delay] versus E[throughput] performance of the optimum algorithm is obtained; this is illustrated in Fig. 7. Finally, it is shown that the maximum average throughput is 0.430 packets/slot and that all the moments of the delay are finite for $\lambda < 0.430$ packets/slot. Section IV contains several observations and extensions, and Section V is the conclusion.

II. Static Binary Tree

A. Presentation of the Tree Algorithm

In this section the tree algorithm is stated, and an example is presented. First, however, some definitions concerning the tree graph are given (see Fig. 1).

Depth of a node: the number of branches between the node and the root node (the root node is at depth zero).

Degree of node: the number of branches that emanate from a node.

Subtree T_{ij}: the rooted subtree whose root node is n_{ij}, where, in a binary rooted tree, j corresponds to the particular one of the 2^i nodes at depth i.

Algorithm Statement: Let each source correspond to a leaf on a binary tree as shown in Fig. 1(a). If the number of sources is infinite, then the tree extends to infinity. This representation of the sources may be considered to be a binary addressing scheme. In Fig 1(a), for example, each source has a four-bit binary address. Also, let T_x and T_y be two rooted subtrees, and assume that no collisions have occurred up to the beginning of the present pair of slots. Then, the binary tree algorithm can be stated as follows.

1) Set $T_x = T_{10}$; $T_y = T_{11}$.
2) Transmit all the packets from sources in T_x in the first slot of the present pair of slots, and transmit all

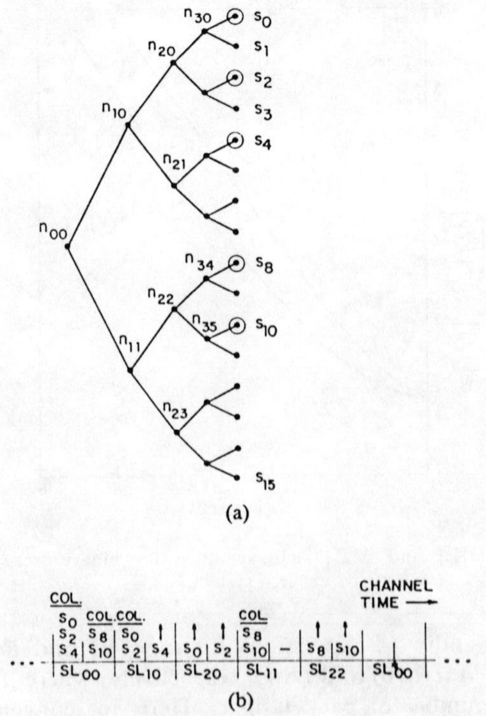

Fig. 1. Example of binary tree algorithm.

the packets from sources in T_y in the second slot. (As has been pointed out above, a source in the Poisson source model can have at most one packet.)

3) If any collisions occur in the preceding step, then
   a) until these collisions are resolved, no new packets are transmitted;
   b) resolve the first collision (if any) before resolving the second (if any).

   A collision in T_x (or T_y) is resolved by dividing T_x (or T_y) into two halves (say A and B), setting $T_x = A$, $T_y = B$, and then repeating steps 2) and 3).

This algorithm is equivalent to the following tree search. Beginning with the root node and at each succeeding node, ask whether there are zero, one, or more than one packet in the subtree stemming from each of the two emanating branches. If the answer at any of the branches is more than one, then proceed to the node at the end of this branch and repeat the question. This continues until all the leaves are separated into sets such that it is known that each set contains at most one packet. Note that the number of slots used in a conflict resolution interval equals twice the number of nodes visited.

Example: Let there be 16 sources $\{S_0, S_1, \cdots, S_{15}\}$, and let each correspond to a leaf on the 16-leaf binary tree as shown in Fig. 1(a). Fig. 1(b) depicts the slotted channel time. Note that the slots are paired and that a slot pair is designated by SL_{ij}. For convenience, the round-trip delay τ_r is taken as zero. If $\tau_r > 0$, then the slot pairs would be separated by τ_r.

Now assume that no collisions have occurred until the beginning of SL_{00}, when sources S_0, S_2, S_4, S_8, and S_{10} each has a packet to transmit. Then beginning with SL_{00} where the first contention arises, the tree algorithm takes the following steps in the indicated slot pairs.

SL_{00} All the sources in T_{10} that have packets to transmit (i.e., S_0, S_2, and S_4) do so in the first slot of SL_{00}, and the corresponding sources in T_{11} do so in the second slot. This results in two collisions, one among S_0, S_2, and S_4, and the other between S_8 and S_{10}. Since there was at least one collision in SL_{00}, any new packets that arrive are not transmitted until that contention is resolved.

SL_{10} Since there was a collision in T_{10}, the sources at T_{10} are divided in half and the packets in T_{20} and T_{21} are transmitted in the first and second slots of SL_{10}, respectively. This results in a collision between S_0 and S_2 and in a successful transmission by S_4.

SL_{20} Since there was a collision in T_{20}, T_{30} and T_{31} transmit their packets in the first and second slots of SL_{20}, respectively. This results in two successful transmissions by S_0 and S_2.

SL_{11} Since there was a collision in T_{11}, T_{22} and T_{23} transmit their packets in succession. This results in a collision between S_8 and S_{10} in the first slot and no transmission in the second.

SL_{22} Since there was a collision in T_{22}, T_{34} and T_{35} transmit. This results in two successful transmissions by S_8 and S_{10}.

SL^*_{00} Since each of the sources participated in at least one collision-free transmission, we know that the original contention has been resolved. Any new packets that may have arrived to T_{10} during this conflict resolution interval are transmitted in the first slot of SL^*_{00}, and packets that arrived to T_{11} are transmitted in the second slot. The process continues, as described above.

In this example, 10 slots were used to transmit 5 packets.

B. Average Delay

The objective of this section is to develop upper and lower bounds to the average delay that are functions of the packet arrival rate. A by-product of this analysis is the determination of the maximum average throughput. The stability analysis of the binary tree algorighm will not be presented here, because it is similar to that of the optimum dynamic tree algorithm given in Section III. However, in [13] we have shown that $E[(\text{delay})^k]$ exists for all k if $\lambda < 1/3$ packets/slot.

The unit of measure for the delay is the time between two successive steps of the algorithm. This is constant, and it equals the round-trip delay plus the transmission time of two packets. Next, several definitions are presented (all lengths are in algorithmic steps, where an algorithmic step is the round-trip delay plus two slots):

epoch The interval beginning with the pair of slots in which a conflict first occurs and ending with the pair of slots where the original conflict is finally resolved. If there is no conflict in the first pair of slots, then the epoch is simply this pair.

ϵ_1 The epoch in which a randomly chosen packet arrives.

ϵ_2 The epoch following ϵ_1.

y_1 The length of ϵ_1. This is a random incidence random variable.

l_j The random variable that equals the length of ϵ_j. The probability that $l_j = L$ is the steady state probability that an epoch of length L occurs. Note that l_j is a discrete state discrete time Markov process.

d_j The time spent by the randomly chosen packet in ϵ_j.

μ_j The average number of packets that arrive in ϵ_j given l_j. Note that $\mu_j = 2\lambda l$.

δ The packet delay, or the time spent in the system by the randomly chosen packet. This equals the sum of d_1 and d_2.

The E[delay] analysis is organized into two parts: first, upper and lower bounds to E[delay] are obtained in terms of $E[l]$ and $E[l]^2$, where

$$E[l] = \lim_{j\to\infty} E[l_j] \qquad E[l^2] = \lim_{j\to\infty} E[l_j^2]; \tag{2.1}$$

second, upper and lower bounds to $E[l]$ and $E[l^2]$ are obtained (as a function of λ) and are substituted into the former bounds to obtain upper and lower bounds on the E[delay] versus arrival rate performance.

1) Bounds to E[delay] versus $E\{l\}$ and $E\{l^2\}$: The packet delay may be written as

$$\delta = d_1 + d_2. \tag{2.2}$$

Therefore,

$$E\{\delta\} = E\{d_1\} + E\{d_2\}. \tag{2.3}$$

Since the slot of epoch ϵ_1 in which the test packet arrives is with equal probability any slot in ϵ_1, we have

$$E\{d_1\} = \frac{1}{2}E\{y_1\}. \tag{2.4}$$

However, from [12, p. 149], the length y_1 of the arrival epoch ϵ_1 is distributed as

$$P_{y_1}(y_1) = \frac{y_1 P_{l_1}(y_1)}{E\{l_1\}}. \tag{2.5}$$

Therefore,

$$E\{d_1\} = \frac{1}{2}\frac{E\{l^2\}}{E\{l\}}. \tag{2.6}$$

Since

$$E\{d_2\} = \sum_{y_1} E\{d_2|y_1\} p_{y_1}(y_1),$$

it follows from (2.3), (2.5), and (2.6) that

$$E\{\delta\} = \frac{1}{2}\frac{E\{l^2\}}{E\{l\}} + \frac{1}{E\{l\}}\sum_{l_1} l_1 E\{d_2|l_1\} p_{l_1}(l_1), \tag{2.7}$$

which is as far as we can go in obtaining a closed-form solution for $E\{\delta\}$. Next we consider bounding $E\{\delta\}$.

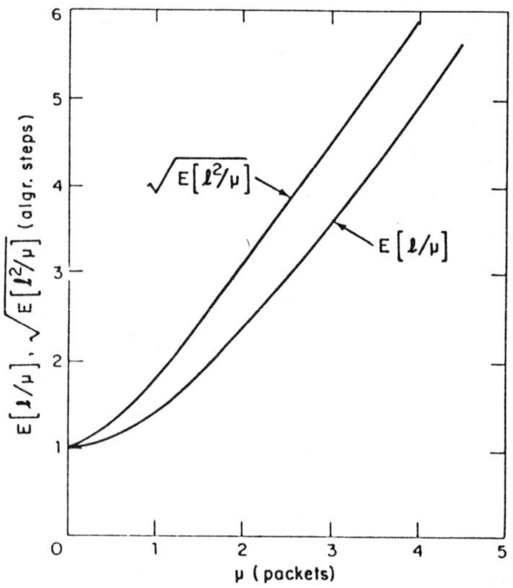

Fig. 2. $E\{l|\mu\}$ and $\sqrt{E\{l|\mu\}}$ versus μ for binary tree/Poisson source system.

In Appendix A, expressions for $E\{d_2|\mu\}$ and $E\{l_2|\mu\}$ are derived ((A.10) and (A.4), respectively) where μ is the average number of packets in ϵ_1. Here, for convenience, the subscript of μ is dropped—these expressions more generally apply to $E\{d_{j+1}|\mu_j\}$ and $E\{l_{j+1}|\mu_j\}$. From (A.4) and (A.10), it can be shown (see [13]) that

$$\tfrac{1}{2}E\{l_2|\mu\} \leqslant E\{d_2|\mu\} \leqslant 0.55E\{l_2|\mu\} + 0.321, \qquad \text{for } \mu \geqslant 0. \tag{2.8}$$

Using (2.8) in (2.7) together with the relation $\mu = 2\lambda l$ and noting that $E^2\{l\} \leqslant E\{l_1 l_2\} \leqslant E\{l^2\}$ (see Appendix B), we obtain the bounds

$$E\{\delta\} \leqslant 1.05\frac{E\{l^2\}}{E\{l\}} + 0.321 \tag{2.9}$$

$$E\{\delta\} \geqslant \frac{1}{2}\frac{E\{l^2\}}{E\{l\}} + \frac{1}{2}E\{l\}. \tag{2.10}$$

This concludes the first part of the E[delay] analysis.

2) Bounds to $E\{l\}$ and $E\{l^2\}$ versus λ: From Appendix A, we have

$$E\{l_{j+1}|\mu\} = 1 + 2D(\mu/2) \tag{2.11}$$

where $D(\cdot)$ is defined by (A.2), and

$$E\{l_{j+1}^2|\mu\} = 1 + 2D(\mu/2) + [2D(\mu/2)]^2 + F(\mu) \tag{2.12}$$

where $F(\cdot)$ is defined by (A.12). The quantities $E\{l_{j+1}|\mu\}$ and $\sqrt{E\{l_{j+1}^2|\mu\}}$ are plotted in Fig. 2.

A. Upper Bound to $E\{l\}$

Theorem 2.1: Let l_j be a positive integer corresponding to the state of a Markov chain after the jth transition. Also assume that, for some constants b and α_u such that $0 \leqslant \alpha_u < 1$,

$$E\{l_{j+1}|l_j\} \leqslant \alpha_u(l_j - 1) + b, \qquad \text{for } l_j \geqslant 1, \tag{2.13}$$

then

$$\lim_{j\to\infty} E\{l_j\} \leq \frac{b-\alpha_u}{1-\alpha_u}. \tag{2.14}$$

Proof: Taking expectations on both sides of (2.13) results in

$$E\{l_{j+1}\} \leq \alpha_u E\{l_j\} + b - \alpha_u. \tag{2.15}$$

Since $0 \leq \alpha < 1$, (2.15) can be solved recursively to obtain (2.14). Q.E.D.

In relating the above theorem to the system parameters, we have from Fig. 2 and (2.13) that

$$\alpha_u = 2.88\lambda \tag{2.16a}$$

and

$$b = E\{l_{j+1}|l_j=1\} = E\{l_{j+1}|\mu=2\lambda\}, \tag{2.16b}$$

so that (2.14) gives

$$E\{l\} \leq \frac{E\{l|\mu=2\lambda\} - 2.88\lambda}{1-2.88\lambda}. \tag{2.17}$$

B. Lower Bound to $E\{l\}$

Theorem 2.2: Let l_j be the state of a Markov chain (with initial state l_0) after the jth transition, and let $f_c(l_j)$ be a positive, convex, and nondecreasing lower bound to $E\{l_{j+1}|l_j\}$. If

$$l_j = f_c(l_j) \tag{2.18}$$

has a unique solution $l_j = l^*$ and if $f_c(l_j) < l_j$ for $l_j > l^*$, then

$$\lim_{j\to\infty} E\{l_j\} \geq l^*. \tag{2.19}$$

If, on the other hand, (2.18) has two solutions, then there exists an initial state l_0 for the Markov chain such that

$$\lim_{j\to\infty} E\{l_j\} = \infty. \tag{2.20}$$

Proof:

$$E\{l_j\} = \sum_{l_j, l_{j-1}} l_j p(l_j|l_{j-1}) p(l_{j-1}) \tag{2.21}$$

$$= \sum_{l_{j-1}} E\{l_j|l_{j-1}\} p(l_{j-1}) \tag{2.22}$$

$$\geq \sum_{l_{j-1}} f_c(l_{j-1}) p(l_{j-1}) \tag{2.23}$$

$$\geq f_c[E\{l_{j-1}\}]. \tag{2.24}$$

Inequality (2.23) follows because $f_c(l_j) \leq E\{l_{j+1}|l_j\}$, and (2.24) follows from the convexity of f_c.

Applying (2.24) to $l_{j-1}, l_{j-2}, \cdots, l_1$ recursively and using the fact that f_c is nondecreasing yields

$$E\{l_j\} \geq f_c^{(j)}(l_0) \tag{2.25}$$

where

$$f_c^{(j)}\{l_0\} \triangleq \underbrace{f_c[f_c[\cdots[f_c(l_0)]]]}_{j \text{ times}}. \tag{2.26}$$

Now assume that $l_j = f_c(l_j)$ at only one point l^* and that

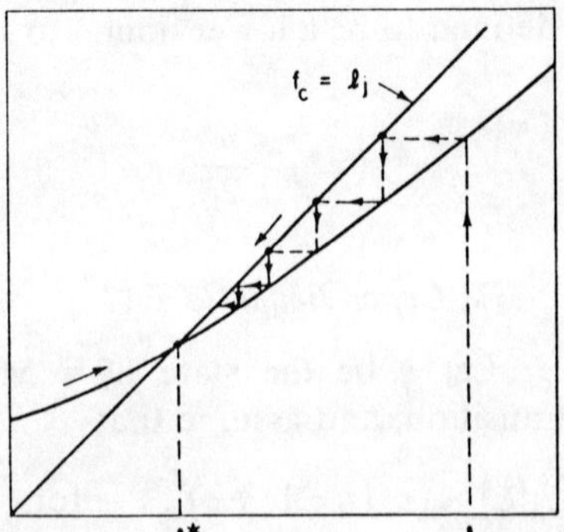

Fig. 3. Illustrating conditions for convergence of lower bound to $\lim_{j\to\infty} E\{l_j\}$.

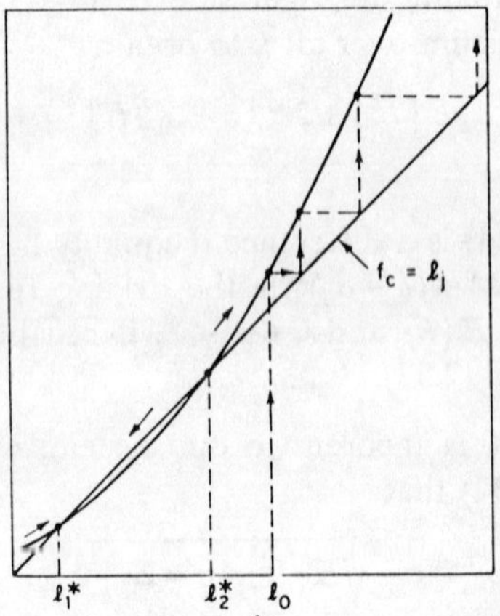

Fig. 4. Illustrating conditions for divergence of lower bound to $\lim_{j\to\infty} E\{l_j\}$.

$f_c(l_j) < l_j$ for $l_j > l^*$. Then $f_c(l_j)$ is as shown in Fig. 3, and since f_c is convex it follows that

$$\lim_{j\to\infty} f_c^{(j)}(l_0) = f_c(l^*) = l^*. \tag{2.27}$$

Now if $l_j = f(l_j)$ has two solutions, l_1^* and l_2^* ($l_1^* > l_2^*$), then $f(l_j)$ and l_j are as shown in Fig. 4. Hence, if $l_0 > l_2$, then

$$\lim_{j\to\infty} f_c^{(j)}(l_0) = \infty. \tag{2.28}$$

Q.E.D.

The next step is to relate this theorem to the system and to obtain a lower bound to $E\{l\}$ as a function of λ.

It follows from (2.11) and Fig. 2 that $f_l(\mu)$ (given below) is a convex increasing positive lower bound to $E\{l_{j+1}|\mu\}$:

$$f_l(\mu) \triangleq 1 + \begin{cases} 2D(\mu/2), & \text{for } \mu \leq 8 \\ 2.88\left(\frac{\mu}{2}-4\right) + 2D(4), & \text{for } \mu > 8. \end{cases} \tag{2.29}$$

Since $\mu = 2\lambda l_j$, it follows from Theorem 2.2, (2.29), and Fig. 2 that

$$\bar{l}_l(\lambda) = \begin{cases} \dfrac{\mu^*}{2\lambda}, & \text{for } \lambda < 0.347 \\ \infty, & \text{for } \lambda \geq 0.347 \end{cases} \tag{2.30}$$

where $\bar{l}_l(\lambda)$ is defined to be a lower bound to $E\{l\}$, and μ^* satisfies

$$f_l(\mu^*) = \frac{\mu^*}{2\lambda}. \tag{2.31}$$

C. Upper Bound to $E\{l^2\}$

Theorem 2.3: Let l_j be the state of a Markov chain after the jth transition, and assume that

$$E\{l_{j+1}^2 | l_j\} \leqslant (\alpha_u(l_j - 1) + c)^2, \qquad \text{for } l_j \geqslant 1. \tag{2.32}$$

Then, for $0 \leqslant \alpha_u < 1$ and $\lim_{j\to\infty} E\{l_j\} = E\{l\}$,

$$\lim_{j\to\infty} E\{l_j^2\} \leqslant \frac{2\alpha_u(c-\alpha_u)E\{l\} + (c-\alpha_u)^2}{1-\alpha_u^2}. \tag{2.33}$$

Proof: Expand the right side of (2.31), multiply by $p(l_j)$, and then sum over all l_j to obtain

$$E\{l_{j+1}^2\} \leqslant \alpha_u^2 E\{l_j^2\} + 2\alpha_u(c-\alpha_u)E\{l_j\} + (c-\alpha_u)^2. \tag{2.34}$$

Inequality (2.34) is a difference inequality in $E\{l_j^2\}$ where $2\alpha_u(c-\alpha_u)E\{l_j\} + (c-\alpha_u)^2$ is the driving function. Since $\lim_{j\to\infty} E\{l_j\} = E\{l\}$ and $0 \leqslant \alpha_u < 1$, it can be used recursively to obtain (2.33). Q.E.D.

In relating this theorem to our system, we have from Fig. 2 and (2.32) that

$$c = \sqrt{E\{l_{j+1}^2 | \mu = 2\lambda\}} \tag{2.35}$$

$$\alpha_u = 2.88. \tag{2.36}$$

Therefore,

$$E\{l^2\} \leqslant \frac{5.76(c-\alpha_u)\lambda E\{l\} + (c-\alpha_u)^2}{1-(2.88\lambda)^2}. \tag{2.37}$$

D. Lower Bound to $E\{l^2\}$

Theorem 2.4: Let l_j be as given in the statement of Theorem 2.3 except that

$$E\{l_{j+1}^2 | l_j\} \geqslant (\alpha_l l_j + c_l)^2. \tag{2.38}$$

Then,

$$\lim_{j\to\infty} E\{l_j^2\} \geqslant \frac{2c_l\alpha_l E\{l\} + c_l^2}{1-\alpha_l^2}. \tag{2.39}$$

Proof: This proof is identical to that of Theorem 2.3 with the inequalities reversed.

Finally, in relating Theorem 2.4 to the system parameters, we have from Fig. 2 and (2.38) that

$$E\{l^2 | \mu\} \geqslant (1.44\mu + 0.25)^2. \tag{2.40}$$

Therefore,

$$E\{l^2\} \geqslant \frac{1.44E\{l\}\lambda + 0.125}{1-(2.88\lambda)^2}. \tag{2.41}$$

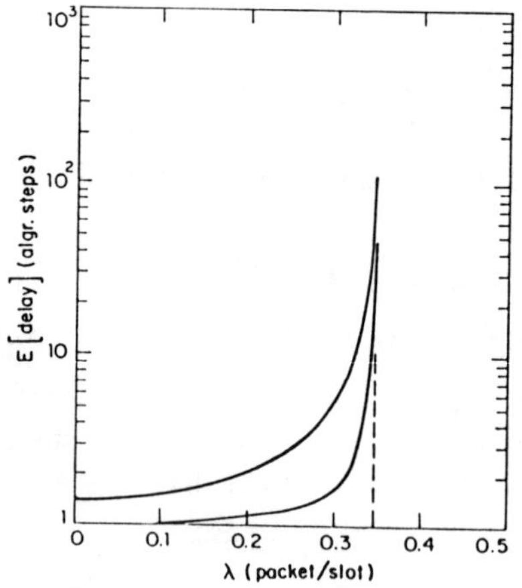

Fig. 5. Upper and lower bounds to average delay versus average arrival rate for binary tree/Poisson source system. Note: An algorithmic step equals one round-trip delay plus two slots.

E. Upper and Lower Bounds to $E\{\delta\}$

From (2.9), (2.30), and (2.37) we have

$$E\{\delta\} \leqslant \frac{6.05\lambda(c-2.88\lambda)}{1-(2.88\lambda)^2} + \frac{1.05(c-2.88\lambda)^2}{(1-(2.88\lambda)^2\bar{l}_l(\lambda)} + 0.321 \tag{2.42}$$

where c and $\bar{l}_l(\lambda)$ are given by (2.35) and (2.30), respectively.

Similarly, a lower bound to $E\{\delta\}$ is given by

$$E\{\delta\} \geqslant \frac{0.72\lambda}{1-(2.88\lambda)^2} + \frac{1}{2}\bar{l}_l(\lambda). \tag{2.43}$$

Since $E\{\delta\} \geqslant \bar{l}_l(\lambda)$, it follows that a tighter bound is

$$E\{\delta\} \geqslant \max\left[\frac{0.72\lambda}{1-(2.88\lambda)^2} + \frac{1}{2}\bar{l}_l(\lambda), \bar{l}_l(\lambda)\right]. \tag{2.44}$$

Inequalities (2.42) and (2.44) (illustrated in Fig. 5) are the desired results.

C. Average Throughput

If $E\{\delta\} < \infty$, then $E\{\text{throughput}\} = \lambda$. Consequently, (2.42) and (2.44), as well as Fig. 5, may be interpreted as the $E\{\text{delay}\}$ versus $E\{\text{throughput}\}$ relation. Therefore, the maximum average throughput for this static algorithm is 0.347 packets/slot.

III. The Optimum Dynamic Tree Algorithm

This section considers a dynamic algorithm in which the tree is allowed to vary from epoch to epoch depending on traffic conditions. Within any one epoch, the tree is held fixed, and the search is similar to that of the static algorithm of Section II.

This section is organized into two parts. The first considers the optimum tree to be used in ϵ_{j+1}. The optimum tree for ϵ_{j+1} is defined to be that tree which minimizes the expected number of slots used in ϵ_{j+1}, given the observation of the transmission process up to the end of ϵ_j. The second part considers the delay, throughput, and stability analysis of an optimal dynamic protocol in which (for ease in analysis) all nodes are constrained to have even degree. (Note that here the tree is fixed within any one epoch. Allowing the tree to vary optimally within an epoch would result in an even faster and higher throughput algorithm.)

A. The Optimum Tree Protocol

Since the number of arrivals in any one slot is a Poisson random variable that is independent of the arrivals in any other slot, it follows that the only quantity from the transmission process that is needed to fully characterize the number of packets to be processed during ϵ_{j+1} is h_j, the number of slots used in ϵ_j. (Note that, given h_j, the number of packets to be processed in ϵ_{j+1} is Poisson distributed will mean $\mu=\lambda h_j$.)

It can be shown, by a procedure similar to that in Appendix A, that the expected number of slots required to process the packets, given μ, is

$$E\{h|\mu\}=g_0+\mu\sum_{i=0}^{\infty}\frac{g_{i+1}}{M_i}\xi(M_i) \tag{3.1}$$

where $\xi(\cdot)$ is defined in (A.3), g_i is the degree of an ith level node, and

$$M_i \triangleq \frac{\mu}{g_0 g_1\cdots g_i}. \tag{3.2}$$

Thus the problem of determining the optimum tree is identical to determining the $\{g_i;\ i=0,1,2,\cdots\}$ which minimize $E\{h|\mu\}$ subject to the constraints that g_i is an integer and

$$g_i \geqslant \begin{cases} 1, & i=0, \\ 2, & i>0. \end{cases} \tag{3.3}$$

It can be shown that the optimum g_i is given by

$$g_0^*=\begin{cases} 1, & \mu\leqslant 1.70 \\ n, & 1.70+1.15(n-2)<\mu\leqslant 1.70+1.15(n-1) \end{cases}$$

$$g_i^*=2, \qquad \text{for } i\geqslant 1, \text{ all } \mu. \tag{3.4}$$

Equation (3.4) is obtained in [13] as follows. First, it is shown that g_i^* equals 2 for sufficiently large i. Second, it is proved that the minimum of $E\{h|\mu\}$ over g_{k-1} and g_k, subject to the condition that $g_i=2$ for $i>k$, occurs at $g_k^*=2$. Next, this result is applied recursively (with the first result as the boundary condition) to prove that $g_i^*=2$ for $i\geqslant 1$. Finally, g_0^* is calculated as a function of μ.

In light of (3.4), the average number of packets per transmission is seen to be close to one. This is an intuitively pleasing result, since a packet is successfully transmitted if, and only if, there is exactly one packet per transmission. Now the optimum dynamic tree algorithm may be stated as follows.

Statement of the Optimum Dynamic Tree Algorithm

1) Observe h_j, the number of slots in the preceding epoch.
2) Calculate $\mu=\lambda h_j$.
3) Determine the optimum tree to be used in the following epoch from (3.4).
4) Execute the tree search in the following epoch.

B. Analysis of the Dynamic Algorithm

The dynamic tree that is analyzed in this section is slightly less optimum than that given in (3.4). Here, all the nodes are restricted to be binary except for the root node whose degree can take on values $2r$ ($r=1,2,3,\cdots,$). As in the preceding section, r is chosen to minimize $E\{h|\mu,r\}$. The optimum r was computed and found to be

$$r^*(\mu)=\begin{cases} 1, & \text{for } \mu\leqslant 3.40 \\ r, & \text{for } 3.40+(r-2)2.30<\mu\leqslant 3.40+(r-1)2.30. \end{cases} \tag{3.5}$$

The tree search is carried out by processing two slots at a time (i.e., two branches transmit in two consecutive slots, and the results of these transmissions are received before two other branches can transmit). Consequently, the number of algorithmic steps per epoch equals r plus the number of nodes, besides the root node, that are visited. Note that the search of a tree in which $g_0=2r$ is equivalent to r binary tree searches where each of the r searches is over a fraction $1/r$ of the sources.

1) Average Delay: In the following two theorems, $E\{l|\mu,r\}$, $E\{l^2|\mu,r\}$, and $E\{d_2|\mu,r\}$ are related to $E\{l|\mu/r,1\}$, $E\{l^2|\mu/r,1\}$, and $E\{d_2|\mu/r,1\}$, respectively, the corresponding quantities for the binary tree.

Theorem 3.1: Let l and μ be as defined above and let the source model be Poisson. Then

$$E\{l|\mu,r\}=rE\left\{l|\frac{\mu}{r},1\right\} \tag{3.6}$$

and

$$E\{l^2|\mu,r\}=r(r-1)E^2\left\{l|\frac{\mu}{r},1\right\}+rE\left\{l^2|\frac{\mu}{r},1\right\}. \tag{3.7}$$

Proof: Divide the $2r$ subtrees that emanate from the root node into r pairs, and let l_i be the number of nodes that are visited in the ith pair. Then,

$$l=\sum_{i=1}^{r}(1+l_i). \tag{3.8}$$

Equation (3.6) follows by taking expectations of (3.8) and noting that $1+l_i$ equals the number of nodes visited in a binary tree in which the mean number of contending packets is μ/r.

Next, square both sides of (3.8) to obtain

$$l^2=2\sum_{i=1}^{r-1}\sum_{j=i+1}^{r}(1+l_i)(1+l_j)+\sum_{i=1}^{r}(1+l_i)^2. \tag{3.9}$$

Taking expectation in (3.9) and noting the independence of l_i and l_j for $i\neq j$, we obtain (3.7). Q.E.D.

Theorem 3.2: Let the source model be Poisson, and let l, d_2, and μ be as defined above. Then

$$E\{d_2|\mu,r\} \leqslant 0.55E\{l|\mu,r\}+0.321 \tag{3.10}$$

and

$$E\{d_2|\mu,r\} \geqslant \frac{1}{2}E\{l|\mu,r\}. \tag{3.11}$$

Proof: Assuming that each of the r subtree pairs that emanate from the root node is processed in sequence, we have

$E\{d_2|\mu,r$, test packet is in the ith pair$\}$

$$=(i-1)E\left\{l|\frac{\mu}{r},1\right\}+E\left\{d_2|\frac{\mu}{r},1\right\}. \tag{3.12}$$

Next multiply by $1/r$ and sum over i to obtain

$$E\{d_2|\mu,r\}=\frac{r-1}{2}E\left\{l|\frac{\mu}{r},1\right\}+E\left\{d_2|\frac{\mu}{r},1\right\}. \tag{3.13}$$

Substituting (2.8) into (3.13) results in

$$E\left\{d_2|\frac{\mu}{r},1\right\} \leqslant \frac{(r+.1)}{2r}rE\left\{l|\frac{\mu}{r},1\right\}+0.321. \tag{3.14}$$

Inequality (3.12) follows by applying Theorem 3.1 to (3.14) and then noting that, for $r \geqslant 1$, $(r+0.1)/2r \leqslant 0.55$.

To obtain (3.11), substitute (2.8) into (3.13) and apply Theorem 3.6. Q.E.D.

By using Theorems 3.1 and 3.2 in the argument that resulted in (2.9) and (2.10), it can be seen that these two equations also hold for the dynamic tree.

Next substitute $r^*(\mu)$ into (3.6) and (3.7) to obtain $E\{l|\mu,r^*(\mu)\}$ and $\sqrt{E\{l^2|\mu,r^*(\mu)\}}$ as shown in Fig. 6. As in Section II, it can be shown that

$$E\{l\} \leqslant 1+\frac{2D(\lambda)}{1-2.325\lambda} \tag{3.15}$$

$$E\{l\} \geqslant l^* \tag{3.16}$$

$$E\{l^2\} \leqslant \frac{4.65a\lambda E\{l\}+a^2}{1-(2.35\lambda)^2} \tag{3.17}$$

$$E\{l^2\} \geqslant \frac{2.79\lambda E\{l\}+0.36}{1-(2.325\lambda)^2} \tag{3.18}$$

where $l^*=f_c(l^*), f_c$ is a convex positive increasing lower bound to $E\{l_{j+1}|l_j,r^*\}$, and $a=\sqrt{E\{l^2|\mu=2\lambda\}-2.325\lambda}$. Finally, substituting (3.15) and (3.18) into (2.9) and (2.10) results in

$$E\{\delta\} \leqslant \frac{4.88a\lambda}{1-(2.325\lambda)^2}+\frac{1.05a^2}{[1-(2.325\lambda)^2]\,\bar{l}_l(\lambda)}+0.321 \tag{3.19}$$

$$E\{\delta\} \geqslant \max\left[\left(\frac{1.395\lambda}{1-(2.325\lambda)^2}+\frac{0.18}{[1-(2.325\lambda)^2]\,\bar{l}_u(\lambda)}+\frac{1}{2}\bar{l}_l(\lambda)\right);\ \bar{l}_l(\lambda)\right]. \tag{3.20}$$

Inequalities (3.19) and (3.20) are illustrated in Fig. 7.

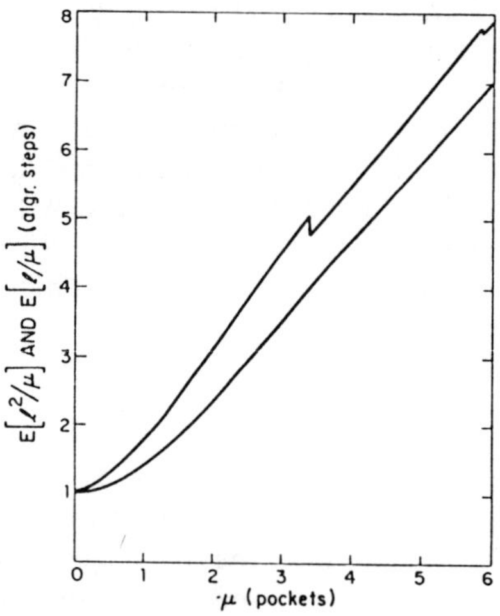

Fig. 6. $E\{l|\mu\}$ and $\sqrt{E\{l^2|\mu\}}$ versus μ for optimum dyanmic tree/Poisson source system.

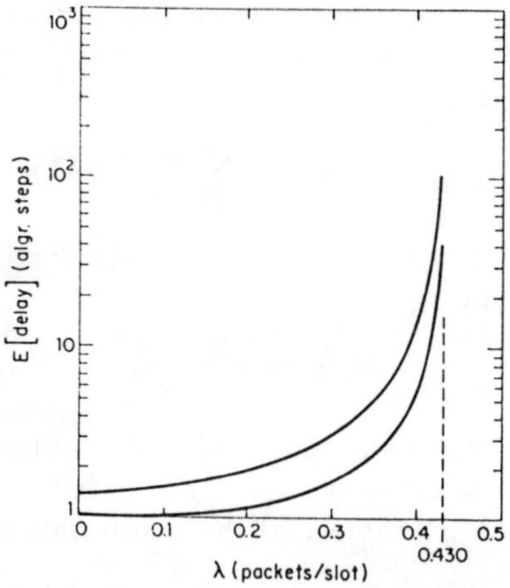

Fig. 7. Upper and lower bounds to the $E\{\text{delay}\}$ versus arrival rate for the optimum dynamic tree/Poisson source system. Note: An algorithmic step equals one round-trip delay plus two slots.

2) Average Throughput: The average delay versus average arrival rate results of (3.19) and (3.20) and Fig. 7 may be interpreted as average delay versus average throughput. Therefore, the maximum average throughput is 0.430 packets/slot.

3) System Stability: Let a dynamic tree be binary everywhere except for the root node which has degree $2r^*(\mu)$ as given by (3.5); then, $E\{(\text{delay})^k\}<\infty$ for all k if $\lambda<0.430$ packets/slot. This follows from the following two theorems.

Theorem 3.3: Let $E_d\{l|\mu\}$ be the expectation of l, given μ, for the dynamic algorithm, and let $G(s,\mu)$ and $G_d(s,\mu)$ be the generating functions of l, given μ, for the static and dynamic tree algorithms. Then

$$G_d(s,\mu)<e^{s[E_d\{l|\mu\}+\mu\epsilon/3.4]} \tag{3.21}$$

for $\mu \geqslant 0$, $0 \leqslant s \leqslant s_0$, $s_0>0$ and any $\epsilon>0$.

Proof: From Appendix C, $G(s,\mu)$ and $\partial G(s,\mu)/\partial s$ are bounded for $s<\ln 2$. Since G is a generating function, it follows that G and $\partial G/\partial s$ are continuous in s. Since $\Pr\{j \text{ packets are active} | \mu\} = \mu^j e^{-\mu}/j!$, it also follows that G and $\partial G/\partial s$ are continuous in μ. Therefore, in any closed bounded region, say $[0 \leqslant s \leqslant \frac{1}{2}\ln 2,\ 0 \leqslant \mu \leqslant 3.4]$, $\partial \ln G(s,\mu)/\partial s$ is uniformly continuous. Thus, for any $\epsilon > 0$, an $s_0 > 0$ exists such that

$$\frac{\partial}{\partial s}\ln G(s,\mu) \leqslant \frac{\partial}{\partial s}\ln G(s,\mu)|_{s=0} + \epsilon, \quad \text{for } 0 \leqslant s \leqslant s_0,\ 0 \leqslant \mu \leqslant 3.4. \tag{3.22}$$

Integrating and noting that $\partial \ln G(s,\mu)/\partial s|_{s=0} = E\{l|\mu\}$ results in

$$G(s,\mu) \leqslant e^{s[E\{l|\mu\}+\epsilon]}, \qquad \text{for } s \leqslant s_0,\ 0 \leqslant \mu \leqslant 3.4. \tag{3.23}$$

But the dynamic tree is identical to $r^*(\mu)$ independent binary trees, each with an average number of μ/r^* contending packets. Therefore,

$$G_d(s,\mu) \leqslant e^{sr^*[E\{l|\mu/r^*\}+\epsilon]}, \qquad \text{for } s \leqslant s_0. \tag{3.24}$$

Equation (3.21) follows by noting that $rE\{l|\mu/r\} = E_d\{l|\mu\}$ (see (3.6)), and $\mu/r^* \leqslant 3.40$ (see (3.5)). Q.E.D.

Theorem 3.4: Let l_j be the state of a Markov chain (with initial state $l_0 < \infty$) after the jth transition. Furthermore, let

$$E\{e^{sl_{j+1}}|l_j\} \leqslant e^{s(al_j+b)} \tag{3.25}$$

where a and b are constant, $0 \leqslant a < 1$. Then,

$$\lim_{j\to\infty} E\{e^{sl_j}\} \leqslant e^{sb/(1-a)}. \tag{3.26}$$

Proof: Taking expectations in (3.25) results in

$$E\{e^{sl_{j+1}}\} \leqslant E\{e^{sal_j}\}e^{sb}.$$

Using the above inequality recursively, we have

$$E\{e^{sl_j}\} \leqslant E\{e^{a^j sl_0}\}e^{sb(a^{j-1}+\cdots+1)}. \tag{3.27}$$

Since $l_0 < \infty$ and $0 \leqslant a < 1$, (3.26) follows by letting $j \to \infty$. Q.E.D.

From Fig. 6, $E_d\{l|\mu\} \leqslant 1 + 1.63\mu$. Since $l = l_{j+1}$ and $\mu = 2\lambda l_j$, we have from Theorems 3.3 and 3.4 that

$$E\{e^{sl}\} \text{ exists if } \lambda \leqslant \frac{1}{2(1.163+\epsilon/3.4)} \text{ and } s \leqslant s_0.$$

But ϵ is arbitrary so $\lambda_{\max} = 0.430$ packets/slot.

Now since $E\{e^{sl}\}$ exists for $0 \leqslant s \leqslant s_0$, $s_0 > 0$, it follows that $E\{l^k\}$ exists for all k. We conclude by observing that, if $E\{l^{k+1}\} < \infty$, then $E\{\delta^k\} < \infty$.

IV. Observations and Extensions

A. Serial and Parallel Tree Searching

There are two basic ways to execute the tree search, serially or in parallel. In the serial search, the one analyzed in this paper, two branches at a time transmit their packets in two consecutive slots. In the parallel search, on the other hand, all the branches at some depth, whose parents have had a collision, transmit their packets in consecutive slots. The number of slots required to process any particular set of contending sources is the same for both schemes. However, when the round-trip delay is nonzero, the parallel search will result in a smaller $E\{\text{delay}\}$. Consequently, the $E\{\text{delay}\}$ results developed here for the serial search may be considered as upper bounds for the parallel search $E\{\text{delay}\}$.

B. Random Addressing

Another variation of the tree algorithm is that in which the source addresses are assigned randomly. With random addressing, the sources are not preassigned to the leaves of a tree. When a collision occurs, then, beginning with the root node, each of the contending sources decides independently and with equal probability which branch to take at each node. The process continues until all the sources have been grouped into sets which have not experienced any collisions. The random addressing scheme has the advantage that, since the addresses are not preassigned, sources can enter or leave the system with much greater ease. Furthermore, note that the tree search may still be either serial or parallel, and that random addressing may be applied to the static, as well as the dynamic, algorithm. If the source model is the Poisson, then both addressing schemes have the same performance. If, on the other hand, the number of sources is finite, deterministic addressing is slightly better. This is because, for a finite source model, deterministic addressing requires only a finite tree whereas random addressing requires an infinite tree.

C. TDMA and the Tree Algorithms

A tree algorithm (for a finite source model), in which the tree has exactly one node with degree equaling the number of sources, is identical to TDMA. Therefore, TDMA may be considered as a special case of the tree algorithm. Consequently, TDMA is at most as efficient as the optimally chosen tree algorithm. Furthermore, it can be shown (see [13]) that, as traffic increases, the optimum dynamic tree for a finite source model adaptively changes from a binary tree to TDMA. This is an intuitively pleasing result.

D. Suboptimum but Efficient Algorithm

A suboptimum dynamic algorithm that is easy to implement is the one in which the root node has degree 2^n and all other nodes are binary (n is chosen so as to minimize the expected number of slots needed). The implementation advantage arises from the observation that the variations in the root node degree may be realized in a single-binary-tree simply by varying the depth of the nodes where the algorithm originates. This algorithm has also been analyzed in [13] where it is shown that $0.420 \leqslant \max E\{\text{throughput}\} \leqslant 0.430$.

E. Higher Throughput Algorithms

Finally, the question arises whether there are multi-accessing algorithms that yield higher $E\{\text{throughput}\}$. There are two restrictions imposed upon tree algorithms considered here whose relaxation will result in $E\{\text{throughput}\}$ greater than 0.430 packets/slot. The first restriction is the one which requires that the packets arriving to nonactive sources in one epoch not be transmitted until the following epoch. In Fig. 1, for example, if S_9 and/or S_{11} receive packets after SL_{00} and before SL_{11}, they can join the ongoing algorithm as if their packets arrived in SL_{00}. The second restriction is the one which requires that the tree be fixed within any one epoch. If the transmission process is at some point within an epoch, there is information (concerning the contending sources) which has been received since the beginning of the present epoch that might be used to restructure[1] the tree optimally for the remaining epoch.

V. Conclusion

The tree algorithms, a general class of multiaccessing algorithms, have been introduced and analyzed here. They are based upon the observation that a channel contention is resolved if, and only if, all the sources are divided into sets such that each set contains at most one source with a packet to transmit. These algorithms were shown to be stable, to have maximum $E\{\text{throughput}\}$ of 0.430 packets/slot, and to have very good average delay properties (see Figs. 5 and 7).

The results developed here are directly applicable to packet switching broadcast channels. However, they may also be applied to a more general system in which a central facility is accessible by a number of independent users. If, for example, the number of users that the facility can handle simultaneously is less than the maximum number that can place requests upon it, then contentions will arise that may be resolved advantageously with the techniques developed here.

Acknowledgment

This research was supervised by Professor R. G. Gallager. His comments and guidance have greatly contributed to the substance, as well as form, of this work.

Appendix A
$E\{l|\mu\}$, $E\{l^2|\mu\}$, $E\{d_2|\mu\}$ of the Binary Tree Algorithm

Only the derivation of $E\{l|\mu\}$ is presented here. The derivations of $E\{l^2|\mu\}$ and $E\{d_2|\mu\}$ are similar (see [13]). We begin with the following.

[1]One restructuring technique, proposed by J. L. Massey, is based upon the observation that an empty slot, immediately following a collision, means that the next slot must give a collision. Thus the corresponding node may be skipped, resulting in maximum $E\{\text{throughput}\}$ of 0.376 for the static algorithm and 0.462 for the optimum dynamic algorithm.

Lemma A.1: A node n_{ij}, $i \neq 0$, is traversed by the binary tree algorithm if, and only if, there are at least two active sources in T_{ij}.

Proof: Let $\{n_{mr}: m=0,1,\cdots,i\}$ be the set of all the nodes that lie on the path from n_{00} to n_{ij}. Since T_{ij} is included in $\{T_{mr}: m=0,1,\cdots,i\}$, it follows that if T_{ij} contains at least two active sources, so does $\{T_{mr}: m<i\}$. Now, if the protocol is at node n_{mr}: $m<i$ and asks for the number of active sources in $T_{m+1,r}$, the answer will be greater than one, and it will move to node $n_{m+1,r}$. Since this holds for at least $m=0,1,\cdots,i-1$, we conclude that node n_{ij} will be traversed by the algorithm. If, on the other hand, there is at most one source in T_{ij} and the algorithm has got at least as far as $n_{i-1,r}$ and asked for the number of active sources in T_{ij}, the answer will be 0 or 1, and it will not continue to n_{ij}. Q.E.D.

Theorem A.2: Let the number of contending sources be a Poisson random variable with mean μ, and let l be the number of nodes traversed by the binary tree algorithm in resolving the conflict. Furthermore, let

$$x_{ij} \triangleq \begin{cases} 1, & \text{if } n_{ij} \text{ is visited} \\ 0, & \text{otherwise} \end{cases} \tag{A.1}$$

$$D(\mu) \triangleq \sum_{i=0}^{\infty} 2^i \xi(\mu/2^i) \tag{A.2}$$

$$\xi(\mu) \triangleq 1 - e^{-\mu} - \mu e^{-\mu}. \tag{A.3}$$

Then

$$E\{l|\mu\} = 1 + 2D(\mu/2). \tag{A.4}$$

Proof: From (A.1) we have that

$$l = \sum_{i=0}^{\infty} \sum_{j=0}^{2^i-1} x_{ij} \tag{A.5}$$

and, from Lemma A.1,

$$\Pr(x_{ij}=1|\mu) = \begin{cases} \Pr(\text{At least 2 active sources in } T_{ij}|\mu), & i \neq 0 \\ 1, & i=0. \end{cases} \tag{A.6}$$

But since the number of active sources is Poisson distributed and the sources are independent of each other, it follows that the number of active sources in T_{ij} is also Poisson distributed with parameter $\mu/2^i$. Therefore,

$$\Pr(x_{ij}=1|\mu) = \xi(\mu/2^i). \tag{A.7}$$

From (A.5), we have

$$E\{l|\mu\} = \sum_{i=0}^{\infty} \sum_{j=0}^{2^i-1} \Pr(x_{ij}=1|\mu) \tag{A.8}$$

and, from (A.7),

$$E\{l|\mu\} = 1 + \sum_{i=1}^{\infty} 2^i \xi(\mu/2^i). \tag{A.9}$$

Similar but somewhat more involved derivations result in

$$E\{d_2|\mu\} = 1 + \sum_{i=1}^{\infty} \left[1 + \tfrac{1}{2} D(\mu/2^i)\right] \xi(\mu/2^i)/(1 - e^{-\mu/2^i}) \tag{A.10}$$

and

$$E\{l^2|\mu\} = 1 + 2D(\mu/2) + [2D(\mu/2)]^2 + F(\mu) \tag{A.11}$$

where

$$F(\mu) \triangleq 2\sum_{i=1}^{\infty} 2^i D(\mu/2^i)[1-\xi(\mu/2^i)] + \sum_{i=1}^{\infty} 2^i \xi^2(\mu/2^i). \tag{A.12}$$

APPENDIX B
$E\{l_1 l_2\} > E\{l_1\}E\{l_2\}$

First, note that by substituting $l = l_2$ and $\mu = 2\lambda l_1$ into (A.4) and then differentiating, it can be shown that $(\partial/\partial l_1)E\{l_2|l_1\} \geq 0$.

Theorem B.1: Let $E\{l_2|l_1\}$ be nondecreasing in l_1. Then, assuming all expectations exist,

$$E\{l_1 l_2\} \geq E\{l_1\}E\{l_2\}. \tag{B.1}$$

Proof: An equivalent statement of (B.1) is

$$E\{(l_1-\bar{l}_1)(l_2-\bar{l}_2)\} \geq 0. \tag{B.2}$$

But

$$E\{(l_1-\bar{l}_1)(l_2-\bar{l}_2)\} = \sum_{l_1}(l_1-\bar{l}_1)[E\{l_2|l_1\}-\bar{l}_2]p(l_1) \tag{B.3}$$

$$= \sum_{l_1<\bar{l}}(\bar{l}_1-l_1)[\bar{l}_2-E\{l_2|l_1\}]p(l_1) + \sum_{l_1>\bar{l}_1}(l_1-\bar{l}_1)[E\{l_2|l_1\}-\bar{l}_2]p(l_1). \tag{B.4}$$

Since $E\{l_2|l_1\}$ is nondecreasing in l_1, we have

$$E\{(l_1-\bar{l}_1)(l_2-\bar{l}_2)\} \geq \sum_{l_1<\bar{l}_1}(\bar{l}_1-l_1)[\bar{l}_2-E\{l_2|\bar{l}_1\}]p(l_1) + \sum_{l_1>\bar{l}_1}(l_1-\bar{l}_1)[E\{l_2|\bar{l}_1\}-\bar{l}_2]p(l_1)=0, \tag{B.5}$$

which completes the proof.

APPENDIX C
UPPER BOUND TO $E\{e^{sl}|\mu\}$

$$G_u(s,\mu) = e^{-\mu}\left(e^s - \left(\frac{2-e^s}{e^s}\right)\right) + \mu e^{-\mu}(e^s-1) + \frac{2-e^s}{e^s}e^{\mu[(e^s/(2-e^s))-1]}. \tag{C.1}$$

Then

$$E\{e^{sl}|\mu\} \leq G_u(s,\mu), \qquad \text{for } s \leq \ln 2 \tag{C.2}$$

$$\frac{\partial}{\partial s}E\{e^{sl}|\mu\} \leq \frac{\partial}{\partial s}G_u(s,\mu), \qquad \text{for } s \leq \ln 2. \tag{C.3}$$

Proof: Let $f_j(s) \triangleq E\{e^{sl}|j \text{ packets to resolve}\}$; then

$$f_j(s) = \sum_{i=1}^{j-1} e^s 2^{-j}\binom{j}{i} f_i(s) f_{j-i}(s) + 2^{-j+1}e^s f_j(s). \tag{C.4}$$

We have divided j into i packets in the upper subtree and $j-i$ packets in the lower. Note that e^s is the contribution of the root node, $2^{-j}\binom{j}{i}$ is the probability of the assumed division, and $f_i(s)$ and $f_{j-i}(s)$ are the generating functions for the number of steps to resolve each subtree. Furthermore, note that $f_0(s)=f_1(s)=1$, because exactly one step is required when there is either zero or one packet to resolve.

Now, solving (C.4) for $f_j(s)$, we have

$$f_j(s) = \frac{\sum_{i=1}^{j-1} e^s 2^{-j}\binom{j}{i} f_i(s) f_{j-i}(s)}{1-2^{-j+1}e^s}, \qquad \text{for } j>1, \tag{C.5}$$

and by induction,

$$f_j(s) \leq \left(\frac{e^s}{2-e^s}\right)^{j-1}, \qquad \text{for } j \geq 1,\ s < \ln 2. \tag{C.6}$$

Taking expectations in (C.6) results in (C.2).

The derivation of (C.3) is similar to that of (C.1) and is not given here.

REFERENCES

[1] N. Abramson, "Packet switching with satellites," in *Proc. National Computer Conference*, vol. 42, pp. 695–702, 1973.

[2] L. G. Roberts, "Aloha packet system with and without slots and capture, ASS note 8, ARPA Network Info. Ctr., Stanford Res. Inst., Stanford, CA, June 26, 1972.

[3] R. M. Metcalf, "Packet communication," Tech. Rep. MAC TR 114, Mass. Inst. Tech., Cambridge, MA, Dec. 1973.

[4] L. Kleinrock and S. S. Lam, "Packet switching in a multiaccess broadcast channel: Performance evaluation," *IEEE Trans. Commun.*, vol. COM-23, pp. 410–423, Apr. 1975.

[5] S. S. Lam and L. Kleinrock, "Packet swsitching in a multiaccess broadcast channel: Dynamic control procedures," *IEEE Trans. Commun.*, vol. COM-23, pp. 891–904, Sept. 1975.

[6] A. B. Carleial and M. B. Hellman, "Bistable behavior of Aloha-type systems," *IEEE Trans. Commun.*, vol. COM-23, pp. 401–410, Apr. 1975.

[7] W. Crowther, R. Rettberg, D. Walden, S. Ornstein, and F. Heart, "A system for broadcast communication: Reservation-Aloha," in *Proc. 6th Hawaii Inter. Conf. Syst. Sci.*, pp. 371–374, Jan. 1973.

[8] L. G. Roberts, "Dynamic allocation of satellite capacity through packet reservation," *Nat. Comp. Conf.*, vol. 42, pp. 711–716, 1973.

[9] R. Binder, "A dynamic packet-switching system for satellite broadcast channels," Tech. Rep. B74-5, University of Hawaii, Honolulu, HI, Aug. 1974.

[10] L. Kleinrock and F. A. Tobagi, "Packet switching in radio channels, Part I," *IEEE Trans. Commun.*, vol. COM-23, pp. 1400–1416, Dec. 1975.

[11] F. A. Tobagi and L. Kleinrock, "Packet switching in radio channels, Part II, *IEEE Trans. Commun.*, vol. COM-23, pp. 1417–1433.

[12] A. W. Drake, *Fundamentals of Applied Probability Theory*. New York: McGraw-Hill, 1967.

[13] J. I. Capetanakis, "The multiple access broadcast channel: Protocol and capacity considerations," Tech. Rep. ESL-R-806, Mass. Inst. of Tech., Cambridge, MA, March 1978.

Packet Transmission in Radio Networks

B. S. TSYBAKOV AND V. L. BAKIROV

Abstract—**The article considers a communication network employing an ALOHA random multiple-access system. A general formulation of the problem is given and three network models are proposed: with random-walk packets, with fixed routes, and a general model. The stability conditions of the network are determined. Optimization of the throughput with respect to the radius of action of the transmitters is performed.**

1. Introduction

COMPUTER and data networks employ random multiple access (RMA) of packets. RMA is employed to organize common use of a broadcast channel by many users with low pulsating traffic.

Most studies dealing with RMA consider communications systems in which all the users transmit packets over a single cable, as in Ethernet [1], or over a single satellite channel, as in the ALOHA system [2]. In practice, however, there are systems that are lacking a single common channel for all users.

Assume, for example, that the users are arranged in such a way that they are at the nodes of a regular rectangular grid with distance d between adjacent nodes. The system employs nondirectional radio communication with a transmitter radius of action equal to d. In this case, the reception zone of each node includes four other nodes of the lattice that are spaced a distance d from it, i.e., the distance of one hop. If the addressee of a packet departing from a node is not one of its four neighbors, the packet will be transmitted through the grid from one node to another with retransmissions (e.g., via the shortest path) until it reaches the addressee.

A difficulty arises in that the network does not exclude conflicts (interference) between packets; specifically, packet conflicts occur at the receiver input of any node on some time interval if at least two of the four adjacent nodes transmit packets in this interval. A node that has sent a packet learns at the end of its transmission whether the packet became involved in a conflict; if it did, the transmission is assumed to be unsuccessful, and the node must subsequently make another attempt to transmit the packet in accordance with the RMA algorithm.

In the network example under consideration, therefore, there is no channel that is common to all users, but rather some transmitting medium in which adjacent nodes may become involved in the conflicts. The aim of this paper is to consider networks with conflicts of this type and to investigate the efficiency of the ALOHA RMA algorithm in them. We will not confine ourselves (as in the example) to networks with nodes on a rectangular grid.

In the extreme particular case in which any network node is a neighbor of any other, the network becomes the network with one common channel (single-hop network) that is usually considered in RMA theory. In this particular case, our results become those of [3].

It is known that in a single-segment network the ALOHA RMA algorithm is unstable under certain conditions. This manifested itself in that the lengths of the packet queues at the nodes increase without limit. Unfortunately, the property of instability is preserved in a multihop network with ALOHA algorithm. In this paper we determine the conditions under which a multihop network operates stably, and the conditions under which stability is lost. The paper consists of two parts. The first part comprises section 3, in which the network is analyzed under a special assumption regarding packet routing. The stability conditions are expressed in terms of the average route length, the number $k^{(i)}$ of nodes located in the zone of the transmission/reception equipment of node i, and the parameter p of the ALOHA algorithm. The second part comprises sections 4–6. Here we offer a very general description of probabilistic packet routing, with the routing of the first part included as a particular case. The stability conditions are expressed in terms of the average number of packets whose route passes through a node, and the average number of packets successfully transmitted by a node.

As the radius of action of the transmitters increases, with the network geography remaining constant, there is a positive effect stemming from the fact that the number of hops required for a packet to reach its destination is reduced, but there is also a negative effect resulting from the increase in the number of nodes with which any given node may become involved in a conflict. Under these conditions, there exists an optimal radius of transmitter action which yields the highest transmission rate in the network. In section 6 we offer an exact formulation of the optimal-radius problem; the problem is solved using examples of symmetrically matched networks.

Multihop networks have been treated in a large number of studies [4]. However, the use of RMA in them has rarely been considered. The ALOHA algorithm was investigated in [5] and [6] in relation to networks. These studies did not consider the incoming streams at the stations or

Translated from *Problemy Peredachi Informatsii*, vol. 21, no. 1, pp. 80–101, January–March, 1985. Original article submitted July 16, 1984.

Reprinted with permission from *Problemy Peredachi Informatsii*, vol. 21, no. 1, pp. 80–101, Jan.-March 1985.

the queues at them. It was assumed that the network geography is random [5] or specified [6], and that all stations are statistically identical. A definition of network stability was lacking.

Some results of this study were published in abridged form in [7] and [8].

2. General Assumptions

Here we offer some general assumptions regarding the structure and action of the network, which will remain valid until the end of the article.

Consider a radio network consisting of $M < \infty$ stations, the stations being numbered $i = 1, \dots, M$. Stations i and j are called neighbor (or adjacent) stations if they lie within the transmitter radii of one another. The topology of the network is arbitrary but specified. The network topology is a nondirectional graph with M vertices. The graph has an edge connecting vertices i and j if and only if stations i and j are neighbor stations. Station i has $k^{(i)}$ neighbors; the set of neighbors of station i is denoted by $B^{(i)}$.

Stations exchange messages which are transmitted in the form of packets. The packets are of equal length α, measured in "slot length" units; it is assumed that $\alpha < 1$.

Assume that station j belonging to $B^{(i)}$ transmits a packet. It is assumed that station i has successfully received a packet (and station j has successfully transmitted it) if this packet is addressed to i as the final addressee or for transit and, during transmission, station i and all stations from $B^{(i)}$ except for j were silent. Let us clarify that if a station receives without conflict a packet that is not addressed to it either as the final addressee or for transit, then this station is not assumed to have successfully received a packet.

A station that has successfully received a packet informs the station that transmitted the packet by sending an acknowledgment. After receiving the acknowledgment, the station that successfully transmitted the packet discards it from its queue.

If a station successfully receives a transit packet or generates a new packet, it places it in its queue. The location of the packet in queue, i.e., the servicing discipline at the station, is unimportant for us, since in what follows we will be concerned only with the queue lengths at the stations and the conditions under which they are finite. However, we will always assume that the servicing discipline at the stations is specified, although not necessarily the same at different stations. If a station successfully receives a packet that is addressed to it as the final addressee, it removes it from the network.

If a station has transmitted a packet but has not received acknowledgment of successful reception, it leaves the packet in queue for subsequent attempts of transmission.

Stations can initiate packet transmission only at times $t \in I_0$. The time interval $[t, t+1)$ is called a slot. If a station transmits a packet in slot $[t, t+1)$, this transmission occurs in the interval $[t, t+\alpha)$. In the remaining interval $[t+\alpha, t+1)$ of slot $[t, t+1)$, a station that has successfully received a packet transmits an acknowledgment. If a station has not successfully received a packet in $[t, t+\alpha)$, it remains silent in $[t+\alpha, t+1)$. Since for any station i that has transmitted a packet in slot $[t, t+1)$, there is not more than one station from $B^{(i)}$ that has successfully received a packet, acknowledgments cannot become involved in conflict.

The ALOHA RMA algorithm in a network means that if the queue at station i is not empty at time t, then, with probability p, station i transmits a packet in slot $[t, t+1)$ and is silent with probability $(1-p)$, where p is independent of t and i.

The queue vector of the network is $\mathbf{N}_t = (n_t^{(1)}, \dots, n_t^{(M)})$, where $n_t^{(i)}$ is the length of the queue of packets at station i at time t, i.e., at the instant when slot $[t, t+1)$ starts. Vector $\mathbf{N}_t$, considered as a function of t, where $t \in I_0$, while $I_j = \{j, j+1, \dots\}$ is the set of integers not less than j, is a random-vector sequence.

The initial state of the network is some set of random variables associated with the network, pertaining to time $t = 0$. In section 3, the initial state will be vector $\mathbf{N}_0$. In section 4, the initial state will be the ordered set of routes of packets that are in queues at network stations at time $t = 0$. A third example of initial state is given in section 5. In all the cases considered here, the queue vector $\mathbf{N}_0$ is necessarily included in the initial state. A given initial state is called finite if $\sum_{i=1}^{m} l_0^{(i)} < \infty$, where $l_0^{(i)} \geq 0$ are the nonrandom queue lengths at network stations for this given initial state, i.e., $n_0^{(i)} = l_0^{(i)}$. For convenience, in the general case, the initial state is denoted by $\mathbf{A}_0$.

Station i is called stable if

$$\lim_{x \to \infty} \liminf_{t \to \infty} \Pr\{n_t^{(i)} < x | \mathbf{A}_0\} = 1 \qquad (2.1)$$

for any given finite initial state $\mathbf{A}_0$. A network is called stable if all its stations are stable. Otherwise a network is called unstable. In a stable network, the sequence $\mathbf{N}_t$ is called stable.

The network description in this section is not complete. It is lacking descriptions of routing and of the input packet traffic. This stems from the fact that different assumptions regarding routing and input traffic are made in sections 3–5.

Single hop network is a particular case of the network under consideration. In such network, $k^{(i)} = M - 1$, $B^{(i)}$ is the set of all stations other than station i.

3. Network With Random-Walk Packets

In this section, we will obtain the stability conditions for a network in which packet routing is performed as follows.

In any transmission, a packet of station i is addressed

to one of its neighbors from $B^{(i)}$; this neighbor is chosen equiprobably among all the $k^{(i)}$ neighbors. Under the condition that the packet has traversed some given path through the network, and has been successfully received by station i by time t, we denote by $\mathscr{P}$ the conditional probability that station i is the addressee of the packet. It is assumed that $\mathscr{P} \geq 1/L$, where L is some constant that is independent of t, i, and the path traversed. Here L has the meaning of the upper bound for the average length of packet path through the network, measured in terms of the number of hops. In what follows, we will also consider the assumption that $\mathscr{P} = 1/L$; all quantities considered under this assumption are marked by an additional superscript, and it is said that they pertain to "model B."

The input traffic to the network is specified as follows. At station i at time $t \in I_0$, a new packet arises with probability $q^{(i)}$; with probability $1 - q^{(i)}$, no new packets arise. Packets that have arisen are placed by the station in its queue.

Construction of the mathematical model of the network is completed by specifying the network topology, routing, input traffic, and RMA algorithm. In what follows, the proposed network model is called Basic Model 1 (BM-1). In BM-1, $\mathbf{N}_0$ is taken to be the initial state of the network, i.e., $\mathbf{A}_0 = \mathbf{N}_0$ in (2.1).

In our mathematical analysis of the network, in addition to BM-1 we will consider auxiliary model A and the above mentioned model B.

Model A. In this model, in the period from the instant at which the queue at the station becomes empty to the instant at which it ceases to be empty, the station is engaged in transmission of dummy packets. Until its first successful transmission, a dummy packet is not included in the number of packets in queue at the station that generated it. Otherwise, dummy packets are in no way different from ordinary ones, and model A coincides with BM-1.

Model B. This model coincides with BM-1 in all respects except for the fact that $\mathscr{P} = 1/L$.

Quantities pertaining to model A and B are indicated by superscripts A and B, respectively.

The following theorem contains a sufficient stability condition for a network with BM-1.

THEOREM 1. If for all $i = 1, \ldots, M$ the following condition is satisfied:

$$q^{(i)} + \frac{L-1}{L} p(1-p)^{k(i)} \sum_{m \in B^{(i)}} \frac{1}{k^{(m)}} < \frac{p}{k^{(i)}} \sum_{m \in B^{(i)}} (1-p)^{k^{(m)}}, \quad (3.1)$$

then a network with BM-1 is stable.

Condition (3.1) admits the following interpretation. The right side of (3.1),

$$s^{(i)} = \frac{p}{k^{(i)}} \sum_{m \in B^{(i)}} (1-p)^{k^{(m)}} \quad (3.2)$$

is the probability of successful packet transmission by station i in model A. Here $s^{(i)}$ is also equal to the average number of packets successfully transmitted by station i in model A in a slot.

The second term of the left side of (3.1),

$$r^{(i)} = \frac{L-1}{L} p(1-p)^{k^{(i)}} \sum_{m \in B^{(i)}} \frac{1}{k^{(m)}} \quad (3.3)$$

is the probability of successful reception of a transit packet by station i in model A. Here $r^{(i)}$ is also equal to the average number of transit packets successfully received by station i in model A in a slot.

Thus, condition (3.1), written in the form $q^{(i)} + r^{(i)} < s^{(i)}$, means that on average in any slot station i successfully transmits more packets than it receives transit and newly arriving packets. It seems that this interpretation of condition (3.1) is in fact a more general form of the sufficient stability condition, which is maintained when the results are extended to networks that do not fit into the BM-1 model.

For a single hop network, $r^{(i)} = 0$ for all i; condition (3.1) becomes $q^{(i)} < p(1-p)^{M-1}$ and coincides with the condition of [3].

Before proceeding to prove Theorem 1, we will give two lemmas that are used in it.

LEMMA 1. For any given i, $1 \leq i \leq M$, the sequence $n_t^{(i)\mathrm{A}}$, $t \in I_0$, is an ergodic Markov chain if and only if condition (3.1) is satisfied for this i.

Proof. Note that the sequence $n_t^{(i)\mathrm{A}}$, $t \in I_0$, is an irreducible aperiodic Markov chain with transition probabilities

$$p_{lm}^{(i)\mathrm{A}} = \Pr\{n_{t+1}^{(i)\mathrm{A}} = l | n_t^{(i)\mathrm{A}} = m\}$$

$$= \begin{cases} s^{(i)}(1-q^{(i)}), & l-m=-1, \\ s^{(i)}q^{(i)} + h^{(i)}(1-q^{(i)}), & l-m=0, \\ r^{(i)}(1-q^{(i)}) + h^{(i)}q^{(i)}, & l-m=1, \\ r^{(i)}q^{(i)}, & l-m=2, \\ 0 \text{ otherwise}, \end{cases} \quad (3.4)$$

for $l \in I_0$, $m \in I_1$, and

$$p_{lm}^{(i)\mathrm{A}} = \begin{cases} h_0^{(i)}(1-q^{(i)}), & l=0, \\ h_0^{(i)}q^{(i)} + r^{(i)}(1-q^{(i)}), & l=1, \\ r^{(i)}q^{(i)}, & l=2 \\ 0 \text{ otherwise}, \end{cases} \quad (3.5)$$

for $l \in I_0$, $m = 0$, where $h^{(i)} = 1 - s^{(i)} - r^{(i)}$, $h_0^{(i)} = 1 - r^{(i)}$.

Such a chain will be ergodic if and only if there exists a

positive solution of the system of linear equations

$$\sum_{m=0}^{\infty} p_{lm}^{(i)A}\pi_m^{(i)} = \pi_l^{(i)}, \quad \sum_{m=0}^{\infty} \pi_m^{(i)} = 1, \quad l \in I_0, \quad (3.6)$$

relative to $\pi_m^{(i)}$, $m \in I_0$, where $\pi_m^{(i)}$ has the meaning of the stationary (as $t \to \infty$) probability of state $n_t^{(i)A} = m$.

We can obtain a solution of system (3.6) by changing over to generator functions $\Phi^{(i)}(x) = \sum_{k=0}^{\infty} x^k \pi_k^{(i)}$ and using the normalization $\lim_{x \to +1} \Phi^{(i)}(x) = 1$. As a result we obtain

$$\pi_0^{(i)} = (s^{(i)} - r^{(i)} - q^{(i)})/s^{(i)}. \quad (3.7)$$

Equation (3.7) shows that $\pi_0^{(i)} > 0$ if and only if

$$s^{(i)} - r^{(i)} - q^{(i)} > 0. \quad (3.8)$$

For system (3.6), the condition $\pi_0^{(i)} > 0$ implies that $\pi_k^{(i)} > 0$, $k \in I_0$ (see, e.g., the Appendix in [3]).

Condition (3.8) coincides with (3.1) for specified i. ▼

LEMMA 2. Assume that n_t and n_t^*, $t \in I_0$, are two random sequences with values from I_0 while A_1 and A_2 are certain random events associated with them. If for any t, x, m, and l from I_0

$$\Pr\{n_{t+1} \geq x | n_t = l, n_0 = m, A_1\} \leq \Pr\{n_{t+1}^* \geq x | n_t^* = l, n_0^* = m, A_2\},$$

$$\Pr\{n_{t+1}^* \geq x | n_t^* = l, n_0^* = m, A_2\} \leq \Pr\{n_{t+1}^* \geq x | n_t^* = l+1, n_0^* = m, A_2\},$$

then

$$\Pr\{n_t^* < x | n_0^* = l, A_2\} \leq \Pr\{n_t < x | n_0 = l, A_1\}$$

for any t, x, m from I_0.

Proof. This lemma differs from Lemma 1 of [3] only in that event A_2 is introduced. Its proof coincides with that of Lemma 1 of [3] and is omitted here. ▼

Proof of Theorem 1. We use Lemma 2, setting $n_t = n_t^{(i)}$, $n_t^* = n_t^{(i)B}$ for an arbitrary given i and assuming that $A_1 = \{\mathbf{N}_0 = \mathbf{l}\}$, where $\mathbf{l} = (l^{(1)}, \ldots, l^{(M)})$ is an arbitrary given finite state, while $A_2 = \{n_0^* = m\}$. We have

$$\Pr\{n_t^{(i)B} < x | n_0^{(i)B} = l^{(i)}\} \leq \Pr\{n_t^{(i)} < x | \mathbf{N}_0 = \mathbf{l}\}. \quad (3.9)$$

Noting that $\mathbf{N}_t^{\mathrm{B}}$ is a Markov chain, and employing (3.9), we obtain that if $\mathbf{N}_t^{\mathrm{B}}$ is an ergodic chain, then $\mathbf{N}_t$ is a stable sequence.

Now we will show that if condition (3.1) is satisfied, then $\mathbf{N}_t^{\mathrm{B}}$ is an ergodic chain. First, setting

$$\sigma_t = \sum_{i=1}^{M} n_t^{(i)},$$

we note that for any t and x from I_0, finite states $\mathbf{m} = (m^{(1)}, \ldots, m^{(M)})$, $\mathbf{l} = (l^{(1)}, \ldots, l^{(M)})$ and events $A_1 = \{\mathbf{N}_0^{\mathrm{B}} = \mathbf{l}\}$, $A_2 = \{\mathbf{N}_0^{\mathrm{A}} = \mathbf{l}\}$ we have

$$\Pr\{\sigma_{t+1}^{\mathrm{B}} \geq x | \sigma_t^{\mathrm{B}} = m, \sigma_0^{\mathrm{B}} = l, A_1\} \leq \Pr\{\sigma_{t+1}^{\mathrm{A}} \geq x | \sigma_t^{\mathrm{A}} = m, \sigma_0^{\mathrm{A}} = l, A_2\} \leq \Pr\{\sigma_{t+1}^{\mathrm{A}} \geq x | \sigma_t^{\mathrm{A}} = m+1, \sigma_0^{\mathrm{A}} = l, A_2\},$$

where

$$m = \sum_{i=1}^{M} m^{(i)}, \qquad l = \sum_{i=1}^{M} l^{(i)}$$

Therefore we can apply Lemma 2 to $n_t = \sigma_t^{\mathrm{B}}$, $n_t^* = \sigma_t^{\mathrm{A}}$, $\mathrm{A}_1 = \{\mathbf{N}_0^{\mathrm{B}} = \mathbf{l}\}$, $A_2 = \{\mathbf{N}_0^{\mathrm{A}} = \mathbf{l}\}$; in accordance with this lemma,

$$\Pr\{\sigma_t^{\mathrm{A}} > x | \mathbf{N}_0^{\mathrm{A}} = \mathbf{l}\} \geq \Pr\{\sigma_t^{\mathrm{B}} > x | \mathbf{N}_0^{\mathrm{B}} = \mathbf{l}\}. \quad (3.10)$$

Now we employ Lemma 1, according to which, if conditions (3.1) is satisfied for all i, then for any $\epsilon > 0$ and any finite l there exists an $x > 0$ such that for all i

$$\lim_{t \to \infty} \Pr\{n_t^{(i)A} > x/M | n_0^{(i)A} = l^{(i)}\} < \epsilon/M. \quad (3.11)$$

Now, employing (3.10) and (3.11), we have

$$\lim_{t \to \infty} \Pr\left\{\bigcup_{i=1}^{M} \{n_t^{(i)B} \leq x\} | \mathbf{N}_0^{\mathrm{B}} = \mathbf{l}\right\} \geq 1 - \lim_{t \to \infty} \Pr\{\sigma_t^{\mathrm{A}} > x | \mathbf{N}_0^{\mathrm{A}} = \mathbf{l}\} \geq 1 - \lim_{t \to \infty} \sum_{i=1}^{M} \Pr\{n_t^{(ii)A} > x/M | n_0^{(i)A} = l^{(i)}\} \geq 1 - \epsilon. \quad (3.12)$$

It is known that if $\mathbf{N}_t^{\mathrm{B}}$ is not an ergodic chain; then for any set of finite states L and any finite initial state $\mathbf{N}_0^{\mathrm{B}} = \mathbf{l}$ we have $\Pr\{\mathbf{N}_t^{\mathrm{B}} \in L | \mathbf{N}_0^{\mathrm{B}} = \mathbf{l}\} \to 0$ as $t \to \infty$. Therefore (3.12) shows that if (3.1) is satisfied, then $\mathbf{N}_t^{\mathrm{B}}$ is an ergodic chain. ▼

In proving Theorem 1, we applied Lemma 2 to the sum of queue lengths σ_t [see (3.10)], not to the queue length $n_t^{(i)}$ at the station. This was done because the conditions of Lemma 2 are not satisfied for $n_t = n_t^{(i)B}$, $n_t^* = n_t^{(i)A}$ without an additional constraint. In the following theorem, this constraint is introduced in the form

$$p \leq 1 - \exp(-1/k^{(i)}), \qquad i = 1, \ldots, M. \quad (3.13)$$

Inequality (3.13) is sufficient for the probability of successful reception of a transit packet

$$\chi_t^{(i)} p (1-p)^{\chi_t^{(i)}} \frac{L-1}{L} \frac{1}{k^{(i)}} \sum_{j \in B^{(i)}} \frac{1}{k^{(j)}}$$

to be maximal with respect to $\chi_t^{(i)}$ for $\chi_t^{(i)} = k^{(i)}$, where $\chi_t^{(i)}$ is the number of stations (among $B^{(i)}$) with nonempty queues at time t. It enables us to obtain a bound for the limiting distribution function of the queue length at the stations. At the same time, this constraint is not excessively burdensome; it is satisfied by the p value for which $s^{(i)} - r^{(i)}$ is maximized in the case $k^{(i)} = k$.

THEOREM 2. If for all i either (3.1) and (3.13) are satisfied or (3.1) and $L = 1$, then for the limiting probability

$$F^{(i)}(x) = \lim_{t\to\infty} \sup \Pr\{n_t^{(i)} > x\}$$

we have the bound

$$F^{(i)}(x) \le \sum_{m:m>x} \pi_m^{(i)}, \qquad i = 1,\ldots, M,$$

where $\pi_0^{(i)}$ is given by (3.7), while

$$\pi_1^{(i)} = \frac{\pi_0^{(i)}[1 - (1 - q^{(i)})h_0^{(i)}]}{s^{(i)}(1 - q^{(i)})}$$

$$\pi_2^{(i)} = \frac{\pi_1^{(i)}[1 - (1 - q^{(i)})h^{(i)} - q^{(i)}s^{(i)}] - \pi_0^{(i)}[q^{(i)}h_0^{(i)} + (1 - q^{(i)})r_0^{(i)}]}{s^{(i)}(1 - q^{(i)})}$$

$$\pi_m^{(i)} = \frac{\pi_{m-1}^{(i)}[1 - (1 - q^{(i)})h^{(i)} - q^{(i)}s^{(i)}] - \pi_{m-2}^{(i)}[q^{(i)}h^{(i)} + (1 - q^{(i)})r^{(i)}] - \pi_{m-3}^{(i)}q^{(i)}r^{(i)}}{s^{(i)}(1 - q^{(i)})}, \qquad m \ge 3, \qquad i = 1,\ldots, M.$$

Proof. Since $\pi_m^{(i)} = \lim_{t\to\infty} \Pr\{n_t^{(i)A} = m\}$, allowing for the fact that under condition (3.13) we have $n_t^{(i)A} \ge n_t^{(i)}$, we obtain

$$F^{(i)}(x) = \lim_{t\to\infty} \sup \Pr\{n_t^{(i)} > x\} \le \lim_{t\to\infty} \Pr\{n_t^{(i)A} > x\} = \sum_{m:m>x} \pi_m^{(i)}$$

The values of $\pi_m^{(i)}$ can be obtained by solving system (3.6) with allowance for (3.4)–(3.5). ▲

The following theorem expresses the condition of lack of network stability under the assumption that $\mathscr{P} = 1/L$.

THEOREM 3. If for all $i = 1,\ldots, M$ the following condition is satisfied:

$$q^{(i)} > s^{(i)} - r^{(i)},$$

then $n_t^{(i)B} \to \infty$ as $t \to \infty$ with probability 1 for all i.

The proof is analogous to that of Theorem 2 from [3], with Eq. (36) of [3] replaced by $\mu_i(h) = e^h s^{(i)}(1 - q^{(i)}) + s^{(i)}q^{(i)} + h(1 - q^{(i)}) + e^{-h}hq^{(i)} + e^{-h}r^{(i)}(1 - q^{(i)}) + e^{2h}r^{(i)}q^{(i)}$, $i = m$, and, furthermore, $\mu_i(0) = 1$, $\mu_1(0) = s^{(i)} - q^{(i)} - r^{(i)} < 0$. ▲

In concluding, we should note that Theorem 1 remains valid if $k^{(i)}$ denotes the upper bound for the number of stations in $B^{(i)}$, rather than the number of stations itself. Theorem 3 is valid if $k^{(i)}$ is the lower bound for the number of stations in $B^{(i)}$. Moreover, Theorem 3 is valid if in BM-1 the condition $\mathscr{P} \ge 1/L$ is replaced by $\mathscr{P} \le 1/L$; here $n_t^{(i)B} \to \infty$ in Theorem 3 is replaced by $n_t^{(i)} \to \infty$.

4. NETWORK WITH FIXED PACKET ROUTES

Here all the assumptions of section 2 remain valid; the routing and input traffic of the network are assumed to be more general.

At the instant of its generation, each new packet is assigned a transmission route through the network from the station where it arises to the station that is its final addressee. For any given packet, the route remains unaltered over the entire residence time of the packet in the network. Over the set of routes that can be assigned to packets arising at station i a probability distribution is defined, so that any given route is assigned to a packet with a probability that is defined by this distribution.

The initial state $\mathbf{A}_0$ of the network is taken to be an ordered set of random variables that represent the numbers of packets in queues at the stations at $t = 0$, and the routes of these packets. The ordering in $\mathbf{A}_0$ reflects the order of packet transmission, corresponding to the servicing discipline at the station.

We introduce some notation in order to complete the description of the routing and input traffic of the network.

Assume that $\mu_{t+1}^{(i)}$ denotes the sum of the number of packets generating at station i in slot $[t, t + 1)$ and of the number of times that routes of packets generating in the network in slot $[t, t + 1)$ intersect station i in transit fashion.

We set

$$\zeta_t^{(i)} = \sum_{j=0}^{t} \mu_j^{(i)}.$$

It is assumed that the following law of large numbers holds for any given finite initial state $\mathbf{A}_0$ of the network:

$$\zeta_t^{(i)}/t \to \rho^{(i)}, \qquad t \to \infty, \qquad i = 1,\ldots, M \qquad (4.1)$$

with probability 1, where $\rho^{(i)} \ge 0$ are given numbers that are independent of $\mathbf{A}_0$.

Note that condition (4.1) is fairly general. For example, it does not involve the introduction of any other assumptions regarding the dependence (or independence) of the input traffic of the queues at stations and routing. Moreover, it is not assumed that only one packet can arise at a station in a slot; the number of packets can be arbitrary. Finally, we should point out that in the case of routes without self-intersections, the limit of the right side of (4.1) has the meaning of the average stationary number of packets that arise at all stations in one slot and whose route begins at or passes through station i. Thus, law (4.1) is a condition only on this average number of packets, and does not impose any constraints on the high-order moments or any other more complete probability characteristics of $\zeta_t^{(i)}$. We should note that condition (4.1) is ensured by the ergodic nature of the input streams in the network.

In what follows, the above network model will be called Basic Model 2 (BM-2).

At first glance, it would seem that, because packets in BM-1 are not assigned routes at the instant they generated, BM-1 is not a particular case of BM-2. However, this is not so. The independence of the generation of packets and of the choice of segments of their routes in BM-1, allows to consider finally established routes, as if they were assigned to the packets at the instant of their generation. Thus, BM-1 is a particular case of BM-2.

In analyzing a network with BM-2, we introduce the additional concept of virtual queue V at station i. All quantities pertaining to this queue will henceforth be designated by a superscript V.

Let us consider some station i. We will assume that there are two queues of packets at it. The first is the queue that arises in BM-2; its length at time t is denoted by $n_t^{(i)}$. The second is called the virtual queue V of packets at station i; its length is denoted by $n_t^{(i)\mathrm{V}}$. Virtual queue V does not have a physical interpretation, and is introduced only as a concept that is required for mathematical analysis of BM-2. In BM-2, a packet leaves the queue immediately after successful transmission. By analogy with this, we will say that a packet in virtual queue V leaves it immediately after successful transmission, although there will not be any real transmission of packets from virtual queue V. Packets arrive at queue $n_t^{(i)\mathrm{V}}$ in the same numbers, with the same routes, and at the same times as they arrive at $n_t^{(i)}$. One can imagine that at the instant of generation or of successful reception of a transit packet at station i this station generates a copy that has the same route as the original packet, and the copy is placed in virtual queue V. The servicing discipline for the packets in virtual queue V is the same as the servicing discipline in BM-2. This means, for example, that if a FIFO discipline is employed for the queue at station i in BM-2, then a FIFO discipline will also be operative for virtual queue V.

If, at time t, virtual queue V becomes empty and, at this instant, no new packet arises at station i and it successfully does not receive a transit packet in BM-2, then the station places a dummy packet in virtual queue V, thus preventing virtual queue V from being empty. The dummy packet is assigned a route of unit length (the length is measured as the number of successful transmissions), the final destination being one of the stations of $B^{(i)}$. The route of a dummy packet remains unaltered over the entire time that this packet is in virtual queue V. In what follows, whenever the concept of virtual queue V is employed, we will specify how the route for a dummy packet is to be chosen from among the $k^{(i)}$ possible routes.

If, prior to successful transmission of a dummy packet, a copy of an ordinary packet is delivered to virtual queue V, servicing of the dummy packet is interrupted at this instant and is renewed only at the first successive instant at which virtual queue V becomes empty. We will agree not to include the dummy packet in $n_t^{(i)\mathrm{V}}$.

Now we will show how packets leave virtual queue V. For this we introduce the concept of virtual instant $t^{(i,j)}$ of successful packet transmission from station i to $j \in B^{(i)}$. In BM-2, in accordance with the ALOHA algorithm, each station that has a packet independently transmits a packet at time t with probability p, and is silent at this time with probability $(1-p)$. The ALOHA algorithm can be described by means of the random logical variable $\Omega_t^{(i)}$. Let us assume that, for any given i and t, $\Omega_t^{(i)}$ independently takes the values 1 and 0 with probabilities p and $(1-p)$, respectively. If $\Omega_t^{(i)} = 1$ and $n_t^{(i)} > 0$, then in BM-2 station i transmits a packet in slot $[t, t+1)$. If $\Omega_t^{(i)} = 0$, then in BM-2 station i is silent in slot $[t, t+1)$. For any given i, j ($j \in B^{(i)}$), the time $t^{(i,j)}$ is any point in time t such that $\Omega_t^{(i)} = 1$, $\Omega_t^{(u)} = 0$ for $u = j$ and all u from $B^{(j)}$ except for $u = i$.

Thus, if a copy of an ordinary packet, that requires transmission to station $j \in B^{(i)}$, to be serviced at station i at time t, it is successfully transmitted and leaves virtual queue V at a time equal to $\min\{t_1 : t_1 = t^{(i,j)}, t_1 \geq t\}$, i.e., the closest virtual instant of successful transmission. A similar rule of successful transmission, but with allowance for interrupts, is valid for dummy packets.

From this point up to the end of this paper, we define that station i is stable if $n_t^{(i)} \nrightarrow \infty$ with probability 1 when $t \to \infty$ for any finite initial state $\mathbf{A}_0$.

Now let us give the stability condition for a network with BM-2, under the additional condition that $k^{(i)} = k$, where k is independent of i.

THEOREM 4. If for all $i = 1, \ldots, M$ (where $M = \infty$ is allowed) the following condition is observed:

$$\rho^{(i)} < p(1-p)^k, \tag{4.2}$$

then an ALOHA network with BM-2 and $k^{(i)} = k$ is stable.

Proof. First we note that, when virtual queue V is employed in this proof, it will be assumed that all dummy packets at a given stations are always assigned the same specified route.

In commencing the proof, we will assume that our BM-2 network is unstable under condition (4.2). This means that for some i and some given finite initial state $\mathbf{A}_0$ there exists an $\epsilon > 0$ such that

$$\Pr\{n_t^{(i)} \to \infty | \mathbf{A}_0\} \geq 2\epsilon \tag{4.3}$$

(Up to the end of this proof, all the probabilities are considered under the above condition $\mathbf{A}_0$; it will be omitted, however, in order to abridge the notation.)

It follows from (4.3) that

$$\lim_{T\to\infty} \inf \Pr\left\{\bigcap_{t=T}^{\infty} \{n_t^{(i)\mathrm{B}} > 0\}\right\} \geq 2\epsilon \tag{4.4}$$

where $n_t^{(i)}$ is replaced by $n_t^{(i)\mathrm{V}}$ in view of the inequality

$$n_t^{(i)\mathrm{V}} \geq n_t^{(i)} \tag{4.5}$$

which is valid for all t and i.

We have the following equivalence of events:

$$\{n_t^{(i)\mathrm{V}} > 0\} = \left\{\hat{\zeta}_t^{(i)} > \theta_t^{(i)\mathrm{V}}\right\} \tag{4.6}$$

where $\hat{\zeta}_t^{(i)}$ denotes the sum of the number of ordinary packets that arise at station i in the interval $[0, t)$ and of the number of successful receptions of transit packets by this station in this same interval; while $\theta_t^{(i)\mathrm{V}}$ denotes the number of ordinary packets that leave virtual queue V in $[0, t)$. The equality $\hat{\zeta}_t^{(i)} = \hat{\zeta}_t^{(i)\mathrm{V}}$ is employed in (4.6).

In addition, we have the following inclusion of events:

$$\bigcap_{t=T}^{\infty} \{\hat{\zeta}_t^{(i)} > \theta_t^{(i)\mathrm{V}}\} \subset \bigcap_{t=T}^{\infty} \{\zeta_t^{(i)} > \theta_T^{(i)\mathrm{V}} + \alpha_{T,t}^{(i)\mathrm{V}}\}, \tag{4.7}$$

where $\alpha_{T,t}^{(i)\mathrm{V}}$ denotes the number of ordinary and dummy packets that leave the virtual queue in the interval $[T, t)$, while $\zeta_t^{(i)}$ is defined in (4.1).

It follows from (4.4), (4.6), (4.7) that

$$\lim_{T\to\infty} \inf \Pr \left\{\bigcap_{t=T}^{\infty} \{\zeta_t^{(i)} > \theta_T^{(i)\mathrm{V}} + \alpha_{T,t}^{(i)\mathrm{V}}\}\right\} \geq 2\epsilon \tag{4.8}$$

or

$$\Pr \left\{\bigcap_{t=T}^{\infty} \{\zeta_t^{(i)} > \theta_T^{(i)\mathrm{V}} + \alpha_{T,t}^{(i)\mathrm{V}}\}\right\} \geq \epsilon$$

for any $T \geq T_1$, where $T_1 < \infty$ is a sufficiently large number.

Allowing for the fact that the probability of intersection of events does not exceed the probability of each of them, we have from (4.9)

$$\lim_{t\to\infty} \inf \Pr \left\{\zeta_t^{(i)} > \alpha_{T_1,t}^{(i)\mathrm{V}}\right\} \geq \epsilon. \tag{4.10}$$

Since $\alpha_{T_1,t}^{(i)\mathrm{V}} = \alpha_t^{(i)\mathrm{V}} - \alpha_{T_1}^{(i)\mathrm{V}}$, where $\alpha_t^{(i)\mathrm{V}} \equiv \alpha_{0,t}^{(i)\mathrm{V}}$, and $\alpha_{T_1}^{(i)\mathrm{B}} \leq T_1$, for any $\Delta > 0$ and all $t > T_2 = T_1/\Delta$ we have the inequality $(\alpha_{T_1,t}^{(i)\mathrm{V}}/t) \geq \alpha_t^{(i)\mathrm{V}}/t - \Delta$. Therefore (4.10) yields

$$\lim_{t\to\infty} \inf \Pr \left\{\frac{\zeta_t^{(i)}}{t} \geq \frac{\alpha_t^{(i)\mathrm{V}}}{t} - \Delta\right\} \geq \epsilon. \tag{4.11}$$

Now let us set up a contradiction with inequality (4.11), on the basis of (4.1) and (4.2). For this purpose, we denote by $\kappa_{t+1}^{(i)\mathrm{V}}$ the number of packets (copies of ordinary and dummy packets) that leave virtual queue V at station i in slot $[t, t+1)$. We have

$$\alpha_t^{(i)\mathrm{V}} = \sum_{j=0}^{t} \kappa_j^{(i)\mathrm{V}}. \tag{4.12}$$

In view of the independence of the random variables $\Omega_t^{(i)}$ that were introduced in defining virtual queue V in this section, the limit of $\alpha_t^{(i)\mathrm{V}}/t$ as $t \to \infty$ exists with probability 1 and, in view of the fact that $k^{(i)} = k$, this limit is equal to $p(1-p)^k$. Allowing for this fact, and also for (4.1) and (4.2), we obtain that for any $\epsilon > 0$, $\Delta > 0$ there exists a $T_3 < \infty$ such that for all $t > T_3$ we have the inequality $\Pr\{(\zeta_t^{(i)}/t) \geq (\alpha_t^{(i)\mathrm{V}}/t) - \Delta\} < \epsilon$ or

$$\lim_{t\to\infty} \inf \Pr \left\{\frac{\zeta_t^{(i)}}{t} \geq \frac{\alpha_t^{(i)\mathrm{V}}}{t} - \Delta\right\} < \epsilon. \tag{4.13}$$

Inequality (4.13) contradicts (4.11), and thus the theorem is proved. ▲

Now let us generalize Theorem 4 to the case of $k^{(i)}$ that are not necessarily equal. This generalization requires the introduction of the following additional assumption, expressed in terms of the $\alpha_t^{(i)\mathrm{V}}$ that were introduced after (4.10).

Let us consider virtual queue V at station i; we will assume that dummy packets choose a route to station $j \in B^{(i)}$ with specified probability $P_j^{(i)}$, $\sum_{j\in B^{(i)}} P_j^{(i)} = 1$. For any given finite initial state $\mathbf{A}_0$, assume that the following condition holds:

$$\frac{\alpha_t^{(i)\mathrm{V}}}{t} \to Q^{(i)}, \qquad t \to \infty, \qquad i = 1, \ldots, M \tag{4.14}$$

with probability 1, where $Q^{(i)} \geq 0$ are specified numbers that are independent of $\mathbf{A}_0$.

In the case $k^{(i)} = k$ condition (4.14) is satisfied and $Q^{(i)} = p(1-p)^k$. In the case of different $k^{(i)}$, (4.14) is a condition on the stream of packets arriving to station i.

THEOREM 5. If condition (4.14) is satisfied for some given probability distribution $P_j^{(i)}$, and $\rho^{(i)} < Q^{(i)}$, $i = 1, \ldots, M$ (where $M = \infty$ is allowed), then an ALOHA network with BM-2 is stable.

The proof is analogous to that of Theorem 4.

Now let us give the condition for a network with BM-2 to be lacking in stability.

THEOREM 6. If $L = 1$; $\mu_t^{(i)}$, $t \in I_0$, $i = 1, \ldots, M$, are independent identically distributed random variables, and for all $i = 1, \ldots, M$ the following condition is satisfied:

$$\rho^{(i)} > Q^{(i)}, \tag{4.15}$$

then a network with BM-2 is unstable.

The proof is the same in many respects as the proof of Theorem 2 of [3] and of Theorem 5 of [9]. Therefore the common parts of these proofs will be summarized.

For any finite vector $\mathbf{d} = (d^{(i)}, \ldots, d^{(M)})$, $\mathbf{d} \in Z_+^M = \{\mathbf{z} = (z^{(i)}, \ldots, z^{(M)}) \colon z^{(i)} \geq 0 \text{ and integers}\}$ we define the set

$$\Gamma(\mathbf{d}) = \bigcup_{i=1}^{M} \{\mathbf{z} \colon z^{(i)} \leq d^{(i)}, \mathbf{z} \in Z_+^M\}.$$

Let us assume that the network is stable under condition (4.15). We will show that this assumption leads to a contradiction which proves the theorem.

Under this assumption, we will first prove that for any $\epsilon > 0$ and $\mathbf{d}$ there exists a finite vector $\mathbf{l} \in Z_+^M$ such that the following inequality is valid:

$$\Pr\left\{\bigcup_{t=0}^{\infty}\{\mathbf{N}_t \in \Gamma(\mathbf{d})\}|\mathbf{N}_0 = \mathbf{l}, \mathbf{A}_0\right\} < \epsilon \quad (4.16)$$

for any specified finite initial network state $\mathbf{A}_0$ that is matched with the equality $\mathbf{N}_0 = \mathbf{l}$, where, as before, $\mathbf{N}_t = (n_t^{(1)}, \ldots, n_t^{(M)})$.

For any i we have

$$n_t^{(i)} = n_{t-1}^{(i)} + \beta_{t-1}^{(i)} - \eta_{t-1}^{(i)}, n_{-1}^{(i)} = 0, \quad (4.17)$$

where $\beta_t^{(i)}$ is the number of packets that arise in slot $[t, t+1)$ at station i; $\eta_t^{(i)}$ is the number of packets successfully transmitted by station i in slot $[t, t+1)$, $t \in I_0$.

We introduce the following quantity:

$$v_t^{(i)} = v_{t-1}^{(i)} + \beta_{t-1}^{(i)} - \kappa_{t-1}^{(i)\mathrm{V}}, v_0^{(i)} = n_0^{(i)} \quad (4.18)$$

for $i = 1, \ldots, M$, where $\kappa_t^{(i)\mathrm{V}}$ was defined just before (4.12).

We have

$$v_t^{(i)} = v_0^{(i)} + \sum_{k=1}^{t} \xi_k^{(i)}, \quad (4.19)$$

where $\xi_t^{(i)} = \beta_t^{(i)} - \kappa_t^{(i)\mathrm{V}}$, $t \in I_1$.

For any $\mathbf{l}, \mathbf{d} \in Z_+^M$ and any specified finite initial state $\mathbf{A}_0$ that is matched with $\mathbf{N}_0 = \mathbf{l}$, we have

$$\Pr\left\{\bigcup_{k=0}^{\infty}\{\mathbf{N}_k \in \Gamma(\mathbf{d})\}|\mathbf{N}_0 = \mathbf{l}, \mathbf{A}_0\right\}$$

$$= \Pr\left\{\bigcup_{k=0}^{\infty}\{\mathbf{V}_k \in \Gamma(\mathbf{d})\}|\mathbf{V}_0 = \mathbf{l}, \mathbf{A}_0\right\}, \quad (4.20)$$

where $\mathbf{V}_k = (v_k^{(1)}, \ldots, v_k^{(M)})$.

Inequalities (4.19) and (4.15) yield, when $t \to \infty$, that $v_t^{(i)} \to \infty$ with probability 1 for any $\mathbf{A}_0$ and $i = 1, \ldots, M$. This means that for any $\epsilon > 0$ and finite $\mathbf{l}$ and $d^{(i)}$ there exists a number $\hat{r}^{(i)}$ such that for all $r^{(i)} > \hat{r}^{(i)}$

$$\Pr\left\{\bigcap_{k=r^{(i)}}^{\infty}\{v_k^{(i)} > d^{(i)}\}|\mathbf{V}_0 = \mathbf{l}\mathbf{A}_0\right\} > 1 - \epsilon/M \quad (4.21)$$

for any $\mathbf{A}_0$ that is matched with $\mathbf{V}_0 = \mathbf{l}$. Consequently, taking accounting of (4.20), we obtain the proof of inequality (4.16).

The assumption of stability also implies that for an arbitrary given finite $\mathbf{d}^*, \mathbf{l}^* \in Z_+^M$

$$\lim_{T\to\infty} \Pr\left\{\bigcup_{k=T}^{\infty}\{\mathbf{N}_k \in \Gamma(\mathbf{d}^*)\}|\mathbf{N}_0 = \mathbf{l}^*, \mathbf{A}_0\right\} = 1 \quad (4.22)$$

for any $\mathbf{A}_0$ that is matched with $\mathbf{N}_0 = \mathbf{l}^*$.

Since for any $p < 1$ (for $p = 1$ the network is trivially unstable) and any i we have $\Pr\{\bigcap_{i=1}^{M}\bigcap_{t=1}^{T}\{\Omega_t^{(i)} = 0\}\} = (1-p)^{TM}$, we obtain, allowing for (4.1), that there exist a $T^* < \infty$ and $\epsilon_1 < 1$ such that for all $T > T^*$, $\mathbf{d}, \mathbf{l}^* \in Z_+^M$

$$\Pr\left\{\bigcap_{k=1}^{T}\{\mathbf{N}_k \in \Gamma(\mathbf{d})\}|\mathbf{N}_0 = \mathbf{l}^*; \mathbf{A}_0\right\} < \epsilon_1 \quad (4.23)$$

for any $\mathbf{A}_0$ that is matched with $\mathbf{N}_0 = \mathbf{l}^*$.

We choose a number $\epsilon_2 > 0$ such that $\epsilon_1 + \epsilon_2 < 1$, and we find an $\mathbf{l}$ such that for $\mathbf{d} = \mathbf{d}^*$ and $\epsilon = \epsilon_2$ inequality (4.16) is satisfied. Then, taking account of (4.16), we obtain that for all $T > T^*$

$$\Pr\left\{\bigcup_{k=T}^{\infty}\{\mathbf{N}_k \in \Gamma(\mathbf{d}^*)\}|\mathbf{N}_0 = \mathbf{l}^*, \mathbf{A}_0\right\}$$

$$= \Pr\left\{\left[\bigcup_{k=T}^{\infty}\{\mathbf{N}_k \in \Gamma(\mathbf{d}^*)\}\right]\right.$$

$$\left.\cap\left[\bigcap_{k=1}^{T}\{\mathbf{N}_k \in \Gamma(\hat{\mathbf{l}})\}\right]|\mathbf{N}_0 = \mathbf{l}^*, \mathbf{A}_0\right\}$$

$$+\Pr\left\{\left[\bigcup_{k=T}^{\infty}\{\mathbf{N}_k \in \Gamma(\mathbf{d}^*)\}\right]\right.$$

$$\left.\cap\left[\bigcup_{k=1}^{T}\{\mathbf{N}_k \notin \Gamma(\hat{\mathbf{l}})\}\right]|\mathbf{N}_0 = \mathbf{l}^*, \mathbf{A}_0\right\} < \epsilon_1 + \epsilon_2 < 1$$

for any $\mathbf{A}_0$ matched with $\mathbf{N}_0 = \mathbf{l}^*$.

The resultant inequality contradicts (4.22), and this proves that the network is unstable under condition (4.15). ▲

5. General Network[1]

In the general approach, we will assume that the network is set up and operates as described in section 2. In addition, we will assume that, via its nodes, the network receives an input traffic of new packets that is completely specified in the probabilistic sense. For each point in time and for each station, the input traffic defines the number of new packets that arise at a station.

It is also assumed that there is a specified rule of motion, in accordance with which, for each attempt at transmission, any packet in the network is unambiguously sent to one of the neighbors of the station that transmits it.

The initial state $\mathbf{A}_0$ of BM-3 is assumed to be an ordered set of random variables that represent the numbers of packets in queues at the stations of the network at time $t = 0$ and the neighbors of the corresponding stations to which the packets are addressed in attempting transmission in slot $[0, 1)$.

Finally, it is assumed that for any packet there is a

[1] Elaborated consideration of this general network was given in the paper B. S. Tsybakov, V. L. Bakirov, "Analysis of Packet Switching Network Stability and Unified Approach to Slotted and Unslotted Packet Aloha Radio Networks," *Probl. Peredachi Inform.*, vol. 24, no. 2, pp. 70–85, 1988.

specified exit rule, according to which a determination is made at each point in time whether or not to remove the packet from the network. If a packet is not removed from the network over its residence time at station i, then, in relation to station i, this packet is called a transit packet over the course of this residence time.

Rule of motion and exit rule can be either deterministic or probabilistic; it is required, however, that all random variables that describe the network form a causal time sequence, i.e., a sequence whose terms are random variables that are conditionally, in some sence, independent of the future of this sequence.

Let us give some examples of specification of the input traffic, rules of motions, and exit rules.

The input traffic can be specified by independent random variables, $\beta_t^{(1)}, \dots, \beta_t^{(M)}$, $t \in I_0$, such that $\beta_t^{(i)}$ denotes the number of new packets that arise at station in slot $[t, t+1)$. Another example that is considered in closed networks is input traffic with $\beta_0^{(1)} \geq 0, \dots, \beta_0^{(M)} \geq 0$ and $\beta_t^{(i)} = 0$ for all $t > 0$, $i > 1, \dots, M$.

A common exit rule is the following: A packet is assigned some station i as the final destination, and, if the packet is successfully received by station i, it is immediately removed from the network. For closed networks, the exit rule is degenerate, and amounts to the fact that packets are never removed from the network. It may occur that a packet is removed from the network, e.g., after visiting all nodes or e.g., after some node is visited twice, or, e.g., after a certain amount of time has elapsed. In section 3 we considered an exit rule in which, after each successful transmission, a packet leaves the network with probability $1/L$, or remains in the network with probability $1 - 1/L$.

The rule of motion is frequently specified as follows. Assume that the exit rule is such that, when a packet arises, it is assigned a final destination that is not subsequently altered. A new packet is also assigned a route through the network that originates at the node at which it arises and terminates at the destination node. The rule of motion involves shifting the packet in accordance with its route. Under these conditions, for specified initial and final node, the route between them need not be specified in advance; rather, it can be chosen as the packet advances through the network, in relation, e.g., to the queue lengths at the nodes. Yet another rule of motion was considered in section 3, where it was assumed that, independently of previous choices, in each transmission, a packet at station i chooses equiprobably a new neighbor (from among the $k^{(i)}$ neighbors) to which it will be sent.

This completes our consideration of the examples; we will now proceed to formulate the constraints on the network. These constraints are introduced under the assumption that the network topology and input traffic are specified, as well as the rule of motion and exit rule.

We denote by $\zeta_t{}^{*(i)}$ the sum of the number of new packets that arise at station in the interval $[0, t)$ and the number of transit packets that are successfully received by station i in $[0, t)$.

The first constraint on the network involves the requirement that the following condition be met:

$$\frac{\zeta_t^{*(i)}}{t} \to \rho^{*(i)}, \qquad i = 1, \dots, M \tag{5.1}$$

with probability 1, where $\rho^{*(i)} \geq 0$ are some numbers.

The second constraint is formulated in terms of virtual queue G, which we will now define. All quantities pertaining to this queue are designated by a superscript G.

In many respects, virtual queue G at station i is defined in the same way as queue V in Section 4. The only reason for the difference is that now ordinary packets and their copies do not have routes that are "set" for their entire residence time in the network. Accordingly, it is assumed that a copy in virtual queue G is sent to the same station to which the ordinary packet corresponding to this copy was directed. If the ordinary packet, in accordance with the exit rule, leaves the queue of station i and the network without successful transmission, its copy leaves virtual queue G at the same instant. A dummy packet chooses a route from station i to stations $j \in B^{(i)}$ in accordance with some given probability distributions, which may depend on the values of certain random variables that define the behavior of the network.

Assume that $\alpha_t^{(i)\mathrm{G}}$ denotes the number of packets (both dummy and copies of ordinary packets) that leave virtual queue G at station i during the interval $[0, t)$.

The second constraint on the network involves the requirement that the following condition be satisfied:

$$\frac{\alpha_t^{(i)\mathrm{G}}}{t} \to Q^{(i)}, \qquad t \to \infty,\ i = 1, \dots, M \tag{5.2}$$

with probability 1, where $Q^{(i)} \geq 0$ are certain numbers.

For convenience, we will denote the above general network model with constraints (5.1) and (5.2) by BM-3.

THEOREM 7. If for all $i = 1, \dots, M$ (where $M = \infty$ is allowed) the condition $\rho^{*(i)} < Q^{(i)}$ is satisfied, then an ALOHA network with BM-3 is stable.

The proof is similar to that of Theorems 4 and 5, with $\hat{\zeta}_t^{(i)}$ replaced by $\zeta_t{}^{*(i)}$ and virtual queue V by G.

The results obtained in this section can readily be extended to conflict-free packet-switched networks, e.g., to wire networks.

6. Capacity and Optimal Transmitter Radius of Uniform Network

In this section we will consider a network with BM-2. As was stated in section 1, of interest is the problem of optimizing the transmitter radius of action, with the aim or maximizing the capacity of the network. The existence of an optimal radius stems from the fact that as the radius increases, on the one hand, the number of necessary hops decreases; on the other, there is an increase in the probability of conflict upon packet transmission.

Let us give an exact formulation of the problem. For this we introduce and refine the necessary concepts.

We call the location of stations in some space V the geography of the network, and we denote it by G. The zone of station i is a region $z^{(i)}$ of space V such that if station j is at $z^{(i)}$ (denoted $j \in z^{(i)}$), then station i and j are directly linked. The number of stations in zone i is equal to $k^{(i)} + 1$. Directly linked stations are called adjacent (or neighboring) stations. It is assumed that if $j \in z^{(i)}$, then $i \in z^{(j)}$. This symmetry is required in order to transmit the positive acknowledgments as provided in section 2.

For a network with specified geography and zones, as in section 2, we define topology T, this being a graph with nodes corresponding to stations, and edges that connect pairs of adjacent stations. A route through the network is a path through its graph T; the length of the route is the number of constituent edges. We denote the ensemble of routes of station i by $\mathscr{M}^{(i)}$ and define it as the set of routes assigned to packets that arise at station i, together with the probability distribution on this set. The average route length for packets arising at station i is the average route length in ensemble $\mathscr{M}^{(i)}$; it is denoted by $L^{(i)}$.

A network is called connected if its graph is connected.

A network is called uniform if $\rho^{(i)} = \rho$, $Q^{(i)} = Q$ for all i, where ρ and Q are independent of i, and $Q^{(i)}$ is obtained under the assumption that some distribution $P_j^{(i)}$ of routes of dummy packets is specified and condition (4.14) is satisfied.

A network is uniform, e.g., in the case in which $k^{(i)} = k$, $L^{(i)} = L$, $\zeta_t^{(i)} = \sum_{j=0}^{t} \mu_j^{(i)}$, where the $\mu_j^{(i)}$, for each specified i, are independent identically distributed variables for all j; $\mathbf{M}\mu_t^{(i)} = \mathbf{M}\mu_t^{(j)}$ for any i and j; $\mathbf{M}\beta_t^{(i)} = \lambda$, where $\beta_t^{(i)}$ is a random variable equal to the number of new packets that arise at station i in slot $[t, t+1)$; the random number of new packets Γ that arise in the network in any slot is independent of the lengths $l_1, \ldots, l_\Gamma$ of the routes assigned to them.

Indeed, in this example, we have

$$\mathbf{M} \sum_{i=1}^{M} \mu_t^{(i)} = \mathbf{M} \sum_{j=1}^{\Gamma} l_j = (\mathbf{M}\Gamma)(\mathbf{M}l_j) = \lambda L M.$$

Therefore $\mathbf{M}\mu_t^{(i)} = \lambda L$. Thus, $(\zeta_t^{(i)}/t) \to \rho = \lambda L$. In addition, for $k^{(i)} = k$ we have $Q^{(i)} = p(1-p)^k$.

For a uniform network with BM-2, stability condition (4.3) has the form $\rho < Q$, while condition (4.16) of lack of stability has the form $\rho > Q$. For a uniform network, therefore, the condition $\rho = Q$ defines the boundary of the capacity region of the network; we will not give an exact definition of this boundary in the general case. For the above example of a homogeneous network, however, the throughput region can be defined as the interval $[0, C_0]$, where C_0 is the upper bound of numbers λ for which the network remains stable. The number MC_0 has the meaning of the largest overall intensity of input traffic which the network transmits with finite delay.

In accordance with the above results, we have $C_0 \geq p(1-p)^k/L$.

Now we can seek the maximum of lower bound $p(1-p)^k/L$ with respect to p, which is attained for $p = (k+1)^{-1}$ and is equal to

$$C = k^k/(k+1)^{k+1}L. \tag{6.1}$$

The number C is called the capacity of the network under consideration.

For a network with given geography G, we assume now that its topology $T(r)$, zones $z^{(i)}(r)$, number of neighbors $k^{(i)}(r)$, ensembles $\mathscr{M}^{(i)}(r)$, and average lengths $L^{(i)}(r)$ depend on some parameter r, $r \in R_+ = [0, \infty)$, such that if $r_1 > r_2$, then $z^{(i)}(r_2) \subset z^{(i)}(r_1)$, this parameter having the meaning of the radius of action of the transmitters of the stations. We denote by r_0 the minimum value of the parameter $r \in R_+$ for which the network is connected. Graph $T(r_0)$, zones $z^{(i)}(r_0)$, and ensembles $\mathscr{M}^{(i)}(r_0)$ are called initial.

In what follows, we will consider only the particular case of the dependence on r in which the following three conditions 1)–3) are satisfied; this is called the symmetrically matched case.

1) Input flows $\beta_t^{(i)}$, $i = 1, \ldots, M$, are assumed to be independent of r. (The definition of $\beta_t^{(i)}$ is the same as that given in the example above.)
2) The dependence on r is assumed to be such that for any r the network remains uniform, i.e., $\rho^{(i)} = \rho(r)$, $Q^{(i)} = p(1-p)^{k(r)}$.
3) It is assumed that for every $r \geq r_0$ ensembles $\mathscr{M}^{(i)}(r)$ are such that packets are transmitted via the shortest path (if there is more than one, any of these may be used.)

For the example under consideration of a uniform network, with given geography G and symmetrically matched dependence on r, the capacity $C(r)$ has the following form for any r, in accordance with (6.1):

$$C(r) = k(r)^{k(r)}/[k(r)+1]^{k(r)+1}L(r). \tag{6.2}$$

For this network the problem of determining the optimal radius r can be formulated as follows: For given geography G, initial graph $T(r_0)$ with zones $z^{(i)}(r_0)$ and ensembles $\mathscr{M}^{(i)}(r_0)$, determine the value r_{opt} for which $C(r)$ is maximized; this is denoted by $C_m = C(r_{opt})$.

In order to determine C_m, we need to know the functions $L(r)$ and $k(r)$, whose form depends on G, $T(r)$, $z^{(i)}(r)$, and $\mathscr{M}^{(i)}(r)$. To determine r_{opt} and C_m, therefore, it is necessary to specify G, $T(r)$, $z^{(i)}(r)$, and $\mathscr{M}^{(i)}(r)$.

Let us consider some examples of uniform networks, and let us obtain r_{opt} for them. Although, strictly speaking, because the number of stations is finite, all our network examples should be considered on a torus, we will describe them as if they lie on a plane. In the first four examples we will assume that the geography G is the same, the stations being arranged on the plane at the nodes of a regular integer-valued lattice with a distance of 1 between nearest nodes. We will specify the ensemble of routes for packets that arise at station (0, 0). For the remaining stations, the ensembles of routes are obtained by shifts parallel to the axes of the lattice. Thus, if station (0, 0) assigns some route to station (x, y) with probability $\bar{p}$, then station (α, b) will assign an analogous route to station $(x + \alpha, y + b)$ with probability $\bar{p}$, where α, b, x, and y are integers.

For the lattice network under consideration, $r_0 = 1$.

Example 1. Zones $z^{(i)}(r)$ are circles of radius r. Packets are addressed to stations (1, 0), (0, 1), (−1, 0), (0, −1), each with probability $\bar{p}/4$ (these stations are shown by open circles in Fig. 1), and to stations (2, 0), (0, 2), (−2, 0), (0, −2), each with probability $(1 - \bar{p})/4$ (these stations are shown by solid circles in Fig. 1). For any $\bar{p}$, $0 \le \bar{p} \le 1$, we have $r_{\text{opt}} = 1$, $k(r_{\text{opt}}) = 4$ and $C_m = 4^4/5^5(2 - \bar{p})$; for $\bar{p} = 0$, $C_m = 0.04096\ldots$.

Example 2. Zones $z^{(i)}(r)$ are circles of radius r. Packets are addressed to stations (1, 0), (0, 1), (−1, 0), (0, −1), each with probability $\bar{p}/4$ (these stations are shown by open circles in Fig. 2) and to stations (1, 1), (1, −1), (−1, 1), (−1, −1), each with probability $(1 - \bar{p})$ (shown by solid circles in Fig. 2). In this case, r_{opt} depends on $\bar{p}$. For $0 \le \bar{p} \le \bar{p}_1$, where $\bar{p}_1 = 2 - (4^4 9^9/5^5 8^8) = 0.1083\ldots$, we have $r_{\text{opt}} = \sqrt{2}$ and $C_m = 8^8/9^9 = 0.0433\ldots$. For $\bar{p}_1 \le \bar{p} \le 1$ we have $r_{\text{opt}} = 1$ and $C_m = 4^4/5^5(2 - \bar{p})$. For $\bar{p} = 1$ we have $C_m = 0.082\ldots$. At point $\bar{p} = \bar{p}_1$ there are two optimal radii, 1 and $\sqrt{2}$. Figure 3 shows C_m plotted against $\bar{p}$.

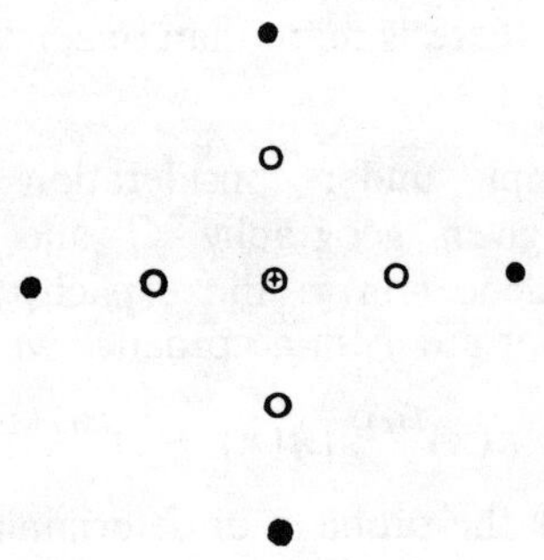

Fig. 1. Addressing of packets for Example 1.

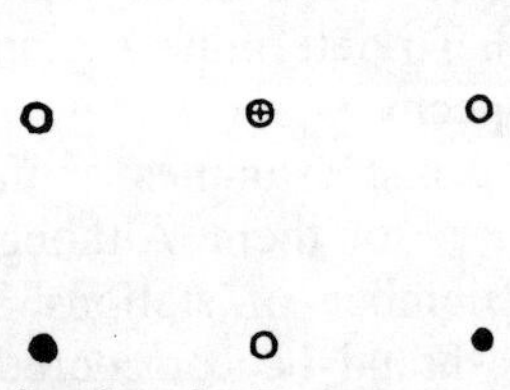

Fig. 2. Addressing of packets for Example 2.

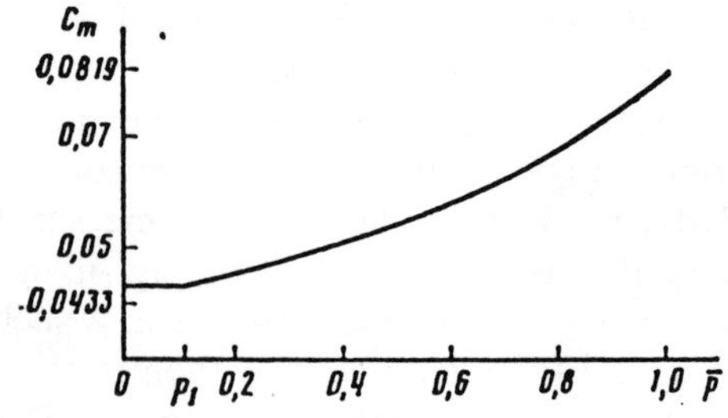

Fig. 3. Graph of C_m plotted against $\bar{p}$.

Example 3. Zones $z^{(i)}(r)$ are four-leaved; they include stations in four directions (forward, backward, right, left) with range r. Thus, the zone for station (0, 0) includes station (0, 1), ..., (0, $[r]$), (1, 0), ..., ($[r]$, 0), (0, −1), ..., (0, −$[r]$), (−1, 0), ..., (−$[r]$, 0), where $[r]$ is the integer part of r. Packets are addressed equiprobably to stations $(n, 0)$, $(0, n)$, $(-n, 0)$, and $(0, -n)$, where n is an integer parameter, $n \ge 1$. In this case $r_{\text{opt}} = n$, $k(r_{\text{opt}}) = 4n$. For example, for $n = 2$ we have $C_m = 0.0433,\ldots$; for $n = 3$, $C_m = 0.0294\ldots$; for $n = 4$, $C_m = 0.0223\ldots$.

Example 4 (for $M = \infty$). Zones $z^{(i)}(r)$ are the same as in Example 3. For any n, $n \in I_1$, packets are addressed to stations $(n, 0)$, $(0, n)$, $(-n, 0)$, $(0, -n)$ with equal probabilities $(1 - 1/L)^{n-1}/4L$, where L has the meaning of the average route length. In this case

$$C(r) = (4r)^{4r} L \cdot \left\{ (4r + 1)^{4r+1} \sum_{i=1}^{\infty} (-[-i/n])(1 - 1/L)^{i-1} \right\}^{-1}.$$

For $L \le 9$, $r_{\text{opt}} = 1$; for $10 \le L \le 21$, $r_{\text{opt}} = 2$; etc.

Example 5. In this example, we assume that the geography G is as shown in Fig. 4. It consists of equilateral triangles with stations at the vertices. We specify the addressing for the stations marked by a cross in Fig. 4. For other stations it can be obtained by shifting, as in the preceding examples. Packets are addressed to stations indicated by open circles with probability $\bar{p}/6$, and to stations indicated by solid circles with probability $(1 - \bar{p})/6$.

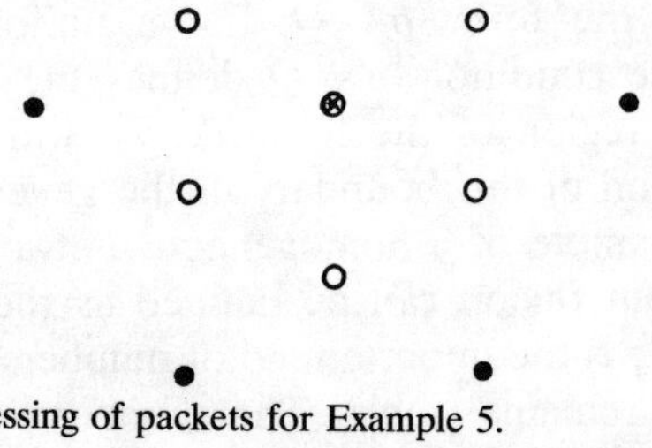

Fig. 4. Addressing of packets for Example 5.

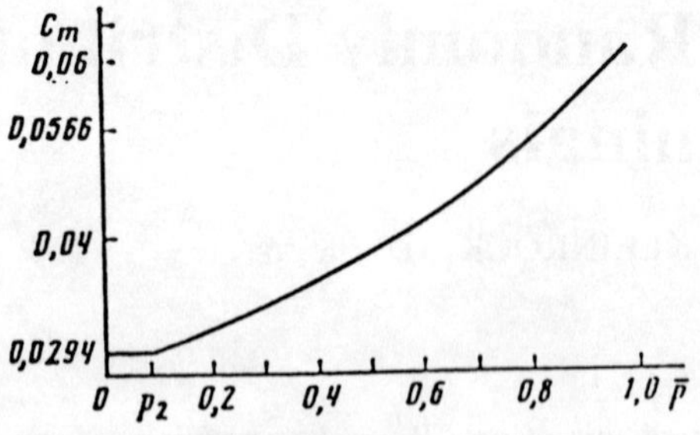

Fig. 5. Graph of C_m plotted against $\bar{p}$.

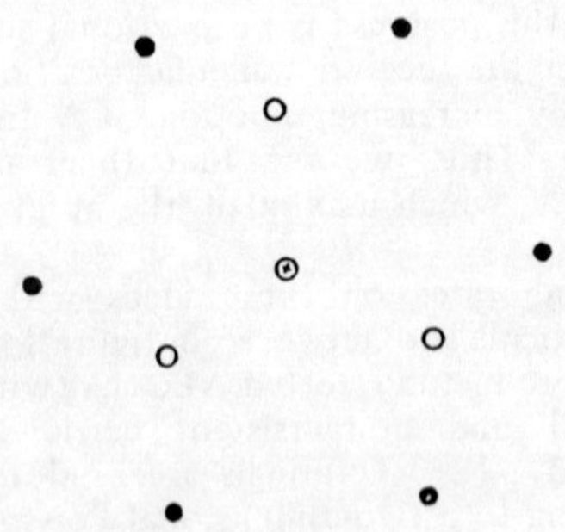

Fig. 6. Addressing of packets for Example 6.

In this case, r_{opt} will depend on $\bar{p}$. If we take the length of a side of a small triangle to be 1, then for $0 \le \bar{p} \le \bar{p}_2$, where $\bar{p}_2 = 2 - 6^6 13^{13}/7^7 12^{12} = 0.0755\ldots$, $r_{\text{opt}} = \sqrt{3}$ and $C_m = 12^{12}/2 \cdot 13^{13} = 0.02943\ldots$. For $\bar{p}_2 \le \bar{p} \le 1$ we have $r_{\text{opt}} = 1$ and $C_m = 6^6/7^7(2 - \bar{p})$. For $p = 1$ we have $C_m = 0.05664\ldots$. At point $\bar{p} = \bar{p}_2$ there are two optimal radii, 1 and $\sqrt{3}$. Figure 5 shows C_m plotted against $\bar{p}$.

Example 6. In this example we assume that the geography G is as shown in Fig. 6. It consists of regular hexagons. We specify the addressing for the station indicated by a cross in Fig. 6. The addressing for the other stations can be obtained as in the preceding example. Packets are addressed to stations indicated by open circles with probability $\bar{p}/3$, and to stations indicated by solid circles with probability $(1 - \bar{p})/6$. In this case r_{opt} is independent of $\bar{p}$ and $r_{\text{opt}} = 1$ (the length of an edge of the hexagon is taken to be 1), $k(r_{\text{opt}}) = 3$ and $C_m = 3^3/4^4(2 - \bar{p})$; for $\bar{p} = 0$, $C_m = 0.05273\ldots$.

References

[1] J. F. Shoch and J. A. Hupp, "Measured performance of an ETHERNET local network," *Commun. ACM*, vol. 23, no. 12, pp. 711–720, 1980.

[2] N. Abramson, "The throughput of packet broadcasting channels," *IEEE Trans. Commun.*, vol. 25, no. 1, pp. 117–128, 1977.

[3] B. S. Tsybakov and V. A. Mikhailov, "Ergodicity of a synchronous ALOHA system," *Probl. Peredachi Inf.*, vol. 15, no. 4, pp. 73–81, 1979.

[4] R. E. Kahn, S. A. Gronemeyer, J. Burchfiel, and R. C. Kunzelman, "Advances in packet-radio technology," *Proc. IEEE*, vol. 66, no. 11, pp. 1468–1496, 1978.

[5] L. Kleinrock and J. A. Silvester, "Optimum transmission radii in packet radio networks or why six is a magic number," in *Proc. Nat. Telecommun. Conf.*, Birmingham, Alabama, 1978.

[6] J. A. Silvester and L. Kleinrock, "On the capacity of multihop slotted ALOHA networks with regular structure," *IEEE Trans. Commun.*, vol. 31, no. 8, pp. 974–982, 1983.

[7] B. S. Tsybakov and V. L. Bakirov, "Communication networks with ALOHA algorithm," in *Packet-Switched Computer Networks* [in Russian], Nauka, Moscow, 1984, pp. 63–66.

[8] B. S. Tsybakov and V. L. Bakirov, "ALOHA algorithm in communications networks," in *Abstracts of Reports to Sixth International Symposium on Information Theory* [in Russian], Part 1, Tashkent, 1984, pp. 199–201.

[9] B. S. Tsybakov and V. L. Bakirov, "Stability of an asynchronous ALOHA system," *Probl. Peredachi Inform.*, vol. 20, no. 1, pp. 82–94, 1984.

Optimal Transmission Ranges for Randomly Distributed Packet Radio Terminals

HIDEAKI TAKAGI, MEMBER, IEEE, AND LEONARD KLEINROCK, FELLOW, IEEE

Abstract—**In multihop packet radio networks with randomly distributed terminals, the optimal transmission radii to maximize the expected progress of packets in desired directions are determined with a variety of transmission protocols and network configurations. It is shown that the FM capture phenomenon with slotted ALOHA greatly improves the expected progress over the system without capture due to the more limited area of possibly interfering terminals around the receiver. The (mini)slotted nonpersistent carrier-sense-multiple-access (CSMA) only slightly outperforms ALOHA, unlike the single-hop case (where a large improvement is available), because of a large area of "hidden" terminals and the long vulnerable period generated by them. As an example of an inhomogeneous terminal distribution, the effect of a gap in an otherwise randomly distributed terminal population on the expected progress of packets crossing the gap is considered. In this case, the disadvantage of using a large transmission radius is demonstrated.**

I. Introduction

One of the key issues in providing efficient and cost-effective multihop packet radio networks is to find an adequate transmission power for each terminal in the network. The environment we have in mind is one in which communicating terminals are geographically distributed and possibly mobile and require multiaccess to a communication channel shared among themselves. It has been shown [7] that the spatial reuse of the channel obtained by reducing the transmission power to such a level that only a few neighbors are within the range gives rise to an improved throughput (the average rate of successful transmissions) for the network. However, since the purpose of transmitting packets in a multihop environment is to advance them towards their destinations, a more appropriate measure of performance is the expected one-hop progress of a packet in the desired direction [4], [7].

The optimal transmission power to maximize the expected progress involves the following tradeoff. (Here we assume every terminal uses the same power.) A short-range transmission is favorable in terms of successful transmission because of its low possibility of collision (the overlapping of packet transmission periods from multiple transmitters) at the receiver. A long-range transmission is favorable because 1) it moves a packet far ahead in one hop if successful, and 2) there is high probability of finding a candidate receiver in the desired direction. Roughly speaking, if we denote by N the average number of terminals within the transmission radius (N is clearly an increasing function of the radius), then the probability of successful transmission is proportional to $1/N$, whereas the progress is proportional to $\sqrt{N}$, and the contribution from the receiver's angular position is expressed as a monotonically increasing function of N from 0 to some asymptotic value. Thus, we see that there must exist an optimal value of N, which maximizes the obtainable expected progress.

This paper elaborates on these ideas with a variety of transmission protocols and network configurations. The protocols considered here include slotted ALOHA (with and without FM capture) [6] and nonpersistent carrier-sense-multiple-access (CSMA) [3], [8]. Terminals are randomly located in the plane according to a two-dimensional Poisson distribution with homogeneous or inhomogeneous density. Each section below begins with the description of the model used in that section, followed by the formulation of the optimization problem. The optimal transmission range is found, and the performance is compared to other models. The results are summarized in the concluding section.

II. Optimal Transmission Radii for Slotted ALOHA

This section is concerned with the optimal transmission radii for randomly distributed terminals using slotted ALOHA as the transmission protocol. The same problem was considered by Kleinrock and Silvester [4], [7] who provided the "magic number" 6 as the optimal number of terminals to be covered by one transmission. However, there appears to be an inconsistency in their treatment. (In evaluating the probability of successful reception [7, eq. (6.7)], the number of terminals around the receiver is confused with that around the transmitter. As a matter of fact, the resultant optimal p, $p^* = 1/N$, could be greater than 1 [inconsistent with slotted ALOHA] for a very small transmission radius.) Therefore, we reconsider their problem and show a different magic number nearly equal to 8. The present section also serves to provide the most basic model among those considered in this paper.

We consider the progress that a given packet makes in the direction towards its final destination for a single (arbitrary) slot only and do not discuss its behavior along the entire path. The basic assumptions and associated parameters used in this section are as follows.

Transmission protocol: slotted ALOHA. The slot length in time is equal to the transmission time of a packet. (All packets are assumed to be of the same length.) The propagation time is ignored (or considered to be included in the slot). We do not take into account the acknowledgment traffic. It is assumed that the successful reception of a packet is immediately made known to the transmitter (e.g., by using a different (free) channel of wide bandwidth).

Transmission probability: p. All terminals are supposed to

Paper approved by the Editor for Computer Communications of the IEEE Communications Society for publication without oral presentation. Manuscript received January 6, 1983; revised May 2, 1983. This work was supported by the Defense Advanced Research Projects Agency of the Department of Defense under Contract MDA 903-77-C-0272.

H. Takagi is with the IBM Japan Science Institute, Tokyo, Japan.

L. Kleinrock is with the Department of Computer Science, University of California, Los Angeles, CA 90024.

Reprinted from *IEEE Trans. Commun.*, vol. COM-32, no. 3, pp. 246–257, March 1984.

have packets at all times (heavy-traffic assumption). For every slot, each terminal transmits a packet with probability p (and does not with probability $1 - p$), where $0 < p \leq 1$.

Transmission radius: R. All terminals use the same transmission radius. This means that terminals within a circle of radius R centered at the transmitter hear the transmission, whereas others do not hear it at all. More than one transmission within a distance R of the receiver in the same slot bring about the collision of all packets at that receiver.

Spatial distribution of terminals: two-dimensional Poisson distribution with the average number of terminals per unit area λ. We assume that a new sample of the spatial distribution is given for every slot.

Distribution of the sources and destinations of packets: two-dimensional isotropic, i.e., uniform over the plane. For every slot, the direction of the final destination for a packet in each terminal is assumed to be uniformly distributed in angle.

Routing strategy: most forward within R (MFR). Each terminal is assumed to know the position of those terminals within a distance R. Given a packet and its final destination, a terminal transmits to the terminal most forward (among those whose positions it knows) in the direction of the final destination. If no terminals are in the forward direction, it transmits to the least backward terminal, if any. (A terminal cannot transmit to itself.) In case there are no terminals in the circle of radius R at all, it does not transmit in that slot. (Note that MFR may not be minimizing the remaining distance to be traveled to the destination; MFR is myopic routing.)

$N \triangleq \lambda \pi R^2$: the average number of terminals within a radius R, and also a measure of connectivity of the network.

In this environment, we have the following two measures of performance.

$S(p, N) \triangleq$ the one-hop throughput, defined as the average number of successful transmissions per slot from a terminal.

$Z(p, N) \triangleq$ the expected progress of a packet in the direction of its final destination per slot from a terminal. The progress x is attained when x is the distance between the transmitter and the receiver projected onto a line drawn towards the final destination and the transmission to that receiver is successful.

Note that $Z(p, N)$ has the dimension of length (e.g., miles). Therefore, $Z(p, N)\sqrt{\lambda}$ may conveniently be used as a dimensionless measure of the expected progress in the number of hopped-over terminals since $1/(2\sqrt{\lambda})$ is the average distance between two nearest terminals. (See (37) below.) We employ $Z(p, N)$ as the objective function for our optimization problem in accordance with the routing strategy MFR. A point in the (p, N) plane which maximizes $Z(p, N)$ is sought. However, the value of $S(p, N)$ at this optimal point is also interesting. It will turn out that the same $p = p^*(N)$ maximizes both $S(p, N)$ and $Z(p, N)$.

In order to evaluate $S(p, N)$, we first note that e^{-N} is the probability of there being no terminals within a distance R of the transmitter. In such a case, no transmission can occur. Under the condition that there is at least one candidate rereceiver within R, let A_i be the event that there are i other terminals (excluding the transmitter P and receiver Q) within a distance R of Q. (See Fig. 1.) Thanks to the memoryless property of the Poisson distribution, the distribution of the number of other terminals within R does not depend on the existence of P and Q. Thus, we have

$$\text{Prob}[A_i] = \frac{N^i}{i!} e^{-N} \qquad i = 0, 1, 2, \cdots. \tag{1}$$

The transmission from P to Q is successful (let this event be

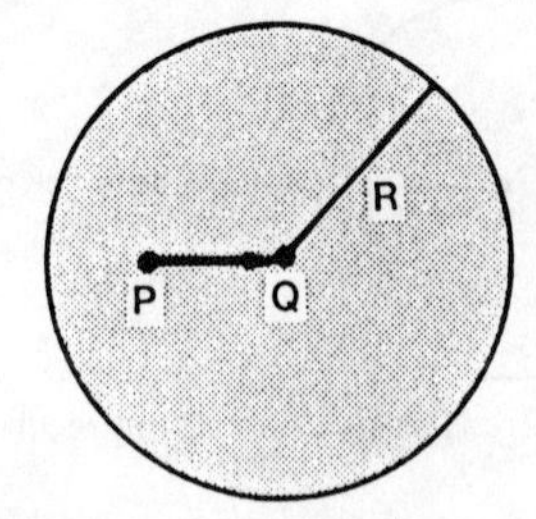

Fig. 1. The area of interference with the transmission $P \rightarrow Q$.

denoted by $P \rightarrow Q$) if none of the terminals within a distance R of Q transmit (including Q itself). Thus,

$$\text{Prob}[P \rightarrow Q \mid A_i] = (1 - p)^{i+1}. \tag{2}$$

It follows that

$$\begin{aligned} S(p, N) &= \text{Prob}[\text{there is at least one terminal within } R] \\ &\quad \cdot \text{Prob}[P \text{ transmits}] \cdot \text{Prob}[P \rightarrow Q] \\ &= (1 - e^{-N})p \sum_{i=0}^{\infty} \text{Prob}[P \rightarrow Q \mid A_i] \, \text{Prob}[A_i] \\ &= p(1 - p)e^{-pN}(1 - e^{-N}) \\ &\cong p(1 - p)N \quad \text{as } N \rightarrow 0, \\ &\cong p(1 - p)e^{-pN} \quad \text{as } N \rightarrow \infty. \end{aligned} \tag{3}$$

Given N, $S(p, N)$ is maximized by

$$\begin{aligned} p = p^*(N) &\triangleq \frac{2}{N + 2 + \sqrt{N^2 + 4}} \\ &\cong 1/2 \quad \text{as } N \rightarrow 0, \\ &\cong 1/N \quad \text{as } N \rightarrow \infty. \end{aligned} \tag{4}$$

The maximum value itself is given by

$$\begin{aligned} S(p^*(N), N) &= \frac{1 - e^{-N}}{2 + \sqrt{N^2 + 4}} \exp\left(-\frac{2N}{N + 2 + \sqrt{N^2 + 4}}\right) \\ &\cong (1/4)N \quad \text{as } N \rightarrow 0, \\ &\cong 1/(Ne) \quad \text{as } N \rightarrow \infty. \end{aligned} \tag{5}$$

In the case $N \rightarrow 0$, at most only pairs of terminals can hear each other. So, they transmit with probability 1/2 (the well-known optimal p for two terminals). The optimized results for large N are the same is in [4] and [7], in which it is stated that they correspond to setting the average traffic load to be equal to one packet per slot within the transmission range. The optimized throughput of N terminals for large N, $1/e$, also conforms to Roberts' result [6] for a single-hop system of an infinite number of terminals.

We now proceed to find $Z(p, N)$. According to the MFR routing, $\tilde{z}$, the progress of a packet per transmission, is not greater than x if there are no terminals in the area A in Fig. 2.

$$\text{Prob}[\tilde{z} \leq x] = e^{-(N/\pi)q(x/R)} \tag{6}$$

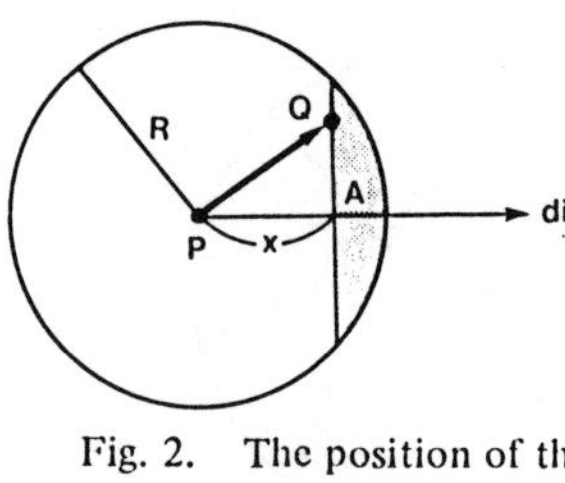

Fig. 2. The position of the receiver Q.

where

$$q(t) \triangleq \cos^{-1}(t) - t\sqrt{1-t^2}. \quad (7)$$

Note that $q(t)$ is the area of A in Fig. 2 when $R = 1$ (unit circle) and $x = t$. Therefore, we have

$$Z(p, N) = \text{Prob}[P \text{ transmits}] \cdot \text{Prob}[P \to Q] \cdot E[\text{progress of a packet}]$$
$$= p(1-p)e^{-pN} \int_{-R}^{R} x \cdot \text{Prob}[x < \tilde{z} \leqslant x + dx]$$
$$= p(1-p)e^{-pN} \sqrt{\frac{N}{\lambda\pi}} \left[1 + e^{-N} - \int_{-1}^{1} e^{-(N/\pi)q(t)}\, dt\right]. \quad (8)$$

Thus, given N, $Z(p, N)$ is also maximized by $p = p^*(N)$ given by (4), and the normalized maximum is given by

$$Z(p^*(N), N)\sqrt{\lambda} = \frac{1}{2+\sqrt{N^2+4}} \exp\left(-\frac{2N}{N+2+\sqrt{N^2+4}}\right) \cdot \sqrt{\frac{N}{\pi}}\left[1 + e^{-N} - \int_{-1}^{1} e^{-(N/\pi)q(t)}\, dt\right] \quad (9)$$
$$\cong \frac{16}{45}\left(\frac{N}{\pi}\right)^{5/2} \quad \text{as } N \to 0,$$
$$\cong \frac{1}{e\sqrt{\pi N}} \quad \text{as } N \to \infty.$$

The functions $p^*(N)$, $S(p^*(N), N)$, and $Z(p^*(N), N)\sqrt{\lambda}$ as given by (4), (5), and (9), respectively, are plotted in Fig. 3. $Z(p^*(N), N)\sqrt{\lambda}$ has its maximum value at

$$N = N^* = 7.72.$$

Thus, we propose a new magic number, 8, as the optimal number of terminals to be covered in the transmission range. In terms of transmission radius, we have

$$R^* = 3.14(1/(2\sqrt{\lambda})).$$

The associated optimal values are

$$p^* \triangleq p^*(N^*) = 0.113$$

$$S^* \triangleq S(p^*, N^*) = 0.0419$$

$$Z^*\sqrt{\lambda} \triangleq Z(p^*, N^*)\sqrt{\lambda} = 0.0431.$$

Therefore, the sketch of optimal transmission is described as follows. Each terminal transmits a packet in every ninth slot on the average ($1/p^* = 8.85$). The probability of success of such a transmission is $S^*/p^* = 0.37$, as slotted ALOHA predicts. It uses a transmission radius to span just about three (3.14) nearest neighbors in linear distance. Then, the expected progress of the packet is $Z^*/p^* = 0.76(1/(2\sqrt{\lambda})) \cong (2/3)(R^*/e)$. Here the factor $1/e$ accounts for the probability of successful transmission, and $(2/3)\ R^*$ represents the effective distance that a packet is advanced by a successful transmission with radius R^*.

III. Optimal Transmission Radii for ALOHA with Capture

The analysis of the preceding section is here extended to the case of a slotted ALOHA system with FM capture. The observation that the capture phenomenon increases the throughput for a single receiver has been investigated by Roberts [6] and Abramson [1]. Fratta and Sant [2] have shown how capture affects the throughput behavior of an ALOHA network which has multiple transmitters and receivers. They did not use the notion of the transmission radius as we have done in Section II. Their work will be the basis of Section VI. In this section, we consider the optimization problem of the expected progress of packets through the MFR (most forward within the transmission radius R) routing in a capture environment. Similar work has been done by Nelson [5] using a different routing strategy. (Specifically, in his routing, one of the [say] k terminals within a half circle [in the forward direction] of radius R is picked as a receiver with probability $1/k$. As a result, his optimized expected progress is somewhat smaller than ours. For example [using the notation defined below], in the case of perfect capture, he gives $Z^*/R^* = 0.0346$, while we give $Z^*/R^* = 0.0393$.)

The basic assumptions and parameters for the model we study here are the same as in Section II, except for the conditions for successful transmission. They include the slotted ALOHA transmission protocol, transmission probability p, transmission radius R, Poisson distribution of terminals with parameter λ, MFR routing, isotropic distribution of source-destination pairs, and $N \triangleq \lambda\pi R^2$.

The concept of FM capture used in this section and Section VI is the same as in the papers cited above, that is, a receiver will correctly receive a packet from a transmitter which is located at a distance r of the receiver if none of the terminals within a distance αr of the receiver transmit simultaneously. The capture parameter α is related to the capture ratio CR in decibels via $\text{CR} = 20 \log_{10}\alpha$, $1 \leqslant \alpha < \infty$. The case $\alpha = 1$ is called perfect capture, whereas the case $\alpha \to \infty$ corresponds to the system without capture (i.e., the case considered in Section II).

Under these circumstances, we evaluate the throughput $S(p, N; \alpha)$ and the expected progress $Z(p, N; \alpha)$. We employ $Z(p, N; \alpha)$ as the objective function of our optimization problem with respect to p and N. [Now the optimum $p^*(N; \alpha)$ for $S(p, N; \alpha)$ is different from that for $Z(p, N; \alpha)$.]

First, we state the conditions for successful transmission of a packet. Since all terminals are using the same transmission radius R, the transmission from the transmitter P to the receiver Q, under the condition that they are a distance r apart, is successful if no other terminals within a distance

$$r' \triangleq \min[\alpha r, R] \quad (10)$$

of Q (including Q itself) transmit at the same time [5], [6]. Fig. 4 shows the area of potential interfering terminals for the transmission from P to Q. Thus, unconditioning on the num-

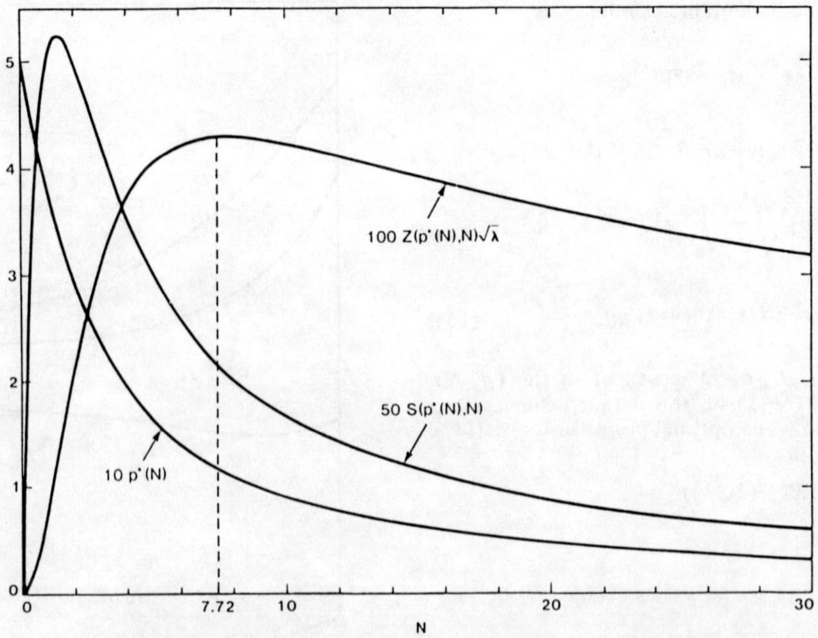

Fig. 3. The optimal transmission for slotted ALOHA networks without capture. (In this and some other figures, for conciseness we have plotted scaled values against a single vertical axis.)

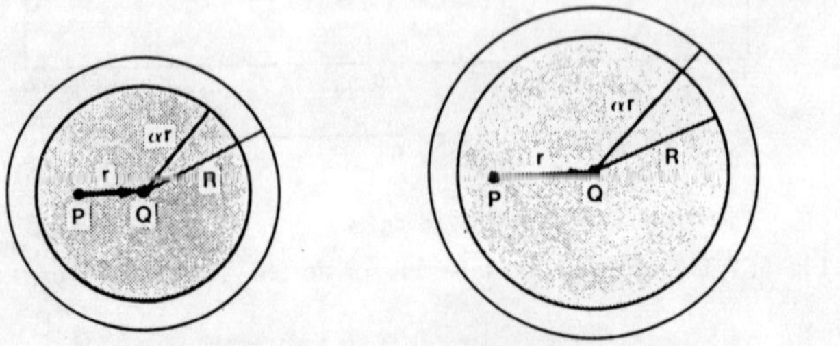

Fig. 4. The area of terminals possibly interfering with the transmission $P \rightarrow Q$. (a) $\alpha r < R$. (b) $\alpha r > R$.

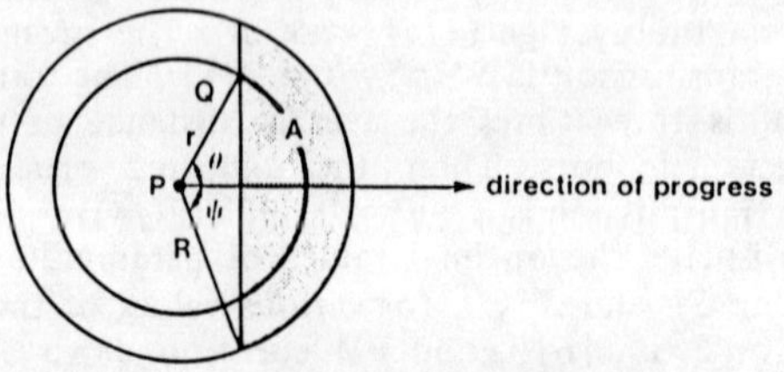

Fig. 5. The position of the receiver Q with respect to the position of the transmitter P.

ber of terminals in the area as in Section II, we have

$$\text{Prob}\,[P \rightarrow Q \mid \tilde{r} = r] = (1-p)e^{-\lambda p \pi r'^2} \tag{11}$$

where $P \rightarrow Q$ represents the event that the transmission from P to Q is successful, and $\tilde{r}$ is the distance between P and Q.

Second, we need the expression for the distribution of the positions of the receiver with respect to the transmitter. Let $(\tilde{r}, \tilde{\theta})$ be the polar coordinates of the position of the receiver Q, where the origin of the coordinates is at the position of the transmitter P, and $\tilde{\theta}$ is measured from the direction in which a packet at P is destined to proceed. See Fig. 5 for the configuration. Let A be the shaded area. Due to the MFR routing, the receiver is located at (r, θ) if and only if there are no terminals in A and there is a terminal at (r, θ). Therefore,

$$\text{Prob}\,[r < \tilde{r} \leqslant r + dr, \theta < \tilde{\theta} \leqslant \theta + d\theta]$$

$$= e^{-\lambda R^2(\psi - \sin\psi\cos\psi)} 2\lambda r\, dr\, d\theta$$

$$= 2\lambda r e^{-(N/\pi)q((r/R)\cos\theta)}\, dr\, d\theta$$

$$0 \leqslant r \leqslant R,\ 0 \leqslant \theta \leqslant \pi \tag{12}$$

where we have used the relation $r \cos\theta = R \cos\psi$ and the definition of $q(t)$ given in (7). Note that this is an expression for an event similar to the event defined for (6).

Thus, we have the throughput [similar to (3)]

$$S(p, N; \alpha) = p \int_0^R \int_0^\pi (1-p)e^{-\lambda\pi p r'^2}$$

$$\cdot\ \text{Prob}\,[r < \tilde{r} \leqslant r + dr, \theta < \tilde{\theta} \leqslant \theta + d\theta]$$

$$= \frac{2}{\pi} pN(1-p) \int_0^1 t e^{-pNt'^2}\, dt$$

$$\cdot \int_0^\pi e^{-(N/\pi)q(t\cos\theta)}\, d\theta \tag{13}$$

where

$$t' = \min\,[\alpha t, 1]. \tag{14}$$

We see that (13) reduces to (3) when $\alpha \rightarrow \infty$. Also, we have

$$S(p, N; \alpha) \cong p(1-p)N$$

for $N \ll 1$ and a moderate value of α, which is again the same as before.

The expected progress can be obtained similarly as

$$Z(p, N; \alpha) = p \int_0^R \int_0^\pi (1-p)e^{-\lambda \pi p r'^2} r \cos\theta$$

$$\cdot \text{Prob}\,[r < \tilde{r} \leqslant r + dr, \theta < \tilde{\theta} \leqslant \theta + d\theta]$$

$$= \frac{2}{\pi} pN(1-p)\sqrt{\frac{N}{\lambda\pi}} \int_0^1 t^2 e^{-pNt'^2}\, dt$$

$$\cdot \int_0^\pi \cos\theta e^{-(N/\pi)q(t\cos\theta)}\, d\theta. \qquad (15)$$

The maximum of $Z(p, N; \alpha)\sqrt{\lambda}$ is sought in the (p, N) plane for given α. Let $p^*(N; \alpha)$ be the p that achieves this maximum for given N and α. The optimal point for $\alpha = 1$ (perfect capture) is found as follows:

$$N^* = 7.1 \quad \text{or} \quad R^* = 3.0(1/(2\sqrt{\lambda}))$$

$$p^* \triangleq p^*(N^*; 1) = 0.17$$

$$S^* \triangleq S(p^*, N^*; 1) = 0.068$$

$$Z^*\sqrt{\lambda} \triangleq Z(p^*, N^*; 1)\sqrt{\lambda} = 0.059.$$

The expected progress is about 36 percent better than the system without capture. The optimal transmission is now sketched as follows. Each terminal transmits a packet in every sixth slot on the average ($1/p^* = 5.88$). The probability of successful transmission is $S^*/p^* = 0.4 > 1/e$. The transmission radius used is three times the average distance between the two nearest neighbors. Then, the expected progress of a packet per transmission is $Z^*/p^* = 0.69(1/(2\sqrt{\lambda}))$.

Fig. 6 displays the optimal values of parameters N and p and resulting S^* and $Z^*\sqrt{\lambda}$ for various values of the capture parameter α. From [6], good FM corresponds to CR = 1.5, while moderate FM corresponds to CR = 3.0 and poor FM corresponds to CR = 6.0. As we noted earlier, the results for $\alpha = \infty$ (no capture) coincide with those in Section II. We first notice that $Z^*\sqrt{\lambda}$ with some capture is always greater than that without capture. Thus, a conclusion here is that the FM capture always helps the progress of packets. The reason for this is that we limit the area of interfering terminals within $\min[\alpha r, R]$, which is always no greater than R for the case without capture. This implies a smaller number of interfering terminals, thus giving higher throughput and greater expected progress. It is also interesting that as α increases, N^* first decreases and then increases to reach its final value. This might be explained as follows. For small α, the limitation of a conflicting area by αr is more effective than that by R, so N^* decreases with more conflict as α increases. On the other hand, for large α, the limitation by R is dominant; so N^* approaches the value without capture.

IV. Optimal Transmission Radii for CSMA

In a single-hop network, another (great) improvement over ALOHA is made possible by CSMA. With this protocol, each terminal utilizes the information about channel status (busy or idle) obtained by listening to the channel. However, the existence of some terminals which are not in line of sight of others causes degradation in performance; this is called the hidden-terminal effect [8], [9]. If we use the CSMA protocol in a multihop network, we expect a similar effect because the hearing range of the receiver is more or less different from the listening range of the transmitter. The purpose of this section is to estimate the effect of hidden terminals associated with CSMA, with the same terminal distribution and with the same packet routing strategy as in the preceding sections. The basic assumptions and parameters carried over from Section II include the Poisson distribution of terminals with parameter λ, transmission radius R, MFR (most forward within R) routing, isotropic distribution of source-destination pairs, and $N \triangleq \lambda\pi R^2$.

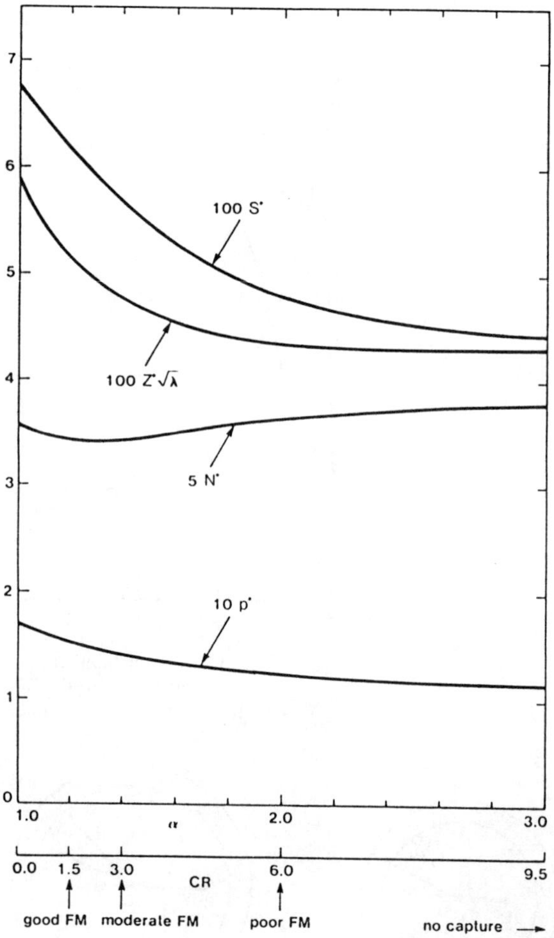

Fig. 6. The optimal transmissions for slotted ALOHA with capture.

We now explain the protocol of slotted nonpersistent CSMA. The constant packet transmission time is chosen as the unit of time, and the length of a (mini)slot, denoted by a, accounts for the signal propagation delay. In the derivation below, $\tau \triangleq 1/a$ is assumed to be an integer. (Propagation delay a is used to imply a time interval long enough for all the terminals in the transmission range to recognize the events that occurred time a before.) See Fig. 7 for the illustration of the channel activity heard at the receiver. We assume that all the terminals within a distance R of the transmitter recognize the transmission in one slot and that they hear the transmission one slot more after the completion of transmission. Assuming that every terminal is ready to transmit at all times, the nonpersistent protocol is described as follows. In every slot, each terminal listens to the channel with probability p (and does not with probability $1 - p$). That is, the channel-sensing behavior in a sequence of slots (except during the transmission) at each terminal constitutes independent Bernoulli trials. The parameter p is the sensing rate per slot. If the channel is sensed idle, it begins transmission in the same slot with probability 1. If the channel is sensed busy, it suppresses the transmission, and stops sensing the channel until the end of the current transmission. When the channel becomes idle, the above sensing procedure is repeated.

It is clear that the events whether an actual transmission occurs or not as a result of channel sensing in a sequence of slots

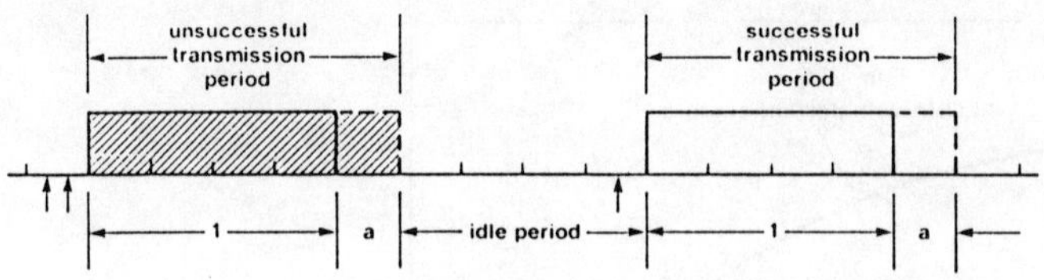

Fig. 7. Slotted nonpersistent CSMA. Transmission and idle periods.

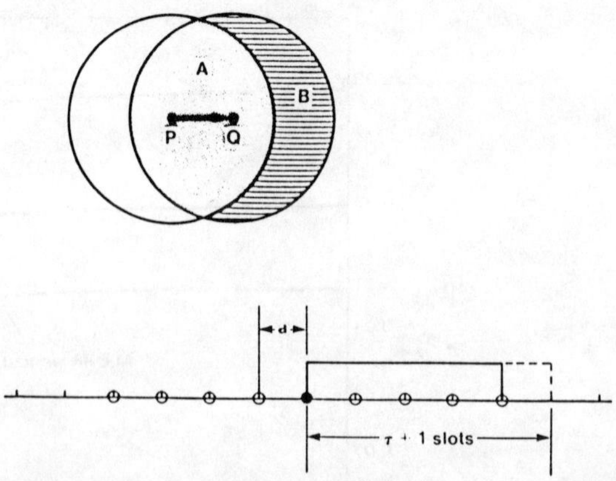

Fig. 8. The period for the transmission P to Q vulnerable to the transmissions from areas A and B. (a) Configuration. (b) Time line (•: vulnerable points to A and B; o: vulnerable points to B).

at each terminal are no longer independent Bernoulli trials. However, we introduce the assumption that they are. That is, for every slot (except during the transmission), each terminal transmits a packet with probability p' (and does not with probability $1 - p'$). A similar assumption was used in [9], and the validity of results obtained was claimed by comparing the throughput values against simulation. The parameter p' is the transmission rate per slot. We leave the determination of p' in terms of p to the Appendix since we formulate our optimization problem with only p'. Under these conditions, we will evaluate the throughput of transmission $S(p', N; a)$, and the expected progress $Z(p', N; a)$. We employ $Z(p', N; a)$ as the objective function of our optimization problem with respect to p' and N. [Again $S(p', N; a)$ and $Z(p', N; a)$ are not optimized by the same $p'(N)$.]

A particular transmission is successful when no other terminals within a distance R of the receiver transmit during the transmission period $1 + a$. Let us consider the conditions for the successful transmission from the transmitter P to receiver Q, referring to Fig. 8. The shaded area A and B shows the area of terminals whose transmission may collide with the transmission from P to Q at Q. Since the terminals in area A recognize the transmission in one slot, a collision will be avoided if they do not begin transmission in the same slot. On the other hand, since the transmissions from the terminals in area B occur independently, it is sufficient that they keep silent throughout the entire vulnerable period of length $2 + a$ or $2\tau + 1$ slots shown in Fig. 8(b) (the first τ slots are included so as to prevent any interference with the ongoing transmissions and the second $\tau + 1$ slots are included not to be interfered with newly started transmissions). (Two packets whose transmissions start with τ slots apart may or may not be received successfully; however, we exclude such a case to pessimistically evaluate the probability of success.) Therefore, if $\tilde{r}$ denotes the distance between P and Q, and $P \to Q$ denotes the successful transmission from P to Q, then we have

$$\begin{aligned}
\text{Prob}\,[P \to Q \mid \tilde{r} = r] &= \text{Prob}\,[Q \text{ does not start transmission in the same slot}] \\
&\quad \cdot \text{Prob}\,[\text{no transmission from } A \text{ during a slot} \mid \tilde{r} = r] \\
&\quad \cdot \text{Prob}\,[\text{no transmission from } B \text{ during } 2\tau + 1 \\
&\qquad \text{slots} \mid \tilde{r} = r].
\end{aligned}$$

Since the area of A is $2R^2 q(r/2R)$ and the area of B is $\pi R^2 - 2R^2 q(r/2R)$, we get

$$\begin{aligned}
\text{Prob}\,[P \to Q \mid \tilde{r} = r] &= (1 - p')e^{-p'\lambda \cdot 2R^2 q(r/2R)} \\
&\quad \cdot e^{-(2\tau+1)p'\lambda \cdot [\pi R^2 - 2R^2 q(r/2R)]} \\
&= (1 - p')e^{-p'N\{1 + 2\tau[1 - (2/\pi)q(r/2R)]\}} \qquad (16)
\end{aligned}$$

where $q(t)$ is defined in (7). An assumption involved here is that an independent sample of terminal distributions is given afresh for every slot throughout the vulnerable period from B. Based on this assumption we have evaluated the probability of success in each slot independently.

Since the assumptions about routing are the same as in Section III, the distribution of the position $(\tilde{r}, \tilde{\theta})$ of the receiver Q with respect to the transmitter P is given by (12). It follows that the one-hop throughput is given by

$$\begin{aligned}
S(p', N; a) &= \frac{p'}{a} \int_0^R \int_0^\pi \text{Prob}\,[P \to Q \mid \tilde{r} = r] \\
&\quad \cdot \text{Prob}\,[r < \tilde{r} \leqslant r + dr, \theta \leqslant \tilde{\theta} < \theta + d\theta] \\
&= \frac{2}{\pi} p'\tau N(1 - p')e^{-p'(2\tau+1)N} \\
&\quad \cdot \int_0^1 t e^{(4p'\tau N/\pi)q(t/2)}\, dt \\
&\quad \cdot \int_0^\pi e^{-(N/\pi)q(t\cos\theta)}\, d\theta \qquad (17)
\end{aligned}$$

and similarly the expected progress is given by

$$\begin{aligned}
Z(p', N; a) &= \frac{2}{\pi} p'\tau N(1 - p')\sqrt{\frac{N}{\lambda\pi}}\, e^{-p'(2\tau+1)N} \\
&\quad \cdot \int_0^1 t^2 e^{(4p'\tau N/\pi)q(t/2)}\, dt \\
&\quad \cdot \int_0^\pi e^{-(N/\pi)q(t\cos\theta)}\, d\theta. \qquad (18)
\end{aligned}$$

(We note that the above $S(p', N; a)$ and $Z(p', N; a)$ are not the long-time average values because we have not taken into account the channel activity cycles (idle and busy) whose duration is variable. Thus, (17) and (18) may be viewed as giving the instantaneous values at transmission start times; note that S and Z in slotted ALOHA cases are overall means and instantaneous values at the same time. Thus, the comparison between CSMA and ALOHA is meaningful.)

The maxima of the function $Z(p', N; a)\sqrt{\lambda}$ are determined numerically in the (p', N) plane for various values of a. The optimal point for $a = 0$ (zero propagation delay) is found as

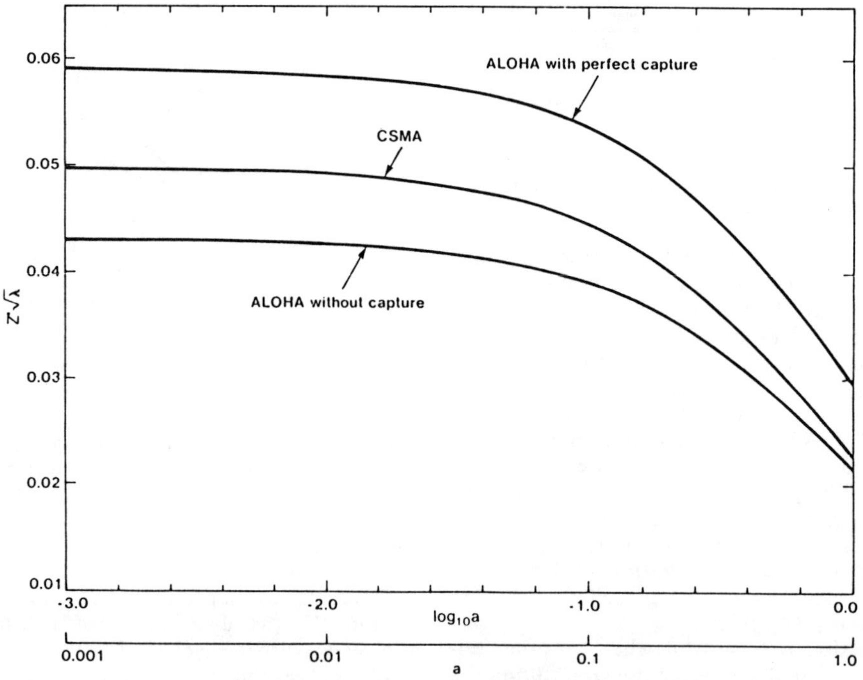

Fig. 9. Comparison of the optimized expected progress among ALOHA with and without capture, and CSMA networks.

follows:

$$N^* = 5.3 \quad \text{or} \quad R^* = 2.6(1/(2\sqrt{\lambda}))$$

$$\lim_{a\to 0} \frac{p^*}{a} = \lim_{\tau\to\infty} \tau p^* = 0.20$$

$$S^* \triangleq S(p^*, N^*; 0) = 0.077$$

$$Z^*\sqrt{\lambda} \triangleq Z(p^*, N^*; 0)\sqrt{\lambda} = 0.050.$$

Therefore, the optimized expected progress is only about 16 percent ($\doteq$(0.050 − 0.0431) × 100/0.0431) better than ALOHA system without capture.

This small improvement in performance, unlike the single-hop case, appears due to the large area of hidden terminals (about half of the hearable range for $N = N^*$) and the long period (twice as long as the packet transmission time) vulnerable to their transmission.

In Fig. 9, the optimized expected progress with CSMA is plotted for various values of a, together with those for ALOHA systems without capture and with perfect capture. (For proper comparison, the optimized expected progress with slotted ALOHA should be divided by $1 + a$ to include the propagation time in a slot.) It is seen that the performance of CSMA lies between ALOHA without capture and ALOHA with perfect capture. With reference to Fig. 6, CSMA's performance turns out comparable to that of ALOHA with capture ratio about 1.5 dB which corresponds to good FM. The degradation of the expected progress with increasing a is due to the longer vulnerable period.

V. Optimal Transmission Radii in an Inhomogeneous Density of Terminals

So far we have considered only the Poisson distribution of terminals with the same spatial density everywhere. However, it is of importance in our multihop packet radio studies to extend the analysis to inhomogeneous structures. For example, how should the transmission power be controlled as one passes from a region of low density terminals to higher density terminals and then back out again to lower density terminals; this corresponds to a kind of geographical bottleneck. Another configuration is what we call the "dumbbell" configuration in which we have high density regions (say, two cities) connected together with an extremely low density region (say, a desert). Here one inquires whether the low density region helps the transmission or not. These are some of the motivations for our study of inhomogeneous configurations of packet radio terminals.

Specifically, the configuration of terminals we consider in this section is a vacant strip of width b in an otherwise Poisson-distributed terminal population with uniform average density λ. Taking the x-axis perpendicular to the gap length, the average density of terminals at x is given by

$$g(x) \triangleq \begin{cases} 0 & 0 \leqslant x \leqslant b \\ \lambda & \text{elsewhere.} \end{cases} \tag{19}$$

We introduce the "intensity" of the gap by

$$\beta \triangleq \lambda b^2 \tag{20}$$

which is the average number of terminals that would be in the gap of length b if it were not for the gap. The dimensionless quantity β will be used as a characteristic parameter below.

In the following we evaluate the expected progress of a packet residing at the terminal P on the left bank ($x = 0$) and destined to cross the gap. See Fig. 10 for the configuration. We assume slotted ALOHA protocol and the transmission radius R ($>b$) for all terminals. For simplicity, we do not optimize the transmission probability p but will use the value $p = 0.113$ which has been found optimal for the case of homogeneous Poisson distribution (see Section II). We recognize the terminals being with a distance R of the receiver as those which may cause conflict with our transmission. Then, our usual procedure yields the probability of successful trans-

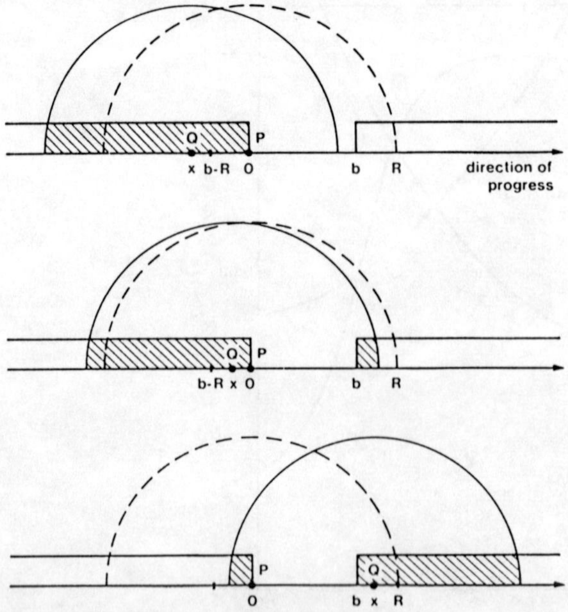

Fig. 10. Three cases of the position of the receiver Q. P: the transmitter, ////: the area of possibly interfering terminals. (a) $x \leq b - R$. (b) $b - R \leq x \leq 0$. (c) $b \leq x \leq R$.

mission to the receiver at x as

$$p(1-p)e^{-pn(x;R)} \tag{21}$$

where $n(x; R)$ is the average number of terminals within a circle of radius R around the receiver at x, given by

$$n(x;R) = \begin{cases} \lambda R^2 \left[\pi - q\left(-\dfrac{x}{R}\right)\right] & -R \leq x \leq b - R \\ \lambda R^2 \left[q\left(\dfrac{b-x}{R}\right) + \pi - q\left(-\dfrac{x}{R}\right)\right] & b - R \leq x \leq 0, \quad b \leq x \leq R \end{cases} \tag{22}$$

where $q(t)$ is defined in (7). The probability distribution function $F(x; R)$ of the position of the receiver is given by

$$F(x;R) \triangleq \text{Prob}\,[\text{no terminal in } (x, R) \mid R] = \begin{cases} \exp\left\{-\lambda R^2 \left[q\left(\dfrac{b}{R}\right) + q\left(\dfrac{x}{R}\right) - \dfrac{\pi}{2}\right]\right\} & -R \leq x \leq 0 \\ \exp\left\{-\lambda R^2 q\left(\dfrac{x}{R}\right)\right\} & b \leq x \leq R. \end{cases} \tag{23}$$

Using these expressions, the expected progress of our packet is calculated as

$$Z(R) = p(1-p) \int_{-R}^{R} e^{-pn(x;R)}\, x \, dF(x;R) \tag{24}$$

or, in a normalized form,

$$\begin{aligned} Z(N;\beta)\sqrt{\lambda} &= 2p(1-p)\left(\frac{N}{\pi}\right)^{3/2} \Bigg\{ \int_r^1 t\sqrt{1-t^2} \\ &\cdot \exp\left\{-\frac{pN}{\pi}[q(r-t) + \pi - q(-t)] - \frac{N}{\pi} q(t)\right\} dt \\ &+ \int_{r-1}^{0} t\sqrt{1-t^2} \exp\left\{-\frac{pN}{\pi}[q(r-t) + \pi - q(-t)] \right. \\ &\left. - \frac{N}{\pi}\left[q(r) + q(t) - \frac{\pi}{2}\right]\right\} dt + \int_{-1}^{r-1} t\sqrt{1-t^2} \\ &\cdot \exp\left\{-\frac{pN}{\pi}[\pi - q(-t)] \right. \\ &\left. - \frac{N}{\pi}\left[q(r) + q(t) - \frac{\pi}{2}\right]\right\} dt \Bigg\} \end{aligned} \tag{25}$$

where

$$N = \lambda \pi R^2; \qquad r = b/R = \sqrt{\pi \beta / N}. \tag{26}$$

Fig. 11 shows the optimal radii N^* and R^* and the expected progress $Z(N; \beta)\sqrt{\lambda}$ for various values of gap intensity β. We see that, for fixed λ

$$N^* \cong \begin{cases} 7.7 & \text{as } \beta \to 0 \\ \infty & \text{as } \beta \to \infty \end{cases} \tag{27}$$

and

$$Z(N^*)\sqrt{\lambda} \cong \begin{cases} 0.043 & \text{as } \beta \to 0 \\ 0.050 \text{ (maximum)} & \text{at about } \beta = 1.0 \\ 0 & \text{as } \beta \to \infty. \end{cases} \tag{28}$$

The results for narrow gaps reduce to the case with homogeneous density. As the gap width increases, the expected progress increases because some of the possibly interfering terminals are removed by the gap. However, for too wide a gap, the transmission radius must be accordingly larger in order to cross it, which causes more conflicts at the receiver; thus, the expected progress decreases. It is interesting that the optimized expected progress achieves its maximum at about $\beta = 1$. We can also see that for fixed b

$$R^*/b \cong \begin{cases} \infty & \text{as } \beta \to 0 \\ 1 & \text{as } \beta \to \infty \end{cases} \tag{29}$$

and

$$Z(R^*)/b \cong \begin{cases} \infty & \text{as } \beta \to 0 \\ 0 & \text{as } \beta \to \infty. \end{cases} \tag{30}$$

Therefore, for large λ, the optimal transmission radius is just large enough to reach the other bank. However, the higher possibility of interference with the terminals behind the receiver (i.e., those in the area $x > b$) diminishes the value of $Z(R^*)/b$. For small λ, since there is almost no inter-

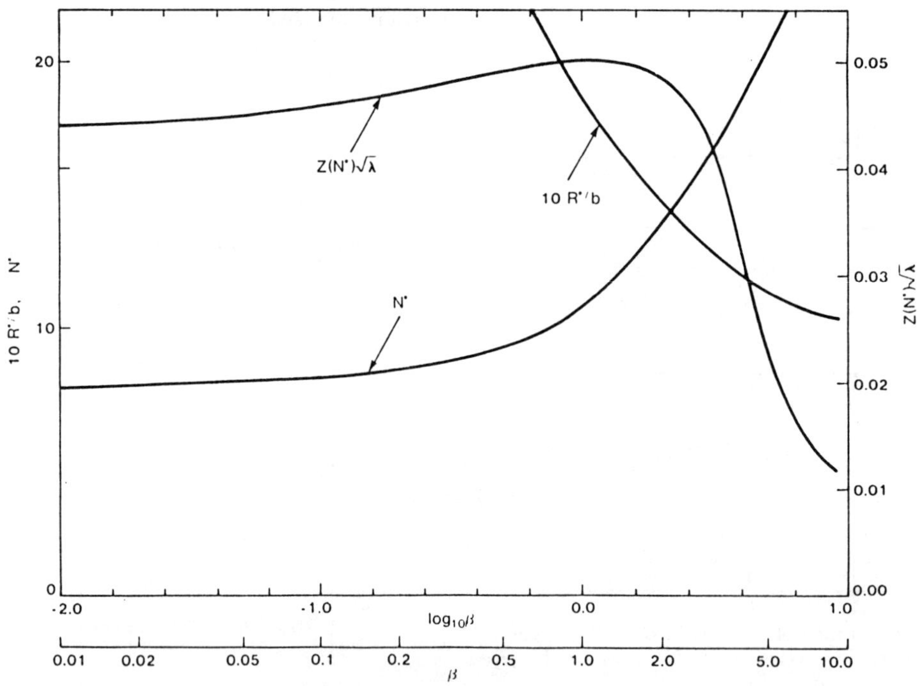

Fig. 11. The optimal transmission radii and expected progress for packet crossing a gap.

ference, the packet can proceed as far as an arbitrarily large transmission radius.

From Fig. 11, we see that the existence of the gap such that $\beta < 2$ helps the transmission. Notice, however, that a gap has an effect on performance only when $\beta \gg 1$. Thus, a conclusion here is that to cross the gap, we should not use a large transmission power with the same channel; rather, we had better use a separate channel (or wire) to avoid possible collisions.

VI. Optimal Number of Neighbors for ALOHA with Capture

In this section, we extend the model of a partially connected packet radio network with capture proposed by Fratta and Sant [2] to the context of our optimization problem. The reason for doing this is that the packet routing algorithm possibly implemented in each terminal is more suitably handled with their model than with the aforementioned MFR which assumes each terminal knows the position of an indefinite number of terminals within a distance R. However, without a notion of transmission radius, their model has a drawback of having an unrealistically wide area of interfering terminals for the case of poor capture (large capture ratio). Therefore, the results obtained here should be applicable only to the case of good capture.

The present model assumes a slotted ALOHA transmission protocol with transmission probability p in each slot, a Poisson distribution of terminals with homogeneous density λ, and an isotropic distribution of source–destination pairs. The concept of capture is described in Section II with capture parameter α. Every terminal is assumed to use the same transmission power. We do not use the notion of transmission radius, which implies that a transmission over a distance r is successful if none of the other terminals within the distance αr of the receiver transmit in the same slot.

We now explain the routing strategy employed here. Each terminal is assumed to know all the positions of its N nearest neighbors. Given a packet and its final destination, a terminal transmits to the most forward terminal in the direction of the final destination among those N neighbors whose positions are known. In case no terminals exist ahead, it transmits to the least backward neighbor. We call this routing MFN (most forward within N). MFN assumes that each terminal keeps the positions of only a fixed number of terminals, lending itself to easy implementation.

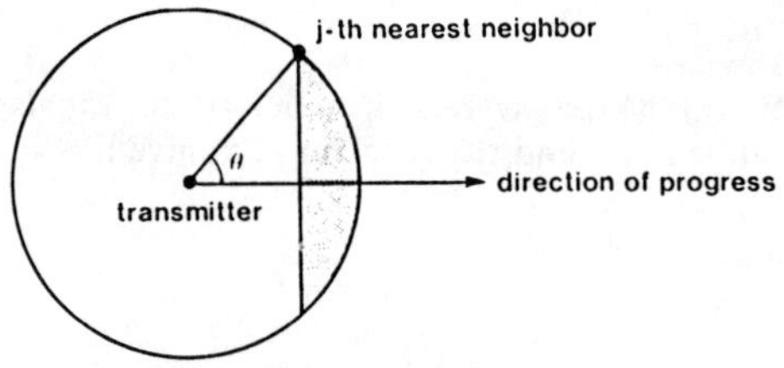

Fig. 12. The angular position of the jth nearest neighbor.

The routing algorithm at each terminal is formally stated as follows:

$j \leftarrow N$

L: Consider j nearest neighbors. If the jth nearest one is the most forward, transmit to it.

Otherwise, $j \leftarrow j - 1$ and go to L.

This algorithm always terminates in at most N cycles. We can evaluate the routing probability $a_j(N)$ that the jth nearest neighbor is chosen as the receiver when considering N neighbors. To this end, let θ be the angular position of the jth nearest neighbor measured from the direction of progress, as shown in Fig. 12. Since the jth nearest neighbor is selected as the receiver only when it is the most forward among j neighbors, its probability is given by

$$\left(1 - \frac{1}{\pi} |\theta - \sin\theta\cos\theta|\right)^{j-1} \qquad -\pi \leqslant \theta \leqslant \pi. \tag{31}$$

(Notice that the distribution of the positions of up to the

TABLE I
THE ROUTING PROBABILITIES $a_j(N)$

j	$N=1$	2	3	4	5	6	7	8	9	10
1	1.0000	0.5000	0.3017	0.1976	0.1359	0.0966	0.0704	0.0523	0.0395	0.0302
2		0.5000	0.3017	0.1976	0.1359	0.0966	0.0704	0.0523	0.0395	0.0302
3			0.3967	0.2598	0.1787	0.1271	0.0926	0.0688	0.0519	0.0397
4				0.3450	0.2373	0.1687	0.1230	0.0913	0.0689	0.0527
5					0.3122	0.2220	0.1618	0.1202	0.0907	0.0693
6						0.2890	0.2106	0.1564	0.1180	0.0902
7							0.2712	0.2015	0.1520	0.1162
8								0.2571	0.1940	0.1483
9									0.2455	0.1876
10										0.2356

$(j-1)$st nearest neighbors is no longer Poisson but uniform since we have specified j.) Unconditioning on θ with the isotropic assumption gives

$c_j \triangleq$ Prob [a terminal transmits to the jth nearest neighbor among j of them]

$$=\frac{1}{\pi}\int_0^{\pi}\left[1-\frac{1}{\pi}(\theta-\sin\theta\cos\theta)\right]^{j-1}d\theta \qquad j=1,2,\cdots,N. \tag{32}$$

Using the above definition of c_j's, we may finally write the routing probability as

$$a_j(N)=c_j\prod_{k=j+1}^{N}(1-c_k) \qquad j=1,2,\cdots,N. \tag{33}$$

Clearly

$$\sum_{j=1}^{N} a_j(N)=1. \tag{34}$$

In Table I, we show some values of $a_j(N)$.

From this point, we follow the derivation in [2]. First, the probability density function of the distance r_j to the jth nearest neighbor is given by

$$P_{r_j}(r)=\frac{2(\lambda\pi r^2)^{j-1}}{r(j-1)!}e^{-\lambda\pi r^2} \qquad r>0 \tag{35}$$

It follows that the mean distance to the jth nearest neighbor is

$$E[r_j]=\frac{(2j-1)!!}{2\sqrt{\lambda}(2j-2)!!} \tag{36}$$

where $(2j-1)!!=(2j-1)\cdot(2j-3)\cdots 3\cdot 1$ and $(2j)!!=(2j)\cdot(2j-2)\cdots 4\cdot 2$. Particularly, the mean distance between the two nearest neighbors is given by

$$E[r_1]=1/(2\sqrt{\lambda}) \tag{37}$$

which (without the factor 1/2) we have used extensively to normalize the expected progress in the preceding sections.

Next, let S_j be the event that a packet transmitted to the jth nearest neighbor is successfully received. As shown in [2],

$$\text{Prob}[S_j \mid r_j=r]=(1-p)(1-pq)^{j-1}e^{-\lambda p\pi r^2(\alpha^2-q)} \tag{38}$$

where

$$q\triangleq\begin{cases}\frac{2}{3}-\frac{\sqrt{3}}{2\pi}\doteq 0.391 & \alpha=1\\ \frac{1}{\pi}\left\{\pi+(\alpha^2-2)\cos^{-1}\left(\frac{\alpha}{2}\right)-\alpha\sqrt{1-\frac{1}{4}\alpha^2}\right\} & 1\leqslant\alpha\leqslant 2\\ 1 & \alpha\geqslant 2.\end{cases} \tag{39}$$

Unconditioning on r_j with (35) gives

$$\text{Prob}[S_j]=\frac{(1-p)(1-pq)^{j-1}}{(1+\alpha^2 p-pq)^j}. \tag{40}$$

It follows that the throughput of transmission is given by

$$S(p,N;\alpha)=p\sum_{j=1}^{N}a_j(N)\,\text{Prob}[S_j]. \tag{41}$$

This completes our quotation from [2]. Similarly, we obtain the expected progress as

$$Z(p,N;\alpha)=\frac{p}{\sqrt{1+\alpha^2 p-pq}}\sum_{j=1}^{N}b_j(N)E[r_j]\,\text{Prob}[S_j] \tag{42}$$

where

$$b_j(N)\triangleq\prod_{k=j+1}^{N}(1-c_k)\cdot\frac{1}{\pi}\int_0^{\pi}\cos\theta\cdot\left[1-\frac{1}{\pi}(\theta-\sin\theta\cos\theta)\right]^{j-1}d\theta \qquad j=1,2,\cdots,N. \tag{43}$$

Our optimization problem is to find the maximum of $Z(p,N;\alpha)\sqrt{\lambda}$ in the (p,N) plane. In Fig. 13, the optimal values of N, N^*, and the maximum values of $Z(p,N;\alpha)\sqrt{\lambda}$, $Z^*\sqrt{\lambda}$, are plotted for various values of capture parameter α. It can be seen that the expected progress decreases rapidly as α increases, which does not agree with the result shown in Fig. 6. This comes from the present assumption that there is no fixed transmission radius. However, for small α, the results that $N^*=7$ and $Z^*\sqrt{\lambda}\cong 0.05$ agree with the previous results. Therefore, we may conclude that 7 is suitable for the number of known terminals when the MFN routing is adopted.

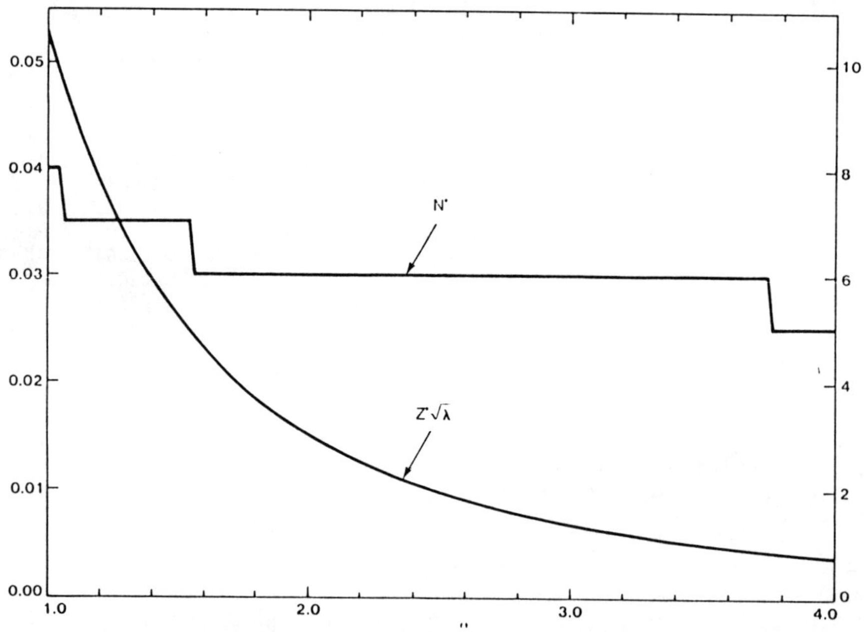

Fig. 13. The optimal transmissions for ALOHA with capture and without transmission radius.

VII. Conclusion

We have solved for the maximum expected progress per hop (Z), provided by the optimal transmission probability (p) and transmission radius (expressed in terms of the number of terminals in the range, N), in some models of randomly distributed packet radio terminals (with average density λ) under the assumption of heavy traffic (all terminals always have ready packets). The quantity $Z\sqrt{\lambda}$ has been used consistently as the dimensionless objective function for optimization problems with respect to p and N. Major conclusions about the performance of each model are as follows.

The optimal transmission with slotted ALOHA without capture is attained by $N = 7.72$ and $p = 0.113$ which gives $Z\sqrt{\lambda} = 0.0431$. Therefore, each terminal transmits once in every nine slots on the average with the transmission radius covering just about eight nearest neighbors in the direction of packet's final destination. The probability of success of such a transmission is nearly equal to $1/e$. The expected progress per transmission is about two thirds of R/e, where R is the optimal transmission radius ($N = \lambda\pi R^2$).

FM capture improves the performance of slotted ALOHA systems due to the more limited area of possibly interfering terminals around the receiver. The expected progress in a system with perfect capture (optimized with $N = 7.1$ and $p = 0.17$) is about 36 percent greater than that in the system without capture. The probability of successful transmission is also higher than $1/e$. A model which is more amenable to implementation (each terminal knows the positions of only a fixed number of its neighbors) has shown similar results.

The slotted nonpersistent CSMA provides a nominal improvement in performance over the ALOHA system (16 percent improvement in the optimized expected progress for the zero propagation delay), which is not as large an improvement as we have obtained in the single-hop case. The reason for this is the large area of "hidden" terminals (about half of the interfering area) which cannot hear the transmission, and the long vulnerable period (twice as long as the packet transmission time) due to those terminals. The performance of (slotted nonpersistent) CSMA is comparable to that of ALOHA with good FM capture (capture ratio about 1.5 dB). The degradation occurs as the ratio of propagation delay to the transmission time increases.

As an example of an inhomogeneous terminal distribution, the effect of a gap of width b in an otherwise uniformly Poisson-distributed terminal population on the optimal transmission has been considered. The expected progress of a packet residing at the terminal on the bank and destined to cross the gap is evaluated with parameter $\beta = \lambda b^2$, called gap intensity. For fixed λ, the existence of the gap helps the progress for $\beta < 2$, because some of the possibly interfering terminals are removed by the gap. The maximum in the optimized expected progress occurs at about $\beta = 1$. Thus, to cross most gaps wider than the average interterminal distance, one had better not use a large transmission radius, but should more sensibly use a separate channel or wire.

Appendix

Determination of Transmission Rate for CSMA

In this Appendix, we derive the relation between transmission rate p' and channel-sensing rate p for slotted nonpersistent CSMA. See Section IV. As in [9], we assume that

$$p' = pP_I$$

where P_I is the probability that the channel is sensed idle. Since the probability of an empty slot is given by $e^{-p'N}$, the expected value of the idle period I (see Fig. 7) is

$$\bar{I} = \sum_{k=1}^{\infty} kae^{-kp'N}(1 - e^{-p'N}) = \frac{ae^{-p'N}}{1 - e^{-p'N}}.$$

On the other hand, the transmission period is $1 + a$. Therefore,

$$P_I = \frac{\bar{I}}{1 + a + \bar{I}} = \frac{ae^{-p'N}}{1 + a - e^{-p'N}}.$$

Thus, we have obtained an equation which determines p' in terms of p:

$$p' = \frac{ape^{-p'N}}{1 + a - e^{-p'N}}.$$

For $a \ll 1$, we have

$$p' = \frac{p}{1 + (p'/a)N}$$

which explicitly gives p'.

Using the optimal values in this case ($N = 5.3$ and $p'/a = 0.20$), we see that the actual transmission rate is 41 percent of the sensing rate.

References

[1] N. Abramson, "The throughput of packet broadcasting channels," *IEEE Trans. Commun.*, vol. COM-25, pp. 117–128, Jan. 1977.

[2] L. Fratta and D. Sant, "Some models of packet radio networks with capture," in *Proc. 5th Int. Conf. Commun.*, Oct. 27–30, 1980, pp. 155–161.

[3] L. Kleinrock and F. A. Tobagi, "Packet switching in radio channels: Part I—Carrier sense multiple-access modes and their throughput-delay characteristics," *IEEE Trans. Commun.*, vol. COM-23, pp. 1400–1416, Dec. 1975.

[4] L. Kleinrock and J. A. Silvester, "Optimum transmission radii for packet radio networks or why six is a magic number," in *Conf. Rec., Nat. Telecommun. Conf.*, Dec. 1978, pp. 4.3.1–4.3.5.

[5] R. Nelson, "Channel access protocols for multi-hop broadcast packet radio networks," Dep. Comput. Sci., School Eng. Appl. Sci., Univ. California, Los Angeles, Eng. Rep. UCLA-ENG-8259, CSD-820731, July 1982.

[6] L. G. Roberts, "ALOHA packet system with and without slots and capture," ARPA Network Inform. Cen., Stanford Res. Inst., Menlo Park, CA, ARPA Satellite Syst. Note 8 (NIC 11290), June 26, 1972; reprinted in ACM SIGCOMM *Comput. Commun. Rev.*, vol. 5, pp. 28–42, Apr. 1975.

[7] J. A. Silvester, "On the spatial capacity of packet radio networks," Dep. Comput. Sci., School Eng. Appl. Sci., Univ. California, Los Angeles, Eng. Rep. UCLA-ENG-8021, May 1980.

[8] F. A. Tobagi, "Random access techniques for data transmission over packet switched radio networks," Dep. Comput. Sci., School Eng. Appl. Sci., Univ. California, Los Angeles, Eng. Rep. UCLA-ENG-7499, Dec. 1974.

[9] F. A. Tobagi and L. Kleinrock, "Packet switching in radio channels: Part II—The hidden terminal problem in carrier sense multiple-access and busy tone solution," *IEEE Trans. Commun.*, vol. COM-23, pp. 1417–1433, Dec. 1975.

SOME NEW APPROACHES TO RANDOM-ACCESS COMMUNICATIONS

James L. MASSEY

Institute for Signal and Information Processing
Swiss Federal Institute of Technology
8092 Zurich, Switzerland

Random-accessing is defined as any technique to accomplish unscheduled seizure of a many-user communications channel; its purpose is to reduce transmission delay below what can be achieved by scheduled-accessing or by channel division. Some general principles regarding channel division, channel seizure, and the effect of feedback are formulated. The "classical" approach to random-accessing, i.e., ALOHA-like techniques, is seen to be subject to instability. A newer approach, collision-resolution algorithms (CRA's), is shown to avoid this problem. The analysis of CRA's has led to bounds on the performance of any random-access system that are briefly discussed. Two new approaches to random-accessing without feedback information are described, viz., protocol sequences for the M-user collision channel and coding for the M-active-out-of-T-user collision channel. Examples are generously used throughout the paper, and some speculations on the practicality of the new approaches are offered.

1. INTRODUCTION

Before describing "new approaches to random-access communications", we should make clear what we mean by "random-accessing" and what we see as its main purpose. To do this, we must first say a few words about "multiple-accessing" in general.

A multiple-access technique is any technique that permits two or more senders to operate on a single communications channel. Time-division multiple-accessing (TDMA), frequency-division multiple-accessing (FDMA) and code-division multiple-accessing(CDMA) are well-known multiple-access schemes of the <u>channel-division</u> type; i.e., they divide the single channel into many "smaller" channels, one for each sender. This division may be fixed, or it may be adjusted from time to time to correspond to the changing needs of the senders as in so-called "demand assignment" schemes. A second class of multiple-access schemes is that of what we shall call the <u>channel-seizure</u> type. In this type of multiple-accessing, a single sender can use the full (time and frequency) resources of the channel for himself alone on some sort of temporary basis. An example of a channel-seizure scheme is a token-ring in which, when the "token" arrives at a sender's station on the ring, that sender can remove the token, send his own message as if he were the only sender on the ring, and then reinsert the token.
A <u>random-access</u> technique can be defined as a multiple-accessing scheme of the channel-seizure type (i) in which it can happen that two or more senders may simultaneously attempt to seize the channel, and (ii) which provides in some way for the recovery from such "access conflicts". In a random-access system, a sender generally "takes a chance" when he attempts to seize the channel, and he relies on the access protocol to repair the damage when he encounters "bad luck".

In some communication scenarios (as we shall see later), access conflicts cannot be avoided. More often, however, it is a matter of choice whether or not to allow access conflicts and hence whether or not to use random-accessing. The obvious question is: why should anyone choose to allow such an obviously bad thing as access conflicts? The answer can be put as a second question: why should anyone demand that a sender always <u>wait for a guarantee</u> of exclusive access before he attempts to seize the channel? When traffic on the channel is

Reprinted with permission from *Performance '87*, pp. 551–569, 1988.

light, the bold sender will be almost sure to succeed in his gamble for access and can thus avoid the delay that a timid sender would incur. The primary purpose of random-accessing is to reduce the delay between the time that a sender obtains an information input and the time that he transmits this information successfully over the channel. Random-accessing is a gamble, but one in which the odds can be on the side of the player rather than on the side of the "house".

In Section 2 of this paper, we show why channel seizure is generally preferable to channel division for multiple-accessing, and we examine the role of channel feedback information. Section 3 describes the ALOHA approach to random-accessing and points out its virtues and defects. In Section 4, we describe one new approach to random-accessing, viz. collision resolution, and we contrast it with the ALOHA approach. Section 5 considers certain general bounds on the throughput of random-access schemes. Section 6 describes two new approaches to random-accessing without feedback. Some concluding remarks are given in Section 7.

2. SOME GENERAL MULTIPLE-ACCESS PRINCIPLES

The simplest multiple-access channel is surely the two-sender binary adder channel (2SBAC) shown in Fig. 1. Each time instant, each sender sends a binary digit (0 or 1) and the received digit is the sum (0, 1 or 2) of these two numbers, i.e.,

$$Y_n = X_{1n} + X_{2n}$$

where X_{1n} and X_{2n} are the binary digits sent by senders 1 and 2, respectively, at time n and Y_n is the received digit. The "wall" shown between the two senders in Fig. 1 signifies that the user on one side is not privy to the information to be sent on the other side, although the two users are allowed in advance to have formulated a common strategy for sending this information.

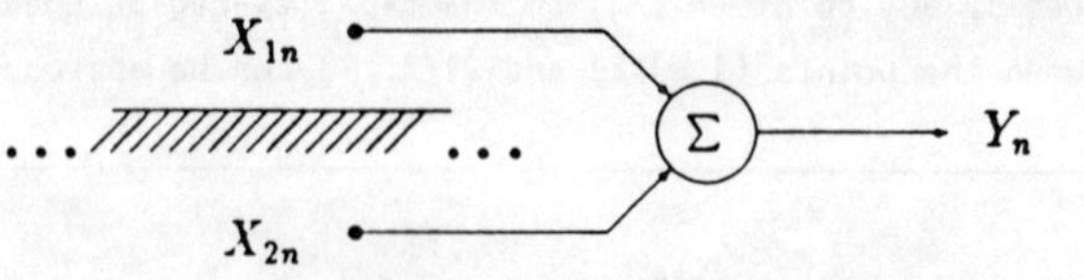

Fig.1 : The two-sender binary adder channel (2SBAC)
$X_{1n} \in \{0,1\}, X_{2n} \in \{0,1\}, Y_n \in \{0,1,2\}$

Fig. 2 shows the pentagonal "capacity region" of the 2SBAC, i.e., the region of rate pairs (R_1, R_2) such that Sender 1 can send data at the rate R_1 (bits per channel use) and Sender 2 can send at the rate R_2, both with arbitrarily small error probability.

It is easy to see how the point $(R_1,R_2) = (1,0)$ on the capacity-region boundary can be achieved. Sender 2 simply always sends 0's (and thus $R_2 = 0$) so that $Y_n = X_{1n}$, and hence Sender 1 can directly send his "raw" information bits over the channel with no need for coding ($R_1 = 1$). The point $(R_1,R_2) = (0,1)$ can be similary achieved. By agreeing to alternate between these two schemes for appropriate periods, Senders 1 and 2 can achieve any point (R_1,R_2) such that $R_1 + R_2 = 1$, i.e., at any point on the "time-sharing line" shown in Fig. 2.

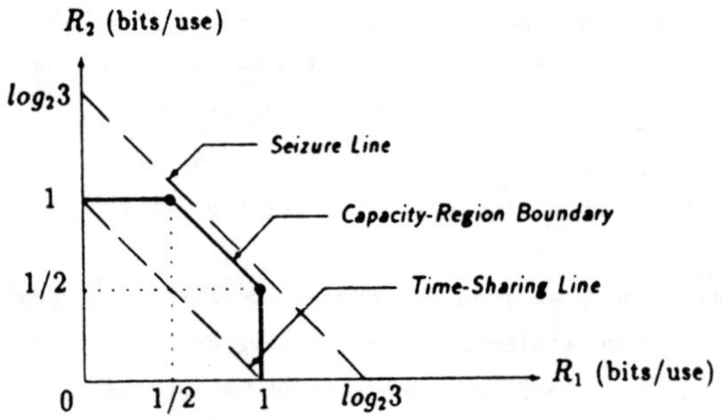

Fig.2 : Capacity Region of the 2SBAC of Fig.1

It is almost as easy to see how the point $(R_1, R_2) = (1, 1/2)$ on the capacity-region boundary can be approached. Sender 1 transmits his raw information bits ($R_1 = 1$). This causes the channel seen by Sender 2 to be that shown in Fig. 3, because, for instance, if Sender 2 should send a 1 then with probability 1/2 Sender 1 will also send a 1 and 2 will be received, while with probability 1/2 Sender 1 will send a 0 and 1 will be received. But the channel of Fig. 3 is the familiar binary erasure channel (in which a received 1 is the "erasure symbol") with erasure probability $\delta = 1/2$ and capacity $C = 1-\delta = 1/2$. Thus, Shannon's noisy coding theorem ensures the existence of a coding scheme that will allow Sender 2 to send information at a rate R_2 arbitrarily close to 1/2 with arbitrarily small error probability. After the receiver has decoded Sender 2's codeword, he can subtract it from the received sequence to obtain the uncoded sequence that was transmitted by Sender 1. The price of making R_2 closer to the capacity 1/2 is an increasingly longer codeword length or, equivalently, a longer delay in recovering the information at the receiver. The point $(R_1, R_2) = (1/2, 1)$ can, of course, be similarly approached. By appropriately alternating between coding schemes, any point (R_1, R_2) on the capacity-region boundary line $R_1 + R_2 = 3/2$ between the points (1, 1/2) and (1/2, 1) can be approached.

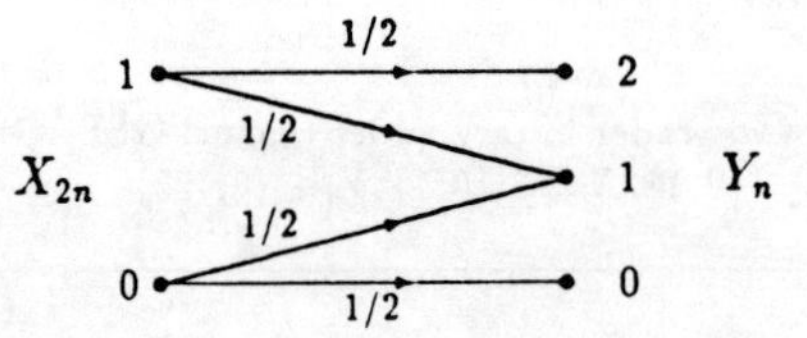

Fig.3 : The binary erasure channel seen by Sender 2 when Sender 1 sends random binary digits over the 2SBAC of Fig.1.

Perhaps the best interpretation of the "wall" shown in Fig. 1 is as a prohibition against seizure of the channel by a single sender. If a single sender is allowed to control both X_{1n} and X_{2n}, then he can by choosing (X_{1n}, X_{2n}) to be (0,0), (0,1) or (1,1) cause Y_n to be 0, 1 or 2, respectively, i.e., he can create a noiseless ternary channel with capacity $\log_2 3$ (bits per use). By alternating appropriately between such seizures, two senders could achieve any point on the "seizure line" shown in Fig. 2 that lies strictly outside the (seizure-prohibited) capacity region.

Suppose now that there is a feedback channel from the receiver to the two senders in Fig. 1 so that each sender learns the value of Y_n immediately after X_{1n} and X_{2n} have been sent. The point $(R_1, R_2) = (1, 1/2)$ can now be achieved with the greatest of ease. Sender 1 still sends his raw information bits ($R_1 = 1$) so that Sender 2 still sees the binary erasure channel of Fig. 3. Sender 2, however, can now (because of the feedback of Y_n) simply send each of his information bits repeatedly until it is received "unerased", i.e., until $Y_n = 0$ when this information bit is a 0 or until $Y_n = 2$ when this information bit is a 1. Because the erasure probability δ is 1/2, Sender 2 will be sending information at the rate $R_2 = 1-\delta = 1/2$ bits/use. Moreover, the average delay between first transmission and successful transmission is only 2-1 = 1 time instant. Something even more remarkable, however, results from the availability of feedback (as was first shown by Gaarder and Wolf [1]): points outside the capacity region of Fig. 2 can be achieved! This was quite surprising when first discovered because it had long been known that feedback could not increase the capacity of a single-sender memoryless channel. The actual capacity region of the 2SBAC with feedback was only recently determined by Willems [2]; it differs from the capacity region without feedback, shown in Fig. 2, in that the boundary line between the points (1, 1/2) and (1/2, 1) is bowed slightly outward (but still well away from the "seizure line").

The simple 2SBAC of Fig. 2 is a rich source of lessons about multiple-accessing. With its help, we have been able to illustrate all of the following general principles of multiple-access communications:

(1) Channel seizure, when possible, is the most effective way to utilize a multiple-access channel.

(2) When channel seizure is prohibited, time sharing (or other types of channel division) generally is still sub-optimum in the sense that it cannot be used to achieve all points in the capacity region.

(3) Feedback, when available, can be exploited to reduce the coding delay and complexity required to achieve a given transmission rate.

(4) When channel seizure is prohibited, feedback can also enlarge the capacity region.

The first of these principles supports the way that computer communications is carried out today. Virtually all newer local area networks (LAN's) operate on a channel seizure basis, sometimes with deterministic access (as in a token ring) and sometimes with random access (as in Ethernet). The third principle suggests that feedback will play an especially crucial role in random-accessing, because some kind of "coding" is absolutely necessary to overcome the losses due to access conflicts.

3. THE ALOHA APPROACH TO RANDOM-ACCESSING

The ALOHA system, devised by Abramson [3] and his colleagues at the University of Hawaii, was the first random-access system; its approach underlies most present-day random-access systems, e.g. Ethernet. To illustrate the ALOHA approach, we now describe the ALOHA system, including the modification of "time slotting" that was introduced by Roberts [4].

Suppose that all data to be sent is in the form of "packets", all of which have the same length (measured in transmission time on the seized channel) that we take to be the unit of time. We define the time interval $(n-1) \leq t < n$ to be the n-th channel "slot". "Time-slotting" means that senders can transmit packets only by beginning transmission at a slot boundary. Thus, transmitted packets from two senders will either overlap completely at the receiver or not at all.

The channel model postulated by Abramson was that, when 2 or more transmitted packets overlap at the receiver, then they mutually destroy one another, but otherwise packets are received error-free. Moreover, there is feedback from the receiver at the end of each slot so that all users learn whether or not a collision occurred (collision/no-collision binary feedback).

The information-generation model postulated by Abramson was that of a very large number (essentially infinite) of identical sources, each with an associated sender, such that the number of new information packets generated during any slot is a Poisson random variable with mean λ (packets/slot), independent of previously generated packets. The essentially infinite number of senders means that access conflicts cannot be entirely avoided, i.e., random-accessing becomes a necessity. [In fact, the original operational ALOHA system had a very small number of transmitters so that random-accessing was a matter of choice, made by Abramson and his colleagues for the express purpose of reducing access delay.]

The random-access protocol devised by Abramson was ingeniously simple. A new packet must be transmitted in the slot immediately following that in which it was generated. When a collision occurs,each "colliding" sender must retransmit in a randomly-selected later slot. Each such sender, of course, independently makes this random selection of retransmission delay.

Abramson's analysis of the ALOHA system was equally ingenious, if not rigorous. He postulated that the retransmission policy could be shaped in such a way that the number of retransmitted packets in any slot would also be a Poisson random variable, independent from slot to slot and independent of the new-packet generation process, with a mean of λ_r (packets/slot). Because the sum of independent Poisson random variables is again Poisson, this implies that the total number of packets transmitted in any slot is also a Poisson random variable with mean $\lambda_t = \lambda + \lambda_r$. Because the <u>throughput</u> τ of successful packets at the receiver is the fraction of slots in which exactly one packet is transmitted, it follows that τ is just the probability that a Poisson random variable with mean λ_t takes on the value 1, i.e.,

$$\tau = \lambda_t e^{-\lambda_t}. \tag{1}$$

Equation (1), which is the so-called <u>throughput</u> equation for slotted-ALOHA, is shown graphically in Fig. 4. It is easy to check from (1) that τ is maximized when $\lambda_t = 1$ (packet/slot), which seems quite natural, and that this maximum is

$$\tau_{max} = e^{-1} \approx .368 \text{ (packets/slot)},$$

which seems quite fundamental. It is common to say that e^{-1} is the "capacity of the slotted-ALOHA channel", but, as we shall see, this description is misleading.

The reader may (and should) be disturbed by the fact that the new-packet arrival rate λ appears nowhere in the throughput equation (1). To bring λ into the picture, one must invoke the <u>equilibrium hypothesis</u> which states that packets are entering and leaving the system at the same rate, i.e.,

$$\tau = \lambda.$$

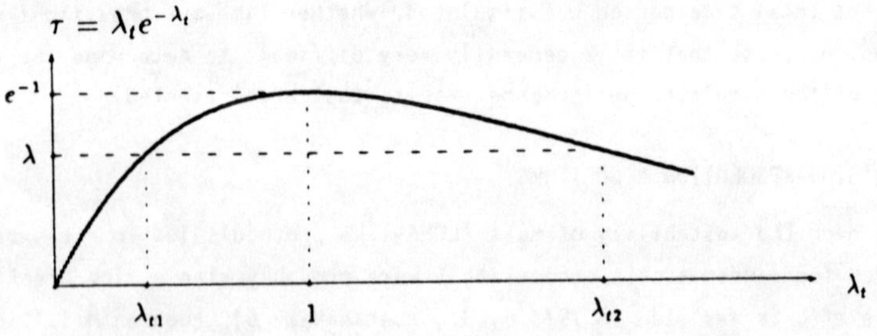

Fig.4 : Graphical depiction of the throughput equation for slotted-ALOHA.

This seems similar to the constancy assumption for the retransmission rate, but in fact neither assumption implies the other.The equilibrium hypothesis is really an expression of the hope that the ALOHA system is stable, i.e., that the queue of packets awaiting retransmission is not steadily growing at a positive rate $\lambda-\tau$ (packets/slot). Such a positive growth rate is not inconsistent with a constant retransmission rate if the retransmission delay is chosen randomly in a way to depend on how many times the given packet has been previously transmitted unsuccessfully. The equilibrium hypothesis should in fact be called the stability hypothesis for ALOHA.

It is easy to argue from Fig. 4 that the ALOHA system cannot be stable for a retransmission policy that does not take into account the number of previous unsuccessful transmissions. Suppose that the arrival rate λ satisfies $\lambda < e^{-1}$ as shown in Fig. 4. If equilibrium prevails, then the traffic rate λ_t will be λ_{t1} as shown in Fig. 4. This is of course an "average" rate and, over any fixed length interval, the actual rate will fluctuate about this mean. If the actual traffic rate moves a little above λ_{t1}, the actual throughput increases a little above λ. Thus, packets leave the system faster than they arrive, which causes the actual traffic rate to move back down to λ_{t1}. Hence, the point $(\tau,\lambda_t) = (\lambda,\lambda_{t1})$ is a conditionally stable point, i.e., it is stable under small fluctuations. But if a large fluctuation causes the actual traffic rate to move to the right of λ_{t2} in Fig. 4, then the actual throughput decreases below λ. Thus, packets leave at a slower rate than they enter, which causes a further increase in the actual traffic rate, a further decrease in actual throughput, etc. The system never returns to the point (λ,λ_{t1}), but rather drifts relentlessly toward the catastrophic "unconditionally stable" point $(\tau,\lambda_t) = (0,\infty)$. The maximum stable throughput of a fixed-policy ALOHA system is 0.

The virtue of the ALOHA approach is its simplicity, its Achilles' heel is its instability. In fact, it is possible to devise retransmission policies that stabilize an ALOHA system, cf. [5], but the protocol then loses its simplicity. Most practical ALOHA-type random-access systems appear in fact to be unstable. These systems incorporate some kind of time-out feature that switches the system to a non-random type of accessing to clear away the backlog of traffic when the channel is jammed with collisions, then switches back again to the basic ALOHA protocol. Again this means some loss of simplicity in the access protocol, as well as some performance anomalies. Published simulations of ALOHA-type systems invariably appear to have been purged of any anomalies, if indeed any occurred. In fact, there is usually very little information provided with such simulations about what total time period was simulated, whether time-out provisions were included, etc., so that it is generally very difficult to determine the real meaning of the simulated performance results that are presented.

4. COLLISION-RESOLUTION ALGORITHMS

Concern over the instability of most ALOHA-like protocols led some researchers to search for random-access schemes that were provably stable. The breakthrough in these efforts was made in 1977 by J. Capetanakis [6], then an M.I.T. doctoral student working with Prof. R. Gallager, and independently achieved shortly thereafter by two Soviet researchers, B. Tsybakov and V. Mikhailov [7]. The essence of their contributions was the "collision-resolution approach" to random-accessing, which we now consider.

The channel model assumed for collision resolution is the same as that for slotted-ALOHA, namely a time-slotted collision-type channel with some form of feedback to the senders at the end of each slot. We will assume the same binary (collision/no-collision) feedback as for slotted-Aloha [as Capetanakis also assumed; Tsybakov and Mikhailov considered ternary (collision/success/idle) feedback]. The information-generation model is the same as for slotted-ALOHA, i.e., essentially-infinitely many identical sources, each with an associated sender, such that the number of new packets generated in each slot is an independent Poisson random variable with mean λ.

A <u>collision-resolution algorithm</u> can be defined as a random-access protocol such that, whenever a collision occurs, then at some later time (provided λ is not too large) all senders will simultaneously learn from the feedback information that all packets involved in that collision have now been successfully transmitted. The crux of collision resolution is the exploitation of the feedback information to control the "random" retransmission process in such a way that chaotic retransmission can never occur. Because there is no upper bound on the number of packets that initially collide, it was not at all obvious that collision-resolution algorithms existed before the first such algorithms were presented by Capetanakis and by Tsybakov and Mikhailov.

As an example of a collision-resolution algorithm, we now describe the <u>binary stack algorithm</u>, which is essentially Capetanakis' binary "tree algorithm", but we prefer the terminology "stack algorithm" introduced by Tsybakov and Mikhailov. The terminology "binary" stems from the fact that every sender is assumed to have a fair "binary coin"(with "0" on one side and "1" on the other) which he flips whenever his packet is involved in a collision. The term "stack" comes from the fact that one can conveniently visualize the operation of the algorithm in terms of the <u>conceptual stack</u> shown in Fig. 5.

Suppose at the outset that the stack is empty, i.e., that $S = 0$. Suppose also that the access gate is opened and some number X of senders then enter the "current sending box", which is just the conceptual location of all senders who will send packets in the current slot. Perhaps $X = 2$ or $X = 0$ or even $X = 6 \times 10^{23}$; for the moment, we assume only that $X \geq 2$ so that a collision ("C") occurs in this slot. Then these colliding senders flip their binary coins, those who flip "0" remain in the "current sending box" (i.e., they again transmit in the next slot) while those who flip "1" are pushed down (conceptually) into level 1 of the stack. The stack size is now S = 1. The general rule is: if "C", then S ← S + 1. One sees that now about X/2 of the original X colliding senders will remain in the "current sending box" while about X/2 will be pushed down into the stack. For the moment, we assume the blocked-access protocol, which states that the "access gate" is closed at the initial collision and remains closed until all senders learn that the original X colliding packets have all been successfully transmitted. The same process of stack growth and concomitant "thinning out" of the "current sending box" continues until the feedback "no-collision" ("NC") occurs, which means that either 1 sender or none had been in the "current sending box". This is the signal for the stack to be pushed upward one level so that senders who were in level 1 of the conceptual stack are now again in the "current sending box" and the stack size is reduced by 1. The general rule is: if "NC" and S > 0, then S ← S - 1. After some time, because the above process will "thin out" crowded levels in the stack, the stack size will again reach S = 0. If now no collision occurs, this means that all of the original X senders must have successfully sent their packets. (If a collision occurs, the previous process continues.) The general rule is: if "NC" and S = 0, then the collision is resolved.

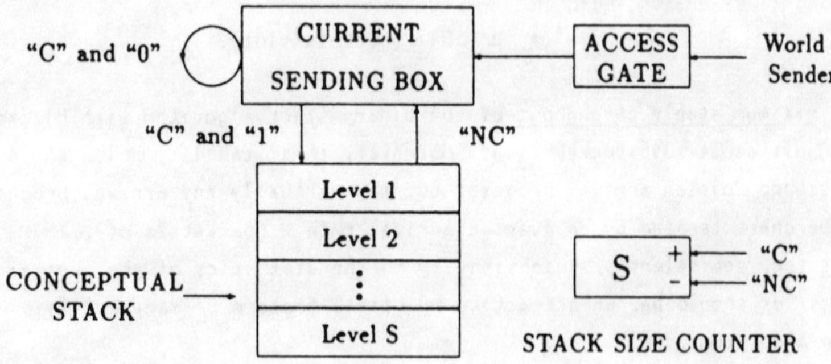

Fig.5 : Conceptual diagram of the "*binary stack algorithm*" for collision resolution.

The binary stack algorithm is clearly simple to implement. When involved in a collision, a sender need only generate a binary random variable ("0" or "1") so a minimum of retransmission "randomization" is required. His only other requirements are to maintain two counters, one of which gives his own position in the stack (if he were a party to the collision) while the other keeps track of the stack size S so that he knows when the collision has been resolved. The big question of course is: how effective is this random-access protocol?

Perhaps the main theoretical advantage of the collision-resolution approach over the ALOHA approach to random-accessing is that the former lends itself to precise (and reasonably simple) analysis as we now demonstrate for the binary stack algorithm with blocked-access. Letting Y be the number of slots needed to resolve the original collision of X senders, then the quantity of principal interest is $L_N = E[Y|X = N]$, the average number of slots needed to resolve a collision of N transmitters. We see that $L_0 = L_1 = 1$ as then there is no initial collision. We see further that

$$L_2 = 1 + \frac{1}{2}(L_0 + L_2) + \frac{1}{2}(L_1 + L_1)$$

as, after the initial collision, with probability 1/2 the two senders will flip the same binary number leaving no one in the "current sending box" and both in level 1 (or vice versa), while with probability 1/2 they will flip different binary numbers leaving 1 sender in the "current sending box" and 1 sender in level 1 of the stack. Solving for L_2 gives

$$L_2 = 5$$

slots required on the average to resolve a collision of 2 packets. It is easy to write the general recursion for L_N and to show that the solution satisfies

$$2.8810 < \frac{L_N}{N} + \frac{1}{N} < 2.8867, \quad N \geq 4; \tag{2}$$

the interested reader is referred to [8] for details of this argument. The conclusion to be drawn from (2) is quite remarkable: whenever the initial collision is moderately large, then just about 2.89 slots will be required to "service" each of the packets involved in this initial collision. This means that the algorithm will be <u>stable</u> (the "server" will not drop hopelessly behind in serving customers) provided only that

$$\lambda \lesssim \frac{1}{2.8867} \approx .346 \text{ (packets/slot)}$$

whereas it will be unstable if

$$\lambda \gtrsim \frac{1}{2.8810} \approx .347 \text{ (packets/slot)}.$$

Thus, the <u>maximum stable throughput</u> of the binary stack algorithm with blocked-access is just about .346 packets/slot. Moreover, this stability holds not only for the assumed Poisson arrival process, but for virtually any arrival process that can be characterized by an average arrival rate λ (packets/slot). This <u>robustness</u> (or, equivalently, insensitivity to the statistics of the arrival process) is, or should be, an attractive practical feature of many collision-resolution algorithms.

To examine the role placed by the "blocked-access protocol" in the above analysis, we first observe that the binary stack algorithm never makes any assumption in advance about the occupancy of the "current sending box" or any of the S stack levels. There could just as well be none or 6×10^{23} senders in any of these locations as far as the binary stack algorithm is concerned. This means that there is no reason to block new senders from entering the "current sending box" at any time. The <u>free-access protocol</u> leaves the "access gate" of Fig. 5 open at all times. Free-access has the practical advantage that senders need to monitor the channel feedback only after they become "active" in the transmitting process. Intuition suggests that free-access should also give a better throughput than blocked-access. Unfortunately, free-access also complicates the analysis, but Mathys and Flajolet have shown that the binary stack algorithm with free-access has a maximum stable throughput of $\lambda = .360$ (packets/slot). More surprisingly, they showed also that the <u>ternary</u> stack algorithm (in which senders flip a fair three-sided coin after a collision, moving into levels 1 or 2 if they flip "1" or "2" while S is increased by 2) with free-access has an even larger maximum stable throughput of

$$\lambda = .401 \text{ (packets/slot)}.$$

(Because this maximum stable throughput exceeds $e^{-1} \approx .368$, we see that it is indefensible to call e^{-1} the "capacity of the slotted-ALOHA channel".) Mathys and Flajolet also showed that this ternary stack algorithm with free-access has a better delay vs. throughput characteristic than does either the binary stack algorithm or the usual ALOHA algorithm (when analyzed assuming optimistically that the equilibrium hypothesis holds. [9]

Many other collision-resolution algorithms have been proposed that have maximum stable throughputs exceeding .401 (packets/slot) -- the current record for binary (collision/no-collision) feedback is .4493 packets/slot [10]. However, the ternary stack algorithm with free-access appears to us to be the best practical choice by virtue of its simplicity, its robustness and its relatively high maximum stable throughput. It also seems to us to be a better practical choice than any ALOHA-like algorithm for random-accessing on the collision channel with feedback, and we wonder why algorithms of the latter type are still being proposed for new random-access systems.

The reader interested in delving more deeply into the mathematics needed for a precise analysis of collision-resolution algorithms will find the recent book [11] by Hofri to be a useful source of information.

5. UPPER BOUNDS ON MAXIMUM STABLE THROUGHPUT

Our discussion of collision-resolution algorithms may have raised the question in the reader's mind: what is the capacity of the collision channel with feedback or, equivalently, what is the largest possible maximum stable throughput that can be achieved for Poisson arrivals? It is known that the answer depends in general on the kind of feedback available so we continue to consider only binary (collision/no-collision) feedback. Some ingenious and complex arguments have been used to obtain upper bounds on the maximum stable throughput. Rather than describing the best bound, we illustrate the general idea by describing a very simple bound due to Kelly [12].

Kelly's bound actually applies only to algorithms (such as the ALOHA algorithm or any collision-resolution algorithm with free-access) with the immediate-first-transmission property that a newly-generated packet must be transmitted in the slot immediately following that in which it was generated. Suppose such an algorithm is operating stably for Poisson new arrivals with a mean of λ (packets/slot). Let p_r be the fraction of slots in which at least one packet is retransmitted. Then, because the number of new senders in any slot is independent of the number of retransmitted packets in that slot, it follows that the fraction of slots with exactly one packet (i.e., the throughput τ) satisfies

$$\tau \leqslant \lambda e^{-\lambda}(1-p_r) + p_r e^{-\lambda}$$

with equality when at most 1 packet is retransmitted in any slot (i.e., perfect scheduling of retransmissions). But stability implies $\tau = \lambda$ and thus

$$\lambda \leqslant \lambda e^{-\lambda}(1-p_r) + p_r{}^{-\lambda} \qquad (3)$$

if the system is stable. It is readily checked that the choice $p_r = 1$ maximizes λ for which (3) can be satisfied. Thus

$$\lambda \leqslant e^{-\lambda} \qquad (4)$$

is required for stability. The largest λ satisfying (4) is Kelly's bound on maximum stable throughput, namely

$$\lambda_{max} \leq .5671 \text{ packets/slot,} \tag{5}$$

that holds for any random-access algorithm with immediate-first-transmission. In fact, our argument has never used the assumption of binary (collision/no-collision) feedback so that Kelly's bound (5) applies to any random-access algorithm with immediate-first-transmission on the collision channel with any kind of feedback.

By using similar but much more intricate arguments, Tsybakov and Likhanov [13] have recently proved that, for any random-access algorithm, the maximum stable throughput on the collision channel with binary (collision/no-collision) feedback satisfies

$$\lambda \leq .5683 \text{ (packets/slots)} \tag{6}$$

for a Poisson new arrival process. This is significantly greater than the largest stable throughput yet achieved, .4493 (packets/slot). The "capacity" of the ALOHA channel with binary (collision/no-collision) feedback lies somewhere between .4493 and .5683 (packets/slot), which is the most that can be said today.

The thoughtful reader may well ask: why are we giving so much attention to maximum stable throughput when the real purpose of random accessing is small (average) delay? The incomplete answer is this. The maximum stable throughput of a random-access algorithm is the smallest throughput where the (average) delay becomes infinite. Thus if one algorithm has a larger maximum stable throughput than another, then it will also have a better delay-throughput characterstic for all sufficiently large throughputs. The complete answer is that one <u>hopes</u> that if the first algorithm is reasonably simple (so that the large maximum stable throughput was not achieved by "trickery" that used high arrival rates to special advantage) then the first algorithm will have a better delay-throughput characteristic for <u>all</u> throughputs. The previous discussion of the ternary stack algorithm shows that there is some justification for this hope.

6. RANDOM-ACCESS WITHOUT FEEDBACK

We now consider some quite recent developments in random-accessing that deal with the situation where there is no feedback to notify senders whether or not their packets have suffered collisions. At first glance, it might seem that random-accessing would be impossible in this situation. The "trick" that makes it possible is for the senders to send <u>redundant packets</u> so that the information packets can still be recovered at the receiver when some of the packets are lost through collisions.

6.1 The M-Sender Collision Channel without Feedback

Again we assume a time-slotted collision channel, but now with no feedback to the senders. Rather than essentially infinite, we suppose there is a given number M ($M \geq 2$) of senders, each with its own information source. We further assume that, although the senders are slot-synchronized, they are unsynchronized at any higher level, i.e., they may all have a different idea of which slot if slot 1. The senders can be thought of as having clocks that "tick" together at clock boundaries but are otherwise unsynchronized. Because there is no feedback from the channel, the senders can never attain any further synchronization of their clocks. The senders each have a "protocol sequence generator" that emits a binary digit at each clock tick; the sender sends a packet if this bit is a 1 and keeps silent in that slot if this bit is a 0. The task is to choose the protocol sequences for the M users in such a way that, by proper coding of the information packets, the receiver can reconstruct the information packets from each user regardless of relative time shifts of the protocol sequences (corresponding to the senders' different understanding of which slot is the first slot). This scenario describes the M-sender collision channel without feedback that was introduced by this writer in 1982 [14]. The capacity C_M is defined as the maximum rate R in packets/slot such that each user can send information packets at a rate at least R/M and the receiver can reconstruct these packets at the receiver with arbitrarily small error probability. In [14], it was shown that

$$C_M = (1 - \frac{1}{M})^{M-1} \quad \text{(packets/slot)} \qquad (7)$$

and moreover that zero-error probability could be achieved at this rate. Note that

$$\lim_{M\to\infty} C_M = e^{-1} \approx .368$$

so we now see that e^{-1} is indeed a capacity, but not that of the slotted-ALOHA channel where feedback is present, but rather that of the M-user collision channel without feedback when M is essentially infinite. Rather than to describe the general argument that leads to (7), we illustrate the main ideas by considering the special case M = 2.

For M = 2 users, capacity is achieved by choosing periodic protocol sequences with period 4 whose first periods are

[1, 1, 0, 0]	Sender 1
[1, 0, 1, 0]	Sender 2.

(Note that Sender 2's protocol sequence actually has least period 2.) The coding scheme is very simple, the Sender simply transmits each packet twice, namely at the two positions where there are 1's in the next period of his protocol sequence. Because Sender 1 always sends these two packets in adjacent slots, it follows that, whatever is the time offset between the two protocol sequences, exactly 1 of these packets will be lost in a collision with a packet from Sender 2. The same conclusion holds for the two packets sent by Sender 2; exactly one is lost in a collision. Moreover, the receiver knows to whom a successfully received packet belongs; if it is adjacent to a collision it came from Sender 1, otherwise it came from Sender 2. Each sender thus sends information error-free at a rate of 1 packet every 4 slots so that the senders achieve $R = 1/4 + 1/4 = C_2 = 1/2$ packet per slot.

For larger M, the scheme to achieve C_M becomes much more complex. In fact, the protocol sequences have period M^M and rather sophisticated coding schemes are required. More interesting and somewhat surprisingly, the capacity is still given by (7) even when the assumption of slot synchronization is removed [14], i.e., when the M senders are completely unsynchronized. The paper by Massey and Mathys [15], which also treats the generalization to the case when the M users wish to send information at different rates, gives full details of the necessary arguments.

It is illuminating to think upon the $M = 2$ user scheme described above as a way of creating a deterministic server for each sender that services information packets at the rate of 1 packet every 4 slots. Thus, if the sender is receiving new packets from his information source at any rate $\lambda/2 < 1/4$ (packet/slot), then the queue at that sender will not grow without bound. Thus, the system will be stable if the total arrival rate λ is equally distributed between the two users and satisfies $\lambda < 1/2 = C_2$ (packets/slot). Conversely, if $\lambda > 1/2 = C_2$, then the system will be unstable. Thus, C_2 is the maximum stable throughput for traffic equally divided between the two users. The similar conclusion holds for all $M \geq 2$.

6.2 The M-out-of-T-Sender Collision Channel without Feedback

A very interesting generalization of the previously described model for random-accessing without feedback was made by Bassalygo and Pinsker [16]. Their model differs from that of the M-Sender collision channel without feedback only in that there is assumed to be a total of T senders, but only at most M of these senders (in advance it is unknown which M) happen to have active information sources. We write C_M^T to denote the capacity of this "M-out-of-T-sender collision channel without feedback".

By using random coding arguments, Bassalygo and Pinsker proved the quite surprising fact that, when M is fairly large, then $C_M^T \approx e^{-1}$. In other words, it costs almost nothing in the achievable maximum stable throughput to have designed the protocol sequences for many more senders than will actually be using the channel actively. The price of increasing T for fixed M is rather an increase in the "complexity" of the necessary protocol sequences (i.e., an increase in the period of these sequences) and a concomitant increase in the complexity of the scheme for coding the information packets. As is typical for random coding arguments, the work of Bassalygo and Pinsker does not provide any specific schemes for achieving any given throughput less than C_M^T, but rather provides a proof of their existence. Thus, it seems especially appropriate here to illustrate the ideas involved in random-accessing on the M-out-of-T-sender channel, by describing a very specific scheme.

For our example, we take $T = 3$ and $M = 2$. We begin by choosing protocol sequences of period $N = 19$ for the 3 possible senders. We number the $N = 19$ locations in the first period from 0 to $N-1 = 18$ and represent the first period by those locations where this binary sequence of length N contains 1's. We choose the first periods of the protocol sequences for senders 1, 2 and 3 to be (0, 1, 8), (0, 2, 16) and (0, 4, 13), respectively. Note that (0, 1, 8), by our convention, denotes the binary sequence of length $N = 19$ with a 1 in positions 0, 1 and 8 only.

We consider next the set of distances between two 1's in the protocol sequence with first period (0, 1, 8) when those two 1's are less than N positions from each other. This set of distances is seen to be {1, 7, 8, 18, 12, 11} as follows from the fact that the 1 in position 0 in one period is distance 8 from the 1 in position 8 of the same period but is distance $N-8 = 19-8 = 11$ from the 1 in position 8 of the previous period, etc. Similarly, the sets of distances between 1's in the protocol sequences with first periods (0, 2, 16) and (0, 4, 13) are {2, 14, 16, 17, 5, 3} and {4, 9, 13, 15, 10, 6 }, respectively. The point to be noticed is that these three sets of distances are pairwise disjoint -- no distance appears in more than one set! This means that, regardless of the synchronization among the 3 protocol sequences, no two users can collide in more than one slot in any span of N consecutive slots; because if two protocol sequences both have a 1 corresponding to some slot, then they cannot both have a 1 in some other slot less than N slots distant from the former.

It follows that if any M = 2 senders are active, then each is guaranteed that at least two of the three packets that he sends within one period of his protocol sequence will be successfully received. Moreover, each sender can send two information packets error-free (assuming that the packet itself is a binary sequence of some fixed length n) in each period simply by sending the information packets in the first two slots where his protocol sequence has 1's and sending the bit-by-bit modulo-two sum of these packets as the "redundant packet" in the remaining slot. If either information packet is lost in a collision, it can be recovered at the receiver by subtracting the other information packet from the redundant packet. It follows that this T = 3 sender protocol-sequence/coding scheme allows any two senders to be active and each to send information packets error-free at a rate of 2/N = 2/19 (packets/slot). The total rate of 4/19 ≈ .211 (packets/slot) is not unreasonably smaller than C_2 = 1/2, the best that can be done with T = M = 2.

6.3 Delay Considerations

It is time to recall once again that the usual purpose of random-accessing is small (average) delay rather than high throughput. Space does not permit us to say much about how the "high-throughput" schemes in sections 6.1 and 6.2 can be modified to reduce delay, but we will give the flavor by considering again the two sequences, [1, 1, 0, 0] and [1, 0, 1, 0], that were used as first periods of periodic protocol sequences in section 6.1. The trick to reducing delay is not to use these sequences periodically but rather to use them only when the corresponding sender actually has a new packet to send, filling 0's into the protocol sequence during idle periods. [In general, one also needs to add some (at most N-1) 0's to the end of each first period of length N to ensure that the scheme still functions correctly, but no such additional 0's are needed for this M = 2 scheme.] The coding scheme is unchanged; each sender still sends each information packet twice. One sees that, if the traffic is light, there will usually be zero transmission delay as the first information packet will get through correctly. When this packet experiences a collision, the transmission delay will be 1 slot if it was from sender 1 and 2 slots if it was from sender 2. Similar delay considerations apply to the M-out-of-T channel coding schemes. One sees, moreover, from these arguments that one can allow the set of active senders to change with time, as long as no more than M are ever active and provided that there is an idle period of at least N - 1 slots between the time that one out of exactly M active senders ends his activity and the next one begins.

6. CONCLUDING REMARKS

We have described some new approaches to random-accessing, with emphasis on the collision-resolution approach for the reason that this approach appears to us to be an eminently practical one but not yet familiar to many practicioners of random-accessing. The reader may wonder why we have said nothing about "carrier-sensing", "collision detection", and many other techniques that are often used in practice to improve the efficiency of random-accessing with feedback. Our reason for avoiding a discussion of these "frills" is that (1) they can be used just as effectively with the collision-resolution approach as with the ALOHA approach to random-accessing, cf. [8], and (2) their introduction tends to obscure the real issue of random-accessing, viz., how effectively does one handle access conflicts. The recent book by Bertsekas and Gallager [17] is "must reading" for anyone who wants to deepen his understanding of the real issues.

We have also described two approaches to random-accessing without feedback. This area is still very much in the research stage, but our guess is that it will also find practical applications. Indeed, when lecturing recently on the material in section 6.1, we were pleasantly surprised by the enthusiastic response of a listener who had wanted to build a random-access system for the remote collection of data from a few sensors in his laboratory, but had previously thought it would be necessary to build a two-way channel so that the sensors could be provided with feedback after their access attempts. There may be other random-access applications where we are now providing feedback out of assumed necessity rather than as a practical choice. If so, the approaches described in section 6 merit some scrutiny by practicioners of random-accessing.

REFERENCES

[1] N.T. Gaarder and J.K. Wolf, "The capacity region of a multiple-access discrete memoryless channel can increase with feedback", IEEE Trans. Info.Th., vol. IT-21, pp. 100-102, Jan. 1975.

[2] F.M.J. Willems, "On multiple access channels with feedback", IEEE Trans. Info. Th., vol. IT-30, pp. 842-845, Nov. 1984.

[3] N. Abramson, "The ALOHA system - Another alternative for computer communications", Proc. Fall Joint Computer Conf., AFIPS Press, vol. 37, pp. 281-285, 1970.

[4] L.G. Roberts, "ALOHA packet system with and without slots and capture", Comp. Comm. Rev., vol. 5, pp. 28-42, April 1975.

[5] B. Hajek, "Stochastic approximation methods for decentralized control of multiaccess communications", IEEE Trans. Info. Th., vol. IT-31, pp. 176-184, March 1985.

[6] J. Capetanakis, "A protocol for resolving conflicts on ALOHA channels", Abstracts of Papers, IEEE Int. Symp. Info. Th., Ithaca, New York, pp. 122-123, Oct. 1977.

[7] B.S. Tsybakov and V.A. Mikhailov, "Slotted multiaccess packet broadcasting feedback channel", Probl. Peredachi Inform., vol. 14, pp. 32-59, Oct. - Dec. 1978.

[8] J.L. Massey, "Collision resolution algorithms and random access communications", pp. 73-137 in Multi-User Communications (Ed. G. Longo), CISM Courses and Lectures, No. 265. Wien and New York: Springer 1981.

[9] P. Mathys and Ph. Flajolet, "Q-ary collision resolution algorithms in random-access systems with free or blocked channel access", IEEE Trans. Info. Th., vol. IT-31, pp. 217-243, March 1985.

[10] P. Studer and H. Pletscher, "Part-and-try algorithms for random-access with the Poisson arrival model", Report of semester project, Inst. Signal & Info. Proc., Swiss Federal Inst. Tech., Zurich, Feb. 1984.

[11] M. Hofri, Probabilistic Analysis of Algorithms. Heidelberg and New York: Springer, 1987.

[12] F.P. Kelly, "Stochastic models of computer communication systems", J.Royal Stat. Soc., vol. 47, 1985.

[13] B.S. Tsybakov and N.B. Likhanov, "An upper bound for capacity of a random multiple-access system", Probl. Peredachi Inform., vol. 23, pp. 64-78, Jul-Sept., 1987.

[14] J.L. Massey, "The capacity of the collision channel without feedback", Abstracts of Papers, IEEE Int. Symp. Info. Th., Les Arcs, France, p. 101, June 1982.

[15] J.L. Massey and P. Mathys, "The collision channel without feedback", IEEE Trans. Info. Th., vol. IT-31, pp. 192-204, March 1985.

[16] L.A. Bassalygo and M.S. Pinsker, "Restricted asynchronous multiple access", Probl. Peredachi Inform., vol. 19, pp. 92-96, Oct.-Dec. 1983.

[17] D. Bertsekas and R.G. Gallager, Data Networks. Englewood Cliffs, NJ: Prentice-Hall, 1987.

VSAT Data Networks

NORMAN ABRAMSON, FELLOW, IEEE

Invited Paper

Two general methods have been used to provide random access packet communications in very small-aperture (VSAT) data networks: spread spectrum (or CDMA) and ALOHA. In this paper we review the use of satellite channels for such networks and discuss certain basic aspects of the architecture of VSAT data networks. Although spread spectrum and ALOHA have different origins and are sometimes represented as competing technologies, they can in fact be characterized as different ways of viewing the same low-dimensional signals in a high-dimensional signal space. After a brief introduction to the architecture of VSAT networks we show how a simple linear transformation of conventional ALOHA packets leads to signals identical in all respects to the most common form of spread-spectrum signals. We call the result of this transformation spread ALOHA.

There are two practical consequences of this theoretical result. First, for the case of small earth stations it is not possible to find an access technique with a higher throughput than spread ALOHA. Second, the use of different spreading sequences for different users in a packet network using spread spectrum is not necessary for user separation.

I. Introduction

A. VSAT Architecture

A communication satellite can be used to provide conventional point-to-point channels for data networks and other applications. But the key characteristic of communication satellites for many applications is the broadcast and multiple access nature of the satellite channel. The physics of the satellite path provides the built-in capability of one-to-many (broadcast) channels and of many-to-one (multiple access) channels to and from any earth station in the satellite footprint. This characteristic of satellite communications can be used to provide new types of service which are simply not practical or economical using conventional ground communications.

The broadcast nature of the satellite channel encouraged the development of satellite communications for video and audio distribution in the 1970s. This same broadcasting capability of satellite channels has been employed for a variety of one-way data distribution networks since the early 1980s. During the past few years we have seen the development of a powerful and flexible network architecture, which combines the capabilities of satellite broadcasting and satellite multiple access for two-way data networks composed of very small-aperture terminals (VSATs).

Manuscript received December 21, 1989; revised January 22, 1990.
The author is with the Department of Electrical Engineering, University of Hawaii, Honolulu, HI 96822, USA.
IEEE Log No. 9037998.

The network architecture used in VSAT data networks is almost always designed around a single large (hub) earth station transmitting data in a broadcast channel to a large number of VSATs (Fig. 1).

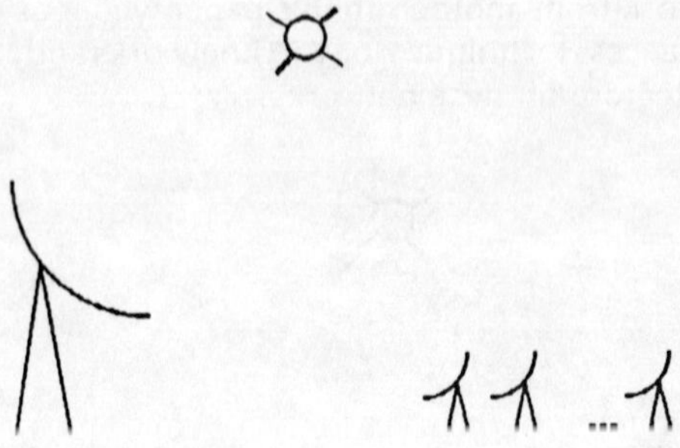

Fig. 1. A VSAT network.

The VSATs in such a network transmit data in packets to the hub station using the multiple-access capability of the satellite channel. Since there are no direct links from one VSAT to another, any VSAT to VSAT traffic must take a two-hop path from its source to its destination. There are two good reasons for such a limitation. First, transmit power and receive power requirements on the large number of VSATs in such a network can be relaxed considerably by the fact that the uplink and downlink in the network each benefit from the higher performance capabilities of the hub station. Second, many of the applications which can benefit from the use of VSAT technology require communications from a large number of users to a central information resource. These applications include large financial networks, reservation networks, and shared data bases.

The link from the hub station of a VSAT network to the VSATs is easily configured using conventional time-division multiplexing (TDM). And although there are differences in data rates, modulation techniques, and transmission formats among different VSAT networks, there is general agreement on the use of TDM for multiplexing from the hub to the terminals (Fig. 2).

The multiple-access link from the VSATs to the hub, however, has been subjected to a greater degree of variation in data networks built during the past decade. And in the overall design of the network, the choice of an access technique for the VSAT-to-hub multiple-access channel is the key deci-

Reprinted from *Proc. IEEE*, vol. 78, no. 7, pp. 1267–1274, July 1990.

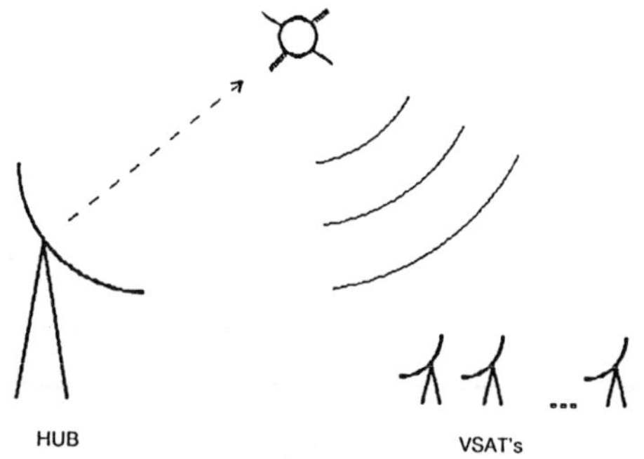

Fig. 2. Broadcast channel: Hub to VSATs.

sion to be made by the network architect. There are still a number of important choices in data rates, modulation techniques, and transmission formats. But while these differences will have a major impact on overall network performance, it is probably accurate to characterize the choice of access technique as the primary feature distinguishing one network from another. In this paper we focus on the choice of access techniques for VSAT networks and analyze the architecture of these networks.

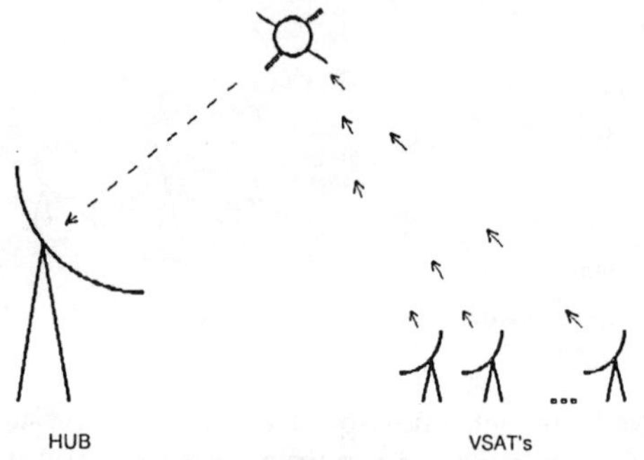

Fig. 3. Multiple-access channel: VSATs to hub.

B. Access Techniques

Frequency-division multiple access (FDMA) and time-division-multiple access (TDMA) are the two primary access techniques used for sharing the high capacity of a typical commercial satellite transponder among several earth stations in a voice network. With either of these techniques the total transponder capacity is divided into frequency or time slices for use in fixed allocation channels of moderate capacity. When the traffic from a single terminal in a network is bursty, the efficiency of this kind of operation drops and a demand-assigned multiple access (DAMA) technique is required in order to allocate capacity in response to fluctuating user demands.

In a DAMA system, network capacity is allocated (either in frequency or time slices) in response to user requests for channel capacity. Note, however, that in one sense the use of a DAMA system merely shifts the problem of allocating capacity in response to random requests to another level. A special channel, usually called the request channel, is assigned to carry requests for capacity assignments in a DAMA network and some access method to allow reasonably efficient sharing of the request channel must be found.

During the past few years it has become practical to build VSAT networks composed of hundreds or thousands of small (typically 1.2 m in diameter) terminals. The traffic in these networks is usually in the form of single data packets originating from interactive users or bursts of data packets originating from some file transfer protocol. In general, as the number of earth stations in a network increases, the more the traffic from a single station will appear to fluctuate due to random user demand. In such a network, of course, the use of fixed-assignment FDMA or TDMA is usually impractical, while the use of DAMA would ordinarily impose an unreasonable amount of overhead in the network. In order to provide multiple access to packet data networks, therefore, two other access techniques have been used—spread-spectrum multiple access (or code-division multiple access—CDMA) [1] and ALOHA [2] (in a so-called unslotted or slotted mode).

Spread spectrum has been defined as communications which uses a "bandwidth well beyond what is required to transmit digital data" [3]. In direct-sequence spread spectrum, the most common form of spread spectrum, each bit to be transmitted is converted into r binary chips, where r is much greater than 1 and the original signal bandwidth is expanded by the same factor of r in order to accommodate the higher transmission rate of the chips. Each transmitter in such a network is usually assigned a distinct spreading sequence, and the receiver must then convert the stream of chips from each transmitter into a (lower data rate) stream of bits by means of what is essentially a digital matched filter, one matched to each distinct transmitter spreading sequence. A frequency-hopping version of this procedure is also possible, but frequency-hopped spread spectrum seems better suited to military rather than commercial applications. (See Fig. 4.)

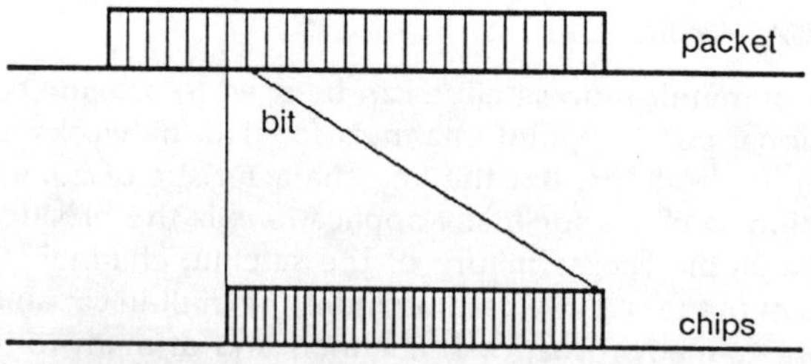

Fig. 4. Spread spectrum: from bits to chips.

In the ALOHA form of multiple access (Fig. 5) the bandwidth of the channel is also required to be greater than the information bandwidth corresponding to the average data rate transmitted from any single user in the channel. Each transmitter in such a multiple access channel will transmit its data packets at the maximum data rate of the channel. Because the channel bandwidth is high compared to the data rate from a single user, the duty cycle of each user is low and the probability that two packets from different users will overlap in the channel will be low. A variety of protocols have been developed [4], [5] to deal with the case of packets that are lost due to contention in the channel. The ratio of the maximum data rate which can be sustained in such a channel to the maximum data rate when the channel is used by a single user in a point-to-point mode is called the maximum throughput of the ALOHA channel, and is equal to $1/2e$ or 0.184 [2]. If the channel is modified so that each transmitter synchronizes the beginning of its packet transmis-

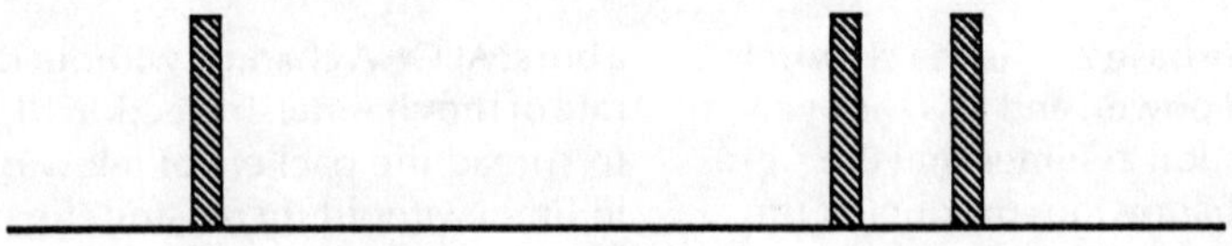

Fig. 5. ALOHA multiple access.

sions to fixed slots, then the maximum throughput of the channel is doubled to 1/e or 0.368 [6].

Note, however, that both the 1/2e and the 1/e maximum throughput results compare the throughput of a low duty cycle (typically 10% or less) ALOHA channel to that of a point-to-point channel operated continuously. The continuous channel, of course, uses considerably more power (typically 10 dB or more) than the bursty ALOHA channel. For the case of the original packet radio ALOHANET [2] where the *peak* power is important such a throughput comparison is appropriate. However in the case of most satellite multi-access channels, and certainly in the case of VSAT channels, where the limiting resource is the *average* power transmitted in the downlink, a different kind of comparison is appropriate. In this case it makes more sense to compare the data rate of the ALOHA channel with the data rate of a point-to-point channel using the same average downlink power as the ALOHA channel. And as might be expected, when proper allowances are made for a 10 dB or more power mismatch, the ALOHA channel throughput looks considerably more attractive. In Section I-D we discuss the throughput of an ALOHA channel with an average rather than a peak power constraint.

C. History

Each of the two multiple-access techniques described above was originally developed for applications other than the satellite channel. Spread spectrum has its origins in a variety of military applications [7] and much of the early work in spread spectrum dealt with the transmission of continuous signals rather than the packet transmissions of interest in most present-day systems. ALOHA channels were originally developed in the early 1970s for use in the ALOHANET, a local-area wireless network (LAWN) at the University of Hawaii [8]. The first demonstration of the use of either of these techniques for satellite communications, however, was in 1973 in the ALOHA system's PACNET network, using the NASA ATS-1 satellite to link VSATs in Hawaii, Alaska, Japan, and Australia.

In 1976 an unslotted ALOHA channel was implemented as the request channel of the COMSAT MARISAT system [9] in a commercial network—the first commercial application of this technology. The first commercial use of spread spectrum in a satellite channel was in 1981 with the introduction of the Equatorial Communications C-100 Micro Earth Station for satellite data distribution. The MARISAT ALOHA channel was a true multiple-access channel, providing many-to-one communications to thousands of small ship stations in a three-ocean system. The first Equatorial Communications terminals were receive-only units in which only the interference rejection capabilities of spread-spectrum signals were utilized for data transmission. In 1984 Equatorial introduced a true multiple-access satellite network using the C-200 Micro Earth Station. By 1985 several other companies had entered the multiple-access VSAT market in the United States.

Ironically, by 1987, Equatorial, which had first demonstrated the commercial viability of the VSAT concept, had run into trouble in its C-200 systems, and had been taken over by ConTel. By 1986 this market had matured so that several other companies had entered the market and by the end of 1989 there were four major suppliers of this equipment in the USA: TRIDOM/AT&T, Hughes Network Systems/GM, GTE/Spacenet, and Equatorial/ConTel. Each of these systems was based on an ALOHA channel (either slotted or unslotted) in some cases supplemented by TDMA for large file transfers, although ConTel also offered its original spread-spectrum multiple-access network as an option. By 1989 there were about 10 000 multiple-access commercial VSAT terminals in operation, not including the large number of ship stations using ALOHA in the INMARSAT (formerly MARISAT) system.

D. Satellite Multiple Access

Although both spread spectrum and ALOHA can provide a multiple-access capability in many different kinds of com munication networks, there are a number of special characteristics of the VSAT packet data multiple-access channel which affect the operation of the overall system. When spread spectrum is used to distinguish among large numbers of intermittent packet transmitters, each using a different spreading sequence, the hub receiving station must be provided with a digital matched filter operating at the high-speed chip rate for each of the possible transmitters. In the Equatorial C-200 system these receivers are called "ear cards." And in a network with thousands of VSATs, the logistics of installing, removing, maintaining, and testing a special piece of hardware at the hub station for each network user is a difficult task.

The throughput of an ALOHA channel is ordinarily calculated on the basis of comparing the average data rate of the burst ALOHA channel to that of a full-time dedicated point-to-point channel between only two users. In the case of ground communications, where the average power radiated by the transmitter is not a major constraint, such a comparison is appropriate. But the bursty ALOHA channel (see Fig. 5) will ordinarily require 10% or less of the average power required by the point-to-point channel in such a comparison.

In the case of a satellite channel, the average power transmitted from the satellite transponder is a limited and expensive resource, so it is appropriate to begin the comparison using Shannon's basic equation for the capacity of the band-limited additive white Gaussian noise channel. In this case R, the data rate of the channel in bits per second, is limited by the channel capacity

$$C_0 = W \log \left(1 + \frac{S}{N}\right) \tag{1}$$

where the logarithm is taken to the base 2, W is the channel bandwidth, S is the average signal power, and N is the average noise power. Shannon's equation assumes that the signal is continuous, but if the time bandwidth product of the packets is large enough we can apply Shannon's equation to the packets during the times of packet transmission. And if the duty cycle of packet transmissions is d ($d \ll 1$), the data rate of the multi-access ALOHA channel must be less than

$$\begin{aligned} C_1 &= dC_0 \\ &= dW \log\left(1 + \frac{S}{N}\right). \end{aligned} \tag{2}$$

It is not difficult to find the corresponding capacity of a point-to-point channel with the same average power as the ALOHA channel. For a given duty cycle d, the average power required by the ALOHA channel is just dS, and the data rate of the continuous channel must therefore be less than

$$C_2 = W \log\left(1 + \frac{dS}{N}\right). \tag{3}$$

Because of possible packet overlap, Eq. (3) is valid for $d \ll 1$ and certain corrections must be made for the more general case. These calculations are provided in [10; section V] where the exact derivation is provided for four separate cases—the slotted and unslotted ALOHA channel, operating with either a linear or a hard limited transponder. In all cases the the basic results are the same. As might be expected, we always have

$$C_1 < C_2. \tag{4}$$

That is, the multi-access ALOHA channel will always have a lower maximum throughput than the point-to-point channel with the same average power. However, for all values of the duty cycle, the ratio of the maximum ALOHA throughput compared to the capacity of the average power limited channel exceeds the ratio of the maximum ALOHA throughput compared to the capacity of the peak-power limited channel by a significant amount. And for the case of a low signal-to-noise ratio in the receiving antenna (for example, VSATs)

$$C_1 \approx C_2. \tag{5}$$

Thus we have the surprising result that for the case of small satellite earth stations operating with a low duty cycle, the ALOHA capacity approaches the Shannon capacity, and it is not possible to find an access method with a higher data throughput [10; section V].

For the kind of ALOHA operation described in the previous paragraph, in order to approach the Shannon channel capacity it is necessary to operate at a low duty cycle, and therefore for interesting values of average power, the peak power of the VSAT transmitter during data bursts can be quite high. And this characteristic of the low duty cycle ALOHA channel could limit the application of this mode of operation in satellite networks composed of large numbers of VSATs. Since the signal detectability depends upon the energy per bit and not the average signal power, it makes sense to spread the transmitted packets in time in order to decrease the peak power requirements of the earth station transmitters while keeping the transmitted energy per bit constant. Note that this strategy can be used in the case of a burst ALOHA channel without decreasing the average data rate of the channel. In Section III, we show how it is possible to spread the packets of a low duty cycle ALOHA channel in time, without increasing the probability of packet overlap out of the channel receiver. But first we use Section II to introduce a notation which will be helpful in the remainder of this paper.

II. Notation

A. Packets and Matched Filters

Consider the case of packet transmission where each packet consists of exactly n bits. Let the value of these bits be d_j, where the data sequence d_j is given by

$$d_j = \pm 1 \qquad j = 0, 1, \cdots, n-1. \tag{6}$$

In this paper we shall be concerned with binary phase-shift keyed (BPSK) channels for transmission of packets and with the representation of these packets in a form that emphasizes their correspondence to conventional direct-sequence spread-spectrum channels [11]. Let $p(t)$ be the bit pulse waveform used in the channel so that a single packet $D(t)$, consisting of n bit transmissions at the times 0, 1, $\cdots$, $n - 1$, can be represented as

$$D(t) = \sum_{j=0}^{n-1} d_j p(t - j). \tag{7}$$

A useful device, which allows us to separate the analog effects of pulse shape from the digital effects of the data sequence, is to represent $D(t)$ as the convolution of the pulse $p(t)$ with a sequence of impulses multiplied by the data sequence values d_j. Then

$$D(t) = \sum_{j=0}^{n-1} d_j \delta(t - j) * p(t). \tag{8}$$

Note that $D(t)$ can be thought of as the output of a time-invariant linear operator with impulse response $p(t)$ (Fig. 6), when the input is

$$d(t) = \sum_{j=0}^{n-1} d_j \delta(t - j). \tag{9}$$

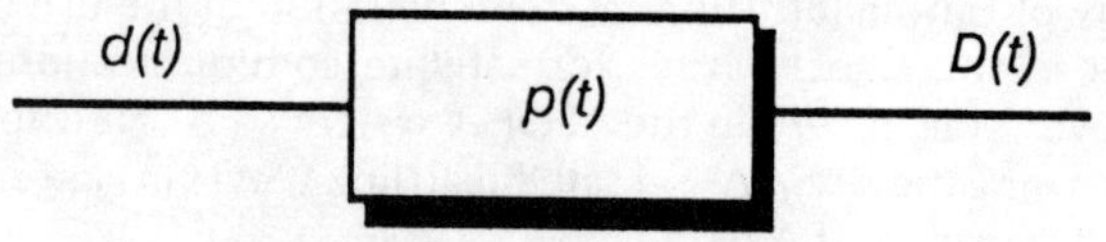

Fig. 6. The packet as output of linear filter.

The reception of a sequence of packets occurring at random times in a multiple-access channel is usually broken down into two stages: a packet detection and synchronization stage (to detect the presence of the packet and to synchronize) and a signal-detection stage (to demodulate the signal and to determine the values of the information bearing elements. In a VSAT network the synchronization problem is made considerably simpler by the fact that the hub station can provide information feedback to the VSAT terminals in order to help them synchronize the phase of their packet transmissions. We therefore need only consider the signal-detection part of the receiver. We assume the packet $D(t)$ is transmitted in a channel with additive white Gaussian noise (AWGN) $n(t)$, so that the received signal $a(t)$ is

$$a(t) = D(t) + n(t). \tag{10}$$

Under a wide variety of assumptions, a key element of the packet receiver [12], [13] is the matched filter—matched to the pulse shape $p(t)$. Ignoring questions of time delays in the receiver, we can represent the impulse response function of the matched filter as $p(-t)$, and the output of the matched filter as $b(t)$ [14], [15] (Fig. 7).

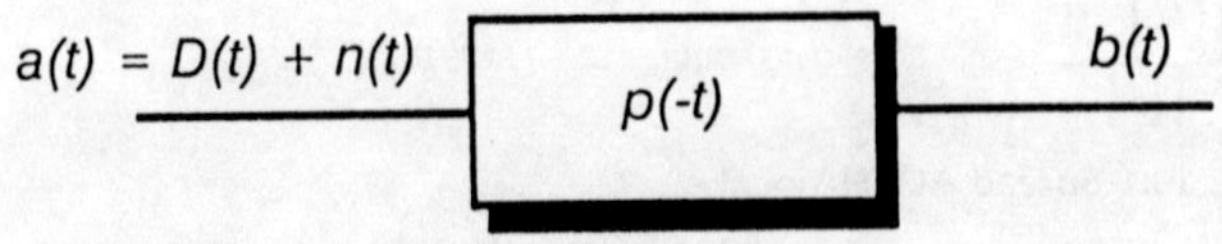

Fig. 7. Input and output of the matched filter.

Then the output of the matched filter is given by

$$\begin{aligned} b(t) &= a(t) * p(-t) \\ &= D(t) * p(-t) + n(t) * p(-t) \\ &= \sum_{j=0}^{n-1} d_j \delta(t-j) * p(t) * p(-t) + n(t) * p(-t). \end{aligned} \tag{11}$$

Define the correlation of the bit waveform as

$$\begin{aligned} \rho(t) &= p(t) * p(-t) \\ &= \int_{-\infty}^{\infty} p(x)p(t-x)\, dx. \end{aligned} \tag{12}$$

Then,

$$\begin{aligned} b(t) &= \sum_{j=0}^{n-1} d_j \delta(t-j) * \rho(t) + n(t) * p(-t) \\ &= \sum_{j=0}^{n-1} d_j \rho(t-j) + n_1(t) \end{aligned} \tag{13}$$

where we define

$$n_1(t) = n(t) * p(-t). \tag{14}$$

Many decision rules to obtain the data sequence d_k from the received signal rely on sample values of $b(t)$ taken at t equal to the bit transmission times $k = 0, 1, 2 \cdots$

$$\begin{aligned} b(k) &= \sum_{j=0}^{n-1} d_j \rho(k-j) + n_1(k) \\ &= d_k \rho(0) + I(k) + n_1(k) \end{aligned} \tag{15}$$

where

$$I(k) = \sum_{\substack{j=0 \\ j \neq k}}^{n-1} d_j \rho(k-j) \tag{16}$$

is the intersymbol interference for the kth symbol in the packet. The energy per bit is

$$E_b = \int_{-\infty}^{\infty} p^2(t)\, dt. \tag{17}$$

Then, if N_0 is the power spectral density of the noise, E_b/N_0 provides a measure of the quality of the signal-detection phase of the receiver [16].

B. Spreading Sequences

In order to achieve the time spreading of the packets discussed in Section I-D we use binary sequences with low autocorrelation properties, such as Barker sequences [17], [18]. Define a binary sequence of length r as

$$s_j = \pm 1 \qquad j = 0, 1, \cdots, r-1 \tag{18}$$

and a time-invariant linear operator (spreader) corresponding to the sequence s_j with an impulse response function

$$s(t) = \sum_{j=0}^{r-1} s_j \delta(t-j). \tag{19}$$

Define the correlation of $s(t)$ as

$$\begin{aligned} \sigma(t) &= s(t) * s(-t) \\ &= \int_{-\infty}^{\infty} s(x)s(x+t)\, dx \\ &= \sum_{j=1-r}^{r-1} s_j * s_j \delta(t-j) \\ &= \sum_{j=1-r}^{r-1} \sigma_j \delta(t-j) \end{aligned} \tag{20}$$

where

$$\sigma_j = s_j * s_j = \sum_{k=0}^{r-1} s_k s_{k-j}. \tag{21}$$

In the case of Barker sequences,

$$\sigma_j = r \qquad \text{for } j = 0 \tag{22}$$

and

$$|\sigma_j| \leq 1 \qquad \text{for } j \neq 0. \tag{23}$$

Barker sequences are known only for the cases of $r = 2$, 3, 4, 5, 7, 11, and 13, but binary sequences of other lengths with good values of σ_j for use in spread ALOHA channels as well as efficient methods of searching for such sequences have been obtained by Li [18].

III. Spread ALOHA

A. Spreading the Packets

As explained in Section I-D, the low duty cycle ALOHA channel shown in Fig. 5 can be used in VSAT networks with one major theoretical advantage (it can achieve the Shannon bound) and one major practical disadvantage (it may require a high value of peak power out of the small aperture earth station). In this section we modify the signals in the low duty cycle ALOHA channel so that the disadvantage of high peak power is eliminated while at the same time E_b/N_0, the measure of signal quality defined in Section II-A, remain unchanged. We call this modification *spread ALOHA*, and our results show that the performance of spread ALOHA is the same as that of low duty cycle ALOHA, while the performance of the latter is the best possible for a given average power and a given channel bandwidth.

The modification we use is a natural one—we take the packets shown in Fig. 5 and spread them in time. Such time spreading can be done in a number of ways, but the method we describe in Section III-B has the property that the spreading leaves E_b/N_0 unchanged and at the same time the spreading does not affect the probability of packet overlap at the output of the detector. If the satellite channel is operated in a linear mode, the increased channel overlap will not affect the operation of the ALOHA channel since the optimum detector consists of a matched filter, which serves to despread the spread packets.

B. Bit Spread ALOHA

Our objective then is simply to spread the packets in time, thus effectively lowering the transmit power requirements of the VSAT terminals, while at the same time keeping the transmitted energy per bit constant, and also by means of the despreading operation at the network hub, keeping the effective overlap silhouette of the transmitted packets constant. Note the distinction made here between the overlap of packets in the physical channel, which is greatly increased by spread ALOHA, and the overlap of packets at the output of the detector, which is not.

For bit spread ALOHA packets we start with Eq. (7) and delay the transmission of each bit by r units of time relative to the previous bit. That is, let the transmitted packet be

$$D_r(t) = \sum_{j=0}^{n-1} d_j p(t - rj) \tag{24}$$

and, following the development of Eqs. (8) and (9)

$$D_r(t) = \sum_{j=0}^{n-1} d_j \delta(t - rj) * p(t). \tag{25}$$

As before, we can represent $D_r(t)$ as the output of a time-invariant linear operator with impulse response $p(t)$ (Fig. 8),

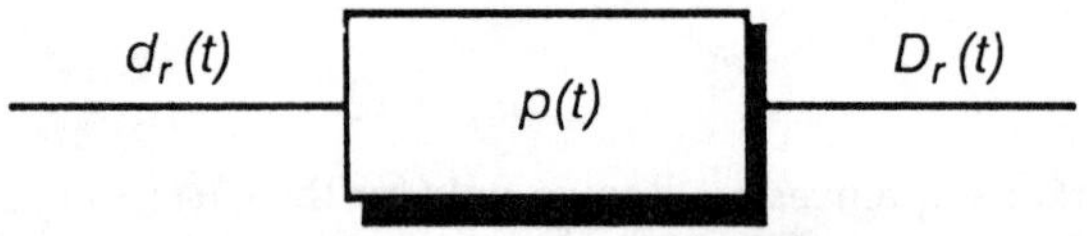

Fig. 8. Stretched packet as output of linear filter.

when the input is

$$d_r(t) = \sum_{j=0}^{n-1} d_j \delta(t - rj). \tag{26}$$

We refer to the packet $D_r(t)$ as the *stretched packet*. And we produce the *spread packet* we want by passing the stretched packet through a spreader, a time-invariant linear filter with impulse response equal to $r^{-1/2}s(t)$, as given in Eq. (19). (See Fig. 9.)

The process of spreading an ALOHA packet for a simplified packet is illustrated in Fig. 10. In order to clarify the process the bits and chips are represented without the pulse-shaping filter $p(t)$. The first line of Fig. 10 depicts a packet composed of 6 bits—110101. The stretched packet of Eqs. (24) and (26) is shown on the second line. The third line illustrates the spread packet for this simple example, where the spreader is taken as the length 7 binary sequence of line 4: 1110010. Of course, more realistic values of both the packet length and the length of the spreading sequence would be in the range from 100 to 10 000.

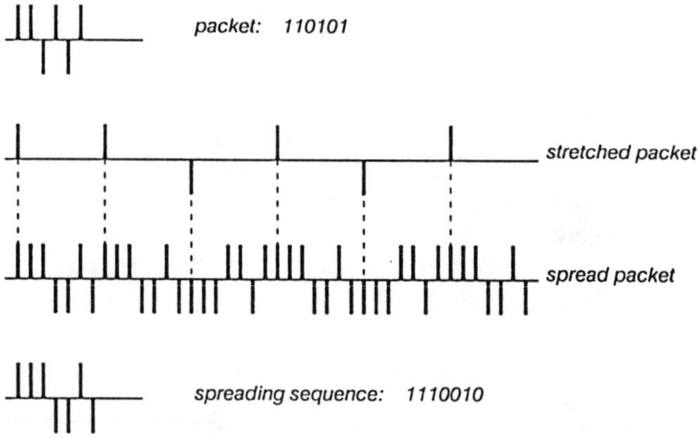

Fig. 10. Spread ALOHA packet.

The form of the spread packet is now easily calculated as

$$\begin{aligned} D_s(t) &= r^{-1/2} d_r(t) * s(t) * p(t) \\ &= \sum_{j=0}^{m-1} c_j \delta(t - j) * p(t) \\ &= \sum_{j=0}^{m-1} c_j p(t - j) \end{aligned} \tag{27}$$

where

$$c_j = r^{-1/2} d_\alpha s_\beta \tag{28}$$

and α is the largest integer less than or equal to j/r and β is equal to j modulo r

$$\alpha = \lfloor j/r \rfloor \tag{29a}$$

$$\beta = j \bmod r. \tag{29b}$$

The binary random variables c_j constitute the "chips" of the spread ALOHA signal; more precisely, c_j is the βth chip of the αth data bit of the spread signal. Since each bit of the original packet is converted to r chips of the spread packet by the spreader, each chip is multiplied by $r^{-1/2}$ in order to maintain the same energy per bit before and after spreading. Thus the peak power requirement of the transmitter is decreased by a factor of r. Time spreading provides a mechanism for reducing the average power of the transmitter to a level consistent with a network of VSAT terminals while maintaining a constant value for the energy per bit.

Note that in contrast to conventional spread spectrum for continuous signals, each packet in the spread ALOHA channel is spread by the same binary sequence s_j. Separation of packets from different users is accomplished by means of the ALOHA contention protocol rather than by the cross-correlation of different spreading sequences. And in contrast to conventional operation of an ALOHA channel, the separation is not limited by the overlap of the transmitted

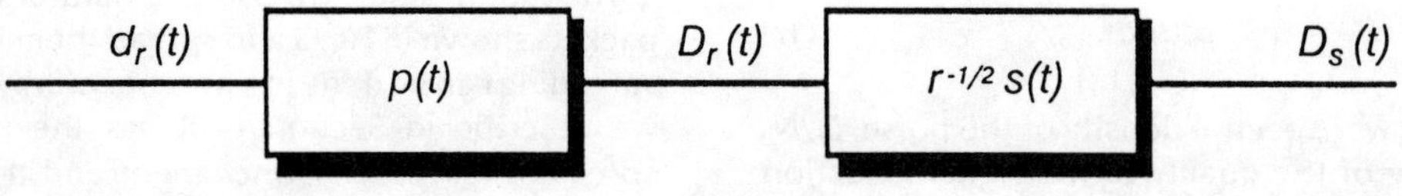

Fig. 9. Bit spread packet as output of two linear filters.

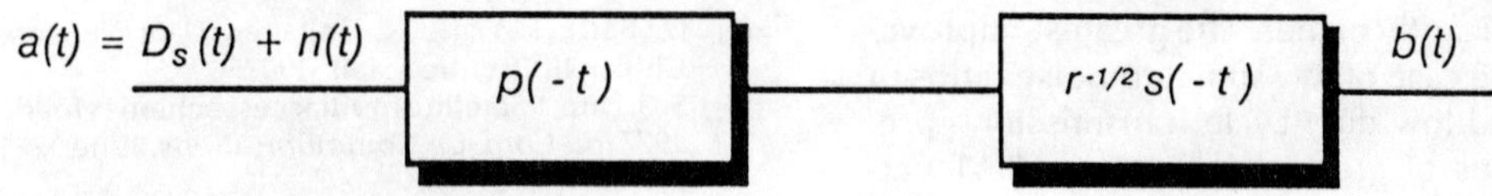

Fig. 11. Input and output of bit spread matched filter.

packets in the channel, but by the overlap of these pulses at the output of a matched filter.

From Fig. 9 we see that the problem of detection of the bits of the spread packet in the presence of AWGN has been reduced to that of the unspread packet of Fig. 2. We need only replace $p(t)$ in Fig. 6 by the convolution of $p(t)$ and $s(t)$ in Fig. 9. Then for the spread case the spread matched filter is shown in Fig. 11.

Now following the development used in Eq. (11) we have

$$\begin{aligned} b(t) &= r^{-1/2}[a(t) * p(-t) * s(-t)] \\ &= r^{-1/2}[D_s(t) * p(-t) * s(-t)] + r^{-1/2}[n(t) * p(-t) * s(-t)] \\ &= r^{-1}[d_r(t) * s(t) * s(-t) * p(t) * p(-t)] \\ &\quad + r^{-1/2}[n(t) * p(-t) * s(-t)] \\ &= r^{-1}[d_r(t) * \sigma(t) * \rho(t)] + n_2(t) \end{aligned} \tag{30}$$

where we define

$$n_2(t) = r^{-1/2}[n(t) * p(-t) * s(-t)]. \tag{31}$$

Using (20) and (26), we can expand $b(t)$, the output of the matched filter in Eq. (30)

$$b(t) = r^{-1} \sum_{j=0}^{n-1} d_j \delta(t - rj) * \sum_{j=1-r}^{r-1} \sigma_j \delta(t - j) * \rho(t) + n_2(t). \tag{32}$$

If we define $d_{-1} = d_n = 0$ and $\sigma_{-r} = 0$, then

$$\begin{aligned} b(t) &= \sum_{j=1-r}^{nr-1} e_j \delta(t - j) * \rho(t) + n_2(t) \\ &= \sum_{j=1-r}^{nr-1} e_j \rho(t - j) + n_2(t) \end{aligned} \tag{33}$$

where

$$e_j = r^{-1}[d_\alpha \sigma_\beta + d_{\alpha+1} \sigma_{r-\beta}] \tag{34}$$

and α is the largest integer less than or equal to j/r and β is equal to j modulo r.

$$\alpha = \lfloor j/r \rfloor \tag{35a}$$

$$\beta = j \bmod r \tag{35b}$$

The output of the matched filter is sampled at the stretched bit transmission times $t = kr$, where $k = 0, 1, 2, \cdots, n - 1$. At these sample points $\beta = 0$ and $\sigma_0 = r$, so that $e_{kr} = d_k$, and we get

$$\begin{aligned} b(kr) &= \sum_{j=1-r}^{nr-1} e_j \rho(kr - j) + n_2(kr) \\ &= d_k \rho(0) + J(kr) + n_2(kr) \end{aligned} \tag{36}$$

where

$$J(kr) = \sum_{\substack{j=1-r \\ j \neq kr}}^{nr-1} e_j \rho(kr - j) \tag{37}$$

is the intersymbol interference for the kth symbol in the packet.

C. Discussion

Note the similarity of the interference term of Eq. (37) to that of Eq. (16). The sample values of the data pulse correlation $\rho(t)$ cause the interference. In this case the interference might properly be termed "interchip interference" rather than intersymbol interference since the stretched packet has separated each bit by r units of time and, for reasonably shaped pulses $p(t)$, this will eliminate the possibility of intersymbol interference. And even in those cases where the sample values of $\rho(t)$ are not negligible, note that the coefficients e_j as given in Eq. (34) decrease as r^{-1}. This means that the interference caused by adjacent chips in bit spread ALOHA will be considerably less than that caused by adjacent bits in unspread packets.

Finally, we note that packet samples utilized by a matched filter detector of bit spread ALOHA packets are taken at intervals of r units of time. In the case of the AWGN channel assumed, packets which overlap in the channel can still be detected correctly as long as all samples of each of the two packets are separated by a single chip interval, taken as a unit of time. Thus the total period of vulnerability in this unslotted spread ALOHA channel is twice the packet length, just as in a conventional unslotted ALOHA channel. The only difference in this case is that the period of vulnerability is broken up into $2n - 1$ small subintervals. In contrast to conventional ALOHA channels, two packets which overlap in the channel do not necessarily result in lost packets due to collision. The spreading and despreading operations effectively separate most of the channel collisions in spread ALOHA signals so that the only overlap of concern is that at the output of the matched filter detector.

The spreading technique we have described is not the only possible method of achieving the practical advantages of spread ALOHA. In [19] a similar dual spreading procedure, chip spread ALOHA, is analyzed and an algebraic coding formulation of spread ALOHA is provided.

D. Conclusions

The arguments of Eqs. (1)–(5) show that the usual analysis of ALOHA throughput for peak power limited channels is not appropriate for the case of average power limited VSAT channels. The well known $1/2e = 0.184$ and $1/e = 0.368$ limits compare the performance of a multiple-access ALOHA channel to that of a point-to-point channel with significantly greater average power in the downlink. Simply put, some of the results originally published by this author are too often used as the right answer to the wrong question.

When the constraint of identical average power in the satellite downlink is taken into account, the maximum value of the throughput of conventional slotted or unslotted ALOHA channels will always be higher than the 1/2e or 1/e limits, and in the case of VSAT networks typical of those available today, significantly higher. The greatest improvement will occur for the case of low signal-to-noise ratios in the VSAT receiver and low duty cycle transmission—precisely those conditions of greatest interest in VSAT networks [10; section V].

For VSAT networks with an average power limitation in the downlink, as the bandwidth used for the multi-access channel is increased, the signal-to-noise ratio in the VSAT receivers and the duty cycle defined in terms of the multi-access bit rate (not the chip rate) both decrease. In the limit the maximum value of the throughput approaches the Shannon bound and it is not possible to find an access technique with a higher throughput.

In order to achieve reasonable values of signal energy per bit using conventional ALOHA operating at a low duty cycle in the high bandwidth channel, it might be necessary to require unreasonably high values of transmitter power from the VSAT terminals. The solution to this problem is to spread the VSAT packets in time to obtain a low-power spread ALOHA signal. In this case the spread ALOHA signal becomes identical in form to a CDMA spread-spectrum signal, and in the limit both types of transmission achieve the Shannon bound. Separation of the signals from different transmitters at the hub station, however, can utilize the ordinary signal separation mechanism of the ALOHA channel rather than the chip code separation of the CDMA channel.

References

[1] E. B. Parker, "Micro earth stations as personal computer accessories," *Proc. IEEE*, vol. 72, pp. 1526–1531, Nov. 1984.

[2] N. Abramson, "Packet switching with satellites," *AFIPS Conf. Proc. of the National Computer Conf.*, New York, June 1973, vol. 42, pp. 695–702.

[3] A. J. Viterbi, "Spread spectrum communications—Myths and realities," *IEEE Commun. Mag.*, vol. 17, no. 3, pp. 11–18, May 1979.

[4] D. Bertsekas and R. Gallager, *Data Networks.* Englewood Cliffs, NJ: Prentice-Hall, 1987.

[5] S. S. Lam, "Satellite multiaccess schemes for data traffic," *Proc. 1977 Int. Conf. on Communications*, June 12–15, 1977, vol. III, pp. 37.1.19–37.1.24.

[6] L. G. Roberts, "ALOHA packet system with and without slots and capture," *Computer Commun. Rev.*, vol. 5, no. 2, pp. 28–42, Apr. 1975.

[7] R. A. Scholtz, "The spread spectrum concept," *IEEE Trans. Commun.*, vol. COM-25, pp. 748–755, Aug. 1977.

[8] N. Abramson, "Development of the ALOHANET," *IEEE Trans. Inform. Theory*, vol. IT-31, no. 2, pp. 119–123, Mar. 1985.

[9] D. W. Lipke *et al.*, "MARISAT—A maritime satellite communication system," COMSAT *Tech. Rev.*, vol. 7, no. 2, Fall 1977.

[10] N. Abramson, "The throughput of packet broadcasting channels," *IEEE Trans. Commun.*, vol. COM-25, pp. 117–128, Jan. 1977.

[11] R. L. Pickholtz, D. L. Schilling, and L. B. Milstein, "Theory of spread-spectrum communications—A tutorial," *IEEE Trans. Commun.*, vol. COM-30, pp. 855–884, May 1982.

[12] M. B. Pursley, D. V. Sarawate, and W. E. Stark, "Error probability for direct-sequence spread-spectrum multiple-access communications—Part I: Upper and lower bounds," *IEEE Trans. Commun.*, vol. COM-30, pp. 975–984, May 1982.

[13] N. F. Krasner, "Optimal detection of digitally modulated signal," *IEEE Trans. Commun.*, vol. COM-30, pp. 885–895, May 1982.

[14] G. L. Turin, "An introduction to matched filters," *IRE Trans. Inform. Theory*, vol. IT-6, pp. 311–329, June 1960.

[15] —, "An introduction to digital matched filters," *Proc. IEEE*, vol. 64, pp. 1092–111, July 1976.

[16] G. R. Cooper and C. D. McGillem, *Modern Communications and Spread Spectrum.* New York: McGraw-Hill, 1986.

[17] R. H. Barker, "Group synchronizing of binary digital systems," in *Communication Theory*, W. Jackson, Ed. Academic Press, 1953, pp. 273–287.

[18] P.-F. Li, "Binary sequences with low autocorrelation," *ALOHA System Tech. Rep. B86-1*, University of Hawaii, Honolulu, May 1986.

[19] N. Abramson, "Spread ALOHA for VSAT's," *ALOHA System Tech. Rep.* B86-4, University of Hawaii, Honolulu, June, 1986.

Part 5
Applications

Ethernet: Distributed Packet Switching for Local Computer Networks

by Robert M. Metcalfe and David R. Boggs

Abstract: Ethernet is a branching broadcast communication system for carrying digital data packets among locally distributed computing stations. The packet transport mechanism provided by Ethernet has been used to build systems which can be viewed as either local computer networks or loosely coupled multiprocessors.

An Ethernet's shared communication facility, its Ether, is a passive broadcast medium with no central control. Coordination of access to the Ether for packet broadcasts is distributed among the contending transmitting stations using controlled statistical arbitration. Switching of packets to their destinations on the Ether is distributed among the receiving stations using packet address recognition.

Design principles and implementation are described based on experience with an operating Ethernet of 100 nodes along a kilometer of coaxial cable. A model for estimating performance under heavy loads and a packet protocol for error-controlled communication are included for completeness.

1. Background

One can characterize distributed computing as a spectrum of activities varying in the degree of decentralization, with one extreme being remote computer networking and the other extreme being multiprocessing. Remote computer networking is the loose interconnection of previously isolated, widely separated, and rather large computing systems. Multiprocessing is the construction of previously monolithic and serial computing systems from increasingly numerous and smaller pieces computing in parallel. Near the middle of this spectrum is local networking, the interconnection of computers to gain the resource sharing of computer networking and the parallelism of multiprocessing.

The separation between computers and the associated bit rate of their communication can be used to divide the distributed computing spectrum into broad activities. The product of separation and bit rate, now about 1 gigabit-meter per second (1 Gbmps), is an indication of the limit of current communication technology and can be expected to increase with time.

Key words and phrases: computer networks, packet switching, multiprocessing, distributed control, distributed computing, broadcast communication, statistical arbitration

Activity	Separation	Bit Rate
Remote Networks	>10 km	<.1 Mbps
Local Networks	10-.1 km	.1-10 Mbps
Multiprocessors	<.1 km	>10 Mbps

1.1 Remote Computer Networking

Computer networking evolved from telecommunications, *terminal-computer* communication, where the object was to connect remote terminals to a central computing facility. As the need for *computer-computer* interconnection grew, computers themselves were used to provide communication [Baran, 1964; Rustin, 1972; Abramson, 1975]. Communication *using* computers as packet switches [Roberts, 1970; Heart, 1970, 1973; Metcalfe 1973b] and communications *among* computers for resource sharing [Crocker, 1972; Thomas, 1973] were both advanced by the development of the Arpa Computer Network.

The Aloha Network at the University of Hawaii was originally developed to apply packet radio techniques for communication between a central computer and its terminals scattered among the Hawaiian Islands [Abramson, 1970, 1975]. Many of the terminals are now minicomputers communicating among themselves using the Aloha Network's Menehune as a packet switch. The Menehune and an Arpanet Imp are now connected providing terminals on the Aloha Network access to computing resources on the U.S. mainland.

Just as computer networks have grown across continents and oceans to interconnect major computing facilities around the world, they are now growing down corridors and between buildings to interconnect minicomputers in offices and laboratories [Ashenhurst, 1975; Willard, 1973; Fraser, 1975; Farber, 1973, 1975].

1.2 Multiprocessing

Multiprocessing first took the form of connecting an I/O controller to a large central computer; IBM's ASP is a classic example [Rustin, 1972]. Next, multiple central processors were connected to a common memory to provide more power for compute-bound applications [Thornton, 1970]. For certain of these applications, more exotic multiprocessor architectures such as Illiac IV were introduced [Barnes, 1968].

More recently minicomputers have been connected in multiprocessor configurations for economy, reliability, and increased system modularity [Wulf, 1972; Ornstein, 1975]. The trend has been toward decentralization for reliability; loosely coupled multiprocessor systems depend less on shared central memory and more on *thin wires* for interprocess communication with increased component isolation [Metcalfe, 1972a, 1973b]. With the continued thinning of interprocessor communication for reliability and the development of distributable applications, multiprocessing is gradually approaching a local form of distributed computing.

1.3 Local Computer Networking

Ethernet shares many objectives with other local networks such as Mitre's Mitrix, Bell Telephone Laboratory's Spider, and U.C. Irvine's Distributed Computing System (DCS) [Willard, 1973; Fraser, 1975; Farber, 1973, 1975]. Prototypes of all four local networking schemes operate at bit rates between one and three megabits per second. Mitrix and Spider have a central minicomputer for switching and bandwidth allocation while DCS and Ethernet use distributed control. Spider and DCS use a ring communication path, Mitrix uses off-the-shelf CATV technology to implement two one-way busses, and our experimental Ethernet uses a branching two-way passive bus. Differences among these systems are due to differences among their intended applications, differences among the cost constraints under which trade-offs were made, and differences of opinion among researchers.

Before going into a detailed description of Ethernet, we offer the following overview (see Figure 1).

2. System Summary

Ethernet is a system for local communication among computing stations. Our experimental Ethernet uses tapped coaxial cables to carry variable-length digital data packets among, for example, personal minicomputers, printing facilities, large file storage devices, magnetic tape backup stations, larger central computers, and longer-haul communication equipment.

The shared communication facility, a branching Ether, is passive. A station's Ethernet interface connects bit-serially through an interface cable to a transceiver which in turn taps into the passing Ether. A packet is broadcast onto the Ether, is heard by all stations, and is copied from the Ether by destinations which select it according to the packet's leading address bits. This is broadcast packet switching and should be distinguished from store-and-forward packet switching in which routing is performed by intermediate processing elements. To handle the demands of growth, an Ethernet can be extended using packet repeaters for signal regeneration, packet filters for traffic localization, and packet gateways for internetwork address extension.

Control is completely distributed among stations with packet transmissions coordinated through statistical arbitration. Transmissions initiated by a station defer to any which may already be in progress. Once started, if interference with other packets is detected, a transmission is aborted and rescheduled by its source station. After a certain period of interference-free transmission, a packet is heard by all stations and will run to completion without interference. Ethernet controllers in colliding stations each generate random retransmission intervals to avoid repeated collisions. The mean of a packet's retransmission intervals is adjusted as a function of collision history to keep Ether utilization near the optimum with changing network load.

Even when transmitted without source-detected interference, a packet may still not reach its destination without error; thus, packets are delivered *only with high probability.* Stations requiring a residual error rate lower than that provided by the bare Ethernet packet transport mechanism must follow mutually agreed upon packet protocols.

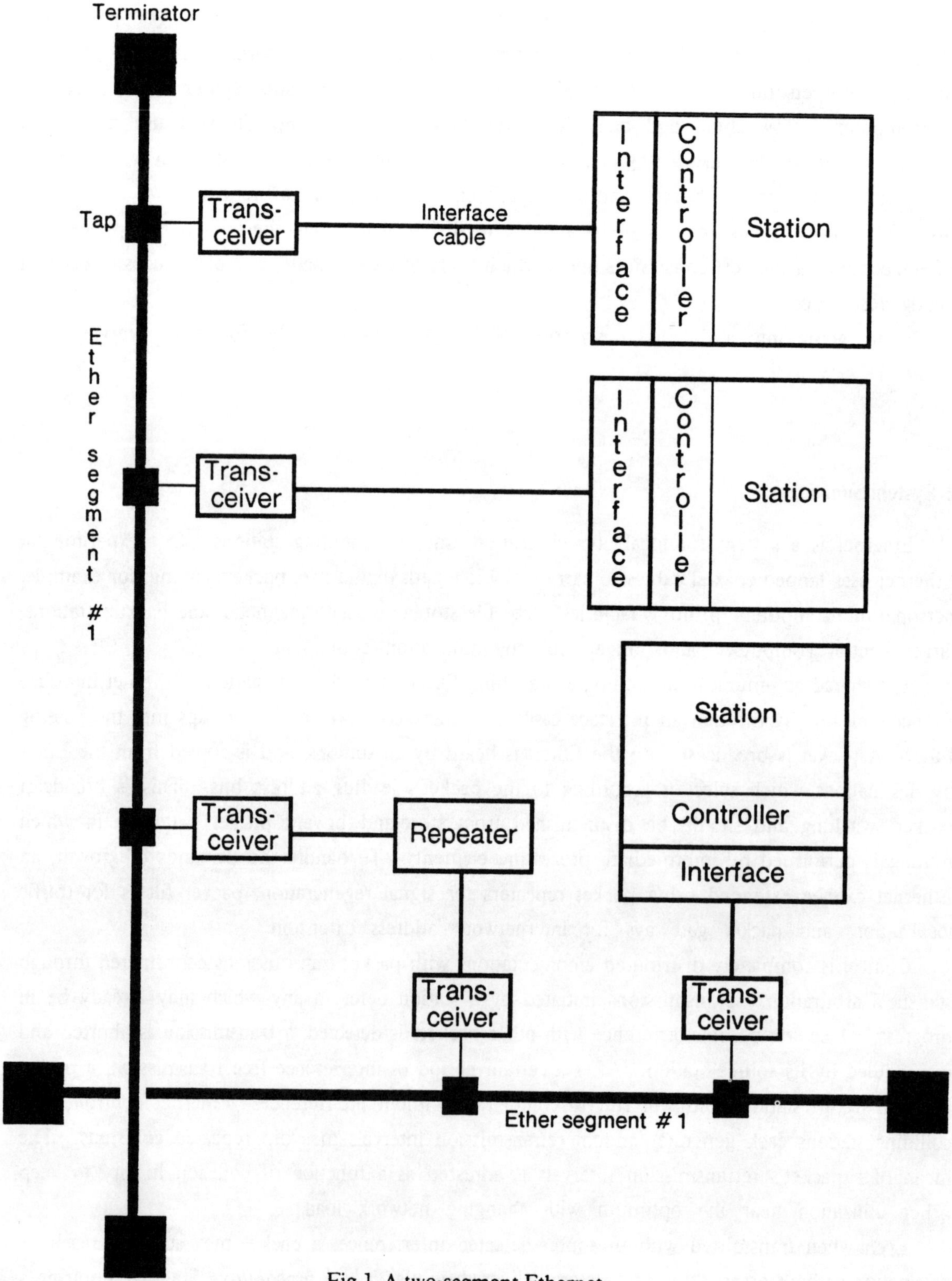

Fig.1. A two-segment Ethernet.

3. Design Principles

Our object is to design a communication system which can grow smoothly to accommodate several buildings full of personal computers and the facilities needed for their support.

Like the computing stations to be connected, the communication system must be inexpensive. We choose to distribute control of the communications facility among the communicating computers to eliminate the reliability problems of an active central controller, to avoid creating a bottleneck in a system rich in parallelism, and to reduce the fixed costs which make small systems uneconomical.

Ethernet design started with the basic idea of packet collision and retransmission developed in the Aloha Network [Abramson, 1970]. We expected that, like the Aloha Network, Ethernets would carry bursty traffic so that conventional synchronous time-division multiplexing (STDM) would be inefficient [Roberts, 1970; Abramson, 1970, 1975; Metcalfe, 1973b]. We saw promise in the Aloha approach to distributed control of radio channel multiplexing and hoped that it could be applied effectively with media suited to local computer communication. With several innovations of our own, the promise is realized.

Ethernet is named for the historical *luminiferous ether* through which electromagnetic radiations were once alleged to propagate. Like an Aloha radio transmitter, an Ethernet transmitter broadcasts completely-addressed transmitter-synchronous bit sequences called packets onto the Ether and hopes that they are heard by the intended receivers. The Ether is a logically passive medium for the propagation of digital signals and can be constructed using any number of media including coaxial cables, twisted pairs, and optical fibers.

3.1 Topology

We cannot afford the redundant connections and dynamic routing of store-and-forward packet switching to assure reliable communication, so we choose to achieve reliability through simplicity. We choose to make the shared communication facility passive so that the failure of an active element will tend to affect the communications of only a single station. The layout and changing needs of office and laboratory buildings leads us to pick a network topology with the potential for convenient incremental extention and reconfiguration with minimal service disruption.

The topology of the Ethernet is that of an unrooted tree. It is a *tree* so that the Ether can branch at the entrance to a building's corridor, yet avoid multipath interference. There must be only one path through the Ether between any source and destination; if more than one path were to exist, a transmission would interfere with itself, repeatedly arriving at its intended destination having travelled by paths of different length. The Ether is *unrooted* because it can be extended from any of its points in any direction. Any station wishing to join an Ethernet taps into the Ether at the nearest convenient point.

Looking at the relationship of interconnection and control, we see that Ethernet is the dual of a star network. Rather than *distributed* interconnection through many separate links and *central* control in a switching node, as in a star network, the Ethernet has *central* interconnection through the Ether and *distributed* control among its stations.

Unlike an Aloha Network which is a star network with an outgoing broadcast channel and an incoming multi-access channel, an Ethernet supports many-to-many communication with a single broadcast multi-access channel.

3.2 Control

Sharing of the Ether is controlled in such a way that it is not only possible but probable that two or more stations will attempt to transmit a packet at roughly the same time. Packets which overlap in time on the Ether are said to *collide*; they interfere so as to be unrecognizable by a receiver. A station recovers from a detected collision by abandoning the attempt and retransmitting the packet after some dynamically-chosen random time period. Arbitration of conflicting transmission demands is both distributed and statistical.

When the Ether is largely unused, a station transmits its packets at will, the packets are received without error, and all is well. As more stations begin to transmit, the rate of packet interference increases. Ethernet controllers in each station are built to adjust the mean retransmission interval in proportion to the frequency of collisions; sharing of the Ether among competing station-station transmissions is thereby kept near the optimum [Metcalfe, 1973a, 1973b].

A degree of cooperation among the stations is required to share the Ether equitably. In demanding applications certain stations might usefully take transmission priority through some systematic violation of equity rules. A station could usurp the Ether by not adjusting its retransmission interval with increasing traffic or by sending very large packets. Both practices are now prohibited by low-level software in each station.

3.3 Addressing

Each packet has a source and destination, both of which are identified in the packet's header. A packet placed on the Ether eventually propagates to all stations. Any station can copy a packet from the Ether into its local memory, but normally only an active destination station matching its address in the packet's header will do so as the packet passes. By convention, a zero destination address is a wildcard and matches all addresses; a packet with a destination of zero is called a *broadcast packet*.

3.4 Reliability

An Ethernet is probabilistic. Packets may be lost due to interference with other packets, impulse noise on the Ether, an inactive receiver at a packet's intended destination, or purposeful discard. Protocols used to communicate through an Ethernet must assume that packets will be received correctly at intended destinations *only with high probability.*

An Ethernet gives its *best efforts* to transmit packets successfully, but it is the responsibility of processes in the source and destination stations to take the precautions necessary to assure reliable communication of the quality they themselves desire [Metcalfe, 1972a, 1973b]. Recognizing the costliness and dangers of promising "error-free" communication, we refrain from guaranteeing reliable delivery of any single packet to get both economy of transmission and high reliability averaged over many packets [Metcalfe, 1973b]. Removing the responsibility for reliable communication from the packet transport mechanism allows us to tailor reliability to the application and to place error recovery where it will do the most good. This policy becomes more important as Ethernets are interconnected in a hierarchy of networks through which packets must travel farther and suffer greater risks.

3.5 Mechanisms

A station connects to the Ether with a *tap* and a *transceiver.* A tap is a device for phyically connecting to the Ether while disturbing its transmission characteristics as little as possible. The design of the transceiver must be an exercise in paranoia. Precautions must be taken to insure that likely failures in the transceiver or station do not result in pollution of the Ether. In particular, removing power from the transceiver should cause it to disconnect from the Ether.

Five mechanisms are provided in our experimental Ethernet for reducing the probability and cost of losing a packet. These are (1) carrier detection, (2) interference detection, (3) packet error detection, (4) truncated packet filtering, and (5) collision consensus enforcement.

3.5.1 Carrier Detection. As a packet's bits are placed on the Ether by a station, they are phase encoded, (like bits on a magnetic tape) which guarantees that there is at least one transition on the Ether during each bit time. The passing of a packet on the Ether can therefore be detected by listening for its transitions. To use a radio analogy, we speak of the presence of *carrier* as a packet passes a transceiver. Because a station can sense the carrier of a passing packet, it can delay sending one of its own until the detected packet passes safely. The Aloha Network does not have carrier detection and consequently suffers a substantially higher collision rate. Without carrier detection, efficient use of the Ether would decrease with increasing packet length. In Section 6 below, we show that with carrier detection, Ether efficiency increases with increasing packet length.

With carrier detection we are able to implement *deference:* no station will start transmitting while hearing carrier. With deference comes *acquisition:* once a packet transmission has been in progress for an Ether end-to-end propagation time, all stations are hearing carrier and are deferring; the Ether has been acquired and the transmission will complete without an interfering collision.

With carrier detection, collisions should occur only when two or more stations find the Ether silent and begin transmitting simultaneously: within an Ether end-to-end propagation time. This will almost always happen immediately after a packet transmission during which two or more stations were deferring. Because stations do not now randomize after deferring, when the transmission terminates, the waiting stations pile on together, collide, randomize, and retransmit.

3.5.2 Interference Detection. Each transceiver has an interference detector. Interference is indicated when the transceiver notices a difference between the value of the bit it is receiving from the Ether and the value of the bit it is attempting to transmit.

Interference detection has three advantages. First, a station detecting a collision knows that its packet has been damaged. The packet can be scheduled for retransmission immediately, avoiding a long acknowledgement timeout. Second, interference periods on the Ether are limited to a maximum of one round trip time. Colliding packets in the the Aloha Network run to completion, but the truncated packets resulting from Ethernet collisions waste only a small fraction of a packet time on the Ether. Third, the frequency of detected interference is used to estimate Ether traffic for adjusting retransmission intervals and optimizing channel efficiency.

3.5.3 Packet Error Detection. As a packet is placed on the Ether, a checksum is computed and appended. As the packet is read from the Ether, the checksum is recomputed. Packets which do not carry a consistent checksum are discarded. In this way transmission errors, impulse noise errors, and errors due to undetected interference are caught at a packet's destination.

3.5.4 Truncated Packet Filtering. Interference detection and deference cause most collisions to result in *truncated packets* of only a few bits; colliding stations detect interference and abort transmission within an Ether round-trip time. To reduce the processing load that the rejection of such obviously damaged packets would place on listening station software, truncated packets are filtered out in hardware.

3.5.5 Collision Consensus Enforcement. When a station determines that its transmission is experiencing interference, it momentarily jams the Ether to insure that all other participants in the collision will detect interference and, because of deference, will be forced to abort. Without this *collision consensus enforcement* mechanism, it is possible that the transmitting station which would otherwise be the last to detect a collision might not do so as the other interfering transmissions successively abort and stop interfering. Although the packet may look good to that last transmitter, different path lengths between the colliding transmitters and the intended receiver will cause the packet to arrive damaged.

4. Implementation

Our choices of 1 kilometer, 3 megabits per second, and 256 stations for the parameters of an experimental Ethernet were based on characteristics of the locally-distributed computer communication environment and our assessments of what would be marginally achievable; they were certainly not hard restrictions essential to the Ethernet concept.

We expected that a reasonable maximum network size would be on the order of 1 kilometer of cable. We used this working number to choose among Ethers of varying signal attenuation and to design transceivers with appropriate power and sensitivity.

The dominant station on our experimental Ethernet is a minicomputer for which 3 megabits per second is a convenient data transfer rate. By keeping the peak rate well below that of the computer's path to main memory, we reduce the need for expensive special-purpose packet buffering in our Ethernet interfaces. By keeping the peak rate as high as is convenient, we provide for larger numbers of stations and more ambitious multiprocessing communications applications.

To expedite low-level packet handling among 256 stations, we allocate the first 8-bit byte of the packet to be the destination address field and the second byte to be the source address field (see Figure 2). 256 is a number small enough to allow each station to get an adequate share of the available bandwidth and approaches the limit of what we can achieve with current techniques for tapping cables. 256 is only a convenient number for the lowest level of protocol; higher levels can accomodate extended address spaces with additional fields inside the packet and software to interpret them.

Our experimental Ethernet implementation has four major parts: the Ether, transceivers, interfaces, and controllers (see Figure 1).

4.1 Ether

We chose to implement our experimental Ether using low-loss coaxial cable with off-the-shelf CATV taps and connectors. It is possible to mix Ethers on a single Ethernet; we use a smaller-diameter coax for convenient connection within station clusters and a larger-diameter coax for low-loss runs between clusters. The cost of coaxial cable Ether is insignificant relative to the cost of the distributed computing systems supported by Ethernet.

4.2 Transceivers

Our experimental transceivers can drive a kilometer of coaxial cable Ether tapped by 256 stations transmitting at 3 megabits per second. The transceivers can *endure* (i.e., work after) sustained direct shorting, improper termination of the Ether, and simultaneous drive by all 256 stations; they can *tolerate* (i.e., work during) ground differentials and everyday electrical noise, from typewriters or electric drills, encountered when stations are separated by as much as a kilometer.

An Ethernet transceiver attaches directly to the Ether which passes by in the ceiling or under the floor. It is powered and controlled through 5 twisted pairs in an interface cable carrying transmit data, receive data, interference detect, and power supply voltages. When unpowered, the transceiver disconnects itself electrically from the Ether. Here is where our fight for reliability is won or lost; a broken transceiver can, but should not, bring down an entire Ethernet. A watchdog timer circuit in each transceiver attempts to prevent pollution of the Ether by shutting down the output stage if it acts suspiciously. For transceiver simplicity we use the Ether's base frequency band, but an Ethernet could be built to use any suitably sized band of a frequency division multiplexed Ether.

Even though our experimental transceivers are very simple and can tolerate only limited signal attenuation, they have proven quite adequate and reliable. A more sophisticated transceiver design might permit passive branching of the Ether and wider station separation.

4.3 Interface

An Ethernet interface serializes and deserializes the parallel data used by its station. There are a number of different stations on our Ethernet; an interface must be built for each kind.

Each interface is equipped with the hardware necessary to compute a 16-bit cyclic redundancy checksum (CRC) on serial data as it is transmitted and received. This checksum only protects against errors in the Ether and specifically not errors in the parallel portions of the interface hardware or station. Higher-level software checksums are recommended for applications in which a higher degree of reliability is required.

A transmitting interface uses a packet buffer address and word count to serialize and phase encode a variable number of 16-bit words which are taken from the station's memory and passed to the transceiver, preceded by a start bit (called SYNC in Figure 2) and followed by the CRC. A receiving interface uses the appearance of carrier to detect the start of a packet and uses the SYNC bit to aquire bit phase. As long as carrier stays on, the interface decodes and deserializes the incoming bit stream depositing 16-bit words in a packet buffer in the station's main memory. When

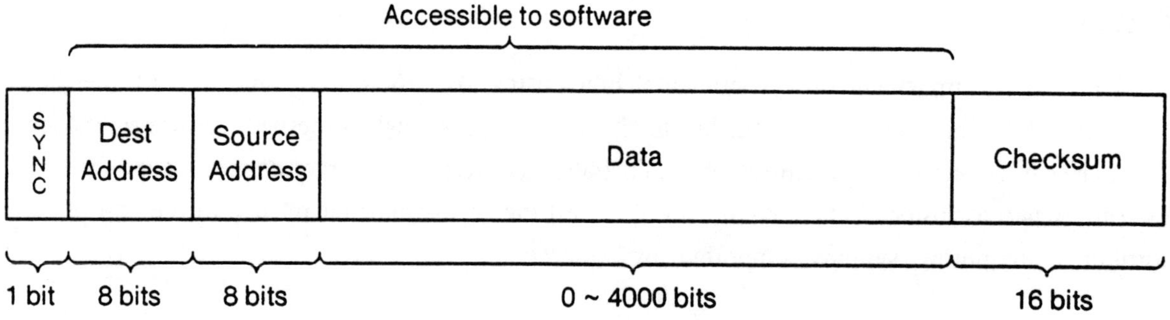

Fig. 2. Ethernet packet layout.

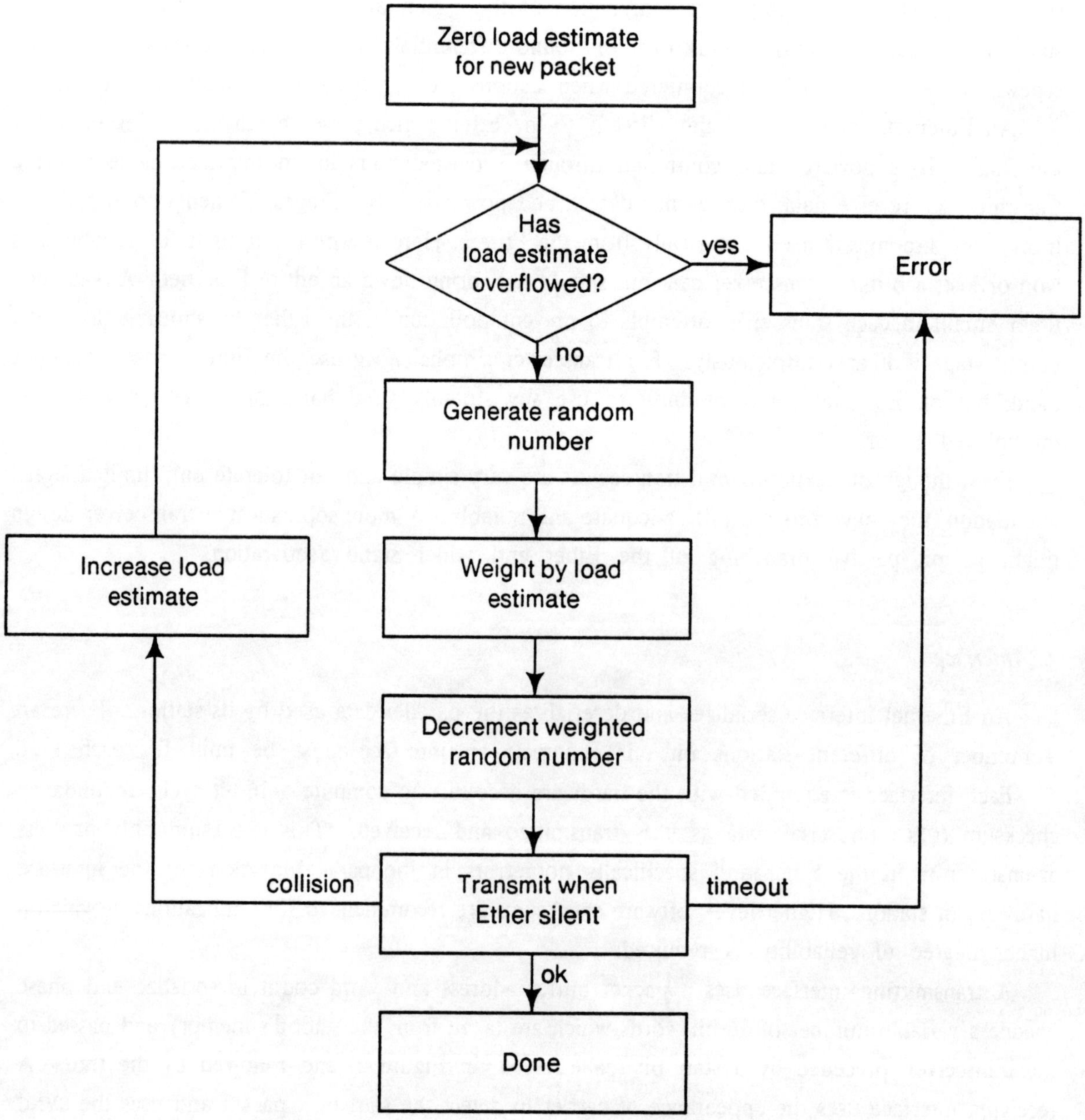

Fig. 3. Collision control algorithm.

carrier goes away, the interface checks that an integral number of 16-bit words has been received and that the CRC is correct. The last word received is assumed to be the CRC and is not copied into the packet buffer.

These interfaces ordinarily include hardware for accepting only those packets with appropriate addresses in their headers. Hardware address filtering helps a station avoid burdensome software packet processing when the Ether is very busy carrying traffic intended for other stations.

4.4 Controller

An Ethernet controller is the station-specific low-level firmware or software for getting packets onto and out of the Ether. When a source-detected collision occurs, it is the source controller's responsibility to generate a new random retransmission interval based on the updated collision count. We have studied a number of algorithms for controlling retransmission rates in stations to maintain Ether efficiency [Metcalfe, 1973a, 1974]. The more practical of these algorithms estimate traffic load using recent collision history.

Retransmission intervals are multiples of a *slot,* the maximum time between starting a transmission and detecting a collision, one end-to-end round trip delay. An Ethernet controller begins transmission of each new packet with a mean retransmission interval of one slot. Each time a transmission attempt ends in collision, the controller delays for an interval of random length with a mean twice that of the previous interval, defers to any passing packet, and then attempts retransmission. This heuristic approximates an algorithm we have called Binary Exponential Backoff (see Figure 3) [Metcalfe, 1974].

When the network is unloaded and collisions are rare, the mean seldom departs from one and retransmissions are prompt. As the traffic load increases, more collisions are experienced, a backlog of packets builds up in the stations, retransmission intervals increase, and retransmission traffic backs off to sustain channel efficiency.

5. Growth

5.1 Signal Cover

One can expand an Ethernet just so far by adding transceivers and Ether. At some point, the transceivers and Ether will be unable to carry the required signals. The *signal cover* can be extended with a simple unbuffered *packet repeater.* In our experimental Ethernet where, because of transceiver simplicity the Ether cannot be branched passively, a simple repeater may join any number of Ether *segments* to enrich the topology while extending the signal cover.

We operate an experimental two-segment packet repeater, but hope to avoid relying on them. In branching the Ether and extending its signal cover, there is a trade-off between using sophisticated transceivers and using repeaters. With increased power and sensitivity, transceivers become more expensive and less reliable. The introduction of repeaters into an Ethernet makes the centrally interconnecting Ether active. The failure of a transceiver will sever the communications of its owner; the failure of a repeater partitions the Ether severing many communications.

5.2 Traffic Cover

One can expand an Ethernet just so far by adding Ether and packet repeaters. At some point the Ether will be so busy that additional stations will just divide more finely the already inadequate bandwidth. The *traffic cover* can be extended with an unbuffered traffic-filtering repeater or *packet filter,* which passes packets from one Ether segment to another only if the destination station is located on the new segment. A packet filter also extends the signal cover.

5.3 Address Cover

One can expand an Ethernet just so far by adding Ether, repeaters, and traffic filters. At some point there will be too many stations to be addressed with the Ethernet's 8-bit addresses. The *address cover* can be extended with *packet gateways* and the software addressing conventions they implement [Cerf, 1974]. Addresses can be expanded in two directions: *down* into the station by adding fields to identify destination ports or processes within a station, and *up* into the internetwork by adding fields to identify destination stations on remote networks. A gateway also extends the traffic and signal covers.

There can be only one repeater or packet filter connecting two Ether segments; a packet repeated onto a segment by multiple repeaters would interfere with itself. However, there is no limit to the number of gateways connecting two segments; a gateway only repeats packets addressed to itself as an intermediary. Failure of the single repeater connecting two segments partitions the network; failure of a gateway need not partition the net if there are paths through other gateways between the segments.

6. Performance

Here is a simple set of formulas with which to characterize the performance expected of an Ethernet when it is heavily loaded. More elaborate analyses and several detailed simulations have been done, but the following simple model has proven very useful in understanding the Ethernet's distributed contention scheme, even when it is loaded beyond expectations [Abramson, 1970; Metcalfe, 1973a, 1973b, 1974; Roberts, 1973; Murthy, 1975].

We develop a simple model of the performance of a loaded Ethernet by examining alternating Ether time periods. The first, called a *transmission interval,* is that during which the Ether has been acquired for a successful packet transmission. The second, called a *contention interval,* is that composed of the retransmission slots of Section 4.4, during which stations attempt to acquire control of the Ether. Because the model's Ethernets are loaded and because stations defer to passing packets before starting transmission, the slots are synchronized by the tail of the preceding acquisition interval. A slot will be empty when no station chooses to attempt transmission in it and it will contain a collision if more than one station attempts to transmit. When a slot contains only one attempted transmission, then the Ether has been acquired for the duration of a packet, the contention interval ends, and a transmission interval begins.

Let P be the number of bits in an Ethernet *packet.* Let C be the peak *capacity* in bits per second, carried on the Ether. Let T be the *time* in seconds of a slot, the number of seconds it takes to detect a collision after starting a transmission. Let us assume that there are Q stations continuously *queued* to transmit a packet; either the acquiring station has a new packet immediately after a successful acquisition or another station comes ready. Note that Q also happens to give the total *offered load* on the network which for this analysis is always 1 or greater. We assume that a queued station attempts to transmit in the current slot with probability $1/Q$, or delays with probability $1-(1/Q)$; this is known to be the optimum statistical decision rule, approximated in Ethernet stations by means of our load-estimating retransmission control algorithms [Metcalfe, 1973a, 1973b].

6.1 Acquisition Probability

We now compute A, the probability that exactly one station attempts a transmission in a slot and therefore *acquires* the Ether. A is $Q*(1/Q)*((1-(1/Q))^{(Q-1)})$; there are Q ways in which one station can choose to transmit (with probability $(1/Q)$) while Q-1 stations choose to wait (with probability $1-(1/Q)$). Simplifying,

$$A = (1-(1/Q))^{(Q-1)}.$$

6.2 Waiting Time

We now compute W, the mean number of slots of *waiting* in a contention interval before a successful acquisition of the Ether by a station's transmission. The probability of waiting no time at all is just A, the probability that one and only one station chooses to transmit in the first slot following a transmission. The probability of waiting 1 slot is $A*(1-A)$; the probability of waiting i slots is $A*((1-A)^i)$. The mean of this geometric distribution is

$$W = (1-A)/A.$$

6.3 Efficiency

We now compute E, that fraction of time the Ether is carrying good packets, the *efficiency*. The Ether's time is divided between transmission intervals and contention intervals. A packet transmission takes P/C seconds. The mean time to acquisition is $W*T$. Therefore, by our simple model,

$$E = (P/C)/((P/C)+(W*T)).$$

Table I presents representative performance figures (i.e., E) for our experimental Ethernet with the indicated packet sizes and number of *continuously queued* stations. The efficiency figures given do not account for inevitable reductions due to headers and control packets nor for losses due to imprecise control of the retransmission parameter $1/Q$; the former is straightforwardly protocol-dependent and the latter requires analysis beyond the scope of this paper. Again, we feel that all of the Ethernets in the table are overloaded; normally loaded Ethernets will usually have a Q much less than 1 and exhibit behavior not covered by this model.

For our calculations, we use a C of 3 megabits per second and a T of 16 microseconds. The slot duration, T, must be long enough to allow a collision to be detected or at least twice the Ether's round trip time. We limit in software the maximum length of our packets to be near 4000 bits to keep the latency of network access down and to permit efficient use of station packet buffer storage.

Table I. Ethernet Efficiency.

Q	$P=4096$	$P=1024$	$P=512$	$P=48$
1	1.0000	1.0000	1.0000	1.0000
2	0.9884	0.9552	0.9143	0.5000
3	0.9857	0.9447	0.8951	0.4444
4	0.9842	0.9396	0.8862	0.4219
5	0.9834	0.9367	0.8810	0.4096
10	0.9818	0.9310	0.8709	0.3874
32	0.9807	0.9272	0.8642	0.3737
64	0.9805	0.9263	0.8627	0.3708
128	0.9804	0.9259	0.8620	0.3693
256	0.9803	0.9257	0.8616	0.3686

For packets whose size is above 4000 bits, the efficiency of our experimental Ethernet stays well above 95 percent. For packets with a size approximating that of a slot, Ethernet efficiency approaches $1/e$, the asymptotic efficiency of a slotted Aloha network [Roberts, 1973].

7. Protocol

There is more to the construction of a viable packet communication system than simply providing the mechanisms for packet transport. Methods for error correction, flow control, process naming, security, and accounting must also be provided through higher level protocols implemented on top of the Ether control protocol decribed in Sections 3 and 4 above [Cerf, 1974; Crocker, 1972; Metcalfe, 1973b; Farber, 1975; Rowe, 1975; Walden, 1972]. Ether control includes packet framing, error detection, addressing and multi-access control; like other line control procedures, Ethernet is used to support numerous network and multiprocessor architectures [SDLC, 1974; SNA, 1975].

Here is a brief description of one simple error-controlling packet protocol. The EFTP (Ethernet File Transfer Protocol) is of interest both because it is relatively easy to understand and implement correctly and because it has dutifully carried many valuable files during the development of more general and efficient protocols.

7.1 General Terminology

In discussing packet protocols, we use the following generally useful terminology. A packet is said to have a *source* and a *destination.* A flow of data is said to have a *sender* and a *receiver,* recognizing that to support a flow of data some packets (typically acknowledgments) will be sourced at the receiver and destined for the sender. A connection is said to have a *listener* and an *initiator* and a service is said to have a *server* and a *user.* It is very useful to treat these as orthogonal descriptors of the participants in a communication. Of course, a server is usually a listener and the source of data-bearing packets is usually the sender.

7.2 EFTP

The first 16 bits of all Ethernet packets contain its interface-interpretable destination and source station addresses, a byte each, in that order (see Figure 2). By software convention, the second 16 bits of all Ethernet packets contain the packet type. Different protocols use disjoint sets of packet types. The EFTP uses 5 packet types: data, ack, abort, end, and endreply. Following the 16-bit type word of an EFTP packet are 16 bits of sequence number, 16 bits of length, optionally some 16-bit data words, and finally a 16-bit software checksum word (see Figure 4). The Ethernet's hardware checksum is present only on the Ether and is not counted at this level of protocol.

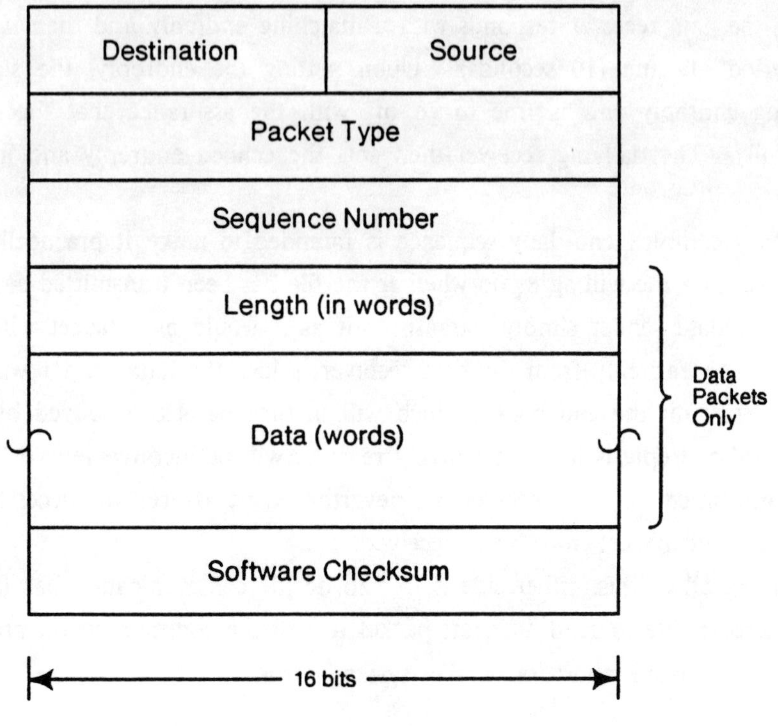

Fig. 4. EFTP packet layout.

It should be obvious that little care has been taken to cram certain fields into just the right number of bits. The emphasis here is on simplicity and ease of programming. Despite this disclaimer, we do feel that it is more advisable to err on the side of spacious fields; try as you may, one field or another will always turn out to be too small.

The software checksum word is used to lower the probability of an undetected error. It serves not only as a backup for the experimental Ethernet's serial hardware 16-bit cyclic redundancy checksum (in Figure 2), but also for protection against failures in parallel data paths within stations which are not checked by the CRC. The checksum used by the EFTP is a 1's complement add and cycle over the entire packet, including header and content data. The checksum can be ignored at the user's peril at either end; the sender may put all 1's (an impossible value) into the checksum word to indicate to the receiver that no checksum was computed.

7.2.1 Data Transfer. The 16-bit words of a file are carried from sending station to receiving station in data packets consecutively numbered from 0. Each data packet is retransmitted periodically by the sender until an ack packet with a matching sequence number is returned from the receiver. The receiver ignores all damaged packets, packets from a station other than the sender, and packets whose sequence number does not match either the expected one or the one preceding. When a packet has the expected sequence number, the packet is acked, its data is accepted as part of the file, and the sequence number is incremented. When a packet arrives with a sequence number one less than that expected, it is acknowledged and discarded; the presumption is that its ack was lost and needs retransmission [Metcalfe, 1973b].

7.2.2 End. When all the data has been transmitted, an end packet is sent with the next consecutive sequence number and then the sender waits for a matching endreply. Having accepted an end packet in sequence, the data receiver responds with a matching endreply and then *dallys* for some reasonably long period of time (10 seconds). Upon getting the endreply, the sending station transmits an echoing endreply and is free to go off with the assurance that the file has been transferred successfully. The dallying receiver then gets the echoed endreply and it too goes off assured.

The comparatively complex end-dally sequence is intended to make it practically certain that the sender and receiver of a file will agree on whether the file has been transmitted correctly. If the end packet is lost, the data sender simply retransmits it as it would any packet with an overdue acknowledgement. If the endreply from the data receiver is lost, the data sender will time out in the same way and retransmit the end packet which will in turn be acknowledged by the dallying receiver. If the echoed endreply is lost, the dallying receiver will be inconvenienced having to wait for it, but when it has timed out, the receiver can nevertheless be assured of successful transfer of the file because the end packet has been received.

At any time during all of this, either side is free to decide communication has failed and just give up; it is considered polite to send an abort packet to end the communication promptly in the event of, say, a user-initiated abort or a file system error.

7.2.3 EFTP Shortcomings. The EFTP has been very useful, but its shortcomings are many. First, the protocol provides only for file transfer from station to station in a single network and specifically not from process to process within stations either on the same network or through a gateway. Second, process rendezvous is degenerate in that there are no mechanisms for finding processes by name or for convenient handling of multiple users by a single server. Third, there is no real flow control. If data arrives at a receiver unable to accept it into its buffers, the data can simply be thrown away with complete assurance that it will be retransmitted eventually. There is no way for a receiver to quench the flow of such wasted transmissions or to expedite retransmission. Fourth, data is transmitted in integral numbers of 16-bit words belonging to unnamed files and thus the EFTP is either terribly restrictive or demands some nested file transfer formats internal to its data words. And fifth, functional generality is lost because the receiver is also the listener and server.

8. Conclusion

Our experience with an operating Ethernet leads us to conclude that our emphasis on distributed control was well placed. By keeping the shared components of the communication system to a minimum and passive, we have achieved a very high level of reliability. Installation and maintenance of our experimental Ethernet has been more than satisfactory. The flexibility of station interconnection provided by broadcast packet switching has encouraged the development of numerous computer networking and multiprocessing applications.

9. Acknowledgements

Our colleagues at the Xerox Palo Alto Research Center, especially Tat C. Lam, Butler W. Lampson, John F. Shoch, and Charles P. Thacker, have contributed in many ways to the evolution of Ethernet ideas and to the construction of the experimental system without which such ideas would be just so much speculation.

10. References

[Abramson, 1970]
N. Abramson, "The Aloha System", *AFIPS Conference Proceedings*, vol. 37, Fall 1970.

[Abramson, 1975]
N. Abramson, F.F. Kuo, *Computer-Communication Networks*, Prentice-Hall, 1975.

[Ashenhurst, 1975]
R.L. Ashenhurst, R.H. Vonderohe, "A Hierarchical Network", *Datamation*, February 1975.

[Baran, 1964]
P. Baran, *On Distributed Communications*, Rand Corporation Memo RM-3420-PR, August 1964.

[Barnes, 1968]
G.H. Barnes, R.M. Brown, M. Kato, D.J. Kuck, D.L. Slotnick, R.A. Stokes, "The Illiac IV Computer", *IEEE Transactions*, C-17, vol. 8, August 1968.

[Binder, 1975]

R. Binder, N. Abramson, F. Kuo, A. Okinaka, D. Wax, "Aloha Packet Broadcasting—A Retrospect", *Proceedings of the National Computer Conference*, May 1975.

[Cerf, 1974]

V.G. Cerf, R.E. Kahn, "A Protocol for Packet Network Intercommunication", *IEEE Transactions on Communications*, vol. COM-22, no. 5, May 1974.

[Computer, 1974a]

"The Shrinking World: Computer Networks and Communications", *Computer*, IEEE Computer Society, February 1974.

[Computer, 1974b]

"Distributed-Function Computer Architectures", *Computer*, IEEE Computer Society, March 1974.

[Crocker, 1972]

S.D. Crocker, J.F. Heafner, R.M. Metcalfe, and J.B. Postel, "Function-Oriented Protocols for the ARPA Computer Network", *AFIPS Conference Proceedings*, vol. 40, May 1972. Reprinted in *Advances in Computer Communications*, edited by W.W. Chu, Artech House Inc., 1974. Reprinted in *Computer Communications*, edited by P.E. Green and R.W. Lucky, IEEE press, 1975.

[Crowther, 1975]

W.R. Crowther, F.E. Heart, A.A. McKenzie, J.M. McQuillan, and D.C. Walden, "Issues in Packet-Switching Network Design", *Proceedings of the National Computer Conference*, May 1975.

[Farber, 1973]

D.J. Farber, et al, "The Distributed Computing System", *Proceedings of the 7th Annual IEEE Computer Society International Conference*, February 1973.

[Farber, 1975]

D.J. Farber, "A Ring Network", *Datamation*, February 1975.

[Fraser, 1975]

A.G. Fraser, "A Virtual Channel Network", *Datamation*, February 1975.

[Heart, 1970]

F.E. Heart, R.E. Kahn, S.M. Ornstein, W.R. Crowther, D.C. Walden, "The Interface Message Processor for the Arpa Computer Network", *AFIPS Conference Proceedings*, vol. 36, May 1970.

[Heart, 1973]

F.E. Heart, S.M. Ornstein, W.R. Crowther, and W.B. Barker, "A New Minicomputer-Multiprocessor for the Arpa Network", *AFIPS Conference Proceedings*, vol. 42, June 1973. Reprinted in *Advances in Computer Communications*, edited by W.W. Chu, Artech House Inc., 1974.

[Kahn, 1975]

R.R. Kahn, "The Organization of Computer Resources into a Packet Radio Network", *Proceedings of the National Computer Conference*, May 1975.

[Metcalfe, 1972a]

R.M. Metcalfe, "Strategies for Interprocess Communication in a Distributed Computing System", *Proceedings of the Symposium on Computer-Communications Networks and Teletraffic*, Polytechnic Press, New York, 1972.

[Metcalfe, 1972b]

R.M. Metcalfe, "Strategies for Operating Systems in Computer Networks", *Proceedings of the ACM National Conference*, August 1972.

[Metcalfe, 1973a]
R.M. Metcalfe, "Steady-State Analysis of a Slotted and Controlled Aloha System with Blocking", *Proceedings of the Sixth Hawaii Conference on System Sciences*, January 1973. Reprinted in the *Sigcomm Review*, January 1975.

[Metcalfe, 1973b]
R.M. Metcalfe, *Packet Communication*, Harvard PhD Thesis, Massachusetts Institute of Technology Project Mac TR-114, December 1973.

[Metcalfe, 1974]
R.M. Metcalfe, "Distributed Algorithms for a Broadcast Queue", talk given at Stanford University in November 1974 and at the University of California at Berkeley in Febraury 1975, paper in preparation.

[Murthy, 1975]
P. Murthy, "Analysis of a Carrier-Sense Random-Access System with Random Packet Lengths", Aloha System Technical Report B75-17, University of Hawaii, May 1975.

[Ornstein, 1975]
S.M. Ornstein, W.R. Crowther, M.F. Kraley, R.D. Bressler, A. Michel, and F.E. Heart, "Pluribus—A Reliable Multiprocessor", *Proceedings of the National Computer Conference*, May 1975.

[Retz, 1975]
D.L. Retz, "Operating System Design Considerations for the Packet Switching Environment", *Proceedings of the National Computer Conference*, May 1975.

[Roberts, 1970]
L. Roberts, B. Wessler, "Computer Network Development to Achieve Resource Sharing", *AFIPS Conference Proceedings*, vol. 36, May 1970.

[Roberts, 1973]
L.G. Roberts, "Capture Effects on Aloha Channels", *Proceedings of the Sixth Hawaii Conference on System Sciences*, January 1973.

[Rowe, 1975]
L.A. Rowe, "The Distributed Computing Operating System", Technical Report Number 66, Department of Information and Computer Science, University of California, Irvine, June 1975.

[Rustin, 1972]
R. Rustin, Editor, *Computer Networks*, Proceedings of the Courant Computer Science Symposium 3, December 1970, Prentice-Hall Series in Automatic Computation, 1972.

[SDLC, 1974]
IBM Synchronous Data Link Control—General Information, IBM Systems Development Division, Publications Center, Department E01, P.O. Box 12195, Research Triangle Park, North Carolina 27709, 1974.

[SNA, 1975]
IBM System Network Architecture—General Infromation, IBM Systems Development Division, Publications Center, Department E01, P.O. Box 12195, Research Triangle Park, North Carolina 27709, 1975.

[Thomas, 1973]
R.H. Thomas, "A Resource Sharing Executive for the Arpanet", *AFIPS Conference Proceedings*, October 1973.

[Thornton, 1970]
J.E. Thornton, *Design of a Computer: the Control Data 6600*, Scott Foresman and Company, 1970.

[Walden, 1972]
D.C. Walden, "A System for Interprocess Communication in a Resource Sharing Computer Network", *Communications of the ACM*, vol. 15, no. 4, April 1972.

[Willard, 1973]
D.G. Willard, "Mitrix: A Sophisticated Digital Cable Communications System", *Proceedings of the National Telecommunications Conference*, November 1973.

[Wulf, 1972]
W. Wulf, R. Levin, "C.mmp—A Multi-Mini-Processor", *AFIPS Conference Proceedings*, Fall 1972.

Measured Capacity of an Ethernet: Myths and Reality

DAVID R. BOGGS, JEFFREY C. MOGUL, AND CHRISTOPHER A. KENT

Abstract—**Ethernet, a 10 Mbit / sec CSMA / CD network, is one of the most successful LAN technologies. Considerable confusion exists as to the actual capacity of an Ethernet, especially since some theoretical studies have examined operating regimes that are not characteristic of actual networks. Based on measurements of an actual implementation, we show that for a wide class of applications, Ethernet is capable of carrying its nominal bandwidth of useful traffic, and allocates the bandwidth fairly. We discuss how implementations can achieve this performance, describe some problems that have arisen in existing implementations, and suggest ways to avoid future problems.**

1. Introduction

Local Area Networks (LANs) have become indispensable in the past few years. Many LAN technologies have been designed, and more than a few of these have made it to market; *Ethernet* [8] is one of the most successful. There are many factors that influence a choice between different technologies, including availability, acceptance as a standard, cost per station and per installation, and ease of maintenance and administration. All of these are difficult or impossible to quantify accurately. Performance characteristics, on the other hand, are easier to quantify, and can thus serve as the basis for religious debates even though the potential performance of a LAN technology may have very little to do with how useful it is.

Considerable confusion exists as to the actual capacity of an Ethernet. This capacity can be determined either by measurement or by analysis. Measurements of intrinsic Ethernet performance (its performance in the limiting case) are fairly meaningless because, at least until recently, most feasible experiments could only measure the performance of host implementations and interfaces, not of the Ethernet per se. If no experiment can "see" the intrinsic performance of the network, then applications probably cannot see it either, and intrinsic performance therefore does not matter. Analyses, on the other hand, have tended to concentrate on the intrinsic performance; first, because software and interface design is too hard to analyze, and second, because Ethernet performance is "interesting" only at very high loads. Therefore, most theoretical studies have examined operating regimes that are not characteristic of actual networks, and their results are also of limited utility in comparing network technologies.

Ethernet works in practice, but allegedly not in theory: some people have sufficiently misunderstood the existing studies of Ethernet performance so as to create a surprisingly resilient mythology. One myth is that an Ethernet is saturated at an offered load of 37%; this is an incorrect reading of the theoretical studies, and is easily disproved in practice. This paper is an attempt to dispel such myths.

We first summarize the theoretical studies relevant to Ethernet, and attempt to extract the important lessons from them. Then, based on measurements of actual implementations, we show that for a wide class of applications, Ethernet is capable of carrying its nominal bandwidth of useful traffic, and allocates the bandwidth fairly. We then discuss how implementations can achieve this performance, describe some problems that have arisen in existing implementations, and suggest ways to avoid future problems.

Reprinted from *Proc. SIGCOMM '88, Symp. Commun. Architectures and Protocols*, pp. 222–234, 1988.

2. What theoretical studies really say

The literature is full of theoretical analyses of carrier-sense multiple access (CSMA), CSMA with collision detection (CSMA/CD), and related access mechanisms. Most of these analyses rely on simplifying assumptions, either for tractability or simply to avoid drowning in the variety of results that would come from exploring the entire parameter space. In spite of the simplifications, these analyses often require mathematical machinery that render them relatively inaccessible to the average practitioner of computer networks.

This inaccessibility, coupled with potential inapplicability due to simplifying assumptions, has led unwary readers into misunderstanding the theoretical studies. As a result, a substantial mythology has arisen to obscure the true properties of the Ethernet.

In this section we examine the theoretical literature pertaining to the Ethernet, and we attempt to dispel some of the mythology by pointing out the applicability of these studies.

2.1. Parameters affecting performance

The performance of an Ethernet depends upon a variety of parameters, in addition to those fixed by the specification. We assume that the reader is familiar with the principles of operation of an Ethernet, specifically the CSMA/CD mechanism [17].

The fixed parameters of the standard Ethernet are described in the specification [8]. These include the following:

- **Bit rate**: The standard Ethernet operates at a bit rate of 10 Mbits/second. Some studies in the literature describe the ''experimental'' Ethernet, which operated at 3 Mbits/second [17]. We will assume the standard Ethernet unless otherwise indicated.
- **Maximum propagation delay**: the maximum round trip delay between any two transmitters is defined to be 464 bit times. This is equivalent to about 5 km of coax cable, much more than a transceiver can drive, but a conceivable delay when repeaters and transceiver multiplexors are used.
- **Maximum jam time**: a transmitter which detects a collision continues to transmit for 32-48 more bit times to insure that the other participants reliably detect the collision.
- **Slot time**: an upper bound on the acquisition time of the network. It must be larger than the sum of the maximum round trip propagation time plus the maximum jam time. This is defined to be 512 bit times (51.2 microseconds).
- **Minimum packet length**: The shortest packet must be no shorter than the slot time, so that if there is a collision it is detectable. This leads to a minimum length of 64 bytes, including a 14-byte header and a 4-byte ''frame check sequence.''
- **Maximum packet length**: This limit, 1518 bytes, bounds the size of buffers that receivers must maintain, and also helps to limit the average access time.
- **Number of hosts**: The Ethernet specification limits the number of hosts per cable segment to 100 (an electrical limit), and the total number of hosts in a multi-segment Ethernet to 1024.

- **Persistence**: The Ethernet is a *1-persistent* CSMA/CD protocol, so-called because a host that becomes ready to transmit when the channel is busy will transmit as soon as the channel is free, with probability 1. Other CSMA/CD protocols have been analyzed; the *non-persistent* protocol waits a random time if the channel is busy. The general case is a *p-persistent* protocol which, after a busy channel becomes free, initiates transmission immediately with probability *p*, and otherwise waits before trying to transmit. Non-persistent protocols lead to higher delays at low loads but perform better at high loads.

In addition to the fixed parameters, performance depends on several parameters determined by the users of the network. These are not entirely independent variables, since intelligent implementors will choose them so as to make best use of the Ethernet. Further, no single choice of these parameters can be viewed as typical; they vary tremendously based on the applications in use.

- **Packet length distribution**: Within the limits of the Ethernet specification, there is substantial freedom to choose packet lengths. In those networks that have been observed, packet length distributions are strongly bimodal, with most packets either near minimum length or near maximum length [20, 11].
- **Actual number of hosts**: The number of hosts on actual Ethernets varies tremendously, and usually is limited for logistical reasons such as the ease with which a single network can be administered. Typical installations have on the order of tens of hosts to a couple of hundred hosts.
- **Arrival rate of packets**: Although the Ethernet allows a host to transmit a minimal length packet every 67.2 microseconds, most hosts are unable to transmit or receive more than a few hundred packets per second; this effectively limits the arrival rate of packets, and it also places a lower bound on the value of average channel access time that actually affects performance.
- **Actual length of cable**: The actual length of an Ethernet may be far shorter than the specification allows. It is not unusual for installations to use a few hundred meters. This means that collisions are detected far sooner than the worst-case propagation delay would imply; channel acquisition may resume as soon as a few microseconds after the beginning of a failed attempt, not the worst-case 51.2 microseconds implied by the specification.

Some of the theoretical analyses make assumptions to provide a simplified model of reality. These may be hard to relate to actual parameters of a real network:

- **Distribution of packet arrivals**: Most theoretical studies assume a simple distribution for the arrival of packets, typically Poisson. Traffic on real networks is seldom so well-behaved; often, there are brief bursts of high load separated by long periods of silence [20].
- **Buffering**: An assumption that hosts have a fixed number of buffers to hold packets waiting for transmission; packets that "arrive" (in the probabilistic sense) for transmission when the buffer is full are discarded. In a real network, flow-control mechanisms limit the arrival rate when the network becomes overloaded; real packets are not discarded, but this is hard to model.

- **Infinite population**: An assumption that even if hosts are blocked waiting for the channel, new packets will arrive for transmission.
- **Slotting**: An assumption that there is a master slot-time clock, with transmissions beginning only at slot boundaries. In a real Ethernet, transmission starts may be separated by less than the slot time; Ethernet is therefore ''unslotted.''
- **''Balanced star'' topology**: An assumption that all hosts are separated from each other by the maximum legal distance; this is essentially impossible in a real installation.
- **Fixed packet size**: It is often much easier to analyze a protocol assuming fixed packet sizes; frequently, the worst-case (minimal packet size) is used.

2.2. How performance is measured

The performance of a network, as with most interesting computer systems, cannot be quantified with a single dimension. Theoretical studies have tended to examine a small set of measures, both for analytic tractability and because it is hard to understand what to measure without knowing what applications are involved.

Performance measures include:

- **Average delay**: the average time it takes to send a packet, measured from the time the host first wishes to acquire the channel.
- **Throughput**: the fraction of the nominal network bandwidth that is actually used for carrying data. Packet headers are considered useful data in calculating this value.
- **Channel capacity**: the maximum achievable throughput for a given set of parameters. Capacity is a function of such parameters as packet length and network length.
- **Fairness**: in a fair network, each host with pending traffic should have an equal probability of acquiring the channel (this is not an equal share of the bandwidth, since hosts use differing packet sizes).
- **Stability**: if the throughput actually drops at high loads then the network is said to be unstable in that region.

2.3. Definitions of offered load

Performance measures are usually described as a function of some parameter of the model; *offered load* is the most common. Several different definitions of offered load are used in the literature.

If one defines the offered load at each host as the fraction of the network bandwidth that the host would use if it had complete access to the network, then G, the offered load on the network as a whole, is simply the sum of the offered loads at each host. Each host's offered load is less than or equal to 1. G can therefore be greater than 1, although the throughput cannot.

Another definition of offered load is the average number Q of hosts waiting to transmit. This concept is what underlies the binary exponential backoff mechanism used to delay retransmission after a collision. If Q hosts are waiting to transmit, one would like each to attempt transmission during a given slot with probability $1/Q$. Essentially, the backoff mechanism is trying to estimate Q.

Although G and Q are both described as the offered load on the network, they are not the same. Q is the offered load as seen by the network, while G is the offered load provided by an idealized customer of the network. Figure 1 depicts this schematically; a *generator* is providing packets at some rate, which are being passed via a *buffer* to the *channel*; if the buffer is full when a packet is generated, the packet is discarded. G is measured at the output of the *generator*, whereas Q is measured at the output of the *buffer*, and so does not count the load contributed by the discarded packets.

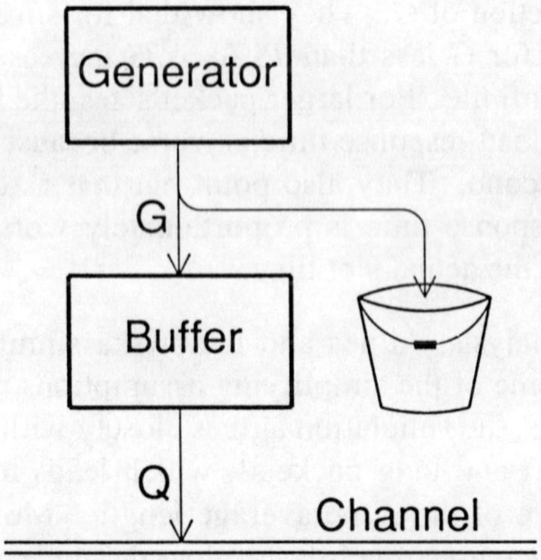

Figure 1: Points for measuring load

While G, at first glance, seems to be a more useful measure from the point of view of the user, in most real systems figure 1 is not an accurate model, because flow-control mechanisms ensure that packets are not discarded in this way. Flow control between the *buffer* and the *generator* is the norm in most applications, except for those, such as packet voice, that are more sensitive to response time than throughput.

Because it is much harder to model the packet arrival distribution of flow-controlled applications, many of the theoretical studies analyze for response time as a function of G, rather than analyzing for throughput as a function of Q.

2.4. A brief guide to the theoretical studies

What follows is a limited survey of some of the theoretical studies relevant to the Ethernet. This is by no means comprehensive, since formal analyses of multiple-access mechanisms are common enough to rate their own section in *IEEE Transactions on Communications*, but we have attempted to include a representative range, listed in order of publication. We have also attempted to extract from these studies information that might be useful in evaluating the performance of the Ethernet for real applications, or for comparing Ethernet with other technologies.

We do not describe the important early work on CSMA without collision detection, such as the papers by Tobagi and Kleinrock [13, 24], and by Metcalfe [15, 16]. The interested reader may refer to these papers for more information.

2.4.1. Metcalfe and Boggs, 1976

In the original paper on the Ethernet [17], Metcalfe and Boggs provide a simple analysis of the 3Mbit/second "experimental" Ethernet invented at Xerox PARC during 1973 and 1974. They compute the throughput of the network as a function of packet length and Q. The throughput remains near 100% for large packets (512 bytes in this case), even with Q as large as 256, but drops to approximately $1/e$ (37%) for minimal-length packets.

2.4.2. Almes and Lazowska, 1979

Almes and Lazowska [1] continue the analysis of the experimental Ethernet. They present values of response time as a function of G. They show that for small packet sizes, the response times stays under 1 millisecond for G less than 75%; as G increases much past that point, the response time is asymptotically infinite. For larger packet sizes, the knee in the curve comes at a higher offered load, but the low-load response time is worse because a single long packet ties up the network for almost a millisecond. They also point out that as the network gets longer, the slot time gets larger, and the response time is proportionately worse. The performance of the network is thus quite sensitive to the actual slot time.

In addition to a theoretical analysis, Almes and Lazowska simulated the performance of an Ethernet as a way of avoiding some of the simplifying assumptions required to keep the analysis tractable. With fixed packet sizes, the simulation agrees closely with the analytical results. They also simulated a mixture of short and long packets[1], which leads to somewhat worse response times than for a fixed packet size of the same average length. More importantly, the standard deviation of the response time increases much faster than the mean with increasing load, which could be a problem for some real-time applications. As with other performance measures, the variance in response time is quite sensitive to the effective slot time.

2.4.3. Tobagi and Hunt, 1980

Tobagi and Hunt [25] provide a more extensive analysis of the throughput-delay characteristics of CSMA/CD networks. Many of the results in this paper are for non-persistent and p-persistent protocols, and so cannot be blindly applied to Ethernets, but it contains a detailed analysis of the effect of bimodal packet-size distributions on throughput, capacity, and delay.

This analysis starts with three parameters: the lengths of the "short" and "long" packets, and the fraction of the packets that are short. The results show that when only a small fraction of the packets are long, the capacity of the network approaches that which can be obtained using a long fixed packet size. Unfortunately, the fraction of the channel capacity available to short packets drops rapidly as the fraction of long packets goes up; this, in turn, increases the average delay for short packets (which are now waiting for long packets to go by, instead of for collisions to be resolved).

---

[1]The "long" packets in this study were only 256 bytes long, although the Pup protocol suite [4] used on the experimental Ethernet used a 576 byte maximum packet.

A large fraction of long packets does allow a somewhat higher total throughput before delays become asymptotically infinite. Further, the delays experienced by the long packets decrease, since fewer total packets are being sent and the amount of time spent resolving collisions is lower.

2.4.4. Bux, 1981

Bux [5] attempts a comparative evaluation of a variety of LAN technologies, including bus systems such as Ethernet, and token-ring systems. Such a comparison has value only if the measure used to distinguish the performance of various systems is relevant to actual users. The measure used in this study is the average packet delay as a function of throughput, and it appears to be biased against Ethernet in that it stresses real-time performance over bulk-transfer performance more than measures used in other studies.

Since throughput is always less than the offered load, curves of delay versus throughput are steeper than those relating delay to offered load, and show knees at lower values of the independent variable. Referring to figure 1, throughput is measured not at the output of the *generator* or the *buffer*, but on the network itself. The curves in [5] show knees for CSMA/CD at throughputs of 0.4, instead of the knees at offered loads of about 0.75. To the unwary, this distinction may not be obvious.

Further, the curves displayed in this study are based on unimodal distributions of packet lengths with means of about 128 bytes. Although Bux has attempted to correct for the effects of analyzing a slotted system, the use only of short packet lengths in this study makes it less applicable to bulk-transfer applications. The comparison favors slotted-ring systems over CSMA/CD systems with equal bandwidth; this is valid only for real-time applications using short packets.

Bux does include an interesting figure that demonstrates the sensitivity of CSMA/CD systems to propagation delay. Packet delays are reasonable even for high throughputs so long as packet lengths are at least two orders of magnitude longer than the propagation delay (expressed in bit-times).

2.4.5. Coyle and Liu, 1983 and 1985

Coyle and Liu [6, 7] refine the analysis of Tobagi and Hunt [25] by using a somewhat more realistic model; specifically, rather than assuming an infinite number of hosts generating an aggregate stream of packets with a Poisson distribution, this analysis assumes a finite number of hosts each generating new packets with an exponential distribution. (They analyze non-persistent CSMA/CD, rather than the 1-persistent Ethernet.)

Coyle and Liu define a stability measure, called ''drift,'' the rate at which the number of backlogged packets is increasing. Drift is a function of the number of users (effectively, the load on the system), and the shape of the drift curve depends upon the retransmission rate. For large numbers of hosts and high retransmission rate, CSMA/CD without backoff is shown to be unstable. In other words, the binary exponential backoff mechanism of Ethernet, which adapts the retransmission rate to the load, is crucial for stability. If the packet length is at least 25 times the collision-detection time, the Ethernet is stable even at extremely high offered loads.

2.4.6. Takagi and Kleinrock, 1985

Takagi and Kleinrock [22] analyze a variety of CSMA protocols under a finite-population assumption. In particular, they provide an analysis of unslotted 1-persistent CSMA/CD. They solve for throughput as a function of G, and one interesting result of their analysis is that not only does the throughput curve depend on the time to abort transmission after detecting a collision, but that for very small values of this parameter, there is a double peak in the curve. In all cases there is a peak at $G = 1$, corresponding to a load exactly matching the capacity of the network. The second peak, present when collisions are detected quickly, comes at offered loads (G) of 100 or more.

2.4.7. Gonsalves and Tobagi, 1986

Gonsalves and Tobagi [10] investigated the effect of several parameters on performance using a simulation of the standard Ethernet. In particular, they examined how the distribution of hosts along the cable affects performance. Previous studies had assumed the balanced star configuration (all hosts separated by the maximum length of the network). This is not a realistic assumption, and so in this simulation several other configurations were examined.

If hosts are uniformly distributed along the cable, or if they are distributed in more than two equal-sized clusters, those hosts in the middle of the cable obtain slightly better throughput and delay than those at the ends. If the clusters are of unequal sizes, hosts in the larger clusters get significantly better service. This is because the hosts within a cluster resolve collisions quickly, whereas the collisions between clusters, which take longer to resolve due to the higher separation, are more likely to be a problem for members of the smaller cluster (since there are more ''distant'' hosts for those hosts than for members of a large cluster).

This is an important result because a number of theoretical analyses explicitly assume that the network is fair by design (for example, that of Apostolopoulos and Protonotarios [2]). Users with real-time applications may find position-dependent unfairness of some unusual Ethernet configurations to be an important effect.

2.4.8. Tasaka, 1986

Tasaka [23] analyzes the effect of buffering on the performance of slotted non-persistent CSMA/CD. As the size of the buffer (measured in packets) increases, fewer packets are dropped due to congestion and thus are retained to use up channel capacity that would otherwise be idle. Increased buffer size leads to higher throughput and average delays at lower offered loads, and causes the knees of these curves to appear at lower offered loads.

2.5. Myths and reality

Although many of the theoretical papers accurately describe the limitations of their approaches, it is still difficult to extract the right lessons from these studies. Thus, many people have learned the wrong lessons, and a variety of myths have been propagated.

Pitfalls for the unwary fall into several categories:

- **Not the same protocol**: the standard Ethernet is an unslotted, 1-persistent, carrier-sense multiple-access method with collision detection and binary exponential back-off. Many analyses assume slotted systems, non-persistent or p-persistent systems, or systems without collision detection. These variations can significantly affect real performance: there are reasons why the Ethernet design is as it is.

- **Unrealistic assumptions**: for analytic tractability, many analyses usually assume balanced-star configurations, infinite populations, unimodal or constant packet lengths, small packet lengths, no buffering, etc. Many (if not most) real-world Ethernets include a relatively small number of hosts, distributed randomly over a cable shorter than the maximum length, with flow-controlled protocols that generate a bimodal distribution of packet lengths.
- **Measuring the wrong dependent variable**: if one is building a real-time system, average packet delay (or perhaps the variance of packet delay) is the critical system parameter. If one is building a distributed file system, throughput may be the critical parameter. Since it is not always possible to optimize for both measures simultaneously, the network designer must keep the intended application clearly in mind.
- **Using the wrong independent variable**: when comparing two studies, it is important to understand what the independent variable is. Most studies use some approximation to offered load, but there are real differences in how offered load is defined.
- **Operating in the wrong regime**: Virtually all of the studies cited in this paper examine the performance of networks at high offered load; that is where the theoretically interesting effects take place. Very few real Ethernets operate in this regime; typical loads are well below 50%, and are often closer to 5%. Even those networks that show high peak loads usually have bursty sources; most of the time the load is much lower. Unless real-time deadlines are important, the capacity of the Ethernet is almost never the bottleneck.

The most well-known myth is that "Ethernets saturate at an offered load of 37%." This is a fair summary of what happens for certain worst-case assumptions, but has very little to do with reality. As we show in section 3, an Ethernet is quite capable of supporting its nominal capacity under realistic conditions. In section 4, we look at how implementors can achieve these conditions.

3. Measurements of a real Ethernet

This section presents measurements of the behavior of an Ethernet under varying combinations of packet lengths, network lengths, and number of hosts.

When many hosts are waiting for a long packet to go by on a busy Ethernet, the instantaneous load on the network routinely exceeds its capacity for short periods. Stressing an Ethernet with an unusual steady overload yields insights into how the network handles the usual momentary overloads encountered in normal operation. Therefore, in these tests we attempt to generate a total offered load that continuously exceeds the capacity of the network.

No real Ethernet should be operated this way ("Don't try this at home"). The Ethernet was designed to handle bursty traffic with a high peak-to-average ratio of bit rates, rather than a continuous high load. Most of the time the Ethernet should be idle, but occasionally the load will spike up, and when this happens, the network must handle it fairly. Fairness is important under heavy load conditions; when the load is light and a host can transmit virtually at will, it is not a concern.

3.1. Previous measurements

Several previous studies have measured the performance of an Ethernet.

Schoch and Hupp [19] reported a number of measurements of the 3 Mbit/second experimental Ethernet. They confirmed that for 512-byte packets, the channel utilization at high offered load stayed near 100% even for 64 hosts, whereas for small packets, the utilization approached $1/e$ for large numbers of hosts. They also found that even for moderately short packets, the network is stable at high offered load: the throughput does not drop as the load increases.

Gonsalves [9] made additional measurements of the experimental Ethernet in order to evaluate its potential performance for packet voice applications. He measured both throughput and delay as a function of offered load; he also studied how load affects the rate at which packets must be discarded because they cannot be delivered before a specific deadline.

3.2. Measurement environment

All of the hosts used in this experiment are Titans, prototype RISC personal computers built by Digital Equipment Corporation's Western Research Laboratory. The cycle time of a Titan is 45 nSec; during these tests the machines were executing about one instruction every 1.5 cycles, or about 15 million instructions per second.

The Titan Ethernet controller was designed by one of the authors. It uses the SEEQ 8023/8003 Ethernet chip set with a custom DMA interface to the Titan's 400 MBit/second ECL backplane. The controller's transmitter sends one packet and then must be serviced by interrupt code before it can send again. This means that the interrupt service latency puts an upper bound on the number of packets that can be generated per second.

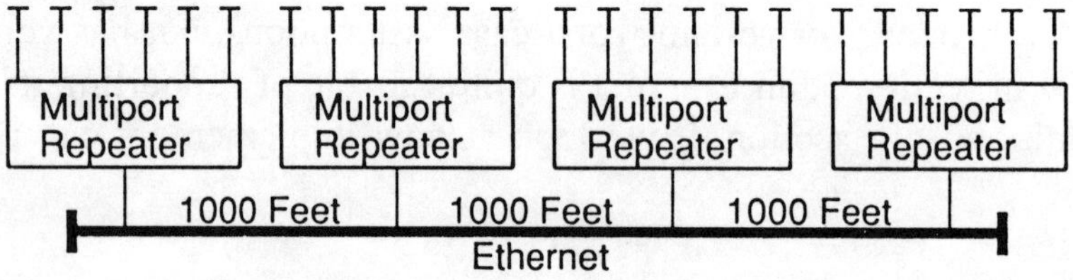

Figure 2: Experimental configuration

During the tests, equal numbers of Titans were connected to one of four DELNI multiport repeaters, whose transceivers were attached to a coaxial cable (see figure 2).

The test software runs in a light-weight operating system written for diagnostics and bootstrap loading. An Ethernet device interrupt takes about 100 microseconds from assertion of interrupt to resumption of user code; this is about 1500 machine instructions.

3.3. Methodology

A full series of tests with N transmitters requires N+1 Titans. One acts as the control machine, coordinating the N generator machines and collecting the data.

Each generator machine listens for a broadcast control packet announcing the beginning of a test and specifying the length of the packets it should generate. After receiving a control packet, it waits 5 seconds, and then transmits packets of the specified length for 20 seconds. During the middle 10 seconds of a test, the Titan measures the performance of its Ethernet transmitter. Next, it waits 5 seconds for the dust to settle and then sends a packet to the control host reporting the statistics gathered during the test. Finally, it goes back and waits for another control packet.

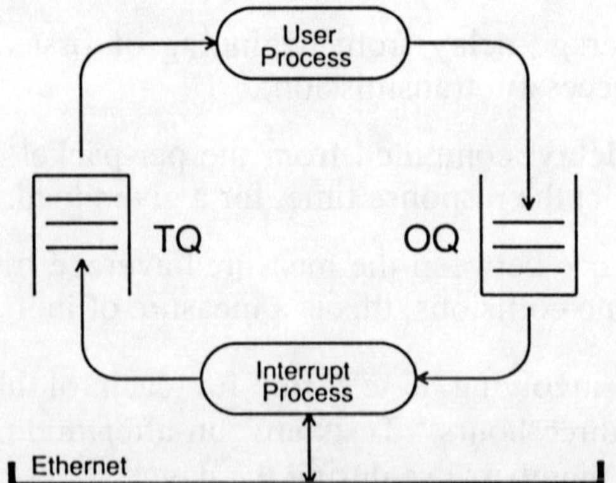

Figure 3: Packet generation process

The program that generates packets for the test load is depicted in figure 3. Each generator Titan allocates 15 packet buffers, builds Ethernet packets in them, and enqueues them on the *transmitted queue* (TQ). A user-level process blocks while the TQ is empty. When a packet buffer appears on the queue, the user process dequeues it, sets the packet length and enqueues it on the Ethernet *output queue* (OQ). If the transmitter is idle, it is started, otherwise nothing more happens until the next transmitter interrupt. When the hardware interrupts at the end of sending a packet, the interrupt routine dequeues the buffer from the OQ and enqueues it on the TQ. Next, if there is another packet buffer on the OQ, it restarts the transmitter. If the time to send a packet is longer than the time to generate one, then the OQ will usually be full, the TQ will usually be empty, and the user process will be idle some of the time. If the time to send a packet is shorter than the time to generate one, then the OQ will usually be empty, the TQ will usually be full, and the user process will be running all of the time.

During a test, each machine counts the number of bits and the number of packets it successfully transmits, and accumulates the sum of the transmission delays and the sum of the delays squared. Each generator machine sends these four numbers to the control machine at the end of a test.

When calculating bit rate, each packet is charged with 24 extra byte times of overhead to account for the 9.6 microsecond interpacket gap (12 byte times), the 64-bit sync preamble (8 bytes), and the 32-bit cyclic redundancy checksum (4 bytes). This method of accounting yields a bit rate of 10.0 MBits/sec when the network is carrying back-to-back packets with no collisions.

The following statistics are computed from the data collected during a test:

- **Bit rate**: total number of useful bits/second (counting overhead), summed over the entire network.

- **Standard deviation of bit rate**: computed from the per-host bit rates, this is a measure of how unfair the network is.
- **Packet rate**: average rate of successful packet generation, summed over the entire network.
- **Standard deviation of packet rate**: computed from the per-host packet rates, this is another measure of unfairness.
- **Transmission delay**: average delay from beginning of first attempt to transmit a packet to the end of its successful transmission.
- **Standard deviation of delay**: computed from the per-packet delays, this indicates how closely one can predict the response time, for a given load.
- **Excess delay**: the difference between the measured average transmission delay and the ideal delay assuming no collisions, this is a measure of inefficiency.

A series of measurements, involving a test run for each of about 200 combinations of parameter values, takes about three hours. Tests are run after midnight because the Titans are the personal computers that lab members use during the day.

All of the graphs presented in this section show performance measures as a function of the number of hosts, N, involved in the test. For large N or for large packet sizes, the network is the bottleneck, and Q (offered load) is approximately equal to N. When N and the packet size are both small, we could not generate packets fast enough to overload the network, so Q is less than one in these cases.

3.4. Maximum rate attainable between a pair of hosts

We first measured the maximum throughput that could be used for a single ''conversation'' between a pair of hosts. This was done by setting up one transmitting host and one receiving host, counting the number of packets sent and received, and verifying that all got through. We found that, for all the packet sizes (and thus packet rates) we tried, the receiver could always keep up with the transmitter. In other words, the rate at which a single host can send packets is what limits the throughput between a single pair of hosts (in these tests).

Because receiver performance was not an issue, during subsequent tests each generator host addressed its packets to itself, and disabled its receiver so that it would not waste cycles receiving its own packets.

3.5. Fixed length packets on a long net

The first set of experiments was conducted with 24 hosts sending fixed length packets on 3000 feet (910 meters) of coaxial cable. Six hosts were connected to each of four multiport repeaters spaced at 1000 foot intervals along the cable. This network configuration is similar to the way many buildings are wired for Ethernet: a long backbone segment with repeaters connecting work groups on short segments. The number of hosts per repeater was kept balanced as the total number of hosts in the test was varied. We did tests with nine different packet lengths, including three that exceed the maximum allowed by the Ethernet specification. The curves are labeled with the packet length in bytes.

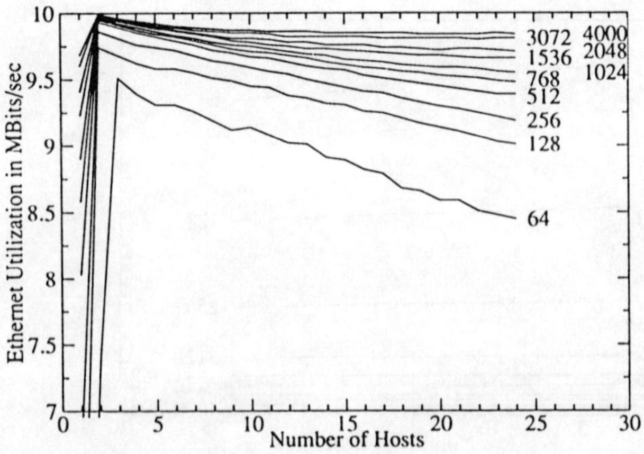

Figure 4: Total bit rate

In figure 4, notice that the bit rate increases with increasing packet size. This is because for larger packets, there are fewer packets per second and so less time is lost at the beginning of packets colliding and backing off. Notice also that the bit rate decreases with increasing number of hosts. This is because for larger numbers of hosts, there are more collisions per packet. (For small numbers of hosts and small packet size, bit rate first increases until the offered load exceeds the network capacity.)

Figure 5 is a measure of the variation in bit rate obtained by individual hosts during a test. (The curves are so hard to distinguish that labelling them would have been pointless.) As the number of hosts increases, the fairness increases. The unfairness for small *N* is due to the intrinsic unfairness of the Ethernet backoff algorithm. As Almes and Lazowska [1] point out, the longer a host has already been waiting, the longer it is likely to delay before attempting to transmit. When $N = 3$, for example, there is a high probability that one host will continually defer to the other two for several collision resolution cycles, and the measured standard deviation becomes a sizeable fraction of the mean bit rate.

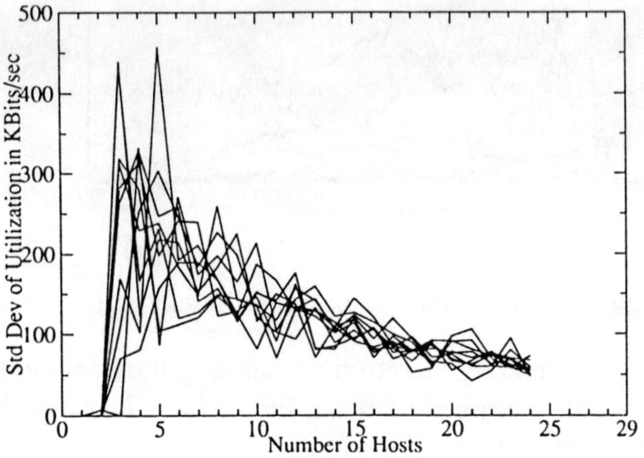

Figure 5: Standard deviation of bit rate

As *N* gets larger, this effect is smoothed out. For 20 hosts, the measured standard deviation is about 20% of mean.

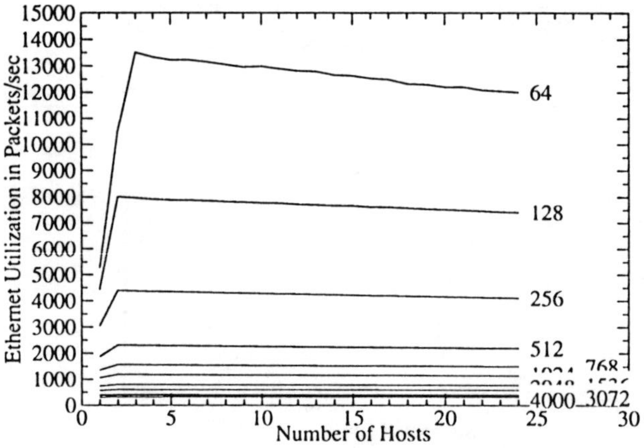

Figure 6: Total packet rate

The theoretical maximum rate for 64 byte packets is about 14,200 packets per second. Figure 6 shows that we were able to achieve about 13,500 packets per second.

The packet rate peaks at two hosts; no collisions occur until at least three hosts are transmitting. There is obviously no contention for the Ethernet when only one host is sending. Because transmitter interrupt service latency is longer than the interpacket gap, two hosts quickly synchronize: each defers to the other and then transmits without contention. When three or more hosts are sending, there are usually two or more hosts ready to transmit, so a collision will occur at the end of each packet.

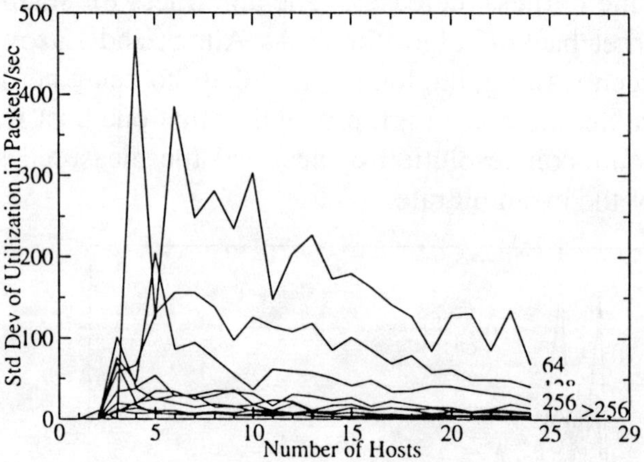

Figure 7: Standard deviation of packet rate

Figure 7 shows the variation in packet generation rate among the hosts in the tests. Note again that the unfairness decreases as the number of hosts increases. The high variance for 64-byte packets may be an experimental artifact.

Figure 8 shows that the average transmission delay increases linearly with increasing number of hosts (i.e. offered load). This contradicts the widely held belief that Ethernet transmission delay increases dramatically when the load on the network exceeds $1/e$ (37%).

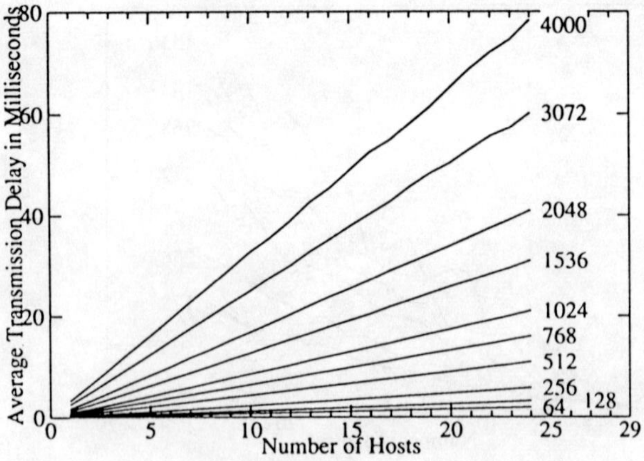

Figure 8: Average transmission delay

The standard deviation of the packet delay, plotted in figure 9 increases linearly with number of hosts. If there are *N* hosts transmitting, then on average for each packet a host sends, it waits while *N-1* other packets are sent. In the worst case, for large packets and many hosts, the standard deviation is about twice the mean.

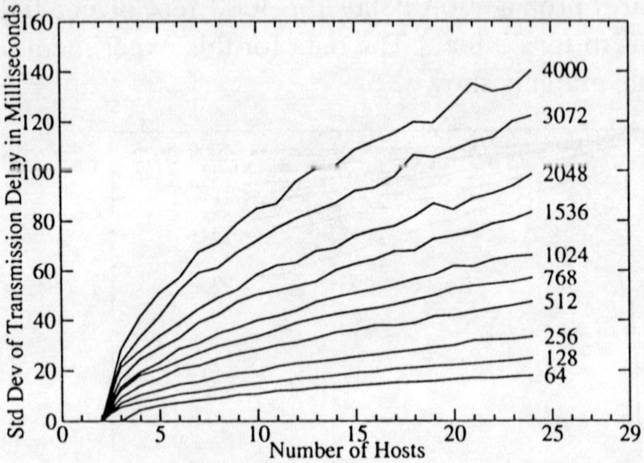

Figure 9: Standard deviation of transmission delay

Figure 10 shows "excess delay", a direct measure of inefficiency. It is derived from the delays plotted in figure 8. The ideal time to send one packet and wait for each other host to send one packet is subtracted from the measured time. The time that remains was lost participating in collisions. Notice that it increases linearly with increasing number of hosts (offered load). When 24 hosts each send 1536-byte packets, it takes about 31 milliseconds for each host to send one packet. Theoretically it should take about 30 mSec; the other 1 mSec (about 3%) is collision overhead. Figure 4 agrees, showing a measured efficiency of about 97% for 1536-byte packets and 24 hosts.

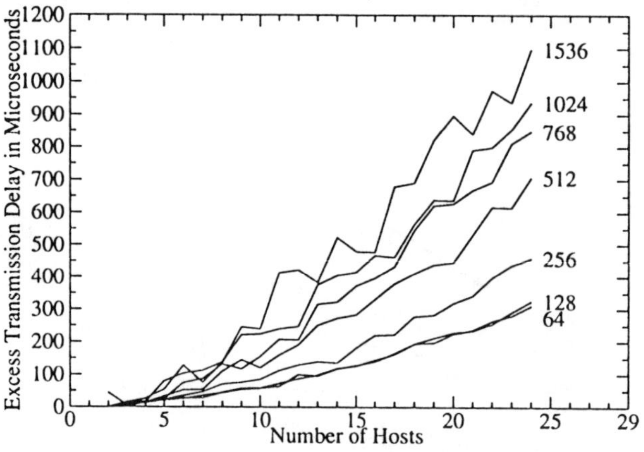

Figure 10: Excess transmission delay

3.6. Fixed length packets on a short net

We also ran a set of tests with 23 hosts on a 20 foot Ethernet. Collision resolution time is a function of round trip propagation delay; the quicker that collisions get resolved, the more efficient the net is. The round trip propagation delay for 3000 feet of coaxial cable is about 10 microseconds; it is essentially zero for 20 feet. The data for this experiment look very similar to the previous set; only the bit rate graph is shown.

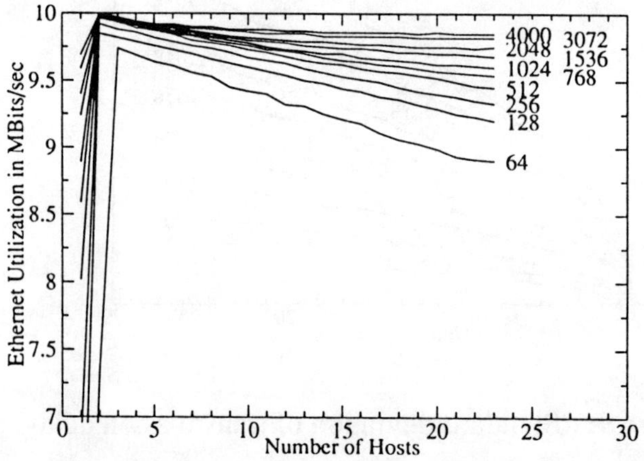

Figure 11: Total bit rate (short net)

Comparing figure 11 with figure 4 shows that network efficiency increases as the collision resolution time decreases. The effect is most pronounced with short packets, where the efficiency drops to only 85% when the packet transmission time is only an order of magnitude larger than the collision resolution time (as in figure 4).

3.7. Bimodal distribution of packet lengths

Because a fixed packet length is not characteristic of most actual Ethernet loads, we repeated our measurements using a number of bimodal length distributions. These distributions were composed of minimal-length (64-byte) and maximal-length (1536-byte) packets in varying ratios, with the packet size chosen randomly while preserving the overall ratio. This approximates real Ethernet traffic, which usually consists of many minimum length packets, some maximum length packets, and a few of intermediate size [20, 11].

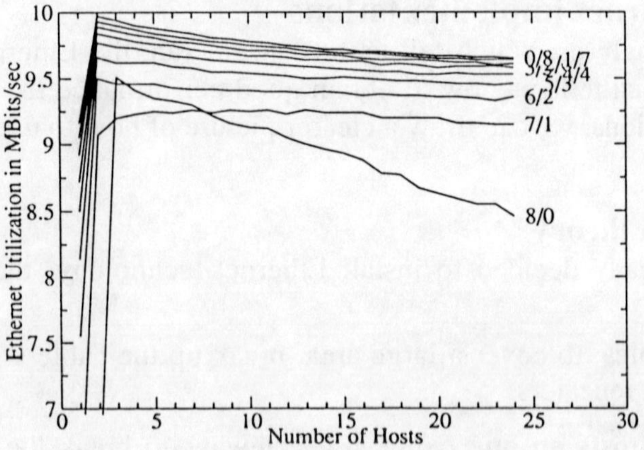

Figure 12: Total bit rate (long net; 4 clusters)

Figure 12 shows the utilizations obtained with bimodal length distributions. The curves are labeled with the ratio of short to long packets; for example, "6/2" means that there were six short packets for every two long packets.) Notice that when only one out of eight packets is long, the utilization is much higher than when all the packets are short. This is as predicted by Tobagi and Hunt [25].

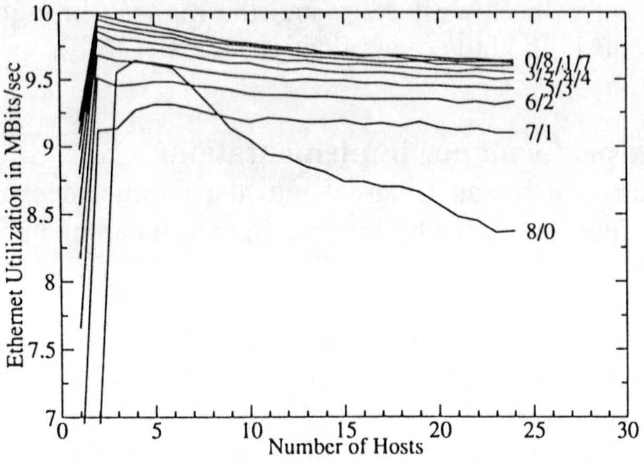

Figure 13: Total bit rate (long net; 2 clusters)

In addition to four groups of hosts at 1000 foot intervals as above, we ran a set of tests with half of the hosts at either end of 3000 feet of cable. The average separation between hosts is greater in this configuration than with four clusters of hosts, so the collision resolution time is increased and the efficiency is decreased. For minimal-length packets and 24 hosts, the average delay is about 1.5% higher, and the total utilization, as shown in figure 13, is about 1.1% lower, in the two-cluster configuration. For maximal-length packets, there is no appreciable difference between the two configurations.

Gonsalves and Tobagi [10] showed, in their simulation, that unequal-sized clusters increase unfairness; we have not yet attempted to measure this effect.

4. Implications for Ethernet implementations

Between the theoretical analyses, which tell us how to operate the Ethernet as a whole, and our practical experience, which tells us how to obtain good performance from actual implementations and on actual applications, we can draw a clearer picture of how to use Ethernets.

4.1. Lessons learned from theory

For the user who has already decided to install Ethernet technology, the theoretical results provide several guidelines:

- **Don't install long cables**: to cover a large area, break up the cable with bridges or gateways (routers), not repeaters.
- **Don't put too many hosts on one cable**: use gateways to break the network into communities of interest, trading higher delay for inter-community traffic for better intra-community response time and throughput.
- **Implement the protocol correctly**: proper collision detection and binary-exponential backoff in interface or host software is essential to good performance.
- **Use the largest possible packet size**: this keeps the packet count down, reducing the likelihood of collision and not incidentally reducing overheads internal to hosts.
- **Don't mix serious real-time and serious bulk-data applications**: it is not possible to simultaneously guarantee the lowest delay and the highest throughput (although for moderate requirements both kinds of applications coexist well).

4.2. Prerequisites for high-performance implementations

It is actually fairly difficult to drive an Ethernet into the regime measured in section 3, let alone into the "unstable" regimes predicted by theory. In fact, it can be hard to get good performance even on a lightly-loaded Ethernet.

Some of the pessimism about Ethernet can be traced to poor implementations. For example, one study [18] found that a typical implementation of an Ethernet interface was unable to provide good performance. This study shows packet rates (for minimal-length packets) of 250 packets per second or less, and throughputs (using large packets) of about 150 Kbytes/second or lower. Although more recent implementations can do better than this, it takes careful design and hard work.

For example, Birrell and Nelson [3] were able to obtain 1 millisecond round-trip times for null Remote Procedure Calls (RPCs) even on a 3 Mbit/second experimental Ethernet. Recently, Van Jacobson has reported achieving 8 Mbits/second using TCP over a standard Ethernet [12]. Unlike the superficially more impressive numbers we show in section 3, these results are for real applications.

Networking performance can be limited by any of several weak links. For example, if the backplane or memory system bandwidth is too low, the bandwidth of the network itself becomes less visible. Also, host processor time is precious; per-packet processing time increases latency and, unless it can be completely overlapped with transmission or reception, decreases throughput. The latency costs of packet processing completely dominate the theoretical channel-access times in the light-load regimes typical of most Ethernet installations. Processor speed often has more influence than network bandwidth on useful throughput (see, for example, Lantz, Nowicki, and Theimer [14].)

When the processor, memory, and software are fast enough to support the full network bandwidth, the network interface can be the weak link. In a high-performance Ethernet interface,

- **Transmitter and receiver performance should be matched**: If the transmitter cannot keep up with the receiver, or vice versa, the ultimate performance suffers, as we found when measuring single-host short-packet rates (see section 3.4).
- **The interface should be able to transmit, and to receive and store, several back-to-back packets without host intervention**: otherwise, bandwidth is wasted on lost packets or on channel idle time.
- **The interface should be able to transmit and receive several packets per interrupt**: interrupt latency is often the bottleneck to efficient handling of packets. The ability to receive back-to-back packets implies the ability to handle them in a batch.

4.3. Problems with certain existing implementations

When Ethernets fail, it is almost never because they have been pushed into instability by a high load of useful traffic. This may be because most implementations are incapable of utilizing the bandwidth of the network. Instead, Ethernets usually fail for one of two reasons: hardware failure (including incorrect designs) or accidental high loads.

Transient hardware failures include open or shorted cables, to which all shared-medium LANs are vulnerable, and ''jabbering'' transmitters (those that send infinitely long packets). Careful hardware maintenance is the only protection against such failures, and it takes considerable preparation and diagnostic skill to locate them; Ethernets are not only ''distributed packet switches'' but ''distributed single points of failure.''

Accidental high loads, usually generated by software bugs, are the most common way that Ethernets are pushed into the overload regime analyzed by theoretical studies. One example of such a bug is distressingly common: a broadcast packet is sent but some of the hosts that receive it mistakenly believe that it should be forwarded, as a broadcast, back onto the cable. The recursive nature of this bug leads to exponential growth in the number of such packets, saturating the network. Unless there is a mechanism for aging packets (such as a hop-count field), it may be impossible to recover the network without shutting down every host.

One design bug that can push a highly-loaded network into overload is to use a linear backoff instead of the exponential backoff specified in the standard. At low loads, a linear backoff will give better service to the host that uses it, especially when set against other hosts that play by the rules. At high loads, an incorrect backoff algorithm can drive the network into instability, making it useless for everyone. For this and other reasons, all implementations should be tested at high offered loads, even if they normally would not be used that way.

Host software that assumes low congestion can also push an Ethernet from high load into overload by retransmitting data too aggressively, especially if acknowledgements are being delayed due to collisions. Retransmission delays in higher protocol levels should be backed off to avoid congestive collapse in any network, not just Ethernets. Hosts should not assume that delays on an Ethernet will always be short; even on a network that is normally lightly-loaded, brief intervals of overload are possible.

4.4. Suitability of Ethernet for specific applications

Both the theoretical studies and our measurements support the use of Ethernet for high-bandwidth applications, especially when response time is not closely constrained. Given that the aggregate bandwidth requirements do not frequently exceed 10 Mbits/second, the Ethernet should not be a bottleneck.

When response time is constrained, then Ethernet may still be a reasonable choice, especially if the number of hosts is relatively small and there are few high-bandwidth applications. For example, Gonsalves [9] found that the experimental (3 Mbit/second) Ethernet was ''capable of supporting about 40 simultaneous 64-Kbps two-way voice conversations with acceptable quality.'' Note that this application can accept occasional lost packets; some real-time applications are more constraining.

The Ethernet may not be the appropriate medium for applications requiring low latency and involving a very large number of simultaneous hosts. One such application is distributed programming with fine-grained remote memory access. Spector [21] was able to obtain reasonable performance on an experimental Ethernet with a small number of hosts, but he suggests that only ring networks are able to support a large amount of parallelism.

5. Summary and conclusions

Our measurements, and a careful reading of the theoretical analyses, show that Ethernet is capable of good performance for typical applications, even at high offered load. The performance of CSMA/CD is sensitive to packet size and the time it takes to detect a collision; users must keep these parameters in mind. At higher bit rates or for longer networks, ring topologies may be the only acceptable approach, but experience with Ethernet proves that at 10 Mbits/second, over a kilometer or so of cable, CSMA/CD is quite successful.

Acknowledgements

John DeTreville provided numerous perceptive comments on an early draft of this paper. Michael Fine, Joel Bartlett and Brian Reid also provided important suggestions. The authors are, of course, solely responsible for any errors in this paper.

We would like to thank the members of the Digital Equipment Corporation Western Research Laboratory (WRL) and Western Software Laboratory (WSL) for their patience. During the process of performing our measurements, we took over 90% of the computing capacity, and on several occasions caused network ''meltdowns'' which made the rest of the computers useless.

Appendix I. All of the Experiment Data

Here are all of the graphs for all of the experiments. Due to space constraints, the conference paper could include only selected graphs.

For completeness, we also include a set of graphs from an experiment on a maximum legal length (45 microsecond round trip delay) network. This experiment was conducted jointly with Robert L. Fink of the Lawrence Berkeley Laboratory. Technical notes describing our experiments with long networks are in preparation by LBL and WRL.

There were five experiements. The first three experiments used fixed length packets on networks with round trip delays of 3, 12 and 45 microseconds. The last two experiments used bimodal length packets on a 12 microsecond net and varied the arrangement of hosts: one experiment with two clusters at the ends of the net, and another with four clusters equally spaced along the net.

Each page contains eight graphs. The X axis of each graph is the number of hosts used to generate the load. The top two graphs are bits per second and standard deviation of bits per second. The upper middle two graphs are packets per second and standard deviation of packets per second. The lower middle two graphs are transmission delay and standard deviation of transmission delay. The lower left graph is the excess transmission delay: the difference between the measured delay and the optimal delay. The lower right graph is the normalized standard deviation of utilization: the standard deviation divided by the mean bit rate.

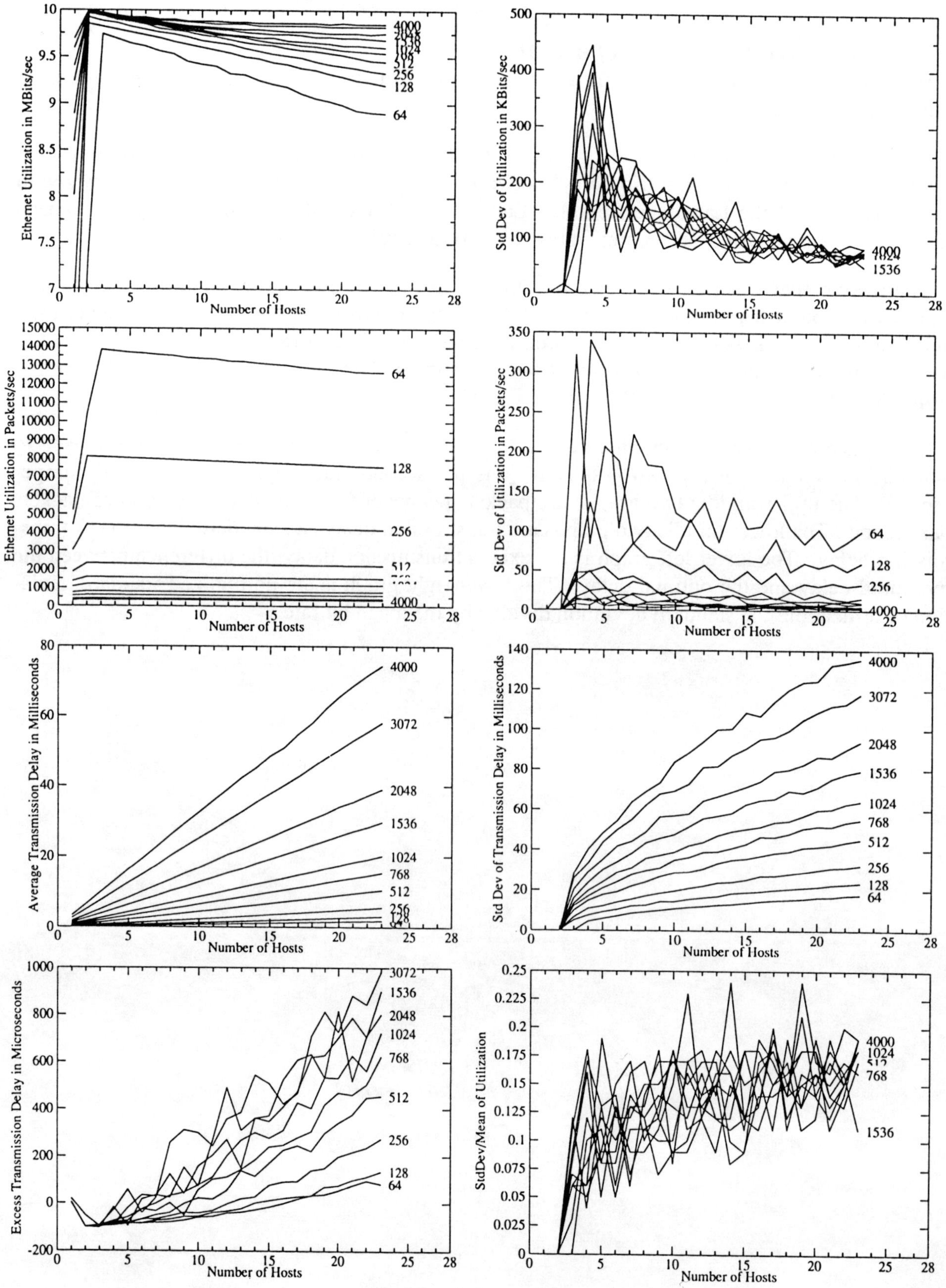

Figure I-1: 3 microsecond net, fixed packet sizes, 1 cluster

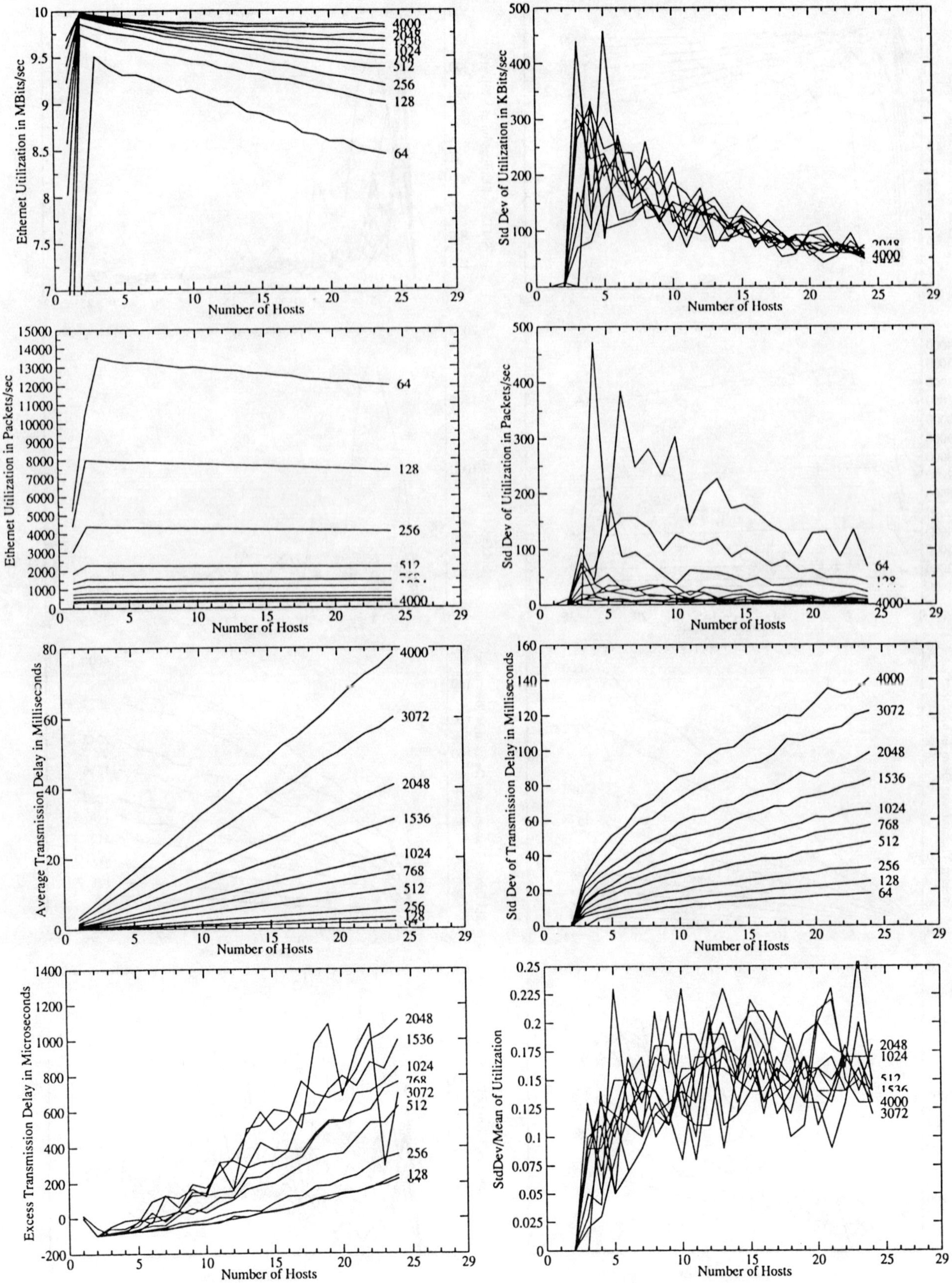

Figure I-2: 12 microsecond net, fixed packet sizes, 4 clusters

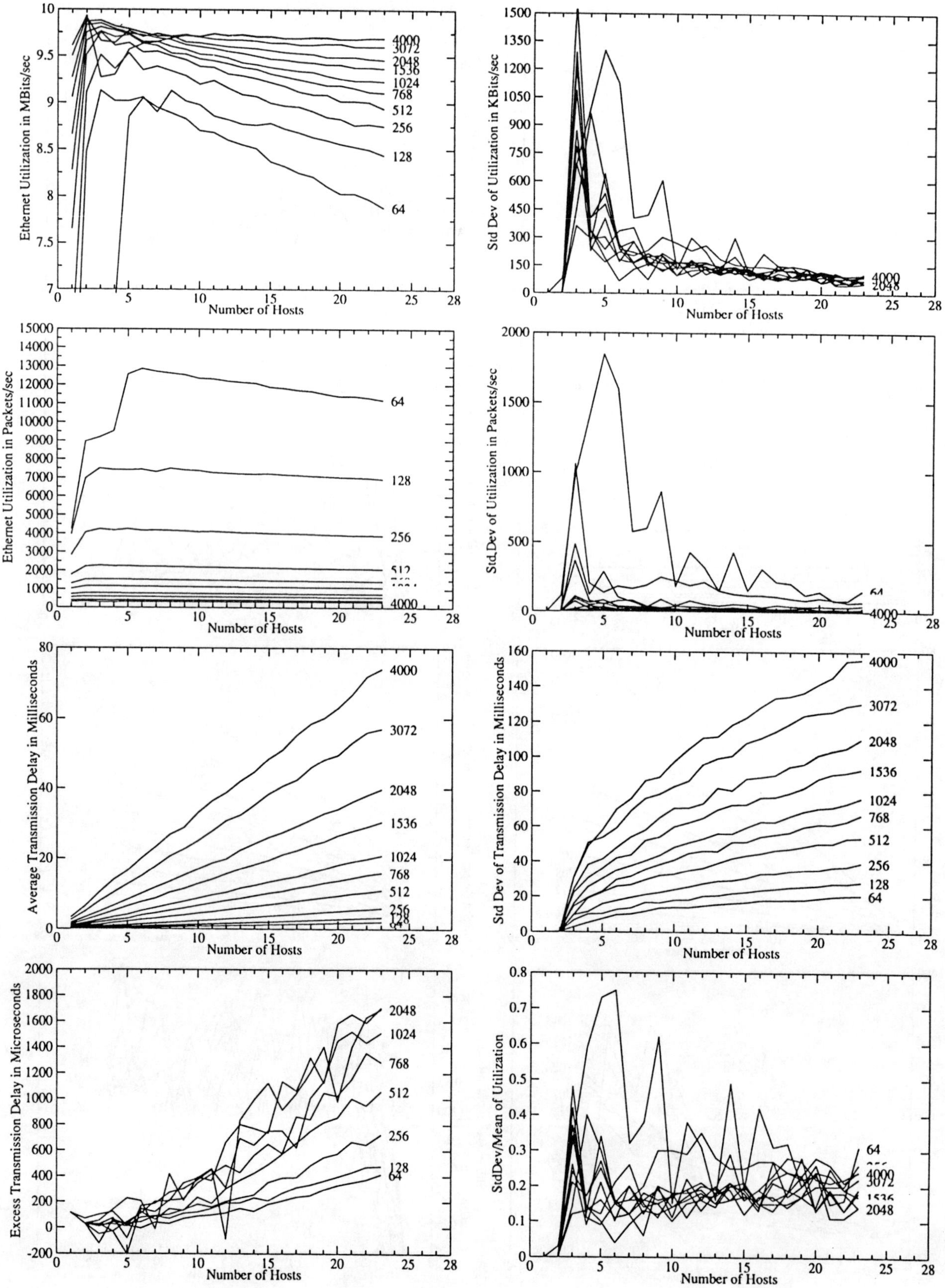

Figure I-3: 45 microsecond net, fixed packet sizes, 2 clusters

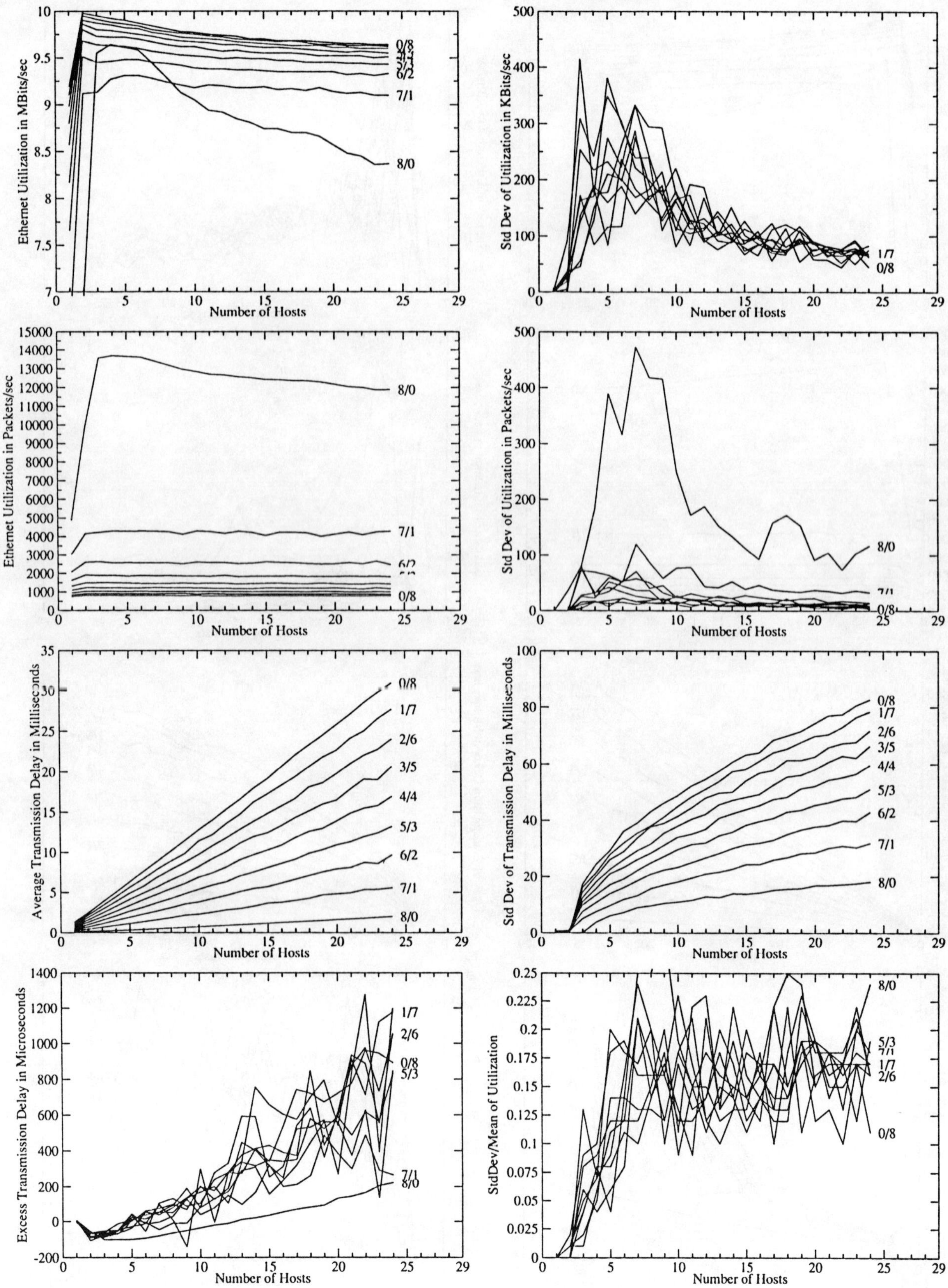

Figure I-4: 12 microsecond net, bimodal packet sizes, 2 clusters

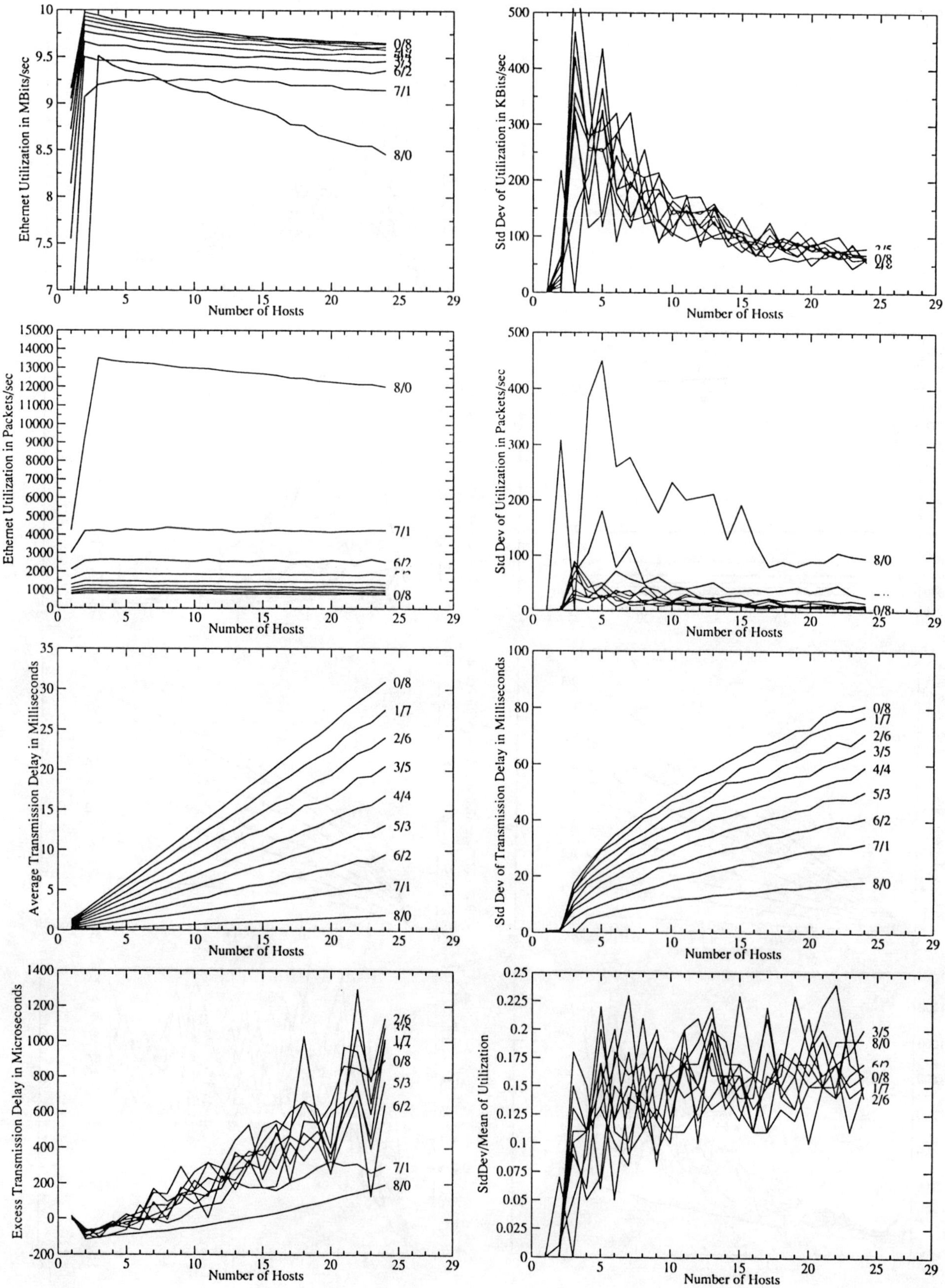

Figure I-5: 12 microsecond net, bimodal packet sizes, 4 clusters

References

[1] Guy T. Almes and Edward D. Lazowska.
The Behaviour of Ethernet-Like Computer Communications Networks.
In *Proceedings of the 7th Symposium on Operating Systems Principles*, pages 66-81. ACM SIGCOMM, Asilomar, California, December, 1979.

[2] Theodore K. Apostolopoulos and Emmanuel N. Protonotarios.
Queueing Analysis of Buffered CSMA/CD Protocols.
IEEE Transactions On Communications COM-34(9):898-905, September, 1986.

[3] Andrew D. Birrell and Bruce Jay Nelson.
Implementing Remote Procedure Calls.
ACM Transactions on Computer Systems 2(1):39-59, February, 1984.

[4] David R. Boggs, John F. Shoch, Edward A. Taft, and Robert M. Metcalfe.
Pup: An internetwork architecture.
IEEE Transactions On Communications COM-28(4):612-624, April, 1980.

[5] Werner Bux.
Local-Area Subnetworks: A Performance Comparison.
IEEE Transactions On Communications COM-29(10):1465-1473, October, 1981.

[6] Edward J. Coyle and Bede Liu.
Finite Population CSMA/CD Networks.
IEEE Transactions On Communications COM-31(11):1247-1251, November, 1983.

[7] Edward J. Coyle and Bede Liu.
A Matrix Representation of CSMA/CD Networks.
IEEE Transactions On Communications COM-33(1):53-64, January, 1985.

[8] *The Ethernet, A Local Area Network: Data Link Layer and Physical Layer Specifications (Version 1.0).*
Digital Equipment Corporation, Intel, Xerox, 1980.

[9] Timothy A. Gonsalves.
Packet-Voice Communications on an Ethernet Local Computer Network: an Experimental Study.
In *Proceedings of SIGCOMM '83*, pages 178-185. ACM SIGCOMM, March, 1983.

[10] Timothy A. Gonsalves and Fouad A. Tobagi.
On The Performance Effects of Station Locations And Access Protocol Parameters In Ethernet Networks.
IEEE Transactions on Communications 36(4):441-449, April, 1988.
Originally published as Stanford University SEL Technical Report 86-292, January, 1986.

[11] Riccardo Gusella.
The Analysis of Diskless Workstation Traffic on an Ethernet.
UCB/CSD 87/379, Computer Science Division, University of California - Berkeley, November, 1987.

[12] Van Jacobson.
Maximum Ethernet Throughput.
Electronic distribution of the TCP-IP Discussion Group, Message-ID <8803100345.AA09833@lbl-csam.arpa>.

[13] Leonard Kleinrock and Fouad A. Tobagi.
Packet Switching in Radio Channels: Part I -- Carrier Sense Multiple Access Modes and their Throughput-delay characteristics.
TRANSCOM COM-23(12):1400-1416, December, 1975.

[14] Keith A. Lantz, William I. Nowicki, and Marvin M. Theimer.
Factors affecting the performance of distributed applications.
In *Proceedings of SIGCOMM '84 Symposium on Communications Architectures and Protocols*, pages 116-123. ACM, June, 1984.

[15] Robert M. Metcalfe.
Steady-State Analysis of a Slotted and Controlled Aloha System with Blocking.
In *Proceedings of the Sixth Hawaii Conference on System Sciences.* January, 1973.
Reprinted in the *SIGCOMM Review*, January, 1975.

[16] Robert M. Metcalfe.
Packet Communication.
PhD thesis, Harvard University, December, 1973.
Massachusetts Institute of Technology Project MAC TR-114.

[17] Robert M. Metcalfe and David R. Boggs.
Ethernet: Distributed Packet Switching for Local Computer Networks.
Communications of the ACM 19(7):395-404, July, 1976.

[18] Jose Nabielsky.
Interfacing To The 10Mbps Ethernet: Observations and Conclusions.
In *Proceedings of SIGCOMM '84*, pages 124-131. ACM SIGCOMM, June, 1984.

[19] John F. Schoch and Jon A. Hupp.
Measured Performance of an Ethernet Local Network.
In *Proceedings of the Local Area Communications Network Symposium.* Mitre/NBS, Boston, May, 1979.
Reprinted in the *Proceedings of the 20th IEEE Computer Society International Conference* (Compcon '80 Spring), San Francisco, February 1980.

[20] John F. Schoch and Jon A. Hupp.
Measured Performance of an Ethernet Local Network.
Communications of the ACM 23(12):711-721, December, 1980.

[21] Alfred Z. Spector.
Multiprocessing Architectures for Local Computer Networks.
Technical Report STAN-CS-81-874, Stanford University, Department of Computer Science, August, 1981.

[22] Hideaki Takagi and Leonard Kleinrock.
Throughput Analysis for Persistent CSMA Systems.
IEEE Transactions On Communications COM-33(7):627-638, July, 1985.

[23] Shuji Tasaka.
Dynamic Behaviour of a CSMA-CD System with a Finite Population of Buffered Users.
IEEE Transactions On Communications COM-34(6):576-586, June, 1986.

[24] Fouad A. Tobagi and Leonard Kleinrock.
Packet Switching in Radio Channels: Part IV -- Stability Considerations and Dynamic Control in Carrier Sense Multiple Access.
IEEE Transactions on Communications COM-23(12):1400-1416, December, 1977.

[25] Fouad A. Tobagi and V. Bruce Hunt.
Performance Analysis of Carrier Sense Multiple Access with Collision Detection.
Computer Networks 4(5):245-259, October/November, 1980.

Channel Access Protocols for Ku-band VSAT Networks: A Comparative Evaluation

D. Raychaudhuri
K. Joseph

In the past few years there has been increasing interest and commercial activity in the area of Very Small Aperture Terminal (VSAT) satellite data networks. VSAT networks employ inexpensive customer-premise earth stations with relatively small antennas (typically less than 2 meters) to provide direct interactive data services to geographically dispersed remote locations. Such satellite based wide area networks potentially offer several economic and operational advantages over conventional terrestrial technologies. This potential can now be realized because of a combination of recent advances in satellite and earth station technology. These include the deployment of high powered Ku-band satellites, the development of low cost solid state power amplifiers and earth station electronics along with advances in shared channel access (multiple access) techniques.

VSAT data networks generally have a star configuration as shown in Fig. 1. In a typical interactive inquiry-response application, the remote stations communicate with a central host computer which is physically located at, or directly connected to, a large hub earth station. Transmissions from the hub to the terminals are sent out in a time division multiplexed (TDM) manner over one or more frequency multiplexed medium to high speed (56 Kbps to 1.544 Mbps) satellite channels. In the terminal to hub direction, channel spacing considerations and transmit power limitations constrain the typical transmission speed to the region of 32 Kbps to 112 Kbps. At these channel speeds, remote stations supporting most interactive applications are very bursty, motivating the use of channel sharing protocols which support a large number of stations on each inbound (terminal to host) frequency. It is clear that reasonably efficient multiaccess is essential for efficient use of satellite bandwidth, relating directly to recurring communications cost. The need for channel sharing efficiency must of course be balanced against the cost of implementing these protocols at the remote stations. Also, it is important to note that unlike other well known scenarios where multiaccess protocols are used (for example, packet radio, local area networks), the overall communication cost per remote station does not vary linearly with multiaccess efficiency. Specifically, in satellite systems, both power and bandwidth are constrained, so that in power limited situations, it may be possible to use a relatively inefficient access protocol without significantly affecting system capacity.

At the time of this writing, shared access of the satellite channel is considered to be the critical path technology item in the development of commercial VSAT data networks. While a considerable body of research on satellite multiaccess has developed during the last decade (see for example [1,2]), there have been few low cost commercial applications. At present, a number of alternative approaches to channel sharing for low cost remote earth stations are simultaneously being developed and tested. The key issues to be considered in selecting a channel sharing protocol for the VSAT scenario are: 1) the efficiency of channel sharing (that is, throughput, the fraction of time useful traffic is carried on the multiaccess channel), 2) the delay characteristics, both in terms of average and peak, 3) the stability properties, relating to

Reprinted from *IEEE Commun. Mag.*, vol. 26, no. 5, pp. 34–44, May 1988.

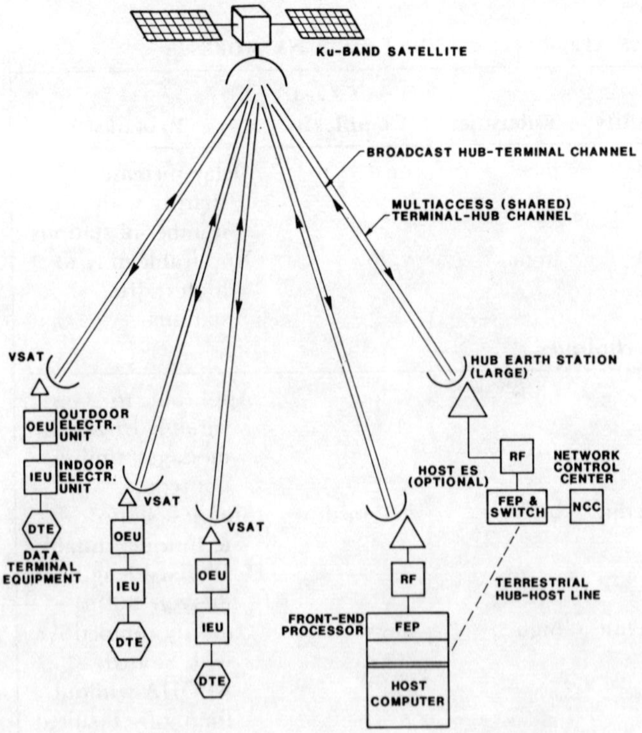

Fig. 1. Star architecture VSAT network.

the possibility of undesirable congestion modes, 4) robustness in the presence of channel errors and equipment failures, 5) operational aspects such as start-up, recovery, and so forth, and 6) the implementation cost/complexity. It may be expected that similar to the evolution of channel access protocols for local area networks, as the VSAT industry matures, a limited number of "standard" protocols for specific application areas will emerge.

In this article, it is our intention to aid this process of selection between alternative protocols by presenting detailed performance comparisons between several candidates for first generation VSAT systems. The performance comparison is carried out over a range of possible traffic, channel and satellite parameters, leading to an understanding of the appropriate regime for each of the protocols under consideration.

Satellite Multiple Access Protocols

Detailed discussions on satellite multiple access can be found in [1-2], so that only a brief introduction is given here. Satellite channel sharing techniques can be classified in terms of the type of message access, which may be either 1) fixed assigned, 2) contention (random access) or 3) reservation (controlled access). Table I provides a summary comparison of a selection of satellite multiaccess protocols belonging to each of these three categories, along with appropriate references for each of the listed protocols.

In fixed assigned systems such as TDMA, a portion of channel time is permanently allocated to each remote station. Since the process of partitioning introduces overhead while resulting in slow service for each station, TDMA is not well suited to applications involving bursty, low volume terminals. Contention techniques provide somewhat more efficient sharing among bursty terminals by permitting stations to access the channel on demand, resorting to appropriate collision resolution procedures when conflicts between transmissions occur. These protocols, of which ALOHA and Tree Algorithm CRA (collision resolution algorithm) are examples, provide low delay and low to moderate efficiency of channel sharing (throughput). The ALOHA protocol also has relatively low implementation complexity. Note that in addition to being used on conventional channels, random access may be employed in conjunction with spread spectrum (CDMA) modulation. When spread spectrum is used, many simultaneous transmissions can be supported on the (wideband) channel, with multiuser conflict manifested in the form of traffic dependent bit error probability. Random access CDMA generally does not provide higher throughput than non-CDMA contention access (when compared on an equal bandwidth basis), but is important in environments where interference considerations require spectrum spreading (such as in the C-band frequencies). Reservation schemes such as Demand Assigned Multiple Access (DAMA) provide higher channel throughput (particularly for long messages) by employing short reservation messages in contention mode to allocate slots for the transmission of longer data messages, using a central or distributed conflict-free scheduling algorithm. Naturally, in the satellite environment, this implies that there will be a high irreducible delay before a message can be transmitted. Also, the associated implementation complexity of reservation schemes tends to be significantly higher than that of contention protocols.

Candidate Protocols for First Generation Ku-band VSAT Systems

We now focus on the major multiaccess alternatives being considered for commercial implementation in Ku-band VSAT networks for interactive data applications. In this environment, typical remote stations generate short messages at a relatively low average rate, but require a rapid response from the host computer. Unless very low bandwidth efficiency can be tolerated, this bursty traffic scenario immediately rules out the application of fixed assigned multiaccess techniques such as TDMA or SCPC (single channel per carrier) FDMA. Accordingly, development of multiaccess protocols for such interactive data applications has centered around random access (ALOHA-type) and demand assigned (DAMA) techniques for which the delay-throughput characteristics are relatively insensitive to the actual number of remote stations from which the traffic is aggregated.

From a survey of current VSAT remote station interface unit products and development prototypes, the primary multiaccess protocol candidates for first generation interactive data Ku-band networks can be identified as: 1) ALOHA, 2) slotted ALOHA, 3) unslotted Selective Reject ALOHA, and 4) Demand Assigned TDMA (DAMA) with Slotted ALOHA reservation access. Note

TABLE I
SUMMARY COMPARISON OF MULTIACCESS PROTOCOLS APPLICABLE TO VSAT DATA NETWORKS

Multiaccess Protocol	Capacity (Max. Throughput)	Delay	Stability	Robustness	VSAT Cost/ Complexity	Remarks
TDMA [3]	0.7–0.8	med-high	good	med	med	Delay increases sharply with number of stations
FDMA/SCPC [4]	0.7–0.8	med	good	high	v. low	Applicable only to high traffic stations
			Fixed Assigned Techniques			
ALOHA [5]	0.13–0.18	low	poor	high	v. low	Applicable to variable length messages; timing not reqd.
Slotted ALOHA [6]	0.25–0.368	low	moderate	high	low-med	Simplest slotted technique; suitable for fixed length message traffic
SREJ-ALOHA [7,8]	0.2–0.3	low	moderate	high	low	Capacity competitive with Slotted ALOHA without timing; well suited to variable length messages
Tree CRA [9,10]	0.43–0.49	med	good	poor	med-high	Suitable for fixed length messages; interleaving delay & deadlock problems
ARRA [11]	0.5–0.6	low	moderate	med	med-high	Side information used to improve Slotted ALOHA type channel performance
Random Access CDMA [12,13]	0.1–0.4	v. low	moderate	high	med	Offers low delay with SSMA advantages; operated with or without timing; complex FEC reqd. for good throughput
Time-of-arrival CRA, FLAP, etc [14,15]	0.4–0.7	low	moderate	high	med	Exploits arrival time inf. in collisions to improve performance of ALOHA channels
			Random Access Techniques			
DAMA/TDMA [16,17]	0.6–0.8	med-high	good	poor	high	Generally attractive for long messages; high latency delay due to reservation
			Controlled Access Techniques			

that in contrast to C-band VSAT applications, spread spectrum approaches have not been generally advocated for Ku-band (which is free from terrestrial interference problems), mainly because of the low bandwidth efficiency characteristic of CDMA. For completeness, we provide a brief review of the operation and qualitative features of each of these access protocols, before we begin a detailed comparison in the next section.

Unslotted ALOHA, [5]

ALOHA is the first and simplest possible contention access technique, and in spite of many subsequent improvements, still has considerable appeal for VSAT applications. The protocol permits users to transmit new messages on the channel as they are generated at remote stations, with colliding messages being retransmitted after a random delay, as shown in Fig. 2a. Note that in the VSAT environment, positive acknowledgement of successful messages is required (since VSATs cannot reliably "hear" their own transmissions); this implies a minimum time-out delay of approximately 0.5 to 0.6 secs before a retransmission can be initiated. The stability of the channel (that is, absence of long term conditions with high congestion due to excessive retransmission traffic), depends upon the selection of the total average retransmission delay. In the satellite scenario, for moderate load levels, the presence of the fixed 0.5–0.6 sec time-out delay permits the use of relatively short additional random delay (typically 0.1 to 0.2 sec).

As one might expect, the simple ALOHA protocol provides relatively low throughput or efficiency of channel use (typically varying between 0.13 to 0.18 depending on the message length distribution). The low throughput is offset by low access delay, the ability to handle variable length message traffic, robust operation and minimal equipment complexity. Moreover, for very short messages (as might be encountered in point-of-sale applications), the throughput supported may be quite competitive with other protocols because of the increasing significance of transmission overheads (which are minimal in ALOHA).

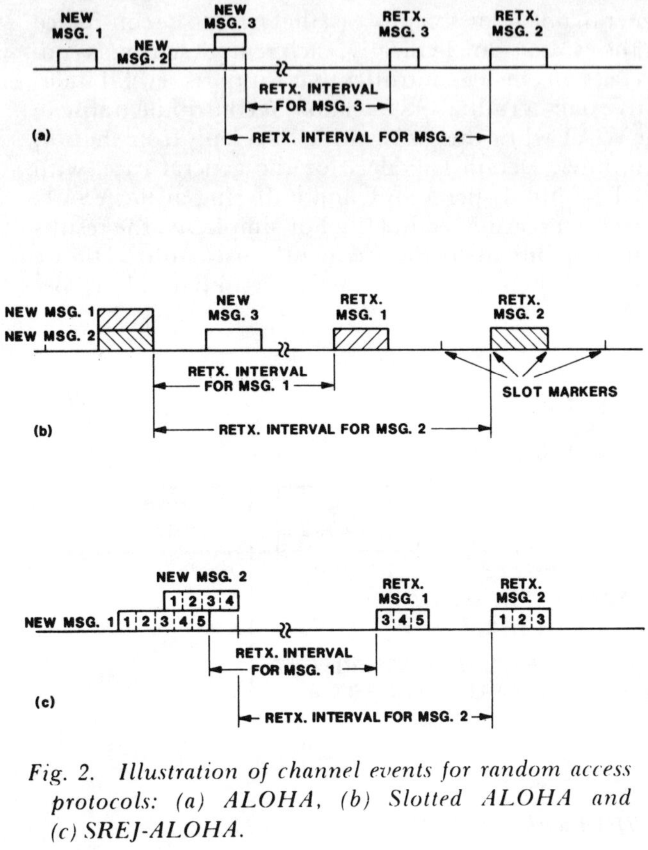

Fig. 2. Illustration of channel events for random access protocols: (a) ALOHA, (b) Slotted ALOHA and (c) SREJ-ALOHA.

Slotted ALOHA, [6]

Slotted ALOHA improves the channel sharing efficiency of ALOHA in an environment with fixed length message transmissions by requiring that all transmissions be synchronized to occupy "time slots" on the channel. As shown in Fig. 2b, after a new message is generated, a station must delay transmission until the start of the next time slot. This reduces the probability of collision relative to unslotted ALOHA, and results in a doubling of the maximum throughput from 0.184 to 0.368 for fixed length messages. This improvement in maximum throughput is somewhat offset by the increased implementation complexity of a slotted, globally synchronous system. Also, the maximum throughput realized in practice will be less than the theoretical value of 0.368 due to both guard time requirements and overhead introduced in converting variable length messages to a fixed packet format. Since short variable length message traffic is typical of most interactive data VSAT applications, these effects must be carefully accounted for when comparing with unslotted ALOHA.

Selective Reject (SREJ) ALOHA, [7,8]

SREJ-ALOHA is an unslotted random access technique in which a maximum throughput comparable to that of Slotted ALOHA is achieved without timing synchronization, using subpacketization of messages in conjunction with a Selective Reject ARQ retransmission strategy. As shown in Fig. 2c, in SREJ-ALOHA, messages are formatted into a contiguous sequence of independently detectable fixed length subpackets, each with its own header and acquisition preamble. The protocol exploits the fact that most collisions in an asynchronous channel result in partial overlaps, so that only the subpackets actually encountering conflict are retransmitted (as in SREJ mode ARQ). The maximum throughput (not accounting for overheads) has been shown to approach 0.368, irrespective of the message length distribution. In practice, the need for acquisition preamble and header in each subpacket limits the maximum useful throughput to the range of 0.2–0.3. Since SREJ-ALOHA is well suited to variable length message scenarios (which are commonly encountered in VSAT applications), it may often outperform Slotted ALOHA, which also involves the additional complexity and overhead associated with timing co-ordination. The critical need for low subpacket overhead places specific hardware constraints on SREJ-ALOHA systems. Specifically, burst modems with short acquisition preamble are a prerequisite for effective SREJ-ALOHA operation.

Demand Assigned TDMA (DAMA), [16-18]

Controlled access or demand assigned strategies are based on allocated access to time slots in a TDMA-like channel. Demand assigned systems typically employ a two level access procedure. The first step is to place a short reservation for message transmission using a fraction of the channel time either in TDMA or contention (such as slotted ALOHA) mode. Once the reservation message is successfully received at the hub station, message slots are allocated in a conflict-free manner by the central controller. Such a scheme can provide higher maximum throughput than contention protocols in an environment in which the data messages are substantially longer than the minimum size of a reservation message. Clearly, DAMA techniques also incur a fixed minimum delay of two satellite round trip times (that is, about 0.5 sec) before a message slot can be allocated, even when traffic is very light. However, the variable component of delay may be expected to increase less rapidly with traffic load when compared with contention access channels.

TDMA type reservation strategies may be used for the message slots only when the DAMA/TDMA channel is shared by a relatively small number of remote stations. This is because both the frame latency and reservation overhead tend to increase with the number of stations supported. If Slotted ALOHA access is used for the reservation packets, the dependence on the number of stations is essentially eliminated. The protocol however, can no longer be considered conflict-free and is potentially subject to the same type of instability problems encountered in ALOHA channels. For the interactive data environment, strong dependence of performance on the station traffic volume is undesirable, so that DAMA/TDMA with slotted ALOHA access is the most popular implementation of controlled access for VSAT applications. An illustration of channel events in such a DAMA/TDMA system is given in Fig. 3. It is clear that although DAMA/TDMA potentially offers higher maximum throughput than contention access, there is a trade-off with ALOHA-type systems in terms of the high irreducible message slot allocation delay, low robustness in presence of channel errors, and high implementation complexity.

Performance Comparison of First Generation Protocols

VSAT Traffic Models

As mentioned earlier, remote stations which support interactive data applications typically generate short, variable length messages with average data rate orders of magnitude lower than the multiaccess (terminal to hub) channel speed. Roughly speaking, the remote station traffic model is described by two attributes: (1) the average rate at which new messages are generated and (2) parameters specifying the message length distribution. Note that (1) and (2) together can be combined to determine the average data rate per station in bits per second. In Fig. 4, approximate regions in the average data rate—average message length plane that are occupied by some common VSAT applications are shown. We observe that the important interactive data scenario is in the region near the origin, corresponding to low average data rate and short messages. Note that although, in Fig. 4, the length distribution has been approximated by a single parameter which is the average value, in general, an accurate evaluation requires the determination of the exact message length distribution for the particular application. It is often reasonable to assume the distribution to be a truncated exponential with specified average and maximum length (typically 256 characters, based on present data communication practice). Since the exponential distribution tends to produce pessimistic results compared with other alternatives, it is a reasonably good choice when the distribution is completely unknown. It should be pointed out here that there are additional subtleties to the station traffic model involving the detailed nature of the arrival process (which is, in general, not a Poisson process) that need to be considered in the evaluation. Typically, each remote station is connected to a cluster controller that supports several interactive data terminals, which do not contribute traffic to the VSAT while they are waiting for a reply from the host computer. Detailed models for the general case, with M>1 terminals per VSAT and buffering at the VSAT, have been considered in [19]. For simplicity, the results here are limited to the frequently used source traffic model with a single interactive terminal (M=1) per VSAT.

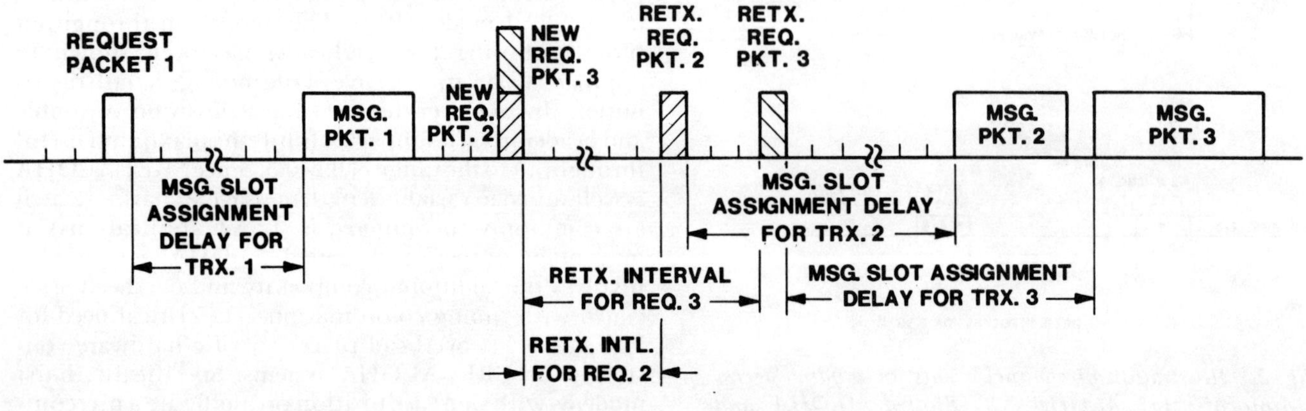

Fig. 3. Illustration of channel events for DAMA/TDMA with Slotted ALOHA reservation.

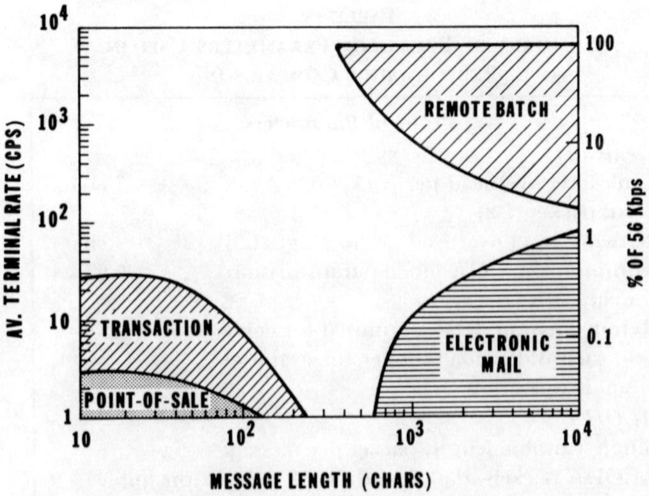

Fig. 4. Traffic source parameter regions for potential VSAT applications.

Performance Evaluation Criteria

The key characteristic of a multiaccess protocol as seen by users of the data network is delay, which is completely described by the average value along with the delay distribution. It is often useful to replace consideration of the distribution of delay by a single "peak" value, such as the 95th percentile of the delay distribution. While delay is the important measure for a user, the number of users sharing the channel (N) is the primary quantity of interest from the system operator's point of view. Since the number of users supported relates directly to the useful data carried or throughput (S), it has been customary to characterize multiaccess protocols in terms of their average delay vs. throughput characteristics. In our presentation of performance results, a more complete characterization will be provided in terms of both average and peak delay vs. throughput (or number of terminals, N), for channels designed to be unconditionally stable with minimum delay. Note that stability of ALOHA type channels with finite, interactive source traffic models, is ensured by selecting an appropriately large average retransmission delay. Since the delay performance worsens with increasing retransmission delay, the minimum value for which the channel is stable should be used at each traffic level. Observe that when DAMA/TDMA employs a slotted ALOHA subchannel for reservation traffic, similar procedures must be carried out to ensure stability of the reservation subchannel.

A second quantity of considerable importance to the system designer is the total channel traffic, G, which is a normalized measure of all traffic, including collisions and overheads (such as reservations), at each level of channel loading, N. The significance of total channel traffic is due to the fact that the average satellite power consumption by the bursty VSAT signals is directly proportional to G. Obviously, G will exceed the throughput, S. For random access, the difference G-S is an indication of the degree of retransmission traffic due to collisions. In reservation-based DAMA systems, the difference between G and S reflects the overhead due to reservation traffic.

While delay and peak delay vs. throughput (or N) are or primary interest, some additional useful secondary characterizations can be identified. Specifically, having used the delay vs. N curves to select an operating point (that is, normal operating load, N*), delay distributions for each protocol at that load, N*, are informative regarding subtle differences in delay properties. A second curve that we have found useful in the variable message length environment under consideration is a plot of the variation of average delay with message length. The above characterizations usually highlight significant qualitative differences between the protocols under consideration.

Performance Evaluation Methodology

It is possible to use analytical methods to obtain average delay vs. throughput curves for the four protocols under consideration. The analytical models generally involve a limited number of assumptions which do not accurately reflect the physical system, but which nevertheless, provide good average delay-throughput results, as borne out by comparison with direct-event stimulation. However, for a more detailed characterization in terms of delay distributions identified above, it is usually necessary to employ detailed simulation models, such as those described in [20]. The stimulation models have been designed to replicate the physical system very closely, so that no significant assumptions are required. Moreover, unlike analytical procedures, the simulation models can easily accommodate variations in the traffic source model, including such details as measured message length distributions. The main drawback of stimulation is the relatively high computation time required, particularly for reliably estimating channel stability. Accordingly, when appropriate, we have used analytical procedures to guide the selection of retransmission delay for system stability, after which stimulation is employed for detailed performance characterization.

Performance Comparison Results

In this section, we present numerical results for a typical interactive data VSAT network scenario. The parameters used are summarized for convenient reference in Tables II–IV. The channel parameters (relating mainly to the modem characteristics) are given in Table II, and are believed to reflect typical VSAT modem capabilities. The source traffic model, given in Table III, is based on

TABLE II
SUMMARY OF CHANNEL PARAMETERS USED IN PERFORMANCE COMPARISON

Channel Parameters	
Channel data rate	= 56 Kbps
Channel symbol rate (rate 1/2 coded)	= 112 Kbps
Minimum E_b/N_0 required	= 8.0 dB
Modem acquisition preamble	= 32 bits (4 characters)
One way propagation delay	= 0.27 sec
Minimum ACK delay (2 hops)	= 0.54 sec

TABLE III
SUMMARY OF TRAFFIC SOURCE PARAMETERS USED IN PERFORMANCE COMPARISON

Source (VSAT) Traffic Model	
Interactive terminals per VSAT	= 1
New message rate per terminal (unblocked mode)	= 250 msg/hr = 0.07 msg/sec
New message rate per terminal (blocked mode)	= 0
New message length distribution	– truncated exponential
Average new message length	= 100 chars (800 bits)
Maximum new message length	= 256 chars (2048 bits)

TABLE IV
SUMMARY OF PROTOCOL PARAMETERS USED IN PERFORMANCE COMPARISON

Protocol Parameters	
General:	
Link level overhead per packet or subpacket (L2)	= 7 chars
Network level overhead per message (L3)	= 8 chars
Minimum ALOHA mode retransmission delay	= 0.67 sec
Retransmission delay optimized for delay at each traffic load, under constraint of stable operation.	
ALOHA:	
Single variable length packet per message	
ALOHA packet—data length + L2 + L3 + preamble	
SREJ-ALOHA:	
Multiple fixed length subpackets per message	
Subpkt size optimized for each average message length (L)	
SREJ subpacket—fixed data length + L2 + preamble	
Subpacket size: S = 40 char for L = 50 char; 50 for 100; 60 for 150	
Slotted ALOHA:	
Single fixed length packet per message	
Slotted ALOHA packet—data length + filler data + L2 + L3 + preamble = 275 chars (fixed)	
Time slot size—packet length + guard time (4 chars) = 279 chars	
DAMA/TDMA (Slotted ALOHA reservation):	
Reservation packet length 26 chars (including preamble)	
Reservation slot length—reservation pkt length + guard time = 30 chars	
DAMA message packet—data length + filler data + L2 + L3 + preamble = 275 chars (fixed)	
DAMA message slot size—message packet length + guard time (4 chars) = 279 chars	
Assignment delay = 0.67 sec	
Fractional allocation of message and reservation slots optimized for minimum delay.	
Message and reservation slots interleaved.	

our understanding of a typical VSAT transaction (inquiry-response) application. Parameters for each of the four protocols under consideration have been selected according to established design principles, and are given in Table IV. Note that as discussed earlier, the key retransmission delay parameter is chosen for minimum stable delay, and is optimized at each level of load (N). For ALOHA, the retransmission delay is the only significant parameter of the protocol. However, for Slotted ALOHA and SREJ-ALOHA, the choice of appropriate packet sizes is important. For Slotted ALOHA, note that we adopt the convenient approach of selecting a message packet size (equal to the slot size less guard time and all overheads) equal to the maximum possible message length. For the present example, referring to the traffic model in Table III, observe that this is assumed equal to 256 characters, as might be the case in many practical systems. For SREJ-ALOHA, subpacket size is an extremely important choice: in general, optimum subpacket size depends upon the acquisition and addressing overhead and is also a load dependent quantity. Fortunately, it can be shown that for the present set of parameters, there is a fairly robust choice of subpacket size (50 characters for an average message length of 100 characters, as given in Table IV), which is close to optimum at moderate to heavy channel load. For DAMA/TDMA, several parameters such as the fraction of time allocated to reservation traffic, the average retransmission delay for colliding reservation messages and the message and reservation slot sizes need to be specified. The design procedure used here is based on optimally allocating a fraction of channel capacity to the reservation channel, noting that total delay is the sum of reservation access delay and message assignment delay. As discussed for the ALOHA channels, the retransmission delay for the Slotted ALOHA reservation subchannel is selected for minimum stable delay. As for the Slotted ALOHA case, the DAMA/TDMA system is based on a message slot size equal to the maximum message length of 256 characters. In general, DAMA efficiency can be improved by using shorter message slots, but this is at the expense of some increase in implementation complexity.

Figure 5 shows the familiar throughput-delay characteristics of the four protocols under consideration. Note that the throughput shown is the useful data throughput, after accounting for overheads such as acquisition preamble in ALOHA, guard time, acquisition time and wasted slot time (due to variable message length) in slotted ALOHA, subpacket overhead in SREJ-ALOHA and reservation channel overhead and wasted message slot time in DAMA/TDMA. The delay values include the effect of slot synchronization latency delays for the slotted protocols. It is observed from the figure that for this message model, Slotted ALOHA performs only slightly

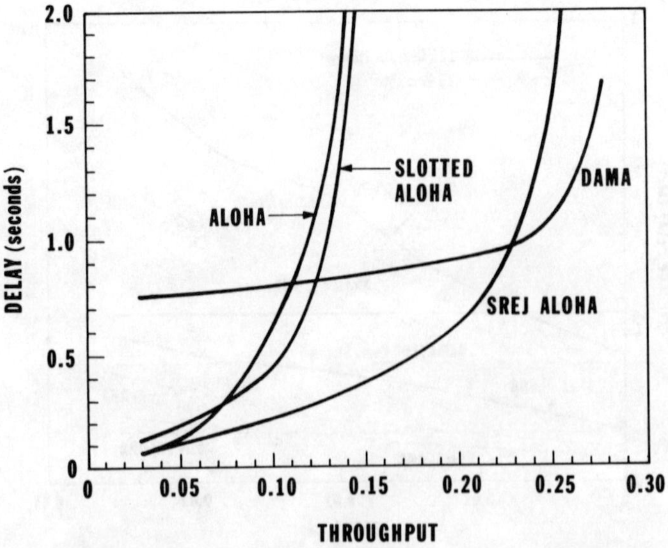

Fig. 5. Delay-throughput characteristics of candidate multiaccess protocols.

better than ALOHA, while both are considerably outperformed by SREJ-ALOHA. Obviously, the relative performance is strongly dependent on message length, an effect which we shall investigate at a later point. Figure 5 also shows that the maximum throughput achieved by the DAMA/TDMA approach is higher than that of the random access protocols, but is achieved with a high irreducible delay of about 0.75 secs.

As we have discussed earlier, a more useful characterization, as in Fig. 6, is provided by plotting curves of average delay and peak (95th percentile) delay as a function of the number of VSAT stations supported on the same channel (N). Since interactive network design specifications must include average as well as peak delay constraints, such a set of curves is a prerequisite for determining the "capacity", that is, the number of remote stations (VSATs), N*, supported on a channel while satisfying the performance requirements. The operating value of N* is determined by the tighter of the

TABLE V
CHANNEL CAPACITIES FOR CANDIDATE PROTOCOLS UNDER ALTERNATIVE DELAY CONSTRAINTS

Multiaccess Protocol	Channel Capacity (Number of VSATS Supported, N*)	
	Delay constraints: 0.75 sec average 2.5 sec peak (95%)	Delay constraints: 1.0 sec average 3.0 sec peak (95%)
ALOHA	75	80
Slotted ALOHA	105	110
SREJ-ALOHA	130	140
DAMA/TDMA	*	230

*Objective not achievable

average and peak delay constraints. From Fig. 6, it can be observed that peak delay for ALOHA tends to rise rather rapidly, while DAMA is characterized by high average delay and low delay variance. Slotted ALOHA and SREJ-ALOHA provide better peak delay properties than ALOHA, but are also characterized by rapidly rising peak delay as the channel becomes overloaded. In Table V, we give a comparison of protocol capacities with delay constraints of 0.75 sec average—2.5 sec peak and 1 sec average—3 sec peak. Observe that the 0.75 average delay case is not achievable with DAMA, but when the average delay requirement is relaxed to 1 sec, DAMA supports many more terminals than the contention access techniques. Comparing the ALOHA protocols, we observe from the table that at both performance levels, SREJ-ALOHA supports about 25 percent and 75 percent more stations than Slotted ALOHA and ALOHA respectively.

Curves of total channel traffic (G), vs. the number of remote stations per channel (N) are given in Fig. 7. As discussed in the previous section, these curves are of some

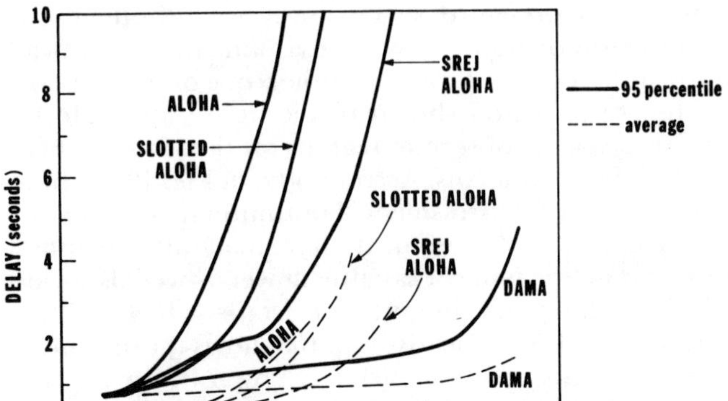

Fig. 6. Average and peak delay vs. number of VSATs for candidate multiaccess protocols.

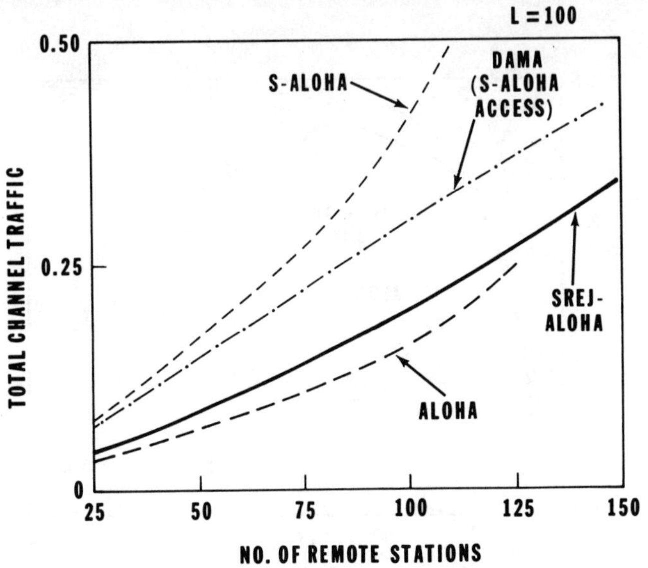

Fig. 7. Variation of total channel traffic with number of VSATs for average message length = 100 char.

ancillary importance because they relate to the "power efficiency" of a protocol, which needs to be considered in power limited situations that frequently characterize VSAT networks. From the figures, observe that at any value of N, ALOHA has the lowest total channel traffic (and hence the best power efficiency), followed by SREJ ALOHA, DAMA and Slotted ALOHA, in that order. Note also that random access systems display a rapid increase in G after reaching their respective congestion points, due to a non-linear increase in retransmission traffic.

Further characterization of the protocols in terms of delay distributions at specified load levels is given in Fig. 8. In the figure, we plot delay distributions for ALOHA, Slotted ALOHA, SREJ-ALOHA and DAMA, designed to operate at an N* corresponding to an average delay of 1 sec (as determined from the curves in Fig. 6). Observe that as expected, DAMA demonstrates relatively low delay variance, while ALOHA is characterized by very high delay variance, mainly due to repeated collisions experienced by long messages. Slotted ALOHA and SREJ-ALOHA display moderate delay variance, although substantially greater than that of DAMA. SREJ-ALOHA appears to have lower "tails" of the delay distribution than the other contention access techniques. A second detailed characterization of interest is the plot of average delay vs. message length shown in Fig. 9. Observe that DAMA and Slotted ALOHA, for which the slot size has been chosen to equal the maximum message length, have delays which are independent of the actual length. In contrast, ALOHA and SREJ-ALOHA accommodate variable length transmissions, so that longer messages tend to have a higher collision probability (and hence delay) than short messages. In ALOHA, the delay penalty experienced by long messages can be rather significant, particularly at high channel load. However, the SREJ mechanism in Selective Reject ALOHA overcomes this problem to a large extent, so that the curves show only a modest increase in delay with message length.

The above results in Figs. 5-9 were based on a specific

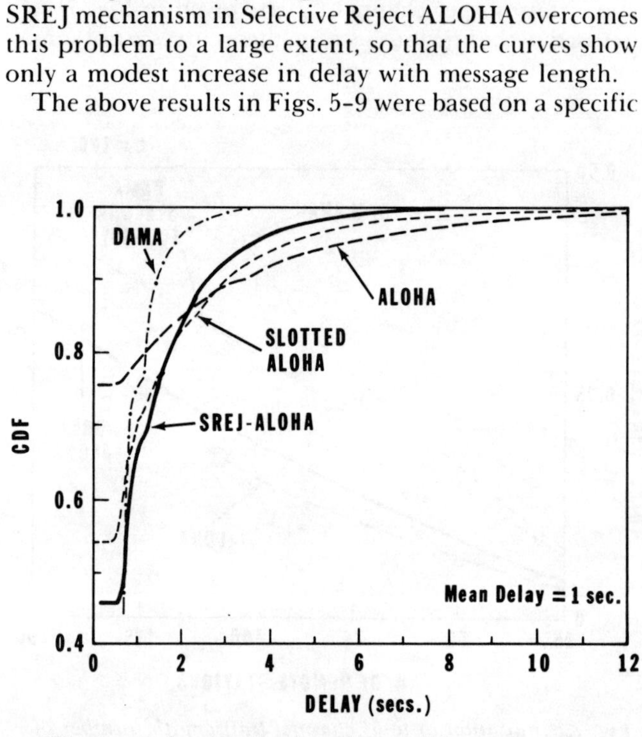

Fig. 8. Delay distributions for candidate protocols at channel load corresponding to an average delay of 1.0 sec.

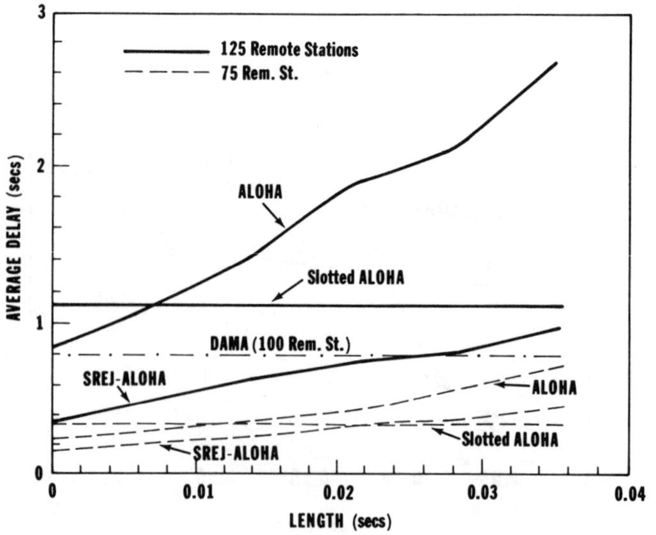

Fig. 9. Variation of average delay with message length for the candidate protocols.

VSAT traffic source model, with an average input rate of 250 messages/hr and truncated exponential message lengths with a maximum of 256 characters and an average value of 100 characters. In order to study the sensitivity of protocol performance with respect to average message length, we introduce Table VI, which shows the number of VSATs supported per channel (based on a 1.0 sec average delay criterion), for average message length equal to 50, 100 and 150 characters respectively (truncated exponential distribution assumed). The results show that as expected, Slotted ALOHA and DAMA have capacities which are insensitive to the message length, while both ALOHA and SREJ-ALOHA exhibit decreasing capacity as message length increases. In general, we find that SREJ-ALOHA has the highest capacity for short to medium length, while for average message lengths of 150 characters or higher, Slotted ALOHA is the preferred protocol.

To complete the discussion, we will attempt to provide a qualitative comparison between multiaccess protocols over the range of traffic source parameters shown in the terminal rate vs. message length diagram shown in Fig. 4. The results presented so far apply to the interactive (transaction or point-of-sale) environment shown near the origin of Fig. 4. If we widen the scope of comparison to the entire region shown in the figure, it would be useful to specify optimum regions for the various alternative VSAT protocols. Accordingly, in Fig. 10, we provide a qualitative assessment of optimum regions for the VSAT protocols discussed. We have made an attempt to incorporate the issue of satellite power as well as bandwidth, without entering into the details, which are discussed in [21]. Specifically, Fig. 10a shows the optimum protocol regions for a moderately power limited example, as might be encountered in a system with small antenna size (1.2 meter), high ratio of hub outbound traffic to inbound traffic and/or high availability requirements. Fig. 10b shows a similar plot for a moderately bandwidth limited system, which may occur in a situation with larger antennas (1.8 meter or greater),

TABLE VI
VARIATION OF CHANNEL CAPACITY
(WITH AVERAGE DELAY CONSTRAINT EQUAL TO 1.0 SEC)
WITH AVERAGE MESSAGE LENGTH

Message Length (Characters)	Channel Capacity (Number of VSATS Supported, N*) ALOHA	Slotted ALOHA	SREJ-ALOHA	DAMA/TDMA
50	180	130	205	250
100	100	130	160	250
150	75	130	115	250

powerful forward-error corrected modems, balanced outbound and inbound traffic and/or low availability requirements. Note that a protocol marked as "optimum" provides maximum system capacity, but does not necessarily provide the highest number of terminals per terminal to hub channel. Note also that we now include TDMA, which is a useful protocol for medium to high volume stations. Additionally, the three ALOHA protocols have been lumped together, with the understanding that the choice between these will depend upon the message length regime of interest along with implementation factors.

Observe from the two figures that, in general, ALOHA-type protocols are optimum for scenarios with short messages and low average data rate per remote station, in agreement with conventional wisdom. DAMA is recommended for higher message lengths, but for low to medium station traffic volume, while TDMA is the choice for systems with high station traffic volume, independent of message length. Observe that there is considerable overlap between optimum regions for the power limited scenario since bandwidth use is relatively non-critical. On the other hand, for the bandwidth limited case, there is little overlap between the optimum regions for each of the protocols, and this can be attributed to the greater importance of the multiaccess efficiency in determining system capacity.

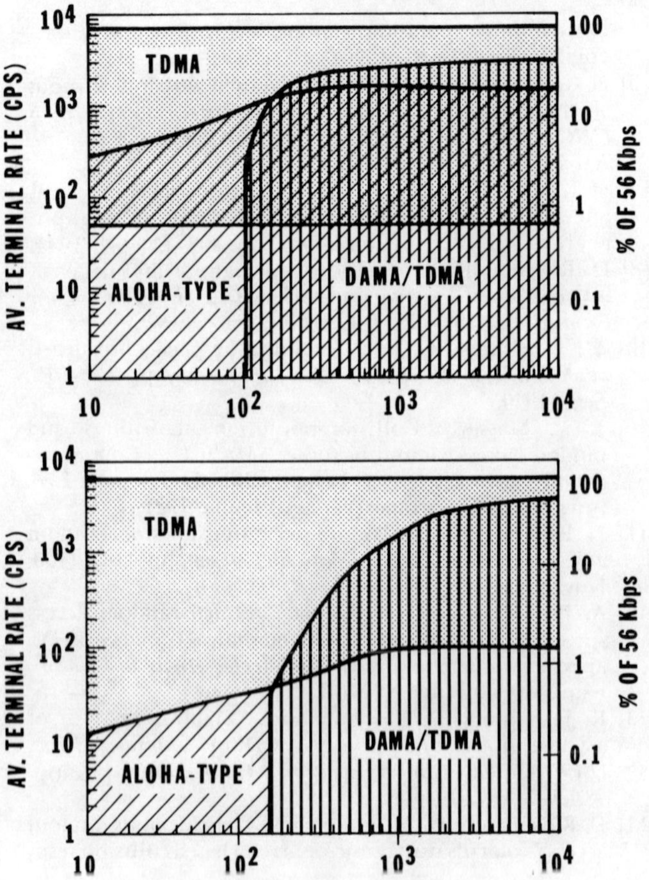

Fig. 10. Optimum protocol regions for typical Ku-band satellite scenario: (a) Moderately power limited case, and (b) Moderately bandwidth limited case.

Concluding Remarks

An overview and performance comparison of first generation satellite channel access protocols for VSAT data networks has been presented. The comparison highlights the strengths and weaknesses of each of the multiaccess protocols considered, and introduces a methodology for detailed performance evaluation. This methodology augments the conventional approach of comparing average delay vs. throughput characteristics, by introducing systematic consideration of delay distribution aspects, which are of significant importance in data network design. The approach developed has been used to identify the optimum VSAT channel access protocol from a performance standpoint over a range of potential traffic parameter scenarios.

It should be emphasized that the numerical results in this paper are intended to be an approximate guide to satellite access protocol selection in the transaction application environment, and should not be interpreted as universally applicable to VSAT network design. On the other hand, it is hoped that the performance evaluation methodology outlined here, which considers many aspects beyond the traditional "throughput-delay" characteristics, will be of some value to system designers as new VSAT network application areas evolve.

References

[1] S. S. Lam, "Satellite packet communication—multiple access protocols and performance," *IEEE Trans. on Comm.*, pp. 1456-1466, Oct. 1979.

[2] D. Raychaudhuri and K. Joseph, "Ku-band satellite data networks using very small aperture terminals—part I: multi-access protocols," *Intl. J. of Satellite Comm.*, pp. 195-212, July-Sept. 1987.

[3] W. G. Schmidt, "Satellite Time Division Multiple Access Systems: Past Present and Future," Telecommun., vol. 7, Aug. 1974, pp. 21-24.

[4] J. W. Schwartz, J. M. Aein and J. Kaiser, "Modulation techniques for multiple access to a hard limiting repeater," *Proc. IEEE*, pp. 763-777, May 1966.

[5] N. Abramson, "The ALOHA system—another alternative

for computer communications," *AFIPS Conf. Proc.*, vol. 37, pp. 281-285, 1970.

[6] L. G. Roberts, "ALOHA packet system with and without slots and capture," *ARPANET Satellite System Note 8 (NIC11290), June 1972; reprinted in Computer Comm. Review*, vol. 5, April 1975.

[7] D. Raychaudhuri, "ALOHA with multipacket messages and ARQ-type retransmission protocols—throughput analysis," *IEEE Trans. on Comm.*, pp. 148-154, Feb. 1984.

[8] D. Raychaudhuri, "Stability, throughput and delay of asynchronous selective reject ALOHA," *IEEE Trans. on Comm.*, pp. 767-772, July 1987.

[9] J. I. Capetanakis, "Tree algorithms for packet broadcasting channels," *IEEE Trans. on Inf. Theory*, pp. 505-515, Sept. 1979.

[10] J. L. Massey, "Collision-resolution algorithms and random-access communications," *Multi-User Communication Systems*, ed. by J. Longo, CISM Courses and Lectures No. 265, Springer-Verlag, Vienna, 1981.

[11] D. Raychaudhuri, "Announced retransmission random access protocols," *IEEE Trans. on Comm.*, pp. 1183-1190, Nov. 1985.

[12] A. Polydoros and J. Silvester, "Slotted random access spread spectrum networks—an analytical framework," to appear in *The IEEE Jour. on Selected Areas in Comm.*, issue on Computer Network Performance.

[13] K. Joseph and D. Raychaudhuri, "Stability analysis of asynchronous random access CDMA systems," *Proc. IEEE Global Communications Conf.*, Houston, pp. 48.1.1-7, Dec. 1986.

[14] D. Raychaudhuri, "High capacity asynchronous random access protocols using time-of-arrival based collision resolution," *Proc. IEEE Global Comm. Conf.*, New Orleans, pp. 48.6.1-5, Dec. 1985.

[15] M. J. Karol and S. C. Schwartz, "The first/last access protocol," *Proc. IEEE Global Comm. Conf.*, pp. 8.4.1-5, Dec. 1985.

[16] I. Jacobs, et al., "CPODA—a demand assigned protocol for satnet," *Proc. 5th Data Comm. Symp.*, Snowbird, Utah, Sept. 1977.

[17] L. Roberts, "Dynamic allocation of satellite capacity through packet reservation," *AFIPS Conf. Proc.*, vol. 42, pp. 711-716, 1973.

[18] S. S. Rappaport and S. Bose, "Demand assigned multiple access systems using collision type request channels: stability and delay considerations," *Proc. IEE*, vol. 128, no. 1, Jan. 1981.

[19] K. Joseph and D. Raychaudhuri, "Evaluation of flow control techniques for ALOHA-type multiaccess channels," *Intl. Conf. on Comm.*, pp. 613-619, 1987.

[20] K. Joseph and D. Raychaudhuri, "Simulation models for performance evaluation of satellite multiaccess protocols," *IEEE Jour. on Selected Areas in Comm.*, p. 210-222, Jan. 1988.

[21] D. Raychaudhuri, Ku-band satellite data networks using very small aperture terminals—part II: system design," *Intl. Jour. of Satellite Comm.*, Oct.-Dec. 1987.

Bibliography

M. El-Amin, L. Chung, B. G. Evans," Access protocols performance for on-board processing business satellites," *Int. Jour. of Satellite Comms.*, vol. 5, no. 2, 1987.

Low Cost VSAT Mesh Network With Time-Spread Aloha Access

M.K. Goel Y.S. Rao

Telecommunications Research Centre
Government of India, Nehru Place
New Delhi 110019
India

Vinod Chandra

Department of Electrical & Computer Engineeering
Florida Atlantic University
Boca Raton, Florida 33431

Abstract

Time spreading of packets involves coded repetition of its data bits on the channel and should theoretically permit reduction in power requirement of VSATs. In practice, use of available sequences with finite auto-correlation sidelobes not only nullifies the enhancement of Eb/No due to spreading but also reduces the network throughput. However, a network operated in a backed-off capacity has fewer simultaneous transmissions which facilitates a part of the spread advantage to be utilised for link design with increased noise. The extent of channel occupancy at different channel throughput values for a VSAT population mesh network, has been obtained by simulation. The results of the simulation have been used to demonstrate the feasibility of using Time-spread ALOHA access to configure a fully connected VSAT mesh network. Possibility of reduction in VSAT size at cost of response time is also discussed.

Introduction

Traditionally VSAT networks have used STAR topology for cost effective design resulting in limitations of 2-hop satellite delay. It is felt that interest in voice/data integration and especially in ISDN could typically force satcom terminals in future to mesh architecture. In addition, messaging and voice traffic tend to follow a wider geographic pattern whereas interactive data tends to funnel into one or more hubs where host computer resides. It is therefore expected that for messaging applications mesh networks would be preferred operationally if costs allow Very Small Aperture Terminals (VSATs) to be built that way.

Time spreading of data packet transmissions has been mainly considered for multi access purposes as a variant of ALOHA channel. However, apart from its multi-access capability it can be used to reduce G/T or EIRP or both in VSAT networks. It is now shown that it is possible to design a mesh link in C-band using VSATs. Such a network has an advantage over the conventional star design in that voice transmission without the disadvantage of double hop delay can take place. Section II of this paper presents the concept of time spreading. It shows that savings in power can be made for networks operated at a backed-off throughput if packets are time spread. An infinite population mesh network was simulated to assess channel occupancy at different throughputs. The results of the simulation indicated that at generally useful throughput values, channel occupancy on Spread ALOHA is such that time spreading advantage for power reduction is still available after collision resolution. Section III gives the details of simulation. A mesh link design for a given traffic requirement is shown in section IV. The last section discusses the importance of the results in respect of mesh configuration of VSATs in C-band.

Time Spreading

Time spreading by low auto-correlation sequences, such as Barker, has been suggested as a means to reduce peak power requirements of VSATs in bursty traffic satellite networks [1]. Bit-spreading of data packets in time basically involves coded repetition of the bits of the data packet while transmitting them on the channel. Since the signal detectability depends upon the energy per bit (Eb/No) and not the average signal power, this repetition provides a spread advantage in the form of enhanced bit energy. For example, if a data bit is repeated 'n' times then for the same average transmitted power it will have 'n' times the bit energy providing 10 log (n) dB spread advantage. It has been argued that the spread advantage can be utilised for power reduction in VSATs.

It has been found now that the claimed spread advantage can come about only for the case of spreading by theoretically zero auto-correlation sidelobe sequences. Practical spreading sequences have finite auto-correlation sidelobes. This limits the spread advantage as well as the available capacity of the ALOHA channel.

Reprinted from *IEEE Proc. 1989 Southeastcon*, vol. 3, pp. 971–974, April 9–12, Columbia, South Carolina, 1989.

Capacity is limited to the extent of [[[n/Ck]-1]/[n/Ck]] where as before 'n' is the length of the spreading sequence and Ck is the absolute value of maximum auto-correlation of the sequence. When Barker sequences are used, the capacity is decreased by (n-1)/n. This capacity limitation can be significant if sequences used have smaller lengths [2]. As far as spread advantage is concerned, practical sequences limit the enhancement of Eb/No of the desired transmission because of the interference-like contribution of their correlation sidelobes as part of other simultaneously present transmissions. In the worst case of a single '1' bit being interfered by 'n' bits of the opposite polarity, the enhancement in bit energy by spreading of the desired data bit is nullified. Therefore, the power advantage envisaged in [1] can come about in practice only when either one uses zero auto-correlation spreading sequences or transmits in such a way that number of simultaneous transmissions is less than the length in bits of the spreading sequence. Since we can do nothing about the first requirement let us analyse the second. It means that if, on a network employing 'n' bit spreading sequence, one can restrict the number of simultaneous transmissions to say '2n/3', these can be resolved with a power-penalty of 10 log (2n/3) dB and the rest of 10 log (n/3) spread advantage can be utilised for power reduction in VSATs. In a real network, restriction of simultaneous transmissions on the random access channel can come about by operating at a backed-off capacity or over-designing the netwrok for capacity. A computer simulation was done to gauge the extent of channel occupancy at different capacity operations of a fully connected mesh network. The simulation demonstrated that at useable throughputs ample power advantage is available to permit the mesh link design of VSATs in C-band.

Simulation

A low bit rate messaging network consisting of an effectively infinite number of terminals has been simulated [4]. In this simulation the terminals communicate to each other directly in a mesh architecture. Channel access is chosen to be Bit-Spread ALOHA with message packet arrival process modelled as a Poisson's process. Message packet formation is according to HDLC procedures and a simple end-to-end protocol is implemented to govern terminal-to-terminal message packet delivery. The protocol is based on positive acknowledgement (POSACK) for every correct reception of message packet and timeout by the source in the event of non-receipt of POSACK. The acknowledgement (ACK) packet is similar to message packet in the number of data bits and format. Timeout results in automatic rescheduling of the transmission of the message packet that is presumed to have failed. Similarly, a failed ACK packet results in automatic retransmission of the originally transmitted message packet.

The network has been studied for throughput at various traffic figures. Fig.1. shows the throughput versus traffic curves obtained for different spreading sequences. Channel occupancy for each throughput value has been shown in fig. 2. Based on this data and the spreading sequence properties table 1 has been drawn to show the collision resolution requirement and the power advantage available for a throughput of 0.1. It shows that for some of the sequences investigated, the full collision resolution capability of the spreading sequence is not utilised simply because there are not enough simultaneous transmissions. From the limited evaluation it comes out that out of the studied sequences, Barker sequence of length 13 is the best spreading sequence available. This sequence provides a power advantage of 7.78 dB at throughput of 0.1. This much power advantage in link design is sufficient to facilitate the configuration of a VSAT mesh network.

Link calculations

Following traffic figures were assumed for the purpose of configuring the network.

No. of messaging terminals	= 10,000
Messages/terminal/day	= 20
No. of words/message	= 30
No. of characters/word	= 6
No. of bits /character	= 8
Overhead bits(incl. 3/4 coding)	= 35%

From the above figures, on an eight hour working day basis the messaging traffic on the network works out to be 13.5 Kbit/s. At a throughput of 0.1 for the channel this requires a burst rate of 135 Kbit/s.

Link Budget:

Uplink:

VSAT Transmit Power	3.0 dBW
VSAT Antenna Diameter	1.8 m
VSAT Antenna Efficiency	55%
Uplink Antenna Gain	38.5 dBi
Pointing and Rain Loss	1.0 dB
Transmit EIRP	41.5 dBW
Uplink Path Loss	200.0 dB
Satellite G/T from EOC	-5.0 dB/K
Uplink C/No	64.1 dBHz

Satellite:

Saturation EIRP	32.0 dBW
Saturation Flux Density from EOC	-92.0 dBW/m^2
Input Backoff	29.5 dB
Output Backoff	21.5 dB
Backed off gain improvement over saturation	8.0 dB
Transfer Gain	169.0 dB

Downlink:

Satellite EIRP	10.5 dBW
Downlink Path Loss	196.0 dB
Rain Attenuation	0.5 dB
Receive Gain	35.0 dBi
VSAT system noise temperature	70 deg K
Downlink C/No	55.2 dBHz

Link Summary:

C/IMo(IM noise at C/N of 18 dB)	70.1 dBHz
C/No available	54.5 dBHz
C/No required (see note 1)	53.5 dBHz
Total operating margin	1.0 dB

Note 1. C/No)reqd. is given by

$$Eb/No + 10 \log Rb + Mi - Gc - Gs$$

where,
Eb/No = 11.4 at BER of 10-7
Rb = burst rate of 135 Kbit/s
Mi = Modem implementation margin, 2.5 dB
Gc = Coding gain for rate 3/4 code, 3.9 dB
Gs = Spread advantage for 13 bit spreading sequence and 0.1 throughput operation

2. Inter-satellite interference is not taken into account.

3. There is no deterioration due to self interference.

The operation of one carrier at 135 Kbit/s in Spread ALOHA requires only 10.5 dBW of transponder power. However, there will be 6 simultaneous transmissions at the considered throughput. This would require a total peak EIRP of 18.3 dBW which amounts to a typical transponder power of 1/23 only. Such a network is therefore very attractive for transponder sharing with bandwidth-limited applications like CDMA.

Discussion

This paper has demonstrated the feasibility of configuring a VSAT mesh network in C-band primarily for messaging applications. However, provision for occasional telephony can be readily incorporated.

Since the network can operate with reduced EIRP or smaller dish per terminal, the costs go down. Moreover, Spread ALOHA, as compared to Spread Spectrum, employs a single spreading sequence for time spreading by all the terminals in the network. This feature of Spread ALOHA can greatly reduce the complexity of the terminals in the network bringing about further cost reductions by economies of scale in production and subsequent maintenance.

Though a 1.8 meter antenna with 2W transmit amplifier has been assumed, the mix can be varied to suit particular situations. For example, a tradeoff reducing the size of the power amplifier in certain situations may enable the terminals to run entirely on Solar power which will be a significant feature for many developing countries.

Table 1 has shown that longer spreading sequences with relatively poorer correlation sidelobes than Barker sequences, do not provide power advantage at useable throughputs. This needs further investigation. The 10% throughput considered here is obtained with typical packet delivery delays of 4 sec or less. However, in messaging applications, one can tolerate larger delays. This implies that the link can tolerate larger number of retransmissions which in turn, means that the link can be designed for lesser number of simultaneous transmissions in the network. The reduced number of simultaneous transmissions permits greater spread advantage to be available for VSAT power reduction. In that case, the longer sequences [3] would also become useful. In general, moderate length sequences with low auto-correlation sidelobes hold the promise of reducing the VSAT power and size further to such levels as to allow configuration of truly low cost portable/mobile satcom terminals.

Acknowledgements:

The authors gratefully acknowledge the improvements brought about by the contributions of Mr. A. K. Ghose and an anonymous colleague.

References:

1. Norman Abramson, "Spread ALOHA for VSATs," Technical Report B86-4, University of Hawaii, June,1986.
2. Rao, Y. S. and Goel, M. K., "Simulation of a Spread ALOHA Protocol for a Rural Messaging Network," To be presented at ICCS '88, Singapore, October 1988.
3. Ping Fai Li, "Binary Sequences with Low Autocorrelation," Technical Report, University of Hawaii, May 1986.
4. Goel, M. K. and Vinod Chandra, "Computer Modelling of a Satellite Based Messaging Network," To be presented at ISMM International Conference, Miami, Florida, December,1988.

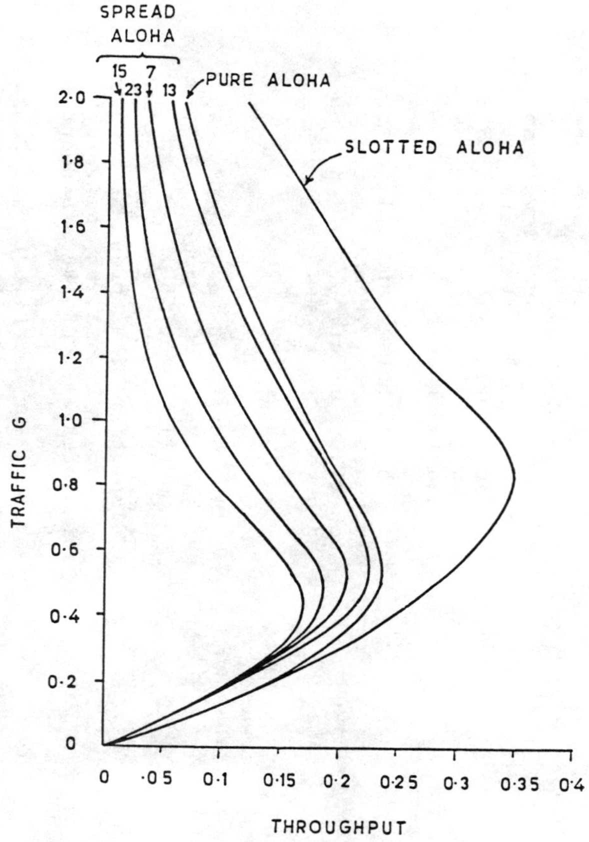

FIG 1

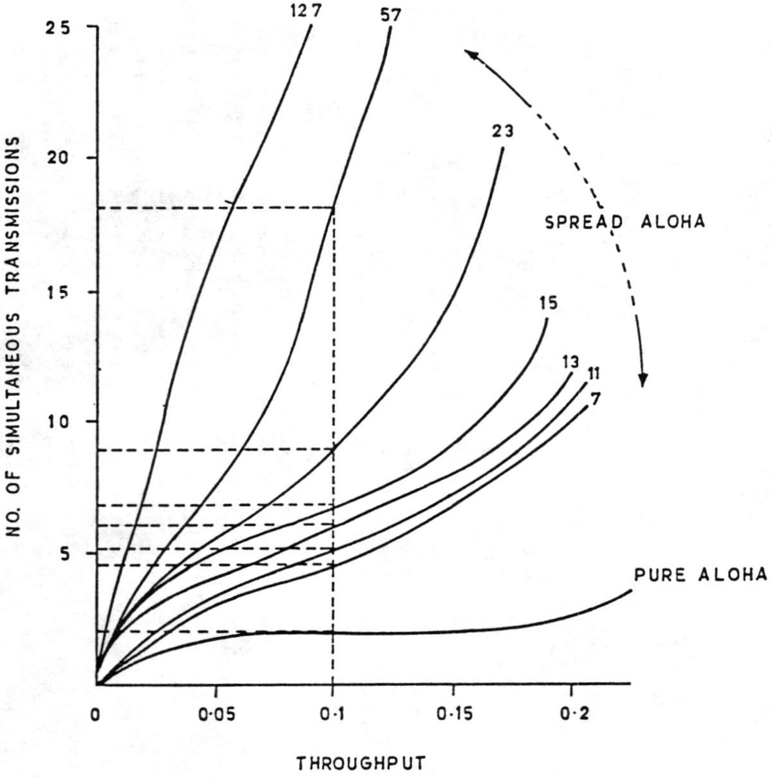

FIG 2

Table 1.

Sequence length n bits	Max. Ck	Collision resolution capability	Channel occupancy at S=0.1	Power advantage available
7(Barker)	1	6	4	3 dB
11(Barker)	1	10	5	6.9 dB
13(Barker)	1	12	6	7.8 dB
15(PRBS)	3	4	7	NIL
23 *	3	7	8	NIL
57 *	5	10	18	NIL
127 *	8	15	31	NIL

* sequences [3].

AA/TDMA - ADAPTIVE SATELLITE ACCESS METHOD FOR MINI-EARTH STATION NETWORKS

A. Fujii, Y. Teshigawara, S. Tejima NEC Corporation, Japan
Y. Matsumoto NEC America, Inc. VA.

ABSTRACT

The newly developed satellite access method called AA/TDMA is presented. This technique is most suitable for star type mini-earth station data communication networks. Short transmission delay can be achieved for transaction type traffic, and at the same time high throughput efficiency can be realized for batch type traffic.

INTRODUCTION

A satellite data communication network using mini-earth stations (1) (2) may be the most attractive way to connect many data application terminals deployed over wide geographically dispersed area with their host computers. The primary applications include point-of-sales (POS), banking/financial, credit card verification, and other computer communications.

The mini-earth stations, easily installed at the user premises, will support user data terminals directly connected to them. The mini-earth station network will replace the conventional terrestrial data network with significant cost savings.

In interactive type applications such as inquiry/response, short response times may be the most important requirement to the user. In this case, since the traffic is usually light and bursty, the throughput efficiency of the satellite channel may be less critical.

The same network is often also used to transfer batch data, such as file transfer or facsimile data. In this case, the throughput efficiency of the channel may be more important than the short response time, since the data volumes are usually higher.

The adaptive satellite access method, AA/TDMA (Adaptive Assignment Time Division Multiple Access), newly developed for NEC's mini-earth station networks called NEXTAR, are presented here. The AA/TDMA scheme offers short response times for the interactive type traffic and at the same time provides high throughput efficiency for the batch type traffic.

MINI-EARTH STATION NETWORK AND SATELLITE ACCESS METHOD

The typical mini-earth station satellite data network consists of many mini-earth stations connected to user data terminals, a satellite, and a hub earth station connected to a host computer system. The user terminals are connected to the host computer through the satellite channels as shown in Fig.1.

The star type network configuration may be the most suitable for the mini-earth station networks. Each mini-earth station can communicate only with the hub earth station which is equipped with a relatively larger antenna. Direct communications between mini-earth stations need more power for both the mini-earth stations and the satellite, thus double hop connections through the hub earth station are practical when necessary.

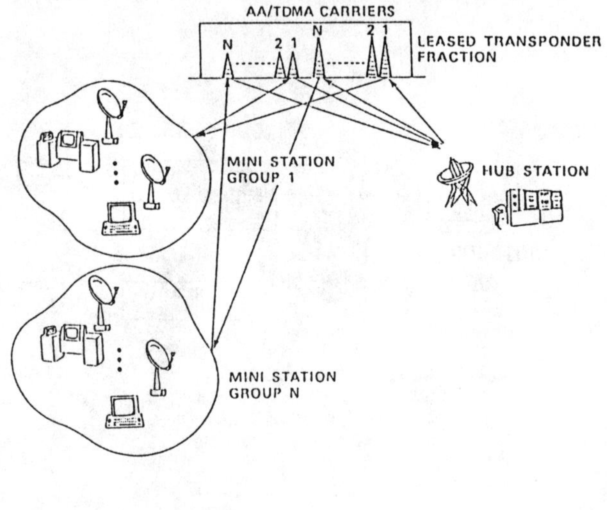

FIG. 1 Mini-Earth Station Network

Reprinted from *Proc. GLOBECOM '86*, pp. 1494–1499, Houston, Texas, December 1986.

Ku-band satellites are particularly suitable for these applications since interference problems are less likely than in C-band satellites. Also, antennas with very small aperture of such as 1.2 meter or 1.8 meter diameter can be used for the mini-earth stations using Ku-band.

The mini-earth stations as shown in Fig.2, consist of an antenna, an outdoor unit (ODU) and an indoor unit (IDU). A block diagram of a typical mini-earth station is shown in Fig.3. These mini-earth stations are installed very easily on roof tops or in the back yard of the user's office buildings where the user data terminals are located. The IDU and the user terminals are usually connected directly using standard interfaces at DCE/DTE points. The baseband processor (BBP), which is integral to the IDU, performs data processing and satellite access control.

FIG. 2 Mini-Earth Stations with 1.2 meter and 1.8 meter Antenna

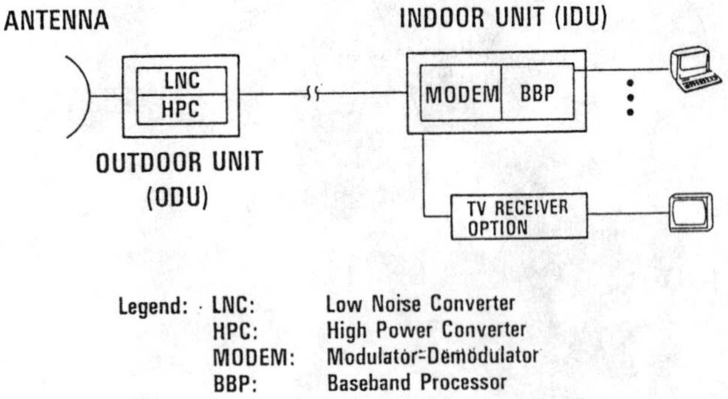

FIG. 3 Block Diagram of the Mini-Earth Station

The hub earth station is typically located near the host computer system. A medium size antenna ranging from 3 meter to 11 meter is used depending upon link budget requirements. The block diagram of a typical hub earth station is shown in Fig.4. Typically, several lines from the front end processor (FEP) of the host computer system are connected to the satellite channel modulators and demodulators through data processing and controller units called Satellite Access Controller (SAC).

Many mini-earth stations can dynamically share one satellite channel since the traffic volume to and from one mini-earth station is usually far less than the transmission capacity of the satellite channel. The satellite channel can therefore be used economically.

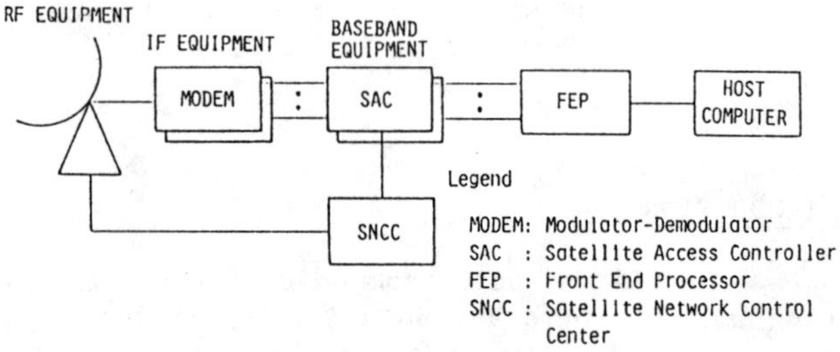

FIG. 4 Block Diagram of the Hub Earth Station

Satellite Access Method

When considering satellite access method for the mini-earth station networks, it should be noted that they are inherently star type. That is, only the hub earth station can receive signals from the mini-earth stations. And, the mini-earth stations can receive only signals from the hub earth station. The mini-earth stations usually cannot receive or monitor signals from the other mini-earth stations.

Various kinds of satellite access method have been proposed to share the satellite channel by many mini-earth stations. Features of some of these conventional methods when applied to the mini-earth station networks are briefly discussed below.

RA/TDMA The so-called "slotted ALOHA" (3) or Random Access TDMA (RA/TDMA) scheme may be suitable when the traffic from each mini-earth station is light and bursty.

In this scheme, each mini-earth station transmits data packets in a burst signal within a TDMA slot any time it is ready. The burst signals are thus transmitted randomly from the participating mini-earth stations, and sometimes they may collide with each other. If the signals collided, they are lost and must be re-transmitted.

This scheme offers short transmission delay when the channel loading is relatively light and thus the probability of the collisions is low. Therefore, this method may be suitable for the transaction type application in which the traffic is light and bursty while the short transmission delay is essential.

However, when the traffic volume become heavy and the probability of collisions become high, the transmission delay increases rapidly. Also, the channel throughput is limited to approximately 37 % of the channel capacity due to the increased collisions.

DA/TDMA The Demand Assignment TDMA (DA/TDMA) scheme may offer high throughput of the satellite channel at a little sacrifice of a transmission delay time. When data packets to be transmitted arrive at a mini-earth station, a request is sent to a control processor at the hub earth station, which replies with TDMA slot assignments. Then the data packets are transmitted in the assigned slots. A delay time of two satellite round trips is spent before the actual data packets are sent.

Therefore, the total delay time is at least three times the satellite round trip delay time. The satellite channel can be used efficiently if this amount of delay can be tolerated.

AA/TDMA - ADAPTIVE ASSIGNMENT TDMA

The Adaptive Assignment TDMA (AA/TDMA) scheme has been developed for the NEXTAR mini-earth station networks which are adaptively applied to both interactive and batch applications. It works essentially the same as the RA/TDMA for interactive data and thus the minimum transmission delay time can be achieved. At the same time, it is also similar to the DA/TDMA to obtain high throughput during batch data transmissions. Co-existence of the interactive and batch data is permitted without sacrificing these critical merits.

The other major feature of the AA/TDMA scheme is that important system parameters such as the TDMA slot length can be changed in order to adapt to the specific applications.

The operating principle of the AA/TDMA scheme is described hereafter.

Outbound and inbound traffic flow is handled in an asymmetric fashion. The outbound channels from the hub station to the mini-earth stations are operated in a continuous TDM mode, and in the inbound channels from the mini-earth stations to the hub earth station the packets of data are transmitted in a burst TDMA mode.

Data packets sent from the hub earth station and destined to one or more mini-earth stations share the same outbound carrier as shown in Fig.5. Each data packet is transmitted successively in a packet multiplex manner, in a first-in first-out basis, asynchronous to the frame timing as soon as it is ready to be sent. The data packets have an address field containing the address information of the destination mini-earth station. All the mini-earth station receive the packets transmitted. However, each mini-earth station checks the address field of the received packets, and the packets with an address field which matches its own address will be accepted, while the other packets will be rejected.

Frame timing and control sequences are periodically inserted in the outbound data. It is used to synchronize the inbound TDMA burst transmitted from the mini-earth stations. It is also used to send acknowledgement for the inbound packets and other control information to the mini-earth stations.

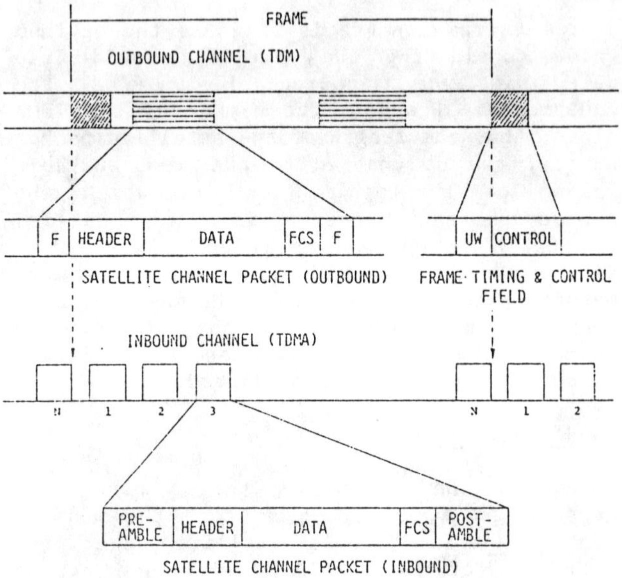

FIG. 5 AA/TDMA Frame Format and Packet Format

The inbound data packets are transmitted in a burst format as also shown in Fig.5. Each burst is transmitted in a TDMA slot. Several TDMA slots are contained in a TDMA frame. A single satellite channel is shared by many mini-earth stations. The timing of the TDMA frame is synchronized to the frame timing signal derived from the frame timing and control sequence in the outbound channel, so that the inbound data bursts transmitted in the same slots from many different mini-earth stations will arrive at the satellite approximately at the same time.

The timing and control sequence in the outbound channel consists of a Unique Word (UW) and a control field. The UW is used to identify the sequence of the other data.

The inbound satellite packet is a burst signal consisting of a preamble, a header, a data portion, a Frame Check Sequence (FCS), and a postamble. The preamble is used for carrier recovery and clock recovery of the hub PSK demodulators. The header contains an address field to identify the transmitting mini-earth stations, and various control information. Since the length of the data field is fixed for particular application, dummy filler bits will be inserted after the user data when the length of the user message is shorter than the field length. On the other hand, if the user message length exceeds the length of the data field, the user message is segmented and transmitted in multiple satellite packets. The FCS is used to detect transmission errors. The postamble is attached for the need of a convolutional error correction circuit in the receiver.

The frame length is fixed to the optimum value considering the satellite round trip delay time and the processing time of the equipments. However, the number of the TDMA slots, thus the length of the satellite packet bursts, can be changed to the most suitable value for the application for which the mini-earth station network is used. That is, the burst length can be selected so that a single satellite packet can just carry a user message packet of average or the most probable length. Then the channel capacity of the inbound channel can be used efficiently adaptive by suit to the application.

Random Access Mode

When the length of the inbound user message is short enough so that the message can be sent in a single satellite packet, the operation of the AA/TDMA is random access quite similar to the slotted ALOHA system. The concept of the random access mode of operation is depicted in Fig.6. Each

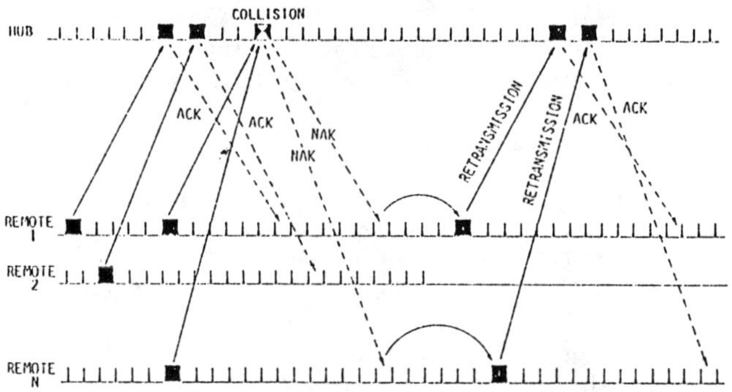

FIG. 6 Random Access Mode Transmission

mini-earth station will transmit its data packets in a TDMA slot as soon as it is ready for transmission. The hub returns an acknowledgement in the frame timing and control sequence in the outbound channel if the packet has been received successfully.

However, since the other mini-earth stations may transmit a burst in the same slot, sometimes collisions may occur. In this case, the data from both the mini-earth stations will be lost, and the mini-earth stations which transmitted the packets will know that since they do not receive the acknowledgement in the expected frame timing and control sequence. Then the mini-earth stations re-transmit the data packets after a randomly chosen time interval to minimize the chance of the second collision.

Under this mode of operation, most of the packets will reach the hub at the first trial if the traffic loading is adequate. In this case, the transmission delay time of the data is the single hop satellite round trip delay time of approximately 270 milliseconds plus processing time. This is the shortest delay time which can be achieved using the satellite transmission.

Even if a collision occurs at the first trial, most of the packets will successfully reach the hub at the second trial. In this case, the transmission delay time will be less than approximately 1 second plus the processing time.

This random access mode operation is most suitable for transmission of delay sensitive transaction data, since usually their message length is short and traffic density is low.

Reservation Mode

When the length of the incoming user data packet is longer than a predetermined value and cannot be accommodated in a single satellite packet, the operation of the AA/TDMA will be the reservation mode as illustrated in Fig.7. In this mode, the mini-earth station transmits the first portion of the user data in a satellite packet with a reservation request for time slots for the remaining portion of the data. This first packet may suffer collision. Therefore, re-transmission may be necessary when collision occurs. If the first packet with the reservation request is successfully received by the hub earth station, the hub satellite access controller (SAC) will assign an appropriate number of time slots to transmit the remaining data, and inform the requesting mini-earth station of the location of the reserved time slots in the frame timing and control sequence.

At the same time, all participating mini-earth stations will also be informed in the frame timing and control sequence that these slots have been reserved and that they are not allowed to transmit packets in these slots. Then the requesting mini-earth station will transmit the remaining portion of the user data in the reserved slots with no possibility of collision. The reservations are made message by message as needed.

The channel capacity is therefore used most efficiently. This reservation mode of operation is suitable for transmission of bulky batch data, since the data can be transmitted very efficiently and relatively fast.

The unique feature of the AA/TDMA system is that it allows the co-existence of the transaction type data and the batch type data. The baseband processor (BBP) of the mini-earth station counts the length of the incoming user messages. If the message length is less than the predetermined value, it is regarded as a short data, and transmitted in the random access mode. If the message length is longer it is regarded as a long data, and transmitted in the reservation mode. The decision will be made message by message. The maximum number of slots in a frame to be used for the reservation is limited to a predetermined value so that a certain number of time slots are always kept free for the random access use.

Computer simulations were performed to check the performance of the AA/TDMA system. An example of the result is shown in Fig.8. This figure shows the delay time vs throughput curve in the case when the short data and the long data are mixed. Total throughput of more than 40 % was achieved with little increase of transmission delay in this case.

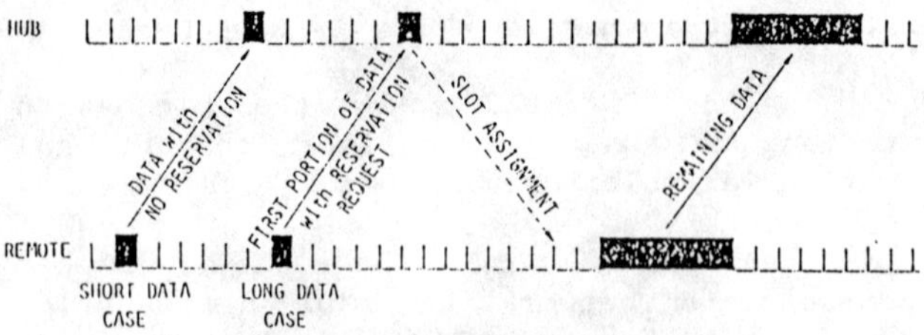

FIG. 7 Reservation Mode Transmission

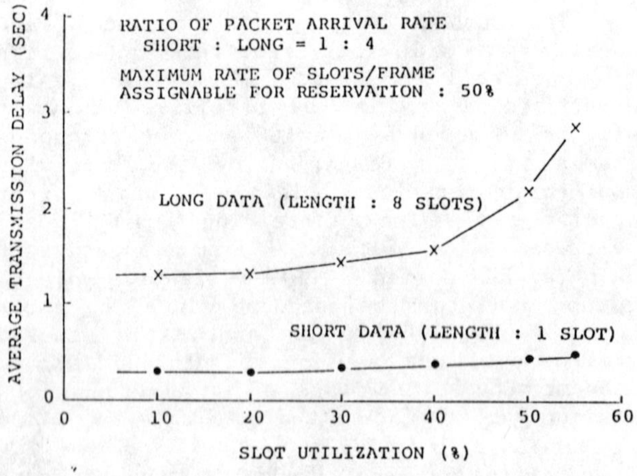

FIG. 8 Delay vs. Throughput Characteristics for Mixed Mode Transmission

ERROR CONTROL, FLOW CONTROL AND PROTOCOL CONVERSION OF AA/TDMA

Basically the proposed mini-earth station satellite data network replaces existing terrestrial data networks with no change of the existing data equipment or their communication protocols. Protocol conversion is performed at the hub earth station and the mini-earth station in order to use the satellite channel efficiently regardless of the satellite delay. The physical level and the link level of the user protocol are converted to the corresponding levels of the satellite protocol and vice versa at both ends of the satellite network.

The SDLC polling from the host is terminated at the hub SAC, and the BBP of the mini-earth station performs local polling to the user terminals. The higher levels of the user protocol are passed in a basically transparent manner within the satellite network, except that the addressing information may be interrogated for the routing purpose. Therefore, the satellite network can be seen as transparent within the user network, while the effect of the satellite transmission delay is compensated.

In the SNA/SDLC case, the mini-earth station network can replace conventional terrestrial multi-drop lines very effectively.

When used in the X.25 packet communication network, the mini-earth station network can form a subnetwork equipped with complete X.25 functions including packet routing.

The outbound channel is continuous and usually with a low bit error rate. HDLC like link protocol is used. If a transmission error occurs within the satellite link, it will be detected by the FCS and the erroneous frames will be discarded by the receiving mini-earth station. Then sequence error occurs at the receiving side and re-transmission of the lost frame is requested using an REJ command. SREJ (selective reject) is not used for the outbound since the error rate is usually very low and little benefit will be obtained by using it. Extended frame sequence numbering is used to increase the maximum number of outstanding frames.

Collisions may occur in the inbound channel. Collided packets are re-transmitted as explained before. This is a kind of the selective reject scheme since only the lost packets are re-transmitted. Inbound Packets may be received out of order and sometimes in duplicate. The hub SAC performs resequencing of the received packets and sends them to the user in a correct order.

Flow Control

The flow control is very important for the inbound channel because instability might be caused when the traffic loading become excessive and the probability of collisions become high.(4) In order to avoid this, elaborate flow control is built into the system. The flow control is performed in two forms. One is a distributed control and the other is a centralized control.

The distributed control operates as follows. When the mini-earth station detects that a packet transmitted from it suffered collision, it regards the system to be congested. Then it suspends transmission of new packets in the random access mode. Only the collided packet can now be transmitted in the random access mode. For the new packets in the queue, the BBP requests slot reservation assignment for them when re-transmitting the collided packet. Then the new packets are transmitted in the reserved slots without collision. The volume of the random access traffic is reduced automatically by this unique algorithm. This mode of operation is canceled automatically once the re-transmitted packet is received by the hub successfully and the acknowledgement for it is received by the mini-earth station. This flow control may be possible only in the AA/TDMA system since the random access and the reservation mode are co-existent in this system.

For the centralized control operation, every inbound packet carries a re-transmission number indicator. The hub SAC can measure the degree of network congestion by counting the indicator of the received packets. If the hub SAC detects excessive network congestion, it first issues a broadcast command to all the participating mini-earth stations to extend the re-transmission interval. This will effectively reduce the traffic which flows into the network. If congestion persists, the hub will issue a broadcast command to stop transmission of new packets.

By using the above mentioned distributed and central flow control scheme, a catastrophic system performance degradation due to excessive traffic flow can be effectively prevented in the AA/TDMA system.

ACKNOWLEDGEMENT

The authors would like to thank members of NEC Corporation and NEC America, Inc. who participated in the development of the mini-earth station systems. They also wish to thank Assistant Prof. Y. Onozato of the University of Electro-Communications for valuable suggestions.

REFERENCES

(1) T. Asai, A. Kikuchi, A. Fujii, "Satellite data communications network by using mini earth stations," in CommunicAsia 86, Conf. Proc., Session 7, pp. 7.26-7.33, April 1986.

(2) A. Fujii, et al, "NEXTAR - The advanced mini-earth station for satellite communications networks," in EASCON 86, Conf. Proc., Session XIII, Sept. 1986.

(3) N. Abramson, "The ALOHA system - Another alternative for computer communications," in Fall Joint Comput. Conf., AFIPS Conf. Proc., 1970, vol.37, pp.281-285.

(4) Y. Onozato, S. Noguchi, "On the thrashing cusp in slotted Aloha systems," IEEE Trans. Commun., vol.COM-33, pp.1171-1182, Nov. 1985.

Packet Radio and Satellite Networks

FOUAD A. TOBAGI, RICHARD BINDER, AND BARRY LEINER

THE FIELD of computer communications has progressed considerably over the past two decades. Based on packet switching, many nationwide and local-area resource-sharing networks exist and operate today. While very successful in supporting conventional computer communications needs, these wire-based networks do not satisfy certain objectives that are often desirable, namely, the provision of high bandwidth for long-haul communications, the support of mobile users, and ease of deployment. Packet radio and packet satellite networks have been introduced to satisfy these objectives. In this paper we describe the structure, protocols, applications, and examples of both packet radio and satellite networks, highlighting the similarities and differences between them.

Introduction

Progress in the field of computer communications began in the early sixties, when we witnessed the problem of connecting remote user terminals to central computer facilities. The solutions to that problem were then based on asynchronous low-speed lines organized in either a star topology with a line dedicated to each terminal, or a tree topology with multidrop lines. By the end of the sixties, the field had made a major leap with the advent of distributed resource-sharing networks. The goal was to interconnect computers and their users at various geographically distributed sites in order to allow the sharing, by all users connected to the network, of hardware and software resources developed at any of the sites. The most prominent example of such a network is ARPANET, which began implementation in 1969 [1].

Such a network comprises:

- the collection of computing resources called *hosts*,
- the collection of users connected to the hosts via terminal access networks, and
- a *communications subnetwork* which interconnects the hosts.

The communications subnet itself consists of two basic components: the relatively high-speed communications channels and the switching nodes. The switching nodes are computers with processing and storage capabilities, which perform what is called *packet switching.* In packet switching, a message to be transmitted between a source and destination is broken into blocks which do not exceed some maximum size. Each block of data is combined with a header containing source and destination addresses as well as control information, and a checksum which is used for error detection and error control purposes, thus constituting a *packet.* With the help of routing and other control protocols implemented in the communications subnet, packets travel independently and asynchronously from source to destination, through (in general) a sequence of intermediate nodes. When a packet reaches some intermediate node, the node receives the entire packet into a buffer, checks its correctness, examines the header information, determines the outgoing channel on which to transmit it, and puts it on the corresponding queue for transmission. The destination node delivers the packet to the appropriate host.

This work was supported in part by the Defense Advanced Research Projects Agency under contracts MDA 903-79-C-0201 and MDA 903-81-C-0638.

Reprinted from *IEEE Commun. Mag.*, vol. 22, no. 11, pp. 24–40, Nov. 1984.

Packet switching is to be contrasted with the more conventional *circuit-switching* technique used in telephony. In the latter, a circuit is set up between the source and destination by the use of special signaling messages before data transmission starts. Once the circuit is established, direct transmission of data from source to destination can take place without any substantial intervention on the part of the communications subnet. Packet switching is favored over circuit switching for computer communications because of the *bursty* nature of computer traffic. Burstiness is a result of the high degree of randomness encountered in the message generation process and message size distribution, combined with a low delay constraint required by the user; that is, the users and devices require the communications resources relatively infrequently, but when they do, they require a relatively rapid response. Thus, bursty users are also characterized by a high peak-to-average data rate requirement. For bursty users, packet switching and the store-and-forward transmission technique offer a more cost-effective solution, since a packet occupies a particular communications link only for the duration of its transmission on that link; the rest of the time it is stored at some intermediate switch, and the link is available for other transmissions. Thus, the communications bandwidth is dynamically allocated, and such allocation is based on elements of fine granularity, namely packets and links in the network.

The success of ARPANET has been evident since its inception, and many similar national and international networks are in operation today (for example, TELENET [2], TYMNET [3], DATAPAC [4], and TRANSPAC [5]). However, experience with ARPANET in the early 1970's has led to the following observations:

- a major component of the cost of a resource-sharing network is incurred by the local collection and dissemination of information, and
- land-based links and the multihop store-and-forward operation do not provide the high bandwidth often required for long-haul communications.

In an attempt to solve the local communications problem, various forms of *local area networks* (*LAN's*) emerged in the early and mid 1970's which circumvent the use of telephone networks. Such networks take the form of rings (exemplified by the Cambridge ring) and broadcast buses (exemplified by Ethernet), or are merely circuit-switched private branch exchanges (PBX's) used for data transmission [6,7]. The transmission media used in these networks range from twisted pairs to coaxial cables, and more recently, optical fibers. While local networks of these types have attained great success in supporting conventional computer communications needs and are becoming standards, there are two often desirable objectives which cannot be satisfied by these solutions; namely, to support mobile users, and to ensure that the network is easy to deploy. A radio-based network can achieve these two objectives [8].

The use of the radio medium for data communications is not new. As early as 1970, a packet-switched radio-based terminal access network called the ALOHA system was built at the University of Hawaii [9,10]. The objective then was primarily to circumvent the use of the telephone network in providing terminal access to a central computer. Given the experience of the ALOHA project, as well as a particular interest in communications and ease of deployment among mobile users, the Defense Advanced Research Projects Agency (DARPA) began (in 1972) to sponsor the design and implementation of a packet radio network with these enhanced features [11–13]. Such a network is intended to serve either as a terminal access network to central computing facilities and long-haul resource sharing networks, or as a standalone network as in tactical applications. Packet radio networks constitute one of the two focuses of this paper.

A solution to the high-bandwidth long-haul communications problem comes with the use of high-bandwidth satellite channels in a packet-switching mode [14]. The satellite transponder is equipped with a global-coverage antenna. Packets are transmitted by the stations on a single uplink channel and are repeated by the transponder on a single downlink channel to reach all stations, thus achieving full connectivity among them. This configuration simplifies the topological design problem considerably, and eliminates the need to store and forward packets many times over long distances.

Packet radio and packet satellite networks have many features in common. They are both packet oriented. They both use shared *multiaccess/broadcast channels;* that is, the communications bandwidth is provided as a single channel to be shared by the many contending users (the radio units and ground stations), offering basically two benefits: a more efficient utilization of the communications bandwidth, and a higher degree of connectivity among the users. Yet these networks differ in many aspects pertaining to network components and their organization, the particular multiaccess protocols used to suit each of the two environments, the various network functions such as routing and flow control, and others.

This paper describes the structure, functions, applications, and some examples of both packet radio and satellite networks; the next section is devoted to packet radio networks, the third section to satellite networks, and the fourth section to a brief comparison of both types, highlighting the similarities and differences between them.

It should be noted that all types of networks (store-and-forward point-to-point networks, LAN's, packet radio networks, and packet satellite networks) have progressed considerably and each type plays an important role of its own. One is not a substitute for another, but instead one complements the other by the added capabilities it can provide. Networks of different types are often interconnected to form a single integrated system, the constitution and the components of which remain transparent to the end users [15].

Packet Radio Networks

Structure

History and Objectives—Pioneering work in packet radio was done at the University of Hawaii as early as 1970. Developed by Abramson and Kuo, the ALOHA system was the first to use the radio medium for data transmission and consisted of a single-hop terminal access network [9,10]. It was also the first to introduce the concepts of *packet broadcasting*, and the use of *random access* in sharing a multiaccess channel. The system comprised two radio channels: a multiaccess channel which the terminals shared to reach the central computer, and a broadcast channel which the computer used to transmit responses back to the terminals. The multiaccess channel was shared according to a "random access" scheme known as *pure* ALOHA. The simplest of its kind, pure ALOHA permits a terminal to transmit any time it desires. However, if a portion of one user's transmission overlaps with another user's transmission, the two collide and "destroy" each other. When the central computer receives a packet successfully, it transmits a positive acknowledgment back to the terminal. If the terminal that has transmitted a packet receives an acknowledgment within a certain appropriate time-out period, then it knows that no conflict occurred. Otherwise, it assumes that a collision occurred and retransmits the packet. To avoid continuously repeated conflicts, the retransmission delay is randomized across the transmitting devices, thus spreading the retransmissions of packets over time. Due to conflicts and idle channel time, the maximum channel efficiency attainable using ALOHA is limited to around 18%. Nevertheless, despite this low channel utilization, pure ALOHA can be shown to be superior to fixed assignment (such as FDMA or TDMA) when the users are bursty and low packet delay is essential [16].

The work done in Hawaii led to the development of a multihop packet radio network called PRNET under the sponsorship of DARPA [11-13]. PRNET is an extension of the ALOHA system in that it permits direct communications among mobile users over wide geographical areas, coexistence with existing users of the frequency band in which it is to operate, protection against multipath effects, protection against jamming, ease of deployment, unattended operation, and reconfigurability. A great deal of research and development work has been undertaken to develop PRNET, leading to a good understanding of how such systems should be implemented. It is the result of this work that we present in this paper.

System Components and Network Organization—The PRNET is a collection of packet-switching nodes, called *packet radio units (PRU's)*, that communicate with each other via broadcast radio. Each PRU consists of a broadcast radio and a digital controller. The broadcast radio acts like the modem

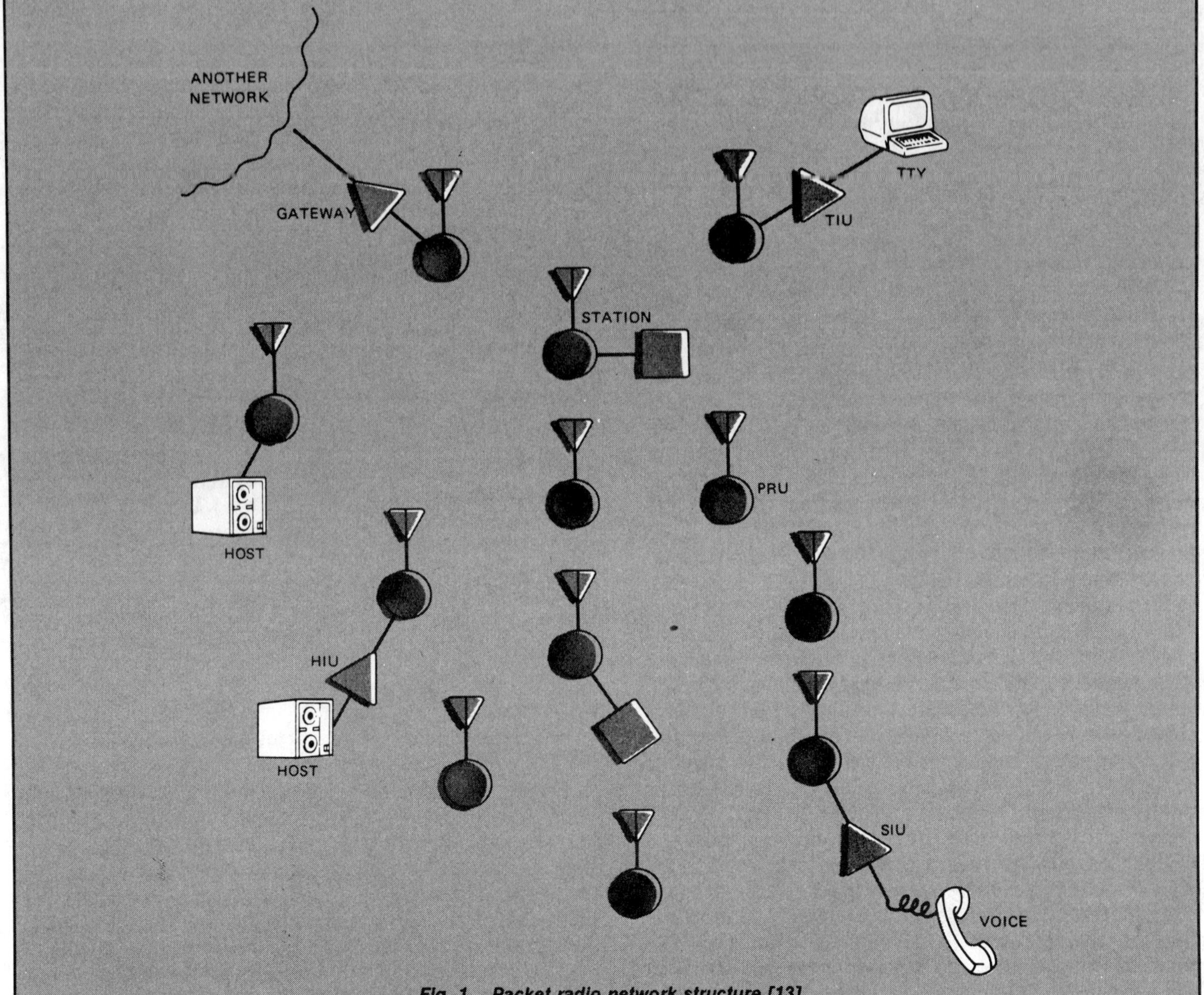

Fig. 1. Packet radio network structure [13].

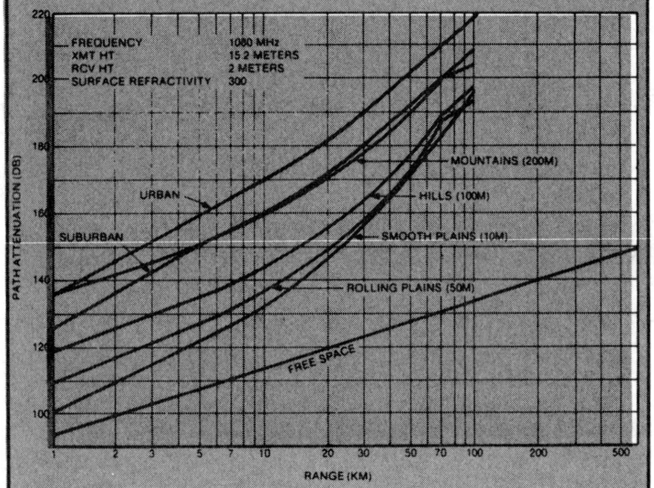

Fig. 2. Path attenuation vs. range for 15-m transmitter height and various terrains [12].

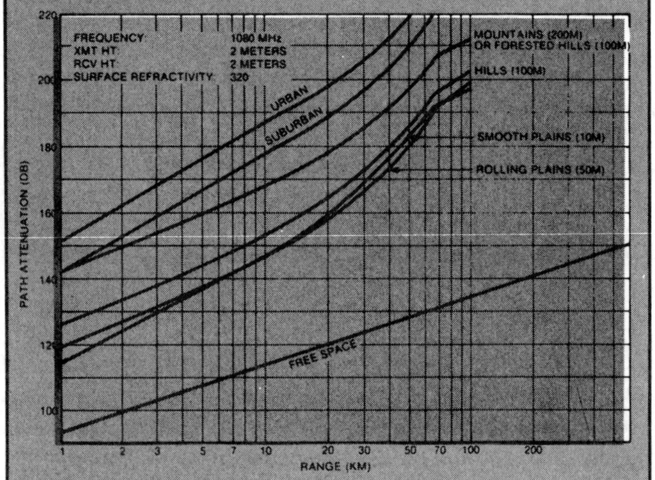

Fig. 3. Path attenuation vs. range for 2-m transmitter height and various terrains [12].

and lines in a terrestrial network, providing the connectivity between neighboring radios. The digital controller provides the packet-switching functions, receiving packets from neighboring radios, making routing decisions, and forwarding packets on to the next radio. In addition, the digital controller implements flow and congestion control mechanisms, assesses connectivity with neighboring radios in a mobile environment, and, in short, implements all the necessary algorithms to permit distributed control of the network in a mobile and potentially hostile environment.

Every PRU in the network is identical. At sources and sinks of information, (that is, user sites), the host computer is plugged into the PRU, and passes packets into the network for delivery at the desired destination. These hosts may include typical service computers, dedicated processors to particular missions, terminal support machines, or internet gateways. In a dense user environment, where there is sufficient connectivity between user sites, repeaters are not necessary as the user radios also serve as repeaters for other user traffic. In certain environments, it is not possible to achieve sufficient connectivity without additional radios. In those cases, standalone PRU's (identical to user PRU's) are used.

Another component in the packet radio network is the *control station*. In the original stages of PRNET development, a station was used to gather connectivity information from the radios, compute suitable routes through the network, and provide these routes dynamically to the radios as new traffic and connectivity patterns developed. In addition, various network control functions are served by the station. Currently, the PRNET uses a multiple-control station algorithm, permitting multiple redundant control of the network. It also uses a stationless fully distributed control algorithm, permitting operation without any control stations at all. Studies and experiments are underway to determine which algorithm best meets the various requirements. A structural model of a packet radio network is shown in Fig. 1.

In order for a packet radio system to support mobile users and be easily deployable, and in order to achieve high bandwidth efficiency, the whole system should operate over a single radio channel centered around one radio frequency, and all PRU's should share this channel using omnidirectional antennas. Accordingly, the radio channel in each locality is at the same time a multiaccess channel and a broadcast channel; the access scheme used by the PRU's is an important element of the network design, and is discussed later in the section titled "Protocols."

The Ground Radio Medium and Its Characteristics—The choice of a frequency band is by and large a function of the propagation characteristics of signals in that band, and the system requirements in terms of bandwidth and connectivity. The higher the bandwidth requirement, the higher the operating frequency must be. On the other hand, a limit on how high the operating frequency can be is dictated by the degree of propagation loss. Indeed, the latter increases with increasing values of the frequency. Beyond 10 GHz, the absorptive losses due to the atmosphere and rain become excessive, considerably reducing the range of PRU's and thus reducing the degree of connectivity. As a result, 10 GHz is considered to be a practical upper limit [12].

Besides having to select the appropriate frequency band, path attenuation and multipath effects in the ground radio environment are two severe problems to be dealt with. Measurements have shown that propagation loss is much more important in real terrains than in free space, and that it is more sensitive to the type of terrain (rolling plains, mountains, suburban, urban) as well as the antenna height than to the frequency band selected. (See Figs. 2 and 3). The implication of this is that the link connectivity among the PRU's is highly variable and may be hard to assess. The system design must take this into account and provide monitoring functions specifically for that purpose.

Multipath is caused by reflections of the original signal leading to a superposition at the receiver of several copies of the signal, delayed with respect to each other. Considering the transmitted signal to be a sequence of symbols drawn from a finite set of waveforms, the effect of multipath is to cause the smearing of individual symbols in the case where the symbol rate is sufficiently low, or to further cause intersymbol interference if the symbol rate is high. This in turn increases the rate of errors in data transmission, an effect which cannot be overcome simply by improving the signal-to-noise ratio. Among the various solutions to combat

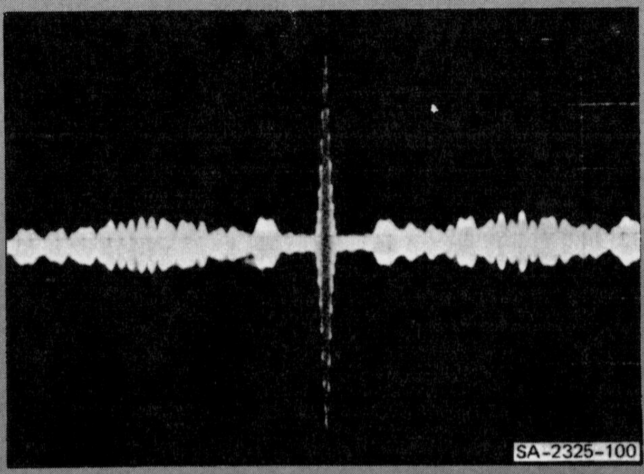

Fig. 4. Typical autocorrelation response of SAWD for a 128 chips/bit PN code [17].

multipath, a good one is to use some form of spread-spectrum signaling. As detailed in the following paragraph, spread spectrum is a combination of bandwidth expansion and coding, whereby the former prevents intersymbol interference from occurring, and the latter allows individual symbols which are smeared by multipath to still be received correctly.

A common way to accomplish spread-spectrum signaling is the use of *direct sequence pseudonoise* (PN) modulation. In its simplest form, PN modulation is as follows. Assume symbols are single bits. A bit of information (say 0) is encoded as a well-specified sequence of n binary bits or "chips"; similarly for information bit 1. It is this n-chip pattern which gets modulated (say by a PSK modulator) and transmitted on the channel. The receiver is equipped with a surface acoustic wave device (SAWD) which knows the pattern to be expected and which performs matched filtering. As the incoming signal is processed, when there is a match, the autocorrelation response of the SAWD will exhibit a main lobe, narrow in width and higher in amplitude than that obtained when there is no match. In the absence of a match, the response is as low as that of noise [17]. (See Fig. 4.) Simple systems may employ fixed-bit-length PN chip patterns used for all bits and by all users. More sophisticated systems may employ bit-by-bit changing PN patterns as a function of time and the users involved, in conjunction with programmable matched filters at the receivers.

The bandwidth expansion factor n is typically one or more orders of magnitude larger than 1. However, the price paid in bandwidth expansion is rewarded by several benefits, the first of which relates to multipath. By keeping the multipath spread to only a fraction of the symbol duration, intersymbol interference is completely suppressed. The delayed signals then create main lobes at slightly different times, allowing the reception of smeared symbols to take place accurately. (See Fig. 5.) Furthermore, these delayed components can be received individually and combined to overcome fading as in diversity systems. The other important benefits beyond combating multipath effects are:

- spread spectrum provides an improvement in the signal-to-signal noise ratio equal to $10 \log(n)$ dB;
- it allows different users and systems to coexist in the same frequency band by providing them with different PN sequences, thus implementing code division multiple access (CDMA); and
- it also offers strong capture properties, allowing a receiver to lock onto a packet in the presence of other interfering packets. This greatly improves the ability of the network to deal with the self-interference inherent to its multiple access nature. (This last topic, which is related to channel access, will be discussed below in "Protocols.")

Another way to achieve spread spectrum is *frequency hopping* (FH). Symbols are modulated and sequentially transmitted on a set of frequencies, hopping from one to another according to some pattern known to the receiver.

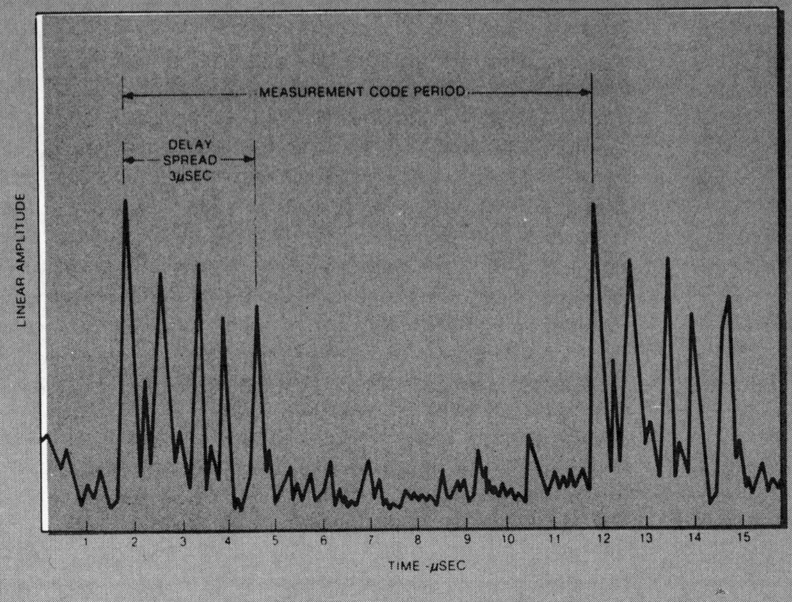

Fig. 5. Channel impulse response in suburban area with multipath [12].

Clearly, symbols transmitted on the same frequency at the same time from two different users are destroyed. To be useful, the FH technique requires forward error correction (FEC), which then allows the entire signal to be recovered despite the presence of lost symbols. In fact, the FEC capability may already have been made available to overcome man-made impulse interference in a radio environment; indeed, observations have shown that such interference could cause a few bit errors in almost every packet transmitted [18]. More details on FH spread spectrum and its benefits in packet radio can be found in [12].

Protocols

Just as with any packet communications network, a packet radio network includes a multitude of protocols ranging from the low-level ones required to transmit a packet across a hop, to higher-level ones required to route a packet from its source to its destination and to provide reliable end-to-end communications. As it is impossible to describe all these protocols in the space available, we shall instead focus on those protocols that are peculiar to packet radio networks. These are: channel access protocols, acknowledgment procedures, routing protocols, and the various network management and monitoring functions necessary to control the operation of such networks.

Channel Access Protocols—Given that a certain communications bandwidth is made available for the network, the question is how to allocate it to the various devices present in the network. In discussing this subject, it is best to differentiate between the two types of signaling; namely, spread-spectrum signaling and (non-spread) narrow-band signaling. One basic difference between the two is that while in narrow-band channels the overlap in time of any two packets reaching the same receiver on the same frequency results in the destruction of both, in spread spectrum there are capture effects which permit overlap in time without necessarily leading to complete loss.

Narrow-Band Channels—At first glance, the solution for systems with a narrow-band channel appears to be one which guarantees orthogonality either in the frequency or the time domain to prevent simultaneous transmissions from destroying each other. Unfortunately, given the bursty character of computer traffic, schemes such as FDMA or synchronous TDMA would lead to low channel utilization, since one would then have to assign enough capacity to each subscriber to meet his peak transmission rates. A more effective approach is to provide a single shareable high-speed channel to all devices, which they would share in some random fashion similar to pure ALOHA. The effectiveness is then due to the strong law of large numbers which guarantees that, with a very high probability, the aggregate demand on the channel at any instant will be approximately equal to the sum of the average demands of the individual users. Moreover, the elimination of complex bandwidth allocation functions simplifies deployment and initialization tasks and improves overall network robustness.

The question then becomes: Can one improve upon pure ALOHA's low efficiency? Recall that the definition we have given for pure ALOHA stated that a device may transmit any time it desires, without regard to channel activity due to other devices. In a packet radio network—where a single channel is available, PRU's are half-duplex, and the transmissions are scheduled randomly in time—it is possible for a PRU to actually be receiving a packet when the scheduled time for a transmission comes up. Requiring that the current reception be aborted to allow the scheduled transmission to take place may result in thrashing. A modification of the above definition would consist of delaying the scheduled transmission until the current reception is complete. This modification has been referred to as *disciplined* ALOHA [13].

Additional improvement over disciplined ALOHA is also possible. Given the data rates achievable (which are limited by the bit error rate required) and the average device range (which is limited by the available power and the propagation loss characteristics), the propagation delay between adjacent units is small compared to the packet transmission time. Under these conditions, it is possible to reduce the possible collisions incurred in a random access scheme by requiring the radio units to sense the channel prior to transmission, and to inhibit transmission if a carrier is detected present. Detecting carrier does not only mean that a packet reception is in progress, it also means that in-band RF energy is present. Known as *carrier sense multiple access* (CSMA), this technique can take different forms depending on the particular action to be taken when the channel is sensed busy. A simple mode, known as *nonpersistent* CSMA, acts as if a collision had occurred, rescheduling the transmission of the packet for a later time. Other variants exist (see [16,19]).

Nonpersistent CSMA offers substantial improvements over ALOHA. In a single-hop fully connected packet radio environment where the ratio *a* of propagation time to transmission time of a packet is about 1%, (for example, 20-mile radius, 100 kb/s data rate, and 1000 bits per packet), nonpersistent CSMA has been shown to achieve a maximum channel utilization equal to 85%. (For $a = 0$, the maximum channel utilization is 100%.) Note, however, that this improvement is attainable only if all nodes are in line-of-sight and within range of each other, and thus can hear each other's transmissions indiscriminately. When this is not the case, unfortunately, some degradation in performance is inevitable. This is known as the *hidden terminal* problem. Studies have shown that if a fraction α of the PRU population equal to 10% is hidden from the rest, all transmitting to a common destination which is in line-of-sight and within range of the entire population, then the channel utilization of nonpersistent CSMA drops to about 35%; it reaches a minimum of about 28% for $\alpha = 0.5$ [16,20].

The hidden terminal problem is intrinsically inherent to multihop packet radio networks. Given that some node *A* is transmitting a packet destined to neighboring node *B*, it is often (if not always) the case that there is some node *C* which is within range of *B* but out-of-range of *A*, and which would thus not hear *A*'s transmission. If *C* had a packet to transmit (no matter to which neighbor) then it would transmit its packet despite the presence of *A*'s transmission, and interfere with the reception of *A*'s transmission at *B*. The performance of CSMA in multihop radio networks has not yet been fully assessed, and work is still in progress [21–24]. Preliminary results obtained for simple multihop network structures have shown that CSMA certainly performs better than ALOHA, but not to the extent that it does in single-hop networks, due to the hidden terminal problem. The fact that collisions occur at the receiving node suggests that a better solution to the access problem could be obtained with the use of a busy tone which is transmitted by a node whenever it is receiving a packet [20,22,25]. Then, any node within range of the receiving node with a packet ready for transmission, sensing the presence of a busy tone, would reschedule the transmission and avoid a collision. Schemes based on the use of a busy tone are called *busy tone multiple access* (BTMA) schemes.

To illustrate how these access schemes compare, we show in Fig. 6 the throughput-delay trade-offs for pure ALOHA, CSMA, and BTMA in a simple network topology consisting of a six-node ring, with uniform traffic among all nodes, infinite buffers available at all repeaters, and $a = 0.01$. These results clearly show the superiority of BTMA over ALOHA and CSMA. Note that, in this figure the cost in bandwidth for the busy-tone channel has not been taken into account. However, it is apparent that the improvement gained in performance outweighs the cost of such a channel. One should also note that the relative performance of these schemes varies with the particular topology examined. Consider, for example, that the network is fully connected; then CSMA and BTMA would become equivalent, with the exception that BTMA again would require additional bandwidth and hardware.

Spread-Spectrum Networks—Spread-spectrum modulation is *not* absolutely essential in packet radio networks. It is, however, motivated by the many benefits listed earlier, pertaining to multipath, processing gain, coexistence, jamming, and self-interference. In this section, we address specifically the self-interference problem and discuss the features spread spectrum offers in handling it. These features are all centered around the resulting property known as *capture*—the ability of a receiver to successfully receive a packet despite the presence of other time-overlapping packets.

The transmission of packets being asynchronous, each packet must be preceded by the transmission of a preamble which the receiver uses to acquire bit and packet synchronization. Typically, the preamble consists of codes which possess a high peak-to-sidelobe autocorrelation property and thus are effective in the bit-sync acquisition process. Having acquired bit synchronization, the receiver can then identify the beginning of the data packet by detecting the end of the preamble. Having successfully processed the preamble of an incoming packet and thus being *locked* onto that packet, the receiver's bit synchronization circuit gates the data detector output through a narrow time window, providing time and multipath discrimination.

For a packet to be correctly received, it is important that its preamble not be completely overlapped in time by another packet's preamble with the same code. More accurately, there is a vulnerable period at the beginning of a packet such that the arrival of another packet during that period would cause the loss of both packets.

Once a packet is locked onto by a receiver, it is important that future overlapping packets do not interfere with the completion of its successful reception. The conditions for such a success are dependent on the ways the PN waveform patterns are selected and used. Consider, for example, that a fixed PN spread-spectrum pattern is used which is identical for each bit transmitted. In this case, an overlapping packet would not interfere with a packet locked onto earlier, as long as its autocorrelation peaks do not coincide with those of the earlier packet; that is, unless the bit periods of the overlapping packets are within a few chip times of each other, causing the peaks to overlap (see Fig. 7). This means that, in addition to the vulnerable period which exists at the beginning of the packet, there is a vulnerable period of a few chip times for each bit during which the arrival of a new packet would cause destruction. Note, however, that the cumulative vulnerable period for a packet remains small compared to its entire transmission time; hence the capture effect.

To minimize the effect of overlapping packets, additional measures can be taken. It is possible to use a PN pattern

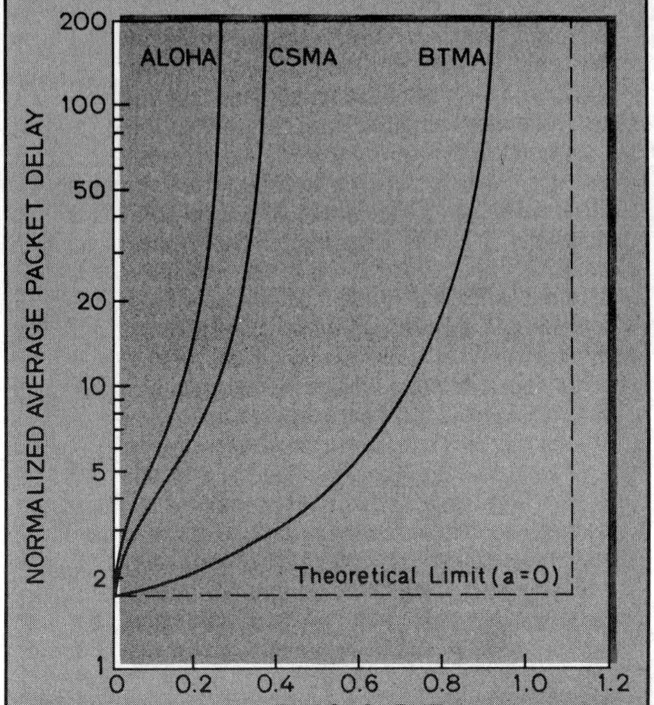

Fig. 6. Throughput-delay trade-offs for ALOHA, CSMA, and BTMA in a ring topology with 6 nodes, infinite transit buffers, uniform traffic requirement, and a = 0.01. Delay is normalized to the packet transmission time. Results are obtained by simulation.

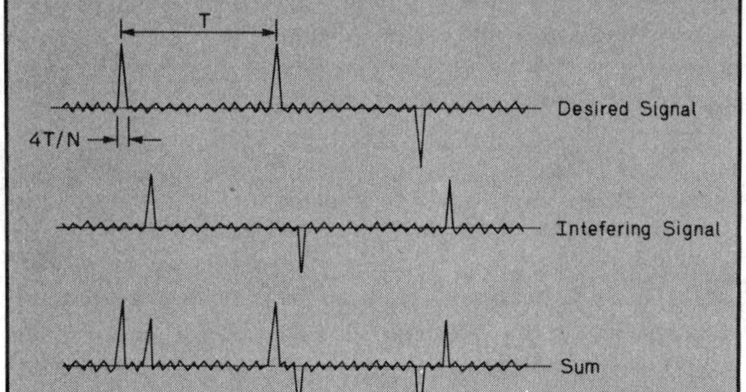

Fig. 7. Effect of interference with no multipath in a PN spread-spectrum network. (T is the bit duration; $N \triangleq 2TW$, where W is the RF bandwidth, and is a measure of the degree of "spread.") [43]

which varies on a bit-by-bit basis, and to equip the receiver with a programmable matched filter which follows the pattern as it varies from bit to bit. If the pattern is long enough so that it does not repeat during the transmission of the longest packet, then any late overlapping packet would not interfere with the packet locked onto earlier, since it would not produce any autocorrelation peak during the entire reception time of the earlier packet. The problem of packet acquisition incurred with arrival times too close to each other would persist, since one would require the PN sequence to start from the same point, known to the receivers.

This last problem could be alleviated by considering orthogonal PN patterns to be associated with different receivers in a given locality. This technique would clearly limit the degree of interference at a receiver to only that caused by the subset of repeaters trying to reach that receiver. It improves both the packet acquisition phase and the protection against late arrivals. The same PN sequence could certainly be associated with two or more PRU's, as long as they are in noninterfering regions of the environment.

Finally, we note that it is possible to combine some form of "carrier sensing" with spread spectrum. Certainly, spread-spectrum networks do not need carrier sensing, as they already exhibit strong capture effects; but the addition of carrier sensing can lead to some additional desirable benefits. Carrier sensing in the spread-spectrum environment cannot simply be an in-band RF energy measurement as with narrow-band channels, since this may result in a high rate of false alarms; instead, similarly to disciplined ALOHA, it is considered to be the indication as to whether or not the receiver has locked onto some packet. If the receiver has locked onto some packet, then presumably the intended destination must also have done so, and it would be advisable to inhibit the transmission. This makes sense particularly if receiver-dependent codes are used, and the transmitting node is performing the carrier-sensing operation using the intended receiver's code. In any case, whenever possible, carrier sensing helps reduce the overall background noise level, a result especially desirable in environments prone to jamming.

Acknowledgment Considerations—Acknowledgments are needed to notify a device of its success in the transmission of a packet. Success or failure is determined by the use of a polynomial-based cyclic redundancy check (CRC). Because of contention, the CRC is typically 32-bits long. As in most networks, two types of acknowledgments exist: the link-level type, also referred to as the hop-by-hop acknowledgment (HBH) type, and the end-to-end (ETE) type. ETE acknowledgments are transmitted as regular packets by the ultimate destination of a packet back to its source, and are part of an end-to-end reliable transport protocol, (such as the DoD standard TCP [15]). HBH acknowledgments are transmitted by every PRU on the route of a packet, to notify its predecessor of the correct reception of the packet. At each hop, the PRU saves a copy of every packet it transmits to the next PRU along the route, awaiting an acknowledgment from the latter. The copy is deleted only when the acknowledgment is received, or when a certain maximum number of tries has been reached.

The two types of acknowledgments appear redundant. However, they are both essential, for the following reasons. ETE acknowledgments guarantee reliable delivery, despite the failures or congestion to which the network is prone. Indeed, with this mechanism, if the source of a packet does not receive an ETE acknowledgment within an appropriate ETE time-out period following transmission of the packet, it resubmits the same packet to the network for delivery to the destination. The new copy may be more successful, perhaps by taking a new route to circumvent congestion and failures. Moreover, in addition to reliability, ETE acknowledgments play an important role in controlling the flow of packets between the source and the destination [26].

HBH acknowledgments are present mainly for efficiency reasons. Without HBH acknowledgments, a failure in the (single) transmission of a packet by some PRU on the packet's route to its intended neighbor would require a restart of the packet from its source, completely wasting the work accomplished prior to the unsuccessful transmission. This is particularly severe in a packet radio network where, due to the random-access nature of transmissions and the environment's propagation characteristics, the probability of a successful transmission over a hop may be low. Studies have shown that in a path 5-hops long, with a probability of success per hop equal to 0.7, the total average delay of a packet is 12.5 packet transmission times for a system with the HBH acknowledgment mechanism, vs. 53 for a system relying totally on the ETE procedure [27].

HBH acknowledgments may be implemented in several ways: One is to have the receiving PRU transmit *explicit* short acknowledgments consisting typically of only the header, since the latter uniquely identifies the packet. The other, due to the broadcast property of a radio channel, consists of considering the relaying of a packet by the next PRU to be the acknowledgment to the current PRU. The latter scheme is called the *Echo acknowledgment scheme*. Clearly, in this scheme, the acknowledgment at the last hop has to be explicit.

While echo acknowledgments might appear to save channel time, as compared to sending explicit acknowledgments, specific problems are associated with them. Consider, for example, that the PRU's are implementing a single first-in-first-out (FIFO) transmit queue with only one outstanding packet awaiting acknowledgment for each neighboring PRU. Let *A* and *B* be two consecutive PRU's on the path for some source-destination pair. PRU *A* transmits the packet at the head of its queue to PRU *B*. When received by PRU *B*, this packet is put at the bottom of its transmit queue. Then, the echo acknowledgment awaited for by PRU *A* will not be received until PRU *B* services that packet, that is, the bottom of its queue. This problem is particularly severe if packets are travelling along a string of repeaters. Another problem associated with echo acknowledgments is that, being as long as the original packet, they are more likely to be interfered with than shorter explicit acknowledgment packets. This may cause severe degradation in performance, especially in highly congested regions, since the original PRU would be required to undertake additional transmissions beyond the first successful one. Thirdly, echo acknowledgments cannot be used in spread-spectrum systems with receiver-dependent codes, since the next PRU would not be using the PN waveform corresponding to the original PRU. Explicit acknowledgments, on the other hand, seem to present benefits which would outweigh the presumably additional channel time that they would require. In addition to being less prone to interference, they can be given priority over regular packets, thus speeding up the freeing of the buffers.

Network Management and Routing [12,13]—Routing consists of the determination and set-up of paths between the sources and destinations of traffic along which packets are forwarded. Were the topology of a packet radio network fixed, then techniques similar to those employed in point-to-point store-and-forward networks would be appropriate. These techniques, typically based on the *sink tree* concept, are implemented by means of routing tables stored at the switching nodes which, for each destination, give the next node to which a packet with that destination is to be transmitted. These routing tables may remain constant, as in static routing, or be dynamically modified to reflect changes in the traffic load and its distribution in the network. But, since in radio networks the topology is constantly changing (due to the mobility of users and network devices, as well as high variations in RF connectivity), routing, to a great extent, must be assisted by various network management functions which constantly track topological changes in the network. This is why routing and network management are closely related and treated jointly here.

In a mobile environment, there are two basic approaches to network management and routing. The first is to have information gathered at a central (or perhaps a few) site(s), have the routing and other management functions performed there, and then have the routes distributed to the various nodes. Each PRU must exchange information with its neighboring radios to assess local connectivity information. This information is then forwarded to the control station where the routing computation is performed. In the case of multiple control stations, each station gathers the information about PRU's in its local area, communicates with other stations to establish global routing decisions, and then provides the information to the PRU's. This approach has the advantage of centralized (or distributed over a limited number of sites) computation, and may reduce overhead further than the more distributed scheme. However, it is inherently less robust, as loss of stations means loss of network management.

The second approach is a more distributed approach, where each PRU exchanges a portion of the global routing data base with its neighboring PRU's, in addition to assessing local connectivity. This data shows each PRU the best next hop on the way to the final destination. Thus, the routing and other network management functions are fully distributed so that loss of any node does not reduce functionality, (although some traffic handling capability may be reduced). This approach may result in more overhead, however, and is the subject of ongoing studies.

In either approach, due to changing connectivity, the current information as to the best route through the network may be incorrect. Thus, alternate routing procedures are necessary. These procedures allow PRU's to ask "help" of their neighboring PRU's in getting the packet to the destination.

Existing Systems

The ALOHA system and PRNET are perhaps the only known packet-radio networks. More than once, public network suppliers have hinted at or even announced future offerings of services based on radio as an alternative to the telephone local loop, but unfortunately no real materialization has yet taken place [8]. The best example is the Xerox Telecommunications Network (XTEN), which aimed at providing "Electronic Message Service." Xerox spent several years of study in this case, but then abandoned the whole project early in 1981. However, with the recent divestiture by AT&T of its operating companies, interest of the private sector in packet radio networks may be resurrected. It is most likely that resulting systems will be privately owned and have a single-hop centralized topology not unlike the ALOHA system.

PRNET remains the only system with all the enhanced features described in this paper. Its design has gone through many revisions, and there are now three experimental networks in operation. A field network is deployed in the San Francisco Bay area and is used for development (Fig. 8). The other two are located at Ft. Bragg, North Carolina, and at the Strategic Air Command (SAC) in Omaha, Nebraska, and are aimed at determining the applicability of this technology to battlefield operations.

It should also be noted that there is packet radio work being carried out by the Royal Signals and Radar Establishment in the United Kingdom [44], and some degree of packet radio investigation being carried out by the amateur radio community.

Packet Satellite

Structure

Broadcast Networks—Existing packet satellite networks have the structure shown in Fig. 9. A single satellite in geosynchronous orbit is used to provide connectivity among an arbitrary number of earth stations. A global-coverage antenna is used on the satellite, allowing each station to communicate with any or all others at any point in time, subject to the protocols discussed in the next section. The region of coverage is typically several thousand miles in diameter.

A single transponder, or a portion thereof, is used on the satellite to repeat received uplink packets on the downlink. All network stations transmit on the same uplink frequency and receive on the same downlink frequency. The uplink packets undergo amplification and frequency translation (for example, 6-GHz uplink to 4-GHz downlink), but are otherwise not subject to processing at the satellite. If only a portion of the transponder is required, FDMA is normally used to provide a channel of the desired bandwidth, with the transponder operated in its linear region to accommodate the multiple simultaneous channels and to minimize intermodulation products. If the entire transponder is used, the transponder can be operated in its saturated region to achieve greater downlink power.

Earth stations can vary greatly in antenna size and overall cost, depending on the particular satellite's downlink power, the frequencies used, the desired data rates, and the required bit error rate (BER). Existing systems range from the use of Intelsat Class A stations with 30-m antennas at 6/4 GHz, to low-cost stations using 2-m antennas at 14/11 GHz. Data rates range from 19.2 kb/s to 3 Mb/s, while BER's range from 10^{-8} to 10^{-3}.

Forward error correction is normally used to improve BER performance, with interleaving also applied when appropriate, for example, in shipboard environments with bursty error sources. Sophisticated modem/codec processors have been developed to allow several combinations of modulation levels and coding rates to be applied within each transmitted burst [28]. This allows optimization of data rate/BER performance for data, voice, and control traffic.

Earth-station processing includes control of packet trans-

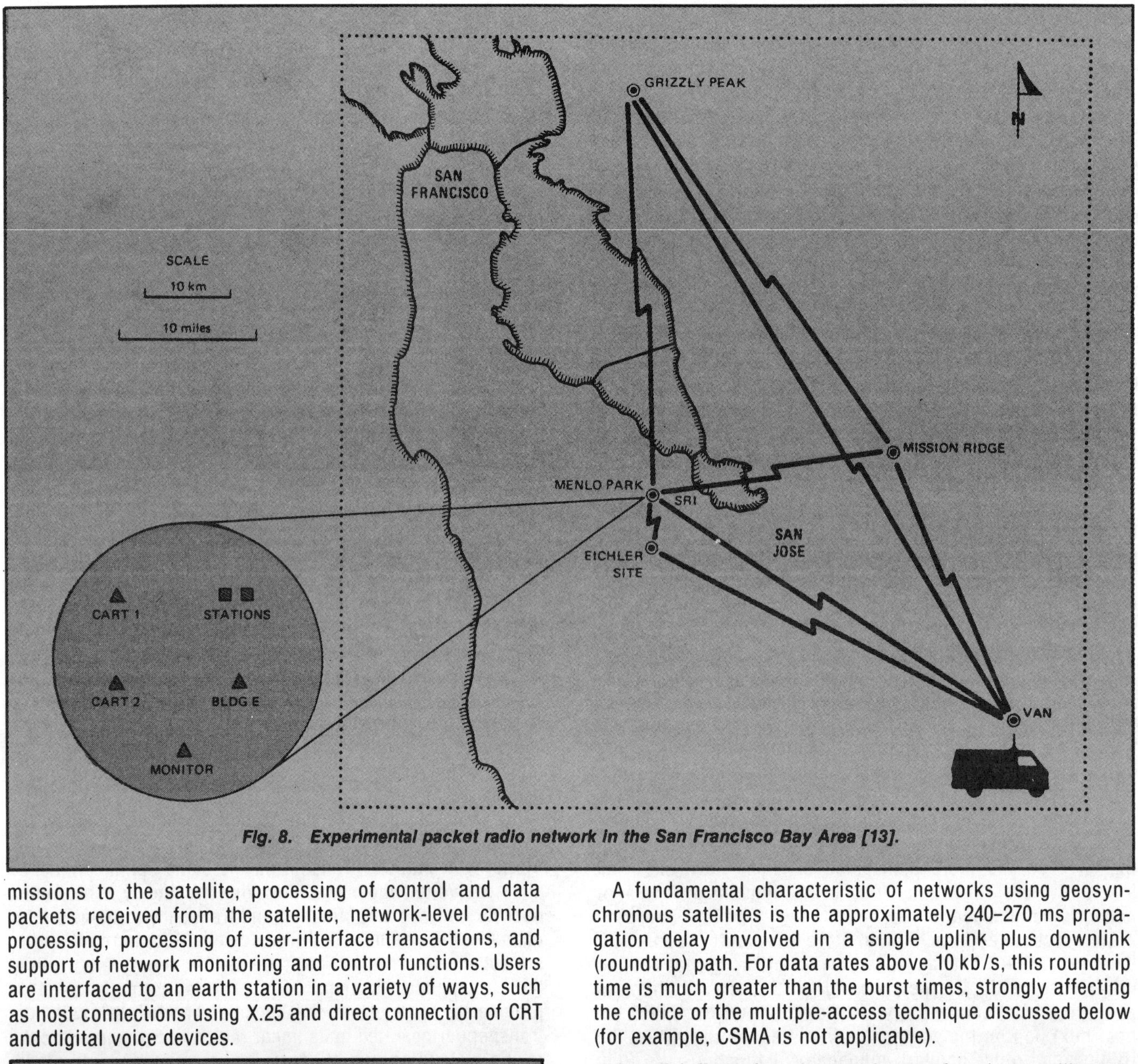

Fig. 8. Experimental packet radio network in the San Francisco Bay Area [13].

missions to the satellite, processing of control and data packets received from the satellite, network-level control processing, processing of user-interface transactions, and support of network monitoring and control functions. Users are interfaced to an earth station in a variety of ways, such as host connections using X.25 and direct connection of CRT and digital voice devices.

Fig. 9. Broadcast satellite network structure.

A fundamental characteristic of networks using geosynchronous satellites is the approximately 240–270 ms propagation delay involved in a single uplink plus downlink (roundtrip) path. For data rates above 10 kb/s, this roundtrip time is much greater than the burst times, strongly affecting the choice of the multiple-access technique discussed below (for example, CSMA is not applicable).

Crosslink Networks—A second type of packet satellite network is shown in Fig. 10. This structure is characterized by the use of two or more satellites with crosslink connections, as well as up/down link connections with the earth stations. The use of several geosynchronous satellites in this manner can provide coverage over most of the world with a single network. If enough satellites are involved, non-geosynchronous orbits can also be used while still providing continuous communications service to a given region. This allows consideration of low-altitude networks with reduced link power requirements and propagation delays.

The communications considerations for a crosslink packet network are in general quite different from the single broadcast network. Because such networks are still in an early stage of development, we shall limit the remaining discussion to the single-satellite broadcast case.

Applications—Packet satellite networks are applicable wherever communication is required among many earth stations spread over a large geographic area. Perhaps their most significant feature is the inherent flexibility in dynami-

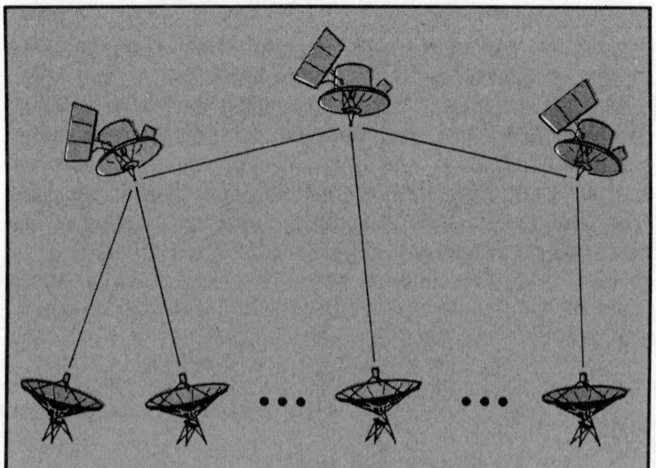

Fig. 10. Crosslink satellite network structure.

cally sharing network bandwidth among the stations. Unlike terrestrial packet networks such as ARPANET, a single station can use the total network bandwidth between itself and one or more other stations if the occasion demands, subject of course to overall traffic priority constraints at the time.

The broadcast networks are also especially well-suited to conferencing and multidestination traffic, since all earth stations can receive every packet transmitted in the network. The large amount of bandwidth potentially available in satellite networks also makes them attractive more generally for integrated voice-data services.

Protocols and Management

Satellite Access Protocols—The major thrust of packet satellite technology has been the development of effective capacity allocation schemes. A central assumption is that traffic requirements among the earth stations have such variability that fixed capacity allocation is inefficient. A second assumption is the absence of terrestrial links between earth stations for passing control information. Thus, the problem is to achieve efficient multiple accessing of the satellite uplink channel, using the uplink plus downlink channels for coordination.

Random Access and Reservation Schemes—Present-day packet satellite systems are a result of work which began about 1972 and proceeded through the decade. An early approach was the use of random access for all data and control packets—in particular, pure and slotted ALOHA [29,30]. In slotted ALOHA, time is divided into slots of duration equal to the transmission time of a single packet (assuming constant-length packets). Each station behaves as in pure ALOHA, with the exception that it synchronizes the start of transmission of its packets to coincide with the slot boundary. Slotting improves the channel utilization of ALOHA by a factor of two, bringing the maximum utilization to 36%.

The direct random-access approaches suffer from relatively limited capacity utilization, however; even with the many variations on the theme which have evolved, practical operating points are limited to about 50% or less [16]. More significantly, they do not readily accommodate heavy flows from one or more stations in the presence of other bursty traffic. The large 270-ms roundtrip propagation delay further aggravates the problem, since each retransmission due to a collision adds at least a roundtrip of delay. This is generally unacceptable for voice traffic, which requires that delays be held as closely as possible to a single roundtrip time.

It was soon realized that, with decreasing processing costs, more sophisticated techniques could be applied to significantly improve performance properties. In particular, explicit reservation schemes were developed in which a short control packet is first sent by stations to request channel time for queued traffic, with subsequent traffic transmission at coordinated (scheduled) times. The relative performance of slotted ALOHA and reservation schemes is shown in Fig. 11.

The first explicit reservation scheme was developed by Roberts [31], in which slotted ALOHA was used for the transmission of reservation packets. Other work at about the same time explored the application of fixed-allocation principles for the reservations and/or portions of the data, as well as the use of implicit reservation schemes in which the data packets constitute reservations [32–34]. In a multicontractor effort sponsored by DARPA, these techniques were synthesized into a reservation system called PODA (Priority-Oriented Demand Assignment), which also extended the reservation concept to include integrated packet voice and data traffic, and priority allocation of capacity [35]. While many choices are possible for the detailed structure of a reservation system, we shall use the PODA scheme to illustrate the principles involved.[1]

PODA—To satisfy the requirements of both data and voice traffic, two distinct reservation mechanisms are used: datagrams and streams. Datagram reservations are made on a

[1] PODA has itself evolved from an early version designed for a 64-kb/s channel to a version accommodating channel rates of several Mb/s. It is the latter's principles that are discussed here.

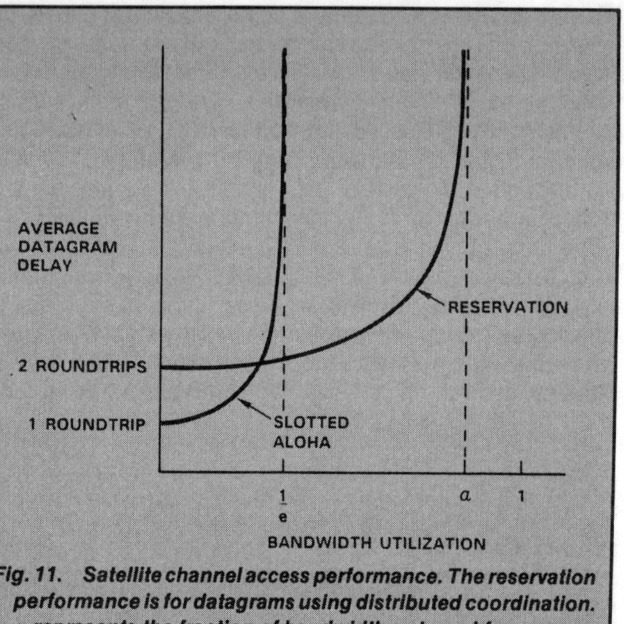

Fig. 11. Satellite channel access performance. The reservation performance is for datagrams using distributed coordination. α represents the fraction of bandwidth not used for reservations, with typical values of 0.7 to 0.9.

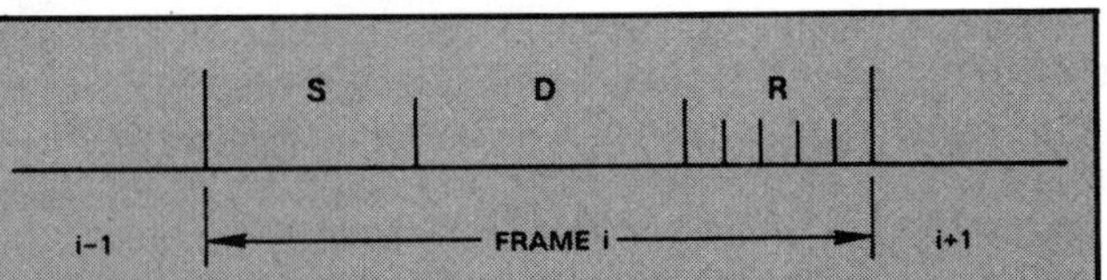

Fig. 12. ***Reservation system channel timeline. S represents the stream subframe; D, the datagram subframe; and R, the reservation subframe.***

per-burst basis, where each reserved burst may contain one or several datagrams. Stream reservations are made on a per-call basis if a single voice call is being supported, and, more generally, as a function of the changing requirements of concentrated voice traffic.

Figure 12 illustrates the channel timeline at the satellite. A fixed repeating frame is used which contains three major subframes. The first subframe contains time currently allocated to streams, the second contains scheduled datagram bursts, and the third contains slots for reservation and other control packets. For a channel data rate of a few Mb/s, the frame size is typically about 20 ms.

In general, the sizes of the subframes vary as a function of time. The stream subframe time can vary from zero when no streams are allocated, to a maximum value which leaves room for a specified minimum datagram and reservation subframes. Similarly, the datagram subframe time can vary from zero when no datagrams are being scheduled, to a maximum which allows minimum values of the other two subframes. The reservation subframe size can also vary as a result of changes in the other subframes.

In the following, we describe the structure of each of the subframes and the associated protocols.

Datagrams

The datagram subframe is shown in more detail in Fig. 13. Each station with an outstanding datagram reservation being honored in the current frame transmits at its scheduled time. (The bursts are separated by a guard time which allows for ranging uncertainties and clock drift.) Bursts vary in size according to their reservation request, containing anywhere from one to *N* distinct datagrams, where *N* is a system parameter which ensures fairness in the use of available datagram capacity. The actual number contained in a burst is a function of the number and priority of datagrams queued in the source station at the time the reservation is made.

The datagram burst structure is shown in Fig. 14. Every burst has a preamble to allow carrier and symbol timing acquisition at the receiver on a per-burst basis. This is followed by a burst control portion which establishes work synchronization, contains the source station address, and contains control fields used for reservation-making and coordination. The individual datagrams also have control information associated with them to identify the intended destination station and host, and to provide error control.

Because an explicit reservation must be made for each datagram burst, the minimum delay for datagrams is either two or three satellite roundtrip times, depending on whether distributed or centralized scheduling coordination is used. Thus, the penalty for achieving the efficient packing of the datagram subframe with different stations' traffic bursts is at least a half-second or more of delay. This not only improves bandwidth efficiency in the presence of both bursty and high-duty-cycle traffic, but also allows flexible priority allocation of capacity among competing stations.

Streams

To provide a delay of only one roundtrip for voice traffic, streams are used. The stream subframe is divided into variable-sized slot allocations, reflecting current stream reservations. The slot for a particular stream can repeat in each successive frame (see Fig. 15), or in selected multiples of a frame. The resulting elapsed time between successive slots is called the stream interval, and reflects the stream's traffic source characteristics (for example, packet voice vocoder rate). The width of each slot allocation, along with the interval, determines the bandwidth allocated to the stream.

A stream might be allocated to a single voice user in a point-to-point conversation, in which case its slots will typically be occupied less than 50% of the time. Because of the inefficient use of channel bandwidth in this case, many voice users are concentrated into a single wider-bandwidth stream at each source host or earth station where possible. For this case, the aggregate statistics of a large number of users allows relatively efficient utilization of each slot, where the slot size is selected to trade-off bandwidth efficiency and talkspurt blocking probability. The stream reservation mechanism allows changes in the slot size to be made while the stream is in progress, accommodating changes in the total number of concentrated users. The stream bursts are structured similarly to the datagram subframe bursts, with one or more individual voice packets contained within each burst.

A single stream can also be allocated to a group of stations for conferencing. In this case, the users participating in the conference coordinate who is to talk at a particular time using a higher-level protocol, perhaps exchanging coordination packets using the datagram subframe (if more than one user speaks at the same time, overlapping packets are sent in the stream's slots with, typically, loss of all the packets). The advantage of this mode is that, at the expense of half-duplex interactions, a single-user stream can be used very efficiently by any number of conference participants. In particular, the broadcast property of the downlink channel

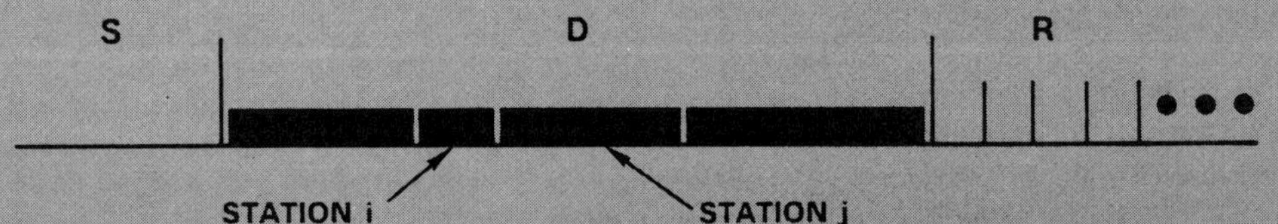

Fig. 13. ***Datagram subframe in PODA.***

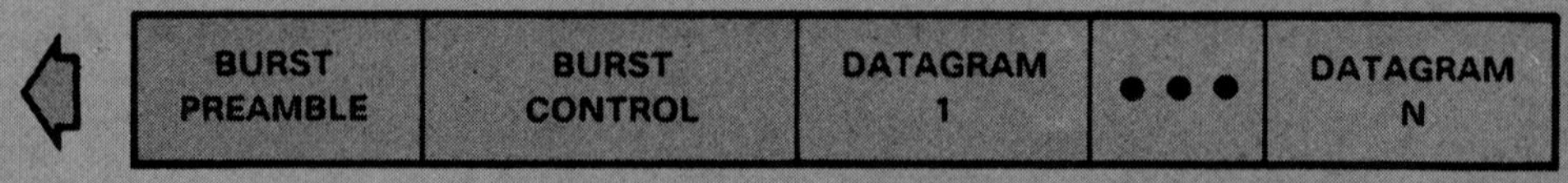

Fig. 14. A datagram burst in PODA.

allows all stations to receive and deliver the packets to all participants.

Stream allocations are also useful for non-voice applications. Certain data applications may require the low delay property, and a stream can allow communication under certain conditions where coordination of datagram scheduling is not possible at a particular station. However, streams impose a system penalty of less flexible allocation changes in response to changing traffic priorities, as well as bandwidth inefficiency for bursty traffic. To accommodate new high-priority streams when capacity is fully utilized, a combination of lower-priority stream preemption and reduction of lower-priority datagram traffic is used.

Reservations

The control subframe can be used in one of two modes: fixed-assignment or contention. In the fixed-assignment mode, a single reservation slot is assigned to each earth station in every frame. In the contention mode, stations use slotted ALOHA to compete for the reservation slots. When the total number of stations in the network is small, for example, less than ten, fixed assignment provides efficient use of reservation subframe capacity while minimizing access delays. For large numbers of stations, contention is used to keep the number of reservation slots small in each frame. In the contention case, the control subframe expands to fill unused frame time, reducing actual contention for the control slots.

A key mechanism which makes the contention mode work well is that reservations can also be made in the headers of scheduled data bursts. Thus, reservations are only sent in the control subframe by a station with no currently scheduled datagrams or streams. When a station has a relatively non-bursty flow of datagram traffic or at least one stream in effect, it does not need a control subframe slot. This technique requires some extra overhead bits in every stream and datagram burst, whether or not actually used for a reservation. This is a reasonably small addition to total burst overhead, and is more than made up in the reduction of control subframe contention.

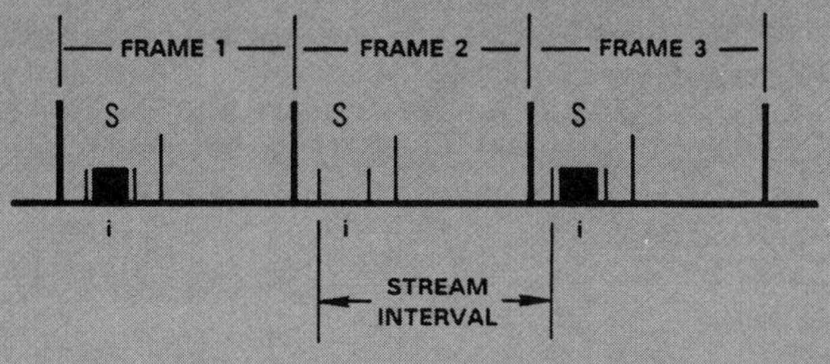

Fig. 15. Stream slots in PODA. (Slot i is allocated to station i in each frame, which uses it to send a stream burst in frames 1 and 3.)

Scheduling Coordination

Two distinct overall control strategies can be applied to the preceding capacity allocation concepts: distributed or centralized. In distributed control, every station executes an identical algorithm to schedule reservations being received on the downlink (Fig. 16). Datagram reservations are placed in a queue and scheduled in the available datagram subframe time. A table is maintained of outstanding stream reservations, and is used to schedule the stream subframe. This approach requires acquisition and maintenance of scheduling synchronization by each station, in addition to the underlying clock and ranging synchronization required of TDMA systems more generally.

Centralized coordination is illustrated in Fig. 17. In this approach, a single station schedules all streams and datagrams, and informs the other stations of their scheduled times. The other stations need be concerned with their own scheduled datagram and transmission times, announced by the central scheduler; no special synchronization is needed beyond the basic TDMA framing. As mentioned earlier, the centralization introduces a third roundtrip delay for datagrams, since each station must wait to be informed about its scheduled datagram times. This additional delay is not significant for streams, since it only impacts the stream set-up time and allocation changes due to priority and concentration factors; the use of individual stream slots still involves only a single roundtrip delay by all stations, whether centralized or distributed coordination is used.

The downlink broadcast property makes distributed control a seemingly attractive goal to pursue, especially so because of the smaller datagram delay and elimination of channel overhead for scheduling announcements. For tactical applications, the robustness inherent in distributed control for dealing with station failures is also important. There are two potential pitfalls to the distributed approach, however: local error rates in downlink control information, and processing/complexity considerations.

Distributed scheduling synchronization is very sensitive to local errors—for a datagram channel rate of 1 Mb/s, performance degrades rapidly for control BER's (after error correction) greater than about 10^{-6} to 10^{-7} (depending on the number of stations and traffic matrix). An inherent difficulty is that, in general, it takes at least one roundtrip time for a station to discover it has lost synchronization due to a missed reservation (accomplished by checking the times of scheduled datagrams received in the downlink against its own schedule). During this time, datagram transmissions made by an out-of-sync station are likely to collide with those made by other stations, due to incorrect scheduling, seriously degrading channel throughput. This effect worsens as channel data rate is increased relative to a given BER, since there is a proportional increase in the number of

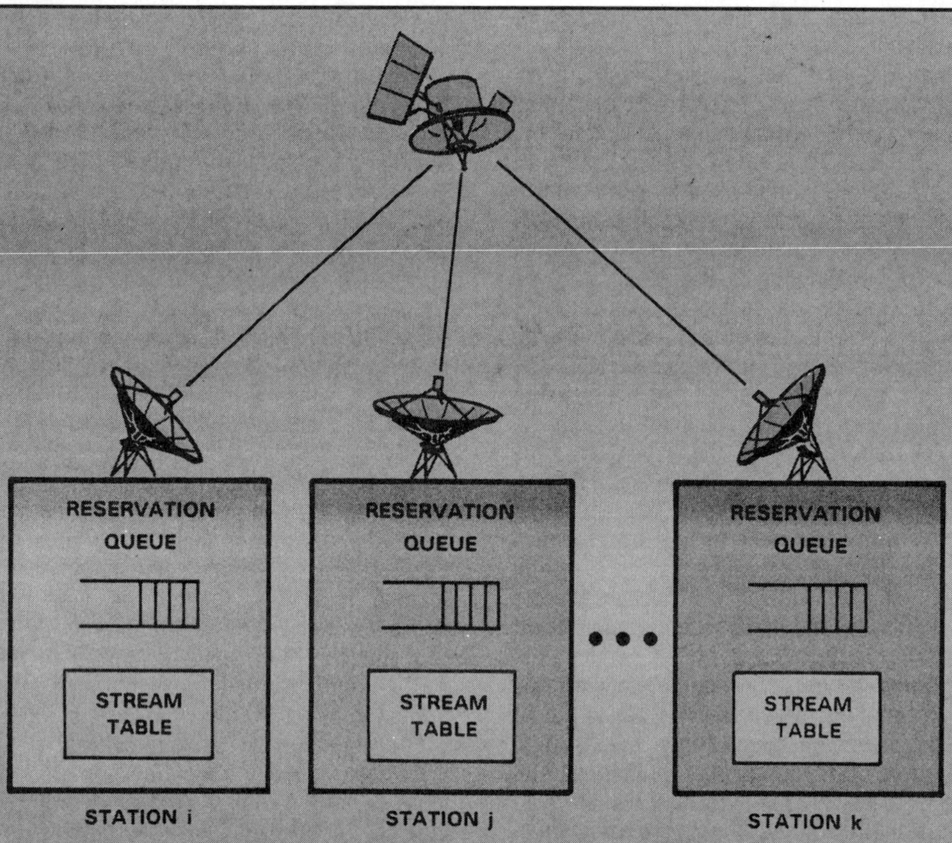

Fig. 16. Distributed scheduling coordination in a satellite network.

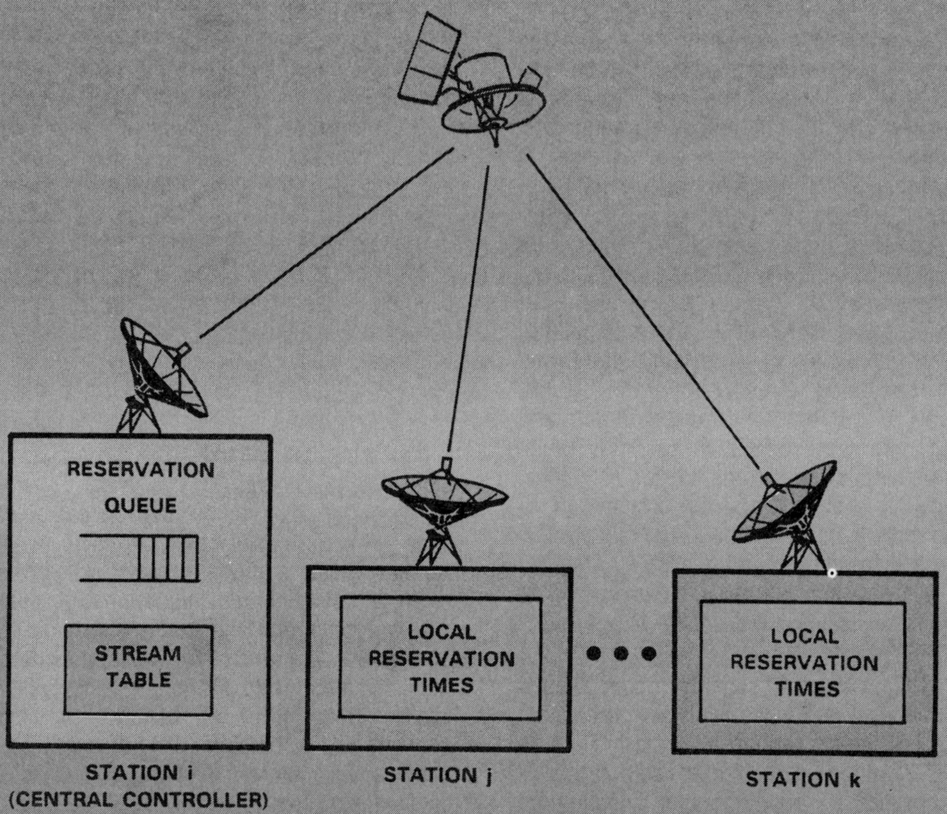

Fig. 17. Centralized scheduling coordination in a satellite network.

reservations per second, and thus more opportunities to miss a reservation transmission.

To achieve the small required BER for distributed control information, multiple forward error correction coding rates are used. For example, rate-1/2 coding might be used for the control portion of each burst, while rate-7/8 or no coding at all is used for the data. A CRC is also used on each control burst to ensure rejection of control information containing uncorrected errors.

In contrast, centralized scheduling is subject only to the point-to-point errors associated with receipt of each station's request at the central site and subsequent receipt of the announced time by the requesting station. An error in the first event requires that the station repeat the request, but does not affect other stations or other transmissions by the requesting station. Missed receipt of the scheduled time results in wasted channel capacity for the unused request, but otherwise imposes only a local penalty on the requesting station. Lower coding rates can also be used for control information in this case, but are not critical to achieve good performance. In particular, centralized control can operate with local control BER's to about 10^{-4} before significant degradation occurs.

Robustness for station failures can be retained in the centralized case by providing the scheduling capability at several or all stations, even though only one station exercises it at any time. Distributed coordination is still required in the form of a baton-pass algorithm, which must reliably hand off control to (exactly) one new station when the current controller fails. This is also a non-trivial problem, but occurs relatively infrequently. Thus, the rapid per-burst scheduling synchronization problem is traded for one with relatively low processing requirements and impact on performance.

The processing bandwidth/complexity problem associated with distributed control mainly affects synchronization. The basic scheduling activity for datagrams and streams is approximately proportional to the channel reservation rate, and imposes reasonable processing requirements even at multi-Mb/s channel rates. Synchronization acquisition processing required due to missed reservations, on the other hand, can be proportional to the square of the reservation rate and involve formidable complexity, depending on the specific design choices made [36].

To reduce this processing bandwidth and complexity, while retaining the lower delay of distributed datagram scheduling and the combined stream/datagram priority allocation of capacity, a combination of distributed and centralized techniques has evolved [37]. By carefully structuring the datagram scheduling and transmission functions, and by using central control for priority preemption of stream/datagram boundary definition, a relatively simple computation can be used to acquire and maintain datagram synchronization. This does not impact the sensitivity to control error rates.

Network Protocols—The absence of repeater nodes results in relatively simple network-level protocols for broadcast satellite networks. In particular, routing algorithms such as those used in ARPANET and terrestrial packet radio networks are not needed. Each station simply inspects the destination address of each received packet to determine if the packet is intended for itself. Logical addresses are used to provide flexibility for host relocations and assignment of temporary conferencing addresses.

The major network-level protocol of significance is congestion control. The broadcast downlink allows a distributed protocol to be used for this function, independently of whether distributed or centralized scheduling control is used. This is because each station can monitor the total channel downlink traffic, throttling its generation of new reservations accordingly. The synchronization problems associated with distributed scheduling do not apply here, since while occasional errors in downlink monitoring may cause a station to send more reservations than it should, correctness of the resulting scheduling by all stations is not affected. Similarly, sending too few reservations unnecessarily reduces the capacity available to the station, but does not otherwise affect global scheduling.

Prioritized congestion control can be achieved by computing current loading separately for each traffic priority, using a priority designator included in each transmitted packet. Each station applies the same set of rules in response to its loading measures, limiting its new reservations to the appropriate priorities. Limitations on the reservations which can currently be made are applied to the control of input traffic from attached hosts at each station, allowing the station to avoid overflowing its local input buffers.

The emphasis in packet satellite networks on flexible prioritized capacity allocation has also motivated development of more flexible host access protocols. To allow more efficient buffer management by both hosts and stations, and improve use of their interface bandwidth, two techniques are used. First, an explicit acceptance/refusal flow control mechanism is used in response to messages sent across the interface. This provides immediate feedback of problems associated with a particular stream or datagram flow. Second, stations routinely inform each host of the traffic priorities currently being accepted, eliminating interface congestion by messages sure to be refused.

These mechanisms prevent the problems associated with simple blocking or discard interfaces. In particular, a blocking interface holds up all traffic because of problems with a particular flow, defeating the benefits of priority and individual stream capacity allocation. A discard interface accepts other traffic, but discards messages associated with problem flows without informing the host. This forces host recovery mechanisms to wait for feedback from the destination host, which, because of the large satellite delays, can result in significant overall performance degradations.

Monitoring and Control—The broadcast channel provides unique opportunities in this area as well. One or a few stations (for redundancy) can be designated as monitoring sites, with terrestrial connections to a network operations center. These stations monitor all transmissions in the satellite downlink for indications of potential satellite or station problems. The use of two or more monitoring stations and reports to a centralized site allows detection of problems with the monitoring stations themselves to be made more easily, along with the benefits of long-term statistical analysis and dispatching of fault-repair personnel.

While the monitoring of channel traffic provides a great deal of information, it is also necessary to continually perform local diagnostics at each station and report problems via the satellite channel if possible. Because of the difficulty in isolating many types of failures in the multiple-access TDMA environment, extensive fault detection/diagnostics are used within both the signal- and packet-processing components of each station [38]. To allow detection of failed stations and gathering of local statistics, each station periodically sends a status packet, whose repeated absence indicates a major failure.

Existing Systems

Three distinct packet satellite networks are presently in operation for experimental purposes. The first to be implemented was SATNET, which uses a 64-kb/s channel on an INTELSAT satellite. Earth stations are located in Virginia, England, Norway, and West Germany, with additional sites planned for the near future. This system provided the first experimental testbed for the original version of PODA and other access protocols [35,39].

A second application consists of a ship-to-ship and ship-to-shore network using a 19.2-kb/s channel on the FLTSAT satellite. This network, called TACNET (aka MATNET), uses the same PODA protocol as SATNET, with the addition of link encryption and error-correction techniques appropriate to the bursty shipboard RF environments [40].

The third implementation utilizes a 3-Mb/s satellite channel on WESTAR, with earth stations located in California, Arizona, Virginia, New Jersey, New York, and Massachusetts. Seven stations are currently installed, with three more expected by Fall 1984. This network implements the principles described in this paper, and has an experimental emphasis on the sharing of a wide-bandwidth channel by many packet voice users [41,42].

Concluding Remarks

As we have indicated, packet satellite networks have been designed for high-bandwidth long-haul communications; packet radio networks have been designed to be easily deployable and to provide communications among ground mobile users. Both types of networks have the multiaccess/broadcast nature in common. They make use of a single radio channel to be shared by many contending users, thus achieving a more efficient utilization of communications bandwidth, a high degree of connectivity among users, and a high degree of flexibility in network organization.

These two types of networks, however, differ in the particular multiaccess protocols used, and in the specific network control and management functions needed to operate the network. These differences stem from two fundamental differences in the physical environment underlying the networks: the *physical connectivity* and the *propagation delay.* A broadcast satellite network is a fully connected star network, allowing any station to reach its destination in only one hop; in a packet radio network, the propagation characteristics of ground radio channels are such that radio connectivity cannot be guaranteed among all radio units, rendering the network an inherently multihop network. In a packet radio network, the propagation delay between neighboring devices is small compared to the transmission time of a packet; in a satellite channel, the round-trip delay is much greater than the packet transmission time.

The impact of the difference in propagation delays is mostly on the multiaccess protocols used. While in packet radio it is possible to base the access scheme on channel status feedback, such as carrier sensing or busy-tone sensing, the long propagation delay in satellite channels renders information from such feedback obsolete and useless. Performance improvement over random access is attained in the satellite case by the use of reservation techniques for both stream traffic and datagrams, with random access techniques used to make reservations.

Another distinction in the multiaccess protocols used by the two types of networks is time synchronization. Due to the full physical connectivity achieved in satellite networks, and the star structure imposed by the satellite transponder, one can relatively easily synchronize all ground stations accurately to a common time clock. This in turn allows a frame-oriented scheme such as PODA to be implemented. On the other hand, in packet radio networks which are of a multihop nature and are intended for mobile users, it is more difficult to achieve accurate systemwide synchronization, and, typically, packet transmissions are asynchronous.

Another implication of physical connectivity is on network management and control. Topology assessment and routing are primary functions in a packet radio network, but are minimal, if not nonexistent, in a satellite network. In satellite networks, the major network-level protocol of significance is congestion control, which, due to the simple broadcast star topology, is readily implemented in a distributed manner. In packet radio networks, special higher-level mechanisms are needed to monitor the connectivity in the local environment, and to relay this information to the control stations as well as to the rest of the network for stationless operation. Based on such collected information, routing algorithms are needed in the PRU's themselves for stationless operation, and in the control stations which then supply the routes to the PRU's for use.

Finally, it should be noted that some of the differences between packet radio and satellite networks will diminish as low-altitude multiple-satellite crosslink networks are considered. The needs to transmit packets over multiple hops and to assess the relative position of constantly moving satellites create networking problems not unlike those encountered in packet radio.

References

[1] L. G. Roberts and B. D. Wessler, "Computer network developments to achieve resource sharing," *1970 Spring Joint Comput. Conf., Proc. AFIPS Conf.*, vol. 36, pp. 543–549, 1970.

[2] H. Opderbeck and R. B. Hovey, "Telenet-network features and interface protocols," *Proc. NTG-Conf. Data Networks*, Baden-Baden, West Germany, Feb. 1976.

[3] L. Tymes, "TYMNET—a terminal oriented communication network," *1971 Spring Joint Comput. Conf., AFIPS Conf. Proc.*, vol. 38, pp. 211–216, 1971.

[4] W. W. Clipshaw and F. Glave, "Datapac network review," *Third Int. Comput. Commun. Conf. (ICCC) Proc.*, pp. 131–136, Aug. 1976.

[5] A. Danet, R. Despres, A. Le Rest, G. Pichon, and S. Ritzenthaler, "The French public packet switching service: the TRANSPAC network," *Third Int. Comput. Commun. Conf. (ICCC) Proc.*, pp. 251–260, Aug. 1976.

[6] D. D. Clark, K. T. Pogran, and D. P. Reed, "An introduction to local area networks," *Proc. IEEE*, vol. 66, no. 11, Nov. 1978.

[7] K. Kummerle and M. Reiser, "Local area communication networks—an overview," *J. Telecommun. Networks*, vol. 1, no. 4, Winter 1982.

[8] D. Nielson, "The role of radio in local area data distribution," *J. Telecommun. Networks*, vol. 1, pp. 39–47, Spring 1982.

[9] N. Abramson, "The Aloha system," *Comput. Commun. Networks*, N. Abramson and F. Kuo, Eds., Englewood Cliffs, NJ: Prentice-Hall, 1973.

[10] R. Binder, N. Abramson, F. Kuo, A. Okinaka, and D. Wax, "ALOHA packet broadcasting—a retrospect," *AFIPS Conf. Proc.*, May 1975.

[11] R. E. Kahn, "The organization of computer resources into a packet radio network," *IEEE Trans. Commun.*, COM-25, no. 1, pp. 169–178, Jan. 1977.

[12] R. E. Kahn, S. A. Gronemeyer, J. Burchfiel, and R. C. Kunzelman, "Advances in packet radio technology," *Proc. IEEE*, vol. 66, pp. 1468–1496, Nov. 1978.

[13] K. Klemba, D. Nielson, J. Tornow, and B. Leiner, "Packet Radio Network Executive Summary," Defense Advanced Research Projects Agency, IPTO-83-7, July 1983.

[14] R. E. Kahn, "The introduction of packet satellite communications," *Proc. 1979 NTC*, Washington, DC, Nov. 1979.

[15] B. Leiner et al., "The DARPA internet protocol suite," submitted to *IEEE Commun. Magazine, Special Issue on Communications Protocols.*

[16] F. A. Tobagi, "Multiaccess protocols in packet communications systems," *IEEE Trans. Commun.*, COM-28, no. 4, pp. 468–488, April 1980.

[17] S. Fralick and J. C. Garrett, "Technological considerations for packet radio networks," *AFIPS Conf. Proc.*, pp. 233–243, 1975.

[18] D. L. Nielson, "Microwave propagation measurements for mobile digital radio applications," *EASCON-77 Proc.*, pp. 14-2A–14-2L, Sept. 1977.

[19] L. Kleinrock and F. A. Tobagi, "Packet switching in radio channels: part I—carrier sense multiple access modes and their throughput delay characteristics," *IEEE Trans. Commun.*, COM-23, pp. 1400–1416, Dec. 1975.

[20] F. Tobagi and L. Kleinrock, "Packet switching in radio channels: part II—the hidden terminal problem in carrier sense multiple access and the busy tone solution," *IEEE Trans. Commun.*, COM-23, pp. 1417–1433, Dec. 1975.

[21] F. A. Tobagi, "Analysis of a two-hop centralized packet radio network—part I: slotted ALOHA,—part II: carrier sense multiple access," *IEEE Trans. Commun., COM-28*, pp. 196–216, Feb. 1980.

[22] F. A. Tobagi and D. H. Shur, "Simulation of busy tone multiple access modes in multihop packet radio networks," *Technical Report No. 217*, Stanford Electronics Laboratories, Aug. 1981.

[23] F. A. Tobagi and J. Brázio, "Throughput analysis of multihop packet radio networks under various channel access schemes," *Proc. INFOCOM '83*, April 1983.

[24] J. M. Brázio and F. A. Tobagi, "Theoretical results in throughput analysis of multihop packet radio networks," *Proc. Int. Conf. Commun., (ICC '84)*, May 1984.

[25] P. Spilling and F. Tobagi, "Activity signalling and improved hop acknowledgements in packet radio systems," in *SRI-International, Packet Radio Temporary Note No. 283*, Menlo Park, CA, Jan. 1980.

[26] M. Gerla and L. Kleinrock, "Flow control: a comparative survey," *IEEE Trans. Commun.*, COM-28, pp. 553–574, April 1980.

[27] H. Frank, I. Gitman, and R. VanSlyke, "Packet radio system—network considerations," *AFIPS Conf. Proc.*, pp. 217–231, May 1975.

[28] I. Gurantz et al., "A high-performance multiple data rate burst modem for satellite packet communication," *Proc. EASCON '81*, New York, Nov. 1981.

[29] N. Abramson, "Packet switching with satellites," *Proc. AFIPS Conf.*, vol. 42, June 1973.

[30] L. Kleinrock and S. Lam, "Packet-switching in a slotted satellite channel," *Proc. AFIPS Conf.*, vol. 42, June 1973.

[31] L. Roberts, "Dynamic allocation of satellite capacity through packet reservation," *Proc., AFIPS Conf.*, vol. 42, June 1973.

[32] W. R. Crowther et al., "A system for broadcast communication: reservation—Aloha," *Proc. 6th Hawaii System. Sci. Conf.*, Jan. 1973.

[33] R. Binder, "A dynamic packet switching system for satellite broadcast channels," *Proc. ICC '75*, San Francisco, CA, June 1975.

[34] R. Weissler et al., "Synchronization and multiple access protocols in the initial satellite IMP," *Proc. Fall COMPCON'78*, Washington, DC, Sept. 1978.

[35] I. M. Jacobs, R. Binder, and E. V. Hoversten, "General purpose packet satellite networks," *Proc. IEEE*, vol. 66, no. 11, Nov. 1978.

[36] N. Hsu and L. Lee, "Channel scheduling synchronization for the PODA protocol," *Proc. ICC '78*, Toronto, Canada, June 1978.

[37] E. Killian and R. Binder, "Control issues in a PODA voice/data satellite network," *Proc. ICC '80*, Seattle, WA, June 1980.

[38] D. A. McNeill et al., "SATNET monitoring and control," *Proc. 1979 NTC*, Washington, DC, Nov. 1979.

[39] W. W. Chu et al., "Experimental results on the packet satellite network," *Proc. 1979 NTC*, Washington, DC, Nov. 1979.

[40] L. N. Evenchik et al., "MATNET, an experimental navy shipboard satellite communications network," *Proc. INFOCOM '82*, Las Vegas, NV, April 1982.

[41] G. Falk et al., "Integration of voice and data in the wideband packet switching satellite network," *IEEE J. Select. Areas in Commun.*, SAC-1, no. 6, Dec. 1983.

[42] H. M. Heggestad and C. J. Weinstein, "Experiments in voice and data communications on a wideband satellite/terrestrial internetwork system," *Proc. ICC '83*, Boston, MA, June 1983.

[43] S. C. Fralick and J. K. Leung, "Study of throughput and delay of spread-spectrum multiple access modes," *SRI-International, Packet Radio Note*, Menlo Park, CA, April 1974.

[44] M. S. Hazell and B. H. Davies, "A fully distributed approach to the design of a 16 kbit/sec VHF packet radio network," *Proc. IEEE MILCOM '83*, pp. 645–649, Washington, DC, 1983.

Analysis of Mobile Radio Slotted ALOHA Networks

CHRISTIAN NAMISLO

Abstract—Given a mobile terminal transmitting a message to some receiver, the level at the receiver will be affected by path loss and by fading phenomena. Transmissions from different terminals will suffer different attenuations on their respective ways to the receiver and will therefore yield different levels at the receiver. Thanks to the capture effect, the receiver will now be able to read the message with the strongest signal, provided that it is sufficiently stronger than its contenders. In a mobile radio ALOHA network this means that a packet collision need not necessarily destroy all packets involved.

A Markov model is developed for slotted ALOHA networks with capture and it is shown that the throughput in such a network can be markedly greater than the famous $1/e$. Perhaps even more important is the result that such networks are very stable under overload.

I. Introduction

THE key feature of all random access protocols to a common transmission channel is the problem that two or more terminals may attempt to use the channel during the same time, or at least overlapping in time. Even though this phenomenon may put a severe limit on the possible throughput achievable in such a network, packet switching using random access procedures still remains attractive and will often be superior to fixed assignment procedures. However, some of those procedures react very sensitively to overload and therefore need careful dimensioning to the expected traffic load.

In an ALOHA network, any terminal which has a packet ready to transmit will simply transmit this packet regardless of other terminals which might already be transmitting themselves. An unsuccessful transmission will be detected by the terminal if, after a certain time, no acknowledgment has been received. The terminal will then assume its packet to be lost. It will determine a certain delay time through a random process, and after this time has elapsed will retransmit the packet. The performance of such a rather anarchic ALOHA network can be improved considerably by imposing two rules: packets are to be of a fixed length and transmissions may only start at discrete points on the time axis. The time interval between two such points is at least as long as the time needed to transmit one packet. The time during which a packet may be transmitted is called a time slot; the network is called a slotted ALOHA network.

Manuscript received February 21, 1983; revised November 10, 1983. Parts of this paper were presented at the Mediterranean Electrotechnical Conference, Athens, Greece, May 24–26, 1983, under the title "ALOHA networks in which collisions do not necessarily destroy all packets involved."

The author is with SIEMENS Central Laboratories for Communication Techniques, Postfach 7000067, 8000 Munich 70, Germany.

Much research has been devoted to slotted ALOHA networks. One of the most prominent results is that not more than $1/e$ (= 0.37) of all time slots can be used for successful packet transmissions. Another one: ALOHA networks tend to break down completely if they are exposed to an overload, even if this overload would have lasted only very shortly.

These investigations usually assumed that the concurrent use of one time slot by more than one terminal (this is called a collision) would certainly result in the destruction of all packets involved. All those packets would have to be retransmitted in later time slots. This is, however, not a realistic assumption in a slotted ALOHA radio network with mobile terminals, and it is therefore the assumption which we are going to drop in this paper.

Apart from a rather complex procedure [1] to cope with collisions, only four papers are known to us where a rather simple physical effect is used to mitigate the severity of packet collisions. If in radio communication the difference in signal strength of two or more colliding packets is beyond a certain value, then a receiver will be able to decode one packet correctly; the other packets will simply cause some noise and get lost. In [2] this is forced to happen by assigning different levels of power to the transmitters. This, of course, means assigning fixed priorities to the terminals.

In radio packet switching with mobile terminals, a somewhat similar situation will arise automatically. Due to different distances between the terminals and the receiver (path loss), and due to shadow fading and Rayleigh fading, the terminals will in fact cause different levels at the receiver. Therefore, the capture effect (capture of the strongest of a number of coinciding signals) will provide for a certain chance that one packet out of a number of colliding ones will be read correctly. The capture effect in slotted ALOHA networks has been considered before [3]–[5], and it has been noticed that it can yield a noticeably higher throughput in the network. All these papers use the "continuous" approach to ALOHA networks as it was introduced by Abramson [6]. We would, however, like to have a discrete Markov model as in [7], [8] which should take care of the capture effect. The Markov model is more

Reprinted from *IEEE J. Selected Areas Commun.*, vol. SAC-2, no. 4, pp. 583–588, July 1984.

desirable because it gives more detailed insight into the network's dynamic behavior.

To this end we shall introduce probabilities q_i that one out of i colliding packets will survive in a collision and the other $i-1$ packets will need retransmission later. Bearing in mind that those q_i need not be zero for $i>1$, we are going to derive a probabilistic mathematical description for such a slotted ALOHA network. We shall assume that positive acknowledgments from the receiver are never lost, so the terminals will not retransmit unnecessarily. This assumption should not be too unrealistic since modern mobile networks usually have to guarantee a 95 percent coverage of the area they serve. With a specific set of q_i, as it would be determined for an (admittedly idealized) network, we then shall illustrate that the classical analysis of slotted ALOHA networks, as in [7], [8], would have fallen rather short of reality.

In the Appendix we shall develop a mathematical model which allows us to actually compute the probabilities q_i.

II. Markov Model

We consider a slotted ALOHA network with N terminals sharing one channel for transmission to one receiver. During any time slot each terminal will be in any of three states: the origination state (O) when a terminal has no packet ready to transmit or if a new packet has just occurred while the terminal is waiting for the next time slot; the transmission state (T) when a terminal is actually transmitting a packet (successfully or not); and the retransmission state (R) after a terminal has unsuccessfully tried to transmit a packet and while it is waiting for retransmission in a future time slot. Depending on its state, a terminal will be referred to as an O-, T-, or R-terminal. It is assumed that the process which generates new packets comes to a halt in a terminal while it has a packet to transmit.

The possible state transitions for an individual terminal are shown in Fig. 1. Changes of state can occur either at the end or beginning of a time slot of the ALOHA network. An O-terminal can either change into a T-terminal at the beginning of a time slot (start transmitting a new packet) or remain in state O (do nothing). The probability for the change from O to T will be denoted p_o (packet generation rate). An R-terminal can either change into a T-terminal at the beginning of a time slot (start retransmis-

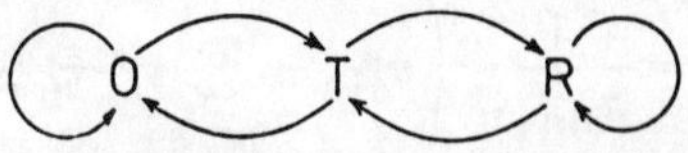

Fig. 1. State transitions for an individual terminal.

sion) or stay in state R (do nothing). The probability for the change from R to T will be called p_r (retransmission rate). After the end of a time slot a T-terminal can either go to O (after a successful transmission) or to R (after an unsuccessful transmission).

Given i $(0 \leqslant i \leqslant N)$ T-terminals transmitting in one time slot, the probability that exactly one of them will go to O and the rest to R will be denoted q_i (probability of success). This is the probability that of i concurrently transmitting and colliding terminals one will be successful. We have $q_o = 0$ and, in an undisturbed channel, $q_1 = 1$. The values of the other q_i, previously assumed to be indiscriminately zero, will depend on specific properties of the network.

The slotted ALOHA network as a whole can now be described as a homogeneous finite Markov process with $N+1$ possible states $(0, 1, \cdots, N)$, corresponding to the number of R-terminals at the end of a time slot [7], [8].

Let $\pi_{n,m}$ denote the state transition probability that the network will move from state n to m in one step (from the end of one time slot to the end of the next one). We distinguish between three types of transitions.

(i) The state decreases by more than one. This is not possible because at most one packet can be transmitted successfully in one time slot.

(ii) The state decreases by one. This will happen with probability q_i iff no O-terminal and i, $1 \leqslant i \leqslant n$, R-terminals change into state T (one packet successful).

(iii) The state moves to $m \geq n$. This will happen either with probability $q_{i+m-n+1}$ iff $m-n+1$ O-terminals and i, $0 \leqslant i \leqslant n$, R-terminals change into state T (one packet successful); or with probability $1-q_{i+m-n}$ iff $m-n$ O-terminals and i, $0 \leqslant i \leqslant n$, R-terminals change into state T (no packet successful).

We assume that the terminals are working independently. Therefore, in state n the numbers of concurrent transmissions of new respective previously collided packets will both be binomially distributed with parameters N-n and p_o, or n and p_r, respectively. We therefore get

$$\pi_{n,m} = \begin{cases} 0 & \text{if } m < n-1 \\ (1-p_o)^{N-n} \cdot \sum_{i=1}^{n} \binom{n}{i} \cdot p_r^i \cdot (1-p_r)^{n-i} \cdot q_i & \text{if } m = n-1 \\ \left\{ \binom{N-n}{N-m} \cdot (1-p_o)^{N-m-1} \cdot p_o^{m-n} \cdot \sum_{i=0}^{n} \binom{n}{i} \cdot (1-p_r)^{n-i} \cdot p_r \right. & \\ \left. \cdot \left[(1-p_o) \cdot (1-q_{i+m-n}) + \frac{N-m}{m-n+1} \cdot p_o \cdot q_{i+m-n+1} \right] \right\} & \text{if } m \geqslant n. \end{cases}$$

From the $\pi_{n,m}$ we can calculate the equilibrium state occupation probabilities π_n. Starting with an arbitrary positive π_o^* we get recursively

$$\pi_n^* = \frac{1}{\pi_{n,n-1}} \cdot \left(\pi_{n-1}^* - \sum_{i=0}^{n-1} \pi_i^* \cdot \pi_{i,n-1} \right)$$

and after normalizing

$$\pi_n = \frac{\pi_n^*}{\sum_{i=0}^{N} \pi_i^*}.$$

The expected throughput of the network in state n, that is, the probability to transmit a packet successfully in a time slot, will be found to be

$$f_n = \sum_{k=0}^{N-n} \binom{N-n}{N-n-k} \cdot p_o^k \cdot (1-p_o)^{N-n-k} \cdot \sum_{i=0}^{n} \binom{n}{i} \cdot p_r^i \cdot (1-p_r)^{n-i} \cdot q_{i+k}.$$

The overall mean throughput, as in [7], is

$$f = \sum_{i=0}^{N} \pi_i \cdot f_i.$$

Let $S = \sum_{i=0}^{N} \pi_i \cdot i$ denote the average state of the system. Although we do not have a formal proof, common sense as well as all our numerical results suggest the simpler formula

$$f = (N-S) \cdot p_o.$$

An indicator of the network's dynamic behavior is its expected drift d_n in state n [7]. For each state n we get [9]

$$d_n = (N-n) \cdot p_o - f_n.$$

This is the difference between the expected number of new packets to be input to the system and the expected number of packets to leave the system within one time slot. System states n with $d_n = 0$ will be equilibrium states of the network. (To be more precise, a state n is an equilibrium state if $d_{n-1} \cdot d_{n+1} < 0$ and if $|d_n| \leqslant \min\{|d_{n-1}|, |d_{n+1}|\}$.)

III. Application

In order to demonstrate the difference between our analysis and [7], [8] we need a set of q_i, the probabilities that one out of i colliding packets will be decoded correctly by the receiver. As will be shown in the Appendix, Fig. 2 gives typical q_i for a specific mobile radio slotted ALOHA.

It is very surprising to observe that the values of the q_i, although monotonously decreasing, do not tend to zero. For the network considered, the limit is approximately 0.45. This already shows that the mobile radio slotted ALOHA network will be stable under overload because even then the throughput cannot drop below 0.45 Erl. (It will, however, be noted in the Appendix that for very large values of i, q_i will have to be zero. So there remains a very slim chance for the network to fail.)

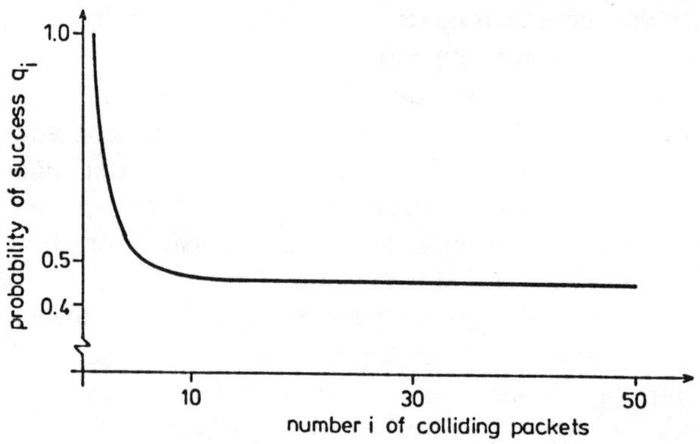

Fig. 2. Probability of success for one out of i colliding packets.

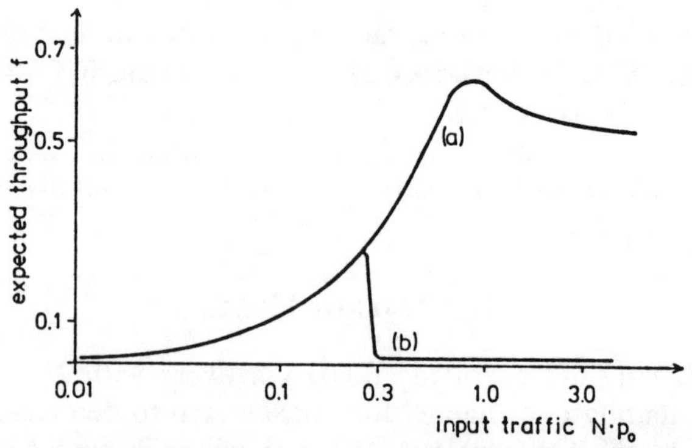

Fig. 3. Expected throughput of a network with $N = 100$ terminals and $p_r = 0.06$. (a) Results of our analysis. (b) Results of [7], [8].

In more detail, the throughput as a function of the input traffic $N \cdot p_o$ is presented in Fig. 3. (Note that the input traffic is different from the traffic load offered to the channel. The latter is, depending on the state n, $(N-n) \cdot p_o + n \cdot p_r$.) The network is assumed to consist of $N = 100$ terminals with a fixed retransmission rate $p_r = 0.06$.

The upper of the two curves gives the throughput of the network when the probabilities q_i are as in Fig. 2. The lower curve would result if one had $q_i = 0$ for $i > 1$ as in former analyses [7], [8].

Our analysis shows that the ALOHA network considered can in fact throughput at most 0.65 packets per time slot (so the famous maximum throughout of $1/e$ will be surpassed by 75 percent). Even for an input traffic of 6.0 Erl the throughput is still as high as 0.5.

Furthermore, Fig. 3 shows that for an input traffic of 0.6 Erl the throughput is still almost 0.6. This indicates that the network can actually carry this much load without much message delay.

The classical analysis of the network would have yielded the lower curve. For the parameters chosen ($N = 100$, $p_r = 0.06$), the network would only have reached a maximum throughput of approximately 0.25 at an input traffic of approximately 0.25 Erl. Even a very slight increase of the input traffic would make the network collapse. Most of the terminals would enter the retransmission state and the throughout would drop almost immediately to its minimum 0.013.

A closer look into the dynamic behavior of the network under different input traffic rates gives Figs. 4 and 5. They display, as functions of the state n, the expected drift d_n

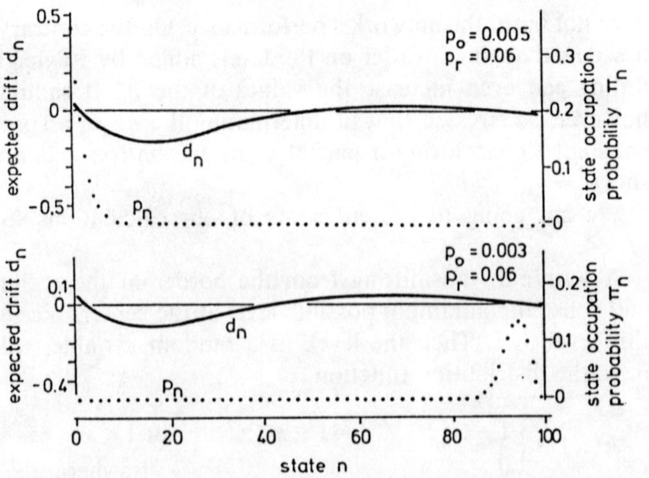

Fig. 4. Expected drift and state occupation probabilities for two packet generation rates in a network with $N = 100$ terminals. Results of [7], [8].

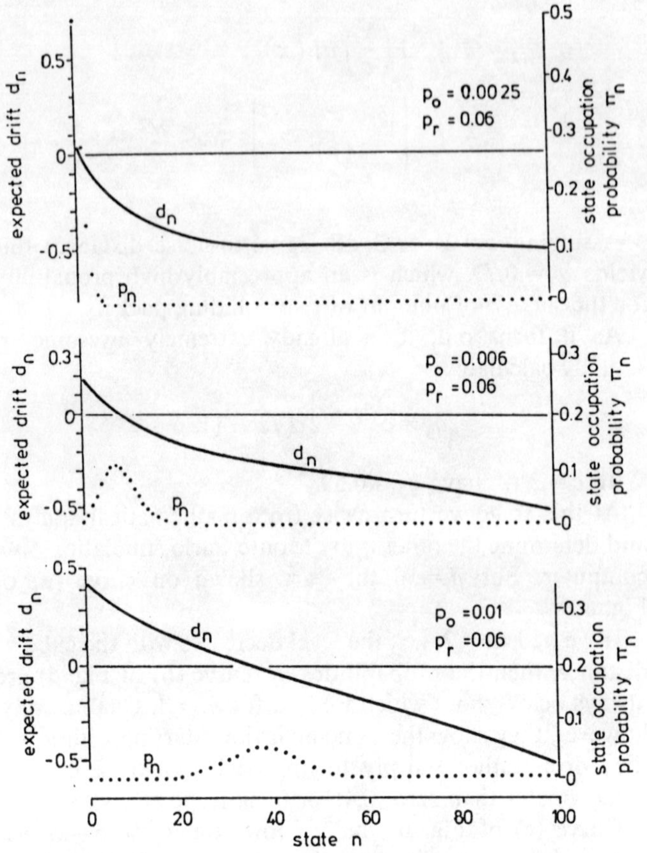

Fig. 5. Expected drift and state occupation probabilities for three packet generation rates in a network with $N = 100$ terminals. Results of our analysis.

and the state occupation probabilities π_n. Fig. 4 contains results which would have been obtained from [7], [8]. Fig. 5 shows the results of our more complex approach.

For an input traffic of 0.25 Erl (corresponding to $p = 0.0025$) the analyses in [7], [8] would have predicted two stable equilibrium states at $n = 1$ and $n = 92$ (zeros of d_n where the sign of the function changes from positive to negative [8]). Even though the steady state occupation probabilities π_n indicate that the network would be found in a state $n \leqslant 5$ with probability 0.943, with probability 1 the network will eventually get stuck in state 92 for some time. If we assume an input traffic of 0.3 Erl, then the approach, as in [7], [8], would suggest the network to have failed. Of the two stable equilibria at $n = 3$ and $n = 95$, a neighborhood of the latter would be inhabited by the network most of the time.

Let us now turn to the results of our analysis as shown in Fig. 5. For any input traffic load considered (low: 0.25 Erl; high: 0.6 Erl; overload: 1.0 Erl), there is only one equilibrium state each. Apart from fairly small neighborhoods of those stable equilibria, the expected drift will be rather big and, of course, will always be directed towards the equilibria. With increasing load, the equilibrium of the network will move *slowly* to higher values, not as abruptly as in Fig. 4. This equilibrium closely corresponds to the mean number of backlogged packets [8, p. 416]. This again proves that not only the network's throughput will degrade rather gracefully under overload, but also that the expected number of backlogged packets, and hence the mean delay time for a packet, will not explode as in Fig. 4.

IV. Conclusion

We have developed a Markov model of a slotted ALOHA network which does not have to assume that packet collisions are necessarily fatal for all colliding packets. With the help of this model we have investigated a mobile radio slotted ALOHA network. We could show that such a network can throughput markedly more traffic than the famous $1/e$ and that it will be very stable under overload.

Appendix
Probabilities of Success

Now that the Markov analysis of slotted ALOHA networks with capture has progressed so smoothly, there remains the task of determining the probabilities q_i that one out of i concurrently transmitting terminals will be successful. This is by no means a trivial part because (ground) radio propagation from/to mobile stations is subject to complicated influences. Assuming identical mobiles, we shall give a (still very general) formula for the q_i, discuss how to make use of it, and finally compute the q_i under fairly idealized conditions.

Let E_k denote the level (of power) at the receiver caused by a transmitting terminal T_k. In the presence of noise (of level S) the receiver is assumed to be able to capture this transmission if $E_k/S \geqslant c$, for some constant $c > 1$. Suppose that there are i terminals $T_1 \cdots T_i$ transmitting concurrently and causing the levels $E_1 \cdots E_i$, respectively. For any k, $1 \leqslant k \leqslant i$, the terminal T_k will succeed with its transmission if its level alone is at least c times the sum of the other levels, that is, if

$$E_k \geqslant c \sum_{\substack{j=1 \\ j \neq k}}^{i} E_j.$$

In order to calculate the probability of this happening,

we view the level E caused by a terminal as a random variable with probability distribution $F(x) = \text{prob}(E < x)$. We assume that this distribution function will be the same for all terminals in the network. Then the level caused by i terminals together will be a random variable whose probability distribution is the i-fold convolution $F_i(x)$ of $F(x)$ with itself.

The probability that the terminal T_k will be successful over its $i-1$ contenders is then

$$\int_0^\infty F_{i-1}\left(\frac{x}{c}\right) dF(x).$$

We are, however, interested in the probability that any one terminal will be successful. This simply is

$$q_i = i \cdot \int_0^\infty F_{i-1}\left(\frac{x}{c}\right) dF(x) \qquad \text{for } i > 1.$$

So far we have transferred the problem of getting hold of the probabilities q_i to the problem of determining the probability distribution $F(x)$. We can actually calculate $F(x)$ if (i) we know how the mobiles are spread over the geographic area covered by the network and (ii) we have a map of that area which shows the mean levels produced at the receiver by a mobile transmitting from different regions of the area. Depending on the mean level caused from a certain region, this region could be shaded more or less heavily on the map. (Using a topographical database of the Federal Republic of Germany, Dr. Lorenz of Fernmeldetechnisches Zentralamt der Deutschen Bundespost has created maps of this kind as a planning tool for the mobile telephone network.) This map will take care of both path loss and shadow fading. If we also want to consider Rayleigh fading, then for each region we get a probability distribution [10] for the level caused from within the region. All that is left is to weigh the different probability distributions with the relative numbers of mobiles expected in the respective regions. The weighed sum of the distributions gives $F(x)$.

Example: We are now going to calculate $F(x)$ for a specific very simple mobile slotted ALOHA network. We assume a number of identical mobile terminals in a circular area of radius 1. The receiver is fixed at the center of the area.

If the terminals are spread evenly over the area of the region, then their distances from the receiver will be random variables with the common distribution function

$$G(x) = \begin{cases} x^2 & \text{if } 0 \leqslant x \leqslant 1 \\ 0 & \text{elsewhere.} \end{cases}$$

Due to path loss, the level of a terminal at the receiver will be proportional to $1/x^m$, where x is the distance between the terminal and the receiver. The exponent m will be a number between 2 (ideal propagation of electromagnetic waves) and about 5, depending very much on the geography of our region (buildings, valleys, forests, etc.). We shall neglect shadow fading and Rayleigh fading here. In fact, first simulation results suggest that Rayleigh fading does not harm the network's performance. On the contrary, it seems that the disorder on the levels added by Rayleigh fading can even increase the values of the q_i. It should, however, be stressed that in determining the q_i we assume constant signal during a packet transmission (packets are short).

We are going to present some of our calculations for $m = 4$.

A terminal transmitting from the border of the region will cause the minimum possible level at the center. We set this level to 1. Then the level, as a random variable, will have the distribution function

$$F(x) = \begin{cases} 1 - G(x^{-1/4}) = 1 - x^{-1/2} & \text{if } 1 \leqslant x \leqslant \infty \\ 0 & \text{elsewhere.} \end{cases}$$

With this result we can begin to calculate the probabilities q_i. We have

$$\begin{aligned} q_2 &= 2 \cdot \int_0^\infty F\left(\frac{x}{c}\right) dF(x) \\ &= 2 \cdot \int_c^\infty \left[1 - \frac{x}{(c)}^{-1/2}\right] \cdot \frac{1}{2} \cdot x^{-3/2} \, dx \\ &= c^{-1/2}. \end{aligned}$$

Assuming $c = 2$ (or 3 dB signal-to-noise distance) this yields $q_2 = 0.71$, which is an appreciably high probability for the success of one out of two colliding packets.

As it turns out, it is already extremely awesome to actually calculate

$$q_3 = c^{-1/2} \cdot 2 \cdot (\sqrt{2} - 1).$$

With $c = 2$ we have $q_3 = 0.59$.

At this stage we turn away from mathematical analysis and determine the other q_i by Monte Carlo simulation on a computer. For $n = 50$ they are shown on curve (a) of Fig. 6.

If we had $m = 3$, i.e., the level decreases with the cube of distance, then the probabilities q_i (curve (b) of Fig. 6) are always below those which we had for $m = 4$. Qualitatively, however, they show the same behavior. Starting with $q_1 = 1$ they drop rather sharply to approach quickly a limiting value greater than zero (0.45 or 0.30, respectively).

Curve (c) of Fig. 6 finally shows the q_i when $m = 2$. Again, they are smaller than those for $m = 3$. But there also is a qualitative difference. The convergence of the q_i is fairly slow and it seems possible that they are actually approaching zero. This could be plausible because in this case the distribution function for the distance from the receiver is of the same order (quadratic) as the function governing the decrease of the level.

In the other two cases the probability for small distances from the receiver decreases quadratically, also. But a terminal very close to the receiver will cause a markedly higher level now. Therefore, a limit for the q_i which is greater than zero appears to be conceivable.

However, as H. Geprägs pointed out to us, in a real mobile ALOHA network the limiting value will have to be

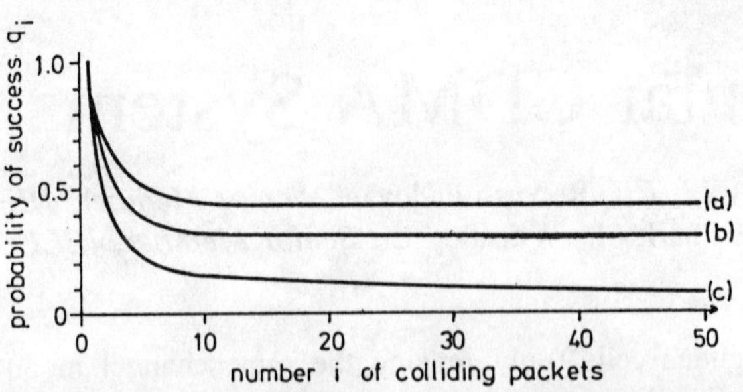

Fig. 6. Probability of success for one out of i colliding packets. Decrease of level with (a) fourth power, (b) cube, and (c) square of the distance.

zero for any function governing the decrease of the level. As the distance between a terminal and the receiver cannot shrink to zero, the level caused by any terminal will be bound from above by some value, say K. If the minimum level in the network is 1, then, with $\lceil K/c \rceil + 2$ (where $\lceil x \rceil$ is the smallest integer greater or equal to x) terminals transmitting concurrently, it will be impossible for any of those terminals to transmit successfully in the same time slot. Therefore, a nonzero probability remains for the network to break down, a very slim one because K will be large.

But, sticking to application, the search for the actual limit of the q_i ought to be of marginal interest for the practitioner. Unless a mobile radio ALOHA network is severely overloaded, the probability of a very great number of terminals transmitting at the same time will have to be negligibly small, anyhow.

References

[1] A. D. Wyner, "Recent results in the Shannon theory," *IEEE Trans. Inform. Theory*, vol. IT-20, pp. 2–10, Jan. 1972.

[2] J. J. Metzner, "On improving utilization in ALOHA networks," *IEEE Trans. Commun.*, vol. COM-24, pp. 447–448, Apr. 1976.

[3] L. G. Roberts, "ALOHA packet system with and without slots and capture," *Comput. Commun. Rev.*, no. 5, pp. 28–42, 1975.

[4] N. Abramson, "The throughput of packet broadcasting channels," *IEEE Trans. Commun.*, vol. COM-25, pp. 117–128, Jan. 1977.

[5] F. Kuperbus and J. Arnbak, "Packet radio in a Rayleigh channel," *Electron. Lett.*, vol. 18, pp. 506–507, June 1982.

[6] N. Abramson, "The ALOHA system—Another alternative for computer communications," in *1970 Fall Joint Comput. Conf., AFIPS Conf. Proc., vol.* 37. Montvale, NJ: AFIPS Press, 1970, pp. 281–285.

[7] A. B. Carleial and M. E. Hellman, "Bistable behavior of ALOHA-type systems," *IEEE Trans. Commun.*, vol. COM-23, pp. 401–410, Apr. 1975.

[8] L. Kleinrock and S. S. Lam, "Packet switching in a multiaccess broadcast channel—Performance evaluation," *IEEE Trans. Commun.*, vol. COM-23, pp. 410–423, Apr. 1975.

[9] Y.-C. Jenq. "On the stability of ALOHA systems," *IEEE Trans. Commun.*, vol. COM-28, pp. 1936–1939, Nov. 1980.

[10] R. C. French, "The effect of fading and shadowing on channel reuse in mobile radio," *IEEE Trans. Veh. Technol.*, vol. VT-28, pp. 171–181, Aug. 1979.

[11] C. Namislo, "ALOHA networks in which collisions do not necessarily destroy all packets involved," in *Proc. Mediterranean Electrotech. Conf.*, May 1983, A2.09, Athens, Greece.

On the Capacity of a Cellular CDMA System

Klein S. Gilhousen, *Senior Member, IEEE,* Irwin M. Jacobs, *Fellow, IEEE,* Roberto Padovani, *Senior Member, IEEE,* Andrew J. Viterbi, *Fellow, IEEE,* Lindsay A. Weaver, Jr., and Charles E. Wheatley III, *Senior Member, IEEE*

Abstract—**The use of spread spectrum or code division techniques for multiple access (CDMA) has long been debated. Certain advantages, such as multipath mitigation and interference suppression are generally accepted, but past comparisons of capacity with other multiple access techniques were not as favorable. This paper shows that, particularly for terrestrial cellular telephony, the interference suppression feature of CDMA can result in a many-fold increase in capacity over analog and even over competing digital techniques.**

I. Introduction

Spread-spectrum techniques, long established for antijam and multipath rejection applications as well as for accurate ranging and tracking, have also been proposed for code division multiple access (CDMA) to support simultaneous digital communication among a large community of relatively uncoordinated users. Yet, as recently as 1985 a straightforward comparison [1] of the capacity of CDMA to that of conventional time division multiple access (TDMA) and frequency division multiple access (FDMA) for satellite applications suggested a reasonable edge in capacity for the latter two more conventional techniques. This edge was shown to be illusory shortly thereafter [2] when it was recognized that since CDMA capacity is only interference limited (unlike FDMA and TDMA capacities which are primarily bandwidth limited), any reduction in interference converts directly and linearly into an increase in capacity. Thus, since voice signals are intermittent with a duty factor of approximately 3/8 [3], capacity can be increased by an amount inversely proportional to this factor by suppressing (or squelching) transmission during the quiet periods of each speaker. Similarly, any spatial isolation through use of multibeamed or multisectored antennas, which reduces interference, also provides a proportional increase in capacity. These two factors, voice activity and spatial isolation, were shown to be sufficient to render CDMA capacity at least double that of FDMA and TDMA under similar assumptions for a mobile satellite application [2].

While previous comparisons primarily applied to satellite systems, CDMA exhibits its greatest advantage over TDMA and FDMA in terrestrial digital cellular systems, for here isolation among cells is provided by path loss, which in terrestrial UHF propagation typically increases with the fourth power of the distance. Consequently, while conventional techniques must provide for different frequency allocation for contiguous cells (only reusing the same channel in one of every 7 cells in present systems), CDMA can reuse the same (entire) spectrum for all cells, thereby increasing capacity by a large percentage of the normal frequency reuse factor. The net improvement in capacity, due to all the above features, of CDMA over digital TDMA or FDMA is on the order of 4 to 6 and over current analog FM/FDMA it is nearly a factor of 20.

The next section deals with a single cell system, such as a hubbed satellite network, and develops the basic expression for capacity. The subsequent two sections derive the corresponding expressions for a multiple cell system and determine the distribution on the number of users supportable per cell. The last section presents conclusions and system comparisons.[1]

II. Single Cell CDMA Capacity

The network to be considered throughout consists of numerous mobile (or personal) subscribers communicating with one or multiple cell sites (or base stations) which are interconnected with a mobile telephony switching office (MTSO), which also serves as a gateway to the public switched telephone network. We begin by considering a single cell system, which can also serve as a model for a satellite system whose "cell site" is a single hub.

Each user of a CDMA system occupies the entire allocated spectrum, employing a direct sequence spread spectrum waveform. Without elaborating on the modulation and spreading waveform, we assume generic CDMA modems at both subscriber units and the cell site with digital baseband processing units as shown in Fig. 1 for the transmitter sides of each. These consist of (digital) forward-error correction (FEC), modulation and (direct sequence) spreading functions, preceding the (analog) amplification and transmission functions. Each of the digital functions can be performed using binary sequences in the subscriber modulator.

At the cell-site transmitter, the spread signals directed to the individual subscribers are added linearly and phase randomness is assured by modulating each signal with independent pseudorandom sequences on each of the two quadrature phases. The weighting factors $\varnothing_1, \varnothing_2, \cdots, \varnothing_N$ can be taken to be equal for the time being, but for the multiple cell case they will provide power control based on considerations to be

Manuscript received May 1, 1990; revised September 14, 1990. This paper was presented at the 1990 IEEE GLOBECOM Conference, San Diego, CA.

The authors are with QUALCOMM, Inc., 10555 Sorrento Valley Road, San Diego, CA 92121-1617.

IEEE Log Number 9144470.

[1] It should be noted that our purpose is not to evaluate or optimize modem performance for the channels under consideration. Rather, assuming an efficient modulation and FEC code for the given channels, we shall establish conditions under which the modems will achieve an acceptable level of performance, particularly in terms of the maximum number of users supportable per cell.

Reprinted from *IEEE Trans. Vehicular Technol.*, vol. 40, no. 2, pp. 303–312, May 1991.

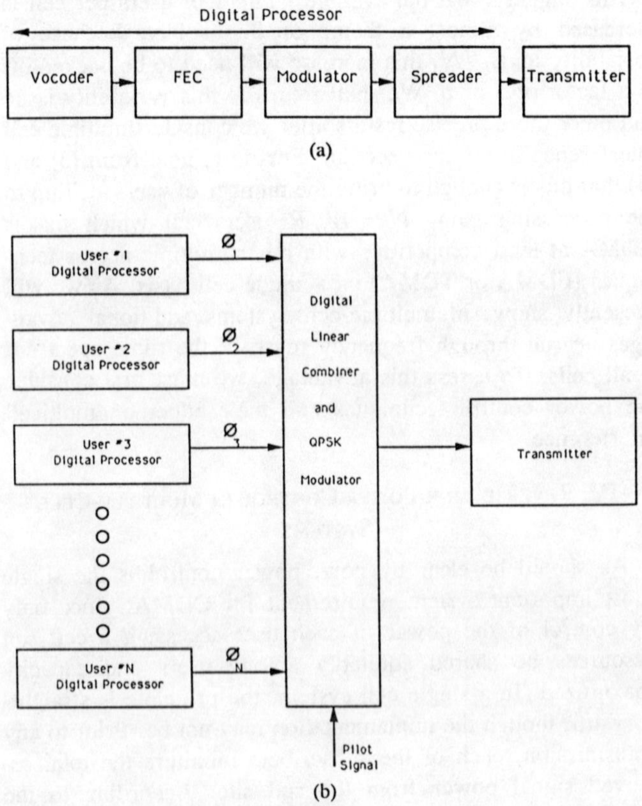

Fig. 1. Cellular system simplified block diagram. (a) Reverse link subscriber processor/transmitter. (b) Forward link cell-site processor/transmitter.

described later. The receiver processors in both subscriber and cell-site receivers provide the inverse baseband functions, which are of course considerably more complex than the transmitter baseband functions.

One other key feature of the cell-site transmitter is the inclusion of a pilot signal in the forward (cell-site-to-subscriber) direction. This provides for acquisition by the mobile terminals, including initial power control by the mobile, which adjusts its output power inversely to the total signal power it receives. Power control is a basic requirement in CDMA and will be expanded on in a later section.

We note also that the pilot signal is used by the subscriber demodulator to provide a coherent reference which is effective even in a fading environment since the desired signal and the pilot fade together. In the subscriber-to-cell-site (reverse) direction, no pilot is used for power efficiency considerations, since unlike the forward case, an independent pilot would be needed for each signal. A modulation consistent with, and relatively efficient for, noncoherent reception is, therefore, used for the reverse direction.

Without elaborating further on the system implementation details, we note that for a single cell site with power control, all reverse link signals (subscribers-to-cell site) are received at the same power level. For N users, each cell-site demodulator processes a composite received waveform containing the desired signal having power S and $(N-1)$ interfering signals each also of power S. Thus the signal-to-noise (interference) power is

$$\text{SNR} = \frac{S}{(N-1)S} = \frac{1}{N-1}.$$

Of greater importance for reliable system operation is the bit energy-to-noise density ratio, whose numerator is obtained by dividing the desired signal power by the information bit rate, R, and dividing the noise (or interference) by the total bandwidth, W. This results in

$$E_b/N_0 = \frac{S/R}{(N-1)S/W} = \frac{W/R}{N-1}. \tag{1}$$

This paper does not explicitly address modulation techniques and their performance. Rather, an E_b/N_0 level is assumed which ensures operation at the level of bit error performance required for digital voice transmission. Among the factors to be considered in establishing the modulation and the resulting required E_b/N_0 level are phase coherence, amplitude fading characteristics and power control techniques and their effectiveness, particularly for the reverse link. One of the lesser considerations, albeit one of the most cited, is the probability distribution of the interfering signals. While Gaussian noise is often assumed, this is not strictly necessary to establish the E_b/N_0 requirements. Nonetheless, the assumption is quite reasonable when powerful forward error-correcting codes are employed, particularly at low code rates, because in such cases decisions are based on long code sequence lengths over which the interfering signal sequence contributions are effectively the sums of a large number of binomial variables, which closely approximate Gaussian random variables.

Equation (1) ignores background noise, η, due to spurious interference as well as thermal noise contained in the total spread bandwidth, W. Including this additive term in the denominator of (1) results in a required

$$E_b/N_0 = \frac{W/R}{(N-1) + (\eta/S)}. \tag{2}$$

This implies that the capacity in terms of number of users supported is

$$N = 1 + \frac{W/R}{E_b/N_0} - \frac{\eta}{S} \tag{3}$$

where W/R is generally referred to as the "processing gain" and E_b/N_0 is the value required for adequate performance of the modem and decoder, which for digital voice transmission implies a BER of 10^{-3} or better. In words, the number of users is reduced by the inverse of the per user signal-to-noise ratio (SNR) in the total system spread bandwidth, W. In a terrestrial system, the per user SNR is limited only by the transmitter's power level. As will be justified below, we shall assume SNR just below unity corresponding to a reduction in capacity equivalent to removing one user. The background noise, therefore establishes the required received signal power at the cell site, which in turn fixes the subscriber's power or the cell radius for a given maximum transmitter power.

For the reverse (subscriber-to-cell-site) direction, noncoherent reception and independent fading of all users is assumed. With dual antenna diversity, the required $E_b/N_0 = 7$ dB for a relatively powerful (constraint length 9, rate 1/3) convolutional code. Since the forward link employs coherent demodulation by the pilot carrier which is being tracked, and since its multiple transmitted signals are synchronously combined, its performance in a single cell system will be much superior to that of the reverse link. For a multiple cell system however, other cell interference will tend to equalize performance in the two directions, as will be described below.

All this leaves us at the point of our previous conclusions [1], only worse because of the Rayleigh fading encountered in terrestrial mobile applications. In the next section we begin to remedy the situation.

III. Augmented Performance Through Sectorization and Voice-Activity Monitoring

Short of reducing E_b/N_0 through improved coding or possibly modulation, which rapidly reaches the point of diminishing returns for increasing complexity (and ultimately the unsurmountable Shannon limit), we can only increase capacity by reducing other user interference and hence the denominator of (1) or (2). This can be achieved in two ways.

The first is the common technique of sectorization, which refers to using directional antennas at the cell site both for receiving and transmitting. For example, with three antennas per cell site, each having 120° effective beamwidths, the interference sources seen by any antenna are approximately one-third of those seen by an omnidirectional antenna. This reduces the $(N-1)$ term in the denominator of (2) by a factor of 3 and consequently, in (3) N is increased by nearly this factor. Henceforth, we shall take N_s to be the number of users per sector and the interference to be that received by one sector's antenna. Using three sectors, the number of users per cell $N = 3N_s$.

Secondly, voice activity can be monitored, a function which virtually already exists in most digital vocoders, and transmission can be suppressed for that user when no voice is present. Extensive studies show that either speaker is active only 35 % to 40 % of the time [3]. We shall assume for this the "voice activity factor," $\propto = 3/8$ throughout. On the average, this reduces the interference term (in the denominator of (2)) from $(N-1)$ to $(N-1)\propto$. Below, we will find through a more careful analysis that the net improvement in capacity due to voice activity is reduced from 8/3 to about 2 due to the fact that with a limited number of calls per sector, there is a nonnegligible probability that an above average number of users are talking at once. We ignore this in this preliminary discussion but include it in the results described below. Thus with sectorization and voice activity monitoring, the average $\overline{E_b}/N_0$, is increased relative to (2) to become[2]

$$\frac{\overline{E_b}}{N_0} = \frac{W/R}{(N_s-1)\propto + (\eta/S)}. \qquad (4)$$

[2] These arguments leading to (4) were first advanced by Cooper and Nettleton [4].

This suggests that the average number of users per cell is increased by almost a factor of 8. In fact, because of variability in E_b/N_0 this increase will need to be backed off to a factor of 5 or 6. We shall return to this variability issue and other more precise results after we consider multiple cell interference in the next section. For now, note from (3) and (4) that this is enough to bring the number of users/cell up to the processing gain, $N \approx W/R$ users/cell which makes CDMA at least competitive with other multiple access techniques (FDMA or TDMA) on a single cell basis. As we will presently show, in multiple-cell systems additional advantages accrue through frequency reuse of the same spectrum in all cells. To assess this advantage, we must first consider the power control techniques and their effect on multicell interference.

IV. Reverse Link Power Control in Multiple-Cell Systems

As should be clear by now, power control is the single most important system requirement for CDMA, since only by control of the power of each user accessing a cell can resources be shared equitably among users and capacity maximized. In a single cell system, the principle is straightforward, though the implementation may not be. Prior to any transmission, each of the subscribers monitors the total received signal power from the cell site. According to the power level it detects, it transmits at an initial level which is as much below (above) a nominal level in decibels as the received pilot power level is above (below) its nominal level. Experience has shown that this may require a dynamic range of control on the order of 80 dB. Further refinements in power level in each subscriber can be commanded by the cell site depending on the power level it receives from the subscriber.

The relatively fast variations associated with Rayleigh fading may at times be too rapid to be tracked by the closed-loop power control but variations in relative path losses and shadowing effects, which are modelled as an attenuation with log-normal distribution, will generally be slow enough to be controlled. Also, while Rayleigh fading may not be the same for forward and reverse links, log-normal shadowing normally will exhibit reciprocity. For the forward link, no power control is required in a single cell system, since for each subscriber any interference caused by other subscriber signals remains at the same level relative to the desired signal; inasmuch as all signals are transmitted together and hence vary together, there are no resulting degradations due to fading assuming the background noise may be neglected.

In multiple-cell CDMA systems, the situation becomes more complicated in both directions. First, for the reverse link, subscribers are power controlled by the base station of their own cell. Even the question of cell membership is not simple. For it is not minimum distance which determines which base station (cell site) the subscriber joins, but rather the maximum pilot power among the cell sites the subscriber receives. In any case the interference level from subscribers in the other cells varies not only according to the attenuation in the path to the subscriber's cell site, but also inversely to the attenuation from the interfering user to his own cell site, which through power control by that cell site may increase, or decrease, the interference to the desired cell site. These issues will be treated in the next section.

As for the forward link for a multiple cell system, interference from neighboring cell sites fade independently of the given cell site and thereby degrade performance for any level of interference. This becomes a particularly serious problem in the region where two or even three cell transmissions are received at nearly equal strengths. Techniques for mitigating this condition are treated in Section VI.

V. Reverse Link Capacity for Multiple Cell CDMA

Recalling that power control to a given mobile is exercised by the cell whose pilot signal power is maximum to that mobile, it follows that if the path loss were only a function of distance from the cell site, then the mobile would be power controlled by the nearest cell site, which is situated at the center of the hexagon in which it lies, as shown in Fig. 2(a) for an idealized placement of cell sites. In fact, the loss is proportional to other effects as well, the most significant being shadowing. The generally accepted model is an attenuation which is the product of the fourth power of the distance and a log-normal random variable whose standard deviation is 8 dB. That is, the path loss between the subscriber and the cell site is proportional to $10^{(\xi/10)}r^{-4}$ where r is distance from subscriber to cell site and ξ is a Gaussian random variable with standard deviation $\sigma = 8$ and zero mean. Fast fading (due largely to multipath) is assumed not to affect the (average) power level.

We note that other propagation exponents can be found in different environments. In fact, within a single cell the propagation may vary from inverse square law very close to the cell antenna to as great as the inverse 5.5 power far from the cell in a very dense urban environment such as Manhattan. The present analysis is primarily concerned with interference from neighboring and distant cells so the assumption of inverse fourth law propagation is a reasonable one.

The interference from transmitter within the given subscriber's cell is treated as before; that is, since each user is power controlled by the same cell site, it arrives with the same power S, when active. Thus given N subscribers per cell, the total interference is never greater than $(N-1)S$, but on the average it is reduced by the voice activity factor, $\propto$. Subscribers in other cells, however, are power controlled by other cell sites (Fig. 2(a)). Consequently, if the interfering subscriber is in another cell and at a distance r_m from its cell site and r_0 from the cell site of the desired user, the other user, when active, produces an interference in the desired user's cell site equal to

$$\frac{I(r_0, r_m)}{S} = \left(\frac{10^{(\xi_0/10)}}{r_0^4}\right)\left(\frac{r_m^4}{10^{(\xi_m/10)}}\right) = \left(\frac{r_m}{r_0}\right)^4 10^{(\xi_0-\xi_m)/10} \leq 1 \tag{5}$$

where the first term is due to the attenuation caused by distance and blockage to the given cell site, while the second term is the effect of power control to compensate for the corresponding attenuation to the cell site of the out-of-cell interferer.[3] Of course ξ_0 and ξ_m are independent so that the difference has zero mean and variance $2\sigma^2$. For all values of the above parameters, the expression is less than unity, for otherwise the subscriber would switch to the cell site which makes it less than unity (i.e., for which the attenuation is minimized).

Then, assuming a uniform density of subscribers, and normalizing the hexagonal cell radius to unity, and since the average number of subscribers/cell is $N = 3N_s$, the density of users is

$$\rho = \frac{2N}{3\sqrt{3}} = \frac{2N_s}{\sqrt{3}} \text{ per unit area.} \tag{6}$$

Consequently, the total other-cell user interference-to-signal ratio is

$$I/S = \iint \psi\left(\frac{r_m}{r_0}\right)^4 \{10^{(\xi_0-\xi_m)/10}\} \cdot \varnothing(\xi_0 - \xi_m, r_0/r_m)\rho\, dA \tag{7}$$

where m is the cell-site index for which

$$r_m^4 10^{-\xi_m} = \min_{k\neq 0} r_k^4 10^{-\xi_k} \tag{8}$$

and

$$\varnothing(\xi_0 - \xi_m, r_0/r_m) = \begin{cases} 1, & \text{if } (r_m/r_0)^4 10^{(\xi_0-\xi_m)/10} \leq 1 \\ & \text{or } \xi_0 - \xi_m \leq 40\log_{10}(r_0/r_m) \\ 0, & \text{otherwise} \end{cases} \tag{9}$$

and ψ is the voice activity variable, which equals 1 with probability $\propto$ and 0 with probability $(1 - \propto)$. To determine the moment statistics of the random variable I, the calculation is much simplified and the results only slightly increased if for m we use the smallest distance rather than the smallest attenuation. Thus (7), with (9), holds as an upper bound if in place of (8) we use that value of m for which

$$r_m = \min_{k\neq 0} r_k. \tag{8'}$$

In Appendix I, it is shown that the mean or first moment, of the random variable I/S is upper bounded (using (8′) rather than (8) for m) by the expression

$$E(I/S) = \propto \iint \frac{r_m^4}{r_0} f\left(\frac{r_m}{r_0}\right)\rho\, dA$$

where

$$f\left(\frac{r_m}{r_0}\right) = \exp\left[(\sigma \ln 10/10)^2\right]\left\{1 - Q\left[\frac{40}{\sqrt{2\sigma^2}} \cdot \log_{10}\left(\frac{r_0}{r_m}\right) - \sqrt{2\sigma^2}\frac{\ln 10}{10}\right]\right\} \tag{10}$$

[3] Cooper and Nettleton [4] employed similar geometric arguments to compute interference, but did not consider log-normal statistical variations due to blockage.

and

$$Q(x) = \int_x^\infty e^{-y^2/2}\, dy / \sqrt{2\Pi}\,.$$

This integral is over the two-dimensional area comprising the totality of all sites in the sector (Fig. 2(a)). The integration, which needs to be evaluated numerically, involves finding for each point in the space the value of r_0, the distance to the desired cell site and r_m, which according to (8′), is the distance to the closest cell site, prior to evaluating at the given point the function (10). The result for $\sigma = 8$ dB is

$$E(I/S) \le 0.247 N_s.$$

Calculation of the second moment, var (I/S) of the random variable requires an additional assumption on the second-order statistics of ξ_0 and ξ_m. While it is clear that the relative attenuations are independent of each other, and that both are identically distributed (i.e., have constant first-order distributions) over the areas, their second-order statistics (spatial correlation functions) are also needed to compute var (I). Based on experimental evidence that blockage statistics vary quite rapidly with spatial displacement in any direction, we shall take the spatial autocorrelation functions of ξ_0 and ξ_m to be extremely narrow in all directions, the two-dimensional spatial equivalent of white noise. With this assumption, we obtain in Appendix I that

$$\text{var}(I/S) \le \iint \left(\frac{r_m}{r_0}\right)^8 \left[\propto g\left(\frac{r_m}{r_0}\right) - \propto^2 f\left(\frac{r_m}{r_0}\right)\right] \rho\, dA$$

where

$$g\left(\frac{r_m}{r_0}\right) = \exp\left[\left(\frac{\sigma \ln 10}{5}\right)^2\right]\left\{1 - Q\left[\frac{40}{\sqrt{2\sigma^2}} \cdot \log_{10}\left(\frac{r_0}{r_m}\right) - \sqrt{2\sigma^2}\left(\frac{\ln 10}{5}\right)\right]\right\}. \quad (11)$$

This integral is also evaluated numerically over the area of Fig. 2(a), with r_m defined at any given point by condition (8′). The result for $\sigma = 8$ dB is var $(I/S) \le 0.078 N_s$. The above argument also suggests that I, as defined by (7), being a linear functional on a two-dimensional white random process, is well modelled as a Gaussian random variable.[4]

We may now proceed to obtain a distribution on the total interference, both from other users in the given cell, and from other-cell users on the desired user's reverse link transmission. With sectorization, variable voice activity and the other-cell interference statistics just determined, the received E_b/N_0 on the reverse link of any desired user becomes the random variable

$$E_b/N_0 = \frac{W/R}{\sum_{i=1}^{N_s-1} \chi_i + (I/S) + (\eta/S)} \quad (12)$$

where N_s is the users/sector and I is the total interference from users outside the desired user's cell. This follows easily

[4] Of course, it can never be negative, but since the ratio of mean-to-standard deviation is approximately $\sqrt{N_s}$, with typical values of $N_s > 30$ the approximating Gaussian distribution is nearly zero for negative values.

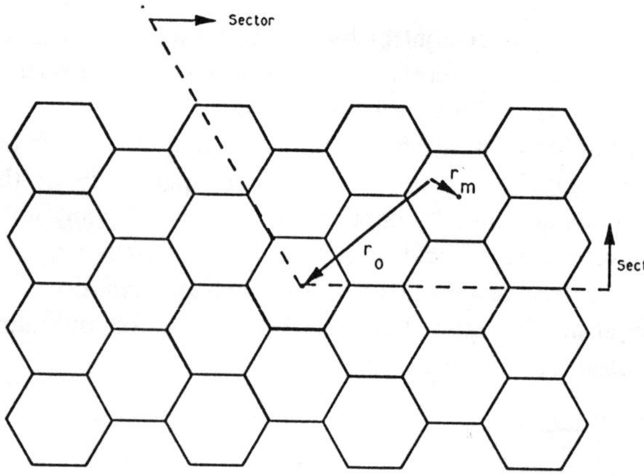

(a)

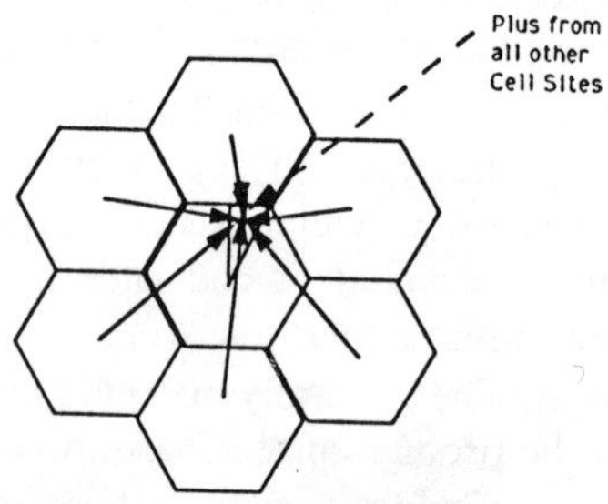

(b)

Fig. 2. Capacity calculation geometrics. (a) Reverse link geometry. (b) Forward link allocation geometry.

from (2) with the recognition that the $N_s - 1$ same sector normalized power users, instead of being unity all the time, now are random variables χ_i with distribution

$$\chi_i = \begin{cases} 1, & \text{with probability } \propto \\ 0, & \text{with probability } 1 - \propto\,. \end{cases} \quad (13)$$

The additional term I represents the other (multiple) cell user interference for which we have evaluated mean and variance,

$$E(I/S) \le 0.247 N_s \quad \text{and} \quad \text{var}(I/S) \le 0.078 N_s \quad (14)$$

and have justified taking it to be a Gaussian random variable. The remaining terms in (12), W/R and S/η, are constants.

As previously stated, with an efficient modem and a powerful convolutional code and two-antenna diversity, adequate performance (BER $< 10^{-3}$) is achievable on the reverse link with $E_b/N_0 \ge 5$ (7 dB). Consequently, the required performance is achieved with probability $P = \Pr(\text{BER} < 10^{-3}) = \Pr(E_b/N_0 \ge 5)$. We may *lower bound* the probability of achieving this level of performance for any desired fraction of users at any given time (e.g., $P = 0.99$) by obtaining an upper bound on its complement, which according to (12), depends on the distribution of χ_i and I, as follows

$$1 - P = \Pr(\text{BER} > 10^{-3}) = \Pr\left(\sum_{i=1}^{N_s} \chi_i + I/S > \delta\right) \quad (15)$$

where

$$\delta = \frac{W/R}{E_b/N_0} - \frac{\eta}{S}, \qquad E_b/N_0 = 5.$$

Since the random variable χ_i has the binomial distribution

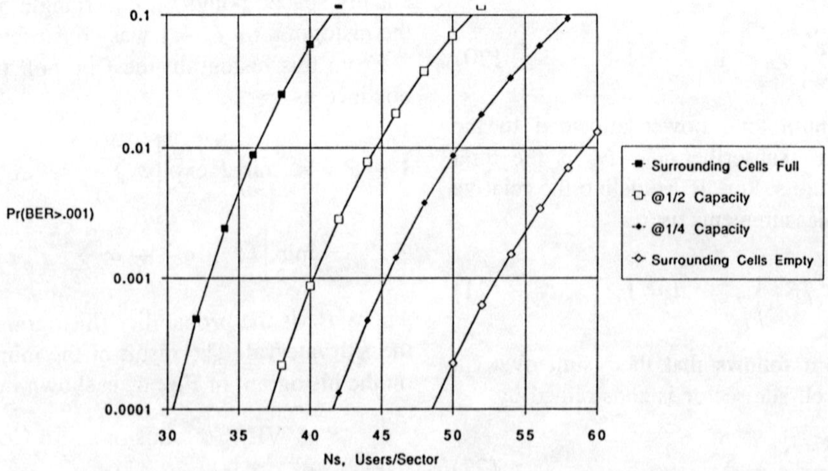

Fig. 3. Reverse link capacity/sector. (W = 1.25 MHz, R = 8 kb/s, voice activity = 3/8).

given by (13) and I/S is a Gaussian variable with mean and variance given by (14) and all variables are mutually independent, (15) is easily calculated to be

$$1 - P = \sum_{k=0}^{N_s-1} \Pr\left(I/S > \delta - k \,\middle|\, \sum x_i = k\right) \Pr\left(\sum x_i = k\right)$$

$$= \sum_{k=0}^{N_s-1} \binom{N_s-1}{k} \propto^k (1-\propto)^{N_s-1-k} \cdot Q\left(\frac{\delta - k - 0.247N_s}{\sqrt{0.078N_s}}\right). \quad (16)$$

This expression is plotted for $\delta = 30$ (a value chosen as discussed in the conclusion) and $\propto = 3/8$, as the leftmost curve of Fig. 3. The rightmost curve applies to a single cell without other cell interference ($I = 0$), while the other intermediate curves assume that all cells other than the desired user's cells are on the average loaded less heavily (with averages of 1/2 and 1/4 of the desired user's cell).

We shall discuss these results further in the concluding section, and now concern ourselves with forward link performance.

VI. Multiple-Cell Forward Link Capacity with Power Allocation

As noted earlier, although with a single cell no power control is required, with multiple cells it becomes important, because near the boundaries of cells considerable interference can be received from other cell-site transmitters fading independently.

For the forward link, power control takes the form of power allocation at the cell-site transmitter according to the needs of individual subscribers in the given cell. This requires measurement by the mobile of its relative SNR, defined as the ratio of the power from its own cell site transmitter to the total power received. Practically, this is done by acquiring (correlating to) the highest power pilot and measuring its energy, and also measuring the total energy received by the mobile's omnidirectional antenna from all cell site transmitters. Both measurements can be transmitted to the selected (largest power) cell site when the mobile starts to transmit. Suppose then that based on these two measurements, the cell site has reasonably accurate estimates of S_{T_1} and $\sum_{i=1}^{K} S_{Ti}$, where

$$S_{T_1} > S_{T_2} > \cdots > S_{T_K} > 0 \quad (17)$$

are the powers received by the given mobile from the cell site sector facing it, assuming all but K (total) received powers are negligible. (We shall assume hereafter that all sites beyond the second ring around a cell contribute negligible received power, so that $K \leq 19$.) Note that the ranking indicated in (17) is not required of the mobile—just the determination of which cell site is largest and hence which is to be designated T_1.

The ith subscriber served by a particular cell site will receive a fraction of S_{T_1} the total power transmitted by its cell site, which by choice and definition (17) is the greatest of all the cell site powers it receives, and all the remainder of S_{T_1} as well as the other cell site powers are received as noise. Thus its received E_b/N_0 can be lower bounded by

$$\left(\frac{E_b}{N_0}\right)_i \geq \frac{\beta \varnothing_i S_{T_1}/R}{\left[\left(\sum_{j=1}^{K} S_{T_j}\right)_i + \eta\right]/W} \quad (18)$$

where S_{T_j} is defined in (17), β is the fraction of the total cell site power devoted to subscribers ($1 - \beta$ is devoted to the pilot) and $\varnothing_i$ is the fraction of this devoted to subscriber i. Because of the importance of the pilot in acquisition and tracking, we shall take $\beta = 0.8$. It is clear that the greater the sum of other cell-site powers relative to S_{T_1}, the larger the fraction $\varnothing_i$ which must be allocated to the ith subscriber to achieve its required E_b/N_0. In fact, from (18) we obtain

$$\varnothing_i \leq \frac{(E_b/N_0)_i}{\beta W/R}\left[1 + \left(\frac{\sum_{j=2}^{K} S_{T_j}}{S_{T_1}}\right)_i + \frac{\eta}{(S_{T_1})_i}\right] \quad (19)$$

where

$$\sum_{i=1}^{N_s} \varnothing_i \le 1 \tag{20}$$

since βS_{T_1} is the maximum total power allocated to the sector containing the given subscriber and N_s is the total number of subscribers in the sector. If we define the relative received cell-site power measurements as

$$f_i \triangleq \left(1 + \sum_{j=2}^{K} S_{T_j}/S_{T_1}\right)_i, \qquad i = 1,\cdots, N_s \tag{21}$$

then from (19) and (20) it follows that their sum over all subscribers of the given cell site sector is constrained by

$$\sum_{i=1}^{N_s} f_i \le \frac{\beta W/R}{E_b/N_0} - \sum_{i=1}^{N_s} \frac{\eta}{S_{T_{1_i}}} \triangleq \delta'. \tag{22}$$

Generally, the background noise is well below the total largest received cell site signal power, so the second sum is almost negligible. Note the similarity to δ in (15) for the reverse link. We shall take $\beta = 0.8$ as noted above to provide 20 % of the transmitted power in the sector to the pilot signal, and the required $E_b/N_0 = 5$ dB to ensure BER $\le 10^{-3}$. This reduction of 2-dB relative to the reverse link is justified by the coherent reception using the pilot as reference, as compared to the noncoherent modem in the reverse link. Note that this is partly offset by the 1-dB loss of power due to the pilot.

Since the desired performance (BER $\le 10^{-3}$) can be achieved with N_s subscribers per sector provided (22) is satisfied with $E_b/N_0 = 5$ dB, capacity is again a random variable whose distribution is obtained from the distribution of variable f_i. That is, the BER can not be achieved for all N_s users/sector if the N_s subscribers combined exceed the total allocation constraint of (22). Then following (15),

$$1 - P = \Pr(\text{BER} > 10^{-3}) = \Pr\left(\sum_{i=1}^{N_s} f_i > \delta'\right). \tag{23}$$

But unlike the reverse link, the distribution of the f_i, which depends on the sum of ratios of ranked log-normal random variables, does not lend itself to analysis. Thus we resorted to Monte Carlo simulation, as follows.

For each of a set of points equally spaced on the triangle shown in Fig. 2(b), the attenuation relative to its own cell center and the 18 other cell centers comprising the first three neighboring rings was simulated. This consisted of the product of the fourth power of the distance and the log-normally distributed attenuation

$$10^{(\xi_k/10)} r_k^{-4}, \qquad k = 0, 1, 2 \cdots, 18.$$

Note that by symmetry, the relative position of users and cell sites is the same throughout as for the triangle of Fig. 2(b). For each sample, the 19 values were ranked to determine the maximum (S_{T_1}), after which the ratio of the sum of all other 18 values to the maximum was computed to obtain $f_i - 1$. This was repeated 10 000 times per point for each of 65 equally spaced points on the triangle of Fig. 2(b). From this, the histogram of $f_i - 1$ was constructed, as shown in Fig. 4.

From this histogram the Chernoff upper bound on (23) is obtained as

$$1 - P = < \min_{s>0} E \exp\left[s \sum_{i=1}^{N_s} f_i - s\delta'\right]$$

$$= \min_{s>0}\left[(1 - \propto) + \propto \sum_k P_k \exp(sf_k)\right]^{N_s} e^{-s\delta'} \tag{24}$$

where P_k is the probability (histogram value) that f_i falls in the kth interval. The result of the minimization over s based on the histogram of Fig. 4, is shown in Fig. 5.

VII. Conclusions and Comparisons

Figs. 3 and 5 summarize performance of reverse and forward links. Both are theoretically pessimistic (upper bounds on probability). Practically, both models assume only moderately accurate power control.

The parameters for both links were chosen for the following reasons. The allocated total spread bandwidth $W = 1.25$ MHz represents ten percent of the total spectral allocation, 12.5 MHz, for cellular telephone service of each service provider. which as will be discussed below, is a reasonable fraction of the band to devote initially to CDMA and also for a gradual incremental transition from analog FM/FDMA to digital CDMA. The bit rate $R = 8$ kb/s is that of an acceptable nearly toll quality vocoder. The voice activity factor, 3/8, and the standard sectorization factor of 3 are used. For the reverse channel, the received SNR per user $S/\eta = -1$ dB reflects a reasonable subscriber transmitter power level. In the forward link, 20 % of each site's power is devoted to the pilot signal for a reduction of 1 dB ($\beta = 0.8$) in the effective processing gain. This ensures each pilot signal (per sector) is at least 5 dB above the maximum subscriber signal power. The role of the pilot, as noted above, is critical to acquisition, power control in both directions and phase tracking as well as for power allocation in the forward link. Hence, the investment of 20 % of total cell site power is well justified. These choices of parameters imply the choices $\delta = 30$ and $\delta' \approx 38$ in (16) and (24) for reverse and forward links, respectively.

With these parameters, according to Fig. 3, the reverse link can support over 36 users/sector or 108 users/cell, with 10^{-3} bit error rates better than 99 % of the time. This number becomes 44 users/sector or 132 users/cell if the neighboring cells are kept to half of this loading. The forward link according to Fig. 5, can do the same or better for 38 users/sector or 114 users/cell.

Clearly, if the entire cellular allocation is devoted to CDMA, these numbers are increased tenfold. Similarly, if a lower bit rate vocoder algorithm is developed, or if narrower sectors are employed, the number of users may be increased further.

Remaining with the parameters assumed, interesting comparisons can be drawn to existing analog FM/FDMA cellular systems as well as other proposed digital systems. First, the former employs 30-kHz channel allocation, and assuming 3

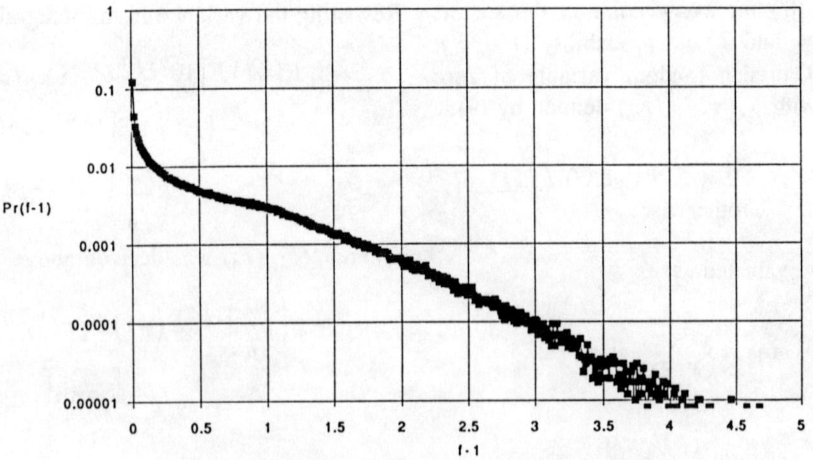

Fig. 4. Histogram of forward power allocation.

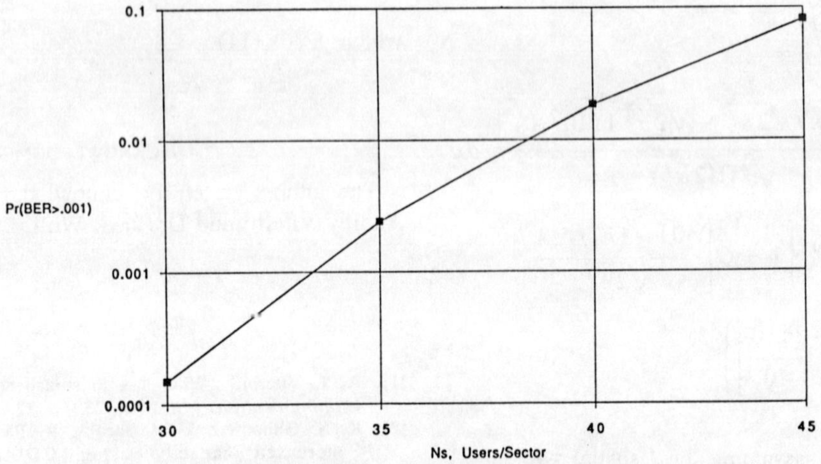

Fig. 5. Forward link capacity/sector. (W = 1.25 MHz, R = 8 kb/s, voice activity = 3/8, pilot power = 20 %).

sectors/cell, requires each of the six contiguous cells in the first ring about a given cell to use a different frequency band. This results in a "frequency reuse factor" of 1/7. Hence, given the above parameters, the number of channels in a 1.25-MHz band is slightly less than 42, and with a frequency reuse factor of 1/7, this results in slightly less than 6 users/cell for a 1.25-MHz band. Thus, CDMA offers at least an eighteenfold increase in capacity. Note further that use of CDMA over just ten percent of the band supports over 108 users/cell whereas analog FM/FDMA supports only 60 users/cell using the entire 12.5 MHz band. Thus by converting *only 10 % of the band* from analog FDMA to digital CDMA, overall capacity is increased *almost threefold*.

Comparisons of CDMA with other digital systems are more speculative. However, straightforward approaches such as narrower frequency channelization with FDMA or multiple time slotting with TDMA can be readily compared to the analog system. The proposed TDMA standard for the U.S. is based on the current 30-kHz channelization but with sharing of channels by three users each of whom is provided one of three TDMA slots. Obviously, this triples the analog capacity but falls over a factor of 6 short of CDMA capacity.

In summary, properly augmented and power-controlled multiple-cell CDMA promises a quantum increase in current cellular capacity. No other proposed scheme appears to even approach this performance. Other advantages of CDMA not treated here include inherent privacy, flexibility in supporting multiple services and multiple voice and data rates, lower average transmit power requirements and soft limit on capacity, since if the bit error rate requirement is relaxed more users can be supported. With all these inherent advantages, CDMA appears to be the logical choice henceforth for all cellular telephone applications.

Appendix I
Reverse Link Outer-Cell Interference

Outer-cell normalized interference, I/S, is a random variable defined by (7), (8), and (9), and upper bounded by replacing (8) by (8′). Then the upper bound on its first moment, taking into account also the voice activity factor of the outer-cell subscribers, α , becomes

$$E(I/S) \leq \iint_{\text{sector}} \left(\frac{r_m}{r_0}\right)^4 E(\psi \cdot 10^{x/10} \varnothing(x, r_0/r_m)) \rho \, dA$$

where r_m is defined by (8′) for every point in the sector, $\psi = 1$ with probability $\propto$ and 0 with probability $(1 - \propto)$, and $\chi = \xi_0 - \xi_m$ is a Gaussian random variable of zero mean and variance $2\sigma^2$ with $\varnothing(\chi, r_0/r_m)$ defined by (9),

$$\varnothing(\chi, r_0/r_m) = \begin{cases} 1, & \text{if } \chi \le 40 \log (r_0/r_m) \\ 0, & \text{otherwise.} \end{cases}$$

The expectation is readily evaluated as

$$\propto f\left(\frac{r_m}{r_0}\right) \triangleq E(\psi) E\left(e^{x \ln 10/10} \varnothing(\chi, r_0/r_m)\right)$$

$$= \propto \int_{-\infty}^{40 \log r_0/r_m} e^{x \ln 10/10} \frac{e^{-x^2/4\sigma^2}}{\sqrt{4\Pi\sigma^2}} dx$$

$$= \propto e^{(\sigma \ln 10/10)^2} \int_{-\infty}^{40 \log(r_0/r_m)} \cdot \frac{\exp\left[-\frac{1}{2}\left(x/\sqrt{2\sigma^2} - \sqrt{2\sigma^2} \ln 10/10\right)^2\right]}{\sqrt{2\Pi(2\sigma^2)}} dx$$

$$= \propto e^{(\sigma \ln 10/10)^2} \left\{1 - Q\left[\frac{40 \log (r_0/r_m)}{\sqrt{2\sigma^2}} - \sqrt{2\sigma^2} \frac{\ln 10}{10}\right]\right\}$$

which yields (10).

To evaluate var (I/S), assuming the "spatial whiteness" of the blockage variable, we have

$$\operatorname{var}(I/S) \le \iint_{\text{sector}} \left(\frac{r_m}{r_0}\right)^8 \cdot \operatorname{var}\left(\psi \cdot 10^{x/10} \varnothing(\chi, r_0/r_m)\right) \rho \, dA.$$

Rewriting the variance in the integral as

$$E(\psi^2) E\left[10^{2x/10} \varnothing^2(\chi, r_0/r_m)\right] - \left\{E(\psi) E\left[10^{x/10} \varnothing(\chi, r_0/r_m)\right]\right\}^2 = \propto g\left(\frac{r_m}{r_0}\right) - \propto^2 f^2\left(\frac{r_m}{r_0}\right)$$

where $f(r_m/r_0)$ was derived above and

$$\propto g\left(\frac{r_m}{r_0}\right) = E(\psi^2) E\left[e^{x \ln 10/5} \varnothing^2(\chi, r_0/r_m)\right]$$

$$= \propto e^{(\sigma \ln 10/5)^2} \left\{1 - Q\left[\frac{40 \log (r_0/r_m)}{\sqrt{2\sigma^2}} - \frac{\sqrt{2\sigma^2} \ln 10}{5}\right]\right\}$$

which yields (11).

Acknowledgment

The authors gratefully acknowledge the contribution of Dr. Audrey Viterbi and Dr. Jack Wolf.

References

[1] A. J. Viterbi, "When not to spread spectrum—A sequel," *IEEE Communications Mag.* vol. 23, pp. 12–17, Apr. 1985.

[2] K. S. Gilhousen, I. M. Jacobs, R. Padovani, and L. A. Weaver, "Increased capacity using CDMA for mobile satellite communications," *IEEE Trans. Select. Areas Commun.*, vol. 8, pp. 503–514, May 1990.

[3] P. T. Brady, "A statistical analysis of on-off patterns in 16 conversations," *Bell Syst. Tech. J.*, vol. 47, pp. 73–91, Jan. 1968.

[4] G. R. Cooper and R. W. Nettleton, "A spread spectrum technique for high capacity mobile communications," *IEEE Trans. Veh. Technol.*, vol. VT-27, pp. 264–275, Nov. 1978.

Twelve Random Access Strategies for Fiber Optic Networks

NICHOLAS F. MAXEMCHUK, SENIOR MEMBER, IEEE

***Abstract*—Twelve random access strategies are described and compared that do not constrain the distance or transmission rate of a network, and can use the capabilities of fiber optic components.**

The twelve strategies consist of three protocols, each of which can use two timing arrangements and two network access devices. The three protocols are the standard ALOHA protocol, LCSMA, and LCSMA-CD. The last two protocols operate on linear unidirectional networks and use local information at the transmitter to increase the throughput of the system. The networks considered in this paper have a common point that all transmitted signals pass through before being received. This makes two timing arrangements possible, a slotted system or an unslotted system. Finally, the taps on the network can be passive or active.

I. Introduction

Random access protocols, such as the ALOHA [1] and carrier sense multiple access with collision detection (CSMA-CD) [2] protocols, provide a simple means for sharing networks. The CSMA-CD protocol makes efficient use of the transmission facility when the propagation delay between sources and receivers is much less than the time it takes to transmit a packet. Typical CSMA-CD networks are a few kilometers long, transmit at a few megabits per second, and have a few kilobits in a packet. Fiber optic networks can operate at transmission rates and distances that are two or three orders of magnitude greater than this. In order to retain the efficiency of the CSMA-CD protocol, the (Packet Size)/(Distance * Transmission Rate) of the system must remain the same. Therefore, either the capabilities of fiber optic components are not used or the packet length must be made extremely long.

CSMA-CD protocols have been implemented on star configured fiber optic networks [3]–[5], however, these networks have the same constraints on transmission rate and distance as conventional CSMA-CD systems. On linear unidirectional networks, such as the "*D*" [6], [7], spiral [8], and double bus [9], information at the transmitter has been used to control access to the network. Access protocols that use information at the transmitter have the advantage that the propagation delay between transmitters and receivers can be greater than the packet transmission time, so that the constraints associated with conventional CSMA-CD protocols can be relaxed; however, protocols that use the information available at the transmitter have the disadvantage that they do not have information that is as complete as that used in CSMA-CD strategies, as explained in Section III-C.

In this paper, random access strategies that use the information at the transmitter in a unidirectional network are described and compared. Three access strategies are considered, ALOHA, LCSMA, and LCSMA-CD. The prefix "L" is used to distinguish strategies that use local information at the transmitter from strategies that use complete information at the receiver. Each of these strategies can be implemented with

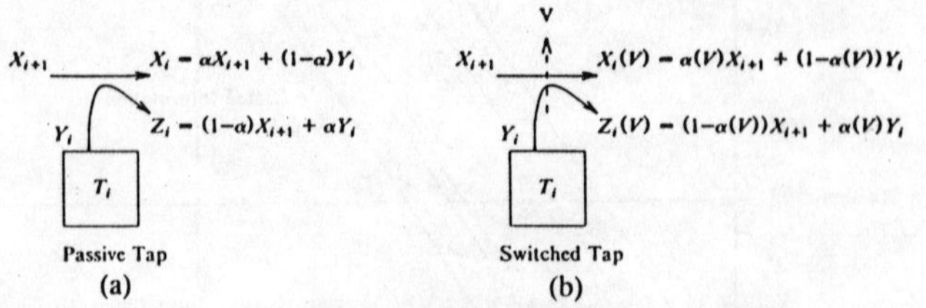

Fig. 1. Fiber optic taps.

passive taps, that transmit the "or" of the locally generated signal and the signal on the network, or active taps, that switch between the locally generated signal and the signal on the network, and can be slotted or unslotted, for a total of twelve possible access strategies. The access strategies are described in Section III, and a quantitative comparison is given in Section VI. The strategies provide a means of trading various types of complexity for throughput. It is shown that fair access strategies can be constructed by combining unfair techniques that give priority to sources at particular positions in the network, and that the throughput of the system can approach the transmission rate.

II. System

A. Fiber Optic Taps

Two types of taps are considered in this paper, passive taps [6], [10], Fig. 1(a), and active taps [11], [12], Fig. 1(b). The taps have two characteristics in common:

1) They are directional. The signal Y_i inserted at terminal T_i propagates in one direction.

2) They are reciprocal. In order to insert a fraction of a signal $(1 - \alpha) Y_i$ onto a fiber, the same fraction $(1 - \alpha) X_{i+1}$ of the signal on the fiber must be removed.

In addition, the active taps act as switches. When a voltage V is applied, the fraction $(1 - \alpha)$ changes from nearly 0 to nearly 1.

Fiber optic taps are not ideal devices. Because of reciprocity and coupling losses, signal amplifiers or regenerators must be placed in the serial data path; and because active taps do not switch completely from zero to one, cross-talk results. Serial regenerators reduce the reliability of a network. In linear networks, such as the "*D*" network discussed in Section II-B, regenerators can be physically located in a common location to make them easier to maintain. In addition, in the protocols described in Section III, signals are not removed from the network unless they are replaced by another signal. Failures that break the signal path can be detected, and failed segments of the network can be automatically bypassed, as is common in loop networks [13]. For the discussions in this paper, it is assumed that the delay caused by regenerators is included in the propagation delays, and that because of proper regeneration, X_i in a passive tap is the "or" of X_{i+1} and Y_i, and X_i in an active tap is either X_{i+1} or Y_i.

B. Topology

The topology considered in this paper is the "*D*" network [6], [10], Fig. 1. In this network, a node T_i transmits a signal

Paper approved by the Editor for Communication Networks of the IEEE Communications Society. Manuscript received July 12, 1987; revised January 6, 1988.

The author is with AT&T Bell Laboratories, Murray Hill, NJ 07974.

IEEE Log Number 8822465.

Reprinted from *IEEE Trans. Commun.*, vol. 36, no. 8, pp. 942–950, August 1988.

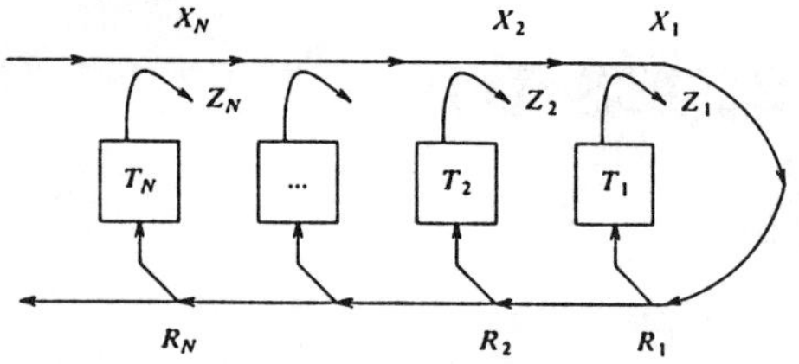

Fig. 2. *D*-network.

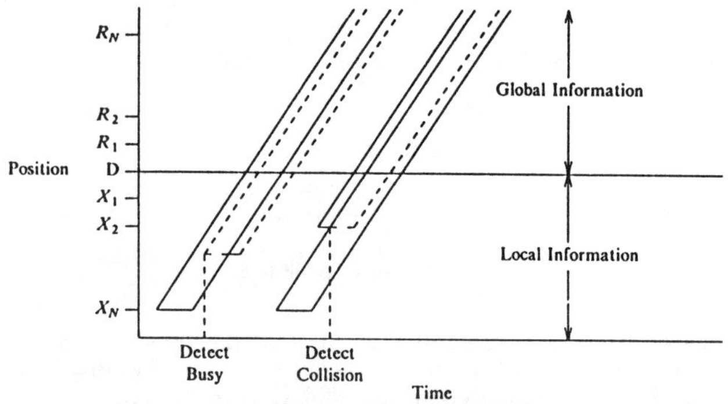

Fig. 3. Global and local information.

at X_i and receives the signal from all of the receivers at R_i. The signal passes the transmitter at each node before passing the receiver at any node. The "*D*" network is well suited to the directional nature of fiber optic taps. In addition, a slotted system can be implemented by generating slot timing pulses at the transmitter that is the furthest upstream and detecting these pulses at the other transmitters, as was done in Fastnet [9].

The protocols that are described in this paper are not dependent upon the signal passing the receivers in sequence and can also be applied to spiral networks [8], and hybrid linear-star networks [14]. In addition, the ALOHA protocols with passive taps do not depend upon the signal passing the transmitters in sequence and can be applied to star networks [15]–[18].

C. Information

At node T_i, information about the state of the network can be extracted at R_i or Z_i. Fig. 3 shows the propagation of signals in the "*D*" network. The duration of packet transmissions is shown as being much less than the propagation time to emphasize the fact that in a high rate, fiber optic network, the signal at R_i may not be useful to determine when to transmit. The signals at X_i and Z_i are at the same position on this diagram.

The signal at R_i is the composite signal from all of the sources that arrive at the bend in the "*D*" network at a particular instant of time, and is referred to as global information. The global signal provides complete information on the state of the network at a particular instant of time; however, as far as the transmitter is concerned, it is out of date and indicates what signals the transmitter would have collided with if it had transmitted a roundtrip propagation time ago. In the access strategies described in this paper, the global signal is used to determine when a retransmission is required. Since the same global signal is received by all of the receivers, when a source receives the signal it transmits, it knows that a collision has not occurred and that a retransmission is not necessary. Using global information in this way does not constrain the transmission rate or span of the network, as it would if this information were used to determine when to transmit in a CSMA-CD strategy.

The signal at Z_i is the composite signal from the sources that are upstream from terminal T_i, and is referred to as local information. The local signal provides partial information about the state of the network at a particular instant of time because downstream terminals are not included; however, as far as the transmitter is concerned, it is current and indicates signals that the transmitter will collide with if it transmits. In the access strategies described in this paper, local information is used to determine when the channel is being used by an upstream terminal. By listening to this signal before transmitting, a source can delay its transmission when an upstream source is using the medium and avoid some collisions. The number of collisions can be further reduced by listening while transmitting, realizing that an upstream source cannot detect transmissions from a downstream source, and aborting transmission if a signal from an upstream source is received.

D. Implementation

In order to implement strategies that use local information the signal from upstream sources must be detected. A separate tap, upstream from the transmit tap, can be avoided because of the reciprocity of passive taps and the cross-talk caused by incomplete switching in active taps. The signal $Z_i = (1 - \alpha(V))X_{i+1}$ can be used to determine if the channel is busy before transmitting. When source T_i is transmitting, a signal from an upstream source can be detected by subtracting the locally generated signal $\alpha(V)Y_i$ from Z_i. Alternatively, a separate fiber from Z_{i+1} to T_i can be installed. This increases the amount of fiber in the network, but has the advantage that collisions can be detected by detecting carrier rather than processing the received signal.

During the time it takes to detect the beginning of a packet the upstream transmission is distorted. A fiber delay can be inserted on the transmit path in the "*D*" network when an upstream tap is used to detect collisions, so that the collision indicator arrives before the packet. Alternatively, a preamble can be transmitted preceding the useful information in a packet so that the distorted portion of the packet is not required at the receiver.

At the receiver it is necessary to determine when collisions have occurred, and when a new packet is preempting a packet transmission that is in progress. This can be done in a system with fixed size packets by using a Manchester code to transmit data, and transmission rule violations to designate the start and end of packet. For instance, if a data one is transmitted as 1, 0, a data zero as 0, 1, the start of packet as three or more 1's, and the end of packet by at least three 0's, and, if all successful packets have the same size, then any combination of collisions and preemptions can be detected in either the "or'ed" or switched system. A description of a framing sequence that can be used to implement all twelve strategies is presented in [19].

III. Access Strategies

In this section, twelve random access strategies are described that do not have the distance or bandwidth constraints of the conventional CSMA-CD protocol. The twelve strategies consist of three protocols, each of which can be slotted or unslotted and use active or passive taps. The three protocols are the standard ALOHA protocol, LCSMA, and LCSMA-CD. LCSMA and LCSMA-CD operate in a manner similar to the conventional CSMA and CSMA-CD protocols, except that they use local rather than global information.

The strategies are referred to as protocol/timing/access where

—the protocol is ALOHA, LCSMA, or LCSMA-CD,
—the timing is unslotted (*U*) or slotted (*S*), and
—the network access is passive (*P*) or active (*A*).

An *X* indicates that all of the values of a parameter are considered.

A. ALOHA

ALOHA/*U*/*P* is the conventional ALOHA protocol [1]. A source transmits as soon as it has a packet, and it retransmits

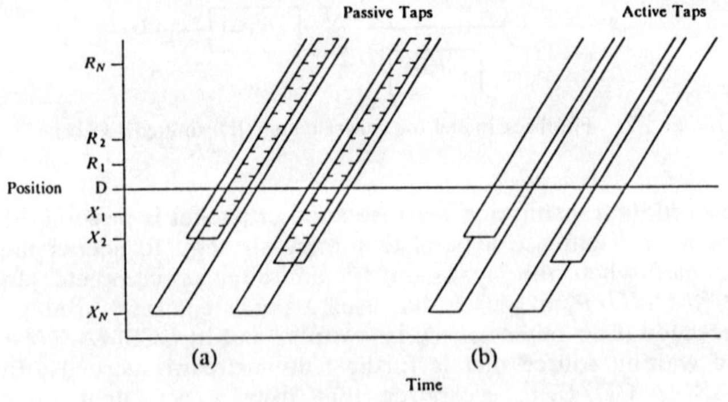

Fig. 4. ALOHA/U/X.

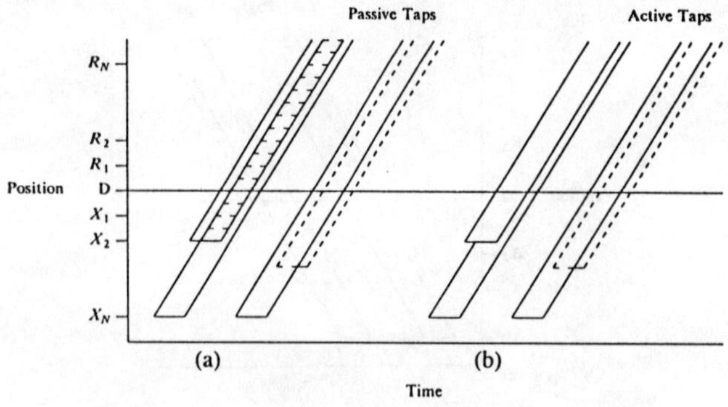

Fig. 5. LCSMA/U/X.

after a random delay if it collides with another terminal. Fig. 4(a) is the time-space diagram for two sources that collide in this system. The shaded area indicates where two packets are transmitted simultaneously.

Acknowledgments are not required in ALOHA/U/P to determine when a signal was received without collision because the same global signal is received by all of the receivers, and a source can determine whether or not it has collided by examining this signal. By contrast, on a bidirectional network, the same composite signal is not seen by all of the sources and acknowledgments must be transmitted. Acknowledgment messages increase the traffic and the probability of collision on the network, and packets must be retransmitted if either the packet or the acknowledgment collides. Therefore, ALOHA/U/P on a unidirectional network requires fewer retransmissions than an ALOHA strategy on a bidirectional network. In the original ALOHA system, there is one destination, acknowledgments are transmitted on a separate channel, and acknowledgments do not experience collisions. ALOHA/U/P on a unidirectional network has the same number of retransmissions as the original ALOHA system even though there are multiple destinations.

ALOHA/U/A uses the same access strategy as ALOHA/U/P except that a source that is transmitting changes α of the tap from 1 to 0. When collisions occur the signal from the downstream source successfully acquires the network. Fig. 4(b) shows the portion of packets from an upstream source that are removed from the system by a downstream source. Switched systems give preference to downstream sources and are analogous to perfect capture radio networks [20], [21] in which the power transmitted by the sources is different enough to guarantee that one succeeds.

In a slotted network, the source furthest from the bend in the "D" network X_N periodically generates timing pulses that signify the start slot. ALOHA/S/P is the conventional slotted ALOHA protocol. Packets that arrive during a packet transmission interval are transmitted at the beginning of the next interval, and a collision occurs if more than one packet arrives during a transmission interval. ALOHA/S/A is similar to ALOHA/S/P except that the switch allows the terminal that is furthest downstream to win when a collision occurs. In ALOHA/S/A, there is a successful transmission in every interval in which one or more packets are transmitted.

B. LCSMA

In LCSMA/X/X, a source listens to the signal from the upstream sources before transmitting. If an upstream source is transmitting, the local source delays its transmission. This strategy gives preference to upstream sources.

In LCSMA/U/P, a source

1) is delayed by packets from upstream sources that are propagating past the transmitter when a packet arrives,

2) collides with packets from upstream sources that arrive at the transmitter during the packet transmission, and

3) collides with downstream sources that are already transmitting when the packet arrives at their transmitter.

The shaded area in Fig. 5(a) shows where two packets collide, and the dashed line indicates a packet that is not transmitted because a source detects a busy channel. There are fewer collisions in this system that in ALOHA/U/P because downstream sources that detect the presence of a transmission from an upstream source do not transmit and avoid a collision. In this system, sources that detect a busy channel can implement a retry strategy immediately instead of waiting for a round trip propagation delay as in ALOHA/X/X.

In LCSMA/U/A, a packet is

1) prevented from entering the system if a packet from an upstream source is passing the transmitter when the packet arrives, and

2) switched off of the system by a downstream source that is transmitting when the packet arrives at its transmitter.

Fig. 5(b) shows an instance in which a portion of a packet from an upstream terminal is switched off the system by a downstream source and an instance in which a downstream source does not transmit because the channel is busy.

LCSMA/U/A combines taps that give priority to downstream sources with a transmission rule that gives priority to upstream sources and results in a fair access strategy. A source successfully transmits a packet if, at the instant that the packet would arrive at the bend in the "D" network if it were transmitted, there are no other packets passing the bend in the "D" network. This is independent of the source's location, and results in a protocol that gives equal access to all sources. There is some unfairness in the protocol because downstream sources know when they have failed, and can retry sooner than upstream sources.

In LCSMA/U/A it is also possible to interrupt an upstream source that is transmitting, depending upon how far into the packet the upstream source has gotten or depending upon the content of the initial portion of the packet. This ability may be useful to establish priority classes of sources. However, no advantage has been found in the performance of the network, and this option is not considered further.

For completeness, LCSMA/S/X systems must be considered, however, these systems perform the same as ALOHA/S/X systems. All sources with packets to transmit at the beginning of the slot, transmit when they receive the start of slot pulse, and packets from an upstream source are not detected in time to delay the transmission. The performance can be improved by defining a slot to be longer than a packet, and starting the transmission from sources at different points in the slot so that downstream sources can detect when an upstream source is transmitting. However, this type of protocol is complex, position dependent, and is not considered further.

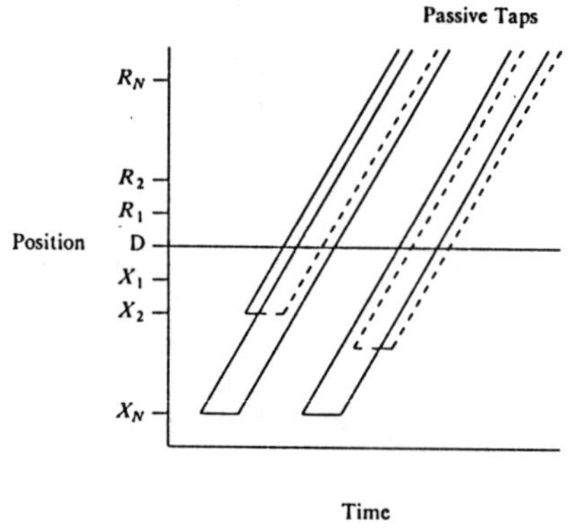

Fig. 6. LCSMA-CD/*U*/*P*.

C. LCSMA-CD

In LCSMA-CD/*X*/*X*, a source listens to the signal both before and during its transmission. A source does not transmit if an upstream source is active, and stops transmitting if an upstream source becomes active. Upstream sources always have preference over downstream sources. While this is unfair, it has the advantage that a source immediately knows whether or not it has successfully accessed the channel and can retry sooner.

In LCSMA-CD/*U*/*P*, a source does not transmit if a packet from an upstream source is passing the local transmitter when a packet arrives, and is preempted if an upstream source transmits a packet that arrives at the local transmitter while a packet is being transmitted. The dashed lines in Fig. 6 show the portions of packets that are not transmitted by a downstream source. In the first instance, a downstream source detects a transmission from an upstream source while it is transmitting and does not transmit the end of its packet, and in the second instance, a busy channel is detected at the beginning of transmission and the packet is not transmitted.

Unlike LCSMA/*S*/*P*, LCSMA-CD/*S*/*P* differs significantly from ALOHA/*S*/*P*. In LCSMA-CD/*S*/*P*, one source wins in every slot in which one or more sources transmit, as in ALOHA/*S*/*A*. However, in LCSMA-CD/*S*/*P*, upstream, rather than downstream, sources are given priority.

Switches in LCSMA-CD/*X*/*A* strategies allow additional choices. In LCSMA-CD/*U*/*A*, a source can defer to an upstream source only if less than a fraction γ of its own packet has been transmitted. As $\gamma \rightarrow 0$, LCSMA-CD/*U*/*A* $\rightarrow$ LCSMA/*U*/*A*, and as $\gamma \rightarrow 1$, LCSMA-CD/*U*/*A* $\rightarrow$ LCSMA-CD/*U*/*P*. In LCSMA-CD/*S*/*A*, a source can defer to the upstream source or replace the data from that terminal with its own data with a specified probability, *P*. As $P \rightarrow 1$, LCSMA-CD/*S*/*A* $\rightarrow$ ALOHA/*S*/*A*, and $P \rightarrow 0$, LCSMA-CD/*S*/*A* $\rightarrow$ LCSMA-CD/*S*/*P*. Both of these strategies approach the same probability of success in a slot, however, one gives priority to upstream terminals and the other to downstream terminals. Adjusting the probability of switching to be different at every node, dependent upon its position, presents the possibility of constructing fairer systems. However, for the remainder of this paper, $\gamma = 1$ and $P = 0$, so that LCSMA-CD/*X*/*A* is equivalent to LCSMA-CD/*X*/*P*.

IV. Retry Strategies

In order to completely define an access strategy, both the access protocol and the retry strategy must be specified. In the analysis in this paper, a random retry strategy will be used, however, other strategies are possible. These strategies are described because they show how this work relates to other work in the area, and indicate possible directions for further work.

In LCSMA/*X*/*X* and LCSMA-CD/*X*/*X*, a source that has delayed its transmission because another packet is passing the transmitter can use a persistent retry strategy to access the channel when the upstream transmission is complete. In LCSMA/*U*/*P*, a source that uses a persistent retry strategy succeeds if no other source is waiting, and in LCSMA/*U*/*A* the waiting source that is furthest downstream succeeds. In LCSMA-CD/*U*/*P*, a source that uses a persistent retry strategy acquires the channel as soon as there are no upstream sources waiting to transmit, and trains of packets, similar to the trains in Expressnet [8], are formed. A difference between this strategy and Expressnet is that an upstream source that becomes active during a train can break into the middle, since the global signal is not used to inhibit new sources from breaking into a train. In LCSMA-CD/*S*/*P*, the preamble at the beginning of the packet has the same function as the busy bit in Fastnet [9]. With a persistent retry strategy, at the beginning of each slot, every source with data to transmit writes a preamble, reads the channel to see if an upstream source is also writing a preamble, and defers to active upstream sources. As in Expressnet and Fastnet, a degree of fairness can be imposed on this access and retry strategy by requiring sources to wait until receiving the end of the train with their packet before transmitting a second packet. This gives downstream sources a chance to transmit.

The LCSMA-CD/*U*/*P* protocol can be made more efficient by adopting a preemptive-resume strategy, in which a source that stops transmitting, transmits the interrupted portion of the packet, rather than the entire packet, when it resumes. This strategy increases the throughput, by not retransmitting data that have gotten through successfully, but significantly increases the complexity of the system.

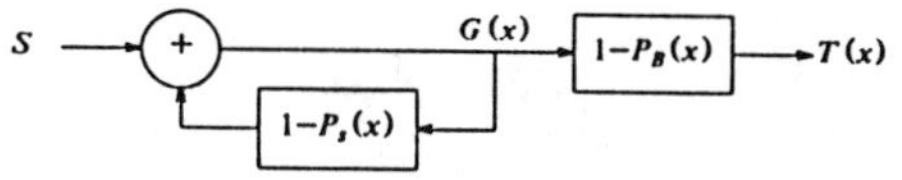

Fig. 7. Feedback model for generated and transmitted packets.

V. Model

In Section VI, the access strategies are compared assuming constant size packets, a Poisson arrival distribution, a random retry strategy, and an infinite user population. The arrival rate for new packets *S* is independent of a source's position in the system. A packet is successfully transmitted with probability $P_s(x)$ that may be dependent upon a sources position in the system. If a packet is not successfully transmitted, a retransmission attempt occurs a random time later and adds to the rate $G(x)$ at which packets attempt to access the system, as shown in Fig. 7. Since the probability of success may be dependent upon a source's position in the network, $G(x)$ may be position dependent, even though *S* is not, and

$$S = P_s(x) * G(x). \tag{5.0.1}$$

In the analysis, $G(x)$ is assumed to be a Poisson process.

In CSMA/*X*/*X* and CSMA-CD/*X*/*X*, a packet encounters a busy channel with probability $P_B(x)$, and is not transmitted. These packets cannot collide with other packets, and $T(x)$, the rate at which packets are transmitted, must be distinguished from $G(x)$, the rate at which packets attempt to access the system.

A. Normalizations

As is common in this type of analysis, time is measured in units of packet transmission times and results in the analysis being independent of the transmission rate. Therefore, *S* is the average number of new arrivals during a packet transmission interval, and *G* is the average number of packet transmissions attempted during a packet transmission interval.

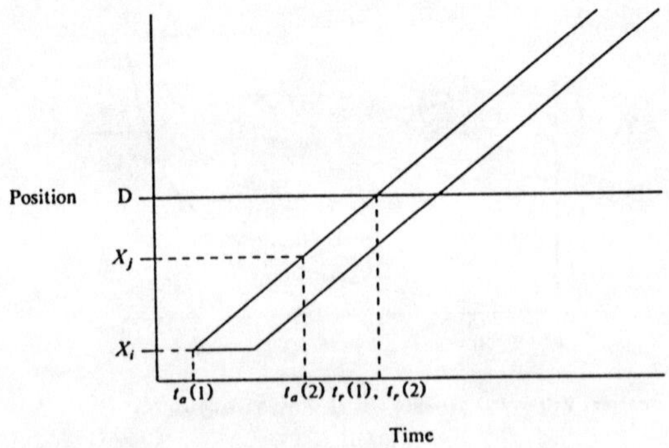

Fig. 8. Relative time.

In addition, the span of the network, from the first source to the bend in the D, is normalized to one. A source is located at position x where $0 \leqslant x \leqslant 1$ and 1 is the bend in the D network. Because the length of the network is normalized, the total arrivals are $S_T = \int_{x=0}^{1} S\,dx = S$.

B. Relative Time

Absolute time is the time at which an event actually occurs. Relative time at each transmitter is defined as the absolute time plus the propagation delay from the transmitter to the bend in the "D" network, and is the absolute time that a signal from this transmitter would arrive at the bend in the "D" network. Relative time is convenient to discuss "D" networks because traffic that is transmitted at a source at its relative time passes a downstream source at that source's same relative time. For instance, in Fig. 8 source X_i transmits a packet at absolute time $t_a(1)$ that passes source X_j at absolute time $t_a(2)$. As far as source X_i is concerned, it transmits the packet at relative time $t_r(1) = t_a(1) + \tau_p(i)$ where $\tau_p(i)$ is the propagation delay from X_i to the common point. As far as source X_j is concerned, the packet passes this source at relative time $t_r(2) = t_a(2) + \tau_p(j)$. Because of the linear structure of the "D" network, the propagation delay from X_i to X_j is $\tau_p(i) - \tau_p(j)$, $t_a(2) = t_a(1) + (\tau_p(i) - \tau_p(j))$, and $t_r(2) = t_r(1)$. In a "D" network, packets that interfere with one another are transmitting at the same relative time, and, in a slotted system, slots arrives at each source at the same relative time. In Section VI, all times are relative times.

In the access strategies that are being discussed, the success or failure of a packet is only dependent upon the relative times that other packets are transmitted, and the relative position of the other sources, whether they are upstream or downstream. The relative time intervals that upstream and downstream sources can interfere with the current source is shown Figs. 9, 12, and 15 for the various access strategies. The effect that the interfering source has on the current source is also shown in these diagrams. An interfering source can collide with the current source, remove it from the network with a switch, delay its transmission, or force it to abort its transmission.

The advantage of using relative time can be seen by comparing the complexity of the analysis in Section VI with the analysis of the positional dependence of sources on a bidirectional CSMA-CD network [22].

C. Delay

The average number of times that a source at position x transmits a data packet is $G(x)/S$, and the average number of retries is

$$\bar{R}(x) = \frac{G(x)}{S} - 1. \tag{5.3.1}$$

Between each retry, a source must wait to determine that a packet has not been successfully transmitted, and then wait for an additional random time. The average delay, $\bar{D}$ is related to the average number of retries $\bar{R}$ as

$$\overline{D(x)} = (\overline{\tau_r} + \tau_p(x)(1 - F(x)) + k)\bar{R}(x) \tag{5.3.2}$$

where

- $\bar{\tau}_r$ is the average time between detecting a collision and retrying,
- $\tau_p(x)$ is propagation delay from the transmitter to the receiver at position x,
- $F(x)$ is the fraction of unsuccessful transmission attempts that are detected at the transmitter rather than at the receiver, and
- k accounts for the average fraction of the packet that passes or is transmitted before an unsuccessful attempt is detected and any processing delays in the electronics.

In ALOHA/X/X systems, none of the unsuccessful attempts are detected at the transmitter, and $F(x) = 0$, in LCSMA-CD/X/X, all unsuccessful attempts are detected at the transmitter, and $F(x) = 1$, and in LCSMA/U/X, $F(x)$ is between 0 and 1 as shown in Figs. 13(b) and 14(b).

D. Limitations of the Analysis

The analysis conducted in Section VI is a first-order approximation to the protocols. In LCSMA/X/X, if a packet from an upstream source arrives a short time before the packet at this source, the hardware may not have time to detect the packet, and a collision may occur. In LCSMA-CD/X/X, packets have a preamble that makes them slightly longer. These effects are ignored in this analysis.

In addition, this analysis does not model all of the detail of practical retry strategies, but is used to approximate randomly distributed retries. The persistent retry strategies are not well approximated by this model, and require a more detailed analysis [23], [24]. Persistent retry strategies can achieve fairness and increase the throughput of the system. The advantages they obtain are above those indicated by this analysis.

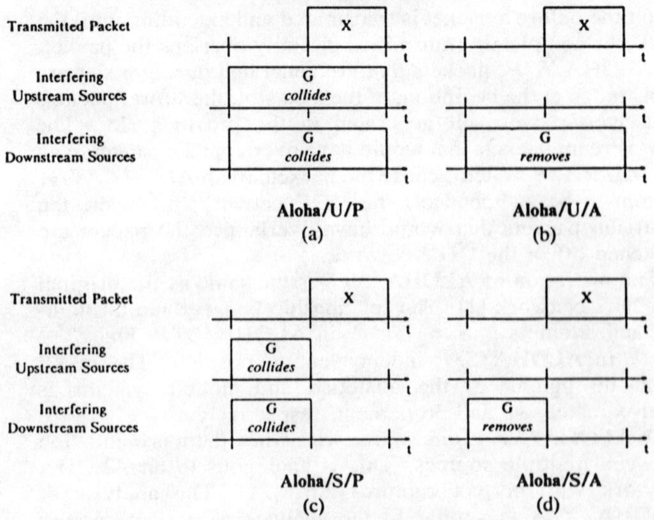

Fig. 9. Interference in ALOHA systems.

VI. Quantitative Evaluation

A. ALOHA/X/X

The upstream and downstream sources that interfere with a packet transmitted by a source in an ALOHA/X/X system are shown in Fig. 9 as a function of the relative time that the packets are transmitted. In ALOHA/X/P, the relative position of the source does not matter. In ALOHA/U/P, any packet generated during an interval starting a packet transmission time before a packet is transmitted and extending until the packet is completely transmitted partially overlaps the packet. In ALOHA/*S*/*P*, packets that are generated during a slot are transmitted at the beginning of the next slot, therefore, packets that overlap were all generated in the previous slot. The downstream packets that would have overlapped a packet in an ALOHA/*X*/*P* system, cause the packets in an ALOHA/*X*/*A* system to be switched off the "*D*" network; however, the upstream packets that would have overlapped the packet are switched off of the "*D*" network.

The operation of ALOHA/*X*/*P* is the same as the original ALOHA Network [1]. The relationship between the throughput and attempts is $S = Ge^{-2G}$ in ALOHA/*U*/*P*, and $S = Ge^{-G}$ in ALOHA/*S*/*P* independent of position. The maximum throughput of the unslotted and slotted systems is approximately 18 and 36 percent, respectively.

In ALOHA/*X*/*A*, one source wins when there is a collision between multiple sources. This is analogous to an ALOHA network with perfect capture [20], [21]. The analysis of ALOHA/*X*/*A* is similar to the analysis of perfect capture systems in [20] except that the transmitters are distributed over a one-dimensional line rather than a two-dimensional plane.

From Fig. 9, the probability of a successful transmission for a source using ALOHA/*U*/*A* at position *x* is the probability that no downstream sources, between positions *x* and 1, start transmitting during the packet interval prior to, or during this source's packet transmission. Therefore,

$$P_s(x) = e^{-2\int_x^1 G(y)\,dy}. \tag{6.1.1}$$

Similarly, from Fig. 9, in ALOHA/*S*/*A*,

$$P_s(x) = e^{-\int_x^1 G(y)\,dy}. \tag{6.1.2}$$

From (5.0.1),

$$S = G(x)e^{-a\int_x^1 G(y)\,dy} \tag{6.1.3}$$

where $a = 1$ for ALOHA/*S*/*A* and $a = 2$ for ALOHA/*U*/*A*. Differentiating (6.1.3) with respect to *x*,

$$G'(x) + aG^2(x) = 0. \tag{6.1.4}$$

At $x = 1$, a source always succeeds when it transmits, therefore, the boundary condition for this equation is $G(1) = S$, and the solution is

$$G(x) = \frac{S}{1 - aS(1-x)}. \tag{6.1.5}$$

As long as $S < 1/a$, $G(x) < \infty$ for $0 \leqslant x \leqslant 1$, and every user has the same throughput. Therefore, the maximum value of *S* for uniform throughput is 1/2 for ALOHA/*U*/*A* and 1 for ALOHA/*S*/*A*. When *S* exceeds these values, the users further from the bend do not get any data through, while those closer to the bend get equal access. This is analogous to the Sisyphus circle Abramson describes in perfect capture networks [20].

The average number of retries in ALOHA/*U*/*A* and ALOHA/*S*/*A* is shown in Figs. 10 and 11. These systems are unfair because sources near the bend in the "*D*" experience fewer retries on the average than sources further away. The effect of position on the average number of retries is shown in Figs. 10 and 11. In Fig. 10, the effect of operating the system at higher offered loads than it is capable of accommodating, $S = 0.75$, is also shown.

B. LCSMA/U/X

The interference diagrams for LCSMA/*U*/*X* systems are shown in Fig. 12. The symbol *G* in Fig. 11 is replaced by *T* because not all of the packets that are generated are transmitted. If the channel is busy because an upstream terminal is transmitting when a packet is generated, the packet is not

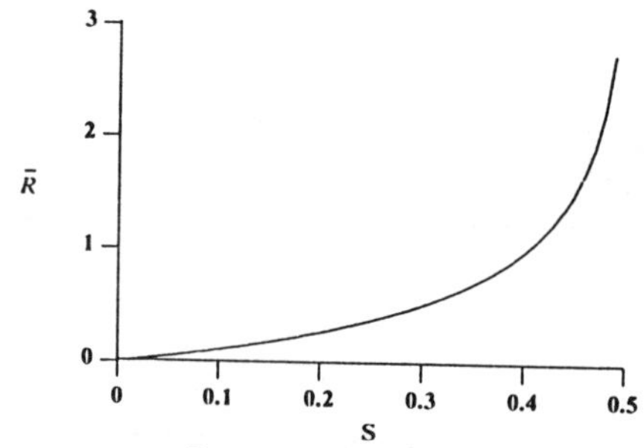

Average Retries ($\bar{R}$) averaged over all *x* vrs. Throughput (S)
(a)

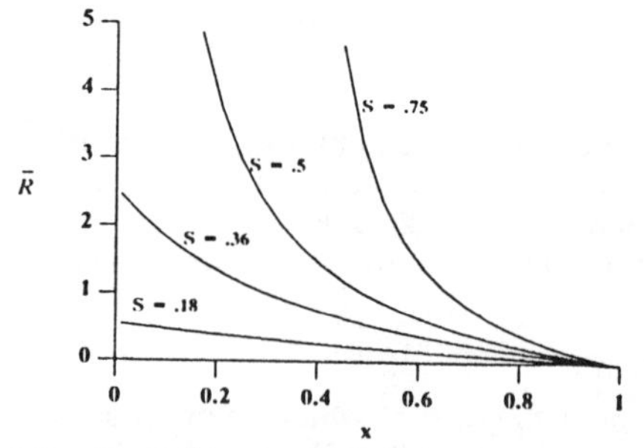

Average Retries ($\bar{R}$) vrs. Position (x) for Several Throughputs (S)
(b)

Fig. 10. Retries in ALOHA/*U*/*A*.

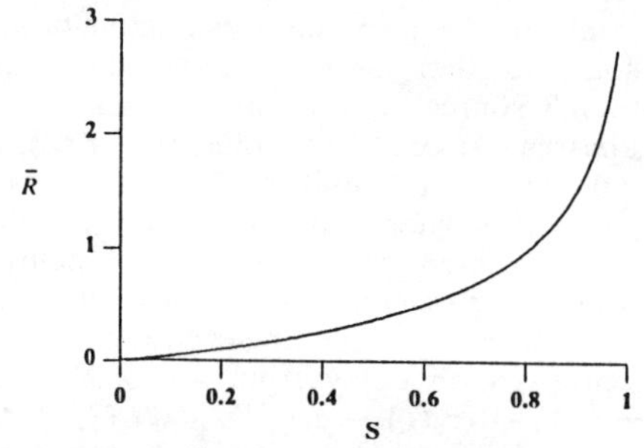

Average Retries ($\bar{R}$) averaged over all *x* vrs. Throughput (S)
(a)

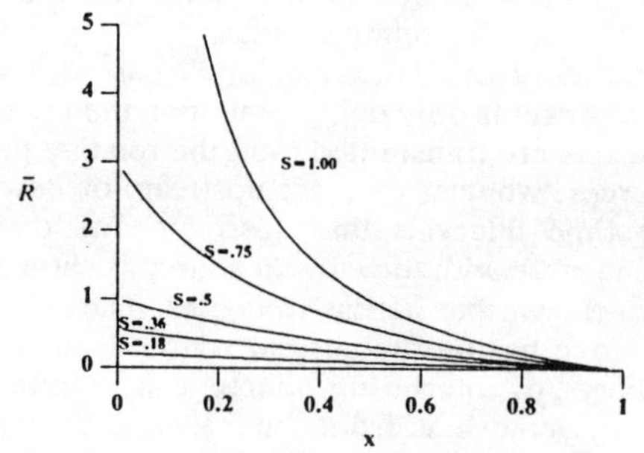

Average Retries ($\bar{R}$) vrs. Position (x) for Several Throughputs (S)
(b)

Fig. 11. Retries in ALOHA/*S*/*A*.

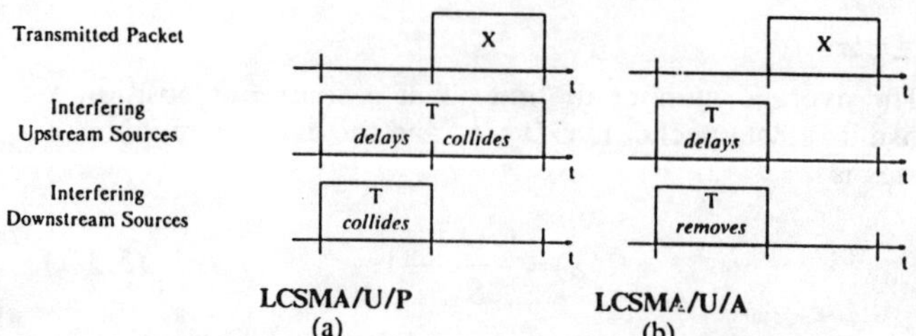

Fig. 12. Interference in LCSMA/*U*/*X* systems.

transmitted and cannot interfere with other packets. LCSMA/S/X systems operate the same as ALOHA/S/X systems, as explained in Section III-B, and are not analyzed.

In LCSMA/U/A, a source fails to transmit a packet if any source is transmitting when the packet arrives, Fig. 12(b). Therefore, $P_s(x)$ is independent of x and

$$P_s = 1 - \rho(1) \tag{6.2.1}$$

where $\rho(1)$ is the utilization of the channel at $x = 1$. From Fig. 7, if $P_s(x)$ is independent of x, then $G(x)$ is also independent of x. The utilization of the channel at $x + \Delta x$ is

$$\rho(x+\Delta x) = \rho(x) + (1-\rho(x))\bar{L}(x)G\Delta x \tag{6.2.2}$$

where

- $G\Delta x$ = is the average number of arrivals per packet transmission time in the interval Δx,
- $1 - \rho(x)$ is the probability that a packet that arrives, finds a clear channel and transmits, and
- $\bar{L}(x)$ is the average fraction of the packet that uses a channel that would have been clear at position x, and thereby adds to the utilization.

At position x, the cumulative arrival rate from upstream sources is

$$G_u(x) = \int_0^x G\,dy = Gx, \tag{6.2.3}$$

since the span of the network is normalized to one, and $G(x)$ is independent of x. Time is normalized so that a packet takes one unit of time to transmit. If, after the start of a packet transmission, the next packet from an upstream source arrives at $t < 1$, then the current transmission adds a fraction t of a packet to the cumulative channel utilization. If the next arrival from an upstream source occurs at $t \geqslant 1$ then the transmission adds one packet to the cumulative channel utilization. Since packet arrivals are governed by a Poisson process, the average fraction of a packet that adds to the cumulative channel utilization at position x is

$$\bar{L}(x) = \int_0^1 Gxte^{-Gxt}\,dt + \int_1^\infty Gxe^{-Gxt}\,dt$$

$$= \frac{1}{Gx}[1-e^{-Gx}]. \tag{6.2.4}$$

Substituting (6.2.4) into (6.2.2), and taking the limit as $\Delta x \to 0$,

$$\frac{d\rho}{1-\rho} = \frac{1}{x}[1-e^{-Gx}]dx. \tag{6.2.5}$$

The solution to this differential equation, with the boundary condition $\rho(0) = 0$, is

$$\rho(x) = 1 - e^{-I(x,G)} \tag{6.2.6}$$

where

$$I(x, G) = \int \frac{1}{x}[1-e^{-Gx}]\,dx$$

$$= \sum_{i=1}^{\infty} (-1)^{i-1}\frac{(Gx)^i}{i*i!}. \tag{6.2.7}$$

The throughput of this system is determined by evaluating (6.2.7) and (6.2.6) at $x = 1$, substituting into (6.2.1), and then into (5.0.1), so that

$$S = Ge^{-I(1,G)}. \tag{6.2.8}$$

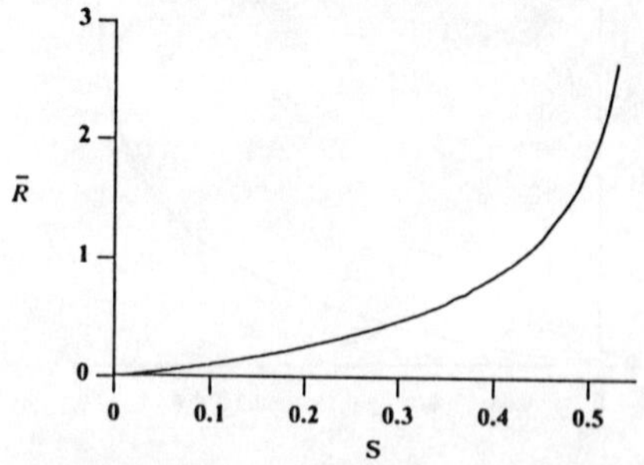

Average Retries ($\bar{R}$) averaged over all x vrs. Throughput (S)

(a)

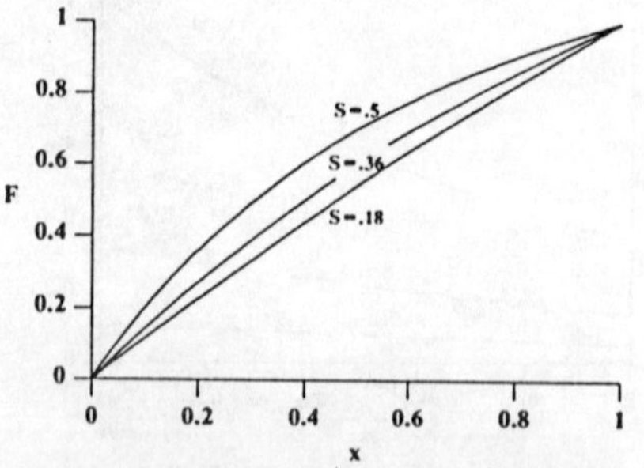

Fraction of the Retries due to CSMA (F) vrs. Position (x) for Several Throughputs (S)

(b)

Fig. 13. Retries in LCSMA/U/A.

The number of retries required to achieve a given throughput is shown in Fig. 13(a). As the throughput approaches 57 percent the number of retries goes to infinity, so that the maximum throughput is about 57 percent.

This system is fair to the extent that every source experiences the same average number of retries. However, sources that detect a busy channel can retry sooner than sources that must wait a roundtrip delay to detect a collision at the receiver. The probability that a source at position x encounters a busy channel is $\rho(x)$, and the probability that a transmission attempt is unsuccessful is $1 - P_s(x)$. Therefore, the fraction of unsuccessful transmissions that are detected at the source is

$$F(x) = \frac{\rho(x)}{1-P_s(x)} \tag{6.2.9}$$

where $\rho(x)$ is defined in (4.2.5) and $P_s(x)$ is defined in (4.2.1). In Fig. 13(b), F as a function of x is plotted for several throughputs.

In LCSMA/U/P, a source fails if any source is transmitting when it attempts to transmit, or if any of the upstream sources start to transmit while it is transmitting, Fig. 12(a). Equation (6.2.1) becomes

$$P_s = (1-\rho(1))(1-e^{-G_u(x)}) \tag{6.2.10}$$

where

$$G_u(x) = \int_{y=0}^{x} G(y)\,dy. \tag{6.2.11}$$

is the total traffic generated by upstream sources.

In this system, P_s is dependent upon x, therefore, G is dependent upon x, and (6.2.1) and (6.2.2) become

$$\rho(x+\Delta x) = \rho(x) + (1-\rho(x))\bar{L}(x)G(x)\Delta x \tag{6.2.12}$$

and, as in the derivation of (6.2.4)

$$\bar{L}(x) = \int_0^1 G_u(x)te^{-G_u(x)t}\,dt + \int_1^\infty G_u(x)e^{-G_u(x)t}\,dt. \tag{6.2.13}$$

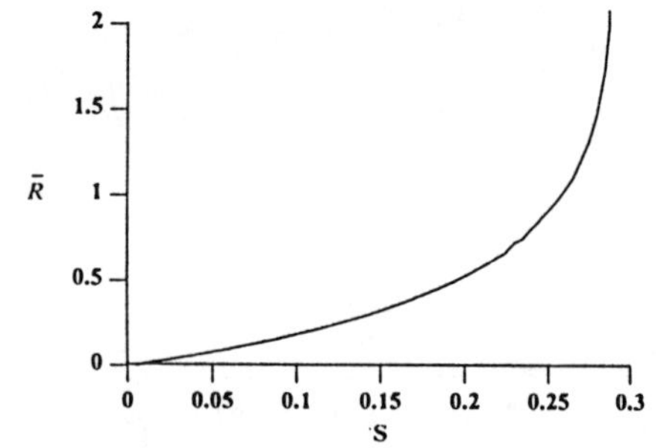

Average Retries ($\bar{R}$) averaged over all x vrs. Throughput (S)
(a)

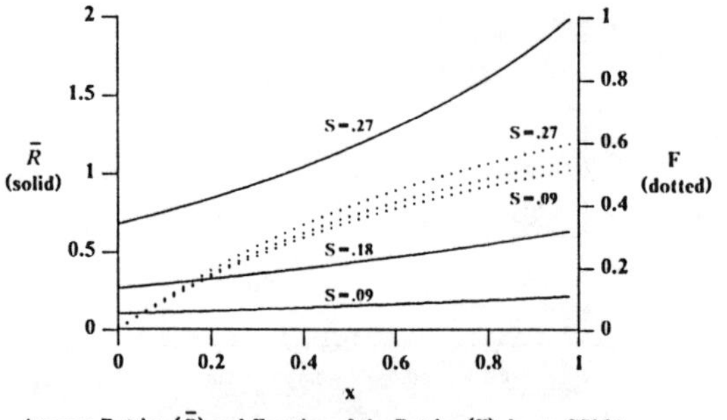

Average Retries ($\bar{R}$) and Fraction of the Retries (F) due to CSMA vrs. Position (x) for Several Throughputs (S)
(b)

Fig. 14. Retries in LCSMA/U/P.

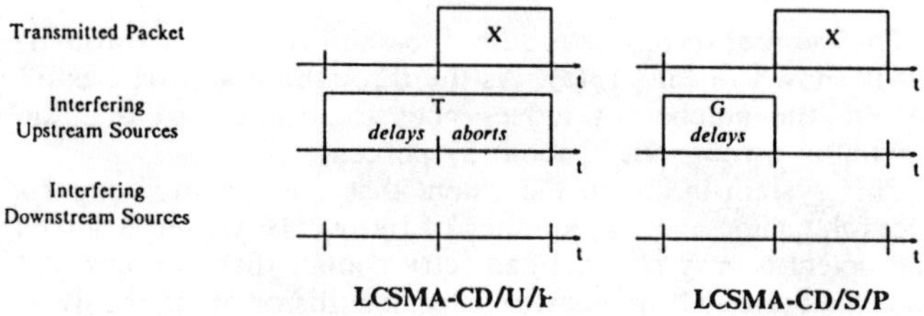

(a) (b)

Fig. 15. Interference in LCSMA-CD/X/P systems.

Equations (5.0.1), (6.2.10), (6.2.11), (6.2.12), and (6.2.13) define the relationship between S and G. Approximate solutions to these equations have been obtained numerically by dividing x into a number of intervals Δx, and adjusting $G(x)$ to obtain the same $S(x)$ in each interval. The number of retries required to achieve a specific throughput is plotted in Fig. 14(a). As the throughput approaches 29 percent the number of retries goes to infinity, so that the maximum throughput for LCSMA/U/P is about 29 percent. Not every source experiences the same delay, and the average number of retries as a function of x, for several throughputs is plotted in Fig. 14(b) along with the fraction of the unsuccessful attempts that are detected at the source. $F(x)$ is calculated as in (6.2.9) with $P_s(x)$ from (6.2.10) and $\rho(x)$ from (6.2.12).

C. LCSMA-CD/X/P

The interference diagram for LCSMA-CD/X/P systems is shown in Fig. 15. Because of the LCSMA component of this protocol, not all of the generated packets are transmitted, and G must be replaced by T in LCSMA-CD/U/P, as in Section VI-B. In LCSMA-CD/S/P, any packet generated by an upstream source during the previous slot interval interferes with a packet transmitted during the current interval. As explained in Section III-C, LCSMA-CD/X/A is equivalent to LCSMA-CD/X/P and is not considered separately.

In LCSMA-CD/U/X, a packet is delayed if an upstream source is transmitting when the packet is generated, and is aborted if an upstream source starts transmitting while the transmission is in progress, as shown in Fig. 15(a). The

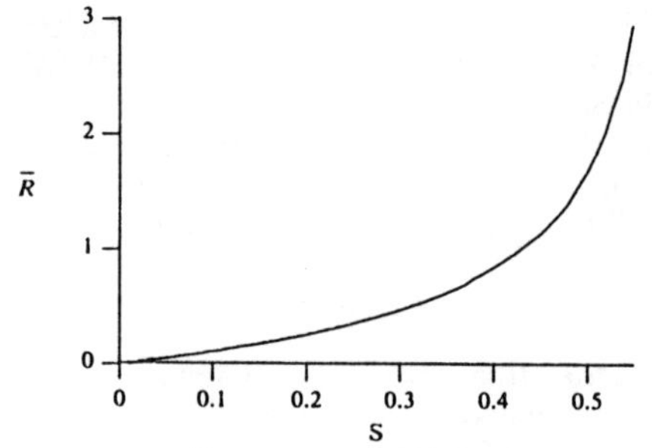

Average Retries ($\bar{R}$) averaged over all x vrs. Throughput (S)
(a)

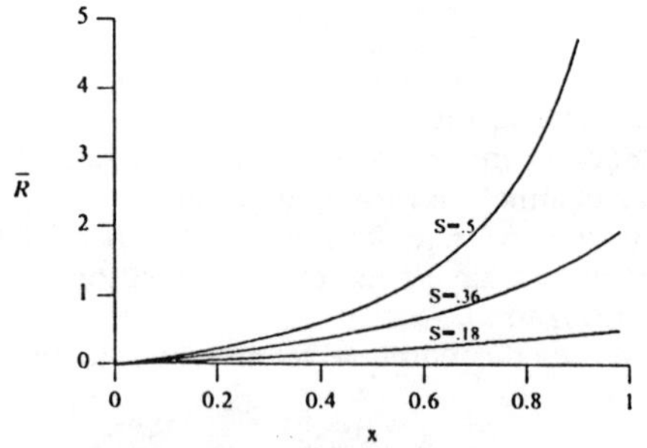

Average Retries ($\bar{R}$) vrs. Position (x) for Several Throughputs (S)
(b)

Fig. 16. Retries in LCSMA-CD/U/P.

probability of successful transmission in this system is

$$P_s = (1-\rho(x))(1-e^{-G_u(x)}) \tag{6.3.1}$$

where $G_u(x)$ is defined as in (6.2.11). The generated traffic is a function of position, and ρ and $\bar{L}(x)$ are defined as in (6.2.12) and (6.2.13). Equations (5.0.1), (6.3.1), (6.2.11), (6.2.12), and (6.2.13) define the relationship between S and G. These equations were solved numerically, as in Section VI-B, the average number of retries is shown Fig. 16(a), and the maximum throughput is about 57 percent. This strategy gives preference to upstream sources, and the average number of retries as a function position is shown in Fig. 16(b).

In LCSMA-CD/S/P, packets are denied access by upstream sources that generate packets in the interval prior to transmission, as shown in Fig. 15(b). The probability of success in this system is

$$P_s(x) = e^{-\int_0^x G(x)\,dx}. \tag{6.3.2}$$

This results in a relationship between S and G that is similar to ALOHA/S/A in Section VI-A, and the solution is

$$G(x) = \frac{S}{1-Sx}. \tag{6.3.3}$$

The maximum throughput in this system is 100 percent.

VII. Conclusion

The maximum throughputs that can be achieved in twelve random access strategies are summarized in Fig. 17. These strategies provide a mechanism for trading three types of complexity, time synchronization, signal processing, and tap structure, for throughput. Time synchronized, or slotted systems, adjust the transmission times of the various sources so that all fixed size packets arrive at a common point in the network at the same instant. Signal processing is used to avoid collisions with upstream sources by examining the channel before or during transmission. Finally, taps that switch a transmitter into the medium are used to allow sources to win in a contention situation.

Slotted systems are always more efficient than unslotted systems, and when synchronization is combined with either

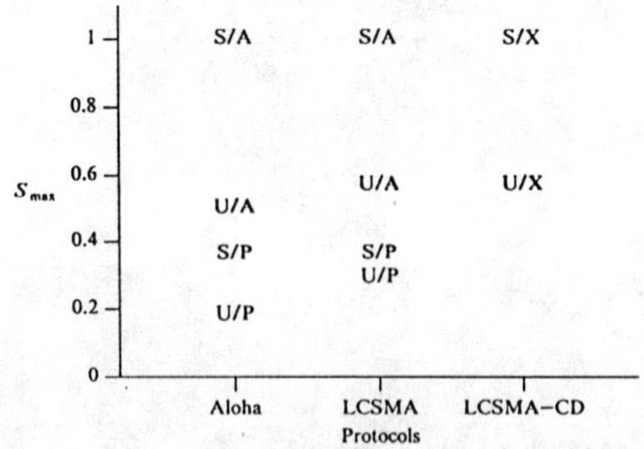

Fig. 17. Twelve maximum throughput in random access strategies.

signal processing or switched taps, these systems can approach a utilization of one. Signal processing in systems with switched taps provides only minor increases in throughput. However, signal processing can be used to make systems with passive taps as efficient as systems with switched taps.

In addition to throughput and delay, fairness must also be considered when selecting a strategy. All of the strategies are fair to the extent that sources achieve the same average throughput. However, the delays experienced by sources can vary from zero to nearly infinity. This type of unfairness is first noted in Section VI-A for ALOHA/X/A. The unfairness in switched systems can be eliminated by signal processing, and in Section VI-B it is shown that in LCSMA/U/A all sources retry the same number of times on the average.

In systems with passive taps, signal processing increases the throughput, but introduces a degree of unfairness that cannot be removed by the access strategy alone. However, when LCSMA-CD is used, a source detects transmission failures immediately, can retry sooner, and thereby, reduce its average delay. In an LCSMA-CD network, a source recognizes immediately if its transmission will succeed, and can use a persistent retry strategy to further reduce its delay.

References

[1] N. Abramson, "The ALOHA system—Another alternative for computer communications," in *Proc. Fall Joint Comput. Conf., AFIPS Conf.*, vol. 37, 1970, pp. 281–285.

[2] R. M. Metcalf and D. R. Boggs, "Ethernet: Distributed packet switching for local computer networks," *Commun. ACM*, vol. 19, no. 7, pp. 395–404, July 1976.

[3] R. V. Schmidt, E. G. Rawson, R. E. Norton, S. B. Jackson, and M. D. Bailley, "Fibernet II: A fiber-optic Ethernet," *IEEE J. Select. Areas Commun.*, vol. SAC-1, pp. 702–711, Nov. 1983.

[4] M. H. Coden and F. W. Scholl, "Fiber optics multiplexing and modulation," in *Proc. SPIE*, vol. 417, Apr. 1983, pp. 48–52.

[5] E. G. Rawson, "Fibernet II Ethernet-compatible fiber-optic LAN," *IEEE J. Lightwave Technol.*, vol. LT-3, pp. 496–501, June 1985.

[6] C. Tseng and B. Chen, "D-Net: A new scheme for high data rate optical local area networks," *IEEE J. Select. Areas Commun.*, vol. SAC-1, Apr. 1983.

[7] N. F. Maxemchuk, "Random access strategies for fiber-optic networks," in *Proc. INFOCOM '87*, San Francisco, CA, Mar. 31–Apr. 2, 1987, pp. 307–311.

[8] L. Fratta, F. Borgonovo, and F. Tobagi, "The Express-Net: A local area communication network integrating voice and data," in *Proc. Int. Conf. Perform. Data Commun. Syst.*, Paris, France, Sept. 1981, pp. 77–88.

[9] J. O. Limb and C. Flores, "Description of Fastnet—A unidirectional local area communications network," *BSTJ*, vol. 61, no. 7, pp. 1413–1440, Sept. 1982.

[10] S. D. Personick, "Protocols for fiber-optic local area networks," *IEEE J. Lightwave Technol.*, vol. LT-3, pp. 426–431, June 1985.

[11] T. Hermes, B. Hoen, J. Saniter, and F. Schmidt, "LOCNET—A local area network using optical switching," *IEEE J. Lightwave Technol.*, vol. LT-3, pp. 467–471, June 1985.

[12] R. C. Alferness, "Waveguide electro-optic modulators," *IEEE Trans. Microwave Theory Tech.*, vol. MTT 30, pp. 1121–1137, Aug. 1982.

[13] H. E. White and N. F. Maxemchuk, "An experimental TDM data loop exchange," presented at Proc. ICC '74.

[14] N. F. Maxemchuk, "Linear, unidirectional networks and random access strategies in fiber-optic local distribution systems," in *Proc. GLOBCOM '86*, Houston, TX, Dec. 1–4, 1986, pp. 29–33.

[15] T. Shibagaki, H. Ibe, and T. Ozeki, "Video transmission characteristics in WDM star networks," *IEEE J. Lightwave Technol.*, vol. LT-3, pp. 490–495, June 1985.

[16] Y. Hakamada and K. Oguchi, "32-Mbit/s star configured optical local area network design and performance," *IEEE J. Lightwave Technol.*, vol. LT-3, pp. 511–524, June 1985.

[17] S. Ohshima, T. Ito, K.-I. Donuma, H. Sugiyama, and Y. Fujii, "Small loss-deviation tapered fiber star coupler for LAN," *IEEE J. Lightwave Technol.*, vol. LT-3, pp. 536–560, June 1985.

[18] D. Navaro, "Couplers, network topologies and access protocols for optical fibre LAN's," presented at Proc. IEEE Colloquium "Optical Fibre Local Networks," Digest 1983/59, June 2, 1983.

[19] U. Mukherji, N. F. Maxemchuk, C. W. Wu, and J. C. Swartzwelder, "Transmission format and receiver structure for random access strategies in a fiber-optic network," presented at Proc. Comput. Net. Symp., Washington, DC, Apr. 11–13, 1988.

[20] N. Abramson, "The throughput of packet broadcasting channels," *IEEE Trans. Commun.*, vol. COM-25, pp. 117–128, Jan. 1977.

[21] D. J. Goodman and A. A. M. Saleh, "Local ALOHA radio communication with capture and packet buffers," in *Proc. GLOBCOM '85*, New Orleans, LA, Dec. 2–5, 1985, pp. 984–990.

[22] K. Sohraby, M. L. Molle, and A. N. Venetsanopoulos, "Why analytical models of Ethernet-like local networks are so pessimistic," in *Proc. IEEE Global Telecommun. Conf.*, Nov. 1984, pp. 659–664.

[23] M. Fine and F. Tobagi, "Demand assignment multiple access schemes in broadcast bus local area network," *IEEE Trans. Comput.*, vol. C-33, pp. 1130–1159, Dec. 1984.

[24] L. Kleinrock and F. A. Tobagi, "Packet switching in radio channels: Part I—Carrier sense multiple-access modes and their throughput-delay characteristics," *IEEE Trans. Commun.*, vol. COM-23, pp. 1400–1416, Dec. 1975.

Spread Spectrum Technology: A Solution to the Personal Communications Services Frequency Allocation Dilemma

Jack T. Taylor, Jim K. Omura, Cylink Corporation

On October 1, 1990, a virtual tidal wave of paper hit the Federal Communications Commission (FCC). That was the due data for comments in the FCC's Notice of Inquiry (NOI) into the establishment of new Personal Communications Services (PCS) (Gen. Docket No. 90-314). Over 100 companies filed comments.

In the NOI, the Commission identified PCS in the broadest of terms to encompass wireless communications services "that free individuals from the constraints of the wireline public switched telephone network and enable them to communicate when they are away from their home or office telephone."

The broad scope of the NOI encouraged comments from almost all sectors of the telecommunications industry. Naturally, each sector set forth a specific view of this broad extensive landscape. However, a constant thread throughout the comments is the importance of an allocation of spectrum in the near term for personal communications. The imperative of keeping the U.S. competitive with the rest of the world (especially Europe) is driving the regulatory process to an early resolution of the spectrum issue. It is apparent from a cursory reading of the comments that waiting for the federal government to transfer spectrum to the FCC is not an acceptable near term option.

If the FCC is going to satisfy the spectrum needs of PCS from spectrum under their jurisdiction, the identification of where in the spectrum and how much spectrum to allocate takes on primary importance. The evolutionary model represented by the FCC's broad view of PCS contains within itself several clues to the future direction of the spectrum allocation for PCS in the U.S. The allocation will have to be large enough to satisfy the long-term residential, office, and public access demands of the marketplace. The need for a large block of spectrum requires that the access technology used must be able to accommodate some initial interference in the band chosen and have the capacity growth potential for the long term. It seems apparent that the initial deployment will be in a shared environment, which will transition over time to an exclusive band. Accordingly, the access technology must be able to function in a shared environment and have the inherent growth potential to satisfy the long-term capacity needs of PCS, as described by the FCC.

In the comments, the size of the allocation recommended depended on whether the commentator was taking the long or short view of the PCS model. The former saw the need for a large allocation (on the order of 100 to 200 MHz) while the latter chose the near term Cordless Telephone second generation (CT2) view, requiring merely a 3 to 4 MHz allocation in the 900 MHz range. In both cases, an early spectrum allocation was seen as critical.

The corollary to the spectrum question is which access technology will dominate the PCS marketplace. This is a critical question and one that is tied inevitably to the spectrum allocation issue. The technology question will be discussed later in this article.

This NOI was the opening shot in a revolution which will, in the next decade, change the way people access the telephone network. Currently, copper wire dominates the access of telephones to the network. There are over 130 million "wired" access lines in the U.S. telephone network. In the last several years, the beginning of a wireless revolution in access has taken hold. Cellular radio and cordless telephones have introduced millions of people to the concept of wireless communications.

Over the next decade, radio technology will first augment and eventually fully complement the wired network. Radio will become as common as wire to connect a majority of people to the telephone network. This type of major overhaul of telephone access methodology will bring about a myriad of social, economic, political, and regulatory considerations, each of which could form the basis of a separate study. This article, however, will confine itself to the spectrum and technology ramifications of the coming wireless revolution.

The technology and the architecture chosen for PCS will be critical to its success. Because of the enormous capacity which will be needed to serve the large numbers of people in a wireless environment, a high degree of frequency reuse will necessitate a microcell approach.

Microcells in their purest form are simply small low-powered radio cells closely spaced to maximize frequency reuse in a high density area. This efficient use of the spectrum will allow the large numbers of PCS users to access the network.

Reprinted from *IEEE Commun. Mag.*, February 1991.

The resulting concept, a large block of spectrum operating interference free in a closely spaced microcell environment, embodies the key PCS issues facing the industry and the regulatory community, spectrum and technology.

Spectrum: Tough Choices

The issue of spectrum for PCS has two dimensions: how much spectrum and where to find it. Determining a precise amount of spectrum for PCS is an elusive goal. Some conclusions, however, can be drawn from the recent experience of licensing Personal Communications Networks (PCNs) in the U.K. Although no specific allocation has been publicly announced, it is generally believed that between 75 to 100 MHz per licensee will be required. Another reference point is the International Radio Consultive Committee Interim Working Party 8/13. During meetings in preparation for WARC '92, they concluded that a minimum of 60 MHz will be necessary for each PCS provider. U.S. telecommunications policy will dictate that at least two service providers will be licensed for each market, which would translate to a minimum of 120 MHz of spectrum needed for a competitive marketplace. Finally the House and Senate versions of spectrum legislation, which are generally believed to be targeted toward the PCS, identify the spectrum needs at 175 to 200 MHz. By inference it can then be safely concluded that the U.S. spectrum allocation for PCS will be between 100 and 200 Mhz of spectrum.

Most commentators agree that the spectrum chosen for personal communications must fall between 500 and 3,000 MHz to be of practical use for PCS. From a technical standpoint, the short distance propagation and microcell architecture could be best supported in this range. In Europe, the U.K. PCN and the more broadly based Digital European Cordless Telephone (DECT), both of which propose a low-powered microcell architecture, are scheduled to use spectrum in the 2 GHz range. Also, the spectrum between 1.7 and 2.3 GHz is being considered for personal communications operations in other parts of the world. Although it would be difficult for the U.S. to exactly match the international allocations, locating within the 1.7 to 2.3 GHz would provide some technological advantages of scale for U.S. telecommunications Research & Development (R&D) firms that are participating in both domestic and international developments of personal communications.

A spectrum allocation between 500 to 3,000 MHz and of the magnitude discussed above will pose tough domestic choices. The first of these is whether the allocation will come from the government or the nongovernment allocations. The next is whether to dislocate existing users or share the band. There are only two sources of spectrum for PCS: discrete or shared. Discrete spectrum would require finding a large block of unoccupied spectrum. As there are no large unoccupied bands between 500 to 3,000 MHz, this approach would require relocating existing users. The sharing approach would necessitate the sharing of a common block of spectrum between PCS and an existing set of users. A discussion of both concepts follows: spectrum sharing and discrete allocations from both sources, government and nongovernment jurisdictions.

Some progress has already been made in the process of identifying a large block of discrete spectrum. Congress has introduced legislation to require the government to identify spectrum presently occupied by government users for reallocation to the FCC for "new and emerging technologies." This is generally believed to be an allocation for PCS. The pending legislation would allow, as an option, sharing on 50% of the identified spectrum.[1]

Within the FCC jurisdiction, clearing a large block of spectrum at the relevant bands is not possible. The procedural delay alone makes it an unacceptable option. Unlike the government option of using appropriated funds to cover the equipment changeout cost, a corresponding FCC procedure would be equally time consuming and much more difficult to implement.

Therefore, if the use of discrete government spectrum is the chosen solution, then government allocation is the best alternative. However, the process of reallocation from government to nongovernment use will be both expensive and protracted. It will entail a two-step process: relocating existing users to another band and assigning licenses by the FCC. The subsequent licensing time frame could be further drawn out because of the existence of competing FCC claimants for spectrum: High Definition Television (HDTV), Land Mobile Radio, etc.

The Notice of Proposed Rulemaking required to effect the PCS licensing will by necessity require consideration of these competing claimants—the result: at least an additional year added to the licensing process.

If discrete spectrum is the path selected, the delay in licensing could easily take seven years,[2] and add fuel to arguments that we are falling behind the Europeans in PCN technology development.

Accordingly, the other option, spectrum sharing, has begun to develop consensus among all the participants in the PCS discussion. In fact, a week before adjournment, a 50% sharing option was added to the legislation discussed above. Sharing could be a much quicker (and less expensive)[3] alternative to dislocation of existing users. If sharing is allowed, then rapid progress could be made toward an allocation by the FCC for PCS.

A case could be made that if shared spectrum was used, the first PCS licenses could be issued in 24 to 30 months. Depending on whether government or nongovernment frequencies were primarily used, a shared spectrum approach could also lead to the eventual migration of existing fixed users as PCS matures and grows from the primary shared band to another shared band.

Government/nongoverment sharing opens up several options for resolving the PCS allocation controversy.

Shared Allocation: A Proposal

The requirements of a shared proposal should meet several criteria. First, the allocation should be sufficient to satisfy the total PCS requirement, i.e., at least 120 MHz. Second, the shared band should be primarily occupied by fixed users to help ameliorate the interference potential of sharing. Third, it should contain a reasonable transition path to allow the eventual migration of the existing fixed users to another band as PCS matures and grows. Finally, the band chosen should fall generally within the 1.7 to 2.3 GHz band which is the preferred international allocation for personal communications.

A survey of available bands between 500 MHz and 3,000 MHz discloses only two bands that meet the above criteria: 1,710 to 1,850 (government allocation) and 1,850 to 1,990 (non-government allocation). Both bands are used primarily by fixed microwave systems. For the reasons discussed above, sharing on the 1,850 to 1,990 nongovernment band appears to be the most desirable option. In fact, several companies (including Cylink Corporation) are currently evaluating the technical characteristics of this band in order to determine the fea-

[1]The Senate Bill died in committee. Similar legislation will be introduced next year.

[2]An Arthur D. Little analyst recently predicted that if discrete spectrum is needed, it will be 1997 before the first PCS license is granted.

[3]The Assistant Secretary of Commerce for Telecommunications testified before Congress that there was $100 billion invested in government radio communications equipment and pointed out that a relocation of any government users would incur a cost in time and dollars. The exact cost and time involved would depend on the amount of spectrum to be reallocated.

sibility of co-primary sharing using Code Division Multiple Access (CDMA)/spread spectrum technology.

This band would yield 140 MHz which would satisfy the 120 MHz minimum discussed earlier. It would also allow sufficient spectrum in the event that a small allocation is needed for CT2 technology trials. Currently, most petitioners are seeking an allocation at 900 MHz for such trials. However, if spectrum is allocated to CT2 and market acceptance is poor or, more likely, the competition from PCS drives it from the marketplace, then the 900 MHz allocation would be stranded. This would not be the case if CT2 was given an allocation contiguous to the PCS allocation whereby the CT2 allocation could subsequently be subsumed into the PCS allocation.

A further advantage to selecting this band is the natural migration path for existing fixed users. Fixed users on this band could be easily and inexpensively migrated to the adjacent government band 1,710 to 1,850 MHz.[4] The government band is lightly loaded and occupied predominantly by government fixed users along with some mobile and space services. The migration could be paced by the growth of PCS and should allow sufficient time for planning by both nongovernment as well as government users of the two bands. Using the 1,710 to 1,850 government band as the band to which nongovernment users would migrate for a permanent allocation, would also satisfy a large portion of any legislative requirement of government reallocation/sharing of up to 175 to 200 MHz.

This proposal would satisfy all the criteria outlined above and, as the 1,850 to 1,990 band is controlled by the FCC, the licensing process could start quickly. It should be noted that a converse proposal could be made to initially share the 1,710 to 1,850 MHz government band. That is less desirable for several reasons.

First, it would invoke a time penalty to negotiate the government/nongovernment sharing agreement. Furthermore, the 1,710 to 1,850 bands is more lightly used than the nongovernment band, making it the ideal location for the migration of the displaced 1,850 to 1,990 MHz private microwave users. (It should be noted that only those users who are in the large metropolitan areas would be migrated off the band and only when the growth of PCS dictated such a move.) Finally, the critical nature of the space services currently conducted on the 1,710 to 1,850 band by the government cast doubt on the feasibility of ever freeing up the band exclusively for PCS.

Technology

If sharing of spectrum between fixed users and PCS licensees is the path chosen by the FCC, then the technology choice for PCS is critical. Most would agree that until the band is completely cleared and exclusively used for PCS, interference between the coprimary users is the primary concern. Accordingly, the technology selected must be robust against interference and have the capacity potential to satisfy the long-term needs of PCS. One such technology is CDMA/Spread Spectrum (SS).

Until recently SS technology was used primarily by the military for the purpose of overcoming intentional interference by hostile jamming radios. Because of its noiselike characteristics it has also been used to hide the transmission signal from detection. These Antijamming (AJ) and Low Probability of Intercept (LPI) properties of SS radios can be very useful in commercial applications for the purpose of:

- Allowing many SS radio users with different spreading codes to share the same channels and rely on the AJ capabilities to overcome mutual interference. Multiple users sharing the same channel are referred to as CDMA as opposed to the more commonly used Frequency Division Multiple Access (FDMA) and Time Division Multiple Access (TDMA) techniques where signals are assigned noninterfering frequency or time slots. With CDMA, users are assigned different spreading codes to minimize mutual interference.
- Minimizing interference to conventional radios sharing the same band due to the wideband noiselike characteristics of SS radios.

These are well-known properties which are backed up by several years of both defense and commercial R&D which validate these properties. In 1985, the FCC issued Part 15.247 which allows for the unlicensed use of SS radios with up to one watt of output power in three shared frequency bands: 902 to 928 MHz, 2.400 to 2.4835 GHz, 5.725 to 5.850 GHz. This ruling has stimulated a wide range of commercial applications for SS radios.

For PCS the number of simultaneous users allowed in a given area is critical to the overall system design. The use of CDMA with SS radios has been a controversial and not well-understood topic. They key claim is that spread spectrum radios have the capability to allow many more active users in a metropolitan area than conventional narrowband analog and digital voice radios using FDMA or TDMA.

In order to compare capacity claims of conventional narrowband modulations with SS modulations, one should look at the number of simultaneous transmissions possible per unit bandwidth, where the narrowband signals occupy the total band in nonoverlapping FDMA or TDMA slots. Making the comparison within a fixed bandwidth with powerful forward error correction codes, using conventional Direct Sequence spread Binary Phase Shift Keying modulation (DS/BPSK), the number of simultaneous CDMA users is approximately the same as with an FDMA/TDMA system.

The primary advantage of SS radios is their ability to tolerate a fair amount of interfering signals compared to conventional radios, which typically cannot tolerate any such interference. From a system design point of view, this can be a considerable advantage.

As a result of the interference tolerance of CDMA/SS, the systems issues of frequency band assignment, adjacent channel interference and adjacent cell interference are greatly simplified when using SS technology. Moreover, flexibility in system design and deployment are greatly improved since some interference to others is not a problem. On the other hand, conventional radios must be carefully assigned a frequency or time slot to make sure that there is no interference with other similar radios, and sophisticated filtering and guard band protection is required with FDMA and TDMA technologies.

Capacity improvements can also result from speech patterns. For example, if each of the SS radios transmits power only when the PCS user is speaking and, on the average, a speaker is talking only about one third of the time, then on the average, the SS radios can tolerate three tmes more users than FDMA/TDMA radios on the same bandwidth. Conventional radios cannot use such a scheme.

Moreover, with CDMA/SS, adjacent microcells can share the same frequencies. The sensitivity of conventional radios to interference generally does not allow for adjacent microcells sharing the same frequencies.

Further capacity gains will result from antenna technology advancements. In order to increase the maximum number of radios handled by a microcell, the use of directional antennas can allow the microcell area to be divided into sectors defined by directional antennas. Again, with SS radios this can be done quite easily compared to conventional radios. Using sectorized antennas is similar in concept to allowing adjacent microcells to use the same frequencies. Each sector, defined by a directional antenna, can use the same frequencies as the other sectors in a microcell. Further, because of the interference rejection properties of SS, adjacent sector interference is not as much a major engineering consideration as with narrowband radios.

---

[4]FCC Chairman Alfred Sikes told the Practicing Law Institute (December 6–7), "If we are to have sufficient spectrum to meet projected new technology needs, some of the incumbents will have to find new homes."

When all of these systems issues are taken into account the overall capacity advantage of using CDMA with SS radios is evident. A detailed analysis with conventional Binary Phase Shift Keying (BPSK) and coded spread spectrum DS/BPSK shows that on a per bandwidth basis CDMA allows about the same total number of simultaneous transmissions as FDMA or TDMA. However, there can be a capacity factor of 3 in favor of CDMA if each radio is powered down when a user is not speaking. Another factor of 7 (over an FM analog cellular reuse pattern) comes in when the adjacent microcells are allowed to use the same frequencies. In a densely packed network of microcells each microcell has 6 adjacent neighbors and in conventional systems adjacent microcells must use different frequencies. If 4 sector antennas are used in each microcell, then another factor of 4 favors CDMA. This theoretical argument, at the system level easily shows a large advantage in a densely packed microcell architecture in the number of active radios with CDMA over conventional FDMA or TDMA.

Another advantage of CDMA/SS is handoff. The handoff from one microcell to another is much more flexible with SS radios. Since interference is allowed, handing off to a busy microcell does not mean that the communication must be terminated. Instead switching to a highly loaded microcell may only mean that the average noise level goes up while communications is maintained. Using conventional technology, busy signals would result.

On balance, CDMA/SS will result in greater overall capacity. This capacity advantage will, however, depend to a large degree on the quality of the specific SS radio and the design of the entire system of microcells. Practical system issues such as power control and network time synchronization will play a critical role in achieving the potential advantages of CDMA/SS.

Conclusions

Wireless access to the public switched network in the form of personal communications services will revolutionize the way people communicate in the decades to come. In response to this change, the FCC has begun a major proceeding to identify the nature and, the spectrum needs of and the technology options availale for these services.

Because of the congested nature of the frequency spectrum in the U.S., the options for spectrum are limited. Although discrete spectrum would be the preferred option for all concerned, the realization that sharing spectrum is the only viable option is becoming apparent.

This approach, sharing spectrum between two users' groups, will mandate the choice of technology. SS technology is an ideal technology for such an environment. The interference tolerance, system flexibility, and capacity advantage of CDMA/SS make it the technology of choice for the coming generation of PCS.

INVITED PAPER *Special Issue on Spread Spectrum Techniques and Its Applications*

Spread Spectrum for Consumer Communications
—— Applications of Spread Spectrum Communications in Japan ——

Masao NAKAGAWA† *and* Takaaki HASEGAWA††

SUMMARY The recent progress in communication devices has reduced the cost of spread spectrum systems that have been mainly used in the military field, enabling the possibility of using spread spectrum systems for nonmilitary use. Nonmilitary communication networks are classified into two categories: public communication networks and consumer communication networks. At the present, the public communication network share is much larger than that of the consumer communication networks. Both of them will increase; however, the consumer communication network share will be comparable to the public one. The progress of consumer communications is due to inexpensiveness (without charge), user privacy, free design from regulations and advantages of local and small zone communications. To overcome fading, power restrictions, interference and interception is very important for its progress. One of the key technologies of consumer communications is the spread spectrum. Consumer communication networks are characterized by user possession of the network itself in addition to the possession of personal terminals. From the viewpoint of analogy between communication networks and transportation networks, "from public to individual" seems to be today's trend. Several spread spectrum applications for consumer communications in Japan have been introduced as follows: a home security system using power line communication as a wire spread spectrum communication system; a data carrier, a radio remote control, low-power radio systems and a clock rate modulation system as radio spread spectrum communication systems. In addition SAW device applications and digital signal processing applications are introduced.

1. Introduction

Spread spectrum (SS) communication systems have many applications mainly in military communications, because of its antijamming and anti-interception properties. However, there have been few nonmilitary applications, because the spread spectrum systems were thought to be too expensive for nonmilitary uses. Today, the progress of communication devices reduces the cost of the systems and enables us the possibility of nonmilitary applications.

Nonmilitary communication networks can be classified into two kinds: public communication networks and consumer communication networks*, both of which are very promising. Figure 1 shows today's share of nonmilitary communication networks; the public communication network share is much larger than that of the consumer communication networks. Figure 2 shows predictions for the future share; the total number of both kinds will become much greater than that of today, and consumer communication networks will be comparable to public communication networks. The progress of consumer communications is due to the inexpensiveness (without charge), user privacy, free design from regulations and advantages for local and small zone communications. Consumer communications will be most fruitful in future communications as mentioned above. However, several problems must be overcome, not only problems of hardware

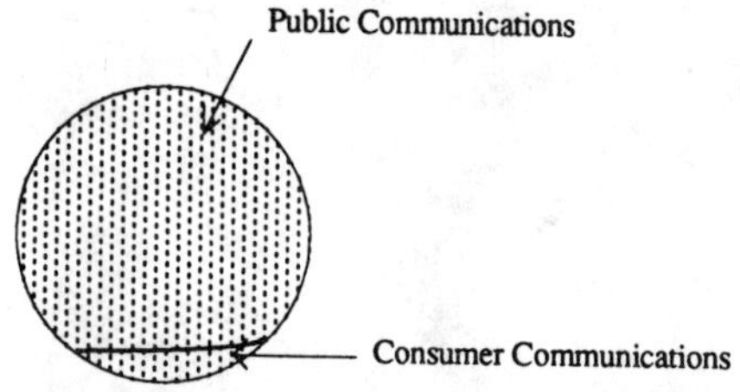

Fig. 1 Today's share.

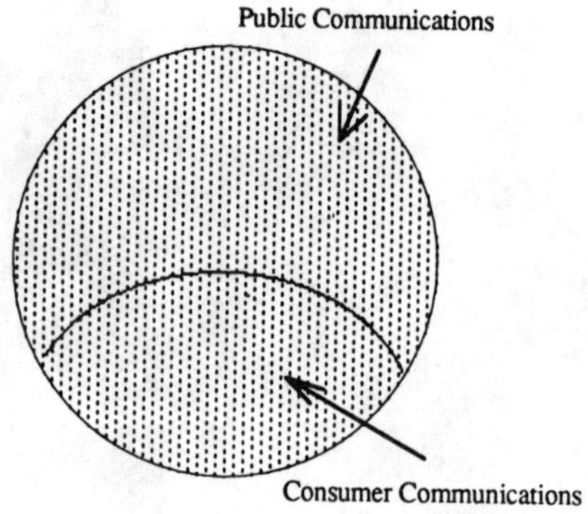

Fig. 2 Future share.

Manuscript received February 6, 1991.

† The author is with the Faculty of Science and Technology, Keio University, Yokohama-shi, 223 Japan.
†† The author is with the Faculty of Engineering, Saitama University, Urawa-shi, 338 Japan.

* LAN, WLAN, indoor communications, cordless telephones, communications for robots, remote control, data carrier, home communications, office communications, factory communications etc.

Reprinted from *IEICE Trans. Commun.*, vol. J74-B-I, no. 5, pp. 1093–1102, May 1991.

but also those of transmission lines, because consumer communications are hampered by poor transmission lines with fading, restrictions of low power transmission, interference and interception in their networks. If we demand good transmission lines, instead of poor ones, we will encounter economic restrictions and severe output power restrictions especially in radio consumer communication networks. Therefore, we must find a method to compensate poor transmission lines to overcome these problems in consumer communication networks.

The key technology to overcome the above problems is the spread spectrum, which is suitable for poor transmission lines with multipath fading, interference and interception. Many researchers are interested in spread spectrum communication technology as one of the key technologies for consumer communications in Japan. This paper presents the growing applications of spread spectrum communications in consumer communications.

2. Consumer† Communications

Communication networks have long been mainly public or military, and have never been owned by consumers; therefore, users have had to share these networks because communication equipment has been so expensive that consumers could not possess their own communication networks. Hovever, due to today's progressing equipment technology, it is now possible for consumers to possess their own communication networks. We can find a good analogy between communication networks and transportation networks. Railway networks appeared at the first stage in the history of transportation in this century, while the transportation network of cars owned by consumers

† Consumers usually mean people who consume energy, services and goods. However, we may add the meaning of consumers being people who consume information.

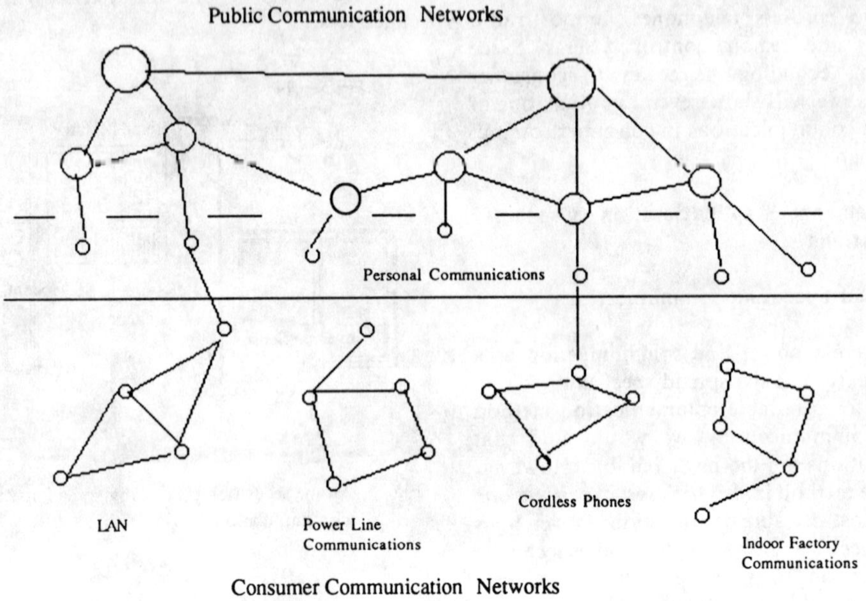

Fig. 3 Public communication networks and consumer communication networks.

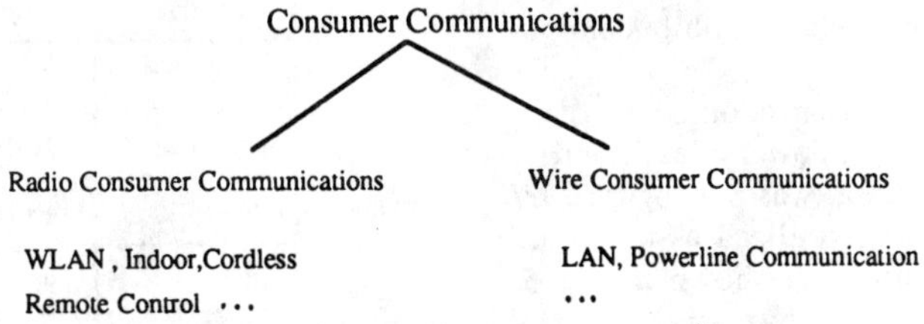

Fig. 4 Classification of consumer communications.

appeared at the second stage. The trend "from the public to the individual" seems to hold true in communication networks as well.

Figure 3 shows the relationship between consumer communication networks and public communication networks, where personal communication networks are a part of the public communication networks. The final goal of the personal communications is that all users possess their own personal terminals in public communication networks. On the contrary, consumer communication networks are not public, but are owned and used by consumers. Public communication networks are designed usually in top-down style; however, the design of consumer communication networks is done in bottom-up style. Such a bottom-up style will bring much competition into the communications industries.

Consumer communications are classified into two parts, wire consumer communications and radio consumer communications, as shown in Fig. 4. Examples of wire consumer communications are LAN and power line communications; those of radio consumer communications are cordless telephones, indoor radio communications and remote control systems. Since spread spectrum technology is a key to consumer communications, we will show several applications of spread spectrum communications in consumer communications in Japan.

3. Spread Spectrum Applications in Consumer Communications

3. 1 Wire Spread Spectrum Communications

We show here a power line communication as a successful application of the spread spectrum.

If we look for a reliable communication method for consumer communications, we would find that wire communications are the most reliable. However, the most reliable method is also the most expensive one because of the cost of cable construction. Power lines which supply electrical power to transceivers can be used as transmission lines between the transceivers without any extra cable construction. However, they have their problems in communication performance, i. e., the fluctuations of the frequency characteristics due to the load changes and the heavy noises generated by motors, regulators and so on. The problems are similar to those in fading channels in mobile communications.

The spread spectrum has been adopted to overcome the above problems in such power lines. The first proposal to adopt the SS method was put forward by NEC in 1983. The power line is utilized as the home bus connected with the terminals, as shown in Fig. 5. NEC Home Electronics Co., Ltd. has developed the spread spectrum home communication system whose specifications are shown in Table 1. Figure 6 shows the bit error performance[(1)-(3)].

One of the most difficult problems is the synchronization of a spreading code in power line communications whose channel fluctuation often makes the system asynchronous. Such instability is decreased by an improved DLL[(4)].

Power line transmission suffers severe bandwidth limitation, where the spreading gain must be reduced in order to increase the bit rates in the limited bandwidth transmission. A parallel spread spectrum communication system was proposed as an effective method to increase the bit rates without large reduction of the spreading gain[(5)]. Figure 7 shows the block diagram and the autocorrelation function of the parallel spread spectrum system.

A code division multiple access method for power line transmission which uses cancellers to cancel the mutual interference between the transmitted spread spectrum signals is proposed[(6)]. Figure 8 shows the system block diagram and Fig. 9 shows the performance comparison of both systems with and without cancellers.

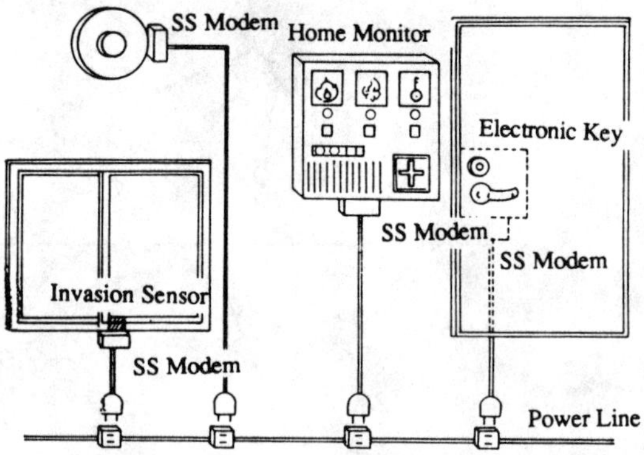

Fig. 5 Home security system using spread spectrum power line communication.

Table 1 SS home bus system.

Term	Specification
Modlation	DS/SS
Processing Gain	31
Freq. Band	10kHz~ 450kHz
Code	Bi-Phase
Bit Rate	960kbs
Aquisition Time	Max 13 msec

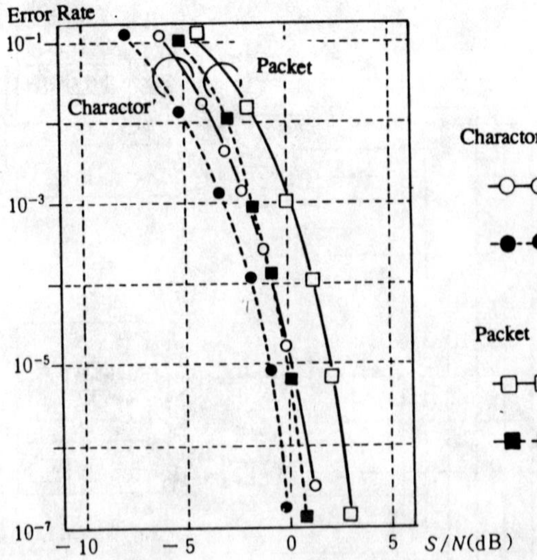

Fig. 6 Error rate in SS power line transmission.

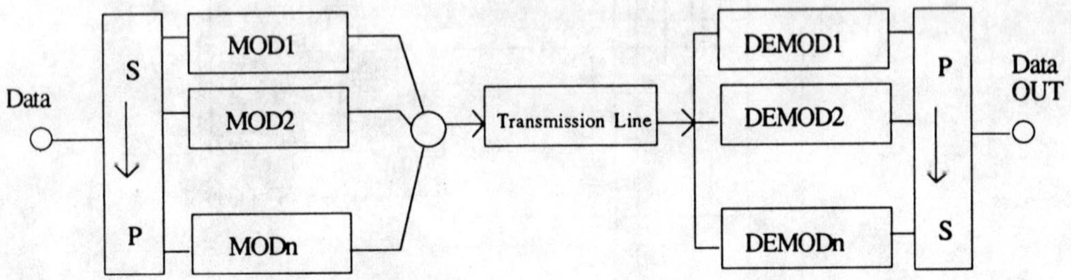

(a) Parallel Spread Spectrum System

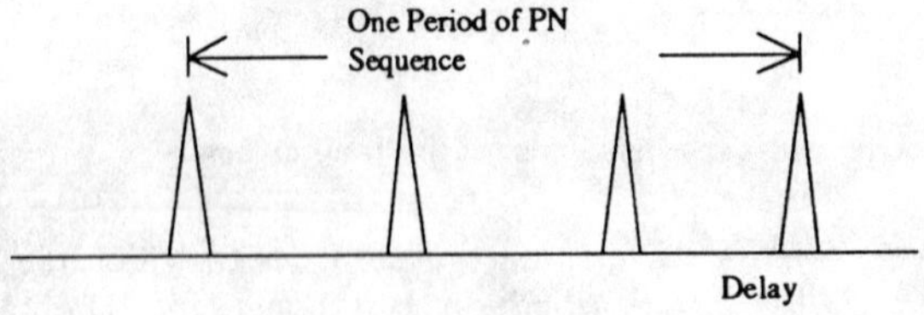

(b) Autocorrelation Function of Parallel Spread Spectrum System

Fig. 7 Parallel spread spectrum communication system.

3.2 Radio Spread Spectrum Communications

The most elegant applications of the spread spectrum method are in radio communications, which are used in the presence of multipath, interference, interception and so on. We show here four examples: a data carrier, a radio control system, low-power radio systems and a clock rate modulation system.

3.2.1 Data Carrier[7]

Figure 10 shows the block diagram of a data carrier system which is mainly used in factories where manufactured objects must be automatically identified. The data carrier system usually consists of a controller unit, a read/write (R/W) head unit and a data carrier unit (tag) ; the controller and the R/W unit are fixed and the data carrier unit is mobile because the

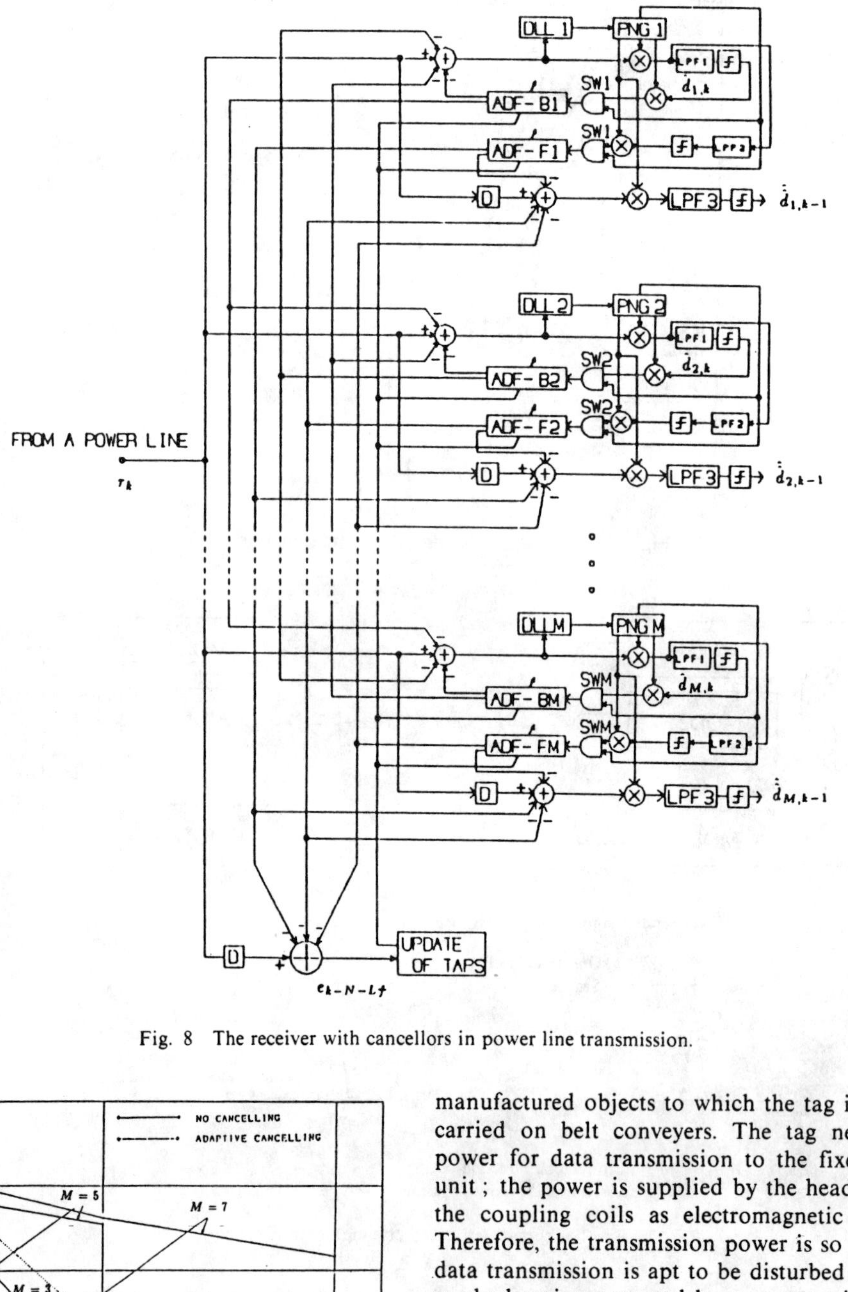

Fig. 8 The receiver with cancellors in power line transmission.

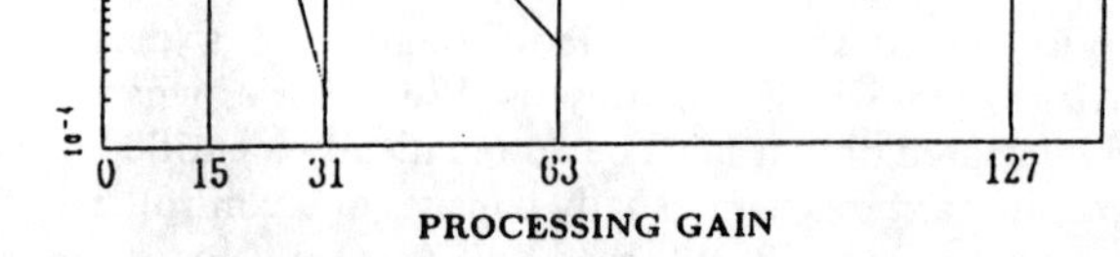

Fig. 9 BER vs. processing gain (M = the number of accessing users).

manufactured objects to which the tag is attached are carried on belt conveyers. The tag needs electrical power for data transmission to the fixed R/W head unit ; the power is supplied by the head unit through the coupling coils as electromagnetic wave power. Therefore, the transmission power is so small that the data transmission is apt to be disturbed by noise such as clock noise generated by processors in the controllers and the R/W units. The spectrum of the clock noise is not white, but a set of line spectra. Spread spectrum modulation is strong against such noise. Figure 11 shows the outside view of the SS data carrier system.

3. 2. 2 Radio Remote Control

Radio remote control systems are widely used in factories, construction sites, offices, homes and so on

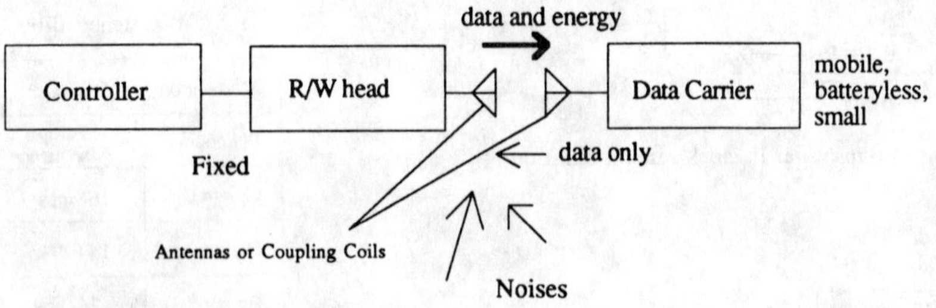

Fig. 10 Data carrier system.

Fig. 11 Photo of SS data carrier system (the big one is R/W head, the small one is data carrier).

Table 2 SS radio remote controller.

Term	Specification
Trans. Power	Low power regulation
FHband width	5MHz
The number of Hopping	15 (RS code)
Hopping Freq.	4.88kHz
DS Processing Gain	32
Modulation	FSK(m=1)
Chip rate	6.4μs/chip
Bit rate	4.88kbps
Error Control	1/2 convolutional code, Viterbi decoding, CRC
DS synch.	64 steps mached filter
FH synch.	DS synch. signal

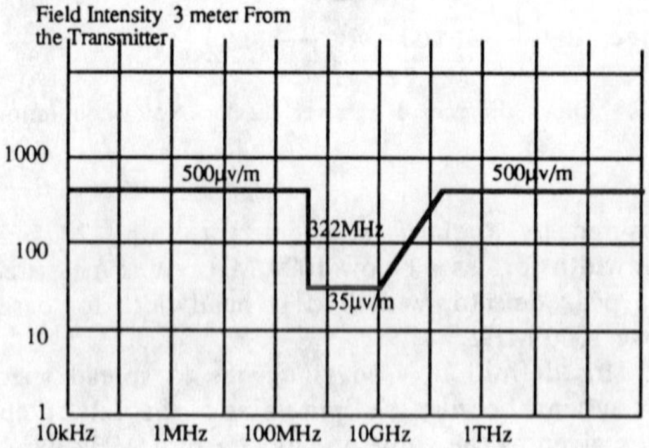

Fig. 12 Field intensity limit for a low power radio station.

Table 3 DS/SS low power experiment.

	No.1	No.2
Transmitted centre freq.	250MHz	310MHz
Received centre freq.	310MHz	250MHz
PN code	M-sequence	
Code length	127chips	
Chip rate	14MHz	
Modulation	DS-FSK	
Radio band width	28MHz	
Trans. antenna	1/4λ whip	
Receiv. antenna	1/2λ DP	

for robots, cranes, computers, etc. These systems are required to have high reliability because a data transmission error may cause a serious accident. As mentioned in the previous sections, multipath fading, interference and noise degrade the reliability of the data transmission in the radio remote control systems. A radio control system using a DS/FH (direct sequence/frequency hopping) hybrid technique was proposed, whose specifications are shown in Table 2[8]. DS/FH, Viterbi decoding and the error correction by CRC (cyclic redundancy check) increase the reliability.

3. 2. 3 Low-Power Radio Communications

Bijaku musen setsubi (BMS: low-power radio equipment) is a low power radio station which is not required to be licensed in Japan. It can use any frequency, if it satisfies the following regulation.

The field intensity 3 meters from the BMS must be below the lines shown in Fig. 12. The intensity is

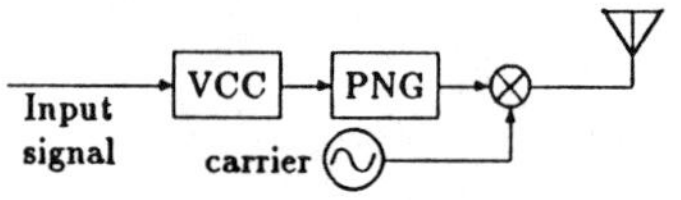

Fig. 13 Block diagram of transmitter in clock rate modulation system.

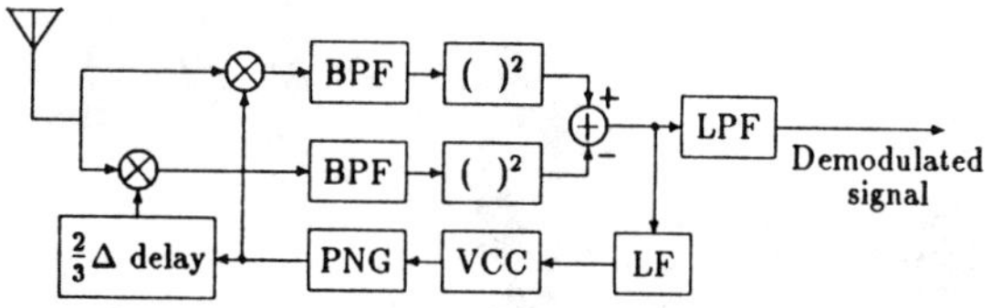

Fig. 14 Block diagram of receiver in clock rate modulation system.

measured by a quasi peak detector with 120 kHz bandwidth for cases below 1000 MHz. It is measured by a peak detector with 1 MHz bandwidth for cases above 1000 MHz.

This definition is advantageous to spread spectrum systems because the limit value of the field intensity is given as the value of 120 kHz or 1 MHz bandwidth and spread spectrum signals usually have bandwidth wider than 120 kHz or 1 MHz.

A group affiliated with Tohoku University and Clarion Co., Ltd. confirmed the above fact through experiments whose specifications are shown in Table 3[2]. In comparison with the conventional FM-BMS, SS-BMS has about ten times the communication range.

3.2.4 Clock Rate Modulation System[9]

The above-mentioned system is for digital signals, while the clock rate modulation is for analog signals. Of course, it is available for digital signals by using FSK. The system uses 2/3 Δ DLL for good demodulation quality[10]. The block diagrams of the transmitter and the receiver are shown in Figs. 13 and 14, respectively. The various performance of the system was evaluated by computer simulations.

4. Saw Device Applications

One of the most difficult problems for spread spectrum communication systems is the synchronization of the local pseudonoise (PN) code to the received PN code at the receivers. The sliding correlation method is the simplest, but it needs long acquisition time. Matched filter methods need very short acquisition time; in addition, the matched filters and the convolvers are available not only for acquisition but also for demodulation. However, they need complicated devices. SAW devices have often been adopted as these devices.

Table 4 SAW matched filter experiment.

Modulation	DS/SS
PN code	127chip Msequence
Chip rate	16Mcps
Centre freq.	144MHz
Bit rate	126kbps

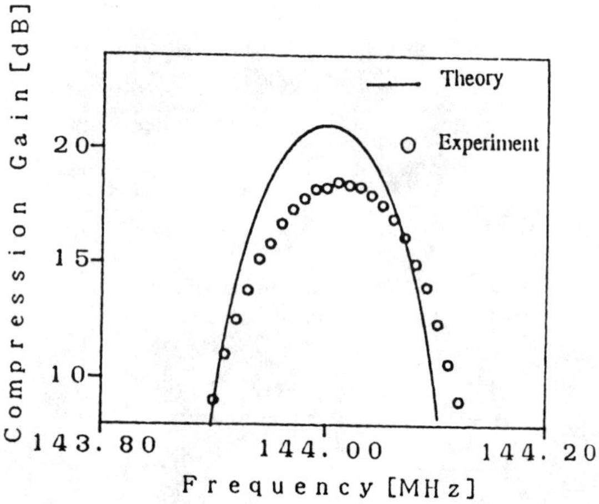

Fig. 15 Effect of frequency offset.

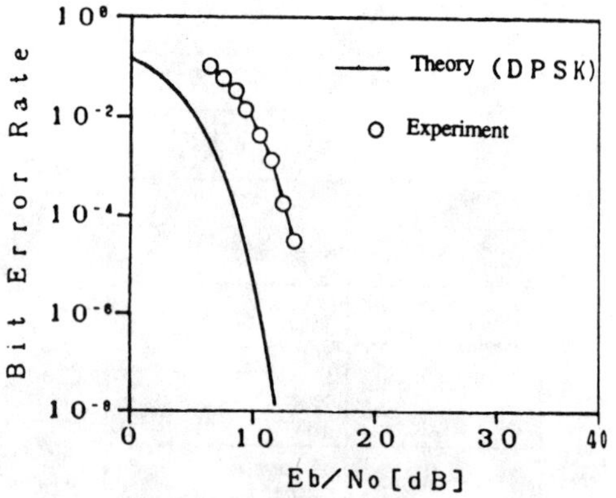

Fig. 16 Bit error rate.

Fig. 17 SAW matched filter.

4. 1 Matched Filter

Reference (11) shows a SAW device matched filter experiment whose specifications are shown in Table 4. Figure 15 shows the performance degradation due to the offset of the carrier frequency. Figure 16 shows BER versus *Eb/No*. Figure 17 shows the matched filter module.

Another topic of the SAW matched filter is the restriction of the data rate. In the case of using only one SAW matched filter in on-off keying or differential phase shift keying, the data rate cannot exceed the sequence rate. To overcome this restriction, the spread spectrum pulse position modulation system was proposed[12].

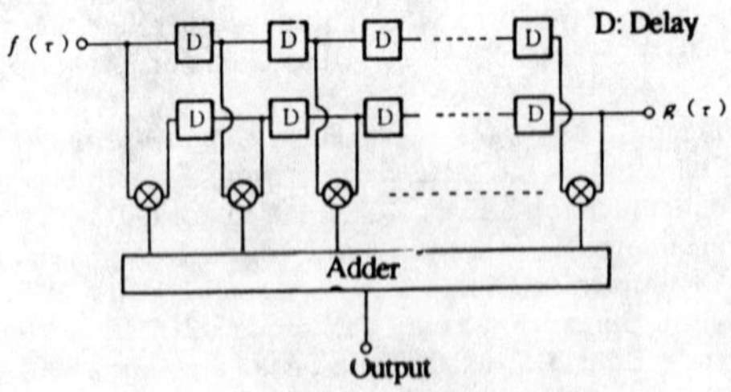

Fig. 18 SAW convolver structure.

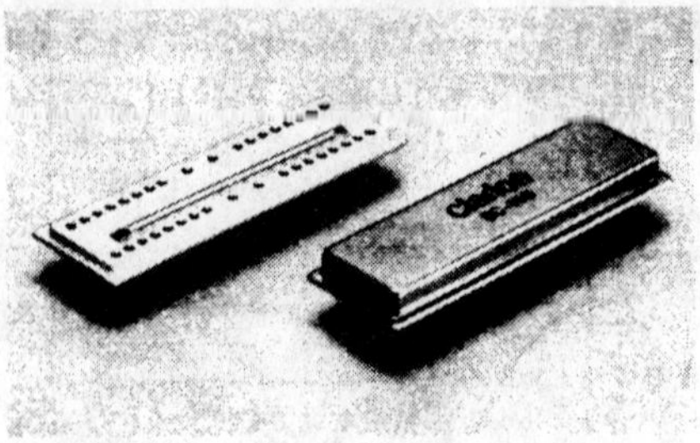

Fig. 19 Exterior of convolver.

Table 5 SAW convolver specifications.

Input 3dB Bandwidth	24MHz
Processing Time	9.0μsec
Time Bandwidth Product	216
Input Centre Freq.	215MHz
Output Centre Freq.	430MHz
Terminal Efficiecy	-50dBm
DC Voltage	+12v, -12v

4. 2 Convolver

Matched filters such as those mentioned above utilize a convolutional effect between the impulse response of the filter and the received pseudonoise (PN) code. However, it is difficult to change one received PN code to another received PN code because the pattern of the impulse response depends on the pattern of the electrodes on the SAW filter.

The structure of a convolver, as shown in Fig. 18, overcomes this problem. We can change one received PN code to another by changing the external reference PN code without any internal structural change. The multipliers in a convolver shown in Fig. 18 which utilize a nonlinear effect in the SAW device are for convolution between the received PN code and the external PN code. The specifications of the spread spectrum convolver are shown in Table 5. Figure 19 shows the exterior view of the convolver[2]. Field tests done by Clarion Co., Limited in Los Angeles area of a spread spectrum wireless modem using a SAW convolver were reported[13].

A spread spectrum signal with a polyphase PN code[14] can be easily detected by the convolver[15]. The polyphase PN code has less side robe than the biphase PN code. The SAW convolver is available for various systems.

5. Digital Signal Processing Applications

Digital signal processors have been adopted in many communications systems because of the stability, the variability of parameters, the flexibility and the capability of new functions. We show here a demodulation method in spread spectrum communications receivers with new functions which can be done only by digital signal processing. Also, other examples are shown.

5. 1 Spread Spectrum Block Demodulator for Packet SS[16]

Spread spectrum techniques have been applied to packet radio networks for interference resistances, code division multiple access, antifading capability and so

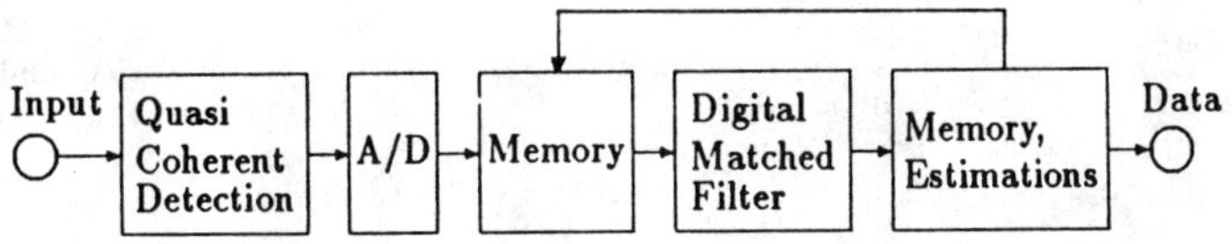

Fig. 20 Spread spectrum block demodulator for packet SS.

on[17]. As mentioned in the previous section, in general, the initial synchronization between the received PN code and the local PN code at the receiver is one of the most difficult problems in SS techniques because it takes a long time. The problem is more serious in packet radio networks. Each SS packet must have a long preamble for initial synchronization, which cannot be used as data and degrades the efficiency of the packet data transmission. Conventional demodulation systems have no solution to the problem because they have no memory function. The SS block demodulator has a memory block to store a quasicoherent-demodulated packet, as shown in Fig. 20. The memorized packet is applied to the digital matched filter (DMF) and then the DMF generates matched pulses at the output. The arctan function detects the frequency and phase offsets (Δf, $\Delta\phi$) of the carrier signal which a matched pulse includes but are not eliminated by the preceeding quasi-coherent demodulator. Thus, both PN code and carrier synchronizations are done and the data are detected. Δf and $\Delta\phi$ are estimated using the least mean squares algorithm and the estimation quality is equal from the packet head to the tail.

On the contrary, the estimation quality is so bad at the packet head in conventional SS demodulators that a preamble part may be required because of its transient pull-in phenomenon.

5. 2 Other Examples

We show two other examples of digital signal applications. One is an acquisition system for power line communications using a microprocessor[18]. The system accomplished quick correlation. The other is an application for sonar using real-valued orthogonal pseudonoise sequences[19]. Although this is not a communication system, we must not forget that spread spectrum systems simultaneously accomplish both communication and measurements.

6. Conclusions

Consumer communications will play an important role and one of the key technologies is the spread spectrum method. Several examples of spread spectrum applications for consumer communications in Japan were shown.

"From military to civil" and "from public to individual" will become true simultaneously.

References

(1) Marubayashi G., Nakagawa M. and Kohno R.: "Research and Developments of Spread Spectrum Communications", J. IEICE, **72**, 5, pp. 580(1989).

(2) ed. Nakagawa M.: "Basic and Applied Spread Spectrum Communications", Triceps(1987).

(3) Hirosaki B., Hasegawa S. and Endo K.: "A Power Line Home Bus System using SS Communication", IEEE ICCE'85(June 1985).

(4) Tachikawa S., Nagase M. and Marubayashi G.: "Lose-lock of Synchronization and the Improvement for the Electric Power Line Data Transmission Using Spread Spectrum Communication Systems—By the Synchronization Monitor Circuit", IEICE Technical Report, **SS88**-3 (April 1988).

(5) Kadoi Y., Sasaki S. and Marubayashi G.: "A Study of Parallel Spread Spectrum Communication System", IEICE Technical Report, **SSTA90**-16(March 1990).

(6) Kohno R., Imai H., Hatori M. and Pasupathy S.: "An Adaptive canceller of Cochannel Interference for Spread Spectrum Multi-Access Communication Network in a Power Line", IEEE J. Sel. Areas Commun., **8**, 4, pp. 691-699(1990).

(7) Harada K., Kajiwara A., Takeuchi K. and Nakagawa M.: "Characteristic of Data-Carrier using Spread Spectrum Communication", IEEE ICCE'90(June 1990).

(8) Yamamoto M., Hoshikuki A., Kohno R. and Imai H.: "An Implementation of R/C System Using DS/FH Hybrid Spread Spectrum Technique", IEICE Technical Report(Spread Spectrum), **SSTA 90**-24(June 1990).

(9) Suzuki Y., Hasegawa T., Okada T. and Hakura Y.: "Characteristics of Clock Rate Modulation and Demodulation System Using 2/3 Δ Type DLL of Manchester Coded PN", IEICE Technical Report, **SSTA90**-38(Oct. 1990).

(10) Hasegawa T., Okada T., Hakura Y. and Haneishi M.: "Performance of a Delay-Lock Loop for Manchester Coded Pseudo Noise Signal", Trans. IEICE, **73-B-I**, 8, pp. 663-665(Aug. 1990).

(11) Takehara K.: "DS-SS Communication Demodulator using SAW Matched Filter", IEICE Technical Report, **SSTA 89**-30(Aug. 1989).

(12) Hasegawa T. and Hakura Y.: "Spread-Spectrum Pulse Position Modulation", 1990 Autumn Natl. Conv. Rec. IEICE, SB-5-4.

(13) Mori M., Takeuchi S., Gochi M. and Tsubouchi K.: "Field Tests of a Spread Spectrum Wireless Modem in USA", IEICE Technical Report, **SSTA 90**-33(June 1990).

(14) Suehiro N. and Hatori M.: "Modulatable Orthogonal Sequences and their Applications to SSMA Systems", IEEE Trans. Inf. Theory, **34**, pp. 93-100(Jan 1988).

(15) Hamatsu M. and Endo M.: "Autocorrelation Property of Polyphase Orthogonal Sequences using SAW Convolver", 1989 Spring Natl. Conv. Rec. IEICE, SB-8-3.

(16) Kajiwara A. and Nakagawa M.: "Spread Spectrum Block Demodulator", International Conference on Spread Spectrum and its Applications, London(Sept. 1990).

(17) Pursley M. B.: "The Role of Spread Spectrum in Packet Radio Networks", Proc. IEEE, **75**, 1(1987).

(18) Kimura N., Shimamura T., Miyashita T. and Sato C.: "Acquisition System for DS-SS Transceiver in Power Line Communication using Microprocessor", 1989 Autumn Natl Conv. Rec. IEICE, B-347(1989).

(19) Tanada Y., Sagasaki Y., Yunokuchi K., Yoshida H. and Nagasawa Y.: "Pulse Compression Multiplexing Sonar Using Real-Valued Orthogonal Pseudonoise Sequences", IEICE Technical Report, **SSTA 90**-37(Oct. 1990).

Perfect-Capture ALOHA for Local Radio Communications

BHASKAR RAMAMURTHI, ADEL A. M. SALEH, FELLOW, IEEE, AND DAVID J. GOODMAN, SENIOR MEMBER, IEEE

Abstract—We consider a set of uncoordinated users employing an ALOHA protocol to transmit packets to a central base station in a local radio environment, e.g., in an urban area or within a building. Because of the different distances between the various users and the base station, the signals received from different users have substantially different power levels. This near/far phenomenon gives rise to a capture effect at the receiver such that when several packets are transmitted simultaneously, the receiver has a good chance of accurately detecting the packet arriving with the highest energy. Moreover, capture can be made near perfect in this environment through the use of properly design spread-spectrum signaling technique that enables the receiver to differentiate among the arriving packets based on both power differences and differing arrival times.

We investigate variations to the conventional ALOHA protocols that take advantage of perfect capture to reduce the delays and increase the throughputs of all users, including the furthest one. Three protocols are introduced and compared, which we call "persistent ALOHA with capture" (PAC), "gated ALOHA with capture" (GAC), and "gated ALOHA with capture/exhausted" (GAC/E). Two goals are considered in the comparison: an equatable delay profile among the users, and a small expected number of simultaneously transmitted packets. The latter is quite relevant for the practical implementation of a perfect-capture receiver. We conclude that the GAC/E protocol is most desirable from the twin points of view of delay profile and receiver implementation.

I. Introduction

A. The ALOHA Technique

THE use of radio for local data and voice communications, e.g., in an urban area or within office buildings, hospitals, factories, warehouses, convention centers or apartment buildings is an attractive proposition. We consider the use of an ALOHA multiple-access packet radio communications technique in such environments. With this technique, many terminals concurrently communicate with a central station over the same radio channel in an uncoordinated fashion. A small building may contain only one central station, while a large building, or a local urban environment, may be divided into several cells with one such station in the center of each cell. We concentrate on communications from the terminals to the local central station. Messages that are generated at the various terminals are transmitted in the form of packets over the common channel to the central station. An error-detecting code is used with each packet to enable the central station to check for the integrity of the received packets. Acknowledgments of accurately received packets are broadcast by the central station to all the terminals over a side channel, which can be made very reliable because of its extremely low data rate requirement. Because we are concerned with local environments having relatively short round-trip propagation delays, it is assumed that acknowledgments are in effect instantaneous.

When two or more terminals transmit their packets simultaneously, it is said that a *collision* is caused at the receiver. In this case, the central station may not receive any packet accurately or it may, at best, be able to receive one packet accurately. In the latter case, the receiver is said to *capture* the packet in spite of the presence of others. The unsuccessful terminals, whose packets have not been acknowledged, retransmit their packets according to some predetermined protocol until they are successfully received. Collisions and subsequent retransmissions are the price of giving many users uncoordinated access to the same channel. They limit the amount of information that can be transmitted through the channel (throughput). They also cause some data packets to suffer delays. These effects are random, and the statistics of the throughput and the delays are the figures of merit that characterize the system performance. In the usual model, an ALOHA system without capture is plagued with low system throughput and high user delays.

Because users in a local environment are at different distances from the central station, their received signals would typically have substantially different power levels [1]-[4]. This near/far phenomenon enhances the capture effect of the receiver because the packet arriving with the highest energy now has a good change of being detected accurately. This effect clearly improves the overall system throughput compared to the case of the standard ALOHA assumption of no capture. Furthermore, the capture effect was shown to improve the throughputs and the delays of all users, including the furthest ones [5]. The improvements become dramatic when capture is (near) *perfect*, i.e., when the central receiver (almost) *always* succeeds in correctly detecting the strongest packet, irrespective of the number of simultaneously transmitted packets.

Manuscript received August 15, 1986; revised January 30, 1987.
B. Ramamurthi is with the Indian Institute of Technology, Madras, India.
A. A. M. Saleh is with AT&T Bell Laboratories, Holmdel, NJ 07733.
D. J. Goodman is with the Philips Research Laboratories, Redhill, England, on leave from AT&T Bell Laboratories, Holmdel, NJ 07733.
IEEE Log Number 8714181.

Reprinted from *IEEE J. Selected Areas Commun.*, vol. SAC-5, no. 5, pp. 806–814, June 1987.

B. Perfect Capture

The realization of a near-perfect-capture receiver based only on differences among arriving signal strengths caused by the near/far phenomenon does not seem feasible. Adding channel coding improves capture somewhat [6], but by no means makes it perfect. On the other hand, the use of an appropriate direct-sequence, spread-spectrum signaling technique [7]–[10] appears to be very promising not only in achieving near-perfect capture, but also in combating the multipath delay speed [11], [12], which characterizes urban and indoor radio propagation [3], [13], [14].

Briefly, in this spread-spectrum technique, the packet transmitted by each user is preceded by a synchronization header code, which is identical for all users. The header is separated by a short silent period from the main body of the packet, which contains the information bits. A small set of quasi-orthogonal, pseudorandom codes, which again is the same for all users, is employed to cyclically modulate the successive information bits. The repetition period of each code is chosen sufficiently large to assure that all of its multipath echoes arriving at the receiver are sufficiently attenuated, which results in virtually eliminating intersymbol interference among nearby bits [7], [9].

We consider a slotted-ALOHA system where the packets arriving at the receiver fall within discrete packet time slots. The packet slots, however, are slightly longer than the actual packet duration to allow for sufficient randomization of the exact packet arrival times. At the receiver, a passive surface-acoustic wave (SAW) correlator that is matched to the header code is used to synchronize the spead-spectrum receiver of the information bits to any desired received packet [8], [10]. The slightly random header arrival times and the silent separation period between the header and the main body of the packet give a logic circuit at the receiver sufficient time to observe the correlation peaks of the SAW device. The locations and amplitudes of these peaks indicate the exact arrival times and the signal strengths of all packets that are received during the current time slot. This information is then used to synchronize the main-body receiver to the packet with the strongest energy. Thus, the receiver differentiates among arriving packets based not only on their potentially vast power differences, but also through their differing arrival times. Hence, interfering packets, either from other users or from multipath echoes of the packet being received, are suppressed by the processing gain of the spread-spectrum code. In this way, perfect capture is very nearly achieved. In fact, one can even conceive of a receiver capable of accurately detecting different arriving packets simultaneously. However, we will not consider this case here.

C. The Purpose of this Paper

In this paper, we consider a slotted-ALOHA system in which uncoordinated terminals communicate with a central station with a perfect-capture receiver. Our main purpose is to study various protocols that exploit perfect capture. All protocols are based on instantaneous feedback information from the receiver indicating the identity of the terminal whose packet is received correctly. In addition, we assume that the terminals are instantaneously informed when no packets arrive at the receiver, i.e., when an empty slot occurs.

The presence of perfect capture induces priority among the terminals because when many terminals transmit packets simultaneously in any slot, only the packet with the highest received energy is successfully decoded, while the others have to be retransmitted. Thus, the statistics of the delay experienced by the packets are different for different terminals. In our application, the induced priorities of the terminals may slowly vary with time, for example, as they are moved about, or if some other terminal becomes active or inactive. Therefore, we assume that the terminals are unaware of their induced priorities.

We consider the average delay experienced by each terminal, particularly the lowest priority one, as a measure of the performance of any particular protocol. Moreover, since the number of colliding packets within any time slot clearly has a bearing on the feasibility of perfect capture, we also consider the statistics of the number of colliding packets when comparing protocols.

Three main protocols are introduced and compared through analysis and/or simulations. A fourth, ideal protocol (first-come-first-served) is also considered, which, however, is not implementable within our system since it requires information flow between the terminals and the receiver that are excluded by our simple acknowledgment model. That protocol is considered only because it provides an idealized lower bound on the averaged delay of the lowest priority terminal.

The results of the analysis for each of the protocols considered in this paper depend on capture being perfect. Thus, if used in practice, these protocols may lead to a deadlock if capture fails in an autocorrelated way in successive time slots. However, temporary modifications can be made to each protocol during these periods by switching to a collision-resolution algorithm [15] until the deadlock is resolved. If this occurs only occasionally, the long-term average performance of the protocol would be only slightly affected.

We adopt the following notation. There are M terminals, denoted by T_i, $i = 1, 2, \cdots, M$, and each one is assumed to have an infinite buffer for queueing its packets. Each terminal generates at most one packet in each slot independently with probability p. The terminals are independent of one another, and the overall throughput is denoted by $S(= Mp)$. Without loss of generality, we will assume that the terminal numbered i has priority i, with priority decreasing from 1 to M. The average delay for a terminal i is denoted $\bar{\tau}_i$, and the number of colliding packets is denoted by n.

II. First-Come-First-Served (FCFS) Protocol

Under this protocol, packets are transmitted in the order in which they are generated by the terminals. Capture is used to resolve ties, i.e., when two packets are generated in the same slot, the one from the higher priority terminal is received first. The average delay $\bar{\tau}_i$ for terminal i is given (from Appendix I) by

$$\bar{\tau}_i = \frac{S^2}{2(1-S)} \frac{M-1}{M} + \frac{i-1}{M} S. \tag{1}$$

When M is large, we obtain for the lowest priority terminal

$$\bar{\tau}_M \approx \frac{S}{2} \frac{2-S}{1-S}. \tag{2}$$

Thus, we see that $\bar{\tau}_M$ has a single pole at $S = 1$. Moreover, the difference in the average delay between the lowest and highest priority terminals is less than S. For S close to unity, this difference is negligible. The FCFS protocol is *work-conserving*, i.e., the terminals are not idle when there are packets to be transmitted nor are packets dropped without being successfully transmitted. Consequently, the Conservation law [16] for work-conserving protocols applies and, in our case, it simplifies to

$$\sum_{i=1}^{M} \bar{\tau}_i = \text{constant} \tag{3}$$

where the constant is the same for *all* work-conserving protocols. Hence the average system delay, given by

$$\frac{1}{M} \sum_{i=1}^{M} \bar{\tau}_i = \frac{M-1}{2M} \frac{S}{1-S} \tag{4}$$

is the same for all such protocols.

If we do not permit packet-dropping, a protocol that does not conserve work can only increase the average delay for the system. The FCFS protocol is thus the optimal protocol for our performance objective: it has the lowest possible average system delay, and it gives approximately equal delays for all the terminals. Since the FCFS protocol is not implementable in our system, our goal is to find an implementable protocol whose performance approaches that of the FCFS protocol in terms of both the system delay and the delay profile among the terminals. We do not restrict ourselves to work-conserving protocols to achieve our goal. While packet dropping is not permitted, the terminals may remain idle when there are packets to be transmitted. This will certainly increase the average system delay, but will hopefully lead to a delay profile that is closer to that of the FCFS protocol.

III. Persistent ALOHA with Capture (PAC)

The first implementable protocol we consider is a simple extension of the conventional ALOHA protocol, which takes full advantage of perfect capture. Under this protocol, each terminal transmits the packet at the head of its buffer persistently until the packet is successfully acknowledged. Since the feedback is instantaneous, the terminal begins transmitting the next packet in the very next slot. Thus, each terminal transmits continuously until its buffer is cleared. This protocol is identical to the head-of-the-line priority queueing discipline in the queueing theory literature [16]. The average delay $\bar{\tau}_i$ for terminal i is given (from Appendix I) by

$$\bar{\tau}_i = S \frac{i-1}{2M} \frac{2 - \frac{i}{M} S}{\left(1 - \frac{i-1}{M} S\right)\left(1 - \frac{i}{M} S\right)}. \tag{5}$$

For large M, the average delay $\bar{\tau}_M$ of the lowest priority terminal is approximately

$$\bar{\tau}_M \approx \frac{S}{2} \frac{2-S}{(1-S)^2}. \tag{6}$$

We see that $\bar{\tau}_M$ has a double pole at $S = 1$, and thus grows very rapidly as S approaches unity. This result is independent of M provided M is sufficiently large. The highest priority terminal T_1 has zero delay, and a large fraction of the terminals have very small delay compared to $\bar{\tau}_M$. Since the PAC protocol is work conserving, the average system delay is given by (4), and is thus the lowest possible. Although this aspect of our performance objective is met, the delay profile is highly nonuniform. This, of course, occurs since, with PAC, a higher priority terminal can at any time preempt a lower priority terminal even if the latter has already been waiting for a long time.

IV. Gated ALOHA with Capture (GAC)

We now propose a new protocol that reduces the delay experienced by the lowest priority terminal considerably compared to PAC. Under the PAC protocol, only the acknowledgment part of the feedback is used by the terminals. Under the GAC protocol, the feedback about empty slots is also used. Indeed, the observation of empty slots by the terminals is the key to the GAC protocol.

The protocol is as follows: Each terminal is permitted to transmit all those packets which were generated prior to or during the most recent empty slot. Packets generated after the most recent empty slot have to wait until the next empty slot in order to get cleared for transmission. Each terminal begins transmitting the cleared packets (in the sequence in which they are buffered) in the slot immediately after the empty slot. The packets are captured by the receiver according to the priority of the originating terminals. When all the cleared packets in the system have been received, an empty slot would occur. The packets that were generated since the previous empty slot are then cleared for transmission and the cycle resumes.

The reason we refer to this protocol as gated ALOHA with capture is that one can visualize a gate in the buffer of each terminal which separates the packets that are cleared for transmission in the current cycle from those that are being held for the next one. This protocol resembles that of roll-call polling described in [17, p. 178], but with gating at the beginning of the polling cycle.

We refer to the period from the end of one empty slot to the end of the next one as a *cycle*. Each cycle consists of full slots (slots with one or more transmitted packets) followed by one empty slot. We refer to the sequence of full slots as a *run*. Thus, the cycle length is one slot longer than the run length. The sequence of run lengths is a random process involving nonnegative numbers, which we denote by R_k, $k = 1, 2, \cdots$.

The statistics of the run-length process determine the statistics of the delays. A fundamental fact about the GAC protocol is that the statistics of R_{k+1} depend only on R_k, i.e., the run-length process is a Markov process. In the steady state, each moment of the run-length distribution tends to a limting value. In particular, we show in Appendix II that the limiting values of the mean and second moment, denoted, respectively, by $\overline{R}$ and $\overline{R^2}$, are given by

$$\overline{R} = \frac{S}{1 - S} \tag{7}$$

$$\overline{R^2} = \frac{S(1 + S + S^2 - p)}{(1 - S)^2 (1 + S)}. \tag{8}$$

Actually, the mean run length $\overline{R}$ is independent of the protocol. To see this, note that if the throughput is S, the fraction of full slots is S and the fraction of empty slots is $(1 - S)$, on the average. The mean run length is simply given by the ratio of the fraction of full slots to the fraction of empty slots (see also [17, eq. (7.4b), p. 184]). The ratio of the variance of the run length, σ_R^2 $(=\overline{R^2} - (\overline{R})^2)$, to the square of the mean is given by

$$\frac{\sigma_R^2}{(\overline{R})^2} = \frac{1 - p}{S(1 + S)}. \tag{9}$$

The average delay $\overline{\tau}_i$ for terminal i, as derived in Appendix II, is

$$\overline{\tau}_i = \frac{1}{2(1 - S)(1 + S)} \left\{ \frac{2i}{M} S(1 + 2S - Sp) + S^2(1 + p) + S(1 - p)(2 - p) - 2p \right\}. \tag{10}$$

For large M, p is very small compared to S, and

$$\overline{\tau}_i \approx \frac{1}{2} \frac{S}{1 - S} \frac{2 + S + (1 + 2S)\frac{2i}{M}}{1 + S}. \tag{11}$$

The corresponding delay $\overline{\tau}_M$ for the lowest priority terminal is

$$\overline{\tau}_M \approx \frac{1}{2} \frac{S}{1 - S} \frac{5S + 4}{1 + S}. \tag{12}$$

Thus, with the GAC protocol, $\overline{\tau}_M$ has only a single pole at $S = 1$, as with the FCFS protocol. Note that the delay is a linear function of the priority. This protocol is not work-conserving since the transmitters are idle in the empty slots even when packets are waiting to be trans mitted. In fact, the average system delay with the GAC protocol is about three times higher with the FCFS or PAC protocol, which is given by (4).

V. Gated ALOHA with Capture/Exhaustive (GAC/E)

This protocol is a variation of GAC, with the gating done differently. Just as in GAC, the packets in each buffer are cleared at the end of the current cycle for transmission in the next cycle. That is, the gates at the heads of all buffers are opened at the end of the current cycle. However, the gate in each buffer is not immediately moved to the end of the buffer as in GAC. Instead, in each buffer, the gate is closed only when the buffer is *exhausted* of all the packets it contains. This type of gating makes this protocol similar to a well-known protocol with desirable characteristics called *loop polling* (LP) [18], [19], which, however, is not implementable in our model.

In the GAC/E protocol, packets that are generated at any terminal during the new cycle are also cleared for transmission *provided the gate in the terminal's buffer is still open when they are generated*. The gate in each buffer is closed as soon as that terminal has been exhaustively "served." If a buffer is empty at the beginning of the new cycle, the gate in that buffer remains open until at least one packet is generated and successfully received. Since the key feature of this protocol is the gating upon exhaustive service we refer to it as *Gated ALOHA with Capture/Exhaustive*.

In contrast to the GAC protocol, the terminals in the GAC/E protocol are not necessarily served in the order of their priority in each cycle. The GAC/E protocol resembles the PAC protocol in that a higher priority terminal captures the receiver from a lower priority terminal when the higher priority terminal generate a new packet. However, GAC/E differs from PAC in that this happens only if the higher priority terminal is not already gated.

We have been unable to analyze the GAC/E protocol for the average delays. Thus, we rely on computer simulations to compare GAC/E to the other protocols.

VI. Comparisons of the Protocols

A. Delay and Throughput

Fig. 1 shows the lowest priority delay, $\overline{\tau}_{100}$ (for $M = 100$), as a function of S for the four protocols. For the FCFS, PAC, and GAC protocols, the analytical results are plotted. For the GAC/E protocol, the results were obtained from computer simulations. The significant differences in performance between the four protocols occur for values of S greater than 0.8. For $S = 0.9$, for example, $\overline{\tau}_{100}$ is about 50 with PAC, 20 with GAC, 14 with GAC/E, and only 5 with FCFS. For S very close to unity, $\overline{\tau}_{100}$ is much lower with GAC or GAC/E than with PAC because of the single pole at $S = 1$ instead of the double pole. When compared with FCFS, $\overline{\tau}_{100}$ with GAC is about 4.5 times larger for S close to unity, while it is only 2.4 times larger with GAC/E. Thus, GAC/E yields a significant improvement over GAC.

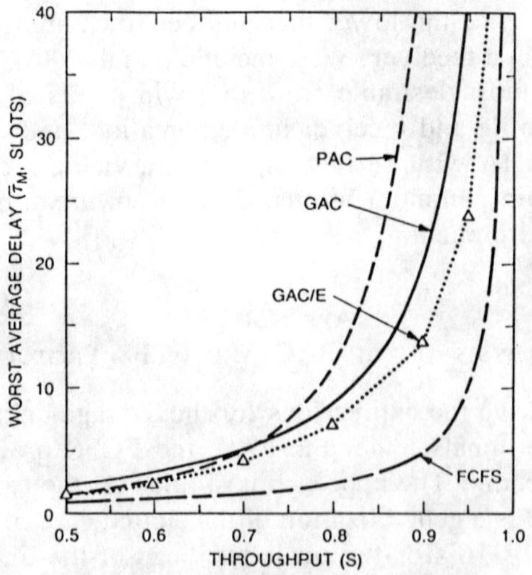

Fig. 1. Lowest priority average delay versus throughput for four protocols. Triangles indicate computer simulation results for GAC / E.

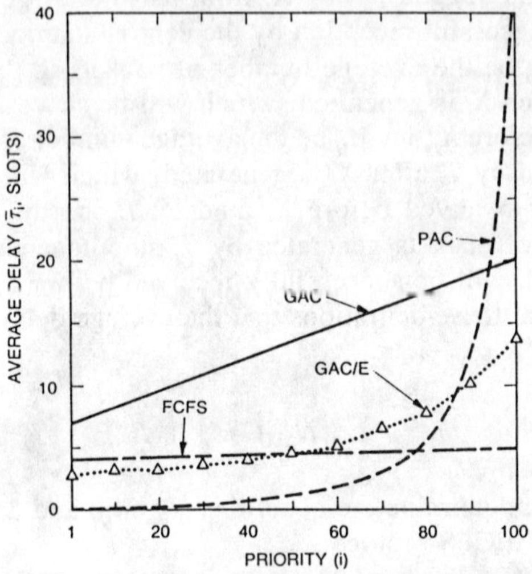

Fig. 2. Average delay versus priority for four protocols, 100 terminals, and throughput $S = 0.9$. Triangles indicate computer simulation results for GAC / E.

The delay $\overline{\tau}_i$ as a function of the priority i when $M = 100$ and $S = 0.9$ is shown in Fig. 2 for the four protocols. Under the PAC protocol, $\overline{\tau}_i$ is small for a large fraction of the terminals but grows rapidly as the priority approaches 100. In fact, most of the terminals have a lower average delay with the PAC protocol than with the FCFS protocol. However, the average delay for the FCFS protocol is nearly constnat for all the terminals. Furthermore, the average system delay is the same for the FCFS and PAC protocols.

Under the GAC protocol, $\overline{\tau}_i$ is a linear function of the priority i. The delay profile is fairer than that of the PAC protocol but not as good as that of the FCFS protocol. The price paid for this improvement over the PAC protocol is a higher average system delay. The average system delay is three times higher for the GAC protocol than for either the PAC or FCFS protocols. With the GAC / E protocol, the average delays are reduced for *all* the terminals when compared to the GAC protocol. Thus, while the delay profile remains fair, the average system delay is reduced for $S = 0.9$ by a factor of about two.

B. Number of Colliding Packets

Finally, we compare the percentile statistics of the number of colliding packets n under the various protocols, with the exception of FCFS, which is not implementable in our model. As mentioned earlier, this comparison is based on computer simulation results. We note that n is a random variable whose statistics change with S. Figs. 3 and 4 show, as a function of S, the values of n not exceeded, respectively, for 99 and 90 percent of the slots. It is clear from these figures that, for $S < 0.8$, both the 90th and 99th percentiles are almost identical for the PAC and GAC / E protocols. The two percentiles for either protocol are less than the corresponding percentiles for the GAC protocol. Furthermore, the two percentiles for the PAC protocol are lower than the corresponding ones for the GAC / E protocol only for $S \geq 0.9$, for which range the GAC protocol suffers from a significantly larger number of collisions.

Although Figs. 3 and 4 do not give all the relevant statistics for the number of collidating packets, they are useful in estimating the number of simultaneous transmissions a perfect-capture receiver must contend with. Assume, for example, that a particular receiver is unable to detect the highest priority packet if the number of simultaneous transmissions exceeds ten. Then we see from Fig. 3 that we can operate GAC / E and PAC successfully for 99 percent of the time with S up to 0.85, while with GAC we are limited to $S \approx 0.75$. For the remaining one percent of the time in these cases, capture may fail.

When capture fails, the three protocols considered here lead to deadlock, unless certain modifications are made to resolve the destructive collision. One approach is to temporarily employ a collision resolution algorithm (see, for example, [15]) whenever capture fails. As soon as the colliding packets are successfully transmitted, the original perfect-capture protocol may be resumed. Many variations are possible, including partial resolution of collisions before resumption of the perfect-capture protocol, in the hope that capture would be perfect with the fewer remaining packets.

Yet another approach to deadlock resolution, which is suited for speech packets, would be to drop the colliding packets at the terminals whenever there is a destructive collision. If this occurs for, say, one percent of the packets, the missing packets could be reconstructed at the receiver using one of various known reconstruction techniques (see, for example, [20]).

Whatever approach is adopted for deadlock resolution, the combination of perfect-capture protocol and collision-resolution method must be further studied in the context of a specific receiver to determine the actual performance of the complete system. The results presented in this paper give us an idea of the performance that can be ex-

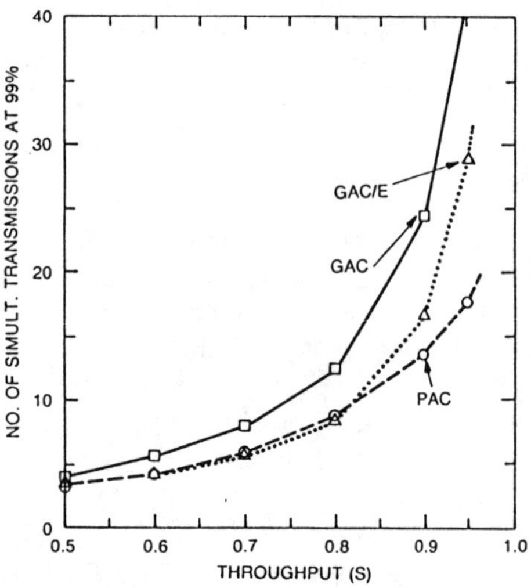

Fig. 3. Simulation results showing the 99th percentile of the number of simultaneous transmissions versus throughput for 100 terminals.

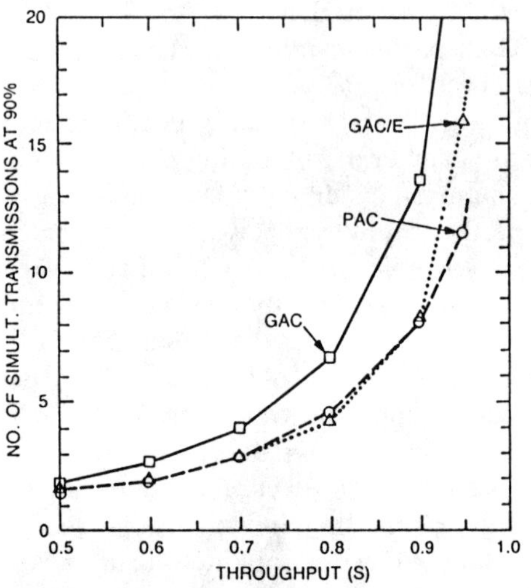

Fig. 4. Simulation results showing the 90th percentile of the number of simultaneous transmissions versus throughput for 100 terminals.

pected in a system with a low rate of capture-failure and a reasonable collision resolution algorithm.

VII. Conclusions

We have argued that the near/far phenomenon, which characterizes local radio environments, combined with appropriate spread-spectrum signaling, can be exploited to realize a receiver with near-perfect capture, which also combats the multipath effect. This makes possible the implementation of high-throughput, low-delay ALOHA protocols for local radio communications. The attractiveness of ALOHA is that it does not require coordination among the various users. Three new protocols, PAC, GAC, and GAC/E, were introduced and compared. Two criteria were used in the comparisons; namely, an equatable delay profile among the users, and a small expected number of simultaneously transmitted packets. The latter is quite relevant for the practical implementation of a perfect-capture receiver since the lower that number, the simpler it is to build such a receiver. We conclude that the GAC/E protocol is most desirable from the twin points of view of delay profile and receiver implementation, and that slotted ALOHA with perfect capture is a viable method for local radio communications, e.g., in an urban or an indoor environment.

Appendix I
Analysis of the PAC and FCFS Protocols

We derive the expressions for the average delays $\{\overline{\tau}_i\}$ of the terminals under the PAC and FCFS protocols in this appendix. The approach is similar for the two protocols, and is a generalization of the slotted case of the derivation in [16] for the unslotted head-of-the-line-priorities queueing discipline. Following [16], we introduce the notation of a *tagged* packet X_j generated by terminal T_j. We will obtain the average delay $\overline{\tau}_j$ for terminal T_j by observing the tagged packet X_j from its time of generation till its successful reception by the central station.

Let N_{ij} be the average number of packets in the buffer of T_i when X_j is generated, which will be successully received before X_j; let M_{ij} be the average number of packets generated by T_i after X_j is generated, which will be successfully received before X_j; and let L_{ij} be the average number of packets generated by T_i simultaneously with X_j, which will be successfully received before X_j. It follows from these definitions that the average delay $\overline{\tau}_j$ of X_j is

$$\overline{\tau}_j = \sum_{i=1}^{M} (N_{ij} + M_{ij} + L_{ij}). \tag{A1.1}$$

Now, we determine the quantities N_{ij}, M_{ij}, and L_{ij} for the PAC and FCFS protocols.

A. PAC Protocol

Under the PAC protocol, X_j has to be successfully received before any packets from terminals of lower priority $i > j$ can be received. Thus, the following equalities follow immediately: for $i = 1, 2, \cdots, M$ and $j = 1, 2, \cdots, M$

$$N_{ij} = 0, \quad i > j \tag{A1.2}$$

$$M_{ij} = 0, \quad i \geq j \tag{A1.3}$$

$$L_{ij} = 0, \quad i \geq j. \tag{A1.4}$$

Further, from Little's result [16], we know that the average number in the buffer of terminal T_i at any time, and in particular at the time when X_j is generated, is given by

$$N_{ij} = p\overline{\tau}_i, \quad i \leq j. \tag{A1.5}$$

Since X_j is delayed on the average by $\overline{\tau}_j$, the average number of packets generated by terminals of higher priority while X_j is held is given by

$$M_{ij} = p\overline{\tau}_j, \quad i < j. \tag{A1.6}$$

Finally, an average of p packets are generated by each terminal of higher priority simultaneously with X_j, i.e.,

$$L_{ij} = p, \; i < j. \qquad (A1.7)$$

Substituting for N_{ij}, M_{ij}, and L_{ij} from (A1.2)-(A1.7) in (A1.1), we obtain

$$\overline{\tau_j} = \sum_{i=1}^{j} p\overline{\tau_i} + \sum_{i=1}^{j-1} p\overline{\tau_j} + \sum_{i=1}^{j-1} p. \qquad (A1.8)$$

Solving for $\overline{\tau_j}$, we get for $j = 1, 2, \cdots, M$

$$\overline{\tau_j} = \frac{p \sum_{i=1}^{j-1} (\overline{\tau_i} + 1)}{1 - jp}. \qquad (A1.9)$$

Starting from $\overline{\tau_1} = 0$, solving these equations iteratively, and substituting $p = S/M$, we obtain (5) in the main text.

B. FCFS Protocol

Under the FCFS protocol $M_{ij} = 0$ for all i because packets generated after X_j will be received only after X_j. Similarly, $N_{ij} = p\overline{\tau_i}$ for all i because packets generated before X_j are received before it. Since priority is used to resolve ties among packets generated simultaneously, L_{ij} will be identical to that of PAC, which is given by (A1.4) and (A1.7).

Substituting for N_{ij}, M_{ij}, and L_{ij} in (A1.1), we obtain for $j = 1, 2, \cdots, M$

$$\overline{\tau_j} = p \sum_{i=1}^{M} \overline{\tau_i} + (j - 1)p. \qquad (A1.11)$$

Adding the M equations for $j = 1, 2, \cdots, M$, simplifying, and substituting $p = S/M$, we get

$$\sum_{j=1}^{M} \overline{\tau_j} = \frac{S}{1 - S} \frac{M - 1}{2}. \qquad (A1.12)$$

From (A1.11) and (A1.12), we obtain (1) in the main text.

We note that the sum in (A1.12) divided by M gives the average system delay for the FCFS protocol. Since the FCFS and PAC protocols are both work-conserving, they have the same average system delay.

Appendix II
Analysis of the GAC Protocol

A. Run-Length Statistics

The run-length process is a first-order Markov process. R_{k+1} is the number of new packets generated during the cycle of duration $1 + R_k$. Thus, conditional on R_k, it has the binomial probability function

$$\Pr(R_{k+1} | R_k) = \binom{M(R_k + 1)}{R_{k+1}} p^{R_{k+1}} (1 - p)^{M(R_k+1) - R_{k+1}}. \qquad (A2.1)$$

The mean and second moment are given by

$$E[R_{k+1} | R_k] = S(R_k + 1), \qquad (A2.2)$$

$$E[R_{k+1}^2 | R_k] = S^2(R_k + 1)^2 + S(R_k + 1)(1 - p) \qquad (A2.3)$$

where we recall that $S = Mp$. Taking expectations over the distribution of R_k, we obtain

$$E[R_{k+1}] = S(E[R_k] + 1) \qquad (A2.4)$$

$$E[R_{k+1}^2] = S^2 E[(R_k + 1)^2] + S(E[R_k] + 1)(1 - p). \qquad (A2.5)$$

In the steady state, each moment tends to a limiting value, which we denote simply by $\overline{R^n}$ where

$$\overline{R^n} = \lim_{k \to \infty} E[R_k^n]. \qquad (A2.6)$$

Hence, we obtain, in the limit, $\overline{R}$ and $\overline{R^2}$ given by (7) and (8) in the main text.

B. Average Delay

Now, we determine the average delay $\overline{\tau_j}$ for terminal T_j, $j = 1, 2, \cdots, M$, under the GAC protocol. Let us assume that terminal T_j generates a packet during slot number $(\tau_g + 1)$ in a cycle of length $(R + 1)$, $(0 \le \tau_g \le R)$. To be successfully received, this packet has to wait first till the end of the cycle, then for all the packets generated by terminals of higher priority, T_i, $i < j$, to be received, and finally for the packets generated by T_j in the slots numbered 1 through τ_g. We determine these three components of the delay separately.

Let us determine the expected delay of the packet in question *given that it is generated in a cycle of length* $(R + 1)$. The average number of packets of higher priority that are generated during the cycle is $(j - 1)p(R + 1)$. Given that the packet in question is generated after τ_g slots, an average of $p\tau_g$ packets are generated by T_j prior to it. Further, the packet has to wait $(R - \tau_g)$ slots for the next cycle to begin. Adding up the three components of the delay, we obtain

$$E[\tau_j | \text{packet gen. in } (R + 1) \text{ after } \tau_g \text{ slots}] = (R - \tau_g) + p\tau_g + (j - 1)p(R + 1). \qquad (A2.7)$$

Since the packet can be generated with equal probability in any slot in the cycle

$$\Pr[\tau_g = i] = \frac{1}{R + 1}, \quad i = 0, 1, \cdots, R. \qquad (A2.8)$$

Averaging over the probability distribution of τ_g, we obtain

$$E[\tau_j \mid \text{packet gen. in } (R+1)]$$

$$= (j-1)p(R+1) + \frac{R}{2}(1+p). \quad \text{(A2.9)}$$

In order to determine $\overline{\tau_j}$, we need to determine the probability that the packet in question is generated in a cycle of length $(R+1)$. Let us consider a very long sequence of, say, N cycles. On the average, the number of runs of length R is N Pr (R). If a packet is generated in a cycle of length $(R+1)$, it may be generated in any of of $N(R+1)$ Pr (R) slots. The total number of slots is, on the average, $N(\overline{R}+1)$. As $N \to \infty$ we obtain the probability of a packet being generated in a cycle of length $(R+1)$ as

$$\Pr[\text{packet gen. in } (R+1)] = \frac{(R+1)\Pr(R)}{\overline{R}+1} \quad \text{(A2.10)}$$

The average delay $\overline{\tau_j}$ is thus given by

$$\overline{\tau_j} = \frac{1}{\overline{R}+1}\sum_{R=0}^{\infty}\Big\{(j-1)p(R+1)^2 \Pr(R) + \frac{1+p}{2}R(R+1)\Pr(R)\Big\}. \quad \text{(A2.11)}$$

Substituting for $\overline{R}$ and $\overline{R^2}$ from (7) and (8), we obtain (10) in the main text.

Acknowledgment

We thank G. J. Foschini for his probing questions, which led eventually to the invention of the GAC/E protocol.

References

[1] S. E. Alexander, "Radio propagation within buildings at 900 MHz," *Electron Lett.*, vol. 18, no. 21, pp. 913–914, Oct., 1982.
[2] S. E. Alexander, "Characterizing buildings for propagation at 900 MHz," *Electron. Lett.*, vol. 19, no. 20, p. 860, Sept., 1983.
[3] R. R. Murray, H. W. Arnold, and D. C. Cox, "815 MHz radio attenuation measured within a commercial building," in *Dig. 1986 IEEE Int. Symp. Antennas and Propagation*, vol. 1, June 8–13, 1986, pp. 209–212.
[4] A. A. M. Saleh and R. A. Valenzuela, "A statistical model for indoor multipath propagation," *IEEE J. Select. Areas Commun.*, vol. SAC-5, no. 2, pp. 128, Feb. 1987.
[5] D. J. Goodman and A. A. M. Saleh, "The near/far effect in local ALOHA radio communications," *IEEE Trans. Veh. Technol.*, vol. VT-36, no. 1, pp. 137, Feb. 1987.
[6] D. J. Goodman and C.-E. W. Sundberg, "The effect of channel coding on the efficiency of cellular radio systems," *IEEE Trans. Commun.*, vol. COM-33, pp. 288–291, Mar. 1985.
[7] R. E. Khan, S. A. Gronemeyer, J. Burchfiel, and R. C. Kunzelman, "Advances in packet radio technology," *Proc. IEEE*, vol. 66, pp. 1468–1496, Nov. 1978.
[8] P. W. Baier, K. Dostert, and M. Pandit, "A novel spread-spectrum receiver synchronization scheme using SAW-tapped delay line," *IEEE Trans. Commun.*, vol. COM-30, pp. 1037–1047, May 1982.
[9] G. J. Turin, "Communications signaling—an antimultipath technique," *IEEE J. Select. Areas Commun.*, vol. SAC-2, pp. 548–562, July 1984.
[10] A. T. Kozlowski, "The capture effect for spread-spectrum signal receivers," in *Dig. 1985 IEEE Global Telecommun. Conf.*, Dec. 2–5, 1985, pp. 980–983.
[11] M. Kavehrad and P. J. McLane, "Performance of low-complexity channel coding and diversity for spread spectrum in indoor, wireless communication," *AT&T Tech. J.*, vol. 64, no. 8, pp. 1927–1965, Oct. 1985.
[12] M. Kavehrad and B. Ramamurthi, "Direct-sequence spread-spectrum with DPSK modulation and diversity for indoor, wireless communications," *IEEE Trans. Commun.*, vol. COM-35, no. 2, pp. 224–236, Feb. 1987.
[13] D. M. J. Devasirvatham, "The delay spread measurements of wideband radio signals within a building," *Electron. Lett.*, vol. 20, no. 23, pp. 950–951, Nov. 1984.
[14] ——, "A comparison of time delay spread measurements within two dissimilar office buildings," in *Dig. 1986 IEEE Int. Conf. Commun.*, vol. 2, June 22–25, 1986, pp. 852–857.
[15] I. Cidon and M. Sidi, "The effect of capture on collision-resolution algorithms," *IEEE Trans. Commun.*, vol. COM-33, no. 4, pp. 317–324, Apr. 1985.
[16] L. Kleinrock, *Queueing Systems, Vol. II: Computer Applications.* New York: Wiley, 1976.
[17] J. F. Hayes, *Modeling and Analysis of Computer Communications Networks.* New York: Plenum, 1984.
[18] A. G. Konheim and B. Meister, "Waiting lines and times in a system with polling," *J. Ass. Comput. Mach.*, vol. 21, no. 3, pp. 470–490, July 1974.
[19] G. B. Swartz, "Polling in a loop system," *J. Ass. Comput. Mach.*, vol. 27, no. 1, pp. 42–59, Jan. 1980.
[20] D. J. Goodman, G. B. Lockhart, O. J. Wasem, and W.-C. Wong, "Waveform substitution techniques for recovering missing speech segments in packet voice communications," *IEEE Trans. Acoust. Speech Signal Processing*, vol. ASSP-34, pp. 1440–1448, Dec. 1986.

Author Index

Subject Index

A

B

C

D

E

R

S

T

U

V

W

Editor's Biography

Norman Abramson received the A.B. in physics from Harvard University, the M.A. in physics from UCLA, and the Ph.D. in electrical engineering from Stanford University. From 1959 to 1965 he was Assistant Professor and Associate Professor in the Electrical Engineering Department at Stanford University. In 1966, he was appointed Professor of Electrical Engineering and Professor of Information Sciences at the University of Hawaii. At the University of Hawaii he has served as a Chairman of the Information and Computer Sciences Department and as Director of THE ALOHA SYSTEM—a university research project concerned with new forms of computer communication networks.

While on visiting appointments, Dr. Abramson has taught communication theory, computer networks and satellite communication courses at Berkeley, Harvard and M.I.T. He has served as a consultant in data networks and satellite communications to government and industrial laboratories in the United States. Dr. Abramson has also served as an expert in computer networks for the ITU/Geneva and for the United Nations in Asia and Europe.

Dr. Abramson is a Fellow of the IEEE and a past Chairman of the Administrative Committee of the IEEE Group on Information Theory. In 1972 he was the recipient of the IEEE Sixth Region Achievement Award. Dr. Abramson has been involved in communication theory, satellite communications, and data network research for more than twenty years. His recent work deals with the use of new forms of data network architecture.